Digital Signal Processing

Principles and Applications

Combining clear explanations of elementary principles, advanced topics and applications with step-by-step mathematical derivations, this textbook provides a comprehensive yet accessible introduction to digital signal processing. All the key topics are covered, including discrete-time Fourier transform, z-transform, discrete Fourier transform and FFT, A/D conversion, and FIR and IIR filtering algorithms, as well as more advanced topics such as multirate systems, the discrete cosine transform and spectral signal processing. Over 600 full-color illustrations, 200 fully worked examples, hundreds of end-of-chapter homework problems and detailed computational examples of DSP algorithms implemented in Matlab® and C aid understanding and help put knowledge into practice. A wealth of supplementary material accompanies the book online, including interactive programs for instructors, a full set of solutions and Matlab® laboratory exercises, making this the ideal text for senior undergraduate and graduate courses on digital signal processing.

Online resources available at www.cambridge.org/holton

- 600+ full-color figures
- Interactive programs for instructors
- A full set of solutions
- Extensive supplementary material, including 100+ pages of text and 70+ illustrations
- Matlab® laboratory exercises

Thomas Holton is Professor of Electrical and Computer Engineering at San Francisco State University with interests in speech and audio processing.

"Professor Holton has done a great service to faculty who teach digital signal processing. The material is developed in a clear and thorough manner with an excellent range of topics, from elementary to advanced and from theoretical to applied. Many insightful analytical and computational examples and homework problems are included, with Matlab intelligently integrated. This textbook is the clear frontrunner in a crowded field."

Howard Weinert, Johns Hopkins University

"... a student-friendly book, making learning DSP a fun journey."

Professor Xiyi Hang, California State University–Northridge

"The *Digital Signal Processing* (DSP) textbook written by Thomas Holton is an excellent textbook for undergraduate as well as graduate students. It is well written, very clearly defined and presents all DSP topics, using many examples including the use of Matlab from chapter 1. In 14 chapters, Dr. Holton covers all necessary materials for a thorough understanding of DSP concepts and practical applications of a subject which is very mathematical. There are 3 appendices, including a tutorial on Matlab. In addition, the text has a website for additional reference materials. It is without any reservations that I strongly endorse and recommend the DSP book by Professor Holton."

Mousavinezhad Hossein, Idaho State University

"The Holton text includes technical materials that a practicing engineer needs to know to prototype a fixed-coefficient DSP system architecture using Matlab. There are many unique features of this textbook, including a full chapter on visualizing frequency response from pole-zero plots, multicolor plots for better comprehension, rigorous derivation of all formulas, and up-to-date hardware- and software-based implementation ideas for the benefit of novice and practicing engineers. I strongly recommend its adoption."

Kalyan Mondal, Fairleigh Dickinson University

DIGITAL SIGNAL PROCESSING

PRINCIPLES AND APPLICATIONS

Thomas Holton
San Francisco State University

CAMBRIDGE
UNIVERSITY PRESS

CAMBRIDGE
UNIVERSITY PRESS

University Printing House, Cambridge CB2 8BS, United Kingdom

One Liberty Plaza, 20th Floor, New York, NY 10006, USA

477 Williamstown Road, Port Melbourne, VIC 3207, Australia

314–321, 3rd Floor, Plot 3, Splendor Forum, Jasola District Centre, New Delhi – 110025, India

79 Anson Road, #06–04/06, Singapore 079906

Cambridge University Press is part of the University of Cambridge.

It furthers the University's mission by disseminating knowledge in the pursuit of education, learning, and research at the highest international levels of excellence.

www.cambridge.org
Information on this title: www.cambridge.org/9781108418447
DOI: 10.1017/9781108290050

First published 2021

Printed in Singapore by Markono Print Media Pte Ltd 2021

A catalog record for this publication is available from the British Library.

ISBN 978-1-108-41844-7 Hardback

Additional resources for this publication at www.cambridge.org/holton.

To my parents, Gerald and Nina Holton.

CONTENTS

PREFACE

Digital signal processing (DSP) is still a comparatively young field. The first textbooks arrived in the 1970s and were relatively scholarly books for graduate students that assumed a level of mathematical sophistication and facility that was appropriate to the intended audience at the time. Over the last 50 years, DSP has evolved from an elective subject aimed at advanced students into a subject that is required in many undergraduate programs. However, I feel that many textbooks still owe their approach and their expectations of the abilities of their audience to books of an earlier era.

The purpose of this book is simple: to lay out the basic principles of digital signal processing, to excite you about the possibility of applying DSP to your own applications, and to give you the tools to do so. The book was originally written for my students, and for the bulk of students like them in engineering programs everywhere. My students have a range of mathematical abilities, from those who could appreciate an advanced book to those who sometimes struggle with basic mathematical material. I suspect this describes the profile of students in many programs around the country and the world. It has been a challenge to engage all of my students and get them to understand the material. This book evolved as my response to that challenge. It started, as many books do, as a series of supplementary notes to an undergraduate DSP course that I have been teaching for a number of years, and has now evolved into a full-featured text that not only covers the basics but also includes coverage of a range of topics to satisfy more advanced students and perhaps even practitioners in the field.

Noteworthy features of the book

Accessible, comprehensive text

In this book, I have striven for three things: *simplicity*, *clarity* and *thoroughness* of explanation. In writing the book, I had in my mind's eye explaining material to an average student, one-on-one, during office hours. Hence, the tone of the book is conversational, with an occasional dash of humor to keep the reader interested (I hope).

The content of the book is comprehensive, starting with patient, clear explanations of elementary principles and proceeding to more advanced topics, some contained in "starred (★) sections" in the text or as supplementary material on the web. A lot of effort went into the early chapters dealing with the impulse response, convolution, the discrete-time Fourier transform and the z-transform. They are particularly detailed, with a lot of examples, because it is essential that the students master this material before trying to move forward. While I have resisted the temptation to cover every single topic in this book, I have included a range of useful material not usually covered, particularly in the later chapters: Chapter 7 (Finite impulse response filters), Chapter 8 (Infinite impulse response filters), Chapter 12 (Discrete cosine transform), Chapter 13 (Multirate systems) and Chapter 14 (Spectral analysis). While some of these topics may be of more interest to advanced readers, my intention has been to make these topics accessible to everyone who has mastered the earlier material.

Good balance of theory and application

In any text such as this, there is also a trade-off between theory and practical discussion. Here, I have followed the famous advice of Yogi Berra[1]: "When you come to a fork in the road, take it." In other words, I have tried to give the reader a good balance of the two. In terms of theory, I have not assumed that I can safely leave steps out of important derivations. So, where necessary, derivations are worked out fully in the text, sometimes using more than one approach. Some of the more algebraically exhausting and tedious derivations have been relegated to the problems, but I have nevertheless tried to structure those problems in such a manner that readers can follow the gist of the derivations even if they do not wish to work them out themselves.

In terms of practical applications, I am mindful of the fact that many (most?) engineering students are likely aiming to get jobs in industry. If they are tasked to use their DSP knowledge, it will most likely be to implement existing algorithms or to use existing design tools to design filters and transforms for specific applications. Accordingly, most chapters have sections that discuss details of hardware or software implementations and applications, including computational examples. Some chapters, for example the supplementary sections of Chapter 6 (A/D and D/A conversion), discuss relevant hardware in considerable detail, something that is often not done in DSP texts. Many students may not see this material in any other course and they ought to in order to understand how DSP theory is put into practice.

Multi-level integration of Matlab®

Matlab is a standard computing language and computing environment for DSP algorithm development and simulation. So, it has become *de rigueur* to include Matlab in any text, sometimes (I feel) gratuitously. I have thoroughly integrated Matlab into the text in several ways:

1. **Language basics**: In early chapters and in Appendix C, the book provides an introduction to the features of the basic Matlab language, and presents a series of examples that indicate features and pitfalls of the language of which the student and practitioner must be aware.
2. **DSP toolbox functions**: While I have resisted cataloging every relevant Matlab function, in later chapters, the text describes the syntax of essential Matlab functions of the signal processing toolbox and gives examples of their use. For example, in Chapter 7 on finite impulse response (FIR) filters, I show how to use Matlab toolbox functions for designing window-based, frequency-sampling, least-square-error and optimum filters, including `rectwin`, `hamming`, `hann`, `hanning`, `blackman`, `kaiserord`, `kaiser`, `window`, `fir1`, `fir2`, `firls`, `firpm` and `firpmord`. But…
3. **DSP algorithms from scratch**: While providing a summary of Matlab's functions is necessary, it isn't adequate for a couple of reasons. First, simply giving the syntax of Matlab's basic filter design functions (such as those listed above) and a quick example or two of their use doesn't give students any clue how to design or implement a filter for themselves, and also weds them to the Matlab toolbox functions. So, the text includes examples of filter algorithms written *from scratch* without those functions. For example, in Chapter 7

[1] Lorenzo Pietro ("Yogi") Berra (1925–2015) was one of the greatest catchers in American baseball history, a winner of 10 World Series Championships in 18 seasons with the New York Yankees, a team he also managed later in his career. Berra was known for his many unintentionally amusing yet wise aphorisms, such as this timeless one, "It ain't over 'till it's over."

(Finite impulse response filters), you will find many Matlab examples of the design of window-based, frequency-sampling and least-square-error designs written in basic Matlab that a student can easily test. In Chapter 8 (Infinite impulse response filters), example code for the design for all filters is provided. In Chapter 12 (Discrete cosine transform), Chapter 13 (Fast Fourier transform) and Chapter 14 (Spectral analysis), there are numerous pieces of code that exemplify every major point.

Providing basic code that does not rely on the Matlab toolboxes is also important because basic Matlab is similar enough to other languages (e.g., Python) that users who understand the basic Matlab code in the chapters can easily port the algorithms to the language of their choice. This is significant, because there are ongoing efforts, particularly in the academic community, to move away from Matlab, which is proprietary and expensive, to open-source languages, either Matlab clones such as Octave, or general-purpose languages such as Python.

4. **End-of-chapter laboratory exercises**: In my DSP course, there are mandatory, weekly, open-ended, project-based laboratory exercises in which the students use Matlab to design, code, debug and test algorithms to solve specific problems related to the material they are learning in lecture. I have adapted a number of these exercises into special sections of Matlab laboratory exercises that appear as supplementary content at www.cambridge.org/holton. I have made an effort to make these laboratory exercises relevant to real applications, including DTMF-to-text decoding and audio compression.

Examples and illustrations

The chapters contain over 200 fully worked theoretical and practical Matlab-based examples, and over 600 illustrations, many in color, designed to complement the text. Each chapter has end-of-chapter problems, ranging from basic exercises to more complex examples that extend the discussion of the text.

Detailed description of the book's chapters

The book has 14 chapters of which the first five comprise the essential, core material upon which everything else depends. The remaining chapters contain a variety of topics, a selection of which would serve for a one-semester course when added to the core chapters. For example, one could add the first sections of Chapter 6 (A/D and D/A conversion), up to and including the discussion of upsampling and downsampling, the first sections of Chapter 7 (Finite impulse response filters) covering window-based filters, the first sections of Chapter 9 (Filter architecture) covering cascade and parallel architectures, much of Chapter 10 (Discrete Fourier transform), and the first section of Chapter 11 (Fast Fourier transform), which deals with the decimation-in-time (DIT) transform. The later sections of many chapters, such as Chapter 7 (Finite impulse response filters), Chapter 8 (Infinite impulse response filters) as well as all of Chapter 12 (Discrete cosine transform), Chapter 13 (Multirate systems) and Chapter 14 (Spectral analysis) contain a variety of more-or-less advanced topics and applications that might provide interesting supplements to the basic course material.

Here is a description of the content of each of the chapters:

Chapter 1 (Discrete-time signals and systems) introduces the fundamental concepts of signals and systems in the discrete-time domain that will be required in the remaining chapters. It starts with a description of the basic signals – impulses, steps, sinusoids and exponentials, both real and complex – and the basic operations upon them, such as flip and shift. Then, using a series of examples, the chapter describes important system properties: linearity, time invariance, causality and stability.

Chapter 2 (Impulse response) defines the notion of the impulse response – the response of the system to an impulse – and shows that for linear time-invariant (LTI) systems, the impulse response is all that is necessary to compute the response of the system to any input through the operation of convolution. The properties of convolution are presented with examples from finite impulse response (FIR) and infinite impulse response (IIR) systems. We demonstrate simple tests for system causality and stability based on the impulse response. Deconvolution is described as well as two methods of convolving infinitely long sequences: the overlap-add and overlap-save methods.

Chapter 3 (Discrete-time Fourier transform) deals with the basic discrete-time Fourier transform (DTFT) material. The chapter describes the response of LTI systems to complex exponential sequences and shows how the orthogonality of these sequences leads to the definition of the DTFT and its inverse. The chapter features a catalog of the important signals and their transforms and a comprehensive treatment of the main properties of the DTFT. This chapter also introduces the idea of linear-phase systems, which are central to the development of FIR filters in Chapter 7.

Chapter 4 (*z*-transform) develops the *z*-transform as a generalization of the DTFT that is designed to handle a larger variety of signals and systems. There is an extended discussion of the visualization of the singularities – poles and zeros – and the region of convergence of the *z*-transform by means of pole-zero plots. The following section of the chapter builds upon the discussion of linear-phase systems started in Chapter 3 and shows that the poles and zeros of linear-phase systems can only occur at constrained locations on the *z*-plane. The properties of the *z*-transform are developed and applied to a number of examples, including the solution of linear constant-coefficient difference equations (LCCDEs). Finally, there is a brief discussion of the unilateral *z*-transform and its application to the solution of LCCDEs with initial conditions.

Chapter 5 (Frequency response) complements the preceding chapter on the *z*-transform. It shows how to visualize the magnitude and phase of the frequency response directly from the *z*-transform in many simple cases, and even design basic filters by inspection. This chapter also introduces allpass filters and minimum-phase systems.

Chapter 6 (A/D and D/A conversion) covers the basic material of analog-to-digital (A/D) and digital-to-analog (D/A) conversion that is required to understand how to apply DSP to real-world analog signals. The chapter first goes over analog sampling and reconstruction and then adds the idea of the "continuous-to-discrete" converter. Anti-aliasing filters on the A/D and anti-imaging filters on the D/A are discussed in the context of a digital recording application. The chapter includes coverage of basic sample-rate conversion: upsampling, downsampling and resampling. One section of the chapter is devoted to the effects of quantization, which is inherent in sampled-data systems, and an explanation of how quantization noise can be reduced by oversampling and noise-shaping A/D converters. Supplementary material to this chapter includes a survey of A/D and D/A conversion hardware.

Chapter 7 (Finite impulse response filters) deals with many varieties of FIR filters, and is one of the longest in the book. It describes a number of methods of designing FIR filters to meet specific design criteria. The first half of the chapter is devoted to window-based filters, which are the most important to understand and implement. The windows (e.g., Hamming, Hann and Kaiser) are necessary not only for filter design, but for windowing data in spectral applications, a topic which will appear again in Chapter 14 (Spectral analysis). The second half of the chapter covers an array of more advanced filtering topics, including frequency-sampling, least-square and optimum (Parks–McClellan) filter design. The chapter is complemented by multiple Matlab programs for the design of filters from scratch.

Chapter 8 (Infinite impulse response filters) describes the design of discrete-time filters that are based on analog filter prototypes. The first half of the chapter describes in detail the theory of design of classical analog Butterworth, Chebyshev, inverse Chebyshev and elliptic filters, complemented by multiple pieces of Matlab code that implement these filters. The chapter then describes several methods of converting from an analog prototype to a discrete-time realization, including the impulse-invariance and bilinear transformations, and shows why the bilinear-transformation method is preferred. The next section of the chapter describes spectral transformations that can be used to convert a prototype discrete-time lowpass filter into a highpass, bandpass or bandstop filter. The last section of the chapter describes how to perform zero-phase filtering to compensate for the nonlinear phase delay of an IIR filter in a non-real-time application.

Chapter 9 (Filter architecture) is concerned with the manner in which the filters described in Chapters 7 and 8 can be implemented in software or hardware. The first section introduces the signal-flow graph, which is an implementation-independent way of representing the basic architecture of FIR and IIR filters. The discussion covers two basic ways of implementing these filters: cascade and parallel configurations in both canonical and transpose variants, as well as lattice and lattice-ladder filters, which are useful in a number of special applications, particularly in speech and audio signal processing. The subsequent section discusses the effect of the quantization of filter coefficients on the filter's response. Finally, the last section considers a number of issues that arise in practical implementations in software and special-purpose hardware implementations.

Chapter 10 (Discrete Fourier transform) begins with the derivation and implementation of the discrete Fourier transform (DFT) from the point of view of the DTFT of periodic sequences. The chapter presents the DFT of basic signals and a description of the fundamental properties of the DFT, with particular attention being paid to circular convolution and to its interpretation as linear convolution followed by time-domain aliasing. Additional topics include the use of the DFT to increase the resolution in the frequency domain by zero-padding in the time domain and the corollary operation of upsampling in the time-domain by zero-padding in the frequency domain.

Chapter 11 (Fast Fourier transform) deals with the fast Fourier transform (FFT), which is the practical implementation of the DFT covered in Chapter 10. The chapter first derives in detail the most common variant of the transform: the radix-2 decimation-in-time (DIT) transform. There follow sections discussing radix-4 and composite (mixed-radix) transforms, the decimation-in-frequency (DIF) transform and transposed forms of the DIT and DIF transforms. The chapter then shows how considerable computational savings can be achieved by exploiting the symmetry of the DFT/FFT in the calculation of real sequences. The chapter concludes with a discussion of recursive and non-recursive strategies for implementing the FFT.

Chapter 12 (Discrete cosine transform) develops the discrete cosine transform (DCT) and its application to sound and image encoders that people use on a daily basis: their cellphones, music players and cameras. The chapter begins with a description of the four main variants of the DCT, and concentrates on the derivation and implementation of the most common variant, the DCT-II. Supplementary material for this chapter also covers the modified windowed DCT (MDCT), which is the main transform used in audio codecs such as the MP3 format. The last two main sections of the chapter give a fairly broad overview of compression of audio by the MP3 codec and compression of images by the JPEG standard of MP3 and JPEG compression. There is quite a bit of non-DSP detail here, such as a description of psychophysical auditory and visual models, color spaces and minimum entropy coding, but I feel that it is important to show how DSP interfaces with all these other disciplines in order to create workable systems.

Chapter 13 (Multirate systems) is somewhat more advanced than the preceding chapters, but polyphase concepts are now so pervasive in filtering and sample-rate conversion that it seemed necessary to create an accessible treatment. To do this, the chapter starts with a derivation of polyphase upsampling, downsampling and resampling from a time-domain perspective, which is perhaps the most intuitive way to introduce the subject. Then, the same material is reexamined using the z-transform, which makes the concepts such as the key multirate identities easier to derive and understand. Since polyphase methods are often paired with multirate filters, multistage filters and filter banks, this chapter is the logical place for that material. There is a section on multistage systems for decimation and interpolation, another section on multistage and multirate filtering and a section on half-band and L-band filters for downsampling and upsampling. Supplementary material associated with this chapter contains a discussion of multirate filter banks including quadrature mirror filters (QMF) and a variety of maximally decimated filter banks.

Chapter 14 (Spectral analysis) covers a wide range of topics relating to spectral analysis, which is the process of measuring, estimating and characterizing the frequency content of signals. The chapter starts with an explanation of two fundamental issues that affect the practical measurement of the frequency spectrum of signals: data windowing and frequency sampling. The following section develops the short-term Fourier transform (STFT), which is useful in many cases where signals are inherently time-varying, such as the analysis of speech and music. The remaining sections of the chapter are concerned with the spectral analysis of random signals. The chapter covers two broad categories of techniques for spectral measurement and estimation of probabilistic systems: nonparametric and parametric. Nonparametric spectral methods do not depend on knowing any specific information of the signal being analyzed. These include several variants of the periodogram (e.g., Bartlett's method and Welch's method). Parametric methods, which are covered next, are based on estimating the parameters of models that have been specifically designed to match the characteristics of the process being analyzed, such as speech. The chapter discusses the autoregressive (AR) model in detail, and derives the

Yule–Walker equations and their solution via the Levinson–Durbin algorithm. The chapter concludes with a discussion of linear prediction and its application to encoding and decoding speech sounds using the source-filter model of speech production.

Appendix A (Linear algebra) provides a brief review of linear algebra used throughout the book, including the solution of systems of linear equations in matrix form and a derivation of the normal equations.

Appendix B (Numeric representations) discusses how integers and floating-point numbers are represented in binary form.

Appendix C (Matlab tutorial) is a brief introduction to Matlab, which highlights features that make Matlab suited to the processing of arrays, and also indicates a number of potential pitfalls for programmers.

Appendix D (Probability and random processes) is a relatively comprehensive review of the topics that are prerequisite for understanding the material in Chapter 14 (Spectral analysis). The appendix covers probability distributions and density functions, basic random processes including discussions of stationary processes, filtered random processes and power spectral density.

Supplementary material In addition to the material published in the book, there are about 100 pages of supplementary material available at www.cambridge.org/holton containing an elaboration and extension of topics discussed in the chapters. Examples of this more advanced material include an extensive discussion of the hardware implementation of A/D and D/A converters (Chapter 6), derivation and examples of the Parks–McClellan algorithm for optimal FIR filtering (Chapter 7), development of the modified discrete cosine transform (MDCT) (Chapter 12) and a discussion of multirate filter banks, including quadrature and conjugate mirror filters, and complex-modulated filter banks (Chapter 13).

1 Discrete-time signals and systems

Introduction

This chapter provides the foundation of our study of digital signal processing. In Sections 1.1–1.3, we start by looking at the two basic paradigms for doing signal processing, analog and digital, and highlight some of the advantages and applications of the digital signal processing paradigm. The remaining sections of the chapter define and explore two fundamental concepts: **signals** and **systems**. These are the basic elements of signal processing in the same sense that vocabulary and grammar are the basic elements of a language. In Sections 1.4 and 1.5, we define the notion of a discrete-time signal and introduce some basic operations that can be performed upon such signals. In Section 1.6, we introduce a collection of important signals – e.g., the impulse, sinusoids and exponentials – combinations of which will be used to create almost all the other signals we use throughout this book. In Section 1.7, we define the notion of a system and in the remaining sections present four key properties of systems – linearity, time invariance, causality and stability.

1.1 Two signal processing paradigms

There are two basic ways of looking at the signal-processing world: the **analog signal processing paradigm** and the **digital signal processing (DSP) paradigm**. In the analog paradigm, shown in **Figure 1.1a**, a circuit – for example, a filter – is constructed using familiar analog elements, both passive (resistors, capacitors, inductors) and active (transistors, operational amplifiers) to perform a specific task. A continuous-time (analog) signal $x(t)$, such as speech or music, is processed by this system to form an analog output $y(t)$.

In contrast, in the DSP paradigm, shown in **Figure 1.1b**, the analog signal $x(t)$ is first **sampled** by a hardware **analog-to-digital (A/D) converter** at a rate of f_s samples/s to form a sequence of numbers $x[n]$. The heart of the DSP system is a mathematical **algorithm**, implemented in software or hardware, that processes the input sequence to form an output sequence $y[n]$. Finally, the output sequence is converted back into an analog output $y(t)$ by a **digital-to-analog (D/A) converter** operating at a rate of f_s samples/s.

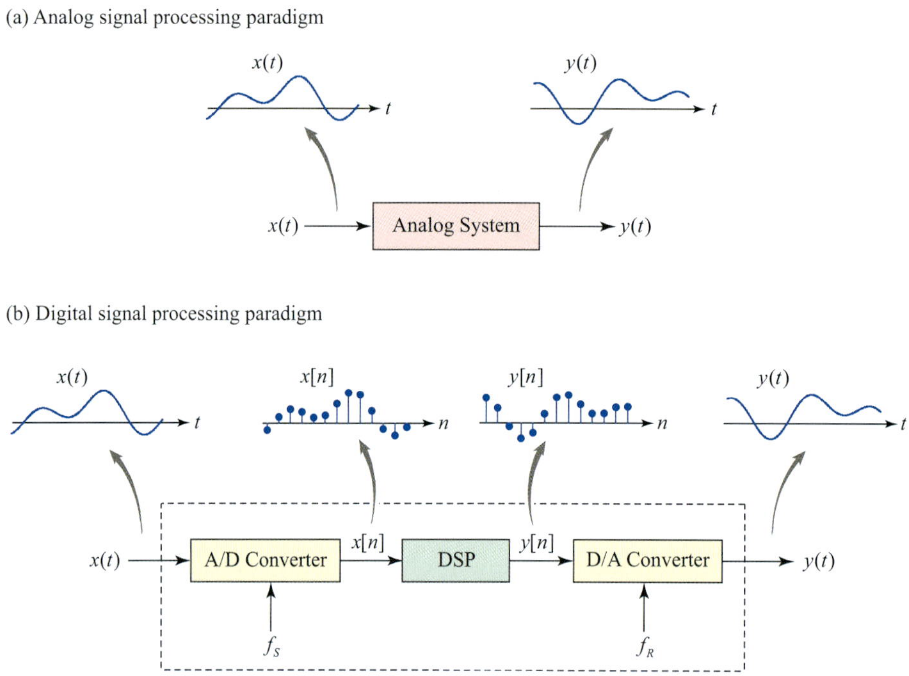

Figure 1.1 Analog and digital signal processing paradigms

Now imagine applying these two paradigms to the problem of creating a simple second-order lowpass filter for an audio application with the following design specifications: a **corner (cutoff) frequency** of $f_c = 1000$ Hz and a **quality factor** of $Q = 0.8$. The left panel of **Figure 1.2** shows this filter implemented using a textbook **Sallen–Key circuit** with discrete components: a single **op-amp** (e.g., the venerable NE5534 op-amp) and some resistors and capacitors chosen from the list of standard values in order to meet our design specifications as closely as possible.

To implement this filter using the DSP paradigm, we might begin by selecting an **audio codec** (e.g., the tiny (5mm square) Analog Devices ADAU1761), a package that includes a 24-bit stereo A/D converter, a 24-bit D/A converter and a programmable processor on which we can implement our filter algorithm. The right panel of **Figure 1.2** shows a DSP algorithm for this same filter – a simple "two-tap" digital filter that has been prototyped as a Matlab function. It is just a few lines of code that accepts the input sequence `x`, along with values of the filter's cutoff frequency `fc` and `Q`, and the sample frequency of the A/D and D/A `fs`. The output of the function, `y`, goes to the D/A converter. While this particular algorithm is written in a high-level language and is not designed for a real-time application, it can be easily ported or cross-compiled to function in a real-time application on the target processor.

Figure 1.3 shows the frequency response that results from the analog and digital filter designs. The analog design has the expected −40 dB/decade roll-off above the corner frequency. The simple

Analog filter

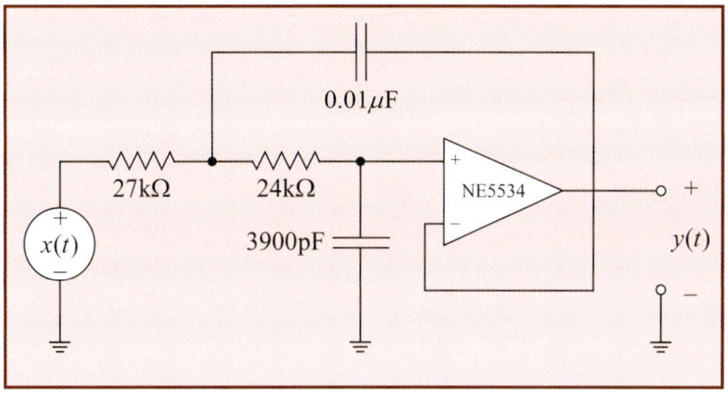

Digital filter

```
function y = lpf(x, fc, Q, fs)
    wt = 2*tan(pi*fc/fs);
    b = wt^2*[1 2 1];
    a = [4+wt^2+2*wt/Q ...
        2*wt^2-8 4-2*wt/Q+wt^2];
    y = filter(b, a, x);
end

>> y = lpf(x, 1000, 0.8, 48000)
```

Figure 1.2 A second-order lowpass analog and digital filter

digital filter matches the frequency response – magnitude and phase – at low frequencies, and actually rolls off faster at higher frequencies, which is not a bad thing.

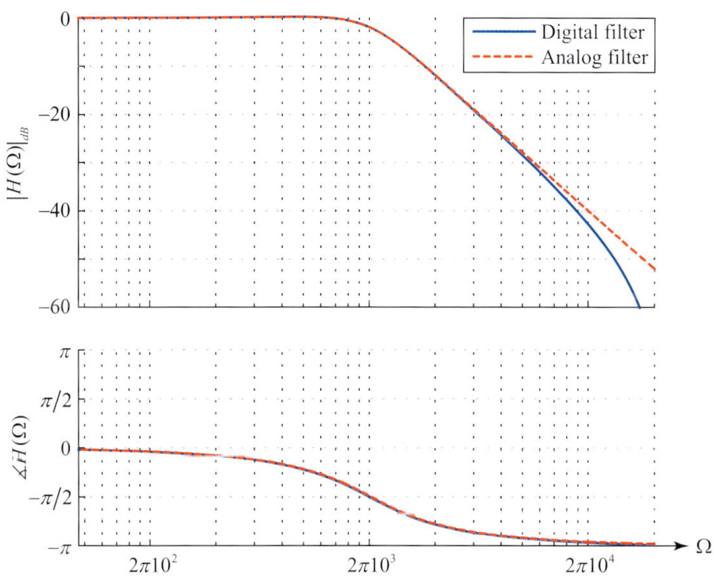

Figure 1.3 Comparison of analog and digital filtering paradigms for a second-order filter

Implementing a signal processing system such as our filter using the DSP paradigm would appear to require a lot more sophisticated parts than the simple analog filter – the A/D converter, the D/A converter, the processor – plus the algorithm. So, why would one want to use this approach? There are many reasons, some of which are itemized in the next section.

1.2 Advantages of digital signal processing

Implementing signal processing tasks using the digital signal processing paradigm has many advantages.

Powerful algorithms. The "heavy-lifting" in digital signal processing tasks is performed by algorithms programmed in software and embedded or compiled into hardware. These algorithms can be very complex and powerful. Signal processing tasks such as echo cancellation, compression, linear-phase and adaptive filtering can be accomplished with ease by digital means whereas they are difficult (or in some cases, impossible) to accomplish by analog means. To be sure, there are important areas – particularly very low-latency, ultra-high-speed circuits and high-power electronics – where purely analog techniques have no substitute. However, as of this writing, the latest generation of high-speed A/D converters can digitize radio-frequency signals directly at frequencies up to 4 GHz. This has enabled purely digital systems to replace analog implementations in applications such as software-defined radio, wireless base stations, radar and spectroscopy.

Simple prototyping. With the analog signal processing paradigm, translating between the circuit design on paper and its realization in silicon or on a circuit board is an art. Careful attention must be paid to the selection and layout of components to prevent stray capacitances and inductance, interference and noise from influencing the performance of the designed circuit. With the DSP paradigm, you can simulate the result of most any DSP algorithm you are developing with a high-level programming language such as Matlab, taking into account effects such as round-off error and coefficient quantization. Once you are satisfied with the results of your simulation, you can even get Matlab and other similar DSP simulators to produce microcode for a range of popular processors. What you see with the simulator is what you will get in the final product.

Zero-tolerance design. With analog design, it is usually necessary to design carefully to accommodate the tolerances of various components, for example resistor and capacitor values. In a design with discrete components, typical off-the-shelf resistors and capacitors have a range of standard values and tolerances of only $\pm5\%$ or $\pm10\%$, respectively. In critical applications, components may need to be individually matched or trimmed on the final circuit boards to insure that products meet the designed specifications. With the DSP paradigm, once the designer specifies the speed and the number of bits of quantization (e.g., 24-bit) of the A/D and D/A converters, the performance of all devices with the same specifications will be effectively identical.

Flexibility. With analog devices, once the circuit design is committed to the silicon chip or circuit board, that's it; there is generally only very limited possibility to change the characteristics of the device, for example the critical frequencies or bandwidths of a filter. Any changes necessary to retrofit devices that are already deployed in the field are likely to be expensive, or perhaps impossible – for example, if your circuit happens to be in space, or part of a medical device that is implanted in a human being. With the DSP paradigm, signal processing is based on algorithms that are coded or microcoded. If the need arises to change the algorithm, in many instances one can just update the code. The DSP program may reside in EPROM or flash memory that can be reprogrammed or replaced in the field. Many consumer devices, even cheap ones like cellphones and MP3 players, enable consumers to download new code from a company's website to update their devices in order to allow better performance or to handle new formats. Devices connected to the Internet can update themselves to fix bugs and add capability.

Multiplatform support: Depending on the application, a DSP algorithm can be implemented in software that runs on a general-purpose microprocessor, or on a special-purpose DSP chip. Even embedded microcontrollers (e.g., ARM Cortex-M4) now include "DSP cores" that feature special instructions to speed up common DSP tasks. In addition to software implementations, DSP algorithms can be "compiled" into custom hardware such as **field-programmable gate arrays (FPGA)** or **application-specific integrated circuits (ASIC)** or **systems-on-chip (SoC)**. The highly parallel architecture of the **graphics processing units (GPU)** found in video cards is well suited to many common DSP tasks, such as convolution. Whereas a central processing unit in an extremely powerful desktop computer may have as many as sixteen cores that can support DSP operations, a large GPU has *thousands* of cores that can simultaneously run *tens of thousands* of threads. Big numbers!

Cost: The cost of DSP processors and A/D and D/A converters has dropped to the point where the DSP paradigm is often the signal processing paradigm of choice for all but the most cost-sensitive applications. As an example, the ADAU1761 we specified for our filter application in Section 1.1 costs about $4 in quantity. A powerful little microcontroller like the ARM Cortex-M4 costs about the same.

1.3 Applications of DSP

Digital signal processing has become ubiquitous in modern electronics. The fact that even cheap microcontrollers natively support sophisticated DSP operations has enabled a range of applications in consumer electronic goods and mobile devices that were essentially unthinkable even a decade ago. Applications include

- **Speech signal processing**: The processing of speech has always been a strong focus of DSP research and application, including such areas as speech recognition, speaker verification, identification, language identification, compression and storage.
- **Audio signal processing**: Another strong area of DSP emphasis is the recording, mixing, equalization, filtering, compression, denoising, storage and playback of music in both professional and consumer applications. The .mp3 format is an example of a highly useful algorithm produced by the marriage of DSP techniques with insights from human auditory psychophysics. We will give an outline of this application in Chapter 12.
- **Image processing**: DSP has made possible a range of sophisticated image processing applications, ranging from image enhancement, transformation and compression to feature extraction and pattern recognition. Many of the basic DSP techniques (e.g., Fourier transformation, convolution) have two-dimensional equivalents that are applicable to the processing of images and video. In Chapter 12, we will show how some of these techniques are behind the JPEG standard for compression of images.
- **Digital communication**: Digital communication includes the modulation, transmission and demodulation of digital data over a physical layer (wires, or a fiber-optic cable) as well as through wireless means. Real-time compression makes applications like video conferencing possible. The cellphone is a prime example of a technology that owes its existence to DSP algorithms such as linear predictive coding, which we will discuss in Chapter 14. High-speed data converters have enabled applications such as software-defined radio, wireless base stations, radar and spectroscopy.

- Medical applications: DSP techniques have enabled the measurement, interpretation and display of data from the electrocardiogram (EKG), the electroencephalogram (EEG), digital X-rays and real-time magnetic resonance imaging systems (MRI). Recently developed **systems-on-chip (SoC)**, which include sensors, processors and high-speed I/O as well as powerful DSP processors on one chip, have enabled the development of low power medical devices such as portable ultrasound systems.

Having given a summary of the multiple advantages and wide applications of DSP, we are now ready to introduce and develop two basic building blocks for our study of DSP: **signals** and **systems**.

1.4 Signals

A signal can be defined as any measurable quantity that conveys information. Examples of signals include electrical signals (e.g., the voltage output of an amplifier); acoustical signals (the pressure measured by a microphone); mechanical signals (the force measured by a strain gauge); biological signals (body temperature, blood pressure, waveforms of EKG, EEG); financial (bank balance, stock price). All these are examples of **one-dimensional signals**, where the **dependent**

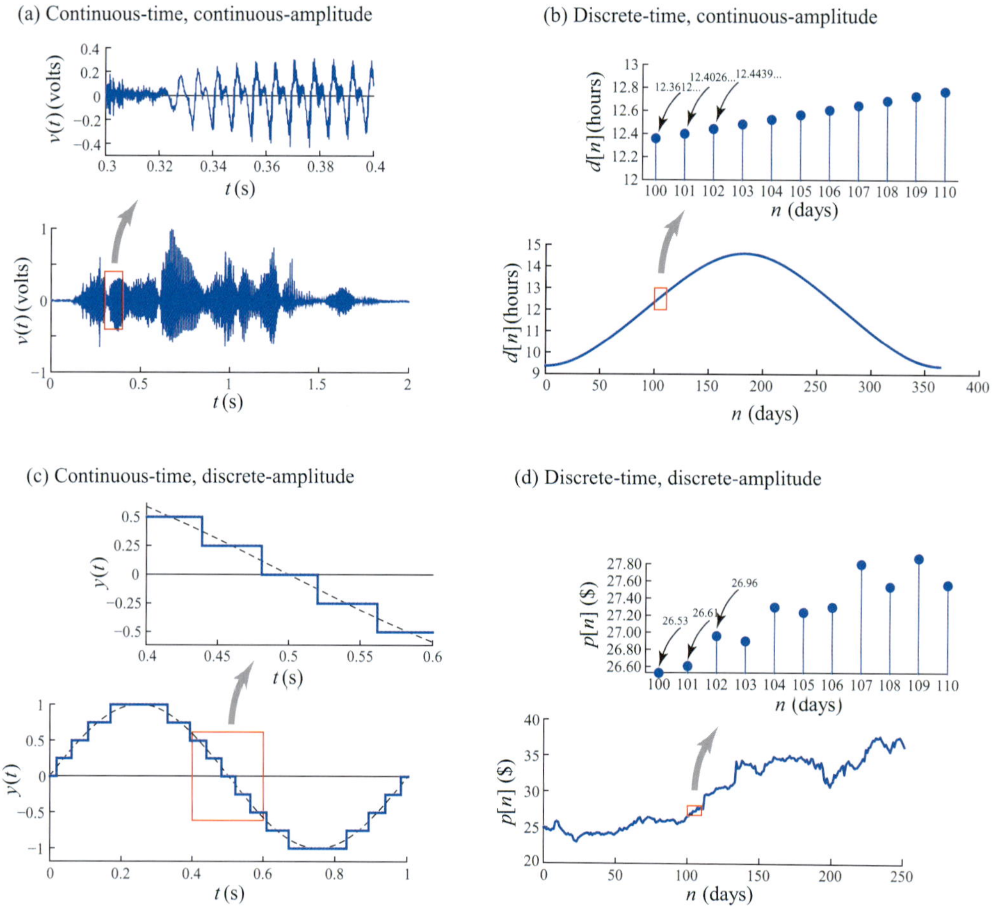

Figure 1.4 Continuous-time and discrete-time signals

variable, which is the measured quantity of interest (e.g., voltage), is a function of the **independent variable** (e.g., time). Signals can also be **multi-dimensional**. An example of a two-dimensional signal is the intensity of a point on a black-and-white display (in lumens) as a function of the x and y coordinates of the point. In this chapter, we will be mostly concerned with the analysis of one-dimensional signals, but the techniques we develop will prove to be extensible to the analysis of multi-dimensional signals as well. For example, in Chapter 13, we will talk about the JPEG image compression standard and show how the techniques developed for the compression of one-dimensional signals such as speech and music can be extended to the compression of images involving three color planes in two dimensions.

1.4.1 Signal classification

Signals can be classified in several ways, the most important of which for our purposes are shown in **Figure 1.4**. A **continuous-time signal**, often simply called an **analog** signal, is one in which the independent variable, which will almost always be time in our examples, is an **unquantized** real number, meaning that the signal has a measurable or definable value at every possible value of the time. For example, the lower plot in **Figure 1.4a** shows the output of a microphone $v(t)$ (in volts) as a function of time t (in seconds) in response to the phrase "she sells seashells." Because the time variable is unquantized, we can find values of $v(t)$ for any value of t in the range $0 \le t \le 2$. The red box highlights the time interval $0.3 \le t \le 0.4$, and the upper plot in the figure shows the voltage for this expanded interval. This is also a continuous curve. In theory, we could continue to zoom in on smaller and smaller portions of the waveform. In this book, the independent variable of a continuous-time signal or function will always be enclosed in round parentheses, e.g., $v(t)$.

A **discrete-time signal**, which is often simply called a **digital signal**, is one in which the independent variable is **quantized**, which means that the signal is only defined at specific values of the independent variable. For example, the bottom panel of **Figure 1.4b** shows sequence $d[n]$, the length of a day (i.e., the amount of daylight, measured in hours) as a function of n, the day of the year, where $n = 0$ is January 1 and $n = 365$ is December 31. The curve looks nice and continuous, but if we select and plot just the 11-day range in the upper panel, $100 \le n \le 110$, you can see that the time axis is quantized in integers. For example, the signal has value on April 11 ($d[100] = 12.3612\cdots$) and value on April 12 ($d[101] = 12.4026\cdots$), but there is no value defined for $d[n]$ for $n = 100.5$. That makes sense because obviously the measured quantity – day length – can only be computed once for each integer day. In this book, we will exclusively consider the case where the independent variable is quantized in integer units.

The signal of **Figure 1.4b** is an example of one that is naturally discrete in time. However, in Chapter 7 we will discuss a key technique of digital signal processing, analog-to-digital (A/D) conversion, which is the process by which we *create* discrete-time signals by sampling continuous-time signals at regular intervals of time.

The signals shown in **Figures 1.4a** and **b** both have the property that their values (the ordinate of the axis) are unquantized real numbers. We will refer to such signals as **continuous-amplitude signals**. **Figure 1.4a** is therefore a continuous-time, continuous-amplitude signal. At any value of t, the value of $v(t)$ can theoretically have any real value. For example, $v(0.3415) = 0.2104\cdots$.

Figure 1.4b is an example of a discrete-time, continuous-amplitude signal. For this signal, even though the independent variable (i.e., the time axis) is quantized in units of days, the ordinate is not quantized. For example, $d[101] = 12.4026\cdots$.

The signals shown in **Figures 1.4c** and **d** have the property that their ordinate is quantized. **Figure 1.4c** shows an example of a continuous-time, **discrete-amplitude** signal

$$y(t) = \text{rnd}(4 \sin 2\pi t)/4,$$

where rnd( ) indicates rounding to the nearest integer. Because this is a continuous-time signal, $y(t)$ has value at every value of t. However, the amplitude has only nine possible discrete levels spanning the range $-1 \le y(t) \le 1$, in increments of 0.25. **Figure 1.4d** shows an example of a discrete-time, discrete-amplitude signal $p[n]$, the closing stock price of Intel (ticker INTC), as a function of day n, for the 252 trading days of the year 2014. The upper panel of the figure shows $p[n]$ for the range of days $100 \le n \le 110$. This is a discrete-time signal, because the closing stock price is only computed once, at the end of each day. It is also a discrete-amplitude signal, because the stock price is quantized in units of one penny. For example, $p[100] = 26.53$.

In this book, we will have cause to discuss signals of each of these four types. But, let us begin with an in-depth look at discrete-time signals.

1.4.2 Discrete-time signals

To restate, a discrete-time signal or **sequence** is essentially an array of numbers $x[n]$ presented as a function of an integer index n. Sequences can be described in several ways.

Table A sequence can be presented as a table or list of numbers:

$$x[n] = \begin{cases} 1, & n = -1 \\ 2, & n = 0 \\ 3, & n = 1 \\ 4, & n = 2 \end{cases}$$

or

n	$x[n]$
−1	1
0	2
1	3
2	4

The values of the sequence are denoted by an ordinate variable, in this case x, described as a function of an integer abscissa variable, in this case n.

Plot A sequence can be graphically represented by a stem plot. **Figure 1.5** shows a plot of $x[n]$:

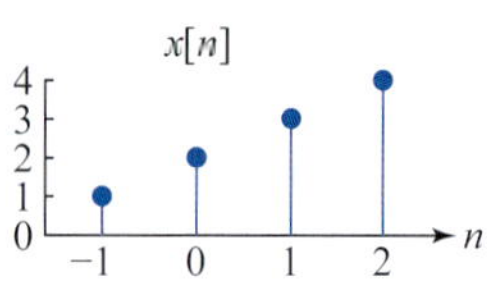

Figure 1.5 Plot of a sequence

Sequences generated by functions Many sequences are described as arithmetic or trigonometric functions of an integer index, for example,

$$y[n] = \cos \pi n/4$$
$$y[n] = \tfrac{1}{2}^n$$
$$y[n] = 1 + n.$$

We will discuss these and other sequences in detail below.

Features of sequences Here are some important points about sequences:

- Since $x[n]$ is a discrete-time sequence, $x[n]$ is only defined for integer values of n. Thus, there is no such thing as $x[1.2]$. However, we will still refer to the index n as "time," even though the values of n can only have integer values.
- Even though $x[n]$ is only defined for integer values of n, the sequence is plotted on a continuous number line in **Figure 1.5**, which we refer to as the **time axis**.
- To emphasize the discrete-time nature of the sequence, we plot each value of $x[n]$ as a little individual "lollipop."
- In **Figure 1.5**, we have let the ordinate values be real integers just for convenience, but in general, the values of $x[n]$ can be real or complex, quantized or unquantized. For the most part, we will concentrate on the analysis of discrete-time, continuous-amplitude signals, but in later chapters, we will also discuss the effects of amplitude quantization.
- The abscissa values n can be both positive and negative. So, it is OK to have $x[-1]=1$. Of course, the ordinate values can also be positive or negative.
- The sequence $x[n]$ is presumed to be defined for *all* n. So, we assume that $x[n]$ is zero for all values of n that are not explicitly stated. In the above example, this means that $x[n]=0$, $n<-1$ or $n>2$.
- $x[n]$ can refer to a single value of the sequence or to the entire sequence. That may seem confusing, but the context almost always makes things clear. If a value of the index n is explicitly specified (e.g., $n=1$), then $x[n]$ is a single value of the sequence, $x[n]=x[1]$. If no value of n is specified, then $x[n]$ refers to the entire sequence.
- In these chapters, we will generally reserve the variables i, j, k, l, m and n to refer to the integer index of the sequence.[1] So, we will write $x[i]$, $x[k]$ and the like.

Sequence representation in Matlab We will be using Matlab extensively throughout this book. If you are unfamiliar with the language, Appendix C provides a brief tutorial introduction. (Even if you are familiar, the tutorial may contain some nuggets of value.) In Matlab, a sequence is denoted by an array of ordinate values enclosed in square brackets, for example,

```
x = [1 2 3 4];
```

However, when an array is defined in Matlab, the index of the first value is always 1; so, in the example above, `x[1]=1`. There is no default way to assign other ordinate values to the index (abscissa) values, n. For example, we would like to have `x[-1]=1`. To produce the plot of **Figure 1.5**, you create an array of index values, using Matlab's colon notation, `n=-1:2`, and use the stem function:

```
x = [1 2 3 4];
n = -1:2;
stem(n,x)
```

[1] This practice stems from the first widely used, general-purpose programming language, FORTRAN, dating from 1954, in which variable names starting with these letters represented integers to the compiler by default.

In the first laboratory exercise accompanying this book and available at www.cambridge.org/holton, you will figure out a clever way of tricking Matlab into being able to assign arbitrary abscissa values to arrays.

1.5 Basic operations on signals

Let us talk about some of the common operations we can perform on sequences.

1.5.1 Shift

The shift is perhaps the most common operation on a single sequence. We define a shifted sequence $y[n]$ as

$$y[n] = x[n - n_0], \tag{1.1}$$

where n_0 is the (integer) amount of shift, and n_0 can be either positive or negative. The shift operation is best explained by a simple example. The center column of **Table 1.1** shows the data, $x[n]$, of the example of **Figure 1.5** in tabular form.

Table 1.1 Shift of sequences

n	$x[n+1]$	$x[n]$	$x[n-1]$
−2	1	0	0
−1	2	1	0
0	3	2	1
1	4	3	2
2	0	4	3
3	0	0	4

Assume that these data represent the number of packages we expect to receive on our doorstep as a function of the day, where $n=0$ represents "today," $n=-1$ is "yesterday," $n=1$ is "tomorrow," and so on. Now, let $n_0=1$ in Equation (1.1), so that $y[n]=x[n-1]$. This states that at any time n, the value of the output sequence $y[n]$ is equal to $x[n-1]$, which is the value of the input sequence at the *previous* time point. So, $y[0]=x[-1]$, $y[1]=x[0]$, and so on. This creates the shifted sequence shown in the right column of **Table 1.1**. In our example, this is what would happen if there was a snowstorm and all our packages were delayed by a day. When $n=0$ (today), instead of getting $x[0]=2$ packages, we only get $y[0]=x[-1]=1$ package. Thus, $y[n]=x[n-n_0]$ represents a delayed version of the sequence $x[n]$. In this example, the amount of delay is *positive*, $n_0>0$, indicating that $y[n]$ is looking *backwards* in time with respect to $x[n]$.

Now, let $n_0=-1$. Then $y[n]=x[n+1]$, as shown in the left data column of **Table 1.1**. This is a shifted sequence in which the value of the output sequence is equal to the value of the input sequence at the *next* time point. So, $y[0]=x[1]$, $y[1]=x[2]$, and so on. In our example, this would happen if we had our packages shipped by air instead of by truck. Now, instead of getting $x[0]=2$ packages today, we get $y[0]=x[+1]=3$ packages. In this example, $y[n]=x[n-n_0]$ again represents a delayed version of the sequence $x[n]$, but in this case the amount of delay is *negative*, $n_0<0$, which indicates that $y[n]$ is looking *forward* in time with respect to $x[n]$.

The center panel of **Figure 1.6** is a plot of the same data, $x[n]$. To help visualize things, we have highlighted in red the point with value $x[0] = 2$. A positive delay (right panel) corresponds to a shift of the entire sequence to the right along the abscissa. A negative delay (left panel) corresponds to a shift to the left.

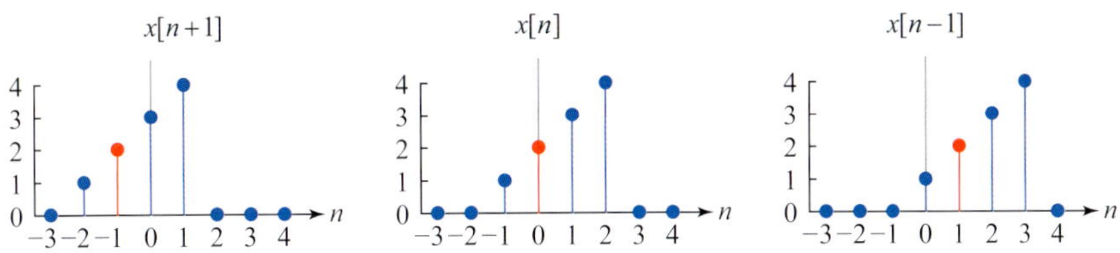

Figure 1.6 Plot of shifted sequence

If we consider again the general relation $y[n] = x[n - n_0]$, a good way to think about the relation between n_0 and the direction that the sequence shifts is to ask the question, "What is the value of n that makes $x[n - n_0] = x[0]$?" Or, in terms of our package example, "When will I receive my $x[0] = 2$ packages?" When $n_0 = 1$, then $y[n] = x[n - 1]$, and the value of n that makes $x[n - 1] = x[0]$ is $n = 1$. So, when we graph the shifted sequence, we place the value $x[0]$ (the "red point") at abscissa value $n = 1$ as on the right panel of **Figure 1.6**. Once we have "pinned" the sequence at one value of n, all the other values of the sequence fall into place. We will discuss the Matlab implementation of shift when we come back to this operation in Section 1.7.4.

1.5.2 Flip

A sequence can be flipped by reversing the abscissa; that is, $y[n] = x[-n]$, as shown in **Table 1.2**.

Table 1.2 Flipped sequence

n	$x[n]$	$x[-n]$
−2	0	4
−1	1	3
0	2	2
1	3	1
2	4	0

This time reversal of $x[n]$ corresponds to a mirror-image reflection of the sequence around the point $n = 0$, as shown in **Figure 1.7**

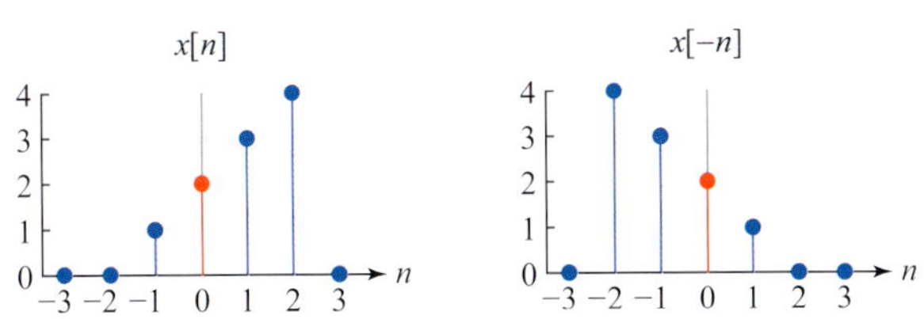

Figure 1.7 Flipped sequence

There are several ways to flip a sequence in Matlab:

```
y = x(end:-1:1);
y = x(length(x):-1:1);
y = fliplr(x);
y = flipud(x);
```

The reserved function `end` is equivalent to the last index in the array, namely `length(x)`. The Matlab functions `fliplr` and `flipud` flip arrays column-wise and row-wise, respectively. However, allow me to make a suggestion. While it is sort-of fun to rifle through the documentation or scurry around the Internet looking for just the *perfect* Matlab function to achieve a given result, using the simplest form (in this case the first line) results in cleaner code that is "self-documenting" and more readable, both by you and by those who may be tasked with understanding what you have done. Also, by using only the set of basic Matlab functions, your code can be more easily ported to other languages.

1.5.3 Flip and shift

We can combine flipping and shifting of sequences. **Table 1.3** shows data for flipped sequences that are shifted, and **Figure 1.8** shows the corresponding plots of these data. Again, the value of $x[0] = 2$ is shown in red.

Table 1.3 Flipped and shifted sequence

n	$x[-n+1]$	$x[-n]$	$x[-n-1]$
−3	0	0	4
−2	0	4	3
−1	4	3	2
0	3	2	1
1	2	1	0
2	1	0	0

A positive delay of the flipped sequence (i.e., $y[n] = x[-n-1]$, right panel) corresponds to a shift of the sequence to the left along the abscissa. A negative delay (i.e., $y[n] = x[-n+1]$, left panel) corresponds to a shift to the right.

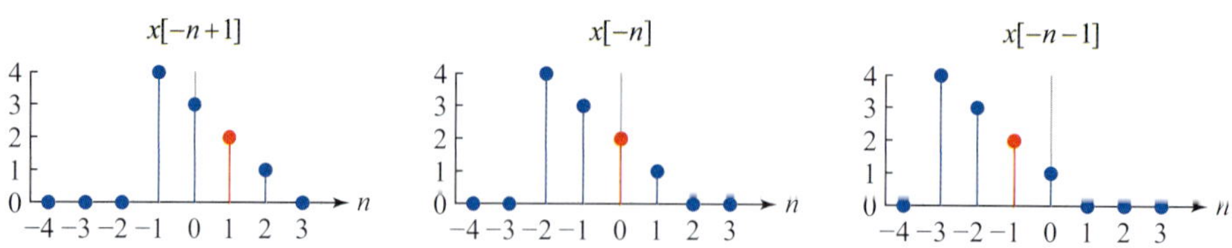

Figure 1.8 Flipped and shifted sequence

Rather than trying to remember which type of delay corresponds to a right or left shift of a flipped sequence, just consider the general relation that describes a flipped and shifted sequence: $y[n] = x[-n-n_0]$. As we discussed above, a good way to think about the relation between n_0

and the direction that the sequence shifts is to ask the question, "What is the value of n that makes $x[-n-n_0]=x[0]$?" For example, when $n_0=1$, then $y[n]=x[-n-1]$, and the value of n that makes $x[-n-1]=x[0]$ is $n=-1$. Thus, in order to graph the flipped and shifted sequence, we first flip $x[n]$ and then shift it so as to place the value $x[0]$ at abscissa value $n=-1$, as shown on the right panel of **Figure 1.8**. Once we have "pinned" the sequence at one value of n, all the other values of the sequence fall into place. The idea of flipping and shifting sequences will be central to our understanding of **convolution** in Chapter 3, which is the mathematical operation behind the filtering of discrete-time sequences.

1.5.4 Time decimation

Time **decimation** or **compression** of a sequence is defined by the equation

$$y[n]=x[nD],$$

where D is the **decimation factor**. An example of decimation is shown in the left column of **Figure 1.9**.

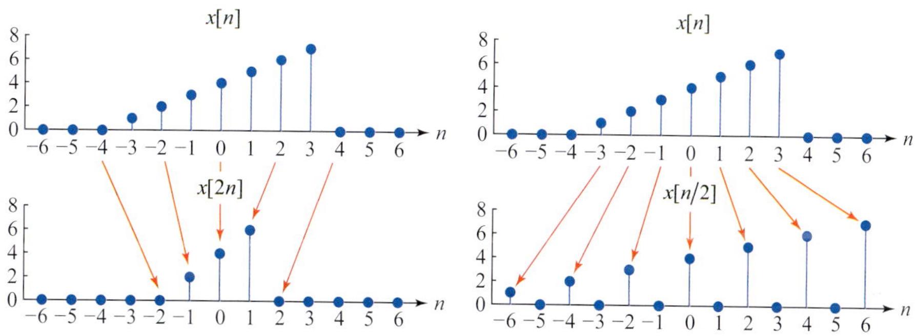

Figure 1.9 Time decimation and expansion

The top row shows an example input sequence $x[n]$. The bottom row shows an output sequence $y[n]=x[2n]$ produced by compressing $x[n]$ by a factor of $D=2$. Hence,

$$\begin{aligned} &\vdots \\ y[-1]&=x[-2] \\ y[0]&=x[0] \\ y[1]&=x[2] \\ &\vdots \end{aligned}$$

In Matlab, we just index every Dth point,

```
y = x(1:D:end);
```

The effect of this decimation is to select every other point of the input sequence. If $D=3$, the output sequence would contain every third point of the input sequence, and so on. Decimation of this sort will become important to us in Chapters 7 and 11, when we discuss what happens when sequences are **downsampled** in the A/D conversion process.

1.5.5 Time expansion

Time **expansion** of a sequence is defined by the equation

$$y[n] = \begin{cases} x[n/U], & n=0, \pm U, \pm 2U, \ldots \\ 0, & \text{otherwise} \end{cases},$$

where U is the **expansion factor**. An example of expansion is shown in the right column of **Figure 1.9** for $U=2$. For this value of U, the even values of $y[n]$ (i.e., $y[n]$, $n=0, \pm 2, \pm 4, \ldots$) select all the points of $x[n]$, $n=0, \pm 1, \pm 2, \ldots$. Thus,

$$\begin{aligned} &\vdots \\ y[-4] &= x[-2] \\ y[-2] &= x[-1] \\ y[0] &= x[0] \\ y[2] &= x[1] \\ y[4] &= x[2] \\ &\vdots \end{aligned}.$$

Because evaluation of the formula $y[n]=x[n/2]$ at odd values of n is not possible, $y[n]$ needs to be defined to be zero at these values. For example, $y[1]$ would be equal to $x[1/2]$, which is undefined. In Matlab, the best way to do this is to create an array with the appropriate number of zeros and then populate it with our data:

```
y = zeros(1, U*length(x));
y(1:U:end) = x;
```

In Chapters 6 and 13, we will explore what happens when we apply time expansion to signals as part of algorithms for upsampling (interpolation).

1.5.6 Operation on multiple sequences

Basic operations on sequences include addition, subtraction and multiplication. If $x_1[n]$ and $x_2[n]$ are two sequences, their sum is $y_s[n]=x_1[n]+x_2[n]$, their difference is $y_d[n]=x_1[n]-x_2[n]$ and their product is $y_p[n]=x_1[n]x_2[n]$. In each case, the operation is performed point-by-point. For example,

$$\begin{aligned} &\vdots \\ y_s[-1] &= x_1[-1]+x_2[-1] \\ y_s[0] &= x_1[0]+x_2[0] \\ y_s[1] &= x_1[1]+x_2[1] \\ &\vdots \end{aligned}.$$

The top row of **Figure 1.10** shows examples of sequences $x_1[n]$ and $x_2[n]$, and the bottom row shows the result of the sum, difference and product operations.

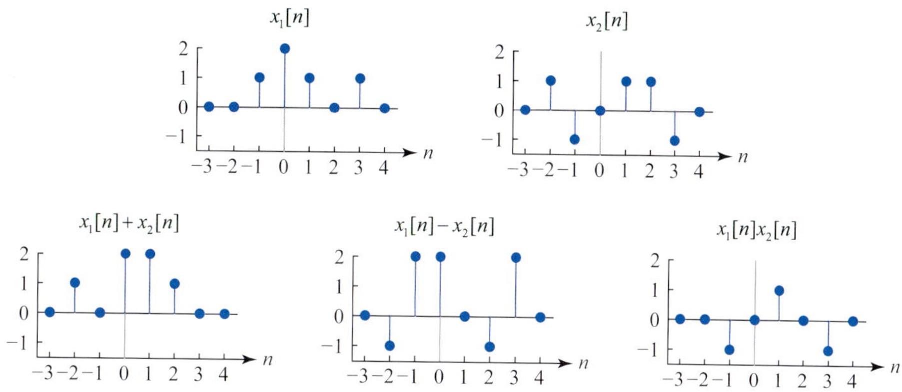

Figure 1.10 Addition, subtraction and multiplication of sequences

We can multiply a sequence by a scalar constant (e.g., $2x[n]$) or a function of n (e.g., $nx[n]$). In Matlab, arithmetic operations on arrays are **vectorized**, meaning that they are carried out point-by-point, which makes them trivial, assuming the sequences are the same length,

```
ys = x1+x2;
yd = x1-x2;
yp = x1.*x2;
```

Point-by-point multiplication of two sequences is performed by the ".*" operator, not the "*" operator, which would indicate a matrix multiplication of two arrays.

1.6 Basic sequences

Now, we will discuss some important sequences that we will be using throughout this book: impulses, steps, sinusoids and exponentials, both real and complex.

1.6.1 Impulse

The most basic sequence is the unit sample or discrete-time **impulse**, which is denoted by the symbol $\delta[n]$, and simply defined as

$$\delta[n] \triangleq \begin{cases} 1, & n=0 \\ 0, & n \neq 0 \end{cases}. \tag{1.2}$$

A plot of the impulse function is shown in **Figure 1.11**. The ellipses (i.e., "...") on either side of the plot are to remind you that $\delta[n]$ is defined to be zero for all $n \neq 0$. It is just a lot of zeros for $n \neq 0$.

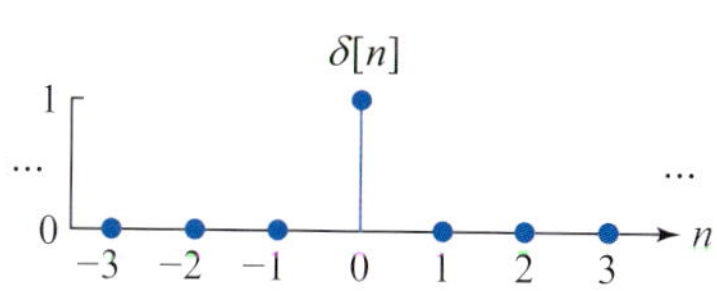

Figure 1.11 Discrete-time impulse function

It is valuable to contrast the discrete-time impulse $\delta[n]$ with the continuous-time impulse $\delta(t)$, which you may remember from your study of analog signal processing as an exceedingly unpleasant function[2] of zero width that goes to ∞ at $t=0$. In contrast, the discrete-time impulse is a simple function that is explicitly defined by Equation (1.2) for all n. A shifted discrete-time impulse is just $\delta[n-k]$, where k is an integer number. The discrete-time impulse is symmetric, that is $\delta[-n]=\delta[n]$, hence, $\delta[n-k]=\delta[-(k-n)]=\delta[k-n]$.

Use of discrete-time impulses in the synthesis and analysis of signals An important use of the impulse is in the synthesis and analysis of arbitrary sequences. Let us first talk about synthesis. A scaled and shifted impulse function $A\delta[n-n_0]$ is non-zero only at $n=n_0$, where it has value A. Using this fact, we can assemble or **synthesize** an arbitrary sequence from the sum

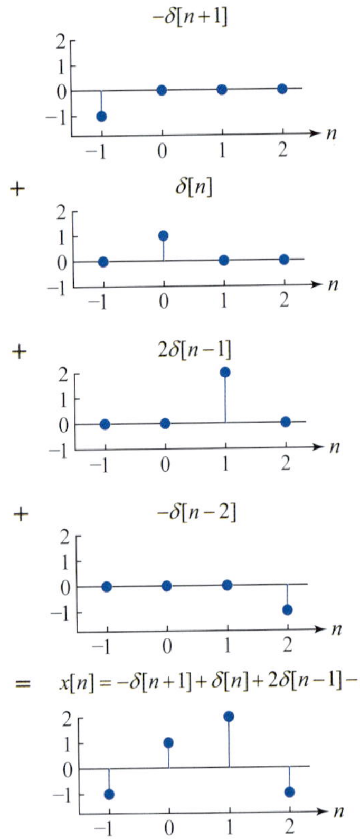

Figure 1.12 Construction of arbitrary sequences using impulse functions

[2] The continuous-time impulse is not a function, but rather a distribution that is defined implicitly by its behavior inside an integral, for example,

$$\delta(t)=0, \quad t\neq 0$$

$$\int_{-\infty}^{\infty}\delta(\tau)d\tau=1.$$

There is an infinite number of distributions that satisfy these conditions.

of scaled and shifted impulse functions. For example, consider an arbitrary sequence $x[n]$ that comprises just four values,

$$x[n] = \begin{cases} -1, & n = -1 \\ 1, & n = 0 \\ 2, & n = 1 \\ -1, & n = 2 \end{cases}. \tag{1.3}$$

Figure 1.12 shows the construction of $x[n]$ as the sum of four scaled and shifted impulses,

$$x[n] = -1 \cdot \delta[n+1] + 1 \cdot \delta[n] + 2 \cdot \delta[n-1] - 1 \cdot \delta[n-2]. \tag{1.4}$$

We can rewrite Equation (1.4) as

$$x[n] = x[-1]\delta[n+1] + x[0]\delta[n] + x[1]\delta[n-1] + x[2]\delta[n-2],$$

where $x[-1]$, $x[0]$, $x[1]$ and $x[2]$ are constants, as specified in Equation (1.3). Now, we generalize the preceding example by stating that any sequence $x[n]$ can be expressed as the sum of scaled and shifted sequences:

$$x[n] = \sum_{k=-\infty}^{\infty} x[k]\delta[n-k].$$

Pay attention to the indices n and k in this expression. For each fixed value of index k in the summation, $x[k]$ is a *value*, a constant scale factor, and $\delta[n-k]$ is a *sequence* on index n; that is, an impulse shifted by amount k. When all the scaled and shifted impulse sequences are summed together, they form $x[n]$:

$$\text{Synthesis:} \quad \underbrace{x[n]}_{\substack{\text{sequence} \\ \text{on } n}} = \sum_{k=-\infty}^{\infty} \underbrace{x[k]}_{\text{value}} \underbrace{\delta[n-k]}_{\substack{\text{sequence on } n \\ \text{shifted by } k}}. \tag{1.5}$$

Equation (1.5) states that an arbitrary time sequence $x[n]$ can be synthesized from the summation of impulse sequences $\delta[n-k]$, each of which is determined by a single parameter, the shift k.

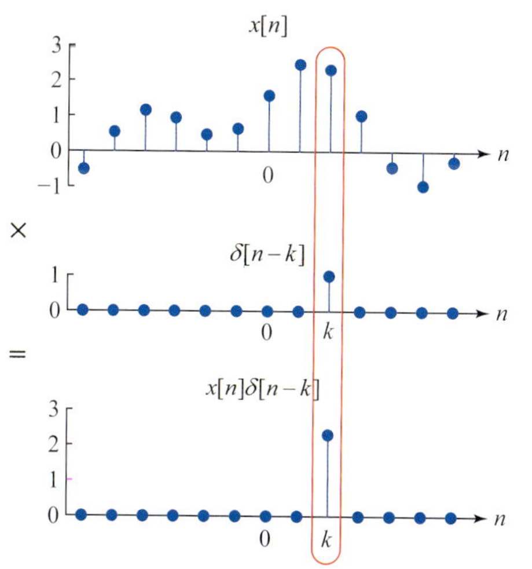

Figure 1.13 Product of a sequence and an impulse

A comparable approach can be used to **analyze** a sequence. **Figure 1.13** shows what happens if an arbitrary sequence $x[n]$ is multiplied by a shifted impulse $\delta[n-k]$.

The product is just a shifted impulse scaled by the value of $x[n]$ at the time $n=k$; that is, $x[n]\delta[n-k]=x[k]\delta[n-k]$. Again, we have made the distinction between $x[n]$, which is a *sequence* defined for all values of n, and $x[k]$, which is a *value* of the sequence at a single point in time, k. If the product $x[n]\delta[n-k]$ is summed, the result is just a single value:

$$\sum_{n=-\infty}^{\infty} x[n]\delta[n-k]=\sum_{n=-\infty}^{\infty} x[k]\delta[n-k]=x[k]\sum_{n=-\infty}^{\infty}\delta[n-k]=x[k],$$

since

$$\sum_{n=-\infty}^{\infty}\delta[n-k]=1.$$

This treatment shows how the shifted impulse can be used to analyze an arbitrary sequence in order to determine the value of the sequence at one time point:

$$\text{Analysis:}\quad \underbrace{x[k]}_{\text{value}}=\sum_{n=-\infty}^{\infty}\underbrace{x[n]}_{\substack{\text{sequence}\\ \text{on } n}}\ \underbrace{\delta[n-k]}_{\substack{\text{sequence on } n\\ \text{shifted by } k}}\ . \tag{1.6}$$

If you look closely, you will see that Equations (1.5) and (1.6) are actually the *same* equation, just with different interpretations. For example, take Equation (1.6) and recognize that the impulse is a symmetrical function so that $\delta[n-k]=\delta[k-n]$. Therefore,

$$x[k]=\sum_{n=-\infty}^{\infty} x[n]\delta[n-k]=\sum_{n=-\infty}^{\infty} x[n]\delta[k-n]. \tag{1.7}$$

In our prior discussions, we said that $x[n]$ is a sequence on time variable n, and $x[k]$ is a value of the sequence at one particular time point $n=k$. However, there is nothing sacred about our choice of variables n and k. We could just as well have said that $x[k]$ is a sequence on time variable k, and $x[n]$ is a value of the sequence at one particular time point $k=n$. Hence, we could rewrite Equation (1.7) as

$$\text{Analysis:}\quad \underbrace{x[n]}_{\text{value}}=\sum_{k=-\infty}^{\infty}\underbrace{x[k]}_{\substack{\text{sequence}\\ \text{on } k}}\ \underbrace{\delta[n-k]}_{\substack{\text{flipped sequence}\\ \text{on } k,\text{ shifted by } n}}\ ,$$

which is identical to Equation (1.5). Only the interpretation is different.

All this might seem like a long, exhausting trip to a destination we do not care about, but soon (Chapter 2) we will see why it is useful to express any sequence in terms of the sum of scaled and shifted impulses. It turns out that computing the response of certain discrete-time systems – **linear time-invariant (LTI) systems** – to scaled and shifted impulses is very easy. We will also revisit the idea of expressing a sequence as the sum of a family of orthogonal sequences in Chapter 3. There, we will discover that powerful analytic tools result when we express a sequence as the sum of another class of orthogonal sequences: the complex exponential sequence, $e^{j\omega n}$.

1.6.2 Unit step

The unit step is denoted by the symbol $u[n]$, and simply defined as

$$u[n] \triangleq \begin{cases} 1, & n \geq 0 \\ 0, & n < 0 \end{cases}.$$

A plot of the step function is shown in **Figure 1.14**. The ellipsis on the right side of the plot is to remind you that $u[n]$ is defined to have value one for all $n > 0$.

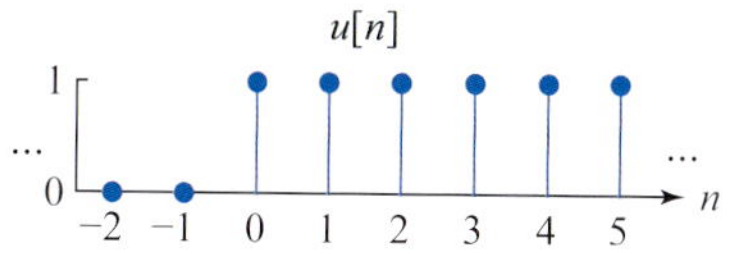

Figure 1.14 Unit step sequence

It is valuable to contrast the discrete-time step $u[n]$ with the continuous-time step $u(t)$ that you will remember from your study of analog signal processing. The continuous-time step is defined as

$$u(t) = \begin{cases} 0, & t < 0 \\ 1, & t > 0 \end{cases}. \tag{1.8}$$

The value of the continuous-time step at $t = 0$ is undetermined by Equation (1.8). In contrast, the value of the discrete-time step at $n = 0$ is completely determined. It is just $u[0] = 1$.

To implement a unit step in Matlab, we can create a simple function:

```
function x = u(n)
    x = double(n>=0);
end
```

Using `double` casts `(n>=0)`, which is a logical value, to a number. You can also implement this as a one-line **anonymous function** that you can put anywhere in your code:

```
u = @(n) double(n>=0);
```

In either event, here is an example of the result:

```
>> u(-2:2)
ans =
     0     0     1     1     1
```

Relation between the step and the impulse The discrete-time unit step can be defined as the sum of discrete-time impulses:

$$u[n] = \delta[n] + \delta[n-1] + \delta[n-2] + \delta[n-3] + \cdots = \sum_{k=0}^{\infty} \delta[n-k].$$

In a similar manner, an impulse can be derived from the difference between a step and a shifted step,

$$\delta[n] = u[n] - u[n-1],$$

as shown in **Figure 1.15**.

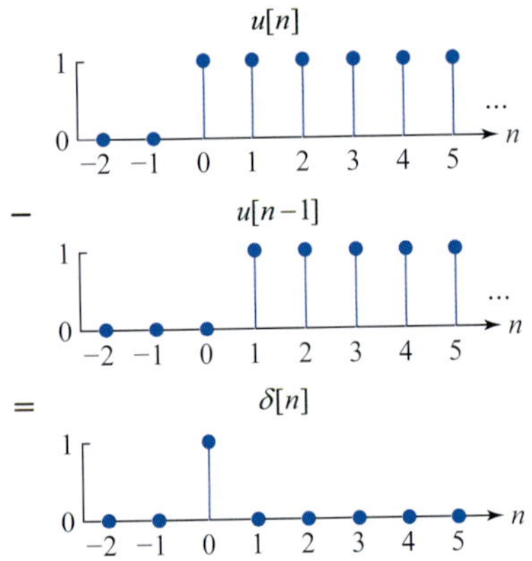

Figure 1.15 Derivation of impulses from steps

Use of the step as a switch One use of the step is as a switch to "turn on" or "turn off" a sequence. For example, given a sequence $x[n]$, we can define a new sequence $y[n]$:

$$y[n] = x[n]u[n] = \begin{cases} x[n], & n \geq 0 \\ 0, & n < 0 \end{cases}.$$

An example is shown in **Figure 1.16**.

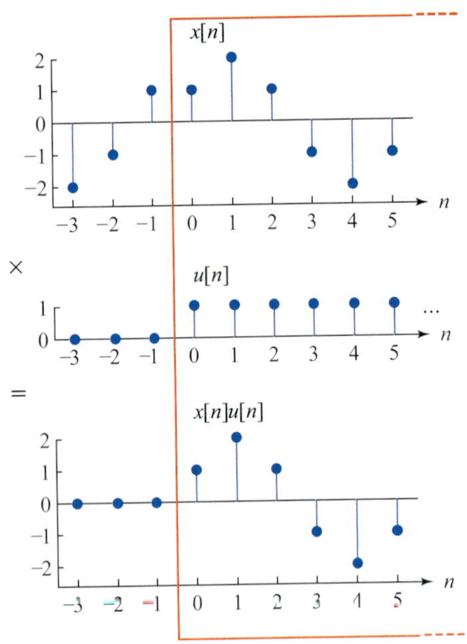

Figure 1.16 Use of step as a switch

Multiplying $x[n]$ by $u[n]$ "turns on" $x[n]$ for $n \geq 0$. Multiplying $x[n]$ by $u[-n]$ "turns off" $x[n]$ for $n > 0$. In Matlab, an example would be the following:

```
>> x = -2:2;
>> x .* u(x)
ans =
        0    0    0    1    2
```

1.6.3 Pulse

We can form a pulse of width N by subtracting two steps that are shifted with respect to each other by N. Define this pulse as $p[n]$:

$$p[n] = u[n] - u[n-N] = \begin{cases} 1, & 0 \le n < N \\ 0, & \text{otherwise} \end{cases}.$$

Figure 1.17 shows an example of a step of width $N = 4$.

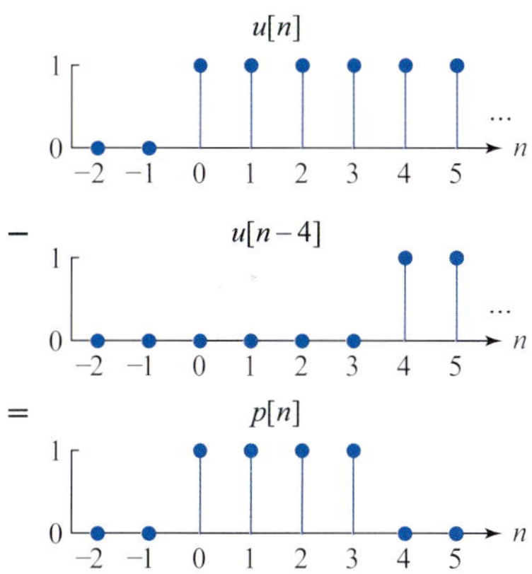

Figure 1.17 Derivation of pulse from steps

We can use this pulse to "select" a portion of a sequence by multiplying a sequence with the pulse. For example, multiplying the sequence $x[n]$ of **Figure 1.16** by the pulse $p[n]$ of **Figure 1.17**, we get the result shown in **Figure 1.18**.

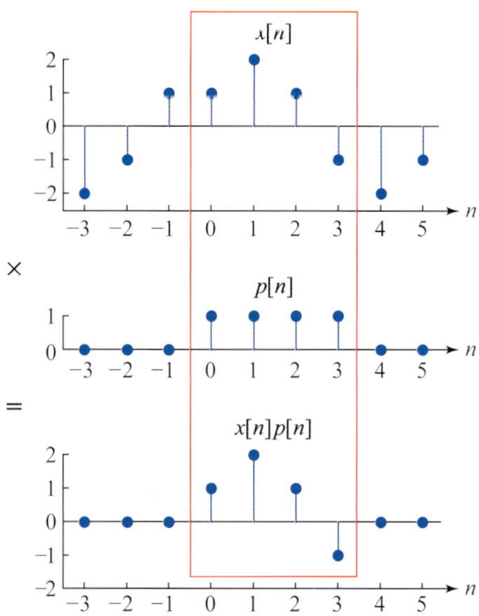

Figure 1.18 Selection of a portion of a sequence with a pulse

1.6.4 Power-law sequences

Power-law sequences will turn out to play a central role in our study of DSP. A real-valued power-law sequence is defined as

$$x[n] = A\alpha^n, \tag{1.9}$$

where both A and α are real. A general real-valued exponential sequence $x[n] = Ae^{\beta n}$ is just a special case of the power-law sequence, where $\alpha = e^{\beta}$.

The behavior of the real-valued power-law sequence depends on the value of α. There are a few cases to be considered, shown in **Table 1.4**.

Table 1.4 Behavior of power-law sequences

Case	α	Sequence characteristic
I	$1<\alpha<\infty$	Monotonically increasing as $n\to\infty$.
II	$\alpha=1$	Constant and has value 1.
III	$0<\alpha<1.$	Monotonically decreasing as $n\to\infty$.
IV	$-1<\alpha<0.$	Alternating with magnitude decreasing as $n\to\infty$.
V	$\alpha=-1$	Alternating with values ±1.
VI	$-\infty<\alpha<-1$	Alternating with magnitude increasing for $n\to\infty$.

Examples of each of these cases are shown in **Figure 1.19**. Running clockwise from the top left panel to the bottom left panel, the power-law sequence α^n is plotted for $\alpha = 2$, 1, 1/2, −1/2, −1 and −2, which represent cases I–VI, respectively.

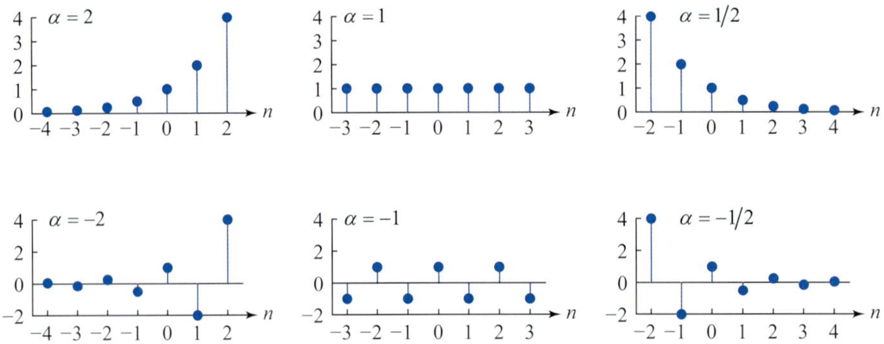

Figure 1.19 Power-law sequences with different values of α

1.6.5 Sinusoidal sequences

Discrete-time sinusoidal sequences deserve a bit of attention. They have properties that are very different from continuous-time sinusoids, and much of the intuition we may have developed for understanding the way that continuous-time sinusoids behave does not serve us well in understanding discrete-time sinusoids. The areas in which discrete-time and continuous-time sinusoids differ are (1) what is meant by "frequency," (2) under what circumstances sinusoids are periodic and (3) when sinusoids of different frequencies are unique. We will now address each of these three points in turn.

Discrete-time frequency is different from continuous-time frequency Consider a continuous-time signal $\cos \Omega_0 t$. Since the argument of the cosine must be in radians, $\Omega_0 t$ must have units of radians, and therefore the continuous-time frequency Ω_0 has units of *radians/sec*; that is, it is an angular frequency.

Now consider a discrete-time sequence $\cos \omega_0 n$. Again, the argument of the cosine must be radians, so $\omega_0 n$ must be in radians. Thus, the discrete-time frequency ω_0 must have units of radians/sample, which is equal to *radians*, since "sample" is a number and therefore dimensionless. That is, discrete-time frequency is actually equivalent to *phase*.

In this book, we will always use capital omega Ω to denote continuous-time frequency and small omega ω to denote discrete-time frequency.

Not all discrete-time sinusoidal sequences are periodic A continuous-time function $x(t)$ is said to be *periodic* if there exists some time T such that $x(t) = x(t+T)$ for all t. The period is defined as the smallest value of T for which this is true. So, a continuous-time sinusoid of frequency Ω_0 will be periodic if we can find a value of T such that $\cos \Omega_0 t = \cos \Omega_0(t+T)$. This requires that

$$\cos \Omega_0(t+T) = \cos(\Omega_0 t + \Omega_0 T) = \cos(\Omega_0 t + 2\pi k) = \cos \Omega_0 t,$$

where k is an integer. Thus, $\cos \Omega_0 t$ is periodic if we can find values of T and k such that $\Omega_0 T = 2\pi k$. Since both Ω_0 and T are real, we will *always* be able to do this. Since the period is defined as the smallest value of T at which this is true, let $k = 1$, which gives the period as

$$T = 2\pi/\Omega_0.$$

To summarize, $\cos \Omega_0 t$ is periodic for *every value* of frequency Ω_0.

In the discrete-time world, *not every sinusoidal sequence is periodic*. A discrete-time sequence will be periodic if there exists an integer N such that

$$x[n] = x[n+N] \tag{1.10}$$

for all n. The period is taken to be the smallest value of N for which this is true. For a discrete-time sinusoid of frequency ω_0 to be periodic, we require that $\cos \omega_0 n = \cos \omega_0(n+N)$. This means that

$$\cos \omega_0(n+N) = \cos(\omega_0 n + \omega_0 N) = \cos(\omega_0 n + 2\pi k) = \cos \omega_0 n,$$

which in turn requires that $\omega_0 N = 2\pi k$ for some integers N and k. This means that N must be

$$N = 2\pi k/\omega_0.$$

However, since N must be an integer, $\cos \omega_0 n$ will be periodic only if there exists an integer k such that $2\pi k/\omega_0$ is an integer. Put another way, $\cos \omega_0 n$ will be periodic if ω_0 is a rational multiple of 2π that can be expressed in the form

$$\omega_0 = 2\pi k/N. \tag{1.11}$$

This is clearly not true for every discrete-time frequency ω_0. Let us evaluate the periodicity of some discrete-time sequences.

Example 1.1

Determine if $\cos \omega_0 n$ is periodic for each of the following values of ω_0:

(a) $\omega_0 = 2$.

(b) $\omega_0 = \pi$.

(c) $\omega_0 = 3\pi/4$.

➤ **Solution:**

(a) $\omega_0 = 2$: This sequence will be periodic when

$$N = 2\pi k/\omega_0 = \cancel{2}\pi k/\cancel{2} = \pi k.$$

Since N is not an integer for *any* choice of k, $\cos 2n$ is not periodic. This example is shown in **Figure 1.20a**. The plot shows the sequence $\cos 2n$. Also, superimposed on the plot is a dashed line that corresponds to the continuous-time sinusoid $\cos 2t$, plotted as if the abscissa were continuous. This continuous-time sinusoid is obviously periodic with a period of π.

(b) $\omega_0 = \pi$: This sequence will be periodic when

$$N = 2\cancel{\pi}k/\cancel{\pi} = 2k.$$

Since N is an integer for *every* choice of k, $\cos \pi n$ is periodic. The period is given as the smallest value of $N = 2k$, so we set $k = 1$, which gives a period of $N = 2$. This example is shown in **Figure 1.20b**. To emphasize the periodicity of this sequence, subsequent periods are plotted in alternating colors. Also, superimposed on the plot is the continuous-time sinusoid $\cos \pi t$, which is periodic with a period of 2 sec.

(c) $\omega_0 = 3\pi/4$: This sequence will be periodic when

$$N = 2\cancel{\pi}k/(3\cancel{\pi}/4) = 8k/3.$$

N will be an integer for $k = 3, 6, \ldots$ So $\cos 3\pi n/4$ is periodic. The period is smallest when $k = 3$, which gives the period $N = 8$. This example is shown in **Figure 1.20c**. Superimposed on the plot is the continuous-time sinusoid $\cos 3\pi t/4$. This sinusoid is periodic with period 8/3, whereas the discrete-time sinusoid, whose period must be an integer, is periodic with period $3 \cdot 8/3 = 8$. Two periods of the discrete-time sinusoid are plotted in different colors.

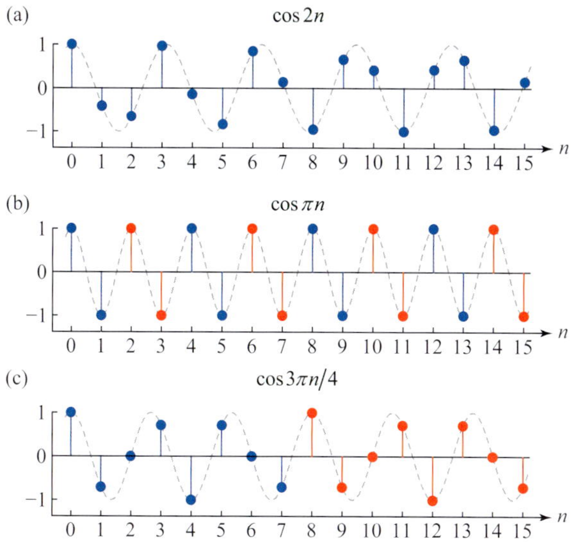

Figure 1.20 Periodicity of discrete-time sinusoidal sequences at different frequencies

Not all discrete-time frequencies are unique In the continuous-time world, sinusoids of different frequencies are unique. That is, given two frequencies Ω_1 and Ω_2, if $\Omega_1 \neq \Omega_2$, then $\cos \Omega_1 t \neq \cos \Omega_2 t$ for all t. In the discrete-time world, sinusoidal sequences of different frequencies are not necessarily distinct. Consider two "different" discrete-time sinusoidal sequences $x_1[n] = \cos \omega_1 n$ and $x_2[n] = \cos \omega_2 n$, where $\omega_1 \neq \omega_2$. We might expect $x_1[n] \neq x_2[n]$ over all n, but that is not always true. Specifically, let $\omega_2 = \omega_1 + 2\pi k$, where k is an integer. As long as $k \neq 0$, then $\omega_1 \neq \omega_2$, but

$$x_2[n] = \cos \omega_2 n = \cos(\omega_1 + 2\pi k) = \cos \omega_1 n = x_1[n].$$

Thus, $x_1[n] = x_2[n]$ for all n. This says that discrete-time sinusoidal sequences whose frequencies ω_1 and ω_2 differ by an integer multiple of 2π are indistinguishable. Of course, that makes sense when you remember that discrete-time frequency is actually phase. **Figure 1.21** shows the equivalence of three discrete-time sinusoidal sequences whose frequencies differ by multiples of 2π.

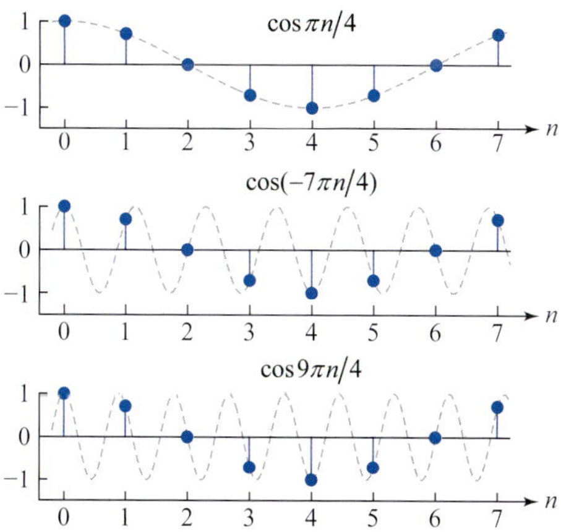

Figure 1.21 Equivalence of discrete-time sinusoidal sequences at selected frequencies

The top panel shows one period of the discrete-time periodic sequence $\cos \pi n/4$. The middle panel shows one period of $\cos(-7\pi n/4)$. This is clearly indistinguishable from $\cos \pi n/4$, which makes sense since

$$\cos(-7\pi n/4) = \cos(-7\pi n/4 + 2\pi) = \cos \pi n/4.$$

The bottom panel shows one period of $\cos 9\pi n/4$, which is also indistinguishable from $\cos \pi n/4$, since

$$\cos 9\pi n/4 = \cos(9\pi n/4 - 2\pi) = \cos \pi/4.$$

Superimposed on each plot is a dotted line corresponding to the continuous-time sinusoids: $\cos \pi t/4$, $\cos(-7\pi t/4)$ and $\cos 9\pi t/4$. These continuous-time sinusoids are obviously different.

The fact that discrete-time sinusoids at frequencies that differ by a factor of 2π are indistinguishable is equivalent to stating that discrete-time frequencies are unique within any contiguous 2π range. Accordingly, in this book we will adopt the convention of expressing all discrete-time frequencies in the range $[-\pi, \pi)$ or $[0, 2\pi)$. A moment's thought will show you why a 2π range is sufficient. If you have a sinusoidal sequence with frequency ω outside this range, you can always find an integer value k such that $-\pi \le \omega - 2\pi k < \pi$. **Figure 1.22** shows this graphically for the examples of **Figure 1.21**. We can view the discrete-time frequency domain as contiguous blocks of range 2π that are centered on multiples of 2π. Any sinusoid with a frequency ω outside of the central range $-\pi \le \omega < \pi$ can be mapped into the central range by addition or subtraction of the appropriate multiple of 2π.

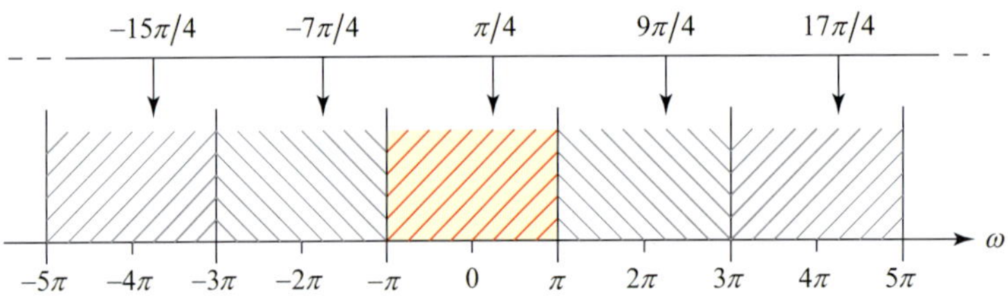

Figure 1.22 Equivalence of discrete-time frequencies

1.6.6 Complex exponential sequences

The complex exponential sequence is a very important case of a power-law sequence. These sequences are going to form the basis of our discussion of the discrete-time Fourier transform in Chapter 4. A general complex exponential sequence is defined by Equation (1.9) where both A and α are complex:

$$\begin{aligned} A &= |A|e^{j\varphi} \\ \alpha &= |\alpha|e^{j\omega_0} \end{aligned} . \tag{1.12}$$

Substituting Equation (1.12) into Equation (1.9), we get

$$\begin{aligned} x[n] &= A\alpha^n = |A|e^{j\varphi}(|\alpha|e^{j\omega_0})^n = |A||\alpha|^n e^{j(\omega_0 n+\varphi)} \\ &= \underbrace{|A||\alpha|^n \cos(\omega_0 n + \varphi)}_{\mathrm{Re}\{x[n]\}} + \underbrace{j|A||\alpha|^n \sin(\omega_0 n + \varphi)}_{\mathrm{Im}\{x[n]\}}. \end{aligned}$$

The complex exponential sequence comprises both a real part $\mathrm{Re}\{x[n]\}$ and an imaginary part $\mathrm{Im}\{x[n]\}$. Both $\mathrm{Re}\{x[n]\}$ and $\mathrm{Im}\{x[n]\}$ are made up of a sinusoidal sequence (sine or cosine) multiplied by a power-law sequence. The sinusoidal sequences are oscillatory with frequency ω_0 and phase φ. These sinusoidal sequences are multiplied by a real power-law sequence $|A||\alpha|^n$ that determines the magnitude of $x[n]$. In fact, this power-law sequence *is* the magnitude of $x[n]$, since

$$|x[n]| = |A\alpha^n| = |A||\alpha|^n.$$

The magnitude of $x[n]$ is monotonically increasing if $|\alpha| > 1$, constant if $|\alpha| = 1$, and monotonically decreasing if $|\alpha| < 1$.

Now, consider the special case of the complex exponential with value of $A = 1$ and $\alpha = 1$. Then,

$$x[n] = e^{j\omega_0 n} = \mathrm{Re}\{e^{j\omega_0 n}\} + j\,\mathrm{Im}\{e^{j\omega_0 n}\} = \cos\omega_0 n + j\sin\omega_0 n.$$

This sequence has a magnitude of one and a phase of $\omega_0 n$. The real and imaginary parts of $x[n]$ are sinusoids of the same frequency ω_0. Specifically, the real part is $\cos\omega_0 n$ and the imaginary part is $\sin\omega_0 n$.

As long as ω_0 is a rational multiple of 2π, each of these sinusoidal sequences will be periodic with the same period $N = 2\pi k/\omega_0$, and therefore $x[n]$ will be periodic with period N. We can directly determine the condition for periodicity of $e^{j\omega_0 n}$ by recalling from Equation (1.10) that a sequence is periodic if there exists an integer value of N such that $x[n+N] = x[n]$. This means that $e^{j\omega_0 n}$ is periodic if

$$e^{j\omega_0(n+N)} = e^{j\omega_0 n}e^{j\omega_0 N} = e^{j\omega_0 n}e^{j(2\pi/N)N} = e^{j\omega_0 n},$$

which occurs when $e^{j\omega_0 N} = 1$. This happens when $\omega_0 N = 2\pi k$, which implies that

$$N = 2\pi k/\omega_0.$$

Not surprisingly, this is exactly the same condition for the periodicity of sinusoidal sequences determined in Equation (1.11).

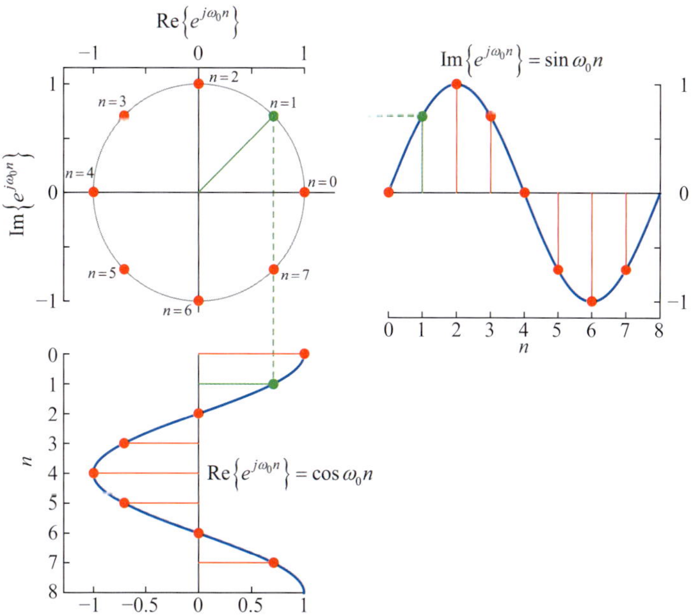

Figure 1.23 Relation between complex exponential and sinusoidal sequences

Figure 1.23 shows the relation between the complex exponential and its real and imaginary parts. The top left panel plots $x[n] = e^{j\omega_0 n}$ as a vector (line) in the complex plane. For this example we have chosen $N = 8$, which implies that $\omega_0 = 2\pi/N = \pi/4$. Each of the eight colored dots corresponds to one of the N distinct values of $x[n]$, $0 \le n \le N-1$. Since $|x[n]| = |e^{j\omega_0 n}| = 1$, all the points lie on a circle of radius one. The nth point, corresponding to $x[n]$, lies at an angle of $n\pi/4$. The top right panel shows $\text{Im}\{e^{j\omega_0 n}\} = \sin\omega_0 n$ and the bottom left panel shows $\text{Re}\{e^{j\omega_0 n}\} = \cos\omega_0 n$. The orientation of the plots makes clear that $\cos\omega_0 n$ is the projection of the vector $x[n] = e^{j\omega n}$ onto the real axis, and $\sin\omega_0 n$ is the projection of the vector $x[n] = e^{j\omega_0 n}$ onto the imaginary axis. The point on each plot corresponding to $n = 1$ is plotted in green.

1.6.7 Sequence classification

In this section, we will investigate the classification of sequences. We will show that sequences can be classified as **energy sequences** or **power sequences** and can be decomposed into real and imaginary parts. Sequences can also be decomposed into even or odd parts. The utility of this kind of decomposition will become very useful in future chapters. Specifically, we will find that powerful algorithms, such as the fast Fourier transform (the FFT) that we will study in Chapter 11, make use of these decompositions.

Energy and power signals For an analog circuit such as that shown in **Figure 1.24**, with $R = 1$, the **instantaneous power** consumed by the circuit at time t is defined as

$$p(t) = v(t)i(t)^* = Ri(t)i(t)^* = |i(t)|^2,$$

or, equivalently,

$$p(t) = v(t)i(t)^* = v(t)\left(\frac{v(t)}{R}\right)^* = |v(t)|^2.$$

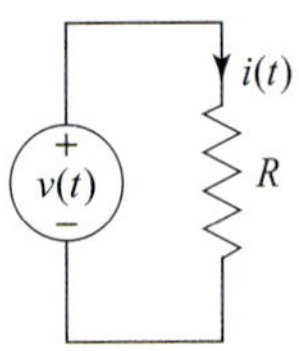

Figure 1.24 Power of a continuous-time signal

Hence, in this analog circuit, power is proportional to the square of the voltage or current in the circuit. The **energy** is the integral of the power over all time,

$$E = \int_{-\infty}^{\infty} p(t)dt = \int_{-\infty}^{\infty} |i(t)^2|dt = \int_{-\infty}^{\infty} |v(t)|^2 dt.$$

Similarly, consider a discrete-time signal $x[n]$, which is possibly complex. By analogy with the analog case, the instantaneous power at time n is defined as

$$x[n]x^*[n] = |x[n]|^2.$$

For a real signal, this simplifies to $|x[n]|^2 = x^2[n]$. The **energy** of this signal is defined as the sum of the power over all time,

$$E_x \triangleq \sum_{n=-\infty}^{\infty} x[n]x^*[n] = \sum_{n=-\infty}^{\infty} |x[n]|^2 = \sum_{n=-\infty}^{\infty} x^2[n].$$

The **power** of this signal over a time period of length $2N+1$ is

$$P_x \triangleq \frac{1}{2N+1} \sum_{n=-N}^{N} |x[n]|^2,$$

and the **average power** is found by sending $N \to \infty$:

$$\bar{P}_x \triangleq \lim_{N\to\infty} \frac{1}{2N+1} \sum_{n=-N}^{N} |x[n]|^2.$$

For a signal that is periodic with period N it suffices to define the power as the average of the energy over one period,

$$\bar{P}_x \triangleq \frac{1}{N} \sum_{n=0}^{N-1} |x[n]|^2.$$

If E_x is finite, then $x[n]$ is termed an **energy signal**, and $\bar{P}_x = 0$. In contrast, if $\bar{P}_x$ is finite (and non-zero), $x[n]$ is termed a **power signal**, and E_x is infinite. A signal can be an energy signal or a power signal, or neither, but not both. Any signal of finite amplitude and length is an energy signal. However, signals of infinite duration can be either power or energy signals, as shown in the following example.

Example 1.2

Determine the energy and average power of each of the following signals:
(a) $x[n] = \cos \pi n/4$.
(b) $x[n] = \frac{1}{2}^n u[n]$.
(c) $x[n] = 2^n$.

➤ Solution:

(a) This is a periodic signal, with period $N = 8$. The average power is

$$\bar{P}_x = \frac{1}{N}\sum_{n=0}^{N-1} \cos^2(\pi n/4) = \frac{1}{N}\sum_{n=0}^{N-1} \left(\tfrac{1}{2} + \tfrac{1}{2}\cos \pi n/2\right) = \frac{1}{N}\sum_{n=0}^{N-1} \tfrac{1}{2} + \frac{1}{2N}\cancel{\left(\sum_{n=0}^{N-1} \cos \pi n/2\right)} = \frac{1}{2}.$$

Since $\bar{P}_x$ is finite, this is a power signal.

(b) Even though this signal has infinite duration, it is absolutely summable. The energy is

$$E_x = \sum_{n=-\infty}^{\infty} |x[n]|^2 = \sum_{n=-\infty}^{\infty} \left(\tfrac{1}{2}^n u[n]\right)^2 = \sum_{n=0}^{\infty} \tfrac{1}{4}^n = \frac{1}{1-\frac{1}{4}} = \frac{4}{3}.$$

Since E_x is finite, this is an energy signal.

(c) Because this signal blows up as $n \to \infty$, both the energy and average power are infinite. Hence, this is neither an energy signal nor a power signal.

Real and imaginary sequences As we saw in Section 1.6.6, a sequence can be complex valued. Hence, we can express a general sequence as the sum of real and imaginary parts,

$$\begin{aligned} x[n] &= \mathrm{Re}\{x[n]\} + j\,\mathrm{Im}\{x[n]\} \\ &= x_r[n] + jx_i[n], \end{aligned}$$

where we have defined $x_r[n] \triangleq \mathrm{Re}\{x[n]\}$ as the real part of $x[n]$ and $x_i[n] \triangleq \mathrm{Im}\{x[n]\}$ as the imaginary part of $x[n]$. Remember, both $x_r[n]$ and $x_i[n]$ are purely *real* sequences. The sequence $jx_i[n]$ is purely imaginary.

The complex conjugate of a sequence has the same real part as $x[n]$ with a negated imaginary part:

$$\begin{aligned} x^*[n] &= \mathrm{Re}\{x[n]\} - j\mathrm{Im}\{x[n]\} \\ &= x_r[n] - jx_i[n]. \end{aligned}$$

If $x[n]$ is purely real, it is equal to its conjugate, $x[n] = x^*[n]$, since its imaginary part is zero. Given a sequence $x[n]$ and its conjugate $x^*[n]$, the real part can be computed as

$$\mathrm{Re}\{x[n]\} = \tfrac{1}{2}(x[n] + x^*[n]).$$

If $x[n]$ is purely imaginary, it is equal to the negative of its conjugate, $x[n] = -x^*[n]$, since its real part is zero. The imaginary part can be computed from the sequence and its complex conjugate as

$$\mathrm{Im}\{x[n]\} = \tfrac{1}{2j}(x[n] - x^*[n]).$$

Example 1.3

Given $x[n] = (2+j)\delta[n+1] + \delta[n] - 3j\delta[n-1]$.

(a) Find $\mathrm{Re}\{x[n]\}$ and $\mathrm{Im}\{x[n]\}$.

(b) For this example, verify that $\mathrm{Re}\{x[n]\} = (x[n] + x^*[n])/2$ and $\mathrm{Im}\{x[n]\} = (x[n] - x^*[n])/2j$.

➤ Solution:

(a)

$$\begin{aligned} \mathrm{Re}\{x[n]\} &= 2\delta[n+1] + \delta[n] \\ \mathrm{Im}\{x[n]\} &= \delta[n+1] - 3\delta[n-1]\cdot \end{aligned}$$

(b) $x^*[n] = (2-j)\delta[n+1] + \delta[n] + 3j\delta[n-1]$ so

$$\begin{aligned}\mathrm{Re}\{x[n]\} &= \tfrac{1}{2}(x[n] + x^*[n]) = \frac{1}{2}\left(+\begin{array}{c}(2+j)\delta[n+1] + \delta[n] - 3j\delta[n-1]\\ (2-j)\delta[n+1] + \delta[n] + 3j\delta[n-1]\end{array}\right)\\ &= \tfrac{1}{2}(4\delta[n+1] + 2\delta[n]) = 2\delta[n+1] + \delta[n]\end{aligned},$$

and

$$\begin{aligned}\mathrm{Im}\{x[n]\} &= \tfrac{1}{2j}(x[n] - x^*[n]) = \frac{1}{2j}\left(-\begin{array}{c}(2+j)\delta[n+1] + \delta[n] - 3j\delta[n-1]\\ (2-j)\delta[n+1] + \delta[n] + 3j\delta[n-1]\end{array}\right)\\ &= (2j\delta[n+1] - 6j\delta[n-1]) = \delta[n+1] - 3\delta[n-1]\end{aligned}.$$

Even and odd sequences An **even** or **symmetric** sequence $x_e[n]$ is defined by the fact that it is symmetric about $n=0$; that is,

$$x_e[n] = x_e[-n]. \tag{1.13a}$$

An **odd** or **antisymmetric** sequence $x_o[n]$ is defined by the fact that it is antisymmetric about $n=0$; that is,

$$x_o[n] = -x_o[-n]. \tag{1.13b}$$

We can express any sequence as the sum of even and odd parts:

$$x[n] = x_e[n] + x_o[n]. \tag{1.14}$$

Appropriate manipulations of Equations (1.13) and (1.14) allow us to find the even and odd parts of $x[n]$:

$$\begin{aligned}x_e[n] &= \tfrac{1}{2}(x[n] + x[-n])\\ x_o[n] &= \tfrac{1}{2}(x[n] - x[-n]).\end{aligned}$$

The notion of the symmetry of sequences will be particularly important in Chapter 7, where we introduce an entire class of filters – linear-phase, finite impulse response (FIR) filters – whose useful properties derive primarily from considerations of the symmetry of the filter's response to an impulse.

Example 1.4

Find and plot $a_e[n]$ and $a_o[n]$, the even and odd parts of the power-law sequence $a[n] = \alpha^n u[n]$, $|\alpha| < 1$.

➤ Solution:

$$\begin{aligned}a_e[n] &= \tfrac{1}{2}(\alpha^n u[n] + \alpha^{-n}u[-n]) = \delta[n] + \tfrac{1}{2}\alpha^{|n|}\\ a_o[n] &= \tfrac{1}{2}(\alpha^n u[n] - \alpha^{-n}u[-n]).\end{aligned}$$

The plots are shown below in **Figure 1.25**.

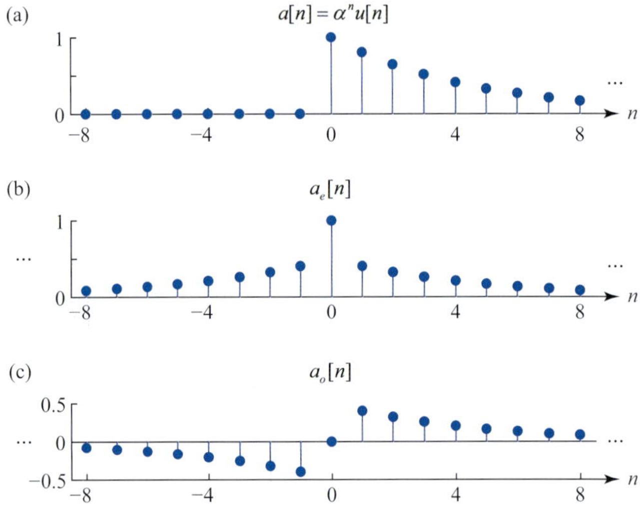

Figure 1.25 Even and odd parts of a power-law sequence

1.7 Systems

A system is generally defined as any group of elements or components that operates on one or more input signals to produce one or more output signals. The idea of systems and their analysis is pervasive not only in engineering and the physical sciences, but also in other disciplines. For example, we are familiar with the non-engineering ideas of a biological system, a social system and an economic system. Implicit in all definitions of systems is the idea that the **inputs** (e.g., force, voltage, body temperature, industrial capacity) can be measured and that the system acts on those inputs to produce **outputs** (e.g., displacement, current, blood pressure, GDP) in a manner that can be modeled and quantified. In digital signal processing, a discrete-time system is a formula, an algorithm, a mathematical operation or a piece of code that transforms an input sequence into an output sequence. We can schematize this relation as shown in **Figure 1.26**, where the notation $T\{\}$ denotes the transformation.

Figure 1.26 Schematic description of a transformation

The idea of transformation will be made clear with an extended series of examples.

1.7.1 Discrete-time scalar multiplier

Consider a discrete-time scalar multiplier defined by the relation

$$y[n] = 2x[n], \tag{1.15}$$

where $x[n]$ is the input sequence and $y[n]$ is the output sequence. **Figure 1.27** shows the relation between a sample input and output sequence for this system.

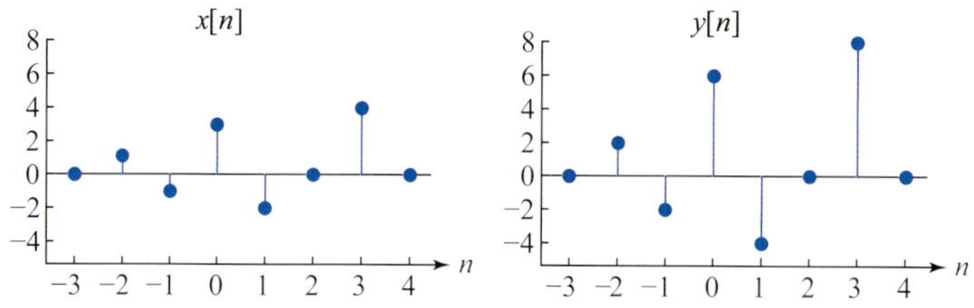

Figure 1.27 Input and output of discrete-time scalar multiplier

The system is basically a discrete-time "amplifier." You can view $x[n]$ and $y[n]$ either as entire *sequences* on the index n, or as a single value of sequences $x[n]$ and $y[n]$ at a specified value of n. If you view $x[n]$ and $y[n]$ as entire sequences, then Equation (1.15) says, "sequence $y[n]$ is twice sequence $x[n]$." This is the way that Matlab treats sequences; it performs vectorized operations that operate on the entire sequence; for example, the previous example would just be accomplished by writing

```
x = [1 -1 3 -2 0 4];
y = 2 * x;
```

If you view $x[n]$ and $y[n]$ as values, then Equation (1.15) says, "the value of the output sequence $y[n]$ at each time point n is just twice the value of the input $x[n]$ at that time point," so,

$$\begin{aligned} &\ldots \\ &y[0] = 2x[0] = 2 \cdot 3 = 6 \\ &y[1] = 2x[1] = 2 \cdot -2 = -4, \\ &\ldots \end{aligned}$$

and so on. This is the way you would program this discrete-time signal processing operation in a lower-level language like assembly (or C, though I guess C programmers may not like to think of themselves as programming in a "lower-level language"). Of course, you could use Matlab to do the same thing, using a `for` loop:

```
x = [1 -1 3 -2 0 4];
y = zeros(1, length(x));
for i = 1:length(x)
    y(i) = 2 * x(i);
end
```

The second line of this little program uses the Matlab function `zeros` to preallocate the output array `y`, as explained in Appendix C. While allocation is necessary in C, it is not strictly necessary in Matlab, though it is probably still good programming practice. This particular discrete-time system is an example of an **instantaneous transformation**; that is, a transformation for which the output value at each time point n depends only on the value of the input sequence at time n. It does not depend on future or past values of the input sequence. This is certainly not true in general for discrete-time systems, as we will shortly see.

1.7.2 Offset

Here is an example of a system that adds an offset of one to the input sequence to form the output:

$$y[n] = x[n] + 1.$$

Figure 1.28 shows the resulting picture, using the same input sequence as **Figure 1.27**:

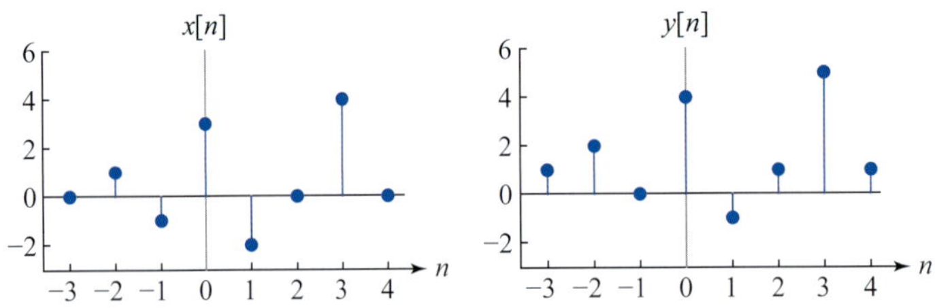

Figure 1.28 Input and output of an offset

As with the previous example, this is an instantaneous transformation: each point in the output sequence is just computed by adding one to the corresponding point in the input sequence. In Matlab, the following adds the constant to each element of array x:

```
y = x + 1;
```

1.7.3 Squarer

Here is the transformation for a squarer:

$$y[n] = x^2[n].$$

Figure 1.29 shows the picture of the input and output for this system.

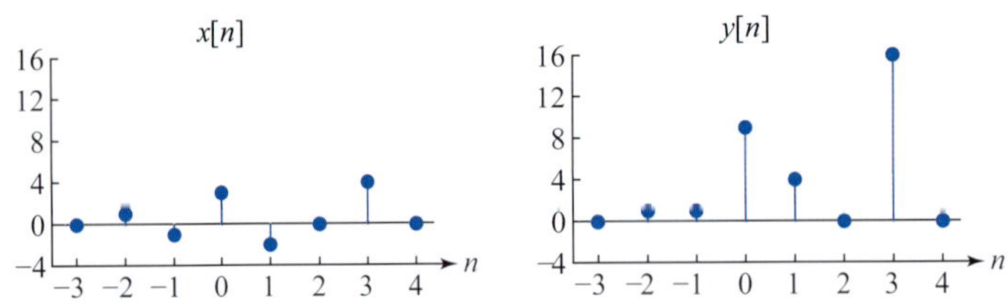

Figure 1.29 Input and output of a squarer

The squarer is another instantaneous transformation in which each output point is just the square of the corresponding input point. It is an example of the use of discrete-time systems to specify precisely a functional relation between an input and an output. This kind of functional specification is very easy to accomplish with high precision using discrete-time methods but more difficult and expensive to accomplish precisely with analog methods, which require skillful design as well as tight component tolerances.

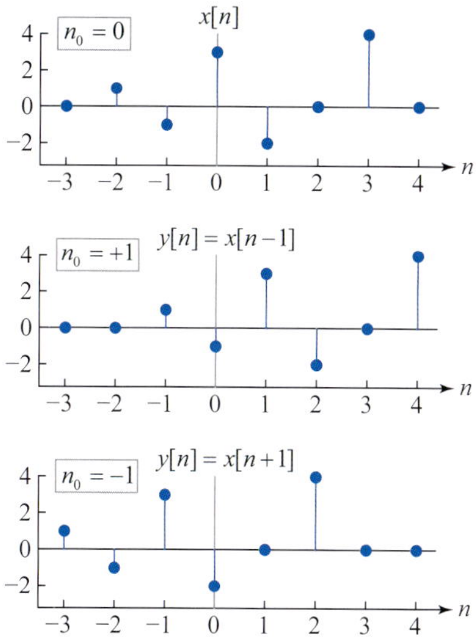

Figure 1.30 Input and output of a delay

In Matlab, you would use the "`.^`" operation to raise an entire array to a power at once,

```
y=x.^2;
```

1.7.4 Shift

As we discussed in Section 1.5.1, the shift is a very common and important signal processing operation, specified by the equation

$$y[n] = x[n - n_0],$$

where n_0 specifies an integer amount of delay that can be positive or negative (or trivially, zero) as shown in the example of **Figure 1.30**. In Matlab, implementing shifting is just a matter of changing the indices that address our data array. Just be careful not to "run out" of data at either end of the array. For example, to emulate the result of **Table 1.1** exactly, we would have to pad one end or the other with zeros when shifting:

```
y = [zeros(1, n0) x zeros(1, -n0)];
```

This is actually a tricky little piece of code. When $n_0 = 1$, we are using the Matlab `zeros` function to pad the front of the array `x` with one zero, and we are padding the back of the array with −1 zeros. What? Here, we are exploiting an undocumented quirk of the `zeros` function. When the argument to the function is negative, it pads with the empty array `[ ]`, which has no effect. You can see how the same trick works on the front of the array when n_0 is negative.

More generally, it is easy to implement a positive delay programmatically or in hardware. Consider the output given a positive delay of $n_0 = 3$. At time point n, $y[n] = x[n - n_0] = x[n - 3]$. To implement this system, we need to store the previous n_0 values of the input. For example, let $n = 0$. Here, we have three previous values of the input stored in memory: $x[-3]$, $x[-2]$ and $x[-1]$. Since $y[n] = x[n - 3]$, we set $y[0] = x[-3]$ and we are done with this time point. Once we have set $y[0]$, we no longer need $x[-3]$ in memory, since we will not use it again when we increment n by one. However, we are going to need $x[0]$, the current value of the input, in the future, so we remove $x[-3]$ from the memory and replace it with $x[0]$. Our memory now contains $x[-2]$, $x[-1]$ and $x[0]$. We increment the time index by one, so that $n = 1$, and are now ready to set the output $y[1] = x[-2]$. We then replace $x[-2]$ with $x[1]$ in our memory, increment n, and repeat. Here is what the general sequence of operations looks like at any value of n:

1:	$y[n] \leftarrow x[n-3]$	Output $y[n]$
2:	$x[n-3] \leftarrow x[n-2]$	Move memory element (or pointer)
3:	$x[n-2] \leftarrow x[n-1]$	Move memory element (or pointer)
4:	$x[n-1] \leftarrow x[n]$	Move current input value into memory element

Depending on the implementation, we may not have to physically move data from one memory element to another, but instead might accomplish the same task by moving pointers, something we will discuss in Section 9.9.

While it is easy to think about implementing a system with a positive delay in a programming language, how can we think about implementing a system with a negative delay, so that the current output is the value of the input at some *future* time? How can a system look into the future? One of the wonderful things about discrete-time systems is that, in many applications, one can store all or part of the input sequence and make the output at any time depend on past, present and/or future stored values. For example, assume you have a file of music which you want to process with a negative-delay transformation $y[n] = x[n + 3]$. At time point $n = 0$, we begin our signal processing operation. Well, $y[0] = x[3]$; that is, the value of the current output point is the value of the input three time points in the future. If we have the entire file, $x[3]$ becomes our first output point. The next output point is $y[1] = x[4]$, which is the next point in the file, and so on. The point here is that in the discrete-time world, it is possible to create systems that have both negative and positive delays.

1.7.5 Moving-window average

A moving-window average is a system in which the output at any time point depends on the input at several adjacent or surrounding time points. For example,

$$y[n] = \tfrac{1}{3}(x[n-1] + x[n] + x[n+1]). \qquad \textbf{(1.16)}$$

There are two ways to think about this relation. First, you can think of $x[n]$ and $y[n]$ as entire *sequences* with a dummy index n. Then, Equation (1.16) can be interpreted as follows: "Sequence $y[n]$ is obtained by averaging three sequences together: sequence $x[n]$, sequence $x[n - 1]$ (which is just sequence $x[n]$ delayed by +1), and sequence $x[n + 1]$ (which is just sequence $x[n]$ delayed by -1)." Alternately, you can think of $x[n]$ and $y[n]$ as *values* of the input and output sequences at a particular time point n. Now, Equation (1.16) can be interpreted as saying, "at a particular time n, the value of the output sequence $y[n]$ is obtained by taking the average of three points: $x[n]$ (which is the input at time n), $x[n-1]$ (which is the input at the previous time) and $x[n+1]$ (which is the value of the input sequence at the next time)."

This is the first example of a discrete-time system we have seen in which the output at any time point depends on values of the input sequence at more than one time point. It is also a system in which the output depends on future values of the input (i.e., $x[n+1]$) as well as past (i.e., $x[n-1]$) and present (i.e., $x[n]$) values. In fact, the moving-window average of Equation (1.16) is a specific example of a **finite impulse response (FIR) filter**, which we will study in considerable detail in Chapter 7. In that chapter, we will learn how to create filters with frequency responses that match our design specifications. But for now, let us just remark that anyone who has ever plotted a bunch of data has probably used this kind of moving-window filter without giving it a fancy name or understanding its properties. For example, assume we had some "raw" data $x[n]$ that we considered to be "noisy" in some way, and we wanted to extract the average trend of the data. One logical thing to do is to replace each point in the original data with the average of that point and the two surrounding points, thus "smoothing the data." This is exactly what Equation (1.16) specifies: each output point is the average (mean) of three input points: the present point, the past point and the future point. **Figure 1.31** shows the moving-window average of Equation (1.16) applied to the sample input sequence of the previous figures. As you can see, this simple moving-window average does a pretty nice job of "smoothing" the data.

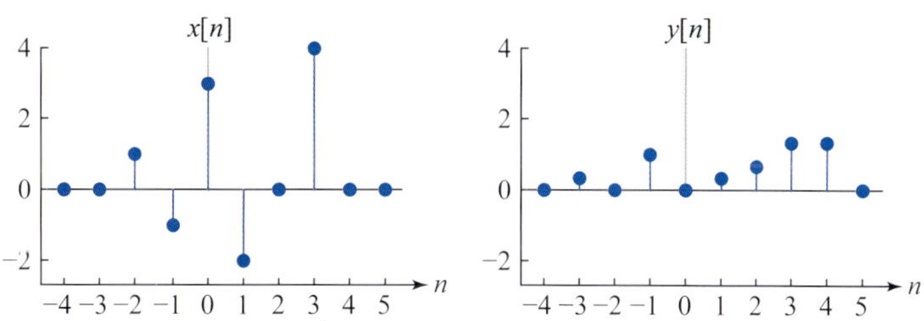

Figure 1.31 Input and output of a moving-window average

1.7.6 Summer

A summer is a system in which the output at any time point n depends on the cumulative sum of all previous values of the input up to and including the present time point,

$$y[n] = \sum_{k=-\infty}^{n} x[k].$$

Figure 1.32 shows the result of applying the sample input sequence to the summer.

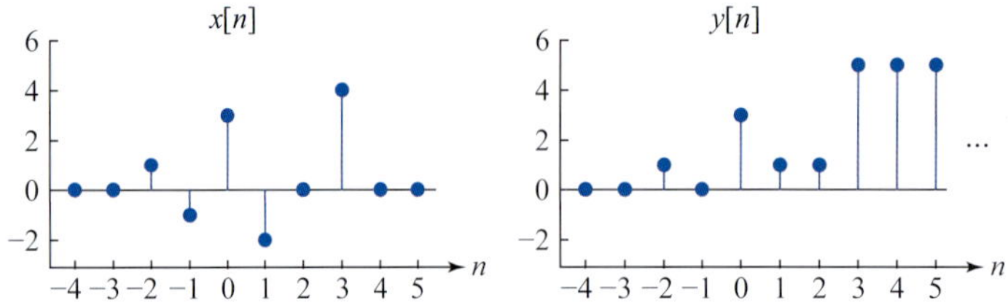

Figure 1.32 Input and output of summer

The output of the summer is the cumulative sum of all the previous inputs. Since the input has a finite number of points, when $n > 3$, the output will be constant. Matlab has a function `cumsum` that can do this for us:

```
x = [0 0 1 -1 3 -2 0 4 0 0];
y = cumsum(x)
y =
    0    0    1    0    3    1    1    5    5    5
```

1.7.7 Switch

In some discrete-time systems, the output depends not only on the input, but also on some measure of absolute time. For example,

$$y[n] = x[n]u[n].$$

The output sequence $y[n]$ is the product of two sequences, an input sequence $x[n]$ and a unit step $u[n]$. As we mentioned in Section 1.6.2, the effect of multiplying the input sequence by the step is to set all the values of the output sequence to zero for time $n < 0$, regardless of the values of the input sequence. For this reason, this system acts as a discrete-time switch. **Figure 1.33** shows the effect of applying the switch to the input sequence of our other examples.

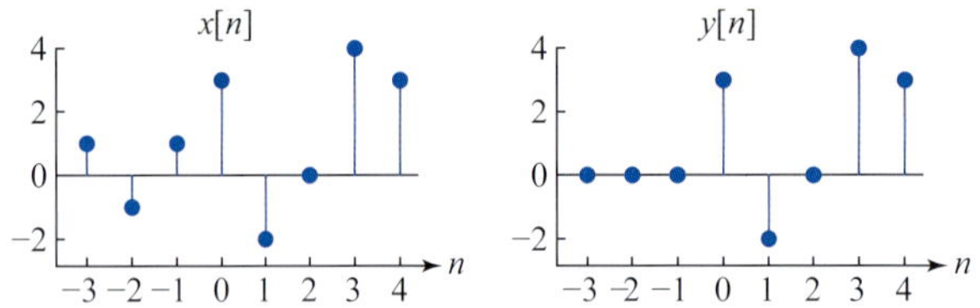

Figure 1.33 Input and output of a switch

1.7.8 Linear constant-coefficient difference equation (LCCDE)

A **linear constant-coefficient difference equation (LCCDE)** is the discrete-time version of a linear constant-coefficient continuous-time differential equation. It is defined by the general relation

$$\sum_{k=0}^{N} a_k y[n-k] = \sum_{k=0}^{M} b_k x[n-k]. \tag{1.17}$$

Equation (1.18) is a simple example of an LCCDE, derived by setting $a_0 = 1$, $a_1 = -1/2$, $b_0 = 1$ and all other a_k and b_k to zero in Equation (1.17):

$$y[n] - \tfrac{1}{2}y[n-1] = x[n]. \quad \textbf{(1.18)}$$

Rewriting Equation (1.18), we get

$$y[n] = \tfrac{1}{2}y[n-1] + x[n]. \quad \textbf{(1.19)}$$

The LCCDE is a fundamentally different system from the systems we have looked at so far. In all the previous systems we have studied, the output sequence is *explicitly* defined solely in terms of the input sequence. That is, the output at any time $y[n]$ depends only on values of the input sequence $x[n]$ at some time or times. As you can see from Equation (1.19), the output of this LCCDE at any time point depends not only upon the present value of the input sequence $x[n]$ but also on a previous value of the output sequence $y[n-1]$. The output sequence depends upon itself as well as the input sequence. This is an *implicit* definition of a system. In words, Equation (1.18) says, "I will not tell you explicitly what $y[n]$ *is*, but I will tell you that if you take $y[n]$ and subtract $0.5y[n-1]$, you will get the input sequence $x[n]$." Systems characterized by LCCDEs form an important class of systems called **infinite impulse response (IIR) filters**, which we will study in Chapter 8.

1.8 Linearity

Linearity is one of the central concepts relating to the analysis of discrete-time systems. Linearity is equivalent to the superposition property you may remember from your study of circuits. In essence, a linear system is one for which the output of a system in response to the sum of a number of inputs is equivalent to the sum of the system's response to each individual input. To formalize the concept of linearity, consider two inputs $x_1[n]$ and $x_2[n]$. We will define the output of the system to each of these separate inputs as $y_1[n]$ and $y_2[n]$, respectively; that is,

$$\begin{aligned} y_1[n] &\triangleq T\{x_1[n]\} \\ y_2[n] &\triangleq T\{x_2[n]\} \end{aligned}. \quad \textbf{(1.20)}$$

A system is said to be linear if it satisfies two properties: **additivity** and **scaling**.

1.8.1 The additivity property

The additivity property states that the transformation of the sum of inputs is equal to the sum of the transformations of the inputs:

$$T\{x_1[n] + x_2[n]\} = T\{x_1[n]\} + T\{x_2[n]\} = y_1[n] + y_2[n]. \quad \textbf{(1.21)}$$

The additivity property can be best understood by considering the result of performing two "experiments," Experiment A and Experiment B, as shown in **Figure 1.34**.

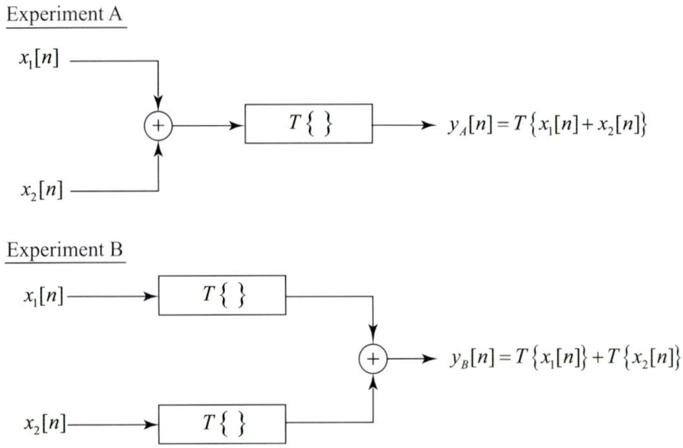

Figure 1.34 Additivity property

In Experiment A, we first add the two inputs $x_1[n]$ and $x_2[n]$, and apply the sum as the input to a discrete-time system, denoted by the transformation $T\{\ \}$. The output of this system is the transformation of the sum of the inputs, which we define as $y_A[n] \triangleq T\{x_1[n] + x_2[n]\}$. In Experiment B, we put the two inputs $x_1[n]$ and $x_2[n]$ into two identical systems and then add the outputs of these systems. The result is the sum of the transformations of the inputs, which we define as $y_B[n] \triangleq T\{x_1[n]\} + T\{x_2[n]\}$. If the results of Experiment A and Experiment B are equal – that is, if the transformation of the sum is equal to the sum of the transformations $T\{x_1[n] + x_2[n]\} = T\{x_1[n]\} + T\{x_2[n]\}$ – then the additivity property is satisfied.

1.8.2 The scaling property

The scaling property states that the transformation of the scaled input is equal to the scaled, transformed input,

$$T\{kx[n]\} = kT\{x_1[n]\}, \tag{1.22}$$

where k is a constant. Again, we can think of the scaling property in terms of two experiments, shown in **Figure 1.35**.

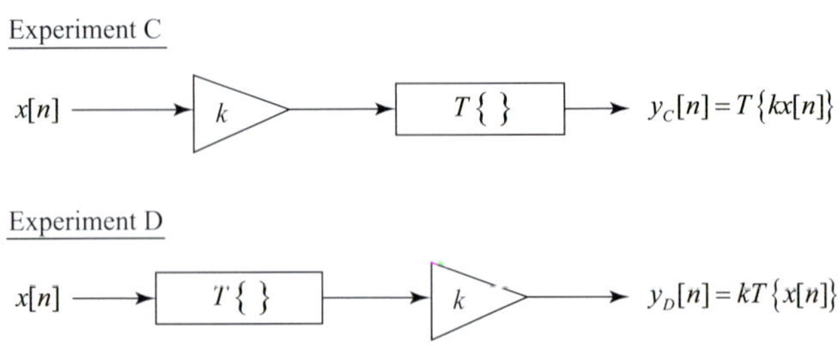

Figure 1.35 Scaling property

In Experiment C, we first scale the input and then transform that scaled input to produce the output $y_C[n] \triangleq T\{kx[n]\}$. In Experiment D, we first transform the input and then scale the output of the transformation, $y_D[n] \triangleq kT\{x[n]\}$. If the result of these two experiments is the same, then the scaling property is satisfied.

The additivity and scaling properties can be combined into one simple statement of linearity:

$$\begin{aligned} T\{ax_1[n] + a_2x_2[n]\} &= T\{a_1x_1[n]\} + T\{a_2x_2[n]\} = a_1T\{x_1[n]\} + a_2T\{x_2[n]\} \\ &= a_1y_1[n] + a_2y_2[n]. \end{aligned} \tag{1.23}$$

That is, the transformation of the scaled sum of inputs is equal to the sum of the scaled transformation of the inputs.

Let us now test the linearity of the systems we introduced in Section 1.7.

1.8.3 Discrete-time scalar multiplier

Additivity property

In this case, the transformation $T\{\ \}$ is equivalent to the statement, "take the input and double it."

Experiment A: Given two inputs $x_1[n]$ and $x_2[n]$, the result of Experiment A is double the sum of the inputs: $T\{x_1[n] + x_2[n]\} = 2(x_1[n] + x_2[n])$.

Experiment B: The result of Experiment B is formed by taking the transformation of the two inputs separately and then adding the results, namely the sum of twice $x_1[n]$ plus twice $x_2[n]$: $T\{x_1[n]\} + T\{x_2[n]\} = 2x_1[n] + 2x_2[n]$.

Since $2(x_1[n] + x_2[n]) = 2x_1[n] + 2x_2[n]$, the results of these two experiments are the same, and the additivity property is satisfied.

Scaling property

Experiment C: The result of Experiment C is the transformation of k times the input: $T\{kx[n]\} = 2(kx[n])$.

Experiment D: The result of Experiment D is k times the transformation of the input: $kT\{x[n]\} = k(2x[n])$.

Since $2(kx[n]) = k(2x[n])$, the results of these two experiments are the same, and the scaling property is satisfied. For this transformation, both the additivity and scaling properties are satisfied; hence, the transformation is linear.

1.8.4 Offset

Additivity property

In this case, the transformation $T\{\ \}$ is equivalent to the statement, "take the input and add one to it."

Experiment A: Given two inputs $x_1[n]$ and $x_2[n]$, the result of Experiment A is the sum of the inputs plus one: $T\{x_1[n] + x_2[n]\} = (x_1[n] + x_2[n]) + 1 = x_1[n] + x_2[n] + 1$.

Experiment B: The result of Experiment B is obtained by taking the transformation of the two inputs separately and then adding the result: $T\{x_1[n]\} + T\{x_2[n]\} = (x_1[n] + 1) + (x_2[n] + 1) = x_1[n] + x_2[n] + 2$.

Clearly the results are *not* the same; hence, the additivity property is not satisfied.

Scaling property

Experiment C: The result of Experiment C is the transformation of k times the input: $T\{kx[n]\} = (kx[n]) + 1 = kx[n] + 1$.

Experiment D: The result of Experiment D is k times the transformation of the input: $kT\{x[n]\} = k(x[n] + 1) = kx[n] + k$.

The results of these two experiments are not the same, so the scaling property is not satisfied either. Since neither the additivity nor the scaling properties are satisfied for this transformation, the transformation is not linear.

1.8.5 Squarer

Additivity property

Experiment A: The result of Experiment A is the square of the sum of the inputs:

$$T\{x_1[n] + x_2[n]\} = (x_1[n] + x_2[n])^2 = x_1^2[n] + 2x_1[n]x_2[n] + x_2^2[n].$$

Experiment B: The result of Experiment B is obtained by taking the square of the two inputs separately and then adding the result

$$T\{x_1[n]\} + T\{x_2[n]\} = x_1^2[n] + x_2^2[n].$$

The results are not the same, so the additivity property is not satisfied.

Scaling property

Experiment C: The result of Experiment C is the transformation of k times the input: $T\{kx[n]\} = (kx[n])^2 = k^2x^2[n]$.

Experiment D: The result of Experiment D is k times the transformation of the input: $kT\{x[n]\} = kx^2[n]$.

The results of these two experiments are not the same, so the scaling property is not satisfied either. Since neither the additivity nor the scaling properties are satisfied, the squarer is a nonlinear transformation.

1.8.6 Shift

Additivity property

Experiment A: $T\{x_1[n] + x_2[n]\} = (x_1[n] + x_2[n]) = (x_1[n - n_0] + x_2[n - n_0])$.

Experiment B: $T\{x_1[n]\} + T\{x_2[n]\} = x_1[n - n_0] + x_2[n - n_0]$.

The results are the same, so the additivity property is satisfied.

Scaling property

Experiment C: $T\{kx[n]\} = kx[n - n_0]$.

Experiment D: $kT\{x[n]\} = kx[n - n_0]$.

The results of Experiments C and D are identical, so the scaling property is satisfied. Since both the additivity and scaling properties are satisfied, the delay is a linear transformation.

1.8.7 Moving-window average

Additivity property

Experiment A: Let $q[n] = x_1[n] + x_2[n]$. Now,

$$T\{x_1[n] + x_2[n]\} = T\{q[n]\} = \tfrac{1}{3}(q[n-1] + q[n] + q[n+1])$$
$$= \frac{1}{3}\left\{\begin{array}{l}(x_1[n-1] + x_2[n-1])\\ +(x_1[n] + x_2[n])\\ +(x_1[n+1] + x_2[n+1])\end{array}\right\} = \frac{1}{3}\left\{\begin{array}{l}(x_1[n-1] + x_1[n] + x_1[n+1])\\ +(x_2[n-1] + x_2[n] + x_2[n+1])\end{array}\right\}.$$

Experiment B:

$$T\{x_1[n]\} + T\{x_2[n]\} = \frac{1}{3}\left\{\begin{array}{l}(x_1[n-1] + x_1[n] + x_1[n+1])\\ +(x_2[n-1] + x_2[n] + x_2[n+1])\end{array}\right\}.$$

The results are the same, so the additivity property is satisfied.

Scaling property

Experiment C: $T\{kx[n]\} = \frac{1}{3}(kx[n-1] + kx[n] + kx[n+1]) = \frac{1}{3}k(x[n-1] + x[n] + x[n+1])$.
Experiment D: $kT\{x[n]\} = k\frac{1}{3}(x[n-1] + x[n] + x[n+1])$.

The results of Experiments C and D are identical, so the scaling property is satisfied. Since both the additivity and scaling properties are satisfied, the moving-window average is a linear transformation.

1.8.8 Summer

Additivity property

Experiment A:

$$T\{x_1[n] + x_2[n]\} = \sum_{k=-\infty}^{n} (x_1[k] + x_2[k]).$$

Experiment B:

$$T\{x_1[n]\} + T\{x_2[n]\} = \sum_{k=-\infty}^{n} x_1[k] + \sum_{k=-\infty}^{n} x_2[k].$$

The results are the same, so the additivity property is satisfied.

Scaling property

Experiment C:

$$T\{kx[n]\} = \sum_{k=-\infty}^{n} kx[k].$$

Experiment D:

$$kT\{x[n]\} = k\sum_{k=-\infty}^{n} x[k].$$

The results of Experiments C and D are identical, so the scaling property is satisfied. Since both the additivity and scaling properties are satisfied, the summer is a linear transformation.

1.8.9 Switch

Additivity property

Experiment A: $T\{x_1[n] + x_2[n]\} = (x_1[k] + x_2[k])u[n]$.

Experiment B: $T\{x_1[n]\} + T\{x_2[n]\} = x_1[k]u[n] + x_2[k]u[n]$.

The results are the same, so the additivity property is satisfied.

Scaling property

Experiment C: $T\{kx[n]\} = (kx[n])u[n]$.

Experiment D: $kT\{x[n]\} = k(x[n]u[n])$.

The results of Experiments C and D are identical, so the scaling property is satisfied. Since both the additivity and scaling properties are satisfied, the switch is a linear transformation.

1.8.10 Linear constant-coefficient difference equation

Systems defined by linear constant-coefficient difference equations are perhaps the most important systems we will encounter in this book. Because the relation between input and output is implicit rather than explicit, the proof that these systems are linear is somewhat different than the proofs in previous examples.

Additivity property

Experiment A: Let $q[n]$ be the sum of the inputs, $q[n] = x_1[n] + x_2[n]$, and let $y_A[n]$ be the output of the system to $q[n]$. So, $T\{x_1[n] + x_2[n]\} = T\{q[n]\}$ implies that $y_A[n]$ satisfies the relation

$$y_A[n] - \tfrac{1}{2}y_A[n-1] = q[n] = x_1[n] + x_2[n].$$

Experiment B: If we define $y_1[n]$ to be the output of the system to $x_1[n]$ and $y_2[n]$ to be the output of the system to $x_2[n]$, then $T\{x_1[n]\}$ and $T\{x_2[n]\}$, respectively, satisfy the relations

$$y_1[n] - \tfrac{1}{2}y_1[n-1] = x_1[n]$$

$$y_2[n] - \tfrac{1}{2}y_2[n-1] = x_2[n].$$

Writing both together and adding gives

$$
\begin{array}{rl}
 & y_1[n] - \frac{1}{2}y_1[n-1] = x_1[n] \\
+ & \underline{y_2[n] - \frac{1}{2}y_2[n-1] = x_2[n]}
\end{array}
$$

$$(y_1[n] + y_2[n]) - \tfrac{1}{2}(y_1[n-1] + y_2[n-1]) = x_1[n] + x_2[n].$$

Since, by definition, $y_B[n] = y_1[n] + y_2[n]$, this equation becomes

$$y_B[n] - \tfrac{1}{2}y_B[n-1] = x_1[n] + x_2[n].$$

Comparing with the result of Experiment A, we see that the right sides of the difference equation are the same $(x_1[n] + x_2[n])$, so the left sides of the equations must also be equal. This implies that $y_A[n] = y_B[n]$. Hence, the output of the system to the sum of inputs $(y_A[n])$ is equal to the sum of the outputs of the system to two independent inputs $(y_B[n] = y_1[n] + y_2[n])$. The additivity property is satisfied.

Scaling property

Experiment C: Define $y_C[n]$ to be the output of the LCCDE system to $T\{kx[n]\}$,

$$y_C[n] - \tfrac{1}{2}y_C[n-1] = kx[n].$$

Experiment D: The output of the LCCDE system to $x[n]$ is just $y[n]$, where $y[n]$ satisfies

$$y[n] - \tfrac{1}{2}y[n-1] = x[n].$$

Multiplying both sides by k gives

$$k(y[n] - \tfrac{1}{2}y[n-1]) = (ky[n]) - \tfrac{1}{2}(ky[n-1]) = kx[n].$$

Letting $y_D[n] = ky[n]$, we see that

$$y_D[n] - \tfrac{1}{2}y_D[n-1] = kx[n].$$

Comparing the results of Experiments C and D, we see that the right sides of the difference equations are the same $(kx[n])$, so the left sides of the equations must also be equal; that is, $y_C[n] = y_D[n]$. The output of the system to the scaled input $(y_C[n])$ is the same as the scaled output of the system $(y_D[n])$, so the scaling property is satisfied. The results of Experiments C and D are identical, so the scaling property is satisfied. Since both the additivity and scaling properties are satisfied, the system defined by the LCCDE is a linear transformation.

This exercise proves that the specific system defined by the LCCDE of Equation (1.19) is linear. But a similar proof can be used to show that any system defined by the general equation for LCCDEs in Equation (1.17) is also linear.

1.8.11 The "zero-in, zero-out" property of linear systems

A necessary but not sufficient condition for a system to be linear is that it satisfies the **zero-in, zero-out** property. This means that *if* the system is linear and the input is $x[n]=0$, *then* the output must be $y[n]=0$. The proof is pretty simple. If a system is linear, the scaling property applies, which guarantees that $T\{kx[n]\}=kT\{x[n]\}$ for any constant k. Letting $k=0$, we obtain $T\{0\}=0$. This means that if a system is linear, then a zero input must yield a zero output. All of our example systems that are linear satisfy this condition (e.g., the multiplier, the delay, the LCCDE). The offset system $y[n]=x[n]+1$ does not satisfy the zero-in, zero-out property and is therefore not linear. But remember that satisfying the zero-in, zero-out property is a necessary but not sufficient condition for linearity. For example, the squarer $y[n]=x^2[n]$ satisfies the zero-in, zero-out property, but is not a linear system.

1.9 Time invariance

Time invariance, also called **shift invariance**, is the second important system property we will discuss in this chapter. A time-invariant system is one for which the output does not depend on the *absolute time* at which the input is presented to the system. To give a prosaic, non-engineering example, the ATM machine outside the bank is a time-invariant system. If you go to the ATM machine at 9 a.m., put in your card, enter your PIN and ask for \$20, you will get \$20 in cash at 9:01 a.m. If you go twelve hours later, at 9 p.m., and put in your card, you will get \$20, just twelve hours later, namely at 9:01 p.m. Since the output of this system (the ATM) does not depend on the absolute time at which the system receives the input (your card), the ATM is a time-invariant system. In contrast, the teller at the same bank is an example of a time-variant system. If you go to the teller window at 9 a.m., hand the teller your card and ask for \$20, you will get \$20 in cash at 9:01 a.m. If you go at 9 p.m., you will find the bank closed and you will get nothing. Since the output of this system (the teller) depends on the absolute time at which the input occurs, this system is time-variant.

To formalize things a bit, consider again a system for which the output $y[n]$ is defined as the result of transforming input signal $x[n]$: $y[n]=T\{x[n]\}$. We can now formulate a procedure to determine if a system is time-invariant by thinking in terms of a couple of experiments, as indicated in **Figure 1.36**.

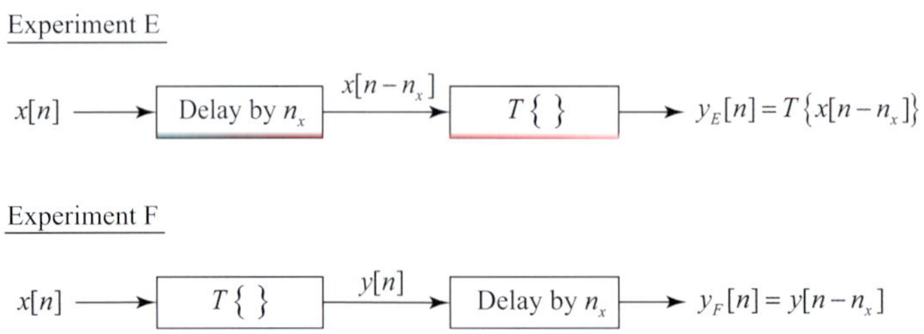

Figure 1.36 Test for time invariance

In Experiment E, we first delay the input $x[n]$ by an amount n_x, and present this delayed input to the system. The result is $y_E[n] = T\{x[n-n_x]\}$. In Experiment F, we input $x[n]$ into the system, producing $y[n]$, and then delay this output by n_x, yielding $y_F[n] = y[n-n_x]$. If the results of these two experiments are the same (i.e., $y_E[n] = y_F[n]$), the system is said to be time-invariant. In words, a time-invariant system is one for which a delayed input to a system produces a delayed output.

Let us now test the time invariance of some of the examples of systems we introduced above.

1.9.1 Discrete-time scalar multiplier

The output of the system is $y[n] = T\{x[n]\} = 2x[n]$.

Experiment E: The output of a system to a delayed input is $y_E[n] = T\{x[n-n_x]\} = 2x[n-n_x]$.

Experiment F: The delayed output of the system is $y[n]$ delayed by n_x, which is found by replacing n by $n-n_x$, resulting in $y_F[n] = y[n-n_x] = 2x[n-n_x]$.

The results of these two experiments are the same, so this system is time-invariant.

1.9.2 Offset

The output of the system is $y[n] = T\{x[n]\} = x[n] + 1$.

Experiment E: The output of a system to a delayed input is $y_E[n] = T\{x[n-n_x]\} = x[n-n_x] + 1$.

Experiment F: The delayed output of the system is $y_F[n] = y[n-n_x] = x[n-n_x] + 1$.

The results of these two experiments are the same, so this system is time-invariant.

1.9.3 Squarer

The output of the system is $y[n]T\{x[n]\} = x^2[n]$.

Experiment E: The output of a system to a delayed input is $y_E[n] = T\{x[n-n_x]\} = x^2[n-n_x]$.

Experiment F: The delayed output of the system is $y_F[n] = y[n-n_x] = x^2[n-n_x]$.

The results of these two experiments are the same, so this system is time-invariant.

1.9.4 Shift

The output of the system is $y[n] = T\{x[n]\} = x[n-n_0]$.

Experiment E: The output of a system to a delayed input is $y_E[n] = T\{x[n-n_x]\} = x[(n-n_0)-n_x]$.

Experiment F: The delayed output of the system is $y_F[n] = y[n-n_x] = x[(n-n_x)-n_0]$.

The results of these two experiments are the same, so this system is time-invariant.

1.9.5 Moving-window average

The output of the system is $y[n] = T\{x[n]\} = \frac{1}{3}(x[n-1] + x[n] + x[n+1])$.

Experiment E. The output of a system to a delayed input is

$$y_E[n] = T\{x[n-n_x]\} = \tfrac{1}{3}(x[(n-n_x)-1] + x[(n-n_x)] + x[(n-n_x)+1]).$$

Experiment F: The delayed output of the system is

$$y_F[n] = y[n - n_x] = \tfrac{1}{3}(x[(n - n_x) - 1] + x[(n - n_x)] + x[(n - n_x) + 1]).$$

The results of these two experiments are the same, so this system is time-invariant.

1.9.6 Summer

The output of the system is

$$y[n] = T\{x[n]\} = \sum_{k=-\infty}^{n} x[k].$$

Experiment E: The output of a system to a delayed input is

$$y_E[n] = T\{x[n - n_x]\} = \sum_{k=-\infty}^{n} x[k - n_x].$$

Let $l = k - n_x$, and recognize that $l = -\infty$ when $k = -\infty$, so

$$y_E[n] = \sum_{k=-\infty}^{n} x[k - n_x] = \sum_{l=-\infty}^{n-n_x} x[l].$$

Experiment F: The delayed output of the system is

$$y_E[n] = y[n - n_x] = \sum_{k=-\infty}^{n-n_x} x[k].$$

The results of these two experiments are the same, so this system is time-invariant.

1.9.7 Switch

The output of the system is $y[n] = T\{x[n]\} = x[n]u[n]$.

Experiment E: The output of a system to a delayed input is

$$y_E[n] = T\{x[n - n_x]\} = x[n - n_x]u[n].$$

Experiment F: The delayed output of the system is

$$y_F[n] - y[n-n_x] = x[n-n_x]\,u[n-n_x].$$

The results of these two experiments are not the same, so this system is *not* time-invariant. We need to look in a bit more detail to understand why this is not a time-invariant system. Let us consider a simple input $x[n] = \delta[n]$. The output of this system is $y[n] = T\{x[n]\} = x[n]u[n] = \delta[n]u[n] = \delta[n]$. The results of our two experiments are:

Experiment E: $y_E[n] = x[n - n_x]u[n] = \delta[n - n_x]u[n]$.
Experiment F: $y_F[n] = x[n - n_x]u[n - n_x] = \delta[n - n_x]u[n - n_x]$.

If we select a value of $n_x = 1$, we get $y_E[n] = \delta[n - 1]u[n] = \delta[n - 1]$ and $y_F[n] = \delta[n - 1]u[n - 1] = \delta[n - 1]$. The results of these two experiments are shown in the top row of **Figure 1.37**.

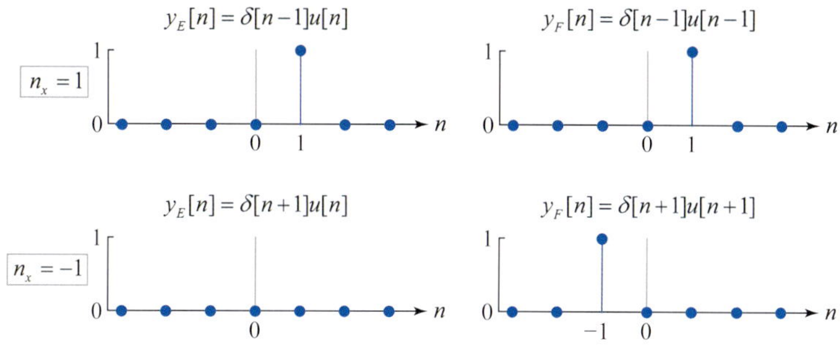

Figure 1.37 Time invariance example

Since we get the same results from our two experiments, this would seem to contradict our conclusion that this system is time-variant! However, if we choose a different value of n_x, the results tell a different story. Select a value of $n_x = -1$. Now we get $y_E[n] = \delta[n+1]u[n] = 0$ and $y_F[n] = \delta[n+1]u[n+1]$. Clearly the results of these two experiments are not the same, as shown in the bottom row of **Figure 1.37**. Hence this system is not time-invariant. This example raises an important point: if a system is not time-invariant for even one value of delay n_x, then it is not time-invariant, period.

1.9.8 Linear constant-coefficient difference equation (LCCDE)

The output of the system is defined by the difference equation $y[n] - \frac{1}{2}y[n-1] = x[n]$. Or we can express the output as

$$y[n] = x[n] + \tfrac{1}{2}y[n-1].$$

Experiment E: The output of a system to a delayed input satisfies

$$y_E[n] - \tfrac{1}{2}y_E[n-1] = x[n-n_x].$$

Experiment F: The delayed output of the system is

$$y_F[n] = y[n-n_x] = x[(n-n_x)] + \tfrac{1}{2}y[(n-n_x)-1] = x[n-n_x] + \tfrac{1}{2}y_F[n-1].$$

Hence,

$$y_F[n] - \tfrac{1}{2}y_F[n-1] = x[n-n_x].$$

Comparing the results of Experiments E and F, we see that the right sides of the difference equation are the same ($x[n-n_x]$), so the left sides of the equations must also be equal. So, $x[n-n_x]$, and we conclude that the LCCDE system is time-invariant. The same conclusion applies to the general LCCDE system described by Equation (1.17).

1.10 Causality

A **causal system** is **non-anticipative**. This means that the output of the system $y[n_x]$ at a particular time point n_x does not depend upon any *future* values of the input, i.e., values of the input for times n greater than n_x. More exactly, a system is causal if, for every

n_x, $y[n_x]$ depends only on values of n for $n \le n_x$ (that is, $x[n]$, $n \le n_x$). For example, if a given output of the system $y[0]$ depends on $x[0]$, $x[-1]$ and $x[-2]$, then the system is causal. However, if $y[0]$ depends on $x[1]$, $x[0]$ and $x[-1]$, the system is non-causal.

In Chapter 2, we will discover a simple, no-fail, test for causality for systems that are both linear and time-invariant. However, for now, it is worth getting an intuitive understanding of what causality means. So, we will run through all the examples from the previous sections.

1.10.1 Discrete-time scalar multiplier

For this system, $y[n_x] = 2x[n_x]$, so the output value at time n_x depends only on the single input value at time n_x. The system is causal.

1.10.2 Offset

Here, $y[n_x] = x[n_x] + 1$, so again the output value at time n_x depends only on the input value at time n_x and the system is causal.

1.10.3 Squarer

This is another system for which the output value at time n_x depends only on the input value at time n_x: $y[n_x] = x^2[n_x]$. The system is causal.

1.10.4 Shift

Here, $y[n_x] = x[n_x - n_0]$. For a positive delay (i.e., $n_0 \ge 0$), it should be clear that the system is causal, because the output of the system depends on a previous value of the input. However, for a negative delay (i.e., $n_0 < 0$), the system is not causal, because the output of the system depends on a future value of the input. For example, if $n_0 = -1$, $y[n_x] = x[n_x + 1]$, then

$$\begin{array}{l} \vdots \\ y[0] = x[1] \\ y[1] = x[0] \\ \vdots \end{array},$$

and so on.

1.10.5 Moving-window average

The output at any time point n_x depends on values of the input both before and after n_x: $y[n_x] = \frac{1}{3}(x[n_x + 1] + x[n_x] + x[n_x - 1])$. This system "looks ahead" (because of the $x[n_x + 1]$ term) and is not causal.

1.10.6 Summer

At time n_x, the output of the summer depends upon all previous values of the input,

$$y[n_x] = \sum_{k=-\infty}^{n_x} x[k],$$

so the system is causal.

1.10.7 Switch

This is another causal system, since the output $y[n_x] = x[n_x]u[n_x]$ depends only on the value of the input at the same time point $x[n_x]$.

1.10.8 Linear constant-coefficient difference equation (LCCDE)

The general relation for the linear constant-coefficient difference equation is given in Equation (1.17). Without loss of generality, we can let $a_0 = 1$, which is equivalent to dividing both sides of the equation by a_0. We can rewrite this equation as

$$y[n] = \sum_{k=1}^{N} a_k y[n-k] + \sum_{k=0}^{M} b_k x[n-k].$$

As long as $k \geq 0$ in the second summation, $y[n_x]$ depends only on the values of $x[n]$, $n \leq n_x$, and this system is causal.

1.11 Stability

Stability is a key, desirable, property of systems. One of the simplest definitions of stability is the **bounded-input, bounded-output (BIBO)** definition. This states that a system is considered stable if every bounded-input signal produces a bounded-output signal. A bounded signal is one for which the maximum absolute amplitude of the entire signal is bounded by some maximum value, $x_{\max}$:

$$|x[n]| < x_{\max}.$$

In practical terms, if a discrete-time system described by an algorithm is unstable, it can produce an arithmetic overflow or saturation, which can lead to unfortunate consequences in the application.

The BIBO definition of stability is conceptually useful, but not practically useful in many cases, since in order to test for stability, one would presumably have to test *every* bounded signal to see if it produced a bounded output. In Chapter 2, we will discover a simple test for the stability of linear time-invariant systems. But for now, let us just assess the stability of our example systems using the BIBO criterion.

1.11.1 Discrete-time scalar multiplier

For this system, $y[n] = 2x[n]$, so the output value is just double the input value. More formally, if $|x[n]| \leq x_{\max}$, then $x[-n-1]$, so any bounded input gives rise to a bounded output and the system is stable.

1.11.2 Offset

Here, $y[n] = x[n] + 1$, so if $|x[n]| \leq x_{\max}$, then $|x[n] + 1| \leq x_{\max} + 1$, which is bounded, so the system is stable.

1.11.3 Squarer

If $|x[n]| \leq x_{\max}$, then $|y[n]| = |x^2[n]| \leq x_{\max}^2$. The system is stable.

1.11.4 Shift

If $|x[n]| \le x_{max}$, then $|x[n-n_0]| \le x_{max}$ because the maximum amplitude of the delayed sequence is the same as that of the original sequence. The system is stable.

1.11.5 Moving-window average

The output of the moving-window average is just the average of the original sequence and two delayed sequences. Using the delay property, we find that $|x[n-1]| \le x_{max}$ and $|x[n+1]| \le x_{max}$. Since $y[n] = \frac{1}{3}(x[n-1] + x[n] + x[n+1])$,

$$\begin{aligned} |y[n]| = |\tfrac{1}{3}(x[n-1] + x[n] + x[n+1])| &\le \tfrac{1}{3}(|x[n-1]| + |x[n]| + |x[n+1]|) \\ &\le \tfrac{1}{3}(x_{max} + x_{max} + x_{max}), \end{aligned}$$

so $|y[n]| < x_{max}$ and the system is stable.

1.11.6 Summer

For the summer,

$$y[n] = \sum_{k=-\infty}^{\infty} x[k].$$

Consider the bounded input $x[n] = u[n]$. Then,

$$y[n] = \sum_{k=-\infty}^{n} u[n] = \begin{Bmatrix} \sum_{k=0}^{n} 1, & n \ge 0 \\ 0, & n < 0 \end{Bmatrix} = nu[n].$$

Clearly, the output is not bounded, but grows without limit as $n \to \infty$. So this system is not stable.

1.11.7 Switch

This is a stable system, since $|y[n]| = |x[n]u[n]| \le |x[n]| \le x_{max}$.

1.11.8 Linear constant-coefficient difference equation (LCCDE)

The stability of this important class of systems cannot be easily determined using the BIBO criterion. We will have to wait until Chapter 2 to find a simple test for stability of these systems. For now, we state that $y[n] - \frac{1}{2}y[n-1] = x[n]$ is the LCCDE of a stable system and $y[n] - 2y[n-1] = x[n]$ is the LCCDE of an unstable system (assuming both systems are causal).

SUMMARY

In this chapter, we have presented some important system properties: linearity, time invariance, causality and stability. We will rely extensively on knowledge of these properties in future chapters.

Systems that are both linear and time-invariant are called **linear time-invariant systems**, or simply **LTI systems**. LTI systems hold a place of particular importance in our study of digital signal processing, both because they are particularly easy to analyze and because many of the most commonly implemented systems are LTI systems. For LTI systems, we will soon be able to develop computationally quick and easy methods of discovering whether they are causal and stable.

It is important to remember that linearity, time invariance, causality and stability are properties of *systems*; they are independent of the signal that is input into the system. It is also important to recognize that these are *independent* properties. For example, knowing that a system is linear (or nonlinear) tells you nothing about whether it is time-variant or -invariant (or causal or stable). Conversely, knowing that a system is time-variant or -invariant tells you nothing about whether it is linear or nonlinear. Just look at the examples of systems we have presented so far in this unit, as summarized in **Table 1.5**.

Table 1.5 Summary of system properties of example systems

System	Equation	Linear	Time-invariant	LTI	Causal	Stable
Multiplier	$y[n]=2x[n]$	✓	✓	✓	✓	✓
Offset	$y[n]=x[n]+1$	✗	✓	✗	✓	✓
Squarer	$y[n]=x^2[n]$	✗	✓	✗	✓	✓
Delay	$y[n]=x[n-n_0]$	✓	✓	✓	$n_0 \geq 0$	✓
Average	$y[n]=(x[n-1]+x[n]+x[n+1])/3$	✓	✓	✓	✗	✓
Summer	$y[n]=\sum_{k=-\infty}^{n} x[k]$	✓	✓	✓	✓	✗
LCCDE	$\sum_{k=0}^{N} a_k y[n-k]=\sum_{k=0}^{M} b_k x[n-k]$	✓	✓	✓	✓	✓ or ✗
Switch	$y[n]=x[n]u[n]$	✓	✗	✗	✓	✓

Systems can be linear and time-invariant (e.g., the LCCDE), linear and not time-invariant (e.g., the switch), nonlinear and time-invariant (e.g., the squarer) or nonlinear and not time-invariant (for example, $y[n]=x^2[n]u[n]$).

PROBLEMS

Problem 1-1

Given $x[n]$ shown in **Figure 1.38**, sketch the following:

(a) $x[n-1]$.

(b) $x[n+1]$.

(c) $x[-n-1]$.

(d) $x[-n+1]$.

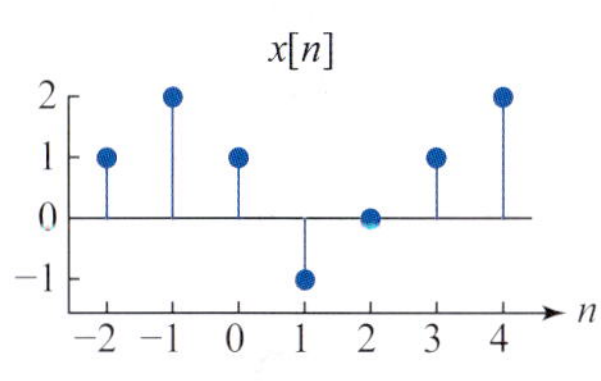

Figure 1.38

Problem 1-2

Given $x[n]$, as shown in **Figure 1.38**, sketch the following:

(a) $x[2n]$.

(b) $x[-2n-1]$.

(c) $x[2n+1]$.

(d) $x[1-2n]$.

Problem 1-3

Given $x[n]$, as shown in **Figure 1.39**, repeat Problem 1-1.

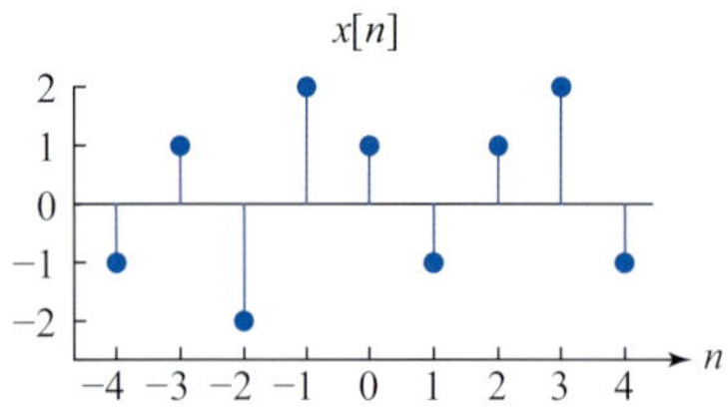

Figure 1.39

Problem 1-4

Given $x[n]$, as shown in Problem 1-3, repeat Problem 1-2.

Problem 1-5

Given $x[n]$ and $y[n]$, as shown in **Figure 1.40**, sketch the following:

(a) $x[1-n]+y[n+1]$.

(b) $x[n-1]-y[2-n]$.

(c) $x[2n+1]+y[2n]$.

(d) $x[n+1]y[n-1]$.

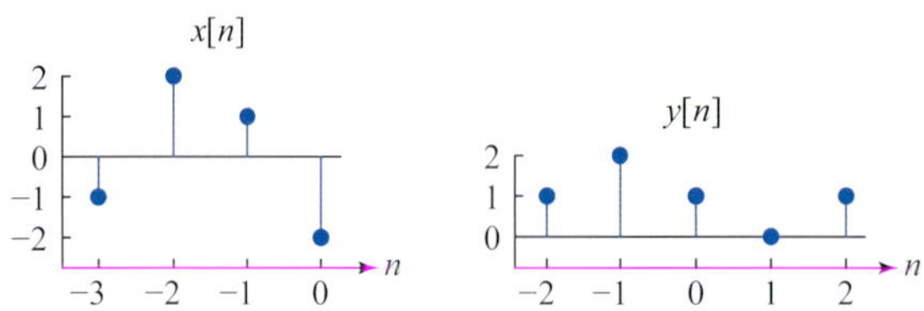

Figure 1.40

Problem 1-6

Show that

$$x_e[n] = \tfrac{1}{2}(x[n] + x[-n])$$
$$x_o[n] = \tfrac{1}{2}(x[n] - x[-n]).$$

Problem 1-7

Show that for any sequence $x[n]$ with even and odd parts $x_e[n]$ and $x_o[n]$, it follows that $x_e[0] = x[0]$ and $x_o[n] = 0$.

Problem 1-8

For each of the sequences shown in **Figure 1.41**, find and plot $x_e[n]$ and $x_o[n]$.

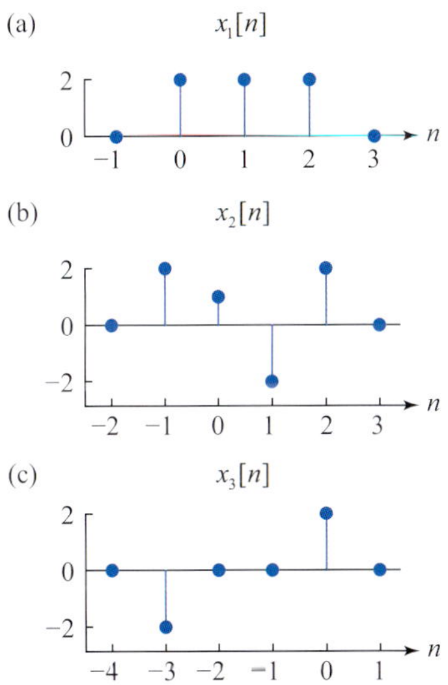

Figure 1.41

Problem 1-9

For each of the sequences shown in **Figure 1.42**, find and plot $x_e[n]$ and $x_o[n]$.

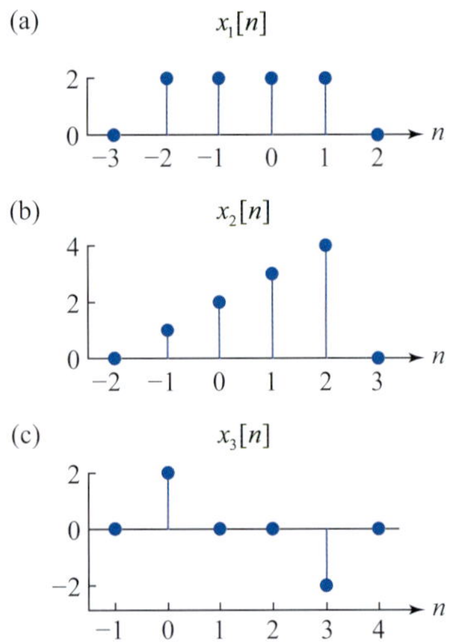

Figure 1.42

Problem 1-10

For each of the sequences shown in **Figure 1.43**, find and plot $x_e[n]$ and $x_o[n]$.

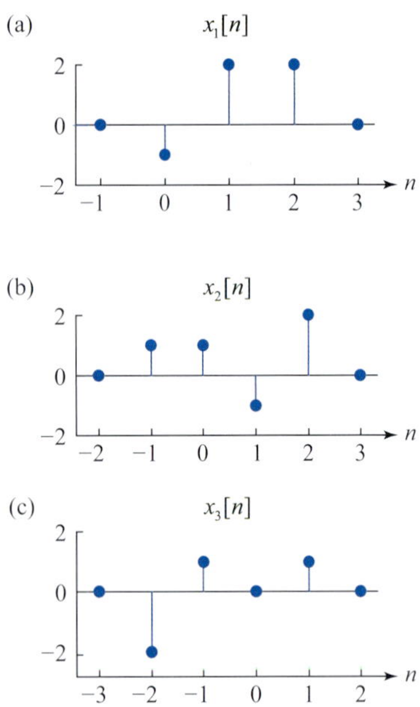

Figure 1.43

Problem 1-11

Given $x[n] = (2 + j)\delta[n + 1] + \delta[n] - 3j\delta[n - 1]$, find $\text{Even}\{x[n]\}$ and $\text{Odd}\{x[n]\}$.

Problem 1-12

Given $x[n] = (2 + j)\delta[n + 1] + \delta[n] - 3j\delta[n - 1]$, find $x_{re}[n]$, $x_{ie}[n]$, $x_{ro}[n]$ and $x_{io}[n]$.

Problem 1-13

Given $x[n] = (1 + j)\delta[n + 2] + 2\delta[n + 1] + (1 - j)\delta[n - 1]$, find

(a) $\mathrm{Re}\{x[n]\}$.
(b) $\mathrm{Im}\{x[n]\}$.
(c) $\mathrm{Even}\{x[n]\}$.
(d) $\mathrm{Odd}\{x[n]\}$.
(e) $\mathrm{Re}\{\mathrm{Even}\{x[n]\}\}$.
(f) $\mathrm{Odd}\{\mathrm{Im}\{x[n]\}\}$.

Problem 1-14

Given $x[n] = (1 - 2j)\delta[n + 2] - 2\delta[n] + (1 + j)\delta[n - 1]$, find

(a) $\mathrm{Re}\{x[n]\}$.
(b) $\mathrm{Im}\{x[n]\}$.
(c) $\mathrm{Even}\{x[n]\}$.
(d) $\mathrm{Odd}\{x[n]\}$.
(e) $\mathrm{Im}\{\mathrm{Even}\{x[n]\}\}$.
(f) $\mathrm{Even}\{\mathrm{Im}\{x[n]\}\}$.

Problem 1-15

(a) Show that $\mathrm{Re}\{\mathrm{Even}\{x[n]\}\} = \mathrm{Even}\{\mathrm{Re}\{x[n]\}\}$.
(b) Show that $\mathrm{Im}\{\mathrm{Odd}\{x[n]\}\} = \mathrm{Odd}\{\mathrm{Im}\{x[n]\}\}$.

Problem 1-16

For each of the following sequences, find $x_e[n]$ and $x_o[n]$.

(a) $x[n] = e^{j\omega_0 n}$.
(b) $x[n] = (j + 1)\delta[n + 1] + (j + 1)\delta[n] + (j - 1)\delta[n - 1]$.
(c) $x[n] = \delta[n] + 2\delta[n - 1] - 2j\delta[n - 2]$.
(d) $x[n] = 2\cos(\omega_0 n + \varphi)$.

Problem 1-17

For each of the following sequences, find $x_e[n]$ and $x_o[n]$.

(a) $x[n] = 1 + e^{j\pi n/4}$.
(b) $x[n] = (1 - j)\delta[n + 1] + j\delta[n] + (j + 1)\delta[n - 1]$.
(c) $x[n] = \delta[n + 1] - 2\delta[n - 1] + 2j\delta[n - 2]$.
(d) $x[n] = 2\sin(\pi n/4 + \pi/3)$.

Problem 1-18

For each of the sequences in Problem 1-16a–c, find $x_{re}[n]$, $x_{ie}[n]$, $x_{ro}[n]$ and $x_{io}[n]$.

Problem 1-19

For each of the sequences in Problem 1-17a–c, find $x_{re}[n]$, $x_{ie}[n]$, $x_{ro}[n]$ and $x_{io}[n]$.

Problem 1-20

By direct substitution, show that

(a) $y[n] = (n+1)u[n]$ satisfies the equation $y[n] - y[n-1] = u[n]$.

(b) $y[n] = 2\delta[n] - \frac{1}{2}^n u[n]$ satisfies the equation $y[n] - \frac{1}{2}y[n-1] = \delta[n] - \delta[n-1]$.

Problem 1-21

Use the result of Problem 1-20a to show that

$$\sum_{k=0}^{\infty} u[n-k] = (n+1)u[n].$$

Problem 1-22

Determine which of the following sequences is periodic. If the sequence is periodic, compute the fundamental period N.

(a) $\cos(0.01\pi n)$.

(b) $\cos(30\pi n/102)$.

(c) $\cos(5\pi n)$.

(d) $\sin(5n)$.

(e) $\sin(62\pi n/10)$.

(f) $\cos(5\pi n/3) + \sin(7\pi n/3)$.

(g) $e^{j\pi/6}$.

(h) $e^{j(2\pi n/6 - \pi)}$.

Problem 1-23

Determine which of the following sequences is periodic. If the sequence is periodic, compute the fundamental period N.

(a) $\cos(3\pi n/11)$.

(b) $\cos(n)$.

(c) $\cos(5\pi n/2 + 2)$.

(d) $\cos(7\pi n/3) + \sin(5\pi n/2)$.

(e) $\cos(7\pi n/3) + \sin(5n)$.

(f) $\cos(7\pi n/3)\sin(5\pi n/2)$.

Problem 1-24

Determine which of the following sequences is periodic. If the sequence is periodic, compute the fundamental period N.

(a) $\cos(3\pi n/13)$.
(b) $\cos(2n)$.
(c) $\cos(5\pi n/3 - 1)$.
(d) $\cos(7\pi n/3) + \sin(5\pi n/7)$.
(e) $\cos(7\pi n/3) + \sin(5n)$.
(f) $\cos(7\pi n/3)\sin(5\pi n/7)$.

Problem 1-25

Determine which of the following sequences is periodic. If the sequence is periodic, compute the fundamental period N.

(a) $\cos(0.2\pi n)$.
(b) $\cos(30\sqrt{2}/101)n$.
(c) $\sin(5\pi n) + 1.23\cos(5\pi n)$.
(d) $\cos(5n)$.
(e) $\sin(6.1\pi n)$.
(f) $e^{jn/6}$.
(g) $e^{j(2\pi n/6 - \pi)}$.

Problem 1-26

Plot $x[n] = \cos(5.2\pi n)$ over two periods.

Problem 1-27

For each of the periodic sequences in Problem 1-23, plot the sequence over two periods.

Problem 1-28

For each of the periodic sequences in Problem 1-24, plot the sequence over two periods.

Problem 1-29

Given

$$x(n) = \begin{cases} |n|, & -3 \le n \le 3 \\ 0, & \text{otherwise} \end{cases},$$

determine and sketch the following sequences:

(a) $y[n] = 2x[n] - 1$.

(b) $y[n] = x[n-3]$.
(c) $y[n] = \frac{1}{3}(x[n+1] + x[n] + x[n-1])$.
(d) $y[n] = \sum_{k=-\infty}^{n} x[k]$.
(e) $y[n] = x[2-n]$.
(f) $y[n] = nx[n]$.

Problem 1-30

Given

$$x(n) = \begin{cases} n, & -3 \le n \le 3 \\ 0, & \text{otherwise} \end{cases},$$

determine and sketch the following sequences:
(a) $y[n] = 2x[n] - 1$.
(b) $y[n] = x[n-3]$.
(c) $y[n] = \frac{1}{3}(x[n+1] + x[n] + x[n-1])$.
(d) $y[n] = \sum_{k=-\infty}^{n} x[k]$.
(e) $y[n] = x[2-n]$.
(f) $y[n] = nx[n]$.

Problem 1-31

Given

$$x(n) = \begin{cases} 4 - |n|, & -3 \le n \le 3 \\ 0, & \text{otherwise} \end{cases},$$

determine and sketch the following sequences:
(a) $y[n] = 2x[n] - 1$.
(b) $y[n] = x[n+2]$.
(c) $y[n] = \frac{1}{3}(x[n+1] + x[n] + x[n-1])$.
(d) $y[n] = \sum_{k=-\infty}^{n} x[k]$.
(e) $y[n] = x[n-1]u[n+1]$.
(f) $y[n] = nx[n]$.

Problem 1-32

Determine whether each of the following systems is linear. For each, show whether additivity and scaling are satisfied.
(a) $y[n] = nx[n]$.
(b) $y[n] = \cos(x[n])$.
(c) $y[n] = e^{j\omega x[n]}u[n]$.
(d) $y[n] = x[2n]$.

Problem 1-33

Determine whether each of the systems in Problem 1-32 is time-invariant.

Problem 1-34

Determine whether each of the following systems is causal, stable, linear and time-invariant.

		Linear	Time-invariant	Causal	Stable
(a)	$y[n] = 2x[n+1] + 1$				
(b)	$y[n+2] = x[n+2] - x[n+1]$				
(c)	$y[n] = nx[n-1]$				
(d)	$y[n] = \alpha^n x[n]u[n], \quad \alpha > 0$				
(e)	$y[n] = \cos(x[n])u[n]$				

Problem 1-35

(a) Show that the transformation $y[n] = T\{x[n]\} = x^*[n]$ satisfies the additivity property but not the scaling property.
➤**Hint:** Assume the scale factor is a complex number.

(b) Show that the transformation $y[n] = T\{x[n]\} = x^2[n]/x[n-1]$ satisfies the scaling property but not the additivity property.

2 Impulse response

Introduction

In the first section of this chapter, we will demonstrate a key result: for linear time-invariant (LTI) systems, the response of the system to an impulse $\delta[n]$ completely characterizes the system. Then, in Section 2.2, we will show that given the impulse response, we can calculate the output of the system to *any* input through the process of convolution. In Sections 2.3 and 2.4, we develop the properties of convolution and show that the stability and causality of LTI systems can be easily determined just by examination of the impulse response. Section 2.6 introduces deconvolution, the reverse operation of convolution. Section 2.7 explains several methods of applying convolution to input sequences of possibly infinite length.

2.1 FIR and IIR systems

Let us start by recalling the relation between the output and input of a system, shown graphically in **Figure 2.1**.

Figure 2.1 Relation of input and output of a system

$x[n]$ is the input to the system, $y[n]$ represents the output and $T\{\ \}$ represents the transformation that is applied to the input to produce the output $y[n] = T\{x[n]\}$. In the particular case where the input is an impulse, $x[n] = \delta[n]$, the output is given a special name, the **impulse response**, and is assigned the variable $h[n]$; that is,

$$h[n] \triangleq T\{\delta[n]\}.$$

There are two types of systems. **Finite impulse response (FIR)** systems have impulse responses that are finite in time duration. **Infinite impulse response (IIR)** systems have impulse responses that are infinite in time duration.

2.1.1 Finite impulse response (FIR) systems

The general input–output relation for an FIR system is

$$y[n] = \sum_{k=M}^{N} b_k x[n-k], \tag{2.1}$$

where b_k, k, N and M are specified constants. The response of this system to an impulse is easy to calculate. We simply set $x[n] = \delta[n]$ in Equation (2.1), which by definition makes the output equal to $h[n]$:

$$h[n] = \sum_{k=M}^{N} b_k \delta[n-k].$$

This equation makes it clear that the impulse response of an FIR system contains a finite number of impulses over a limited range of time; namely $M \leq n \leq N$.

Example 2.1

We have already seen examples of FIR systems in Chapter 1, for example, the moving-window average system defined by

$$y[n] = \tfrac{1}{3}(x[n-1] + x[n] + x[n+1]).$$

This is just an example of Equation (2.1) with

$$M = -1$$

$$N = +1$$

$$b_k = \begin{cases} \tfrac{1}{3}, & k = -1, 0, 1 \\ 0, & \text{otherwise} \end{cases}.$$

The impulse response of this system is obtained by setting $x[n] = \delta[n]$,

$$h[n] = \tfrac{1}{3}(\delta[n-1] + \delta[n] + \delta[n+1]).$$

This is shown in **Figure 2.2**.

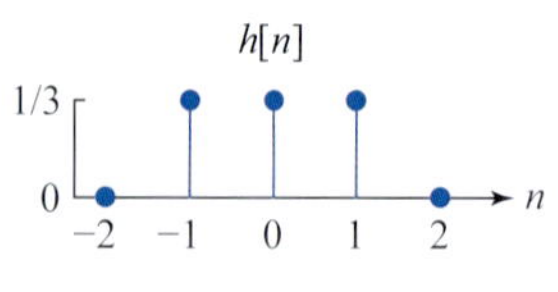

Figure 2.2 Impulse response of a moving-window average filter

The output of a causal FIR system depends only on the present and past values of the input. So for these systems, Equation (2.1) becomes

$$y[n] = \sum_{k=0}^{N} b_k x[n-k]. \tag{2.2}$$

2.1.2 Infinite impulse response (IIR) systems

The general input–output relation for a causal IIR system is defined by the linear constant-coefficient difference equation (LCCDE),

$$\sum_{k=0}^{N} a_k y[n-k] = \sum_{k=0}^{M} b_k x[n-k]. \tag{2.3}$$

Without loss of generality, let $a_0 = 1$. Then, rewriting this equation to solve for the output $y[n]$, we get

$$y[n] = \sum_{k=0}^{M} b_k x[n-k] - \sum_{k=1}^{N} a_k y[n-k]. \tag{2.4}$$

Comparing Equation (2.4) with Equation (2.2) reveals the major difference between FIR and IIR systems. For an FIR system, the output depends only on past and present values of the *input*; for an IIR system, the output also depends on past values of the *output*. Formal solution of LCCDEs such as Equation (2.4) requires analytical techniques similar to those used to solve linear constant-coefficient differential equations in the continuous-time domain. We will delve into these techniques in Chapter 4, but for now, let us just calculate the impulse response of a simple IIR system by inspection.

Example 2.2

A first-order IIR system is described by the LCCDE

$$y[n] - \tfrac{1}{2} y[n-1] = x[n].$$

This equation is obtained from Equation (2.3) by setting $a_0 = 1$, $a_1 = -1/2$, $b_0 = 1$ and all other a_k and b_k to zero. Find the output of this system $y[n]$ for all time when the input $x[n]$ is an impulse, assuming the initial condition that the output has been zero before we applied the impulse to the input.

➤ Solution:

The problem assumptions are

$$\begin{aligned} x[n] &= \delta[n], \\ y[n] &= 0, \quad n < 0. \end{aligned} \tag{2.5}$$

Starting with $n = 0$, we calculate

$$y[0] = \tfrac{1}{2} y[-1] + x[0] = \tfrac{1}{2} \cdot 0 + 1 = 1,$$

where we note from Equation (2.5) that $x[0] = \delta[0] = 1$ and $y[-1] = 0$. Proceeding with $n = 1$,

$$y[1] = \tfrac{1}{2} y[0] + x[1] = \tfrac{1}{2} \cdot 1 + 0 = \tfrac{1}{2}.$$

Note the recursive nature of this calculation. Each value of the output $y[n]$ requires that we have already calculated the previous value of the output $y[n-1]$. After calculating a few more time points in a similar fashion, we have the following table of values:

Table 2.1 Values of input and output for an LCCDE

n	$x[n]$	$y[n-1]$	$y[n] = \frac{1}{2}y[n-1] + x[n]$
<0	0	0	0
0	1	0	1
1	0	1	$\frac{1}{2}$
2	0	$\frac{1}{2}$	$\frac{1}{4}$
3	0	$\frac{1}{4}$	$\frac{1}{8}$
$\vdots$	$\vdots$	$\vdots$	$\vdots$
n		$\frac{1}{2}^{n-1}$	$\frac{1}{2}^{n}$

From this table, we see by inspection that the solution of the LCCDE is

$$\begin{aligned} y[n] &= \begin{cases} 0, & n < 0 \\ \frac{1}{2}^{n}, & n \geq 0 \end{cases} \\ &= \frac{1}{2}^{n} u[n]. \end{aligned} \tag{2.6}$$

Here, we have used the notion of the switch to express $y[n]$ simply as the product of a power-law sequence and a step. **Figure 2.3** shows the output of the system described by this LCCDE to an impulse. As you can see from the figure and from Equation (2.6), the response to the impulse is infinite in duration. That is why this is called an infinite impulse response (IIR) system.

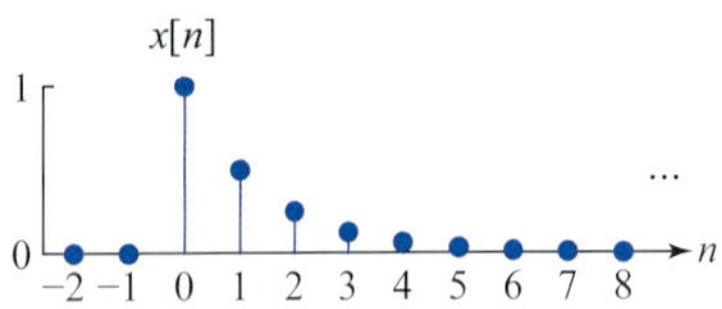

Figure 2.3 Response of an LCCDE system to an impulse

It is pretty unsatisfying to have to find the impulse response by inspection like this. In Chapter 4, we will show how to derive impulse responses of IIR systems analytically.

2.1.3 Response of a system to a flipped and shifted impulse

In Chapter 2, we introduced the notion of linear time-invariant (LTI) systems and claimed that they were particularly easy to analyze. In this section, we will show why this is so. First, let us consider the response of a general system (which may or may not be linear and/or time-invariant) to a shifted and scaled impulse.

Response of a system to a shifted impulse Define $h_k[n]$ as the response of the system to $\delta[n-k]$, an impulse that has been shifted by k samples:

$$h_k[n] \triangleq T\{\delta[n-k]\}.$$

What is the response of this system to a shifted impulse? In general, we cannot say. The result depends on the detailed specification of the system transformation. For example, consider the *time-variant* system defined by

$$y[n] = nx[n].$$

The impulse response of this system (the response of the system to an impulse at $n = 0$) is

$$h_0[n] = T\{\delta[n]\} = n\delta[n] = 0.$$

The output of the system to a shifted impulse $\delta[n-k]$ is

$$h_k[n] = T\{\delta[n-k]\} = n\delta[n-k].$$

Clearly, for this system, $h_k[n] \neq h[n-k]$. The response of the system to the impulse at $n = 0$ provides absolutely no information that could help us to predict the output of the system to a shifted impulse at $k \neq 0$. In general, this is true for systems that are not time-invariant, such as this one. However, now consider what happens if the system is *time-invariant*. Then it follows from the definition of time invariance that the response to a shifted impulse is the shifted impulse response; that is,

$$h_k[n] = T\{\delta[n-k]\} = h[n-k]. \tag{2.7}$$

So, if a system is time-invariant, the response of the system to a shifted impulse is *completely predicted* from the response of the system to an impulse at $n = 0$, namely the impulse response $h[n]$. In order to specify completely the system's response to a shifted impulse, we need know nothing about the details of the transformation $T\{\,\}$ except $h[n]$, the response of the system to the impulse.

Response to a scaled impulse What is the response of our general system to a scaled impulse, that is $x[n] = k\delta[n]$, where k is a constant? Again, in general we cannot say without a detailed specification of the system transformation $T\{\,\}$. As an example, consider the *nonlinear* system defined by the equation

$$y[n] = x[n] - \delta[n].$$

The impulse response of this system is

$$h[n] = T\{\delta[n]\} = \delta[n] - \delta[n] = 0,$$

and the output of the system to a scaled impulse $A\delta[n]$ is

$$T\{A\delta[n]\} = A\delta[n] - \delta[n] = (A-1)\delta[n].$$

For this system, the impulse response $h[n]$ obviously provides no information that would enable us to predict the output of the system to a scaled impulse. However, if the system is *linear*, then it follows immediately from the scaling property that the response of the system to a scaled impulse is the scaled impulse response; namely

$$T\{A\delta[n]\} = AT\{\delta[n]\} = Ah[n].$$

Thus, for a linear system, the impulse response carries all the information necessary to predict the response of the system to scaled impulses.

Impulse response of a linear time-invariant system Now, consider what happens if a system is *both* linear and time-invariant. Then, the results of the previous paragraphs show that the response of the system to a scaled and shifted impulse is the scaled and shifted impulse response. That is,

$$T\{A\delta[n-k]\} = Ah[n-k].$$

2.2 Convolution

We are now in a position to see why systems that are both linear and time-invariant are so important and easy to analyze. Recall from Chapter 1 that we can express any sequence $x[n]$ as the sum of scaled and shifted impulses:

$$\underbrace{x[n]}_{\text{sequence}} = \sum_{k=-\infty}^{\infty} \underbrace{x[k]}_{\text{value}} \underbrace{\delta[n-k]}_{\text{shifted sequence}}.$$

For each value of k in the sum, we shift $\delta[n]$ by k samples to form $\delta[n-k]$ and scale it by a value $x[k]$. Then we add all the scaled and shifted impulses together to produce $x[n]$. Now, the response of a general system described by transformation $T\{\ \}$ to input $x[n]$ is

$$y[n] = T\{x[n]\} = T\left\{\sum_{k=-\infty}^{\infty} x[k]\delta[n-k]\right\}.$$

If the system is linear and time-invariant, we can easily find the response of the system to $x[n]$. Given that the system is linear, the additivity property tells us that the transformation of the sum is equal to the sum of the transformations, so

$$y[n] = T\left\{\sum_{k=-\infty}^{\infty} x[k]\delta[n-k]\right\} = \sum_{k=-\infty}^{\infty} T\{x[k]\delta[n-k]\}. \tag{2.8}$$

The scaling property tells us that multiplying the input by a constant and then transforming it is equivalent to first transforming the input and then multiplying the result by the constant. For every k in the summation of Equation (2.8), $x[k]$ is a constant, so

$$y[n] = \sum_{k=-\infty}^{\infty} T\{x[k]\delta[n-k]\} = \sum_{k=-\infty}^{\infty} x[k]T\{\delta[n-k]\}.$$

Finally, if the system is time-invariant, then applying Equation (2.7) yields

$$y[n] = \sum_{k=-\infty}^{\infty} x[k]h[n-k]. \tag{2.9}$$

This is a significant result. It says that if you know $h[n]$, the response of a linear time-invariant system to an impulse $\delta[n]$, then you can predict the response of the system to *any other input* $x[n]$ without knowing anything else about the system. To put this in anthropomorphic terms, if a person were a linear time-invariant system, you could ask that person a *single* question and from their response to this one question you could predict their response to *any other question*.

The summation of Equation (2.9) is called the **convolution sum** and the mathematical operation of applying the summation to two sequences $x[n]$ and $h[n]$ to yield an output $y[n]$ is called **convolution**. To compute the response of an LTI system to an arbitrary input, take the impulse response, shift it by k to form $h[n-k]$, where k is the amount of the shift, scale it by the appropriate value of the input sequence $x[k]$ and then sum all the scaled and shifted impulse responses to form $y[n]$.

Convolution is a mathematical operation in the same sense that addition or multiplication is an operation on two sequences. The convolution operator is given the symbol "$*$," which is somewhat unfortunate, since many computer languages often use this symbol to denote multiplication. So, we write Equation (2.9) as

$$y[n] = x[n] * h[n] \triangleq \sum_{k=-\infty}^{\infty} x[k]h[n-k]. \tag{2.10}$$

In words, we say, "$y[n]$ equals $x[n]$ convolved with (not 'convoluted with') $h[n]$." Schematically we draw this convolution as in **Figure 2.4**.

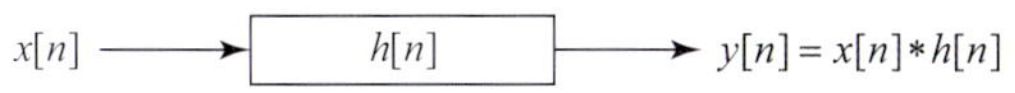

Figure 2.4 Convolution

By putting $h[n]$ in the box that represents an LTI system, this schematic reinforces the main point that the response of an LTI system is completely characterized by the impulse response and that all you need to determine the output of the system $y[n]$ to an arbitrary input $x[n]$ is the impulse response $h[n]$ and Equation (2.10).

There are two graphical ways of understanding the convolution sum formula of Equation (2.10): the **direct-summation method** and the **flip-and-shift method**.

2.2.1 Direct-summation method

To understand the direct-summation method, let us embellish Equation (2.10) with a few annotations:

$$\underbrace{y[n]}_{\substack{\text{sequence}\\ \text{on } n}} = \sum_{k=-\infty}^{\infty} \underbrace{x[k]}_{\text{value}}\, \underbrace{h[n-k]}_{\substack{\text{sequence}\\ \text{shifted by } k}}. \tag{2.11}$$

In this interpretation of convolution, $x[k]$ is a *value* and $h[n-k]$ is a *sequence* on index n shifted by an amount k for each value of k in the summation. The convolution sum says, for each k, shift the impulse response by k samples to form $h[n-k]$, and scale it by $x[k]$, the kth sample of the input sequence, thereby forming a scaled and shifted impulse response $x[k]h[n-k]$. The entire output sequence $y[n]$ is then computed as the sum of the scaled and shifted impulse responses for all values of k.

Example 2.3

Let $x[n]$ and $h[n]$ be a couple of simple sequences, shown in **Figure 2.5**. Find the convolution $y[n] = x[n] * h[n]$ by the direct-summation method.

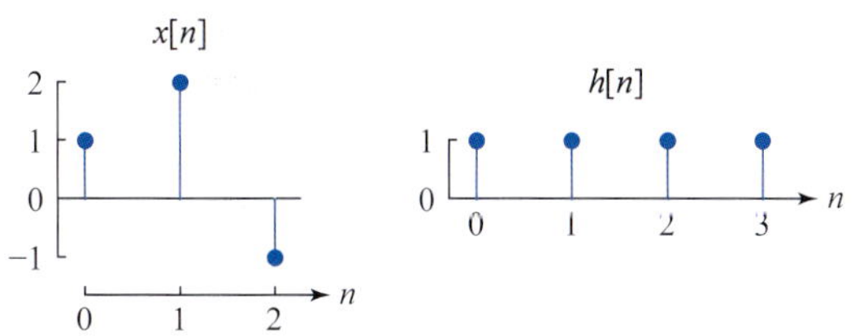

Figure 2.5 Sequences $x[n]$ and $h[n]$

➤ **Solution:**

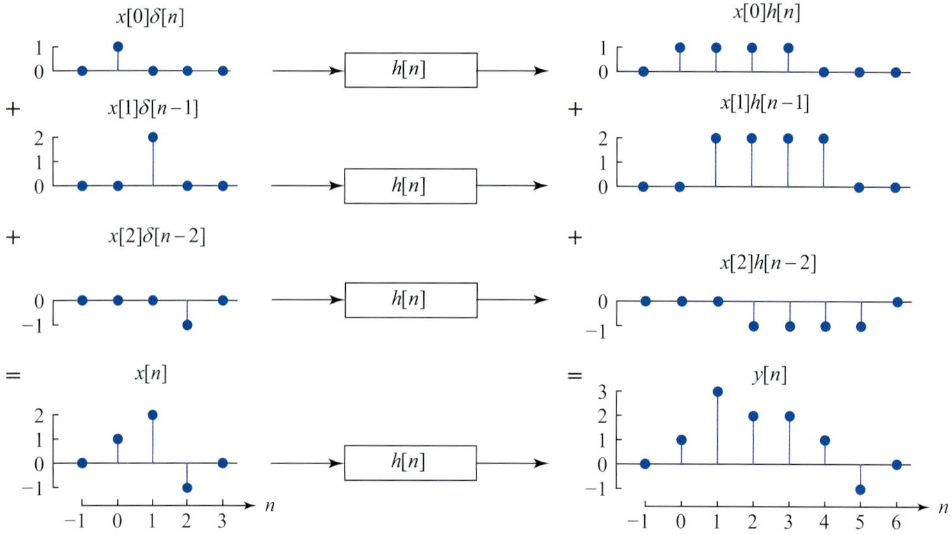

Figure 2.6 Convolution of $x[n]$ and $h[n]$ using the direct-summation method

First, write $x[n]$ as the sum of impulses,

$$\begin{aligned} x[n] &= \underbrace{\cdots}_{\text{all zero}} + x[0]\delta[n] + x[1]\delta[n-1] + x[2]\delta[n-2] + \underbrace{\cdots}_{\text{all zero}} \\ &= 1\cdot\delta[n] + 2\cdot\delta[n-1] - 1\cdot\delta[n-2]. \end{aligned} \tag{2.12}$$

When convolved with the impulse response $h[n]$, each scaled and shifted impulse in Equation (2.12) produces a scaled and shifted impulse response, so the first few terms of the convolution sum of Equation (2.11) are

$$\begin{aligned} y[n] &= \sum_{k=-\infty}^{\infty} x[k]h[n-k] \\ &= \underbrace{\cdots}_{\text{all zero}} + x[0]h[n] + x[1]h[n-1] + x[2]h[n-2] + \underbrace{\cdots}_{\text{all zero}} \\ &= 1\cdot h[n] + 2\cdot h[n-1] - 1\cdot h[n-2]. \end{aligned}$$

All terms of $x[n]$ for $n<0$ and $n>2$ are zero, so $y[n]$ just consists of three scaled and shifted impulse response terms. This is schematized in **Figure 2.6**.

The first three rows of the left column of the figure show $\delta[n]$ shifted by 0, 1 and 2 samples and multiplied by values $x[0]=1$, $x[1]=2$ and $x[2]=-1$ to form $x[0]\delta[n]$, $x[1]\delta[n-1]$ and $x[2]\delta[n-2]$, respectively. When summed (bottom panel), these scaled and shifted impulses add up to the input $x[n]$. The first three rows of the right column show $h[n]$ shifted by 0, 1 and 2 samples and multiplied by $x[0]=1$, $x[1]=2$ and $x[2]=-1$, respectively. When summed (bottom panel), these scaled and shifted impulse responses add up to the output $y[n]$.

The visualization of **Figure 2.6** shows that convolution is just the sum of scaled and shifted impulse responses. In the direct-summation method, we construct the entire output sequence $y[n]$ all at once by adding together these scaled and shifted impulse responses. In this example, we had to sum three sequences to get output $y[n]$.

2.2.2 Flip-and-shift method

The flip-and-shift method is another way of looking at convolution. In the flip-and-shift method, we interpret the same convolution sum formula in a different way,

$$\underbrace{y[n]}_{\substack{\text{value of } y\\ \text{at a given}\\ \text{value of } n}} = \sum_{k=-\infty}^{\infty} \underbrace{x[k]}_{\substack{\text{sequence}\\ \text{on } k}} \; \underbrace{h[n-k]}_{\substack{\text{sequence on } k,\\ \text{flipped and}\\ \text{shifted by } n}} .$$

In this interpretation, we compute the output sequence $y[n]$ *point-by-point*. $x[k]$ is interpreted as a sequence on index k, and $h[n-k]$ as a sequence on k that has been flipped to form $h[-k]$ and then shifted by value n to form $h[n-k]$. These two sequences on k are then multiplied together to form the product sequence $x[k]h[n-k]$. All the values of this product sequence are then summed on k to give a single value, which corresponds to $y[n]$ at the specified value of n. Then we choose the next value of n, compute the sum again to form the next value of $y[n]$ and so on.

Figure 2.7 shows the computation of convolution sum using the flip-and-shift method. This figure comprises six panels, each of which has four axes. The top left panel shows the computation of the sum for $n=0$. The first axis in the panel shows the sequence $x[k]$ as a function of k. The second axis shows $h[n-k]$, the flipped and shifted impulse response of the filter, again as a function of k. For this example, the amount of shift is $n=0$, so the sequence is just flipped. The red point indicates $h[0]$. The third axis shows the product of the two sequences, $x[k]h[n-k]=x[k]h[-k]$, which has only one non-zero element. The last axis shows all the values of $y[n]$ as a function of n up to $n=0$. At any value of n, $y[n]$ is the sum of all the points of the product $x[k]h[n-k]$. So, at $n=0$,

$$y[0]=\sum_{k=-\infty}^{\infty} x[k]h[0-k]=1\cdot 1=1.$$

This value is circled in the panel. The top middle panel shows the computation of the next point, $y[1]$. Again, the first axis shows $x[k]$, which is the same as in the previous panel. The second axis shows $h[1-k]$, which is the impulse response flipped and shifted to the right by one sample. The point with value $h[0]$ has shifted to the right to $k=1$. The next axis shows the product of these two sequences, $x[k]h[1-k]$, which has two non-zero elements. The sum of this product sequence is $y[1]$,

$$y[1]=\sum_{k=-\infty}^{\infty} x[k]h[1-k]=x[0]h[1]+x[1]h[0]=(1\cdot 1)+(2\cdot 1)=3.$$

The last axis shows the values of $y[n]$ up to $n=1$. The remaining panels of this figure show the computation of the remaining points of $y[n]$. Comparing the last (bottom right) panel of this figure with **Figure 2.6**, which was computed using the direct-summation method, you can see that the results are the same.

The essential difference between the direct-summation method and the flip-and-shift method is this: in the direct-summation method, all the points of the output sequence $y[n]$ are computed at once from the sum of scaled sequences. In the flip-and-shift method, the output sequence is

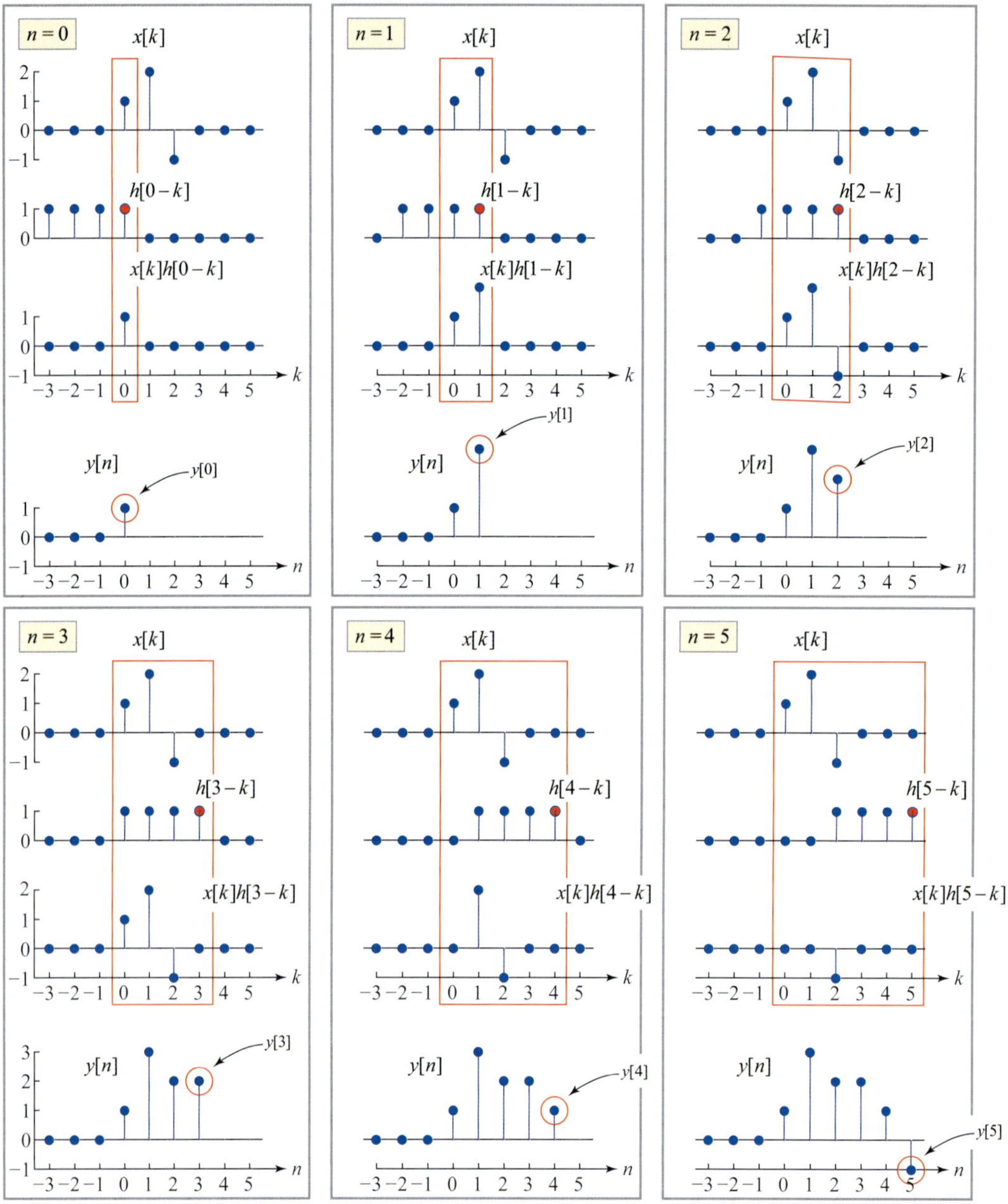

Figure 2.7 Convolution of $x[n]$ and $h[n]$ using the flip-and-shift method

computed one point at a time by multiplying the input sequence by the scaled and shifted output sequence and summing the product to get a single value of $y[n]$.

2.2.3 Convolution examples

In the previous examples described in this section, both the input sequence $x[n]$ and the impulse response $h[n]$ were of finite length. To gain further insight into convolution, let us do some examples in which the impulse response is of infinite duration and the input is of either finite or infinite duration.

Example 2.4

Consider the IIR system of Example 2.2,

$$y[n] - \frac{1}{2}y[n-1] = x[n].$$

The impulse response of this system was found to be

$$h[n] = \frac{1}{2}^n u[n]. \tag{2.13}$$

Find the response of this system to the finite-duration input $x[n] = \delta[n] + 2\delta[n-1] - \delta[n-2]$, shown in Figure 2.5.

➤ Solution:

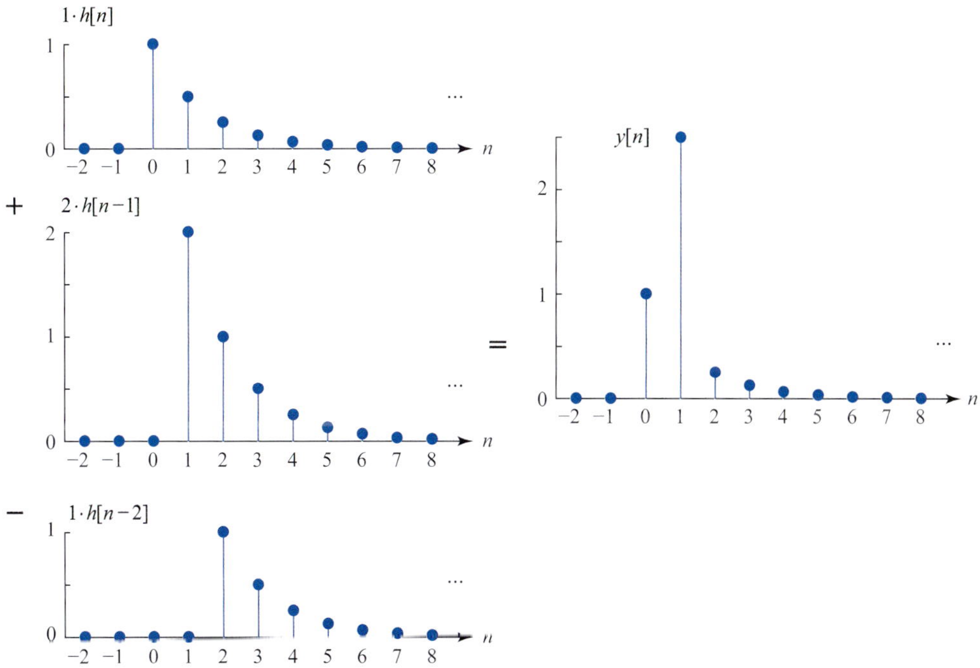

Figure 2.8 Response of an IIR system to a finite-length input

By the direct-summation method, the answer is

$$\begin{aligned} x[n]*h[n] &= \sum_{k=-\infty}^{\infty} x[k]h[n-k] = x[0]h[n] + x[1]h[n-1] + x[2]h[n-2] \\ &= 1 \cdot \frac{1}{2}^n u[n] + 2 \cdot \frac{1}{2}^{n-1} u[n-1] - 1 \cdot \frac{1}{2}^{n-2} u[n-2] \end{aligned}$$

The response is shown in Example 2.8. The second term can be written as

$$2 \cdot \frac{1}{2}^{n-1} u[n-1] = 2\delta[n-1] + \frac{1}{2}^{n-2} u[n-2],$$

so the sum of the terms simplifies to

$$y[n] = \frac{1}{2}^n u[n] + 2\delta[n-1].$$

Step response We now analyze the response of this IIR system to an infinitely long input, a step, $x[n] = u[n]$.

Example 2.5

Find the response of the IIR system of Example 2.2 to a step, $x[n] = u[n]$.

➤ Solution:

Given $x[n] = u[n]$, and $h[n]$ of Equation (2.13), the convolution formula yields

$$y[n] = \sum_{k=-\infty}^{\infty} x[k]h[n-k] = \sum_{k=-\infty}^{\infty} \underbrace{u[k]}_{\substack{\text{affects}\\ \text{bottom}\\ \text{limit}}} \left(\frac{1}{2}^{n-k} \overbrace{u[n-k]}^{\substack{\text{affects}\\ \text{top limit}}} \right). \tag{2.14}$$

The effect of the term $u[k]$ is to set to zero all terms in the summation for $k < 0$, which is equivalent to changing the lower limit of the sum to $k = 0$. We can say that $u[k]$ "turns on" the summation for $k \geq 0$. The effect of the term $u[n-k]$ is to set to zero all terms in the summation for $n - k < 0$, which is equivalent to $k > n$, so we can say that $u[n-k]$ "turns off" the summation for $k > n$. This is equivalent to changing the upper limit of the sum to $k = n$. Hence, we can rewrite Equation (2.14) as

$$y[n] = \frac{1}{2}^{n} \sum_{k=0}^{n} \frac{1}{2}^{-k} = \frac{1}{2}^{n} \frac{1 - \frac{1}{2}^{-(n+1)}}{1 - \frac{1}{2}^{-1}} = 2 - \frac{1}{2}^{n}. \tag{2.15}$$

This is *almost* correct, but not quite. To see what is wrong, consider what happens to the summation of Equation (2.14) when we evaluate $y[n]$ for $n < 0$. The term $u[k]$ turns on the summation for $k \geq 0$ and the term $u[n-k]$ turns off the summation for $k > n$. But if $n < 0$, then the summation is turned off for $k < 0$. Because the product $u[k]u[n-k]$ is zero for all k, the summation is zero. Therefore, $y[n]$ must also be zero for all $n < 0$. Put another way, the occurrence of the product $u[k]u[n-k]$ inside the summation does two things: (a) it sets the limits of the summation to $0 \leq k < n$ and (b) it sets the entire summation to zero if the value of the top limit of the summation (i.e., n) is less than the bottom limit of the summation (i.e., 0). Equation (2.15) does not reflect this latter condition. The summation is non-zero for all n, and is therefore incorrect. Adding this condition, we state:

$$y[n] = \begin{cases} 2 - \frac{1}{2}^{n}, & n \geq 0 \\ 0, & n < 0 \end{cases}.$$

The condition that $y[n]$ is non-zero for $n \geq 0$ and zero for $n < 0$ is economically expressed using the step function:

$$y[n] = \left(2 - \frac{1}{2}^{n}\right)u[n]. \tag{2.16}$$

The response of a system to a step is, unsurprisingly, called the **step response**. It is shown in **Figure 2.9**.

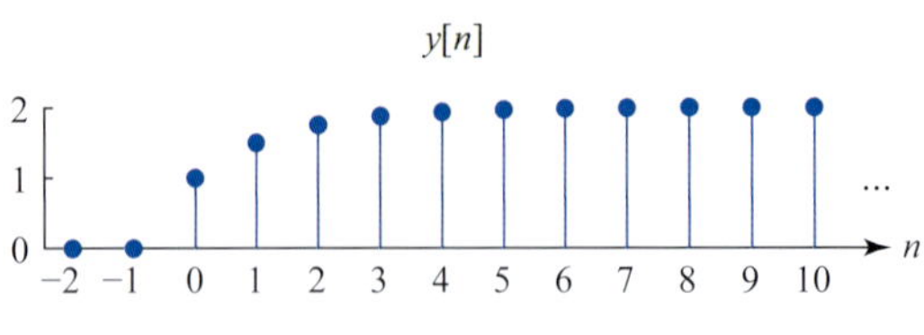

Figure 2.9 Response of an IIR system to a step

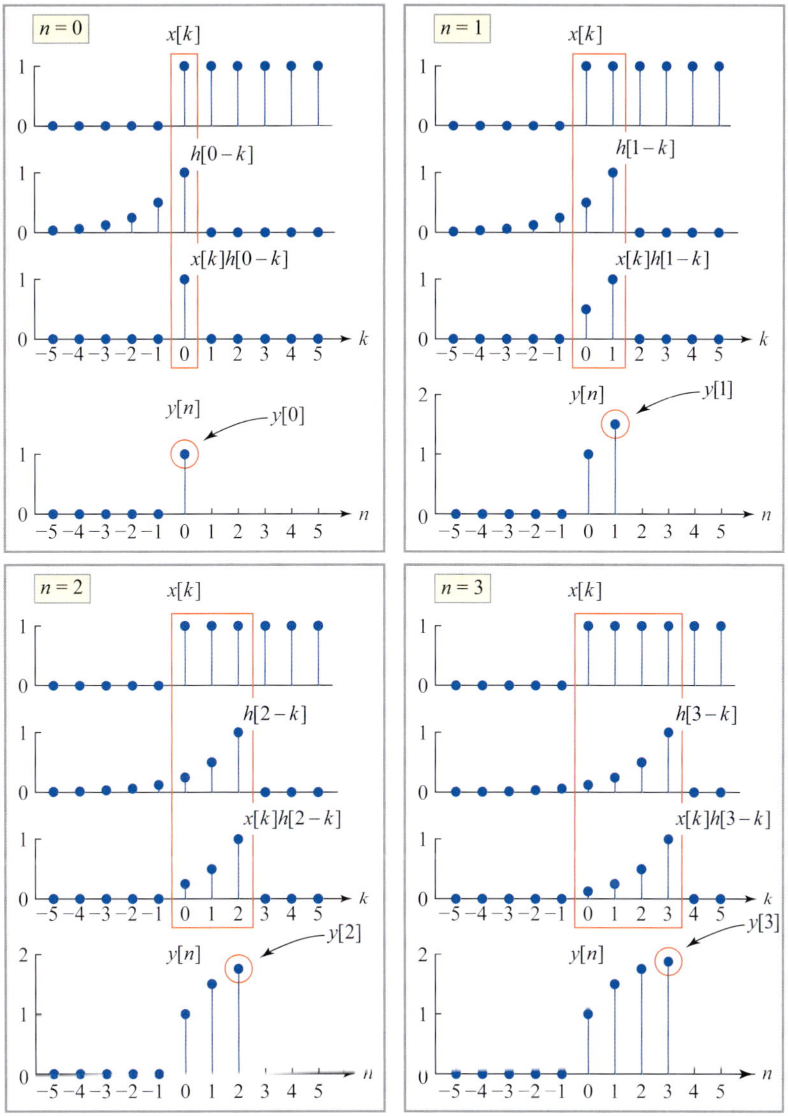

Figure 2.10 Step response of IIR system visualized with the flip-and-shift method

We can visualize the convolution of the step with the impulse response using either the direct-summation method or the flip-and-shift method. For example, **Figure 2.10** shows the first four points of the impulse response visualized with the flip-and-shift method.

Figure 2.11 shows the step response of the same system visualized using the direct-summation method. The top row of the figure shows an impulse $\delta[n]$ in the left column and the response of the system to an impulse $h[n]$ in the right column. Subsequent rows show the responses to shifted impulses $\delta[n-1]$, $\delta[n-2]$ and

$\delta[n-3]$, which are, of course, the shifted impulse responses $h[n-1]$, $h[n-2]$ and $h[n-3]$. The left column of the bottom row shows that the sum of all the shifted impulses forms a step, and the right column shows that the sum of all the shifted impulse responses forms the step response.

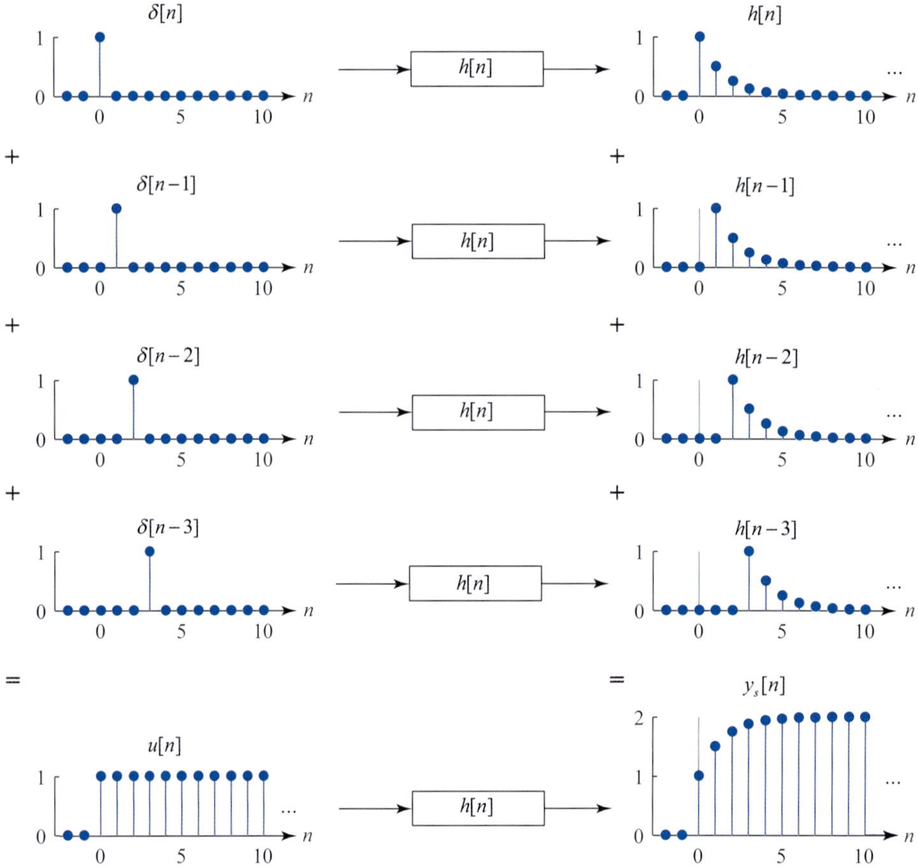

Figure 2.11 Step response of IIR system visualized with the direct-summation method

Response to a pulse The response of a system to a pulse is easy to compute given the step response.

Example 2.6

Analyze the response of the IIR system of Example 2.2 to a finite duration pulse of length N.

➤ Solution:

We can construct a pulse of length N as the difference of a step and a shifted step,

$$x[n] = u[n] - u[n-N]. \tag{2.17}$$

The construction of the pulse is shown in **Figure 2.12** for $N = 4$.

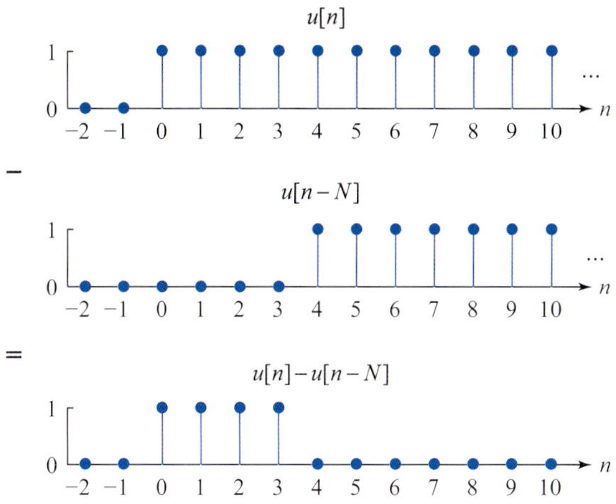

Figure 2.12 Construction of a pulse from two steps

There are several ways to solve this problem. We could obtain the solution by brute-force by plugging Equation (2.17) into the convolution summation (see Problem 2-9a). Alternately, recognize that since the pulse can be expressed as the difference of a step and a step shifted by N, $u[n]-u[n-N]$, therefore the response of the LTI system can be computed as the difference of the response to a step (the step response) and the step response shifted by N.

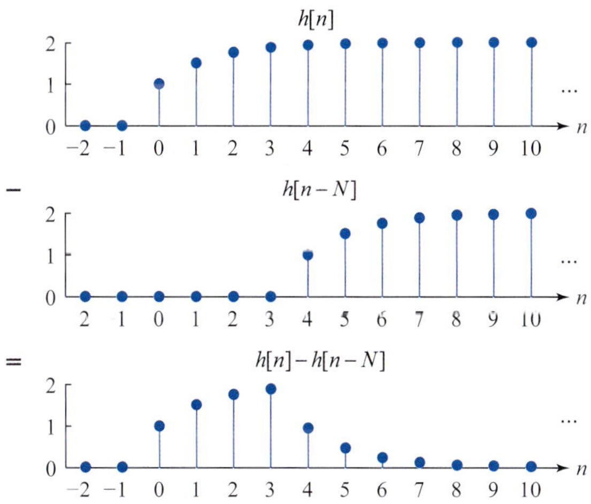

Figure 2.13 Response of an IIR filter to a pulse

Equation (2.16) gives us the response of the system to a step,

$$y[n]=\left(2-\tfrac{1}{2}^{n}\right)u[n],$$

so the response to a pulse is

$$y[n]=T\{u[n]\}-T\{u[n-N]\}=\left(2-\tfrac{1}{2}^{n}\right)u[n]-\left(2-\tfrac{1}{2}^{(n-N)}\right)u[n-N].$$

This result is shown in **Figure 2.13**.

2.3 Properties of convolution

Convolution is an operation that satisfies **commutative**, **associative** and **distributive** properties. Understanding these properties is not just a matter of theoretical interest; application of these properties has a number of useful practical consequences, including allowing more efficient algorithms to be created to perform common tasks.

2.3.1 The commutative property

The commutative property says that the order of convolution does not matter. That is,

$$x[n] * h[n] = h[n] * x[n].$$

The proof is simple. From the definition of convolution,

$$x[n] * h[n] = \sum_{k=-\infty}^{\infty} x[k]h[n-k].$$

Now let $m = n - k$, and substitute:

$$x[n] * h[n] = \sum_{m=\infty}^{-\infty} x[n-m]h[m] = \sum_{m=-\infty}^{\infty} h[m]x[n-m] = h[n] * x[n],$$

where we have recognized that $m \to -\infty$ as $k \to \infty$ and that the order of the summation does not matter (i.e., it does not matter if we sum from $m = -\infty \to +\infty$ or $m = +\infty \to -\infty$).

The commutative property can be understood graphically in terms of two experiments, as shown in **Figure 2.14**.

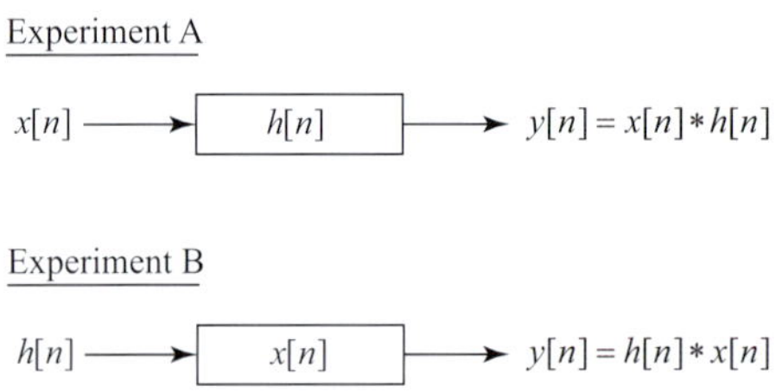

Figure 2.14 Commutative property of convolution

In Experiment A, we pass an input $x[n]$ through a system characterized by impulse response $h[n]$, yielding $y[n] = x[n] * h[n]$. In the second experiment, we convolve $h[n]$ with $x[n]$, which is equivalent to passing input $h[n]$ through a system characterized by impulse response $x[n]$, yielding $y[n] = h[n] * x[n]$. The commutative property says that the results of these two experiments are the same: $x[n] * h[n] = h[n] * x[n]$.

Example 2.7

Given $x[n]$ and $h[n]$, the sequences shown in **Figure 2.5**, show that the commutative property is satisfied: $x[n] * h[n] = h[n] * x[n]$.

➤ Solution:

Figure 2.15 shows the convolution of $h[n]$ and $x[n]$ using the data of **Figure 2.5** and the direct-summation method. As you can see, the results are the same as the convolution of $x[n]$ and $h[n]$ shown in **Figure 2.6**. You can obtain the same result using the flip-and-shift method (see Problem 2-35).

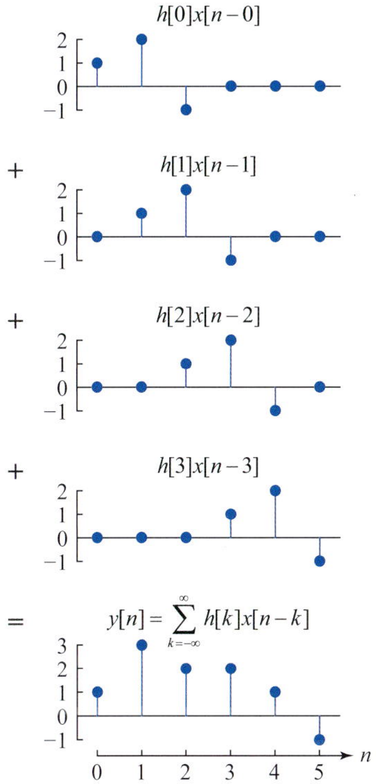

Figure 2.15 Convolution of $h[n]$ and $x[n]$ using direct-summation method

The commutative property can be used to aid in the computation of simple FIR filtering problems. When both the input $x[n]$ and the impulse response $h[n]$ start at $n = 0$, the output $y[n]$ also starts at $n = 0$. When either the input $x[n]$ or the impulse response $h[n]$ are offset from the origin by non-zero amounts, the commutative property provides a good way to understand what happens.

Example 2.8

Given $x[n]$ and $h[n]$ shown in **Figure 2.16**,

(a) Find $x[n] * h[n]$.

(b) Find $x[n-1] * h[n]$.

(c) Find $x[n] * h[n-1]$.

(d) Find $x[n-1] * h[n-1]$.

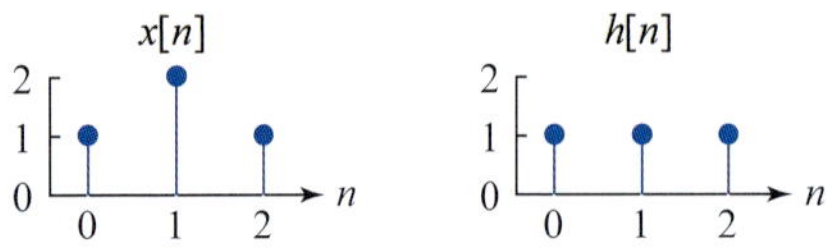

Figure 2.16

➤ Solution:

(a) $x[n]$, $h[n]$ and $y[n] = x[n] * h[n]$ are shown in **Figure 2.17a**.

(b) $x[n-1]$, $h[n]$ and the result of the convolution, $x[n-1] * h[n]$, are shown in **Figure 2.17b**. By time-invariance, $x[n-1] * h[n]$ is equivalent to $y[n-1]$.

(c) $x[n]$, $h[n-1]$ and the result of the convolution, $x[n] * h[n-1]$, are shown in **Figure 2.17c**. By the commutative property, $y[n] = x[n] * h[n] = h[n] * x[n]$. Hence, $x[n] * h[n-1] = h[n-1] * x[n]$, which by time-invariance is equivalent to $y[n-1]$.

(d) $x[n-1]$, $h[n-1]$ and the result of the convolution, $x[n-1] * h[n-1]$, are shown in **Figure 2.17d**. The output is equivalent to applying a delayed input to the previous example, hence $y[n-2]$.

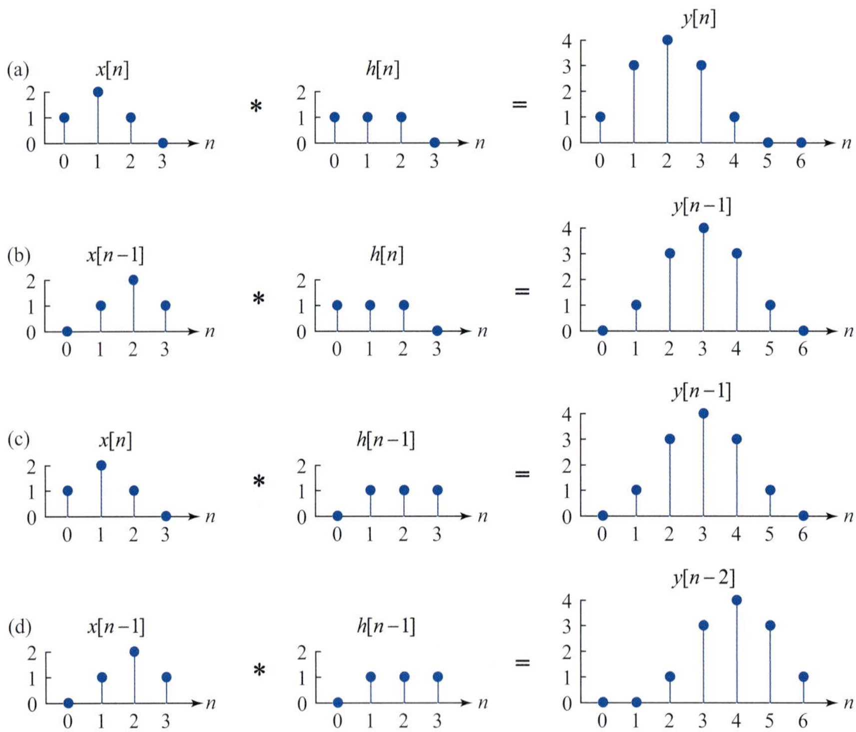

Figure 2.17

In general, if $x[n] * h[n] = y[n]$, then $x[n-n_x] * h[n-n_y] = y[n-n_x-n_y]$.

2.3.2 The associative property

The **associative property** says the convolution of an input with two filters having impulse responses $h_1[n]$ and $h_2[n]$, respectively, arranged in a **cascade** or **series** configuration is equivalent to the convolution of the input with a single filter whose impulse response is $h_1[n] * h_2[n]$. This can be understood as the result of two experiments, as shown in **Figure 2.18**.

In Experiment C, the input $x[n]$ is first passed through a filter with impulse response $h_1[n]$ to form output $w[n]$. Then $w[n]$ is input to a second filter with impulse response $h_2[n]$ to form output $y[n]$.

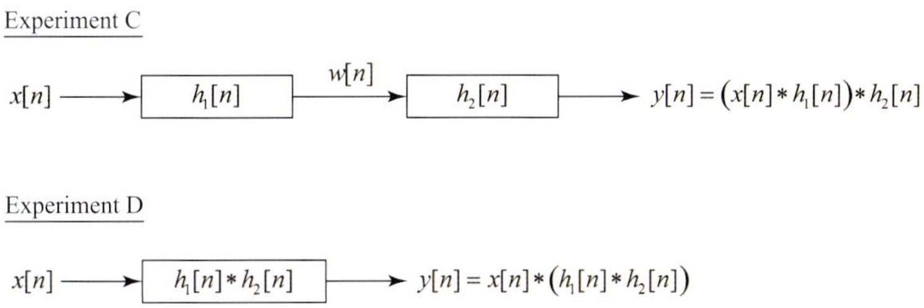

Figure 2.18 Associative property of convolution

Expressing $y[n]$ in terms of convolution, we get

$$y[n] = w[n] * h_2[n] = (x[n] * h_1[n]) * h_2[n]. \tag{2.18}$$

Writing out the convolution sum for $w[n]$ and $y[n]$:

$$w[n] = x[n]*h_1[n] = \sum_{l=-\infty}^{\infty} x[l]h_1[n-l],$$

and

$$\begin{aligned} y[n] &= w[n]*h_2[n] = \sum_{k=-\infty}^{\infty} w[k]h_2[n-k] = \sum_{k=-\infty}^{\infty} \left(\sum_{l=-\infty}^{\infty} x[l]h_1[k-l] \right) h_2[n-k] \\ &= \sum_{l=-\infty}^{\infty} x[l] \sum_{k=-\infty}^{\infty} h_1[k-l]h_2[n-k]. \end{aligned}$$

Let $m = k - l$ in the inner summation, and note that $m \to \infty$ as $k \to \infty$. So,

$$y[n] = \sum_{l=-\infty}^{\infty} x[l] \sum_{m=-\infty}^{\infty} h_1[m]h_2[n-(m+l)] = \sum_{l=-\infty}^{\infty} x[l] \underbrace{\sum_{m=-\infty}^{\infty} h_1[m]h_2[(n-l)-m]}_{h[n-l]}. \tag{2.19}$$

If we define $h[n]$ to be

$$h[n] \triangleq h_1[n]*h_2[n] = \sum_{m=-\infty}^{\infty} h_1[m]h_2[n-m],$$

then the inner summation in Equation (2.19) is just $h[n-l]$, so this equation becomes

$$y[n] = \sum_{l=-\infty}^{\infty} x[l]h[n-l] = x[n]*h[n] = x[n]*(h_1[n]*h_2[n]). \tag{2.20}$$

Comparing Equations (2.18) and (2.20), we see that

$$y[n] = (x[n] * h_1[n]) * h_2[n] = x[n] * (h_1[n] * h_2[n]), \tag{2.21}$$

which establishes the associative property.

Equation (2.21) states that the series convolution of $x[n]$ through a cascade of two filters in series with impulse responses $h_1[n]$ and $h_2[n]$, as shown in Experiment C, is equivalent to the convolution of $x[n]$ with a single filter with impulse response $h[n] = h_1[n] * h_2[n]$, as shown in Experiment D.

The associative property has practical consequences. Frequently, one is called upon to perform a cascade filtering operation, as shown in Experiment C of **Figure 2.18**. For example, we might wish to bandpass a signal $x[n]$ by passing it first through a lowpass filter characterized by impulse response $h_1[n]$, and then pass the output of this lowpass filter through a highpass filter characterized by impulse response $h_2[n]$, to create output $y[n]$. Processing $x[n]$ in this manner requires the implementation of *two* discrete-time filters in series and hence two convolution operations. Convolution is a computationally demanding operation, so it is desirable to do everything we can to reduce the number of convolutions we perform. The associative property tells us that we can implement the same bandpass filter by processing $x[n]$ through a single filter whose impulse response is given by $h[n] = h_1[n] * h_2[n]$. Processing $x[n]$ in this manner requires only one filter and hence one convolution operation. Use of the associative property allows us to cut our computational burden in half. Of course, calculating the impulse response of the effective bandpass filter $h[n]$ also requires a convolution, but this convolution only need happen *once* when we design the filtering system. In Chapter 9, we will discuss a variety of cascade filter architectures that derive from the associative property.

2.3.3 The distributive property

The **distributive property** says the convolution of an input with two filters whose impulse responses are $h_1[n]$ and $h_2[n]$ arranged in a **parallel** configuration is equivalent to the convolution of the input with a single filter whose impulse response is $h_1[n] + h_2[n]$. This can again be understood as the result of two experiments, as shown in **Figure 2.19**.

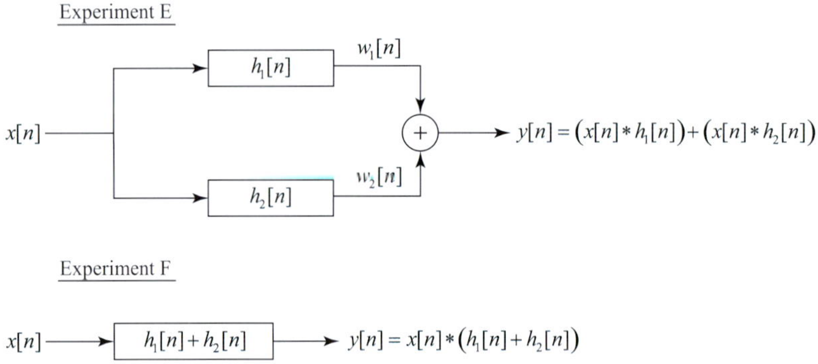

Figure 2.19 Distributive property of convolution

In Experiment E, the input $x[n]$ is passed through two filters in parallel. The output of the first filter, with impulse response $h_1[n]$, is $w_1[n]$. The output of the second filter, with impulse response $h_2[n]$, is $w_2[n]$. The output of these two filters is then added to form $y[n]$,

$$y[n] = x[n] * h_1[n] + x[n] * h_2[n]. \tag{2.22}$$

Expressing $w_1[n]$ and $w_2[n]$ in terms of the convolution sum, we have

$$w_1[n] = x[n] * h_1[n] = \sum_{k=-\infty}^{\infty} x[k]h_1[n-k],$$

and

$$w_2[n] = x[n] * h_2[n] = \sum_{k=-\infty}^{\infty} x[k]h_2[n-k].$$

So,

$$\begin{aligned} y[n] = w_1[n] + w_2[n] &= \sum_{k=-\infty}^{\infty} x[k]h_1[n-k] + \sum_{k=-\infty}^{\infty} x[k]h_2[n-k] \\ &= \sum_{k=-\infty}^{\infty} x[k]\underbrace{(h_1[n-k] + h_2[n-k])}_{h[n-k]}. \end{aligned} \tag{2.23}$$

If we define $h[n]$ to be

$$h[n] \triangleq h_1[n] + h_2[n],$$

then Equation (2.23) becomes

$$y[n] = \sum_{k=-\infty}^{\infty} x[k]h[n-k] = x[n] * h[n] = x[n] * (h_1[n] + h_2[n]). \tag{2.24}$$

Comparing Equations (2.22) and (2.24) we get

$$y[n] = x[n] * h_1[n] + x[n] * h_2[n] = x[n] * (h_1[n] + h_2[n]), \tag{2.25}$$

which establishes the distributive property. Equation (2.25) states that the sum of the convolution of $x[n]$ by two filters with impulse responses $h_1[n]$ and $h_2[n]$, arranged in parallel as shown in Experiment E, is equivalent to the convolution of $x[n]$ with a single filter with impulse response $h[n] = h_1[n] + h_2[n]$, shown in Experiment F.

The distributive property underlies a number of practical applications. In Chapter 9, we will discuss a variety of parallel filter architectures that derive from the distributive property. Also, a number of DSP applications call for implementation of parallel multi-channel **filter banks** – for example a graphic equalizer – of which Experiment E of **Figure 2.19** is a simple two-channel example. If there are N filters in a filter bank with impulse responses $h_1[n]$, $h_2[n]$, ..., $h_N[n]$, the distributive property tells us that the filter bank is equivalent to a single filter with impulse response $h[n] = (h_1[n] + h_2[n] + \cdots + h_N[n])$. We discuss some of these applications in Chapter 13.

2.4 Stability and causality

Having now defined the properties of LTI systems and introduced the notion of the impulse response, we are in a position to revisit the issues of stability and causality. We will show that it is easy to evaluate the stability and causality of LTI systems based solely on the impulse response.

2.4.1 Stability

In Chapter 1, we defined stability using the **bounded-input, bounded-output (BIBO)** criterion. A system was defined to be stable if every amplitude-bounded input gives rise to an amplitude-bounded output. In our discussion, we remarked that the BIBO criterion was not really very helpful as a practical test of stability, since it seemed to require that we test every possible bounded input to see if one could cause the output to become unbounded. Fortunately, for LTI systems, there is a simple test for the stability that is based on the impulse response: an LTI system is stable if and only if the impulse response is **absolutely summable**; that is, if the summation of $|h[n]|$ is finite:

$$\sum_{k=-\infty}^{\infty} |h[k]| < \infty. \tag{2.26}$$

To prove this powerful result, we need to show that absolute summability of the impulse response is both a *necessary* and a *sufficient* condition for stability. The sufficiency condition says that *if* the impulse response is absolutely summable, *then* the system is stable. For a system of **Figure 2.4**,

$$y[n] = x[n] * h[n] = \sum_{k=-\infty}^{\infty} x[k]h[n-k] = \sum_{k=-\infty}^{\infty} h[k]x[n-k]. \tag{2.27}$$

Stability in the BIBO sense means that the output is bounded in amplitude for every bounded input. That is, the system is stable if $|y[n]| < \infty$ for every $|x[n]| < \infty$. Equation (2.27) says that if we require $|y[n]| < \infty$, then

$$\left| \sum_{k=-\infty}^{\infty} h[k]x[n-k] \right| < \infty.$$

However, since the sum of absolute values is always greater than or equal to the absolute value of the sum,

$$\left| \sum_{k=-\infty}^{\infty} h[k]x[n-k] \right| \le \sum_{k=-\infty}^{\infty} |h[k]x[n-k]|.$$

The absolute value of the product is the product of the absolute values, so

$$\sum_{k=-\infty}^{\infty} |h[k]x[n-k]| = \sum_{k=-\infty}^{\infty} |h[k]||x[n-k]|.$$

Given a bounded input, then $|x[n-k]| \le x_{\max}$, for all n and k, where $x_{\max}$ is the maximum absolute value of the input. Thus,

$$\left| \sum_{k=-\infty}^{\infty} h[k]x[n-k] \right| \le \sum_{k=-\infty}^{\infty} |h[k]||x[n-k]| \le \sum_{k=-\infty}^{\infty} |h[k]|x_{\max}.$$

So, if the input $x[n]$ is bounded, then

$$\sum_{k=-\infty}^{\infty} |h[k]| x_{\max} < \infty, \tag{2.28}$$

and certainly

$$\left| \sum_{k=-\infty}^{\infty} h[k] x[n-k] \right| < \infty,$$

which means that $|y[n]| < \infty$ and the system will be stable. Dividing both sides of Equation (2.28) by $x_{\max}$ gives Equation (2.26), thereby showing that absolute summability of the impulse response is a sufficient condition for stability. To show that Equation (2.26) is a necessary condition for stability, we need to show that *if* the system is stable, *then* the impulse response is absolutely summable. It is easier to show the contrapositive (which is logically equivalent), namely that if the impulse response is *not* absolutely summable then the system is *not* stable in the BIBO sense. Given an impulse response $h[n]$ that is not absolutely summable, the necessary condition for stability will be satisfied if we can find any bounded input $x[n]$ that causes the output $y[n] = x[n] * h[n]$ to become unbounded.

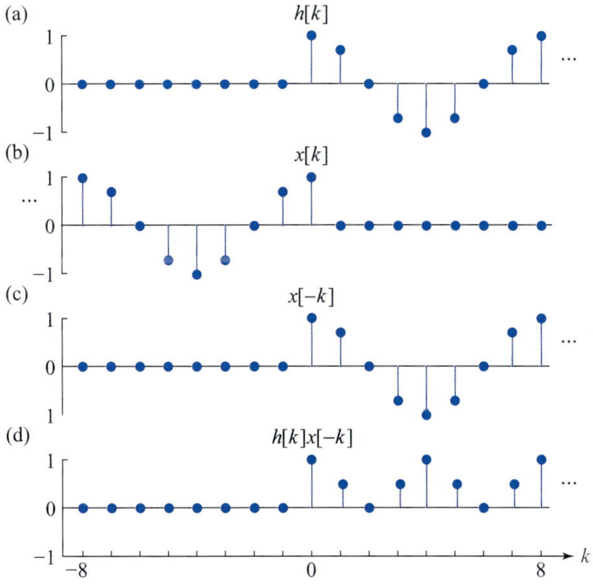

Figure 2.20 Necessary condition for BIBO stability

Consider a (bounded) impulse response of infinite duration, $h[n] = \cos(n\pi/4)\, u[n]$, a small portion of which is shown in **Figure 2.20a**. Clearly, the response is not absolutely summable since

$$\sum_{n=-\infty}^{\infty} |h[k]| = \infty.$$

Now, let the input to the system be the anticausal sequence $x[n] = \cos(-n\pi/4)\, u[-n]$, shown in **Figure 2.20b**. The output of the system is given by the convolution, Equation (2.27). Consider just one time point of the output, $y[0]$. From Equation (2.27),

$$y[0] = \sum_{k=-\infty}^{\infty} h[k]x[-k] = \sum_{k=-\infty}^{\infty} h^2[k] = \sum_{k=0}^{\infty} \cos^2 \pi n/4 = \infty.$$

We have found a bounded input $x[n]$ that makes the output of this system unbounded (infinite) at least at one time point, $n = 0$. This proves the necessity condition for BIBO stability.

Stability of FIR systems The impulse response of a general FIR system can be written as the sum of a finite number of scaled and shifted impulses:

$$h[n] = \sum_{k=M}^{N} h[k]\delta[n-k].$$

Since $h[n]$ is non-zero only over a limited range of time, namely $M \le n \le N$, the sum of these weighted impulses must be finite as long as each of the individual coefficients $h[k]$ is finite. Thus, *every FIR system is stable*. This is a very important conclusion and is one of the chief attractions of FIR filters. Chapter 7 will be devoted to the design of these filters.

Stability of IIR systems With Equation (2.3), we introduced the linear constant-coefficient difference equation (LCCDE) that defines the input–output relation for a causal IIR system:

$$\sum_{k=0}^{N} a_k y[n-k] = \sum_{k=0}^{M} b_k x[n-k].$$

We will defer the solution of these difference equations to Chapter 4, where we will discover that system stability depends upon the coefficients of the LCCDE, specifically the value of a_k. Nevertheless, it is worth considering a couple of specific examples now.

Example 2.9

Let us take a simple LCCDE, derived by setting $a_0 = 1$, $a_1 = -\alpha$, $b_0 = 1$, and all other a_k and b_k to zero in Equation (2.3),

$$y[n] - \alpha y[n-1] = x[n].$$

We can solve this LCCDE by an ad hoc method identical to that shown in Example 2.2, yielding

$$h[n] = \alpha^n u[n]. \tag{2.29}$$

To determine whether the system with this impulse response is stable, substitute Equation (2.29) into Equation (2.26),

$$\sum_{k=-\infty}^{\infty} |h[k]| = \sum_{k=-\infty}^{\infty} |\alpha^n u[n]| = \sum_{k=0}^{\infty} |\alpha|^n. \tag{2.30}$$

The conclusion is that the system may or may not be absolutely summable, depending on the value of α.

Case I, $|\alpha| < 1$

When $|\alpha| < 1$, $|\alpha|^k$ is a convergent series, so Equation (2.30) is absolutely summable,

$$\sum_{k=-\infty}^{\infty} |h[k]| = \sum_{k=0}^{\infty} |\alpha|^k = \frac{1}{1-|\alpha|}.$$

Hence, the system defined by Equation (2.3) is *stable*. We have already seen an example of a stable system in Example 2.2, where $\alpha = 1/2$.

Case II, $|\alpha| > 1$

When $|\alpha| > 1$, $|\alpha|^k$ is a divergent series, so the summation of Equation (2.30) is infinite and the system is not stable. In this case, it is obvious that the system is *not stable* by inspection of the impulse response.

Now consider the system defined by setting $a_0 = 1$, $a_1 = +1/2$, $b_0 = 1$, and all other a_k and b_k to zero in Equation (2.3),

$$y[n] + \tfrac{1}{2}y[n-1] = x[n].$$

This is identical to Equation (2.3), except that $a_1 = +1/2$ instead of $-1/2$.

Case III, $|\alpha| = 1$

When $|\alpha| = 1$, $|\alpha|^k = 1^k = 1$ is neither divergent nor convergent. Consider what happens if $\alpha = 1$. Then, the system has impulse response $h[n] = u[n]$. This impulse response is not absolutely summable, hence the system is technically unstable. However, it is easy to find inputs for which the output is bounded, for example $x[n] = \delta[n]$ or $x[n] = \delta[n] - \delta[n-1]$. Such systems are referred to as **quasi-stable**.

This example points out the practical value of the absolute summability test for stability over the BIBO test we discussed previously. The absolute summability test gives us an exact answer for the value of α that yields a stable system.

It is important to emphasize that stability is a property of the *system*, independent of any particular *signal* input into that system. If a system is stable, you are guaranteed that its output will never "blow up" (i.e., grow without bound), for *any* input. However, as we showed in the previous example, if a system is unstable, it is not necessarily true that its output will blow up for every input.

Example 2.10

Determine whether each of the following linear time invariant systems is stable or not.

(a) Discrete-time scalar multiplier: $y[n] = 2x[n]$.

(b) Delay: $y[n] = x[n - n_0]$.

(c) Moving-window average: $y[n] = \frac{1}{3}(x[n+1] + x[n] + x[n-1])$.

(d) Summer:

$$y[n] = \sum_{k=-\infty}^{n} x[k].$$

➤ Solution:

(a) Discrete-time scalar multiplier: The impulse response is $h[n] = 2\delta[n]$. Clearly this is absolutely summable, so the system is stable.

(b) Delay: The impulse response is $h[n] = \delta[n - n_0]$. This is absolutely summable, so the system is stable.

(c) Moving-window average: The impulse response is $h[n] = \frac{1}{3}(\delta[n+1] + \delta[n] + \delta[n-1])$. This is absolutely summable, so the system is stable.

(d) Summer: The impulse response is $h[n]=u[n]$. Since

$$\sum_{n=-\infty}^{\infty}|u[n]|=\sum_{n=0}^{\infty}1=\infty,$$

the system is not stable.

2.4.2 Causality

Formally, a system is defined to be **causal** if, at any time n_0, the output $y[n_0]$ depends on values of the input $x[n]$ for $n \le n_0$. A causal system is said to be **non-anticipative**, meaning that at any moment, the output of the system depends only on present and past values of the input, not on future values. Causality is particularly easy to establish for a linear time-invariant system. For an LTI system, we can cast the definition of causality in terms of the impulse response as follows. At time n_0, the output of the LTI system, $y[n_0]$, is

$$\begin{aligned} y[n_0] &= x[n_0] * h[n_0] = \sum_{k=-\infty}^{\infty} h[k]x[n_0-k] \\ &= \cdots h[-2]x[n_0+2]+h[-1]x[n_0+1]+h[0]x[n_0]+h[1]x[n_0-1]+h[2]x[n_0-2]\cdots. \end{aligned}$$

For values of $k<0$, the summation includes terms that depend upon future values of $x[n]$; that is, $x[n]$, $n>n_0$. The only way to assure that the system will be causal is to guarantee that terms of the form $h[k]x[n_0-k]$ are zero for $k<0$. Since we have no control of the input, the only way to prevent these terms from contributing to the summation is to make sure that $h[k]=0$, $k<0$. Thus, a linear time-invariant system will be causal if the impulse response is zero for negative time. This makes some intuitive sense: if the impulse response contains any impulses for $n<0$, then the output of the system to a single impulse at $n=0$ also contains impulse(s) for $n<0$, and the output to any arbitrary input must contain impulses that precede the input.

Example 2.11

Determine whether each of the following linear time-invariant systems is causal or not.

(a) Discrete-time scalar multiplier: $y[n]=2x[n]$.

(b) Delay: $y[n]=x[n-n_0]$.

(c) Moving-window average: $y[n]=\frac{1}{3}(x[n+1]+x[n]+x[n-1])$.

(d) Summer:

$$y[n]=\sum_{k=-\infty}^{n}x[k].$$

➤ Solution:

(a) Discrete-time scalar multiplier: The impulse response is $h[n]=2\delta[n]$. Clearly this is causal.

(b) Delay: The impulse response is $h[n]=\delta[n-n_0]$. This is causal for $n_0 \ge 0$, but not causal for $n_0<0$.

(c) Moving-window average: The impulse response is $h[n]=\frac{1}{3}(\delta[n+1]+\delta[n]+\delta[n-1])$. This is not causal due to the $\delta[n+1]$ term.

(d) Summer: The impulse response is $h[n]=u[n]$, hence, the system is causal.

2.5 ★ Convolution reinterpreted

In Section 2.2, we introduced two ways of looking at convolution, the direct-summation method and the flip-and-shift method. In this section, we revisit convolution and show that these two methods are just two sides of the same coin.

2.5.1 Convolution as polynomial multiplication

Start by expressing both $x[n]$ and $h[n]$ in terms of component impulses:

$$y[n] = x[n] * h[n] = \left(\sum_{k=-\infty}^{\infty} x[k]\delta[n-k]\right) * \left(\sum_{l=-\infty}^{\infty} h[l]\delta[n-l]\right)$$

$$= \begin{array}{c} (\cdots + x[-1]\delta[n+1] + x[0]\delta[n] + x[1]\delta[n-1] + \cdots) \\ * \\ (\cdots + h[-1]\delta[n+1] + h[0]\delta[n] + h[1]\delta[n-1] + \cdots) \end{array}$$

Each impulse term of $x[n]$ convolves each term of $h[n]$. In order to see this clearly, we can write this convolution in a manner similar to the way we multiply polynomials:

$$y[n] = \begin{Bmatrix} x[n] \\ * \\ h[n] \end{Bmatrix} = \begin{Bmatrix} \sum_{k=-\infty}^{\infty} x[k]\delta[n-k] \\ * \\ \sum_{l=-\infty}^{\infty} h[l]\delta[n-l] \end{Bmatrix} = \begin{Bmatrix} (\cdots + & x[-1]\delta[n+1] + & x[0]\delta[n] + & x[1]\delta[n-1] & + \cdots) \\ & & * & & \\ (\cdots + & h[-1]\delta[n+1] + & h[0]\delta[n] + & h[1]\delta[n-1] & + \cdots) \end{Bmatrix}$$

$$\vdots$$

$$\begin{array}{rrrrrrl}
 & & + \; x[-1]h[1]\delta[n] & + \; x[0]h[1]\delta[n-1] & + \; x[1]h[1]\delta[n-2] & + \cdots = & h[1]x[n-1] \\
 & + \; x[-1]h[0]\delta[n+1] & + \; x[0]h[0]\delta[n] & + \; x[1]h[0]\delta[n-1] & & + \cdots = & h[0]x[n] \\
\cdots + x[-1]h[-1]\delta[n+2] & + \; x[0]h[-1]\delta[n+1] & + \; x[1]h[-1]\delta[n] & \vdots & & + \cdots = & h[-1]x[n+1] \\
\hline
\cdots + \; y[-2]\delta[n+2] & + \; y[-1]\delta[n+1] & + \; y[0]\delta[n] & + \; y[1]\delta[n-1] & + \; y[2]\delta[n-2] & + \cdots = & y[n]
\end{array}$$

$$h[0]x[n] = h[0]\sum_{k=-\infty}^{\infty} x[k]\delta[n-k] \qquad y[0]\delta[n] = \left(\sum_{k=-\infty}^{\infty} x[k]h[0-k]\right)\delta[n] \qquad y[n] = \sum_{k=-\infty}^{\infty} h[k]x[n-k]$$

Each row of the convolution represents the input sequence shifted by k and scaled by $h[k]$; that is, $h[k]x[n-k]$ at a single value of k. For example, the row outlined in red corresponds to the value $k=0$, which is $(h[0]\delta[n]) * x[n] = h[0]x[n]$. The sum of all the rows, shown in green, is the sum of scaled and shifted input sequences for all n. This is equivalent to using the direct-summation method to evaluate the convolution sum,

$$y[n] = \sum_{k=-\infty}^{\infty} h[k]x[n-k],$$

at all values of n simultaneously. In contrast, each column of the convolution reflects the value of $y[n]$ at a single value of n. For example, the column outlined in blue is $y[0]\delta[n]$. The sum of all the elements of the column is equivalent to using the flip-and-shift method to evaluate the convolution sum,

$$y[n] = \sum_{k=-\infty}^{\infty} x[k]h[n-k],$$

at a single value of n, in this case, $n=0$.

This approach is very reminiscent of polynomial multiplication, and in fact we can use this idea as a shortcut to calculate the convolution of two sequences quickly. Consider the two sequences shown in **Figure 2.5**. **Table 2.2** shows this convolution interpreted as a sort of polynomial multiplication. The first two rows show the values of $x[n]$ and $h[n]$ arrayed as a function of n. The next four rows show $h[0]x[n]$, $h[1]x[n-1]$, $h[2]x[n-2]$ and $h[3]x[n-3]$. Shifting $x[n]$ to form $x[n-1]$ corresponds to moving all numbers to the right one column in the table. The last row of the table is the sum of the preceding four rows, which yields $y[n]$.

Table 2.2 Convolution example

n	0	1	2	3	4	5
$x[n]$	1	2	−1			
$h[n]$	1	1	1	1		
$h[0]x[n]$	1	2	−1			
$h[1]x[n-1]$		1	2	−1		
$h[2]x[n-2]$			1	2	−1	
$h[3]x[n-3]$				1	2	−1
$y[n]$	1	3	2	2	1	−1

Dispensing with the formality of the table, we write

$$
\begin{array}{rrrrrrr}
x[n]: & 1 & 2 & -1 & & & \\
h[n]: & 1 & 1 & 1 & 1 & & \\
\hline
 & 1 & 2 & -1 & & & \\
 & & 1 & 2 & -1 & & \\
 & & & 1 & 2 & -1 & \\
 & & & & 1 & 2 & -1 \\
\hline
y[n]: & 1 & 3 & 2 & 2 & 1 & -1.
\end{array}
$$

Since both $x[n]$ and $h[n]$ start at $n=0$, we know that the first point of $y[n]$ is also at $n=0$.

Convolution is commutative, so the convolution of $h[n]*x[n]$ yields the same results as the convolution of $x[n]*h[n]$:

$$
\begin{array}{rrrrrrr}
h[n]: & 1 & 1 & 1 & 1 & & \\
x[n]: & 1 & 2 & -1 & & & \\
\hline
 & 1 & 1 & 1 & 1 & & \\
 & & 2 & 2 & 2 & 2 & \\
 & & & -1 & -1 & -1 & -1 \\
\hline
y[n]: & 1 & 3 & 2 & 2 & 1 & -1.
\end{array}
$$

The first of these short-hand convolutions consists of four rows and six columns and the second one has three rows and six columns.

2.5.2 Convolution using Matlab

Matlab's `conv` function implements convolution of arrays.

Example 2.12

(a) Find the convolution of the two sequences $x[n]$ and $h[n]$, shown in **Figure 2.21**.

(b) Show that $x[n]*h[n]=h[n]*x[n]$.

(c) Check the results of the convolution with Matlab.

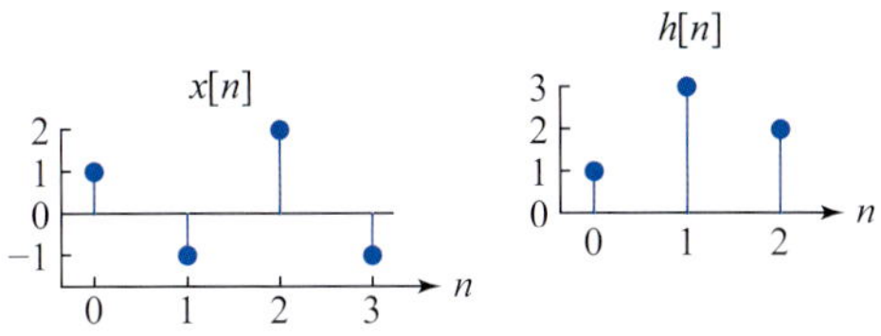

Figure 2.21 Two sequences

➤ **Solution:**

(a)

$h[n]$:	1	3	2			
$x[n]$:	1	−1	2	−1		
	1	3	2			
		−1	−3	−2		
			2	6	4	
				−1	−3	−2
$y[n]$:	1	2	1	3	1	−2.

(b)

$x[n]$:	1	−1	2	−1		
$h[n]$:	1	3	2			
	1	−1	2	−1		
		3	−3	6	−3	
			2	−2	4	−2
$y[n]$:	1	2	1	3	1	−2.

(c)

```
>> conv( [1 3 2], [1 -1 2 -1])
ans =
     1  2  1  3  1 -2
>> conv( [1 -1 2 1], [1 3 2])
ans =
     1  2  1  3  1  2
```

2.5.3 Convolution as matrix multiplication

Convolution can be expressed compactly in terms of matrix multiplication. To do so, express the input sequence $x[n]$ and output sequence $y[n]$ as row vectors,[1]

$$\begin{aligned}\mathbf{x}^T &= [y[n-1]\ y[n-2]\ \cdots\ y[n-N]]\\ \mathbf{y}^T &= [y[n-1]\ y[n-2]\ \cdots\ y[n-N]].\end{aligned}$$

Then convolution is expressed as a matrix multiplication, $\mathbf{y}^T = \mathbf{x}^T\mathbf{H}$,

[1] Here, we follow standard matrix notational convention, wherein $\mathbf{x}$ represents a column vector and $\mathbf{x}^T$ a row vector.

$$\underbrace{[\cdots y[-2]\, y[-1]\, y[0]\, y[1]\, y[2] \cdots]}_{\mathbf{y}^T} = \underbrace{[\cdots x[-1]\, x[0]\, x[1] \cdots]}_{\mathbf{x}^T} \underbrace{\begin{bmatrix} & & \vdots & & & \\ & h[-1] & h[0] & h[1] & h[2] & h[3] & \\ \cdots & h[-2] & h[-1] & h[0] & h[1] & h[2] & \cdots \\ & h[-3] & h[-2] & h[-1] & h[0] & h[1] & \\ & & \vdots & & & \end{bmatrix}}_{\mathbf{H}}.$$

$\mathbf{H}$ is a **Toeplitz matrix**, which has the property that each diagonal of the matrix descending from left to right has a constant value. The kth row of $\mathbf{H}$ corresponds to $h[n-k]$. We can interpret this matrix multiplication in a couple of different ways. One way is to multiply each row of $\mathbf{H}$ (i.e., $h[n-k]$) by the appropriate element of $\mathbf{x}^T$(i.e., $x[k]$), so that each row of the resulting matrix is equal to $x[k]h[n-k]$. Then sum all the rows to give $\mathbf{y}^T$:

$$\mathbf{y}^T = \mathbf{x}^T\mathbf{H} = \begin{array}{ccccccc} & & & \vdots & & & \\ +\cdots & x[-1]h[-1] & x[-1]h[0] & x[-1]h[1] & x[-1]h[2] & x[-1]h[3] & \cdots \\ +\cdots & x[0]h[-2] & x[0]h[-1] & x[0]h[0] & x[0]h[1] & x[0]h[2] & \cdots \\ +\cdots & x[1]h[-3] & x[1]h[-2] & x[1]h[-1] & x[1]h[0] & x[1]h[1] & \cdots \\ & & & \vdots & & & \\ \hline \cdots & y[-2] & y[-1] & y[0] & y[1] & y[2] & \end{array}$$

This way of interpreting matrix multiplication is exactly equivalent to the direct-summation method of convolution. The other, more usual, method of thinking about matrix multiplication is to calculate each element of the row vector $\mathbf{y}^T$ separately: the nth element of $\mathbf{y}^T$(i.e., $y[n]$) is calculated by multiplying the row vector $\mathbf{x}^T$ by the nth column of $\mathbf{H}$ and summing:

$$\begin{aligned} &\vdots \\ y[-2] &= \sum_{k=-\infty}^{\infty} x[k]h[-2-k] = \cdots + x[-1]h[-1] + x[0]h[-2] + x[1]h[-3] + \cdots \\ y[-1] &= \sum_{k=-\infty}^{\infty} x[k]h[-1-k] = \cdots + x[-1]h[0] + x[0]h[-1] + x[1]h[-2] + \cdots \\ y[0] &= \sum_{k=-\infty}^{\infty} x[k]h[0-k] = \cdots + x[-1]h[1] + x[0]h[0] + x[1]h[-1] + \cdots \\ y[1] &= \sum_{k=-\infty}^{\infty} x[k]h[1-k] = \cdots + x[-1]h[2] + x[0]h[1] + x[1]h[0] + \cdots \\ y[2] &= \sum_{k=-\infty}^{\infty} x[k]h[2-k] = \cdots + x[-1]h[3] + x[0]h[2] + x[1]h[1] + \cdots \\ &\vdots \end{aligned}$$

Each column of $\mathbf{H}$ corresponds to $h[n-k]$, which can be interpreted as $h[k]$ flipped and shifted by different values of n. Thus, this interpretation of matrix multiplication is exactly equal to the flip-and-shift method.

Example 2.13

(a) Perform the convolution of the two sequences $x[n]$ and $h[n]$ shown in **Figure 2.21** using matrix multiplication: $\mathbf{y}^T = \mathbf{x}^T\mathbf{H}$.

(b) Find $\mathbf{y}^T = \mathbf{h}^T\mathbf{X}$, showing that the commutative property applies.

➤ **Solution:**

(a)

$$\mathbf{y}^T = \mathbf{x}^T\mathbf{H} = [1\ -1\ 2\ -1]\begin{bmatrix} 1 & 3 & 2 & 0 & 0 & 0 \\ 0 & 1 & 3 & 2 & 0 & 0 \\ 0 & 0 & 1 & 3 & 2 & 0 \\ 0 & 0 & 0 & 1 & 3 & 2 \end{bmatrix} = [1\ 2\ 1\ 3\ 1\ -2].$$

In Matlab, we can use the `toeplitz` function to set up the matrix, **H**:

```
y = x*toeplitz([1 zeros(1,length(x)-1)], [h zeros(1,length(x)-1)])
```

(b)

$$\mathbf{y}^T = \mathbf{h}^T\mathbf{X} = [1\ 3\ 2]\begin{bmatrix} 1 & -1 & 2 & -1 & 0 & 0 \\ 0 & 1 & -1 & 2 & -1 & 0 \\ 0 & 0 & 1 & -1 & 2 & -1 \end{bmatrix} = [1\ 2\ 1\ 3\ 1\ -2],$$

or in Matlab:

```
y = h*toeplitz([1 zeros(1,length(h)-1)], [x zeros(1,length(h)-1)])
```

Given finite-length sequences $x[n]$ and $h[n]$, the commutative property of convolution assures us that $x[n] * h[n]$ is theoretically identical to $h[n] * x[n]$. However in practical terms, the number of operations necessary to achieve these two convolutions can differ. Notice that in Example 2.13a, the number of rows of **H** is equal to the length of $\mathbf{x}^T$, while the number of columns is equal to the length of the output $\mathbf{y}^T$, which is the length of $\mathbf{x}^T$ plus the length of $\mathbf{x}^T$ minus one. In Example 2.13b, the number of rows of **X** is equal to the length of $\mathbf{h}^T$, while the number of columns is the same as **H**. This means that it takes 24 multiplications and 18 additions to compute $\mathbf{y}^T = \mathbf{x}^T\mathbf{H}$, but only 18 multiplications and 12 additions to compute $\mathbf{y}^T = \mathbf{h}^T\mathbf{X}$.

2.6 ★ Deconvolution

In the previous sections, we have discussed at length the meaning and mechanics of convolution: for a linear time-invariant system with input $x[n]$ and impulse response $h[n]$, the output $y[n]$ is given by the convolution $y[n] = x[n] * h[n]$. Now we ask a related question: if we are given the output and the impulse response, can we find the input that produced this output? The answer is "yes," and in Chapters 3 and 4, we will derive the frequency-domain techniques for solving this question in a general manner. For now, let us look at a few simple examples.

For simplicity, let $y[n]$ and $h[n]$ start at $n = 0$. Then we can write,

$$\begin{aligned} y[n] &= \sum_{k=-\infty}^{\infty} h[k]x[n-k] = \sum_{k=0}^{\infty} h[k]x[n-k] = h[0]x[n] + h[1]x[n-1] + \cdots \\ &= h[0]x[0]\delta[n] + h[0]x[1]\delta[n-1] + h[0]x[2]\delta[n-2] + h[0]x[3]\delta[n-3] + \cdots \\ &\qquad + h[1]x[0]\delta[n-1] + h[1]x[1]\delta[n-2] + h[1]x[2]\delta[n-3] + \cdots \\ &\qquad\qquad h[2]x[0]\delta[n-2] + h[2]x[1]\delta[n-3] + \cdots \\ &\qquad\qquad\qquad + h[3]x[0]\delta[n-3] + \cdots \\ \hline &\quad y[0]\delta[n] + \quad y[1]\delta[n-1] + \quad y[2]\delta[n-2] + \quad y[3]\delta[n-3] + \cdots \end{aligned}$$

Adding the columns, we have

$$\begin{aligned}
y[0] &= x[0]h[0] \\
y[1] &= x[0]h[1] + x[1]h[0] \\
y[2] &= x[0]h[2] + x[1]h[1] + x[2]h[0] \\
y[3] &= x[0]h[3] + x[1]h[2] + x[2]h[1] + x[3]h[0]\,. \\
&\vdots
\end{aligned}$$

The first equation gives $x[0]$ directly in terms of known values $x[0]$ and $h[0]$. Substituting this value into the second equation allows us to solve for $x[1]$. Substituting the values of $x[0]$ and $x[1]$ into the third equation gives $x[2]$, and so on. This deconvolution is an example of a recursive process called **back substitution** in which the value of $x[n]$ at a particular value of n depends on the values of $h[k]$ for $k < n$ as well as all the previous values of $x[k]$ for $k < n$:

$$\begin{aligned}
x[0] &= \frac{y[0]}{h[0]} \\
x[1] &= \frac{y[1] - x[0]h[1]}{h[0]} = \left(\frac{y[1] - h[1]\dfrac{y[0]}{h[0]}}{h[0]} \right) \\
&\vdots \\
x[n] &= \frac{y[n] - \sum_{k=0}^{n-1} x[k]h[n-k]}{h[0]}\,.
\end{aligned} \tag{2.31}$$

Deconvolution as polynomial division Although it may not be obvious, deconvolution can be viewed as a sort of long division, where the divisor is $h[n]$ and the dividend is $y[n]$:

$$\begin{array}{r|l}
 & \dfrac{y[0]}{h[0]}\delta[n] \quad + \left(\dfrac{y[1] - h[1]\dfrac{y[0]}{h[0]}}{h[0]} \right) \delta[n-1] + \cdots \\ \hline
h[0]\delta[n] + h[1]\delta[n-1] + h[2]\delta[n-2] + \cdots & y[0]\delta[n] \quad + \quad y[1]\delta[n-1] + \cdots \\
 & y[0]\delta[n] + \quad h[1]\dfrac{y[0]}{h[0]}\delta[n-1] + \cdots \\ \hline
 & \left(y[1] - h[1]\dfrac{y[0]}{h[0]} \right) \delta[n-1] + \cdots
\end{array} \tag{2.32}$$

This is really not a wonderful notation. We will fix this when we discuss the z-transform in Chapter 4.

Deconvolution using Matlab Matlab can be used to deconvolve arrays. The syntax is

```
[x,r] = deconv(y,h),
```

where the function deconvolves `y` and `h` to produce `x` and a remainder vector `r` of the same length as `y`, such that `y = conv(x, h) + r`.

Deconvolution via matrix inversion If we choose to express $x[n]$, $h[n]$ and $y[n]$ as matrices, we have $\mathbf{y}^T = \mathbf{x}^T\mathbf{H}$. Deconvolution of $\mathbf{y}^T$ to obtain $\mathbf{x}^T$ is equivalent to a matrix inversion: $\mathbf{x}^T = \mathbf{y}^T\mathbf{H}^{-1}$. In fact, we can compute the inverse using exactly the same equations shown in Equation (2.31). The example, below, should make all this a bit clearer.

Example 2.14

Given the sequences $h[n]$ and $y[n]$ shown in **Figure 2.22**, find the input $x[n]$ by the following methods:

(a) the recursive method of Equation (2.31)

(b) "long division" as in Equation (2.32)

(c) matrix inversion

(d) using Matlab's `deconv` function.

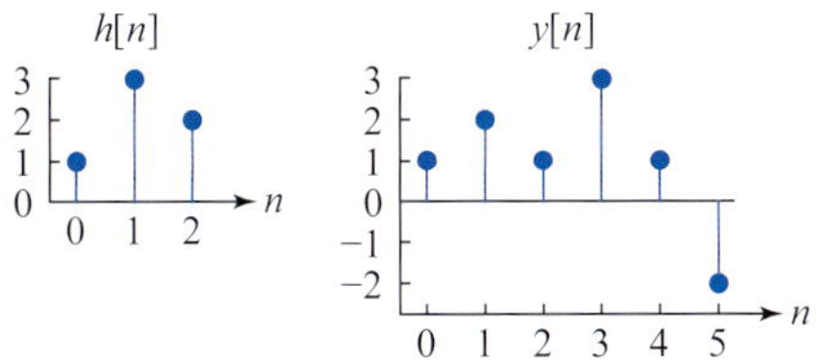

Figure 2.22 $h[n]$ and $y[n]$

➤ Solution:

(a) For the convolution of two finite-length sequences, length($y[n]$) = length($x[n]$) + length($h[n]$) − 1. Here, length($x[n]$) = length($y[n]$) − length($h[n]$) + 1 = 6 − 3 + 1 = 4. The first point of $x[n]$ must be at $n = 0$, because both $h[n]$ and $y[n]$ start at $n = 0$. So we can write $x[n]$ as

$$x[n] = x[0]\delta[n] + x[1]\delta[n-1] + x[2]\delta[n-2] + x[3]\delta[n-3].$$

Then,

$$\begin{aligned} y[n] = \sum_{k=-\infty}^{\infty} h[k]x[n-k] &= \sum_{k=0}^{\infty} h[k]x[n-k] = h[0]x[n] + h[1]x[n-1] + \cdots \\ &= h[0]x[0]\delta[n] + h[0]x[1]\delta[n-1] + h[0]x[2]\delta[n-2] + h[0]x[3]\delta[n-3] + \cdots \\ &\quad + h[1]x[0]\delta[n-1] + h[1]x[1]\delta[n-2] + h[1]x[2]\delta[n-3] + \cdots \\ &\quad + h[2]x[0]\delta[n-2] + h[2]x[1]\delta[n-3] + \cdots \\ &\quad + h[3]x[0]\delta[n-3] + \cdots \\ \hline &\quad y[0]\delta[n] + \quad y[1]\delta[n-1] + \quad y[2]\delta[n-2] + \quad y[3]\delta[n-3] + \cdots \end{aligned}$$

So,

$$\begin{array}{lclllllll} y[0]=1 & = & x[0] \\ y[1]=2 & = & 3x[0] & + & x[1] \\ y[2]=1 & = & 2x[0] & + & 3x[1] & + & x[2] \\ y[3]=3 & = & & & 2x[1] & + & 3x[2] & + & x[3] \\ y[4]=1 & = & & & & & 2x[2] & + & 3x[3] \\ y[5]=-2 & = & & & & & & & 2x[3]. \end{array} \tag{2.33}$$

You could solve this system of simultaneous equations by standard methods (e.g., Kramer's rule,

Gaussian elimination), but these equations are most easily solved by iterative back-substitution. The first equation gives $x[0] = 1$ directly. Substituting this value back into the second equation gives $x[1] = y[1] - 3x[0] = 2 - 3 = -1$. Substituting the values of $x[0]$ and $x[1]$ into the third equation gives $x[2] = y[2] - 2x[0] - 3x[1] = 2$. Finally, the fourth equation gives $x[3] = y[3] - 2x[1] - 3x[2] = -1$. This is the same $x[n]$ seen in **Figure 2.21**.

Because $x[n]$ only has four values, we only need to solve four unique equations. In the example above, we chose the first four equations, but we could have chosen any four rows.

(b) Let us also see this result by a "long division" shortcut:

$$
\begin{array}{rrrrrrrrr}
 & & & 1 & -1 & 2 & -1 & & \\
\cline{4-9}
1 & 3 & 2 &)1 & 2 & 1 & 3 & 1 & -2 \\
 & & & 1 & 3 & 2 & & & \\
\cline{4-6}
 & & & & -1 & -1 & 3 & & \\
 & & & & -1 & -3 & -2 & & \\
\cline{5-7}
 & & & & & 2 & 5 & 1 & \\
 & & & & & 2 & 6 & 4 & \\
\cline{6-8}
 & & & & & & -1 & -3 & -2.
\end{array}
$$

(c) Now, let us do the same deconvolution by matrix inversion. Because $h[n]$ is of length three and $y[n]$ is of length six, therefore $x[n]$ must be of length four and $\mathbf{x}^T$ must therefore be a 1×4 vector. So, if $\mathbf{x}^T = \mathbf{y}^T\mathbf{H}^{-1}$, $\mathbf{y}$ must be a 4×4 vector and $\mathbf{H}$ must be a 4×4 square matrix so that its inverse is also 4×4. We arbitrarily chose the first four values of $\mathbf{y}^T$ to be the first four rows of Equation (2.33), and form a 4×4 submatrix of $\mathbf{H}$: $\hat{\mathbf{y}}^T = \mathbf{x}^T\hat{\mathbf{H}}$, where $\hat{\mathbf{y}}^T = [1\ \ 2\ \ 1\ \ 3]$ and

$$
\hat{\mathbf{H}} = \begin{bmatrix} 1 & 3 & 2 & 0 \\ 0 & 1 & 3 & 2 \\ 0 & 0 & 1 & 3 \\ 0 & 0 & 0 & 1 \end{bmatrix}.
$$

The matrix $\hat{\mathbf{H}}$ in this example is called an **upper-triangular matrix**; all the elements of this matrix below the main diagonal are zero. The inverse of an upper-triangular matrix is particularly simple; it is another upper-triangular matrix, which can be efficiently computed using exactly the same back-substitution technique described in the context of Equation (2.33). Then,

$$
\mathbf{x}^T = \hat{\mathbf{y}}^T\hat{\mathbf{H}}^{-1} = [1\ \ 2\ \ 1\ \ 3]\begin{bmatrix} 1 & -3 & 7 & -15 \\ 0 & 1 & -3 & 7 \\ 0 & 0 & 1 & -3 \\ 0 & 0 & 0 & 1 \end{bmatrix} = [1\ \ -1\ \ 2\ \ -1].
$$

We could have chosen any four rows of Equation (2.33) and found the deconvolution by the appropriate choice of $\hat{\mathbf{y}}^T$ and $\hat{\mathbf{H}}$. For example, choosing the row corresponding to $y[0]$, $y[2]$, $y[4]$ and $y[5]$ gives

$$
\mathbf{x} = \hat{\mathbf{y}}\hat{\mathbf{H}}^{-1} = [1\ \ 1\ \ 1\ \ -2]\begin{bmatrix} 1 & 2 & 0 & 0 \\ 0 & 3 & 0 & 0 \\ 0 & 1 & 2 & 0 \\ 0 & 0 & 3 & 2 \end{bmatrix}^{-1} = \frac{1}{12}[1\ \ 1\ \ 1\ \ -2]\begin{bmatrix} 12 & -8 & 0 & 0 \\ 0 & 4 & 0 & 0 \\ 0 & -2 & 6 & 0 \\ 0 & 3 & -9 & 6 \end{bmatrix} = [1\ \ -1\ \ 2\ \ -1],
$$

which is the same as the previous result.

(d) Finally, using deconv:

```
>> [x, r] = deconv(y, h)
x =
    1    -1     2    -1
r =
    0     0     0     0     0     0
```

The fact that the vector r is all zeros indicates that this deconvolution had no remainder.

We have used deconvolution to determine the unknown input signal $x[n]$ that gave rise to a known output $y[n]$ in a system with known impulse response $h[n]$. By the commutative property of convolution, the roles of $x[n]$ and $h[n]$ can be reversed. Accordingly, we can use the deconvolution techniques we have just developed to determine an unknown $h[n]$ given known $x[n]$ and $y[n]$.

2.7 ★ Convolution of long sequences

In Section 2.2, we considered the convolution of two sequences of finite length. We have implicitly assumed that, in practice, the lengths of both the input $x[n]$ and the impulse response $h[n]$ are small enough that the computation can be carried out within the limitations of conventional processors and memory. We now consider the case where the length of the input sequence is very large (possibly infinite), or unknown. This reflects what would happen, for example, in a real-time discrete-time filter where the input to the filter is continuously provided by a data-acquisition system.

Two methods have been developed to deal with infinite-length sequences: the **overlap-add** and the **overlap-save** methods. These methods essentially involve segmenting the input sequence into finite-length input subsequences (called blocks, frames or "chunks" of data), convolving each subsequence with the impulse response to form an output subsequence, and then stitching the output subsequences together to form the complete output.

2.7.1 Overlap-add method

Assume we want to convolve a long sequence $x[n]$ with impulse response $h[n]$, where $h[n]$ is assumed to be of finite length, as shown in **Figure 2.23**.

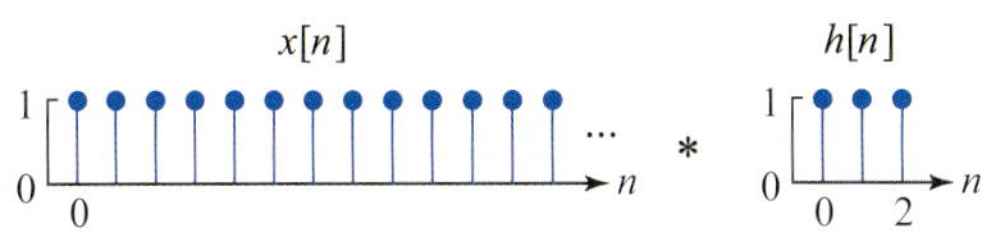

Figure 2.23 Convolution of finite-length $h[n]$ with infinite-length $x[n]$

In the overlap-add method, diagrammed in **Figure 2.24a**, the infinite-length input signal $x[n]$ is cut into finite, *non-overlapping* chunks, $x_1[n]$, $x_2[n]$, $x_3[n]$, . . . so that

$$x[n] = x_1[n] + x_2[n] + x_2[n] + \cdots.$$

Each chunk is then convolved with the impulse response $h[n]$ to form the output chunks shown in **Figure 2.24b**,

$$y_1[n] = x_1[n]*h[n]$$
$$y_2[n] = x_2[n]*h[n]$$
$$\vdots$$

By the distributive property of convolution, when these output chunks are added together, they yield $y[n]$:

$$\begin{aligned} y[n] &= x[n]*h[n] = (x_1[n] + x_2[n] + x_3[n] + \cdots)*h[n] \\ &= x_1[n]*h[n] + x_2[n]*h[n] + x_3[n]*h[n] + \\ &= y_1[n] + y_2[n] + y_3[n] + \cdots. \end{aligned}$$

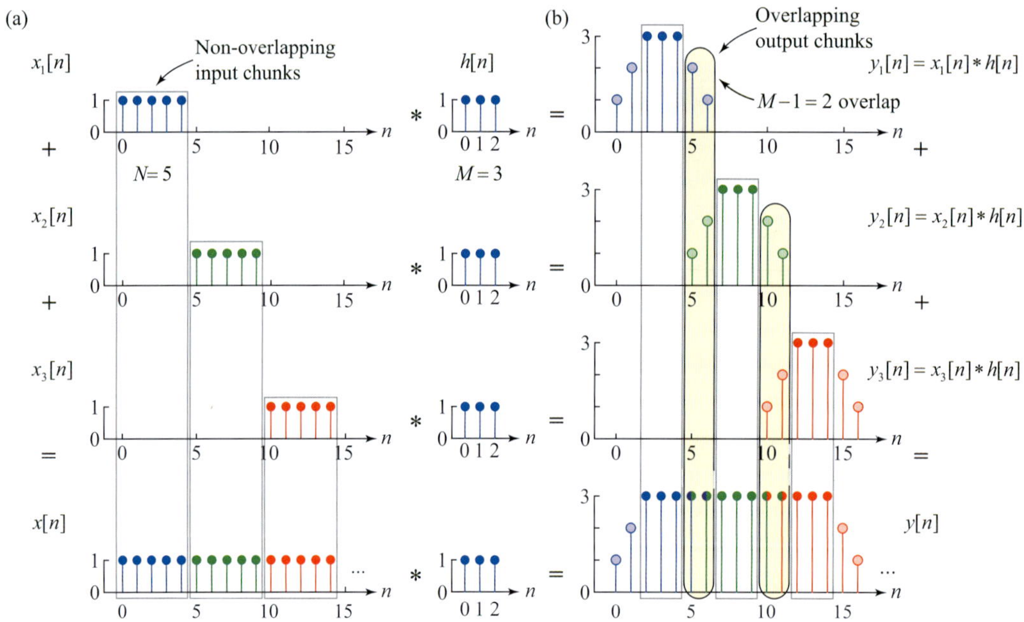

Figure 2.24 The overlap-add method

If each input chunk is of length N, and the impulse response is of length M, then each output chunk is of length $N + M - 1$. In the example of **Figure 2.24**, $N = 5$ and $M = 3$. Whereas the input sequences are non-overlapping in time, the output subsequences overlap by $M - 1 = 2$ points, as indicated by the filled yellow ovals. As a result, the center $N - M + 1$ points of each output chunk (denoted by the thin rectangular boxes) derive from a single output chunk, while the last $M - 1$ points of each chunk must be added to the beginning $M - 1$ points of the subsequent chunk to form points in the output sequence $y[n]$. For example, in **Figure 2.24**, $y[4] = y_1[4]$ derives solely from the first output chunk, but $y[5] = y_1[5] + y_2[5]$ derives from the output of the first two chunks. If the input subsequences are adequately long with respect to the length of $h[n]$, no point in $y[n]$ requires the summation of data from more than two output chunks.

It can be shown (Problem 2-36) that if each output chunk is efficiently computed using the convolution sum formula, Equation (2.9), then each point in $y[n]$, $n \geq M$, requires M multiplications and $M-1$ additions. This suggests that the size of the input chunk, N, does not matter, at least in terms of the number of arithmetic operations.

2.7.2 Overlap-save method

Figure 2.25 shows the basic scheme of the overlap-save method.

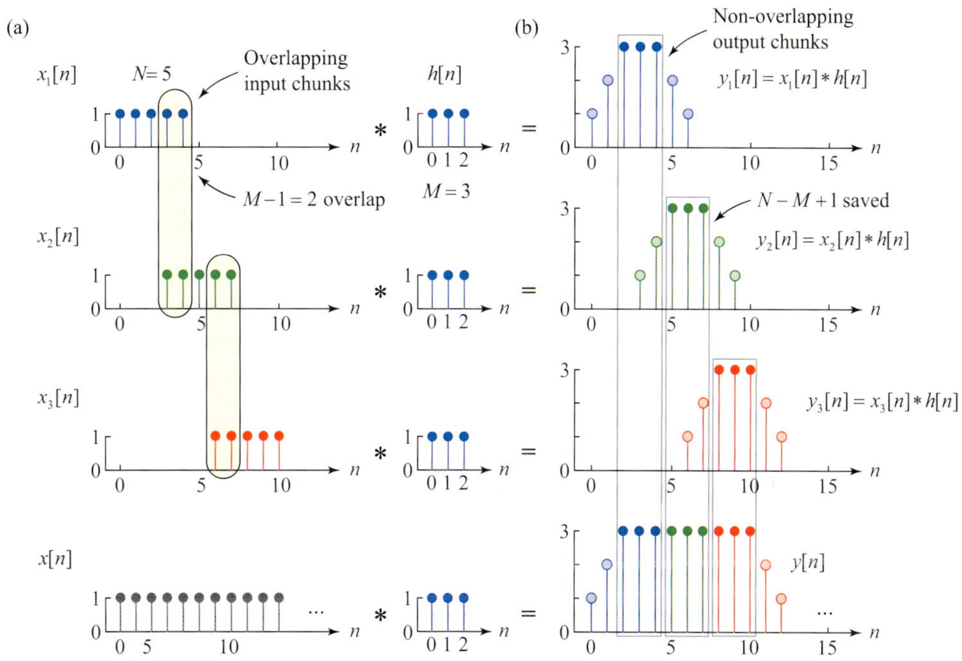

Figure 2.25 Overlap-save method

First, the signal $x[n]$ is segmented into overlapping input chunks $x_1[n]$, $x_2[n]$, . . ., as shown in **Figure 2.25a**. However, unlike the segmentation of the overlap-add method, the points of each chunk overlap with the following subsequence by $M-1$ points, as shown with yellow ovals. Each input chunk is then convolved with $h[n]$ to form output subsequences $y_1[n]$, $y_2[n]$, . . . (as shown in **Figure 2.25b**), where

$$y_1[n] = x_1[n] * h[n]$$
$$y_2[n] = x_2[n] * h[n]$$
$$\vdots$$

In contrast with the overlap-add method, here we only choose the points from the center region of the output subsequences; that is, values of $y[n]$ for which the convolution of $h[n]$ and the corresponding input subsequence (e.g., $x_1[n]$) is exactly the same as the convolution of $h[n]$ and the original sequence $x[n]$. For example, consider the convolution $y_1[n] = x_1[n] * h[n]$ in the first chunk. When $n = 2$, we have

$$y_1[2] = x_1[2]h[0] + x_1[1]h[1] + x_1[0]h[2].$$

Since $x_1[2]=x[2]$, $x_1[1]=x[1]$ and $x_1[0]=x[0]$, thus $y_1[2]=y[2]$. The same conclusion applies to the next two output points from this chunk: $y_1[3]=y[3]$ and $y_1[4]=y[4]$. The final two points of this chunk are not usable since $y_1[5]\neq y[5]$ and $y_1[6]\neq y[6]$. However, looking at the second output chunk, you can see that $y_2[5]=y[5]$, $y_2[6]=y[6]$ and $y_2[7]=y[7]$. Therefore, we can construct a complete output sequence, $y[n]$, by taking only the center $N-M+1$ points of each output chunk, as shown by the thin rectangular boxes in the figure.

Even though the overlap-save method seems simpler to implement and more efficient – after all, we do not have to do all those pesky additions of the head and tail of subsequent chunks – it can be shown (Problem 2-36) that if the convolution sum is only computed for the center region of each output chunk, then each point in $y[n]$, $n\geq M$, requires only M multiplications and $M-1$ additions, which is exactly the same as the overlap-add method. This suggests that there is no theoretical advantage of one method over the other, at least in terms of the number of operations if these methods are implemented as direct convolution in the time domain. In Chapter 11, we will discover how the overlap-add and overlap-save methods can be used to implement **block convolution** using the fast Fourier transform (FFT).

2.8 Implementation issues

Implementing convolution using the direct-summation or matrix-multiplication methods of Section 2.2 and 2.5 is not practical. These methods assume that the input sequence $x[n]$ is of finite duration and is available in its entirety. In real-time filtering applications, the sequence may be of infinite duration and you may get the input one point at a time, for example as the result of an interrupt from an A/D conversion occurring at a constant rate. For each point of the input, the application may be required to produce one point of the output $y[n]$. Alternately, you may get a frame of input data at a time – a contiguous block of data points – and be required to produce a frame of output sequence, as we discussed in the previous section. Even if the input sequence is of finite duration and available all at once, the direct-summation or matrix multiplication methods are not very practical. Sufficient memory must exist so that multiple shifted copies of the sequence can be maintained in memory so that they can be scaled and summed. If $x[n]$ is of length N and the impulse response $h[n]$ is of length M, we still require $(N+M-1)\min(N,M)$ memory elements.

To do a real-time convolution, we need to write an efficient routine that does the minimum number of multiplications and additions and requires the minimum number of storage elements and data moves. Performing all these steps in software on a general-purpose serial processor requires multiple lines of code and takes a lot of CPU cycles. However, special-purpose digital signal processors (and DSP cores in modern general-purpose processors and microcontrollers) exist. These processors have hardware and instructions that are optimized to handle convolution. In Chapters 10 and 11, we go into more detail on aspects of software and hardware implementations of convolution using multiplications of Fourier transforms.

SUMMARY

This chapter has introduced the basics of signal processing in the discrete-time domain. We first derived a key result, namely that linear time-invariant systems are completely characterized by their response to an impulse. Knowing the impulse response is theoretically all you need to determine the response of an LTI system to any input by the operation of convolution. We described in detail two methods of understanding convolution, the direct-summation method and the "flip-and-shift" method. Deconvolution can be understood as the process of determining the input given the output and impulse response. (We will see convolution and deconvolution again in the next chapter, where we introduce frequency-domain methods of characterizing and analyzing systems.) Finally in this chapter, we discussed the method of processing inputs of potentially infinite length using the overlap-add and overlap-save methods. These methods will make a reappearance in Chapter 11, when we discuss their practical implementation using the fast Fourier transform.

PROBLEMS

Problem 2-1

Use convolution to find the response to an IIR system whose impulse response is $h[n]=\frac{1}{2}^{n}u[n]$.

(a) $x[n]=\frac{1}{4}^{n}u[n]$.

(b) $x[n]=u[n]-u[n-4]$.

Problem 2-2

Given $h[n]=\delta[n]-\delta[n-1]$, find $y[n]=x[n]*h[n]$ for each of the following $x[n]$:

(a) $x[n]=\cos\omega_0 n$. Express your answer as a single cosine of the form $A\cos(\omega_0 n+\phi)$.

(b) $x[n]=\frac{1}{2}^{n}u[n]$.

(c) $x[n]=u[n+1]-u[n-2]$.

Problem 2-3

Given $x[n]=\cos\pi n/4$ and $h[n]=\frac{\sqrt{2}}{2}\delta[n]-\delta[n-1]$, find $y[n]=x[n]*h[n]$. Express your answer as a single cosine of the form $A\cos(\omega_0 n+\varphi)$.

Problem 2-4

Given $x[n]=\sin\pi n/4$ and $h[n]=\delta[n]-\sqrt{2}\delta[n-1]$, find $y[n]=x[n]*h[n]$. Express your answer as a single cosine of the form $A\cos(\omega_0 n+\varphi)$.

Problem 2-5

For each of the pairs of $x[n]$ and $h[n]$ in **Figure 2.26**, plot $y[n]=x[n]*h[n]$.

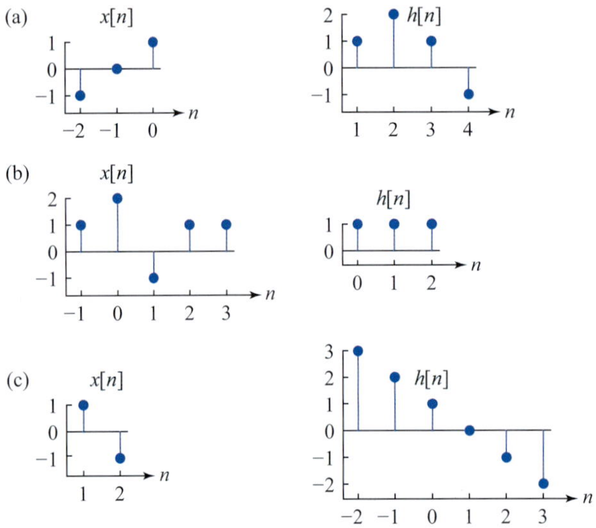

Figure 2.26

Problem 2-6

For each of the pairs of $x[n]$ and $h[n]$ in **Figure 2.27**, plot $y[n] = x[n] * h[n]$.

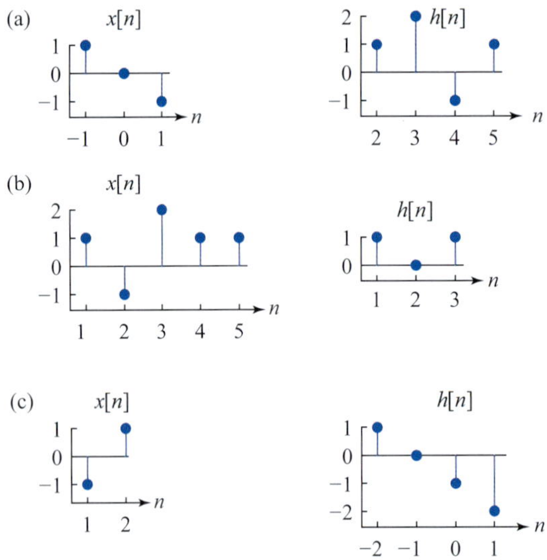

Figure 2.27

Problem 2-7

Given a filter with impulse response $h[n]$ and output $y[n]$, we observe that when the input is $x[n] = \delta[n] - \delta[n-1] + \delta[n-2]$, the output is $y[n] = \delta[n] + \delta[n-1] - 2\delta[n-2] + 3\delta[n-3] - \delta[n-4]$. Find the output when $x[n] = \delta[n] + \delta[n-1] + \delta[n-2]$.

Problem 2-8

Given a filter with impulse response $h[n]$ and output $y[n]$, we observe that when the input is $x[n] = \delta[n] + 2\delta[n-1] + \delta[n-2]$, the output is $y[n] = \delta[n] + 3\delta[n-1] + 2\delta[n-2] - \delta[n-3] - \delta[n-4]$. Find the output when $x[n] = \delta[n] + \delta[n-1] + \delta[n-2]$.

Problem 2-9

Given that the impulse response of an LTI system is $h[n] = \alpha^n u[n]$, show that the response to an input $x[n] = u[n] - u[n-N]$ is

$$y[n] = \frac{1-\alpha^{n+1}}{1-\alpha} u[n] - \frac{1-\alpha^{n-N+1}}{1-\alpha} u[n-N].$$

(a) Find the solution in closed form by evaluating both convolution terms,

$$y[n] = x[n] * h[n] = (u[n] - u[n-N]) * h[n] = u[n] * h[n] - u[n-N] * h[n].$$

(b) Find the solution by computing just $u[n] * h[n]$ and using the fact that the system is linear and time-invariant.

Problem 2-10

Given the system shown in **Figure 2.28** with

$$h_1[n] = \delta[n] + \delta[n-1] + \delta[n-2]$$
$$h_2[n] = \delta[n+1] - \delta[n-1]$$
$$h_3[n] = \delta[n-3].$$

(a) Find the equivalent impulse response of the system, $h[n]$.
(b) Find the response of the system when $x[n] = \delta[n+1] + 2\delta[n] + \delta[n-1]$.

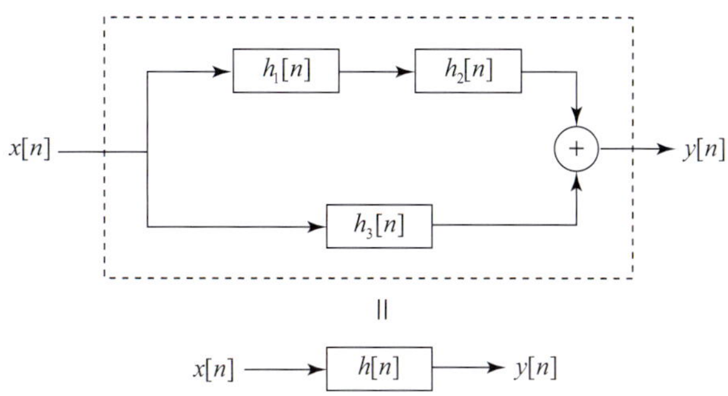

Figure 2.28

Problem 2-11

Repeat Problem 2-10 with

$$h_1[n] = \delta[n] - \delta[n-1] + 2\delta[n-2]$$
$$h_2[n] = \delta[n+1] + \delta[n-1]$$
$$h_3[n] = \delta[n-2] - 2\delta[n-3].$$

Problem 2-12

Repeat Problem 2-10 when $h_1[n] = h_2[n] = h_3[n] = \delta[n+1] - \delta[n-1]$ and $x[n] = u[n] - u[n-6]$.

Problem 2-13

Given the system shown in **Figure 2.28** with

$$h_1[n] = \delta[n] + \delta[n-1] + 2\delta[n-2]$$
$$h_3[n] = -\delta[n-1] + \delta[n-2] - 2\delta[n-3]$$
$$h[n] = \delta[n+1] + \delta[n-1],$$

find $h_2[n]$.

Problem 2-14

Given the system shown in **Figure 2.28** with

$$h_1[n] = \delta[n-1] + 2\delta[n-2] + \delta[n-3]$$
$$h_2[n] = -\delta[n+3] + \delta[n+1],$$

find $h_3[n]$ such that $h[n] = -2\delta[n+2] + 2\delta[n-2]$.

Problem 2-15

Given the system shown in **Figure 2.28**, with $h_1[n]$, $h_2[n]$, $h_3[n]$ and $h[n]$ as shown in **Figure 2.29**, find the values of all the unknowns: a, b, c, d, e, f and n.

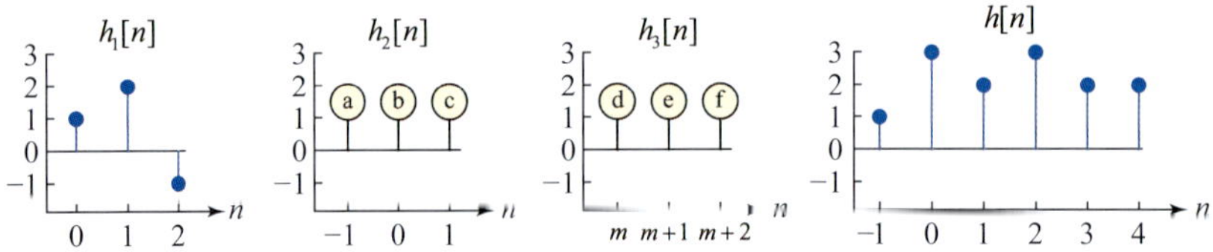

Figure 2.29

Problem 2-16

Given the system shown in **Figure 2.28**, with

$$h_1[n] = \delta[n+3]$$
$$h_2[n] = \tfrac{1}{2}^{n} u[n]$$
$$h_3[n] = \delta[n-3].$$

(a) Find the equivalent impulse response of the system, $h[n]$.
(b) Find the response of the system when $x[n] = u[n] - u[n-6]$.

Problem 2-17

You are given the system shown in **Figure 2.28**, with

$$h_1[n] = \delta[n+n_0]$$
$$h_2[n] = \delta[n+1] + \delta[n] + \delta[n-1]$$
$$h_3[n] = \delta[n-3],$$

where n_0 is unknown. The input $x[n]$ and output $y[n]$ are shown in **Figure 2.30**. Find n_0.

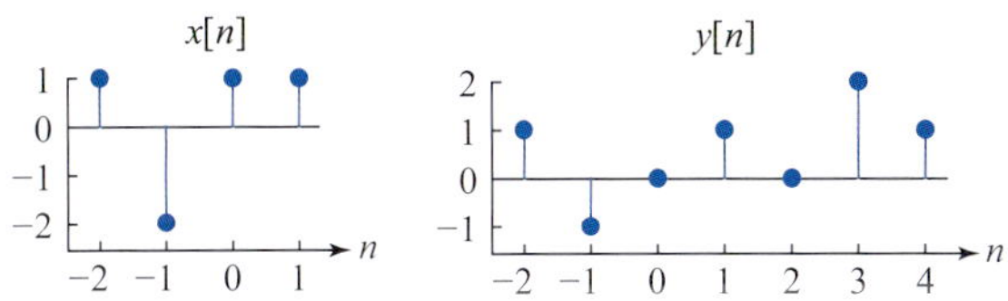

Figure 2.30

Problem 2-18

We are given the system shown in **Figure 2.28**. The input is shown in **Figure 2.31a**. All that is known about the impulse responses of the three subsystems $h_1[n]$, $h_2[n]$ and $h_3[n]$ is that each one is chosen uniquely from the impulse responses $g_1[n]$, $g_2[n]$ and $g_3[n]$, shown in **Figures 2.31b**, **c** and **d**.

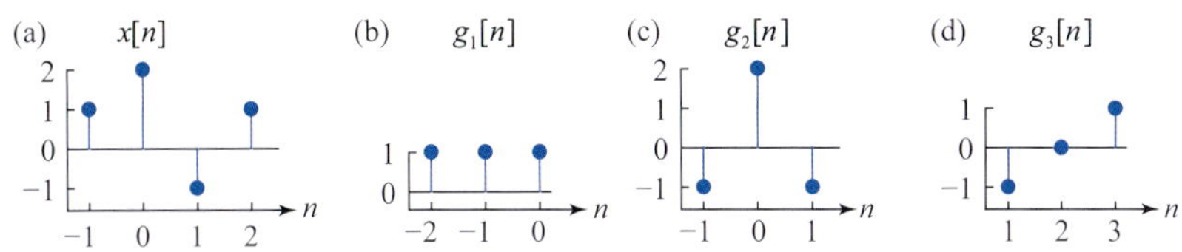

Figure 2.31

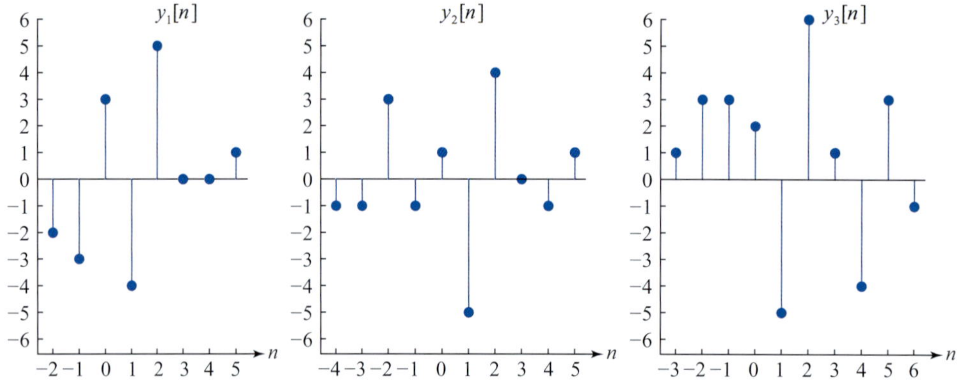

Figure 2.32

(a) Given that the output of the system is $y_1[n]$, as shown in **Figure 2.32a**, determine all possible ways of assigning $g_1[n]$, $g_2[n]$ and $g_3[n]$ to $h_1[n]$, $h_2[n]$ and $h_3[n]$. It is *not* necessary to perform any convolutions to answer this question!

(b) Repeat part (a), given that the output of the system is $y_2[n]$, as shown in **Figure 2.32b**.

(c) Repeat part (a), given that the output of the system is $y_3[n]$, as shown in **Figure 2.32c**.

Problem 2-19

Given the system shown in **Figure 2.33**, with

$$h_1[n] = \delta[n] + \delta[n-1] + \delta[n-2]$$
$$h_2[n] = \delta[n+1] - \delta[n-1]$$
$$h_3[n] = \delta[n+1] - \delta[n].$$

(a) Find the equivalent impulse response of the system $h[n]$.

(b) Find the response of the system when $x[n] = 2\delta[n] - \delta[n-1]$.

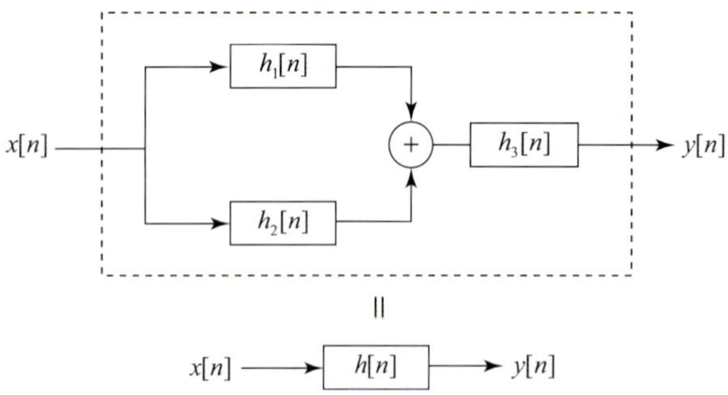

Figure 2.33

Problem 2-20

Given the system shown in **Figure 2.33**, with

$$h_1[n] = \delta[n] + \delta[n-1]$$
$$h_2[n] = \delta[n-1] - \delta[n-2]$$
$$h_3[n] = \delta[n+1] - \delta[n-1].$$

(a) Find the equivalent impulse response of the system, $h[n]$.
(b) Find the response of the system when $x[n] = \delta[n+1] - 2\delta[n]$.

Problem 2-21

Given the system shown in **Figure 2.33**, with

$$h_1[n] = \delta[n+1] - \delta[n] - 2\delta[n-1]$$
$$h_3[n] = \delta[n] - \delta[n-1]$$
$$h[n] = \delta[n+1] - \delta[n] - \delta[n-1] + 2\delta[n-2] - \delta[n-3],$$

find $h_2[n]$.

Problem 2-22

Given the system shown in **Figure 2.34**, with

$$x[n] = \delta[n] + \delta[n-1]$$
$$h_1[n] = \delta[n-1] - \delta[n]$$
$$y[n] = 2\delta[n] + 4\delta[n-1] - \delta[n-2] - 3\delta[n-3],$$

find $h_2[n]$.

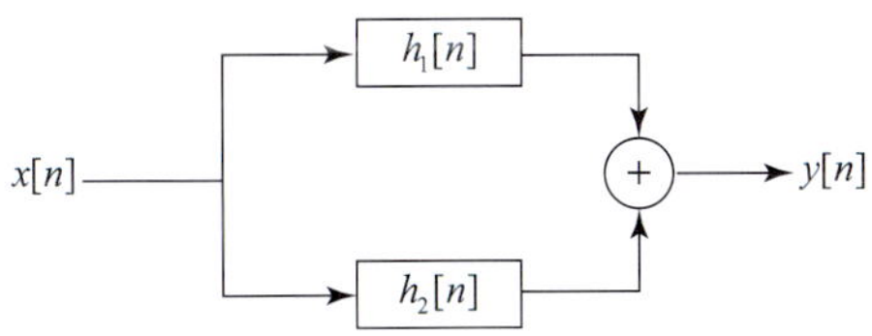

Figure 2.34

Problem 2-23

For the system shown in **Figure 2.34**, you are given that $h_2[n] = \delta[n] + 2\delta[n-1] + \delta[n-2]$. You are also told that when $x[n] = \delta[n] - \delta[n-1]$, then $y[n] = \delta[n-1] - \delta[n-3] + \delta[n-4] - \delta[n-5]$. Find $h_1[n]$.

Problem 2-24

Given the system shown in **Figure 2.33**, with

$$h_1[n] = \delta[n] + \delta[n-1]$$
$$h_2[n] = \delta[n-1] - \delta[n-2]$$
$$h_3[n] = \delta[n+1] - \delta[n-1].$$

(a) Find the equivalent impulse response of the system, $h[n]$.
(b) Find the response of the system when $x[n] = \delta[n+1] - 2\delta[n]$.

Problem 2-25

Given the system shown in **Figure 2.33**, with

$$h_1[n] = \delta[n] - \delta[n-1]$$
$$h_2[n] = 2\delta[n-1] + \delta[n+1]$$
$$h_3[n] = \delta[n+1] - \delta[n-1].$$

(a) Find the equivalent impulse response of the system, $h[n]$.
(b) Find the response of the system when $x[n] = \delta[n+1] - \delta[n]$.

Problem 2-26

Given the system shown in **Figure 2.33**, with

$$h_1[n] = \delta[n+3]$$
$$h_2[n] = \frac{1}{2}^{n} u[n]$$
$$h_3[n] = \delta[n-3].$$

(a) Find the equivalent impulse response of the system, $h[n]$.
(b) Find the response of the system when $x[n] = u[n] - u[n-6]$.

Problem 2-27

Given the system shown in **Figure 2.33**, with $h[n] = \delta[n+1] + 2\delta[n] - 3\delta[n-1] + \delta[n-2] - \delta[n-3]$, and

$$h_1[n] = \delta[n] - \delta[n-1] + \delta[n-2]$$
$h_2[n]$ unknown
$$h_3[n] = \delta[n] - \delta[n-1],$$

find $h_2[n]$.

➤**Hint:** You might use the fact that $\delta[n] = (\delta[n] - \delta[n-1]) * u[n]$.

Problem 2-28

A digital system has two filters, with impulse responses $h_1[n]$ and $h_2[n]$, connected in series as shown in **Figure 2.35**. This is equivalent to a single filter with impulse response $h[n]$.

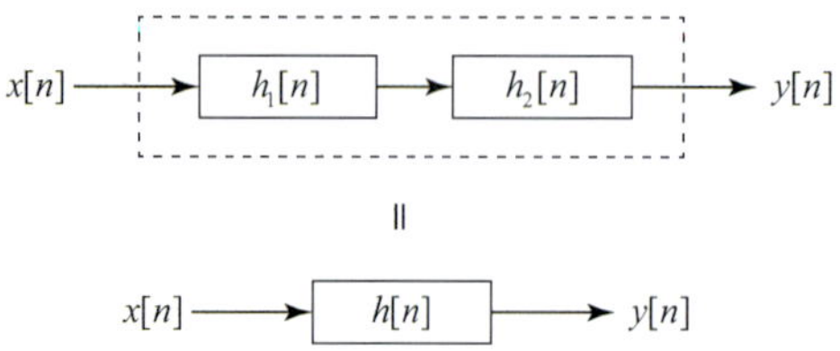

Figure 2.35

(a) Given input $x[n] = \delta[n] + 2\delta[n-1] - \delta[n-2]$, and output $y[n] = \delta[n-1] + 2\delta[n-2] - 2\delta[n-3] - 2\delta[n-4] + \delta[n-5]$, find $h[n]$.

(b) Given $h_1[n] = \delta[n] - \delta[n-1]$, find $h_2[n]$.

Problem 2-29

Find and sketch the convolution of the following sequences:

(a)

$$x[n] = \begin{cases} 1, & n = -2, 0, 1 \\ 2, & n = -1 \end{cases}$$

$$h[n] = \delta[n] - \delta[n-1] + \delta[n-4].$$

(b)

$$x[n] = u[n+1] - u[n-4] - \delta[n-5]$$
$$h[n] = (3 - |n|)(u[n+3] - u[n-4]).$$

(c) (A single expression not containing sums of sines would be nice.)

$$x[n] = \sin \pi n/4$$
$$h[n] = \delta[n+1] - \delta[n-1].$$

(d)

$$x[n] = [1\ \ 4\ \ 3\ \ 5\ \ 2\ \ 5\ \ 1\ \ 4], \text{offset} = 0$$
$$h[n] = [1\ \ -2\ \ 1], \text{offset} = 0.$$

Problem 2-30

We observe that when $x[n] = [1\_\_2\ 1]$ and $h[n] = [1\ \_\_\ 1]$, then $y[n] = x[n] * h[n] =$ [______4 4 __]. Fill in all the blanks, assuming all sequences start at $n = 0$.

Problem 2-31

We observe that when $x[n] = [1\_\_2\ 1]$ and $h[n] = [\ 1\ \_\_\ \_\_]$, then $y[n] = x[n] * h[n] =$ [__ 2 0____ 1]. Find all possible outputs, assuming all sequences start at $n = 0$.

Problem 2-32

We observe that when $x[n] = [1\_\_2\ 1]$ and $h[n] = [\_\_\ \_\_\ \_\_]$, then $y[n] = x[n] * h[n] =$ [__0__ 5 4 1]. Find all possible outputs, assuming all sequences start at $n = 0$.

Problem 2-33

The input to a linear time-invariant system, $x[n]$, is shown in **Figure 2.36**. The output $y[n]$ is shown in **Figure 2.36b**. Find the impulse response of the system, $h[n]$.

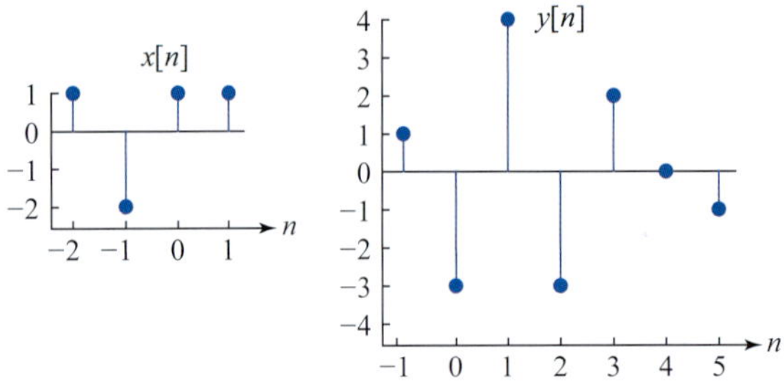

Figure 2.36

Problem 2-34

An input $x[n]$ and impulse response $h[n]$ are shown in **Figure 2.37**. Find $y[1]$, $y[6]$ and $y[10]$.

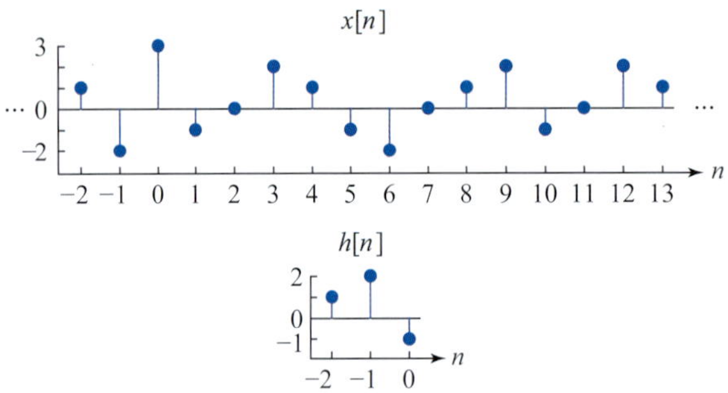

Figure 2.37

Problem 2-35

In **Figure 2.7**, we used the flip-and-shift method to find the convolution $y[n]=x[n]*h[n]$ using the sequence in **Figure 2.5**. Find the convolution $y[n]=h[n]*x[n]$ using the flip-and-shift method and plot the results as in **Figure 2.7** for values of shift, $0\le n\le 6$, thereby showing that the commutative property is satisfied: $x[n]*h[n]=h[n]*x[n]$.

Problem 2-36

In this problem we consider the number of arithmetic operations – two-operand multiplications and additions – that are necessary in the overlap-add and overlap-save methods. Each method segments the input into a number of input chunks, each of which is convolved with the impulse response to form an output chunk. **Figure 2.38** shows the convolution of an input $x[n]$ of length N with impulse response $h[n]$ of length M to form output chunk $y[n]$ for the case $N=5$ and $M=3$. The output of the convolution comprises a central region $N-M+1$ points in length, shown with solid dots bounded by a rectangular box in the figure. Computation of each point in this region requires M values of $x[n]$ as well as the M values of $h[n]$. On either side of the central region are "ramp-up" and "ramp-down" regions $M - 1$ points in length, shown with open dots. Computation of each point in these regions requires fewer than M values of $x[n]$ or $h[n]$.

In the overlap-add method, the $N - M + 1$ points of the central region directly contribute points in the final output sequence. A further $M - 1$ points in the final output sequence are formed from the summation of the ramp-down region of one output chunk with the ramp-up region of the subsequent chunk.

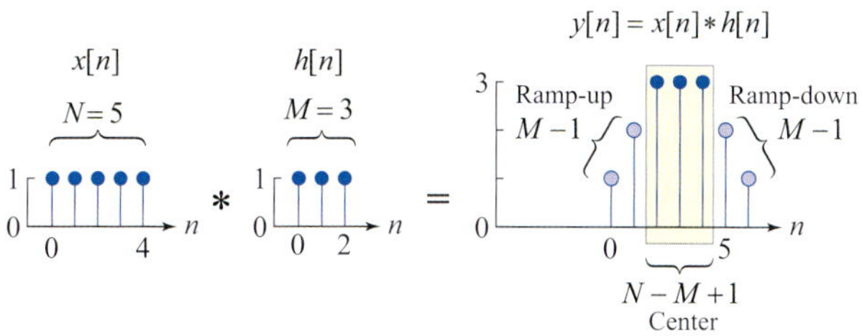

Figure 2.38

(a) Show that the number of multiplications required for the convolution of the points in the center region is $(N - M + 1)M$ and the number of additions is $(N - M + 1)(M - 1)$.

(b) Show that the number of multiplications required for the convolution of the points in either the ramp-up or the ramp-down region is $(M - 1)M/2$ and the number of additions is $(M - 2)(M - 1)/2$.

(c) Show that for the overlap-add method, each output point requires M multiplications and $M - 1$ additions.

(d) Show that for the overlap-save method also, each output point requires M multiplications and $M - 1$ additions.

3 Discrete-time Fourier transform

Introduction

In Chapter 2, we showed that a linear time-invariant (LTI) system is completely characterized in the time domain by its impulse response, and that the response of the system to any input can be expressed as the sum of scaled and shifted impulse responses. In this chapter, we introduce the notion of the **frequency domain** and show that we can express an arbitrary sequence in terms of the integral of complex exponential sequences. In Section 3.1, we explore the response of LTI systems to complex exponential and sinusoidal sequences. In Section 3.2, this will lead us to develop the **discrete-time Fourier transform (DTFT)**, which is a powerful tool for analyzing LTI systems. In Section 3.3, we show how the DTFT can be graphically represented in plots of magnitude and phase. Section 3.5 discusses symmetry properties of the DTFT, some of which will become practically important in later chapters. Section 3.7 introduces linear-phase systems, which will form the basis of an entire class of filters – **finite impulse response (FIR) filters** – which we discuss in Chapter 7. The remaining sections of the chapter detail important properties of the DTFT.

Pretty much all of the material in this chapter (except perhaps the starred sections) is essential to master. Some topics, especially the properties of the DTFT, underlie everything that is to follow in later chapters.

3.1 Complex exponentials and sinusoids

Complex exponentials and sinusoids are two basic building blocks that we will use to express signals and systems in the frequency domain.

3.1.1 Response of LTI systems to complex exponentials

Complex exponential sequences are sequences of the form $e^{j\omega_0 n}$, where ω_0 is the frequency. We will show that *any* sequence can be expressed in terms of the sum of complex exponential sequences, and demonstrate why that is a valuable thing to do. In this section, we will explain what makes complex exponential sequences special and show that the response of linear time-invariant systems to complex exponentials is particularly easy to analyze.

Figure 3.1 A discrete-time system

Complex exponentials and the system function Let $x[n]$ and $y[n]$ be respectively the input and output of a system with transformation $T\{\}$, as shown in **Figure 3.1**.

As we discussed in Chapter 2, a linear time-invariant system is completely characterized by its impulse response $h[n]$, and the output of the system is related to the input through convolution with the impulse response:

$$y[n] = h[n] * x[n] = \sum_{k=-\infty}^{\infty} h[k]x[n-k].$$

When the input to the system is a complex exponential sequence $x[n] = e^{j\omega_0 n}$, the output is

$$y[n] = \sum_{k=-\infty}^{\infty} h[k]e^{j\omega_0(n-k)} = \underbrace{\left(\sum_{k=-\infty}^{\infty} h[k]e^{-j\omega_0 k}\right)}_{H(\omega_0)} e^{j\omega_0 n} = H(\omega_0)e^{j\omega_0 n} = H(\omega_0)x[n].$$

The function

$$H(\omega) = \sum_{n=-\infty}^{\infty} h[n]e^{-j\omega n} \tag{3.1}$$

is termed the **system function** or **frequency response**. $H(\omega)$ is, in general, a complex function of real variable ω. As we shall soon show, $H(\omega)$ characterizes a linear time-invariant system completely as a function of frequency ω, just as the impulse response $h[n]$ characterizes the system as a function of time n. For now, just note that at any particular value of frequency, say $\omega = \omega_0$, $H(\omega_0)$ is a (possibly complex) constant. Hence, the output of the LTI system is identical to the input (i.e., a complex exponential sequence at the same frequency as the input), just multiplied by the constant $H(\omega_0)$. The complex exponential sequence $x[n] = e^{j\omega_0 n}$ is termed an **eigenfunction** of an LTI system, because the output $y[n] = T\{x[n]\} = H(\omega_0)x[n]$ has the same functional dependence on n as the input, just scaled by a constant $H(\omega_0)$, which is termed the **eigenvalue** of the system.[1] This is schematized in **Figure 3.2**.

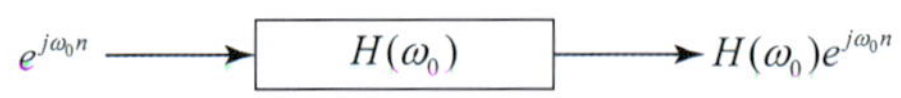

Figure 3.2 Response of a linear time-invariant system to a complex exponential

[1] You may have run across these terms in a physics or mathematics course. The root of the words, the German *eigen*, means "own" or "self" and indicates that these functions have the special self-referential property that the output of the system to an eigenfunction input is just a scaled replica of the same eigenfunction.

A few concrete examples will show how easy it is to analyze the response of LTI systems to complex exponential inputs.

Example 3.1

Consider a simple, non-causal moving-window average filter defined by its impulse response

$$h[n] = \tfrac{1}{3}(\delta[n+1] + \delta[n] + \delta[n-1]).$$

Let the input to the system be a complex exponential at frequency ω_0: $x[n] = e^{j\omega_0 n}$.

(a) Find $H(\omega_0)$.
(b) Find the output of this system, $y[n]$, when the input is $x[n] = e^{j\pi/3}$.

➤ Solution:

(a) Equation (3.1) gives

$$\begin{aligned} H(\omega_0) &= \sum_{n=-\infty}^{\infty} h[n]e^{-j\omega_0 n} = \tfrac{1}{3}\sum_{n=-\infty}^{\infty}(\delta[n+1] + \delta[n] + \delta[n-1])e^{-j\omega_0 n} = \tfrac{1}{3}(e^{j\omega_0} + 1 + e^{-j\omega_0}) \\ &= \tfrac{1}{3}(1 + 2\cos\omega_0). \end{aligned}$$

(b) When $x[n] = e^{j\pi/3}$, then $\omega_0 = \pi/3$, so

$$H(\pi/3) = \tfrac{1}{3}(1 + 2\cos\pi/3) = \tfrac{2}{3},$$

and

$$y[n] = T\{x[n]\} = H(\omega_0)e^{j\omega n} = H(\pi/3)e^{j\pi n/3} = \tfrac{2}{3}e^{j\pi n/3}.$$

For this system, the constant $H(\omega_0)$ is real, so $y[n]$ and $x[n]$ differ only by a real constant 2/3 for all n. However, $H(\omega_0)$ is not always real, as the next example shows.

Example 3.2

Given a system defined by

$$h[n] = \delta[n] - \delta[n-1].$$

(a) Find $H(\omega_0)$.
(b) Find $y[n]$, the output of this system when $x[n] = e^{j\pi/3}$.
(c) Find $y[n]$, the output of this system when $x[n] = e^{j\pi/2}$.

➤ Solution:

(a) Equation (3.1) gives

$$\begin{aligned} H(\omega_0) &= \sum_{n=-\infty}^{\infty} h[n]e^{-j\omega_0 n} = \sum_{n=-\infty}^{\infty}(\delta[n] - \delta[n-1])e^{-j\omega_0 n} = 1 - e^{-j\omega_0} = e^{-j\omega_0/2}(e^{j\omega_0/2} - e^{-j\omega_0/2}) \\ &= 2je^{-j\omega_0/2}\sin\omega_0/2. \end{aligned}$$

(b) For the specific example, $x[n] = e^{j\pi n/3}$, we have $\omega_0 = \pi/3$. So,

$$H(\pi/3) = 2je^{-j\pi/6}\sin\pi/6 = e^{j\pi/2}e^{-j\pi/6} = e^{j\pi/3}.$$

In this case, the value of $H(\omega_0)$ is complex and

$$y[n] = H(\pi/3)x[n] = e^{j\pi/3}e^{j\pi n/3} = e^{j(\pi n/3 + \pi/3)}.$$

(c) Here, $x[n] = e^{j\pi n/2}$, we have $\omega_0 = \pi/2$ and

$$H(\pi/2) = 2je^{-j\pi/4}\sin\pi/4 = \sqrt{2}e^{j\pi/4}.$$

So,

$$y[n] = H(\pi/2)x[n] = \sqrt{2}e^{j\pi/4}e^{j\pi n/2} = \sqrt{2}e^{j(\pi n/2+\pi/4)}.$$

Complex exponentials are eigenfunctions of linear time-invariant systems. However, they are not, in general, eigenfunctions of systems that are nonlinear or not time-invariant. Here is an example:

Example 3.3

Show that $e^{j\omega_0 n}$ is not an eigenfunction of the nonlinear and time-invariant system defined by

$$y[n] = x^2[n].$$

➤ Solution:

If $x[n] = e^{j\omega_0 n}$, then $y[n] = x^2[n] = e^{j2\omega_0 n}$. Hence, for all n, the output is not simply a constant times the input.

Example 3.4

Show that $e^{j\omega_0 n}$ is not an eigenfunction of the linear and time-variant system defined by

$$y[n] = x[n]u[n].$$

➤ Solution:

If $x[n] = e^{j\omega_0 n}$, then $y[n] = x[n]u[n] = e^{j\omega_0 n}u[n]$. Again, for all n, the output is not a constant times the input.

3.1.2 Response of linear time-invariant systems to sinusoids

Determining the output of linear time-invariant systems to complex exponentials is not only theoretically nice, but it also leads us directly to an understanding of the response of these systems to sinusoidal signals, which are of practical interest.

Example 3.5

Determine the output of the system of Example 3.1 to the input $x[n] = \cos \pi n/3$.

➤ Solution:

A cosine of frequency ω_0 can be expressed as the sum of two complex exponentials scaled by 1/2,

$$x[n] = \cos \omega_0 n = \tfrac{1}{2} e^{j\omega_0 n} + \tfrac{1}{2} e^{-j\omega_0 n}.$$

Since the system defined by $h[n]$ is linear and time-invariant, the output of the system to the sum of complex exponentials is the sum of the response to the individual terms:

$$\begin{aligned} y[n] &= T\{x[n]\} = T\{\tfrac{1}{2} e^{j\omega_0 n} + \tfrac{1}{2} e^{-j\omega_0 n}\} = \tfrac{1}{2} T\{e^{j\omega_0 n}\} + \tfrac{1}{2} T\{e^{-j\omega_0 n}\} \\ &= \tfrac{1}{2} H(\omega_0) e^{j\omega_0 n} + \tfrac{1}{2} H(-\omega_0) e^{-j\omega_0 n}. \end{aligned} \tag{3.2}$$

Let $\omega_0 = \pi/3$, so that $x[n] = \cos \pi n/3$. In Example 3.1, we showed that $H(\pi/3) = 2/3$. Similarly,

$$H(-\pi/3) = \tfrac{1}{3}(1 + 2\cos(-\pi/3)) = \tfrac{2}{3},$$

so,

$$\begin{aligned} y[n] &= \tfrac{1}{2} H(\pi/3) e^{j\pi n/3} + \tfrac{1}{2} H(-\pi/3) e^{-j\pi n/3} = \tfrac{1}{2} \cdot \tfrac{2}{3} e^{j\pi n/3} + \tfrac{1}{2} \cdot \tfrac{2}{3} e^{-j\pi n/3} = \tfrac{2}{3}(\tfrac{1}{2} e^{j\pi n/3} + \tfrac{1}{2} e^{-j\pi n/3}) \\ &= \tfrac{2}{3} \cos \pi n/3. \end{aligned}$$

You can see from the preceding example that $\cos \omega_0 n$ is also an eigenfunction of this *particular* system, since the input and output differ only by a multiplicative constant. However, $\cos \omega_0 n$ is *not* an eigenfunction of *every* LTI system, as the next example shows.

Example 3.6

Determine the output of the system of Example 3.2 to the input $x[n] = \cos \pi n/2$.

➤ Solution:

From Equation (3.2),

$$y[n] = \tfrac{1}{2} H(\omega_0) e^{j\omega_0 n} + \tfrac{1}{2} H(-\omega_0) e^{-j\omega_0 n}.$$

Let $\omega_0 = \pi/2$, so that $x[n] = \cos \pi n/2$. We showed in Example 3.2(a) that $H(\pi/2) = \sqrt{2} e^{j\pi/4}$. Similarly,

$$H(-\pi/2) = 2j e^{j\pi/4} \sin(-\pi/4) = \sqrt{2} e^{-j\pi/4}.$$

So,

$$\begin{aligned} y[n] &= \tfrac{1}{2} H(\pi/2) e^{j\pi n/2} + \tfrac{1}{2} H(-\pi/2) e^{-j\pi n/2} = \tfrac{1}{2} \cdot \sqrt{2} e^{j\pi/4} e^{j\pi n/2} + \tfrac{1}{2} \cdot \sqrt{2} e^{-j\pi/4} e^{-j\pi n/2} \\ &= \sqrt{2} \left(\frac{e^{j(\pi n/2 + \pi/4)} + e^{-j(\pi n/2 + \pi/4)}}{2} \right) = \sqrt{2} \cos(\pi n/2 + \pi/4). \end{aligned}$$

In this example, $\cos\omega_0 n$ is *not* an eigenvalue of this system, since the output $y[n]$ is not simply expressible as a constant times $\cos\omega_0 n$. However, the output *is* a cosine of the same frequency as the input cosine, just with a different magnitude and phase. Furthermore, $H(\omega)$ contains all the information necessary to predict the magnitude and phase of the output to a cosine of frequency ω_0. Specifically, the magnitude of the output is $|H(\omega_0)| = \sqrt{2}$ and the phase is $\measuredangle H(\omega_0) = \pi/4$:

$$y[n] = |H(\omega_0)| \cos(\omega_0 n + \measuredangle H(\omega_0)).$$

In this example, the frequency response $H(\omega)$ tells you everything that you need to know about the response of the system to an input cosine of frequency ω_0. The output is also a cosine of the same frequency, whose magnitude is scaled by $|H(\omega_0)|$ and whose phase is changed by $\measuredangle H(\omega_0)$.

3.2 Discrete-time Fourier transform (DTFT)

The discrete-time Fourier transform (DTFT) is the single most valuable tool we have to describe and analyze the frequency response of both simple and complex stimuli and systems. In this section, we will define it and present its most essential properties.

3.2.1 Orthogonality of complex exponential sequences

A key reason for our interest in complex exponential sequences of the form $e^{j\omega n}$ is that they are **orthogonal**. In the next section, we will show how this property enables us to express an arbitrary time function in terms of the sum of complex exponentials. Orthogonality is defined to mean that

$$\frac{1}{2\pi}\int_{-\pi}^{\pi} e^{j\omega n} e^{-j\omega k} d\omega = \begin{cases} 1, & n = k \\ 0, & n \neq k \end{cases}. \tag{3.3}$$

Here is the proof:

$$\frac{1}{2\pi}\int_{-\pi}^{\pi} e^{j\omega n} e^{-j\omega k} d\omega = \frac{1}{2\pi}\int_{-\pi}^{\pi} e^{j\omega(n-k)} d\omega = \frac{1}{2\pi j(n-k)} e^{j\omega(n-k)}\Big|_{-\pi}^{\pi} = \frac{1}{\pi(n-k)} \frac{e^{j\pi(n-k)} - e^{-j\pi(n-k)}}{2j}$$
$$= \frac{\sin\pi(n-k)}{\pi(n-k)} = \operatorname{sinc}\pi(n-k).$$

The **sinc function** is one of the most important functions in signal processing. We shall run into it again and again in our study. In this example, the numerator of the sinc function, $\sin\pi(n-k)$, is zero for all integer n and k, so the sinc function is also zero for all integer n and k, except when $n = k$, in which case both the numerator and denominator are zero. Then, the ratio is indeterminate, but we can easily evaluate the integral of Equation (3.3) directly with $n - k = 0$ to give

$$\left(\frac{1}{2\pi}\int_{-\pi}^{\pi} e^{j\omega(n-k)} d\omega\right)\Bigg|_{n-k=0} = \frac{1}{2\pi}\int_{-\pi}^{\pi} e^{j\omega 0} d\omega = 1.$$

Therefore,

$$\frac{1}{2\pi}\int_{-\pi}^{\pi} e^{j\omega(n-k)}d\omega = \operatorname{sinc}\pi(n-k) = \begin{cases} 1, & n=k \\ 0, & n\neq k \end{cases}.$$

This integral defines a function that is one when $n=k$ and zero when $n\neq k$. Wait a minute! We have seen this before. It is exactly equivalent to the sequence $\delta[n-k]$, as shown in **Figure 3.3**.

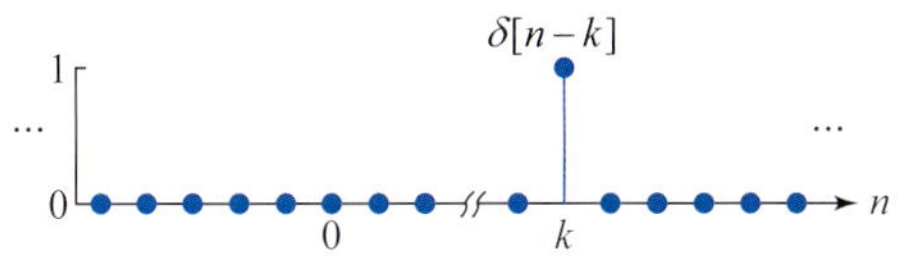

Figure 3.3 The sequence $\delta[n-k]$

Therefore,

$$\delta[n-k] = \frac{1}{2\pi}\int_{-\pi}^{\pi} e^{j\omega(n-k)}d\omega. \quad \textbf{(3.4)}$$

This is a very useful result. Not only are complex exponential sequences orthogonal, but we can express the discrete-time impulse as the integral of complex exponentials of all frequencies, $-\pi \leq \omega < \pi$, and unit amplitude. That is,

$$\delta[n] = \frac{1}{2\pi}\int_{-\pi}^{\pi} e^{j\omega n}d\omega.$$

In the next section, we will use this important result to show that *any* sequence can be expressed in terms of complex exponentials. This will lead us directly to the discrete-time Fourier transform (DTFT).

3.2.2 Definition and derivation

In Chapter 2, we showed that we could express an arbitrary sequence $x[n]$ in terms of the sum of scaled and shifted impulses,

$$x[n] = \sum_{k=-\infty}^{\infty} x[k]\delta[n-k], \quad \textbf{(3.5)}$$

where for any k, $x[k]$ is a scale factor and $\delta[n-k]$ is a sequence on time index n shifted by k. In the preceding section, we just showed that we could express $\delta[n]$ in terms of the integral of complex exponentials. Substituting Equation (3.4) into Equation (3.5), we get

$$x[n] = \sum_{k=-\infty}^{\infty} x[k]\delta[n-k] = \sum_{k=-\infty}^{\infty} x[k]\left(\frac{1}{2\pi}\int_{-\pi}^{\pi} e^{j\omega(n-k)}d\omega\right) = \frac{1}{2\pi}\int_{-\pi}^{\pi}\underbrace{\left(\sum_{k=-\infty}^{\infty} x[k]e^{-j\omega k}\right)}_{X(\omega)} e^{j\omega n}d\omega$$
$$= \frac{1}{2\pi}\int_{-\pi}^{\pi} X(\omega)e^{j\omega n}d\omega.$$

The summation inside the integral, $X(\omega)$, is strictly a function of ω, so we define it as

$$X(\omega) \triangleq \sum_{n=-\infty}^{\infty} x[n]e^{-j\omega n}. \quad \textbf{(3.6)}$$

Hence,

$$x[n] = \frac{1}{2\pi}\int_{-\pi}^{\pi} X(\omega)e^{j\omega n}d\omega. \quad \textbf{(3.7)}$$

Equation (3.6) is termed the **discrete-time Fourier transform (DTFT)** or, more properly, the **forward DTFT**, or most often, just "the DTFT." The DTFT describes how to *analyze* a given $x[n]$ to determine $X(\omega)$. Equation (3.7) is termed the **inverse DTFT**. It describes how to *synthesize* an arbitrary sequence $x[n]$ from the integral of scaled complex exponential sequences of different frequencies:

$$\begin{array}{rlr}
 & 00101_2 \longrightarrow & 5_{10} \\
+ & 00111_2 \longrightarrow & 7_{10} \\ \hline
 & \boxed{01}100_2 \longrightarrow & -4_{10} \\
+ & 01000_2 \longrightarrow & -8_{10} \\ \hline
 & \boxed{00}100_2 \longrightarrow & 4_{10}
\end{array}$$

✗ Intermediate result overflows!

✓ Final result O.K.

At each value of frequency ω, $X(\omega)$ is a (possibly complex) constant that scales a complex exponential sequence $e^{j\omega n}$. Then all the scaled complex exponential sequences are integrated as a function of ω to form $x[n]$.

Equation (3.7) can be thought of as a "recipe" for the synthesis of $x[n]$ from scaled orthogonal complex exponential sequences $e^{j\omega n}$ of different frequencies ω. The $e^{j\omega n}$ sequences constitute the entire list of possible "ingredients" out of which any $x[n]$ can be made. The $X(\omega)$ constitute the "proportions" of those ingredients at different frequencies, ω, that are necessary to make a particular $x[n]$. All $x[n]$ are made of these same basic ingredients – that is, the same $e^{j\omega n}$, for $-\pi \le \omega < \pi$. What distinguishes one $x[n]$ from another is the proportion of these ingredients – that is, $X(\omega)$ as a function of ω.

3.2.3 Notation of the DTFT

The forward and inverse DTFT of Equations (3.6) and (3.7) form a unique and invertible pair. Given $x[n]$, there is one and only one $X(\omega)$. Given $X(\omega)$, there is one and only one $x[n]$. We can denote this relationship schematically with a double arrow,

$$x[n] \leftrightarrow X(\omega).$$

Also, we will frequently use the following notation:

$$\mathfrak{F}\{x[n]\} \text{ means "the forward DTFT of } x[n]\text{"}$$
$$\mathfrak{F}^{-1}\{X(\omega)\} \text{ means "the inverse DTFT of } X(\omega)\text{."}$$

So,

$$\mathfrak{F}\{x[n]\} = X(\omega)$$
$$\mathfrak{F}^{-1}\{X(\omega)\} = x[n].$$

3.2.4 Existence of the DTFT

Here is an important question: can we compute the DTFT for every sequence $x[n]$? A *sufficient* condition for the existence of the DTFT is that $X(\omega)$ does not "blow up" for any value of ω; namely, that $|X(\omega)| < \infty$ for all ω. From Equation (3.6), we can argue that

$$|X(\omega)| = \left|\sum_{n=-\infty}^{\infty} x[n]e^{-j\omega n}\right| \leq \sum_{n=-\infty}^{\infty} \left|x[n]e^{-j\omega n}\right|. \tag{3.8}$$

The inequality in the last relation reflects the fact that the absolute value of the sum of elements is always less than or equal to the sum of the absolute value of the elements, because the sum of $|x[n]e^{-j\omega n}|$ is always positive, whereas the sum of $x[n]e^{-j\omega n}$ may not be. Now,

$$\sum_{n=-\infty}^{\infty} \left|x[n]e^{-j\omega n}\right| = \sum_{n=-\infty}^{\infty} |x[n]|\left|e^{-j\omega n}\right| = \sum_{n=-\infty}^{\infty} |x[n]|, \tag{3.9}$$

because the absolute value of the product is the product of the absolute values and $|e^{-j\omega n}| = 1$. Putting Equations (3.8) and (3.9) together, we get

$$|X(\omega)| \leq \sum_{n=-\infty}^{\infty} |x[n]|.$$

Thus, if

$$\sum_{n=-\infty}^{\infty} |x[n]| < \infty, \tag{3.10}$$

then $|X(\omega)| < \infty$, which implies that the DTFT exists. A sequence $x[n]$ that satisfies Equation (3.10) was defined in Section 2.4.1 as being **absolutely summable**. Equation (3.10) states that if the sequence is absolutely summable, then the DTFT exists. The absolute summability of a sequence is a *sufficient* condition for the existence of the DTFT, not a *necessary* condition. So, we cannot state the converse, that if $X(\omega)$ exists then $x[n]$ is absolutely summable. In fact, we will soon discover a number of important sequences – for example, constants and periodic sequences – that are not absolutely summable, but for which we can still calculate the DTFT. But beyond that, there are clearly sequences such as $\alpha^n u[n]$ in Example 3.12, below, for which the DTFT exists for certain values of α and not for others. In fact, in Chapter 4, we will introduce the z-transform, a generalization of the DTFT that can handle this sequence and many other sequences of this sort.

3.2.5 The system function (again)

We previously showed that complex exponentials are eigenfunctions of linear time-invariant systems, and that when the input to the system is $x[n]=e^{j\omega_0 n}$, the output is $y[n]=H(\omega_0)x[n]$, where $H(\omega)$ is the system function given by Equation (3.1),

$$H(\omega)=\sum_{n=-\infty}^{\infty} h[n]e^{-j\omega n}.$$

This equation is exactly the same as the DTFT of $h[n]$. Hence, the system function is equivalent to the DTFT of the impulse response:

$$h[n]\leftrightarrow H(\omega).$$

Using the results of the preceding section, we state that a sufficient condition for the frequency response to exist is that the impulse response be absolutely summable. So, if

$$\sum_{n=-\infty}^{\infty} |h[n]| < \infty,$$

then the system function exists. However, remember that in Section 2.4, we showed that this same criterion is both a necessary and a sufficient condition for the system to be stable. Hence, if a system is stable, then its system function exists.

3.2.6 Periodicity of the DTFT

Complex exponential sequences like $e^{j\omega_0 n}$ are periodic with a period of 2π. That is,

$$e^{j(\omega_0+2\pi k)n}=e^{j\omega_0 n}\cancel{e^{j2\pi kn}}=e^{j\omega_0 n}.$$

This fact guarantees that the DTFT $X(\omega)$ is also periodic with a period of 2π, namely, $X(\omega+2\pi k)=X(\omega)$. This is easy to show:

$$X(\omega+2\pi k)=\sum_{n=-\infty}^{\infty} x[n]e^{-j(\omega+2\pi k)n}=\sum_{n=-\infty}^{\infty} x[n]e^{-j\omega n}=X(\omega).$$

Since $X(\omega)$ is periodic with period 2π, $X(\omega)$ is only unique over a continuous 2π range of ω, for example the range $-\pi\le\omega<\pi$. We shall term this to be the range of **principal values** of $X(\omega)$. There is nothing magic about this particular range of ω; we could just as well have chosen $0\le\omega<2\pi$ or any other continuous 2π range. Having selected a range, we can then derive the value of $X(\omega)$ at any other frequency outside of this range by adding or subtracting the appropriate multiple of 2π from ω.

3.2.7 DTFT of finite-length sequences

The DTFT of a finite-length sequence can be expressed as a finite sum,

$$X(\omega)=\sum_{n=n_0}^{n_1} x[n]e^{-j\omega n},$$

where n_0 and n_1 are respectively the lowest and highest time indices of the sequence. Since the number of terms in the sequence is finite, a finite-length sequence is guaranteed to be absolutely summable,

$$\sum_{n=n_0}^{n_1} |x[n]| < \infty,$$

so the DTFT of any finite-length sequence is also guaranteed to exist.

Let us compute the DTFT of a few simple examples of finite-length sequences. These include an impulse and pulses of various lengths.

Example 3.7

Let $x[n] = \delta[n]$. Find $X(\omega)$.

➤ Solution:

Just plug into Equation (3.6):

$$X(\omega) = \sum_{n=-\infty}^{\infty} x[n]e^{j\omega n} = \sum_{n=-\infty}^{\infty} \delta[n]e^{j\omega n} = 1.$$

Remember that when the delta function $\delta[n]$ occurs inside the summation, it "extracts" the value of the function it multiplies (in this case $e^{j\omega n}$) at $n = 0$. Using our shorthand notation, we write

$$\mathfrak{F}\{\delta[n]\} = 1.$$

In this case, $X(\omega)$ is trivially periodic, since it is constant for all ω. Plots of $x[n] = \delta[n]$ and $X(\omega) = 1$ are shown in **Figure 3.4**.

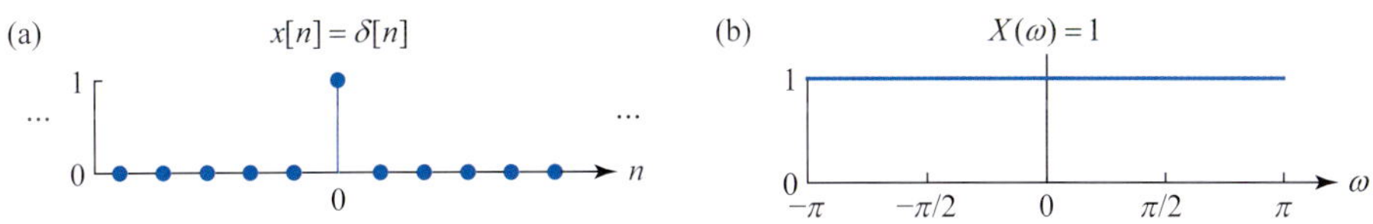

Figure 3.4 DTFT of an impulse and a constant

Example 3.8

Consider the impulse response of the moving-window filter, $h[n] = \delta[n+1] + \delta[n] + \delta[n-1]$, as shown below in **Figure 3.5a**. Find $H(\omega)$.

➤ Solution:

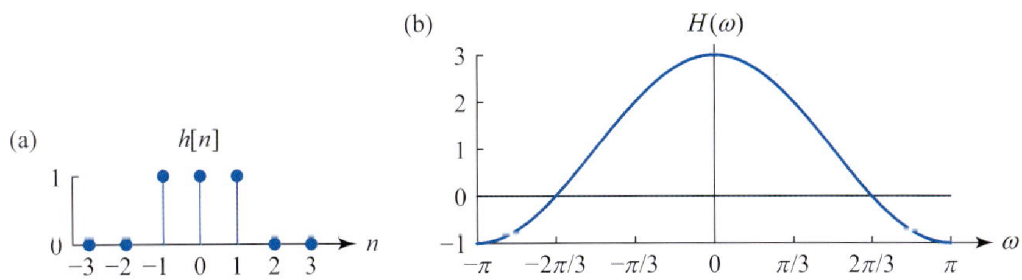

Figure 3.5 DTFT of a pulse

$$H(\omega) = \sum_{n=-\infty}^{\infty} h[n]e^{-j\omega n} = \sum_{n=-\infty}^{\infty} (\delta[n+1] + \delta[n] + \delta[n-1])e^{-j\omega n} = e^{j\omega} + 1 + e^{-j\omega} = 1 + 2\cos\omega. \quad \textbf{(3.11)}$$

In this case, $H(\omega)$ is purely real, so we can plot it on a single axis of amplitude vs. frequency, as shown in **Figure 3.5b**.

Example 3.9

Now consider a generalization of the above example. Find the DTFT of a centered symmetric pulse of odd length N,

$$h_N[n] = \begin{cases} 1, & |n| \leq (N-1)/2 \\ 0, & \text{otherwise.} \end{cases}$$

This is shown in **Figure 3.6a** for several values of N.

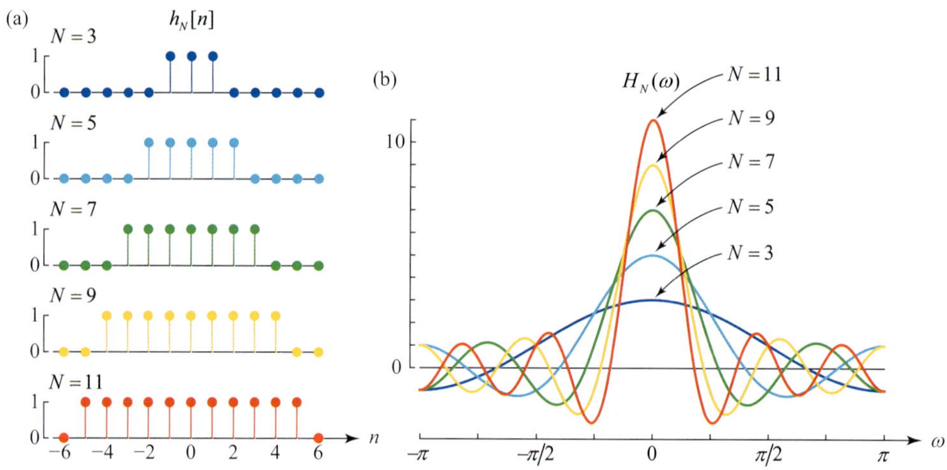

Figure 3.6 DTFT of pulses of varying widths

➤ Solution:

The DTFT of this family of pulses is

$$H_N(\omega) = \mathfrak{F}\{h_N[n]\} = \sum_{n=-\infty}^{\infty} h_N[n]e^{-j\omega n} = \sum_{n=-(N-1)/2}^{(N-1)/2} e^{-j\omega n}. \quad \textbf{(3.12)}$$

Let $m = n + (N-1)/2$, so

$$
\begin{aligned}
H_N(\omega) &= \sum_{n=-(N-1)/2}^{(N-1)/2} e^{-j\omega n} = e^{j\omega(N-1)/2} \sum_{m=0}^{N-1} e^{-j\omega m} = e^{j\omega(N-1)/2} \frac{1-e^{-j\omega N}}{1-e^{-j\omega}} \\
&= \cancel{e^{j\omega(N-1)/2}} \frac{\cancel{e^{-j\omega N/2}}}{\cancel{e^{-j\omega/2}}} \frac{e^{j\omega N/2} - e^{-j\omega N/2}}{e^{j\omega/2} - e^{-j\omega/2}} = \frac{\sin \omega N/2}{\sin \omega/2}.
\end{aligned}
\tag{3.13}
$$

This function is called the **Dirichlet kernel**, which you may have seen in your study of the Fourier series.[2] We can also rewrite Equation (3.13) as the ratio of sinc functions,

$$H_N(\omega) = \frac{\sin \omega N/2}{\sin \omega/2} = \frac{(\omega N/2)\,\mathrm{sinc}\,\omega N/2}{(\omega/2)\,\mathrm{sinc}\,\omega/2} = N \frac{\mathrm{sinc}\,\omega N/2}{\mathrm{sinc}\,\omega/2}. \tag{3.14}$$

When expressed this way, we call this transform the **periodic sinc function**, and we will refer to it often in the chapters that follow. There are a couple of things to note about this function. First, it is periodic with period 2π, as are all DTFTs. Second, at any value of N, the transform is maximum at $\omega = 0$ and has value

$$H_N(0) = N \frac{\mathrm{sinc}(0)}{\mathrm{sinc}(0)} = N.$$

(Recall that $\mathrm{sinc}(0) = 1$.) We could also determine $H_N(0)$ directly from Equation (3.12) by substitution of $\omega = 0$,

$$H_N(0) = \sum_{n=-(N-1)/2}^{(N-1)/2} 1 \cdot e^{-j0n} = N.$$

Figure 3.6b shows the DTFT of pulses of different lengths N. As N increases, the maximum amplitude of $H_N(\omega)$ increases in direct proportion to N. This figure also shows that $H_N(\omega)$ oscillates with a frequency proportional to N. This might be expected since the numerator of $H_N(\omega)$ is sinc $\omega N/2$, which in turn varies as N. Hence, the main "lobe" of $H_N(\omega)$, centered at $\omega = 0$, has zero-crossings at values of $\omega = 2\pi k/N$, where k is an integer. The frequency domain characteristics of these pulses (e.g., the width of the main lobe) will become important to us in Chapter 7, where we design FIR filters (see Example 3.34 for a preview), and in Chapter 14, where we discuss the spectral analysis of data that is limited in time duration (see Section 3.10.11 for a preview).

Any sequence or impulse response $h[n]$ can be expanded as an infinite sum of delayed impulses,

$$h[n] = \sum_{k=-\infty}^{\infty} h[k]\delta[n-k], \tag{3.15}$$

[2] Peter Gustav Lejeune Dirichlet (1805–1859) was a prolific German mathematician (and successor to Carl Friedrich Gauss at the University of Göttingen) who established fundamental results in mathematical analysis and number theory. Of particular interest to the study of signal processing, Dirichlet showed in 1829 that the Fourier series approximation of a function could be described by the convolution of the function with the Dirichlet kernel and he determined the conditions (known as the **Dirichlet conditions**) under which the Fourier series converges.

where $\delta[n-k]$ is just an impulse shifted by k and $h[k]$ is a constant for any particular value of k. $H(\omega)$ can also be expanded as an infinite sum of complex exponential sequences $e^{-j\omega k}$,

$$H(\omega)=\sum_{k=-\infty}^{\infty} h[k]e^{-j\omega k}, \tag{3.16}$$

where, again, $h[k]$ is a constant for any particular value of k. Note the homology between Equations (3.15) and (3.16). There is a one-to-one correspondence between each term in the series that makes up $h[n]$ and each term in $H(\omega)$:

$$\begin{array}{lllllll} h[n] = \sum_{k=-\infty}^{\infty} h[k]\delta[n-k] & = \cdots h[-1]\delta[n+1] & +h[0]\delta[n] & +h[1]\delta[n-1]+ & \cdots & +h[k]\delta[n-k]+ & \cdots \\ & \updownarrow & \updownarrow & \updownarrow & & \updownarrow & \\ H(\omega) = \sum_{k=-\infty}^{\infty} h[k]e^{-j\omega k} & = \cdots h[-1]e^{-j\omega(-1)} & +h[0]e^{-j\omega 0} & +h[1]e^{-j\omega 1} & +\cdots & +h[k]e^{-j\omega k} & +\cdots \end{array}$$

In general,

$$\mathfrak{F}\{\delta[n-k]\}=e^{-j\omega k}.$$

The DTFT of a shifted impulse is a single complex exponential. When $h[n]$ is a sequence of length N, the summations in Equations (3.15) and (3.16) can each be truncated to N terms, and $H(\omega)$ is evaluated explicitly from the summation, as shown in the next example.

Example 3.10

Find the DTFT of $x[n]=\delta[n+1]+\delta[n]-2\delta[n-1]+3\delta[n-2]$.

➤ Solution:

$$\begin{array}{ccccccccc} x[n] & = & \delta[n+1] & + & \delta[n] & - & 2\delta[n-1] & + & 3\delta[n-2] \\ \updownarrow & & \updownarrow & & \updownarrow & & \updownarrow & & \updownarrow \\ X(\omega) & = & e^{j\omega} & + & 1 & - & 2e^{-j\omega} & + & 3e^{-j2\omega}. \end{array}$$

If we wish, we can express any finite-length sequence in terms of the sum of trigonometric function by use of the Euler relation,

$$e^{-j\omega n}=\cos\omega n-j\sin\omega n.$$

So, this example becomes

$$\begin{aligned} X(\omega)&=e^{j\omega}+1-2e^{-j\omega}+3e^{-j2\omega}\\ &=(1+\cos\omega-2\cos\omega+3\cos 2\omega)+j(\sin\omega+2\sin\omega-3\sin 2\omega). \end{aligned}$$

Of course, given the DTFT of a finite-length sequence, we can find the inverse DTFT by inspection, as shown in the next example.

Example 3.11

Find the inverse DTFT of $X(\omega) = 4\cos\omega - 2\sin 2\omega$.

➤ Solution:

$$\begin{array}{ccccccccccc} X(\omega) & = & 4\cos\omega - 2\sin 2\omega & = & je^{j2\omega} & + & 2e^{j\omega} & + & 2e^{-j\omega} & - & je^{-j2\omega} \\ \updownarrow & & & & \updownarrow & & \updownarrow & & \updownarrow & & \updownarrow \\ x[n] & = & & & j\delta[n+2] & + & 2\delta[n+1] & + & 2\delta[n-1] & - & j\delta[n-2]. \end{array}$$

3.2.8 DTFT of infinite-length sequences

Infinite-length sequences include constants and the impulse responses of infinite impulse response (IIR) systems. The existence of the DTFT for infinite-length sequences is not guaranteed, since infinite-length sequences may not be absolutely summable. We will consider two important examples: a power-law sequence, which is absolutely summable under certain conditions, and a constant, which is not absolutely summable, but whose transform we can nonetheless obtain by trickery.

Example 3.12

Find the DTFT of a power-law sequence, $h[n] = \alpha^n u[n]$, and determine the range of α for which the DTFT exists. You may remember this sequence as the impulse response of the difference equation, $y[n] - \alpha y[n-1] = x[n]$.

➤ Solution:

$$H(\omega) = \sum_{n=-\infty}^{\infty} h[n]e^{-j\omega n} = \sum_{n=-\infty}^{\infty} \alpha^n u[n]e^{-j\omega n} = \sum_{n=0}^{\infty} (\alpha e^{-j\omega})^n.$$

The summation has the form of an infinite geometric series. This series is convergent and can be reduced to closed form if $|\alpha e^{-j\omega}| < 1$. But $|\alpha e^{-j\omega}| = |\alpha|$. So, if $|\alpha| < 1$, then $H(\omega)$ exists and can be expressed as

$$H(\omega) = \frac{1}{1 - \alpha e^{-j\omega}}.$$

We will refer to this result again and again. However, if $|\alpha| > 1$, then the summation does not converge and $H(\omega)$ does not exist.

Example 3.13

Let $x[n] = 1$. Find $X(\omega)$.

➤ Solution:

We should immediately recognize that we are sort-of in trouble because, in this example, $x[n]$ is clearly not absolutely summable. In fact,

$$\sum_{n=-\infty}^{\infty} |x[n]| = \infty.$$

If we simply plug $x[n]=1$ into Equation (3.6), we discover that the infinite sum of $e^{j\omega n}$ is difficult to compute:

$$X(\omega) = \sum_{n=-\infty}^{\infty} x[n]e^{-j\omega n} = \sum_{n=-\infty}^{\infty} e^{-j\omega n} = ???$$

However, we can guess an appropriate solution by looking at Equation (3.7) and asking, "What does $X(\omega)$ have to be in order to make $x[n]=1$?" A candidate solution is $X(\omega)=2\pi\delta(\omega)$. We verify this by plugging it into Equation (3.7):

$$x[n] = \frac{1}{2\pi}\int_{-\pi}^{\pi} X(\omega)e^{j\omega n}d\omega = \frac{1}{\cancel{2\pi}}\int_{-\pi}^{\pi} \cancel{2\pi}\delta(\omega)e^{j\omega n}d\omega = 1. \tag{3.17}$$

Hence the DTFT of a constant sequence is an impulse in frequency. This makes some intuitive sense. Equation (3.17) says that to make a constant sequence $x[n]=1$ out of complex exponentials, you only need a *single* complex exponential at frequency $\omega=0$, which is, of course, a constant.

Unfortunately, $2\pi\delta(\omega)$ is not quite the correct solution for $X(\omega)$. Why? Because, $X(\omega)$ must not only satisfy Equation (3.7), but must also be periodic with a period of 2π. $X(\omega)=2\pi\delta(\omega)$ is not periodic. The periodic solution for $X(\omega)$ that satisfies Equation (3.17) is

$$X(\omega) = 2\pi\sum_{k=-\infty}^{\infty} \delta(\omega+2\pi k). \tag{3.18}$$

This is just a series of impulses at multiples of 2π. Substitution of Equation (3.18) into Equation (3.7) yields the same result as Equation (3.17), because only the impulse at $\omega=0$ lies within the frequency range of the integral, namely $-\pi\le\omega<\pi$. To summarize, using our shorthand notation, we write

$$\mathfrak{F}\{1\} = 2\pi\sum_{k=-\infty}^{\infty} \delta(\omega+2\pi k).$$

The picture is shown in **Figure 3.7**. There is only one impulse within the principal value range of $X(\omega)$, namely $-\pi\le\omega<\pi$, which is shaded in yellow.

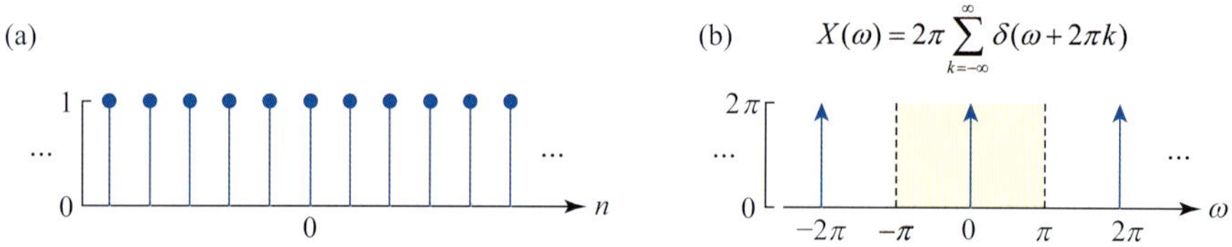

Figure 3.7 DTFT of a constant

This important example demonstrates that, although $x[n]$ is not absolutely summable, the DTFT may nonetheless exist. This should emphasize the point we made earlier that the absolute summability criterion is a sufficient, but not necessary, condition for the existence of the DTFT. Note also the "duality" of the results of the preceding examples:

- The DTFT of an impulse in time is a constant in frequency.
- The DTFT of a constant in time is impulses in frequency.

$h[n]$	$H(\omega)$
$\delta[n]$	1
1	$2\pi\sum_{k=-\infty}^{\infty}\delta(\omega+2\pi k)$

3.3 Magnitude and phase description of the DTFT

In the previous section, we considered several examples of the DTFT that were purely real. In order to visualize these DTFTs, all we needed to do was to graph the amplitude of $X(\omega)$ as a function of ω (see **Figure 3.6b**, for example). However, in general, DTFTs are complex. To visualize these DTFTs, the simple graph of amplitude as a function of frequency will not do. In this section, we discuss the general visual representation of complex DTFTs using separate graphs of magnitude and phase as a function of frequency ω.

3.3.1 Magnitude and phase of the DTFT of an impulse

Let us start by reconsidering the DTFT of an impulse, $x[n]=\delta[n]$. As discussed in Example 3.7, we found that the DTFT is a constant, $X(\omega)=\mathfrak{F}\{\delta[n]\}=1$. Because $X(\omega)$ is positive real for all ω, it can be displayed on a single graph, shown in **Figure 3.4b**. In general, $X(\omega)$ will be complex, and must be graphically displayed as magnitude and phase. The magnitude and phase of the DTFT of $x[n]=\delta[n]$ are shown in **Figures 3.8b** and **c**.

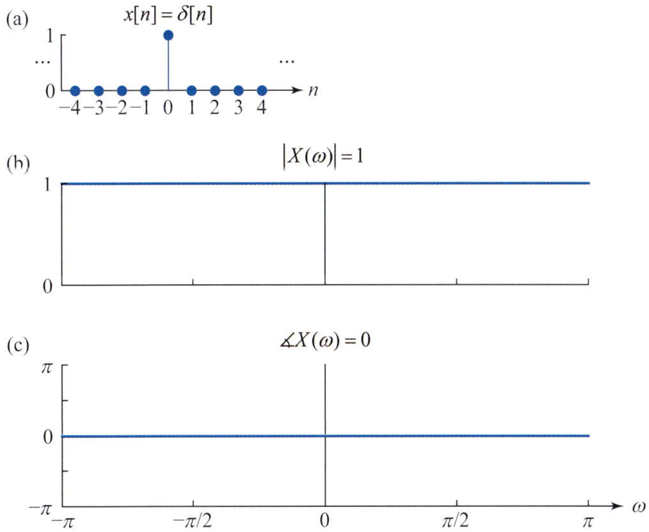

Figure 3.8 Magnitude and phase plot of the DTFT of an impulse

Since $X(\omega) = 1 = 1 \cdot e^{j0}$, therefore, $|X(\omega)| = 1$ and $\measuredangle X(\omega) = 0$ for all ω. This is a pretty uninteresting plot. However, now consider an impulse that is shifted by n_0 samples: $y[n] = \delta[n - n_0]$. The DTFT of the shifted impulse is

$$Y(\omega) = \sum_{n=-\infty}^{\infty} \delta[n - n_0] e^{-j\omega n} = e^{-j\omega n_0}.$$

This is a complex quantity whose magnitude and phase are

$$|Y(\omega)| = 1$$
$$\measuredangle Y(\omega) = -\omega n_0 .$$

The graph is shown in **Figure 3.9** for $n_0 = 1$. The magnitude of the DTFT of the shifted impulse, $|Y(\omega)|$, is the same as that of the unshifted impulse, $|X(\omega)|$, shown in **Figure 3.8**. The phase of the shifted impulse, $\measuredangle Y(\omega)$, has a constant slope of -1 when plotted vs. ω.

Figure 3.10 shows the graph of the magnitude and phase of the DTFT of an impulse shifted by $n_0 = 2$. As expected, the magnitude is again one. The phase, shown with the dotted red line, has a slope of -2. Recall that the phase of a complex quantity can always be expressed in the principal value range $-\pi \leq \measuredangle Y(\omega) < \pi$. Any phase angle outside this range can be mapped inside the range by addition or subtraction of the appropriate multiple of 2π since

$$e^{j(\measuredangle Y(\omega) + 2\pi k)} = e^{j\measuredangle Y(\omega)} e^{j2\pi k} = e^{j\measuredangle Y(\omega)},$$

where k is an integer. Accordingly, we have subtracted 2π from the red line for the range of frequencies $-\pi \leq \omega < -\pi/2$, and added 2π to the curve for the range of frequencies $\pi/2 \leq \omega < \pi$, in order to make all the phase values lie within the range $-\pi \leq \measuredangle Y(\omega) < \pi$, as shown in the blue trace in **Figure 3.10c**.

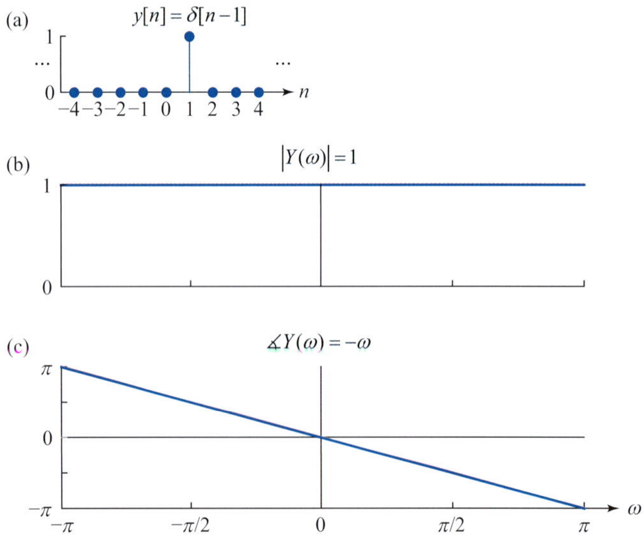

Figure 3.9 Magnitude and phase of the DTFT of a shifted impulse, $n_0 = 1$

The process of replotting phase data to fall within the range $-\pi \le \measuredangle Y(\omega) < \pi$ is termed **phase wrapping**. Phase wrapping is a strictly "cosmetic" operation whose purpose is to allow us to plot the phase in the range of principal values. Although phase wrapping produces **plotting discontinuities** in the phase plot, it does not add **essential phase discontinuities** to the plot. The 2π discontinuities at $\omega = \pm\pi/2$ in the blue plot of **Figure 3.10c** are plotting discontinuities, not essential phase discontinuities. What is an essential phase discontinuity? Stay tuned for the next example.

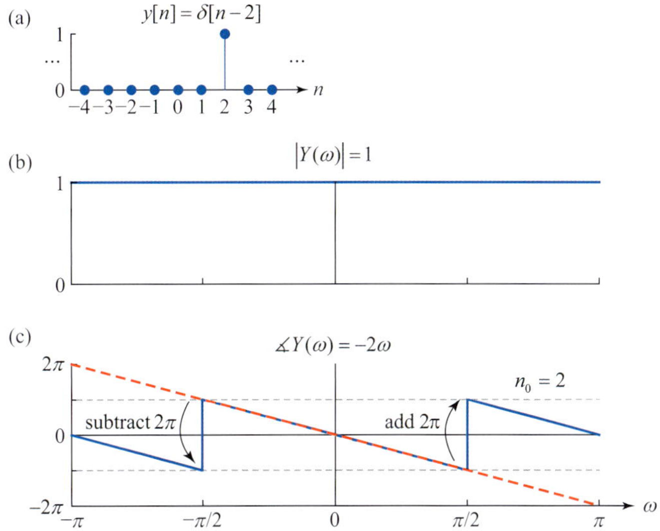

Figure 3.10 Magnitude and phase of the DTFT of a shifted impulse, $n_0 = 2$

3.3.2 Essential phase discontinuities

Let us reconsider the DTFT of Example 3.1, $X(\omega) = 1 + 2\cos\omega$, the graph of which is shown in **Figure 3.5b**. This is a graph of amplitude $X(\omega)$ vs. frequency ω. This is clearly not the same as a graph of magnitude $|X(\omega)|$ vs. ω, even when $X(\omega)$ is real, as it is in this example. For a real quantity, $X(\omega)$ can be less than zero, whereas $|X(\omega)|$ must, by definition, be greater than or equal to zero for all ω. For consistency, even when $X(\omega)$ is real, we will express it in polar form,

$$X(\omega) = |X(\omega)|e^{j\measuredangle X(\omega)}.$$

The separate plots of $|X(\omega)|$ and $\measuredangle X(\omega)$ are shown in **Figure 3.11**. For the frequency range $-2\pi/3 \le \omega < 2\pi/3$, $X(\omega)$ is strictly positive, $X(\omega) > 0$, so we can express $X(\omega) = |X(\omega)|e^{j0}$. For this range of frequencies, $|X(\omega)| = X(\omega)$ and $\measuredangle X(\omega) = 0$, as shown in **Figure 3.11**. For the frequency range $2\pi/3 < |\omega| < \pi$, $X(\omega)$ is strictly negative, $X(\omega) < 0$, so we can express $X(\omega) = |X(\omega)|e^{j\pi} = |X(\omega)|e^{-j\pi} = -X(\omega)$. For this range of frequencies, $|X(\omega)| = -X(\omega)$ and $\measuredangle X(\omega)$ can be either $+\pi$ or $-\pi$. We have chosen to let $\measuredangle X(\omega) = \pi$ for $-\pi \le \omega < -2\pi/3$ and $\measuredangle X(\omega) = -\pi$ for $2\pi/3 \le \omega < \pi$. This choice makes the phase of $X(\omega)$ an odd function of ω; that is, $\measuredangle X(\omega) = -\measuredangle X(-\omega)$. Also, you will notice that the magnitude of $X(\omega)$ is an even function of ω, $|X(\omega)| = |X(-\omega)|$. This symmetry is not a coincidence. In Section 3.5, we will show that if $x[n]$

is a real sequence, then $|X(\omega)|$ is *always* an even function of ω, and $\measuredangle X(\omega)$ is *always* an odd function of ω. The phase discontinuities of π at $\omega = \pm 2\pi/3$ are not cosmetic plotting discontinuities but, rather, they are **essential phase discontinuities** that reflect the change of sign in $X(\omega)$ at those frequencies.

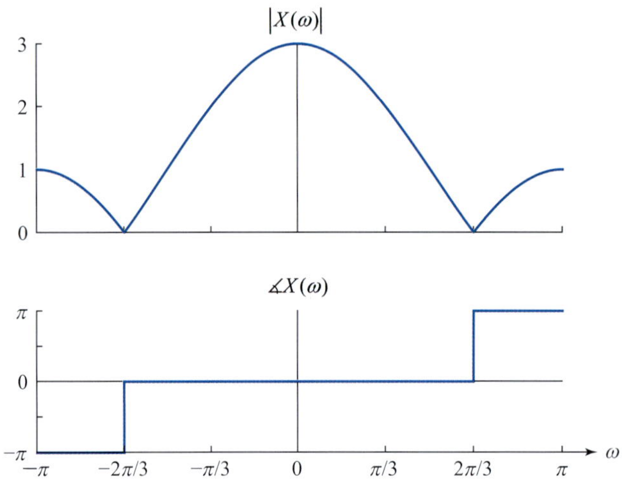

Figure 3.11 Magnitude and phase of the DTFT

Example 3.14

Find and plot the DTFT of the pulse of Example 3.8 shifted by two samples,

$$y[n] = x[n-2] = \delta[n-1] + \delta[n-2] + \delta[n-3],$$

as shown below in **Figure 3.12**.

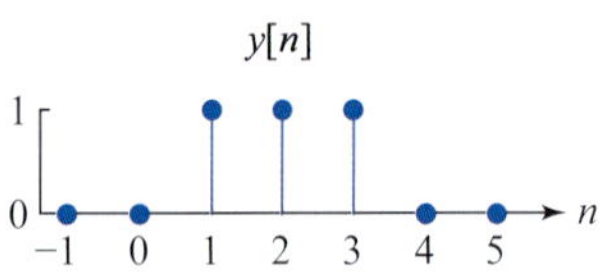

Figure 3.12 Shifted pulse

➤ Solution:

The DTFT is

$$\begin{aligned} Y(\omega) &= \sum_{n=-\infty}^{\infty} y[n]e^{-j\omega n} = \sum_{n=-\infty}^{\infty} (\delta[n-1] + \delta[n-2] + \delta[n-3])e^{-j\omega n} \\ &= e^{-j\omega} + e^{-j2\omega} + e^{-j3\omega} = e^{-j2\omega}(e^{j\omega} + 1 + e^{-j\omega}) = e^{-j2\omega}(1 + 2\cos\omega) = e^{-j2\omega} X(\omega)\,, \end{aligned} \tag{3.19}$$

where $X(\omega)$ is the DTFT of the symmetric pulse given in Equation (3.11). $Y(\omega)$ is complex and can be expressed in polar form as

$$Y(\omega) = |Y(\omega)|e^{j\measuredangle Y(\omega)}.$$

From Equation (3.19),

$$Y(\omega) = e^{-j2\omega}X(\omega) = e^{-j2\omega}\left(|X(\omega)|e^{j\measuredangle X(\omega)}\right) = |X(\omega)|e^{j(\measuredangle X(\omega)-2\omega)}.$$

So, the magnitude $|Y(\omega)|$ is clearly equal to the magnitude $|X(\omega)|$, and the phase $\measuredangle Y(\omega)$ is the sum of two components: $\measuredangle X(\omega)$, and a so-called **linear-phase term**, -2ω:

$$\begin{aligned} |Y(\omega)| &= |X(\omega)| \\ \measuredangle Y(\omega) &= \measuredangle X(\omega) - 2\omega\,. \end{aligned}$$

The two phase terms are shown individually in **Figure 3.13a**.

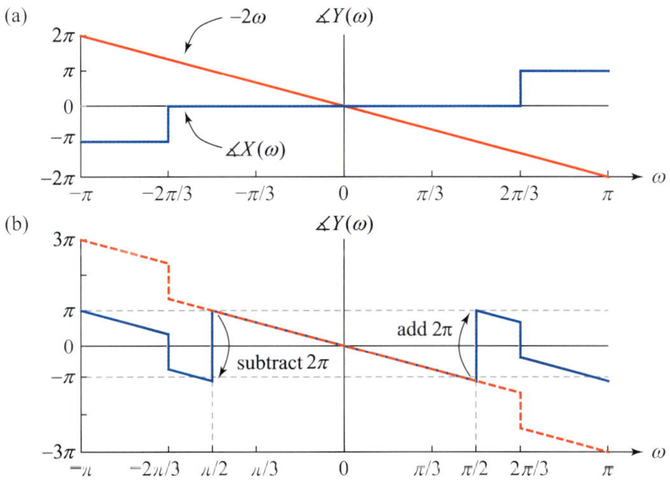

Figure 3.13 Phase wrapping

The blue trace is the same $\measuredangle X(\omega)$ shown in **Figure 3.11**. The red trace is the linear-phase term -2ω, a straight line with a slope of -2. The sum of the two terms, $\measuredangle Y(\omega)$, is shown as the dashed red line in **Figure 3.13b**. The sum is greater than π for $\omega < -\pi/2$ and less than $-\pi$ for $\omega > \pi/2$. However, phase can always be expressed in the principal value range $-\pi \le \measuredangle Y(\omega) < \pi$ by adding or subtracting an appropriate multiple of 2π from phase values that lie outside this range. So, we subtract 2π from the red trace for $-\pi \le \omega < -\pi/2$ and add 2π to the trace for $\pi/2 \le \omega < \pi$, thereby producing phase values within the range $-\pi < \measuredangle Y(\omega) < \pi$, as shown in the blue trace in **Figure 3.13b**. The 2π discontinuities at $\omega = \pm\pi/2$ in the plot are non-essential plotting discontinuities; the π discontinuities at $\omega = \pm 2\pi/3$ in the blue trace of **Figure 3.13b** are essential phase discontinuities. **Figure 3.14** shows the final magnitude and phase plot for this pulse.

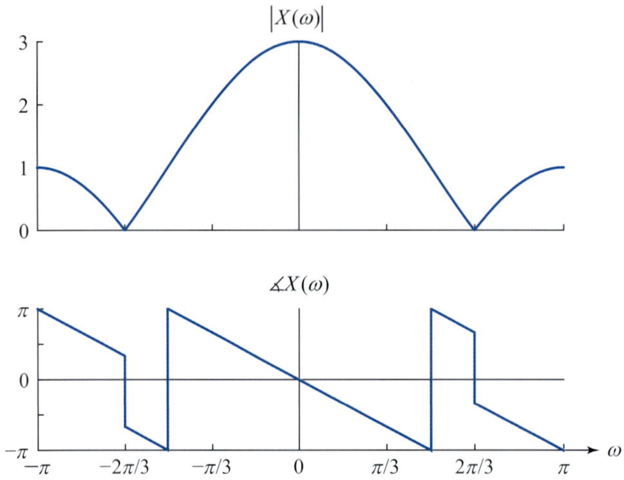

Figure 3.14 DTFT of shifted pulse

Example 3.15

Find and plot the DTFT of $x[n] = \delta[n+1] - \delta[n-1]$, as shown in **Figure 3.15**.

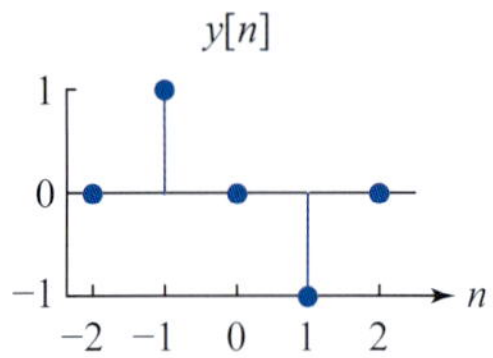

Figure 3.15 Asymmetric pulse

➤ Solution:

The DTFT is

$$Y(\omega) = \sum_{n=-\infty}^{\infty} y[n]e^{-j\omega n} = \sum_{n=-\infty}^{\infty} (\delta[n+1] - \delta[n-1])e^{-j\omega n} = e^{j\omega} - e^{-j\omega} = 2j\sin\omega.$$

This transform is purely imaginary. The magnitude is

$$|Y(\omega)| = |2j\sin\omega| = |2j||\sin\omega| = 2|\sin\omega|,$$

and the phase is

$$\measuredangle Y(\omega) = \measuredangle 2j + \measuredangle \sin\omega = \pi/2 + \measuredangle \sin\omega,$$

where $\measuredangle \sin\omega$ is the phase of the (real) quantity $\sin\omega$ which could be either 0 or $\pm\pi$. The plot of magnitude and phase is shown in **Figure 3.16**. Once again, the magnitude is an even function of ω and the phase is an odd function. An essential π phase transition occurs at $\omega = 0$, reflecting the change in the sign of $\sin\omega$ at $\omega = 0$. For $0 \le \omega < \pi$, $\sin\omega > 0$ so $\measuredangle \sin\omega = 0$. Therefore $\measuredangle Y(\omega) = \pi/2 + 0 = \pi/2$. For $-\pi \le \omega < 0$, $\sin\omega < 0$ so $\measuredangle \sin\omega = \pm\pi$. Therefore, $\measuredangle Y(\omega) = \pi/2 \pm \pi$. We choose to subtract π, making $\measuredangle Y(\omega) = \pi/2 - \pi = -\pi/2$. This keeps the phase in the range $-\pi \le \omega < \pi$.

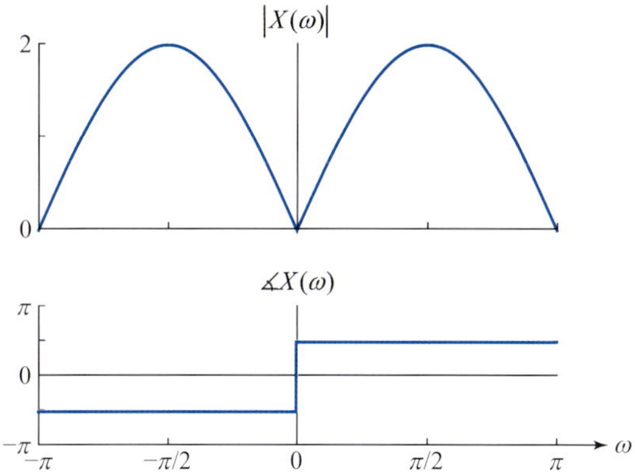

Figure 3.16 DTFT of an asymmetric pulse

3.4 Important sequences and their transforms

There are a number of important sequences whose transforms we will require throughout our study of digital signal processing. We have already seen a few – the impulse, constant and pulse. **Table 3.1** lists a few more. Proofs are mostly left as problems.

Table 3.1 Discrete-time Fourier transform of several important sequences

$x[n]$	$X(\omega)$
$\delta[n]$	1
1	$2\pi \sum_{k=-\infty}^{\infty} \delta(\omega - 2\pi k)$
$\alpha^n u[n],\ \lvert\alpha\rvert < 1$	$\dfrac{1}{1 - \alpha e^{-j\omega}}$
$\alpha^n \cos(\omega_0 n)\, u[n],\ \lvert\alpha\rvert < 1$	$\dfrac{1 - \alpha e^{-j\omega} \cos \omega_0}{1 - 2\alpha e^{-j\omega} \cos \omega_0 + \alpha^2 e^{-2j\omega}}$
$\alpha^n \sin(\omega_0 n)\, u[n],\ \lvert\alpha\rvert < 1$	$\dfrac{\alpha e^{-j\omega} \sin \omega_\theta}{1 - 2\alpha e^{-j\omega} \cos \omega_\theta + \alpha^2 e^{-2j\omega}}$
$\cos \omega_0 n$	$\pi \sum_{k=-\infty}^{\infty} \delta(\omega - \omega_0 - 2\pi k) + \pi \sum_{k=-\infty}^{\infty} \delta(\omega + \omega_0 - 2\pi k)$
$\sin \omega_0 n$	$-\pi j \sum_{k=-\infty}^{\infty} \delta(\omega - \omega_0 - 2\pi k) + \pi j \sum_{k=-\infty}^{\infty} \delta(\omega + \omega_0 - 2\pi k)$
$\cos(\omega_0 n + \varphi)$	$\pi e^{j\omega_0} \sum_{k=-\infty}^{\infty} \delta(\omega - \omega_0 - 2\pi k) + \pi e^{-j\omega_0} \sum_{k=-\infty}^{\infty} \delta(\omega + \omega_0 - 2\pi k)$

3.5 Symmetry properties of the DTFT

In all the examples of the DTFT of real sequences in the preceding section, the magnitude of the transform is an even function of ω and the phase is an odd function of ω. As we will see in this section, this result follows from the so-called **symmetry properties** of the DTFT. We can derive a great deal of useful information about the DTFT from these symmetry properties. We will also use these properties to help us derive powerful practical algorithms for discrete-time filtering in Chapter 7 and for the fast computation of the Fourier transform in Chapter 11.

3.5.1 Time reversal

Consider a sequence $x[n]$, with DTFT $X(\omega)$. The time-reversal property states that the DTFT of a time-reversed (flipped) sequence $x[-n]$ is the frequency-reversed (flipped) transform $X(-\omega)$; that is,

if $x[n] \leftrightarrow X(\omega)$, then $x[-n] \leftrightarrow X(-\omega)$.

The proof is simple:

$$\mathfrak{F}\{x[-n]\} = \sum_{n=-\infty}^{\infty} x[-n]e^{-j\omega n} = \sum_{-n=\infty}^{-\infty} x[-n]e^{j\omega(-n)} = \sum_{m=\infty}^{-\infty} x[m]e^{-j(-\omega)m} = X(-\omega),$$

where we have let $m = -n$ in the last summation.

There are some interesting consequences of the time-reversal property, which reveal themselves when sequences have even or odd symmetry.

3.5.2 Conjugate symmetry and antisymmetry

Assume that $x[n]$ is a complex sequence. Its complex conjugate is $x^*[n]$. The DTFT of the complex conjugate is

$$\mathfrak{F}\{x^*[n]\} = \sum_{n=-\infty}^{\infty} x^*[n]e^{-j\omega n} = \sum_{n=-\infty}^{\infty} (x[n]e^{j\omega n})^* = \left(\sum_{n=-\infty}^{\infty} x[n]e^{-j(-\omega)n}\right)^* = X^*(-\omega), \tag{3.20}$$

where we have used the fact that the conjugate of the product is the product of the conjugates: $x^*[n]e^{-j\omega n} = (x[n]e^{j\omega n})^*$. We have also used the fact that the conjugate of the sum is the sum of the conjugates.

Real sequences Most (but not all) of the sequences that are of interest to us are real, which means that $x[n] = x^*[n]$. Therefore,

$$\mathfrak{F}\{x[n]\} = \mathfrak{F}\{x^*[n]\}.$$

By Equation (3.20), $\mathfrak{F}\{x^*[n]\} = X^*(-\omega)$, so for a real sequence,

$$X(\omega) = X^*(-\omega). \tag{3.21}$$

A function that satisfies Equation (3.21) is said to exhibit **conjugate symmetry** (also known as **Hermitian symmetry**). The fact that the transform of a real sequence is conjugate-symmetric has a number of important consequences. Expressing $X^*(-\omega)$ in polar form, we get

$$X^*(-\omega) = |X^*(-\omega)|e^{j\measuredangle X^*(-\omega)} = |X(-\omega)|e^{-j\measuredangle X(-\omega)}, \tag{3.22}$$

since

$$\begin{aligned}|X^*(-\omega)| &= |X(-\omega)|\\ \measuredangle X^*(-\omega) &= -\measuredangle X(\omega).\end{aligned}$$

Putting Equation (3.21) in polar form, with the help of Equation (3.22), we get

$$|X(\omega)|e^{j\measuredangle X(\omega)} = |X(-\omega)|e^{-j\measuredangle X(-\omega)}.$$

Hence,

$$\begin{aligned}|X(\omega)| &= |X(-\omega)|\\ \measuredangle X(\omega) &= -\measuredangle X(-\omega).\end{aligned} \quad \textbf{(3.23)}$$

In other words, if $x[n]$ is real, then the magnitude of $X(\omega)$ is an *even* function of ω and the phase of $X(\omega)$ is an *odd* function of ω. If you look back over all the plots in this chapter, you will see that this is always so, and now you know why!

Equation (3.23) says that for a real sequence, the DTFT is completely characterized by $X(\omega)$ over the range $0 \le \omega < \pi$. We can then derive $X(\omega)$ over the entire range of $-\pi \le \omega < \pi$ from Equation (3.23) if we wish. This has practical relevance. For example, in Chapters 10 and 11, we will compute the Fourier transform of $x[n]$ algorithmically at discrete values of ω using the discrete Fourier transform (DFT), and its practical realization, the fast Fourier transform (FFT). For real sequences, we only need to compute and store the DTFT over the range $0 \le \omega < \pi$. Furthermore, when we display the magnitude and phase of the transform of a real sequence, our plots only need to show the range $0 \le \omega < \pi$.

We can also write $X(\omega)$ and $X^*(\omega)$ in Cartesian form:

$$\begin{aligned}X(\omega) &= \mathrm{Re}\{X(\omega)\} + j\ \mathrm{Im}\{X(\omega)\}\\ X^*(\omega) &= \mathrm{Re}\{X(\omega)\} - j\ \mathrm{Im}\{X(\omega)\}.\end{aligned}$$

If $x[n]$ is real, then, as indicated in Equation (3.21), $X(\omega) = X^*(-\omega)$, so

$$\begin{aligned}\mathrm{Re}\{X(\omega)\} &= \mathrm{Re}\{X(-\omega)\}\\ \mathrm{Im}\{X(\omega)\} &= -\mathrm{Im}\{X(-\omega)\}\cdot\end{aligned} \quad \textbf{(3.24)}$$

In other words, if $x[n]$ is real, then the real part of $X(\omega)$ is an even function of ω and the imaginary part of $X(\omega)$ is an odd function of ω. Again, for a real sequence we only need to compute $\mathrm{Re}\{X(\omega)\}$ and $\mathrm{Im}\{X(\omega)\}$ over the range $0 \le \omega < \pi$. We can then derive $\mathrm{Re}\{X(\omega)\}$ and $\mathrm{Im}\{X(\omega)\}$ over the entire range of frequency, $-\pi \le \omega < \pi$, from Equation (3.24) if we wish.

★ **Imaginary sequences** Most of the time, the sequences we deal with will be real, but we will occasionally run into sequences that are imaginary, particularly in applications having to do with discrete-time communications (for example, the Hilbert transformer discussed in Chapter 7). For imaginary sequences, $x[n] = -x^*[n]$. Therefore,

$$\mathfrak{F}\{x[n]\} = \mathfrak{F}\{-x^*[n]\} = -\mathfrak{F}\{x^*[n]\}.$$

By Equation (3.20), $\mathfrak{F}\{x^*[n]\} = X^*(-\omega)$, so for an imaginary sequence,

$$X(\omega) = -X^*(-\omega). \tag{3.25}$$

A function that satisfies Equation (3.25) is said to be **conjugate-antisymmetric**. A number of consequences follow from this.

$$\begin{aligned} |X(\omega)| &= |-X^*(-\omega)| = |X(-\omega)| \\ \measuredangle X(\omega) &= \measuredangle(-X^*(-\omega)) = \measuredangle X^*(-\omega) \pm \pi = -\measuredangle X(-\omega) \pm \pi. \end{aligned} \tag{3.26}$$

In other words, if $x[n]$ is imaginary, then the magnitude of $X(\omega)$ is an even function of ω (just as it was when $x[n]$ was real) and the phase of $X(\omega)$ is *neither* an even nor an odd function of ω. However, you will note again that the DTFT of an imaginary sequence is completely characterized by $X(\omega)$ over the range $0 \le \omega < \pi$. We can then derive $X(\omega)$ over the entire range of $-\pi \le \omega < \pi$ by exploiting Equation (3.25).

3.5.3 Even and odd symmetry

Even sequences If $x[n]$ is an even sequence, then, by definition, $x[n] = x[-n]$. Therefore, $\mathfrak{F}\{x[n]\} = \mathfrak{F}\{x[-n]\}$. However, by the time-reversal property, $\mathfrak{F}\{x[-n]\} = X(-\omega)$. So, for an even sequence,

$$X(\omega) = X(-\omega). \tag{3.27}$$

In other words, if $x[n]$ is even, then $X(\omega)$ is also even. Furthermore, expressing both sides of Equation (3.27) in polar form, we have

$$|X(\omega)|e^{j\measuredangle X(\omega)} = |X(-\omega)|e^{j\measuredangle X(-\omega)},$$

so

$$\begin{aligned} |X(\omega)| &= |X(-\omega)| \\ \measuredangle X(\omega) &= \measuredangle X(-\omega)\cdot \end{aligned} \tag{3.28}$$

So, if $x[n]$ is even, then both the magnitude and phase of $X(\omega)$ are even functions of ω.

Odd sequences If $x[n]$ is an odd sequence, then $x[n] = -x[-n]$. Therefore, $\mathfrak{F}\{x[n]\} = \mathfrak{F}\{-x[-n]\} = -\mathfrak{F}\{x[-n]\}$. Again, by the time-reversal property, $\mathfrak{F}\{x[-n]\} = X(-\omega)$. So, for an odd sequence,

$$X(\omega) = -X(-\omega). \tag{3.29}$$

Therefore, if $x[n]$ is odd, then $X(\omega)$ is also odd. Expressing both sides in polar form:

$$|X(\omega)|e^{j\measuredangle X(\omega)} = |-X(-\omega)|e^{j(-\measuredangle X(-\omega))},$$

so

$$\begin{aligned} |X(\omega)| &= |-X(-\omega)| = |X(-\omega)| \\ \measuredangle X(\omega) &= \measuredangle X(-\omega) \pm \pi. \end{aligned}$$

Hence, if $x[n]$ is odd, then the magnitude of $X(\omega)$ is an even function of ω, and the phase is offset by $\pm\pi$.

Here is a bit of a puzzlement: Equation (3.28) says if $x[n]$ is an even (though not necessarily real) sequence, then the phase of $X(\omega)$ will be an even function of ω: $\measuredangle X(\omega) = \measuredangle X(-\omega)$. Equation (3.23) says that if $x[n]$ is a real sequence, then the phase of $X(\omega)$ will be an odd function of ω: $\measuredangle X(\omega) = -\measuredangle X(-\omega)$. So if a sequence is both real and even, how can both be true? (The next section has the answer.)

3.5.4 Consequences of symmetry

Putting the results of the previous paragraphs together yields some interesting and important results.

Real-and-even sequences If a sequence $x[n]$ is real, then Equation (3.21) tells us that $X(\omega)$ is conjugate-symmetric; that is, $X(\omega) = X^*(-\omega)$. If $x[n]$ is even, then Equation (3.27) tells us that $X(\omega)$ is even; that is, $X(\omega) = X(-\omega)$. So, if $x[n]$ is both *real* and *even*, then $X(\omega) = X^*(-\omega)$ and $X(\omega) = X(-\omega)$, from which we derive $X(\omega) = X^*(\omega)$. This means that $X(\omega)$ is purely real, too. To summarize,

If $x[n]$ is real and even, then $X(\omega)$ is real and even.

Since $X(\omega)$ is real, the phase is either 0 (when $X(\omega) > 0$) or $\pm\pi$ (when $X(\omega) < 0$), so we can write $\measuredangle X(\omega) = \pi k(\omega)$, where k is an integer function of ω. This result resolves the puzzle we mentioned above; namely how can a real-and-even sequence have both odd phase, $\measuredangle X(\omega) = -\measuredangle X(-\omega)$ (because it is a real sequence), and also even phase, $\measuredangle X(\omega) = \measuredangle X(-\omega)$ (because it is an even sequence)? The answer is that the only permissible phases for a real-and-even sequence are $\measuredangle X(\omega) = 0$ and $\measuredangle X(\omega) = \pm\pi$, which satisfy both conditions. The former is obvious, and the latter follows from the fact that a phase of $-\pi = -\pi + 2\pi = \pi$. (See Problem 3-52.)

Real-and-odd sequences If $x[n]$ is odd, then Equation (3.29) tells us that $X(\omega)$ is also odd; that is, $X(\omega) = -X(-\omega)$. So, if $x[n]$ is both *real* and *odd*, then $X(\omega) = X^*(-\omega)$ and $X(\omega) = -X(-\omega)$, from which we derive $X(\omega) = -X^*(\omega)$. This means that $X(\omega)$ is purely imaginary. To summarize,

If $x[n]$ is real and odd, then $X(\omega)$ is imaginary and odd.

Since $X(\omega)$ is purely imaginary, the phase is either $\pm\pi/2$, so we can write $\measuredangle X(\omega) = \pi/2 + \pi k(\omega)$, where k is an integer function of ω. **Figure 3.17** shows examples of real-and-even and real-and-odd sequences with their transforms.

In summary, the transforms of real-and-even and real-and-odd sequences can be written as

$$X(\omega) = |X(\omega)| e^{j(\beta + \pi k(\omega))},$$

where $\beta = 0$ for a real-and-even sequence, $\beta = \pi/2$ for a real-and-odd sequence and k has a value of 0 or ± 1.

In Chapter 7, we will show that an important class of finite impulse response filters, called **linear-phase filters**, derive their interesting properties and design methodologies from the symmetry properties we have just discussed.

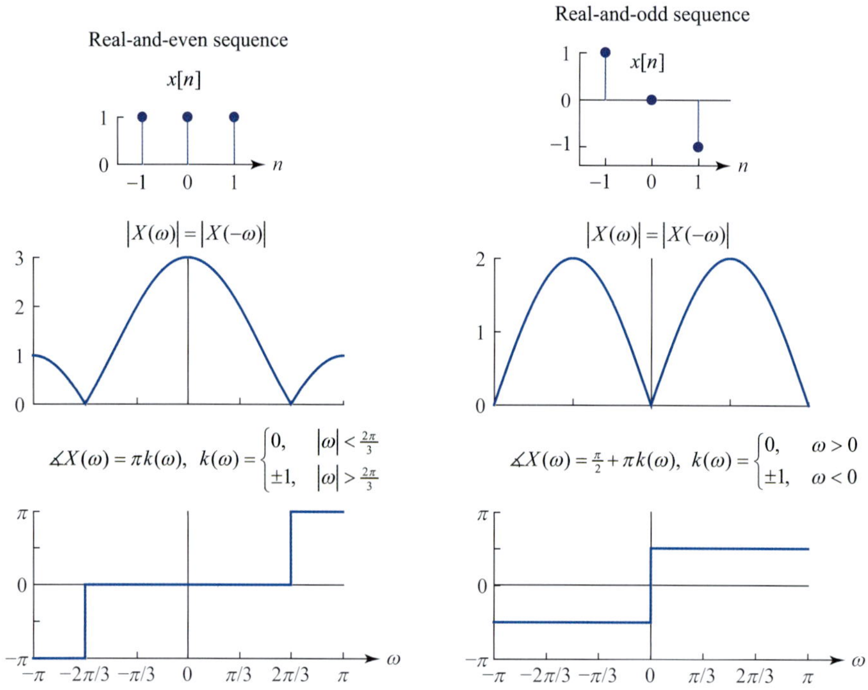

Figure 3.17 Real even and odd sequences

★ **Imaginary sequences** Using the same line of argument as above, it is easy to show that

> If $x[n]$ is imaginary and even, then $X(\omega)$ is imaginary and even.

and

> If $x[n]$ is imaginary and odd, then $X(\omega)$ is real and odd.

3.5.5 ★ Complex sequences

Any sequence $x[n]$ can be expressed as the sum of real and imaginary parts, each of which can be further split into even and odd parts:

$$\begin{aligned} x[n] &= \mathrm{Re}\{x[n]\} + j\mathrm{Im}\{x[n]\} = \mathrm{Re}\{x_e[n] + x_o[n]\} + j\mathrm{Im}\{x_e[n] + x_o[n]\} \\ &= (x_{re}[n] + x_{ro}[n]) + j(x_{ie}[n] + x_{io}[n]). \end{aligned}$$

Here, $x_{re}[n]$ and $x_{ro}[n]$ represent respectively the even and odd parts of the real part of $x[n]$. Both of these sequences are real; $x_{ie}[n]$ and $x_{io}[n]$ represent the even and odd parts of the imaginary part of $x[n]$. Both of these sequences are *real*, too. It is the sequences $jx_{ie}[n]$ and $jx_{io}[n]$ that are purely imaginary. That can be a bit confusing.

We can also split $X(\omega)$ into real and imaginary parts, each of which can be further split into even and odd functions of ω:

$$\begin{aligned} X(\omega) &= \mathrm{Re}\{X(\omega)\} + j\mathrm{Im}\{X(\omega)\} \\ &= (X_{re}(\omega) + X_{ro}(\omega)) + j(X_{ie}(\omega) + X_{io}(\omega)), \end{aligned}$$

where $X_{re}(\omega)$, $X_{ro}(\omega)$, $X_{ie}(\omega)$ and $X_{io}(\omega)$ represent the real-and-even, real-and-odd, imaginary-and-even and imaginary-and-odd parts of $X(\omega)$. Again, all these are real functions of ω. From the discussion in Section 3.5.4, we can now show that each part of $x[n]$ corresponds to one part of $X(\omega)$. Since $x_{re}[n]$ is a real-and-even sequence, its transform is also real and even. Hence,

$$\mathfrak{F}\{x_{re}[n]\} = X_{re}(\omega).$$

Since $x_{ro}[n]$ is an imaginary-and-odd sequence, its transform is imaginary and odd. Accordingly, we write

$$\mathfrak{F}\{x_{ro}[n]\} = jX_{io}(\omega).$$

The sequence $jx_{ie}[n]$ is imaginary and even. (Again, $x_{ie}[n]$ is real, so $jx_{ie}[n]$ is imaginary.) Hence, its transform is also imaginary and even. Since $X_{ie}(\omega)$ is defined to be real, therefore $jX_{ie}(\omega)$ is imaginary and we can write $\mathfrak{F}\{jx_{ie}[n]\} = jX_{ie}(\omega)$. Since j is a scalar multiplier on both sides, we can factor it out, yielding

$$\mathfrak{F}\{x_{ie}[n]\} = X_{ie}(\omega).$$

Finally, the sequence $jx_{io}[n]$ is imaginary and odd, so its transform, $X(\omega)$, is real and odd. So we write $\mathfrak{F}\{jx_{io}[n]\} = X_{ro}(\omega)$. Multiplying both sides by j yields

$$\mathfrak{F}\{x_{io}[n]\} = -jX_{ro}(\omega).$$

To summarize:

$$\begin{aligned}
\mathfrak{F}\{x_{re}[n]\} &= X_{re}(\omega)\\
\mathfrak{F}\{x_{ro}[n]\} &= jX_{io}(\omega)\\
\mathfrak{F}\{x_{ie}[n]\} &= X_{ie}(\omega)\\
\mathfrak{F}\{x_{io}[n]\} &= -jX_{ro}(\omega).
\end{aligned} \tag{3.29}$$

Example 3.16

Consider a simple real sequence, $x[n] = \delta[n] + 2\delta[n-1]$. Show that $x[n]$ can be split into $x_{re}[n]$ and $x_{ro}[n]$. Further show that $X(\omega)$ can be split into $X_{re}(\omega)$ and $X_{io}(\omega)$, and that $\mathfrak{F}\{x_{re}[n]\} = X_{re}(\omega)$ and $\mathfrak{F}\{x_{ro}[n]\} = jX_{io}(\omega)$.

➤ Solution:

$x[n]$ is neither even nor odd, but can be expressed as the sum of even and odd parts, $x[n] = x_e[n] + x_0[n]$, where

$$\begin{aligned}
x_e[n] &= \tfrac{1}{2}(x[n] + x[-n]) = \delta[n+1] + \delta[n] + \delta[n-1]\\
x_0[n] &= \tfrac{1}{2}(x[n] - x[-n]) = -\delta[n+1] + \delta[n-1].
\end{aligned}$$

Since both $x_e[n]$ and $x_o[n]$ are real, we can also write that $x[n] = x_{re}[n] + x_{ro}[n]$.

Finding the DTFT of $x[n]$:

$$\begin{aligned}
X(\omega) &= \mathfrak{F}\{\delta[n] + 2\delta[n-1]\} = \mathfrak{F}\{\delta[n]\} + 2\mathfrak{F}\{\delta[n-1]\} = 1 + 2e^{-j\omega} = 1 + 2\cos\omega - j2\sin\omega\\
&= \underbrace{(1 + 2\cos\omega)}_{X_r(\omega)} + j\underbrace{(-2\sin\omega)}_{X_i(\omega)},
\end{aligned}$$

where $X_r(\omega)$ and $X_o(\omega)$ are the real and imaginary parts. $X_r(\omega)$ is even and $X_o(\omega)$ is odd, so for this DTFT we can also write

$$X(\omega) = \underbrace{(1 + 2\cos\omega)}_{X_{re}(\omega)} + j\underbrace{(-2\sin\omega)}_{X_{io}(\omega)}.$$

Taking the DTFT of $x_{re}[n]$ and $x_{ro}[n]$, we get

$$\mathfrak{F}\{x_{re}[n]\} = \mathfrak{F}\{\delta[n+1] + \delta[n] + \delta[n-1]\} = e^{j\omega} + 1 + e^{-j\omega} = 1 + 2\cos\omega = X_{re}(\omega)$$

$$\mathfrak{F}\{x_{io}[n]\} = \mathfrak{F}\{-\delta[n+1] + \delta[n-1]\} = -e^{j\omega} + e^{-j\omega} = j(-2\sin\omega) = jX_{io}(\omega).$$

3.5.6 Symmetry summary

Table 3.2 presents a complete summary of the symmetry relations for the DTFT.

Table 3.2 Symmetry relations for the DTFT

$x[n]$		$X(\omega)$	
Real $x[n]$	$x[n] = x^*[n]$ $\mathrm{Re}\{x[n]\} = x[n]$ $\mathrm{Im}\{x[n]\} = 0$	Conjugate-symmetric $X(\omega)$	$X(\omega) = X^*(-\omega)$ $\mathrm{Re}\{X(\omega)\} = \mathrm{Re}\{X(-\omega)\}$ $\mathrm{Im}\{X(\omega)\} = -\mathrm{Im}\{X(-\omega)\}$ $\lvert X(\omega)\rvert = \lvert X(-\omega)\rvert$ $\measuredangle X(\omega) = -\measuredangle X(-\omega)$
Imaginary $x[n]$	$x[n] = -x^*[n]$ $\mathrm{Re}\{x[n]\} = 0$ $\mathrm{Im}\{x[n]\} = x[n]$	Conjugate-antisymmetric $X(\omega)$	$X(\omega) = -X^*(-\omega)$ $\mathrm{Re}\{X(\omega)\} = -\mathrm{Re}\{X(-\omega)\}$ $\mathrm{Im}\{X(\omega)\} = \mathrm{Im}\{X(-\omega)\}$ $\lvert X(\omega)\rvert = \lvert X(-\omega)\rvert$ $\measuredangle X(\omega) = -\measuredangle X(-\omega) \pm \pi$
Conjugate-symmetric $x[n]$	$x[n] = x^*[-n]$ $\mathrm{Re}\{x[n]\} = \mathrm{Re}\{x[-n]\}$ $\mathrm{Im}\{x[n]\} = -\mathrm{Im}\{x[-n]\}$	Real $X(\omega)$	$X(\omega) = X^*(\omega)$ $\mathrm{Re}\{X(\omega)\} = X(\omega)$ $\mathrm{Im}\{X(\omega)\} = 0$ $\measuredangle X(\omega) = 0, \pm\pi$
Conjugate-antisymmetric $x[n]$	$x[n] = -x^*[-n]$ $\mathrm{Re}\{x[n]\} = -\mathrm{Re}\{x[-n]\}$ $\mathrm{Im}\{x[n]\} = \mathrm{Im}\{x[-n]\}$	Imaginary $X(\omega)$	$X(\omega) = -X^*(\omega)$ $\mathrm{Re}\{X(\omega)\} = 0$ $\mathrm{Im}\{X(\omega)\} = X(\omega)$ $\measuredangle X(\omega) = \pm\pi/2$
Even $x[n]$	$x[n] = x[-n]$	Even $X(\omega)$	$X(\omega) = X(-\omega)$
Odd $x[n]$	$x[n] = -x[-n]$	Odd $X(\omega)$	$X(\omega) = -X(-\omega)$
Real and even $x[n]$	$x[n] = x_{re}[n]$	Real and even $X(\omega)$	$X(\omega) = X_{re}(\omega)$
Real and odd $x[n]$	$x[n] = x_{ro}[n]$	Imaginary and odd $X(\omega)$	$X(\omega) = jX_{io}(\omega)$
Imaginary and even $x[n]$	$x[n] = x_{ie}[n]$	Imaginary and even $X(\omega)$	$X(\omega) = jX_{ie}(\omega)$
Imaginary and odd $x[n]$	$x[n] = x_{io}[n]$	Real and odd $X(\omega)$	$X(\omega) = X_{ro}(\omega)$

From these relations we can derive some other symmetry relations, for example the DTFT of the real part of $x[n]$ is the conjugate-symmetric part of $X(\omega)$. A list is given in **Table 3.3**.

Table 3.3 Further symmetry relations for the DTFT

$x[n]$		$X(\omega)$	
$\mathrm{Re}\{x[n]\}$	$\frac{1}{2}(x[n]+x^*[n])$	$\mathrm{ConjSym}\{X(\omega)\}$	$\frac{1}{2}(X(\omega)+X^*(-\omega))$
$\mathrm{Im}\{x[n]\}$	$\frac{1}{2j}(x[n]-x^*[n])$	$\mathrm{ConjAntisym}\{X(\omega)\}$	$\frac{1}{2j}(X(\omega)-X^*(-\omega))$
$\mathrm{ConjSym}\{x[n]\}$	$\frac{1}{2}(x[n]+x^*[-n])$	$\mathrm{Re}\{X(\omega)\}$	$\frac{1}{2}(X(\omega)+X^*(\omega))$
$\mathrm{ConjAntisym}\{x[n]\}$	$\frac{1}{2}(x[n]+x^*[-n])$	$\mathrm{Im}\{X(\omega)\}$	$\frac{1}{2j}(X(\omega)-X^*(\omega))$
$\mathrm{Even}\{x[n]\}$	$x[n]=x[-n]$	$\mathrm{Even}\{X(\omega)\}$	$\frac{1}{2}(X(\omega)+X(-\omega))$
$\mathrm{Odd}\{x[n]\}$	$x[n]=-x[-n]$	$\mathrm{Odd}\{X(\omega)\}$	$\frac{1}{2}(X(\omega)-X(-\omega))$

3.6 Response of a system to sinusoidal input

In the introduction, we showed that complex exponentials are eigenfunctions of LTI systems. Specifically, the response of an LTI system characterized by an impulse response $h[n]$ and DTFT $H(\omega)$ to a complex exponential input $e^{j\omega_0 n}$ is $H(\omega_0)e^{j\omega_0 n}$. We used this fact to find the response of a system to a few simple systems with cosine inputs. Let us now revisit the response to sinusoidal stimuli in light of our understanding of the DTFT.

A general sinusoidal input $x[n]$, with frequency ω_0, magnitude $|X|$ and phase $\measuredangle X$, has the form

$$\begin{aligned} x[n] &= |X|\cos(\omega_0 n+\measuredangle X)=\mathrm{Re}\left\{|X|e^{j(\omega_0 n+\measuredangle X)}\right\}=\mathrm{Re}\{|X|e^{j\measuredangle X}e^{j\omega_0 n}\}=\mathrm{Re}\left\{Xe^{j\omega_0 n}\right\} \\ &= \tfrac{1}{2}Xe^{j\omega_0 n}+\tfrac{1}{2}(Xe^{j\omega_0 n})^*=\tfrac{1}{2}Xe^{j\omega_0 n}+\tfrac{1}{2}X^*e^{-j\omega_0 n}, \end{aligned}$$

where

$$X=|X|e^{j\measuredangle X}.$$

The output of a system characterized by system function $H(\omega)$ to a complex exponential input $Xe^{j\omega_0 n}$ is $XH(\omega_0)e^{j\omega_0 n}$. Similarly, the output of the system to $X^*e^{-j\omega_0 n}$ is

$$X^*H(-\omega_0)e^{-j\omega_0 n}=X^*H^*(\omega_0)e^{-j\omega_0 n}=(XH(\omega_0)e^{j\omega_0 n})^*,$$

where we have used the fact that for a system with real impulse response $h[n]=h^*[n]$, the DTFT is conjugate-symmetric, $H(-\omega)=H^*(\omega)$ (Equation (3.21)). Then the response to $x[n]=\frac{1}{2}Xe^{j\omega_0 n}+\frac{1}{2}X^*e^{-j\omega_0 n}$ is

$$\begin{aligned} y[n] &= \tfrac{1}{2}XH(\omega_0)e^{j\omega_0 n}+\tfrac{1}{2}(XH(\omega_0)e^{j\omega_0 n})^*=\mathrm{Re}\{XH(\omega_0)e^{j\omega_0 n}\} \\ &= \mathrm{Re}\left\{|X|e^{j\measuredangle X}|H(\omega_0)|e^{j\measuredangle H(\omega_0)}e^{j\omega_0 n}\right\}=|X||H(\omega_0)|\mathrm{Re}\left\{e^{j\left(\omega_0 n+\measuredangle X+\measuredangle H(\omega_0)\right)}\right\} \\ &= |X||H(\omega_0)|\cos(\omega_0 n+\measuredangle X+\measuredangle H(\omega_0)). \end{aligned} \tag{3.31}$$

So, the effect of the filter is to change the magnitude and phase of the input. The output sinusoid has the same frequency as the input, with magnitude scaled by $|H(\omega_0)|$ and phase shifted by $\measuredangle H(\omega_0)$.

Example 3.17

Consider a system with impulse response $h[n] = \delta[n] + 2\delta[n-1] + 4\delta[n-2] + 2\delta[n-3] + \delta[n-4]$, shown in **Figure 3.18a**. Find the response of this system to the input $x[n] = \cos\omega_0 n$, for four frequencies:

(a) $\omega_0 = \pi/3$.
(b) $\omega_0 = \pi/2$.
(c) $\omega_0 = 2\pi/3$.
(d) $\omega_0 = \pi$.

➤ Solution:

For this input, $|X| = 1$ and $\measuredangle X = 0$. $H(\omega)$, the DTFT of $h[n]$, is

$$\begin{aligned} H(\omega) &= 1 + 2e^{-j\omega} + 4e^{-j2\omega} + 2e^{-j3\omega} + e^{-j4\omega} = e^{-j2\omega}(e^{2j\omega} + 2e^{-j\omega} + 4 + 2e^{j\omega} + e^{2j\omega}) \\ &= e^{-j2\omega}(4 + 4\cos\omega + 2\cos 2\omega) \end{aligned},$$

as shown in **Figure 3.18b**.

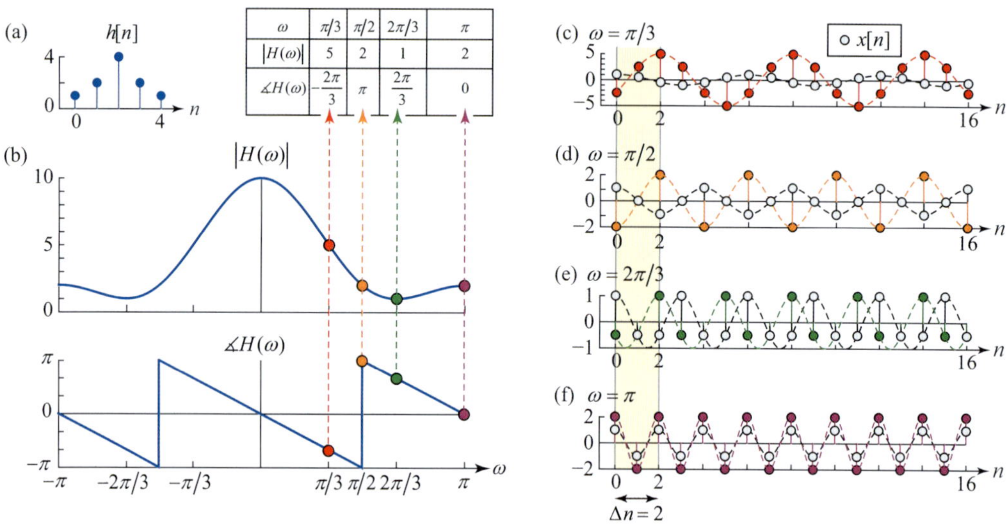

Figure 3.18 Response of an LTI system to sinusoids at different frequencies

The values of $|H(\omega)|$ and $\measuredangle H(\omega)$ for the four values of ω given in the problem are:

ω	$\pi/3$	$\pi/2$	$2\pi/3$	π
$\|H(\omega)\|$	5	2	1	2
$\measuredangle H(\omega)$	$-2\pi/3$	π	$2\pi/3$	0

Therefore, from Equation (3.31), the outputs of the filter are

(a) $\omega_0 = \pi/3$: $y[n] = 5\cos(\pi n/3 - 2\pi/3) = 5\cos\pi(n-2)/3$, shown in red in **Figure 3.18c**.
(b) $\omega_0 = \pi/2$: $y[n] = 2\cos(\pi n/2 + \pi) = 2\cos(\pi n/2 - \pi) = 2\cos\pi(n-2)/2$, shown in orange in **Figure 3.18d**.

(c) $\omega_0 = 2\pi/3$: $y[n] = \cos(2\pi n/3 + 2\pi/3) = \cos(2\pi n/3 - 4\pi/3) = \cos 2\pi(n-2)/3$, shown in green in **Figure 3.18e**.

(d) $\omega_0 = \pi$: $y[n] = 2\cos \pi n = 2\cos \pi(n-2)$, shown in magenta in **Figure 3.18f**.

You will notice that even though each of the output cosines $y[n]$ (colored symbols in **Figures 3.18c–f**) has a different phase, the peak of each has been delayed with respect to the input $x[n]$ (grey symbols) by exactly the same amount of time, $\Delta n = 2$ points, indicated by the shaded band. This is not an accident. It is a characteristic of a system that has **linear phase**, a concept we will discuss in the next section.

3.7 Linear-phase systems

The symmetric and antisymmetric sequences of the previous sections were all of odd length and centered at $n=0$. We now extend the concept of symmetry and antisymmetry to the causal sequences of either even or odd length, such as those shown in **Figure 3.19**. We shall show that such sequences have the important property that the phase of their DTFT is linear.

Linear-phase systems play an important part in a number of areas of DSP. FIR filters having these types of impulse responses are termed **linear-phase filters**. In Chapter 7, we will introduce a number of the most practical and frequently employed of these filters. In Chapter 12, we will discover an entire class of transforms – the **discrete cosine transform** and its close relative the **discrete sine transform** – that is based on creating causal symmetric or antisymmetric sequences from arbitrary sequences. In Chapter 14, we will show that symmetric windows have an important role in spectral signal processing.

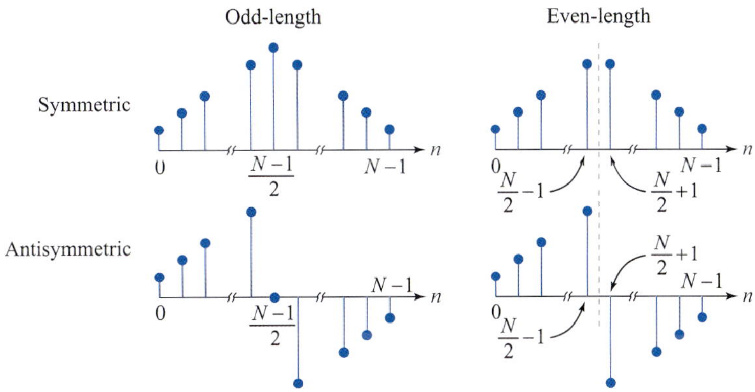

Figure 3.19 Causal symmetric and antisymmetric sequences

3.7.1 Causal symmetric sequences

A real causal symmetric sequence (for example, an impulse response) of length N is defined by

$$h[n] = h[N-1-n]. \tag{3.32}$$

Such sequences can be of either even length or odd length, as shown in the top row of **Figure 3.19**. In Chapter 7, we will study a whole class of important lowpass FIR filters, such as the Hamming, Hann and Kaiser filters, that have such symmetric impulse responses. These filters are particularly appealing because they have **linear phase**. To understand what that means, consider the Fourier transform of $h[N-1-n]$,

$$\mathfrak{F}\{h[N-1-n]\} = \sum_{n=0}^{N-1} h[N-1-n]e^{-j\omega n}.$$

Substituting $m = N-1-n$,

$$\sum_{m=N-1}^{0} h[m]e^{-j\omega(N-1-m)} = e^{-j\omega(N-1)} \sum_{m=0}^{N-1} h[m]e^{j\omega m}.$$

If $h[n]$ is real, then $h[n] = h^*[n]$, so

$$\begin{aligned}\mathfrak{F}\{h[N-1-n]\} &= e^{-j\omega(N-1)}\left(\sum_{m=0}^{N-1} h^*[m]e^{j\omega m}\right) = e^{-j\omega(N-1)}\left(\sum_{m=0}^{N-1} h[m]e^{-j\omega m}\right)^* \\ &= e^{-j\omega(N-1)}H^*(\omega).\end{aligned}$$

For a real symmetric sequence, the transform of Equation (3.32) gives $\mathfrak{F}\{h[n]\} = \mathfrak{F}\{h[N-1-n]\}$. Hence,

$$H(\omega) = e^{-j\omega(N-1)}H^*(\omega). \tag{3.33}$$

Taking the phase of both sides gives

$$\measuredangle H(\omega) = \measuredangle(e^{-j\omega(N-1)}H^*(\omega)) = -\omega(N-1) + \measuredangle H^*(\omega) = -\omega(N-1) - \measuredangle H(\omega).$$

So,

$$2\measuredangle H(\omega) = -\omega(N-1),$$

or

$$\measuredangle H(\omega) = -\omega(N-1)/2. \tag{3.34}$$

This result shows that the phase of a symmetric sequence is a linear function of ω with a slope of $-(N-1)/2$. Odd-length symmetric sequences (i.e., N odd) have slopes that are integer multiples of ω (e.g., ..., $-1, 0, 1$, ...). Even-length symmetric sequences (i.e., N even) have slopes that are non-integer multiples of ω (e.g., ..., -1.5, -0.5, 0.5, 1.5, ...). However, we have to be a bit careful with this conclusion, as the next examples show.

Figure 3.20a shows an example of an odd-length ($N = 3$) symmetric sequence $h[n]$, and its transform $H(\omega)$. The phase $\measuredangle H(\omega)$ (blue line), is generally linear with a slope of $(N-1)/2 = -1$, as predicted by Equation (3.34). However, it has discontinuities of $\pm\pi$ at $\omega = \pm 2\pi/3$, which correspond to the frequencies at which $H(\omega) = 0$. To understand these discontinuities, note that since $H(\omega) = e^{-j\omega}(1 + 2\cos\omega)$, the phase of the response is

$$\measuredangle H(\omega) = \measuredangle(e^{-j\omega}(1 + 2\cos\omega)) = \measuredangle e^{-j\omega} + \measuredangle(1 + 2\cos\omega) = -\omega + \measuredangle(1 + 2\cos\omega).$$

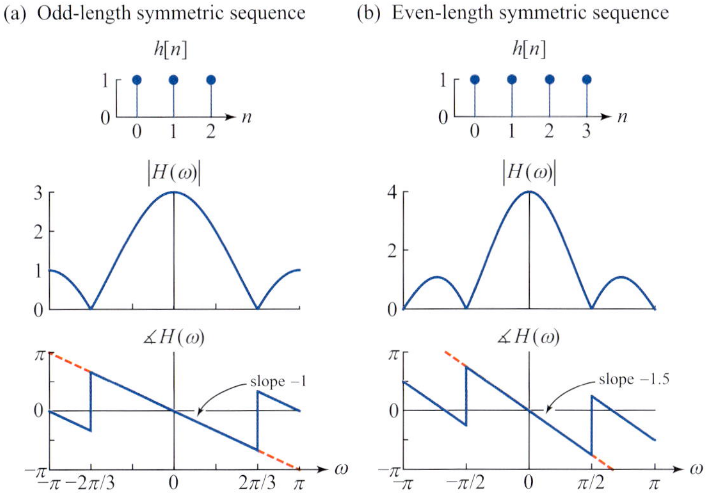

Figure 3.20 Examples of odd-length and even-length causal symmetric sequences

The first term, $-\omega$, is the linear phase we expect from Equation (3.34). The second term, $\measuredangle(1+2\cos\omega)$, is zero where $|H(\omega)|>0$ (i.e., for $|\omega|<2\pi/3$) and becomes $\pm\pi$ where $|H(\omega)|<0$ (i.e., for $|\omega|>2\pi/3$). Hence, there are essential phase discontinuities at frequencies $\omega=\pm 2\pi/3$. **Figure 3.20b** shows an example of an even-length ($N=4$) sequence and its transform. Here, Equation (3.34) predicts a slope of -1.5, but again there are essential phase discontinuities of $\pm\pi$ at frequencies $\omega=\pm\pi/2$, which is where $H(\omega)=0$. We define linear phase in this broader sense to mean that the slope is linear with possible discontinuities of $\pm\pi$ at points where the magnitude goes to zero.

3.7.2 Causal antisymmetric sequences

A real causal antisymmetric sequence of length n is defined by

$$h[n] = -h[N-1-n].$$

Antisymmetric sequences of even length or odd length are shown in the bottom row of **Figure 3.19**. As we will see in Chapter 7, highpass and bandpass filters based on window design generally have antisymmetric impulse responses. The transforms of antisymmetric sequences also have linear phase, with the same caveats we noted with respect to the transforms of symmetric sequences. Following a similar derivation that led to Equation (3.33), we get

$$H(\omega) = -e^{-j\omega(N-1)}H^*(\omega). \tag{3.35}$$

Taking the phase of both sides yields

$$\begin{aligned}\measuredangle H(\omega) &= \measuredangle(-e^{-j\omega(N-1)}\,H^*(\omega)) = \measuredangle(-1) - \omega(N-1) + \measuredangle H^*(\omega)\\ &= \pm\pi - \omega(N-1) - \measuredangle\, H(\omega).\end{aligned}$$

Solving for $\measuredangle H(\omega)$ gives

$$\measuredangle H(\omega) = \pm\pi/2 - \omega(N-1)/2. \tag{3.36}$$

Figures 3.21a and **b** show examples of odd-length ($N=3$) and even-length ($N=4$) antisymmetric sequences $h[n]$, and their transforms $H(\omega)$.

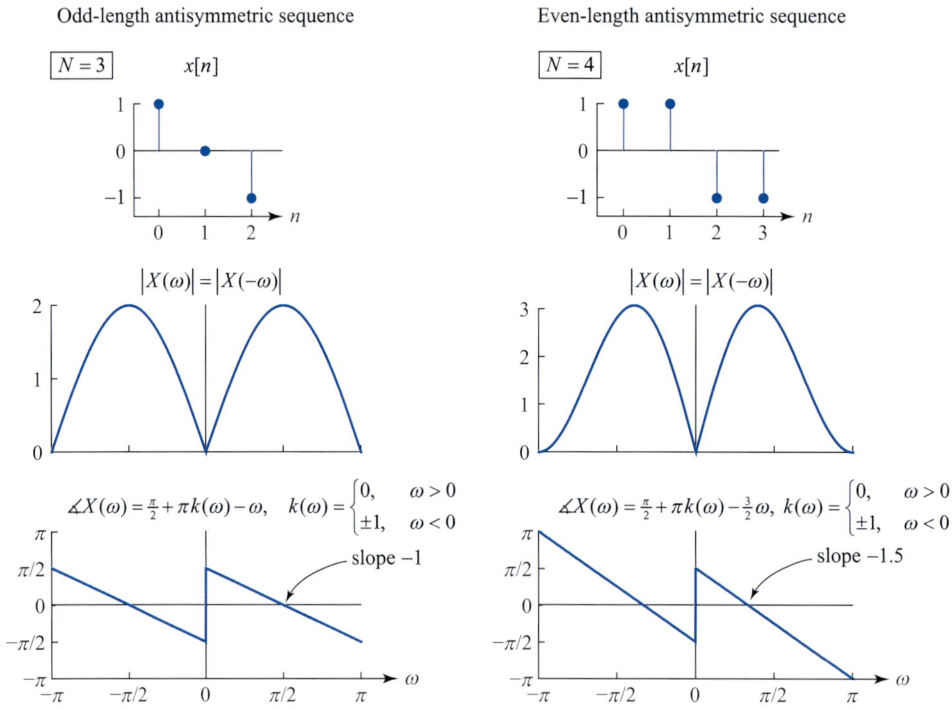

Figure 3.21 Examples of odd-length and even-length causal antisymmetric sequences

The frequency response of an antisymmetric sequence is guaranteed to be zero at $\omega = 0$, as shown in these examples, since from Equation (3.35), $H(0) = -H^*(0)$, which implies that $H(0) = 0$. The transform of an antisymmetric sequence is also guaranteed to have a phase discontinuity of π at $\omega = 0$, as shown in the examples. That is because at $\omega = 0^+$ (i.e., just a little above $\omega = 0$), Equation (3.36) gives $\measuredangle H(0^+) = \pm\pi/2$. For a real sequence, the phase is an odd function of ω, so $\measuredangle H(0^-) = \mp\pi/2$. Hence, there is always a phase discontinuity of $H(0^+) - H(0^-) = \pm\pi/2 - (\mp\pi/2) = \pm\pi$ at $\omega = 0$.

3.7.3 Time delay and group delay

A key characteristic of linear-phase filters, indicated by Equations (3.34) and (3.35), is that each frequency component at the input of the filter is effectively delayed by the same amount of time at the output. As an example, consider the response of a linear-phase filter to an input cosine at frequency ω_0, $x[n] = \cos\omega_0 n$. As we showed in Section 3.6, the response $y[n]$ is a cosine of the same frequency, ω_0, with magnitude and phase determined by $H(\omega_0)$,

$$y[n] = |H(\omega_0)| \cos(\omega_0 n + \measuredangle H(\omega_0)).$$

For a filter with linear phase, we have $\measuredangle H(\omega_0) = -\omega_0(N-1)/2$, so

$$\begin{aligned} y[n] &= |H(\omega_0)| \cos(\omega_0 n - \omega_0(N-1)/2) = |H(\omega_0)| \cos(\omega_0(n-(N-1)/2)) \\ &= |H(\omega_0)| \cos(\omega_0(n - \Delta n)), \end{aligned} \quad (3.37)$$

where

$$\Delta n \triangleq (N-1)/2.$$

So, at any frequency ω_0, the magnitude of an output cosine is $|H(\omega_0)|$, and the phase is delayed by $\omega_0(N-1)/2$, which corresponds to an absolute time delay of $\Delta n = (N-1)/2$ samples that is *independent* of ω_0. As an example, return to the system shown in Example 3.17. The impulse response $h[n]$ in **Figure 3.18a** is a symmetric, linear-phase filter of length $N=5$. Hence, $\Delta n = 2$, which means that the output $y[n] = |H(\omega_0)| \cos \omega_0(n-2)$ at each frequency ω_0 is delayed by two samples with respect to the input, as shown at each of the four frequencies, $\omega_0 = \pi/3$, $\pi/2$, $2\pi/3$ and π.

The main point of this example is that the linear-phase filter delays all frequency components of the input by the same absolute amount of *time*, $\Delta n = (N-1)/2$, and therefore maintains the phase alignment of all the output components. In order to demonstrate the importance of this feature of linear-phase systems, **Figure 3.22** compares the response of two filters to an identical input: a linear-phase FIR filter and an IIR filter that does not have linear phase. The red curves in **Figures 3.22a** and **b** show the magnitude and phase, respectively, of a linear-phase FIR lowpass filter (a Kaiser filter, which we will discuss in Chapter 7) of length $N=41$. The magnitude of the frequency response is essentially flat for frequencies $0 \le \omega < 0.45\pi$, and the phase is linear with a slope of -20. The blue curves in the figure show the response of an IIR filter (a Butterworth filter, which we will discuss in Chapter 8) designed to have an almost identical magnitude response over the same frequency range. However, the phase of this filter is a nonlinear function of ω.

Figure 3.22c shows two ways of quantifying the phase characteristics of these filters as a function of frequency, **time delay** and **group delay**.

Time delay **Time delay** is defined as the phase at frequency ω, divided by frequency,

$$\tau(\omega) \triangleq \frac{\measuredangle H(\omega)}{\omega}.$$

It is simply a measure of the amount of time to which a given amount of phase corresponds at a given frequency. For the linear-phase filter, $\measuredangle H(\omega)$ is a linear function of ω, as specified by Equations (3.34) and (3.36), so the time delay is a constant, $\tau(\omega) = \Delta n = (N-1)/2$, independent of ω, as shown by the red line of value $(41-1)/2 = 20$ in the figure. In contrast, the time delay of the IIR filter is a nonlinear, monotonically increasing function of ω, shown by the dashed blue line.

Group delay **Group delay** $D(\omega)$ is defined as the rate of change of phase at a given frequency ω, as a function of frequency,

$$D(\omega) = -\frac{d\measuredangle H(\omega)}{d\omega}.$$

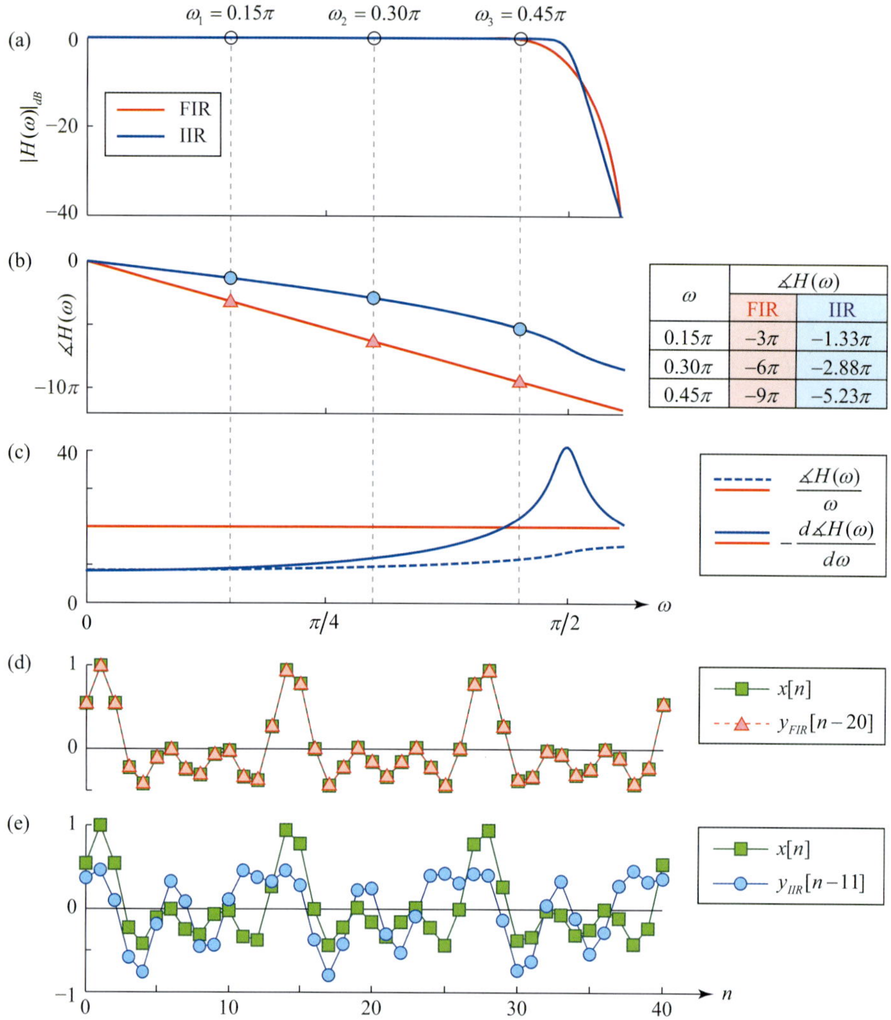

ω	$\measuredangle H(\omega)$	
	FIR	IIR
0.15π	-3π	-1.33π
0.30π	-6π	-2.88π
0.45π	-9π	-5.23π

Figure 3.22 Response of a linear-phase FIR filter and an IIR filter to tones

The group delay can be thought of as a local measure of the time delay of a small range of spectral components of a filter's output around frequency ω. For the linear-phase filter, $\measuredangle H(\omega)$ is a linear function of frequency; hence, the derivative is constant and $D(\omega)$ is a constant that is equal to the time delay, $D(\omega) = \tau(\omega) = \Delta n = (N-1)/2$, independent of frequency. However, for the IIR filter of our example, the group delay is roughly constant only at very low frequencies, and reaches a peak near the corner frequency of $\omega_c = \pi/2$. In classical communication theory, the group delay is often used to characterize the output of a filter whose input is a sinusoid of frequency ω modulated by an envelope function $m[n]$: $x[n] = m[n]\cos\omega n$. For a general filter, the output is roughly equal to the cosine delayed by $\tau(\omega)$ and multiplied by the envelope delayed by $D(\omega)$: $y[n] \simeq m[n - D(\omega)]\cos\omega(n - \tau(\omega))$. That is, the envelope and the cosine experience unequal delays that depend upon frequency. However, for the linear-phase filter, the group delay and time-delay are both constant with frequency and equal to Δn, so the output of the filter is roughly $y[n] \simeq m[n-\Delta n]\cos\omega(n-\Delta n)$. In this instance, the envelope experiences a delay equal to that of the sinusoid, independent of frequency.

Returning to the example of **Figure 3.22a**, consider the response of our FIR and IIR filters to a signal comprising the sum of three cosines at frequencies $\omega_1 = 0.15\pi$, $\omega_2 = 0.3\pi$ and $\omega_3 = 0.45\pi$, so that

$$x[n] = \cos\omega_1 n + \cos\omega_2 n + \cos\omega_3 n = \cos(0.15\pi n) + \cos(0.3\pi n) + \cos(0.45\pi n).$$

The three frequencies are indicated by the circles in **Figure 3.22a**. When this input signal is passed through the FIR and IIR filters, the output is of the form

$$y[n] = |H(\omega_1)|\cos(\omega_1 n + \measuredangle H(\omega_1)) + |H(\omega_2)|\cos(\omega_2 n + \measuredangle H(\omega_2)) + |H(\omega_3)|\cos(\omega_3 n + \measuredangle H(\omega_3)).$$

For the linear-phase FIR filter, $|H(\omega_1)| \cong |H(\omega_2)| \cong |H(\omega_3)| \cong 1$ and $\measuredangle H(\omega_1) = -3\pi$, $\measuredangle H(\omega_2) = -6\pi$, $\measuredangle H(\omega_3) = -9\pi$, so the output of this filter is

$$\begin{aligned} y_{FIR}[n] &= \cos(0.15\pi n - 3\pi) + \cos(0.30\pi n - 6\pi) + \cos(0.45\pi n - 9.\pi) \\ &= \cos(0.15\pi(n-20)) + \cos(0.30\pi(n-20)) + \cos(0.45\pi(n-20)) \\ &= x[n-20]. \end{aligned}$$

Each of the three frequency components has a different phase delay, but each phase delay corresponds to the same time delay of $\Delta n = (N-1)/2 = 20$ samples. **Figure 3.22d** shows (in red) the time response $y_{FIR}[n]$ delayed by 20 samples, superimposed upon the input $x[n]$. You can see that $y_{FIR}[n-20]$ is essentially equal to $x[n]$.

For the IIR filter, again $|H(\omega_1)| \cong |H(\omega_2)| \cong |H(\omega_3)| \cong 1$, but the phase is a nonlinear function of frequency: $\measuredangle H(\omega_1) = -1.33\pi$, $\measuredangle H(\omega_2) = -2.88\pi$, $\measuredangle H(\omega_3) = -5.23\pi$, so the output of the filter is

$$\begin{aligned} y_{IIR}[n] &= \cos(0.15\pi n - 1.33\pi) + \cos(0.30\pi n - 2.88\pi) + \cos(0.45\pi n - 5.23.\pi) \\ &= \cos(0.15\pi(n-8.87)) + \cos(0.30\pi(n-9.60)) + \cos(0.45\pi(n-11.62)). \end{aligned}$$

Here, the phase delays of the three components do not correspond to the same time delay, which results in **phase distortion** of the output waveform. The blue trace in **Figure 3.22e** shows that the shape of $y_{IIR}[n]$ does not match $x[n]$ very closely, even when $y_{IIR}[n]$ is shifted by 11 samples, an amount calculated to minimize the mean-square error difference between $y_{IIR}[n]$ and $x[n]$.

This figure demonstrates an essential advantage of linear-phase FIR filters over IIR filters. Because each spectral component in the output of the linear-phase FIR filter is delayed by the same amount of time, there is no phase distortion, even though the amplitudes of different frequency terms are differently affected by the filter. This advantage is critical for certain applications, for example in digital communication, where maintaining the time alignment of spectral components can be important. This is illustrated by the example of **Figure 3.23**, which shows two periods of the steady-state response of the FIR filter, $y_{FIR}[n]$ (top panel), and the IIR filter, $y_{IIR}[n]$ (bottom panel), of **Figure 3.22** to a periodic square-wave $x[n]$ (shown in green). Both filters attenuate the higher frequencies of the input, but all the spectral components of the FIR-filtered output remain time-aligned with the input, so the response appears relatively symmetric, whereas the output of the IIR filter shows significant phase distortion and is asymmetric.

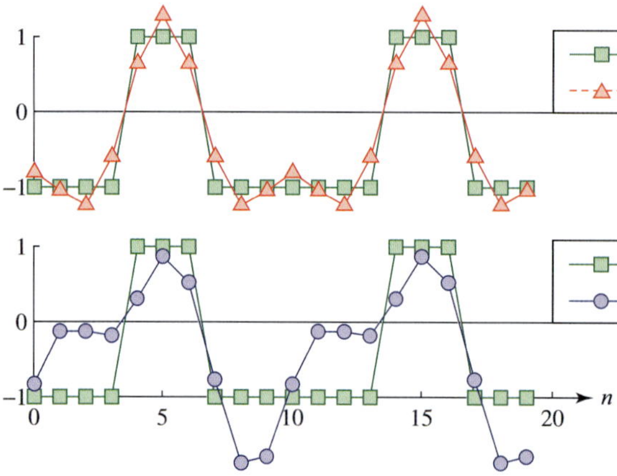

Figure 3.23 Response of a linear-phase FIR filter and an IIR filter to pulses

We will continue our discussion of linear-phase filters in Chapter 7.

3.8 The inverse discrete-time Fourier transform

As defined in Equation (3.7), the inverse DTFT involves taking the integral of a transform multiplied by a complex exponential:

$$x[n] = \frac{1}{2\pi}\int_{-\pi}^{\pi} X(\omega)e^{j\omega n}d\omega.$$

The following is an important example, and one in which it is trivial to evaluate the integral of the inverse DTFT directly.

Example 3.18

The frequency response of the **ideal lowpass filter**, $H(\omega)$, of bandwidth ω_b is shown in **Figure 3.24b** for the case of $\omega_b = \pi/4$. The frequency response of this filter is defined by

$$H(\omega) = \begin{cases} 1, & |\omega| < \omega_b \\ 0, & \omega_b < |\omega| < \pi \end{cases}.$$

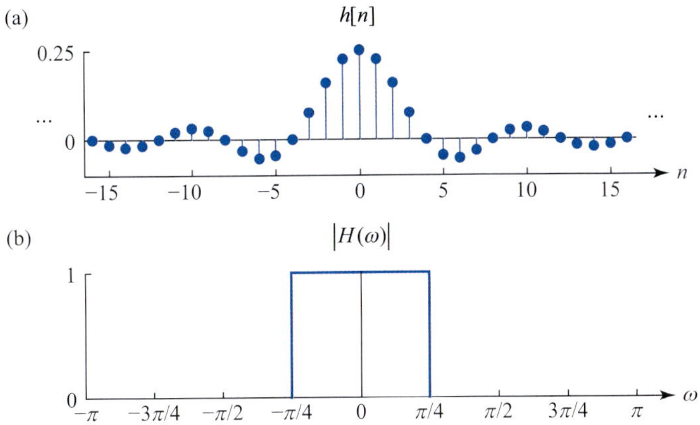

Figure 3.24 Ideal lowpass filter

The impulse response corresponding to this filter's frequency response is found by direct computation,

$$h[n] = \frac{1}{2\pi}\int_{-\pi}^{\pi} H(\omega)e^{j\omega n}d\omega = \frac{1}{2\pi}\int_{-\omega_b}^{\omega_b} e^{j\omega n}d\omega = \frac{1}{2\pi}\frac{1}{jn}(e^{j\omega_b n} - e^{-j\omega_b n}) = \frac{\sin\omega_b n}{\pi n} = \frac{\omega_b}{\pi}\operatorname{sinc}\omega_b n.$$

Clearly, $h[n]$ is a non-causal sequence that is infinite in time duration. For this reason, the ideal lowpass filter cannot be used as a practical filter. However, as we shall see in Chapter 7, it is possible to implement a large and important family of practical FIR filters by shaping the infinite impulse response of the ideal lowpass filter by multiplying it by a variety of finite-length window sequences.

While one can always compute the inverse DTFT by performing the integral of Equation (3.7), there are many cases of practical interest for which we will simply recognize the inverse transform by inspection. Let us discuss finite- and infinite-length sequences separately.

Finite-length sequences The inverse transforms of finite-length sequences are particularly simple. When a finite-length sequence represents the impulse response of a system, then we are, of course, dealing with a finite impulse-response (FIR) system. A finite-length sequence is defined by the fact that it is non-zero only within some range of n, namely $n_{min} \le n \le n_{max}$, where n_{min} and n_{max} represent respectively the minimum and maximum values of n for which the sequence is non-zero (that is, $x[n]=0$, $n < n_{min}$ or $n > n_{max}$). Accordingly, a finite-length sequence $x[n]$ can be written as

$$\begin{aligned} x[n] &= \sum_{k=n_{min}}^{n_{max}} x[k]\delta[n-k] \\ &= x[n_{min}]\delta[n-n_{min}] + x[n_{min}+1]\delta[n-(n_{min}+1)] + \cdots \\ &\quad + x[n_{max}-1]\delta[n-(n_{max}-1)] + x[n_{max}]\delta[n-n_{max}]. \end{aligned}$$

The transform of this sequence is just a finite-length polynomial in powers of $e^{j\omega n}$,

$$\begin{aligned} X(\omega) &= \sum_{n=n_{min}}^{n_{max}} x[n]e^{-j\omega n} \\ &= x[n_{min}]e^{-j\omega n_{min}} + x[n_{min}+1]e^{-j\omega(n_{min}+1)} + \ldots \\ &\quad + x[n_{max}-1]e^{-j\omega(n_{max}-1)} + x[n_{max}]e^{-j\omega n_{max}}, \end{aligned}$$

where $x[n]$ is the coefficient of the polynomial term $e^{-j\omega n}$. So, given a DTFT that can be expressed in the form of a finite-length polynomial in powers of $e^{-j\omega n}$, it is easy to derive a finite-length sequence corresponding to the inverse DTFT by inspection.

Example 3.19

Find the inverse DTFT of the transform $X(\omega) = 4\cos 2\omega + 2\cos 4\omega$.

➤ Solution

First, express $X(\omega)$ as the sum of complex exponentials,

$$X(\omega) = 4\cos 2\omega + 2\cos 4\omega = 2(e^{j2\omega} + e^{-j2\omega}) + (e^{j4\omega} + e^{-j4\omega}) = e^{j4\omega} + 2e^{j2\omega} + 2e^{-j2\omega} + e^{-j4\omega}.$$

The inverse DTFT follows by inspection:

$$\begin{array}{ccccccccc} X(\omega) & = & e^{j4\omega} & + & 2e^{j2\omega} & + & 2e^{-j2\omega} & + & e^{-j2\omega} \\ \updownarrow & & \updownarrow & & \updownarrow & & \updownarrow & & \updownarrow \\ x[n] & = & \delta[n+4] & + & 2\delta[n+2] & + & 2\delta[n-2] & + & \delta[n-4]. \end{array}$$

Infinite-length sequences Performing the inverse transform of infinite-length sequences formally requires performing the integral of Equation (3.7), but it is often possible to recognize the inverse transform by inspection. Some of the most important sequences and their transforms are tabulated in **Table 3.1**.

Beyond the simple transform pairs of **Table 3.1**, there is a large class of transforms whose inverse can be found by inspection after a bit of algebraic manipulation, as we will discuss in considerable detail in Chapter 4. We shall show that the transform of many of the impulse responses of IIR systems are expressible as the ratio of polynomials in $e^{j\omega}$:

$$H(\omega) = \frac{\sum_{m=0}^{M-1} b_m e^{-j\omega m}}{\sum_{n=0}^{N-1} a_n e^{-j\omega n}}. \tag{3.38}$$

For now, let us consider only the case where the roots of the denominator polynomial, λ_k, $0 \le k \le M-1$, are distinct (i.e., there are no repeated roots). In these cases, it is possible to use simple algebraic manipulations to express $H(\omega)$ as the sum of terms:

$$H(\omega) = \frac{\sum_{m=0}^{M-1} b_m e^{-j\omega m}}{\sum_{n=0}^{N-1} a_n e^{-j\omega n}} = \sum_{l=0}^{M-N} C_l e^{-jl\omega} + \sum_{k=0}^{N-1} \frac{B_k}{1-\lambda_k e^{-jk\omega}}.$$

Then we can recognize the inverse transform by inspection:

$$h[n] = \sum_{l=0}^{M-N} C_l \delta[n-l] + \sum_{k=0}^{N-1} B_k \lambda^k u[n].$$

Let us clarify things with a couple of specific examples.

Example 3.20

Find the inverse DTFT of the transform

$$X(\omega) = \frac{7 - e^{-j\omega}}{1 - \frac{3}{4}e^{-j\omega} + \frac{1}{8}e^{-j2\omega}}.$$

➤ Solution:

$$X(\omega) = \frac{7 - e^{-j\omega}}{1 - \frac{3}{4}e^{-j\omega} + \frac{1}{8}e^{-j2\omega}} = \frac{7 - e^{-j\omega}}{(1 - \frac{1}{4}e^{-j\omega})(1 - \frac{1}{2}e^{-j\omega})} = \frac{-3}{1 - \frac{1}{4}e^{-j\omega}} + \frac{10}{1 - \frac{1}{2}e^{-j\omega}}.$$

The denominator can be factored into the product of two terms. This product is then split into the sum of two terms using partial fraction expansion, each term corresponding to one of the two roots. From **Table 3.1**, we recognize that

$$\mathfrak{F}\{\alpha^n u[n]\} = \frac{1}{1 - \alpha e^{-j\omega}},$$

so we then find the inverse transform by inspection,

$$x[n] = -3 \cdot \frac{1}{4}^{n} u[n] + 10 \cdot \frac{1}{2}^{n} u[n].$$

Example 3.21

Find the inverse DTFT of the transform,

$$X(\omega) = \frac{1 - 5e^{-j\omega} + e^{-j2\omega}}{1 - \frac{3}{4}e^{-j\omega} + \frac{1}{8}e^{-j2\omega}}.$$

➤ Solution:

$$X(\omega) = \frac{1 - 5e^{-j\omega} + e^{-j2\omega}}{1 - \frac{3}{4}e^{-j\omega} + \frac{1}{8}e^{-j2\omega}} = 8 - \frac{7 - e^{-j\omega}}{(1 - \frac{1}{4}e^{-j\omega})(1 - \frac{1}{2}e^{-j\omega})} = 8 + \frac{3}{1 - \frac{1}{4}e^{-j\omega}} - \frac{10}{1 - \frac{1}{2}e^{-j\omega}}.$$

In this example, because the orders of the numerator and denominator polynomials are equal, we cannot immediately perform partial-fraction expansion. It is first necessary to perform a long division; then the residual expression is expanded into the sum of two terms, as discussed in Example 3.19. Again, we find the inverse transform by inspection:

$$x[n] = 8\delta[n] + 3 \cdot \frac{1}{4}^{n} u[n] - 10 \cdot \frac{1}{2}^{n} u[n].$$

We will have much more to say about the process of finding the inverse transform in Chapter 4, in which we cover the z-transform, which can be viewed as a generalization of the DTFT. For now, just understand that in many cases of practical interest, it is easy to obtain the inverse DTFT by inspection without having to calculate the integral of Equation (3.7).

3.9 Using Matlab to compute and plot the DTFT

Matlab has a number of functions that can be used to compute and visualize the DTFT. If you have the Signal Processing Toolbox, you can use the `freqz` function,

```
[x, w] = freqz(b, [a], [N]),
```

to compute an array of transform values `x`, and frequency values `w`, for `N` frequencies (default 512) linearly spaced between $0 \leq \omega < \pi$. The transform is specified by arrays `b` and `a` (default, `a=1`), which represent the numerator and denominator polynomials in Equation (3.38). If you specify one or more output arguments (i.e., `x`), then `freqz` returns values, but does not plot. In the absence of output arguments, `freqz` produces a magnitude and phase plot. For example, the following would plot the transform of Example 3.21:

```
freqz([1 -5 1], [1 -3/4 1/8]).
```

It's easy to implement an equivalent routine without the Signal Processing Toolbox by using basic Matlab function with a line or two of your own code. For example, assume that you want to use Matlab to calculate the DTFT of an FIR system, Equation (3.6), specified by a column vector, `h`, of length N at frequencies given by a column vector, `w`, of length M. One line will do it:

```
H=exp(-1j*w*(0:length(h)-1))*h;
```
(3.39)

Here, we are effectively exploiting the matrix multiplication properties of Matlab. Define vector quantities

$$\begin{aligned}\mathbf{h} &= [h_0\ h_1 \cdots h_{N-1}]\\ \boldsymbol{\omega} &= [\omega_0\ \omega_1 \cdots \omega_{M-1}].\\ \mathbf{n} &= [0\ 1 \cdots N-1]^T\end{aligned}$$

Then, the DTFT in matrix form is

$$\begin{aligned}\mathbf{H} = \exp(-j\boldsymbol{\omega}\mathbf{n})\mathbf{h} &= \exp\left(-j\begin{bmatrix}\omega_0\\ \omega_1\\ \vdots\\ \omega_{M-1}\end{bmatrix}[0\ 1\ \cdots\ N-1]\right)\begin{bmatrix}h_0\\ h_1\\ \vdots\\ h_{N-1}\end{bmatrix}\\ &= \exp\left(-j\begin{bmatrix}0 & \omega_0 & \cdots & (N-1)\omega_0\\ 0 & \omega_1 & \cdots & (N-1)\omega_1\\ \vdots & \vdots & \ddots & \vdots\\ 0 & \omega_1 & \cdots & (N-1)\omega_{M-1}\end{bmatrix}\right)\begin{bmatrix}h_0\\ h_1\\ \vdots\\ h_{N-1}\end{bmatrix}.\end{aligned}$$

To compute the response of a general IIR system, specified by Equation (3.38), you can form the ratio of two expressions of the form of Equation (3.39). The following piece of code evaluates the DTFT and either returns the values or plots them.

```
function [x, w] = dtft(b, a)
    N = 512;
    w = pi * linspace(0, 1-1/N, N)';
    X = (exp(-1j*w*(0:length(b)-1))*b(:)) ./ ...
        (exp(-1j*w*(0:length(a)-1))*a(:));
    if (~nargout)
        subplot(2, 1, 1)
        plot(w/pi, abs(X));
        title('|X(\omega)|')
        subplot(2, 1, 2)
        plot(w/pi, angle(X)/pi)
        title('\angleX(\omega)')
        figure(gcf)
    else
        out = X;
    end
end
```

The main calculation of this function is done using the ratio of terms of the form of Equation (3.39). Matlab's `nargout` function checks the number of output arguments. In the absence of output arguments, our function plots the DTFT, otherwise it returns values of the DTFT and the frequencies at which it was computed. The `title` command in the plotting section uses Matlab's symbols for ω and $\measuredangle$. As an example, the plots in **Figure 3.14** can be produced with `dtft([0 1 1 1], 1)`.

3.10 DTFT properties

In this section, we look at some important properties of the DTFT, such as linearity, shifting, complex modulation and others. These properties not only are of theoretical interest, but also form the basis for understanding major topics of DSP, such as filtering, spectral analysis and digital communication. These properties also open the door to practical shortcuts to the computation of complicated transforms.

3.10.1 Linearity

The linearity property states that if $X_1(\omega)$ is the DTFT of sequence $x_1[n]$, and $X_2(\omega)$ is the DTFT of sequence $x_2[n]$, then the DTFT of the sum of scaled sequences is the sum of the scaled DTFTs. That is,

> If $x_1[n] \longleftrightarrow X_1(\omega)$ and $x_2[n] \longleftrightarrow X_2(\omega)$, then $a_1x_1[n]+a_2x_2[n] \longleftrightarrow a_1X_1(\omega)+a_2X_2(\omega)$.

The proof of this just follows from the linearity of the summation operator:

$$\begin{aligned}\mathfrak{F}\{a_1x_1[n]+a_2x_2[n]\} &= \sum_{n=-\infty}^{\infty}(a_1x_1[n]+a_2x_2[n])e^{-j\omega n} = a_1\underbrace{\sum_{n=-\infty}^{\infty}x_1[n]e^{-j\omega n}}_{X_1(\omega)} + a_2\underbrace{\sum_{n=-\infty}^{\infty}x_2[n]e^{-j\omega n}}_{X_2(\omega)}\\ &= a_1X_1(\omega)+a_2X_2(\omega).\end{aligned}$$

This is a useful property of DTFT that we will use constantly. In fact, we have already used it in several examples in this chapter.

3.10.2 Delay (shifting) property

Consider a system that is an ideal delay. The output sequence $y[n]=x[n-n_0]$ is just sequence $x[n]$ delayed by n_0 samples, where n_0 could be either a positive or a negative integer. The shifting property states that

$$\text{If } x[n] \longleftrightarrow X(\omega), \text{ then } x[n-n_0] \longleftrightarrow X(\omega)e^{-j\omega n_0}.$$

The proof is simple:

$$\mathfrak{F}\{y[n]\} = \mathfrak{F}\{x[n-n_0]\} = \sum_{n=-\infty}^{\infty} x[n-n_0]e^{-j\omega n}.$$

Let $m=n-n_0$, so that $n=m+n_0$, and we get

$$Y(\omega) = \mathfrak{F}\{y[n]\} = \sum_{m+n_0=-\infty}^{\infty} x[m]e^{-j\omega(m+n_0)} = e^{-j\omega n_0}\sum_{m=-\infty}^{\infty} x[m]e^{-j\omega m} = e^{-j\omega n_0}X(\omega),$$

where we note that for finite values of n_0, the lower limit of the summation, from $m+n_0=-\infty$, is equivalent to $m=-\infty$.

Some interesting conclusions follow from this result. Since $|e^{-j\omega n_0}|=1$, the magnitude of $Y(\omega)$ is

$$|Y(\omega)| = |X(\omega)e^{-j\omega n_0}| = |X(\omega)|\cancel{|e^{-j\omega n_0}|} = |X(\omega)|,$$

and the phase of $Y(\omega)$ is

$$\measuredangle Y(\omega) = \measuredangle(X(\omega)e^{-j\omega n_0}) = \measuredangle X(\omega) + \measuredangle e^{-j\omega n_0} = \measuredangle X(\omega) - \omega n_0.$$

In words, the magnitude of the DTFT of the shifted sequence is identical to that of the original sequence. The phase of the DTFT of the shifted sequence is equal to that of the original sequence plus a linear-phase term that is directly proportional to n_0, the amount of the shift. We have seen several examples of the shifting property applied to the impulse and pulse functions in Section 3.3.

Figure 3.25a shows the DTFT of a shifted impulse $\delta[n-n_0]$ for various positive and negative values of n_0. Positive values of n_0 correspond to a positive delay; that is, the output lags the input by n_0 samples. Negative values of n_0 correspond to a negative delay; the output leads the input by n_0 samples.

As we have discussed in Section 3.3, the magnitude of the DTFT of a shifted impulse $\delta[n-n_0]$ is unity. The phase of the DTFT, shown in **Figure 3.25b** is a linear function of ω, with a slope equal to $-\omega n_0$. Thus, delay in time corresponds to the addition of a linear-phase term to the DTFT.

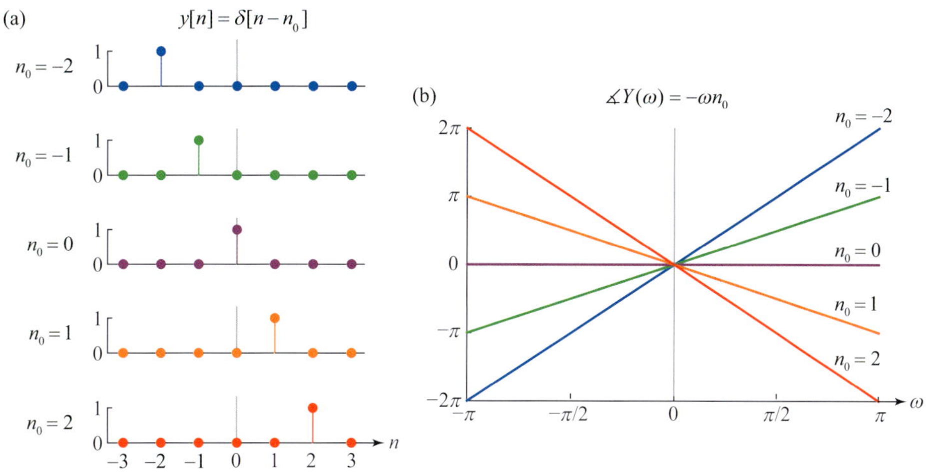

Figure 3.25 DTFT of a shifted impulse

3.10.3 Complex modulation (frequency shift) property

Modulation is a fancy word for multiplication. Multiplication of signals is one of the most important operations in digital signal processing, underlying such areas as digital communication. There are several properties that pertain to multiplication. To perform complex modulation, multiply input sequence $x[n]$ by a complex exponential sequence $e^{j\omega_0 n}$ to form output sequence $y[n] = x[n]e^{j\omega_0 n}$. The complex modulation property states that

$$\text{If } y[n] = x[n]e^{j\omega_0 n}, \text{ then } Y(\omega) = X(\omega - \omega_0).$$

The proof is a one-liner:

$$Y(\omega) = \sum_{n=-\infty}^{\infty} (x[n]e^{j\omega_0 n})e^{-j\omega n} = \sum_{n=-\infty}^{\infty} x[n]e^{-j(\omega-\omega_0)n} = X(\omega - \omega_0).$$

This says that multiplication of a sequence by a complex exponential sequence of frequency ω_0 causes the entire frequency spectrum to shift by ω_0. Hence, we could also call the complex modulation property the **frequency shift** or **frequency translation** property. This property is the dual of the time shift (delay) property. Note the similarity in structure of the two properties:

Delay (Time shift)	Shift sequence in time $y[n] = x[n - n_0]$	$\longleftrightarrow$	Multiply transform by $e^{-j\omega n_0}$ $Y(\omega) = X(\omega)e^{-j\omega n_0}$
Complex modulation (Frequency shift)	Multiply sequence by $e^{j\omega_0 n}$ $y[n] = x[n]e^{j\omega_0 n}$	$\longleftrightarrow$	Shift transform in frequency $Y(\omega) = X(\omega - \omega_0)$

We can use the complex modulation property to find the DTFT of useful and interesting signals, as shown in the following examples.

Example 3.22

Use the complex modulation property to find the DTFT of a cosine, $x[n] = \cos\omega_0 n$.

➤ Solution:

We write

$$x[n] = \cos\omega_0 n = \tfrac{1}{2}e^{j\omega_0 n} + \tfrac{1}{2}e^{-j\omega_0 n}.$$

Each of these terms is a constant (1/2) multiplied by a complex exponential. Now,

$$\mathfrak{F}\{\tfrac{1}{2}\} = \pi \sum_{k=-\infty}^{\infty} \delta(\omega - 2\pi k).$$

As shown graphically in **Figure 3.26a**, this transform is a train of impulses in frequency at multiples of 2π. Using the complex modulation property,

$$\mathfrak{F}\{\tfrac{1}{2}e^{j\omega_0 n}\} = \pi \sum_{k=-\infty}^{\infty} \delta(\omega - \omega_0 - 2\pi k).$$

This is shown in **Figure 3.26b**. All the impulses have been shifted in frequency by ω_0. Similarly,

$$\mathfrak{F}\{\tfrac{1}{2}e^{-j\omega_0 n}\} = \pi \sum_{k=-\infty}^{\infty} \delta(\omega + \omega_0 - 2\pi k),$$

which is shown in **Figure 3.26c**. All the impulses have been shifted in frequency by $-\omega_0$. Now, using the linearity property,

$$\begin{aligned}\mathfrak{F}\{\cos\omega_0 n\} &= \mathfrak{F}\{\tfrac{1}{2}e^{j\omega_0 n} + \tfrac{1}{2}e^{-j\omega_0 n}\} = \mathfrak{F}\{\tfrac{1}{2}e^{j\omega_0 n}\} + \mathfrak{F}\{\tfrac{1}{2}e^{-j\omega_0 n}\}\\ &= \pi \sum_{k=-\infty}^{\infty} \delta(\omega - \omega_0 - 2\pi k) + \pi \sum_{k=-\infty}^{\infty} \delta(\omega + \omega_0 - 2\pi k)\\ &= \pi \sum_{k=-\infty}^{\infty} (\delta(\omega - \omega_0 - 2\pi k) + \delta(\omega + \omega_0 - 2\pi k)).\end{aligned}$$

Visually, the DTFT of a cosine comprises pairs of impulses centered on multiples of 2π. If we concentrate on the principal region of frequency from $-\pi \le \omega < \pi$ (the shaded region in **Figure 3.26d**), the DTFT of a cosine comprises just a single pair of impulses at frequencies $\omega = \pm\omega_0$,

$$X(\omega) = \pi\delta(\omega - \omega_0) + \pi\delta(\omega + \omega_0). \tag{3.40}$$

If you think about it, this makes intuitive sense. The Fourier transform description essentially asks, "How do I express $x[n] = \cos\omega_0 n$ as the sum of complex exponentials of the form $e^{j\omega n}$ over the frequency range $-\pi \le \omega < \pi$?" That is, how do I express $x[n]$ in the form

$$x[n] = \frac{1}{2\pi}\int_{\omega=-\pi}^{\pi} X(\omega)e^{j\omega n}d\omega? \tag{3.41}$$

The answer is that $\cos\omega_0 n$ is *already* in the form of the sum of complex exponentials:

$$\cos \omega_0 n = \tfrac{1}{2}e^{j\omega_0 n} + \tfrac{1}{2}e^{-j\omega_0 n}.$$

You can see by inspection of this formula that you only need two complex exponential sequences, at frequencies ω_0 and $-\omega_0$, to make up $\cos \omega_0 n$. To reinforce this notion, substitute Equation (3.40) into Equation (3.41):

$$\begin{aligned} x[n] &= \frac{1}{2\pi}\int_{\omega=-\pi}^{\pi} X(\omega)e^{j\omega n}d\omega = \frac{1}{2\pi}\int_{\omega=-\pi}^{\pi} (\pi\delta(\omega-\omega_0)+\pi\delta(\omega+\omega_0))e^{j\omega n}d\omega \\ &= \frac{1}{2\not{\pi}}\int_{\omega=-\pi}^{\pi} \not{\pi}\delta(\omega-\omega_0)e^{j\omega n}d\omega + \frac{1}{2\not{\pi}}\int_{\omega=-\pi}^{\pi} \not{\pi}\delta(\omega+\omega_0)e^{j\omega n}d\omega \\ &= \tfrac{1}{2}e^{j\omega_0 n} + \tfrac{1}{2}e^{-j\omega_0 n} = \cos \omega_0 n. \end{aligned}$$

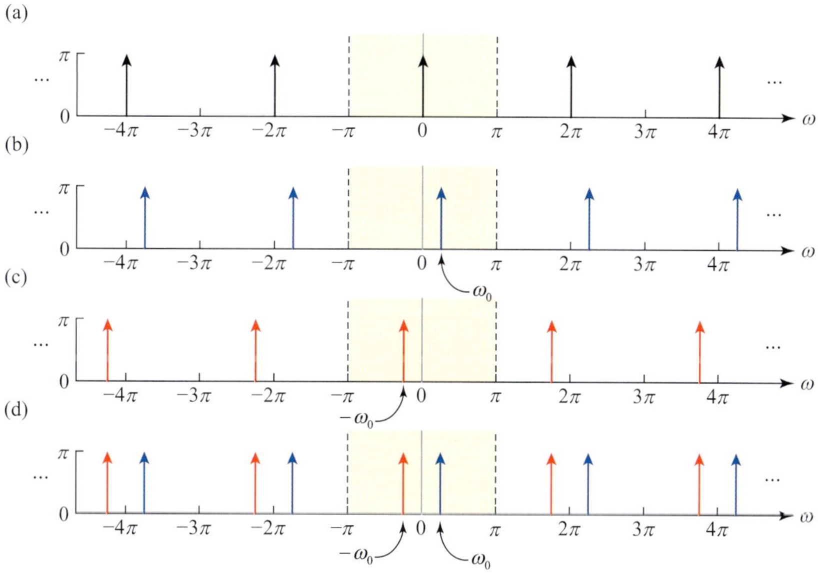

Figure 3.26 DTFT of a cosine

Figure 3.27 offers a graphical way of understanding discrete-time frequency. The left side of the figure shows the time domain: the sequences, $\cos \omega_0 n$, as a function of ω_0. The right column shows the frequency domain: the DTFTs of the sequences.

First look at the left column. As ω_0 increases from $\pi/4$ to π (**Figures 3.27a–d**), the sequence goes from a low frequency of oscillation (a period of eight samples when $\omega_0 = \pi/4$) to a high frequency (a period of two when $\omega_0 = \pi$). So far, so good. However, as ω_0 increases further from π to 2π, as shown in **Figures 3.27e–h**, the sequence goes from a high frequency *back* to a low frequency (a period of eight when $\omega_0 = \pi/4$) and then to a constant (i.e., *zero* frequency) when $\omega_0 = 2\pi$. Of course, this behavior is no mystery. We have already discussed it extensively in Chapter 1. All discrete-time frequencies are periodic modulo 2π, so, for example, $\cos 3\pi/4 = \cos(2\pi - 3\pi/4) = \cos(-5\pi/4)$. In addition, cosine is an even function, so $\cos(-5\pi/4) = \cos 5\pi/4$. That is why the plots of **Figures 3.27c** and **e** look the same.

The plots of DTFTs in the right column of **Figure 3.27** provide a particularly revealing and intuitive explanation for this paradoxical behavior. The shaded region of each frequency plot corresponds to the principal frequency range $-\pi \le \omega < \pi$. At a low frequency (e.g., $\omega_0 = \pi/4$), the DTFT comprises pairs of impulses, each centered around an integer multiple of 2π. We have attempted to keep things visually clear by plotting each pair of impulses in a different color. As ω_0 increases from $\pi/4$ to π, as shown in **Figures 3.27a–d**, the impulses within each pair move apart from one another, while the pair continues to be centered at "its" multiple of 2π. So, for example, within the principal frequency range, the impulses move apart as we expect. The left impulse heads for lower (more negative) frequency and the right impulse heads for higher frequency. At $\omega_0 = \pi$, the impulses are exactly at $\omega = \pm\pi$. Now, as ω_0 increases further from π to 2π, as shown in **Figures 3.27e–h**, the impulses originally centered at $\omega = 0$ move outside the shaded region, and one impulse from each of the adjoining pairs of the DTFT moves into the shaded region. To be specific, look at the frequency plot in **Figure 3.27e**, corresponding to $\omega_0 = 5\pi/4$. Here, the pair of black impulses centered at 0 are located at $\omega = \pm 5\pi/4$, which is outside the principal range. The two "blue" impulses, centered at -2π, are now at $\omega = -2\pi \pm 5\pi/4$; the left impulse of the pair is at $\omega = -11\pi/4$ and the right impulse is

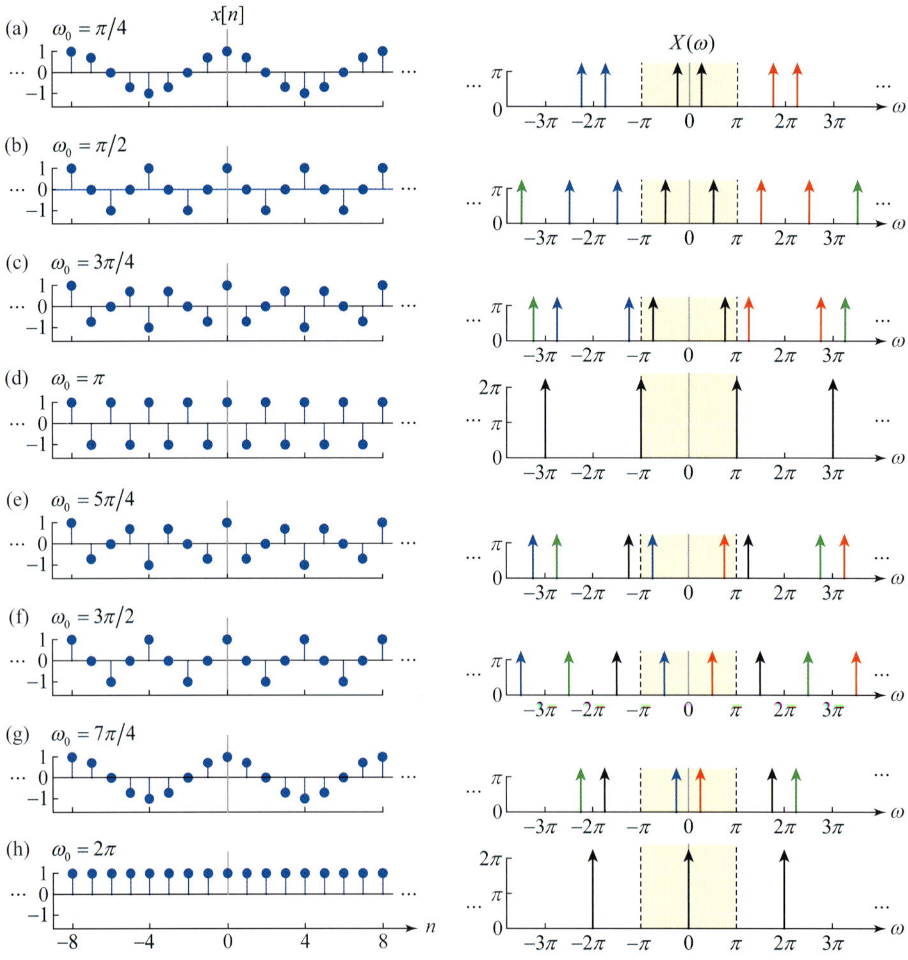

Figure 3.27 DTFT of a cosine of varying frequency

at $\omega = -3\pi/4$; thus, one of this pair now lies within the principal range. Similarly, for the two "red" impulses centered at 2π, the right impulse of the pair is at $\omega = 11\pi/4$ and the left impulse is at $\omega = 3\pi/4$, which lies within the principal range. Hence, when $\omega_0 = 5\pi/4$, there are exactly two impulses within the principal range, and they lie at $\omega = \pm 3\pi/4$. This corresponds to a cosine of frequency $\omega = 3\pi/4$. If you look, you will see that the time and frequency domain plots of **Figure 3.27e** for $\omega_0 = 5\pi/4$ are identical to those of **Figure 3.27c** for $\omega_0 = 3\pi/4$. Similarly, the plots of **Figure 3.27f** for $\omega_0 = 3\pi/2$ are identical to those of **Figure 3.27b** for $\omega_0 = \pi/2$.

One additional question before we leave this topic: look again at the impulses of the DTFT of the cosine. In **Figures 3.27a–c** and **e–g**, there is a pair of impulses within any contiguous 2π range of frequency (e.g., the principal frequency range, $-\pi \le \omega < \pi$), and each impulse has area π. However, for $\omega_0 = \pi$ and $\omega_0 = 2\pi$, there is only one impulse within any range and it has area 2π. Why?

Example 3.23

(a) Use the complex modulation property to find the DTFT of a cosine, $x[n] = A\cos(\omega_0 n + \varphi)$.

(b) Plot the sequence and DTFT for $A = 1$, $\omega_0 = \pi/4$ and for three values of φ: $\varphi = 0$, $-\pi/2$ and $\pi/2$.

➤ Solution:

(a)
$$X(\omega) = \mathfrak{F}\{A\cos(\omega_0 n + \varphi)\} = \mathfrak{F}\left\{\tfrac{1}{2}Ae^{j(\omega_0 n + \varphi)} + \tfrac{1}{2}Ae^{-j(\omega_0 n + \varphi)}\right\} = \mathfrak{F}\left\{\tfrac{1}{2}Ae^{j\varphi}e^{j\omega_0}\right\} + \mathfrak{F}\left\{\tfrac{1}{2}Ae^{-j\varphi}e^{-j\omega_0}\right\}$$
$$= \pi Ae^{j\varphi}\sum_{k=-\infty}^{\infty}\delta(\omega - \omega_0 - 2\pi k) + \pi Ae^{-j\varphi}\sum_{k=-\infty}^{\infty}\delta(\omega + \omega_0 - 2\pi k).$$

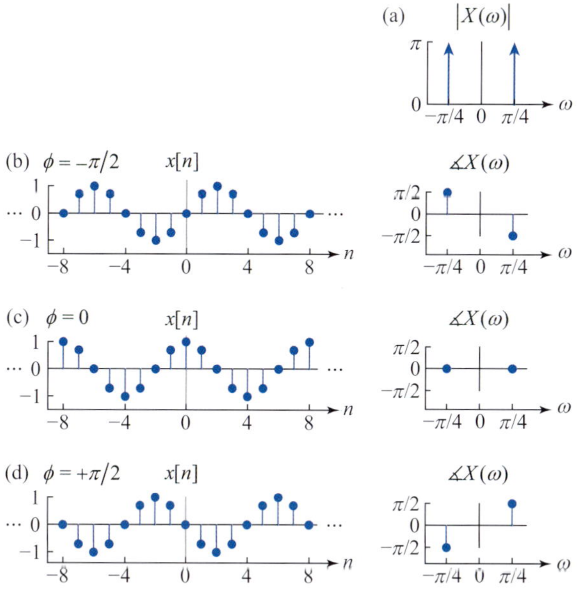

Figure 3.28 DTFT of $\cos(\pi n/4 + \varphi)$ for $\varphi = -\pi/2$, 0 and $\pi/2$.

So, the transform comprises impulses with complex amplitude. Within the principal range of frequency $-\pi \le \omega < \pi$, we can write

$$X(\omega) = \pi A e^{j\varphi}\delta(\omega - \omega_0) + \pi A e^{-j\varphi}\delta(\omega + \omega_0),$$

which means

$$\begin{aligned} |X(\omega)| &= \pi A\delta(\omega - \omega_0) + \pi A\delta(\omega + \omega_0) \\ \measuredangle X(\omega) &= \begin{cases} \varphi, & \omega = \omega_0 \\ -\varphi, & \omega = -\omega_0 \end{cases}. \end{aligned} \tag{3.42}$$

(b) The left panels of **Figures 3.28b**, **c** and **d** show the sequence $x[n] = A\cos(\omega_0 n + \varphi)$ for $A = 1$, $\omega_0 = \pi/4$ and $\varphi = -\pi/2$, 0 and $\pi/2$ (**Figures 3.28b–d**, respectively). The magnitude of the transform, $|X(\omega)|$, shown in **Figure 3.28a**, is the same for all values of phase φ; it comprises a pair of impulses at frequencies $\omega_0 = \pm\pi/4$. The remaining panels show the phase $\measuredangle X(\omega)$ corresponding to $\varphi = 0$, $-\pi/2$ and $\pi/2$. The phase is only defined at frequencies $\omega = \pm\omega_0$.

Modulation by $(-1)^n$ A particularly interesting example of complex modulation occurs when a sequence is multiplied by $e^{j\omega_0 n}$, where $\omega_0 = \pm\pi$:

$$y[n] = x[n]e^{\pm j\pi n} = x[n](-1)^n.$$

In the time domain, modulation by $(-1)^n$ is equivalent to reversing the sign of every other point. In the frequency domain, it corresponds to translating all the frequencies in $X(\omega)$ by $\pm\pi$. The next example shows this.

Example 3.24

Given the pulse $x[n]$ shown in **Figure 3.29a**, find $y[n] = x[n](-1)^n$ and plot the DTFT. Show that $Y(\omega) = X(\omega \pm \pi)$.

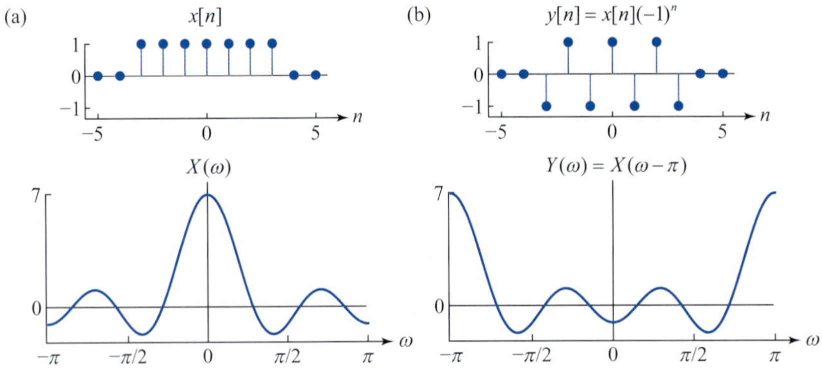

Figure 3.29 Modulation of a pulse by $(-1)^n$

➤ Solution:

The original sequence $x[n]$ is shown in **Figure 3.29a**, along with its DTFT $X(\omega)$. Since $x[n]$ is real and even, $X(\omega)$ is purely real and even and its amplitude can be plotted on a single axis. The modulated sequence $y[n] = x[n](-1)^n$ is shown in **Figure 3.29b**. You can see that its DTFT $Y(\omega)$ is exactly equivalent to $X(\omega)$ translated in frequency by π or $-\pi$. That is, $Y(\omega) = X(\omega \pm \pi)$. In Chapter 7, we will use this trick of modulation by $(-1)^n$ to transform a lowpass filter into a highpass filter.

Now consider a sequence whose DTFT requires us to display both magnitude and phase plots.

Example 3.25

(a) Given a power-law sequence $x[n] = (1/2)^n u[n]$, find and plot $X(\omega)$.

(b) Find $y[n] = x[n](-1)^n$ and $Y(\omega)$. Show analytically that $Y(\omega) = X(\omega \pm \pi)$.

(c) Show this graphically, so that

$$\begin{aligned} |Y(\omega)| &= |X(\omega \pm \pi)| \\ \measuredangle Y(\omega) &= \measuredangle X(\omega \pm \pi) \end{aligned}.$$

➤ Solution:

(a) The power-law sequence $x[n]$ and its DTFT $X(\omega)$ are shown in **Figure 3.30a**. Because $X(\omega)$ is complex, both magnitude and phase are shown. Analytically, the DTFT of $x[n]$ is

$$X(\omega) = \sum_{n=-\infty}^{\infty} x[n]e^{-j\omega n} = \sum_{n=-\infty}^{\infty} \tfrac{1}{2}^n u[n]e^{-j\omega n} = \sum_{n=0}^{\infty} \left(\tfrac{1}{2}e^{-j\omega}\right)^n = \frac{1}{1-\frac{1}{2}e^{-j\omega}}.$$

(b) The modulated power-law sequence $y[n]$ and its DTFT $Y(\omega)$ are shown in **Figure 3.30b**. The modulated sequence is

$$y[n] = x[n](-1)^n = \tfrac{1}{2}^n u[n](-1)^n = (-\tfrac{1}{2})^n u[n].$$

Its transform is

$$Y(\omega) = \frac{1}{1+\frac{1}{2}e^{-j\omega}}.$$

The proof that $Y(\omega) = X(\omega \pm \pi)$:

$$X(\omega \pm \pi) = \frac{1}{1-\frac{1}{2}e^{-j(\omega \pm \pi)}} = \frac{1}{1-\frac{1}{2}e^{\mp \pi}e^{-j\omega}} = \frac{1}{1+\frac{1}{2}e^{-j\omega}} = Y(\omega).$$

(c) Graphically, as you can see from **Figure 3.30**, $|Y(\omega)| = |X(\omega \pm \pi)|$ and $\measuredangle Y(\omega) = \measuredangle X(\omega \pm \pi)$. Both the magnitude and the phase of the transform are shifted by π.

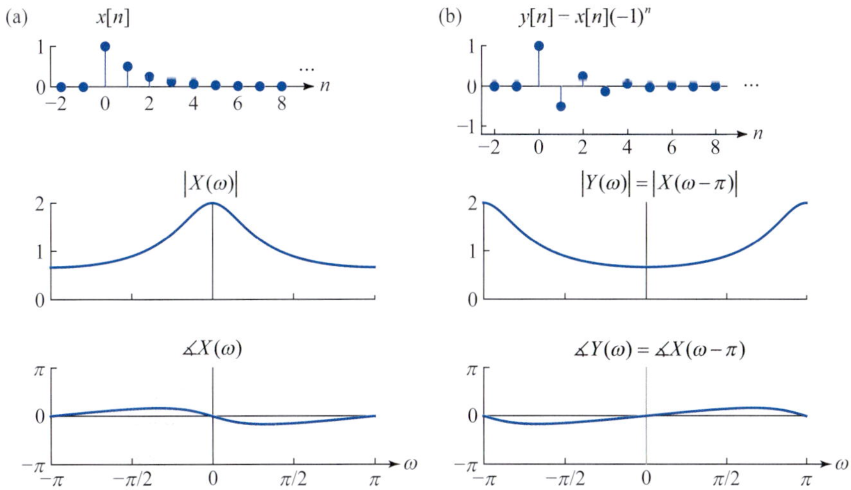

Figure 3.30 Modulation of a power-law sequence by $(-1)^n$

3.10.4 Convolution

The convolution property forms the theoretical basis for understanding three major signal processing operations: **filtering**, **deconvolution** and **system identification**. Filtering is a central application of digital signal processing. In this section, we will spend considerable time discussing and applying this key property. In Section 3.10.6, we apply the insights developed in this section to deconvolution and system identification.

Start by recalling the convolution operation schematized in **Figure 3.31**.

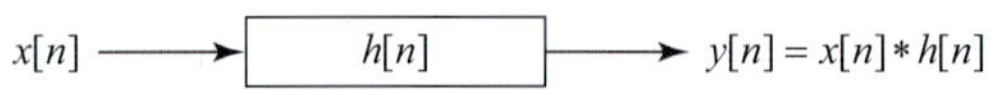

Figure 3.31 Convolution

We have

$$y[n] = x[n] * h[n] = \sum_{k=-\infty}^{\infty} x[k]h[n-k]. \tag{3.43}$$

Taking the DTFT:

$$Y(\omega) = \mathfrak{F}\{x[n] * h[n]\} = \mathfrak{F}\left\{ \sum_{k=-\infty}^{\infty} x[k]h[n-k] \right\}.$$

For each k in the summation, $x[k]$ is a constant. Hence, by linearity,

$$Y(\omega) = \mathfrak{F}\left\{ \sum_{k=-\infty}^{\infty} x[k]h[n-k] \right\} = \sum_{k=-\infty}^{\infty} x[k]\mathfrak{F}\{h[n-k]\}. \tag{3.44}$$

Using the shifting property, the DTFT of $h[n-k]$ is

$$\mathfrak{F}\{h[n-k]\} = \mathfrak{F}\{h[n]\}e^{-j\omega k} = H(\omega)e^{-j\omega k}, \tag{3.45}$$

where we have recognized that the DTFT of $h[n]$ is the system function $\mathfrak{F}\{h[n]\} = H(\omega)$. Substituting Equation (3.45) into Equation (3.44), we get

$$\begin{aligned} Y(\omega) &= \sum_{k=-\infty}^{\infty} x[k]\,\mathfrak{F}\{h[n-k]\} = \sum_{k=-\infty}^{\infty} x[k]H(\omega)e^{-j\omega k} = \underbrace{\left(\sum_{k=-\infty}^{\infty} x[k]e^{-j\omega k} \right)}_{X(\omega)} H(\omega) \\ &= X(\omega)H(\omega). \end{aligned} \tag{3.46}$$

To summarize, the convolution property states the following:

> If $x[n] \longleftrightarrow X(\omega)$ and $h[n] \longleftrightarrow H(\omega)$, then $x[n] * h[n] \longleftrightarrow X(\omega)H(\omega)$.

In words, filtering is the convolution of two sequences in the time domain, which corresponds to multiplication of their DTFTs in the frequency domain.

Expressing Equation (3.46) in polar form allows us to relate the magnitudes and phases of $X(\omega)$, $H(\omega)$ and $Y(\omega)$:

$$|Y(\omega)|e^{\measuredangle Y(\omega)} = |X(\omega)|e^{\measuredangle X(\omega)}|H(\omega)|e^{\measuredangle H(\omega)},$$

so

$$\begin{aligned} |Y(\omega)| &= |X(\omega)||H(\omega)| \\ \measuredangle Y(\omega) &= \measuredangle X(\omega) + \measuredangle H(\omega). \end{aligned} \tag{3.47}$$

3.10.5 Using the convolution property of the DTFT to do filtering

One key reason that the convolution property is important is that it gives us an alternative method to perform filtering.

Method #1: direct time-domain convolution This is one of the methods we know and love (or at least respect): the direct-summation or flip-and-shift methods we have studied in Chapter 2. Direct evaluation of the convolution sum using Equation (3.43) is computationally demanding. If we take the example where $x[n]$ and $h[n]$ are both finite-length sequences that are N points long, then it is easy to show that to compute all $2N-1$ points in the output sequence $y[n]$ efficiently would require on the order of N^2 multiplies.

Method #2: convolution by multiplication of transforms This process is shown pictorially below:

$$\begin{array}{ccccc} x[n] & * & h[n] & = & y[n] \\ \mathfrak{F}\downarrow & & \mathfrak{F}\downarrow & & \mathfrak{F}^{-1}\uparrow \\ X(\omega) & \cdot & H(\omega) & = & Y(\omega). \end{array} \tag{3.48}$$

The method has three steps:

(1) Take the DTFT of $x[n]$ and $h[n]$:

$$\begin{aligned} X(\omega) &= \mathfrak{F}\{x[n]\} \\ H(\omega) &= \mathfrak{F}\{h[n]\}. \end{aligned}$$

(2) Multiply the DTFTs together to form $Y(\omega)$: $Y(\omega) = X(\omega)H(\omega)$.

(3) Take the inverse DTFT to form $y[n]$: $y[n] = \mathfrak{F}^{-1}\{Y(\omega)\}$.

At first glance, this method seems ridiculously complicated. You have to take two DTFTs, multiply them together and then take the inverse DTFT of the product to get $y[n]$. How easy is all that? It turns out that this process can actually be much less computationally demanding than implementing convolution by the direct time-domain method. The reason is that there exist very fast and computationally efficient methods of performing the Fourier transform, such as the **fast Fourier transform (FFT)**. Whereas convolving two N-point sequences such as $x[n]$ or $h[n]$ via the direct method requires on the order of N^2 multiplications, convolving these sequences using the transform-multiplication method only requires on the order of $N\log_2 N$ multiplications. This is a substantial reduction in computational complexity. For example, performing the convolution of two $N = 1000$-point sequences requires on the order of $N^2 = 10^6$ multiplications using the direct time-domain method and on the order of $N\log_2 N \simeq 10^4$ multiplications using the transform-multiplication method. That is over two orders of magnitude faster! We will discuss this method in detail when we cover the discrete Fourier transform (DFT) and the FFT in Chapters 10 and 11.

Example 3.26

Find the convolution of the two sequences $x[n]$ and $h[n]$ shown in **Figure 3.32**, below, by (a) a time-domain convolution method (i.e., direct-summation method) and (b) the transform-multiplication method.

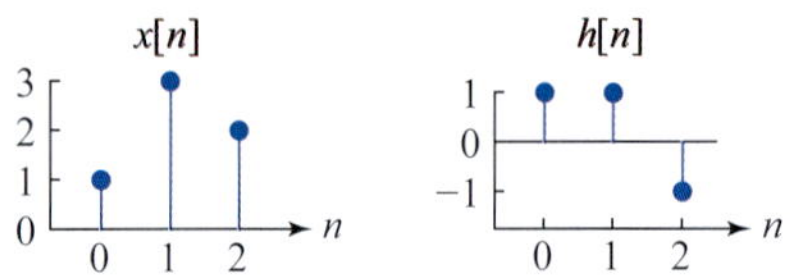

Figure 3.32 Two sequences

➤ Solution:

(a) Time-domain convolution (direct-summation method):

$$y[n] = \sum_{k=-\infty}^{\infty} h[k]x[n-k] = h[0]x[n] + h[1]x[n-1] + h[2]x[n-2]$$

$$\begin{aligned}
&= \delta[n] + 3\delta[n-1] + 2\delta[n-2] \\
&+ \delta[n-1] + 3\delta[n-2] + 2\delta[n-3] \\
&+ \; - \delta[n-2] - 3\delta[n-3] - 2\delta[n-4] \\
\hline
&\delta[n] + 4\delta[n-1] + 4\delta[n-2] - \delta[n-3] - 2\delta[n-4]
\end{aligned}$$

(b) Transform-multiplication method:

$$
\begin{array}{rrrrrrr}
X(\omega)=\sum_{n-\infty}^{\infty}(\delta[n]+3\delta[n-1]+2\delta[n-2])e^{-j\omega n}= & & & 1+ & 3e^{-j\omega} & +2e^{-2j\omega} & \\
H(\omega)=\sum_{n-\infty}^{\infty}(\delta[n]+\delta[n-1]-\delta[n-2])e^{-j\omega n}= & & & 1+ & e^{-j\omega} & -\;e^{-2j\omega} & \\
\hline
 & & & -\,e^{-2j\omega} & -\,3e^{-3j\omega} & -\,2e^{-4j\omega} & \\
 & & +e^{-j\omega} & +3e^{-2j\omega} & +2e^{-3j\omega} & & \\
 & 1+3e^{-j\omega} & +2e^{-2j\omega} & & & & \\
\hline
X(\omega)H(\omega)= & 1+4e^{-j\omega} & +4e^{-2j\omega} & -\;e^{-3j\omega} & -\,2e^{-4j\omega}. & &
\end{array}
$$

$X(\omega)$ and $H(\omega)$, the transforms of finite-length sequences such as $x[n]$ and $h[n]$, are just finite-length series expansions in powers of $e^{j\omega}$. Because each term in the transform of the form $e^{-j\omega n_0}$ corresponds to a single shifted impulse $\delta[n-n_0]$ in the sequence

$$\mathfrak{F}^{-1}\left\{e^{-j\omega n_0}\right\}=\delta[n-n_0],$$

we can derive the inverse transform by inspection using linearity:

$$
\begin{array}{ccccccccccc}
Y(\omega) & = & 1 & + & 4e^{-j\omega} & + & 4e^{-j2\omega} & - & e^{-j3\omega} & - & 2e^{-j4\omega} \\
\updownarrow & & \updownarrow & & \updownarrow & & \updownarrow & & \updownarrow & & \updownarrow \\
y[n] & = & \delta[n] & + & 4\delta[n-1] & + & 4\delta[n-2] & - & \delta[n-3] & - & 2\delta[n-4].
\end{array}
$$

This is the same result found by direct convolution. Of course, if you examine what we just did, you will see that both the direct-summation method and the transform-multiplication method are essentially equivalent to polynomial multiplication. In the direct-summation method, we group terms so as to add coefficients for impulses with the same amount of shift. In the transform-multiplication method, we group terms to add coefficients for the same power of $e^{-j\omega}$.

Another way to think of the transform-multiplication method of convolution is to view it as a natural result of the shifting and linearity properties we have already studied. Consider again the sequence

$$x[n]=\delta[n]+3\delta[n-1]+2\delta[n-2],$$

whose DTFT is

$$X(\omega)=1+3e^{-j\omega}+2e^{-j2\omega}.$$

By the shifting property, if we multiply $X(\omega)$ by $e^{-j\omega}$, this is equivalent to the transform of the time-shifted sequence:

$$X(\omega)e^{-j\omega}-(1+3e^{-j\omega}+2e^{-j2\omega})e^{-j\omega}-e^{-j\omega}+3e^{-j2\omega}+2e^{-j3\omega},$$

which implies that

$$\underbrace{\mathfrak{F}^{-1}\{X(\omega)e^{-j\omega}\}}_{x[n-1]} = \underbrace{\mathfrak{F}^{-1}\{e^{-j\omega}\}}_{\delta[n-1]} + \underbrace{3\mathfrak{F}^{-1}\{e^{-j2\omega}\}}_{3\delta[n-2]} + \underbrace{2\mathfrak{F}^{-1}\{e^{-j3\omega}\}}_{2\delta[n-3]}.$$

Now,

$$y[n] = h[0]x[n] + h[1]x[n-1] + h[2]x[n-2].$$

Taking the transform:

$$\begin{array}{rlllll}
Y(\omega) = \mathfrak{F}\{y[n]\} = h[0]\cdot\mathfrak{F}\{x[n]\} & \rightarrow 1\cdot(1 + 3e^{-j\omega} + 2e^{-j2\omega}) & & & & \\
+\, h[1]\cdot\mathfrak{F}\{x[n-1]\} & \rightarrow & 1\cdot(e^{-j\omega} + 3e^{-j2\omega} & + 2e^{-j3\omega}) & & \\
+\, h[2]\cdot\mathfrak{F}\{x[n-2]\} & \rightarrow & & -1\cdot(e^{-j2\omega} & + 3e^{-j3\omega} & + 2e^{-j4\omega}) \\
\hline
 & 1 + 4e^{-j\omega} + & 4e^{-j2\omega} & - \; e^{-j3\omega} & - 2e^{-j4\omega}. &
\end{array}$$

Taking the inverse transform by inspection yields the result previously obtained:

$$y[n] = \delta[n] + 4\delta[n-1] + 4\delta[n-2] - \delta[n-3] - 2\delta[n-4].$$

You can use a "short-hand" method to do convolution by the transform-multiplication method by writing each transform as an array with just the coefficients of the $e^{-j\omega}$ terms and multiplying them as if they were polynomials:

$$\begin{array}{lrrrrr}
X(\omega): & & & [1 & 3 & 2] \\
H(\omega): & & & [1 & 1 & -1] \\
 & & & -1 & -3 & -2 \\
 & & 1 & 3 & 2 & \\
 & 1 & 3 & 2 & & \\
\hline
Y(\omega): & [1 & 4 & 4 & -1 & -2]
\end{array}$$

3.10.6 Deconvolution and system identification using the convolution property

In the previous section, we used the convolution property, $Y(\omega) = X(\omega)H(\omega)$, to do filtering. However, the same property can be used to accomplish the two other important tasks: deconvolution and system identification.

Deconvolution **Deconvolution** is the process of finding the input $x[n]$ given the output $y[n]$ and the impulse response $h[n]$. In Chapter 2, we performed deconvolution in the time domain, essentially by long division of sequences. Deconvolution can be accomplished in the frequency domain by recognizing that Equation (3.46) can be rewritten as

$$X(\omega) = \frac{Y(\omega)}{H(\omega)}.$$

So, given $y[n]$ and $h[n]$, compute $Y(\omega)$ and $H(\omega)$ and take the inverse transform of their ratio to get $x[n]$.

Example 3.27

Given a system with impulse response $h[n] = \delta[n] - \frac{1}{2}\delta[n-1]$ and output $y[n] = \delta[n] - \delta[n-1]$, find the input $x[n]$.

➤ Solution:

$$X(\omega) = \frac{Y(\omega)}{H(\omega)} = \frac{1 - e^{-j\omega}}{1 - \frac{1}{2}e^{-j\omega}} = 2 - \frac{1}{1 - \frac{1}{2}e^{-j\omega}}.$$

Taking the inverse DTFT gives $x[n] = 2\delta[n] - \frac{1}{2}^n u[n]$.

System identification In a similar manner, we can find the impulse response of the system $h[n]$ given an input $x[n]$ and output $y[n]$, an operation known as **system identification**. Rearranging Equation (3.46) gives

$$H(\omega) = \frac{Y(\omega)}{X(\omega)},$$

from which $h[n]$ can be computed in a manner similar to that shown in Example 3.27.

Example 3.28

Given a system with input $x[n] = (2 - \frac{1}{2}^n)u[n]$ and output $y[n] = \frac{1}{2}^n u[n]$, find $h[n]$.

➤ Solution:

$$X(\omega) = \frac{2}{1 - e^{-j\omega}} - \frac{1}{1 - \frac{1}{2}e^{-j\omega}} = \frac{1}{1 - \frac{3}{2}e^{-j\omega} + \frac{1}{2}e^{-j2\omega}},$$

and

$$Y(\omega) = \frac{1}{1 - \frac{1}{2}e^{-j\omega}}$$

$$H(\omega) = \frac{Y(\omega)}{X(\omega)} = \frac{1 - \frac{3}{2}e^{-j\omega} + \frac{1}{2}e^{-j2\omega}}{1 - \frac{1}{2}e^{-j\omega}} = \frac{\cancel{(1 - \frac{1}{2}e^{-j\omega})}(1 - e^{-j\omega})}{\cancel{1 - \frac{1}{2}e^{-j\omega}}} = 1 - e^{-j\omega}.$$

Taking the inverse DTFT gives $h[n] = \delta[n] - \delta[n-1]$.

3.10.7 Convolution properties

In Chapter 2, we showed that the convolution operator satisfies commutative, associative and distributive properties. For example, convolution is commutative, so that

$$y[n] = x[n] * h[n] = h[n] * x[n].$$

This implies that multiplication of transforms is also commutative:

$$\begin{array}{ccccccccc} y[n] & = & x[n] & * & h[n] & = & h[n] & * & x[n] \\ \mathfrak{F}\downarrow & & \mathfrak{F}\downarrow & & \mathfrak{F}\downarrow & & \mathfrak{F}\downarrow & & \mathfrak{F}\downarrow \\ Y(\omega) & = & X(\omega) & \cdot & H(\omega) & = & H(\omega) & \cdot & X(\omega). \end{array}$$

But we already knew that, because multiplication is itself a commutative operator. A full list of the properties of convolution and their corresponding transform properties is shown in **Table 3.4**. The remaining proofs are left as exercises.

Table 3.4 Correspondence between time-domain and frequency-domain properties of convolution

Property	Time domain	Frequency domain
Convolution	$y[n] = x[n] * h[n]$	$Y(\omega) = X(\omega)H(\omega)$
Commutivity	$x[n] * h[n] = h[n] * x[n]$	$X(\omega)H(\omega) = H(\omega)X(\omega)$
Associativity	$(x[n]*h_1[n])*h_2[n] = x[n]*\underbrace{(h_1[n]*h_2[n])}_{h[n]}$	$(X(\omega)H_1(\omega))H_2(\omega) = X(\omega)\underbrace{(H_1(\omega)H_2(\omega))}_{H(\omega)}$
Distributivity	$(x[n]*h_1[n]) + (x[n]*h_2[n]) = x[n]*\underbrace{(h_1[n] + h_2[n])}_{h[n]}$	$X(\omega)H_1(\omega) + X(\omega)H_2(\omega) = X(\omega)\underbrace{(H_1(\omega) + H_2(\omega))}_{H(\omega)}$

The transform-multiplication method has proven to be an immensely powerful practical technique for implementing convolution of finite length sequences on a computer or microprocessor. This technique is clearly applicable when both of the sequences $x[n]$ and $h[n]$ are of finite length. Practical application of the transform-multiplication method is also possible when only one of the two sequences is of finite length, using the overlap-add and overlap-save methods discussed in Chapters 2 and 11. When $x[n]$ and $h[n]$ are both of infinite length, it is not possible to use the transform-multiplication method in a practical implementation of convolution, but it is still of theoretical value in obtaining transform results.

Example 3.29

A system comprises the cascade of two filters with impulse responses $h_1[n] = (1/2)^n u[n]$ and $h_2[n] = (1/4)^n u[n]$, as shown in **Figure 3.33**.

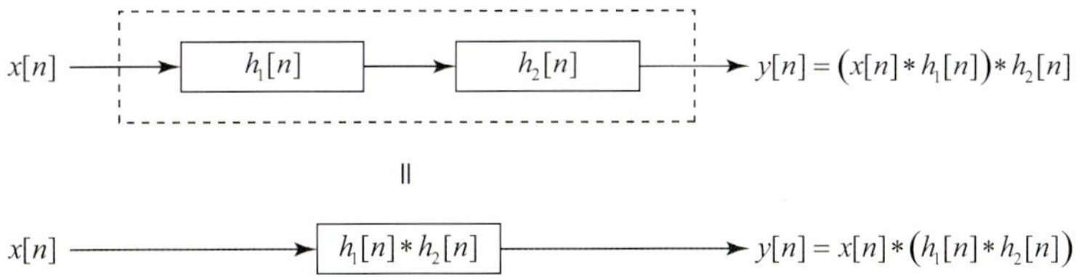

Figure 3.33 Cascade of two filters

Use the time-domain convolution and the transform-multiplication property of convolution to find the impulse response of a single filter with impulse response $h[n]$ that is equivalent to the cascade of these two filters.

➤ Solution:

By the associative property, the cascade of two filters $h_1[n]$ and $h_2[n]$ is equivalent to a single filter with impulse response $h[n] = h_1[n] * h_2[n]$. By the transform method,

$$H(\omega) = H_1(\omega)H_2(\omega) = \frac{1}{1 - \frac{1}{2}e^{-j\omega}} \cdot \frac{1}{1 - \frac{1}{4}e^{-j\omega}} = \frac{2}{1 - \frac{1}{2}e^{-j\omega}} - \frac{1}{1 - \frac{1}{4}e^{-j\omega}},$$

where we have used partial fraction decomposition to split the product of terms into the sum of terms. We derive the inverse transform by inspection of the individual terms:

$$h[n] = \mathfrak{F}^{-1}\left\{\frac{2}{1 - \frac{1}{2}e^{-j\omega}}\right\} - \mathfrak{F}^{-1}\left\{\frac{1}{1 - \frac{1}{4}e^{-j\omega}}\right\} = 2 \cdot \frac{1}{2}^n u[n] - \frac{1}{4}^n u[n].$$

By direct convolution,

$$\begin{aligned} h[n] &= \sum_{k=-\infty}^{\infty} h_1[k]h_2[n-k] = \sum_{k=-\infty}^{\infty} \frac{1}{2}^k u[k] \frac{1}{4}^{n-k} u[n-k] = \left(\frac{1}{4}^n \sum_{k=0}^{n} 2^k\right) u[n] \\ &= \left(\frac{1}{4}\right)^n \frac{1 - 2^{n+1}}{1 - 2} u[n] = \frac{1}{4}^n (2^{n+1} - 1)u[n] = 2 \cdot \frac{1}{2}^n u[n] - \frac{1}{4}^n u[n], \end{aligned}$$

which is the same result.

3.10.8 Understanding filtering in the frequency domain

One of the most important uses of the convolution property is to enable us to understand the relation between filtering in the time and frequency domains, as schematized in **Figure 3.34**. In the time domain, filtering is described by convolution. The output of the filter, $y[n]$, is the convolution of the input $x[n]$ with the filter's impulse response $h[n]$: $y[n] = x[n] * h[n]$. In the frequency domain, convolution is equivalent to multiplication of transforms: $Y(\omega) = X(\omega)H(\omega)$.

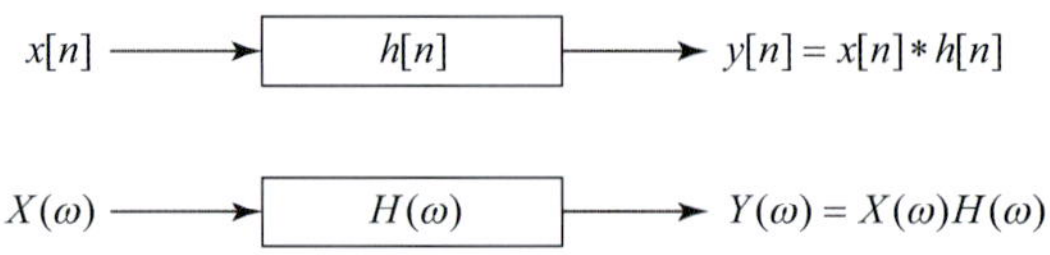

Figure 3.34 Filtering time and frequency domain

The following are examples of the filter's role in amplifying or attenuating components in the signal.

Example 3.30

As a simple example, use the frequency-domain solution technique of multiplying the transforms of input and system functions to find $y[n]$, the response of an ideal lowpass filter with bandwidth $\omega_x = \pi/2$, to an input comprising the sum of cosines at two frequencies, $x[n] = \cos \pi n/4 + \cos 3\pi n/4$.

➤ Solution:

This is a problem that cries out for interpretation in the frequency domain. The left column of **Figure 3.35** shows the input sequence $x[n]$ (top panel) and its transform $X(\omega)$ (bottom panel).

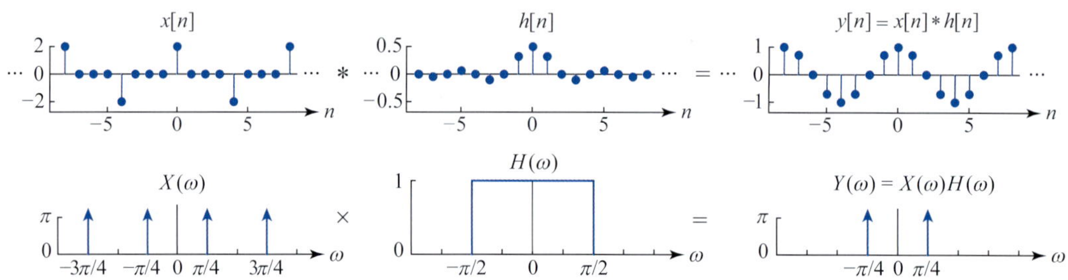

Figure 3.35 Filtering of the sum of cosines by an ideal lowpass filter

Since the input comprises only the sum of two cosines, the DTFT consists of a pair of impulses at $\omega = \pm\pi/4$ and another pair at $\omega = \pm 3\pi/4$:

$$X(\omega) = \mathfrak{F}[x[n]] = \pi(\delta(\omega - \pi/4) + \delta(\omega + \pi/4)) + \pi(\delta(\omega - 3\pi/4) + \delta(\omega + 3\pi/4)).$$

The center column of **Figure 3.35** shows the impulse response of the ideal lowpass filter, $h[n] = \frac{1}{2} \operatorname{sinc} \pi n/2$ (top panel), and its transform (bottom panel), which is the frequency response of this filter,

$$H(\omega) = \begin{cases} 1, & |\omega| < \pi/2 \\ 0, & \pi/2 < |\omega| < \pi \end{cases}.$$

In the frequency domain, the transform of the output is the product of the transform of the input and frequency response of the filter: $Y(\omega) = X(\omega)H(\omega)$. In this example, the impulses in $X(\omega)$ at $\omega = \pm\pi/4$ are multiplied by one and the impulses at $\omega = \pm 3\pi/4$ are multiplied by zero. Hence, the output is exactly $Y(\omega) = \pi(\delta(\omega - \pi/4) + \delta(\omega + \pi/4))$, shown in the right panel of **Figure 3.35** (bottom panel), which corresponds to output sequence $y[n] = \cos \pi n/4$ (top panel). This example demonstrates that the effect of this lowpass filter is most clearly understood in the frequency domain: the filter removes the components of the input signal at the higher frequency, $\omega = \pm 3\pi/4$, and retains the components at the lower frequency, $\omega = \pm\pi/4$.

While the ideal lowpass filter in this example is theoretically valuable, making realizable FIR and IIR filters requires us to consider both the magnitude and phase of response, as shown in the next couple of examples.

Example 3.31

Find the response of an IIR system defined by impulse response $h[n] = \frac{1}{2}^n u[n]$ to a cosine $x[n] = \cos \omega n$, directly using convolution in the time domain, and by multiplication of transforms in the frequency domain.

➤ Solution:

Time-domain solution

The time-domain response of a system to a cosine is shown in **Figure 3.36a**.

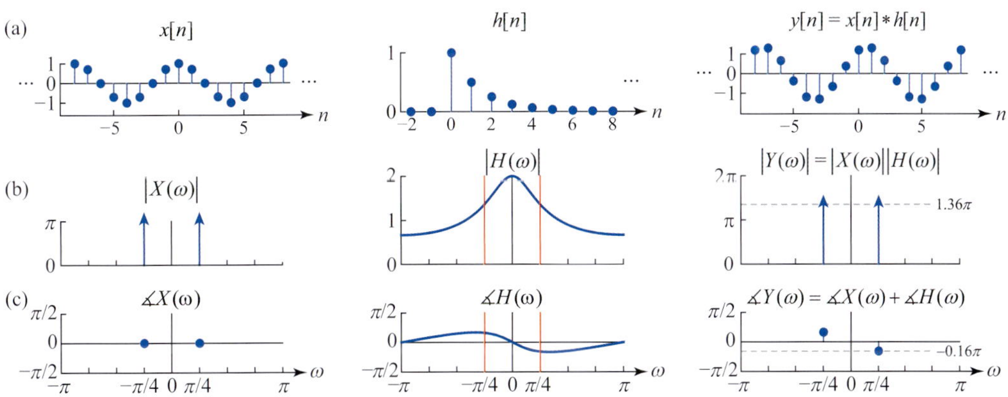

Figure 3.36 Filtering of $\cos \pi n/4$

As we discussed in Section 3.6, the response of a linear time-invariant system to $x[n] = \cos \omega_0 n$, a cosine of frequency ω_0, is a cosine of the same frequency with magnitude $|H(\omega_0)|$ and phase $\measuredangle H(\omega_0)$ that are determined solely by the magnitude and phase of the frequency response of the filter at frequency ω_0,

$$y[n] = |H(\omega_0)| \cos(\omega_0 n + \measuredangle H(\omega_0)). \tag{3.49}$$

In this example, the frequency response is

$$H(\omega)=\mathfrak{F}\{h[n]\}=|H(\omega)|e^{j\measuredangle H(\omega)}=\frac{1}{1-\frac{1}{2}e^{-j\omega}}=\frac{1}{1-\frac{1}{2}(\cos(-\omega)+j\sin(-\omega))}=\frac{1}{1-\frac{1}{2}\cos\omega+\frac{1}{2}j\sin\omega}$$

$$=\frac{1}{\sqrt{(1-\frac{1}{2}\cos\omega)^2+(\frac{1}{2}\sin\omega)^2}\,e^{j\tan^{-1}(\frac{\sin\omega}{2-\cos\omega})}}=\frac{1}{\sqrt{1-\cos\omega+\frac{1}{4}\cos^2\omega+\frac{1}{4}\sin^2\omega}\,e^{j\tan^{-1}(\frac{\sin\omega}{2-\cos\omega})}}\cdot$$

$$\frac{1}{\sqrt{\frac{5}{4}-\cos\omega}}e^{-j\tan^{-1}(\frac{\sin\omega}{2-\cos\omega})}$$

So when $\omega=\omega_0=\pi/4$, $H(\omega_0)=H(\pi/4)\simeq 1.36e^{-j0.159\pi}$, which means $y[n]\simeq 1.36\cos(\pi n/4-0.159\pi)=1.36\cos((\pi/4)(n-0.637))$. Looking at the plot of the sequence $y[n]$ in **Figure 3.36a**, you can see that the output cosine is, in fact, somewhat larger in amplitude than the input and appears to lag the input by not quite one sample.

Frequency-domain solution

A cosine of frequency ω_0 can be expressed solely as the sum of two complex exponentials at frequencies $\pm\omega_0$,

$$x[n]=\cos\omega_0 n=\tfrac{1}{2}e^{j\omega_0 n}+\tfrac{1}{2}e^{-j\omega_0 n}.$$

Thus, the frequency-domain representation of a cosine of frequency ω_0 is

$$X(\omega)=\pi(\delta(\omega-\omega_0)+\delta(\omega+\omega_0)).$$

The magnitude and phase of $X(\omega)$ are

$$|X(\omega)|=\pi(\delta(\omega-\omega_0)+\delta(\omega+\omega_0))$$
$$\measuredangle X(\omega)=0$$

From Equation (3.46), filtering in the frequency domain is represented by multiplying the transform of the input, $X(\omega)$, by the frequency response of the system, $H(\omega)$, so

$$Y(\omega)=X(\omega)H(\omega)=\pi(\delta(\omega-\omega_0)+\delta(\omega+\omega_0))H(\omega)=\pi(H(\omega_0)\delta(\omega-\omega_0)+H(-\omega_0)\delta(\omega+\omega_0)).$$

In general, $H(\omega)$ is complex, so we can write

$$H(\omega)=|H(\omega)|e^{j\measuredangle H(\omega)},$$

which gives,

$$\begin{aligned}Y(\omega)&=\pi(|H(\omega_0)|e^{j\measuredangle H(\omega_0)}\delta(\omega-\omega_0)+|H(-\omega_0)|e^{j\measuredangle H(-\omega_0)}\delta(\omega+\omega_0))\\&=\pi(|H(\omega_0)|e^{j\measuredangle H(\omega_0)}\delta(\omega-\omega_0)+|H(\omega_0)|e^{-j\measuredangle H(\omega_0)}\delta(\omega+\omega_0)).\end{aligned}$$

Here, we have used the fact that for a real $h[n]$, the magnitude of the transform is even, $|H(-\omega_0)|=|H(\omega_0)|$, and the phase is odd, $\measuredangle H(-\omega_0)=-\measuredangle H(\omega_0)$. The magnitude and phase are readily seen to be

$$|Y(\omega)|=\pi|H(\omega_0)|(\delta(\omega-\omega_0)+\delta(\omega+\omega_0))$$

$$\measuredangle Y(\omega)=\begin{cases}\measuredangle H(\omega_0), & \omega=\omega_0\\ -\measuredangle H(\omega_0), & \omega=-\omega_0\end{cases}$$

Comparing this with Equation (3.42), we recognize that the inverse transform,

$$y[n] = \mathfrak{F}^{-1}\{Y(\omega)\} = |H(\omega_0)|\cos(\omega_0 n + \measuredangle H(\omega_0)),$$

is the same result we obtained with the time-domain convolution method, Equation (3.49).

The relation between the magnitude and phase of $X(\omega)$, $H(\omega)$ and $Y(\omega)$ is given by Equation (3.47),

$$|Y(\omega)| = |X(\omega)||H(\omega)|$$
$$\measuredangle Y(\omega) = \measuredangle X(\omega) + \measuredangle H(\omega),$$

and is plotted in **Figures 3.36b** and **c**. This plot illustrates the central idea of filtering in the frequency domain: the effect of the filter is to multiply the magnitude of the input by $|H(\omega)|$ and add $\measuredangle H(\omega)$ to the phase. In this example, the input, $X(\omega)$, comprises impulses at only two frequencies, $\pm\omega_0$, which have area π and phase 0. Thus, the filter only has effect at these two frequencies, indicated by red lines on the plots. The filter multiplies the area of the impulses by $|H(\omega_0)|$ and adds $\pm\measuredangle H(\omega_0)$ to their phase.

If you like, you can think of the filtration of a cosine this way: a cosine at frequency ω_0 comprises a pair of impulses that "probe" the frequency response of the filter at frequencies $\omega = \pm\omega_0$. The DTFT of the output provides information about the magnitude and phase of the filter at exactly these frequencies and nowhere else: $H(\omega_0)$ and $H(-\omega_0)$. The corresponding time-domain response of the filter also contains information at exactly these frequencies.

Example 3.32

Find the response of the same system as Example 3.31, defined by the impulse response $h[n] = \frac{1}{2}^n u[n]$, to an impulse $x[n] = \delta[n]$, using time-domain and frequency-domain solution techniques.

➤ Solution:

By definition, the response of the system to an impulse is, of course, the impulse response:

$$y[n] = x[n] * h[n] = \delta[n] * h[n] = h[n].$$

So $y[n] = h[n]$, as shown in **Figure 3.37a**.

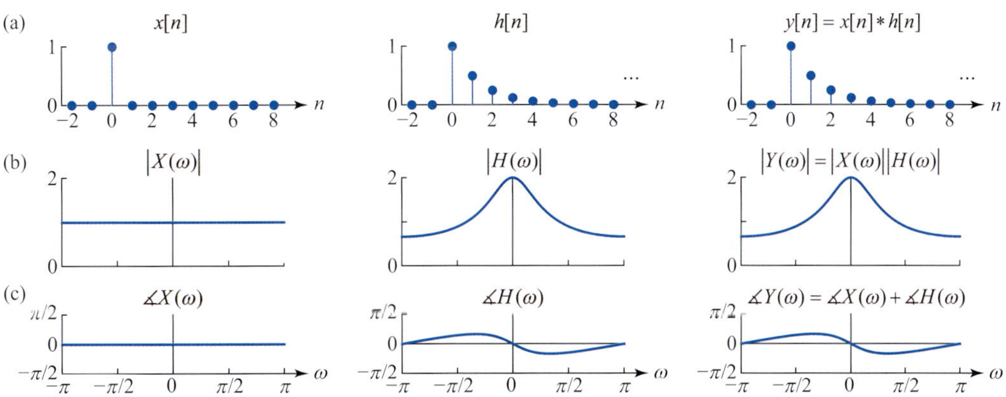

Figure 3.37 Filtering an impulse

The frequency-domain interpretation of filtering is shown graphically in **Figures 3.37b** and **c**. The DTFT of the input is a constant, $X(\omega) = \mathfrak{F}\{\delta[n]\} = 1$, which means that $|X(\omega)| = 1$ and $\measuredangle X(\omega) = 0$. Filtering in the frequency domain is understood as multiplication of $X(\omega)$ by system function $H(\omega)$, which is defined as the DTFT of $h[n]$. In this case,

$$Y(\omega) = X(\omega)H(\omega) = 1 \cdot H(\omega) = H(\omega),$$

which means that

$$|Y(\omega)| = |X(\omega)||H(\omega)| = 1 \cdot |H(\omega)| = |H(\omega)|$$
$$\measuredangle Y(\omega) = \measuredangle X(\omega) + \measuredangle H(\omega) = 0 + \measuredangle H(\omega) = \measuredangle H(\omega).$$

This section can be summarized as follows: the main purpose of a filter is to alter the frequency components of the input signal. The filter may increase (amplify) or decrease (attenuate) the magnitude of a particular frequency component or components in the signal, and/or it may change the phase of the components. The effect of filtering is usually most clearly understood in the frequency domain using the multiplication of transforms.

3.10.9 Multiplication (windowing) property

The multiplication property is the "dual" of the convolution property. This property underlies many important DSP tasks, including the construction of FIR filters using windows, which we will discuss in Chapter 7. It also gives the theoretical basis for understanding the sampling theorem, which we will study in Chapter 6.

The multiplication property states that if $X(\omega)$ is the DTFT of sequence $x[n]$, and $W(\omega)$ is the DTFT of sequence $w[n]$, then the DTFT of the product of $x[n]$ and $w[n]$ is the convolution of their DTFTs. That is,

$$\text{If } x[n] \longleftrightarrow X(\omega) \text{ and } w[n] \longleftrightarrow W(\omega), \text{ then } x[n]w[n] \longleftrightarrow \tfrac{1}{2\pi} X(\omega) * W(\omega).$$

In words, multiplication of sequences in the time domain corresponds to convolution of their DTFTs in the frequency domain. The proof of this follows:

$$Y(\omega) = \sum_{n=-\infty}^{\infty} y[n]e^{-j\omega n} = \sum_{n=-\infty}^{\infty} x[n]w[n]e^{-j\omega n}.$$

But,

$$x[n] = \frac{1}{2\pi} \int_{\xi=-\pi}^{\pi} X(\xi)e^{j\xi n} d\xi,$$

so

$$Y(\omega) = \sum_{n=-\infty}^{\infty}\left(\frac{1}{2\pi}\int_{\xi=-\pi}^{\pi} X(\xi)e^{j\xi n}d\xi\right)w[n]e^{-j\omega n} = \frac{1}{2\pi}\int_{\xi=-\pi}^{\pi} X(\xi)\left(\sum_{n=-\infty}^{\infty} w[n]e^{-j(\omega-\xi)n}\right)d\xi$$
$$= \frac{1}{2\pi}\int_{\xi=-\pi}^{\pi} X(\xi)W(\omega-\xi)d\xi \triangleq \frac{1}{2\pi}X(\omega)*W(\omega). \tag{3.50}$$

We have to define exactly what we mean by convolution in the discrete-time frequency domain. The convolution integral of Equation (3.50) differs from a standard convolution in the continuous-time frequency domain in that ξ only ranges from $-\pi \leq \xi < \pi$ instead of $-\infty \leq \xi < \infty$. If the convolution integral were evaluated over the range $-\infty \leq \xi < \infty$, it would not be bounded because both $X(\xi)$ and $W(\xi)$ are periodic in ξ. Accordingly, we *define* convolution in the discrete-time frequency domain to span just one period of $X(\xi)$ and $W(\xi)$, that is, from $-\pi \leq \xi < \pi$.

One of the most common uses of windowing is to truncate an infinite-length sequence $x[n]$ by **windowing** or multiplying it by a finite-length sequence $w[n]$. The multiplication property allows us to derive the DTFT of the windowed sequence from that of the infinite-length sequence. Let us start with a very simple example.

Example 3.33

Use the windowing property to find the DTFT of a centered symmetric pulse of odd length N,

$$y[n] = \begin{cases} 1, & |n| \leq (N-1)/2 \\ 0, & \text{otherwise} \end{cases}.$$

➤ Solution:

Let $x[n]=1$ be a constant, infinite-length sequence, as shown in **Figure 3.38a**. The constant is multiplied by $w[n]$, a symmetric window of length $N=11$, yielding $y[n]=x[n]w[n]$. Since $x[n]-1$, then $y[n]-x[n]w[n]=1\cdot w[n]=w[n]$, so the output is equal to the input.

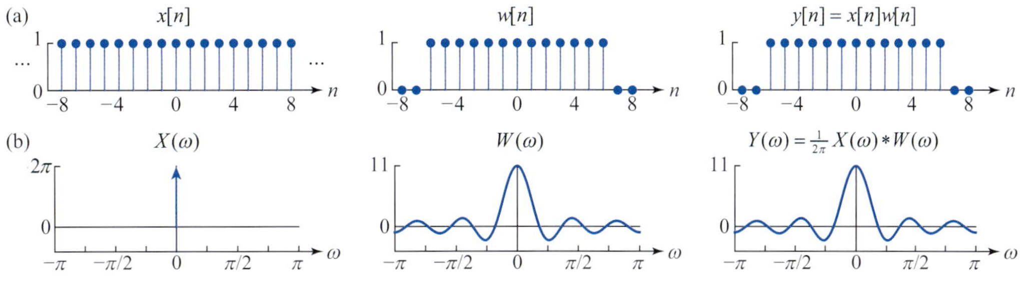

Figure 3.38 Windowing of a constant sequence

Figure 3.38b shows the effect of windowing in the frequency domain. The DTFT of a constant is an impulse in frequency centered at $\omega=0$: $X(\omega)=2\pi\delta(\omega)$, where the plot shows only the principal frequency range $-\pi \le \omega < \pi$, and the DTFT of the $w[n]$ pulse is the periodic sinc function as discussed in connection with Example 3.32 and derived in Equation (3.14),

$$W(\omega) = N\frac{\operatorname{sinc} N\omega/2}{\operatorname{sinc}\omega/2}.$$

The DTFT of the output is the convolution in frequency of the impulse and the periodic sinc function, which yields the periodic sinc function again. The main conclusion is that the act of windowing in the time domain has the effect of "spreading" the spectrum in the frequency domain. The spectrum of the constant input is an impulse at a single frequency, $\omega = 0$. When the input is windowed, this impulse is convolved with a function that has energy at all frequencies $-\pi \le \omega < \pi$; thus, the output contains energy at all frequencies.

One important use of the multiplication property is in a practical application: using windowing to derive an FIR lowpass filter from the infinite impulse response of an ideal lowpass filter. We have previously noted that an ideal lowpass filter is unrealizable because its impulse response is of infinite time duration. One fruitful approach to generating a practical lowpass filter is to truncate the infinite impulse response by multiplying it by a specially designed window function of finite duration. This is a topic that will be covered in detail in Chapter 7. The following example gives a preview of how this approach works.

Example 3.34

Design a symmetric FIR lowpass filter with **cutoff frequency** $\omega_c = \pi/2$ by windowing the impulse response of an ideal lowpass filter with a rectangular window of length $N = 13$.

➤ Solution:

The impulse response of an ideal lowpass filter with cutoff frequency ω_c is $x[n] = (\omega_c/\pi)\operatorname{sinc}\omega_x n$, as shown in the top panel of **Figure 3.39a** for $\omega_c = \pi/2$. The response is symmetric about $n = 0$ and of infinite time duration. To create a finite response filter, we truncate the infinite response by multiplying by a symmetric rectangular window of length $N = 13$, shown in **Figure 3.39b**, yielding the response shown in **Figure 3.39c**.

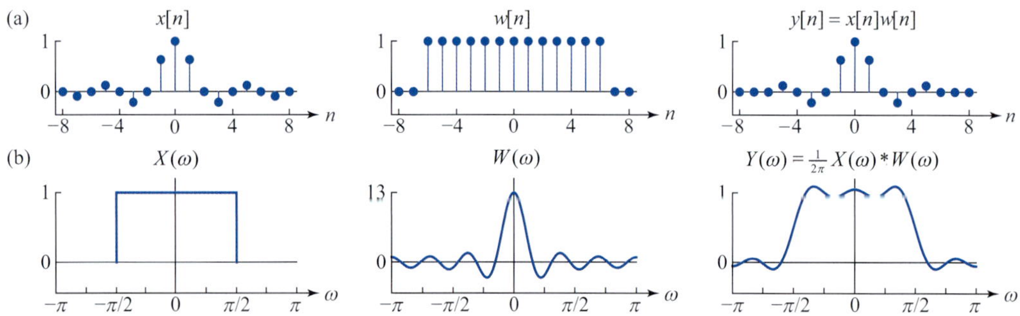

Figure 3.39 Windowing of the impulse response of an ideal lowpass filter

As shown in the bottom panels of **Figure 3.39**, multiplication of $x[n]$ by $w[n]$ yielding $y[n] = x[n]w[n]$ corresponds to the convolution of $X(\omega)$ with $W(\omega)$ yielding $Y(\omega) = (1/2\pi)X(\omega) * H(\omega)$. The frequency response of the ideal lowpass filter is

$$X(\omega) = \begin{cases} 1, & |\omega| < \omega_c \\ 0, & \omega_c < |\omega| < \pi \end{cases}.$$

The fact that this filter is ideal means that (a) the magnitude of the response in the **passband** $|\omega| < \omega_c$ is exactly one; (b) the magnitude of the response in the **stopband** $\omega_c < |\omega| < \pi$ is exactly zero; and (c) the slope of the response at the cutoff frequency ω_c is infinitely steep.

The DTFT of a rectangular window of length N is the periodic sinc function of Equation (3.14). The frequency response of the truncated window, $Y(\omega)$, represents an ideal filter, $X(\omega)$, whose ideal characteristics have been modified by convolution with $W(\omega)$. The response of filter $Y(\omega)$ is non-ideal because (a) the magnitude of the response in the passband is no longer exactly one; (b) the magnitude of the response in the stopband is non-zero; and (c) the slope of the response at the cutoff frequency is finite.

The filter generated by this windowing procedure is clearly a lowpass FIR filter. It turns out that it is not a very *good* lowpass FIR filter, but in Chapter 7 we will discuss a number of ways to use this same windowing approach to design a much better FIR filter; specifically, we investigate what happens when we change the shape and length of the windowing sequence $w[n]$.

3.10.10 ★ Time- and band-limited systems

The preceding two examples of the multiplication property now lead us to make some very fundamental points about the relation between the time and frequency domains of discrete-time systems. First, we need to define a couple of concepts:

- A **time-limited system** is one for which the impulse response is finite in time. That is, there exists some time n_{max}, which represents the maximum absolute time at which $h[n]$ has non-zero value, so $h[n] = 0,\ |n| > n_{max}$.
- A **band-limited system** is one for which there exists some frequency ω_h that represents the maximum absolute frequency at which $H(\omega)$ has non-zero energy,[3] so $H(\omega) = 0$, $\omega_h < |\omega| < \pi$.

Using the modulation and convolution properties, we can now show the truth of the following two statements:

> A time-limited system cannot be band-limited.
> A band-limited system cannot be time-limited.

A time-limited system cannot be band-limited In any real-world application, we turn on our system at some time and turn it off some time later; hence, every signal has a finite duration. We will now show that any signal with a finite duration cannot be band-limited; it is guaranteed to have energy at all discrete-time frequencies $-\pi \le \omega < \pi$. The proof follows directly from the multiplication property. Assume that there exists a system with impulse response $h[n]$ and frequency response $H(\omega)$, which is both time-limited *and* band-limited as shown in **Figure 3.40a** (time and frequency domains on the top and bottom panels, respectively). Since $h[n]$ is time-limited, it has non-zero values restricted to some range of n, say $|n| \le n_{max}$.

[3] We could also make the same arguments by assuming that $H(\omega)$ has a highpass characteristic, whereby there exists some frequency ω_l, such that $H(\omega) = 0$ for $|\omega| < \omega_l$.

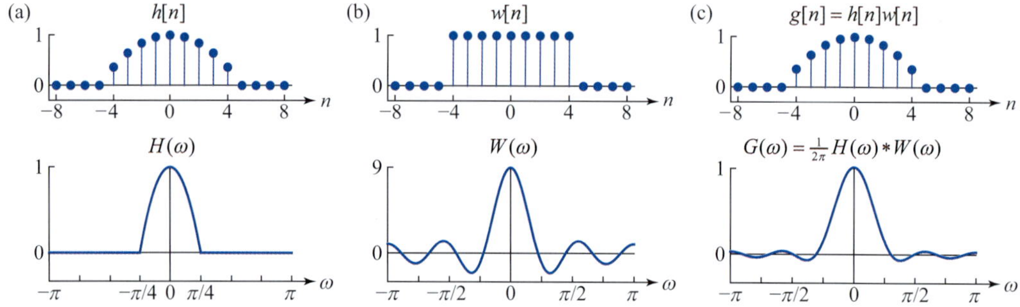

Figure 3.40 A time-limited system cannot be band-limited

If we multiply $h[n]$ by a window $w[n]$ (top panel of **Figure 3.40b**), which has value one in the same range of n as $h[n]$ and zero outside, the result $g[n]$ (top panel of **Figure 3.40c**), is identical to $h[n]$: $g[n] = h[n]w[n] = h[n]$. That means that in the frequency domain, $G(\omega)$, the DTFT of the output, should be equal to $H(\omega)$. It is clearly not. Why not? By the multiplication property, multiplying $h[n]$ by the window $w[n]$ corresponds in the frequency domain to convolving $H(\omega)$ (bottom panel of **Figure 3.40a**), which we have assumed to be band-limited, with $W(\omega)$ (bottom panel of **Figure 3.40b**), which is a periodic sinc function of the form of Equation (3.14). Since $W(\omega)$ is not band-limited, the convolution with $H(\omega)$, which is band-limited, yields a function $G(\omega)$ (bottom panel of **Figure 3.40c**), which is not band-limited. Hence, our initial assumption that $h[n]$ is both time-limited and band-limited is incorrect: a time-limited system cannot be band-limited as well.

A band-limited system cannot be time-limited In an ideal world, we would like to create sharp filters that allow signals to pass in a well-defined range of frequencies and completely attenuate signals in another range of frequencies. For example, the frequency response of an ideal lowpass filter with bandwidth ω_x is zero outside the range $|\omega| < \omega_c$, where ω_c represents the cutoff frequency. However, it is easy to show that the impulse response of any band-limited system cannot be time-limited; in fact, the impulse response is guaranteed to have infinite duration, $-\infty < n < \infty$, and is therefore unrealizable. The proof follows from the convolution property. Again, assume that there exists a system with impulse response $h[n]$ (top panel of **Figure 3.41a**), which is time-limited to $|n| \leq n_{\max}$, and frequency response $H(\omega)$ (bottom panel of **Figure 3.41a**), which is band-limited to $|\omega| < \omega_c$.

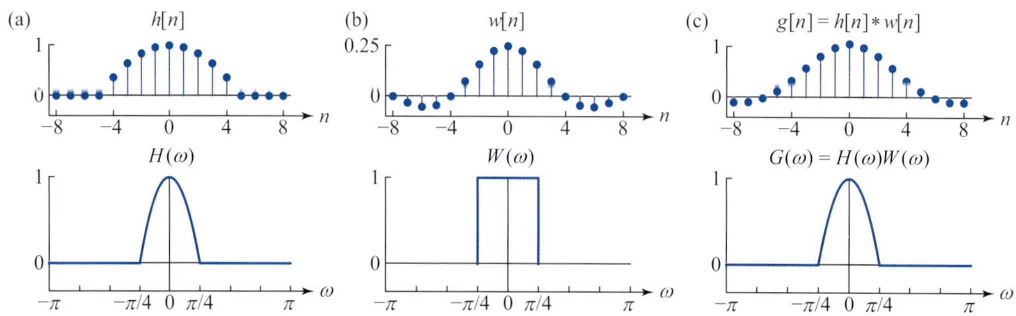

Figure 3.41 A band-limited system cannot be time-limited

Multiplying $H(\omega)$ by the frequency response of an ideal lowpass filter $W(\omega)$ (bottom panel of **Figure 3.41b**) results in response $G(\omega)=H(\omega)W(\omega)$ (bottom panel of **Figure 3.41c**). Because the bandwidth of $G(\omega)$ is equal to ω_c, this multiplication should have no effect, and $G(\omega)$ should be identical to $H(\omega)$: $G(\omega)=H(\omega)$. In the time domain, this should mean that $g[n]$ is equal to $h[n]$. It is not. Multiplication of DTFTs corresponds to convolution of $h[n]$ with $w[n]$ (top panel of **Figure 3.41b**), yielding $g[n]$ (top panel of **Figure 3.41c**): $g[n]=h[n]*w[n]$. However, the convolution of a sequence $h[n]$, which we have assumed to be time-limited, with one that is not time-limited, $w[n]$, results in a sequence $g[n]$ that is not time-limited. Accordingly, it is not possible for a band-limited function to be time-limited as well.

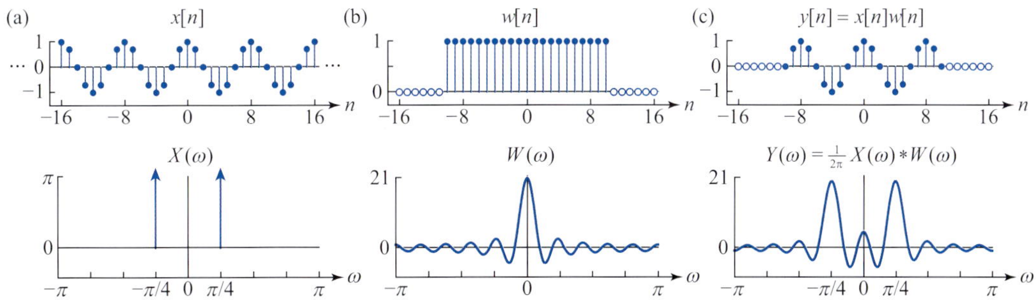

Figure 3.42 Spectral ambiguity resulting from the windowing of a cosine

3.10.11 ★ Spectral and temporal ambiguity

Spectral analysis is one of the most important applications of DSP, one that we will discuss in detail in Chapter 14. It is the process of measuring, estimating and characterizing the frequency content of signals, and finds application in fields as diverse as astrophysics, medicine, chemistry and oil exploration. Spectral analysis also provides an example of the trade-off between spectral and temporal resolution that occurs in the practical measurements of spectra in the real world.

The top panel of **Figure 3.42a** shows a cosine of infinite duration, $x[n]=\cos \pi n/4$, which might represent the signal whose spectrum we are trying to measure. This signal is completely unlocalized in time, meaning that it extends from $-\infty<n<\infty$; however, its spectrum $X(\omega)$ (bottom panel of **Figure 3.42a**) is highly localized to exactly two frequencies, $\omega=\pm\pi/4$. In a practical spectral measurement, we can only measure a finite-length piece of this signal. Accordingly, we will model this finite-length signal as the multiplication of $x[n]$ by a data window of finite duration, $w[n]$. The top panel of **Figure 3.42b** shows a 21-point window $w[n]$, whose spectrum $W(\omega)$ (bottom panel) is a periodic sinc. The top panel of **Figure 3.42c** shows the resulting windowed, finite-duration cosine, $y[n]=x[n]w[n]$. Its spectrum $Y(\omega)$ (bottom panel), is obtained by convolving the two impulses of $X(\omega)$ with the periodic sinc $W(\omega)$,

$$Y(\omega)=\frac{1}{2\pi}X(\omega)*W(\omega).$$

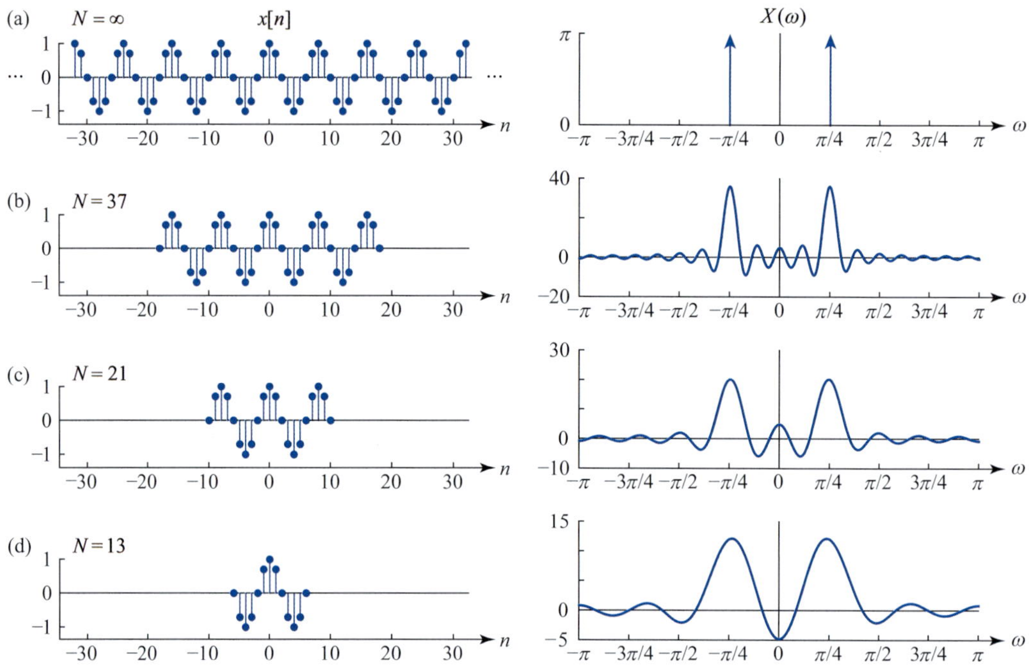

Figure 3.43 Windowing of cosines

The result of this convolution is a signal that is localized in time, but whose spectrum is "smeared out" in frequency; it has two relatively broad peaks and energy spread over all frequencies. **Figure 3.43** shows the consequence of windowing the ideal cosine $x[n]$ with narrower and narrower windows. The conclusion of these two figures is that there is a trade-off between resolution in the time and the frequency domains; as we increase the localization of the signal in the time domain, we reduce the localization in the frequency domain. In Chapter 14, we will discuss this trade-off in detail and see how spectral measurements are affected by factors such as the size and shape of the data windows and the presence of noise.

3.10.12 Time-reversal property

If a sequence $x[n]$ has transform

$$X(\omega) = \sum_{n=-\infty}^{\infty} x[n]e^{-j\omega n},$$

then the sequence $x[-n]$ has transform $X(-\omega)$:

$$\sum_{n=-\infty}^{\infty} x[-n]e^{-j\omega n} = \sum_{n=\infty}^{-\infty} x[n]e^{-j(-\omega)n} = X(-\omega).$$

Example 3.35

A Bartlett (triangular) window, $b_{2N-1}[n]$, of length $2N-1$ is defined as

$$b_{2N-1}[n] = \begin{cases} \dfrac{N-|n|}{N}, & 0 \le |n| \le N-1 \\ 0, & \text{otherwise} \end{cases}.$$

As shown in **Figure 3.44a**, this window can be created by convolving two identical rectangular windows, $w_N[n]$, of length N,

$$b_{2N-1}[n] = \frac{1}{N} w_N[n] * w_N[-n],$$

where

$$w_N[n] = \begin{cases} 1, & 0 \le n \le N-1 \\ 0, & \text{otherwise} \end{cases}.$$

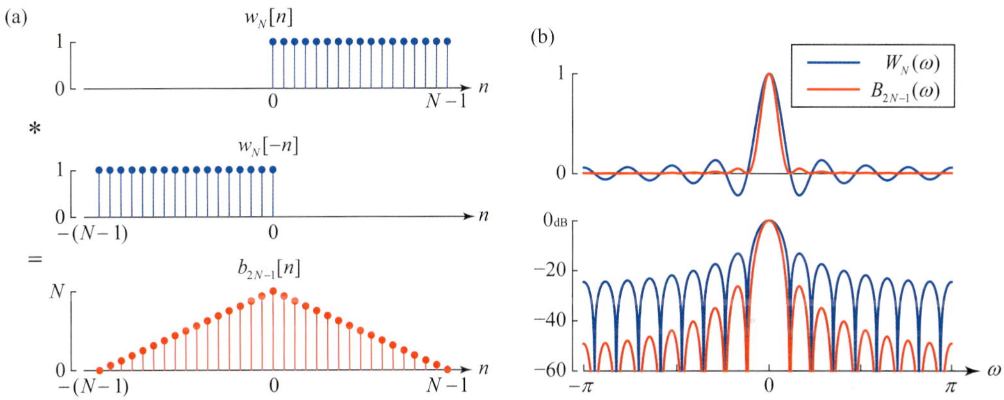

Figure 3.44 Construction of the Bartlett window

Find $B_{2N-1}(\omega)$, the DTFT of $b_{2N-1}[n]$.

➤ Solution:

The DTFT of $w_N[n]$ is

$$W_N(\omega) = \sum_{n=0}^{N-1} e^{-j\omega n} = \frac{1-e^{-j\omega N}}{1-e^{-j\omega}} = e^{-j\omega N/2}\left(\frac{\sin \omega N/2}{\sin \omega/2}\right).$$

By the multiplication and time-reversal properties,

$$\begin{aligned} B_{2N-1}(\omega) &= \mathfrak{F}\{b_{2N-1}[n]\} = \mathfrak{F}\left\{\frac{1}{N} w_N[n] * w_N[-n]\right\} = \frac{1}{N}\mathfrak{F}\{w_N[n]\}\cdot\mathfrak{F}\{w_N[-n]\} = \frac{1}{N} W_N(\omega) W_N(-\omega) \\ &= \frac{1}{N}\left(\frac{\sin \omega N/2}{\sin \omega/2}\right)^2. \end{aligned}$$

Figure 3.44b shows the relation between $B_{2N-1}(\omega)$ and $W_N(\omega)$ on both a linear scale (top) and a dB scale (bottom). See Problem 3-23 for another approach to this solution using other DTFT properties.

3.10.13 Differentiation property

If a sequence $x[n]$ has transform

$$X(\omega) = \sum_{n=-\infty}^{\infty} x[n]e^{-j\omega n},$$

then the sequence $nx[n]$ has transform

$$j\frac{dX(\omega)}{d\omega} = \sum_{n=-\infty}^{\infty} nx[n]e^{-j\omega n}.$$

Example 3.36

Find the DTFT of a ramp of length N,

$$r_N[n] = \begin{cases} n, & 0 \le n \le N-1 \\ 0, & \text{otherwise} \end{cases}.$$

➤ **Solution:**

Let $w_N[n]$ be a rectangular pulse of length N, as shown in the top panel of **Figure 3.44a**. Then the ramp $r_N[n] = nw_N[n]$ has transform

$$R_N(\omega) = j\frac{dW_N(\omega)}{d\omega} = -\frac{Ne^{-j\omega N}(1-e^{-j\omega}) - e^{-j\omega}(1-e^{-j\omega N})}{(1-e^{-j\omega})^2} = \frac{Ne^{-j\omega(N-1/2)}\sin\omega/2 - e^{-j\omega N/2}\sin\omega N/2}{2j\sin^2\omega/2}, \quad \omega \neq 0$$

$$R_N(0) = \sum_{n=0}^{N-1} ne^{-j0n} = \sum_{n=0}^{N-1} n = \frac{N(N-1)}{2}, \quad \omega = 0.$$

3.10.14 Parseval's theorem

We have shown that we can represent a sequence $x[n]$ as the integral of scaled orthogonal frequency components,

$$x[n] = \frac{1}{2\pi}\int_{-\pi}^{\pi} X(\omega)e^{j\omega n}d\omega. \tag{3.51}$$

The orthogonality of the $e^{j\omega n}$ functions makes it possible to relate the energy in a time signal to the energy of its frequency components. The energy of the time signal is defined as

$$E_x = \sum_{n=-\infty}^{\infty} x[n]x^*[n] = \sum_{n=-\infty}^{\infty} |x[n]|^2.$$

Substituting Equation (3.51) for $x[n]$ gives

$$\begin{aligned} E_x &= \sum_{n=-\infty}^{\infty} x[n]x^*[n] = \sum_{n=-\infty}^{\infty} x[n]\left(\frac{1}{2\pi}\int_{-\pi}^{\pi} X(\omega)e^{j\omega n}d\omega\right)^* = \sum_{n=-\infty}^{\infty} x[n]\left(\frac{1}{2\pi}\int_{-\pi}^{\pi} X^*(\omega)e^{-j\omega n}d\omega\right) \\ &= \frac{1}{2\pi}\int_{-\pi}^{\pi} X^*(\omega)\left(\sum_{n=-\infty}^{\infty} x[n]e^{-j\omega n}\right)d\omega = \frac{1}{2\pi}\int_{-\pi}^{\pi} X^*(\omega)X(\omega)d\omega = \frac{1}{2\pi}\int_{-\pi}^{\pi} |X(\omega)|^2 d\omega. \end{aligned}$$

Hence, we arrive at Parseval's theorem,

$$E_x = \sum_{n=-\infty}^{\infty} |x[n]|^2 = \frac{1}{2\pi} \int_{-\pi}^{\pi} |X(\omega)|^2 d\omega,$$

which states that the sum of the energy in the time sequence is equal to the integral of the energy in all the frequency components. We will have occasion to use this relation when we discuss the least-square FIR filter in Chapter 7.

3.10.15 DC- and π-value properties

The DC-value property states that the DC component of the DTFT is the sum of all the values of the sequence. The DC component of the DTFT is $X(0)$, the value at $\omega = 0$ (i.e., DC):

$$X(0) = X(\omega)|_{\omega=0} = \sum_{n=-\infty}^{\infty} x[n]e^{-j0n} = \sum_{n=-\infty}^{\infty} x[n].$$

The π-value property states that

$$X(\pm\pi) = \sum_{n=-\infty}^{\infty} x[n]e^{\mp j\pi n} = \sum_{n=-\infty}^{\infty} x[n](-1)^n.$$

These two properties have some interesting consequences that we will use when we come to the design of the FIR filter in Chapter 7. If a sequence is odd, then both $X(0)$ and $X(\pi)$ must both be zero. If you look back at Example 3.15, you will see that this is true.

3.10.16 ★ Using the DTFT to solve linear constant-coefficient difference equations

The properties of the DTFT can be applied to the solution of linear constant-coefficient difference equations (LCCDEs). While we will most often favor using the z-transform, which we will introduce in Chapter 4 to solve these equations, you can also use the DTFT, as shown here.

Example 3.37

Consider a system whose input–output relation is defined by the first-order difference equation

$$y[n] - \alpha y[n-1] = x[n],$$

where a is a real constant, which may be either positive or negative. Find the impulse response of this system, $h[n]$.

➤ Solution:

Taking the DTFT of both sides of the equation:

$$\mathfrak{F}\{y[n] - \alpha y[n-1]\} = \mathfrak{F}\{x[n]\}.$$

By the linearity property,

$$\mathfrak{F}\{y[n]\} - \alpha\mathfrak{F}\{y[n-1]\} = \mathfrak{F}\{x[n]\}.$$

Now, $\mathfrak{F}\{y[n]\} = Y(\omega)$ and, by the shifting property, $\mathfrak{F}\{y[n-1]\} = Y(\omega)e^{-j\omega}$, so

$$Y(\omega) - \alpha Y(\omega)e^{-j\omega} = X(\omega).$$

Collecting terms on the left-hand side gives the relation between $Y(\omega)$ and $X(\omega)$:

$$Y(\omega)(1 - \alpha e^{-j\omega}) = X(\omega). \tag{3.52}$$

For a linear time-invariant system with input $x[n]$, impulse response $h[n]$ and output $y[n]$, convolution gives $y[n] = x[n] * h[n]$. The convolution property, Equation (3.46), gives the relation between the DTFTs: $Y(\omega) = X(\omega)H(\omega)$. Therefore, $H(\omega)$ is obtainable as the ratio of $Y(\omega)$ and $X(\omega)$:

$$H(\omega) = \frac{Y(\omega)}{X(\omega)}. \tag{3.53}$$

Combining Equations (3.52) and (3.53) gives

$$H(\omega) = \frac{Y(\omega)}{X(\omega)} = \frac{1}{1 - \alpha e^{-j\omega}}. \tag{3.54}$$

We have seen this transform before. It is the transform of the impulse response of the first-order lowpass filter. We state the inverse transform by inspection:

$$h[n] = \mathfrak{F}^{-1}\{H(\omega)\} = \alpha^n u[n].$$

We can use this method to find the impulse response of arbitrarily complicated LCCDEs; however, we will wait until we do the z-transform in Chapter 4. For now, here is another example.

Example 3.38

Find the impulse response of a system described by

$$y[n] - \tfrac{3}{4}y[n-1] + \tfrac{1}{8}y[n-2] = 2x[n].$$

➤ Solution:

Taking the transform of both sides, using the linearity property and recognizing that $\mathfrak{F}\{y[n-1]\} = Y(\omega)e^{-j\omega}$ and $\mathfrak{F}\{y[n-2]\} = Y(\omega)e^{-j2\omega}$, we get

$$Y(\omega) - \tfrac{3}{4}Y(\omega)e^{-j\omega} + \tfrac{1}{8}Y(\omega)e^{-j2\omega} = 2X(\omega)$$
$$Y(\omega)(1 - \tfrac{3}{4}e^{-j\omega} + \tfrac{1}{8}e^{-j2\omega}) = 2X(\omega).$$

Taking the ratio of $Y(\omega)$ and $X(\omega)$ gives the frequency response

$$H(\omega) = \frac{Y(\omega)}{X(\omega)} = \frac{2}{1 - \tfrac{3}{4}e^{-j\omega} + \tfrac{1}{8}e^{-j2\omega}}.$$

The denominator is a polynomial in powers of $e^{-j\omega}$, so we can factor it into the product of two factors and then separate the sum of two terms using the partial fraction technique:

$$H(\omega) = \frac{2}{1-\frac{3}{4}e^{-j\omega}+\frac{1}{8}e^{-j2\omega}} = \frac{2}{(1-\frac{1}{2}e^{-j\omega})(1-\frac{1}{4}e^{-j\omega})} = \frac{4}{(1-\frac{1}{2}e^{-j\omega})} - \frac{2}{(1-\frac{1}{4}e^{-j\omega})}.$$

Each of these two terms should be familiar as the transform of a sequence of the form $\alpha^n u[n]$. So, taking the inverse transform yields

$$h[n] = \mathfrak{F}^{-1}\{H(\omega)\} = \mathfrak{F}^{-1}\left\{\frac{4}{1-\frac{1}{2}e^{-j\omega}}\right\} - \mathfrak{F}^{-1}\left\{\frac{2}{1-\frac{1}{4}e^{-j\omega}}\right\} = 4\cdot\tfrac{1}{2}^n u[n] - 2\cdot\tfrac{1}{4}^n u[n].$$

3.10.17 Summary of DTFT properties

Table 3.5 gives a summary of the main properties of the DTFT.

Table 3.5 Main properties of the DTFT

Property	Time domain	Frequency domain
Transform	$x[n] = \frac{1}{2\pi}\int_{\omega=-\pi}^{\pi} X(\omega)e^{j\omega n}d\omega$	$X(\omega) = \sum_{n=-\infty}^{\infty} x[n]e^{-j\omega n}$
Linearity	$y[n] = a_1x_1[n] + a_2x_2[n]$	$Y(\omega) = a_1X_1(\omega) + a_2X_2(\omega)$
Time shift	$y[n] = x[n-n_0]$	$Y(\omega) = X(\omega)e^{-j\omega n_0}$ $\lvert Y(\omega)\rvert = \lvert X(\omega)\rvert$ $\measuredangle Y(\omega) = \measuredangle X(\omega) - \omega n_0$
Complex modulation (Frequency shift)	$y[n] = x[n]e^{j\omega_0 n}$	$Y(\omega) = X(\omega-\omega_0)$
Modulation	$y[n] = x[n]\cos\omega_0 n$	$Y(\omega) = \frac{1}{2}X(\omega-\omega_0) + \frac{1}{2}X(\omega+\omega_0)$
	$y[n] = x[n]\sin\omega_0 n$	$Y(\omega) = \frac{1}{2j}X(\omega-\omega_0) - \frac{1}{2j}X(\omega+\omega_0)$
	$y[n] = x[n]\cos(\omega_0 n+\varphi)$	$Y(\omega) = \frac{1}{2}e^{j\varphi}X(\omega-\omega_0) + \frac{1}{2}e^{-j\varphi}X(\omega+\omega_0)$
Convolution	$y[n] = x[n] * h[n]$	$Y(\omega) = X(\omega)H(\omega)$
Multiplication (Windowing)	$y[n] = x[n]w[n]$	$Y(\omega) = \frac{1}{2\pi}X(\omega) * W(\omega)$

3.11 ★ The relation between the DTFT and the Fourier series

For those who remember their continuous-time signal processing, we would be remiss not to point out the correspondence between the DTFT and the Fourier series. Consider a continuous-time function $x(t)$ that is periodic with period T. This function can be expressed as the sum of continuous-time complex exponential basis functions $e^{jn\Omega_0 t}$, whose frequencies are multiples of the fundamental frequency $\Omega_0 = 2\pi/T$:

$$x(t) = \sum_{n=-\infty}^{\infty} x_n e^{jn\Omega_0 t}.$$

That is the synthesis relation of the Fourier series, where the Fourier series coefficients x_n are computed from $x(t)$ by the analysis integral,

$$x_n = \frac{1}{T} \int_{t=0}^{T} x(t) e^{-jn\Omega_0 t} dt.$$

These equations are completely analogous to those that define the DTFT, as shown in **Table 3.6**.

Table 3.6 Comparison of Fourier series and DTFT

Fourier series	DTFT
$x(t) = \sum_{n=-\infty}^{\infty} x_n e^{jn\Omega_0 t}$	$X(\omega) \triangleq \sum_{n=-\infty}^{\infty} x[n] e^{-j\omega n}$
$x_n = \frac{1}{T} \int_{t=0}^{T} x(t) e^{-jn\Omega_0 t} dt$	$x[n] = \frac{1}{2\pi} \int_{-\pi}^{\pi} X(\omega) e^{j\omega n} d\omega$

The continuous-time variable t is analogous to the discrete-time variable ω. The Fourier series coefficients x_n are analogous to the sequence $x[n]$. The time interval T is analogous to the discrete-frequency interval 2π, so the fundamental frequency $\Omega_0 = 2\pi/T$ in the expression for the Fourier series becomes $2\pi/2\pi = 1$ in the discrete-time expressions.

SUMMARY

The DTFT is the essential tool for understanding how signals and systems are represented in the frequency domain. This chapter has laid out how the DTFT is computed and how it can be plotted. The properties of the DTFT, such as the convolution property, are essential to understand filtering, deconvolution and system identification. The symmetry properties of the DTFT will provide the basis for our design of linear-phase FIR filters in Chapter 7. Symmetry considerations will also enable the design of fast Fourier transform algorithms in Chapter 11.

PROBLEMS

Problem 3-1

Given $x[n] = \sin \frac{\pi}{4} n$ and $h[n] = \delta[n] - \sqrt{2}\delta[n-1]$, find $y[n] = x[n] * h[n]$ using the properties of the DTFT. Express your answer as a single cosine of the form $y[n] = |Y| \cos(n + \measuredangle Y)$.

Problem 3-2

Given $x[n] = \cos\omega_0 n$, with ω_0 specified in each part below and $h[n] = \delta[n] + \delta[n-1] + \delta[n-2]$, find $y[n] = x[n] * h[n]$. Express your answer as a single cosine of the form $y[n] = A\cos(\omega_0 n + \theta)$.

(a) $\omega_0 = 0$.
(b) $\omega_0 = \pi/3$.
(c) $\omega_0 = \pi/2$.
(d) $\omega_0 = 2\pi/3$.
(e) $\omega_0 = \pi$.

Problem 3-3

Given $x[n] = \cos\omega_0 n$, with ω_0 specified in each part below and $h[n] = \delta[n] + \delta[n-2]$, find $y[n] = x[n] * h[n]$. Express your answer as a single cosine of the form $y[n] = A\cos(\omega_0 n + \theta)$.

(a) $\omega_0 = 0$.
(b) $\omega_0 = \pi/3$.
(c) $\omega_0 = \pi/2$.
(d) $\omega_0 = 2\pi/3$.
(e) $\omega_0 = \pi$.

Problem 3-4

Find and plot $|Y(\omega)|$ and $\measuredangle Y(\omega)$, the magnitude and phase of the DTFT of the pulse $y[n]$ shown in **Figure 3.45**. Make sure your plot of $\measuredangle Y(\omega)$ is appropriately phase wrapped.

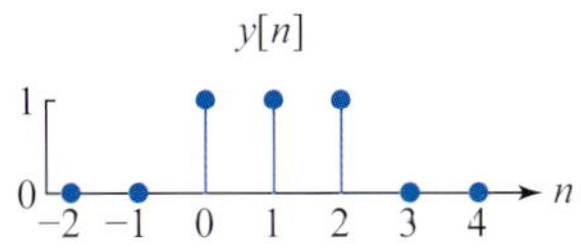

Figure 3.45

Problem 3-5

Find and plot $|Y(\omega)|$ and $\measuredangle Y(\omega)$, the magnitude and phase of the DTFT of the pulse $y[n]$ shown in **Figure 3.46**. Make sure your plot of $\measuredangle Y(\omega)$ is appropriately phase wrapped.

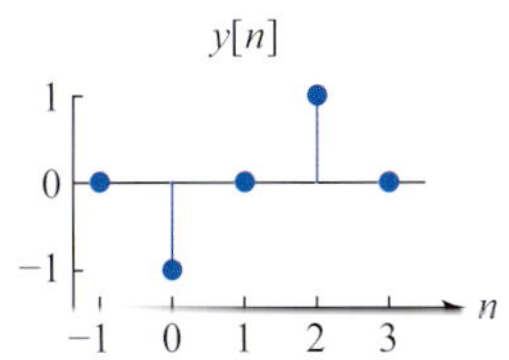

Figure 3.46

Problem 3-6

Find and plot $|H(\omega)|$ and $\measuredangle H(\omega)$, the magnitude and phase of the DTFT of the pulse $h[n]$ shown in **Figure 3.47**. Make sure your plot of $\measuredangle H(\omega)$ is appropriately phase wrapped.

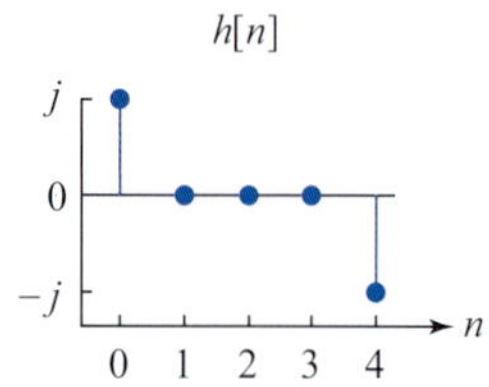

Figure 3.47

Problem 3-7

A system has a linear constant-coefficient difference equation given by

$$y[n] - y[n-1] = x[n].$$

This system is sometimes called quasi-stable because for certain inputs the output will be bounded, and for others it will explode. Find the output of this system when the input is

(a) $x[n] = \sin \pi n/2$.

(b) $x[n] = 1$ (i.e., $x[n] = \cos(0n)$).

Problem 3-8

Find and plot $|H(\omega)|$ and $\measuredangle H(\omega)$, the magnitude and phase of the DTFT of the pulse $h[n]$ shown in **Figure 3.48**.

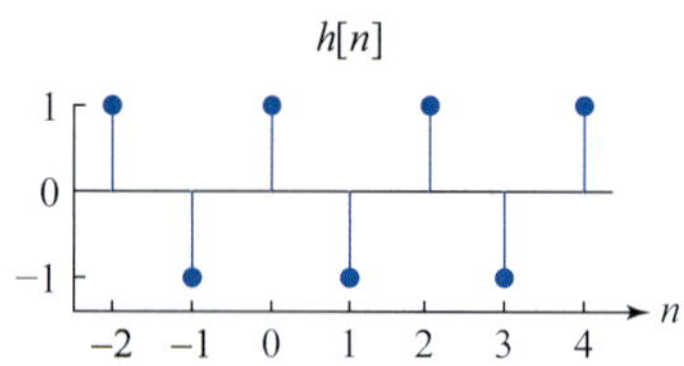

Figure 3.48

Make sure your plot of $\measuredangle H(\omega)$ is appropriately phase wrapped.

Problem 3-9

Find and plot $|H(\omega)|$ and $\measuredangle H(\omega)$, the magnitude and phase of the DTFT of the sequence $h[n]$ shown in **Figure 3.49**.

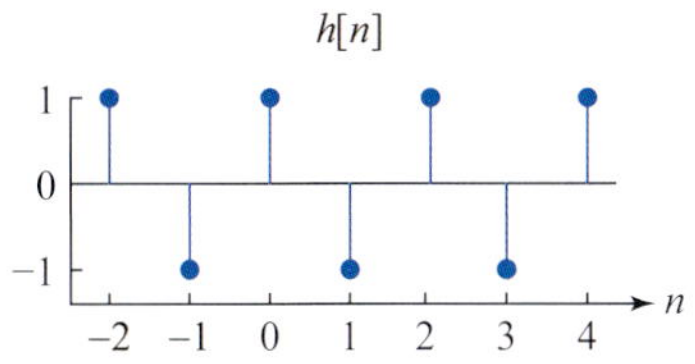

Figure 3.49

Make sure your plot of $\measuredangle H(\omega)$ is appropriately phase wrapped.

Problem 3-10

Find and plot $|H(\omega)|$ and $\measuredangle H(\omega)$, the magnitude and phase of the DTFT of the sequence $h[n] = -\delta[n+1] - \delta[n-3]$. Make sure your plot of $\measuredangle H(\omega)$ is appropriately phase wrapped.

Problem 3-11

Given a system with impulse response $h[n] = \delta[n] - \delta[n-1] + \delta[n-2]$, find the response of the system when $x[n] = \cos 2\pi n/3$.

Problem 3-12

Plot the magnitude and phase of $x[n] = j\delta[n] + j\delta[n-1] + j\delta[n-2]$.

Problem 3-13

Given the sequence shown in **Figure 3.50**, find the range of values of A such that the phase of the DTFT is zero for all ω.

➤**Hint:** If $x[n]$ is *real* and *even*, what can you say about $H(\omega)$ and, specifically, what can you say about the phase of the DTFT?

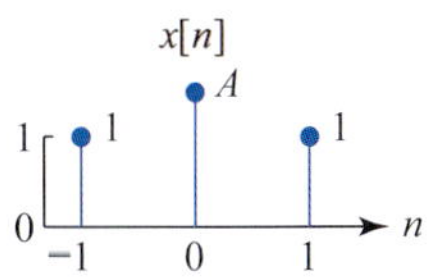

Figure 3.50

Problem 3-14

Repeat Problem 3-13 for the sequence shown in **Figure 3.51**.

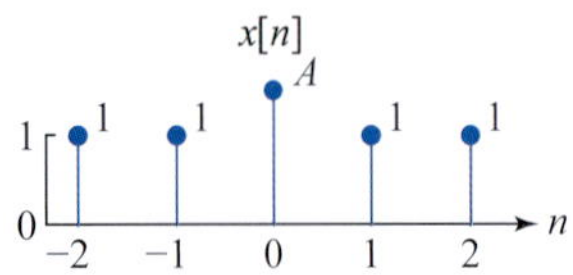

Figure 3.51

Problem 3-15

A system has a linear constant-coefficient difference equation given by $y[n] = x[n] - 2x[n-1] + x[n-2]$.

(a) Find $H(\omega)$.

(b) Find and make fully labeled plots of $|H(\omega)|$ and $\measuredangle H(\omega)$.

Problem 3-16

A system has a linear constant-coefficient difference equation given by $y[n] = -x[n] - x[n-2]$.

(a) Find $H(\omega)$.

(b) Find and make fully labeled plots of $|H(\omega)|$ and $\measuredangle H(\omega)$.

Problem 3-17

Let $x[n]$ be a real sequence, and let $y[n] = jx[n]$. Show that

$$|Y(\omega)| = |X(\omega)|$$
$$\measuredangle Y(\omega) = \measuredangle X(\omega) + \pi/2.$$

Problem 3-18

A system is defined by the linear constant coefficient differential equation $y[n] - (1/2)y[n-1] = x[n-1]$. Find the output, $y[n]$, when the input is $x[n] = (1/4)^n u[n]$.

Problem 3-19

When the input to a system is $x[n] = \frac{1}{4}^{n-1} u[n-1]$, the output is $y[n] = 4(\frac{1}{2}^n - \frac{1}{4}^n)u[n]$.

(a) Find the impulse response $h[n]$.

(b) Find the linear constant-coefficient difference equation that relates $x[n]$ to $y[n]$.

Problem 3-20

For each of the $H(\omega)$ in **Figure 3.52**, find $h[n]$. Hints for the three parts:

(a) $H(\omega)$ is of the form $A \sin a\omega$.

(b) $|H(\omega)|$ is of the form of a raised cosine.
(c) $|H(\omega)|$ is of the form of a periodic sinc function.

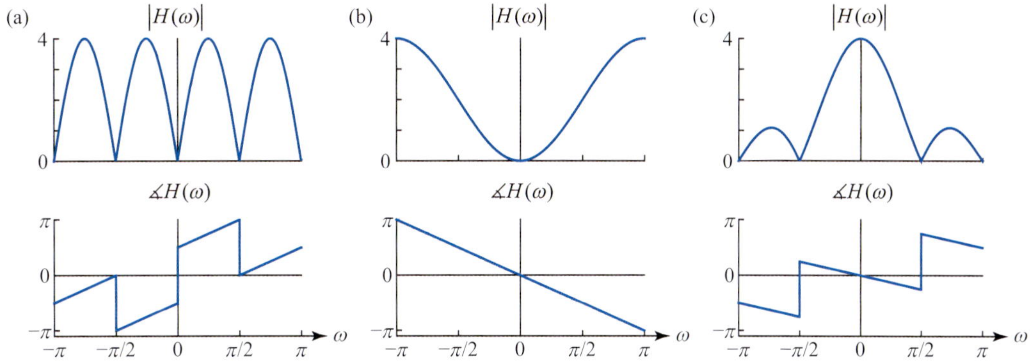

Figure 3.52

Problem 3-21

Consider the sequence from Example 1.4, $x[n]=(2+j)\delta[n+1]+\delta[n]-3j\delta[n-1]$.

(a) Find $X(\omega)$, $X_r(\omega)$, $X_i(\omega)$, $X_{re}(\omega)$, $X_{ro}(\omega)$, $X_{ie}(\omega)$ and $X_{io}(\omega)$.
(b) Show that the four relations of Equation (3.30) apply.

Problem 3-22

Given $h[n]=\cos(\pi n/4)(u[n+4]-u[n-4])$, find $H(\omega)$ by

(a) direct computation of the DTFT.
(b) using the modulation property of the DTFT.

Problem 3-23

Find $B_{2N-1}(\omega)$, the DTFT of a $(2N-1)$-point Bartlett window defined in Example 3.35.

(a) Show that

$$B_{2N-1}(\omega)=\sum_{-(N-1)}^{N-1} e^{-j\omega n}-\frac{1}{N}(R_N(\omega)+R_N(\omega)), \tag{3.55}$$

where $R_N(\omega)$ is the DTFT of a ramp of length N, given in Example 3.36,

$$R_N(\omega)=\begin{cases}\dfrac{Ne^{-j\omega(N-1/2)}\sin\omega/2-e^{-j\omega N/2}\sin\omega N/2}{2j\sin^2\omega/2}, & \omega\neq 0\\[2ex] \dfrac{N(N-1)}{2}, & \omega=0.\end{cases} \tag{3.56}$$

(b) By substituting Equation (3.56) into Equation (3.55), show that

$$B_{2N-1}(\omega)=\left(\frac{\sin\omega N/2}{\sin\omega/2}\right)^2.$$

Problem 3-24

(a) Given the system in **Figure 3.53**, find $H(\omega)$.
(b) Use your answer in (a) to find $y[n]$ in the form $y[n] = A\cos(\omega n + \theta)$, when $x[n] = \cos \pi n/2$.
(c) Use your answer in (a) to find $y[n]$ in the form $y[n] = A\cos(\omega n + \theta)$, when $x[n] = \cos \pi n/4$.

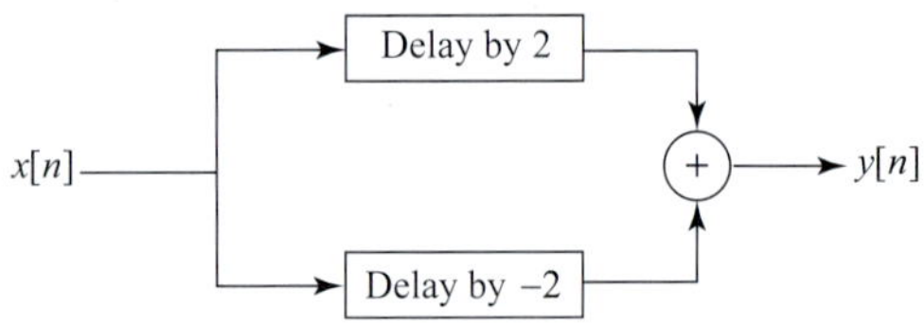

Figure 3.53

Problem 3-25

Given the system in **Figure 3.54** with

$$h_1[n] = \tfrac{1}{4}^{n} u[n]$$
$$h_2[n] = \tfrac{1}{2}^{n} u[n],$$

(a) find $H(\omega)$.
(b) find the difference equation that relates $x[n]$ and $y[n]$.

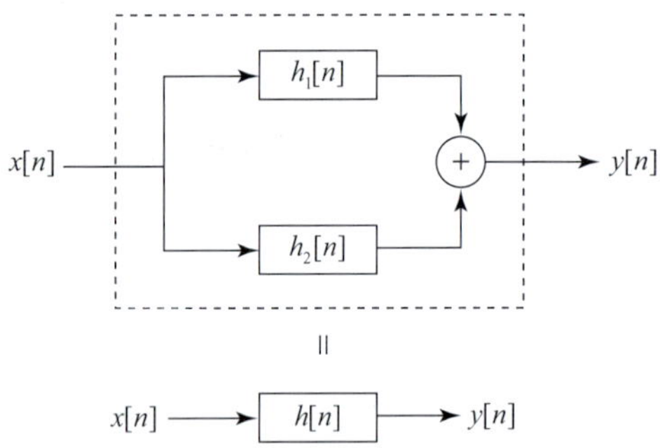

Figure 3.54

Problem 3-26

The system in **Figure 3.54** is defined by the difference equation

$$y[n] - \tfrac{3}{4}y[n-1] + \tfrac{1}{8}y[n-2] = -\tfrac{1}{4}x[n-1].$$

Given that

$$h_2[n] = -\frac{1}{2}^n u[n],$$

find $h_1[n]$.

Problem 3-27

Consider the system of **Figure 3.55** with input $x[n] = \cos \pi n/4$, impulse response $h[n] = \delta[n] - \delta[n - n_0]$ and output $y[n] = A\cos \pi(n + 1)/4$. Find A and n_0.

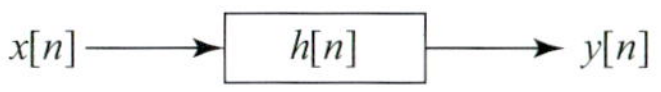

Figure 3.55

Problem 3-28

Consider the system of **Figure 3.55** with input $x[n] = \cos \pi n/4$ and impulse response $h[n] = \delta[n] - \delta[n - n_0]$.

(a) Find $y[n]$ if $n_0 = 2$.

(b) Find all possible values of n_0 such that $y[n] = 0$.

Problem 3-29

In the system of **Figure 3.55** with impulse response $h[n] = \delta[n] + \delta[n - 2]$,

(a) find output $y[n]$ when the input is $x[n] = \cos \pi n/3$.

(b) find output $y[n]$ when the input is $x[n] = \sqrt{2}\cos \pi n/4$.

Problem 3-30

Consider the system of **Figure 3.55** with input $x[n] = \delta[n] - \frac{1}{2}\delta[n - 1]$ and output $y[n] = -\delta[n] + \frac{1}{4}^n u[n]$. Find $h[n]$.

Problem 3-31

Given the system shown in **Figure 3.56** with

$$h_1[n] = \text{unknown}$$
$$h_2[n] = \delta[n] - \delta[n - 1]$$
$$h_3[n] = \frac{1}{2}^n u[n],$$

find $h_1[n]$ such that $y[n] = x[n]$.

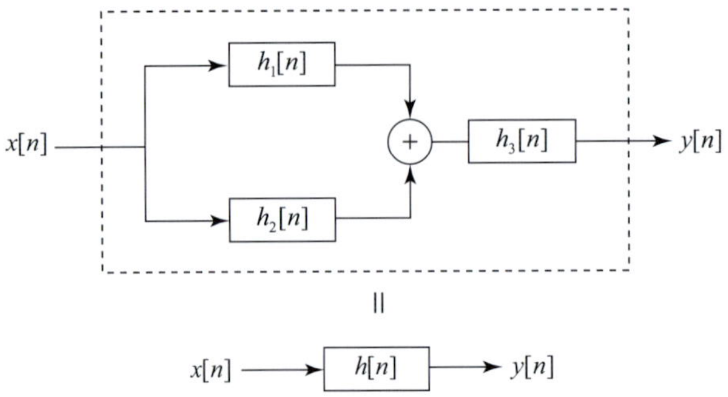

Figure 3.56

Problem 3-32

The system shown in **Figure 3.57** is characterized by the following difference equations:

$$y[n] - 0.5y[n-1] = x[n]$$
$$w[n] - 0.7w[n-1] + 0.1w[n-2] = x[n].$$

(a) Find $h[n]$.
(b) Find $h_2[n]$.

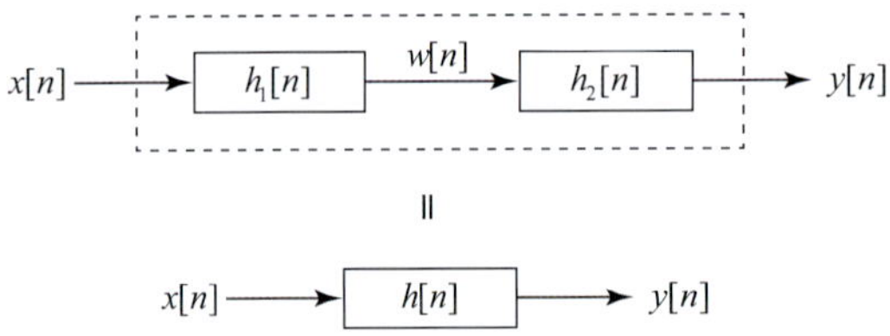

Figure 3.57

Problem 3-33

The system shown in **Figure 3.57** is characterized by the following difference equations:

$$y[n] - \tfrac{3}{4}y[n-1] + \tfrac{1}{8}y[n-1] = x[n-1]$$
$$y[n] - \tfrac{1}{4}y[n-1] = w[n] - \tfrac{1}{3}w[n-1].$$

Find $h_1[n]$.

Problem 3-34

Find $x[n]$, the input to a system defined by the LCCDE $y[n] - \frac{1}{2}y[n-1] = x[n]$, such that the output is $y[n] = \frac{1}{4}^n u[n]$.

Problem 3-35

Given a system defined by the linear constant-coefficient difference equation $y[n] - \frac{3}{4}y[n-1] + \frac{1}{8}y[n-2] = x[n-1]$, find $h[n]$.

Problem 3-36

Given a system defined by the linear constant-coefficient difference equation $y[n] - \frac{3}{4}y[n-1] + \frac{1}{8}y[n-2] = 2x[n+1]$,

(a) find $h[n]$.

(b) Is the system linear? Time-invariant? Causal? Stable?

Problem 3-37

Given the system shown in **Figure 3.58**, with

$$h_1[n] = \frac{1}{2}^{n} u[n]$$

$$h_3[n] = \delta[n] + \delta[n-1] + \delta[n-2],$$

find $h_2[n]$ such that the impulse response of the entire system is $h[n] = \delta[n]$.

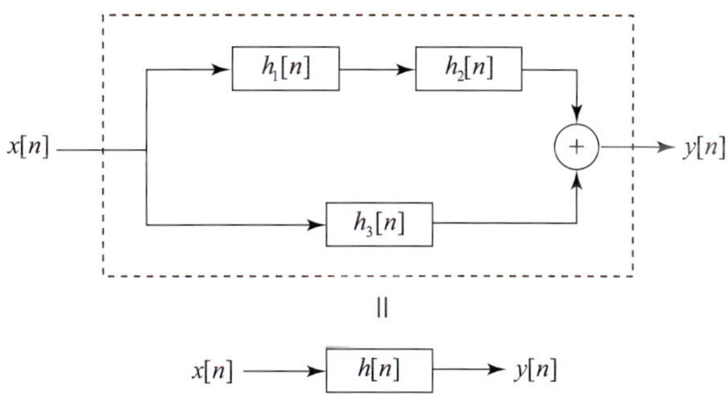

Figure 3.58

Problem 3-38

The system shown in **Figure 3.59** is an example of a simplified comb filter.

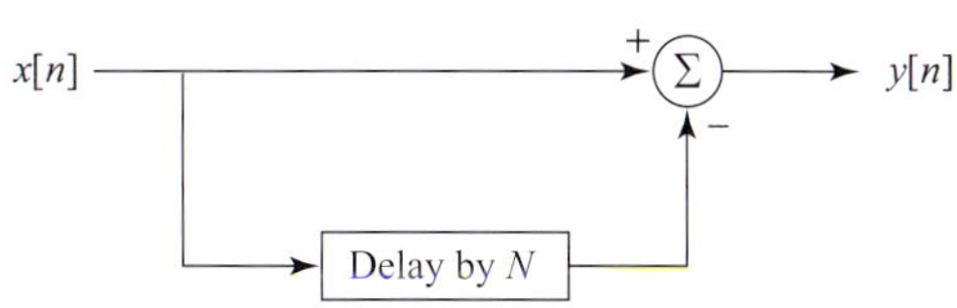

Figure 3.59

(a) Find $H(\omega)$ as a function of N.
(b) Sketch $|H(\omega)|$ and $\measuredangle H(\omega)$ for $N=2$, $N=4$ and $N=8$. Why is this called a comb filter?
(c) Find $y[n]$ when $N=2$ and $x[n]=\cos(\pi n/4)+\cos(\pi n/2)$.
(d) Find $y[n]$ when $N=4$ and $x[n]=\cos(\pi n/4)+\cos(\pi n/2)$.
(e) Find $y[n]$ when $N=8$ and $x[n]=\cos(\pi n/4)+\cos(\pi n/2)$.

Problem 3-39

For the system shown in **Figure 3.60**, assume that $x[n]=\cos\pi n/8$.

(a) Make a fully labeled sketch of $X(\omega)$, $Q(\omega)$, $R(\omega)$ and $Y(\omega)$.
(b) Find $q[n]$, $r[n]$ and $y[n]$.

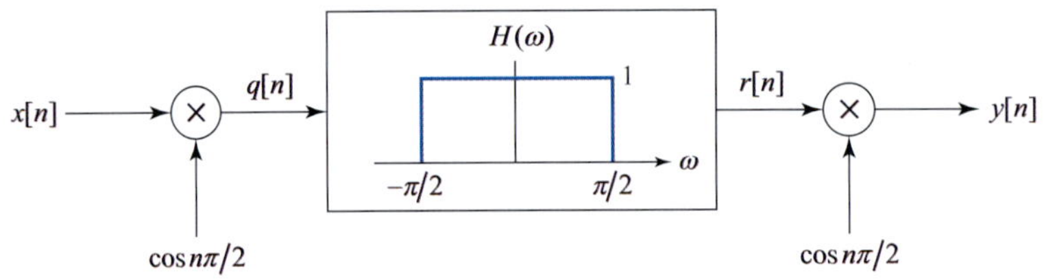

Figure 3.60

Problem 3-40

Repeat Problem 3-39 assuming that $x[n]=\sin\pi n/8$.

Problem 3-41

(a) Repeat Problem 3-39 assuming that $x[n]$ has the spectrum shown in **Figure 3.61**.
(b) Find $y[n]$ in terms of $x[n]$.

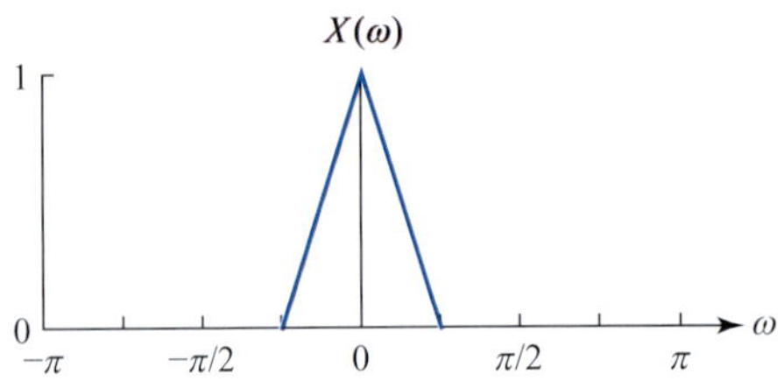

Figure 3.61

Problem 3-42

Find the convolution in **Figure 3.62** of $x[n]$ and $h[n]$ by
(a) direct convolution.
(b) using the convolution property of Fourier transforms.

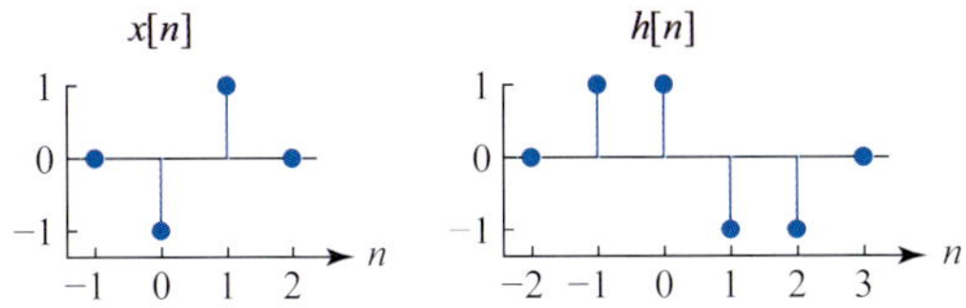

Figure 3.62

Problem 3-43

Consider the periodic sinc function,

$$H_N(\omega) = \frac{\sin \omega N/2}{\sin \omega/2} = N \frac{\operatorname{sinc} \omega N/2}{\operatorname{sinc} \omega/2}.$$

(a) Show that $H_N(\omega)$ is periodic with period 2π, when N is odd: $H_N(\omega + 2\pi k) = H_N(\omega)$.

(b) Show that $H_N(\omega)$ is antiperiodic with period 2π, when N is even: $H_N(\omega + 2\pi k) = -H_N(\omega)$.

(c) Show that $H_N(\omega)$ is periodic with period 4π, when N is even: $H_N(\omega + 4\pi k) = H_N(\omega)$.

Problem 3-44★

Consider a pulse of half-width N,

$$x_N[n] = \begin{cases} 1, & |n| \leq N \\ 0, & \text{otherwise} \end{cases}.$$

Clearly, $x_N[n]$ approaches a constant, one, as $N \to \infty$:

$$\lim_{N\to\infty} x_N[n] = 1.$$

In this problem, we show that the DTFT of $x_N[n]$ approaches the Fourier transform of a constant, which is a series of impulses at multiples of 2π:

$$\lim_{N\to\infty} X_N(\omega) \triangleq \lim_{N\to\infty} \mathfrak{F}\{x_N[n]\} = 2\pi \sum_{k=-\infty}^{\infty} \delta(\omega - 2\pi k).$$

(a) Show that $X(\omega)$ can be expressed as

$$\begin{aligned} X(\omega) &= 1 + 2\sum_{n=1}^{\infty} \cos \omega n \\ &= 1 + 2G(\omega), \end{aligned}$$

where $G(\omega)$ is defined as a summation,

$$G(\omega) = \sum_{n=1}^{\infty} \cos \omega n. \tag{3.57}$$

Then substitute the Chebyshev trigonometric recursion relation,

$$\cos \omega n = 2\cos\omega\cos\omega(n-1) - \cos\omega(n-2),$$

into Equation (3.57) and solve, showing that

$$G(\omega) = \begin{cases} \infty, & \omega = 2\pi k \\ -1/2, & \omega \neq 2\pi k \end{cases},$$

and hence

$$X(\omega) = \begin{cases} \infty, & \omega = 2\pi k \\ 0, & \omega \neq 2\pi k \end{cases}.$$

(b) Use the results from the preceding part to argue that

$$X(\omega) = A \sum_{k=-\infty}^{\infty} \delta(\omega - 2\pi k),$$

where A is a constant. Substitute $X(\omega)$ and $x[n]$ into the inverse DTFT relation to find the value of A.

Problem 3-45★

Use calculus to show that the functions $\cos \omega n$ form an orthogonal set such that

$$\frac{1}{2\pi}\int_{-\pi}^{\pi} \cos\omega k \cos\omega m \, d\omega = \begin{cases} 1, & k=m=0 \\ 1/2, & k=m\neq 0 \\ 0, & k\neq m \end{cases}.$$

➤**Hint:** $\cos\omega k \cos\omega m = \frac{1}{2}(\cos\omega(k+m) + \cos\omega(k-m))$.

Problem 3-46★

Derive the result of Problem 3-45 using properties of the DTFT. Specifically, let $G(\omega) = \cos\omega k \cos\omega m$.

(a) Find the inverse DTFT of $G(\omega)$,

$$g[n] = \frac{1}{2\pi}\int_{-\pi}^{\pi} G(\omega) e^{j\omega n} d\omega.$$

➤**Hint:** Use the convolution property of the DTFT.

(b) Sketch $g[n]$ for three cases: $k \neq m$, $k = m \neq 0$ and $k = m = 0$.

(c) Use the DTFT to demonstrate the result of Problem 3-45.

➤**Hint:** $g[0]$.

Problem 3-47★

Show that $\cos \omega n$ is an eigenfunction of an LTI system that has an even impulse response; that is, $h[n] = h[-n]$.

Problem 3-48

(a) Show that if $x[n]$ is conjugate-symmetric (i.e., $x[n]=x^*[-n]$), then $X(\omega)$ is purely real.

(b) Show that if $x[n]$ is conjugate-antisymmetric (i.e., $x[n]=-x^*[-n]$), then $X(\omega)$ is purely imaginary.

Problem 3-49

For any system, we can express the impulse response $h[n]$ as a sum of an even part $h_e[n]$ and an odd part $h_o[n]$, such that $h[n]=h_e[n]+h_o[n]$. Denote $\mathfrak{F}\{h_e[n]\}\triangleq H_e(\omega)$ and $\mathfrak{F}\{h_o[n]\}\triangleq H_o(\omega)$. Using the results of Problems 3-47 and 3-48, show that

$$T\{\cos\omega n\}=H_e(\omega)\cos\omega n+jH_o(\omega)\sin\omega n$$

and

$$T\{\sin\omega n\}=H_e(\omega)\sin\omega n-jH_o(\omega)\cos\omega n,$$

so that

$$T\{e^{j\omega n}\}=(H_e(\omega)+H_o(\omega))e^{j\omega n}=H(\omega)e^{j\omega n}.$$

Problem 3-50

Show that $\sin\omega n$ is an eigenfunction of an LTI system that has an even impulse response $h[n]=h[-n]$.

Problem 3-51

(a) Show that if $x[n]$ is imaginary and even, then $X(\omega)$ is imaginary and even.

(b) Show that if $x[n]$ is imaginary and odd, then $X(\omega)$ is real and odd.

Problem 3-52★

Plot an example of the DTFT of a real-and-even sequence whose phase can be either as an even function of ω or as an odd function of ω or as neither even nor odd.

Problem 3-53

For each $x[n]$ below, place a check in the table for each property that applies to its transform, $X(\omega)$. More than one property can apply to each transform.

(a) $x[n]=\delta[n]+\delta[n-2]$.

(b) $x[n]=\delta[n+2]+\delta[n-2]$.

(c) $x[n]=\delta[n+2]-\delta[n-2]$.

(d) $x[n]=(1+j)\delta[n-1]+(1-j)\delta[n+1]$.

(e) $x[n]=(1+j)\delta[n-1]-(1-j)\delta[n+1]$.

(f) $x[n]=(1+j)\delta[n-1]+(1+j)\delta[n+1]$.

(g) $x[n]=(1+j)\delta[n-1]-(1+j)\delta[n+1]$.

(h) $x[n] = j\delta[n] - j\delta[n-1]$.
(i) $x[n] = j\delta[n-1] + j\delta[n+1]$.
(j) $x[n] = -j\delta[n+2] + j\delta[n-2]$.

$x[n]$	$X(\omega)$ real	$X(\omega)$ imaginary	$X(\omega)$ conjugate symmetric	$X(\omega)$ conjugate antisymmetric	$X(\omega)$ even	$X(\omega)$ odd
(a)						
(b)						
(c)						
(d)						
(e)						
(f)						
(g)						
(h)						
(i)						
(j)						

Problem 3-54

For each of the sequences of Problem 3-53, place a check in the table for each property that applies to the transform, $X(\omega)$. More than one property can apply to each transform.

$x[n]$	$\lvert X(\omega)\rvert$ even	$\measuredangle X(\omega)$ even	$\measuredangle X(\omega)$ odd	$\measuredangle X(\omega)$ only 0 or $\pm\pi$	$\measuredangle X(\omega)$ only $\pm\frac{\pi}{2}$
(a)					
(b)					
(c)					
(d)					
(e)					
(f)					
(g)					
(h)					
(i)					
(j)					

Problem 3-55

Table 3.7 is a list of 12 sequences $x[n]$. The magnitude of each of these sequences is one of the four magnitude plots, $|M_1(\omega)|$, $|M_2(\omega)|$, $|M_3(\omega)|$ and $|M_4(\omega)|$, shown in **Figure 3.63**. Using only your knowledge of the symmetry properties of the DTFT, select the proper magnitude plot for each sequence.

Table 3.7

$x[n]$	M
$j(\delta[n+1]+\delta[n-1])$	
$j(\delta[n]-\delta[n-2])$	
$\delta[n+1]e^{-j\pi/4}+\delta[n-1]e^{j\pi/4})$	
$j(\delta[n+1]-\delta[n-1])$	
$\delta[n+1]+\delta[n-1]$	
$j(\delta[n]+\delta[n-2])$	
$\delta[n]e^{j\pi/4}+\delta[n-2]e^{-j\pi/4}$	
$\delta[n]+\delta[n-2]$	
$\delta[n+1]e^{-j\pi/4}-\delta[n-1]e^{j\pi/4}$	
$\delta[n+1]-\delta[n-1]$	
$\delta[n]e^{j\pi/4}+\delta[n-2]e^{-j\pi/4}$	
$\delta[n]+\delta[n-2]$	

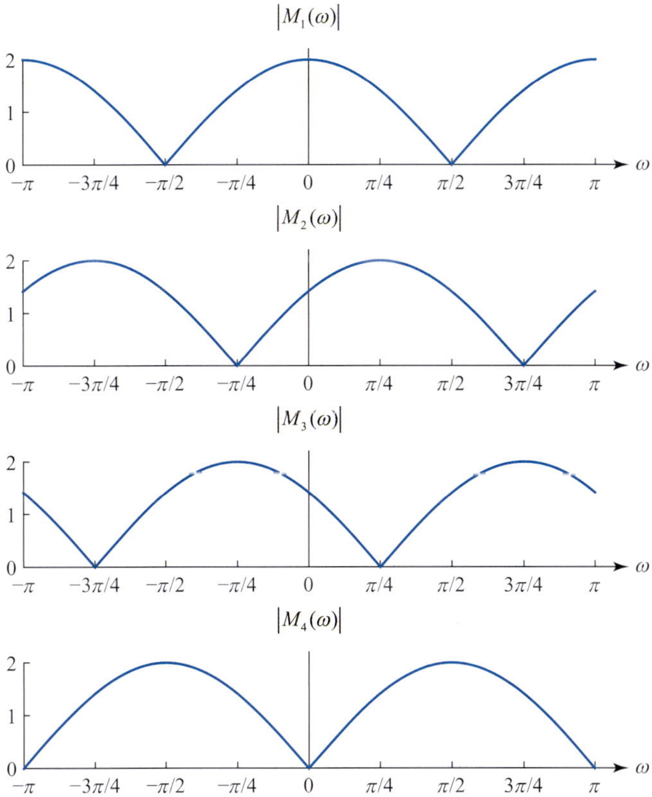

Figure 3.63

Problem 3-56

For each of the 12 sequences $x[n]$ in **Table 3.7**, determine the appropriate phase, $\measuredangle P_1(\omega)$–$\measuredangle P_{12}(\omega)$, shown in **Figure 3.64**.

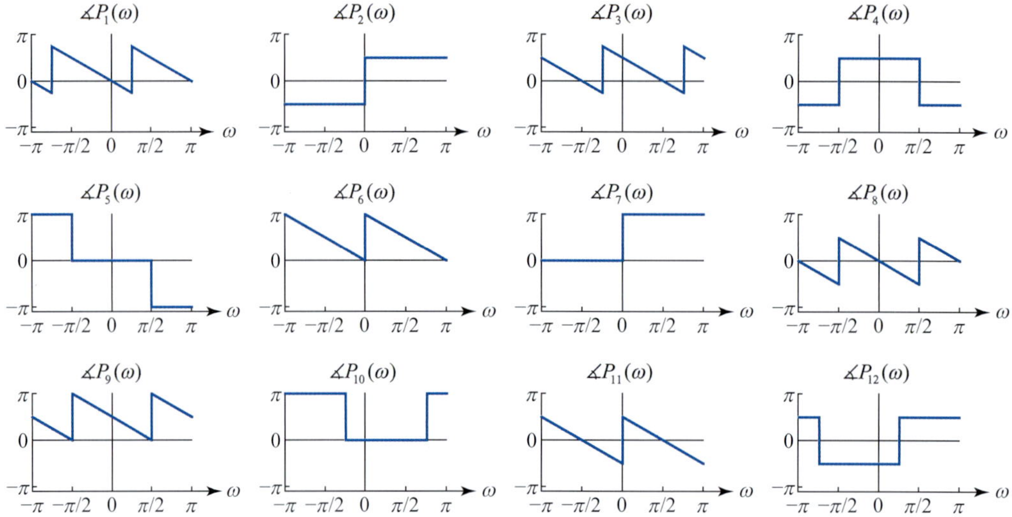

Figure 3.64

Problem 3-57

Suppose that you know $x[n]$, the sequence whose transform is shown in **Figure 3.65**.

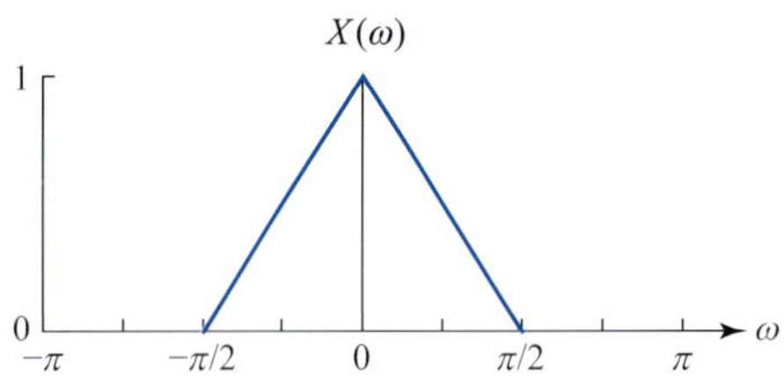

Figure 3.65

Sketch the DTFT (magnitude and phase) of the following sequences:

(a) $y_1[n] = x[n-1]$.

(b) $y_2[n] = x[n] * \cos \pi(n-1)/4$ (i.e., $x[n]$ convolved with a cosine).

(c) $y_3[n] = x[n] \sin \pi n/2$.

(d) $y_4[n] = x[n] * \sin \pi n/4$.

Problem 3-58

Suppose that you know $x[n]$, the sequence whose transform is shown in **Figure 3.65**. Given the DTFTs $Y_1(\omega)$, $Y_2(\omega)$, $Y_3(\omega)$ and $Y_4(\omega)$ shown in **Figure 3.66**, find the sequences $y_1[n]$, $y_2[n]$, $y_3[n]$ and $y_4[n]$ in terms of $x[n]$. You should not have to compute explicitly any transforms or inverse transforms.

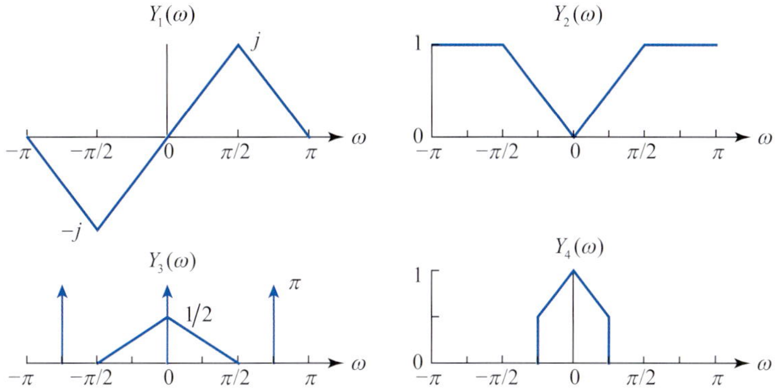

Figure 3.66

Problem 3-59

Show that if a sequence, $x[n]$, is odd, then both $X(0)$ and $X(\pi)$ must be zero.

Problem 3-60

Show that if a sequence $x[n]$ is even, then $X(0)$ will be zero if

$$X[0] = -2\sum_{n-1}^{\infty} x[n].$$

Problem 3-61

(a) Find $A(\omega)$, the DTFT of $a[n] = \alpha^n u[n]$, $|\alpha| < 1$, and show that it can be expressed as

$$A(\omega) = A_{re}(\omega) + jA_{io}(\omega),$$

where $A_{re}(\omega)$ and $A_{io}(\omega)$ are respectively the real and even and imaginary and odd parts of $A(\omega)$.

(b) Show that $A_{re}(\omega) = \mathfrak{F}\{a_{re}[n]\}$ and $A_{io}(\omega) = j\mathfrak{F}\{a_{ro}[n]\}$, where $a_{re}[n]$ and $a_{ro}[n]$ are respectively the real-and-even and real-and-odd parts of $a[n]$.

Problem 3-62

(a) Show that the DTFT of a causal symmetric sequence of even length N can be expressed as

$$X(\omega) = 2e^{-j\omega(N-1)/2}\sum_{n-0}^{N/2-1} x[n]\cos\omega(n \quad (N-1)/2).$$

(b) Show that the DTFT of a causal symmetric sequence of odd length N can be expressed as

$$X(\omega) = x[(N-1)/2]e^{-j\omega(N-1)/2} + 2e^{-j\omega(N-1)/2} \sum_{n-0}^{(N-3)/2} x[n] \cos \omega(n-(N-1)/2).$$

(c) Show that the DTFT of a causal antisymmetric sequence of even length N can be expressed as

$$X(\omega) = 2je^{-j\omega(N-1)/2} \sum_{n-0}^{N/2-1} x[n] \sin \omega(n-(N-1)/2).$$

(d) Show that the DTFT of a causal antisymmetric sequence of odd length N can be expressed as

$$X(\omega) = 2je^{-j\omega(N-1)/2} \sum_{n-0}^{(N-3)/2} x[n] \sin \omega(n-(N-1)/2).$$

Problem 3-63

Find the Fourier transform of the following $h[n]$:

(a) $(-3)^n u[n]$.

(b) $\cos((n-1/2)\pi/4)$.

(c) $2^n \sin(\pi n/3)\, u[n]$.

(d) $2^n u[n] - (-2)^{n-1} u[n-1]$.

Problem 3-64

For a system characterized by the frequency response $H(\omega)$, the *group delay* is defined as

$$D(\omega) = -\frac{d\measuredangle H(\omega)}{d\omega}.$$

(a) Show that

$$D(\omega) = \mathrm{Re}\left\{ j \frac{\left(\dfrac{dH(\omega)}{d\omega}\right)}{H(\omega)} \right\}.$$

➤**Hint:** Express $H(\omega) = |H(\omega)|e^{-j\omega n}$ and use the chain rule to take the derivative.

(b) Show that for an FIR system characterized by impulse response $h[n]$,

$$\frac{dH(\omega)}{d\omega} = -j\mathfrak{F}\{nh[n]\},$$

so that

$$D(\omega) = \mathrm{Re}\left\{\frac{\mathfrak{F}\{nh[n]\}}{\mathfrak{F}\{h[n]\}}\right\}.$$

(c) Show that, for a system characterized by

$$H(\omega) = \frac{B(\omega)}{A(\omega)} = \frac{\sum_{n-0}^{N} b_n e^{-j\omega n}}{\sum_{m-0}^{M} a_n e^{-j\omega n}},$$

we have

$$\begin{aligned}(\omega) &= \mathrm{Re}\left\{j\left(\frac{\frac{dB(\omega)}{d\omega}}{B(\omega)}\right)\right\} - \mathrm{Re}\left\{j\left(\frac{\frac{dA(\omega)}{d\omega}}{A(\omega)}\right)\right\} \\ &= \mathrm{Re}\left\{\frac{\mathfrak{F}\{nb[n]\}}{\mathfrak{F}\{b[n]\}}\right\} - \mathrm{Re}\left\{\frac{\mathfrak{F}\{na[n]\}}{\mathfrak{F}\{a[n]\}}\right\}.\end{aligned}$$

4 z-Transform

Introduction

In Chapter 3, we showed how the discrete-time Fourier transform allowed us to gain insight into the analysis of signals and systems in the frequency domain. However, there are a large number of important signals and systems that do not easily lend themselves to Fourier analysis. Given a sequence $h[n]$, we are guaranteed that the DTFT $H(\omega)$ exists only if $h[n]$ is absolutely summable; that is,

$$\sum_{n=-\infty}^{\infty} |h[n]| < \infty.$$

Consider a simple system defined by the linear constant-coefficient difference equation (LCCDE)

$$y[n] - \alpha y[n-1] = x[n].$$

In Chapter 3, we showed that this system has impulse response

$$h[n] = \alpha^n u[n]. \tag{4.1}$$

This response has different behavior depending on the value of α as shown in **Figure 4.1**.

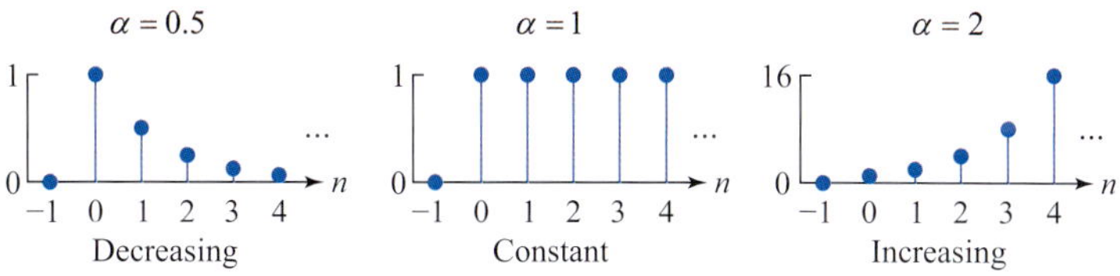

Figure 4.1 Monotonic right-sided sequences

The response is convergent (monotonically decreasing) for $0 < \alpha < 1$; the response is the unit step (i.e., constant for $n \geq 0$) for $\alpha = 1$; and the response is divergent (monotonically increasing) for $\alpha > 1$.

We have shown that the DTFT of $h[n]$ can be easily computed for $0 < \alpha < 1$,

$$H(\omega) = \sum_{n=0}^{\infty} h[n]e^{-j\omega n} = \frac{1}{1 - \alpha e^{-j\omega}},$$

because the summation has a closed-form solution. However, there is no obvious way to compute the DTFT of $h[n]$ for $\alpha \geq 1$, because the summation that defines the DTFT no longer has a closed-form solution. This is a significant problem. We would like to be able to use the *same* frequency-domain methods to analyze the system defined by LCCDEs such as Equation (4.1) regardless of our particular choice of parameter α. The solution to this problem is to modify the Fourier transform by providing an additional degree of freedom to handle cases of divergent sequences. The result is the **z-transform**.

As a look forward, in Section 4.1 we define the z-transform, and in Section 4.2 we show how the z-transforms of the systems of most interest to us have a simple graphical representation: the pole-zero plot. Section 4.3 picks up the discussion of linear-phase systems from Chapter 3 and shows that the poles and zeros of these systems can only occur in certain places. Section 4.4 introduces the inverse z-transform. Section 4.5 provides an important catalog of the properties of the transform and Sections 4.6 and 4.7 show how the z-transform can be used to solve linear constant-coefficient difference equations.

4.1 The z-transform

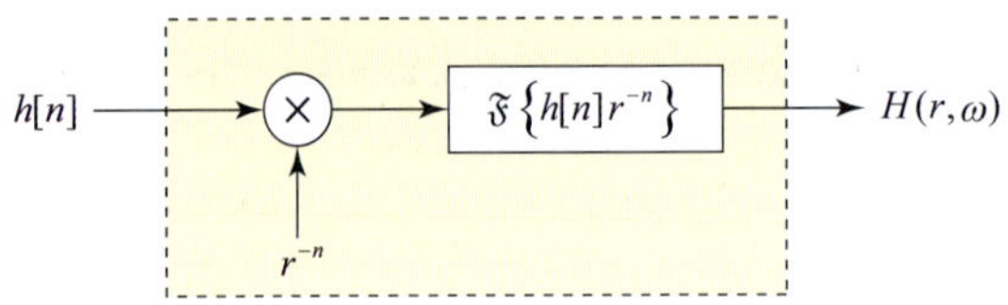

Figure 4.2

The essential modification we need to make is to pre-multiply $h[n]$ by a sequence r^{-n} before taking the Fourier transform, as schematized in **Figure 4.2**, where r is a positive real constant that we control. Depending upon our choice of r, we can make the product $h[n]r^{-n}$ absolutely summable for any value of α in our example. As an example, consider the divergent sequence $h[n] = 2^n u[n]$ shown in **Figure 4.3a**.

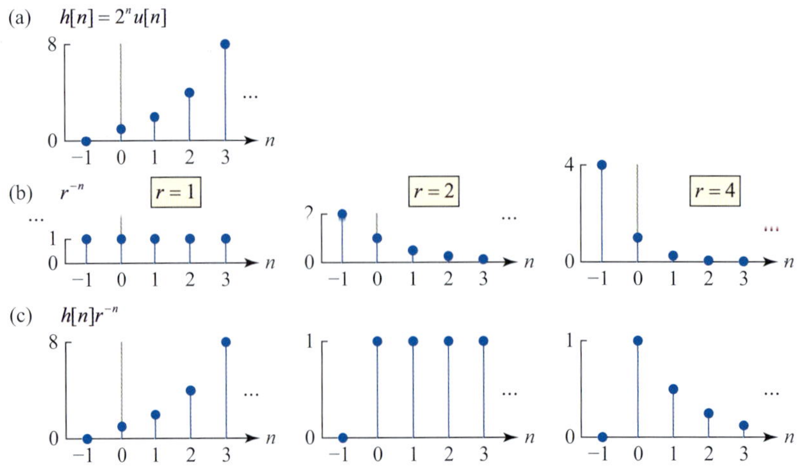

Figure 4.3

When multiplied by r^{-n} with $r = 1$ (left panel of **Figure 4.3b**), the result is just $h[n]$ (left panel of **Figure 4.3c**). When multiplied by r^{-n} with $r = 2$ (middle panel of **Figure 4.3b**), the result is a step (middle panel of **Figure 4.3c**). You should convince yourself that the sequence $r^{-n}\alpha^n u[n]$ will converge and be absolutely summable for any choice of $r > 2$. For example, if we choose $r = 4$ (right panel of **Figure 4.3b**), the result of multiplying $h[n]$ by r^{-n} is a decreasing power-law sequence, $(1/2)^n u[n]$ (right panel of **Figure 4.3c**).

To formalize this observation for an arbitrary $h[n]$, define the DTFT of $h[n]r^{-n}u[n]$ to be $H(r, \omega)$:

$$H(r,\omega) \triangleq \mathfrak{F}\{h[n]r^{-n}\} = \sum_{n=-\infty}^{\infty} (h[n]r^{-n})e^{-j\omega n} = \sum_{n=-\infty}^{\infty} h[n](re^{j\omega})^{-n}. \tag{4.2}$$

This transform depends on two real parameters, r and ω. The Fourier transform of the pre-multiplied sequence $h[n]r^{-n}$ is equivalent to expressing $h[n]$ in terms of basis functions $re^{j\omega}$ instead of $e^{j\omega}$. To proceed, let us define a complex variable $z \triangleq re^{j\omega}$. Then, Equation (4.2) becomes

$$H(z) = \mathfrak{Z}\{h[n]\} \triangleq \sum_{n=-\infty}^{\infty} h[n]\underbrace{(re^{j\omega})}_{z}{}^{-n} = \sum_{n=-\infty}^{\infty} h[n]z^{-n}. \tag{4.3}$$

$H(z)$ is called the **bilateral z-transform** of $h[n]$, to which we give the shorthand notation $\mathfrak{Z}\{\}$. Bilateral means that the summation in Equation (4.3) extends from $n = -\infty$ to ∞. In Section 4.7, we will consider the **unilateral z-transform**, a refinement of the z-transform for cases in which $h[n]$ is causal; i.e. $h[n] = 0$, $n < 0$.

Consider the previous example,

$$H(z) \triangleq \sum_{n=-\infty}^{\infty} h[n]z^{-n} = \sum_{n=-\infty}^{\infty} \alpha^n u[n]z^{-n} = \sum_{n=0}^{\infty} (\alpha z^{-1})^n = \frac{1}{1-\alpha z^{-1}}, \quad |z| > |\alpha|. \tag{4.4}$$

$H(z)$ is an infinite geometric sum that has a closed-form solution if $|\alpha z^{-1}| < 1$, which implies that the values of z must satisfy $|z| > |\alpha|$. Thus, the z-transform of $h[n] = \alpha^n u[n]$ exists (that is, the summation can be computed in closed form) for a given value of α as long as we compute the summation for values of $|z| > |\alpha|$. Since z is just a complex number, every value of z can be placed on the complex plane, which we call the **z-plane**. A contiguous locus of values of z for which $H(z)$ exists is called the **region of convergence (ROC)** of $H(z)$.

4.2 The singularities of $H(z)$

In most cases of interest to us, the z-transform $H(z)$ can be expressed as a ratio of two polynomials: a numerator polynomial $N(z)$, and a denominator polynomial $D(z)$:

$$H(z) = \frac{N(z)}{D(z)} = \frac{\sum_{k=0}^{M} b_k z^{-k}}{\sum_{k=0}^{N} a_k z^{-k}}. \tag{4.5}$$

Depending on what we are trying to accomplish, we will choose to express $H(z)$ either as powers of z^{-1}, as in Equation (4.5), or equivalently in powers of z, which is done by multiplying both $N(z)$ and $D(z)$ by $z^{\max(M,N)}$. In the following sections, where we discuss the singularities of $H(z)$, it will be most convenient to express $H(z)$ as powers of z. In Section 4.4, where we discuss the inverse z-transform, we will want to express $H(z)$ in powers of z^{-1}. The reason for these two equivalent representations will become clear as we proceed.

For z-transforms that can be expressed using Equation (4.5), there is an elegant, essentially graphical way of representing $H(z)$. In order to do so, find the roots of the numerator and denominator polynomial $N(z)$ and $D(z)$, expressed as powers of z. To take a simple example, multiply both the numerator and denominator of Equation (4.4) by z to produce numerator and denominator polynomials that are both functions of z,

$$H(z) = \frac{1}{1-\alpha z^{-1}} = \frac{z}{z-\alpha}, \quad |z| > |\alpha|. \tag{4.6}$$

Now, we need to introduce some terminology:

Zeros: The values of z in the finite z-plane for which $H(z)=0$ are called the **zeros** of $H(z)$.

In the finite z-plane, the zeros of $H(z)$ are the roots of $N(z)$, that is, the values of z at which the numerator polynomial $N(z)=0$. In our example $N(z)=z$, so we say, "There is a zero at $z=0$." Though we are generally only interested in the zeros of $H(z)$ in the finite z-plane, zeros at $z=\pm\infty$ can also occur at values of z at which $D(z)\to\infty$. For example, given the z-transform

$$H(z) = \frac{z}{z^2+\alpha z+\beta},$$

there is a zero at $z=0$, which corresponds to the value of z at which $N(z)=0$. But, there are also zeros at $z=\pm\infty$, which also send $H(z)\to 0$.

Poles: The values of z in the finite z-plane for which $H(z)\to\infty$ are called the **poles** of $H(z)$.

In the finite z-plane, the poles of $H(z)$ are the roots of $D(z)$, that is, values of z at which the denominator polynomial $D(z)=0$. In our example, $D(z)=z-\alpha$, so we say, "There is a pole at $z=\alpha$." Again, we are generally only interested in finite poles of $H(z)$, which are the roots of $D(z)$.

4.2.1 Pole-zero plots

The poles and zeros are collectively termed the **singularities** of $H(z)$. We can indicate the singularities of $H(z)$ graphically on the z-plane as a **pole-zero plot**, denoting the poles with an × and zeros with an O. For example, the pole-zero plot of $H(z)$ of Equation (4.6) is shown in **Figure 4.4** for values of $\alpha = 1/2$, 1 and 2.

Figure 4.4a corresponds to the z-transform

$$H(z) = \frac{z}{z-\frac{1}{2}}, \quad |z| > \tfrac{1}{2}.$$

There is a pole at $z = 1/2$ and a zero at $z = 0$. For any z-transform, we can plot the ROC on the complex z-plane along with the poles and zeros. In this example, the equation for the ROC, $|z| > 1/2$, is equivalent to $|z| = |re^{j\omega}| = r > 1/2$. The ROC is the exterior of a circle of radius 1/2. We have colored the ROC in blue. In addition to the ROC, we have also plotted a reference circle at $|z| = r = 1$, which we call the **unit circle**.

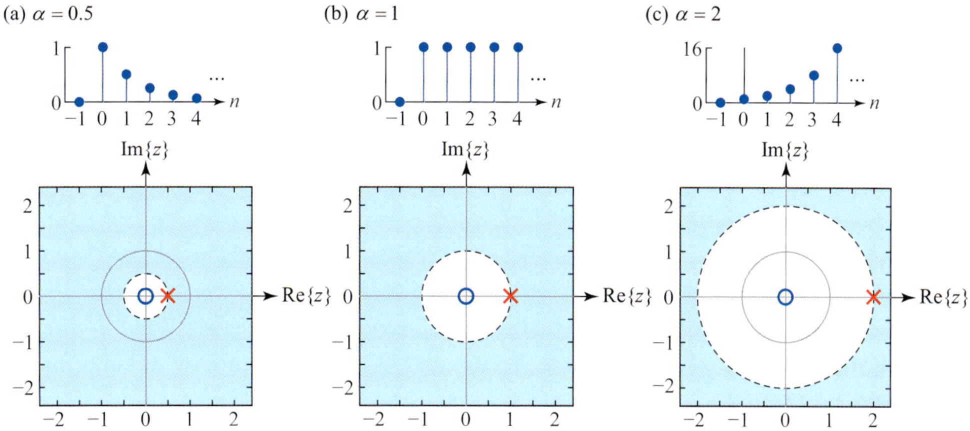

Figure 4.4 Sequences and pole-zero plots for monotonic right-sided sequences

The pole-zero plot for the unit step $h[n] = u[n]$ is shown in **Figure 4.4b**. The pole is at $z = 1$ and the ROC is the exterior of a circle of radius one. **Figure 4.4c** shows the pole-zero plot of our unsummable sequence $h[n] = 2^n u[n]$. The pole is at $z = 2$ and the ROC is the exterior of a circle of radius two.

Sequences such as those shown in **Figure 4.1** are examples of **right-sided sequences**. A sequence $h[n]$ is said to be right-sided if there exists some time n_0 such that $h[n] = 0$, $n < n_0$. The ROC of a right-sided sequence is always the exterior of a circle bounded by a pole. To see why this is so, let us say the sequence $h[n]$ just converges when multiplied by ${r_0}^{-n}$. Then it will also converge for any sequence ${r_1}^{-n}$, where $r_1 > r_0$, and will, in fact, converge faster. Thus the ROC of $H(z)$ has to be the *exterior* of a circle of radius r_0, since this region includes all values of $r > r_0$. Notice also that if the ROC includes the point $z_0 = r_0 e^{j\omega_0}$, then it will also include the entire locus of points with the same magnitude, $|z_0| = |r_0 e^{j\omega_0}| = r_0$, namely a circle of radius r_0. All the sequences shown in **Figure 4.1** of the form $h[n] = \alpha^n u[n]$ are right-sided. As shown in **Figure 4.4**, there is a single pole at $z = \alpha$, so the region of convergence is the exterior of a circle of radius α.

Sequences of the form $h[n] = \alpha^n u[n]$ with $\alpha > 0$, shown in **Figure 4.4**, are examples of monotonic sequences. **Figure 4.5** shows examples of alternating (i.e., non-monotonic) sequences, which occur when $\alpha < 0$.

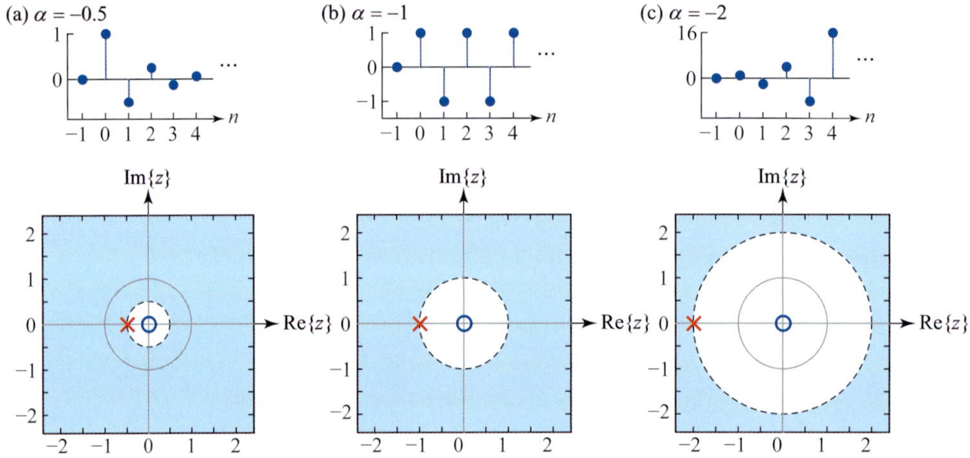

Figure 4.5 Sequences and pole-zero plots for alternating right-sided sequences

The response is alternating-decreasing for $-1 < \alpha < 0$ (e.g., **Figure 4.5a**); it alternates between -1 and $+1$ for $\alpha = -1$ (**Figure 4.5b**); and it is alternating-increasing for $\alpha < -1$ (e.g., **Figure 4.5c**).

The figure also shows pole-zero plots corresponding to the z-transforms of these sequences. Since $\alpha < 0$, the poles are all in the left-half of the z-plane. However, the region of convergence of the z-transforms of each of these alternating sequences is still $|z| > |\alpha|$, which is the exterior of a circle of radius α.

Table 4.1 summarizes the relation between the pole position and the nature of the right-sided sequences shown in **Figures 4.4** and **4.5**.

Table 4.1 Properties of right-sided sequences

α	Monotonic	Alternating	Convergent	Constant	Divergent
$\alpha > 1$	√				√
$\alpha = 1$	√			√	
$0 < \alpha < 1$	√		√		
$-1 < \alpha < 0$		√	√		
$\alpha = -1$		√		√	
$\alpha < -1$		√			√

For $H(z)$ that only have real poles in the right-half of the z-plane (i.e., $\alpha > 0$), the sequence is monotonic; if all the poles are in the left-half of the z-plane (i.e., $\alpha < 0$), the sequence is alternating. If the poles are outside the unit circle (i.e., $|\alpha| > 1$), the response is divergent (since the ROC does not include the unit circle); if the poles are inside the unit circle (i.e., $|\alpha| < 1$), the response is convergent; and if the poles lie exactly on the unit circle (i.e., $|z| = 1$), the response is neither convergent nor divergent.

4.2.2 Left-sided sequences

The top panel of **Figure 4.6** shows several sequences corresponding to the impulse response

$$h[n] = -\alpha^n u[-n-1]$$

for different values of α.

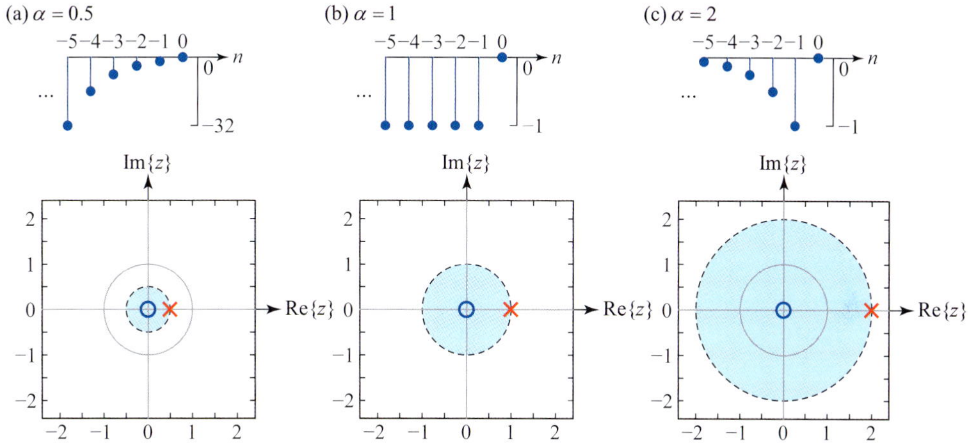

Figure 4.6 Sequences and pole-zero plots for monotonic left-sided sequences

These are said to be **left-sided sequences** since they are anti-causal; that is, there exists some time n_0 such that $h[n] = 0$, $n > n_0$. A left-sided impulse response might appear to be a mathematical curiosity since it would seem to be unrealizable. However, towards the end of this chapter (Example 4.28), we show how a filter with a left-sided impulse response will enable us to perform a practical inverse-filtering task.

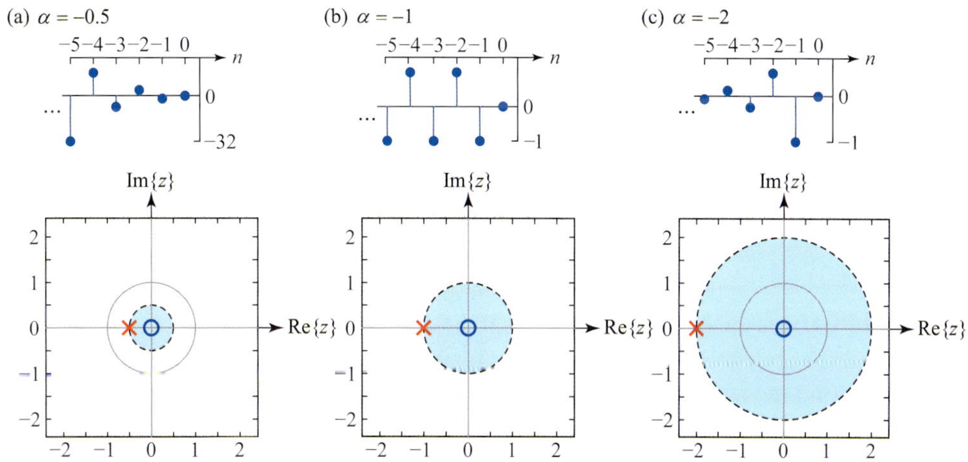

Figure 4.7 Sequences and pole-zero plots for alternating left-sided sequences

For the examples shown in **Figure 4.6**, $n_0 = -1$. For $0 < \alpha < 1$ (**Figure 4.6a**), the response becomes increasingly negative as $n \to -\infty$; for $\alpha = 1$ (**Figure 4.6b**), the response is constant for $n < 0$ and for $\alpha > 1$ (**Figure 4.6c**), the response is monotonically decreasing as $n \to -\infty$. The z-transform of $h[n]$ is

$$\begin{aligned} H(z) &= \sum_{n=-\infty}^{\infty} h[n]z^{-n} = -\sum_{n=-\infty}^{\infty} \alpha^n u[-n-1]z^{-n} = -\sum_{n=-\infty}^{-1} (\alpha z^{-1})^n \\ &= 1 - \sum_{n=-\infty}^{0} (\alpha z^{-1})^n = 1 - \sum_{n=0}^{\infty} (\alpha^{-1} z)^n. \end{aligned}$$

The summation only converges when $|\alpha^{-1}z| < 1$, which implies $|z| < |\alpha|$. In this range of z, the z-transform has a closed-form solution,

$$H(z) = 1 - \frac{1}{1-\alpha^{-1}z} = \frac{-\alpha^{-1}z}{1-\alpha^{-1}z} = \frac{1}{1-\alpha z^{-1}} = \frac{z}{z-\alpha}, \quad |z| < |\alpha| \cdot \tag{4.7}$$

Comparing Equations (4.6) and (4.7), you can see that the z-transforms of the right-sided sequence $h[n] = \alpha^n u[n]$ and the left-sided sequence $h[n] = -\alpha^n u[-n-1]$ are exactly the same except for one thing: the region of convergence. The poles and zeros are in the same places as in **Figure 4.4**, but the region of convergence is now $|z| < |\alpha|$, which is the interior of a circle bounded by the pole. This is a very important lesson. The polynomial expression for $H(z)$ is insufficient by itself to describe the transform of a sequence completely. You also need to know the region of convergence.

Figure 4.7 shows pole-zero plots for left-sided alternating sequences. **Figure 4.7a** shows the plot for $\alpha = -1/2$, which results in a divergent alternating sequence. **Figure 4.7b** shows the plot for $\alpha = -1$, which yields an alternating sequence that is neither convergent nor divergent. **Figure 4.7c** shows the plot for $\alpha = 0.5$, which results in a convergent alternating sequence. **Table 4.2** summarizes the relation between the pole position and the characteristics of left-sided sequences.

Table 4.2 Properties of left-sided sequences

α	Monotonic	Alternating	Convergent	Constant	Divergent
$\alpha > 1$	√		√		
$\alpha = 1$	√			√	
$0 < \alpha < 1$	√				√
$-1 < \alpha < 0$		√			√
$\alpha = -1$		√		√	
$\alpha < -1$		√	√		

4.2.3 Relation between the z-transform and DTFT

The ROC is defined as the values of z for which the z-transform exists. Since $z = |z|e^{j\omega} = re^{j\omega}$, evaluating the z-transform at the locus of points $|z| = r$ for a constant value of r is equivalent to evaluating it on a circle of radius r in the z-plane. The DTFT is just the z-transform evaluated at $z = e^{j\omega}$, which is the locus of points in the z-plane corresponding to the unit circle, $|z| = r = 1$,

$$H(z)|_{z=e^{j\omega}} = \left(\sum_{n=-\infty}^{\infty} h[n]z^{-n} \right)\Bigg|_{z=e^{j\omega}} = \sum_{n=-\infty}^{\infty} h[n]e^{-j\omega n} = H(\omega).$$

In words: the DTFT is just the z-transform evaluated for values of z on the unit circle. Clearly, the DTFT cannot exist unless the region of convergence of the z-transform includes the locus of points $|z| = 1$, that is, the unit circle. Hence, we can state that

> If the ROC of $H(z)$ includes the unit circle, then $H(\omega)$ exists.

In our examples of right-sided sequences, the ROC includes the unit circle only for those sequences whose pole-zero plot has a pole inside the unit circle; that is, $h[n] = (1/2)^n u[n]$ (**Figure 4.4a**) and $h[n] = (-1/2)^n u[n]$ (**Figure 4.5a**). Hence, only for those sequences is the DTFT guaranteed to exist. In our examples of left-sided sequences, the ROC includes the unit circle only for those sequences whose pole-zero plot has a pole outside the unit circle; that is, $h[n] = -2^n u[-n-1]$ (**Figure 4.6c**) and $h[n] = -(-2)^n u[-n-1]$ (**Figure 4.7c**).

One other important result follows from this. We know from the discussion of stability of LTI systems in Chapter 3 that if $H(\omega)$ exists, then the system is stable. Thus,

> If the ROC includes the unit circle, then the system is stable.

4.2.4 Multiple poles and zeros

Let us move on to examples of sequences whose z-transforms have multiple poles and zeros. First, we will look at the simplest example of multiple poles and zeros: the shifted impulse. Then we will look at the sequences formed from the sum of power-law terms in order to explore the more general connection between the placement of the poles and the location of the ROC.

Shifted impulse Given an ideal delay,

$$h[n] = \delta[n - n_0],$$

the z-transform is

$$H(z) = \sum_{n=-\infty}^{\infty} h[n]z^{-n} = \sum_{n=-\infty}^{\infty} \delta[n - n_0]z^{-n} = z^{-n_0}.$$

Clearly, when $n_0 = 0$, we have $h[n] = \delta[n]$ and the z-transform is $H(z) = 1$, as shown in **Figure 4.8a**. There are no poles or zeros, so the z-transform converges for all values of z; that is, the ROC is the entire z-plane. When $n_0 = 1$, we have $h[n] = \delta[n-1]$ and $H(z) = z^{-1} = 1/z$. There is a single pole at $z = 0$ and the ROC is the entire z-plane except the point $z = 0$ (**Figure 4.8b**). Similarly, for $n_0 > 0$, $h[n] = \delta[n - n_0]$ and there are n_0 poles at $z = 0$ (**Figure 4.8c**). When $n_0 = -1$, $h[n] = \delta[n+1]$ and $H(z) = z$. There is a single zero at $z = 0$ and the ROC is the entire z-plane (**Figure 4.8d**). More generally, for $n_0 < 0$, then $h[n] = \delta[n + n_0]$ and there are n_0 zeros at $z = 0$ (**Figure 4.8e**).

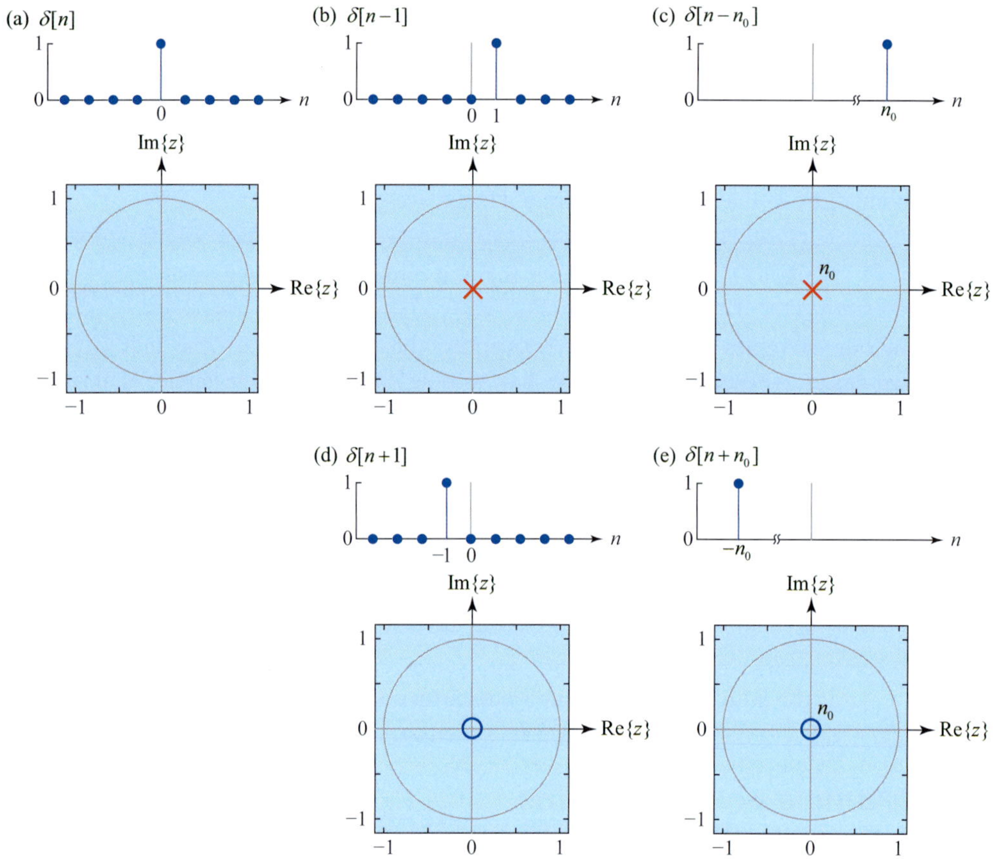

Figure 4.8 Pole-zero plot of shifted impulses

Right-sided sequence Consider the sequence $x_A[n]$, which is formed from the scaled sum of two right-sided sequences, a monotonically decreasing sequence $x_1[n]$ and a monotonically increasing sequence $x_2[n]$, each of which is of the form $\alpha^n u[n]$,

$$x_A[n] = x_1[n] + 2x_2[n] = \underbrace{\tfrac{1}{2}^n u[n]}_{x_1[n]} + 2 \cdot \underbrace{2^n u[n]}_{x_2[n]}.$$

The behavior of the sum is dominated by the increasing term, as shown in **Figure 4.9a**.

$X_A(z)$, the z-transform of $x_A[n]$, is the sum of two terms, each of which converges to closed form for an appropriate range of z:

$$X_A(z) = \sum_{n=-\infty}^{\infty} x_A[n]z^{-n} = \sum_{n=-\infty}^{\infty} x_1[n]z^{-n} + 2\sum_{n=-\infty}^{\infty} x_2[n]z^{-n} = \sum_{n=0}^{\infty} \tfrac{1}{2}^n z^{-n} + 2\sum_{n=0}^{\infty} 2^n z^{-n}$$

$$= \sum_{n=0}^{\infty} \left(\tfrac{1}{2}z^{-1}\right)^n + 2\sum_{n=0}^{\infty} (2z^{-1})^n = \underbrace{\frac{1}{1-\frac{1}{2}z^{-1}}}_{|\frac{1}{2}z^{-1}|<1} + \underbrace{\frac{2}{1-2z^{-1}}}_{\cap\ |2z^{-1}|<1}$$

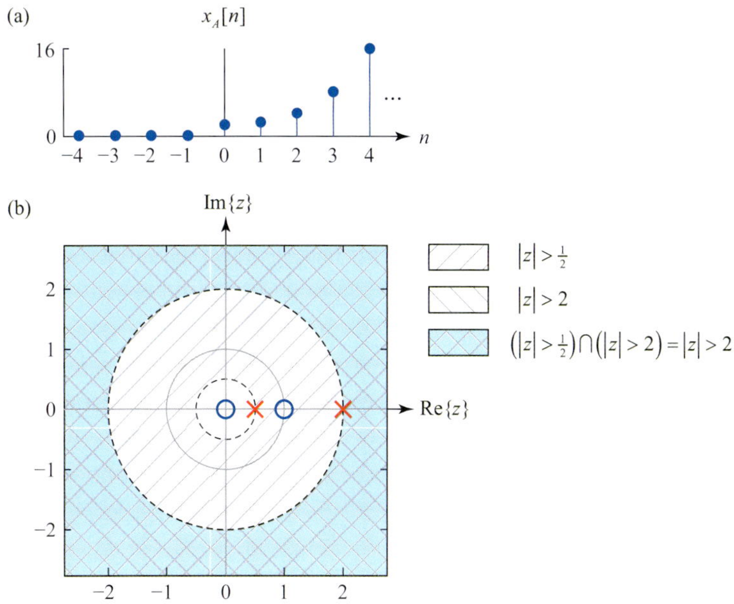

Figure 4.9 Pole-zero plot of right-sided sequence with multiple poles

The critical step is the replacement of each of the infinite summations with closed-form expressions for the infinite sum. The condition for the convergence of the first summation is $|(1/2)z^{-1}| < 1$ or, equivalently, $|z| > 1/2$. The condition for the convergence of the second summation is $|2z^{-1}| < 1$, or $|z| > 2$. The ROCs of these two summations are shown under each term. In order for $X_A(z)$ to exist in closed form, *both* of the summations must converge. If either one (or both) fails to converge, then $X_A(z)$ will blow up. Hence, the condition for the existence of $X_A(z)$ is $|z| > 1/2$ *and* $|z| > 2$. The ROC of $X_A(z)$ is therefore the intersection of the ROCs of the two terms: $(|z| > 1/2) \cap (|z| > 2) = |z| > 2$. Define

$$X_1(z) \triangleq \mathcal{Z}\{x_1[n]\} = \frac{1}{1 - \frac{1}{2}z^{-1}}, \quad |z| > \tfrac{1}{2},$$

and

$$X_2(z) \triangleq \mathcal{Z}\{x_2[n]\} = \frac{1}{1 - 2z^{-1}}, \quad |z| > 2.$$

The complete z-transform is

$$X_A(z) = X_1(z) + 2X_2(z) = \underbrace{\frac{1}{1 - \frac{1}{2}z^{-1}}}_{|z|>\frac{1}{2}} + \underbrace{\frac{2}{1 - 2z^{-1}}}_{|z|>2} = \frac{3(1 - z^{-1})}{(1 - \frac{1}{2}z^{-1})(1 - 2z^{-1})}, \quad |z| > 2.$$

$$= \frac{3z(z-1)}{(z - \frac{1}{2})(z - 2)}$$

We will leave the denominator factored so the roots (i.e., the poles) are obvious. Because the denominator of $X_A(z)$ is the product of the denominators of the two terms $X_1(z)$ and $X_2(z)$, the poles of $X_A(z)$ are just the poles of $X_1(z)$ and $X_2(z)$. The pole at $z = 1/2$ comes from $X_1(z)$; the pole at $z = 2$ comes from $X_2(z)$. However, the zeros of $X_A(z)$ are determined by the cross-multiplication of the denominators and numerators of $X_1(z)$ and $X_2(z)$. The sequences $x_1[n]$ and $x_2[n]$ are both right-sided, so the ROC of each of their transforms, $X_1(z)$ and $X_2(z)$, is the exterior of a circle bounded by the appropriate pole ($z = 1/2$ and $z = 2$, respectively). In **Figure 4.9b**, we show the region of convergence of $X_1(z)$, $|z| > 1/2$ (hatched with), and the region of convergence of $X_2(z)$, $|z| > 2$ (hatched with). The ROC of $X_A(z)$ is the intersection of the ROCs of $X_1(z)$ and $X_2(z)$, namely $(|z| > 1/2) \cap (|z| > 2) = |z| > 2$ (hatched with). This ROC is the exterior of a circle bounded by the "outermost pole," that is, the pole with the greatest absolute value, namely $z = 2$. Accordingly we make the general statement:

> The ROC of the z-transform of a right-sided sequence extends from the outermost pole to infinity.

Since the ROC of $X_A(z)$ does not include the unit circle, the DTFT of this sequence does not exist.

A few other general points follow from this example:

> If a right-sided system is stable, all the poles must be inside the unit circle.

Since the ROC of the z-transform of a right-sided sequence extends from the outermost pole to infinity, the outermost pole must be inside the unit circle in order for the ROC to include the unit circle.

> The ROC cannot include a pole.

That should be obvious, because by definition, the ROC is the range of z for which the z-transform converges, and a pole is a value of z at which the z-transform goes to infinity.

> The ROC can include a zero.

A zero is, by definition, a value of z at which the z-transform goes to zero. You cannot get more convergent than that. However, not every zero is in the ROC. For example, $X_A(z)$ has zeros at $z = 0$ and $z = 1$, neither of which is in the ROC. The reason is that we defined the ROC to be a *contiguous* locus of points, not just a singular value, at which the transform converges. Having a contiguous ROC is important in computing the inverse z-transform using contour integration, where the integration path must describe a continuous closed contour in the z-plane.

> A right-sided system is causal if the number of poles is greater than or equal to the number of zeros.

The proof follows by looking at the limit of $H(z)$ when it is expanded in series form,

$$H(z) = \sum_{n=-\infty}^{\infty} h[n]z^{-n} = \cdots + h[-2]z^2 + h[-1]z + h[0] + h[1]z^{-1} + h[2]z^{-2} + \cdots$$

Now consider the limit as $z \to \infty$,

$$\lim_{z\to\infty} H(z) = \lim_{z\to\infty} (\cdots + h[-2]z^2 + h[-1]z + h[0] + h[1]z^{-1} + h[2]z^{-2} + \cdots).$$

Causality implies that $h[n] = 0$, $n < 0$. Hence, if a system is causal, all the terms in the summation for $n < 0$ must be zero. Furthermore, all the terms in the summation for $n > 0$ must also be zero, since $z^{-n} \to 0$ in the limit as $z \to \infty$ for all $n > 0$. Hence, for a causal system the only term of the summation that survives as $z \to \infty$ is a constant, $h[0]$,

$$\lim_{z\to\infty} H(z) = h[0]. \tag{4.8}$$

Now, look at $H(z)$ in terms of poles and zeros,

$$H(z) = A \frac{\prod_{k=1}^{M} (z - b_k)}{\prod_{k=1}^{N} (z - a_k)},$$

where a_k and b_k are the locations of the N poles and M zeros respectively and A is a constant. Again, investigate what happens to $H(z)$ as we send $z \to \infty$:

$$\lim_{z\to\infty} H(z) = \lim_{z\to\infty} A \frac{\prod_{k=1}^{M} (z - b_k)}{\prod_{k=1}^{N} (z - a_k)} = A\frac{z^M}{z^N} = Az^{M-N}.$$

If the number of poles is less than the number of zeros, $N < M$, then $H(z)$ blows up as $z \to \infty$, but if $N \geq M$, then $H(z)$ approaches a constant given by Equation (4.8),

$$\lim_{z\to\infty} H(z) = \begin{cases} h[0], & N = M \\ 0, & N > M \end{cases}.$$

In other words, $h[n]$ is causal if the number of poles is greater than or equal to the number of zeros.

Before leaving this example, note the correspondence between the time and frequency domains. Just as the ROC of $X_A(z)$ is controlled by the pole with the largest absolute value, so the behavior of $x_A[n]$ is determined by the term with the largest absolute value of α (i.e., $2^n u[n]$).

Left-sided sequence Now consider the sequence $x_B[n]$, which is formed from the scaled sum of two left-sided sequences $x_3[n]$ and $x_4[n]$, each of the form $-\alpha^n u[-n-1]$,

$$x_B[n] = x_3[n] + 2x_4[n] = \underbrace{-\tfrac{1}{2}^n u[-n-1]}_{x_3[n]} - 2 \cdot \underbrace{2^n u[-n-1]}_{x_4[n]}.$$

The sequence $x_B[n]$ and the pole-zero plot and ROC of $X_B(z)$ are shown in **Figure 4.10**. Even though $x_4[n]$ converges monotonically as $n \to -\infty$, the sum is dominated by $x_3[n]$, which decreases monotonically as $n \to -\infty$. Define

$$X_3(z) \triangleq \mathfrak{Z}\{x_3[n]\} = \frac{1}{1 - \frac{1}{2}z^{-1}}, \quad |z| < \tfrac{1}{2},$$

and

$$X_4(z) \triangleq \mathcal{Z}\{x_4[n]\} = \frac{1}{1-2z^{-1}}, \quad |z| < 2.$$

We can now express $X_B(z)$, the z-transform of $x_B[n]$, as the sum of these two terms, each of which converges to closed form for an appropriate range of z,

$$X_B(z) = X_3(z) + 2X_4(z) = \underbrace{\frac{1}{1-\frac{1}{2}z^{-1}}}_{|z|<\frac{1}{2}} \cap + \underbrace{\frac{2}{1-2z^{-1}}}_{|z|<2} = \frac{3(1-z^{-1})}{\left(1-\frac{1}{2}z^{-1}\right)(1-2z^{-1})}$$

$$= \frac{3z(z-1)}{\left(z-\frac{1}{2}\right)(z-2)}, \quad |z| < \frac{1}{2}.$$

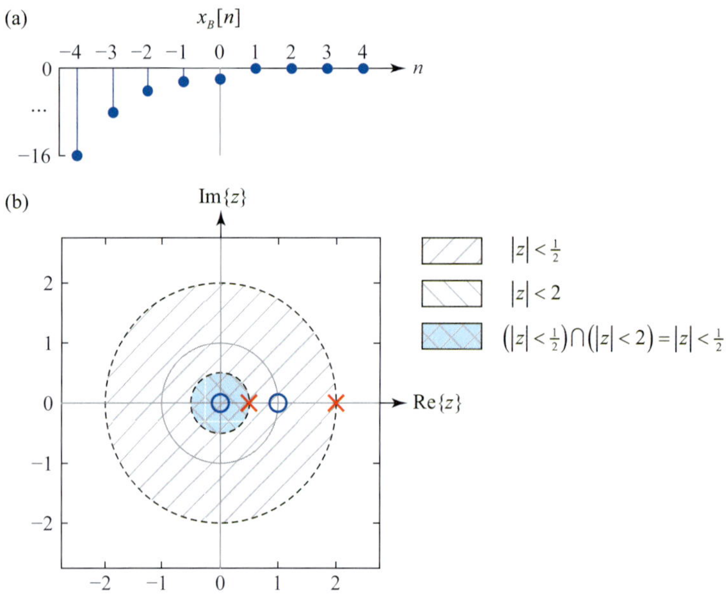

Figure 4.10 Pole-zero plot of left-sided sequence with multiple poles

Both $x_3[n]$ and $x_4[n]$ are left-sided, so the ROCs of their transforms $X_3(z)$ and $X_4(z)$ are the interior of circles bounded by the appropriate pole: $z = 1/2$ for $X_3(z)$ and $z = 2$ for $X_4(z)$. In **Figure 4.10** we show the region of convergence of $X_3(z)$, $|z| < 1/2$ (hatched with), and the region of convergence of $X_4(z)$, $|z| < 2$ (hatched with). The ROC of $X_B(z)$ is the intersection of the ROCs of the two terms $(|z| < 1/2) \cap (|z| < 2) = |z| < 1/2$, shown hatched with). This ROC is the interior of a circle bounded by the "innermost pole," that is, the pole with the smallest absolute value, namely $z = 1/2$. Accordingly we make the following general statement:

> The ROC of the z-transform of a left-sided sequence extends from the innermost pole to zero.

Since the ROC of $X_B(z)$ does not include the unit circle, the DTFT of this left-sided sequence does not exist.

Double-sided convergent sequence Now consider the sequence $x_C[n]$, which is formed from the scaled sum of a right-sided sequence $x_1[n]$ and a left-sided sequence $x_4[n]$, as shown in **Figure 4.11a**,

$$x_C[n] = x_1[n] + 2x_4[n] = \underbrace{\tfrac{1}{2}^n u[n]}_{x_1[n]} - 2\cdot\underbrace{2^n u[-n-1]}_{x_4[n]}.$$

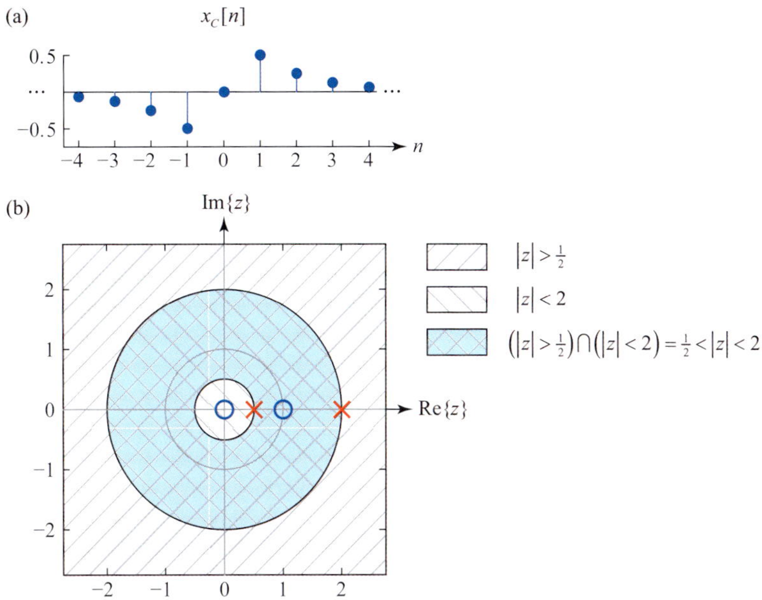

Figure 4.11 Pole-zero plot of double-sided convergent sequence

This is termed a **double-sided sequence**, since it extends from $-\infty < n < \infty$. This double-sided sequence $x_C[n]$ converges both as $n \to \infty$ and as $n \to -\infty$. The z-transform of $x_C[n]$ comprises two terms, each of which converges to closed form for an appropriate range of z,

$$X_C(z) = X_1(z) + 2X_4(z) = \underbrace{\frac{1}{1 - \frac{1}{2}z^{-1}}}_{|z| > \frac{1}{2}} \cap \underbrace{+ \frac{2}{1 - 2z^{-1}}}_{|z| < 2} = \frac{3(1 - z^{-1})}{\left(1 - \frac{1}{2}z^{-1}\right)(1 - 2z^{-1})}, \quad \frac{1}{2} < |z| < 2.$$

$$= \frac{3z(z-1)}{\left(z - \frac{1}{2}\right)(z-2)}$$

As we have already seen, the sequence $x_1[n]$ is right-sided, so its transform $X_1(z)$ has an ROC ($|z| > 1/2$) that is the exterior of a circle bounded by the pole at $z = 1/2$; the sequence $x_4[n]$ is left-sided, so its transform $X_4(z)$ has an ROC ($|z| < 2$) that is the interior of a circle bounded by the pole at $z = 2$. The ROC of $X_C(z)$ is the intersection of the ROCs of two terms: $(|z| > 1/2) \cap (|z| < 2) = 1/2 < |z| < 2$. As shown in **Figure 4.11b**, this ROC is in the shape of an annulus (a doughnut) bounded by the two poles. Since an ROC can never contain a pole, we can make the following general statement:

> The ROC of the z-transform of a two-sided sequence is an annulus bounded by two adjacent poles.

Since the ROC of $X_C(z)$ includes the unit circle, the DTFT of this double-sided sequence exists.

Double-sided, non-convergent sequence Finally, consider the sequence $x_D[n]$, which is formed from the scaled sum of the monotonically increasing right-sided sequence $x_2[n]$ and the monotonically decreasing left-sided sequence $x_3[n]$, as shown in **Figure 4.12a**,

$$x_D[n] = x_2[n] + 2x_3[n] = \underbrace{2^n u[n]}_{x_2[n]} - \underbrace{2 \cdot \tfrac{1}{2}^n u[-n-1]}_{x_3[n]}.$$

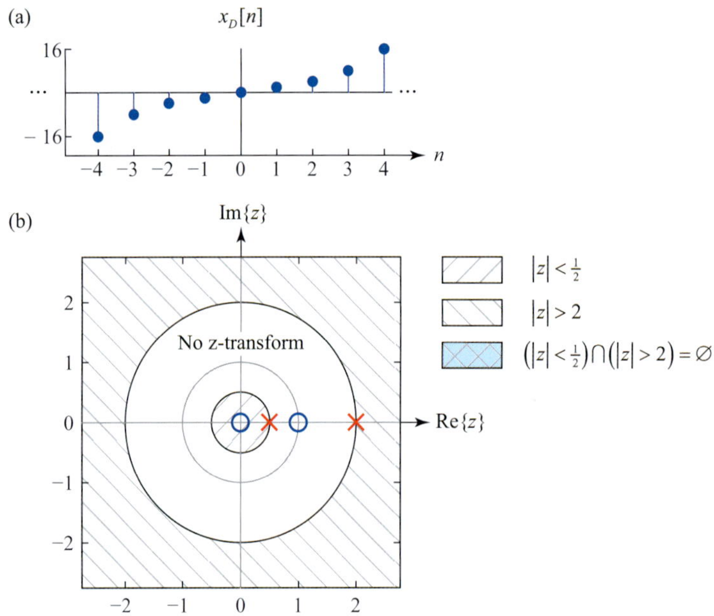

Figure 4.12 Pole-zero plot of a double-sided non-convergent sequence

The sequence $x_D[n]$ is double-sided and non-convergent. The z-transform of $x_D[n]$ comprises two terms, each of which separately converges to closed form for an appropriate range of z,

$$X_D(z) = X_2(z) + 2X_3(z) = \underbrace{\frac{1}{1 - \frac{1}{2}z^{-1}}}_{|z|<\frac{1}{2}} \;\cap\; + \underbrace{\frac{2}{1 - 2z^{-1}}}_{|z|>2} = \varnothing.$$

However, since $X_2(z)$ is the transform of a left-sided sequence whose ROC is the interior of a circle bounded by the pole at $z = 1/2$ and $X_3(z)$ is the transform of a right-sided sequence whose ROC is the exterior of a circle bounded by the pole at $z = 2$, there is no overlap between the ROCs of $X_2(z)$ and $X_3(z)$ and therefore no range of z for which the z-transform of $x_D[n]$ exists.

4.2.5 Finding the z-transform from the pole-zero plot

The pole-zero plot provides a clear graphic summary of the poles and zeros of the z-transform and its ROC. Given the pole-zero plot plus the ROC you can reconstruct the z-transform to within a multiplicative constant. For example, given the pole-zero plot and ROC of **Figure 4.4a**, we can state that the z-transform must be of the form

$$H(z) = A\frac{z}{z-\frac{1}{2}}, \quad |z| > \tfrac{1}{2},$$

where A is an unknown constant. The constant does not affect the location of the pole or the zero and hence cannot be determined from the plot. In order to determine $H(z)$ unambiguously, we also need to specify $H(z_0)$, the value of the transform at a non-singular value of $z = z_0$. For example, if we were given that $H(1) = 4$, we would conclude that

$$H(1) = A\frac{1}{1-\frac{1}{2}} = 4,$$

from which we find that $A = 2$.

4.2.6 Complex poles and zeros

So far, all our examples have been of systems with real poles and zeros. Consider the sequence

$$h[n] = \alpha^n \cos \omega_o n \, u[n].$$

This sequence comprises an oscillatory term, $\cos\omega_0 n$, of frequency ω_0 modulated (i.e., multiplied) by a decaying-exponential envelope, $\alpha^n u[n]$. As we will see a bit later, $h[n]$ is actually the impulse response of a system described by a second-order LCCDE. Writing $h[n]$ as the sum of two terms:

$$h[n] = \alpha^n \cos \omega_o n \, u[n] = \alpha^n\left(\tfrac{1}{2}e^{j\omega_0 n} + \tfrac{1}{2}e^{-j\omega_0 n}\right)u[n] = \tfrac{1}{2}(\alpha e^{j\omega_0})^n u[n] + \tfrac{1}{2}(\alpha e^{-j\omega_0})^n u[n].$$

Each term is a right-sided sequence of the form $\beta^n u[n]$, so application of Equation (4.3) gives the z-transform

$$H(z) = 3\left\{\tfrac{1}{2}(\alpha e^{j\omega_0})^n u[n]\right\} + 3\left\{\tfrac{1}{2}(\alpha e^{-j\omega_0})^n u[n]\right\} = \frac{1}{2}\underbrace{\frac{1}{1-\alpha e^{j\omega_0}z^{-1}}}_{|\alpha e^{j\omega_0}z^{-1}|<1} + \frac{1}{2}\underbrace{\frac{1}{1-\alpha e^{-j\omega_0}z^{-1}}}_{|\alpha e^{-j\omega_0}z^{-1}|<1}.$$

$$|\alpha e^{j\omega_0}z^{-1}|<1 \;\cap\; |\alpha e^{-j\omega_0}z^{-1}|<1$$

Since $|\alpha e^{j\omega_o}z^{-1}| = |\alpha e^{-j\omega_o}z^{-1}| = |\alpha z^{-1}|$, the ROC of these two terms is identical, i.e., $|\alpha z^{-1}| < 1$ or, equivalently, $|z| > |\alpha|$. Hence, the z-transform reduces to

$$\begin{aligned} H(z) &= \frac{1}{2}\frac{1}{1-\alpha e^{j\omega_0}z^{-1}} + \frac{1}{2}\frac{1}{1-\alpha e^{-j\omega_0}z^{-1}} = \frac{1-\alpha\left(\frac{1}{2}e^{j\omega_0} + \frac{1}{2}e^{-j\omega_0}\right)z^{-1}}{(1-\alpha e^{j\omega_0}z^{-1})(1-\alpha e^{-j\omega_0}z^{-1})} \\ &= \frac{z(z-\alpha\cos\omega_o)}{(z-\alpha e^{j\omega_0})(z-\alpha e^{-j\omega_0})}, \quad |z| > |\alpha|, \end{aligned} \tag{4.9}$$

where we have again left the denominator in factored form so the roots are obvious. There are two real zeros plus two poles that are complex conjugates of each other, located at a distance α from the origin at angles $\pm\omega_0$ with respect to the real axis, as shown in **Figure 4.13**.

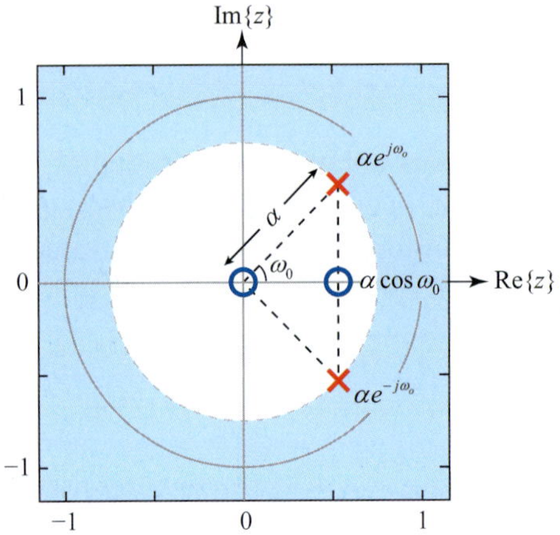

Figure 4.13 Pole-zero plot of a second-order system

It is worth understanding how the positions of the poles and zeros of the z-transform of $h[n] = \alpha^n \cos\omega_0 n u[n]$ depend on our choices of α and ω_0. First, fix the value of ω_0 and vary α. **Figures 4.14a–f** show the poles and zeros of the z-transform of $h[n]$ for values of $\alpha = 0$, 0.25, 0.5, 0.75, 1 and 1.25, at a fixed value of $\omega_0 = \pi/4$.

When $\alpha = 0$ (**Figure 4.14a**),

$$h[n] = 0^n \cos(\pi n/4)\, u[n] = \begin{Bmatrix} 1, & n = 0 \\ 0, & n \neq 0 \end{Bmatrix} = \delta[n].$$

In this degenerate case, Equation (4.9) reduces to $H(z) = 1$, so there are no poles and zeros, and the ROC is the entire z-plane. For all values of $\alpha > 0$, the pole-zero plot comprises two poles at complex-conjugate positions, $z = \pm\alpha e^{j\omega_0}$. As α increases, the poles move out a distance α from the center of the z-plane (i.e., from the point $z = 0$) along lines at angles $\pm\omega_0$ with respect to the real axis. One of the zeros stays at zero, the other one "tracks" the position of the two poles, being located at the position of the projection of either pole onto the real axis. When α is small (e.g., $\alpha = 0.25$, **Figure 4.14b**), the sequence is a highly overdamped cosine. As α approaches one (e.g., $\alpha = 0.75$, **Figure 4.14d**), the damping of the sequence decreases and the oscillations of the cosine become more evident. When $\alpha = 1$ (**Figure 4.14e**), the poles lie on the unit circle and the sequence is an undamped cosine. Finally, as α increases beyond one (e.g., $\alpha = 1.25$, **Figure 4.14f**), the poles are outside of the unit circle, and the impulse response is oscillatory with an exponentially increasing envelope. Even though the damping changes with α, the frequency of the modulated cosine stays constant.

Since $h[n]$ in this example is a right-sided sequence, the ROC extends from the poles to ∞. As long as $\alpha < 1$, the poles are inside the unit circle (at the same distance from the center of

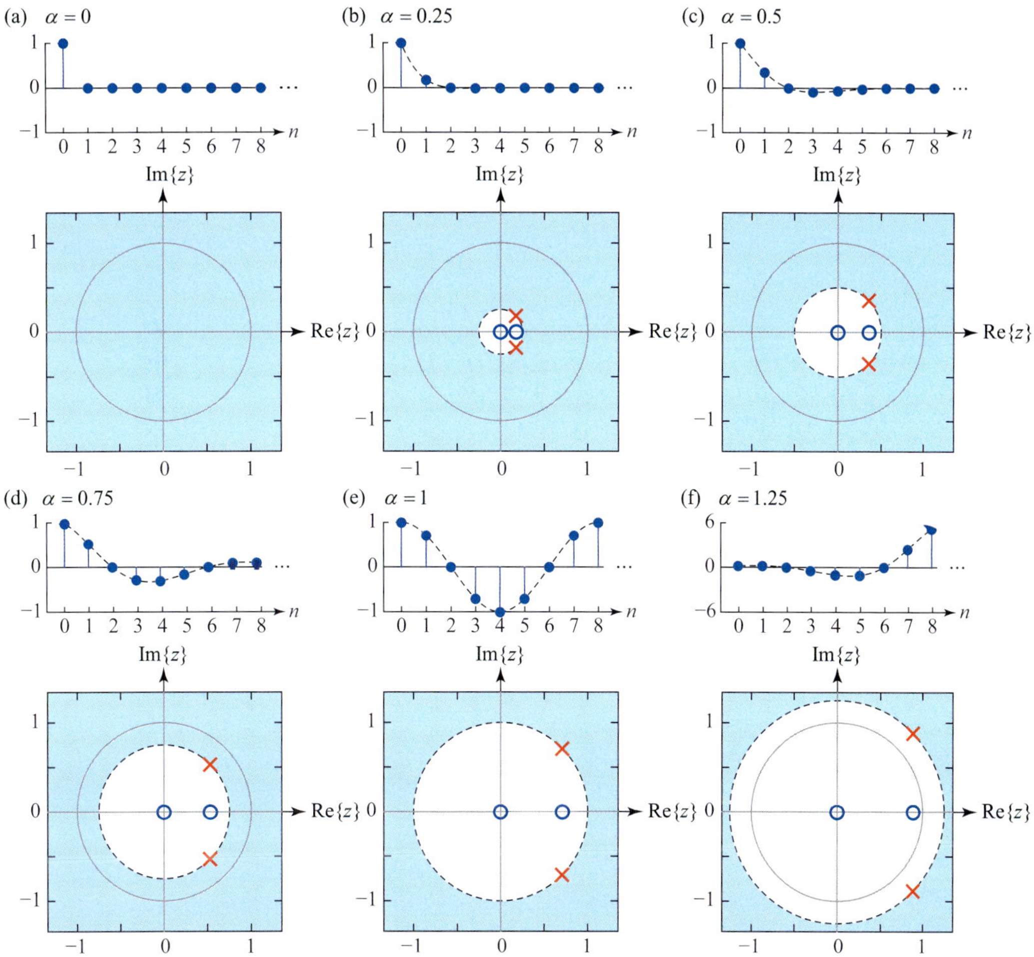

Figure 4.14 Poles and zeros of a second-order system at a fixed value of ω_0

the z-plane), so the ROC includes the unit circle. Thus $H(\omega)$ exists (or, equivalently, $h[n]$ converges) for any choice of ω_0. When $\alpha > 1$, the poles are outside the unit circle. The response is divergent and $H(\omega)$ no longer exists.

Now, let us fix α and vary ω_0. **Figures 4.15a–e** show the poles and zeros of the z-transform of $h[n]$ for values of $\omega = 0$, $\pi/4$, $\pi/2$, $3\pi/4$ and π, at a fixed value of $\alpha = 0.75$.

When $\omega_0 = 0$ (**Figure 4.15a**), the sequence $h[n] = \alpha^n \cos \omega_0 n u[n]$ becomes $h[n] = \alpha^n u[n]$, which is just a decaying power-law sequence. Equation (4.9) then reduces to

$$H(z) = \frac{z\cancel{\left(z - \frac{1}{2}\right)}}{\left(z - \frac{1}{2}\right)^2} = \frac{z}{z - \frac{1}{2}}.$$

A zero at $z = 1/2$ has canceled one of the poles at $z = 1/2$, leaving a single real pole and a single zero.

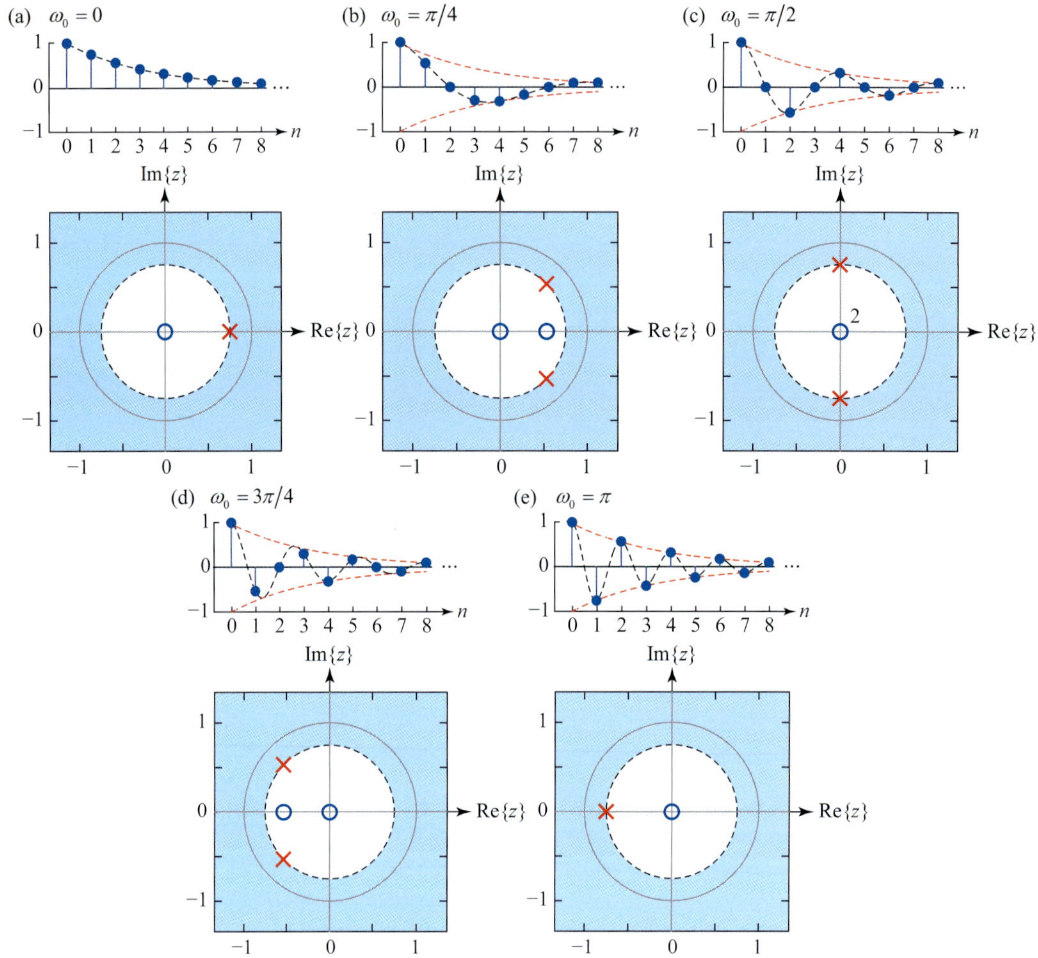

Figure 4.15 Poles and zeros of a second-order system at a fixed value of α

For non-zero values of $0 < \omega_0 < \pi$, the sequence is a cosine of frequency ω_0 that oscillates within the envelope delimited by $\alpha^n u[n]$ and $-\alpha^n u[n]$ (the dotted red curves in **Figure 4.15**). Looking at the pole-zero plot, you see that $H(z)$ has two complex-conjugate poles located at a distance α from the center of the z-plane at angles $\pm\omega_0$, with respect to the real axis. Increasing ω_0 corresponds to moving the poles around a circle of radius α: one pole moves counterclockwise around the top half of the circle from $\omega_0 = 0$ to π; the other (complex-conjugate) pole moves clockwise around the bottom half of the circle from $\omega_0 = 0$ to $-\pi$.

When $\omega_0 = \pi$ (or $\omega_0 = -\pi$), then the sequence becomes $h[n] = \alpha^n \cos \pi n u[n] = \alpha^n(-1)^n u[n] = (-\alpha)^n u[n]$, and $H(z)$ reduces to

$$H(z) = \frac{z\cancel{\left(z+\frac{1}{2}\right)}}{\left(z-\frac{1}{2}\right)\cancel{\left(z+\frac{1}{2}\right)}} = \frac{z}{z+\frac{1}{2}}.$$

Once again, we have a zero (this time at $z = -\alpha$) that cancels one of two poles and leaves $H(z)$ with one pole and one zero. The sequence $h[n]$ is exactly what we would expect from this transform: a decaying, alternating, convergent sequence. Since this is a right-sided sequence, the ROC extends from the poles to ∞. As long as $\alpha < 1$, the poles are inside the unit circle (at the same distance from the center of the z-plane), so the ROC includes the unit circle. $H(\omega)$ exists and $h[n]$ converges for any choice of ω_0.

4.2.7 Some important transforms

Table 4.3 presents some important z-transforms, some of which we have discussed above. The rest are exercises for you. In all cases, note we have to specify both the expression for $H(z)$ and the region of convergence.

Table 4.3 Important transforms

$h[n]$	$H(z)$	ROC
$\delta[n]$	1	Entire z-plane
$\delta[n-n_0]$	z^{-n_0}	Entire z-plane (except $z=0,\ n_0>0$)
$\alpha^n u[n]$	$\dfrac{1}{1-\alpha z^{-1}}$	$\|z\|>\|\alpha\|$
$-\alpha^n u[-n-1]$	$\dfrac{1}{1-\alpha z^{-1}}$	$\|z\|<\|\alpha\|$
$n\alpha^n u[n]$	$\dfrac{-\alpha z^{-1}}{(1-\alpha z^{-1})^2}$	$\|z\|>\|\alpha\|$
$(n+1)\alpha^n u[n]$	$\dfrac{1}{(1-\alpha z^{-1})^2}$	$\|z\|>\|\alpha\|$
$\dfrac{(n+1)(n+2)\cdots(n+N-1)}{(N-1)!}\alpha^n u[n]$	$\dfrac{1}{(1-\alpha z^{-1})^N}$	$\|z\|>\|\alpha\|$
$\alpha^n \cos\omega_o n\, u[n]$	$\dfrac{1-\alpha\cos\omega_o z^{-1}}{(1-\alpha e^{j\omega_0}z^{-1})(1-\alpha e^{-j\omega_0}z^{-1})} = \dfrac{z(z-\alpha\cos\omega_o)}{z^2-2\alpha\cos\omega_0 z+\alpha^2}$	$\|z\|>\|\alpha\|$
$\alpha^n \cos(\omega_o n+\varphi)\, u[n]$	$\dfrac{\cos\varphi-\alpha\cos(\omega_o-\varphi)z^{-1}}{(1-\alpha e^{j\omega_0}z^{-1})(1-\alpha e^{-j\omega_0}z^{-1})} = \dfrac{z(z\cos\varphi-\alpha\cos(\omega_o-\varphi))}{z^2-2\alpha\cos\omega_0 z+\alpha^2}$	$\|z\|>\|\alpha\|$
$\alpha^n \sin\omega_o n\, u[n]$	$\dfrac{\alpha\sin\omega_0 z^{-1}}{(1-\alpha e^{j\omega_0}z^{-1})(1-\alpha e^{-j\omega_0}z^{-1})} = \dfrac{\alpha z\sin\omega_0}{z^2-2\alpha\cos\omega_0 z+\alpha^2}$	$\|z\|>\|\alpha\|$
$\alpha^n \sin(\omega_o n+\varphi)\, u[n]$	$\dfrac{\sin\varphi+\alpha\sin(\omega_0-\varphi)z^{-1}}{(1-\alpha e^{j\omega_0}z^{-1})(1-\alpha e^{-j\omega_0}z^{-1})} = \dfrac{z(z\sin\varphi+\alpha\sin(\omega_0-\varphi))}{z^2-2\alpha\cos\omega_0 z+\alpha^2}$	$\|z\|>\|\alpha\|$

4.2.8 Finite-length sequences

Understanding the z-transform of finite-length sequences is particularly important. For example, we frequently want to take the z-transform of the impulse response of an FIR system, which is, by definition, the sum of a finite number of scaled and shifted impulses. In general, we can express an N-point impulse response as

$$h[n] = \sum_{k=0}^{N-1} b_k \delta[n - n_0 - k],$$

where n_0 represents the "leftmost" point of the impulse response. Hence,

$$\begin{aligned} H(z) &= \sum_{n=-\infty}^{\infty} h[n]z^{-n} = \sum_{n=-\infty}^{\infty} \left(\sum_{k=0}^{N-1} b_k \delta[n - n_0 - k] \right) z^{-n} = \sum_{k=0}^{N-1} b_k \sum_{n=-\infty}^{\infty} \delta[n - n_0 - k] z^{-n} \\ &= \sum_{k=0}^{N-1} b_k z^{-(n_0+k)} = z^{-n_0} \sum_{k=0}^{N-1} b_k z^{-k}. \end{aligned}$$

Expanding this expression:

$$\begin{aligned} H(z) &= z^{-n_0}(b_0 + b_1 z^{-1} + \cdots + b_{N-1} z^{N-1}) = z^{-(n_0+N-1)}(b_0 z^{N-1} + b_1 z^{N-2} + \cdots + b_{N-1}) \\ &= \frac{b_0 z^{N-1} + b_1 z^{N-2} + \cdots + b_{N-1}}{z^{n_0+N-1}}. \end{aligned} \quad \textbf{(4.10)}$$

Expressed as powers of z, you can see that the numerator of $H(z)$ is a polynomial of order $N-1$. The roots of this polynomial are the $N-1$ zeros of $H(z)$. The denominator of $H(z)$ has $n_0 + N - 1$ poles, all of which are at $z = 0$. Because the poles of a finite-length sequence (such as the impulse response of an FIR system) can *only* occur at $z = 0$, an FIR system is sometimes called an **all-zero system**, though it would perhaps be more accurate to call it an "all-zero-except-for-possible-poles-at-$z = 0$" system.

A couple of important related points about FIR systems follow from the above results. Since the only poles of an FIR system are located at zero, the ROC is the entire z-plane (except perhaps the point $z = 0$, if there are poles there). Since the ROC of every FIR system includes the unit circle, then *every FIR system is stable*. This is one of the most compelling features of FIR systems, for example discrete-time FIR filters: they can never be unstable. Here are some examples.

Example 4.1

Given the impulse response of an FIR system $h[n]$ shown in **Figure 4.16a**, find $H(z)$ and plot the pole-zero plot.

➤ Solution:

For this sequence, the transform is

$$H(z) = 1 + 3z^{-1} + 2z^{-2} = \frac{z^2 + 3z + 2}{z^2} = \frac{(z+1)(z+2)}{z^2}, \quad |z| > 0.$$

There are zeros at $z = -1$ and $z = -2$, and a double pole at $z = 0$. **Figure 4.16b** shows the pole-zero plot.

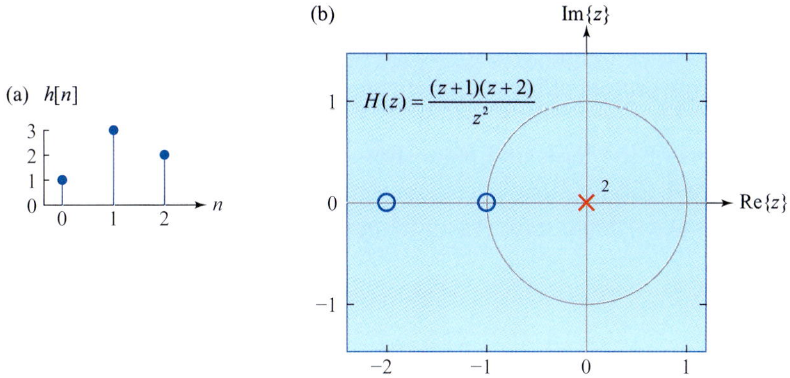

Figure 4.16 Pole-zero plot for an FIR system

Example 4.2

Find the z-transform and pole-zero plot for pulse $h[n]$ of length N shown in **Figure 4.17a**.

➤ Solution:

$h[n]$ is a sequence of N impulses,

$$h[n] = \begin{Bmatrix} 1, & 0 \le n < N \\ 0, & \text{otherwise} \end{Bmatrix} = u[n] - u[n-N].$$

So,

$$H(z) = \sum_{n=-\infty}^{\infty} h[n]z^{-n} = \sum_{n=0}^{N-1} z^{-n} = \frac{1-z^{-N}}{1-z^{-1}} = \frac{1}{z^{N-1}} \cdot \frac{z^N - 1}{(z-1)}, \quad |z| \neq 0.$$

The zeros of $H(z)$ are the N roots of the numerator $z^N - 1$. Writing "1" as a complex number $e^{j2\pi k}$, the roots are the solutions of $z^N - e^{jk2\pi} = 0$, which gives $z = e^{j2\pi k/N}$, $0 \le k < N$. So the N zeros of $H(z)$ are spaced around the unit circle at multiples of $2\pi/N$, starting at $z = e^{j0} = 1$. The denominator of $H(z)$ also has N roots: $N-1$ poles at $z = 0$, plus one pole at $z = 1$ that exactly cancels the zero at $z = e^{j0} = 1$. **Figure 4.17b** shows the pole-zero plot for a pulse of length $N = 8$.

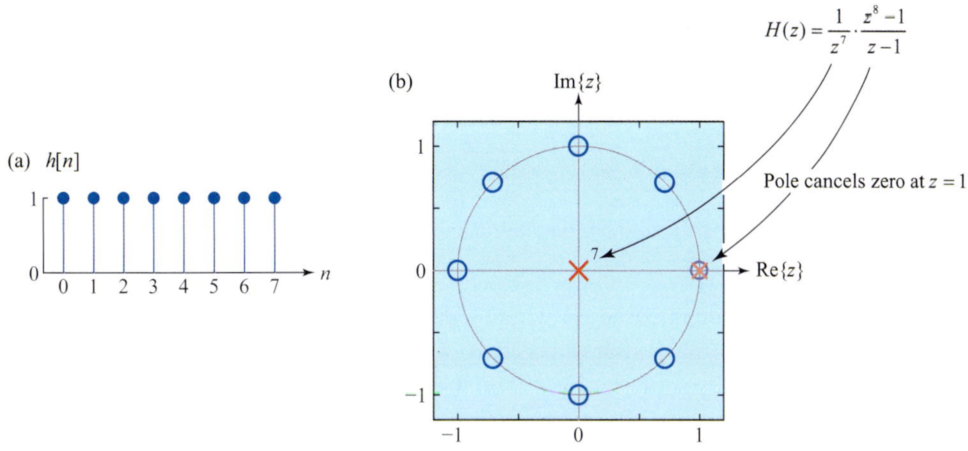

Figure 4.17 Pole-zero plot of a pulse

4.2.9 Plotting pole-zero plots with Matlab

Matlab provides the `zplane` function to plot poles and zeros. The syntax is `zplane(b,a)`, where `b` and `a` are row vectors of the coefficients b_k and a_k of $H(z)$ expressed in powers of z^{-1}, as in Equation (4.5). Alternately, the function can be called using `zplane(z,p)`, where `z` and `p` are column vectors of the zeros and poles. For example, consider producing the pole-zero plot shown in **Figure 4.15b** for $h[n] = \alpha^n \cos \omega_0 n\, u[n]$ with $\alpha = 0.75$ and $\omega_0 = \pi/4$. From Equation (4.9) there are zeros at $z = 0$ and $\alpha \cos \omega_0$, and a pair of complex-conjugate poles at $z = \alpha e^{\pm j\omega_0}$. So:

```
alpha = 0.75;
w0 = pi/4;
z = [0; alpha*cos(w0)];
p = [alpha*exp(1j*w0*[1;-1])];
zplane(z, p)
```

`zplane` does not plot the ROC.

4.3 ★ Linear-phase FIR systems

In Chapter 3, we introduced linear-phase FIR systems and suggested why they are important. Recall that linear-phase systems can have either symmetric or antisymmetric impulse responses, which can be of either even or odd length, as shown by the examples in **Figure 4.18**. A causal symmetric sequence of length N is defined by $h[n] = h[N-1-n]$ and a causal antisymmetric sequence of length N is defined by $h[n] = -h[N-1-n]$.

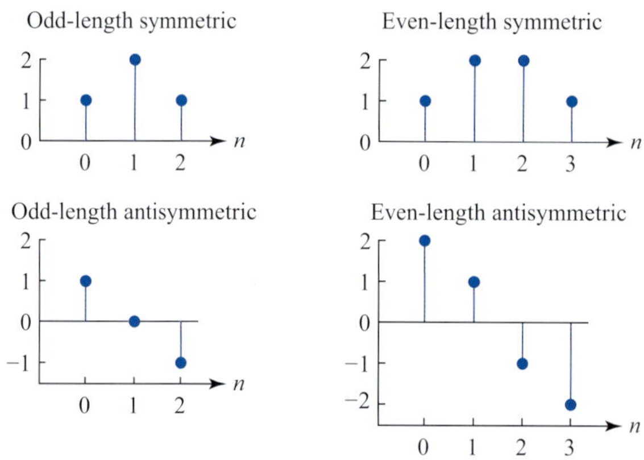

Figure 4.18 Symmetric and antisymmetric impulse responses

All the lowpass filters based on window designs that we will study in Chapter 7, such as the Hamming, Hann and Kaiser filters, have symmetric impulse responses, as do the transformed highpass and bandpass designs based on these windows. Antisymmetric impulse responses yield highpass and bandpass filters. In this section, we will look at the z-transform of linear-phase systems. In particular, we will show that the singularities of the z-transforms of these systems – that is, the poles and zeros of $H(z)$ – can only occur at particular places in the z-plane. The restrictions on their positions result from the following three constraints:

(1) Poles can only occur at the origin. Zeros can only be real or occur in complex conjugate pairs. The z-transform of any real finite-length causal sequence can be written as

$$
\begin{aligned}
H(z) &= \sum_{n=0}^{N-1} h[n]z^{-n} = A(1 + b_1 z^{-1} \cdots + b_{N-2} z^{-(N-2)} + b_{N-1} z^{-(N-1)}) \\
&= A z^{-(N-1)}(z^{N-1} + b_1 z^{N-2} + \cdots + b_{N-2} z + b_{N-1}) = A z^{-(N-1)} \prod_{k=1}^{N-1} (z - z_k) \\
&= A \frac{\prod_{k=1}^{N-1}(z - z_k)}{z^{N-1}},
\end{aligned} \tag{4.11}
$$

where $A = h[0]$ and $b_n = h[n]/h[0]$ are real coefficients. There are $N-1$ poles at the origin and $N-1$ zeros, z_k. The fact that the coefficients of the numerator polynomial, $b_1, \ldots, b_{N-1}$, are real imposes the restriction that the zeros can only (1) be real or (2) occur in complex-conjugate pairs.

(2) Zeros must occur at conjugate-reciprocal positions in the *z*-plane. The symmetry or antisymmetry of $h[n]$ imposes further important restrictions on the locations of the zeros of $H(z)$. For a symmetric sequence, the coefficients satisfy $b_n = b_{N-1-n}$, so Equation (4.11) becomes

$$H(z) = A z^{-(N-1)}(z^{N-1} + b_1 z^{N-2} + \cdots + b_1 z + 1). \tag{4.12}$$

The zeros of $H(z)$ are the roots of the polynomial $z^{N-1} + b_1 z^{N-2} + \cdots + b_1 z + 1$. Since $h[n]$ is symmetric, we have $h[n] = h[N-1-n]$, and

$$
\begin{aligned}
H(z) &= \mathfrak{Z}\{h[N-1-n]\} = \sum_{n=0}^{N-1} \underbrace{h[N-1-n]}_{m} z^{-n} = \sum_{m=N-1}^{0} h[m] z^{-((N-1)-m)} = z^{-(N-1)} \sum_{m=0}^{N-1} h[m] z^{m} \\
&= z^{-(N-1)} H(z^{-1}),
\end{aligned} \tag{4.13}
$$

or, equivalently,

$$H(z^{-1}) = z^{(N-1)} H(z) = A(z^{N-1} + b_1 z^{N-2} + \cdots + b_1 z + 1).$$

For an antisymmetric sequence $b_n = -b_{N-1-n}$, so Equation (4.12) becomes

$$H(z) = A z^{-(N-1)}(z^{N-1} + b_1 z^{N-2} + \cdots - b_1 z - 1). \tag{4.14}$$

The zeros of $H(z)$ are the roots of the polynomial $z^{N-1} + b_1 z^{N-2} + \cdots - b_1 z - 1$. Since $h[n]$ is antisymmetric, we have $h[n] = -h[N-1-n]$, and a derivation similar to that of Equation (4.13) yields

$$H(z^{-1}) = -z^{-(N-1)} H(z) = -A(z^{N-1} + b_1 z^{N-2} + \cdots - b_1 z - 1).$$

In either case – symmetric or antisymmetric – the zeros of $H(z^{-1})$ are the same as the zeros of $H(z)$. This means that if $z = re^{j\theta}$ is a zero of the system, there must be a zero at the **conjugate-reciprocal** position, $z^{-1} = (1/r)e^{-j\theta}$; this zero has a reciprocal magnitude and negative angle.

(3) The product of the zeros is either +1 or −1. There is one final set of conditions that further constrains the position of zeros in symmetric and antisymmetric systems. From Equation (4.11),

$$H(0) = A\prod_{k=1}^{N-1}(-z_k) = A(-1)^{N-1}\prod_{k=1}^{N-1} z_k.$$

For a symmetric sequence, Equation (4.12) says $H(0) = A$, so

$$\prod_{k=1}^{N-1} z_k = (-1)^{N-1}, \tag{4.15a}$$

and for an antisymmetric sequence, Equation (4.14) says $H(0) = -A$, so

$$\prod_{k=1}^{N-1} z_k = (-1)^{N}. \tag{4.15b}$$

Taken together, these three constraints mean that the zeros of a linear-phase FIR system can occur in the z-plane in only four possible configurations, as shown in **Figure 4.19**:

1. Real zeros at $z = +1$ and/or $z = -1$.
2. Real zeros located at reciprocal locations on the real axis.
3. Pairs of complex-conjugate zeros located on the unit circle.
4. "Quartets" comprising two pairs of complex-conjugate zeros located off the unit circle.

We will now dig into these possibilities a bit.

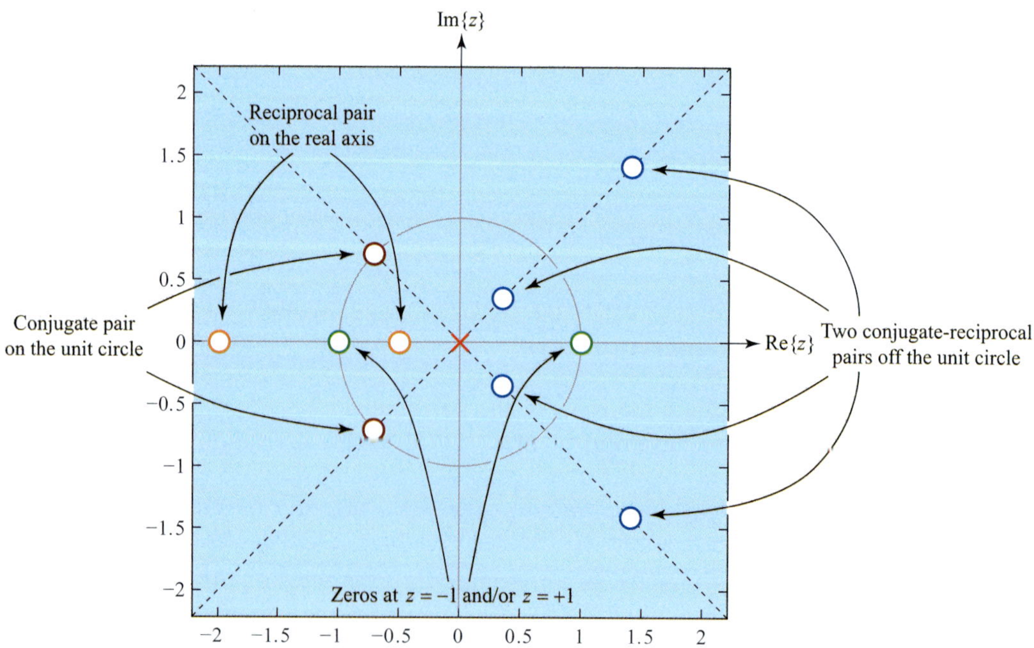

Figure 4.19 Position of zeros for linear-phase FIR systems

4.3.1 Complex zeros

Because complex roots of $H(z)$ must occur in complex-conjugate pairs, if $H(z)$ has a complex zero at $z = re^{j\theta}$ (i.e., $\theta \neq 0$), then there also must be another zero at the conjugate position $z = re^{-j\theta}$. If $r = 1$, this pair of zeros lies on the unit circle at $z = e^{\pm j\theta}$, as shown in the figure with ○ for the example of $\theta = 3\pi/4$. Since $1/z = 1/e^{\pm j\theta} = e^{\mp j\theta}$, these same zeros on the unit circle also satisfy our requirement that there be zeros at the conjugate-reciprocal position. However, if a pair of complex-conjugate zeros is located off the unit circle (i.e., $r \neq 1$), then there has to be *another* pair of zeros at the conjugate-reciprocal positions for a total of *four* zeros: the original conjugate pair at $z = re^{\pm j\theta}$, plus a pair at the reciprocal positions $z = (1/r)e^{\mp j\theta}$, as shown by ○ in the figure for $r = 1/2$ and $\theta = \pi/4$. One pair of this quartet of zeros lies inside the unit circle and the other pair lies outside.

4.3.2 Real zeros

If $H(z)$ has a real zero at $z = r$ (i.e., $\theta = 0$), there are a couple of possibilities. If $r \neq 1$, there must be a second zero on the real axis at the reciprocal position $1/z = 1/r$. One zero will be inside the unit circle and the other outside. An example is shown with ○ in **Figure 4.19** for $z = -1/2$ and $z = -2$. If $H(z)$ has a real zero on the unit circle ($r = 1$), then that zero is at either $z = +1$ or $z = -1$, as shown with ○ in the figure. Since $1/r = r = 1$, the reciprocal position is identical, which satisfies our requirement that there must be a zero at the reciprocal position – it is just the same zero.

Table 4.4 Number of permissible zeros at $z = \pm 1$ for symmetric and antisymmetric sequences

Symmetry	N	Number of zeros $z = +1$	Number of zeros $z = -1$
Symmetric	Odd	Even	Even
Symmetric	Even	Even	Odd
Antisymmetric	Odd	Odd	Odd
Antisymmetric	Even	Odd	Even

Equation (4.15a) additionally determines the number of real zeros that can exist at $z = +1$ or $z = -1$. For example, consider an odd-length symmetric sequence. The transform $H(z)$ then has an *even* number of zeros whose product must therefore be $+1$. If some of these zeros are complex, they must occur either in conjugate pairs ($z = e^{\pm j\theta}$) or in quartets ($z = re^{\pm j\theta}$, $(1/r)e^{\pm j\theta}$) whose product is $+1$. If some of the zeros are real and occur in reciprocal pairs at $r \neq 1$, then their product is also $+1$. This means that if $H(z)$ has a single zero located at $z = +1$, there must be a second zero at $z = +1$ in order that the number of zeros remains even and that the product be $+1$. Similarly, if $H(z)$ has a zero at $z = -1$, there must be a double zero at $z = -1$. Therefore, for an odd-length symmetric sequence, there must be an even number of zeros (i.e., $0, 2, 4, \ldots$) at $z = \pm 1$. By a similar line of reasoning, one can determine the number of permissible zeros at $z = \pm 1$ for sequences that are symmetric or antisymmetric, and of even or odd length, as shown in **Table 4.4**. Note, for example, that an even-length symmetric system must have an odd number of zeros (at least one) at $z = -1$ as well as an even number of zeros at $z = +1$. All this might seem like a pleasant academic exercise (which it also is), but it has important practical consequences. For example, the requirement that there be at least one zero at $z = -1$ means that the frequency response of this filter must go to zero at $\omega = \pi$. Hence, all such systems must be either lowpass or bandpass. Similarly, an antisymmetric system of even length must have at least one zero at $z = +1$, corresponding to $\omega = 0$. Hence, such a system must be either highpass or bandpass. We shall return to this topic when we come to design linear-phase filters in Chapter 7.

Here are a few examples of symmetric and antisymmetric sequences of various lengths N.

Example 4.3

Determine where in the z-plane the zeros can be located for a symmetric sequence with $N = 3$.

➤ Solution:

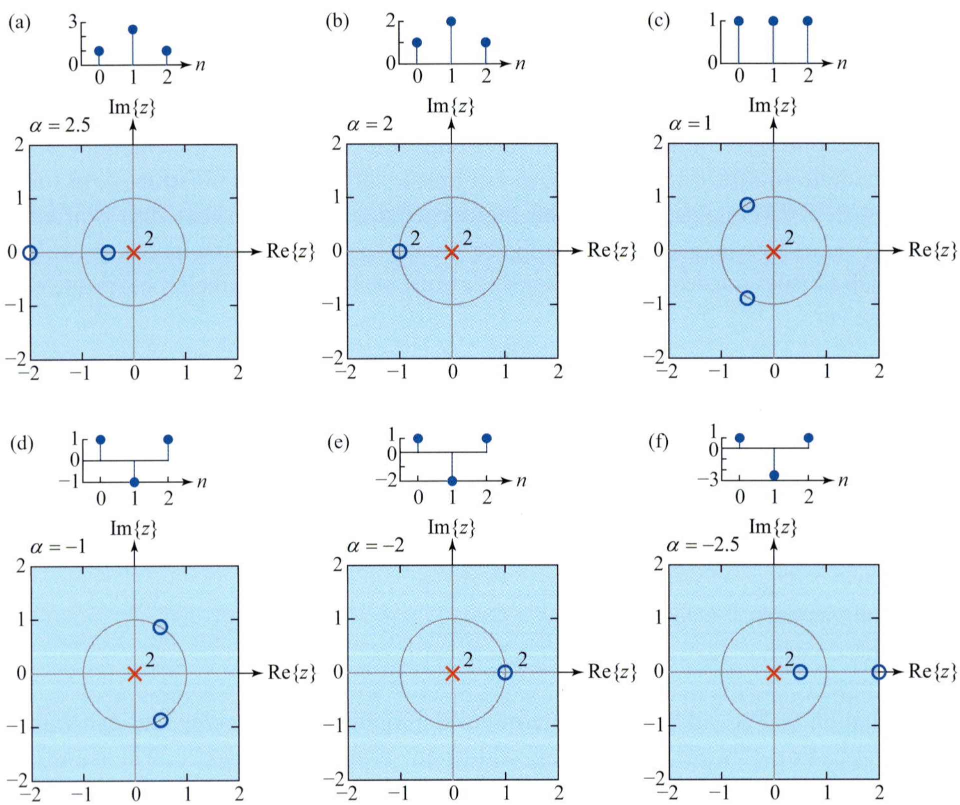

Figure 4.20 Pole-zero plot of an odd-length symmetric sequence of length $N = 3$.

There are two zeros, z_1 and z_2. Equation (4.15a) requires that $z_1 z_2 = 1$, which gives only three possibilities for the position of the zeros: (1) two real zeros located at reciprocal positions on the real axis, $z_1 = r$ and $z_2 = 1/r$; (2) a complex conjugate pair of zeros located on the unit circle, $z_1 = e^{j\theta}$ and $z_2 = e^{-j\theta}$, or (3) a double zero located at either $z_1 = z_2 = -1$ or $z_1 = z_2 = +1$. To verify this result, consider the odd-length symmetric sequence $h[n] = A(\delta[n] + \alpha\delta[n-1] + \delta[n-2])$, where A and α are real constants. The two zeros, z_1 and z_2, are roots of the quadratic

$$z^2 + \alpha z + 1 = (z - z_1)(z - z_2) = z^2 - \underbrace{(z_1 + z_2)}_{-\alpha} z + \underbrace{z_1 z_2}_{1},$$

where the monic form of the quadratic formula gives

$$z_{1,2} = \tfrac{\alpha}{2} \pm \sqrt{\left(\tfrac{\alpha}{2}\right)^2 - 1}.$$

Depending on the value of α, we again find that there are three possibilities for the position of the zeros:

(1) When $|\alpha| > 2$, the discriminant is positive and there are two distinct real zeros. Because $z_1z_2 = 1$, z_1 and z_2 have to be reciprocals of each other (i.e., $z_2 = 1/z_2$). Hence, one zero will be on the real axis inside the unit circle and the other zero will be at the reciprocal position on the real axis outside the unit circle. The zeros are either both negative or both positive. For example, when $\alpha = 2.5$, the two zeros lie on the real axis at -2 and $-1/2$, as shown in **Figure 4.20a**. When $\alpha = -2.5$, the two zeros lie on the real axis at 2 and 1/2, as shown in **Figure 4.20f**.

(2) When $\alpha = 2$, the discriminant is zero and there is a double real zero. Because $z_1z_2 = 1$, the double zero can only occur at either $z = +1$ or $z = -1$, as shown in **Figures 4.20b** and **e**.

(3) When $|\alpha| < 2$, the discriminant is negative and z_1 and z_2 are a pair of complex-conjugate roots that can be expressed in polar form, $z_1 = re^{j\varphi}$ and $z_2 = re^{-j\varphi}$. Because $z_1z_2 = re^{j\varphi} \cdot re^{-j\varphi} = r^2 = 1$, therefore $r = 1$, and these complex-conjugate roots lie on the unit circle, as shown in **Figures 4.20c** and **d** for values of $\alpha = 1$ and -1.

Figure 4.20 shows what happens to the position of the zeros for various values of α.

As a farewell example for this section, **Figure 4.21** shows the plot of a Kaiser FIR lowpass filter, a filter that we will study in Chapter 7, which has a cutoff frequency of $\omega_c = \pi/2$.

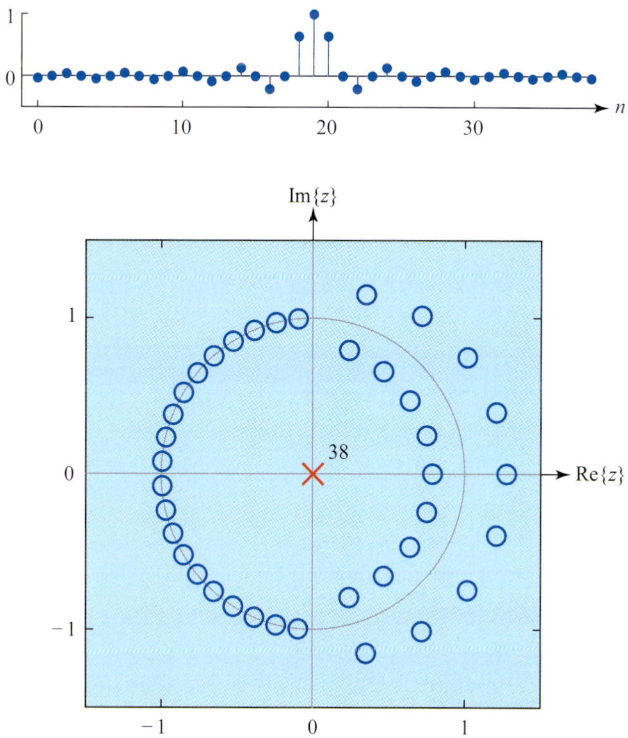

Figure 4.21 Pole-zero plot of a Kaiser lowpass filter

The impulse response is an odd-length ($N = 39$) symmetric sequence. There is a pair of zeros on the real axis at reciprocal positions, four quartets of four zeros at conjugate-reciprocal positions and ten conjugate pairs of zeros on the unit circle in the left half-plane.

4.4 The inverse z-transform

Our justification for developing the z-transform in the first place was that it enabled us to find the transform of a larger class of sequences than the discrete-time Fourier transform would allow. Of course, it would be pointless to be able to take the z-transform of a sequence, $H(z) = \mathcal{Z}\{h[n]\}$, unless there was a unique inverse z-transform, $h[n] = \mathcal{Z}^{-1}\{H(z)\}$, which allowed us to recover the original sequence from the transform.

The inverse z-transform, derived through the use of contour integration, is

$$h[n] = \frac{1}{2\pi j}\oint H(z)z^{n-1}dz.$$

Fortunately, in almost every case of practical interest to us, we do not actually have to evaluate this integral to get the inverse transform. As long as the transform can be expressed as in Equation (4.5) as a ratio of simple polynomial functions of z,

$$H(z) = \frac{N(z)}{D(z)},$$

then we can use techniques such as partial fraction expansion to express $H(z)$ as the sum of terms whose inverse transform can be found by inspection, for example using our list of useful transform pairs in **Table 4.3**. In what follows, we will consider the inverse z-transform of a variety of systems, including those with all zeros, distinct real poles, complex poles and multiple poles.

4.4.1 All-zero systems

Finding the inverse transform of an all-zero system, one expressed by Equation (4.10), is particularly simple. It essentially amounts to reversing the steps of Example 4.1. The result will be a finite-length sequence.

Example 4.4

Find $h[n]$ given

$$H(z) = \frac{z^3 + 2z + 1}{z^2}.$$

➤ **Solution:**

Because $H(z)$ only has poles at $z = 0$, we know that this is a finite-length system. To find the inverse transform, the first step is to divide $N(z)$ by $D(z)$,

$$H(z) = z + 2z^{-1} + z^{-2}.$$

Now, take the inverse transform of each term by inspection using the identity $\mathcal{Z}^{-1}\{z^{-n_0}\} = \delta[n - n_0]$,

$$h[n] = \delta[n+1] + 2\delta[n-1] + \delta[n-2].$$

4.4.2 Distinct real poles

We will first consider cases where $H(z)$ is a proper rational function of z^{-1}, that is, when $H(z)$ can be expressed as in Equation (4.5), where the order of the numerator polynomial $N(z)$, expressed in powers of z^{-1}, is less than the order of the denominator polynomial $D(z)$.

Consider the case where the poles of $H(z)$ (i.e., p_k, the roots of $D(z)$) are real and also distinct, meaning that there are no repeated (multiple) poles, $p_k \neq p_m$, $k \neq m$. Then $D(z)$ can be factored into the product of N terms, each of which corresponds to one pole,

$$H(z) = \frac{N(z)}{D(z)} = \frac{N(z)}{\sum_{k=0}^{N-1} a_k z^{-k}} = \frac{N(z)}{\prod_{k=1}^{N}(1-p_k z^{-1})}. \tag{4.16}$$

The essential idea of partial fraction expansion is to express $H(z)$ as the *sum* of terms, each of which corresponds to one of the distinct poles. Given N distinct real poles, we would express $H(z)$ as

$$H(z) = \frac{B_1}{1-p_1 z^{-1}} + \frac{B_2}{1-p_2 z^{-1}} + \cdots + \frac{B_N}{1-p_N z^{-1}} = \sum_{k=1}^{N} \frac{B_k}{1-p_k z^{-1}}, \tag{4.17}$$

where p_k are the values of the poles and the coefficients B_k are called the **residues** of $H(z)$. Once the values of the residues are determined, the inverse transform of $H(z)$ is just the sum of the inverse transforms of the individual terms, such that the intersection of the ROCs of these terms equals that of $H(z)$.

To find the value of any residue B_k, multiply both sides of Equation (4.17) by the associated pole term $(1 - p_k z^{-1})$, and evaluate at pole $z = p_k$:

$$\begin{aligned} H(z)(1-p_k z^{-1})\big|_{z=p_k} &= \frac{B_1(1-p_k z^{-1})}{1-p_1 z^{-1}}\bigg|_{z=p_k} + \cdots + \frac{B_k\cancel{(1-p_k z^{-1})}}{\cancel{1-p_k z^{-1}}}\bigg|_{z=p_k} + \cdots + \frac{B_N(1-p_k z^{-1})}{1-p_N z^{-1}}\bigg|_{z=p_k} \\ &= 0 + \cdots + B_k + \cdots + 0. \end{aligned}$$

Because the roots are distinct, all terms of the summation except the kth term are set to zero, and the kth term becomes B_k:

$$B_k = H(z)(1-p_k z^{-1})\big|_{z=p_k} = \frac{N(z)(1-p_k z^{-1})}{\prod_{m=1}^{N}(1-p_m z^{-1})}\Bigg|_{z=p_k} = \frac{N(z)}{\prod_{m=1,\, m\neq k}^{N}(1-p_m z^{-1})}. \tag{4.18}$$

A couple of examples should make things clearer.

Example 4.5

Find $h[n]$ given

$$H(z) = \frac{3z^2 - z}{z^2 - 0.75z + 0.125}, \quad |z| > 0.5.$$

➤ Solution:

The first step in the solution of any inverse transform problem with non-trivial poles is to express $H(z)$ in powers of z^{-1}, which is accomplished by dividing both $N(z)$ and $D(z)$ by z^m, where m is the highest power of z in either $N(z)$ or $D(z)$. So, in this case, dividing by z^2 yields

$$H(z) = \frac{3 - z^{-1}}{1 - 0.75z^{-1} + 0.125z^{-2}}.$$

The order of the numerator expressed in powers of z^{-1} is less than that of the denominator, so this is a proper rational fraction of polynomials and we can proceed directly with partial fraction expansion. Factoring the denominator gives

$$H(z) = \frac{3 - z^{-1}}{(1 - 0.5z^{-1})(1 - 0.25z^{-1})}.$$

In this example, there are two distinct poles and our job is to find the residues B_1 and B_2 such that

$$H(z) = \frac{3 - z^{-1}}{(1 - 0.5z^{-1})(1 - 0.25z^{-1})} = \frac{B_1}{1 - 0.5z^{-1}} + \frac{B_2}{1 - 0.25z^{-1}}. \qquad \textbf{(4.19)}$$

Find the value of each coefficient separately using the approach of Equation (4.18). For example, to find B_1, multiply both expressions for $H(z)$ in Equation (4.19) by $1 - 0.5z^{-1}$ and evaluate both expressions at the value of the pole $z = 0.5$ or, equivalently at $z^{-1} = 2$:

$$\left.\frac{3 - z^{-1}}{\cancel{(1 - 0.5z^{-1})}(1 - 0.25z^{-1})}\cancel{(1 - 0.5z^{-1})}\right|_{z^{-1}=2} = \frac{B_1}{\cancel{1 - 0.5z^{-1}}}\cancel{(1 - 0.5z^{-1})}$$
$$\left. + \cancel{\frac{B_2}{1 - 0.25z^{-1}}(1 - 0.5z^{-1})}\right|_{z^{-1}=2} = B_1.$$

The first term of the summation just becomes B_1. The second term, B_2, goes away when evaluated at $z^{-2} = 2$ because the numerator evaluates to zero while the denominator is non-zero. Hence,

$$B_1 = \left.\frac{3 - z^{-1}}{\cancel{(1 - 0.5z^{-1})}(1 - 0.25z^{-1})}\cancel{(1 - 0.5z^{-1})}\right|_{z^{-1}=2} = \frac{1}{0.5} = 2.$$

In similar fashion, to find B_2 multiply both sides of Equation (4.19) by the denominator term, $1 - 0.25z^{-1}$, and evaluate the result at $z^{-1} = 4$. In this case, the B_1 term goes away and

$$B_2 = \left.\frac{3 - z^{-1}}{(1 - 0.5z^{-1})\cancel{(1 - 0.25z^{-1})}}\cancel{(1 - 0.25z^{-1})}\right|_{z^{-1}=4} = \frac{-1}{-1} = 1.$$

This method is sometimes called the "cover-up" method, because the effect of multiplying the left side of Equation (4.19) by one of the denominator terms is equivalent to covering up that term in the denominator and evaluating the remaining parts of the expression at the value of z^{-1} that makes the covered term zero. For example, to calculate the B_1 term, cover up $(1 - 0.5z^{-1})$ in the denominator and evaluate at $z^{-1} = 2$:

$$B_1 = \frac{3-z^{-1}}{\cancel{(1-0.5z^{-1})}(1-0.25z^{-1})}\cancel{(1-0.5z^{-1})}\Bigg|_{z^{-1}=2} = \frac{3-z^{-1}}{(1-0.25z^{-1})}\Bigg|_{z^{-1}=2} = 2.$$

So far, we have

$$H(z) = \frac{3-z^{-1}}{(1-0.5z^{-1})(1-0.25z^{-1})} = \frac{2}{1-0.5z^{-1}} + \frac{1}{1-0.25z^{-1}}.$$

$H(z)$ is the sum of two terms, each of which is of the form

$$\frac{1}{1-\alpha z^{-1}}.$$

The inverse transform of each such term is either $\alpha^n u[n]$ or $-\alpha^n u[-n-1]$, depending on whether the ROC is the exterior or interior of a circle of radius α (i.e., $|z|>|\alpha|$ or $|z|<|\alpha|$). As we have discussed in Section 4.2.4, the ROC of the sum of two terms is the intersection of the ROCs of the individual terms. Each term of $H(z)$ has two possible ROCs, but the intersection of the ROCs of these two individual ROCs must be equal to the ROC of $H(z)$ given in the problem statement: ROC$\{H(z)\} = (|z|>0.5)$.

$$\begin{array}{llllll} H(z) & = & \dfrac{2}{1-0.5z^{-1}} & + & \dfrac{1}{1-0.25z^{-1}} & \\ \text{ROC:} & & \left\{\begin{array}{c}|z|>0.5\\ \text{or}\\ |z|<0.5\end{array}\right\} & \cap & \left\{\begin{array}{c}|z|>0.25\\ \text{or}\\ |z|<0.25\end{array}\right\} & \Rightarrow |z|>0.5. \end{array}$$

Here are the possible ROCs formed as the intersection of the ROCs of these two terms:

Right-sided sequence:	$(\lvert z\rvert>0.5)\cap(\lvert z\rvert>0.25)$	$\Rightarrow$	$\lvert z\rvert>0.5$
Left-sided sequence:	$(\lvert z\rvert<0.5)\cap(\lvert z\rvert<0.25)$	$\Rightarrow$	$\lvert z\rvert<0.25$
Two-sided sequence:	$(\lvert z\rvert<0.5)\cap(\lvert z\rvert>0.25)$	$\Rightarrow$	$0.25<\lvert z\rvert<0.5$
No sequence:	$(\lvert z\rvert>0.5)\cap(\lvert z\rvert<0.25)$	$\Rightarrow$	$\varnothing$.

Only the ROC of the right-sided sequence matches the ROC given in the problem statement, $|z|>0.5$, hence,

$$\begin{array}{llllll} H(z) & = & \dfrac{2}{1-0.5z^{-1}} & + & \dfrac{1}{1-0.25z^{-1}} & \\ \text{ROC:} & & |z|>0.5 & \cap & |z|>0.25 & \Rightarrow |z|>0.5. \end{array}$$

Each term has an ROC that is the exterior of a circle; hence each corresponds to a right-sided sequence. Given the functional form of each term plus its ROC, we can now derive a unique inverse transform by inspection,

$$h[n] = 2\cdot\tfrac{1}{2}^n u[n] + \tfrac{1}{4}^n u[n].$$

Example 4.6

Given the same $H(z)$ as in Example 4.4, but with ROC $0.25<|z|<0.5$, find $h[n]$.

➤ Solution:

The residues of $H(z)$ are the same as in the previous example, irrespective of the ROC. Again, we select the ROC of the individual terms so that their intersection is equal to the ROC of $H(z)$:

$$\begin{array}{llcll} H(z) & = \dfrac{2}{1-0.5z^{-1}} & + & \dfrac{1}{1-0.25z^{-1}} & \\ \text{ROC:} & |z|<0.5 & \cap & |z|>0.25 & \Rightarrow 0.25<|z|<0.5. \end{array}$$

Hence, the first term corresponds to a left-sided sequence and the second term to a right-sided sequence,

$$h[n] = -2 \cdot \tfrac{1}{2}^n u[-n-1] + \tfrac{1}{4}^n u[n].$$

4.4.3 Complex poles

Everything in the preceding discussion of the partial fraction expansion method for systems with real poles applies to systems with complex poles as well. If the coefficients of polynomial $D(z)$, namely the a_k in Equation (4.16), are real, as is the case for systems that are of practical interest, then $D(z)$ must be factorable into the product of terms such that complex poles occur in conjugate pairs; that is,

$$H(z) = \frac{N(z)}{\cdots(1-p_k z^{-1})(1-p_k^* z^{-1})\cdots}.$$

This has important consequences for the residues when $H(z)$ is expanded using partial fraction expansion.

Example 4.7

Given a stable system with z-transform

$$H(z) = \frac{3-2z^{-1}+0.5z^{-2}}{1-1.5z^{-1}+z^{-2}-0.25z^{-3}},$$

find the inverse transform $h[n]$.

➤ Solution:

The denominator has three roots: a real root, $p_1 = 0.5$, and a pair of complex-conjugate roots, $p_2 = 0.5+0.5j = (\sqrt{2}/2)e^{j\pi/4}$ and $p_3 = 0.5-0.5j = (\sqrt{2}/2)e^{-j\pi/4}$. Because $p_3 = p_2^*$ we can write the partial fraction expansion as follows:

$$\begin{aligned} H(z) &= \frac{3-2z^{-1}+0.5z^{-2}}{1-1.5z^{-1}+z^{-2}-0.25z^{-3}} = \frac{3-2z^{-1}+0.5z^{-2}}{(1-p_1z^{-1})(1-p_2z^{-1})(1-p_3z^{-1})} \\ &= \frac{B_1}{1-p_1z^{-1}} + \frac{B_2}{1-p_2z^{-1}} + \frac{B_3}{1-p_2^*z^{-1}}. \end{aligned} \tag{4.20}$$

Applying Equation (4.18) for each residue gives

$$\begin{aligned} B_1 &= 1 \\ B_2 &= 1-j = \sqrt{2}e^{-j\pi/4} \\ B_3 &= 1+j = \sqrt{2}e^{j\pi/4}. \end{aligned}$$

Because all the poles are inside the unit circle, $|p_k| < 1$, and we are given that the system is stable, the inverse transform of each term of the summation in Equation (4.20) must be right-sided:

$$
\begin{aligned}
h[n] &= (B_1(p_1)^n + B_2(p_2)^n + B_3(p_3)^n)u[n] \\
&= \left((0.5)^n + \sqrt{2}e^{-j\pi/4}\left(\tfrac{\sqrt{2}}{2}e^{j\pi/4}\right)^n + \sqrt{2}e^{j\pi/4}\left(\tfrac{\sqrt{2}}{2}e^{-j\pi/4}\right)^n\right)u[n] \\
&= \left(\tfrac{1}{2}^n + \sqrt{2}\left(\tfrac{\sqrt{2}}{2}\right)^n (e^{jn\pi/4}e^{-j\pi/4} + e^{-jn\pi/4}e^{j\pi/4})\right)u[n] \\
&= \left(\tfrac{1}{2}^n + 2\sqrt{2}\left(\tfrac{\sqrt{2}}{2}\right)^n \cos(\pi n/4 - \pi/4)\right)u[n].
\end{aligned}
$$

In the preceding example, the residues B_2 and B_3 associated with the complex-conjugate poles p_2 and p_3 are also complex conjugates: $B_3 = B_2^*$. That is not an accident. You can show (Problem 4-36) that the residues of the pair of terms of a partial fraction expansion that are associated with a pair of complex-conjugate roots are also complex conjugates; namely that

$$
H(z) = \frac{N(z)}{\cdots(1 - p_k z^{-1})(1 - p_k^* z^{-1})\cdots} = \cdots + \frac{B_k}{1 - p_k z^{-1}} + \frac{B_k^*}{1 - p_k^* z^{-1}} + \cdots.
$$

Using this result, you can show (Problem 4-37) that the inverse transform of the sum of the two terms corresponding to a pair of complex-conjugate poles (assuming right-sided) is a damped cosine,

$$
\mathcal{Z}^{-1}\left\{\frac{B_k}{1 - p_k z^{-1}} + \frac{B_k^*}{1 - p_k^* z^{-1}}\right\} = 2|B_k||p_k|^n \cos(n\measuredangle p_k + \measuredangle B_k)u[n]. \tag{4.21}
$$

You can verify that this formula applies in Example 4.7.

4.4.4 Multiple (repeated) poles

When the denominator of $H(z)$ contains multiple repeated roots, the partial fraction expansion of each of the multiple poles requires more than one term. A pole of order (multiplicity) M at $z = p_k$ generates M terms in the partial fraction expansion as follows:

$$
H(z) = \frac{N(z)}{\hat{D}(z)(1 - p_k z^{-1})^M} = \cdots + \underbrace{\left(\frac{B_{k1}}{1 - p_k z^{-1}} + \frac{B_{k2}}{(1 - p_k z^{-1})^2} + \cdots + \frac{B_{kM}}{(1 - p_k z^{-1})^M}\right)}_{M \text{ terms}} + \cdots,
$$

where $\hat{D}(z)(z)(1 - p_k z^{-1})^M$ represents the terms of the denominator with poles not equal to p_k. The calculation of the residues of the poles in $\hat{D}(z)$ is identical to that described above. Calculation of the residues $B_{k1}, B_{k2}, \ldots, B_{kM}$ requires a revision of the procedure established for single poles. Specifically,

$$
B_{km} = \frac{1}{(M - m)!(-p_k)^{M-m}} \frac{d^{M-m}}{d(z^{-1})^{M-m}} \left(H(z)(1 - p_k z^{-1})^M\right)\Bigg|_{z = p_k}. \tag{4.22}
$$

Sounds complicated. We need an example.

Example 4.8

Given a stable system with z-transform

$$H(z) = \frac{1}{1 - 1.25z^{-1} + 0.5z^{-2} - 0.0625z^{-3}} = \frac{1}{(1 - 0.25z^{-1})(1 - 0.5z^{-1})^2},$$

find the inverse transform $h[n]$.

➤ Solution:

$H(z)$ has a single pole at $z = 0.25$ and a double pole at $z = 0.5$. Hence, the partial fraction expansion has three terms, one for the single pole and two for the double pole,

$$H(z) = \frac{1}{(1 - 0.25z^{-1})(1 - 0.5z^{-1})^2} = \frac{B_1}{1 - 0.25z^{-1}} + \underbrace{\frac{B_{21}}{1 - 0.5z^{-1}} + \frac{B_{22}}{(1 - 0.5z^{-1})^2}}_{\text{Terms for the double pole}}.$$

The calculation of residue of the single pole, B_1, is identical to that discussed in the preceding section,

$$B_1 = H(z)(1 - 0.25z^{-1})\big|_{z=p_1} = \left.\frac{1}{(1 - 0.5z^{-1})^2}\right|_{z=4} = 1.$$

The calculation of B_{22} is also easy since by Equation (4.22) this is equivalent to multiplying both sides by $(1 - 0.5z^{-1})^2$ and evaluating at $z = 2$,

$$B_{22} = \left.\frac{1}{(2-2)!(-0.5)^0}\frac{d^0}{d(z^{-1})^0}\left(H(z)(1 - 0.5z^{-1})^2\right)\right|_{z=2} = H(z)(1 - 0.5z^{-1})^2\big|_{z=2} = \left.\frac{1}{1 - 0.25z^{-1}}\right|_{z=2} = 2.$$

The calculation of the residue B_{21} requires taking a derivative,

$$B_{21} = \left.\frac{1}{(2-1)!(-0.5)}\frac{d}{d(z^{-1})}\left(\frac{1}{1 - 0.25z^{-1}}\right)\right|_{z=2} = -2\left.\left(\frac{0.25}{(1 - 0.25z^{-1})^2}\right)\right|_{z=2} = -2.$$

Unfortunately, the math can get ugly really fast for higher multiples of poles. For an Nth-order pole with residues $B_{x1}, B_{x2}, \cdots, B_{xN}$ it is easy to evaluate B_{x1} and B_{xN} as we did in this example, but evaluation of the remaining $N-2$ residues requires application of Equation (4.22). That is why we would use Matlab's `residuez` function (see Section 4.4.6) to find the residues of any complicated inverse transform.

After computing the residues, we can take the inverse transform of the individual terms using the identities in **Table 4.3**. In the case of a stable system where all the poles are inside the unit circle, each term corresponds to a right-sided sequence:

$$\begin{array}{lcccccccc} H(z) & = & \dfrac{1}{1 - 0.25z^{-1}} & - & \dfrac{2}{1 - 0.5z^{-1}} & + & \dfrac{2}{(1 - 0.5z^{-1})^2} & & \\ \text{ROC:} & & |z| > 0.25 & \cap & |z| > 0.5 & \cap & |z| > 0.5 & \Rightarrow |z| > 0.5 & \\ h[n] & = & (0.25^n & - & 2(0.5)^n & + & 2(n+1)0.5^n) & u[n]. & \end{array}$$

4.4.5 Improper rational functions

The preceding discussion of the partial fraction method assumes that $H(z)$ is a **proper rational function** of z^{-1} described by Equation (4.5), where the order M of the numerator polynomial $N(z)$, expressed in powers of z^{-1}, is less than the order N of the denominator polynomial $D(z)$. When $M > N$, we can always apply long division to express $H(z)$ as the sum of the **direct terms** of a simple polynomial in z^{-1} plus a proper rational function:

$$H(z) = \underbrace{\sum_{k=0}^{M-N} c_k z^{-l}}_{\text{Polynomial in } z^{-1}} + \underbrace{\frac{\hat{N}(z)}{D(z)}}_{\text{Proper rational function}}. \tag{4.23}$$

After the long division, the order of the new numerator polynomial $\hat{N}(z)$ is less than that of the denominator polynomial, so we can apply the partial fraction method.

Example 4.9

Given a stable system with z-transform

$$H(z) = \frac{1 + 0.4z^{-1} - 2.2z^{-2} + 0.8z^{-3}}{1 - 1.3z^{-1} + 0.4z^{-2}},$$

find the inverse transform $h[n]$.

➤ Solution:

In this case, the order of the numerator, expressed as powers of z^{-1}, is greater than the order of the denominator, so we perform long division to produce the direct terms plus the remainder that we can factor using partial-fraction expansion. There are two ways of doing this long division. It is important to choose the right one.

I. "Wrong" way:

$$\begin{array}{r l}
 & \quad 1 \\
1 - 1.3z^{-1} + 0.4z^{-2} & \overline{)\,1 + 0.4z^{-1} - 2.2z^{-2} + 0.8z^{-3}} \\
 & \underline{1 - 1.3z^{-1} + 0.4z^{-2}} \\
 & \qquad 1.7z^{-1} - 2.6z^{-2} + 0.8z^{-3}.
\end{array}$$

We now have

$$H(z) = 1 + \frac{1.7z^{-1} - 2.6z^{-2} + 0.8z^{-3}}{1 - 1.3z^{-1} + 0.4z^{-2}}.$$

This is clearly not heading in the right direction. The order of the numerator of the remainder is still greater than that of the denominator, so we have not made any progress. Continuing with long division does not improve things:

$$
\begin{array}{r}
1+1.7z^{-1} \qquad\qquad\qquad\qquad \\
1-1.3z^{-1}+0.4z^{-2}\ \overline{)1+0.4z^{-1}-2.2z^{-2}+0.8z^{-3}} \\
\underline{1-1.3z^{-1}+0.4z^{-2}} \qquad\qquad \\
1.7z^{-1}-2.6z^{-2}+0.8z^{-3} \\
\underline{1.7z^{-1}-2.21z^{-2}+0.68z^{-3}} \\
-0.39z^{-2}+0.12z^{-3}.
\end{array}
$$

Now we have

$$H(z)=1+1.7z^{-1}+\frac{-0.39z^{-2}+0.12z^{-3}}{1-1.3z^{-1}+0.4z^{-2}}.$$

Things just are not getting better (and never will). The order of the numerator of the remainder term is increasing. What we are actually doing is producing an infinite series expansion of $H(z)$ in terms of powers of z^{-1},

$$H(z)=1+1.7z^{-1}-0.39z^{-2}-0.387z^{-3}+\cdots \tag{4.24}$$

That is not terrible, but our goal is to express $H(z)$ as in Equation (4.23) as the sum of a direct polynomial term plus a proper rational function of z^{-1} that we can factor.

II. "Right" way:

First express both the numerator and denominator of $H(z)$ as *decreasing* powers of z^{-1} and then do the long division:

$$
\begin{array}{r}
2z^{-1}+1 \qquad\qquad\qquad\qquad \\
0.4z^{-2}-1.3z^{-1}+1\ \overline{)0.8z^{-3}-2.2z^{-2}+0.4z^{-1}+1} \\
\underline{0.8z^{-3}-2.6z^{-2}+2z^{-1}} \qquad \\
0.4z^{-2}-1.6z^{-1}+1 \\
\underline{0.4z^{-2}-1.3z^{-1}+1} \\
-0.3z^{-1}.
\end{array}
$$

Now we have the desired result: the sum of a polynomial in z^{-1} plus a proper rational function in z^{-1},

$$H(z)=\underbrace{1+2z^{-1}}_{\text{Polynomial in } z^{-1}}+\underbrace{\frac{-0.3z^{-1}}{1-1.3z^{-1}+0.4z^{-2}}}_{\text{Proper rational function}}.$$

Splitting the second term by partial fractions gives

$$
\begin{array}{lcccccccc}
H(z) & = & 1 & + & 2z^{-1} & + & \dfrac{0.5}{1-0.5z^{-1}} & - & \dfrac{0.8}{1-0.8z^{-1}} \\
\text{ROC:} & & \text{all} & \cap & |z|>0 & \cap & |z|>0.5 & \cap & |z|>0.8 \quad \Rightarrow |z|>0.8.
\end{array}
$$

Hence,

$$h[n]=\delta[n]+2\delta[n-1]+(0.5^{n+1}-0.8^{n+1})u[n].$$

By evaluating the first few terms of $h[n]$, you can convince yourself that this expression is identical to the inverse transform of the power series expansion of $H(z)$ we derived by our first attempt at long division, in Equation (4.24).

Example 4.10

Given the pole-zero plot and ROC shown in **Figure 4.22** and the information that $H(0) = 8$, find $h[n]$.

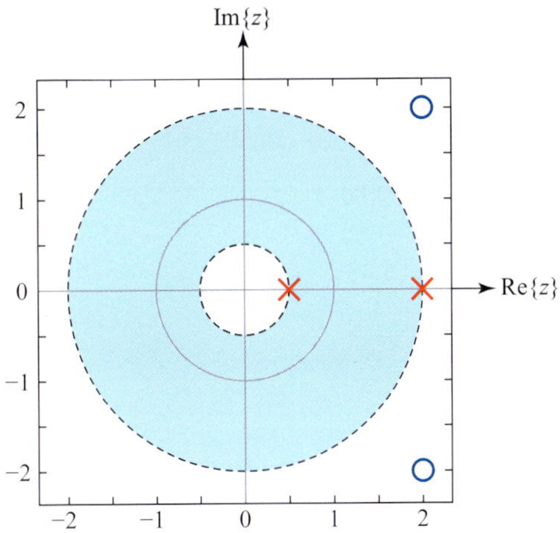

Figure 4.22

➤ Solution:

Poles are at $z = 0.5$ and $z = 2$. The zeros are at $2 \pm 2j$. The ROC is $0.5 < |z| < 2$. Hence, the z-transform has the form

$$H(z) = A\frac{(z-(2+2j))(z-(2-2j))}{(z-\frac{1}{2})(z-2)} = A\frac{z^2-4z+8}{z^2-\frac{5}{2}z+1}, \qquad \tfrac{1}{2} < |z| < 2.$$

To evaluate the constant A, plug $z = 0$ into the equation and use the fact that $H(0) = A \cdot 8 = 8$, so $A = 1$. Note that $H(z)$ is completely specified by three pieces of information:

(1) the location of the poles and zeros on the pole-zero plot,
(2) the ROC,
(3) the value of $H(z)$ at one non-singular value of z (i.e., not on a pole or zero).

This last condition is essential. There is an infinite number of $H(z)$, all of which have the same poles and zeros and the same ROC, all related to each other by a multiplicative constant. Specifying $H(z)$ at a single value of z resolves this ambiguity and specifies a unique $H(z)$.

To find the inverse transform using the partial fraction method, we have to specify $H(z)$ in powers of z^{-1} (not z). So, divide both the numerator and the denominator of $H(z)$ by z^2,

$$H(z) = \frac{1-4z^{-1}+8z^{-2}}{1-\frac{3}{2}z^{-1}+z^{-2}}.$$

Since the orders of the numerator and denominator expressed as powers of z^{-1} are the same, we have to perform long division to reduce the order of the numerator, remembering to express the numerator and denominator as decreasing powers of z^{-1}:

$$\begin{array}{r|l} & \quad 8 \\ z^{-2}-\frac{5}{2}z^{-1}+1 & \overline{)8z^{-2}-4z^{-1}\;\;+1} \\ & \underline{8z^{-2}-20z^{-1}\;+8} \\ & \qquad 16z^{-1}-7 \end{array}.$$

The order of the remainder has been reduced by one, so we have

$$H(z)=8+\frac{-7+16z^{-1}}{1-\frac{5}{2}z^{-1}+z^{-2}}=8+\frac{-7+16z^{-1}}{(1-\frac{1}{2}z^{-1})(1-2z^{-1})}, \qquad \tfrac{1}{2}<|z|<2.$$

Splitting the second term by partial fractions gives

$$\begin{array}{lcccccc} H(z) & = & 8 & + & \dfrac{-\frac{25}{3}}{(1-\frac{1}{2}z^{-1})} & + & \dfrac{\frac{4}{3}}{(1-2z^{-1})} \\ \text{ROC:} & & \mathit{all} & \cap & \left\{\begin{array}{c}|z|>\frac{1}{2}\\ \text{or}\\ |z|<\frac{1}{2}\end{array}\right\} & \cap & \left\{\begin{array}{c}|z|>2\\ \text{or}\\ |z|<2\end{array}\right\} \stackrel{?}{\Rightarrow} \tfrac{1}{2}<|z|<2. \end{array}$$

The ROC of the constant term is the entire z-plane. Each of the other two terms has two possible ROCs. All we now need is to figure out the proper ROC for each term so that the intersection comes out to $1/2<|z|<2$, as indicated in the pole-zero plot of the problem statement. The only possible solution is that the term with the pole at $z=1/2$ must have an ROC of $|z|>1/2$ and the term with the pole at $z=2$ must have an ROC of $|z|<2$:

$$\begin{array}{lcccccc} H(z) & = & 8 & + & \dfrac{-\frac{25}{3}}{(1-\frac{1}{2}z^{-1})} & + & \dfrac{\frac{4}{3}}{(1-2z^{-1})} \\ \text{ROC:} & & \mathit{all} & \cap & |z|>\frac{1}{2} & \cap & |z|<2 \quad \Rightarrow \tfrac{1}{2}<|z|<2. \end{array}$$

By inspection, the inverse transform is

$$h[n]=8\,\delta[n]-\tfrac{25}{3}\cdot\tfrac{1^n}{2}u[n]-\tfrac{4}{3}\cdot 2^n u[-n-1].$$

The term with the pole at $z=1/2$ corresponds to a right-sided sequence, the term with the pole at $z=2$ to a left-sided sequence, so $h[n]$ is two-sided. Furthermore, $h[n]$ is convergent, as we would expect since the ROC includes the unit circle.

4.4.6 Using Matlab to compute the inverse *z*-transform

There are several ways to get Matlab's help with calculating and plotting the inverse z-transform of a system defined by $H(z)$. Matlab's `residuez` function calculates the residues of $H(z)$

and allows you to express them as a sum of a series of terms. This function handles both proper and improper polynomial functions for $H(z)$. The syntax is

```
[r, p, k] = residuez(b, a),
```

where `b` and `a` are the coefficients of $H(z)$ in powers of z^{-1} given in Equation (4.5), `r` are the residues, `p` are the poles and `k` are the coefficients of the direct term that correspond to the coefficients of the polynomial in z^{-1} resulting from the long division of an improper $H(z)$. Here is an example:

Example 4.11

Use Matlab's `residuez` function to compute the residues of Examples 4.8 and 4.9.

➤ Solution:

For the residues of Example 4.8:

```
>> [r, p, k] = residuez(1, [1 -1.25 0.5 -0.0625])
r =
   -2.0000
    2.0000
    1.0000
p =
    0.5000
    0.5000
    0.2500
k =
    []
```

The values of `r` correspond to B_{21}, B_{22} and B_1 in that order. The fact that `k = []` means that $H(z)$ is a proper polynomial function with no direct terms.

For the residues of Example 4.9:

```
>> [r, p, k] = residuez([1 0.4 -2.2 0.8], [1 -1.3 0.4])
r =
   -1.0000
    1.0000
p =
    0.8000
    0.5000
k =
    1   2
```

You can also use the `residuez` function "in reverse" to give the coefficients of $H(z)$, namely b_k and a_k in Equation (4.5), given the residues, poles and direct terms:

```
[b, a] = residuez(r, p, k).
```

Example 4.12

Use Matlab's `residuez` function to compute the values of `b` and `a` given `r`, `p` and `k` in Example 4.8.

➤ Solution:

```
>> [b, a] = residuez(r, p, k)
 b =
      1.0000      0.0000     -0.0000
 a =
      1.0000     -1.2500      0.5000     -0.0625
```

Finally, let us show how to use Matlab to compute and plot the impulse response of a relatively complicated example – a two-sided system with real, complex and multiple poles – by a couple of methods.

Example 4.13

Use Matlab to compute and plot the impulse response $h[n]$ over the range $-20 \le n \le 20$ for a stable system described by

$$H(z) = \frac{5 - 26z^{-1} + 39.5z^{-2} - 23.5z^{-3} + 8z^{-4} - 2z^{-5}}{1 - 5z^{-1} + 8.5z^{-2} - 6z^{-3} + 2z^{-4}}.$$

➤ Solution:

First, compute the residues:

```
>> [r, p, k] = residuez([5 -26 39.5 -23.5 8 -2], [1 -5 8.5 -6 2])
r =
    1.0000 + 0.0000i
   -1.0000 - 0.0000i
    2.0000 + 0.0000i
    2.0000 - 0.0000i
p =
    2.0000 + 0.0000i
    2.0000 - 0.0000i
    0.5000 + 0.5000i
    0.5000 - 0.5000i
k =
     1    -1
```

There is a pair of complex-conjugate poles inside the unit circle at $z = 0.5 \pm 0.5j = (\sqrt{2}/2)e^{\pm j\pi/4}$ and a double pole outside the unit circle at $z = 2$. Hence,

$$H(z) = 1 - z^{-1} + \underbrace{\frac{1}{1-2z^{-1}} - \frac{1}{(1-2z^{-1})^2}}_{\text{left-sided sequence}} + \underbrace{\frac{2}{1-\frac{\sqrt{2}}{2}e^{j\pi/4}z^{-1}} + \frac{2}{1-\frac{\sqrt{2}}{2}e^{-j\pi/4}z^{-1}}}_{\text{right-sided sequence}}.$$

Since the system is stable, the terms of the expansion of the double poles correspond to left-sided sequences and the other terms correspond to right-sided sequences. There are a couple of ways to proceed.

I. Let Matlab compute the residues and use them to help us derive an expression for *h*[*n*] manually.

The inverse transform of the direct terms of the expansion is given by `k`:

$$\mathfrak{Z}^{-1}\{1 - z^{-1}\} = \delta[n] - \delta[n-1].$$

From Equation (4.21), the two terms for the complex-conjugate poles can be combined to give

$$\mathfrak{Z}^{-1}\left\{\frac{2}{1-\frac{\sqrt{2}}{2}e^{j\pi/4}z^{-1}} + \frac{2}{1-\frac{\sqrt{2}}{2}e^{-j\pi/4}z^{-1}}\right\} = 4\left(\tfrac{\sqrt{2}}{2}\right)^n \cos(n\pi/4)\,u[n].$$

The remaining two terms correspond to left-sided sequences. The inverse of the first-order term is

$$\mathfrak{Z}^{-1}\left\{\frac{1}{1-2z^{-1}}\right\} = -2^n u[-n-1].$$

The inverse of the second-order term (see Problem 4-33) is

$$\mathfrak{Z}^{-1}\left\{\frac{1}{(1-2z^{-1})^2}\right\} = -(n+1)2^n u[-n-1].$$

Putting it all together gives

$$h[n] = \delta[n] - \delta[n-1] + 4\left(\tfrac{\sqrt{2}}{2}\right)^n \cos(n\pi/4)u[n] + 2^n u[-n-1].$$

Plot it with Matlab:

```
n = 10;
rhs = 4*(sqrt(2)/2).^(0:n).*cos((0:n)*pi/4);      % right-sided terms
rhs([1 2]) = rhs([1 2]) + [1 -1];                 % add direct terms
lhs = (-n:-1).*2.^(-n:-1);                        % left-sided terms
stem(-n:n, [lhs rhs])
```

The result is shown in **Figure 4.23**. There is nothing wrong with this approach, except that we have only used Matlab to compute the residues, after which we had to use our brain-power to compute the inverse transforms of several right-sided and left-sided terms before having Matlab plot them. The more poles, the more terms, and the more brain-power required.

II. Let Matlab do all the work.

Recognize there are at most two kinds of terms in the partial fraction expansion of the inverse transform of $H(z)$. Since the system comprises both left-sided and right-sided terms and is also stable, the ROC is an annulus that includes the value of $|z| = 1$. The right-sided terms of $H(z)$ comprise all the direct terms plus those terms whose poles have absolute values less than one. The left-sided terms comprise the terms whose poles have absolute values greater than one. We can use Matlab to collect each of these sets of terms into a system whose impulse response we can determine computationally using `residuez`. In this example, we have both right-sided and left-sided terms. Here is the code:

```
N = 10;
ROCval = 1; [r, p, k] = residuez( [5 -26 39.5 -23.5 8 -2], [1 -5 8.5 -6 2 ]);
indx = (abs(p)< ROCval);                         % find right-sided terms
[br, ar] = residuez(r(indx), p(indx), k);        % find right-sided coeffs
rhs = filter(br, ar, [1 zeros(1, N-1)]);         % right-side
[bl, al] = residuez(r(~indx), p(~indx), []);     % find left-sided coeffs
lhs = fliplr(filter(bl(end:-1:1), al(end:-1:1), [1 zeros(1, N-2)]));
stem(-n:n, [lhs rhs])
```

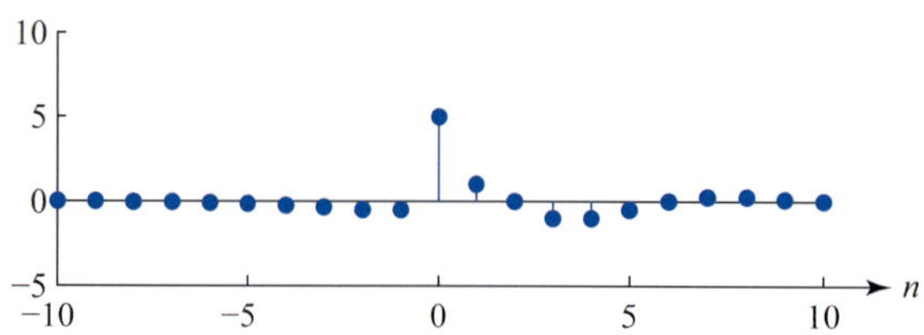

Figure 4.23

As indicated above, we choose `ROCval`, a value of z that lies within the ROC, to be one, namely on the unit circle. After computing all the residues, we select (with `indx`) those poles with absolute values less than one and use `residuez` to combine those terms plus all the direct terms into a single filter with coefficients `br` and `ar`. Then we use `filter` to compute the first ten terms of the right-sided impulse response. We combine the remaining residue terms (selected with `~indx`) to produce a filter with coefficients `bl` and `al`. Now use the time-reversal property (described in Equation (4.25) of the next section) to "flip" the coefficient arrays for `br` and `ar` in order to create the filter required to compute the left-sided component of the impulse response. The output of `filter` corresponds to points $n = -1, -2, \ldots -10$. Now use `fliplr` to reverse the array to form `lhs` and then stitch `lhs` together with `rhs` to form the entire two-sided impulse response and plot it. Less brain-power required.

Matlab has another function, `iztrans`, which works with the Symbolic Toolbox to return the inverse z-transform given a specification of $H(z)$ as a string of characters.

4.5 Properties of the *z*-transform

The properties of the z-transform follow the same pattern as the properties we have already derived for the Fourier transform. The main thing that is different – and it is an important difference – is that we must consider the ROC.

4.5.1 Linearity

Assume we have two sequences, $x_1[n]$ with z-transform $X_1(z)$ and $\text{ROC}\{X_1(z)\}$, and $x_2[n]$ with z-transform $X_2(z)$ and $\text{ROC}\{X_2(z)\}$. If we form a new sequence as the scaled sum of $x_1[n]$ and $x_2[n]$, $y[n] = a_1x_1[n] + a_2x_2[n]$, then

$$\begin{array}{ccccc} Y(z) & = & a_1X_1(z) & + & a_2X_2(z) \\ \text{ROC}\{Y(z)\} & \supseteq & \text{ROC}\{X_1(z)\} & \cap & \text{ROC}\{X_2(z)\}. \end{array}$$

The symbol "$\supseteq$" means that the ROC of $Y(z)$ is "at least as large as" the intersection of the ROCs of $X_1(z)$ and $X_2(z)$. For the most part, the ROC of the sum of terms will be equal to the intersection of the ROCs of the terms, but there can be cases where pole-zero cancelation leads to a circumstance in which the ROC of the sum will be greater than the intersection (see Example 4.19). At any rate, we have already tacitly used this property in many examples in the previous sections, so we will move along.

4.5.2 Shifting property

Assume $x[n]$ has z-transform $X(z)$. Let the sequence $y[n]$ be a shifted version of $x[n]$: $y[n] = x[n - n_0]$. Then,

$$Y(z) = \sum_{n=-\infty}^{\infty} x[\underbrace{n - n_0}_{m}]z^{-n}.$$

Let $m = n - n_0$. Then,

$$Y(z) = \sum_{m=-\infty}^{\infty} x[m]z^{-(m+n_0)} = z^{-n_0} \sum_{m=-\infty}^{\infty} x[m]z^{-m} = X(z)z^{-n_0}.$$

Thus, $Y(z)$ has the same poles and zeros as $X(z)$ in the finite z-plane except at $z = 0$. Hence, $\text{ROC}\{Y(z)\} = \text{ROC}\{X(z)\}$, except perhaps at $z = 0$. An example should make things clear.

Example 4.14

Let $x[n] = (1/2)^n u[n]$. Find the pole-zero plot and ROC for $y[n] = x[n - n_0]$ with $n_0 = 0$, $n_0 = 1$ and $n_0 = 2$.

➤ **Solution:**

$$X(z) = \frac{1}{1 - \frac{1}{2}z^{-1}}, \quad \text{ROC}\{X(z)\} = |z| > \tfrac{1}{2},$$

and

$$Y(z) = X(z)z^{-n_0} = \frac{z^{-n_0}}{1 - \frac{1}{2}z^{-1}} = \frac{z}{z - \frac{1}{2}} \cdot z^{-n_0}, \quad |z| > \tfrac{1}{2}.$$

Figure 4.24 shows the pole-zero plot of $Y(z)$ for $n_0 = 0$, $n_0 = 1$ and $n_0 = 2$.

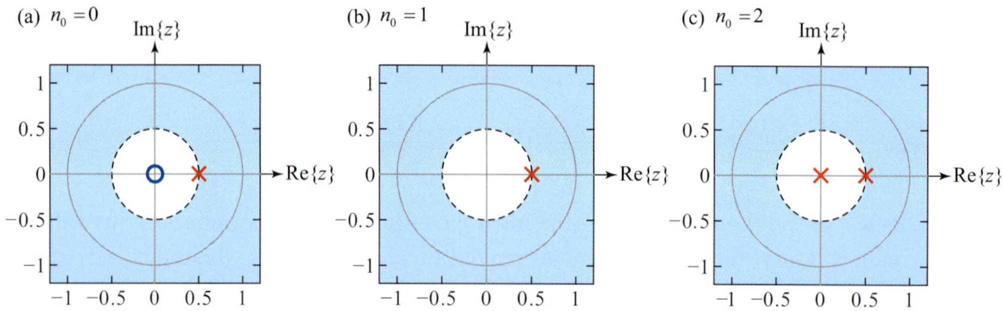

Figure 4.24 Pole-zero plot of a shifted IIR response

When $n_0 = 0$ (**Figure 4.24a**),

$$Y(z) = \frac{z}{z - \frac{1}{2}}.$$

There is a pole at $z = 1/2$ and a zero at $z = 0$. When $n_0 = 1$ (**Figure 4.24b**), shifting $x[n]$ one to the right introduces a pole at $z = 0$ that cancels the existing zero in $X(z)$. Further shifts to the right (e.g., $n_0 = 2$, **Figure 4.24c**) just put more poles at $z = 0$. In this example, $\text{ROC}\{Y(z)\} = \text{ROC}\{X(z)\}$ regardless of the value of n_0 since the ROC of this right-sided sequence is determined by the outermost pole, which is at $z = 1/2$.

4.5.3 Differentiation property

Let $y[n] = nx[n]$. If $\mathfrak{Z}\{x[n]\} = X(z)$ with $\text{ROC}\{X(z)\}$, then

$$\mathfrak{Z}\{y[n]\} = \mathfrak{Z}\{nx[n]\} = -z\frac{dX(z)}{dz}, \quad \text{ROC}\{Y(z)\} = \text{ROC}\{X(z)\}.$$

Proof:

$$-z\frac{dX(z)}{dz} = -z\frac{d}{dz}\sum_{n=-\infty}^{\infty} x[n]z^{-n} = -z\sum_{n=-\infty}^{\infty} x[n](-n)z^{-n-1} = \sum_{n=-\infty}^{\infty} (nx[n])z^{-n} = \mathfrak{Z}\{nx[n]\}.$$

Example 4.15

Use the differentiation property to find the z-transform of $y[n]=n\alpha^n u[n]$.

➤ **Solution:**

Let $x[n]=\alpha^n u[n]$. Then $y[n]=n(\alpha^n u[n])=nx[n]$. So,

$$\mathfrak{Z}\{y[n]\}=-z\frac{dX(z)}{dz}=-z\frac{d}{dz}\left(\frac{1}{1-\alpha z^{-1}}\right)=-z\frac{-\alpha z^{-2}}{(1-\alpha z^{-1})^2}=\frac{\alpha z^{-1}}{(1-\alpha z^{-1})^2}.$$

4.5.4 Time-reversal property

If $\mathfrak{Z}\{x[n]\}=X(z)$ with $\mathrm{ROC}\{X(z)\}$, then $\mathfrak{Z}\{x[-n]\}=X(z^{-1})$ with $\mathrm{ROC}\{X(z^{-1})\}$. Proof:

$$\mathfrak{Z}\{x[-n]\}=\sum_{n=-\infty}^{\infty}x[-n]z^{-n}=\sum_{-n=\infty}^{\infty}x[(-n)]z^{(-n)}=\sum_{m=\infty}^{-\infty}x[m]z^{m}=\sum_{m=\infty}^{-\infty}x[m](z^{-1})^{-m}=X(z^{-1}).$$

For the ROC of the time-reversed signal, all powers of z are replaced with z^{-1}. So, for example, if $\mathrm{ROC}\{X(z)\}$ is $|z|>\alpha$, then $\mathrm{ROC}\{X(1/z)\}$ is the reciprocal, namely $|1/z|>\alpha$, or, equivalently, $|z|<1/\alpha$. The time-reversal property comes in handy for computing the response of left-sided impulse responses, as shown in the next example.

Example 4.16

Given $x[n]=\alpha^n u[n]$ and its z-transform,

$$X(z)=\mathfrak{Z}\{\alpha^n u[n]\}=\frac{1}{1-\alpha z^{-1}},\quad |z|>\alpha,$$

use the time-reversal property to show that the z-transform of $y[n]=-\alpha^n u[-n-1]$ is

$$Y(z)=\mathfrak{Z}\{-\alpha^n u[-n-1]\}=\frac{1}{1-\alpha z^{-1}},\quad |z|<\alpha.$$

➤ **Solution:**

Start with $x[-n]=\alpha^{-n}u[-n]$. By the time-reversal property, the z-transform of $x[-n]$ is

$$X(z^{-1})=\frac{1}{1-\alpha z}=\frac{1}{-\alpha z}\left(\frac{1}{1-\alpha^{-1}z^{-1}}\right),\quad |z|<\alpha.$$

Let $W(z)=-\alpha zX(z^{-1})$. Then,

$$W(z)=\frac{1}{1-\alpha^{-1}z^{-1}},\quad |z|<\alpha.$$

So, $w[n]=-\alpha x[-(n+1)]=-\alpha(\alpha^{-(n+1)}u[-(n+1)])=-\alpha^{-n}u[-n-1]$. Notice that $y[n]=-\alpha^n u[-n-1]$ is the same as $w[n]$ with α instead of α^{-1}. Hence, $Y(z)$ is the same as $W(z)$ with α instead of α^{-1}.

Given a right-sided signal $h[n]$ whose z-transform $H(z)$ is written as a ratio of two polynomials in z^{-1} as in Equation (4.5),

$$H(z) = \frac{N(z)}{D(z)} = \frac{\sum_{k=0}^{M} b_k z^{-1}}{\sum_{k=0}^{N} a_k z^{-1}}, \quad \mathrm{ROC}\{H(z)\},$$

the z-transform of $h[-n]$ is

$$H(z^{-1}) = \frac{\sum_{k=0}^{M} b_k z^k}{\sum_{k=0}^{N} a_k z^k} = \frac{b_0 + b_1 z + \cdots b_M z^M}{a_0 + a_1 z + \cdots a_M z^N} = \left(\frac{z^M}{z^N}\right) \frac{b_M + b_{M-1} z^{-1} + \cdots b_0 z^{-M}}{a_N + a_{N-1} z^{-1} + \cdots a_0 z^{-N}}$$

$$= z^{M-N} \frac{\sum_{k=0}^{M} b_{M-k} z^{-k}}{\sum_{k=0}^{N} a_{N-k} z^{-k}}, \quad \mathrm{ROC}\{H(z^{-1})\}. \tag{4.25}$$

The effect of time reversal on the transform is to "flip" the coefficients b_k and a_k so that they are applied to z^{-k} in reverse order. Another key result is that the singularities of $H(z^{-1})$ occur at the conjugate-reciprocal positions of those of $H(z)$. To see this, write $H(z)$ as the product of terms

$$H(z^{-1}) = C \frac{\prod_{k=0}^{M-1} (z - z_k)}{\prod_{k=0}^{N-1} (z - p_k)},$$

where C is a constant. Then,

$$H(z^{-1}) = C \frac{\prod_{k=0}^{M-1} (z^{-1} - z_k)}{\prod_{k=0}^{N-1} (z^{-1} - p_k)} = \underbrace{C \left(\frac{\prod_{k=0}^{M-1} (-z_k)}{\prod_{k=0}^{N-1} (-p_k)} \right)}_{\hat{C}} \left(\frac{\prod_{k=0}^{M-1} z^{-1}}{\prod_{k=0}^{N-1} z^{-1}} \right) \cdot \frac{\prod_{k=0}^{M-1} (z - 1/z_k)}{\prod_{k=0}^{N-1} (z - 1/z_k)}$$

$$= \hat{C} z^{N-M} \frac{\prod_{k=0}^{M-1} (z - 1/z_k)}{\prod_{k=0}^{N-1} (z - 1/z_k)}, \tag{4.26}$$

where $\hat{C}$ is a constant. Note the factor z^{N-M} in Equation (4.26). It means that poles or zeros at $z = 0$ could either appear or disappear in $H(z^{-1})$ depending on the values of M or N. Here is an example.

Example 4.17

Find the z-transform of

$$h[n] = \left(7\tfrac{1}{4}^{n} - 6\tfrac{1}{2}^{n}\right)u[n],$$

and the z-transform of the time-reversed sequence $h[-n]$.

➤ **Solution:**

$$H(z) = \frac{7}{1 - \frac{1}{4}z^{-1}} - \frac{6}{1 - \frac{1}{2}z^{-1}} = \frac{1 - 2z^{-1}}{1 - \frac{3}{4}z^{-1} + \frac{1}{8}z^{-2}} = \frac{z^2 - 2z}{z^2 - \frac{3}{4}z + \frac{1}{8}} = \frac{z(z-2)}{\left(z - \frac{1}{4}\right)\left(z - \frac{1}{2}\right)}, \quad |z| > \tfrac{1}{2}.$$

The transform has two poles and two zeros, one of which is at $z = 0$ .The transform of $h[-n]$ is

$$H(z^{-1}) = \frac{1 - 2z}{1 - \frac{3}{4}z + \frac{1}{8}z^2} = z^{-1}\frac{-2 + z^{-1}}{\frac{1}{8} - \frac{3}{4}z^{-1} + z^{-2}} = -16\frac{z - \frac{1}{2}}{z^2 - 6z + 8} = -16\frac{z - \frac{1}{2}}{(z-4)(z-2)}, \quad |z| < 2.$$

The poles and zeros of $H(z^{-1})$ are at reciprocal positions of $H(z)$; they are reflected with respect to the unit circle. The zero in $H(z)$ at $z = 0$ effectively "disappeared" in $H(z^{-1})$ by being sent to its reciprocal position, $z = \infty$.

4.5.5 Convolution property

Convolution is the basis for understanding most signal processing operations of linear time-invariant systems. The output of the system, $y[n]$, is the convolution of the input $x[n]$ with the impulse response $h[n]$, as shown in **Figure 4.25**

Figure 4.25 Convolution

The z-transform of the output is

$$Y(z) = \sum_{n=-\infty}^{\infty} y[n]z^{-n} = \sum_{n=-\infty}^{\infty}\left(\sum_{k=-\infty}^{\infty} x[k]h[n-k]\right)z^{-n} = \sum_{n=-\infty}^{\infty} x[k]\sum_{n=-\infty}^{\infty} h[\underbrace{n-k}_{m}]z^{-n}.$$

Letting $m = n - k$, this becomes

$$Y(z) = \sum_{n=-\infty}^{\infty} x[k] \sum_{m=-\infty}^{\infty} h[m] z^{-(m+k)} = \sum_{k=-\infty}^{\infty} x[k] z^{-k} \cdot \sum_{m=-\infty}^{\infty} h[m] z^{-m} = X(z)H(z).$$

In order for $Y(z)$ to exist, both the summations must converge, which means that the region of convergence of $Y(z)$ must be at least the intersection of the regions of convergence of $X(z)$ and $H(z)$:

$$\begin{array}{ccccc} Y(z) & = & X(z) & \cdot & H(z) \\ \mathrm{ROC}\{Y(z)\} & \supseteq & \mathrm{ROC}\{X(z)\} & \cap & \mathrm{ROC}\{H(z)\} \end{array}.$$

Here are some examples.

Example 4.18

Let $x[n] = u[n]$ and $h[n] = (1/2)^n u[n]$. Find $y[n] = x[n] * h[n]$ using the convolution property of z-transforms.

➤ Solution:

$$X(z) = \frac{1}{1 - z^{-1}}, \quad |z| > 1,$$

and

$$H(z) = \frac{1}{1 - \frac{1}{2} z^{-1}}, \quad |z| > \tfrac{1}{2}.$$

So,

$$\begin{aligned} Y(z) = X(z)H(z) &= \frac{1}{1 - z^{-1}} \cdot \frac{1}{1 - \frac{1}{2} z^{-1}} \\ \mathrm{ROC}\{Y(z)\} &= (|z| > 1) \cap (|z| > \tfrac{1}{2}) \Rightarrow |z| > 1. \end{aligned}$$

In this case, the ROC of $Y(z)$ is exactly equal to the intersection of the ROCs of $X(z)$ and $H(z)$, as shown in **Figure 4.26**.

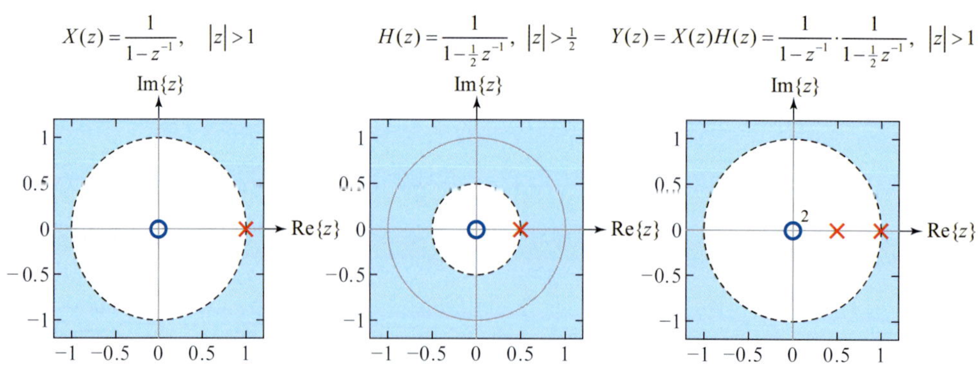

Figure 4.26

To find $y[n]$, first decompose $Y(z)$ into the sum of two terms, being careful to select the proper ROC for each term such that the sum will have the desired ROC, $|z| > 1$.

$$\begin{aligned} Y(z) \; &= \; \frac{2}{1 - z^{-1}} \quad - \quad \frac{1}{1 - \frac{1}{2}z^{-1}} \\ &\quad (|z| > 1) \quad \cap \quad (|z| > \tfrac{1}{2}) \quad \Rightarrow \quad |z| > 1. \end{aligned}$$

Now the inverse z-transform yields $y[n] = 2u[n] - (1/2)^n u[n]$.

Example 4.19

Repeat the previous example with $x[n] = \frac{1}{2}^n u[n]$ and $h[n] = \delta[n] - \frac{1}{4}^n u[n-1]$.

➤ Solution:

$$X(z) = \frac{1}{1 - \frac{1}{2}z^{-1}}, \quad |z| > \tfrac{1}{2}.$$

Write $h[n] = \delta[n] - \frac{1}{4}^n u[n-1] = \delta[n] - \frac{1}{4}\cdot\frac{1}{4}^{n-1} u[n-1]$, so we can use the shifting property of z-transforms,

$$H(z) = 1 - \frac{1}{4}\frac{z^{-1}}{1 - \frac{1}{4}z^{-1}} = \frac{1 - \frac{1}{2}z^{-1}}{1 - \frac{1}{4}z^{-1}}, \quad |z| > \frac{1}{4}.$$

Now

$$\begin{aligned} Y(z) = X(z)H(z) \quad &= \quad \frac{1}{1 - \frac{1}{2}z^{-1}} \quad \cdot \quad \frac{1 - \frac{1}{2}z^{-1}}{1 - \frac{1}{4}z^{-1}} \quad = \quad \frac{1}{1 - \frac{1}{4}z^{-1}} \\ &\quad (|z| > \tfrac{1}{2}) \quad \cap \quad (|z| > \tfrac{1}{4}) \quad \Rightarrow \quad |z| > \tfrac{1}{2}. \end{aligned}$$

Here is a case where ROC$\{Y(z)\}$ is *not* simply equal to the intersection of the ROC$\{X(z)\}$ and ROC$\{H(z)\}$. In this case, there is a pole-zero cancelation which affects the ROC, as shown in **Figure 4.27**. The zero at $z = 1/2$ in $H(z)$ cancels the pole at $z = 1/2$ in $X(z)$. $H(z)$ also puts a pole at $z = 1/4$, which now "controls" the ROC. So the correct ROC is $|z| > 1/4$. This is why we use the "$\supseteq$" symbol, and write

$$\text{ROC}\{Y(z)\} \supseteq \text{ROC}\{X(z)\} \cap \text{ROC}\{H(z)\}.$$

The notation says that the ROC of $Y(z)$ is at least as large as (i.e., *contains*) the intersection of the ROCs of $X(z)$ and $H(z)$. Hence, $h[n] = \frac{1}{4}^n u[n]$.

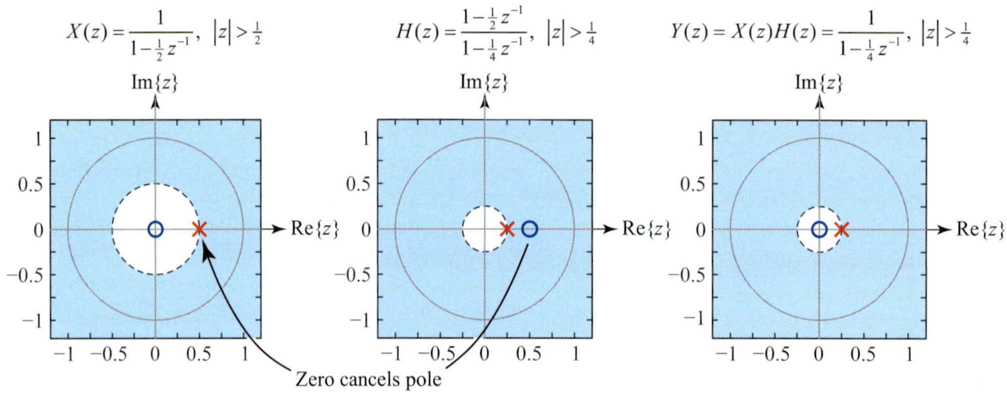

Figure 4.27

4.5.6 Applications of convolution

As we found in Chapter 3, the convolution property of z-transforms allows us to solve three basic problems of linear systems:

(1) **Filtering**. If you know $x[n]$ and $h[n]$, you can obtain $y[n]$ by multiplication of the transforms, $Y(z)=X(z)H(z)$, from which $y[n]=\mathcal{Z}^{-1}\{Y(z)\}$.

(2) **System identification**. If you know $x[n]$ and $y[n]$, you can obtain $h[n]$ by division of transforms,

$$H(z)=\frac{Y(z)}{X(z)},$$

from which $h[n]=\mathcal{Z}^{-1}\{H(z)\}$.

(3) **Deconvolution**. If you know $y[n]$ and $h[n]$, you can obtain $x[n]$ by division of transforms,

$$X(z)=\frac{Y(z)}{H(z)},$$

from which $x[n]=\mathcal{Z}^{-1}\{X(z)\}$.

We have examined several examples of the filtering property to find $y[n]$ (e.g., Examples 4.18 and 4.19, above). Here is an example of using the z-transform to determine $h[n]$ and $x[n]$.

Example 4.20

Given $x[n]=\frac{1}{2}^n u[n]$ and $y[n]=\frac{1}{4}^n u[n]$, find $h[n]$ such that $y[n]=x[n]*h[n]$.

➤ Solution:

$$H(z)=\frac{Y(z)}{X(z)}=\frac{\dfrac{1}{1-\frac{1}{4}z^{-1}}}{\dfrac{1}{1-\frac{1}{2}z^{-1}}}=\frac{1-\frac{1}{2}z^{-1}}{1-\frac{1}{4}z^{-1}}.$$

What is the ROC of $H(z)$? Is it $|z| > 1/4$ or $|z| < 1/4$? To figure out the correct ROC, remember that ROC$\{Y(z)\} \supset$ ROC$\{X(z)\} \cap$ ROC$\{H(z)\}$. Since ROC$\{X(z)\} = (|z| > 1/2)$ and ROC$\{Y(z)\} = (|z| > 1/4)$, the only possibility of the ROC of $H(z)$ is $|z| > 1/4$. Taking the inverse z-transform of $H(z)$ gives $h[n] = 2\delta[n] - \frac{1}{4}^n u[n]$.

Example 4.21

Given $y[n] = \frac{1}{2}^n u[n]$ and $h[n] = \delta[n] - \delta[n-1]$, find $x[n]$ such that $y[n] = x[n] * h[n]$.

➤ Solution:

$$X(z) = \frac{Y(z)}{H(z)} = \frac{\dfrac{1}{1-\frac{1}{2}z^{-1}}}{1-z^{-1}} = \frac{1}{(1-z^{-1})(1-\frac{1}{2}z^{-1})}.$$

There are three possibilities for the ROC of $X(z)$: $|z| < 1/2$, $|z| > 1$ or $1/2 < |z| < 1$. However, only one of these three options will satisfy the requirement that ROC$\{Y(z)\} \supseteq$ ROC$\{X(z)\} \cap$ ROC$\{H(z)\}$. The answer is ROC$\{X(z)\} = (|z| > 1)$. Note here that ROC$\{Y(z)\} = (|z| > 1/2)$, which is larger than simply ROC$\{X(z)\} \cap$ ROC$\{H(z)\}$. That is because $H(z)$ has a zero that cancels the pole in $X(z)$ at $z = 1$. The remaining pole of $X(z)$, at $z = 1/2$, then controls the ROC of $Y(z)$.

To find $x[n]$, apply partial fractions to decompose the product of $X(z)$ into the sum of two terms,

$$X(z) = \frac{1}{(1-z^{-1})(1-\frac{1}{2}z^{-1})} = \underbrace{\frac{2}{1-z^{-1}}}_{|z|>1} \cap \underbrace{-\frac{1}{1-\frac{1}{2}z^{-1}}}_{|z|>\frac{1}{2}}, \quad |z| > 1.$$

The ROCs of the two terms are chosen such that their intersection is $|z| > 1$. The inverse transform yields $x[n] = (2 - \frac{1}{2}^n)u[n]$.

4.5.7 Initial-value theorem

If a sequence $h[n]$ is causal, then

$$\lim_{z\to\infty} H(z) = h[0].$$

We showed this was true in the discussion leading to Equation (4.8).

4.5.8 Final-value theorem

If $H(z)$ is the transform of a causal stable system, then $h[n]$ approaches a constant, $h(\infty)$, in response to a step as $n \to \infty$,

$$\lim_{z\to 1} (z-1)H(z) - h[\infty].$$

To show this, express $h[n]$ as the sum of the constant causal sequence $h[\infty]u[n]$, plus another causal sequence $g[n]$ that goes to zero as $n \to \infty$: $h[n] = h[\infty]u[n] + g[n]$. The transform of this is

$$H(z) = h[\infty]\frac{1}{1 - z^{-1}} + G(z) = h[\infty]\frac{z}{z - 1} + G(z).$$

Multiplying both sides by $z - 1$ and taking the limit as $z \to 1$ gives

$$\lim_{z \to 1}(z - 1)H(z) = \lim_{z \to 1} h[\infty]z + \cancel{\lim_{z \to 1}(z - 1)G(z)} = h[\infty].$$

One use of the final-value theorem is to show that the final value, $y[\infty]$, of a causal stable system defined by Equation (4.5) in response to a step $x[n] = u[n]$ can be determined quite easily from the coefficients b_k and a_k. By the convolution property,

$$Y(z) = X(z)H(z) = \frac{1}{1 - z^{-1}}H(z) = \frac{z}{z - 1}\frac{\sum_{k=0}^{M} b_k z^{-k}}{\sum_{k=0}^{N} a_k z^{-k}}.$$

Then, by the final-value theorem,

$$y[\infty] = \lim_{z \to 1}(z - 1)Y(z) = \lim_{z \to 1}\cancel{(z - 1)}\frac{z}{\cancel{z - 1}}\frac{\sum_{k=0}^{M} b_k z^{-k}}{\sum_{k=0}^{N} a_k z^{-k}} = \frac{\sum_{k=0}^{M} b_k}{\sum_{k=0}^{N} a_k}.$$

Example 4.22

A system has a single pole at $z = 0.75$ and a single zero at $z = -1.5$,

$$H(z) = A\frac{z + 1.5}{z - 0.75}.$$

Determine the value of the constant A such that the steady-state response of the system in response to a step $x[n] = u[n]$ is $y[\infty] = 1$.

➤ Solution:

$$y[\infty] = A\frac{1 + 1.5}{1 - 0.75} = 10A,$$

hence, $A = 0.1$.

4.6 Linear constant-coefficient difference equations (LCCDE)

Pretty much every system in which we are interested can be described by a linear constant-coefficient difference equation (LCCDE) that relates an input of an LTI system, $x[n]$, to the output $y[n]$. The general form of the difference equation is

$$\sum_{k=0}^{N-1} a_k y[n-k] = \sum_{k=0}^{M-1} b_k x[n-k]. \quad \textbf{(4.27)}$$

We can further assume that $a_0 = 1$. (If it is not, divide both sides of Equation (4.27) by a_0, thereby creating a new set of normalized coefficients, $\hat{a}_k = a_k/a_0$ and $\hat{b}_k = b_k/b_0$, such that $\hat{a}_0 = 1$.) Reorganizing the equation yields

$$y[n] = \sum_{k=0}^{M-1} b_k x[n-k] - \sum_{k=1}^{N-1} a_k y[n-k].$$

In words, this equation states that the present value of the output $y[n]$ depends on both present and past values of the input $x[n-k]$, and on past values of the output $y[n-k]$, scaled by the appropriate constants b_k and a_k, respectively.

4.6.1 LCCDE of FIR systems

First, let us consider the case where all the a_i constants in Equation (4.27) are zero. Then the output at time $n=0$ is strictly a function of past and present values of the input scaled by constants b_k:

$$y[n] = \sum_{k=0}^{M-1} b_k x[n-k].$$

We have seen this before. For any LTI system, the output is related to the input via convolution,

$$y[n] = x[n] * h[n] = \sum_{k=-\infty}^{\infty} h[k]x[n-k].$$

If the system is FIR with a causal impulse response $h[n]$ of length M, then $h[n]$ is non-zero only over the interval $0 \le n < M$, and the convolution sum becomes

$$y[n] = x[n] * h[n] = \sum_{k=0}^{M-1} h[k]x[n-k]. \quad \textbf{(4.28)}$$

Comparing Equation (4.28) with Equation (4.27), you can see that the convolution sum is just a linear constant-coefficient difference equation with all the a_k's set to zero and the b_k's set to the impulse response: $b_k = h[k]$. We have already covered the convolution property of FIR systems fairly extensively so maybe we will skip the examples and move on.

4.6.2 LCCDE of IIR systems

When the a_k constants in Equation (4.27) are non-zero, the output of the system at time n depends on past and present values of both the input *and* the output and the system is generally IIR. In arriving at a general solution of these LCCDEs, there are two cases to consider.

Zero-input response If the system has zero input (i.e., $x[n]=0$ for all n), the difference equation becomes

$$y[n]=-\sum_{k=1}^{N-1} a_k y[n-k].$$

Even if $x[n]$ is zero, the output of the system $y[n]$ can still be non-zero for $n \geq 0$ because it depends on the N previous values of the output, which may themselves be non-zero. In particular, the first output point, $y[0]$, will depend on the values of $y[-1]$, $y[-2]$, ..., $y[-(N-1)]$. These previous values are termed the **initial conditions**. They determine the **initial state** of the system at time $n=0$. The response of the system with zero input and non-zero initial state is termed the **zero-input response**.

Zero-state response If the initial state of the system is zero (i.e., $y[-1]=y[-2]=\cdots=y[-(N-1)]=0$), the system is said to be at **initial rest**. Then the system's output depends solely on the input $x[n]$. This is termed the **zero-state response**.

Total response If the system has both an input and non-zero initial conditions, then the total response is the superposition (sum) of the zero-input response and the zero-state response. In this section, we will consider the zero-state solution, which is appropriate for the majority of applications that have zero initial conditions. In Section 4.7, we will introduce the unilateral z-transform, which will allow us to tackle problems with non-zero initial conditions.

4.6.3 Relation between the LCCDE and *H*(*z*)

Take the z-transform of both sides of Equation (4.27),

$$\mathcal{Z}\left\{\sum_{k=0}^{N-1} a_k y[n-k]\right\}=\mathcal{Z}\left\{\sum_{k=0}^{M-1} b_k x[n-k]\right\}$$

$$\sum_{k=0}^{N-1} \mathcal{Z}\{a_k y[n-k]\}=\sum_{k=0}^{M-1} \mathcal{Z}\{b_k x[n-k]\}$$

$$\sum_{k=0}^{N-1} a_k \mathcal{Z}\{y[n-k]\}=\sum_{k=0}^{M-1} b_k \mathcal{Z}\{x[n-k]\}$$

$$Y(z)\sum_{k=0}^{N-1} a_k z^{-k}=X(k)\sum_{k=0}^{M-1} b_k z^{-k},$$

where we have used the shifting property to express $\mathcal{Z}\{y[n-k]\}=Y(z)z^{-k}$ and $\mathcal{Z}\{x[n-k]\}=X(z)z^{-k}$. Rearranging terms and using the convolution property to recognize that $H(z)=Y(z)/X(z)$,

$$H(z) = \frac{Y(z)}{X(z)} = \frac{\sum_{k=0}^{M-1} b_k z^{-k}}{\sum_{k=0}^{N-1} a_k z^{-k}},$$

which is exactly the same as Equation (4.27).

4.6.4 Using Matlab to solve LCCDEs

Matlab's `filter` function can be used to produce the output of an LCCDE to a given input. The syntax is

```
y = filter(b,a,x),
```

where `x` is the input sequence $x[n]$, `y` is the output sequence $y[n]$ and `b` and `a` are the coefficients b_k and a_k in Equation (4.27).

Example 4.23

A stable system is characterized by the LCCDE $y[n] - \frac{1}{4}y[n-1] = x[n] - \frac{1}{2}x[n-1]$.

(a) Find the impulse response $h[n]$.

(b) Verify the answer to part (a) with Matlab by finding the first five points of the impulse response.

➤ Solution:

(a) Find $H(z)$:

$$H(z) = \frac{1 - \frac{1}{2}z^{-1}}{1 - \frac{1}{4}z^{-1}} = 2 - \frac{1}{1 - \frac{1}{4}z^{-1}}.$$

The inverse transform is $2\delta[n] - \frac{1}{4}^n u[n]$.

(b) The trick to using `filter` to calculate an impulse response of length N is to create an array with a single one followed by $N-1$ zeros.

```
b = [1 -0.5];
a = [1 -0.25];
N = 5;
x = [1 zeros(1, N-1)];      % create an impulse followed by four zeros
h_filt = filter(b, a, x);   % use 'filter' to find h[n]
h_comp = [1 -0.25.^(1:4)]; % calculate h[n]
out = [(0:4);h_filt;h_comp];
disp(['n         h [n]'])
fprintf(1, '%d  %+5.4f  %+5.4f\n', out(:))

n         h [n]
0  +1.0000  +1.0000
1  -0.2500  -0.2500
2  -0.0625  -0.0625
3  -0.0156  -0.0156
4  -0.0039  -0.0039
```

Example 4.24

We observe that when the input to a system is $x[n]=\delta[n]-\frac{1}{2}\delta[n-1]$, the output is $y[n]=2\frac{1}{3}^n u[n]$. Find the LCCDE that describes the system.

➤ **Solution:**

$$X(z)=1-\tfrac{1}{2}z^{-1}$$

$$Y(z)=\frac{2}{1-\frac{1}{3}z^{-1}}$$

$$H(z)=\frac{Y(z)}{X(z)}=\frac{2}{\left(1-\frac{1}{2}z^{-1}\right)\left(1-\frac{1}{3}z^{-1}\right)}=\frac{2}{1-\frac{5}{6}z^{-1}+\frac{1}{6}z^{-2}}.$$

Hence,

$$y[n]-\tfrac{5}{6}y[n-1]+\tfrac{1}{6}y[n-2]=2x[n].$$

4.6.5 Inverse filter

The **inverse filter** is a good example of the use of z-transform properties. Assume that we pass an input $x[n]$ through a filter with impulse response $h[n]$ yielding output $y[n]=x[n]*h[n]$. Now, we wish to design a filter with impulse response $h_I[n]$ such that when we pass $y[n]$ through this inverse filter, the output $z[n]$ is the same as the input $x[n]$, as schematized in **Figure 4.28**. This is called the inverse filter.

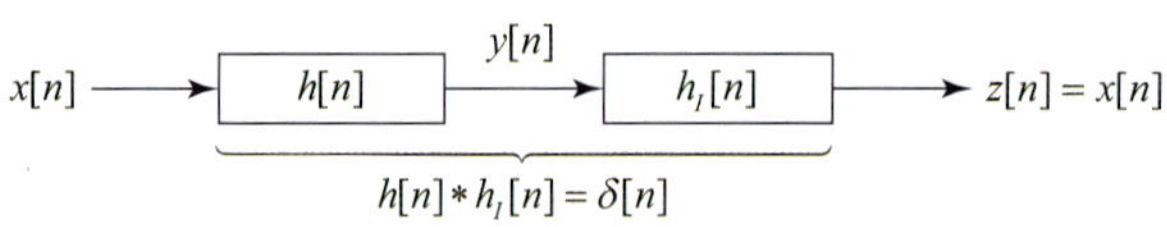

Figure 4.28 Inverse filter

We have $z[n]=x[n]*h[n]*h_I[n]=x[n]$. The only way this can happen is if $h[n]*h_I[n]=\delta[n]$, which implies $H(z)\cdot H_I(z)=1$. Hence,

$$H_I(z)=\frac{1}{H(z)}.$$

That is why it is called the inverse filter. Because $H_I(z)$ and $H(z)$ are inverses, the poles of $H_I(z)$ become the zeros of $H(z)$ and vice versa. This has important consequences for the causality of the inverse filter. We will explore the inverse filter using a series of examples.

Example 4.25

Given a lowpass filter with impulse response $h[n]=\frac{1}{4}\cdot\frac{3^n}{4}u[n]$, find the impulse response of the inverse highpass filter $h_I[n]$, and show the pole-zero plot.

➤ Solution:

Since

$$H(z)=\frac{\frac{1}{4}}{1-\frac{3}{4}z^{-1}}=\frac{\frac{1}{4}z}{z-\frac{3}{4}},\qquad |z|>\frac{3}{4},$$

the inverse filter has transform $H_I(z)=4(1-\frac{3}{4}z^{-1})$. So, $h_I[n]=4\delta[n]-3\delta[n-1]$. You can verify that $h[n]*h_I[n]=\delta[n]$. The plots of $h[n]$ and $H(z)$ are shown in **Figure 4.29a**. $H(z)$ is a lowpass filter that has a pole at $z=3/4$ and a zero at $z=0$. The plots of $h_I[n]$ and $H_I(z)$ are shown in **Figure 4.29b**. The inverse filter $H_I(z)$ is a highpass filter that has a pole at $z=0$ and a zero at $z=3/4$. The ROC is the entire z-plane (except $z=0$).

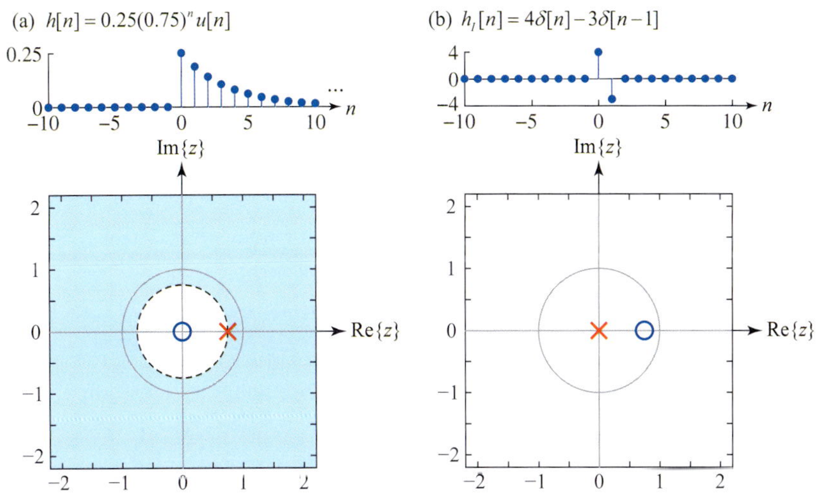

Figure 4.29 Causal and stable system with a causal and stable inverse

In this example, $H(z)$ is causal and stable and so is $H_I(z)$. However, not every causal and stable system has a causal and stable inverse, as we will now show in the following examples.

Example 4.26

Repeat Example 4.25 given $h[n] = -0.2\delta[n] + 0.3(0.75)^n u[n]$.

➤ Solution:

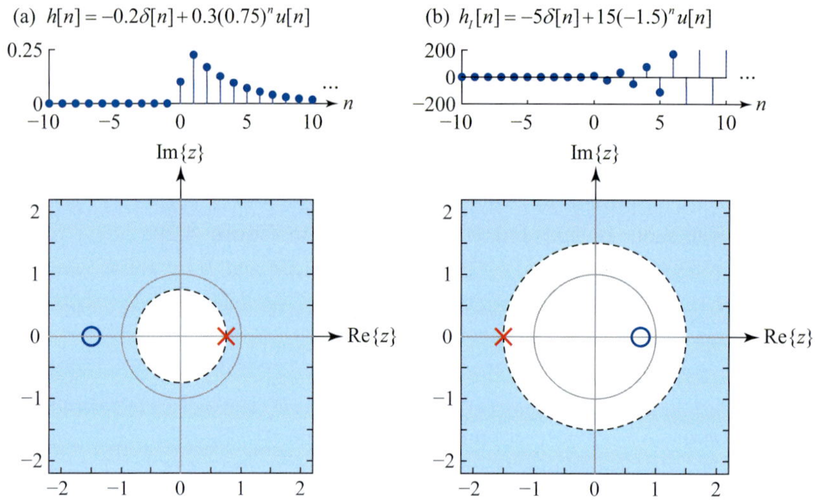

Figure 4.30 Causal and stable system with a causal unstable inverse

Here,

$$H(z) = -0.2 + \frac{0.3}{1 - 0.75z^{-1}} = 0.1\frac{1 + 1.5z^{-1}}{1 - 0.75z^{-1}} = 0.1\frac{z + 1.5}{z - 0.75}, \quad |z| > 0.75. \tag{4.29}$$

There is a pole at $z = 0.75$ and a zero at $z = -1.5$, as shown in **Figure 4.30a**. The causal system is stable, because the ROC, $|z| > 0.75$, includes the unit circle. The inverse filter has transform

$$H_I(z) = \frac{1}{H(z)} = 10\frac{1 - 0.75z^{-1}}{1 + 1.5z^{-1}} = 10\frac{z - 0.75}{z + 1.5}, \quad |z| > 1.5. \tag{4.30}$$

The pole is at $z = -1.5$ and the zero is at $z = 0.75$, as shown in **Figure 4.30b**. The causal inverse system is unstable because the ROC $|z| > 1.5$ does not include the unit circle. The impulse response is $h_I[n] = -5\delta[n] + 15(-1.5)^n u[n]$ (see Problem 4-38), which blows up as $n \to \infty$.

The problem with the system in Example 4.26 is that the zeros in $H(z)$ that lie outside the unit circle become poles in $H_I(z)$. If we require the inverse filter to have a right-sided impulse response, the ROC will extend from the outermost pole to infinity. But, if the outermost pole is outside the unit circle, the inverse filter will be unstable. In Example 4.25, all the zeros (well, there was only one) were inside the unit circle, so the inverse system was also stable.

Example 4.27

Repeat Example 4.25 given $h[n] = 0.25(0.75)^{n-1}u[n-1]$.

➤ Solution:

This system has the same impulse response as that in Example 4.25, except delayed by one. The shift by one introduces a pole at $z = 0$ that cancels the zero in **Figure 4.29a**. The resulting pole-zero plot is shown in **Figure 4.31a**. The inverse has a single zero, so it is stable. But the number of zeros (one) exceeds the number of poles (zero), so the system is non-causal.

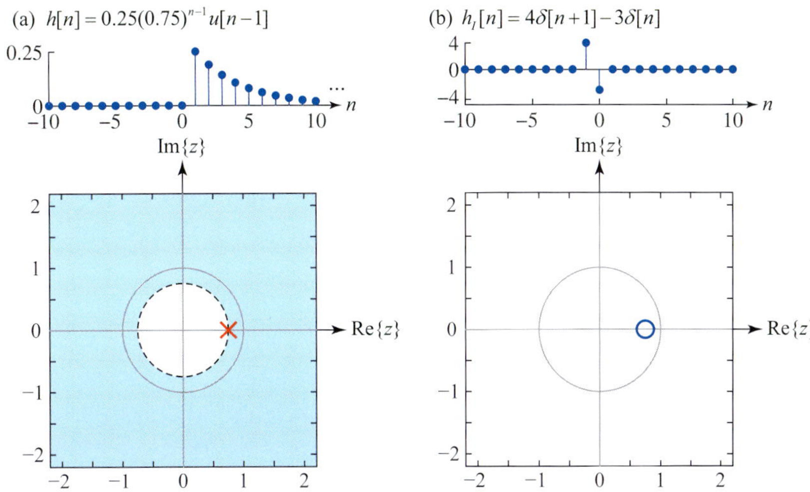

Figure 4.31 Causal and stable system with a non-causal stable inverse

The lesson of the previous three examples can be summarized as follows:

> A causal and stable system, $H(z)$, can have a causal and stable inverse only if (1) all the poles and zeros of $H(z)$ lie inside the unit circle and (2) the number of poles and zeros are equal.

The first condition is necessary to assure that the inverse system will be stable; the second condition is required to make the system causal.

Before leaving the subject of inverse filters, let us return to Example 4.26 for a moment. Is there anything we can do to find a stable inverse for a causal stable system such as that pictured in **Figure 4.30a**, which has the misfortune to have a zero located outside the unit circle? Yes, there are actually a couple of options if we are willing to dispense with the requirement that the processing of the inverse filter's response to an input be done in real time.

Example 4.28

Find a stable inverse for the system shown in **Figure 4.30a**.

➤ Solution:

Almost every $H(z)$ with poles and zeros in the finite z-plane has a stable inverse. It is just a matter of finding the ROC that includes the unit circle. (The exception would be systems with zeros located on the unit circle.) The causal inverse system we found in **Figure 4.30a** was unstable because the ROC $|z| > 1.5$ did not include the unit circle. A stable system with the same pole and zeros has ROC $|z| < 1.5$, as shown in the lower panel of **Figure 4.32a**. Because the ROC extends from the innermost (and only) pole to $z = 0$, this system is stable, though anti-causal, and has impulse response $h_I[n] = -5\delta[n] - 15(-1.5)^n u[-n-1]$. What possible use can such a system be? Read on ...

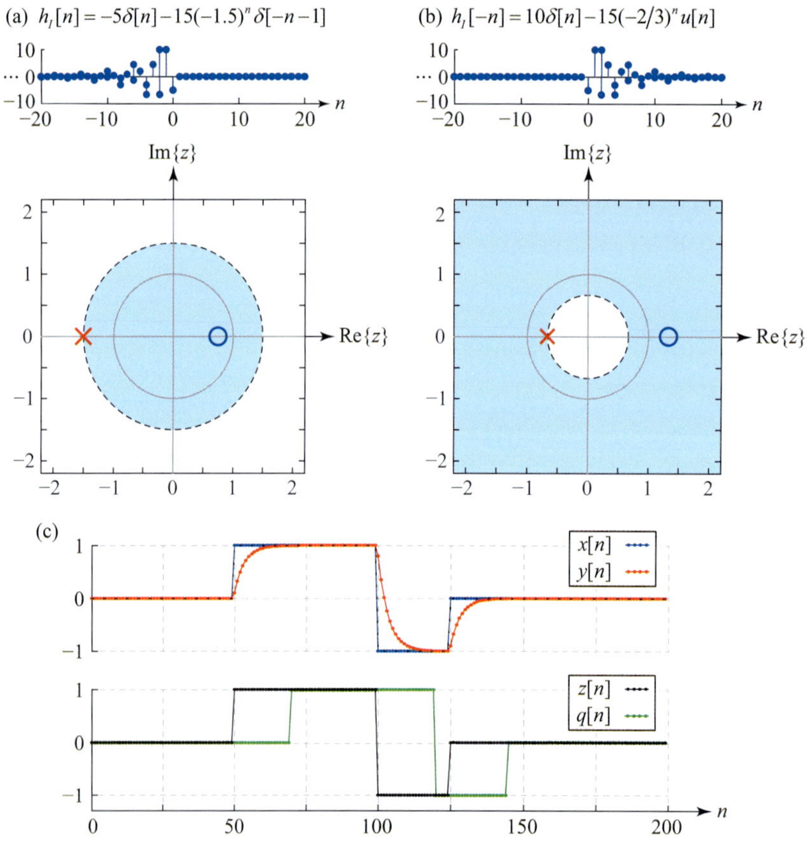

Figure 4.32 An anti-causal stable inverse filter

Assume we have a signal $x[n]$ that has been lowpass filtered by the system in **Figure 4.30a** with impulse response $h[n]$ to form an output $y[n] = x[n] * h[n]$. We would now like to filter $y[n]$ with $h_I[n]$ to recover the input:

$$z[n] = y[n] * h_I[n] = (x[n] * h[n]) * h_I[n] = x[n] * (h[n] * h_I[n]) = x[n] * \delta[n] = x[n]. \tag{4.31}$$

Since the inverse system, $h_I[n]$, is anti-causal, we cannot do this filtering. However, what we *can* do is hinted at by substituting $-n$ into Equation (4.31):

$$z[-n] = y[-n] * h_I[-n] = x[-n]. \tag{4.32}$$

If we have the entire signal $y[n]$ at our disposal, for example as a data file, then we can easily time-reverse it to form $y[-n]$. Moreover, we can also easily find the system that corresponds to $h_I[-n]$ by applying the time-reversal property to Equation (4.30):

$$\mathcal{Z}\{h_I[-n]\} = H_I(z^{-1}) = 10\frac{-0.75 + z^{-1}}{1.5 + z^{-1}} = -5\frac{z - 4/3}{z + 2/3}, \quad |z| > 2/3. \tag{4.33}$$

The poles and zeros of $H_I(z^{-1})$ are reflected with respect to the unit circle; there is a single pole inside the unit circle at $z = -2/3$ and a single zero at $z = 4/3$, as shown in the lower panel of **Figure 4.32b**. Hence, this represents the response of a right-sided, causal and stable system. Good news! To find the impulse response, $h_I[-n]$, do the partial fraction expansion

$$H_I(z^{-1}) = -5\frac{1 - (4/3)z^{-1}}{1 + (2/3)z^{-1}} = 10 - \frac{15}{1 + (2/3)z^{-1}},$$

from which we get $h_I[-n] = 10\delta[n] - 15(-2/3)^n u[n]$, shown in the top panel of **Figure 4.32b**. You should satisfy yourself that this expression is equal to time-reversing the impulse response $h_I[n] = -5\delta[n] - 15(-1.5)^n u[-n-1]$. Now, according to Equation (4.32), we can filter $y[-n]$ with $h_I[-n]$ to form $z[-n]$ and then time-reverse the result to form $z[n] = x[n]$.

Figure 4.32c shows an example of the application of an inverse filter to cancel the effect of a lowpass filter on some data. The input data $x[n]$ (blue trace) is 200 points of a pulse. The output of the lowpass filter is $y[n]$ (red trace). The output of the inverse filter, $z[n]$ (black trace), is identical to $x[n]$. We will describe the remaining trace $q[n]$ in a moment. But first, it is worth looking at the code that generated this example:

```
% Create an input pulse, x[n]
N = 200;
n = 0:N-1;
x = [zeros(1, N/4) ones(1, N/4) -ones(1, N/8) zeros(1, 3*N/8)];

% Create a lowpass filter with zero outside unit circle and filter input
% to form y[n]
b = [1 1.5];
a = [1 -0.75];
A = sum(a) / sum(b); % filter gain is 0.1 so that H(z=1) = 1;
b = b * A;
y = filter(b, a, x);

% Create inverse filter and apply to y[-n] to form z[-n]
zflipped = filter(a(end:-1:1), b(end:-1:1), y(end:-1:1));
z = zflipped(end:-1:1);

plot(n, x, 'bo-', n, y, 'ro-', n, z, 'ko ')
```

First, we create the input pulse `x`. Next, this pulse is lowpass filtered using `filter` to produce output `y`. The coefficients of the filter, `b` and `a`, correspond to the values $b_0 = 1$, $b_1 = 1.5$, $a_0 = 1$ and $a_1 = -0.75$, given in Equation (4.29). The next two lines create scale factor `A`, which is chosen so that the response of the lowpass filter to a step input, $x[n]=u[n]$, will approach a value of $y[\infty]=1$ as $n\to\infty$. We showed how this is done using the final-value theorem in Example 4.22. This scale factor is applied to coefficient `b`.

The next lines of the code create the causal flipped filter of Equation (4.33) with impulse response $h_I[-n]$. As shown in Equation (4.30), the poles and zeros of the anti-causal filter $H_I(z)=1/H(z)$ are interchanged with respect to those of $H(z)$, so `b` and `a` vectors we use in `filter` need to be interchanged; the numerator gets the `a` coefficients and the denominator gets the `b` coefficients. Furthermore, by the time-reversal property, Equation (4.25), it is also necessary to reverse the *order* of the coefficients, as you can verify in this example by comparing Equations (4.30) and (4.33). We then apply time-reversed $y[-n]$ to the causal inverse filter to get $z[-n]$, named `zflipped` in the code. Finally, we time-reverse $z[-n]$ to get $z[n]$. Phew!

This technique is pretty neat, but what can we do if we do not have the luxury of time-reversing the entire signal (e.g., by forming $y[-n]$ from a recording), and processing it with the time-reversed IIR inverse filter? In some cases, we can still make a reasonable (though not perfect) N-point FIR filter by taking the first N points of the time-reversed IIR inverse filter and applying them to $y[n]$. Here is how:

```
N = 21;
hfir = fliplr(filter(a(end:-1:1), b(end:-1:1), [1 zeros(1, N-1)]));
q = filter(hfir, 1, y);
plot(n, q, 'go-')
```

This code uses `filter` to calculate an array of the first N points of the impulse response $h_I[-n]$ and then time-reverses (flips) the array to form `hfir`, the impulse response of an N-point FIR filter. The result of filtering `y` with `hfir` is `q`, labeled $q[n]$ in **Figure 4.32c** (green trace). $q[n]$ is effectively equivalent to input $x[n]$, but is time-delayed by N points. This time delay might be OK in some applications, but may not be acceptable in real-time applications such as digital control where response latency is important.

4.7 ★ The unilateral *z*-transform

The unilateral z-transform is defined as:

$$\mathcal{Z}_U\{x[n]\}\triangleq\sum_{n=-\infty}^{\infty}(x[n]u[n])z^{-n}=\sum_{n=0}^{\infty}x[n]z^{-n}.$$

The definition of the unilateral z-transform is the same as that of the bilateral z-transform, Equation (4.3), except that the limits of the summation go from $n=0$ to ∞. This transform is particularly useful for analyzing the solution of LCCDEs of causal systems with non-zero initial conditions. In supplementary material available at www.cambridge.org/holton, this topic is discussed in detail, with examples of the solution of LCCDEs with initial conditions.

SUMMARY

The z-transform and its inverse are the most useful tools we have to analyze signals and systems in the frequency domain. They permit us to characterize both stable and unstable systems, right-sided, left-sided and two-sided responses. The pole-zero plot is a simple graphical representation of both the singularities of the z-transform and the region of convergence. The properties of the transform provide a framework for understanding convolution, deconvolution and system identification of linear systems. The bilateral z-transform and its relative the unilateral transform can be used to solve linear constant-coefficient difference equations, including those with non-zero initial conditions.

PROBLEMS

Problem 4-1

A system has input $x[n] = \frac{1}{2}^n u[n]$ and output $y[n] = \frac{1}{4}^n u[n-1]$.

(a) Plot $H(z)$, including region of convergence.

(b) Find $h[n]$.

(c) Find the difference equation that relates the output to the input.

Problem 4-2

For each of the systems shown in **Figure 4.33**, the ROC is not specified. Check (√) for each of the following statements that is *always* true.

Statement	a	b	c	d	e	f
The system is FIR.						
The system is IIR.						
The system is (or could be) causal.						
The system is (or could be) stable.						
If the system is causal, it is stable.						
If the system is stable, it is causal.						

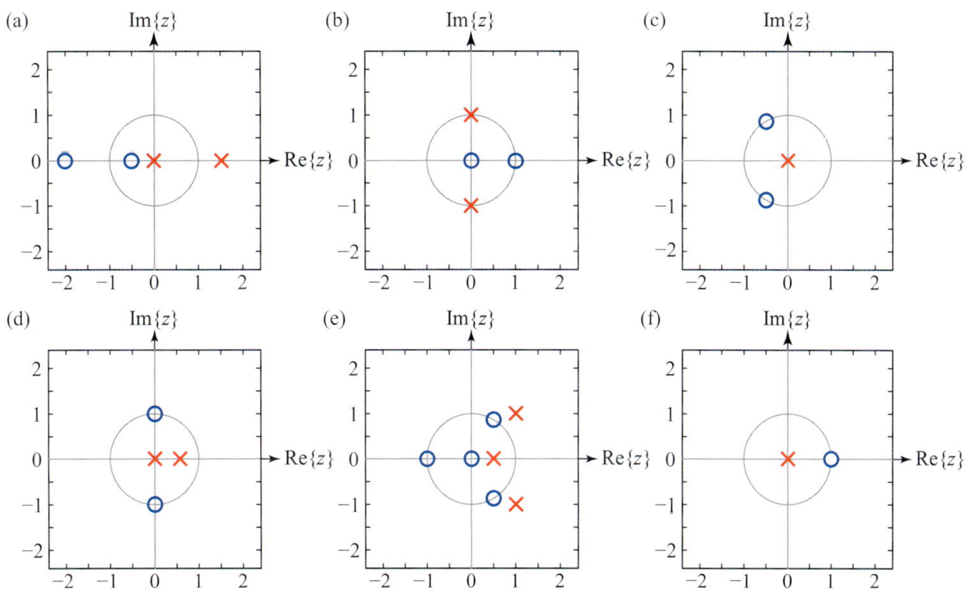

Figure 4.33

Problem 4-3

The stable system shown in **Figure 4.34** has $h_1[n]=\frac{1}{6}^n u[n]$, $h_2[n]=\frac{1}{6}\delta[n-1]$ and $h_3[n]=\frac{2}{3}^n u[n]$.

(a) Find $H(z)$, including the region of convergence.

(b) Find the difference equation that relates the output $y[n]$ to the input $x[n]$.

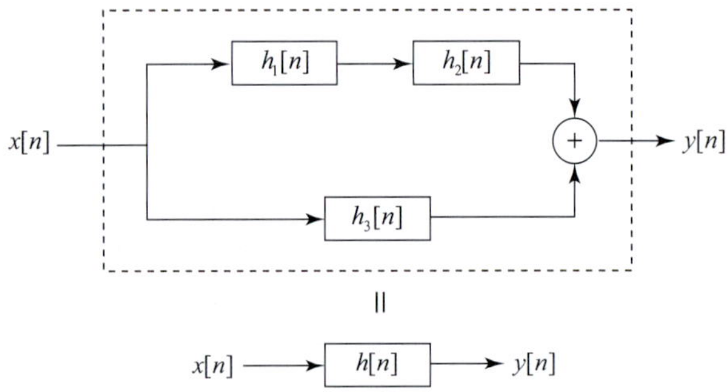

Figure 4.34

Problem 4-4

Find the transform of $h[n]=\alpha^n \sin \omega_0 u[n]$ and sketch the pole-zero plot for values $\alpha=0.75$ and $\omega_0=\pi/4$.

Problem 4-5

When the input to a system is $x[n]=(1-\frac{1}{2}^n)u[n]$, the output of the system is $y[n]=\frac{1}{4}^n u[n]$.

(a) Plot the pole-zero plot for this system, with ROC.

(b) Find the LCCDE of this system.

(c) Find the impulse response $h[n]$.

Problem 4-6

Given the LCCDE $y[n]-\frac{1}{4}y[n-2]=x[n]-\frac{1}{2}x[n-2]$,

(a) find $h[n]$.

(b) find the input $x[n]$ that makes the output $y[n]=\frac{1}{2}^n u[n]$.

Problem 4-7

Given a system with

$$H(z)=\frac{2+2z^{-2}}{1-1.5z^{-1}-z^{-2}},$$

(a) find $h[n]$ given that the system is causal.

(b) find $h[n]$ given that the system is stable.

Problem 4-8

A stable system is shown in **Figure 4.35**, comprising a cascade of two discrete-time filters such that $z[n] = x[n]$.

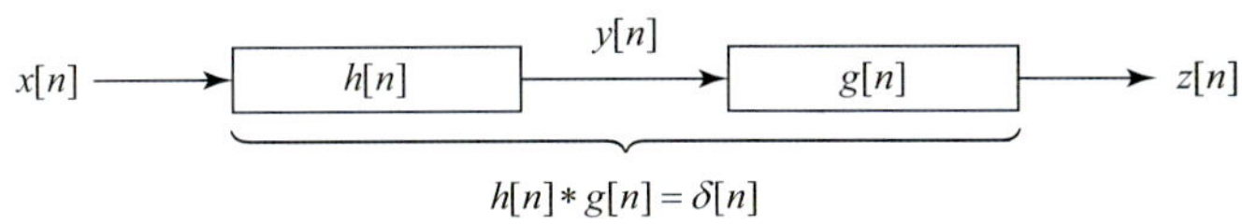

Figure 4.35

The first filter has unknown impulse response $h[n]$. The second filter is defined by the difference equation

$$z[n] = y[n] - \tfrac{3}{4}y[n-1] + \tfrac{1}{8}y[n-2].$$

Find $h[n]$.

Problem 4-9

A stable system comprises a cascade of two discrete-time filters, as shown in **Figure 4.35**. The first filter, with impulse response $h[n]$, is defined by the pole-zero plot for $H(z)$ shown in **Figure 4.36**, along with the information that $h[0] = 2$.

(a) Find $h[n]$.

(b) Find the difference equation of a second system, characterized by $g[n]$, such that $z[n] = x[n]$.

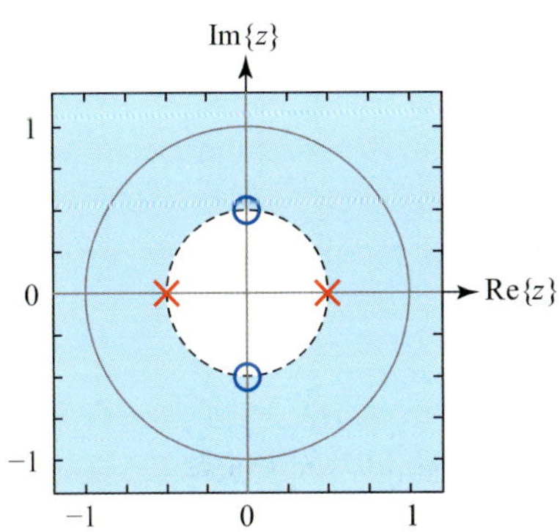

Figure 4.36

Problem 4-10

Repeat Problem 4-9 with $H(z)$ shown in **Figure 4.37**.

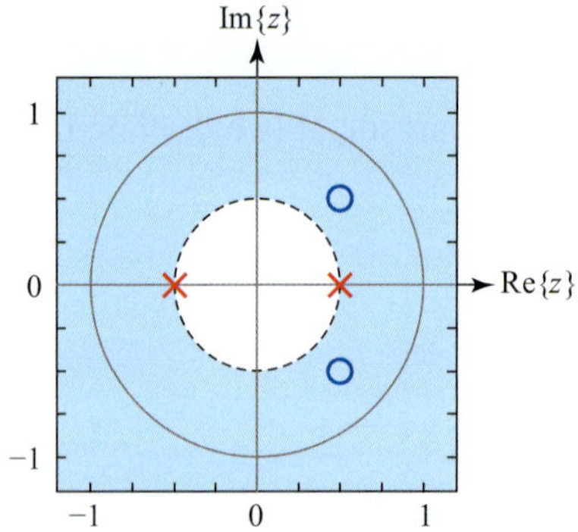

Figure 4.37

Problem 4-11

A stable system is characterized by the difference equation $y[n]-\frac{3}{2}y[n-1]-y[n-2]=x[n]$.

(a) Plot $H(z)$, including the region of convergence.

(b) Find $h[n]$.

Problem 4-12

An input $x[n]$ is filtered by a system comprising two stable subsystems, with impulse responses $f[n]$ and $g[n]$, whose outputs are subtracted to form the output $y[n]$, as shown in **Figure 4.38**.

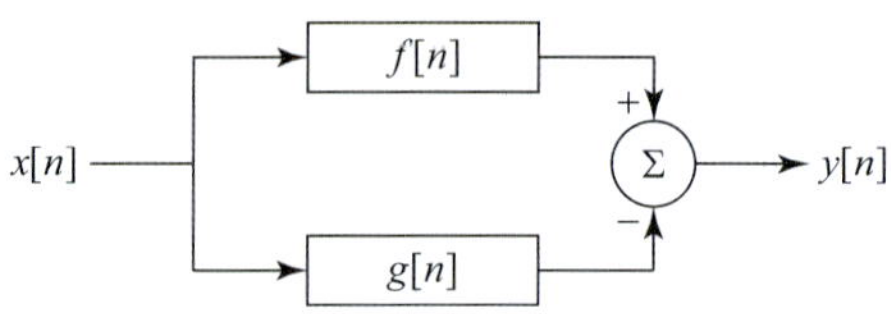

Figure 4.38

The difference equation for the entire system is

$$y[n]-\tfrac{1}{4}y[n-2]=2(x[n]-x[n-2]).$$

You are given that $F(z)$ has a single zero at $z=0.8$ and a single pole at $z=0.5$, and that $F(0)=5$. Find $g[n]$.

Problem 4-13

Given the system of Problem 4-12 with

$$F(z)=\frac{5z-4}{z-\frac{1}{2}},$$

and $g[n]=3\cdot(-\frac{1}{2})^n u[n]$,

(a) find the difference equation relating input $x[n]$ and output $y[n]$.
(b) find the impulse response of the entire system.

Problem 4-14

Given a system with pole-zero plot shown in **Figure 4.39** and the fact that $H(0) = -5$,
(a) find $h[n]$ given that the system is causal.

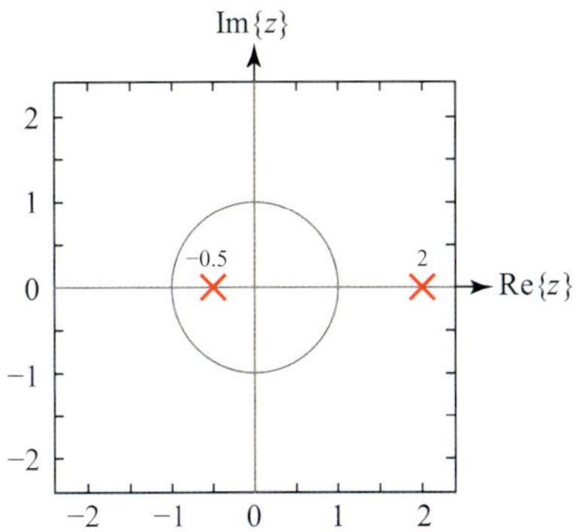

Figure 4.39

(b) find $h[n]$ given that the system is stable.

Problem 4-15

We are given that each $H(z)$ below corresponds to a *stable* system. For each one, determine which of the following attributes apply (more than one may apply to each system), and place a check (√) in the appropriate column(s). If the system is IIR, determine whether the impulse response oscillates or is monotonic.

	$H(z)$	FIR	IIR	Causal	Non-causal	$h[n]$ oscillates	$h[n]$ monotonic
(a)	$\frac{3}{4}z/\left(z-\frac{3}{4}\right)$						
(b)	$(z+2)/z$						
(c)	$\frac{12}{7}\left(z+\frac{3}{4}\right)/z$						
(d)	$\frac{12}{7}\left(z-\frac{3}{4}\right)$						
(e)	$\frac{3}{4}z/\left(z+\frac{3}{4}\right)$						
(f)	$\frac{9}{4}(z-1)/\left(z-\frac{1}{2}\right)$						
(g)	$\frac{9}{7}\left(z-\frac{4}{3}\right)$						
(h)	$z/\left(z-\frac{4}{3}\right)$						

Problem 4-16

We are given that each $H(z)$ in Problem 4-15 corresponds to a *right-sided* system. For each one, determine the impulse response $h[n]$.

Problem 4-17

We are given that each $H(z)$ in Problem 4-15 corresponds to a *right-sided* system. Determine which systems are stable.

Problem 4-18

Each $H(z)$ below corresponds to a *stable* system. For each one, determine which of the following attributes apply (more than one may apply to each system), and place a check (√) in the appropriate column(s).

	$H(z)$	FIR	IIR	Causal	Non-causal	$h[n]$ oscillates	$h[n]$ monotonic
(a)	$H(z)=\frac{1}{3}(z-2)$						
(b)	$H(z)=\frac{2}{3}(z+\frac{1}{2})$						
(c)	$H(z)=1/(z-2)$						
(d)	$H(z)=z/(2z+1)$						
(e)	$H(z)=\frac{1}{2}(z-1)$						

Problem 4-19

For each part below, circle T if the statement is *always* true. Otherwise, circle F.

T or F: You can uniquely determine $h[n]$ if you know the location of the poles and zeros of $H(z)$, and the value of $H(z_0)$ at a single, non-singular value z_0.

T or F: If a system is FIR, then $H(z)$ has no poles, except perhaps at $z=0$.

T or F: The inverse of a causal IIR system is always stable.

T or F: If a causal system has all its poles and all zeros inside the unit circle, its inverse is stable.

T or F: The inverse filter of an FIR system is always stable.

T or F: If you know $H(z)$, and whether a single point in the z-plane is inside or outside the ROC, then you can completely specify the ROC.

T or F: If $H(z)$ has more poles than zeros, the system is causal.

T or F: If $H(z)$ has all its poles and zeros inside the unit circle, it is stable.

Problem 4-20

Given

$$x[n] = -\left(-\tfrac{1}{2}\right)^n u[n-1]$$

$$h[n] = \delta[n] + \tfrac{1}{2}^{n-1} u[n-1],$$

find $y[n] = x[n] * h[n]$.

Problem 4-21

Given

$$y[n] = \delta[n] - \tfrac{1}{2}^{n-1} u[n-1]$$

$$x[n] = \delta[n] - 3\delta[n-1] + \tfrac{9}{4}\delta[n-2],$$

(a) find $h[n]$.

(b) find the difference equation relating $x[n]$ to $y[n]$.

Problem 4-22

Given a system, $H(z)$, with four poles and two zeros whose pole-zero plot is shown in **Figure 4.40**, find how many responses of the given type exist.

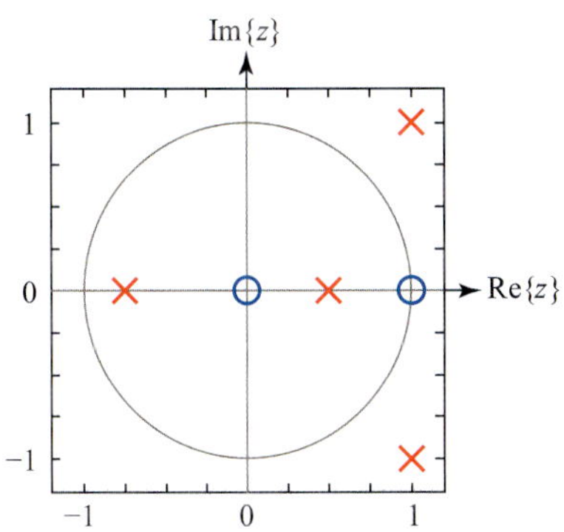

Figure 4.40

Right-sided	Left-sided	Two-sided	Stable	Non-stable	How many?

Problem 4-23

Given a system $H(z)$, whose pole-zero plot is shown in **Figure 4.41**, find how many responses of the given type exist.

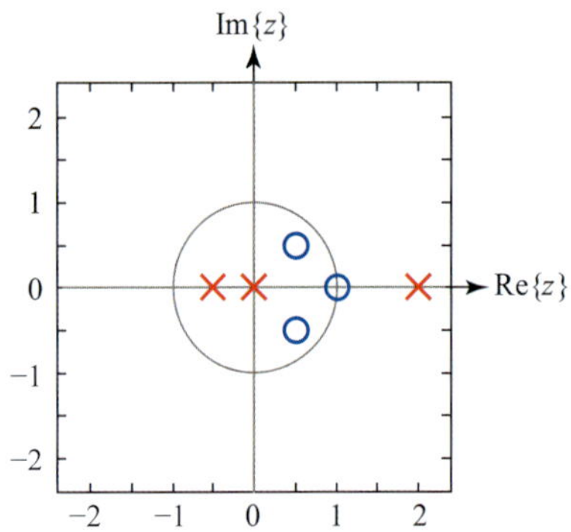

Figure 4.41

Right-sided	Left-sided	Two-sided	Stable	Non-stable	How many?

Problem 4-24

Given a system with

$$H(z) = \frac{3 + \frac{3}{2}z^{-1}}{1 - \frac{5}{2}z^{-1} + z^{-2}},$$

(a) find $h[n]$ given that the system is causal.

(b) find $h[n]$ given that the system is stable.

Problem 4-25

When the input to a system is

$$x[n] = 2u[n] - (-\tfrac{1}{2})^n u[n],$$

the output is

$$y[n] = \tfrac{1}{4}^n u[n].$$

(a) Find $H(z)$.
(b) Find $h[n]$.

Problem 4-26

When the input to a system with impulse response $h[n]$ is

$$x[n] = \alpha^n u[n],$$

the output is

$$y[n] = \beta^n u[n].$$

Find $h[n]$.

Problem 4-27

The step response of a given system is $y[n] = \frac{1}{2}^n u[n]$. Find the input $x[n]$ that produces the output $y[n] = 4\delta[n] - 3 \cdot \frac{1}{4}^n u[n]$.

Problem 4-28

Given $h[n] = \delta[n] - \frac{5}{2}\delta[n-1] + \frac{21}{4}\delta[n-2] - \frac{5}{2}\delta[n-3] + \delta[n-4]$, find $H(z)$ and plot the pole-zero plot.

Problem 4-29

Given $h[n] = 2\delta[n] - 3\delta[n-1] - 1\delta[n-2] - 3\delta[n-3] + 2\delta[n-4]$, find $H(z)$ and plot the pole-zero plot.

Problem 4-30

(a) Determine the pole-zero plot of the z-transform of the odd-length antisymmetric sequence $h[n] = \delta[n] - \delta[n-1] + \delta[n-3] - \delta[n-4]$.
(b) Determine the pole-zero plot of the z-transform of the even-length antisymmetric sequence $h[n] = \delta[n] - \delta[n-2] + \delta[n-3] - \delta[n-5]$.

Problem 4-31

Given a system $H(z)$ whose pole-zero plot is shown in **Figure 4.42**, and the fact that $H(1) = 2$,

(a) find $H(z)$. There should be no complex quantities in your final expression.

(b) find $h[n]$.

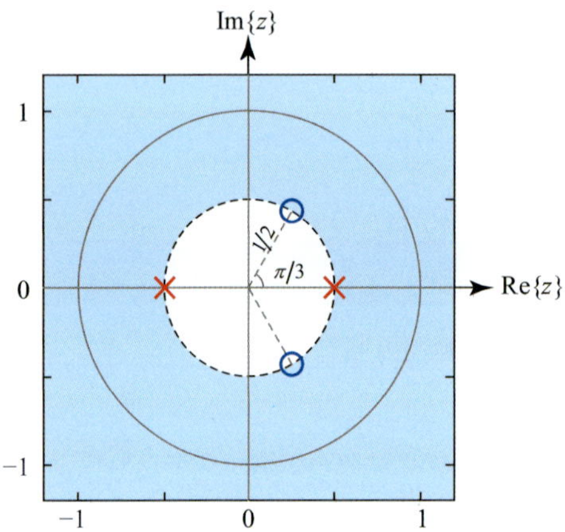

Figure 4.42

Problem 4-32

Given a system $H(z)$, whose pole-zero plot is shown in **Figure 4.43**, and the fact that $H(1) = -12$,

(a) find $h[n]$ given that the system is causal.

(b) find $h[n]$ given that the system is stable.

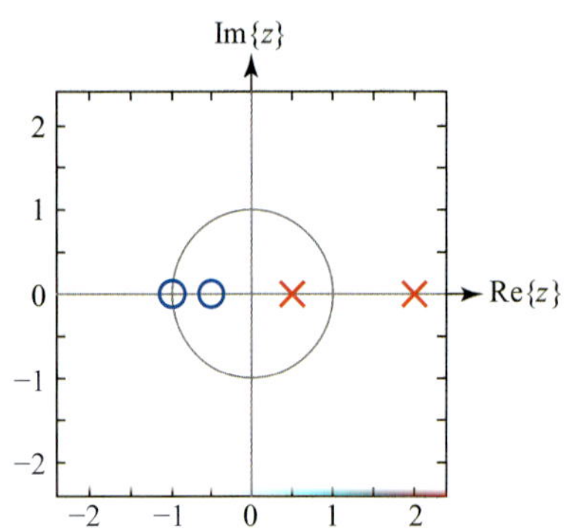

Figure 4.43

Problem 4-33

A system has z-transform

$$W(z) = \frac{1}{(1 - \alpha z^{-1})^2}.$$

(a) Show that $W(z)$ is the transform of a right-sided sequence, $w[n] = (n+1)\alpha^n u[n]$.
(b) Show that $W(z)$ is the transform of a left-sided sequence, $w[n] = -(n+1)\alpha^n u[-n-1]$.

➤Hint: Start with $y[n] = n\alpha^n u[n]$ and use the differentiation property.

Problem 4-34

Use the results of Problem 4-33 and induction to show that the inverse transform of the right-sided sequence corresponding to

$$W_N(z) = \frac{1}{(1-\alpha z^{-1})^N}, \quad N > 1,$$

is

$$w_N[n] = \frac{(n+1)(n+2)\cdots(n+N-1)}{(N-1)!}\alpha^n u[n] = \frac{1}{(N-1)!}\left(\prod_{k=1}^{N-1}(n+k)\right)\alpha^n u[n].$$

➤Hint: Start with $W_N(z)$, apply the differentiation property and relate the result to $W_{N+1}(z)$.

Problem 4-35

Use the differentiation property and induction to show that the z-transform of $y_N[n] = n^N x[n]$ is

$$Y_N(z) = (-z)^N \frac{d^N}{dz^N} X(z).$$

Problem 4-36

Consider the case where $H(z)$ has a pair of complex-conjugate poles p_k and p_k^* whose residues B_k and $\hat{B}_k$ we wish to find:

$$H(z) = \frac{N(z)}{\hat{D}(z)(1-p_k z^{-1})(1-p_k^* z^{-1})} = \cdots + \frac{B_k}{1-p_k z^{-1}} + \frac{\hat{B}_k}{1-p_k^* z^{-1}}.$$

$H(z)$ may also contain other real poles and pairs of complex-conjugate poles. $\hat{D}(z)$ is a polynomial that can be factored to give these other poles. Because these other poles are either real or occur in complex-conjugate pairs, $\hat{D}(z)$ is a polynomial with real coefficients. Show that the residues associated with the complex-conjugate poles are complex conjugates, namely that $\hat{B}_k = B_k^*$.

Problem 4-37

Use the result of Problem 4-36 to simplify the partial fraction expansion of expressions with complex-conjugate poles. For example, if the partial fraction expansion of $H(z)$ has a pair of complex-conjugate poles,

$$H(z) = \cdots + \frac{B_k}{1 - p_k z^{-1}} + \frac{B_k^*}{1 - p_k^* z^{-1}} + \cdots$$

Express both the residues and poles in polar form, $B_k = |B_k|e^{j\measuredangle B_k}$ and $p_k = |p_k|e^{j\measuredangle p_k}$, and show that the inverse transform of the sum of the two terms (assuming a right-sided impulse response) is

$$2|B_k||p_k|^n \cos(n \measuredangle p_k + \measuredangle B_k).$$

Problem 4-38

A causal stable system has impulse response $h[n] = -0.2\delta[n] + 0.3(0.75)^n u[n]$. Find the impulse response of the causal (but unstable) inverse $h_I[n]$.

Problem 4-39

Determine and plot the possible positions in the z-plane where zeros can be located for a symmetric sequence of length $N = 5$.

Problem 4-40

Find the sequence corresponding to the pole-zero plot in **Figure 4.44**, given that $H(1) = 6/5$.

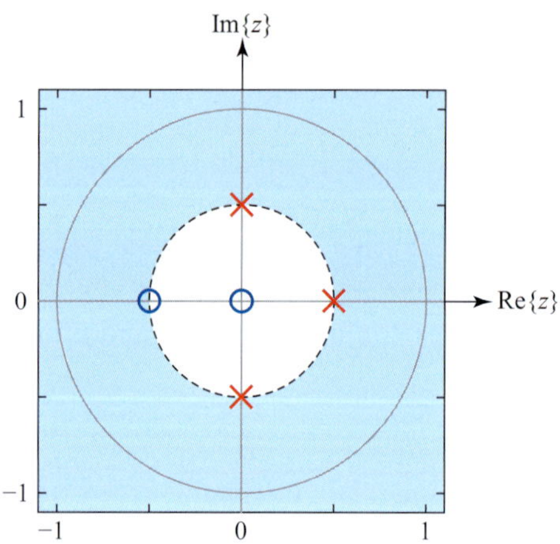

Figure 4.44

Problem 4-41

Given $x_1[n] = \frac{1}{2}^n u[n]$ and $x_2[n] = \frac{1}{4}^n u[n]$, find the z-transform of $y[n] = 2x_1[n] - x_2[n]$.

Problem 4-42

This is a very contrived example of a situation in which ROC$\{Y(z)\}$ can be larger than the intersection of ROC$\{X_1(z)\}$ and ROC$\{X_2(z)\}$. Given $x_1[n] = \frac{1}{2}^n u[n]$ and $x_2[n] = \frac{1}{2}^n u[n-1]$, find the z-transform of $y[n] = x_1[n] - x_2[n]$.

5 Frequency response

Introduction

In Chapter 4, we discussed the relation between the impulse response $h[n]$, the z-transform $H(z)$ and the linear constant-coefficient difference equation (LCCDE) of a linear time-invariant system. In this chapter we will extend our discussion to the frequency response $H(\omega)$. Specifically, we will show that, in many cases, the frequency response can be easily visualized by inspection of the pole-zero plot.

5.1 The computation of $H(\omega)$ from $H(z)$

The relation between the discrete-time Fourier transform $H(\omega)$ and the z-transform $H(z)$ is simple. $H(\omega)$ is just $H(z)$ evaluated for values of z on the unit circle, $z = e^{j\omega}$,

$$H(\omega) = H(z)|_{z = e^{j\omega}}.$$

We will investigate this relation by way of a number of simple examples. First, we will look at systems with a single zero or a single pole and then consider systems with both poles and zeros.

5.2 Systems with a single real zero

Let us start with a simple system which only has a single zero, $H(z) = z - a$, where a is a real constant. **Figure 5.1** shows the case for $a = 1/2$.

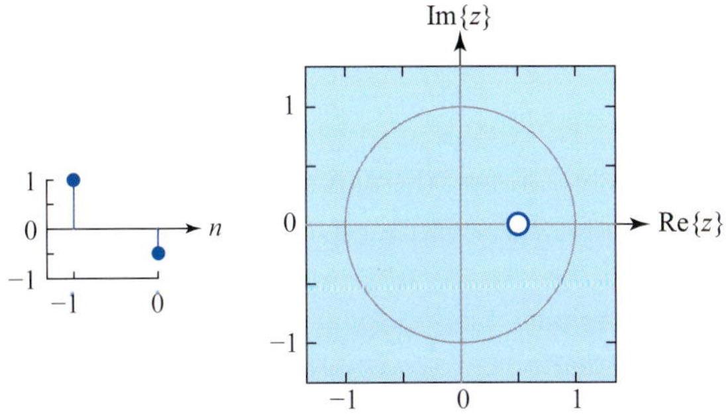

Figure 5.1 Example of a system with one zero

In this case, the Fourier transform is $H(\omega)=e^{j\omega}-a$. We will explore the magnitude $|H(\omega)|$ and phase $\measuredangle H(\omega)$ in two different ways, first by direct computation and then by a graphic visualization method.

5.2.1 Direct computation

Direct computation of $|H(\omega)|$ and $\measuredangle H(\omega)$ produces a lot of algebra:

$$
\begin{aligned}
|H(\omega)| &= |e^{j\omega}-a| = |\cos\omega + j\sin\omega - a| = \sqrt{(\cos\omega-a)^2+\sin^2\omega} \\
&= \sqrt{\cos^2\omega+\sin^2\omega+a^2-2a\cos\omega} = \sqrt{1+a^2-2a\cos\omega}
\end{aligned}
$$

and

$$
\measuredangle H(\omega) = \tan^{-1}\left(\frac{\sin\omega}{\cos\omega - a}\right).
$$

Looking at the expression for the magnitude, you can see that when $\omega=0$, $|H(0)|=\sqrt{1+a^2-2a}=1-a$, and when $\omega=\pi$, $|H(\pi)|=\sqrt{1+a^2+2a}=1+a$. When ω is between 0 and π, $|H(\omega)|$ will vary monotonically between $1-a$ and $1+a$. Hence, if a is positive, you can argue that this is a highpass filter and if a is negative, it is a lowpass filter. However, the visualization of $|H(\omega)|$ and $\measuredangle H(\omega)$ from these formulas is not exactly intuitive, would you say?

5.2.2 Graphical method

Consider a point on the unit circle, as shown by the green dot in **Figure 5.2a**.

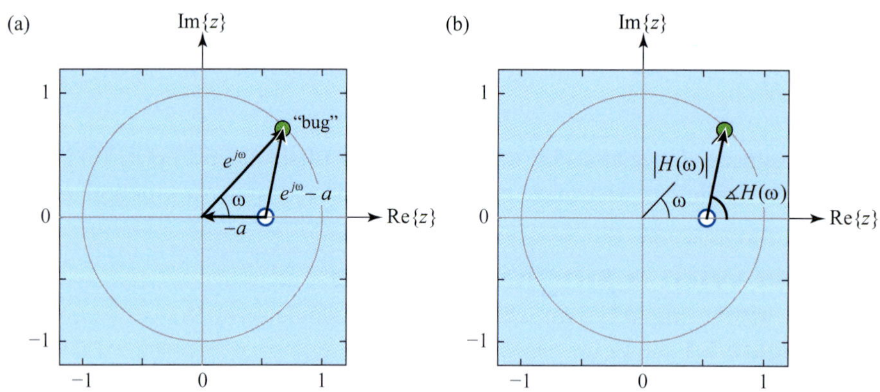

Figure 5.2 Graphical method for deriving $H(\omega)$ from $H(z)$ for a system with a single zero

You can imagine that this is a "bug" crawling around the unit circle. The position of this bug on the circle is completely specified by angle ω. The expression $H(\omega)=e^{j\omega}-a$ is the sum of two vectors, $e^{j\omega}$ and $-a$, as shown in **Figure 5.2a**. At any angle ω, $e^{j\omega}$ is a vector drawn from the center of the z-plane (the point $z=0$) to the bug. The vector $-a$ can be drawn from the zero to the center of the z-plane, so by vector summation $e^{j\omega}-a$ is just a vector drawn from the zero to the bug. If you wish, you can imagine this vector, $H(\omega)=e^{j\omega}-a$, as an elastic string stretching from the zero to the bug. $|H(\omega)|$ is the length of this vector (string) as a function of angle ω with respect to the real axis and $\measuredangle H(\omega)$ is the angle of the vector (string) with respect to the real axis as a function of a, as shown in **Figure 5.2b**.

Here are examples of systems with a single real zero.

Example 5.1

Relate the pole-zero plot to $H(\omega)$ for the case of a single zero, $H(z) = z - 1/2$.

➤ Solution:

$H(\omega) = e^{j\omega} - 1/2$. Start with $\omega = 0$ (**Figure 5.3a**). Here, the bug is located on the unit circle at the intersection of the real axis. The length of the vector drawn from the zero to the bug is $|H(0)| = 1/2$ and the angle is $\measuredangle H(0) = 0$. Now let $\omega = \pi$, the maximum angle to which the bug can travel (**Figure 5.3c**). Here, the length of the vector is $|H(\pi)| = 3/2$ and the angle is $\measuredangle H(\pi) = \pi$. Between the extremes $0 < \omega < \pi$, you can see that both $|H(\omega)|$ and $\measuredangle H(\omega)$ will increase monotonically. Hence, this is a highpass filter. This is the same result at which we arrived by the direct algebraic method, but it seems more intuitive via the graphical method.

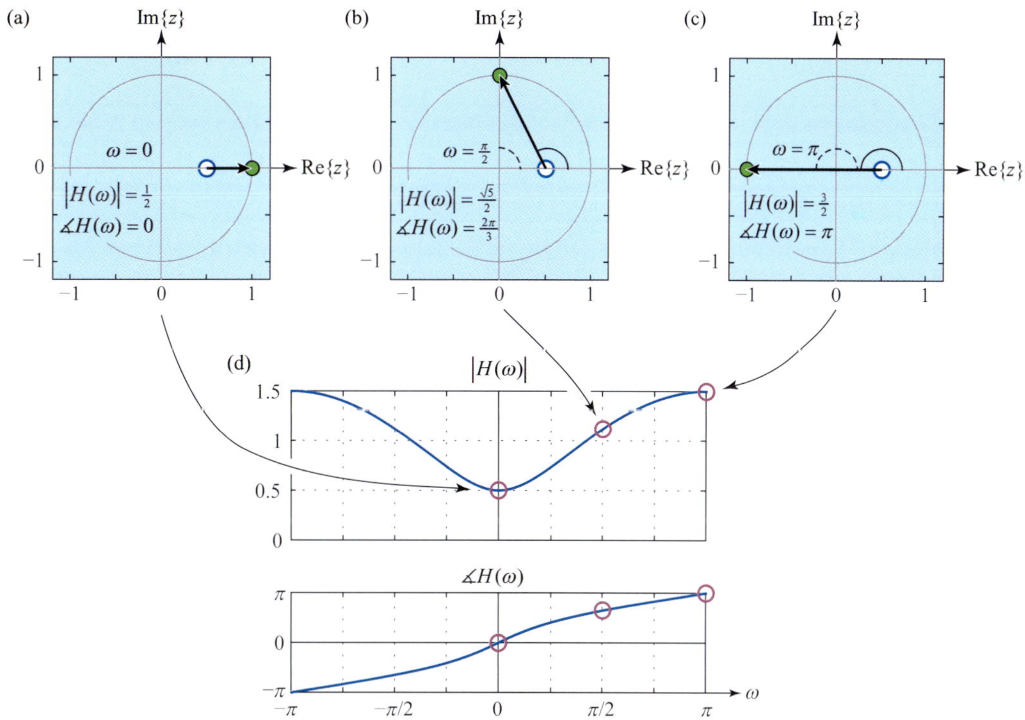

Figure 5.3 Visualization of $H(\omega) = e^{j\omega} - 1/2$

We can gain quite a lot of further insight about this system from the graphic method without a lot of algebra. For example, you can argue that $|H(\omega)|$ will increase faster in the range $0 \le \omega < \pi/2$ than in the range $\pi/2 \le \omega < \pi$. When $\omega = \pi/2$ (**Figure 5.3b**), simple trigonometry gives $|H(\pi/2)| = \sqrt{5}/2$ and

$\measuredangle H(\pi/2) = 2\pi/3$. So, when ω is halfway between 0 and π, $|H(\omega)|$ has increased more than half of the way between its minimum value, 1/2, and its maximum value, 3/2.

With respect to the phase, if you look carefully at **Figure 5.2b**, you can see that at any angle ω, the phase $\measuredangle H(\omega)$ must be greater than ω. That is because the zero is in the right half of the z-plane, so $\measuredangle H(\omega)$ will be more obtuse (i.e., greater) than ω: $\measuredangle H(\omega) > \omega$. Hence at $\omega = \pi/2$, $\measuredangle H(\pi/2) = 2\pi/3$ has gone through more than one-half of the transition between its minimum value, 0, and its maximum value, π.

Example 5.2

Relate the pole-zero plot to $H(\omega)$ for the case of $H(z) = z - 3/2$.

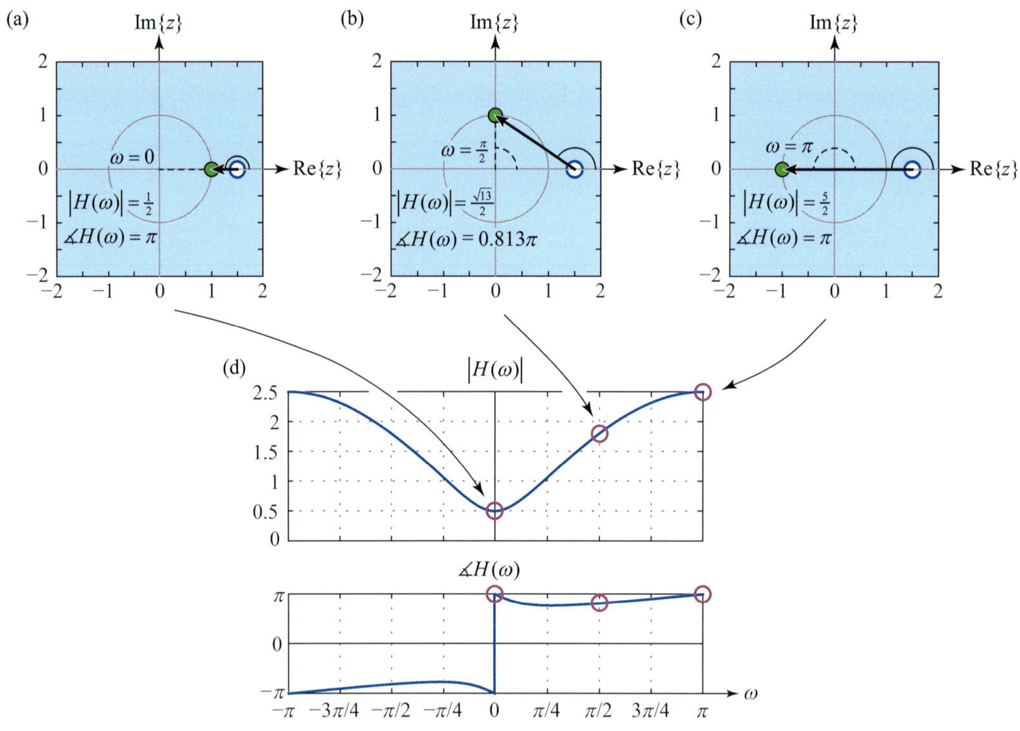

Figure 5.4 Visualization of $H(\omega) = e^{j\omega} - 3/2$

➤ Solution:

Here, the zero is outside the unit circle. At $\omega = 0$ (**Figure 5.4a**), $|H(0)| = 1/2$ and $\measuredangle H(0) = \pi$. As ω increases towards π (**Figure 5.4b**), $|H(\omega)|$ increases monotonically, while the phase first decreases to a minimum value ($\measuredangle H(0.268\pi) = 0.768$) and then increases again, reaching $\measuredangle H(\pi) = \pi$ again when $\omega = \pi$ (**Figure 5.4c**). This is another highpass filter. There is no essential phase discontinuity at $\omega = 0$ in this response. What you see is a plotting discontinuity of 2π due to the fact that we have chosen to wrap $\measuredangle H(\omega)$ into the range $-\pi \leq \omega < \pi$.

Example 5.3

Relate the pole-zero plot to $H(\omega)$ for the case of $H(z) = z - 1$.

➤ Solution:

At $\omega = 0$ (**Figure 5.5a**), $|H(0)| = 0$ and $\measuredangle H(0) = 0$. As ω increases towards π (**Figure 5.5b**), $|H(\omega)|$ increases monotonically towards the value $|H(\pi)| = 2$. Again, this is a highpass filter. The phase has an essential discontinuity of π at $\omega = 0$. When ω is slightly greater than 0 (e.g., $\omega = +\varepsilon$, **Figure 5.5e**), the "bug" is essentially walking directly above the zero, so the phase is $\measuredangle H(\varepsilon) = +\pi/2$. When ω is slightly less than 0 (e.g., $\omega = -\varepsilon$, **Figure 5.5f**), the "bug" is heading directly below the zero, and the phase is $\measuredangle H(-\varepsilon) = -\pi/2$. For values of frequency $0 < |\omega| < \pi$, the phase increases linearly, a result you can prove using analytic geometry if you wish.

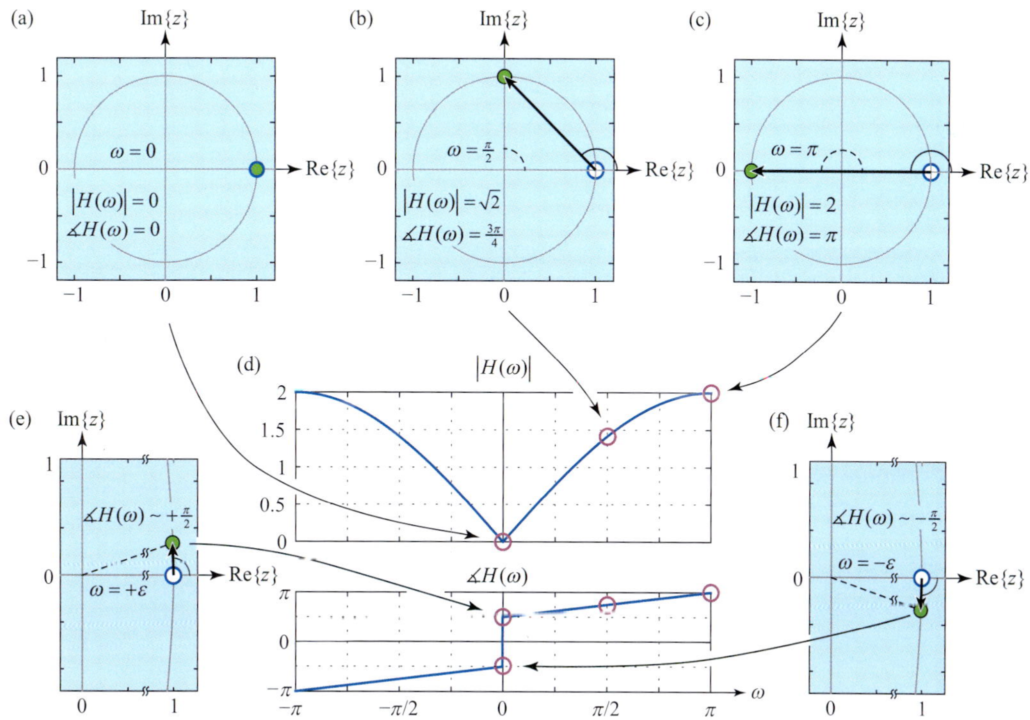

Figure 5.5 Visualization of $H(\omega) = e^{j\omega} - 1$

Example 5.4

Relate the pole-zero plot to $H(\omega)$ for the case of a zero at $z = 0$; namely $H(z) = z$.

➤ Solution:

Clearly, $|H(\omega)| = 1$ is independent of ω, and $\measuredangle H(\omega) = \omega$, as shown in **Figure 5.6d**. This "filter" is just a negative delay of one sample, corresponding to $h[n] = \delta[n+1]$.

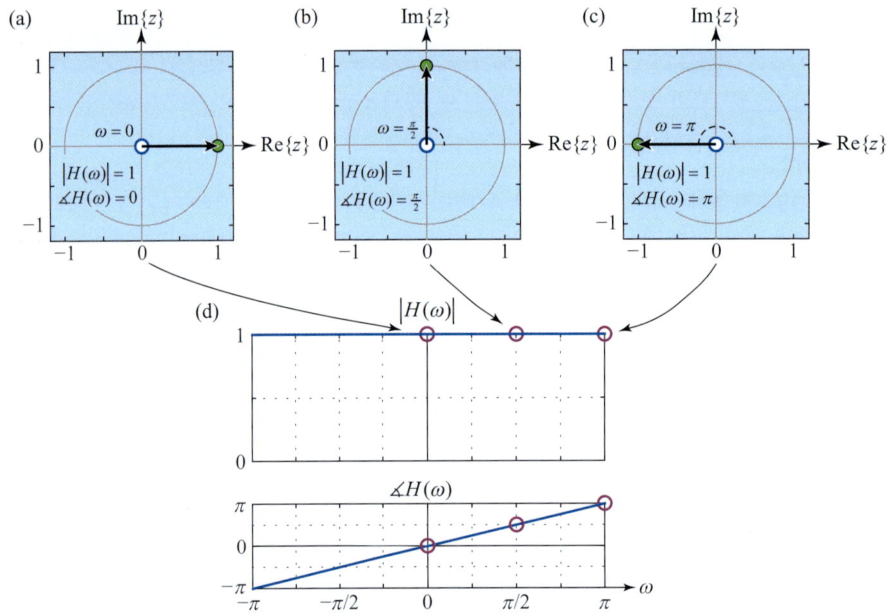

Figure 5.6 Visualization of $H(\omega) = e^{j\omega}$

Example 5.5

Relate the pole-zero plot to $H(\omega)$ for the case of $H(z) = z + 1/2$.

➤ Solution:

Now the zero is in the left half-plane. At $\omega = 0$ (**Figure 5.7a**), $|H(0) = 3/2|$ and $\measuredangle H(0) = 0$. As ω increases (**Figure 5.7b**), $|H(\omega)|$ decreases and $\measuredangle H(\omega)$ increases monotonically, so that as $\omega \to \pi$ (**Figure 5.7c**), $|H(\pi)| \to 1/2$ and $\measuredangle H(\pi) \to \pi$. This is a lowpass filter.

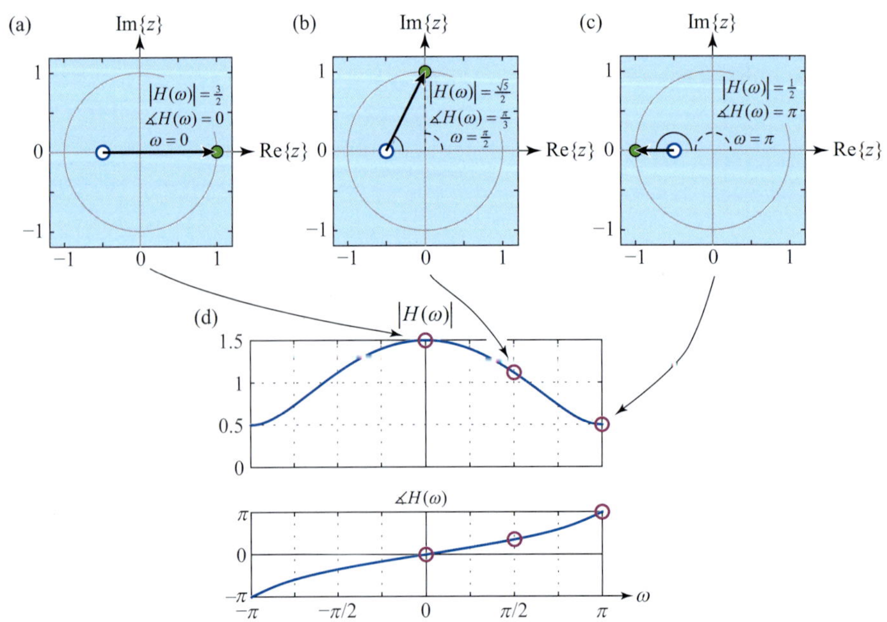

Figure 5.7 Visualization of $H(\omega) = e^{j\omega} + 1/2$

Example 5.6

Relate the pole-zero plot to $H(\omega)$ for the case of $H(z) = z + 1$.

➤ Solution:

At $\omega = 0$ (**Figure 5.8a**), $|H(0)| = 2$ and $\measuredangle H(0) = 0$. As ω increases (**Figure 5.8b**), $|H(\omega)|$ again decreases monotonically and $\measuredangle H(\omega)$ increases monotonically (linearly in fact). As ω approaches π (**Figure 5.8c**), $|H(\omega)| \to 0$ and $\measuredangle H(\omega) \to \pi/2$. $|H(\omega)|$ must be 0 at $\omega = \pi$ since there is a zero on the unit circle at $z = -1$. This is another lowpass filter.

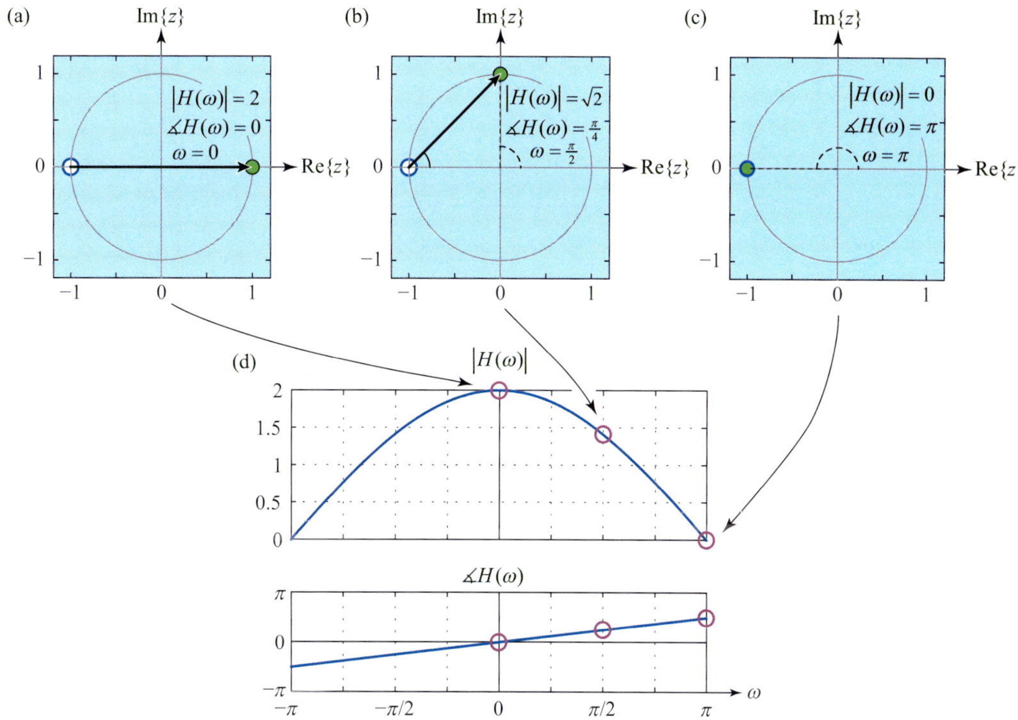

Figure 5.8 Visualization of $H(\omega) = e^{j\omega} + 1$

Example 5.7

Relate the pole-zero plot to $H(\omega)$ for the case of $H(z) = z + 3/2$.

➤ Solution:

At $\omega = 0$ (**Figure 5.9a**), $|H(0)| = 5/2$ and $\measuredangle H(0) = 0$. As ω increases (**Figure 5.9b**), $|H(\omega)|$ again decreases monotonically. $\measuredangle H(\omega)$ increases to a maximum value of $\measuredangle H(0.732\pi) = 0.232\pi$ and then, as ω approaches π (**Figure 5.9c**), $|H(\omega)| \to 1/2$ and $\measuredangle H(\omega)$ returns to 0. This is another lowpass filter.

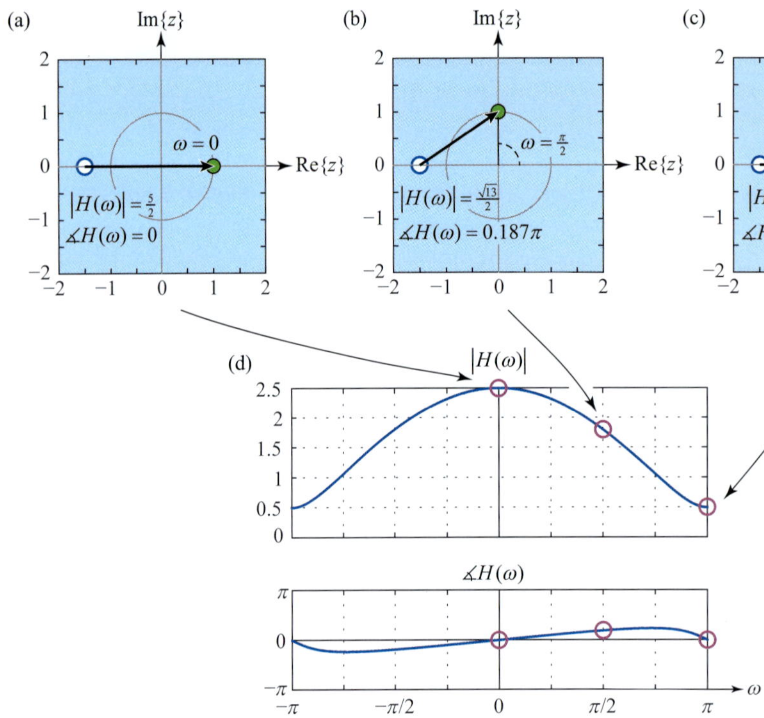

Figure 5.9 Visualization of $H(\omega) = e^{j\omega} + 3/2$

By way of summary, **Figure 5.10** shows $|H(\omega)|$ and $\measuredangle H(\omega)$ for a single real zero derived from $H(z) = z - a$ for values of a ranging from -1.5 to 1.5.

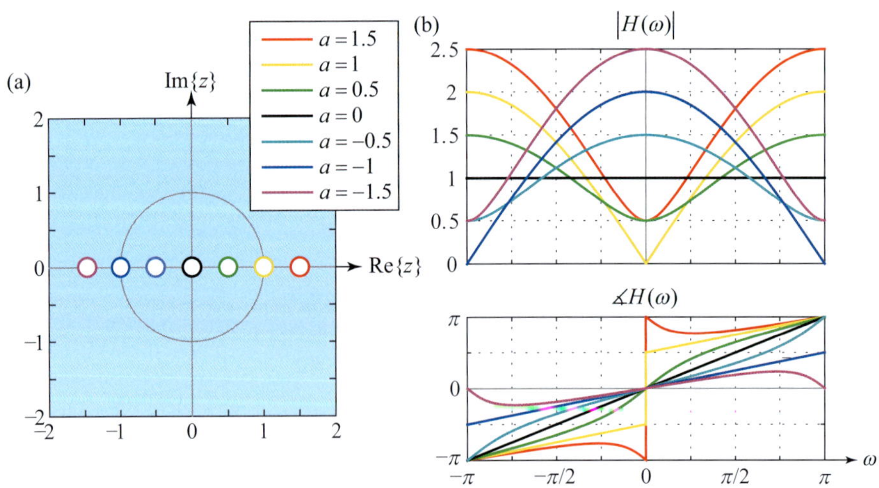

Figure 5.10 Summary of $H(\omega)$ for a single real zero

When the zero is in the right half-plane, $H(z)$ corresponds to a highpass filter; when it is in the left half-plane, $H(z)$ corresponds to a lowpass filter.

5.3 Systems with a single real pole

For a system with a single real pole,

$$H(z) = \frac{1}{z - a}.$$

Visualization of the frequency response is as easy as the case of a single zero. Here,

$$H(\omega) = \frac{1}{e^{j\omega} - a}.$$

The magnitude of the frequency response of the pole,

$$|H(\omega)| = \frac{1}{|e^{j\omega} - a|},$$

is just the reciprocal of the magnitude of the frequency response of the zero; namely, the reciprocal of the length of a vector drawn from the pole to a point on the unit circle (i.e., the "bug") at an angle ω with respect to the real axis, as diagrammed in **Figure 5.11**.

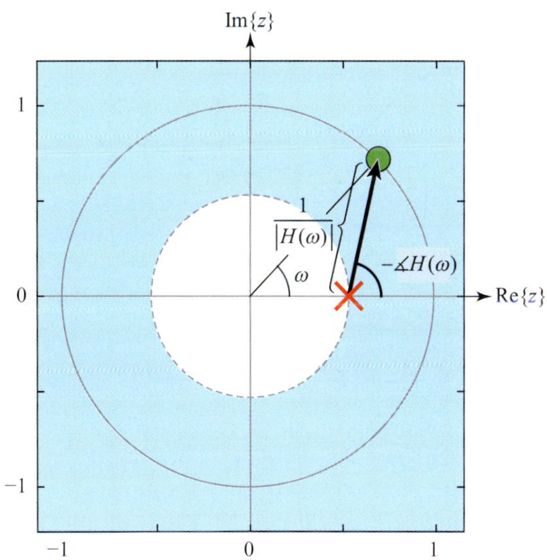

Figure 5.11 Graphical method for deriving $H(\omega)$ from $H(z)$ for a system with a single pole

The phase of the frequency response of a single pole, $\measuredangle H(\omega) = -\measuredangle(e^{j\omega} - a)$, is just the negative of the phase of the frequency response of a zero; namely, the negative of the angle of a vector drawn from the pole to a point located on the unit circle at an angle ω with respect to the real axis.

Here are some examples of systems with a single real pole.

Example 5.8

Relate the pole-zero plot to $H(\omega)$ for the case of a single pole at $z = 1/2$,

$$H(z) = \frac{1}{z - \frac{1}{2}}.$$

➤ Solution:

This is the reciprocal of Example 5.1. At $\omega = 0$ (**Figure 5.12a**), $|H(0)|$ is the reciprocal of the length of the vector drawn from the pole to the bug located on the unit circle, so $|H(0)| = 1/(1/2) = 2$ and $\measuredangle H(0) = 0$. As ω increases towards π, $|H(\omega)|$ and $\measuredangle H(\omega)$ both decrease monotonically, reaching values of $|H(\pi)| = 2/3$ and $\measuredangle H(\pi) = -\pi$ again when $\omega = \pi$. This is a lowpass filter.

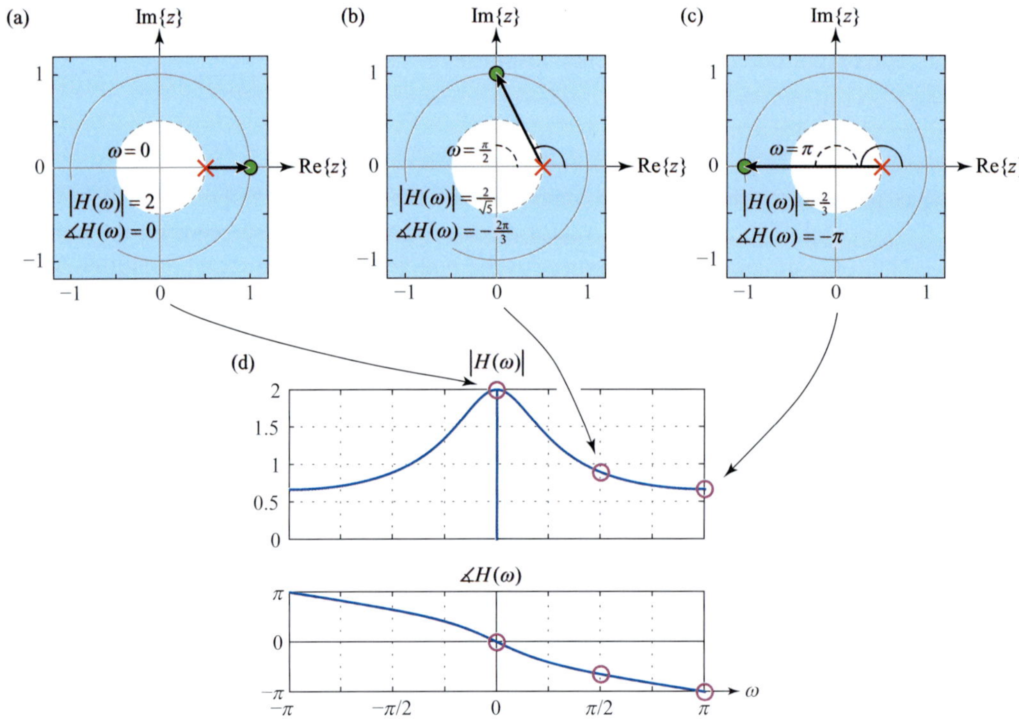

Figure 5.12 Visualization of $H(\omega) = 1/(e^{j\omega} - 1/2)$

The relation between the frequency response of a single pole and that of a single zero is best seen by plotting the magnitudes on a dB scale, as shown in **Figure 5.13**. Here, we consider the case of the frequency response of a single zero, $H_z(\omega)$ (**Figure 5.13a**), and a single pole, $H_p(\omega)$ (**Figure 5.13b**), at $z = 1/2$. The magnitudes of the frequency responses of the zero and pole are

$$|H_z(\omega)| = e^{j\omega} - \tfrac{1}{2}$$
$$|H_p(\omega)| = \frac{1}{e^{j\omega} - \frac{1}{2}}.$$

Expressed on a dB scale, these become

$$|H_z(\omega)|_{dB} = 20\log_{10}|e^{j\omega} - \tfrac{1}{2}|$$
$$|H_p(\omega)|_{dB} = 20\log_{10}\frac{1}{|e^{j\omega} - \frac{1}{2}|} = -20\log_{10}|e^{j\omega} - \tfrac{1}{2}| = -|H_z(\omega)|_{dB}.$$

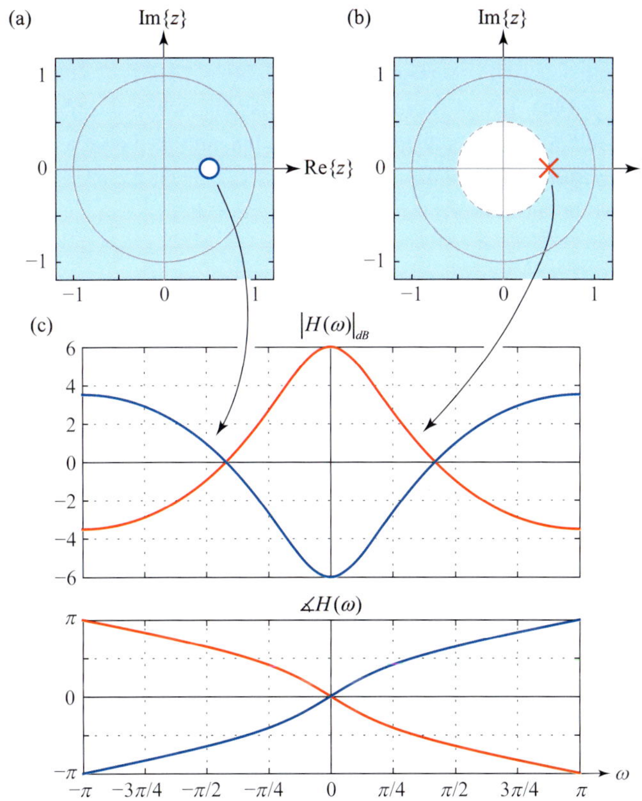

Figure 5.13 Frequency response of a single zero and pole on a dB scale

The phases of the frequency responses of the zero and pole are

$$\measuredangle H_z(\omega) = \tan^{-1}\left(\frac{\sin\omega}{\cos\omega - \frac{1}{2}}\right)$$
$$\measuredangle H_p(\omega) = -\tan^{-1}\left(\frac{\sin\omega}{\cos\omega - \frac{1}{2}}\right) = -\measuredangle H_z(\omega).$$

Hence, on a dB scale, the magnitude of the frequency response of the pole is the negative of the magnitude of the zero. The phase of the pole is also the negative of the zero:

$$|H_p(\omega)|_{dB} = -|H_z(\omega)|_{dB}$$
$$\measuredangle H_p(\omega) = -\measuredangle H_z(\omega).$$

Further examples follow.

Example 5.9

Relate the pole-zero plot to $H(\omega)$ for the case of

$$H(z) = \frac{1}{z - \frac{3}{2}}.$$

➤ Solution:

This is the reciprocal of Example 5.2. At $\omega = 0$ (**Figure 5.14a**), $|H(0)| = 1/(1/2) = 2$ and $\measuredangle H(0) = -\pi$. As ω increases towards π (**Figure 5.14b**), $|H(\omega)|$ decreases monotonically, while the phase first increases (i.e., becomes less negative) to a minimum value, $\measuredangle H(0.268\pi) = -0.768$, and then decreases again, reaching $\measuredangle H(\pi) = -\pi$ again, where $\omega = \pi$ (**Figure 5.14c**). This is a lowpass filter. There is a discontinuity of 2π in the phase plot, which results from wrapping $\measuredangle H(\omega)$ into the range $-\pi \leq \omega < \pi$.

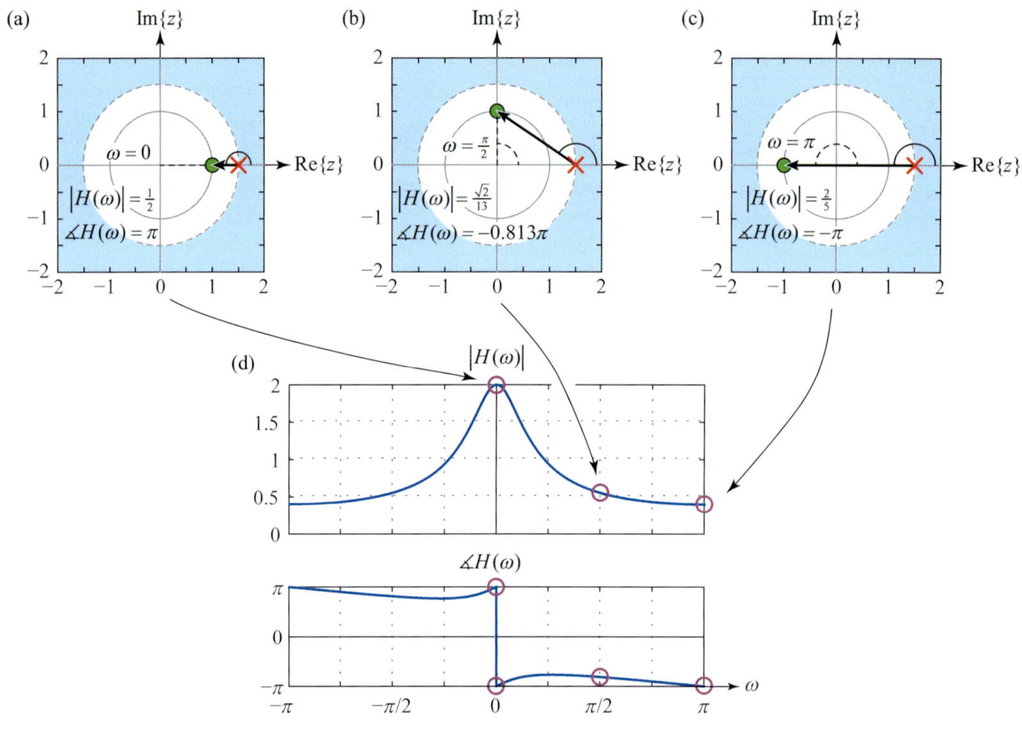

Figure 5.14 Visualization of $H(\omega) = 1/(e^{j\omega} - 3/2)$

Example 5.10

Relate the pole-zero plot to $H(\omega)$ for the case of

$$H(z) = \frac{1}{z - 1}.$$

➤ Solution:

This is the reciprocal of Example 5.3. At $\omega = 0$ (**Figure 5.15a**), $|H(0)| = 1/0 = \infty$. As ω increases towards π (**Figure 5.15b**), $|H(\omega)|$ decreases monotonically towards the value $|H(\pi)| = 1/2$. Again, this is a lowpass filter. The phase has an essential discontinuity of π at $\omega = 0$. When ω is slightly greater than 0 (e.g., $\omega = +\varepsilon$, **Figure 5.15e**), the phase is $\measuredangle H(\varepsilon) = -\pi/2$. When ω is slightly less than 0 (e.g., $\omega = -\varepsilon$, **Figure 5.15f**), the phase is $\measuredangle H(-\varepsilon) = +\pi/2$. For values of frequency $0 < |\omega| < \pi$, the phase changes linearly.

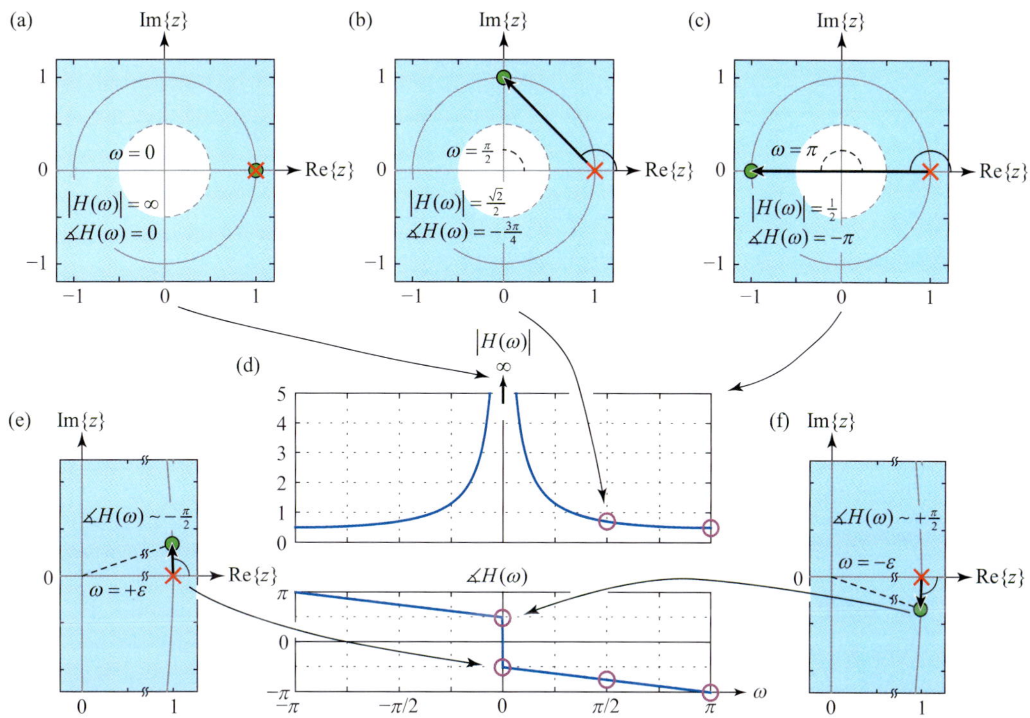

Figure 5.15 Visualization of $H(\omega) = 1/(e^{j\omega} - 1)$

Example 5.11

Relate the pole-zero plot to $H(\omega)$ for the case of a single pole at $z = 0$,

$$H(z) = \frac{1}{z}.$$

➤ Solution:

This is the reciprocal of Example 5.4. $|H(\omega)| = 1$, independent of ω, and $\measuredangle H(\omega) = -\omega$, as shown in **Figure 5.16d**. This "filter" is just a positive delay, corresponding to $h[n] = \delta[n-1]$.

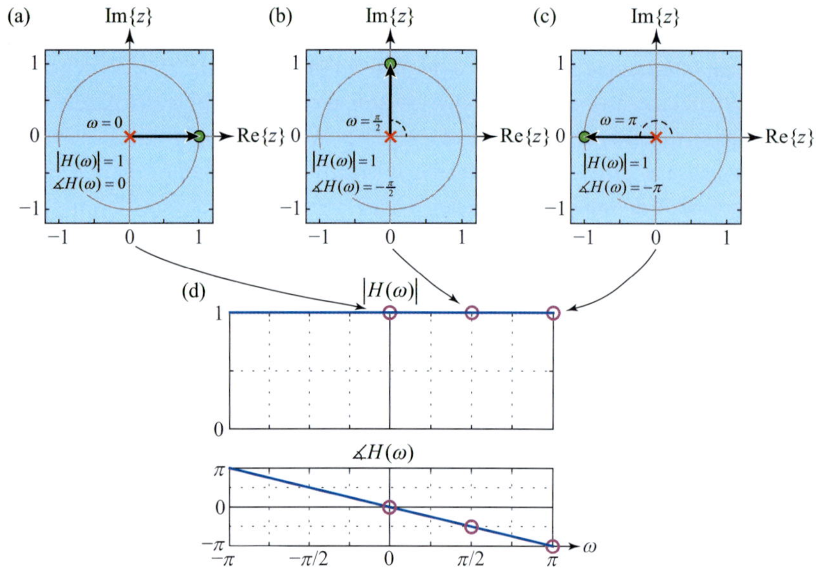

Figure 5.16 Visualization of $H(\omega) = 1/e^{j\omega}$

Example 5.12

Relate the pole-zero plot to $H(\omega)$ for the case of a single pole at $z = -1/2$,

$$H(z) = \frac{1}{z + \frac{1}{2}}.$$

➤ Solution:

This is the reciprocal of Example 5.5. The pole is in the left half-plane. At $\omega = 0$ (**Figure 5.17a**), $|H(0)| = 2/3$ and $\measuredangle H(0) = 0$. As ω increases (**Figure 5.17b**), $|H(\omega)|$ increases and $\measuredangle H(\omega)$ decreases monotonically, so that as $\omega \to \pi$ (**Figure 5.17c**), $|H(\pi)| \to 2$ and $\measuredangle H(\pi) \to -\pi$. This is a highpass filter.

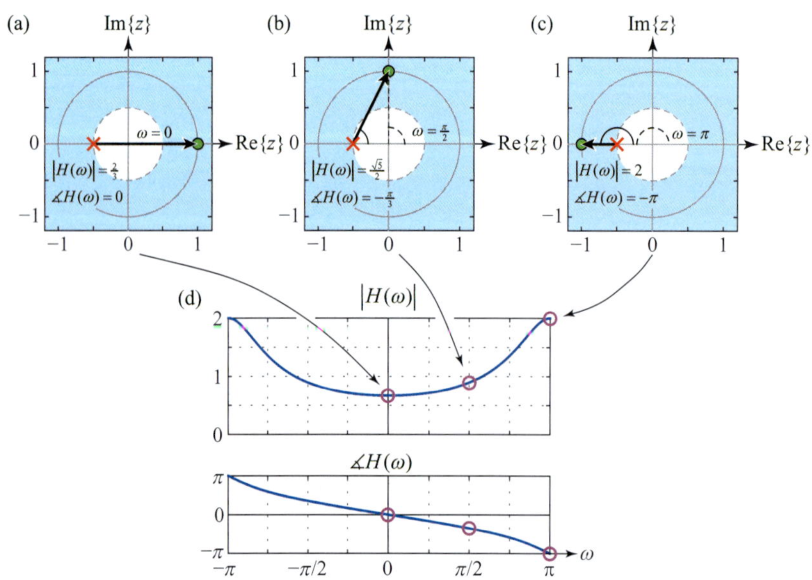

Figure 5.17 Visualization of $H(\omega) = 1/(e^{j\omega} + 1/2)$

Example 5.13

Relate the pole-zero plot to $H(\omega)$ for the case of a single pole at $z=-1$,

$$H(z)=\frac{1}{z+1}.$$

➤ Solution:

This is the reciprocal of Example 5.6. At $\omega=0$ (**Figure 5.18a**), $|H(0)|=1/2$ and $\measuredangle H(0)=0$. As ω increases (**Figure 5.18b**), $|H(\omega)|$ again increases monotonically and $\measuredangle H(\omega)$ decreases monotonically (linearly in fact). As ω approaches π (**Figure 5.18c**), $|H(\omega)|\to\infty$ and $\measuredangle H(\omega)\to-\pi/2$. This is another highpass filter.

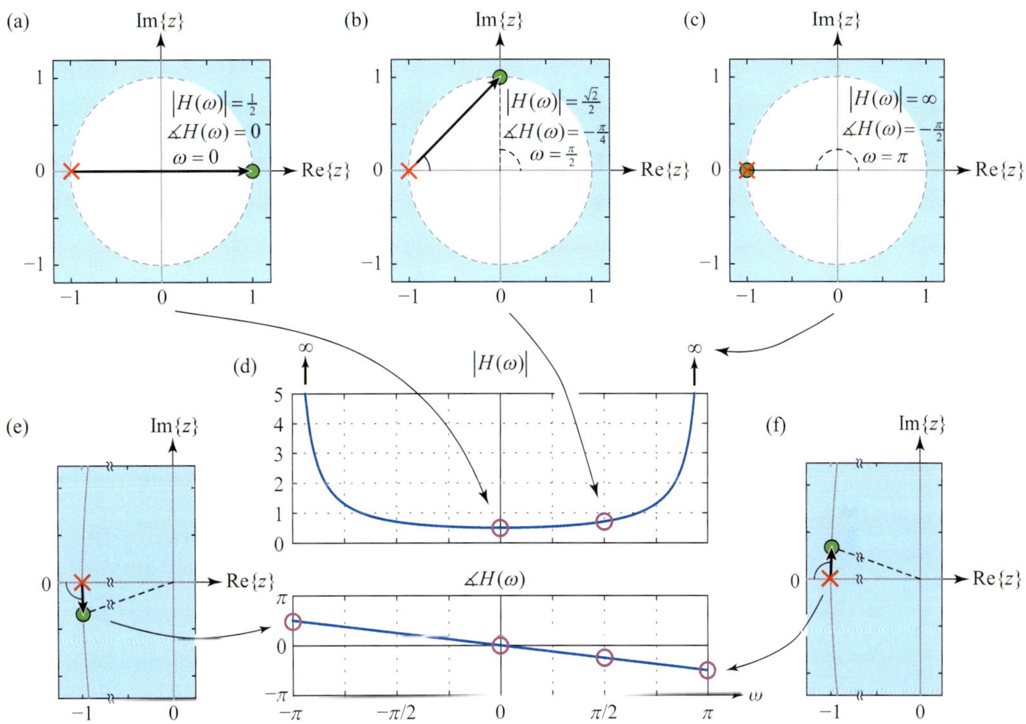

Figure 5.18 Visualization of $H(\omega)=1/(e^{j\omega}+1)$

Example 5.14

Relate the pole-zero plot to $H(\omega)$ for the case of a single pole at $z=-3/2$,

$$H(z)=\frac{1}{z+\frac{3}{2}}.$$

➤ Solution:

This is the reciprocal of Example 5.7. At $\omega=0$ (**Figure 5.9a**), $|H(0)|=2/5$ and $\measuredangle H(0)=0$. As ω increases (**Figure 5.19b**), $|H(\omega)|$ increases monotonically. $\measuredangle H(\omega)$ decreases to a minimum value of $\measuredangle H(0.732\pi)=-0.232\pi$. As ω approaches π (**Figure 5.9c**), $|H(\omega)|\to 2$ and $\measuredangle H(\omega)$ returns to 0. This is another highpass filter.

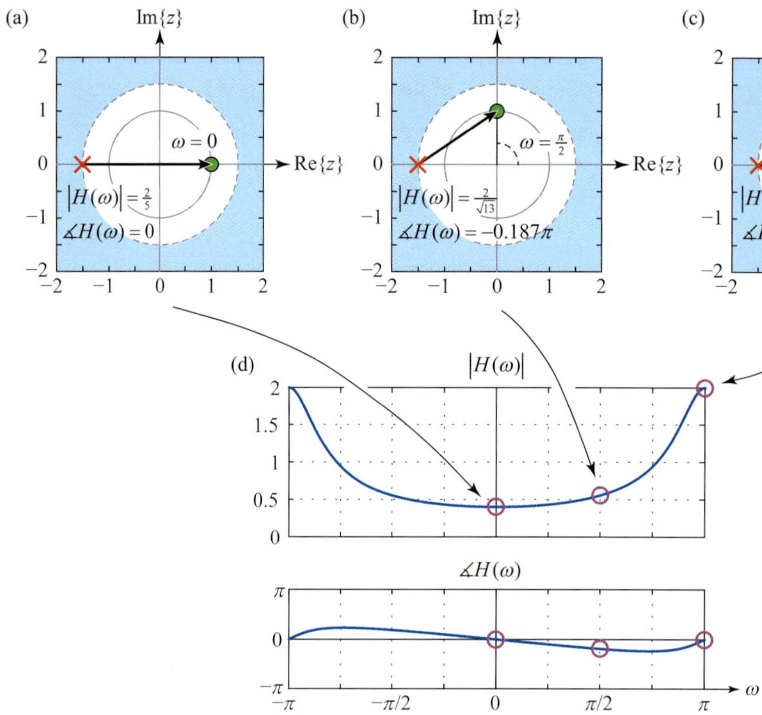

Figure 5.19 Visualization of $H(\omega) = 1/(e^{j\omega} + 3/2)$

By way of summary, **Figure 5.20** shows $|H(\omega)|$ and $\measuredangle H(\omega)$ for a single real pole derived from $H(z) = 1/(z - a)$ for values of a ranging from -1.5 to 1.5.

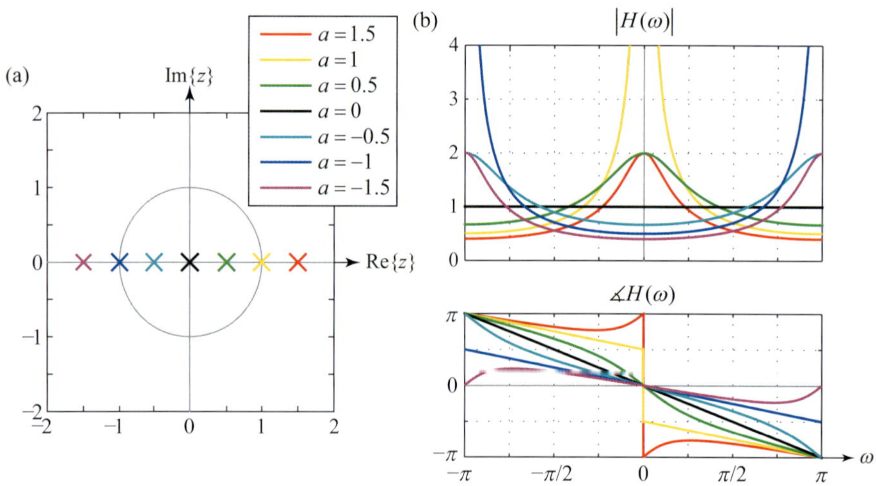

Figure 5.20 Summary of $H(\omega)$ for a single real pole

When the pole is in the right half-plane, $H(z)$ corresponds to a lowpass filter; when it is in the left half-plane, $H(z)$ corresponds to a highpass filter.

5.4 Multiple real poles and zeros

We can extend the results of the previous sections to the case where $H(z)$ has multiple poles and/or zeros on the real axis. In most cases of interest to us, $H(z)$ will be expressed as the product of N poles and M zeros:

$$H(z) = \frac{\sum_{k=0}^{M} b_k z^{-k}}{\sum_{k=0}^{N} a_k z^{-k}} = \frac{\prod_{k=1}^{M} (1 - c_k z^{-1})}{\prod_{k=1}^{N} (1 - d_k z^{-1})} = z^{M-N} \frac{\prod_{k=1}^{M} (z - c_k)}{\prod_{k=1}^{N} (z - d_k)},$$

where we have assumed that $a_0 = b_0 = 1$. Some examples will help make the necessary points.

Example 5.15

Analyze a simple FIR system described by

$$h[n] = \delta[n] - \left(\tfrac{1}{2}\right)\delta[n-1].$$

➤ Solution:

This system has z-transform

$$H(z) = 1 - \tfrac{1}{2}z^{-1} = \frac{z - \frac{1}{2}}{z},$$

which has a zero at $z = 1/2$ and a pole at $z = 0$. To understand the frequency response of this system, recognize that $H(z)$ can be expressed as the product of two factors,

$$H(z) = \frac{z - \frac{1}{2}}{z} = \underbrace{\left(z - \tfrac{1}{2}\right)}_{H_Z(z)} \cdot \underbrace{\left(\frac{1}{z}\right)}_{H_P(z)} = H_Z(z) \cdot H_P(z).$$

$H_Z(z)$ is the system with a single zero that we have already described in Example 5.1. $H_P(z)$ is the single-pole system we analyzed in Example 5.11. The magnitude of $H(\omega)$ is the product of the magnitudes of $H_Z(\omega)$ and $H_P(\omega)$: $|H(\omega)| = |H_Z(\omega)||H_P(\omega)|$. Graphically, at any value of ω, $|H(\omega)|$ is the ratio of the length of vectors drawn from the zero to a point ("bug") on the unit circle divided by the length of a vector drawn from the pole to the same point, as shown in **Figure 5.21**.

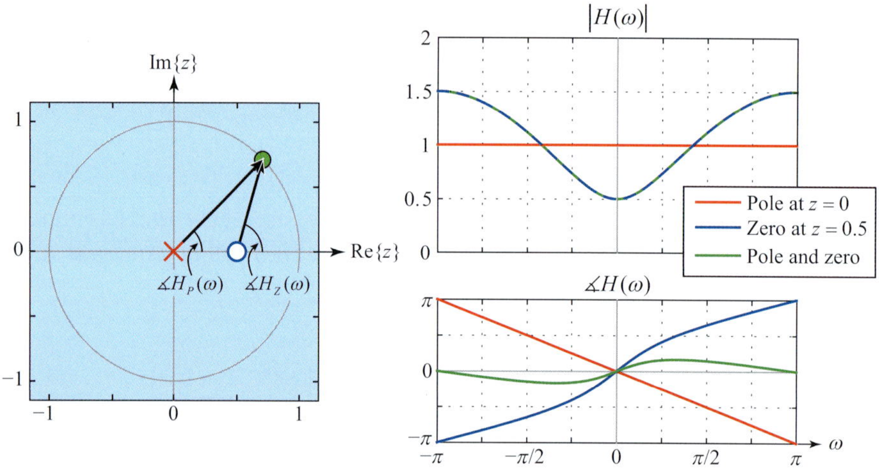

Figure 5.21 Visualization of $H(\omega)$ from $H(z)$ for a system with a pole and a zero

In this case, since the pole is located at $z=0$, the magnitude term is a constant, $|H_P(\omega)|=1$, for all ω. Therefore, the magnitude of the pole-zero combination is the same as the magnitude of the zero term: $|H(\omega)|=|H_Z(\omega)|$. The phase of the pole-zero combination is the sum of the phase of the zero term plus the linear-phase term due to the pole: $\measuredangle H(\omega)=\measuredangle H_Z(\omega)+\measuredangle H_P(\omega)$. Graphically, at any value of ω, $\measuredangle H(\omega)$ is the difference between the angles of vectors drawn from the zero and the pole to a point on the unit circle. The graph shows that the phase of the zero increases monotonically from 0 at $\omega=0$ to π at $\omega=\pi$, while the phase of the pole decreases linearly from 0 at $\omega=0$ to $-\pi$ at $\omega=\pi$. Hence, the total phase starts at $0-0=0$ when $\omega=0$ and ends at $\pi-\pi=0$ at $\omega=\pi$. For any value of $0<\omega<\pi$, $\measuredangle H(\omega)$ is always greater than 0, since $\measuredangle H_Z(\omega)$, the angle drawn from the zero, is always more obtuse (i.e., greater) than $\measuredangle H_P(\omega)$, the angle drawn from the pole.

Example 5.16

Analyze a simple IIR system described by

$$h[n]=\delta[n]+\left(-\tfrac{1}{2}\right)^n u[n-1].$$

➤ **Solution:**

This system has z-transform

$$H(z)=1+\frac{1}{2}\frac{z^{-1}}{1+\frac{1}{2}z^{-1}}=\frac{1+z^{-1}}{1+\frac{1}{2}z^{-1}}=\frac{z+1}{z+\frac{1}{2}},$$

which has a zero at $z=-1$ and a pole at $z=-1/2$. The frequency response is shown in **Figure 5.22**.

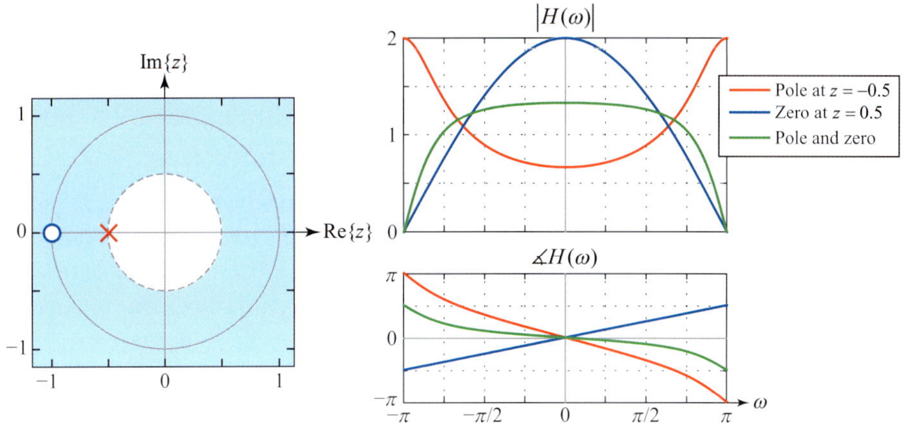

Figure 5.22 Visualization of $H(\omega)=(e^{j\omega}+1)/(e^{j\omega}+1/2)$

This is a broad lowpass filter. The pole flattens the response of the zero at low frequencies near $\omega=0$, and sharpens the response at higher frequencies near $\omega=\pm\pi$. Note also that the zero is exactly on the unit circle at $z=-1$, so the response is guaranteed to go to 0 at $\omega=\pm\pi$, since

$$H(z)|_{z=-1}=H(\omega)|_{\omega=\pm\pi}=0.$$

We can do simple filter design just by visualizing the frequency response we get from placements of poles and zeros, as shown in the next example.

Example 5.17

Assume we have a cosine signal with a DC level added, $x[n]=1+\cos n\pi/8$, as shown in blue in **Figure 5.23c**. Design a filter to remove the DC level.

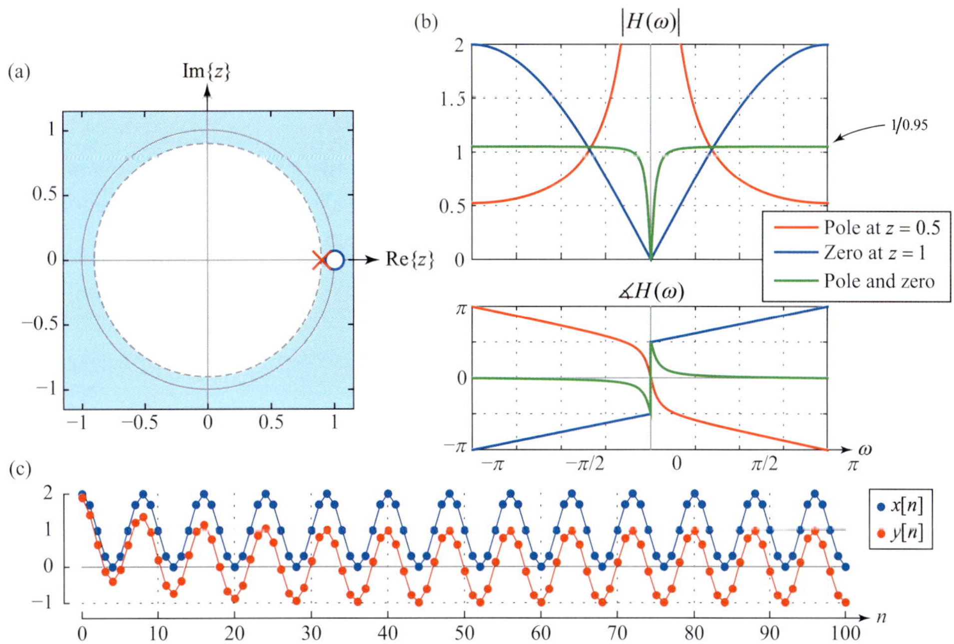

Figure 5.23 DC stop filter

➤Solution:

In order to remove the DC level, we need a filter that makes $H(\omega)=0$ at $\omega=0$. By extension of the previous example, this means we want a zero at $z=1$, since

$$H(z)|_{z=1}=H(\omega)|_{\omega=0}=0.$$

By itself, a filter with a single zero at $z=1$, $H(z)=z-1$, has too broad a response, as shown by the blue curve in **Figure 5.23b**. However, if we put a pole very close to the zero, at $z=0.9$, this does not affect the response at $\omega=0$, but flattens the response at higher frequencies. If you think about this plot in terms of the "bug" going around the unit circle, this result makes sense. At $\omega=0$, the bug is exactly sitting on the zero, so the response must be 0, irrespective of the location of the pole. However, as the bug moves around the unit circle away from the zero, the distance of the bug to the zero is roughly the same as the distance to the pole because the pole is so close to the zero. Hence, the pole and zero effectively cancel each other and the response is almost unity for frequencies away from $\omega=0$. The net result is a sharp DC notch filter, shown by the green curve in **Figure 5.23b**. The z-transform corresponding to this pole-zero plot is

$$H(z)=A\frac{z-1}{z-0.9}=A\frac{1-z^{-1}}{1-0.9z^{-1}}.$$

To find the unknown constant A, note that at $z=-1$, $H(z=-1)=A/0.95$. In order to make the response have roughly unity magnitude, we set $A=0.95$. The difference equation for this system is

$$y[n]-0.9y[n-1]=0.95(x[n]-x[n-1]).$$

The input to this filter $x[n]$ and the output $y[n]$ are shown in **Figure 5.23c**.

5.5 Complex poles and zeros

Most z-transforms of interest to us can be expressed as the ratio of rational polynomials in z with real coefficients. For these transforms, the singularities (i.e., poles and zeros) of $H(z)$ appear on the real axis of the z-plane, and/or in complex-conjugate pairs symmetrically disposed about the real axis, as discussed in Chapter 4. Visualizing the frequency response of systems with complex singularities can be handled in exactly the same way as those with real singularities. Though it is often harder to visualize the shape of the frequency response plots exactly, you can often figure out whether the system is lowpass, highpass, bandpass, etc., just by looking at the position of the poles and zeros in the z-plane.

Let us consider some simple examples.

Example 5.18

Find the pole-zero plot of a system with complex poles given by

$$H(z)=\frac{z^2}{(z-0.8e^{j\pi/4})(z-0.8e^{-j\pi/4})}.$$

➤ Solution:

The pole-zero plot and frequency response are shown in **Figure 5.24**.

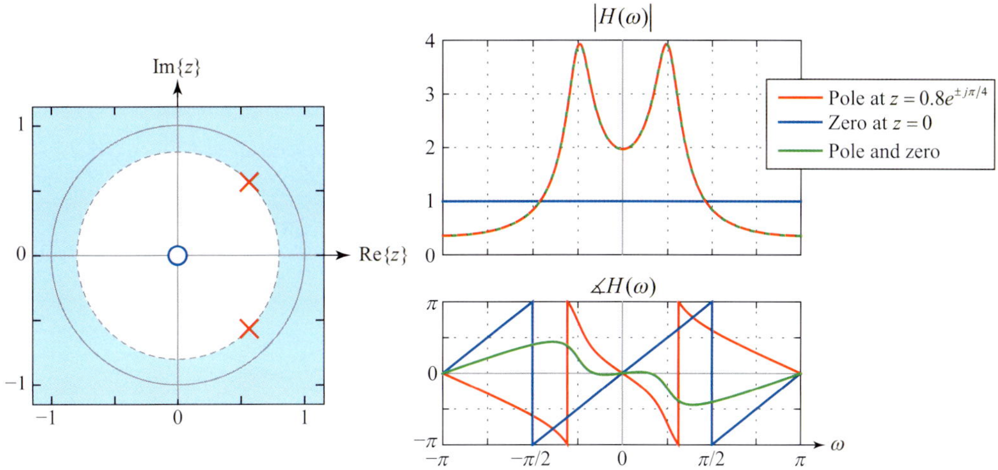

Figure 5.24 Pole-zero plot and frequency response for a system with complex poles

The zero at $z = 0$ has unity magnitude and linear phase. With respect to the poles, if you imagine the "bug" moving around the unit circle, it should be clear that when it is nearest either one of the poles (i.e., $\omega \sim \pm\pi/4$), the distance to that pole reaches a minimum. Therefore the reciprocal of the distance becomes large, and $|H(\omega)|$ peaks around these frequencies. This is a bandpass filter.

In Example 5.17, we designed a simple notch filter to remove the DC level from a signal. With complex poles and zeros, it is easy to extend this idea to create notch filters that selectively remove particular frequency components from a signal.

Example 5.19

Consider an input signal $x[n] = (1 + \cos n\pi/2)u[n]$. Design a notch filter to remove selectively the cosine signal component.

➤ Solution:

Recall that the z-transform of $\cos(\omega_0 n)u[n]$ has poles at $z = e^{\pm j\omega_0}$. So, taking our cue from Example 5.17, we design the filter to have complex zeros on the unit circle at $e^{\pm j\pi/2} = \pm j$, in order that

$$H(z)|_{z=\pm j} = H(\omega)|_{\omega=\pi/2} = 0.$$

We put a pair of complex poles near the unit circle at the same angle, but slightly inboard, $0.9e^{\pm j\pi/2} = \pm 0.9j$, as shown in **Figure 5.25a**.

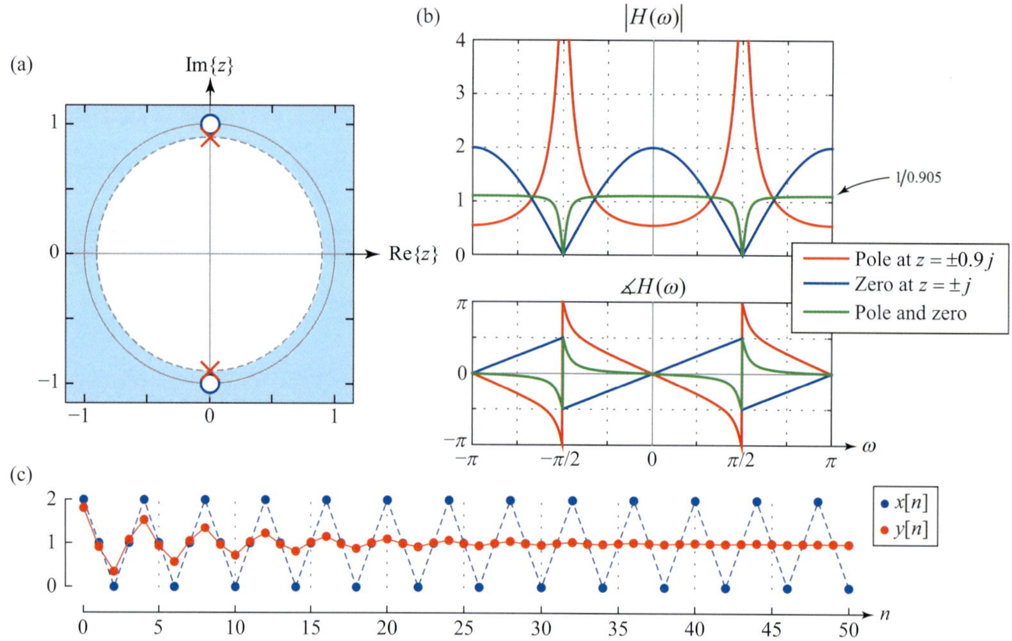

Figure 5.25 A notch filter

The z-transform corresponding to this pole-zero plot is

$$H(z) = A\frac{(z+j)(z-j)}{(z-0.9j)(z+0.9j)} = A\frac{1+z^{-2}}{1+0.81z^{-2}}.$$

In order to make the response have roughly unity magnitude at $\omega = 0$, we can set the unknown constant A so that $H(z=1) = A(2/1.81) = 1$. So, $A = 0.905$. The difference equation for this system is

$$y[n] - 0.81y[n-2] = 0.905(x[n] - x[n-2]).$$

The input to this filter $x[n]$ and the output $y[n]$ are shown in **Figure 5.25c**.

5.6 ★3-D visualization of $H(\omega)$ from $H(z)$

In the previous sections, we developed a graphical approach of determining the magnitude and phase of the DTFT from the pole-zero plot by exploiting the fact that the DTFT is just the z-transform evaluated for values of z on the unit circle. To further our "graphical intuition," it is instructive to visualize the magnitude and phase of $H(z)$ for all values of z, not just those on the unit circle. As an example, **Figure 5.26a** shows a plot of the magnitude of $H(z) = z - 3/4$, a highpass filter which has a single zero at $z = 3/4$.

The magnitude function $|H(z)| = |z - 3/4|$ is a surface that has the shape of a shallow funnel. By definition, the magnitude is 0 at $z = 3/4$. It rises as the distance from the zero increases. The solid green trace superimposed on the surface marks the values of $|H(z)|$ for $z = e^{j\omega}$; that is, the locus of the point corresponding to the unit circle. You can view this three-dimensional picture as a sort of topographic map where the magnitude of the DTFT, shown in **Figure 5.26b**, is the height of the surface along the green path that corresponds to the unit circle. **Figure 5.26c** shows the pole-zero plot. A green dot denotes the "bug" at one frequency, $\omega = \pi/4$, which is also shown in **Figures 5.26a** and **b**.

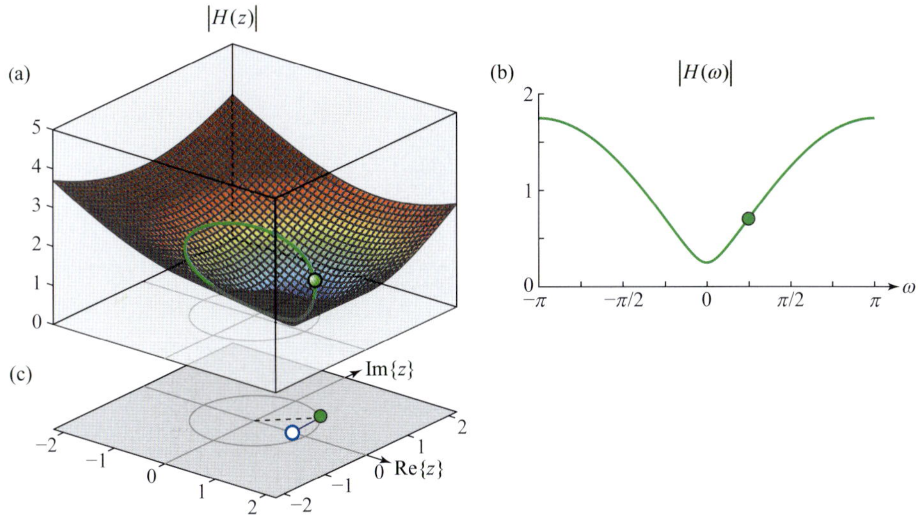

Figure 5.26 Plot of $|H(z)|$ for a single-zero system

Figure 5.27 shows the three-dimensional visualization of the magnitude of $H(z)$ for a lowpass filter comprising a single pole,

$$H(z) = \frac{1}{z - \frac{3}{4}}.$$

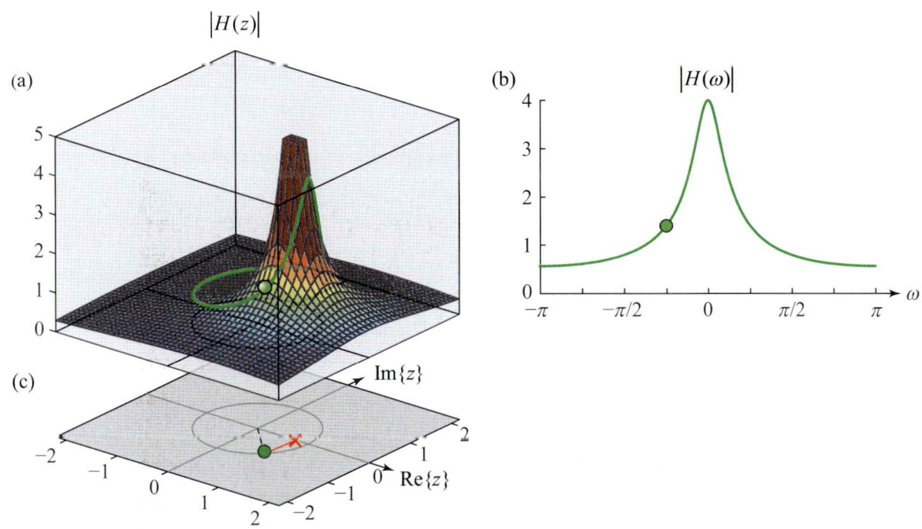

Figure 5.27 Plot of $|H(z)|$ for a single-pole system

Here the surface appears like a volcano that is "pushed up" towards infinity at the value of $z = 3/4$ and decreases in magnitude as z gets farther from the pole. The path corresponding to the unit circle, $z = e^{j\omega}$, rings the volcano's peak and climbs steeply up the flank of the volcano closest to the pole.

As a final example, **Figure 5.28** shows the three-dimensional visualization of $|H(z)|$ for a system with two complex poles and two real zeros, corresponding to a bandpass filter.

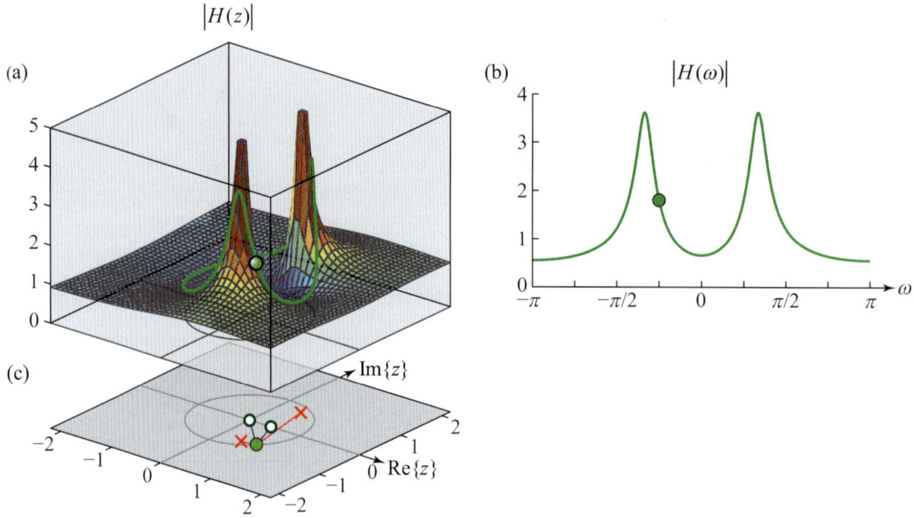

Figure 5.28 Plot of |*H*(*z*)| for a two-pole system

Here the surface is pushed towards infinity at the values of $z = 0.8e^{\pm j\pi/3}$, corresponding to the poles, and is pushed towards 0 at the values of $z = 0$ and $z = 0.5$, corresponding to the zeros. The path corresponding to the unit circle rings rises steeply closest to the poles.

5.7 Allpass filter

A first-order allpass filter is a system defined by the general relation

$$H_{ap}(z) = \frac{z^{-1} - a^*}{1 - az^{-1}} = \frac{1 - a^* z}{z - a} = -a^* \frac{z - 1/a^*}{z - a},$$

where a is a constant. There is a pole at $z = a$ and a zero at the conjugate-reciprocal position $z = 1/a^*$.

To understand why it is called an allpass filter, look at the frequency response,

$$H_{ap}(\omega) = H(z)|_{z=e^{j\omega}} = \frac{1 - a^* e^{j\omega}}{e^{j\omega} - a} = e^{j\omega} \frac{e^{-j\omega} - a^*}{e^{j\omega} - a} = e^{j\omega} \frac{(e^{j\omega} - a)^*}{e^{j\omega} - a}. \quad \text{(5.1)}$$

The magnitude of the response is unity for all ω:

$$|H_{ap}(\omega)| = |e^{j\omega}| \left| \frac{(e^{j\omega} - a)^*}{e^{j\omega} - a} \right| = 1,$$

since $|e^{j\omega}| = 1$ and $|(e^{j\omega} - a)^*| = |e^{j\omega} - a|$. From Equation (5.1), the phase of the allpass filter (Problem 5-2) is

$$\measuredangle H_{ap}(\omega) = \omega - 2\tan^{-1}\left(\frac{\sin\omega - |a| \sin \measuredangle a}{\cos\omega - |a| \cos \measuredangle a} \right).$$

Let us look at a few examples of both real and complex allpass filters.

5.7.1 Real allpass filter

If a is real, then $a = a^*$. Hence, the pole of the first-order allpass filter is on the real axis. The zero is also real and located at the reciprocal position on the real axis, $z = 1/a^* = 1/a$:

$$H_{ap}(z) = \frac{z^{-1} - a}{1 - az^{-1}} = \frac{1 - az}{z - a} = -a\frac{z - 1/a}{z - a}.$$

That means if the pole is located inside the unit circle (i.e., $|a| < 1$), then the zero is outside, and vice versa. For example, **Figures 5.29a** and **b** show the pole-zero plot and frequency response of allpass filters with $a = +0.5$ and $a = -0.5$, respectively.

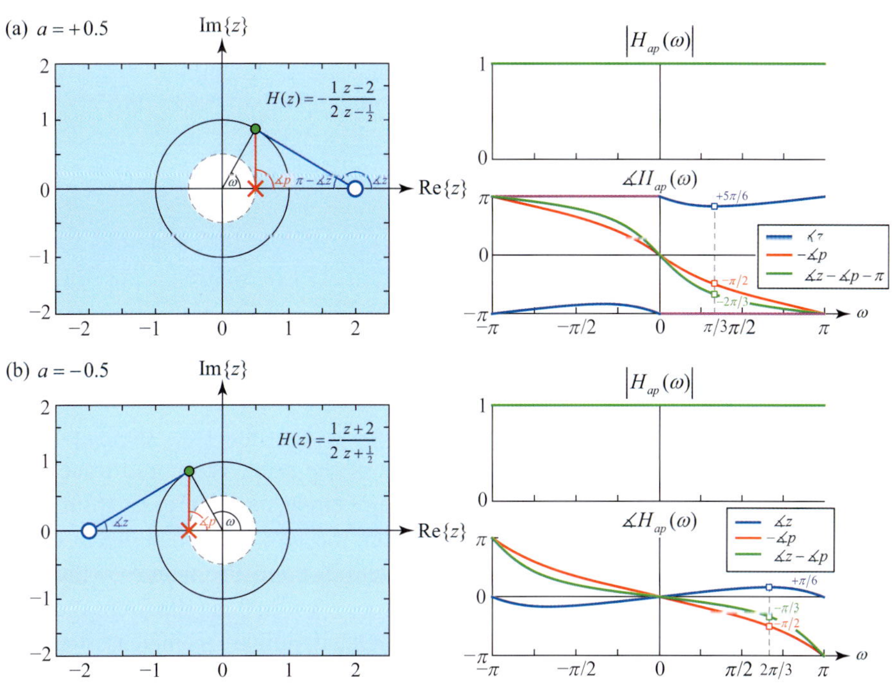

Figure 5.29 Allpass filter with real poles

Because the pole of each filter is inside the unit circle, the ROC of the z-transform includes the unit circle, which means that the filter is stable. Furthermore, because the number of poles is equal to the number of zeros, each filter is causal as well.

For an allpass filter with a real pole at $z = a$, the phase (Problem 5-2) reduces to

$$\measuredangle H_{ap}(\omega) = \omega - 2\tan^{-1}\left(\frac{\sin\omega}{\cos\omega - a}\right),$$

as shown with the green trace in the figure. For any real allpass system with $|a| < 1$, you can show that $\measuredangle H_{ap}(\omega)$ is always 0 for $\omega = 0$ and decreases monotonically as $\omega \to \pi$ (Problem 5-3). This means that an allpass filter always *adds phase-lag* (delay) to a signal at every frequency. While it is easy to prove this result mathematically, you can also appreciate it intuitively by examining the pole-zero plots in **Figure 5.29**. From Equation (5.1), recognize that the phase of the filter is the sum of three components: the phase of the constant, plus the phase of the zero, minus the phase of the pole:

$$\measuredangle H_{ap}(\omega) = \measuredangle\left(-a\frac{e^{j\omega} - 1/a}{e^{j\omega} - a}\right) = \underbrace{\measuredangle(-a)}_{\measuredangle\text{constant}} + \underbrace{\measuredangle(e^{j\omega} - 1/a)}_{\measuredangle\text{zero}} - \underbrace{\measuredangle(e^{j\omega} - a)}_{\measuredangle\text{pole}}.$$

The phase of the constant, $\measuredangle(-a)$, is either 0 if $a < 0$ or $\pm\pi$ if $a > 0$. At any value of ω, the phase of the zero, $\measuredangle(e^{j\omega} - 1/a)$, is the angle with respect to the real axis of a vector drawn from the zero to a point on the unit circle at angle ω, as discussed in Section 5.2. Similarly, the phase of the pole, $\measuredangle(e^{j\omega} - a)$, is the angle of a vector drawn from the pole to the same point on the unit circle, as discussed in Section 5.3. As an example, consider the pole-zero plot in **Figure 5.29b**. Here, $a = -0.5$, so

$$H_{ap}(z) = 0.5\frac{z+2}{z+0.5}$$

and

$$H_{ap}(\omega) = 0.5\frac{e^{j\omega}+2}{e^{j\omega}+0.5}.$$

For this filter, the phase is

$$\measuredangle H_{ap}(\omega) = \underbrace{\measuredangle(0.5)}_{\measuredangle\text{constant}} + \underbrace{\measuredangle(e^{j\omega}+2)}_{\measuredangle\text{zero}} - \underbrace{\measuredangle(e^{j\omega}+0.5)}_{\measuredangle\text{pole}} = 0 + \measuredangle z - \measuredangle p.$$

The phase of the constant is 0. The angle of the zero vector (blue trace on the pole-zero plot and in the plot of $\measuredangle H_{ap}(\omega)$) is denoted $\measuredangle z$; the angle of the pole vector (red trace) is denoted $\measuredangle p$. Because the zero is outside and to the left of the unit circle, as ω increases from 0 to π, the angle of the zero goes from 0 to a maximum value when the zero vector is tangent to the unit circle and then returns to 0. You can show (see Problem 5-4 and Problem 5-5) that the maximum value of $\measuredangle z = -\sin^{-1}(-a) = \pi/6$ occurs when $\omega = \cos^{-1}a = 2\pi/3$, as shown in the figure. Because the pole is inside and to the left of the unit circle, its angle $\measuredangle p$ increases monotonically as $0 \le \omega < \pi$. Furthermore, because the pole is closer to the origin than the zero (that is, $|a| < |1/a|$), the angle of the pole is always more obtuse (greater) than the angle of the zero ($\measuredangle p \geq \measuredangle z$) at any value of ω. Since the phase of this filter is the difference between these angles, $\measuredangle H_{ap}(\omega) = \measuredangle z - \measuredangle p$, the phase of the filter will always be negative and monotonically decreasing.

You can make the same arguments for positive values of a ($0 < a < 1$). For example, consider the filter whose pole-zero plot and frequency response are shown in **Figure 5.29a**. Here, $a = 0.5$, so

$$\measuredangle H_{ap}(\omega) = \underbrace{\measuredangle(-0.5)}_{\measuredangle\text{constant}} + \underbrace{\measuredangle(e^{j\omega} - 2)}_{\measuredangle\text{zero}} - \underbrace{\measuredangle(e^{j\omega} - 0.5)}_{\measuredangle\text{pole}} = -\pi + \measuredangle z - \measuredangle p. \tag{5.2}$$

In this case, the angle of the constant is $-\pi$. The angle of the zero, $\measuredangle z$, starts at π when $\omega = 0$, reaches a minimum value of $\pi - \sin^{-1} a = 5\pi/6$ when the zero vector is tangent to the unit circle at $\omega = \pi/3$, and then increases back to π when $\omega = \pi$. Because the pole is again inside the unit circle, its angle $\measuredangle p$ is positive and increases monotonically from 0 to π. To understand why the phase of this filter is negative and decreases monotonically as ω goes from 0 to π, it is useful to consider the angle $\pi - \measuredangle z$, shown in the figure. This angle is always positive and always more acute (smaller) than $\measuredangle p$; it increases from 0 to $\pi/6$ and then back to 0 as $0 \leq \omega < \pi$. Now, write Equation (5.2) as

$$\measuredangle H_{ap}(\omega) = -\pi + \measuredangle z - \measuredangle p = -(\pi - \measuredangle z) - \measuredangle p.$$

Both terms, $-(\pi - \measuredangle z)$ and $-\measuredangle p$, are monotonic negative and so is their sum.

5.7.2 Multiple poles and zeros

An allpass filter can have multiple poles and zeros. Given two allpass filters $H_1(z)$ and $H_2(z)$, the product $H(z) = H_1(z)H_2(z)$ is also an allpass filter because $|H_1(\omega)| = 1$ and $|H_2(\omega)| = 1$. So,

$$|H(\omega)| = |H_1(\omega)H_2(\omega)| = |H_1(\omega)||H_2(\omega)| = 1.$$

For example, consider the allpass filter whose pole-zero plot is shown in **Figure 5.30**, with real poles at $z = \pm 0.5$ and corresponding zeros at $z = \pm 2$,

$$H(z) = H_1(z)H_2(z) = \underbrace{-0.5\frac{z-2}{z-0.5}}_{H_1(z)} \cdot \underbrace{0.5\frac{z+2}{z+0.5}}_{H_2(z)} = -0.25\frac{z^2-4}{z^2-0.25}.$$

The pole-zero pair in the right half-plane, $H_1(z)$, is the same allpass filter shown in **Figure 5.29**. Its phase is shown in red. The pole-zero pair in the left half-plane, $H_2(z)$, is also an allpass filter whose phase is shown in blue. The product is an allpass filter with unity magnitude and a phase (shown in green) that is the sum of the phases: $\measuredangle H(\omega) = \measuredangle H_1(\omega) + \measuredangle H_2(\omega)$. In the case of a two-pole allpass filter, the phase decreases monotonically and goes through a phase shift of 2π as ω goes from $0 \leq \omega \leq \pi$.

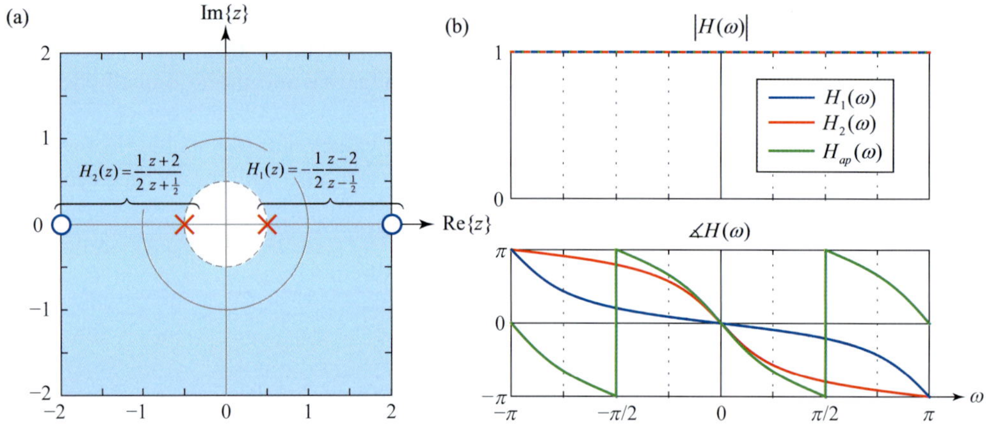

Figure 5.30 Allpass filter with multiple real singularities

5.7.3 Allpass filters with complex poles and zeros

An allpass filter can have complex poles and zeros. Consider an allpass filter,

$$H_1(z) = -a^* \frac{z - 1/a^*}{z - a} = \frac{1 - a^* z}{z - a},$$

where a is complex. The pole lies at complex position $z = a = |a|e^{j\measuredangle a}$, and the zero lies at the conjugate-reciprocal position

$$z = \frac{1}{a^*} = \frac{1}{|a|e^{-j\measuredangle a}} = \frac{1}{|a|}e^{j\measuredangle a}.$$

Thus, the pole and zero lie along a line connected to the origin at $z = 0$, and have reciprocal magnitudes. For example, if the pole is at $a = 0.5e^{j\pi/4}$, then the zero is at $1/a^* = 2e^{j\pi/4}$ and both the pole and zero lie on a line at an angle $\pi/4$ radians from the real axis, as shown in blue in **Figure 5.31a**.

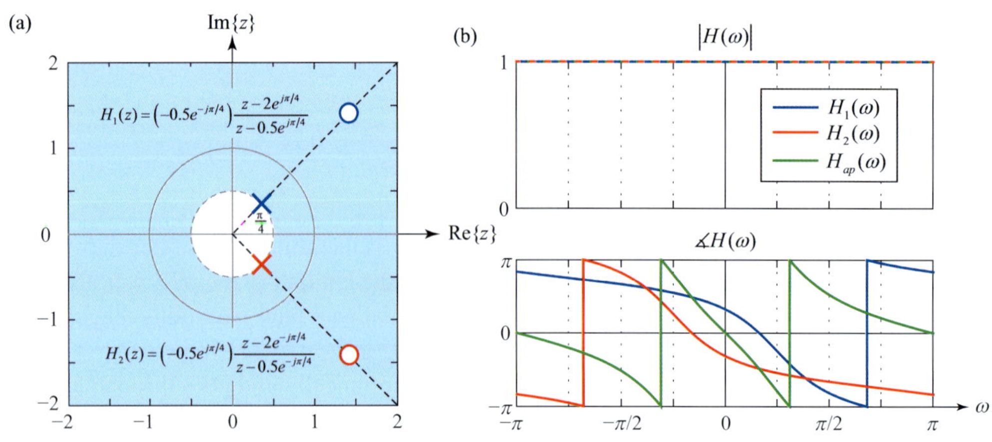

Figure 5.31 Allpass filter with complex poles

The magnitude of this allpass filter is one and its phase is shown in blue in **Figure 5.31b**. The z-transform of this pole-zero pair is

$$H_1(z) = -a^* \frac{z - 1/a^*}{z - a} = (-|a|e^{-j\measuredangle a}) \frac{z - \frac{1}{|a|}e^{j\measuredangle a}}{z - |a|e^{j\measuredangle a}} = \left(-0.5e^{-j\pi/4}\right) \frac{z - 2e^{j\pi/4}}{z - 0.5e^{j\pi/4}}.$$

Of course, in order for a system to be realizable (i.e., for its difference equation to have real coefficients), an allpass filter must be the ratio of polynomials in z that have only *real* coefficients. Hence, if a realizable allpass filter has a pole and zero at complex locations a and $1/a^*$, respectively (shown in blue in **Figure 5.31a**), these singularities must be matched by poles and zeros at the conjugate positions, namely a^* and $1/a$, as shown in red in the figure. The z-transform of this conjugate pole-zero pair is

$$H_2(z) = -a \frac{z - 1/a}{z - a^*} = (-|a|e^{j\measuredangle a}) \frac{z - \frac{1}{|a|}e^{-j\measuredangle a}}{z - |a|e^{-j\measuredangle a}} = \left(-0.5e^{j\pi/4}\right) \frac{z - 2e^{-j\pi/4}}{z - 0.5e^{-j\pi/4}}.$$

The product of $H_1(z)$ and $H_2(z)$ is an allpass filter $H_{ap}(z)$ with real coefficients,

$$\begin{aligned} H_{ap}(z) = H_1(z)H_2(z) &= \left(-a^* \frac{z - 1/a^*}{z - a}\right)\left(-a \frac{z - 1/a}{z - a^*}\right) = |a|^2 \frac{z^2 - \frac{2}{|a|^2}\mathrm{Re}\{a\}z + \frac{1}{|a|^2}}{z^2 - 2\mathrm{Re}\{a\}z + |a|^2} \\ &= \frac{|a|^2 z^2 - 2\cos(\measuredangle a)z + 1}{z^2 - 2|a|\cos(\measuredangle a)z + |a|^2}. \end{aligned} \tag{5.3}$$

Equation (5.3) has the general form,

$$H(z) = \frac{\beta_2 + \beta_1 z^{-1} + z^{-2}}{1 + \beta_1 z^{-1} + \beta_2 z^{-2}} = \frac{\beta_2 z^2 + \beta_1 z + 1}{z^2 + \beta_1 z + \beta_2}. \tag{5.4}$$

In the example of **Figure 5.31**,

$$H(z) = \frac{1}{4} \frac{z^2 - 2\sqrt{2}z + 4}{z^2 - \frac{\sqrt{2}}{2}z + \frac{1}{4}} = \frac{z^2 - 2\sqrt{2}z + 4}{4z^2 - 2\sqrt{2}z + 1}.$$

Since both poles are inside the unit circle, this is a stable filter. The phase of the frequency response $\measuredangle H(\omega)$ (shown in green in **Figure 5.31b**) is the sum of the phases: $\measuredangle H(\omega) = \measuredangle H_1(\omega) + \measuredangle H_2(\omega)$ (see Problem 5-6).

As in the case of the real two-pole allpass filter discussed previously, the phase of the complex allpass filter decreases monotonically and goes through a phase shift of 2π as ω goes from 0 to π.

5.7.4 General allpass filter

The z-transform of an Nth-order real allpass filter is a generalized extrapolation of Equation (5.4),

$$H_{ap}(z) = \frac{N(z)}{D(z)} = \frac{\beta_N + \beta_{N-1}z^{-1} + \cdots + \beta_1 z^{-(N-1)} + z^{-N}}{1 + \beta_1 z^{-1} + \cdots + \beta_{N-1}z^{-(N-1)} + \beta_N z^{-N}}, \tag{5.5}$$

where the coefficients β_k, $1 \le k \le N$, are real. Let the denominator polynomial of Equation (5.5) be

$$D(z) = 1 + \beta_1 z^{-1} + \cdots + \beta_{N-1} z^{-(N-1)} + \beta_N z^{-N}.$$

Then, the numerator polynomial $N(z)$ can be written as

$$\begin{aligned} N(z) &= \beta_N + \beta_{N-1}z^{-1} + \cdots + \beta_1 z^{-(N-1)} + z^{-N} = z^{-N}(1 + \beta_1 z + \cdots + \beta_{N-1} z^{(N-1)} + \beta_N z^N) \\ &= z^{-N} D(z^{-1}), \end{aligned}$$

so,

$$H_{ap}(z) = \frac{N(z)}{D(z)} = z^{-N}\frac{D(z^{-1})}{D(z)}. \tag{5.6}$$

Equation (5.6) can be viewed as the definition of a general allpass filter. To verify this, note that

$$H_{ap}(z)H_{ap}(z^{-1}) = z^{-N}\frac{D(z^{-1})}{D(z)} \cdot z^N \frac{D(z)}{D(z^{-1})} = 1,$$

or, equivalently in the frequency domain,

$$H_{ap}(\omega)H_{ap}(-\omega) = H_{ap}(\omega)H^*_{ap}(\omega) = |H_{ap}(\omega)|^2 = 1,$$

where we have recognized that for a filter with real coefficients, $H_{ap}(-\omega) = H^*_{ap}(\omega)$.

The poles and zeros of $H_{ap}(z)$ given by Equation (5.5) are located at conjugate-reciprocal positions. To see this, note from Equations (5.6) and (5.7) that $H_{ap}(z^{-1}) = 1/H(z) = D(z)/N(z)$. In Section 4.5.4, we showed that singularities of $H_{ap}(z^{-1})$ occur at the conjugate-reciprocal positions from those of $H_{ap}(z)$, so the roots of $D(z)$ (the poles) are located at conjugate-reciprocal positions from the roots of $N(z)$ (the zeros).

In Chapter 9, we will discover an efficient way to implement allpass filters using a **lattice-ladder** architecture.

5.7.5 Systems with the same magnitude

Allpass filters are interesting for both theoretical and practical reasons. Theoretically, the allpass filter allows us to address an important question: is the frequency-response magnitude *unique* to a system with a given z-transform, or are there multiple systems that have the same magnitude?

Figure 5.32a shows the pole-zero plot and frequency response of a system that is characterized by a single zero at $z = 1/2$: $H_1(z) = z - 1/2$. We now create a second system $H_2(z)$ by multiplying the z-transform of $H_1(z)$ by an allpass filter $H_{ap}(z)$,

$$H_2(z) = H_1(z)H_{ap}(z), \tag{5.7}$$

where $H_{ap}(z)$ is designed to have a pole at $z = 1/2$ and a zero at $z = 2$,

$$H_{ap}(z) = -\frac{1}{2}\frac{z-2}{z-\frac{1}{2}},$$

as shown in **Figure 5.32b**. The result is that

$$H_2(z) = H_1(z)H_{ap}(z) = \left(z-\tfrac{1}{2}\right)\left(-\frac{1}{2}\frac{z-2}{z-\frac{1}{2}}\right) = 1-\tfrac{1}{2}z,$$

shown in **Figure 5.32c**.

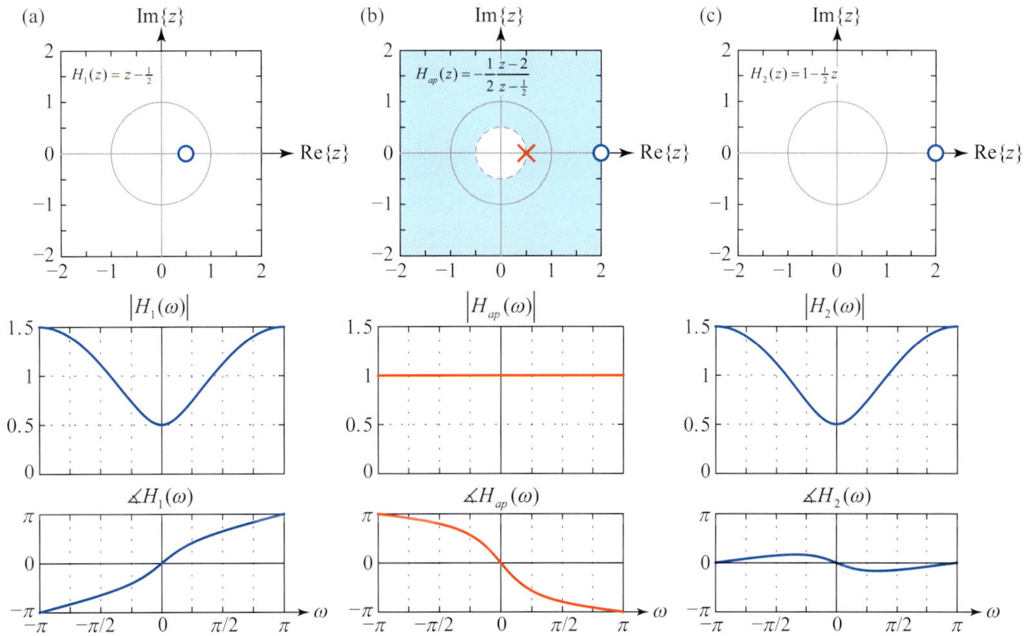

Figure 5.32 Two systems with the same frequency-response magnitude

Because $|H_{ap}(\omega)| = 1$, it follows that $|H_2(\omega)| = |H_1(\omega)H_{ap}(\omega)| = |H_1(\omega)||H_{ap}(\omega)| = |H_1(\omega)|$. The frequency-response magnitudes of these two systems are the same, though their phases are clearly different. The pole of the allpass filter at $z = 1/2$ cancels the zero in $H_1(z)$ and creates a new zero at the conjugate-reciprocal position, $z = 2$. If you wish, you can think of the effect of the allpass filter as "moving" the zero from $z = 1/2$ to its conjugate-reciprocal position at $z = 2$.

Because there is an infinite number of allpass filters, all with magnitude $|H_{ap}(\omega)| = 1$, there is an infinite number of systems $H_2(\omega) = H_1(\omega)H_{ap}(\omega)$ that can have the same magnitude as $H_1(\omega)$. However, if we restrict our consideration only to systems that have the same number of singularities as $H_1(z)$ (i.e., only one zero), then the $H_2(\omega)$ shown in **Figure 5.32c** is the only other system that has the same magnitude as $H_1(\omega)$.

As another example, **Figure 5.33a** shows a system with a pair of complex-conjugate zeros at $z = 0.5e^{\pm j\pi/4}$.

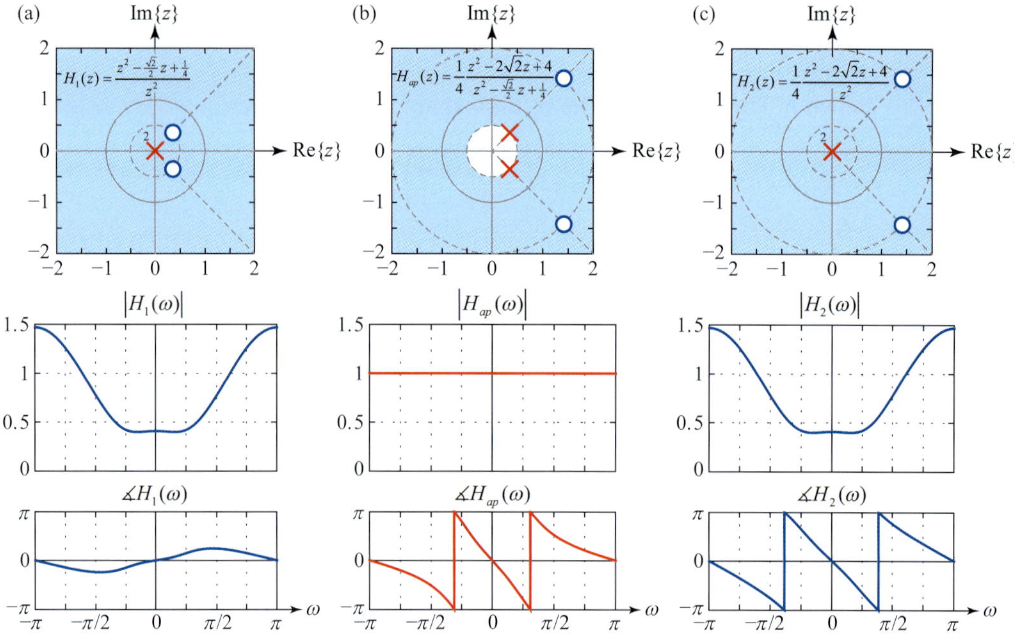

Figure 5.33 Two systems with complex zeros

This system has transform

$$H_1(z) = \frac{(z - \frac{1}{2}e^{j\pi/4})(z - \frac{1}{2}e^{-j\pi/4})}{z^2} = \frac{z^2 - \frac{\sqrt{2}}{2}z + \frac{1}{4}}{z^2}.$$

Figure 5.33b shows an allpass system $H_{ap}(z)$, which has a pair of complex poles at $z = 0.5e^{\pm j\pi/4}$ and a pair of complex zeros at the conjugate-reciprocal positions, $z = 2e^{\pm j\pi/4}$,

$$H_{ap}(z) = \frac{1}{4}\frac{\left(z - 2e^{j\pi/4}\right)\left(z - 2e^{-j\pi/4}\right)}{\left(z - \frac{1}{2}e^{j\pi/4}\right)\left(z - \frac{1}{2}e^{-j\pi/4}\right)} = \frac{1}{4}\frac{z^2 - 2\sqrt{2}z + 4}{z^2 - \frac{\sqrt{2}}{2}z + \frac{1}{4}}.$$

You can verify that $|H_{ap}(\omega)| = 1$. Now create a second system,

$$H_2(z) = H_1(z)H_{ap}(z) = \frac{1}{4}\frac{z^2 - 2\sqrt{2}z + 4}{z^2},$$

whose pole-zero plot and frequency response are shown in **Figure 5.32c**. Since $|H_2(\omega)| = |H_1(\omega)||H_{ap}(\omega)| = |H_1(\omega)|$, these two systems have the same frequency-response magnitude. Once again, you can see that the allpass filter effectively reflects the pair of conjugate zeros of $H_1(z)$ from their positions inside the unit circle to conjugate-reciprocal positions outside the unit circle. Although it would theoretically be possible to design an allpass filter to move only one of the zeros or the other outside the unit circle, such an allpass filter would have complex coefficients, and would therefore be practically unrealizable. In order for the allpass filter to be realizable, both zeros must move as a pair. Hence, $H_2(z)$ is the only other system with only two zeros that has the same frequency response magnitude as $H_1(z)$.

As a final example in this section, consider the system $H_1(z)$, whose pole-zero plot is shown in **Figure 5.34a**.

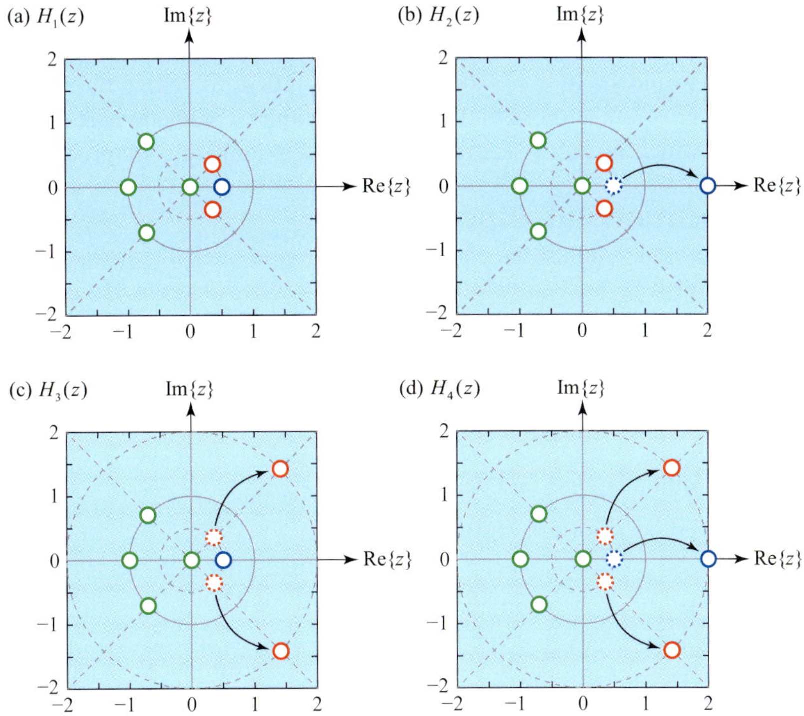

Figure 5.34 Allpass filters

This system has seven zeros: three real zeros located at $z = 0$, $z = 1/2$ and $z = -1$ and four complex zeros located at $z = 0.5e^{j\pi/4}$ and $z = e^{j3\pi/4}$. How many other systems, excluding $H_1(z)$, having seven zeros have the same frequency-response magnitude as $H_1(\omega)$? There are only three. As shown in **Figure 5.34b**, the zero on the real axis at $z = 1/2$ (shown in blue) can be reflected to its reciprocal position at $z = 2$ by application of an allpass filter with transform

$$H_{ap}(z) = -\frac{1}{2}\frac{z-2}{z-\frac{1}{2}},$$

just as we discussed in **Figure 5.32**. As shown in **Figure 5.34c**, the pair of complex zeros located at $z = 0.5e^{\pm j\pi/4}$ (shown in red) can be reflected to their conjugate-reciprocal positions at $z = 2e^{\pm j\pi/4}$ by application of an allpass filter with transform

$$H_{ap}(z) = \frac{1}{4}\frac{z^2 - 2\sqrt{2}z + 4}{z^2 - \frac{\sqrt{2}}{2}z + \frac{1}{4}},$$

as discussed in conjunction with **Figure 5.33**. Finally, as shown in **Figure 5.34d**, both the real zero at $z = 1/2$ and the complex pair at $z = 0.5e^{\pm j\pi/4}$ can be reflected outside the unit circle by applying a third-order allpass filter whose transform is just the product of the transforms of the two allpass filters we just discussed:

$$H_{ap}(z) = -\frac{1}{2}\frac{z-2}{z-\frac{1}{2}} \cdot \frac{1}{4}\frac{z^2 - 2\sqrt{2}z + 4}{z - \frac{\sqrt{2}}{2}z + \frac{1}{4}} = -\frac{1}{8}\frac{z^3 - 2(1+\sqrt{2})z^2 + 4(1+\sqrt{2})z - 8}{z^3 - \frac{1}{2}(1+\sqrt{2})z^2 + \frac{1}{4}(1+\sqrt{2})z - \frac{1}{8}}.$$

How about the other zeros, shown in green in **Figure 5.34**? They cannot be moved using an allpass filter. The singularities at $z = 0$ cannot be moved because the conjugate-reciprocal position would be at $z = \pm\infty$. It is also pointless to reflect the singularities that lie on the real axis at $z = \pm 1$ or those that occur in a complex-conjugate pair on the unit circle. These singularities, which lie at $|z| = 1$, are already at their conjugate-reciprocal positions.

5.7.6 Practical applications of allpass filters

Allpass filters can be used to adjust the phase of systems without changing the magnitude of the response. They can also be used to help make unstable systems stable. An example will help explain this.

Suppose we have an input signal $x[n]$ that has been passed through a highpass FIR filter characterized by the impulse response

$$h[n] = \tfrac{1}{4}\delta[n] - \tfrac{\sqrt{2}}{2}\delta[n-1] + \delta[n-2],$$

to form an output $y[n]$, as shown in **Figure 5.35**.

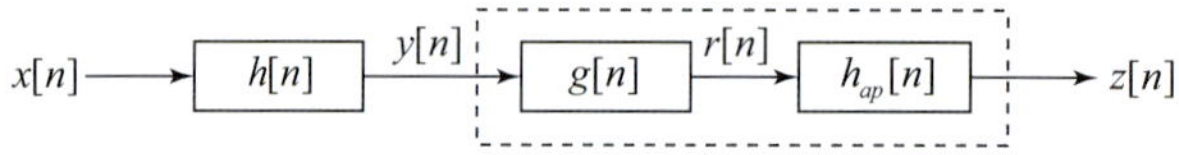

Figure 5.35 Diagram of inverse filter system

This filter has transform

$$H(z) = \frac{1}{4}\frac{z^2 - 2\sqrt{2}z + 4}{z^2}.$$

The pole-zero plot and frequency response of this system are shown in **Figure 5.36a**.

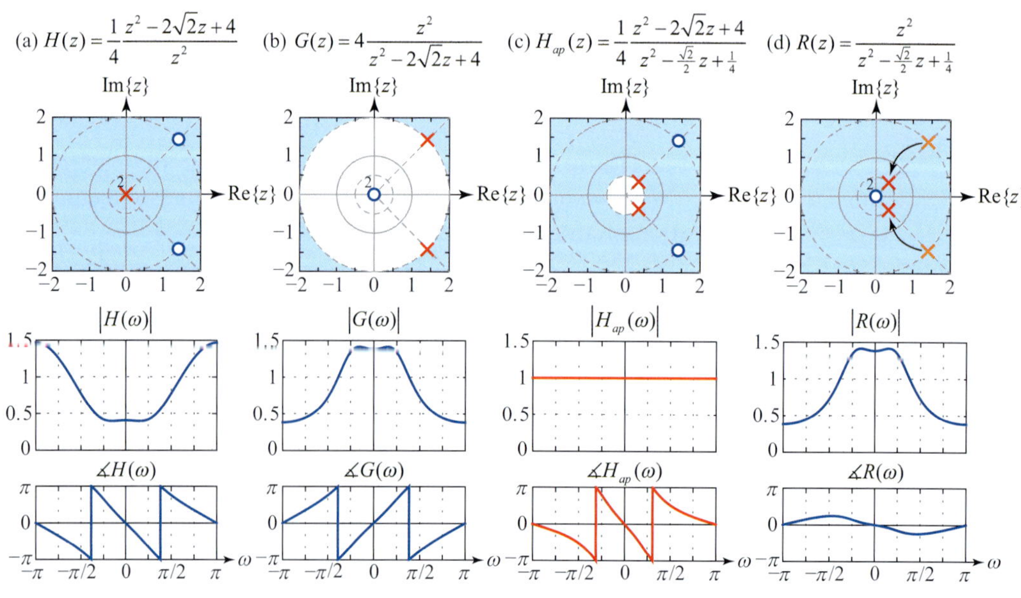

Figure 5.36 Application of allpass filter to inverse filtering

We wish to design an inverse filter to compensate for the effect of $H(z)$ such that $z[n] = x[n]$. First, we attempt to create the inverse filter with impulse response $g[n]$, such that

$$G(z) = \frac{1}{H(z)} = 4\frac{z^2}{z^2 - 2\sqrt{2}z + 4},$$

as shown in **Figure 5.36b**. Unfortunately, a causal inverse filter will not be stable because the poles are outside the unit circle. Instead, we will design a filter $R(z)$, which will have the same frequency-response magnitude as the inverse filter $|R(\omega)| = |G(\omega)|$, and the same number of singularities, but will be both causal and stable. We do this by applying the allpass filter discussed in **Figure 5.33a**, which has transform

$$H_{ap}(z) = \frac{1}{4}\frac{z^2 - 2\sqrt{2}z + 4}{z^2 - \frac{\sqrt{2}}{2}z + \frac{1}{4}}.$$

The effect of this allpass filter is to reflect the poles of $G(z)$ inside the unit circle, as shown in **Figure 5.36d**. The resulting causal stable inverse filter has transform

$$R(z) = G(z)H_{ap}(z) = \frac{z^2}{z^2 - \frac{\sqrt{2}}{2}z + \frac{1}{4}}.$$

This filter exactly compensates for the magnitude of $H(\omega)$, but not for the phase.

5.8 Minimum-phase-lag systems

Of all the possible systems which have the same magnitude of the frequency response (but different phases), the system with all its poles and zeros inside the unit circle is termed the **minimum-phase-lag system**, often just abbreviated **minimum-phase system**. The name says it all: the minimum-phase-lag system is the system which has the minimum phase-lag at any frequency. A system with all its zeros outside the unit circle is the **maximum-phase-lag** (or just **maximum-phase**) system.

A few of the examples we have already studied in the previous section should help clarify the definition. Consider the systems with a single zero, $H_1(z)$ and $H_2(z)$, shown in **Figures 5.32a** and **c**. These two systems have the same frequency-response magnitude, but different phases. $H_1(z)$ has its zero inside the unit circle while $H_2(z)$ has its zero outside. At any value of frequency $0 < \omega \leq \pi$, $\measuredangle H_1(\omega)$ is more positive (i.e., has less phase-lag) than $\measuredangle H_2(\omega)$. Therefore, this is the minimum-phase-lag system. These two systems are related through the application of the allpass filter, which has a pole at $z = 1/2$ and a zero at $z = 2$: $\measuredangle H_2(\omega) = \measuredangle H_1(\omega) + \measuredangle H_{ap}(\omega)$. As we showed in Section 5.7.1, any real allpass system with $|a| < 1$ always *adds* phase-lag (delay) to a signal at every frequency. Hence, moving a zero outside the unit circle always results in increasing the phase-lag of the frequency response.

As another example, **Figure 5.33a** shows a system $H_1(z)$ with two complex zeros. Since both zeros are inside the unit circle, it is the minimum-phase-lag system. The system $H_2(z)$, shown in **Figure 5.33c**, has both zeros outside the unit circle, and is the maximum-phase-lag system.

Minimum-phase-lag systems have a number of nice properties. For one, a causal stable minimum-phase-lag system always has a causal stable inverse, which is also minimum-phase-lag. That is because if a causal system is stable, all its poles are inside the unit circle, and if the system is minimum-phase-lag, all the zeros are inside as well. Hence the inverse – which has zeros and poles swapped – also has all poles and zeros inside the unit circle. Minimum-phase-lag systems also have the nice property that more of the energy in the impulse response is concentrated at the beginning of the sequence than is the case for non-minimum-phase systems. To help understand what this means, look at the simple example shown in **Figure 5.37**.

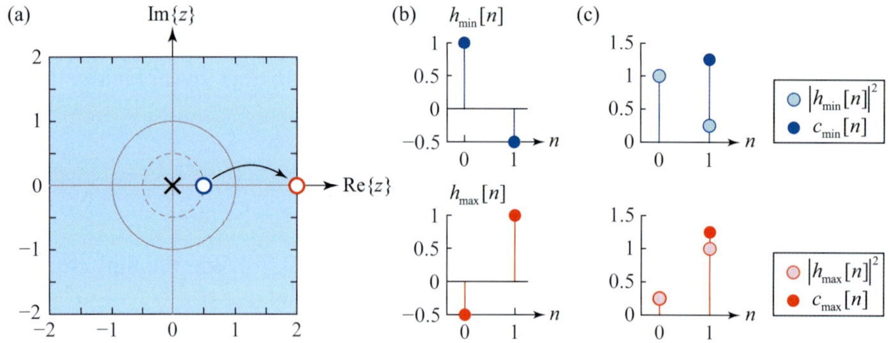

Figure 5.37 Energy of minimum- and maximum-phase-lag systems with one zero

Figure 5.37a shows the pole-zero plot of two causal single-zero FIR systems. The minimum-phase-lag system $H_{min}(z)$ has a zero at $z = 1/2$ (shown in blue) and a pole at $z = 0$,

$$H_{min}(z) = \frac{z - \frac{1}{2}}{z} = 1 - \tfrac{1}{2}z^{-1}.$$

The impulse response of the minimum-phase-lag system,

$$h_{min}[n] = \delta[n] - \tfrac{1}{2}\delta[n-1],$$

is shown in blue in the upper panel of **Figure 5.37b**. The energy of this impulse response, defined as $|h_{min}[n]|^2$, is shown in light blue (●) in **Figure 5.37c**. In this example, most of the energy in the impulse response of the minimum-phase-lag system is in the first sample. We can further quantify this effect by defining the cumulative energy $c_{min}[n]$, shown in dark blue (●) in the upper panel of **Figure 5.37c**, as the sum of the energy in the impulse response up to and including time n,

$$c_{min}[n] \triangleq \sum_{m=0}^{n} |h_{min}[m]|^2.$$

Since $|h_{min}[n]|^2$ is always positive, the cumulative energy increases monotonically with increasing n. For the simple minimum-phase-lag system of this example, 80% of the energy of the impulse response, $(1/(1+1/4) = 4/5)$, is in the first sample.

The maximum-phase-lag system $H_{max}(z)$ has its zero outside the unit circle, produced by cascading $H_{min}(z)$ with an allpass filter that reflects the zero (shown in red) outside the unit circle to $z = 2$,

$$H_{max}(z) = -\frac{1}{2}\frac{z-2}{z} = -\tfrac{1}{2} + z^{-1}.$$

The impulse response of the maximum-phase-lag system is shown in red in the lower panel of **Figure 5.37b**. The energy $|h_{max}[n]|^2$(●) and the cumulative energy $c_{max}[n]$ (●) are shown in the lower panel of **Figure 5.37c**. For this system, most (80%) of the energy in the impulse response is in the last sample.

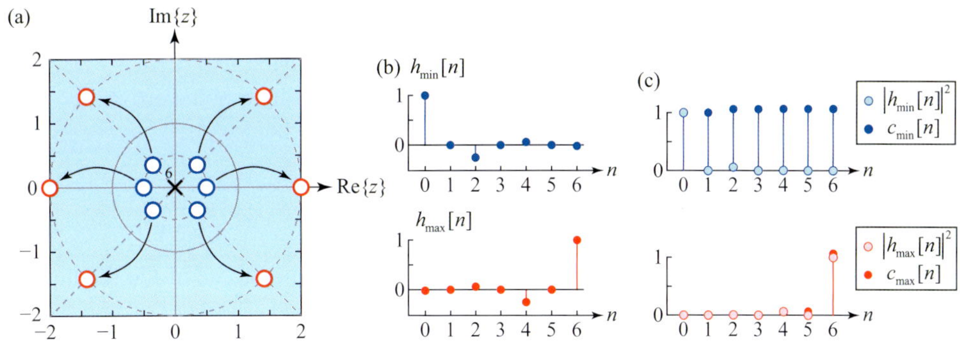

Figure 5.38 Energy of minimum- and maximum-phase-lag systems with multiple zeros

Figure 5.38 shows another example: minimum- and maximum-phase-lag FIR systems with multiple zeros. Here the impulse responses each comprise seven samples.

Again, most of the energy of the minimum-phase-lag system is compacted at the front of the impulse response. In fact, almost 94% of the cumulative energy occurs in the first sample and over 99% in the first three samples. In contrast, only about 0.02% of the energy of the maximum-phase-lag system occurs in the first sample and less than 0.4% in the first three samples. We can prove (see Problem 5-1) that the cumulative energy of the minimum-phase-lag system, $h_{min}[n]$, at time n is always greater than that of the impulse response of any system, $h[n]$, with the same frequency response magnitude:

$$\sum_{m=0}^{n} |h_{min}[m]|^2 > \sum_{m=0}^{n} |h[m]|^2.$$

SUMMARY

In this chapter, we have related the frequency response of a system, $H(\omega)$, to its z-transform, $H(z)$, and have shown that in many instances that frequency response can be easily visualized by inspection of the pole-zero plot. These concepts will be important throughout the rest of this book, particularly when we come to designing FIR and IIR filters, in Chapters 7 and 8, respectively.

PROBLEMS

Problem 5-1

In our discussion of minimum-phase-lag systems, we stated that the cumulative energy of the minimum-phase-lag system $h_{min}[n]$ at time n is always greater than that of the impulse response of any system, $h[n]$, with the same frequency-response magnitude:

$$\sum_{m=0}^{n} |h_{\min}[m]|^2 > \sum_{m=0}^{n} |h[m]|^2.$$

Your task is to prove this. In order to do so, recall that the transforms of these other systems are related to the minimum-phase-lag system through Equation (5.7): $H(z) = H_{\min}(z)H_{ap}(z)$. For simplicity, consider the case where the allpass filter is a single pole-zero pair,

$$H_{ap}(z) = \frac{1 - a^* z}{z - a} = \frac{-a^* + z^{-1}}{1 - az^{-1}} = \underbrace{\frac{1}{1 - az^{-1}}}_{Q(z)} \cdot \underbrace{(-a^* + z^{-1})}_{R(z)},$$

where a represents the position of the pole, which is assumed to be inside the unit circle, so that $|a| < 1$. To aid in the analysis, represent $H_{ap}(z)$ as a product (cascade) of two sections, $H_{ap}(z) = R(z) \cdot Q(z)$, as shown in **Figure 5.39**, where the first section,

$$Q(z) \triangleq \frac{Y(z)}{H_{\min}(z)} = \frac{1}{1 - az^{-1}},$$

is defined in the time domain by the equation $y[n] - ay[n-1] = h_{\min}[n]$ and the second section,

$$R(z) \triangleq \frac{H(z)}{Y(z)} = -a^* + z^{-1},$$

is defined in the time domain by the equation $h[n] = -a^* y[n] + y[n-1]$.

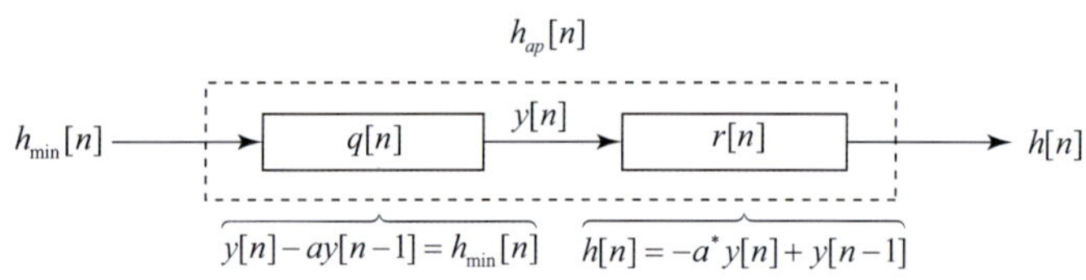

Figure 5.39

Problem 5-2

A causal stable first-order allpass filter has frequency response

$$H_{ap}(\omega) = \frac{1 - a^* e^{j\omega}}{e^{j\omega} - a},$$

where $|a| < 1$.

(a) Show that the phase of the frequency response, $\measuredangle H_{ap}(\omega)$, is

$$\measuredangle H_{ap}(\omega) = \omega - 2\measuredangle a - 2\tan^{-1}\left(\frac{\sin(\omega - \measuredangle a)}{\cos(\omega - \measuredangle a) - |a|}\right).$$

(b) Show that the phase can also be written as

$$\measuredangle H_{ap}(\omega) = \omega - 2\tan^{-1}\left(\frac{\sin\omega - |a|\sin\measuredangle a}{\cos\omega - |a|\cos\measuredangle a}\right).$$

(c) For a real, show that

$$\measuredangle H_{ap}(\omega) = \omega - 2\tan^{-1}\left(\frac{\sin\omega}{\cos\omega - a}\right).$$

Problem 5-3

A causal stable first-order allpass filter has frequency response

$$H_{ap}(\omega) = \frac{1 - ae^{j\omega}}{e^{j\omega} - a},$$

where a is real (and $|a| < 1$, since the filter is stable).

(a) Show that $\measuredangle H_{ap}(\omega = 0) = 0$.

(b) Using the result of Problem 5-2c, show that the derivative of $\measuredangle H_{ap}(\omega)$ with respect to ω is

$$\frac{d\measuredangle H_{ap}(\omega)}{d\omega} = \frac{a^2 - 1}{a^2 - 2a\cos\omega + 1}.$$

(c) Use the result of part (b) to show that $\measuredangle H_{ap}(\omega)$ is monotonically negative for $0 \leq \omega < \pi$.

Problem 5-4

An allpass filter has transform

$$H_{ap}(z) = -a\frac{z - 1/a}{z - a},$$

where $|a| < 1$.

(a) Show that the derivative of the phase of the zero is

$$\frac{d}{d\omega}\measuredangle(e^{j\omega} - 1/a) = \frac{1 + (1/a)\cos\omega}{1 + (2/a)\cos\omega + (1/a)^2}.$$

(b) Set the derivative to zero to show that the value of ω at which the phase of the zero is either maximum or minimum is

$$\omega_x = \cos^{-1}a.$$

(c) Show that for values of $a < 0$, ω_x corresponds to the maximum phase of the zero, $\measuredangle z = -\sin^{-1} a$, and for $a > 0$, ω_x corresponds to the minimum phase of the zero, $\measuredangle z = \pi - \sin^{-1} a$.

Problem 5-5

Derive the results of Problem 5-4b, c for $0 < a < 1$ using analytic geometry. That is, using **Figure 5.29** as a guide, show that the maximum phase of the zero is $\measuredangle z = -\sin^{-1} a$ and occurs at $\omega_x = \cos^{-1} a$.

Problem 5-6

Show that an allpass filter $H_{ap}(z)$ with two complex poles at $z = |a|e^{\pm j\measuredangle a}$ and two zeros at the conjugate-reciprocal position, $z = (1/|a|)e^{\pm j\measuredangle a}$, has phase

$$\begin{aligned}\measuredangle H_{ap}(\omega) &= \left(\omega - 2\measuredangle a - 2\tan^{-1}\left(\frac{\sin(\omega - \measuredangle a)}{\cos(\omega - \measuredangle a) - |a|}\right)\right) + \left(\omega + 2\measuredangle a - 2\tan^{-1}\left(\frac{\sin(\omega + \measuredangle a)}{\cos(\omega + \measuredangle a) - |a|}\right)\right)\\ &\vdots\\ &= 2\tan^{-1}\left(\frac{(a^2 - 1)\sin\omega}{(a^2 + 1)\cos\omega - 2|a|\cos\measuredangle a}\right).\end{aligned}$$

Problem 5-7

For each $H(z)$ in the table below, select the matching magnitude and phase picture from the plots shown in **Figure 5.40**. It is possible that some pictures may apply to more than one system, and some pictures will not apply to any system. The magnitude pictures and phase pictures are not necessarily placed in pairs (i.e., M1 is not necessarily from the same system as P1).

	$H(z)$	Magnitude	Phase
(a)	$z - 2$		
(b)	$2\left(z + \frac{1}{2}\right)$		
(c)	$3z/(2z + 1)$		
(d)	$\frac{3}{2}(z + 1)/(2z - 1)$		
(e)	$\frac{3}{2}(z - 1)$		
(f)	$(z + 2)/z$		
(g)	$2\left(z - \frac{1}{2}\right)$		
(h)	$z + 2$		

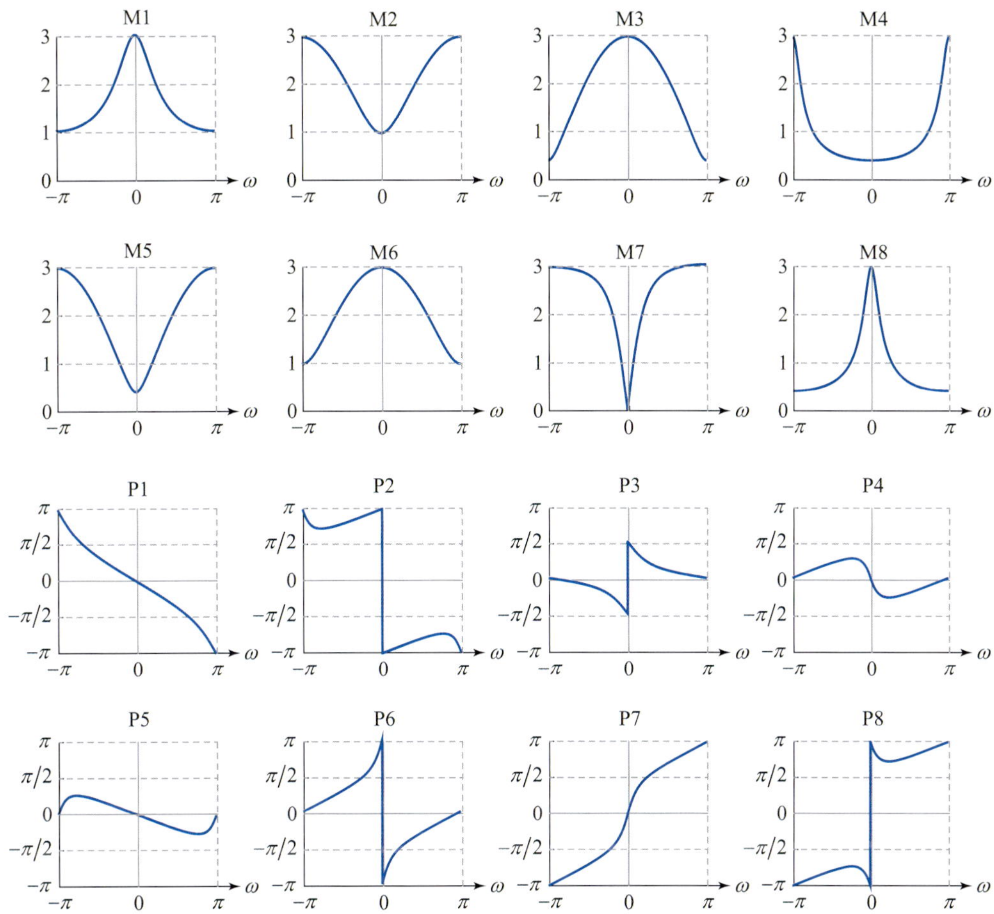

Figure 5.40

Problem 5-8

Plot the pole-zero plot for each $H(z)$ in Problem 5-7.

Problem 5-9

Repeat Problem 5-7 for each $H(z)$ in the table below.

	$H(z)$	Magnitude	Phase
(a)	$\frac{3}{4}z/\left(z-\frac{3}{4}\right)$		
(b)	$(z+2)/z$		
(c)	$\frac{12}{7}\left(z+\frac{3}{4}\right)/z$		
(d)	$\frac{12}{7}\left(z-\frac{3}{4}\right)$		
(e)	$\frac{3}{4}z/\left(z+\frac{3}{4}\right)$		
(f)	$\frac{9}{4}(z-1)/\left(z-\frac{1}{2}\right)$		
(g)	$\frac{9}{7}\left(z-\frac{4}{3}\right)$		
(h)	$\frac{3}{2}z/\left(z-\frac{1}{2}\right)$		

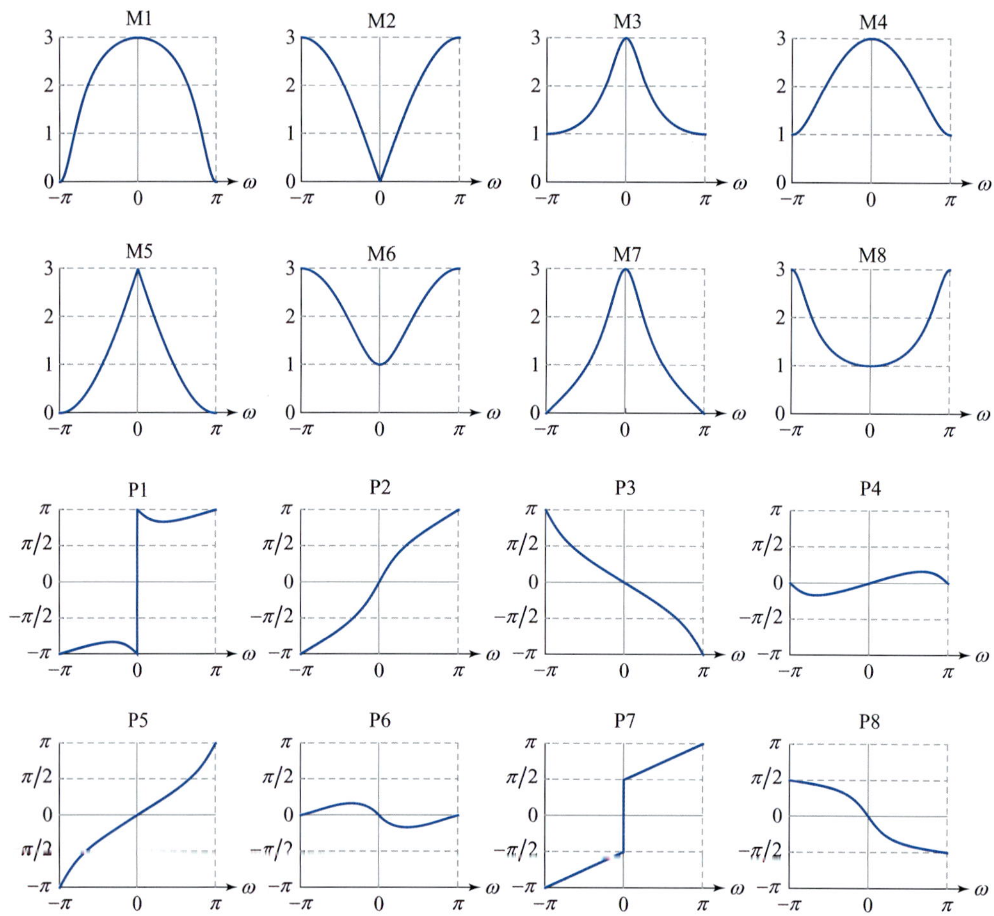

Figure 5.41

Problem 5-10

Plot the pole-zero plot for each $H(z)$ in Problem 5-9.

Problem 5-11

A stable system has impulse response

$$h[n] = \delta[n] + 3(\tfrac{1}{2})^n u[n-1].$$

(a) Plot the pole-zero plot for this system, including the ROC.
(b) Sketch the frequency response of this system (both magnitude and phase).

6 A/D and D/A conversion

Introduction

We live in an analog world. Most of the stimuli we experience in daily life – sounds, sights, tactile stimuli – are continuous in time, that is, analog. Hence, it is not surprising that many of the most important applications of digital signal processing involve processing analog signals, for example: recording, storing, manipulating and playing back audio and video. Many other applications involve sampling signals from transducers that measure information about the natural or constructed world such as barometric pressure, air temperature, the movement of the earth, the electrocardiogram (EKG) or electroencephalogram (EEG) of a patient, the speed of a car, and so on.

In this chapter, we will analyze in detail how analog-to-digital (A/D) and digital-to-analog (D/A) converters are used to convert analog signals into digital signals and back. After a brief overview in Section 6.1, Sections 6.2 through 6.5 cover the theory of A/D and D/A conversion, including such topics as upsampling and downsampling. In Section 6.7, we will discuss quantization, which is a key feature of sampled data systems. Finally, in Sections 6.8 and 6.9, as well as in supplementary material available at www.cambridge.org/holton, we will take a look at how practical A/D and D/A converters are implemented in hardware.

6.1 Overview of A/D and D/A conversion

In Chapter 1, we introduced the notion of the DSP paradigm – the approach of using discrete-time signal processing to process analog signals – that is shown again in **Figure 6.1**.

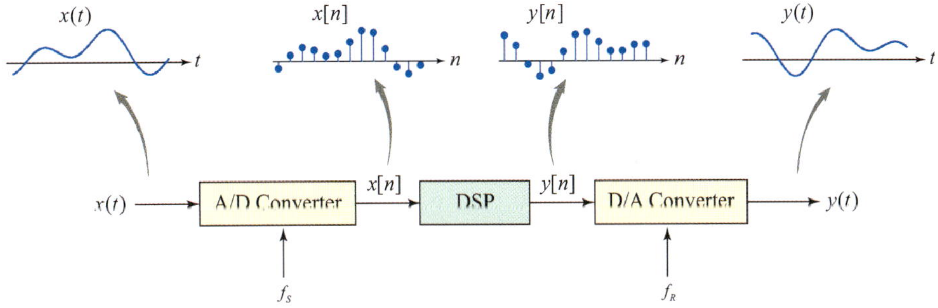

Figure 6.1 Discrete-time processing of analog signals

In the DSP paradigm, an analog signal $x(t)$ is sampled by a hardware device, an **analog-to-digital (A/D) converter**, to form a discrete-time sequence $x[n]$. This input sequence can be processed through a DSP system, for example a discrete-time filter implemented in software or hardware, to form an output sequence $y[n]$. The output sequence is then converted back into the analog signal $y(t)$ by a hardware **digital-to-analog (D/A) converter**.

We will make several passes in order to describe the analog-to-digital and digital-to-analog conversion processes in detail. We begin with an overview of the ideal analog-to-digital and digital-to-analog conversion process, and then progress to discuss some of the more practical aspects of conversion. Finally, we will take a brief tour of some of the dominant technologies used to implement practical A/D and D/A converters. Understanding the theory of sampling and reconstruction presented in this chapter depends heavily on understanding the continuous-time Fourier transform and its properties. So if you are a bit rusty on this material, you might consider reviewing it first.

Figure 6.2 gives a more detailed overview of the digital signal processing paradigm.

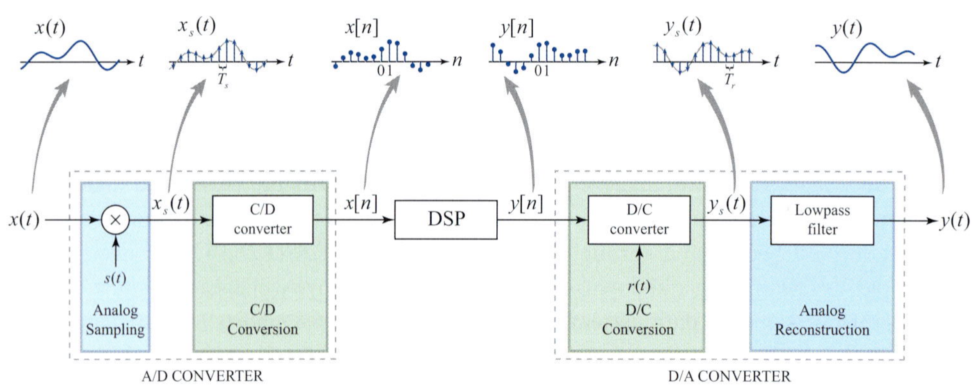

Figure 6.2 Model of the digital signal processing paradigm

In order to understand the A/D process, we divide the ideal A/D converter into two conceptually separate stages, an **analog sampling stage** (shown in blue) and a **continuous-to-discrete (C/D) conversion stage** (shown in green). In the analog sampling stage, a continuous-time analog signal $x(t)$ is sampled by multiplying it by a periodic impulse train $s(t)$, which has a period of T, to form a modulated impulse train $x_s(t)$. In the continuous-to-discrete conversion stage, this modulated impulse train is converted into a discrete-time sample sequence $x[n]$. Each value of the sequence represents one instant of the original input sequence, $x[n] = x(nT)$. It would seem that a lot of information has been thrown out of $x(t)$ in producing $x[n]$; however, it is a remarkable fact that subject to certain conditions, $x[n]$ theoretically contains all the information necessary to completely reconstruct $x(t)$! Following the A/D converter is a DSP operation, such as discrete-time filtering, that takes input sequence $x[n]$ and produces output sequence $y[n]$.

Finally, the output discrete-time sequence $y[n]$ is converted back into an analog signal $y(t)$ by the D/A converter.

We also divide the ideal D/A converter into two conceptually separate stages, a **discrete-to-continuous (D/C) conversion stage** (shown in green) and an **analog reconstruction stage** (shown in blue). In the D/C converter, a continuous-time impulse train $y_s(t)$ is created. The area of each successive impulse of $y_s(t)$ is obtained from successive values of $y[n]$. Then, $y_s(t)$ is filtered to form the output analog signal $y(t)$.

In the following sections we will analyze each of the blocks of the discrete-time signal processing paradigm of **Figure 6.2**. To simplify our analysis and concentrate our attention just on the A/D and D/A process, we will replace the DSP operation with the identity transformation $y[n]=x[n]$. Even this simple system has some practical value. In fact, it is a simplified model of the digital recording and reproduction process used in CDs and other music reproduction technologies. In a digital recording application, $x(t)$ would represent the analog input from a microphone or mixer in a recording studio during a musical performance. The output of the A/D converter is a sequence $x[n]$ that is encoded as a series of pits on a CD (or processed into a computer file such as an MP3 file, as described in Chapter 12). When the CD is placed on a CD player (assuming anyone still has a CD player), the series of pits is read by a laser system and used to create the discrete sequence $y[n]$, which is identical to $x[n]$. Then $y[n]$ is processed by the D/A converter in the player to reconstitute the music, $y(t)$.[1]

6.2 Analog sampling and reconstruction

As a first step in understanding the complete discrete-time signal processing paradigm, we will reduce the A/D and D/A conversion process to its most essential elements by considering the simpler, purely analog, problem of sampling and reconstruction shown in **Figure 6.3**.

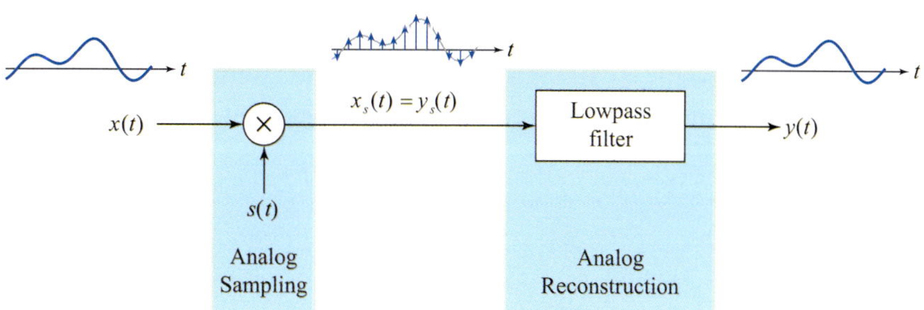

Figure 6.3 Model of analog sampling and reconstruction

[1] It would probably be a bit more accurate to model the digital signal processing operation of the CD player as an ideal delay, $y[n]=x[n-n_0]$, where n_0 represents the delay (in samples) between the time that the music was recorded and the time it is played back.

This system comprises just the outer ("blue") portions of the complete system shown in **Figure 6.2** – the analog sampling stage and the analog reconstruction stage – strung together. In the analog sampling stage, the continuous-time analog signal $x(t)$ is multiplied by the periodic impulse train $s(t)$ to form the modulated impulse train $x_s(t)$. In the analog reconstruction stage, this impulse train is filtered to recover (we hope) the original signal. In the following paragraphs, we will consider sampling and reconstruction separately.

6.2.1 Analog sampling

Figure 6.4 shows a simplified representation of the sampling of an analog signal by a periodic impulse train.

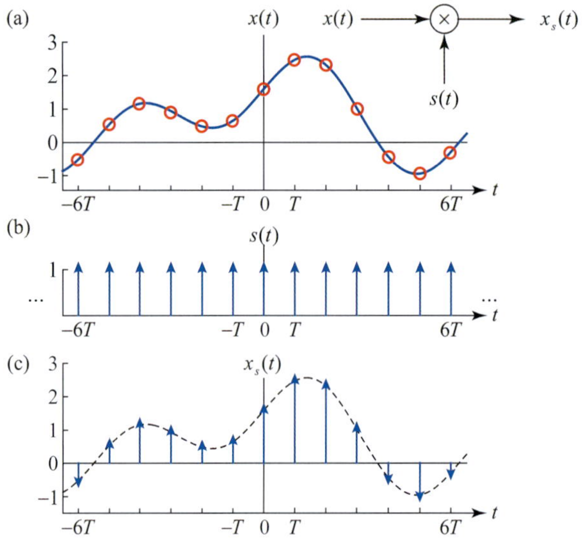

Figure 6.4 Sampling an analog signal

The goal of analog sampling is to select values of a continuous signal $x(t)$ at instants of time that are integer multiples of a **sampling period** T, as shown by the red circles in **Figure 6.4a**. This selection effectively "discards" the rest of the waveform between the samples. Sampling can be modeled theoretically as the multiplication of the continuous-time signal $x(t)$ by an **impulse train** $s(t)$, an infinitely long train of impulses that is periodic with period T (**Figure 6.4b**),

$$s(t) = \sum_{n=-\infty}^{\infty} \delta(t - nT).$$

The result of this multiplication is the sampled input signal $x_s(t)$ shown in **Figure 6.4c**,

$$x_s(t) = x(t) \cdot s(t). \tag{6.1}$$

Analog sampling is a form of modulation, where the input $x(t)$ modulates the impulse train $s(t)$ to form $x_s(t)$. The areas of the impulses of $x_s(t)$ are scaled by the values of $x(t)$ at the instants at which the impulses in $s(t)$ occur. This modulation thereby "encodes" the amplitude of the original analog signal $x(t)$ at discrete moments in time, $t = nT$, into the area of the impulses of $x_s(t)$. All values of the original signal at times other than at multiples of T are lost in $x_s(t)$. The sampled signal $x_s(t)$ is still a continuous-time signal, but it is more "sparse" than the input signal in the sense that it is completely characterized by only a finite number of samples per second. Why would we want to sample a signal in such a manner? Because both $x(t)$ and $x_s(t)$ contain *identical* information, so that the original signal $x(t)$ can be completely recovered from its samples $x_s(t)$ subject to certain conditions, as we will now show.

6.2.2 The sampling theorem

The **sampling theorem**[2] states the following remarkable fact: in theory, a continuous signal $x(t)$ can be perfectly reconstructed from its samples $x_s(t)$. The two conditions for perfect reconstruction are (1) that the signal be **bandlimited** and (2) that we sample "fast enough." A bandlimited signal is one whose spectrum $X(\Omega)$ only has energy in a finite frequency range $-\Omega_b < \Omega < \Omega_b$, where Ω_b represents the highest frequency in $X(\Omega)$. We will explore what "fast enough" means in a moment.

The sampling theorem in the frequency domain While it is not easy to see how this reconstruction is possible from the time-domain pictures of **Figure 6.4**, it will be very clear when we examine the frequency-domain representations of $x(t)$, $s(t)$ and $x_s(t)$. The left column of **Figure 6.5** again shows $x(t)$, $s(t)$ and $x_s(t)$ from **Figure 6.4**.

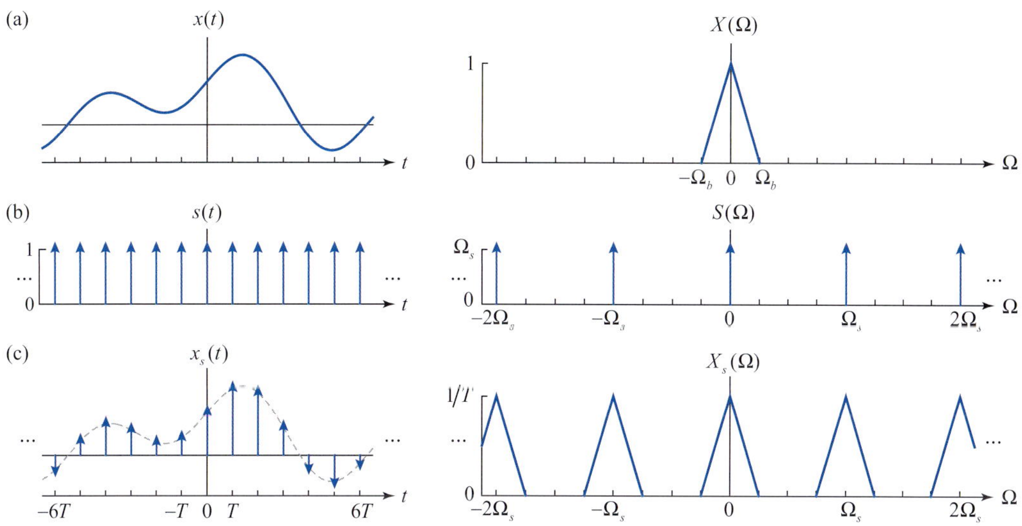

Figure 6.5 Analog sampling

The right panel shows the continuous-time Fourier transforms of these signals: $X(\Omega)$, $S(\Omega)$ and $X_s(\Omega)$, respectively. $X(\Omega)$, which is schematically depicted with a "tent" shape, is assumed to be

[2]The sampling theorem is often called the *Nyquist sampling theorem* or the *Nyquist–Shannon sampling theorem* in honor of two pioneering electrical engineers and mathematicians, Harry Nyquist (1889–1976) and Claude Shannon (1916–2001), who contributed to its formulation.

bandlimited to the frequency range $-\Omega_b < \Omega < \Omega_b$ centered around frequency $\Omega = 0$. Taking the Fourier transform[3] of Equation (6.1), we get

$$X_s(\Omega) = \mathfrak{F}\{x_s(t)\} = \mathfrak{F}\{x(t)s(t)\} = \frac{1}{2\pi}X(\Omega)*S(\Omega). \tag{6.2}$$

The Fourier transform of an impulse train in time $s(t)$ can be shown to be an impulse train in frequency, as depicted in **Figure 6.5b**,

$$S(\Omega) = \mathfrak{F}\{s(t)\} = \frac{2\pi}{T}\sum_{k=-\infty}^{\infty}\delta(t - 2\pi k/T) = \Omega_s\sum_{k=-\infty}^{\infty}\delta(t - k\Omega_s),$$

where

$$\Omega_s \triangleq 2\pi/T$$

is defined as the **sampling frequency**, which has units of rad/s. An alternate, equivalent definition of sampling frequency is

$$f_s \triangleq 1/T,$$

which has units of Hertz (Hz). The relation between the two definitions is

$$\Omega_s = 2\pi f_s.$$

In the example of **Figure 6.5**, we have chosen $\Omega_s = 4\Omega_b$. Substituting $S(\Omega)$ into Equation (6.2) gives

$$\begin{aligned} X_s(\Omega) &= \frac{1}{2\pi}X(\Omega)*S(\Omega) = \frac{1}{2\pi}X(\Omega)*\Omega_s\sum_{k=-\infty}^{\infty}\delta(\Omega - k\Omega_s) = \frac{1}{\cancel{2\pi}}\frac{\cancel{2\pi}}{T}\sum_{k=-\infty}^{\infty}X(\Omega)*\delta(\Omega - k\Omega_s) \\ &= \frac{1}{T}\sum_{k=-\infty}^{\infty}X(\Omega - k\Omega_s). \end{aligned}$$

The convolution of $X(\Omega)$ with $\delta(\Omega - k\Omega_s)$ is $X(\Omega - k\Omega_s)$, which is the spectrum of the input $X(\Omega)$ translated in frequency by $k\Omega_s$. Hence, as shown in **Figure 6.5c**, $X_s(\Omega)$ comprises replicas of $X(\Omega)$, each centered at a multiple of the sampling frequency, $k\Omega_s$, and scaled by the sampling frequency in Hz, $f_s = 1/T$. The replica centered at $\Omega = 0$ is called the **baseband component**. The other replicas are termed **image replicas**. Comparing the frequency-domain pictures of **Figures 6.5a** and **c**, you can see that $X_s(\Omega)$ comprises a baseband component centered at $\Omega = 0$ that is identical to $X(\Omega)$, plus image replicas centered at multiples of $\Omega = k\Omega_s$, $k \neq 0$.

The question, "How do we recover $x(t)$ from $x_s(t)$?" is equivalent to asking, "How do we recover $X(\Omega)$ from $X_s(\Omega)$?" In the frequency domain, the answer is clear: we must do something that preserves the baseband replica of $X_s(\Omega)$ while removing all the image replicas. That "something" is lowpass filtering. Specifically, ideal recovery of a signal from its samples can be accomplished by filtering the sampled signal through a lowpass **reconstruction filter** with frequency response $H_r(\Omega)$, as shown in **Figure 6.6**.

[3] In this chapter, we shall use the notations $\mathfrak{F}\{\}$ and $\mathfrak{F}^{-1}\{\}$ for both continuous-time and discrete-time Fourier transforms. The usage will be unambiguous since the arguments will be either discrete or continuous.

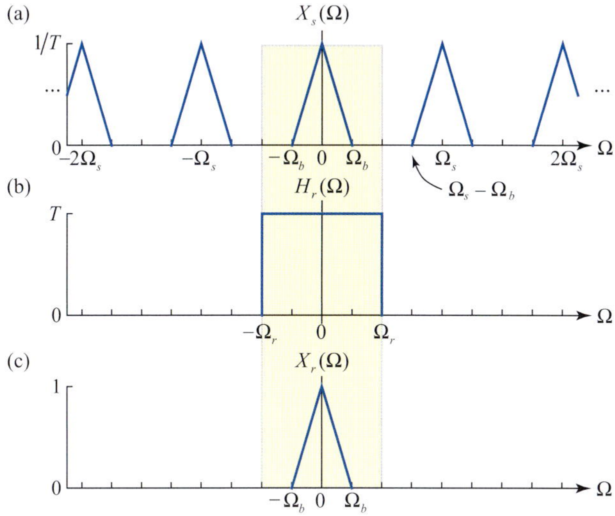

Figure 6.6 Recovery of $X_r(\Omega)$ from $X_s(\Omega)$ by lowpass filtering

Figure 6.6a again shows the spectrum of the sampled signal $X_s(\Omega)$. In the frequency domain, the spectrum of the reconstructed signal at the output of the lowpass filter is given by

$$X_r(\Omega) \triangleq X_s(\Omega)H_r(\Omega). \tag{6.3}$$

In order for the reconstruction of the signal from its samples to be perfect, we require that $X_r(\Omega) = X(\Omega)$. This means that the reconstruction filter needs to pass the baseband of $X_s(\Omega)$ by multiplying it by a constant T in at least the frequency range $-\Omega_b < \Omega < \Omega_b$, and needs to remove all the image replicas by multiplying them by zero. For this reason, the reconstruction filter is also often called the **anti-imaging filter**.

Figure 6.6b shows a candidate reconstruction filter that will accomplish this goal, an ideal lowpass filter with a bandwidth of Ω_r,

$$H_r(\Omega) = \begin{cases} T, & |\Omega| < \Omega_r \\ 0, & \text{otherwise} \end{cases}. \tag{6.4}$$

As long as the bandwidth of the reconstruction filter is greater than the highest frequency in the baseband, Ω_b, but smaller than the lowest frequency of the first image replica, $\Omega_s - \Omega_b$, the filter will effectively separate the baseband component from the surrounding images. Hence, the spectrum of the reconstructed signal, $X_r(\Omega)$, will be equal to $X(\Omega)$, as shown in **Figure 6.6c** and $x_r(t)$ will be equal to $x(t)$. In other words, the original signal will be recovered completely from its samples. This is the essence of the sampling theorem.

The sampling theorem in the time domain It is easy to see how the sampling theorem works in the frequency domain: the reconstruction filter selects the baseband from the spectrum of the sampled signal and removes all the images. In order to understand how lowpass filtering works in the time domain, let $x_r(t)$ be the signal that is recovered by lowpass filtering $x_s(t)$. In the time domain, $x_r(t)$ is the inverse Fourier transform of Equation (6.3), namely

$$x_r(t)=\mathfrak{F}^{-1}\{X_r(\Omega)\}=\mathfrak{F}^{-1}\{X_s(\Omega)H_r(\Omega)\}=x_s(t)*h_r(t),$$

where $h_r(t)$ is the impulse response of the ideal lowpass filter whose frequency response is given by Equation (6.4). Hence,

$$h_r(t)=\mathfrak{F}^{-1}\{H_r(\Omega)\}=\frac{1}{2\pi}\int_{-\Omega_r}^{\Omega_r}Te^{j\Omega t}d\Omega=\frac{1}{\Omega_s}\frac{1}{jt}(e^{j\Omega_r t}-e^{-j\Omega_r t})=\frac{2}{\Omega_s}\frac{\sin\Omega_r t}{t}=2\frac{\Omega_r}{\Omega_s}\operatorname{sinc}\Omega_r t.$$

So,

$$x_r(t)=x_s(t)*h_r(t)=\left(\sum_{n=-\infty}^{\infty}x(nT)\delta(t-nT)\right)*h_r(t)=2\frac{\Omega_r}{\Omega_s}\sum_{n=-\infty}^{\infty}x(nT)\operatorname{sinc}\frac{\Omega_s}{2}(t-nT). \tag{6.5}$$

For simplicity in the discussion that follows, let us set the bandwidth of the reconstruction filter equal to half the sampling frequency, $\Omega_r=\Omega_s/2$. Then, $h_r(t)=\operatorname{sinc}\Omega_r t$ and Equation (6.5) becomes

$$x_r(t)=\sum_{n=-\infty}^{\infty}x(nT)\operatorname{sinc}\frac{\Omega_s}{2}(t-nT).$$

Many other lowpass filters could be used in reconstruction, as we will show below, but this ideal lowpass filter is particularly easy to analyze and understand, so we will tackle it first.

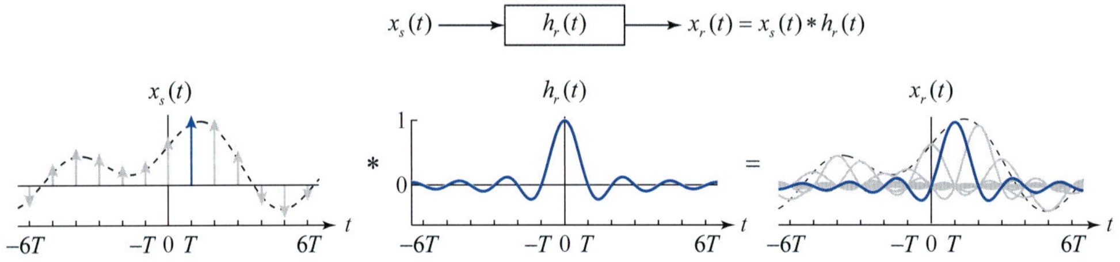

Figure 6.7 Filtering of one impulse of $x_s(t)$

The sampled signal $x_s(t)$ consists of an infinite sum of impulses, each one shifted by an integer multiple of the sampling period, $t=nT$, with area scaled by the value of $x(nT)$, namely $x(nT)\delta(t-nT)$. When each impulse of the sampled signal is filtered by the reconstruction filter whose impulse response is $h_r(t)$, the result is a shifted impulse response scaled by a constant $x(nT)$,

$$(x(nT)\delta(t-nT))*h_r(t)=x(nT)h_r(t-nT)=x(nT)\operatorname{sinc}\frac{\Omega_s}{2}(t-nT),$$

as shown in **Figure 6.7**. The left panel of the figure shows one impulse highlighted in blue, namely $x(T)\delta(n-T)$. When this impulse is filtered by $h_r(t)$, the result is the shifted, scaled impulse response $x(T)h_r(n-T)$, shown in the right panel.

When the sum of all the scaled and shifted impulses that comprise $x_s(t)$ is passed through the reconstruction filter, the output $x_r(t)$ is the sum of sinc functions, each one shifted by an integer multiple of the sampling period, $t = nT$, and scaled by the value of $x(nT)$, as shown in the right column of **Figure 6.8**.

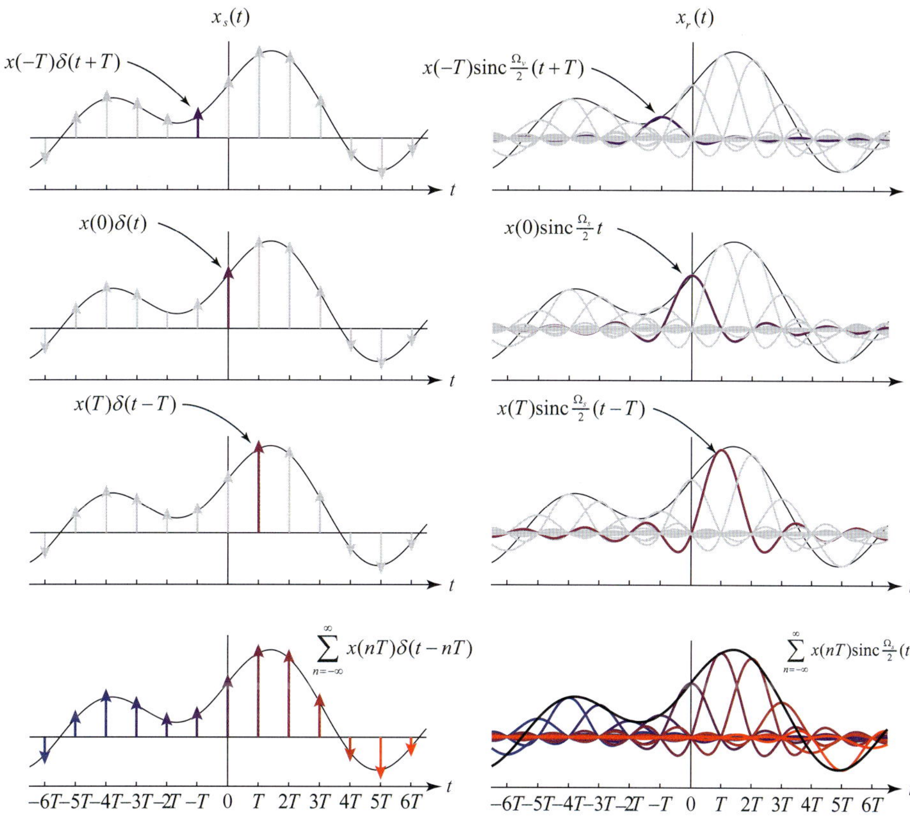

Figure 6.8 Time-domain reconstruction of $x_r(t)$ from $x_s(t)$ using an ideal lowpass filter

Equation (6.5) guarantees that the value of the reconstructed signal is exactly equal to the input signal at values of t that are multiples of the sampling period, namely $x_r(t) = x(t)$, $t = nT$. That is because a sinc function centered at $t = nT$ is equal to one for $t = nT$ and is equal to zero for values of t that are other multiples of the sampling period,

$$\mathrm{sinc}\frac{\Omega_s}{2}(t-nT)\bigg|_{t=kT} = \mathrm{sinc}\frac{\Omega_s T}{2}(k-n) = \mathrm{sinc}\,\pi(k-n) = \begin{Bmatrix} 1, & k=n \\ 0, & k \neq n \end{Bmatrix} = \delta[k-n].$$

Hence, from Equation (6.5),

$$x_r(kT) = \sum_{n=-\infty}^{\infty} x(nT)\,\mathrm{sinc}\frac{\Omega_s}{2}(kT-nT) = \sum_{n=-\infty}^{\infty} x(nT)\delta[k-n] = x(kT). \tag{6.6}$$

This point should be made more clearly in **Figure 6.9**. **Figure 6.9a** shows $X_s(\Omega)$ for the case $\Omega_s = 4\Omega_b$, with the baseband highlighted. The dotted rectangle indicates the reconstruction filter, with bandwidth $\Omega_r = \Omega_s/2 = 2\Omega_b$. **Figure 6.9b** shows $x_r(t)$ from the bottom right panel of **Figure 6.8**, comprising the sum of scaled and shifted impulse responses. A portion of the plot, highlighted in yellow, is replotted in **Figure 6.9c** on an expanded time scale of about one period. At time $t = 0$, Equation (6.6) gives

$$\begin{aligned} x_r(0) &= \sum_{n=-\infty}^{\infty} x(nT)\,\mathrm{sinc}\left(-n\frac{\Omega_s T}{2}\right) = \sum_{n=-\infty}^{\infty} x(nT)\,\mathrm{sinc}(-n\pi) \\ &= \cdots \cancel{x(-T)\,\mathrm{sinc}(\pi)} + x(0)\,\mathrm{sinc}\,0 + \cancel{x(T)\,\mathrm{sinc}(-\pi)} + \cdots \\ &= x(0). \end{aligned}$$

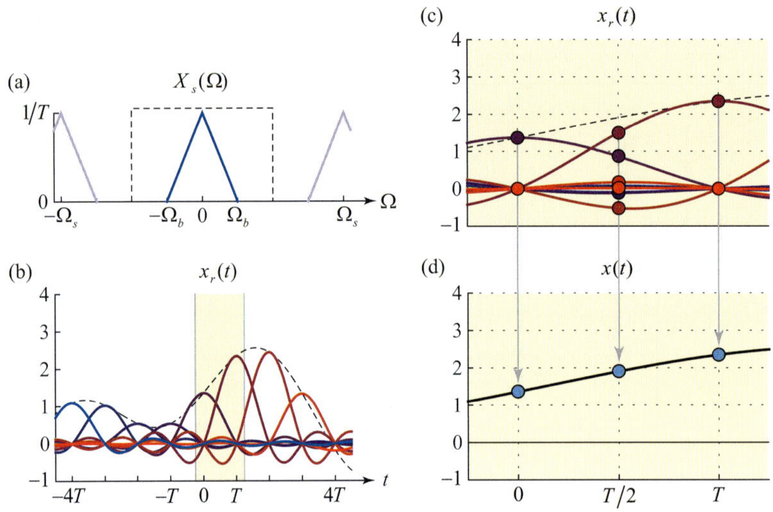

Figure 6.9 Detail of time-domain reconstruction of $x_r(t)$ from $x_s(t)$

So at $t = 0$, only the sinc function $h_r(t)$, which is centered at $t = 0$, has non-zero value; the sinc functions centered at other multiples of T all have zero crossings at $t = 0$ and contribute nothing to the sum. Similarly, at time $t = T$, only the sinc function $h_r(t - T)$, which is centered at $t = T$, contributes a non-zero value to the response.

Figure 6.9c also shows what happens when we consider values of t that are *not* an integer multiple of the sampling period, for example $t = T/2$. Here, the sinc functions centered at *every* multiple of T contribute to the sum. The final result, indicated in **Figure 6.9d**, is that the sum of all the sinc functions adds up to the input signal $x(t)$, not only at values of t that are multiples of the sampling period, but for all t.

The impulse response of the ideal lowpass reconstruction filter, $h_r(t)$, acts as an ideal interpolating function. Each impulse in $x_s(t)$ excites a scaled and shifted version of $h_r(t)$ that extends from $-\infty < t < \infty$, and therefore adds value to $x_r(t)$ for *all* t. Put another way, because $x_r(t)$ is the sum of all these impulse responses, in general, *every* point in the reconstructed signal $x_r(t)$ depends upon *every* impulse in the sampled signal $x_s(t)$, as indicated by Equation (6.5).

Reconstruction filters of different bandwidths So far, the reconstruction filter we have used is an ideal lowpass filter with a bandwidth equal to exactly half the sample frequency,

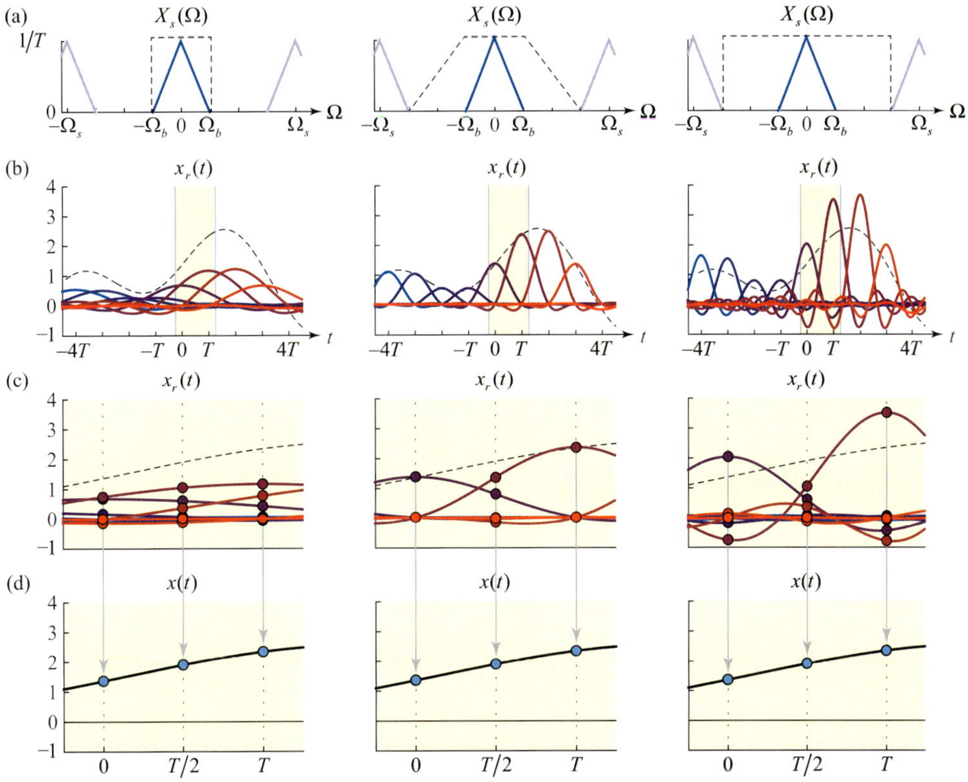

Figure 6.10 Reconstruction of $x_r(t)$ from $x_s(t)$ using different ideal lowpass filters

namely $\Omega_s/2$. This is a particularly easy case to analyze and visualize. However, it is not the only filter, and not even the only ideal lowpass filter, that can be used to recover a signal from its samples. The essential characteristic of an acceptable reconstruction filter is that it preserves the baseband spectrum and removes the image replicas. As long as the reconstruction filter meets these criteria, the original signal can be recovered. This is demonstrated graphically in **Figure 6.10**, which shows what happens when we reconstruct a signal $x_r(t)$ with lowpass filters that have different characteristics, some ideal, some not.

The left column of **Figure 6.10** shows the reconstruction of the signal using an ideal lowpass filter whose bandwidth is just equal to the highest frequency in the baseband, $\Omega_r = \Omega_b$. The right column of the figure shows the case of an ideal lowpass filter whose bandwidth extends to the lowest frequency of the first image replica, $\Omega_r = \Omega_s - \Omega_b$. The center column shows the response of a non-ideal filter whose magnitude is flat in the passband $|\Omega| < \Omega_b$, and tapers to zero at the lowest frequency of the first image replica. Even though the impulse responses of the three filters differ dramatically, the sum of the scaled impulse responses is nevertheless equal to the input signal in all cases.

Summary of the reconstruction process **Figure 6.11** shows a summary of the process of reconstructing a signal from its samples. The left column shows the time domain and the right column shows the frequency domain.

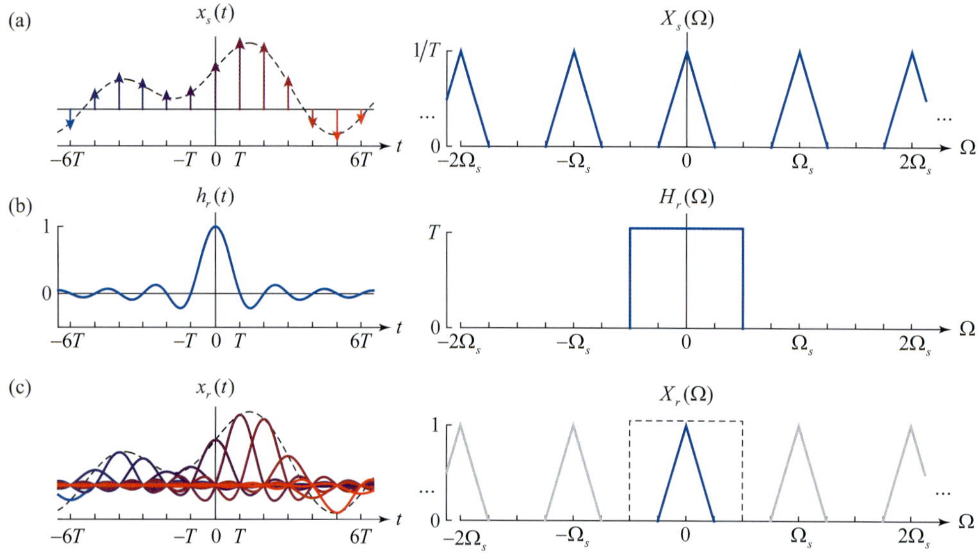

Figure 6.11 Summary of reconstruction of $x_r(t)$ from $x_s(t)$ using an ideal lowpass filter

6.2.3 The Nyquist sampling criterion

In order for a signal $x(t)$ to be reconstructed from its samples, we said that the signal must be bandlimited and that we must sample "fast enough." What does "fast enough" mean? Specifically, is there is a minimum (or maximum) sampling frequency that is required in order for us to reconstruct a signal from its samples? In order to quantify what "fast enough" means, look at the spectrum of $X_s(\Omega)$ in **Figure 6.12**.

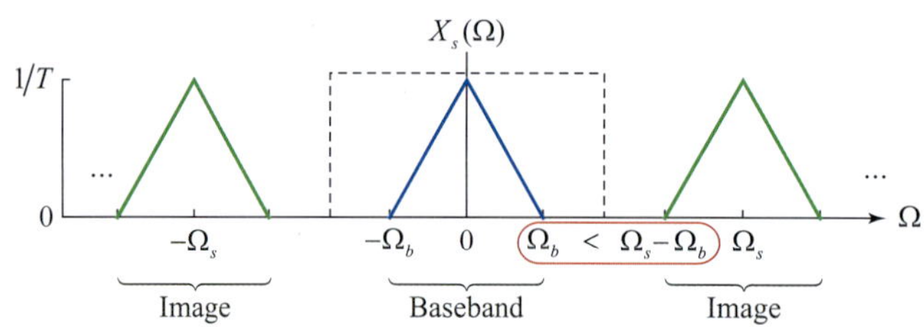

Figure 6.12 Spectral condition for the recovery of a signal from its samples

Assume again that we are sampling a signal with maximum bandwidth Ω_b at frequency Ω_s. If $x(t)$ is to be recovered from $x_s(t)$, we must be able to recover $X(\Omega)$ from $X_s(\Omega)$. This means that the baseband component of $X_s(\Omega)$ (shown in blue) must be distinct from the image components centered at multiples of Ω_s (shown in green); the components cannot overlap. Hence, the lower edge of the replica centered at $\Omega = \Omega_s$, which occurs at frequency $\Omega = \Omega_s - \Omega_b$, must be greater than the upper edge of the baseband, which occurs at frequency Ω_b. That is, $\Omega_s - \Omega_b > \Omega_b$, from which we conclude that $\Omega_s > 2\Omega_b$. This criterion for recovering a signal from its samples is called the **Nyquist sampling criterion** and is the fundamental rule that governs the choice of sample rate in analog-to-digital conversion.

To summarize the preceding sections, the sampling theorem states that a signal can be recovered from its samples if it is bandlimited to frequency Ω_b, and is sampled at a rate Ω_s that is at least twice Ω_b: $\Omega_s > 2\Omega_b$.

6.2.4 Oversampling, undersampling and critical sampling

The Nyquist sampling criterion sets the theoretical minimum sample rate for recovering a signal from its samples. Let us now investigate what happens when a signal $x(t)$ with maximum bandwidth Ω_b is sampled at different sample rates Ω_s, above and below and exactly at the Nyquist rate.

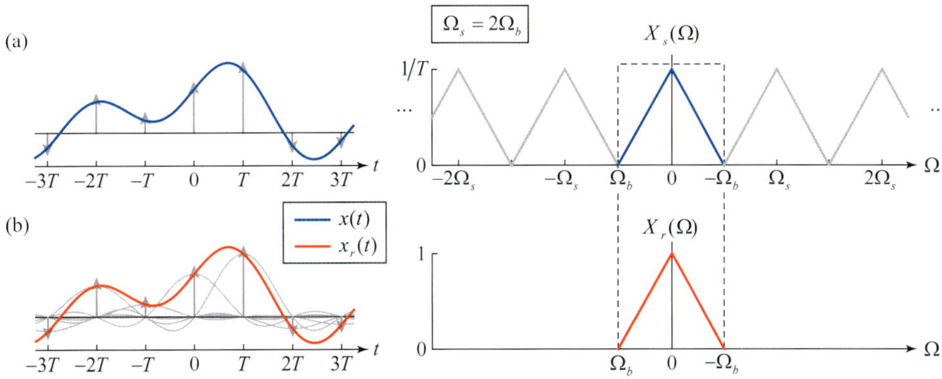

Figure 6.13 Critical sampling: reconstruction of $x(t)$ from $x_s(t)$ with $\Omega_S = 2\Omega_V$

Critical sampling **Figure 6.13** shows what happens if we sample at exactly the Nyquist rate, $\Omega_s = 2\Omega_b$, which we will term **critical sampling**. The left panel of **Figure 6.13a** shows the time domain representation of an input signal $x(t)$ (in blue), as well as the sampled signal $x_s(t)$ (the black impulses). The right panel shows the spectrum of the sampled signal, $X_s(\Omega)$, comprising replicas of the spectrum of the input signal, $X(\Omega)$, at multiples of the sampling frequency. Since the sampling frequency is exactly twice the highest frequency in the input signal, Ω_b, the baseband replica of $X_s(\Omega)$ (denoted in blue) and the image replicas (in grey) adjoin each other. $X_s(\Omega)$ can only be separated from the other replicas by an ideal lowpass filter. The filter must have a bandwidth of exactly $\Omega_s/2$, a gain of T in the passband ($|\Omega| < \Omega_b$), a gain of zero in the stopband ($|\Omega| > \Omega_b$) and an infinite slope at $\Omega = \Omega_b = \Omega_s/2$. The right panel of **Figure 6.13b** shows the spectrum of the reconstructed signal $X_r(\Omega)$, and the left panel shows the reconstructed time signal $x_r(t)$ (red), made up of summed impulse responses (grey) of the reconstruction filter. In any event, the ideal lowpass filter is unrealizable in practice – the impulse response is non-causal, and the filter is unstable to boot (since the impulse response is not absolutely summable) – hence it is not possible to sample at exactly at the Nyquist rate in practical applications. There is another reason we were lucky with this particular example: there was no energy in the input signal at exactly the frequency $\Omega = \Omega_b - \Omega_s/2$. If there had been energy at this frequency we could not have reconstructed the signal from its samples, even with an ideal lowpass filter (see Section 6.2.5 for further discussion of this point).

Oversampling In all practical A/D converters, we need to sample above the Nyquist rate, either by a little (perhaps 10%) or by a lot (an order of magnitude or more, for reasons we will discuss in Section 6.5.3). Sampling above the Nyquist rate is termed **oversampling**. As an example, **Figure 6.14** shows what happens when a signal is sampled at twice the Nyquist rate, that is, $\Omega_s = 4\Omega_b$. We can refer to this as "2× oversampling."

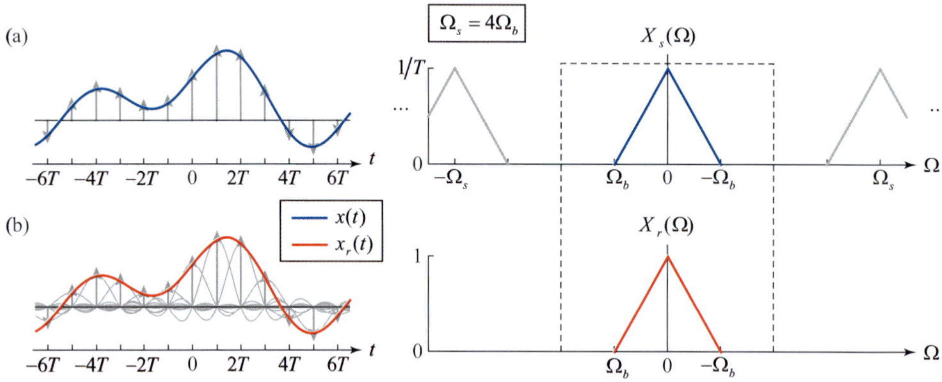

Figure 6.14 Oversampling: reconstruction of $x(t)$ from $x_s(t)$ with $\Omega_S = 4\Omega_V$

The left panel of **Figure 6.14a** shows the signal $x(t)$ (blue trace) and its samples (grey impulses). The right panel of **Figure 6.14a** shows the spectrum of the sampled signal, which consists of a baseband component plus image replicas, each of which has a bandwidth of $2\Omega_b$. Since the sampling frequency is four times the highest frequency in the signal, the replicas of $X_s(\Omega)$ are completely distinct from one another; they do not overlap at all. In particular, the baseband replica of $X_s(\Omega)$ is distinct from the replicas at other multiples of Ω_s and is equal to $X(\Omega)$. So in this case, we can certainly recover the input signal from the sampled signal by filtering the sampled signal with a lowpass reconstruction filter. The right panel of **Figure 6.14b** shows the spectrum of the reconstructed signal $X_r(\Omega)$ and the left panel shows the reconstructed signal $x_r(t)$ (red), made up of summed impulse responses (grey) of the reconstruction filter. When $\Omega_s = 4\Omega_b$, the reconstructed signal is equal to the input signal.

An important benefit of oversampling is that it makes practical reconstruction filters possible. The reconstruction filter needs to pass the baseband of $X_s(\Omega)$ and extinguish the image replicas. So it has to have a gain of T for frequencies up to the highest baseband frequency, $|\Omega| < \Omega_b$, and a gain as close to zero as practicable for all frequencies above the lowest image frequency, $|\Omega| > \Omega_s - \Omega_b$. The complexity of the design of the analog reconstruction filter that lives in the D/A converter – that is, the order of the filter – depends in large measure on the sharpness of the filter's cutoff or, equivalently, the width of the transition zone of the filter between Ω_b and $\Omega_s - \Omega_b$. The further apart the baseband and image replicas are, the lower the slope of the filter that is required and the easier it is to design and implement. For instance, the middle column of **Figure 6.10** shows a non-ideal reconstruction filter whose response has a linear slope between Ω_b and $\Omega_s - \Omega_b$. As the sampling frequency increases, the image replicas move further away from the baseband and the slope of the filter can decrease, making it less and less complex to design. So, from the point of view of the design of the reconstruction filter, the higher the sampling frequency, the better. Of course, from the point of view of data storage and processing, the lower the sampling frequency the better, since a lower sampling frequency means fewer data need to be stored and/or processed per second. We will soon see how clever engineers have managed to reconcile with these opposing goals.

Undersampling Sampling at frequencies below the Nyquist rate ($\Omega_s < 2\Omega_b$) is termed **undersampling**. **Figure 6.15** shows an example of what happens when we lower the sample rate to $\Omega_s = 1.5\Omega_b$, which is only 75% of the Nyquist rate. Recall that $X_s(\Omega)$ is formed from the sum of shifted replicas of $X(\Omega)$:

$$X_s(\Omega) = \frac{1}{T}\sum_{k=-\infty}^{\infty} X(\Omega - k\Omega_s).$$

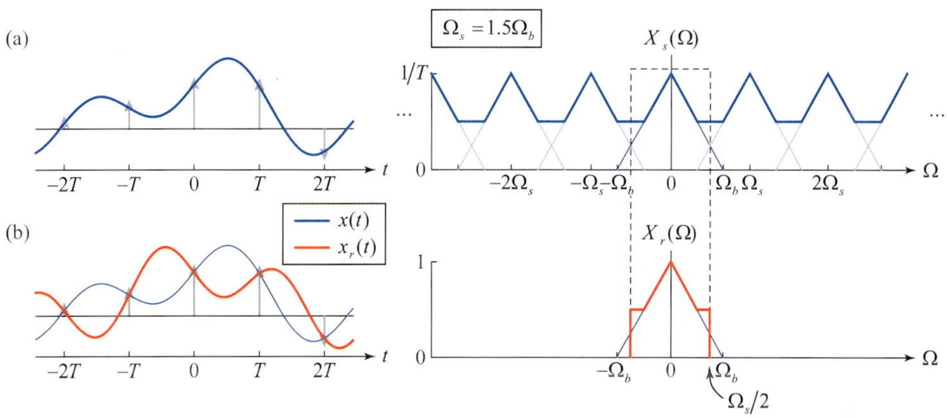

Figure 6.15 Undersampling: reconstruction of $x(t)$ from $x_s(t)$ with $\Omega_s = 1.5\Omega_v$

As long as Ω_s is high enough (i.e., $\Omega_s > 2\Omega_b$), the replicas of $X(\Omega)$ are non-overlapping. However, when $X(\Omega)$ is undersampled ($\Omega_s < 2\Omega_b$), $X_s(\Omega)$ (right panel, blue trace) is the sum of *overlapping* replicas of the spectrum of the input signal (grey traces). There exists no lowpass filter, ideal or otherwise, that could be designed such that we can recover this $X(\Omega)$ from $X_s(\Omega)$. If we choose an ideal lowpass reconstruction filter with bandwidth $\Omega_s/2$, the resulting spectrum of $X_r(\Omega)$, shown in red on the right panel of **Figure 6.15b**, is clearly not equivalent to the spectrum of the original input signal $X(\Omega)$, shown with a thin line blue. The left panel of **Figure 6.15b** shows the time-domain waveforms of the reconstructed signal $x_r(t)$ in red, and the original input signal $x(t)$ in blue. They are not the same either. Actually, Equation (6.5) guarantees that $x(t)$ and $x_r(t)$ *will* be equal for times that are integer multiples of the sampling period (as you can see in **Figure 6.15b**), but they are clearly completely different at other times. This phenomenon – in which the reconstructed signal is distorted as a consequence of undersampling – is termed **aliasing**, and the output is said to be **aliased**.[4] When a signal is

[4]In colloquial speech, the word alias (from the Latin, *alius* = "otherwise") means "an assumed name or identity." In the context of sampling, the word alias is very appropriate.

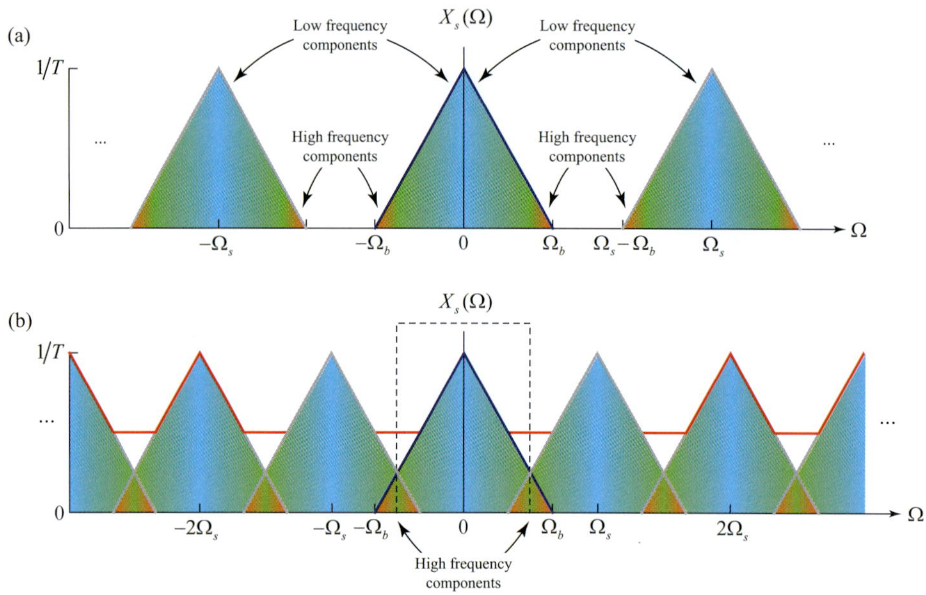

Figure 6.16 Low and high frequencies of $X_s(\Omega)$

undersampled, high-frequency components of images of $X(\Omega)$ are added into the baseband spectrum, becoming indistinguishable from lower-frequency components.

That last statement might need a bit of clarification. To understand aliasing better, let us remind ourselves what "low-frequency" and "high-frequency" components mean in the context of sampling. **Figure 6.16a** shows the spectrum $X_s(\Omega)$ of a signal that has been sampled above the Nyquist rate. It comprises distinct replicas of $X(\Omega)$ at multiples of the sampling frequency, $\Omega = k\Omega_s$. For each replica, the low-frequency components of the input signal are centered at the middle of the replica, the area denoted in blue around multiples of the sampling frequency, $\Omega = k\Omega_s$. The high-frequency components are located at the edges of each replica at frequencies near $|\Omega - k\Omega_s| = \Omega_b$, denoted in red. This means that, for example, the portion of the spectrum in the left edge of the replica at frequency near $\Omega = \Omega_s - \Omega_b$ is considered high-frequency.

Figure 6.16b shows the spectrum of the same signal which has been undersampled. High-frequency components of image replicas of $X(\Omega)$ adjacent to the baseband replica have now been added to the baseband replica. When filtered (for example, by an ideal lowpass filter with bandwidth $\Omega_s/2$), these high-frequency components add energy to the resulting spectrum of the input signal that is indistinguishable from the energy of low-frequency components in the original spectrum.

Aliasing is sometimes called **frequency folding** because you can visualize the frequency overlap due to aliasing by taking all the frequency components in $X(\Omega)$ for $|\Omega| > \Omega_s/2$, "folding" them towards $\Omega = 0$ around lines drawn at $|\Omega| = \Omega_s/2$, and adding them to the lower frequency components, as shown in **Figure 6.17**.

It is important to understand that aliasing is an *unrecoverable* artifact of undersampling. In general, a signal that has been undersampled cannot be uniquely recovered from its samples by

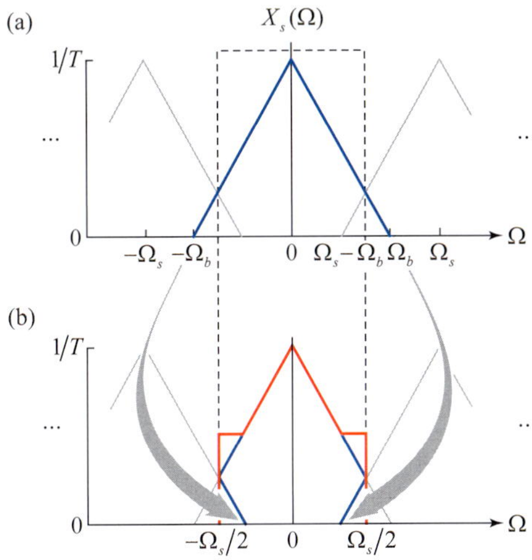

Figure 6.17 Aliasing and frequency folding

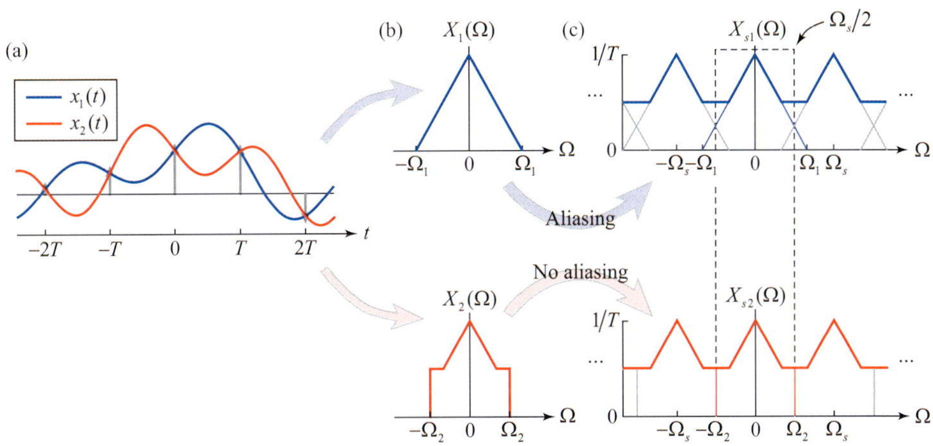

Figure 6.18 Two signals with equal samples

any signal processing magic. As a simple example of this, **Figure 6.18a** shows portions of two input signals, $x_1(t)$ (blue trace) and $x_2(t)$ (red trace). These signals are obviously different, but when sampled at frequency $\Omega_s = 2\pi/T$ (grey impulses at multiples of period T), they have identical values at the sample instants $t = nT$. So, how can unique signals be reconstructed from the samples? Simple answer: they can't. To explain things a bit further, **Figure 6.18b** shows the transform of these two signals, $X_1(\Omega)$ (upper panel, blue trace) and $X_2(\Omega)$ (lower panel, red trace). $X_1(\Omega)$ has a bandwidth greater than half the Nyquist rate, $\Omega_1 > \Omega_s/2$, so when it is undersampled at Ω_s, the sampled signal $X_{s1}(\Omega)$ shows aliasing (**Figure 6.18c**, solid blue trace). In contrast, $X_2(\Omega)$ has a bandwidth of exactly $\Omega_1 = \Omega_s/2$, so, when it is critically sampled at Ω_s, the sampled signal $X_{s2}(\Omega)$ shows no aliasing (**Figure 6.18c**, solid red trace). **Figure 6.18d** shows the transforms $X_{r1}(\Omega)$ and $X_{r2}(\Omega)$ of signals reconstructed by filtering $X_{s1}(\Omega)$ and $X_{s2}(\Omega)$ at half the sampling rate, $\Omega_s/2$, with an ideal lowpass filter. The resulting transforms are identical, $X_{r1}(\Omega) = X_{r2}(\Omega)$, so it would be impossible to tell whether the analog signal reconstructed from these transforms was $x_1(t)$ or $x_2(t)$. An extension of this result is that there is an *infinite* number of input signals of bandwidth $\Omega_1 > \Omega_s/2$ that when undersampled at rate Ω_s and reconstructed by lowpass filtering yield an output signal that is indistinguishable from a signal properly sampled at or above the Nyquist rate. For example, the red trace in **Figure 6.19** is $x_2(t)$, the same signal as in **Figure 6.18a**. It has a bandwidth of $\Omega_s/2$ and is being critically sampled at period T (grey impulses), corresponding to the Nyquist rate $\Omega_s = 2\pi/T$. The remaining traces are different signals with bandwidth greater than $\Omega_s/2$. But, when these signals are sampled at period T, they have identical values at the sample instants. So, a unique signal cannot be reconstructed from the samples of these signals.

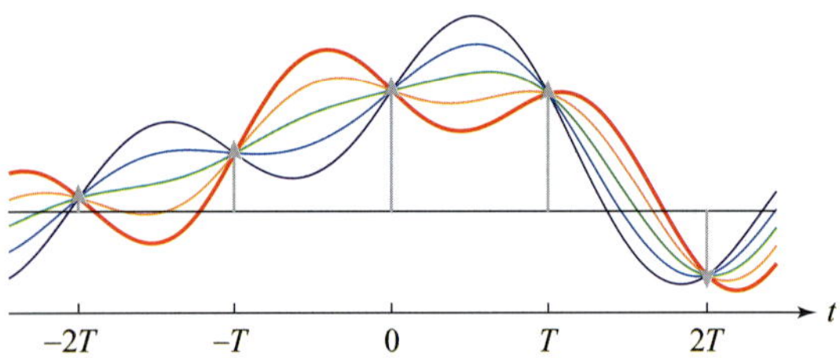

Figure 6.19 Signals with equivalent samples

6.2.5 ★ Sampling a cosine

To further our understanding of the concepts of oversampling, undersampling and critical sampling, let us now consider some simple examples in which we sample and reconstruct a cosine at different sample rates. Our input signal in all cases will be a 1-Hz cosine, $x(t) = \cos(2\pi t)$, shown plotted in blue in the left panel of **Figure 6.20a**.

Oversampling We first sample this signal at a rate of 4 Hz, which is twice the Nyquist frequency, to form $x_s(t)$, shown by the black impulses in the left panel. Since the frequency of the cosine is 1 Hz, there are four samples per cycle. Recall that the spectrum of a continuous-time cosine of frequency 2π is a pair of impulses in the frequency domain of area π located at $\Omega = \pm 2\pi$: $X(\Omega) = \pi(\delta(\Omega + 2\pi) + \delta(\Omega - 2\pi))$. The center panel shows the spectrum of the sampled signal, $X_s(\Omega)$. This spectrum comprises the sum of replicas of $X(\Omega)$ at multiples of the sampling frequency, in this case $k\Omega_s = k(2\pi \cdot 4) = 8\pi k$. In order to clarify this picture and those that follow, the baseband replica (corresponding to $k=0$) centered about $\Omega = 0$ is plotted in blue, and its two impulses at $\Omega = \pm 2\pi$ have been circled in blue to indicate that they belong together. The image replica corresponding to $k=1$ is centered at 8π, with impulses at $\Omega = 8\pi \pm 2\pi$. This replica is plotted in red and circled. Same story with the replica at $k = -1$. Recall from Equation (6.2) that all replicas are scaled by the sampling frequency, in this case by $1/T = 4$, so each impulse has an area of 4π.

To recover the signal, $x_s(t)$ is passed through the ideal lowpass reconstruction filter with a bandwidth of $\Omega_s/2 = 4\pi$ and a gain of $T = 1/4$, denoted in the center plot by a yellow region. In the frequency domain, the effect of this filter is to pass only the "blue" impulses at $\Omega = \pm 2\pi$, which correspond to the $k=0$ replica. The spectrum of the output $X_r(\Omega)$ is shown in the right panel of the figure. The spectrum of the reconstructed signal is equal to the spectrum of the input signal, and therefore $x_r(t)$ is equal to $x(t)$. That is exactly as we expect: a signal can be recovered exactly from its samples if it has been sampled above the Nyquist rate.

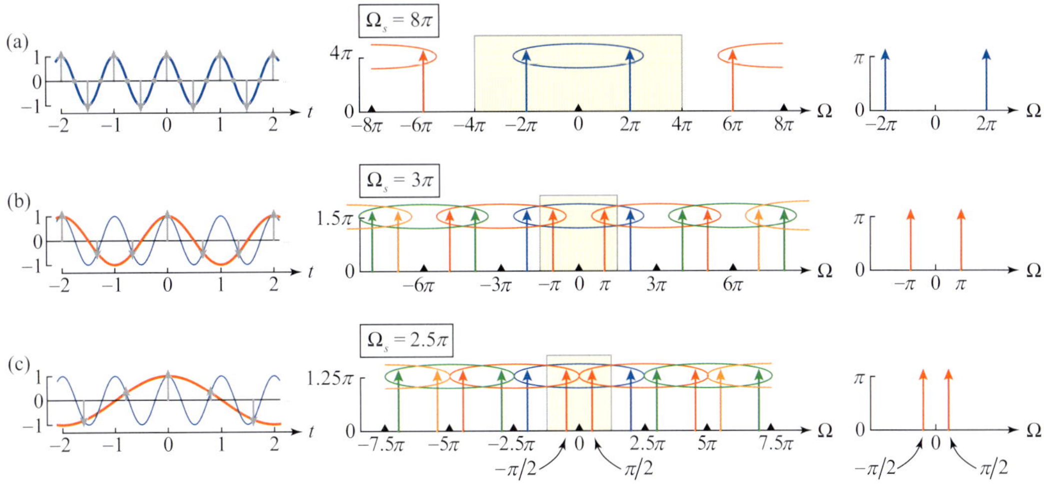

Figure 6.20 Sampling a cosine at 4, 1.5 and 1.25 Hz

Undersampling **Figure 6.20b** shows what happens when we sample the 1-Hz cosine at a frequency of 1.5 Hz, which is only 75% of the Nyquist rate. There are only three samples for every two cycles of the waveform. We expect to see aliasing! The thin blue trace in the right panel is the input signal again, $x(t) = \cos 2\pi t$. The spectrum of the sampled signal $X_s(\Omega)$ (center panel) again comprises the sum of replicas of $X(\Omega)$ at multiples of the sampling frequency, in this case, $\Omega = k\Omega_s = k(2\pi \cdot 1.5) = 3\pi k$. The baseband ($k = 0$) replica, circled in blue, consists of two impulses located at $\Omega = \pm 2\pi$. The $k = 1$ replica (circled in red) is centered at 3π, so its impulses are located at $\Omega = 3\pi \pm 2\pi = \pi, 5\pi$. The $k = -1$ replica is centered at -3π, so its impulses are at $\Omega = -3\pi \pm 2\pi = -\pi, -5\pi$. Since the sampling frequency is below the Nyquist rate, the replicas clearly overlap. Now, when the sampled signal $x_s(t)$ is passed through the reconstruction filter with a bandwidth of half the sampling frequency, $\Omega_s/2 = 3\pi$, *neither* of the impulses that constitute the baseband replica (at $\Omega = \pm 2\pi$) lie within the passband of the filter. However, the impulse at $\Omega = 3\pi - 2\pi = \pi$ from the $k = 1$ replica and the impulse at $\Omega = -3\pi + 2\pi = -\pi$ from the $k = -1$ replica *are* within the passband of the filter. Thus the spectrum of the output of the filter comprises a pair of impulses at $\Omega = \pm\pi$, shown in red in the right panel. This corresponds to a 0.5-Hz cosine, as shown in red in the left panel. So, when the input to the A/D is a 1-Hz cosine, the output of the D/A is a 0.5-Hz cosine. This example points out that aliasing is a *nonlinear* signal distortion, since it is capable of producing frequencies in the output that were not in the input. That is a big problem.

Figure 6.20c shows what happens when we sample the 1-Hz cosine at a frequency of 1.25 Hz, which is less than 63% of the Nyquist rate. Again in this case, when the sampled signal $x_s(t)$ is passed through the reconstruction filter with a bandwidth of half the sampling frequency, $\Omega_s/2 = 2.5\pi$, neither of the impulses that constitute the baseband replica lie within the passband of the filter. However, the impulse at $\Omega = 2.5\pi - 2\pi = 0.5\pi$ from the $k = 1$ replica and the impulse at $\Omega = -2.5\pi + 2\pi = -0.5\pi$ from $k = -1$ replica are within the passband of the filter. Thus the spectrum of the output of the filter comprises a pair of impulses at $\Omega = \pm 0.5\pi$, as shown in red in the right panel. This corresponds to a 0.25-Hz cosine, as shown in red in the left panel. Can you predict what would happen if we sampled this 1-Hz cosine at a frequency of exactly the Nyquist rate, 2 Hz? Let us find out.

Critical sampling In the discussion of critical sampling in conjunction with **Figure 6.13**, we showed that, at least theoretically, a signal could be reconstructed from its samples if the sampling frequency Ω_s is exactly at the Nyquist frequency, as long as we had an ideal lowpass filter. However, we will now show that even if we have an ideal lowpass filter, we cannot reconstruct the signal from its samples uniquely if there is energy in the input signal at the Nyquist frequency. To do so, we will consider the case of a cosine sampled at exactly the Nyquist frequency.

Figure 6.21 shows what happens if we sample a 1-Hz cosine with different phases at exactly the Nyquist frequency, namely 2 Hz. **Figure 6.21a** shows the result of sampling a cosine with zero phase, that is, $x(t) = \cos 2\pi t$ (blue trace in left panel). The spectrum of the sampled signal (middle panel) comprises replicas of the spectrum of the cosine at multiples of the sampling frequency, $\Omega_s = 4\pi$, scaled by two. Since the sampling frequency is exactly twice the frequency of the input, the impulses of the replicas overlap completely and their areas add together: $2\pi + 2\pi = 4\pi$. When the sampled signal is filtered by an ideal lowpass reconstruction filter with a bandwidth of 2π, the area of the impulses at $\Omega = \pm 2\pi$ is exactly cut in half. Furthermore, the gain of the filter is 0.5; hence, in the frequency domain the output of the filter is a pair of impulses at $\Omega = \pm 2\pi$, each with area π. This corresponds to the output time signal $\cos 2\pi t$(red trace), which is the same as the input. Looking at this example, you would be justified in concluding that a signal *can* be reconstructed from its samples if the sampling frequency Ω_s is exactly equal to the Nyquist frequency, even if the signal has energy at frequency $\Omega_s/2$. Not so fast!

Now look at what happens if we sample a cosine of the same frequency, but phase-shifted by a quarter-cycle: $x(t) = \cos(2\pi t - \pi/4)$ (left panel of **Figure 6.21b**, thin blue trace). When this cosine is sampled, this pair of impulses is replicated at multiples of the sampling frequency. The impulses add, but the resulting area is less than 4π. How come?

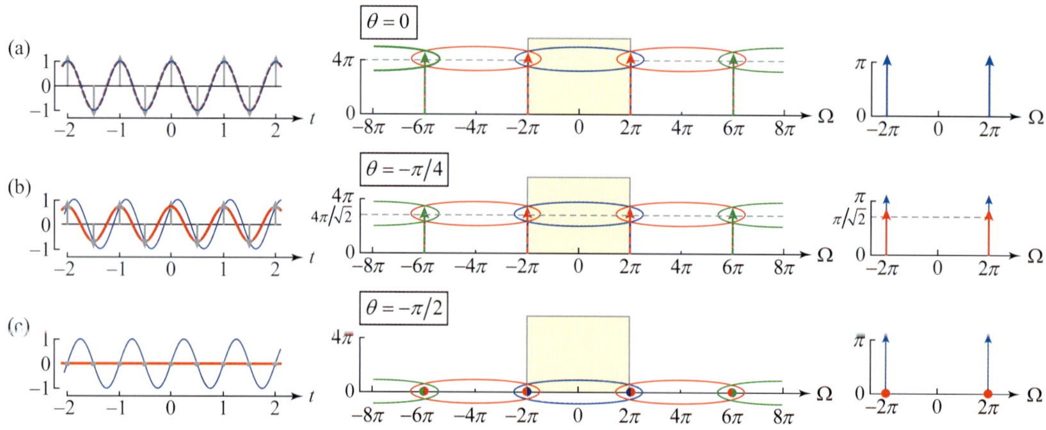

Figure 6.21 Sampling a 1-Hz cosine with different phases at the Nyquist frequency

Figure 6.22 shows in more detail what happens when overlapping impulses are added together. The transform of a cosine with a phase of $\pi/4$ is $X(\Omega) = \pi e^{j\pi/4}\delta(\Omega + 2\pi) + \pi e^{-j\pi/4}\delta(\Omega - 2\pi)$. When the cosine is sampled at a frequency of 2 Hz, each replica is scaled by two. **Figure 6.22a**

shows the baseband ($k = 0$) replica of $X_s(\Omega)$ in blue; **Figure 6.22b** shows the adjacent ($k = \pm 1$) image replicas in red. Concentrating our attention on just the frequency $\Omega = 2\pi$, the sum of the overlapping impulses from the $k = 0$ ("blue") and $k = 1$ ("red") replicas is $2\pi e^{-j\pi/4} + 2\pi e^{j\pi/4} = 4\pi \cos \pi/4 = 4\pi/\sqrt{2}$. The same calculation applies to all other impulses, yielding the picture of $X_s(\Omega)$ shown in **Figure 6.22c**, which is the same as the middle panel of **Figure 6.21b**. When $X_s(\Omega)$ is filtered with the ideal lowpass filter with bandwidth 2π, the result is two impulses in the frequency domain with area $\pi/\sqrt{2} \simeq 0.707\pi$, which corresponds to the time waveform $0.707 \cos 2\pi t$ shown in the left panel (red trace) of **Figure 6.21b**. Neither the magnitude nor the phase of the output is the same as the input, so sampling at exactly the Nyquist rate *is not*, in general, acceptable. Another way to look at this is to note that when $\cos(2\pi t - \pi/4)$ and $0.707 \cos 2\pi t$ are sampled at 1 Hz, the sampled signals are identical, since these two signals are indistinguishable when sampled.

Figure 6.21c shows the ultimate example of aliasing. It shows what happens if we sample a 1-Hz sine at exactly the Nyquist rate. Here, you can see that the samples of the input occur at the zero-crossing of the waveform. The output of the reconstruction filter is zero for all time; hence, it is not possible to tell from the output of the filter whether the input was $x(t) = \cos(2\pi t - \pi/2) = \sin 2\pi t$ or $x(t) = 0$. In general, when a 1-Hz cosine with phase ϕ, $x(t) = \cos(2\pi t - \phi)$, is sampled at exactly the Nyquist rate, the signal reconstructed by an ideal lowpass filter with a 1-Hz bandwidth is $\cos\phi \cos 2\pi t$. Thus, when we sample a cosine at exactly the Nyquist rate, the magnitude and phase of the input signal are altered and the signal cannot be reconstructed unambiguously. The general conclusion of this exercise is that signals need to be sampled at a rate *greater than* the Nyquist frequency, not *greater than or equal to* the Nyquist frequency. As we mentioned above, in practice, sample rates must generally be at least 10% greater than the Nyquist rate.

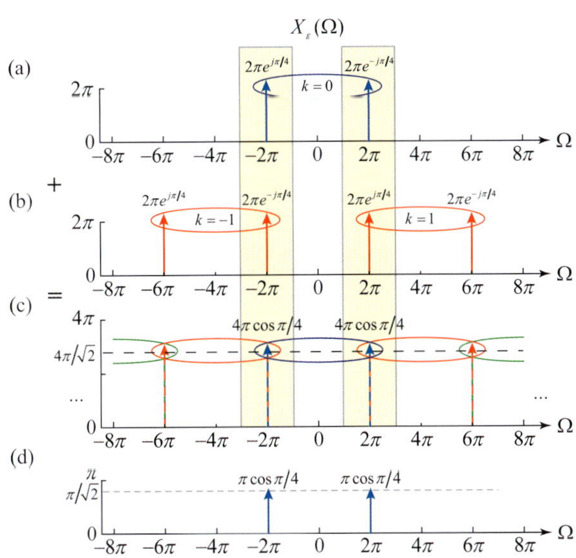

Figure 6.22 Sampling $\cos(2\pi t - \pi/4)$

6.3 Conversion from continuous time to discrete time and back

In the preceding sections, we have described the theoretical basis of analog sampling and reconstruction, as schematized in **Figure 6.3**. The process we have described to this point has been purely analog: all the signals (i.e., $x(t)$, $x_s(t)$, $x_r(t)$ and $y(t)$) have been continuous-time signals. We will now extend these results to complete our understanding of the theory of the analog-to-digital and digital-to-analog conversion process, as schematized in the block diagram of **Figure 6.2**.

6.3.1 The continuous-to-discrete (C/D) converter

The analog sampling stage is the first part of the complete ideal A/D converter that we have discussed above. Here, an analog input signal $x(t)$ modulates an analog impulse train $s(t)$ to yield an analog sampled signal $x_s(t)$. Each impulse in $x_s(t)$ represents the value of $x(t)$ at a time that is an integer multiple of the sampling period, $t=nT$. The second part of the ideal A/D converter is the "continuous-to-discrete" (C/D) converter. This conceptual device produces a discrete-time sequence $x[n]$ whose nth value is the area of $x_s(t)$ at $t=nT$. That is, $x[0]=x(0)$, $x[1]=x(T)$, $x[2]=x(2T)$, …, so,

$$x[n]=x(nT). \tag{6.7}$$

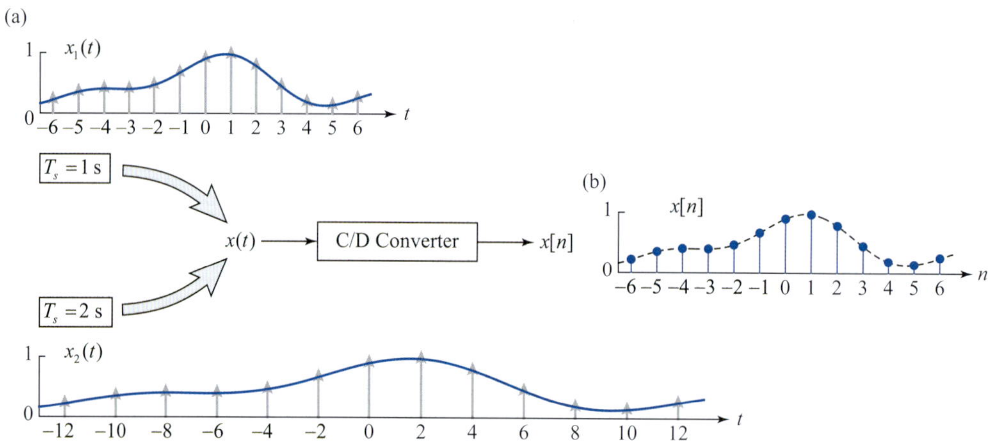

Figure 6.23 Time normalization in the C/D converter

This simple relation describes the C/D converter in the time domain. Hence, $x[n]$ can be viewed as equivalent to a time-normalized version of $x(t)$. In going from the analog signal $x(t)$ to the discrete-time sequence $x[n]$, the value of the period T is effectively normalized out, and in a sense lost. The only thing the index of the sequence, n, indicates is the order of each sample in the sequence. The index gives no indication of the absolute time at which the sample occurred, nor the time that has elapsed between samples (i.e., T). This is illustrated in **Figure 6.23**.

Figure 6.23a shows two different analog signals, $x_1(t)$ and $x_2(t)=x_1(t/2)$, that have been sampled at periods of $T=1$ and $T=2$ s, respectively. While these two signals are clearly different, the resulting discrete-time sequence $x[n]$ shown in **Figure 6.23b** is the same.

Accordingly, without some information about the value of T, it is not possible to know which of the two signals, $x_1(t)$ or $x_2(t)$, were input to the A/D converter.

6.3.2 Spectrum of the discrete-time sequence

The Fourier transform of the sampled analog signal $x_s(t)$ is

$$X_s(\Omega) = \mathfrak{F}\{x_s(t)\} = \int_{t=-\infty}^{\infty} x_s(t)e^{-j\Omega t}dt = \int_{t=-\infty}^{\infty} \left(\sum_{n=-\infty}^{\infty} x(nT)\delta(t-nT)\right)e^{-j\Omega t}dt$$

$$= \sum_{n=-\infty}^{\infty} x(nT)\left(\int_{t=-\infty}^{\infty} \delta(t-nT)e^{-j\Omega t}dt\right) = \sum_{n=-\infty}^{\infty} x(nT)e^{-j\Omega Tn}.$$

From Equation (6.7), we recognize that $x[n] = x(nT)$. Hence,

$$X_s(\Omega) = \sum_{n=-\infty}^{\infty} x[n]e^{-j\Omega Tn}. \tag{6.8}$$

But the DTFT of $x[n]$, the discrete-time sequence that is the output of the C/D converter, is

$$X(\omega) = \mathfrak{F}\{x[n]\} = \sum_{n=-\infty}^{\infty} x[n]e^{-j\omega n}. \tag{6.9}$$

Comparing Equations (6.8) and (6.9), we see that the two expressions are identical with the substitution $\omega = \Omega T$. Thus, $X(\omega)$ is simply $X_s(\Omega)$ evaluated at $\Omega = \omega/T$:

$$X(\omega) = X_s(\Omega)|_{\Omega=\omega/T}.$$

Just as $x[n]$ can be viewed as equivalent to $x_s(t)$ with the time axis normalized by the sample period T, so $X(\omega)$ can be viewed as equivalent to $X_s(\Omega)$ with frequency normalized by the sampling frequency Ω_s. Since the sampling rate of the analog signal is $\Omega_s = 2\pi/T$, we can write the frequency normalization factor as

$$\omega = \Omega T = \Omega(2\pi/\Omega_s) = 2\pi(\Omega/\Omega_s).$$

In words, the discrete-time frequency ω is equal to the continuous-time frequency Ω normalized by the sample rate Ω_s and multiplied by 2π.

Figure 6.24 shows an example of the C/D conversion process for the example of **Figure 6.5**: a signal with a maximum bandwidth of 5 Hz (i.e., $\Omega_b = 2\pi \cdot 5$) sampled at a rate of 20 Hz ($\Omega_s = 2\pi \cdot 20$).

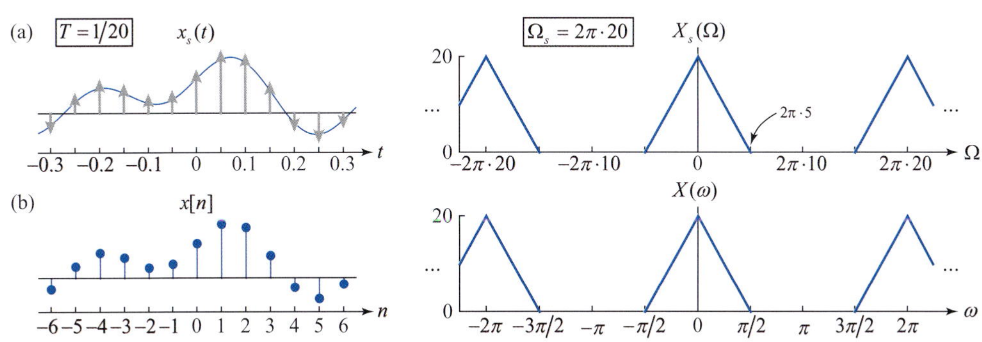

Figure 6.24 Analog-to-digital conversion, $\Omega_s = 2\pi \cdot 20$

Figure 6.24a shows the sampled analog signal $x_s(t)$ (left panel), and its spectrum $X_s(\Omega)$ (right panel). Since this signal is sampled above the Nyquist rate, the image replicas are well separated from the baseband replica. **Figure 6.24b** shows the corresponding sequence $x[n]$ (left panel), and its spectrum $X(\omega)$ (right panel). The relation between $X_s(\Omega)$ and $X(\omega)$ is simply a frequency scaling, $\omega = 2\pi(\Omega/\Omega_s)$. The sampling frequency in this example, $\Omega_s = 2\pi \cdot 20$, maps to $\omega = 2\pi(\Omega_s/\Omega_s) = 2\pi$, and the highest frequency in the baseband, $\Omega_b = 2\pi \cdot 5$, maps to $2\pi(\Omega_b/\Omega_s) = 2\pi(5/20) = \pi/2$.

Because $X_s(\Omega)$ is periodic with period Ω_s, $X(\omega)$ is periodic with period $\omega = 2\pi(\Omega_s/\Omega_s) = 2\pi$. Of course, we already knew that because $X(\omega)$ is the discrete Fourier transform of a discrete-time sequence, which is always periodic with period 2π. Because $X_s(\Omega)$ is periodic, it is only unique in the range $-\Omega_s/2 \le \Omega < \Omega_s/2$ (or an equivalent range spanning Ω_s in frequency). Similarly, $X(\omega)$ is unique in the range $-\pi \le \Omega < \pi$ (or an equivalent range spanning 2π in frequency). The amplitude scale of $X(\omega)$ is identical to the amplitude scale of $X_s(\Omega)$.

Figure 6.25 shows an example of the relation between $X_s(\Omega)$ and $X(\omega)$ for a critically sampled signal. Again, we choose a signal with maximum bandwidth 5 Hz (i.e., $\Omega_b = 2\pi \cdot 5$), which we sample at exactly the Nyquist frequency, namely 10 Hz ($\Omega_s = 2\pi \cdot 10$). Since $\Omega_s = 2\Omega_b$, the replicas of $X_s(\Omega)$ are just touching, as shown in the right panel of **Figure 6.25a**. Comparing $X(\Omega)$ with $X(\omega)$, we note that the sampling frequency, now $\Omega_s = 2\pi \cdot 10$, still maps to $\omega = 2\pi(\Omega_s/\Omega_s) = 2\pi$, but the highest frequency in the baseband, $\Omega_b = 2\pi \cdot 5$, now maps to $2\pi(\Omega_b/\Omega_s) = 2\pi(5/10) = \pi$. Note also that the amplitude of both $X_s(\Omega)$ and $X(\omega)$ is exactly equal to the sampling frequency in Hz, namely 10 Hz.

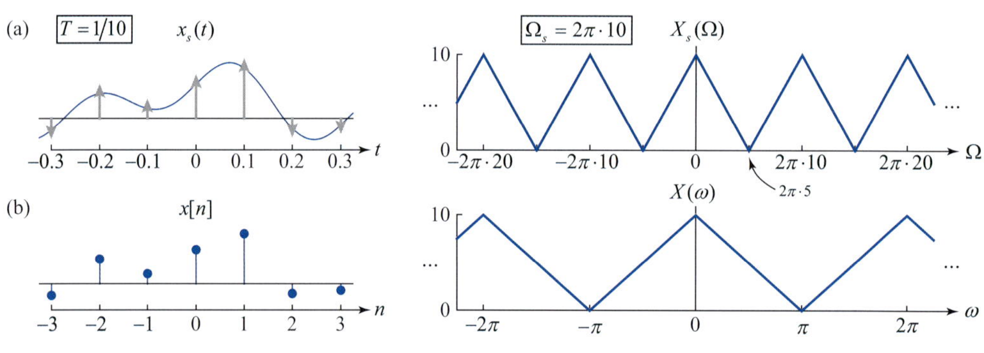

Figure 6.25 Analog-to-digital conversion, $\Omega_s = 2\pi \cdot 10$

Figure 6.26 shows the relation between $X_s(\Omega)$ and $X(\omega)$ for an undersampled signal. If we sample a signal with bandwidth 5 Hz (i.e., $\Omega_b = 2\pi \cdot 5$) at a sampling frequency of 7.5 Hz ($\Omega_s = 2\pi \cdot 7.5$), the sampling frequency is only 1.5 times the highest frequency in the signal ($\Omega_s = 1.5\Omega_b$), so the replicas of $X_s(\Omega)$ overlap. Looking at $X(\omega)$, we see that the sampling frequency again maps to 2π, and the highest frequency in the baseband, $\Omega_b = 2\pi \cdot 5$, now maps to $2\pi(\Omega_b/\Omega_s) = 2\pi(5/7.5) = 4\pi/3$. Any frequency in $X_s(\Omega)$ that is greater than half the Nyquist frequency corresponds to frequencies in the baseband of $X(\omega)$ above π, and will cause aliasing. The amplitude of both $X_s(\Omega)$ and $X(\omega)$ is again equal to the sampling frequency in Hz, namely 7.5 Hz.

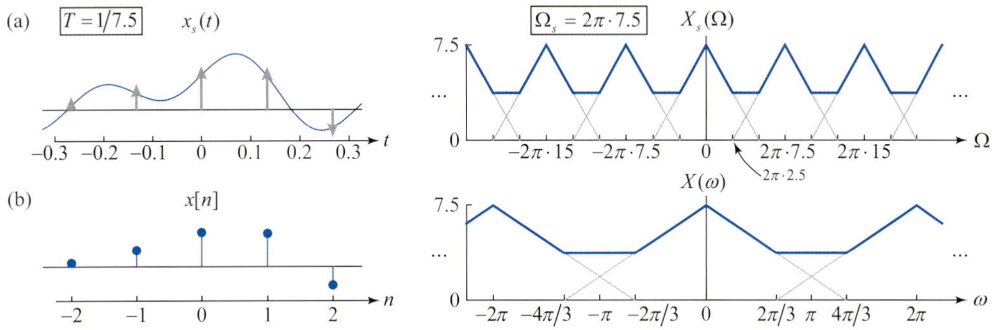

Figure 6.26 Analog-to-digital conversion, $\Omega_s = 2\pi \cdot 7.5$

6.3.3 The discrete-to-continuous (D/C) converter

The discrete-to-continuous (D/C) converter is the first conceptual element of the D/A shown in **Figure 6.2**. The D/C converter takes a sequence $y[n]$ and converts it into a continuous-time impulse-train $y_s(t)$, with impulses spaced at multiples of the reconstruction period T_r and scaled by the values of the sequence $y[n]$. That is,

$$y_s(t) = \sum_{n=-\infty}^{\infty} y[n]\delta(t - nT_r). \tag{6.10}$$

Each continuous-time impulse in $y_s(t)$ for $t = nT_r$ has an area equal to the value of the discrete-time impulse in $y[n]$. $y[n]$ is just a list of ordered values without any indication of the absolute time at which the sample occurred. Hence, in going from the discrete-time impulse train to the continuous-time impulse train, we must provide a value of the reconstruction period T_r. Different values of T_r will produce completely different $y_s(t)$ from the same $y[n]$. This is illustrated in **Figure 6.27**.

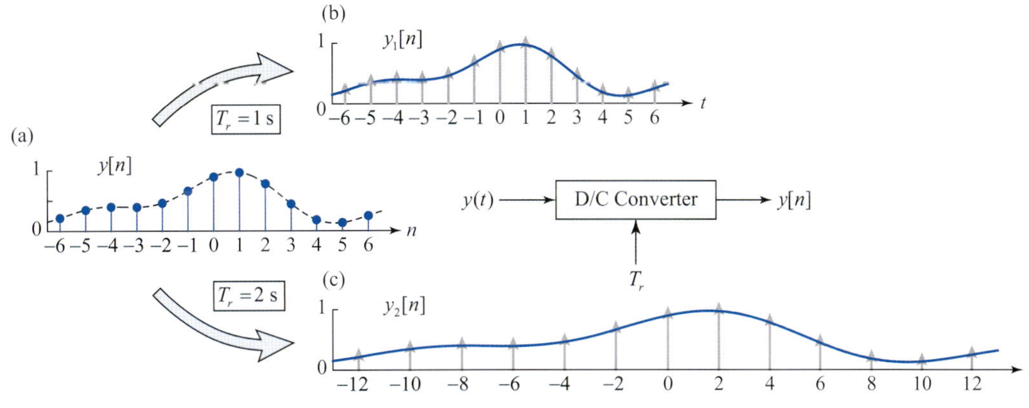

Figure 6.27 Discrete-to-continuous (D/C) converter

Figure 6.27b is a continuous-time impulse train produced by the D/C converter from the discrete-time sequence $y[n]$ shown in **Figure 6.27a** at a value of $T_r = 1$ s. **Figure 6.27c** is a continuous-time impulse train produced by the D/C converter from the same sequence at a value of $T_r = 2$ s.

The spectrum of the continuous-time impulse train $Y_s(\Omega)$ is equivalent to the spectrum of the sequence $y[n]$ with the frequency axis scaled by $1/T_r$. From Equation (6.10),

$$Y_s(\Omega) = \mathfrak{F}\{y_s(t)\} = \int_{t=-\infty}^{\infty} \left(\sum_{n=-\infty}^{\infty} y[n]\delta(t - nT_r) \right) e^{-j\Omega t} dt = \sum_{n=-\infty}^{\infty} y[n] \int_{t=-\infty}^{\infty} \delta(t - nT_r) e^{-j\Omega t} dt$$
$$= \sum_{n=-\infty}^{\infty} y[n] e^{-j(\Omega T_r)n} = Y(\omega)\big|_{\omega=\Omega T_r}.$$

If we define the reconstruction frequency to be $\Omega_r = 2\pi/T_r$, we can express the frequency scaling as

$$\Omega = \omega/T_r = \Omega_r \cdot \omega/2\pi.$$

That is, the continuous-time frequency Ω is equal to the discrete-time frequency ω "de-normalized" by 2π and then multiplied by the reconstruction rate Ω_r. To summarize, $Y_s(\Omega)$ is equal to $Y(\omega)$ evaluated at $\omega = \Omega T_r$:

$$Y_s(\Omega) = Y(\omega)\big|_{\omega=\Omega T_r}.$$

6.3.4 Summary

Figure 6.28 presents a summary of the entire A/D and D/A conversion process as it applies to a simple application in which an analog signal $x(t)$ is sampled at rate Ω_s to form sequence $x[n]$, and then reconstructed at the same rate, $\Omega_r = \Omega_s$, thus yielding analog output signal $y(t)$, which is theoretically identical to the original input signal $x(t)$.

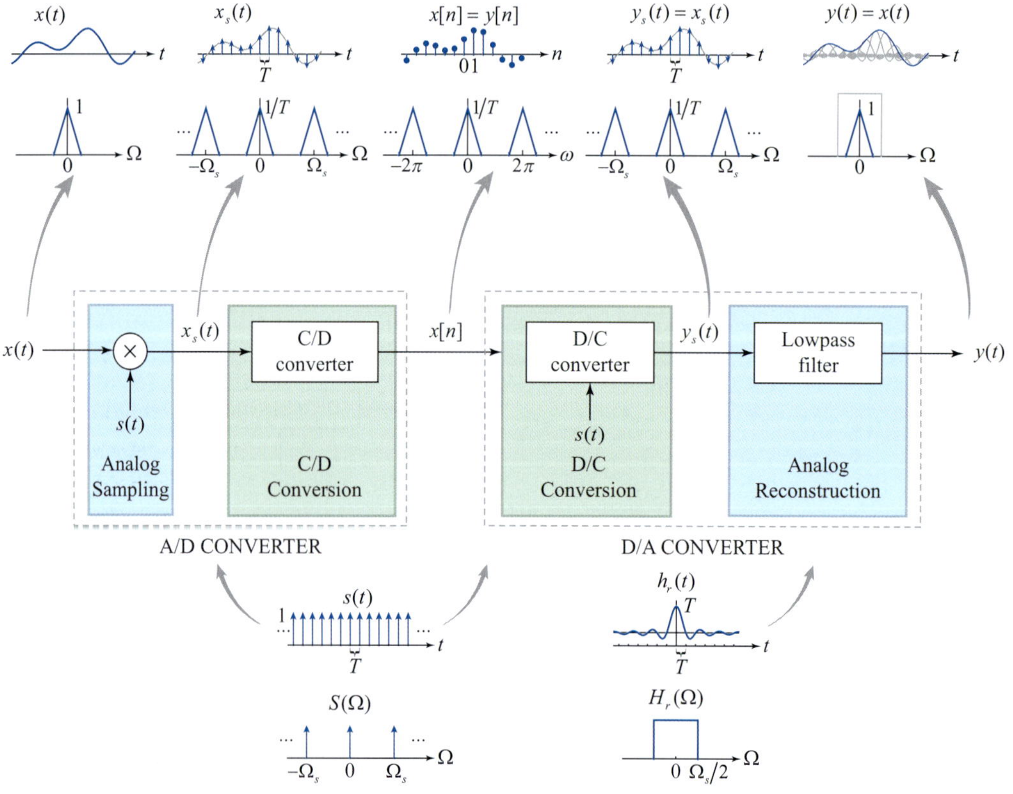

Figure 6.28 Summary of A/D and D/A conversion

6.4 Anti-aliasing and reconstruction filters

6.4.1 The anti-aliasing filter

In our derivation of the sampling theorem, we stated that a signal could be perfectly reconstructed from its samples without aliasing if the signal was of finite bandwidth Ω_b, and we sampled at a rate Ω_s above the Nyquist frequency, that is, $\Omega_s > 2\Omega_b$. That means that to prevent aliasing, we have to make sure that we sample at a rate twice the highest frequency in the input signal or, equivalently, that the maximum bandwidth of the signal we are sampling is less than one-half the sample rate. Every signal of practical interest is of finite duration, and you may remember from our discussion of the DTFT in Chapter 3 that a signal cannot be simultaneously time-limited and bandlimited, which means that, at least theoretically, every signal of finite duration has infinite bandwidth! Since we are constrained to sample at a finite rate, aliasing is guaranteed to happen to some extent in every situation of practical interest. But, the amount of aliasing can be minimized by interposing an analog **anti-aliasing filter** in the signal path between the signal source (e.g., the microphone preamplifier) and the input to the A/D converter, as shown in **Figure 6.29**.

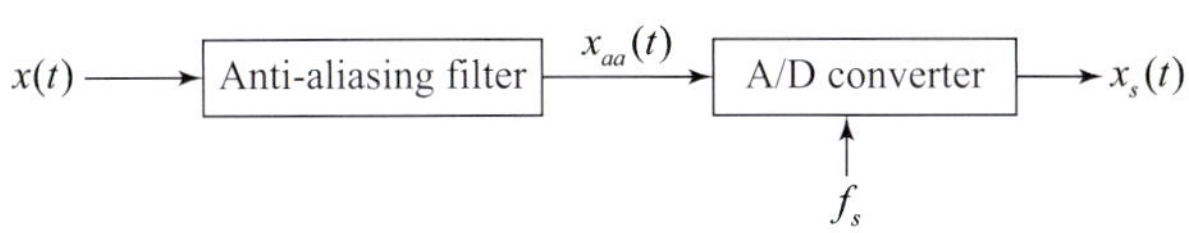

Figure 6.29 Anti-aliasing filter

The purpose of the anti-aliasing filter is to limit the effective bandwidth of the input signal to less than half the sample rate. Let us look at the function of the anti-aliasing filter, first an ideal filter and then a realizable filter that could be used in a practical application.

Ideal anti-aliasing filter Theoretically, the "perfect" anti-aliasing filter is an ideal analog lowpass filter with a bandwidth of half the sample rate. **Figure 6.30** shows the function of this ideal anti-aliasing filter. **Figure 6.30a** is essentially a summary of **Figure 6.15**. It shows what happens when a signal $x(t)$ of bandwidth Ω_b is sampled below the Nyquist rate, $\Omega_s < 2\Omega_b$, without an anti-aliasing filter. The thin blue traces in the left and center panels show, respectively, $x(t)$ and its spectrum $X(\Omega)$. The right panel shows the spectrum of the sampled signal, $X_s(\Omega)$ (solid red line). Because $\Omega_s < 2\Omega_b$, $X_s(\Omega)$ (red trace) comprises the sum of overlapping replicas of $X(\Omega)$ (thin blue lines). When $X_s(\Omega)$ is reconstructed with an ideal lowpass reconstruction filter $H_r(\Omega)$ with a bandwidth of half the sampling rate, $\Omega_s/2$ (dashed black line), the resulting spectrum (solid red trace in the center panel) displays aliasing. As a consequence, the time waveform (solid red trace in the left panel) is markedly different from the original input signal.

Figure 6.30b shows what happens when an anti-aliasing filter is interposed in the signal path before sampling takes place. In order to prevent aliasing, any spectral energy in the input signal $x(t)$ above one-half the sampling frequency, $|\Omega| > \Omega_s/2$, must be removed. That is the job of the

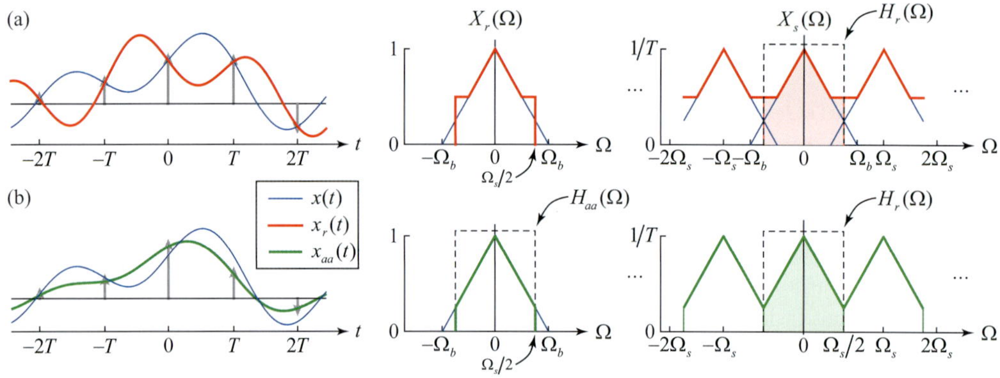

Figure 6.30 Anti-aliasing with an ideal lowpass filter

anti-aliasing filter with ideal frequency response $H_{aa}(\Omega)$ (dotted black line, center panel). The green trace in the center panel shows $X_{aa}(\Omega)$, the spectrum of the anti-aliasing-filtered input. All energy in the input for $|\Omega| > \Omega_s/2$ has been removed. The anti-aliasing-filtered time waveform $x_{aa}(t)$ shown in the left panel is a lowpass filtered version of the original input signal. When $x_{aa}(t)$ is sampled at frequency Ω_s, the replicas of $X_{aa}(\Omega)$ do not overlap (right panel). Hence, when $X_s(\Omega)$ is filtered with an ideal lowpass with a reconstruction filter $H_r(\Omega)$ with a bandwidth of half the sampling rate, $\Omega_s/2$, the result is identical to $X_{aa}(\Omega)$; there is no aliasing. Of course, the imposition of an anti-aliasing filter results in an input signal that is no longer identical to the original input, but this lowpass-filtered version is a well-characterized linear transformation of the original signal, which is generally far preferable to allowing aliasing, which is an ill-characterized, non-linear transformation of the signal.

Non-ideal anti-aliasing filter While it is not possible to create an ideal lowpass anti-aliasing filter, practical analog anti-aliasing filters are employed as the front-end of almost all sampled data systems to mitigate the effect of aliasing. We shall discuss the basic principles of analog filter design in much more detail in Chapter 8, but for the purposes of the present discussion, **Figure 6.31** shows the schematic frequency response of a non-ideal anti-aliasing filter.

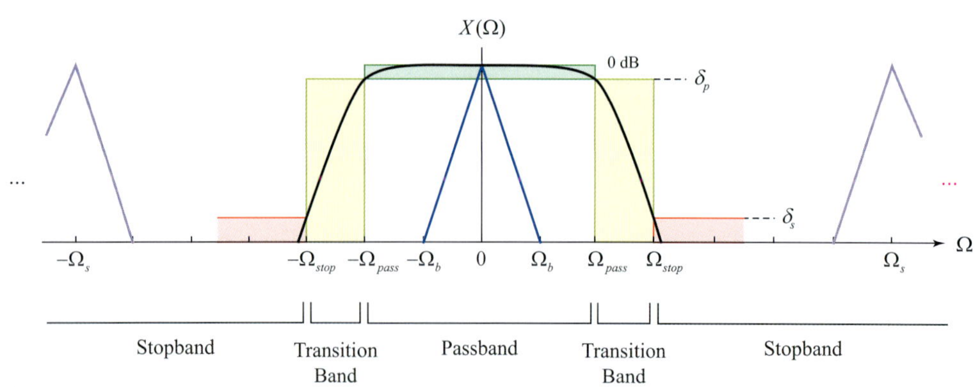

Figure 6.31 Non-ideal anti-aliasing filter

A general lowpass filter is characterized by its **passband** and **stopband**. The passband is the frequency region below the **passband frequency** Ω_{pass} in which the filter is designed to have a reasonably flat frequency response and a gain close to unity (i.e., 0 dB). The filter designer specifies the magnitude of the frequency response of the filter in the passband to be between 0 dB and a **passband attenuation** δ_p, which is usually specified in dB. Thus, for frequencies in the passband $|\Omega| < \Omega_{pass}$ the magnitude of the frequency response is $\delta_p < |H(\Omega)|_{dB} < 0$. For an ideal lowpass filter, $\delta_p = 0$.

The stopband is the frequency region above the **stopband frequency** Ω_{stop} in which the filter is designed to be highly attenuating. A practical filter cannot have a gain of exactly zero throughout the stopband because a perfectly bandlimited filter would have an unlimited impulse response in the time domain. So, in practice the magnitude of the frequency response of the filter is specified to be less than the **stopband gain** δ_s, that is, $|H(\Omega)|_{dB} < \delta_s$, $|\Omega| > \Omega_{stop}$.

The range of frequencies between the passband and stopband, $\Omega_{pass} < |\Omega| < \Omega_{stop}$, is termed the **transition band** of the filter. For an ideal lowpass filter $\Omega_{stop} = \Omega_{pass}$, so the transition band has zero width. For any practical filter, the transition band has a finite width. For an analog filter, the narrower the transition band, the sharper the filter's frequency selectivity but the more components it requires to implement. This is a key concern in implementing a practical anti-aliasing filter. We will now explore that trade-off in the context of a particular application.

6.4.2 A digital recording application

As an example of the use of the anti-aliasing filter, let us look again at the problem of recording sound for an audio application, such as a CD (remember them?). Recording a CD is essentially just an exercise in A/D conversion: an analog signal $x(t)$, for example from a microphone, is sampled to form a sequence $x[n]$, which is then encoded into a data file. The frequency range of practical interest for music recording extends to about 20 kHz, which corresponds to the maximum frequency range of human hearing. The standards selected for CD recording call for a sample frequency of 44.1 kHz, which is just greater than the Nyquist frequency (i.e., 40 kHz). So, if the recorded sound or anything in the recording signal path contributes substantial energy above half the sampling rate (i.e., above 22.05 kHz), there will be aliasing. It is the job of the anti-aliasing filter in the recorder's analog front end to substantially attenuate energy significantly above 22.05 kHz before it is sampled by the A/D converter.

Figure 6.32 shows several possible implementations of an anti-aliasing filter for a digital recording application. Creating a good analog anti-aliasing filter presents a real design challenge: we want the passband of the filter to extend to the highest frequency of interest in the signal (i.e., 20 kHz), and we also want the stopband to start at one half the sample rate (i.e., 22.05 kHz), as shown in **Figure 6.32a**. That means the width of the transition band needs to be less than one-seventh of an octave wide ($\log_2(22.05/20) = 0.14$ octave). If we were to design the filter to attenuate the signal by at least 90 dB at 22.05 kHz, it would need to have a cutoff slope of at least −639 dB/octave! In fact, a conventional analog filter designed to meet these nominal specifications ($\delta_p = -1$ dB, $\delta_s = -90$ dB) would require a filter of high order (e.g., a 27th-order Chebyshev filter). Filters of this kind are conventionally called "brick-wall" filters. They are hard to design and expensive.

The easiest way to lessen the demands on the anti-aliasing filter is to **oversample**, that is to increase Ω_s (or, equivalently, f_s) in order to sample the input signal faster than required by the Nyquist criterion. **Figure 6.32b** shows what would happen if we were to sample the signal, which still has a bandwidth of 20 kHz, at eight times the original sample rate, namely $f_s = 44.1 \times 8 = 352.8$ kHz. At this faster sample rate, the image replicas are more widely separated from the baseband. The anti-aliasing filter can take advantage of this separation and be designed with a wider transition band, so that it meets the design criteria with a lower-order filter. For example, let us design the anti-aliasing filter to attenuate the signal by 90 dB at a frequency corresponding to half the sample rate. The passband frequency is still 20 kHz, but the stopband frequency will now be 176.4 kHz. The transition band is about three octaves wide ($\log_2(176.4/20) = 3.14$ octave), and the slope of the anti-aliasing filter will have a slope of −29 dB/octave, requiring only a 5th-order Chebyshev filter, which seems much more acceptable. The real benefit of such oversampling is that it makes the design of the analog anti-aliasing filter easier. Of course, as the sample rate increases, the size of the data file produced – and thus the size of the potential CD – increases proportionally, which is not desirable. We will address that problem in Section 6.5 below.

If we knew that there was negligible energy in the input signal above 20 kHz, it would be sufficient to increase the stopband frequency of the anti-aliasing filter so that it just excludes the image replica centered at the sampling frequency. In the case of sampling frequency $f_s = 352.8$ kHz, the stopband frequency of the filter can move to $352.8 - 20 = 332.8$ kHz, as shown in **Figure 6.32c**. The transition band would now be about four octaves wide ($\log_2(332.8/20) = 4.06$ octave), and the slope of the resulting anti-aliasing filter would be −22 dB/octave, which could be implemented with a four-pole analog filter.

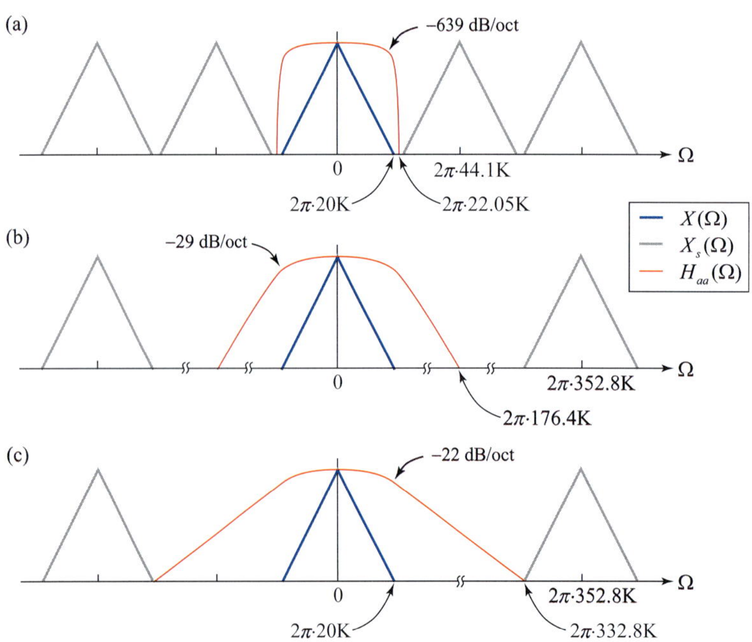

Figure 6.32 Anti-aliasing filter in an audio application

6.4.3 Reconstruction filter

The problem of designing the reconstruction filter in the D/A converter is entirely homologous to the problem of designing the anti-aliasing filter in the A/D converter. To put a practical face on the issue, consider again the problem of playing a CD in which the recorded signal with a maximum bandwidth of 20 kHz had been sampled at 44.1 kHz. The spectrum of the reconstructed signal $x_r(t)$ is shown in **Figure 6.33**.

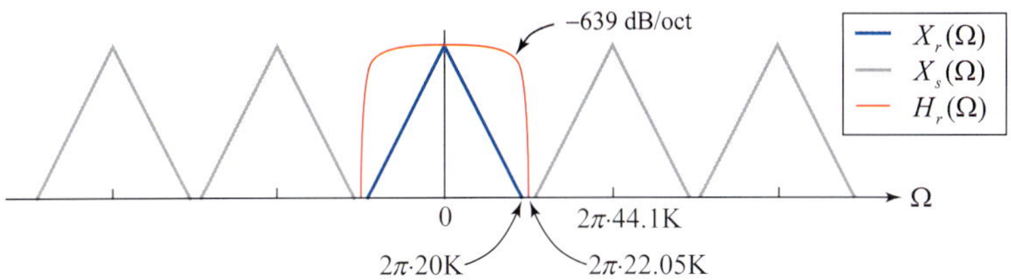

Figure 6.33 Reconstruction of a critically sampled signal

This figure looks exactly the same as **Figure 6.32a**, with $X_r(\Omega)$ and $H_r(\Omega)$ replacing $X(\Omega)$ and $H_{aa}(\Omega)$ respectively. The analog reconstruction filter at the back end of the D/A converter must separate the baseband replica from the surrounding image replicas, and therefore needs to have the same very sharp frequency-selective characteristics as the anti-aliasing filter we just finished discussing. Because the CD was encoded at a fixed sample rate, it looks like we are stuck with designing a very sharp filter, and in fact the first generation of CD players had to have exactly such sharp filters.[5] However, as we shall show in Section 6.5.6, the sample rate of the recording can be effectively increased by a clever software interpolation approach called **upsampling** or **oversampling**. By upsampling the digital data $x[n]$ by a factor of $U = 16$ (so-called "16× oversampling"), this algorithmic approach creates a discrete-time data stream that is equivalent to what would have happened if the input to the A/D, $x(t)$, had been sampled at an effective rate of $16 \times 44.1\,\text{kHz} = 705.6\,\text{kHz}$. The required lowpass reconstruction filter can then be very simple.

6.4.4 Revised model of D/A conversion

So far, we have been describing D/A conversion as a two-step process involving a theoretically ideal discrete-to-continuous (D/C) converter, which converted discrete-time impulses into continuous-time impulses, followed by an ideal lowpass reconstruction filter. Of course, in practice there is no ideal D/C converter and no ideal lowpass filter. What, then?

The top panel of **Figure 6.34** shows a somewhat more tenable model of practical D/A conversion. The remaining panels of the figure show the waveforms of signals (left panels) and their transforms (right panels) at various stages in the conversion, labeled Ⓐ–Ⓕ. D/A conversion starts with the discrete-time sequence $y[n]$ (panel Ⓐ), whose transform $Y(\omega)$ is assumed to be bandlimited to $|\omega| < \omega_B$. At each time n, a practical D/A converter converts

[5] The first commercially available CD player was the Sony CDP-101. It cost \$730 dollars when it was introduced in late 1982, equivalent to more than \$1800 today. It had a single 16-bit D/A converter with a 24th-order lowpass filter multiplexed between the two stereo channels. Because the channels were served sequentially by the D/A, there was an 11.3 μs delay between the channels.

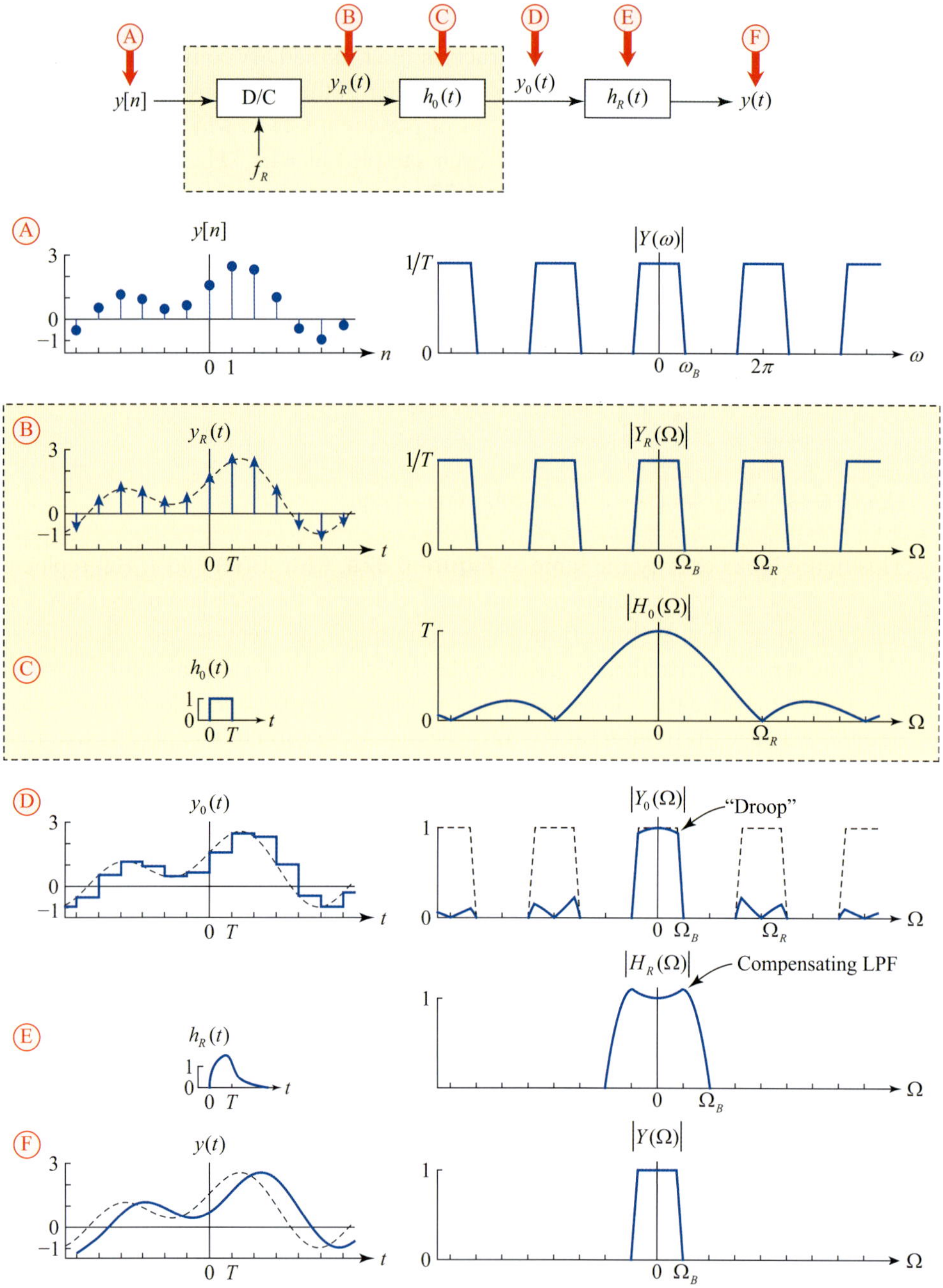

Figure 6.34 D/A conversion with zero-order hold

each sample value not into an impulse, but into a constant voltage that lasts for the reconstruction period T_R (which we assume is the same as the sample period, T). The result is the output waveform $y_0(t)$ (panel Ⓓ), which resembles a staircase in which each trend, $y_0(t)$, $nT \le t < (n+1)T$, has a value given by $y[n]$. The spectrum of $y_0(t)$ can be obtained by recognizing that $y_0(t)$ is theoretically equivalent to the convolution of a continuous-time impulse train $y_R(t)$ (panel Ⓑ),

$$y_R(t) = \sum_{n=-\infty}^{\infty} y[n]\delta(t - nT),$$

with impulse response $h_0(t)$ (Ⓒ),

$$h_0(t) = u(t) - u(t - T) = \begin{cases} 1, & 0 \le t < T \\ 0, & \text{otherwise} \end{cases}.$$

$h_0(t)$ is called a **zero-order hold**. Its function is to hold the value of each impulse, $y[n]$, for a period of length T. Thus,

$$y_0(t) = y_R(t) * h_0(t) = \sum_{n=-\infty}^{\infty} y[n]h_0(t - nT) = \sum_{n=-\infty}^{\infty} y[n](u(t - nT) - u(t - (n+1)T)).$$

As we have discussed in Section 6.3.3, $Y_R(\Omega)$, the spectrum of $y_R(t)$, is equal to $Y(\omega)$ with frequency scale mapped such that $\omega = 2\pi$ corresponds to $\Omega_R = 2\pi/T$ as shown on the right side of panel Ⓑ. The transform of the impulse response of the zero-order hold of length T is a sinc function with zero-crossings at multiples of $\Omega_R = 2\pi/T$, as indicated on the right side of panel Ⓒ. The transform of the staircase waveform $y_0(t) = y_R(t) * h_0(t)$ is the product $Y_0(\Omega) = Y_R(\Omega)H_0(\Omega)$, the magnitude of which is shown on the right side of panel Ⓓ,

$$|H_0(\Omega)| = T\left|\operatorname{sinc}\frac{\Omega T}{2}\right|.$$

The zero-order hold basically does two things to $Y_0(\Omega)$: it causes some amplitude reduction or "droop" in the higher frequencies of the baseband replica and it also severely attenuates all the image replicas, since the zero-crossings of the sinc function of $H_0(\Omega)$ all occur at multiples of $\Omega_R = 2\pi/T$. The final job of the D/A is to clean up this spectral mess by compensating for the droop in the baseband replica and by removing the remains of the image replicas. All this can be accomplished by a single compensating reconstruction filter[6] with impulse response $h_R(t)$ and frequency response $H_R(\Omega)$, as shown in panel Ⓔ. We have already noted that the higher the reconstruction rate of the D/A converter (i.e., the higher f_R), the greater the separation of the image replicas from the baseband, which makes the job of the lowpass reconstruction filter easier. A higher reconstruction rate also has the benefit of flattening the response of $H_0(\Omega)$ in the range of baseband frequencies, which makes the job of reducing the droop in the baseband of $Y_0(\Omega)$ easier, or even unnecessary. For example, if the oversampling factor is $U = 16$, the maximum droop at frequency Ω_B is only about 0.6%. If $U = 64$, the maximum droop is 0.04%, which is not worth worrying about.

The final output of the D/A converter is the reconstructed analog waveform $y(t)$ shown in panel Ⓕ. Any real analog reconstruction filter must, of necessity, be causal, which means that the output will be somewhat delayed with respect to the theoretical response discussed in Section 6.2, which was derived with an ideal non-causal lowpass reconstruction filter.

[6] It is also possible to compensate first for the droop by slightly emphasizing the higher frequencies of $Y(\omega)$ with a discrete-time filter interposed before the D/A converter. Then the reconstruction filter would just be a standard lowpass filter.

6.5 Downsampling and upsampling

One of the most commonly performed digital signal processing operations is changing the effective sampling rate of a signal, either up (upsampling) or down (downsampling) or some combination of up and down (resampling).

6.5.1 Downsampling

Let us first tackle downsampling, which is reducing the effective sample rate of the signal. As a way of fixing ideas, consider the following concrete example. Say that we have obtained a discrete-time sequence $x[n]$ by sampling an analog signal, which we know has a maximum bandwidth of 12 kHz, at a frequency that is double the Nyquist rate, namely $f_s = 48$ kHz. After 8 hours of sampling, we have 1.38 GB of data (assuming each sample is one byte). We want to distribute the data to a client on a very, very cheap flash drive that has a maximum capacity of 1 GB. Bad news! Now, if we had originally sampled the input signal half as fast (i.e., at the Nyquist rate 24 kHz), we would have had only 0.69 GB of data to store. What can we do?

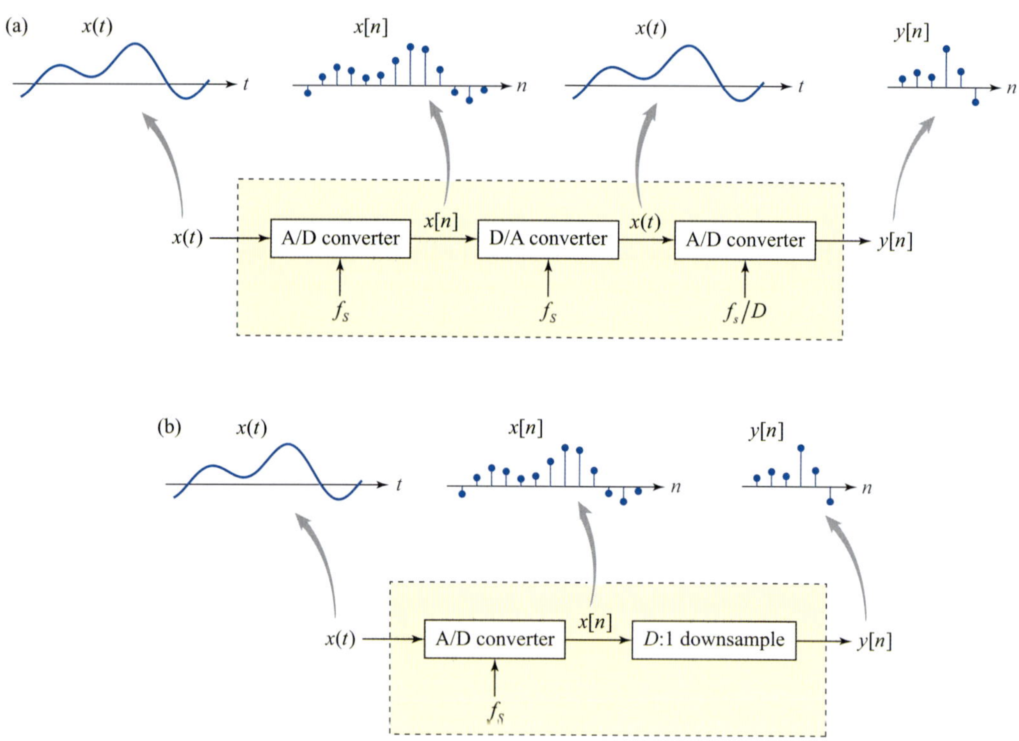

Figure 6.35 Two approaches for downsampling a signal

Approach 1 (bad approach): Here is a truly rotten idea: we will just convert $x[n]$ back to an analog signal and then resample that analog signal at the desired, lower, rate, $f_s/D = 24$ kHz, where $D = 2$ is the **downsample factor**. The result is the output sequence $x_d[n]$ schematized in **Figure 6.35a**. This is a poor approach because it requires both an additional A/D converter and an additional D/A converter to implement.

Approach 2 (better approach): Since $x[n]$ contains all the information of the original analog signal, it should be possible to design a discrete-time operation (namely, an algorithm) that generates the reduced-size output sequence $y[n]$ *directly* from $x[n]$ without going back into the analog domain. We call this operation **downsampling**, and denote it in **Figure 6.35b** with a box with the inscription "D:1 downsample," where D is again the downsample factor. To arrive at the architecture of this downsampler we start with a simpler problem, **decimation**. In its simplest form, you can imagine that the D:1 downsampler would just select every Dth point of the input and pass it to the output,

$$y[n] = x[nD]. \tag{6.11}$$

Equation (6.11) formally defines decimation. **Figure 6.36** shows the result of decimating $x[n]$ by a factor of $D = 2$.

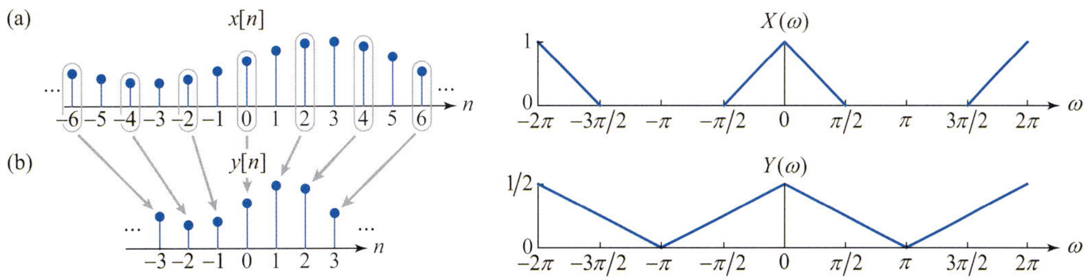

Figure 6.36 Decimation by a factor of two

The left panel of **Figure 6.36a** shows a sequence $x[n]$ derived by sampling the continuous-time signal $x(t)$ at a sample rate f_s, which is twice the Nyquist rate. The spectrum of this sequence, $X(\omega)$, shown in the right panel, therefore has a bandwidth of $\pi/2$. If $x[n]$ is decimated by a factor of $D = 2$, the result is to select every other point of $x[n]$, thereby producing the decimated sequence $y[n]$ shown in **Figure 6.36b**. A bit of thought ought to convince you that the resulting sequence $y[n]$ is *exactly equivalent* to what we would have obtained if we had sampled $x(t)$ half as fast in the first place, namely at $f_s/2$. So, $Y(\omega)$, the spectrum of $y[n]$, has a bandwidth that is double (and magnitude that is half) that of $X(\omega)$. The conclusion is that an A/D converter sampling at rate f_s followed by a D:1 decimation is exactly equivalent to an A/D converter sampling at rate f_s/D, as schematized in **Figure 6.35b**.

Now that we understand intuitively what the decimation does, let us derive the exact mathematical relation between $X(\omega)$ and $Y(\omega)$. **Figure 6.37a** again shows a segment of the sequence $x[n]$ (left panel) and its spectrum $X(\omega)$ (right panel), derived by sampling the continuous-time signal at rate f_s, which is twice the Nyquist frequency. Mathematically, decimating $x[n]$ by a factor of D can be viewed as a two-step process. First, multiply $x[n]$ by the discrete-time impulse-train sequence $s[n]$ shown in the left panel of **Figure 6.37b**. $s[n]$ has a value of one at integer multiples of D and is zero elsewhere,

$$s[n] = \sum_{k=-\infty}^{\infty} \delta[n-kD] = \begin{cases} 1, & n = 0, \pm D, \pm 2D, \ldots \\ 0, & \text{otherwise} \end{cases}.$$

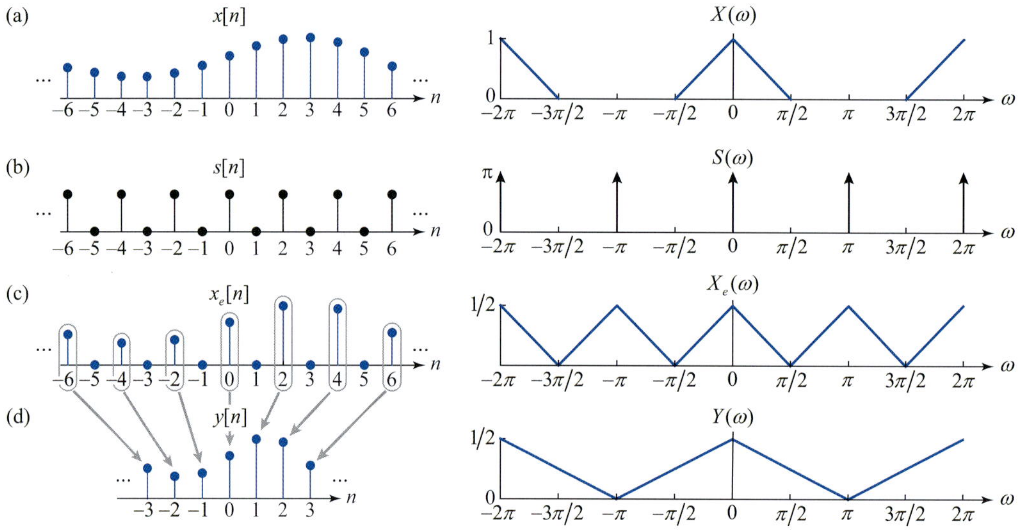

Figure 6.37 Decimation by a factor of two

This impulse train can also be economically expressed as the sum of D complex-exponential sequences with frequencies that are multiples of $2\pi/D$,

$$s[n] = \frac{1}{D}\sum_{k=0}^{D-1} e^{j2\pi kn/D} = \frac{1}{D}\frac{1-e^{j2\pi n}}{1-e^{j2\pi n/D}} = \begin{cases} 1, & n=0,\ \pm D,\ \pm 2D,\ \dots \\ 0, & \text{otherwise} \end{cases}.$$

Then the DTFT of this impulse train is just

$$S(\omega) = \mathfrak{F}\{s[n]\} = \frac{1}{D}\sum_{k=0}^{D-1}\mathfrak{F}\{e^{j2\pi kn/D}\} = \frac{2\pi}{D}\sum_{k=0}^{D-1}\delta(\omega - 2\pi k/D).$$

$S(\omega)$ is an impulse-train in the frequency domain, with impulses of area $2\pi/D$ spaced at multiples of $2\pi/D$, as shown in the right panel of **Figure 6.37b** for a value of $D = 2$.

The left panel of **Figure 6.37c** shows $x_e[n]$, defined as the product of $x[n]$ and $s[n]$; $x_e[n]$ corresponds to setting every Dth value to $x[n]$ and setting the rest to zero:

$$x_e[n] = x[n]s[n] = \sum_{k=-\infty}^{\infty} x[n]\delta[n-kD] = \begin{cases} x[n], & n=0,\ \pm D,\ \pm 2D,\ \cdots \\ 0, & \text{otherwise} \end{cases}.$$

The spectrum of $x_e[n]$ is derived using the convolution property of the DTFT,

$$\begin{aligned} X_e(\omega) &= \mathfrak{F}\{x_e[n]\} = \mathfrak{F}\{x[n]s[n]\} = \frac{1}{2\pi}\{X(\omega)*S(\omega)\} = \frac{1}{2\pi}\left\{X(\omega)*\frac{2\pi}{D}\sum_{k=0}^{D-1}\delta(\omega-2\pi k/D)\right\} \\ &= \frac{1}{\cancel{2\pi}}\frac{\cancel{2\pi}}{D}\sum_{k=0}^{D-1}X(\omega)*\delta(\omega-2\pi k/D) = \frac{1}{D}\sum_{k=0}^{D-1}X(\omega-2\pi k/D). \end{aligned}$$

(6.12)

The result is that $X_e(\omega)$ is the sum of D replicas of $X(\omega)$, each centered at a different multiple of $2\pi/D$ and scaled by $1/D$, as shown in the right panel of **Figure 6.37c** for $D = 2$.

Now, let us relate $y[n]$ to $x_e[n]$:

$$x_e[n] = \sum_{k=-\infty}^{\infty} y[k]\delta[n-kD] = \begin{cases} y[n/D], & n = 0, \pm D, \pm 2D, \ldots \\ 0, & \text{otherwise} \end{cases}. \tag{6.13}$$

Note how this equation works. $x_e[n]$ is expressed as the summation of the product of two sequences, $y[n]$ and $\delta[n-kD]$. For any value of n that is not an integer multiple of D, the expression $\delta[n-kD]$ will be zero for all values of k. Hence the sum will be zero as well, as will $x_e[n]$. But, if n is an integer multiple of D, say $n = rD$, the expression $\delta[n-kD] = \delta[rD-kD] = \delta[(r-k)D]$ will be one when $k = r$ and zero for any other value of k. Hence, the summation picks up a single value of $y[k] = y[r] = x_e[rD]$. $x_e[n]$ is termed the **expanded sequence** of $y[n]$:

$$\begin{aligned} &\vdots \\ x_e[-D] &= y[-1] \\ x_e[0] &= y[0] \\ x_e[D] &= y[1] \\ &\vdots \\ x_e[nD] &= y[n] \\ &\vdots \end{aligned}.$$

Given this relation between $y[n]$ and $x_e[n]$, we can now easily derive the relation between $Y(\omega)$ and $X_e(\omega)$:

$$\begin{aligned} X_e(\omega) &= \mathfrak{F}\{x_e[n]\} = \mathfrak{F}\left\{\sum_{k=-\infty}^{\infty} y[k]\delta[n-kD]\right\} = \sum_{k=-\infty}^{\infty} y[k]\mathfrak{F}\{\delta[n-kD]\} = \sum_{k=-\infty}^{\infty} y[k]e^{-j\omega(kD)} \\ &= \sum_{k=-\infty}^{\infty} y[k]e^{-jk(\omega D)} = Y(\omega D). \end{aligned} \tag{6.14}$$

$Y(\omega)$ and $X_e(\omega)$ are related through a simple scaling of the ω axis by a factor D, as you can see in the right panel of **Figure 6.37d**. Finally, combine Equations (6.12) and (6.14) to yield

$$Y(\omega) = X_e(\omega/D) = \frac{1}{D}\sum_{k=0}^{D-1} X(\omega/D - 2\pi k/D) = \frac{1}{D}\sum_{k=0}^{D-1} X((\omega - 2\pi k)/D).$$

The spectrum of the decimated signal, $Y(\omega)$, is the sum of D replicas of $X(\omega)$, each shifted by a multiple of $2\pi/D$, scaled in frequency by D and scaled in magnitude by $1/D$. The effect of all this manipulation is a spectrum that is exactly the same as the spectrum of $x[n]$ obtained by sampling $x(t)$ at a frequency of Ω_s/D, which is what we wanted to achieve in the first place.

6.5.2 Decimation and aliasing

Because decimating a discrete-time sequence by a factor of D is identical to sampling the original continuous-time signal at a rate of f_s/D, there is the danger that decimation can lead to aliasing. To see this, consider the example of the previous section, but with $D=3$, as shown in **Figure 6.38**.

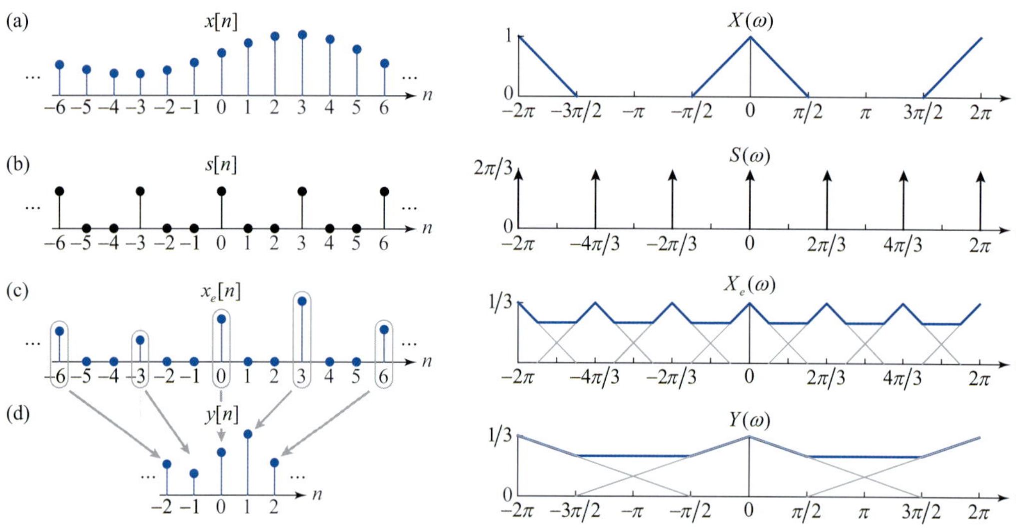

Figure 6.38 Downsampling by a factor of three

Now, $X_e(\omega)$ is the sum of replicas of $X(\omega)$, shifted in frequency by 0, $2\pi/3$ and $4\pi/3$, and scaled by a factor of 1/3. Since the bandwidth of $X(\omega)$ is $\pi/2$, the sum of the replicas overlap and add (thick blue trace), resulting in aliasing. Even though the original signal $x(t)$ was sampled at a value of f_s equivalent to twice the Nyquist rate, decimating $x[n]$ results in aliasing, exactly as if $x(t)$ had been sampled at $f_s/3$, which is 2/3 of the Nyquist rate.

The preceding example shows that we have to exercise care in decimation. We cannot decimate a signal using any value of D and expect that all will be well. Given an analog signal $x(t)$ with a bandwidth of $\Omega_b = 2\pi f_b$ that is sampled at rate $\Omega_s = 2\pi f_s$, the resulting discrete-time signal $x[n]$ has bandwidth $\omega_b = 2\pi(\Omega_b/\Omega_s) = 2\pi(f_b/f_s)$. To ensure that there will no aliasing, we must sample above the Nyquist rate, $\Omega_s > 2\Omega_b$, which means that the bandwidth of the discrete-time signal must be $\omega_b < \pi$ or there will be aliasing. Now, if $x[n]$ is decimated by a factor of D, this is equivalent to sampling $x(t)$ at a rate of Ω_s/D, which is therefore equivalent to "spreading" the discrete-time spectrum by a factor of D, yielding a discrete-time bandwidth of $D\omega_b$, which might exceed π. Thus, the way to prevent aliasing in decimating by a factor of D is to make sure that the maximum discrete-time bandwidth of the input signal does not exceed π, i.e., $D\omega_b < \pi$, which implies that

$$\omega_b < \pi/D. \tag{6.15}$$

This condition is assured when $2\pi f_b/f_s < \pi/D$, which again gives $f_b < f_s/2D$, or equivalently $f_s > 2Df_b$.

Example 6.1

An input signal $x(t)$ has bandwidth $f_b = 4$ kHz. The signal is sampled at a rate f_s and then decimated by a factor of D. Find the maximum D possible without causing aliasing for each of the following sampling rates:

(a) $f_s = 12$ kHz.
(b) $f_s = 16$ kHz.
(c) $f_s = 20$ kHz.

➤ Solution:
To prevent aliasing, the maximum bandwidth of the sampled signal, $\omega_b = 2\pi(\Omega_b/\Omega_s) = 2\pi(f_b/f_s)$, cannot exceed π, i.e., $\omega_b < \pi/D$. Hence the maximum value of $D < \pi/\omega_b$ or $D < f_s/2f_b$. Also, D must be an integer.

(a) $f_s/2f_b = 12/8 = 1.5$ so $D = 1$.
(b) $f_s/2f_b = 16/8 = 2$ so $D = 1$. "Wait a minute," you say. "Why 1?" Because $D < f_s/2f_b$, not $D \leq f_s/2f_b$.
(c) $f_s/2f_b = 20/8 = 2.5$ so $D = 2$.

Example 6.2

An input signal $x(t)$ has bandwidth f_b. The signal is sampled at a rate $f_s = 16$ kHz and then decimated by a factor of D, specified below. For each value of D, find the maximum possible value of f_b such that the signal can be downsampled without aliasing:

(a) $D = 2$.
(b) $D = 3$.
(c) $D = 4$.

➤ Solution:
$f_b < f_s/2D$, so

(a) $f_s/2D = 16/4$, so $f_b < 4$.
(b) $f_s/2D = 16/6$, so $f_b < 2.6667$.
(c) $f_s/2D = 16/8$, so $f_b < 2$.

Unfortunately, we cannot always control the bandwidth or the sampling rate of the analog signal, so we cannot assure that aliasing will not occur when decimating. The only sure way to prevent aliasing is make sure that the bandwidth of the discrete-time sequence $x[n]$ before decimation by a factor of D satisfies Equation (6.15) and is no larger than π/D. This can be accomplished by preceding decimation with a **discrete-time anti-aliasing filter**, a lowpass digital filter that restricts the bandwidth of the input to $\omega_b = \pi/D$. If the bandwidth of $X(\omega)$ is already less than π/D, the lowpass filter will have no effect. However, if the bandwidth of $X(\omega)$ is greater than π/D, the discrete-time filter will have the same effect as an analog anti-aliasing filter with a bandwidth set to $\Omega_s/2D$.

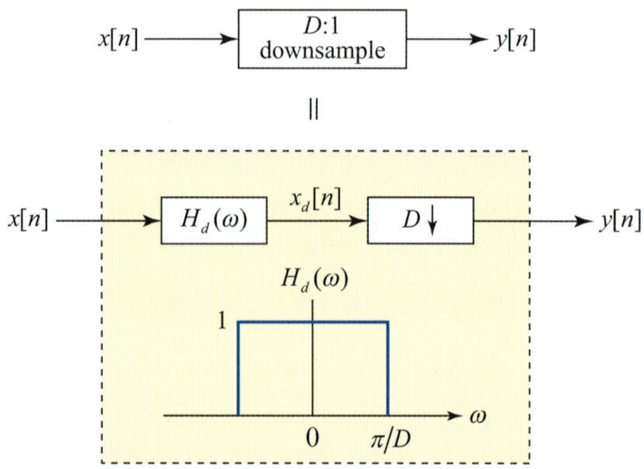

Figure 6.39 Downsampler with a discrete-time anti-aliasing filter

Figure 6.39 shows the architecture of the complete downsampler. The first stage is the discrete-time anti-aliasing filter, labeled $H_d(\omega)$, which is represented here as an ideal lowpass filter with bandwidth π/D and magnitude one. The filtered input sequence $x_d[n]$ is then decimated by a factor of D so that $y[n] = x_d[nD]$, a process schematized by the box with the downsample factor and down-arrow, $D\downarrow$. We will formally use the term **downsampling** to indicate this cascade of lowpass anti-alias filtering followed by decimation.

Figure 6.40 shows the effect of downsampling $x[n]$, the same sequence shown in **Figure 6.38**, by a factor of $D = 3$ by a system that performs a discrete-time anti-aliasing filtering before decimation.

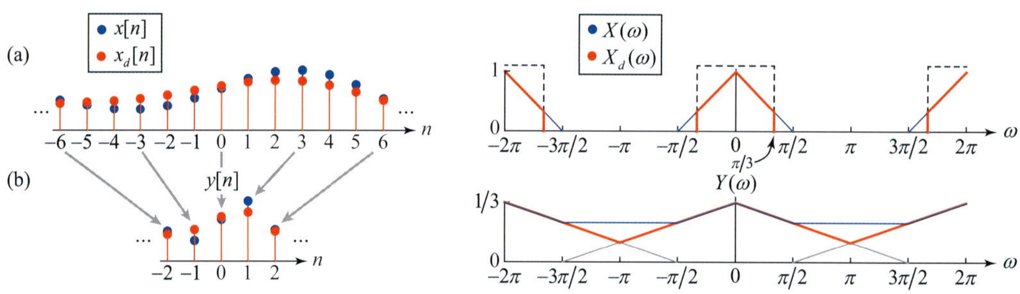

Figure 6.40 Downsampling by a factor of three with a discrete-time anti-aliasing filter

The blue symbols in the left panel of **Figure 6.40a** show the input sequence $x[n]$. The bandwidth of $X(\omega)$ is $\pi/2$, as shown by the thin blue lines in the right panel of the figure. As we discovered in **Figure 6.38**, this exceeds the maximum bandwidth allowed for the downsampler, which is $\pi/D = \pi/3$. Hence, we first filter $x[n]$ with a discrete-time anti-aliasing filter, a lowpass that has a bandwidth of $\pi/3$, as indicated by the dotted lines in the right panel. The spectrum of the output of the filter, $X_d(\omega)$ (solid red trace), has no frequency components above $\pi/3$. The output of the filter, $x_d[n]$, is shown in red symbols in the left panel. When $x_d[n]$ is decimated (**Figure 6.40b**, left panel), the spectrum of the result (solid red trace, right panel) shows no aliasing. The end result of this process is equivalent to having applied an analog anti-aliasing filter with a bandwidth of $\Omega_s/2D$ to the original analog signal and then having sampled it at frequency Ω_s.

6.5.3 Oversampling A/D converter in a digital recording application

Let us briefly return to the digital recording application we discussed in Section 6.4, where the challenge was to design an anti-aliasing filter that will preserve information in the passband, below 20 kHz, and substantially attenuate frequencies above half the Nyquist rate (i.e., 22.05 kHz). **Figure 6.41** shows two possible approaches to designing such a system. **Figure 6.41a** shows the brute-force analog solution, which we described in conjunction with **Figure 6.32a**. The A/D converter samples at 44.1 kHz, which is just greater than the Nyquist rate. We found that if we assume the stopband of the anti-aliasing filter needs to be half the sampling frequency, $2\pi \cdot 44.1/2 = 2\pi \cdot 22.05$ kHz, with a desired attenuation of −90 dB, we need to precede the A/D converter with an extremely sharp analog anti-aliasing filter.

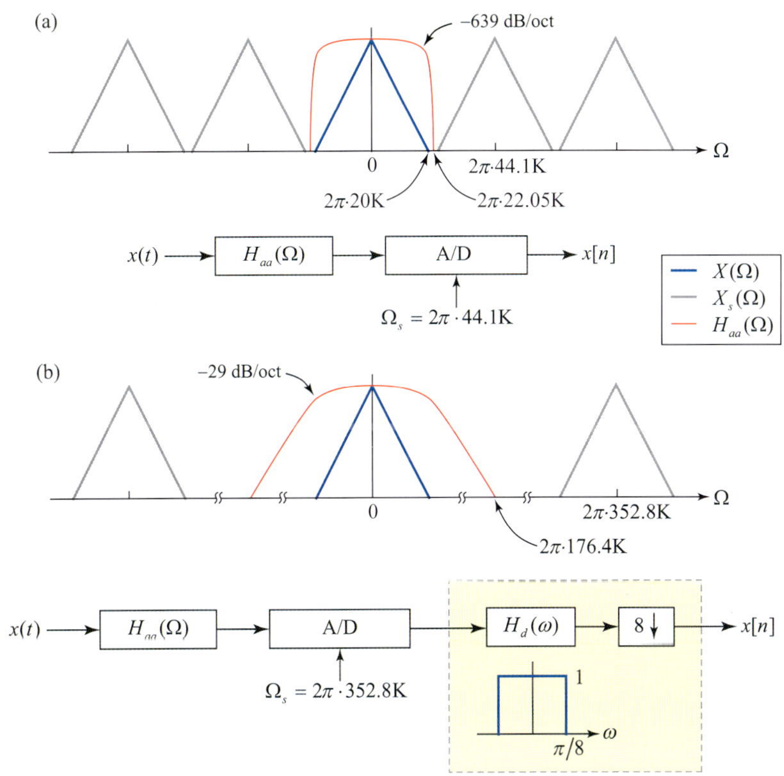

Figure 6.41 Use of downsampling in an A/D converter

Figure 6.41b shows a cleverer solution that utilizes both oversampling and downsampling. We start by oversampling the analog signal at $f_s = 352.8$ kHz, which is eight times the previous sampling rate of 44.1 kHz. Since the faster sampling rate is eight times the Nyquist rate, we can get away with the relatively mediocre anti-aliasing filter, as we described in connection with **Figure 6.32b**. However, even though we have now "fixed" the problem of anti-aliasing filter design, we have also generated eight times as much data as we would have had we sampled at 44.1 kHz. But we can fix that, too. After the signal is oversampled, the next step is to downsample the resulting discrete-time sequence by a factor of eight, which reduces the effective sample rate to 44.1 kHz. In accordance with our discussion in Section 6.5.2, we use an 8:1 downsampler, which includes a discrete-time anti-aliasing filter with a cutoff of $\omega_b = \pi/8$ followed by decimation by a factor of $D = 8$. The $\pi/8$ cutoff corresponds to an analog cutoff of $354.8 \cdot (\pi/8)/2\pi = 22.05$ kHz. So, the entire system appears equivalent to the analog anti-aliasing filter in **Figure 6.41a**. The required discrete-time filter $H_d(\omega)$ can be designed to be very sharp – equivalent to 100s of dB/octave in an analog filter, using methods we will discuss in Chapter 7 (FIR filters) and/or Chapter 8 (IIR filters). The key point here is that the downsampling A/D converter allows us to trade a difficult analog design problem – designing a sharp analog anti-aliasing filter at the input to the A/D converter – for a relatively easier discrete-time problem – designing a sharp discrete-time anti-aliasing filter in the downsampler.

6.5.4 Upsampling

Upsampling is the complementary problem to downsampling. Upsampling is an **interpolation** method that effectively increases the sampling rate of an analog signal by means of digital signal processing techniques. As a way to motivate this section, let us say we have a sequence $x[n]$ derived from sampling an analog signal bandlimited to $f_b = 11.9$ kHz at a sample rate of $f_s = 24$ kHz, which is just above the Nyquist rate, so that bandwidth of $x[n]$ is $\omega_b = 2\pi(f_b/f_s) = 0.99\pi$. Now, assume that we want to distribute this sequence to a client whose D/A converter has a fixed reconstruction rate of $f_r = 48$ kHz. How unfortunate. We wish we had sampled the signal twice as fast in order to match the reconstruction rate of the client's D/A converter, but it is too late now. What can we do? In the same spirit as our two approaches to the problem of downsampling (**Figure 6.35**), we propose two approaches to upsampling.

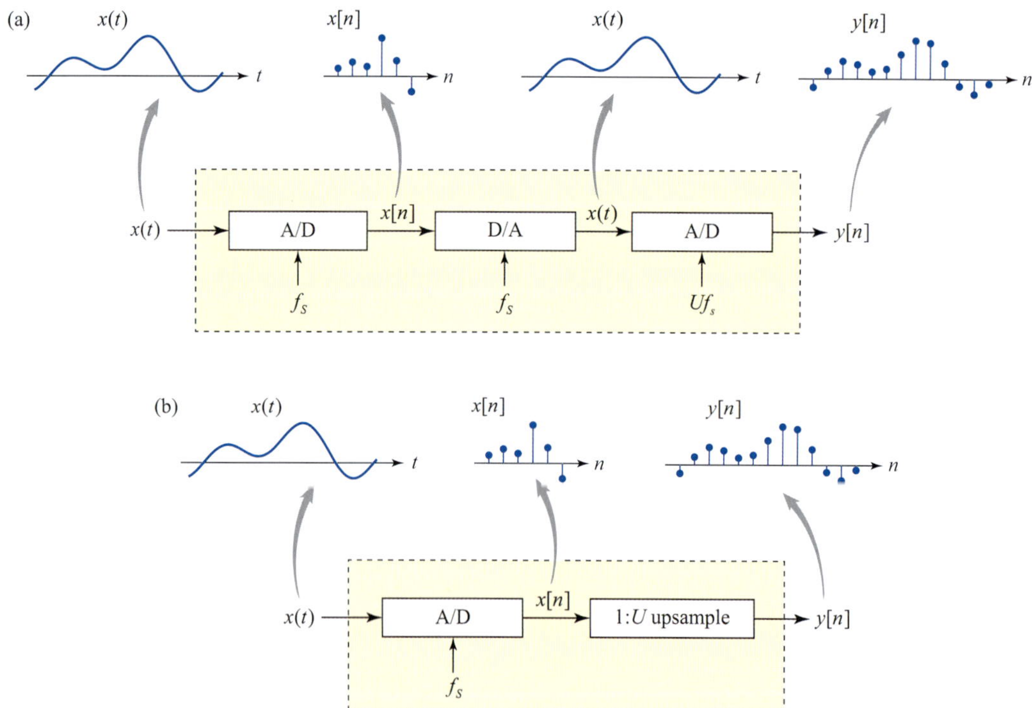

Figure 6.42 Two approaches to upsampling a signal

Approach 1 (bad approach): **Figure 6.42a** shows a not-very-good approach to resampling the signal: we take the original sequence $x[n]$, convert it back into an analog signal $x(t)$ in our laboratory and then resample that analog signal at the desired (higher) sample rate $2f_s = 48$ kHz to form a new output sequence $y[n]$. As bad as this approach might be, it does indicate one thing: all the information necessary to create $y[n]$ must already be present in $x[n]$. Therefore, it should be possible to compute $y[n]$ directly from $x[n]$ without going through the intermediate step of converting $x[n]$ back to an analog signal and then resampling it.

Approach 2 (better approach): This is a purely algorithmic way of creating the sequence $y[n]$ from the original sequence $x[n]$. The algorithm for upsampling is denoted in **Figure 6.42b** with a box inscribed with "1:U upsample," where U is called the **upsample factor**. The method for creating sequence $y[n]$ is best shown by the example in **Figure 6.43**. The left panel of **Figure 6.43a** shows the sequence $x[n]$, and the right panel shows its DTFT. Since the original analog signal was sampled just above the Nyquist rate, the maximum spectral frequency of $X(\omega)$, namely ω_b, is almost π.

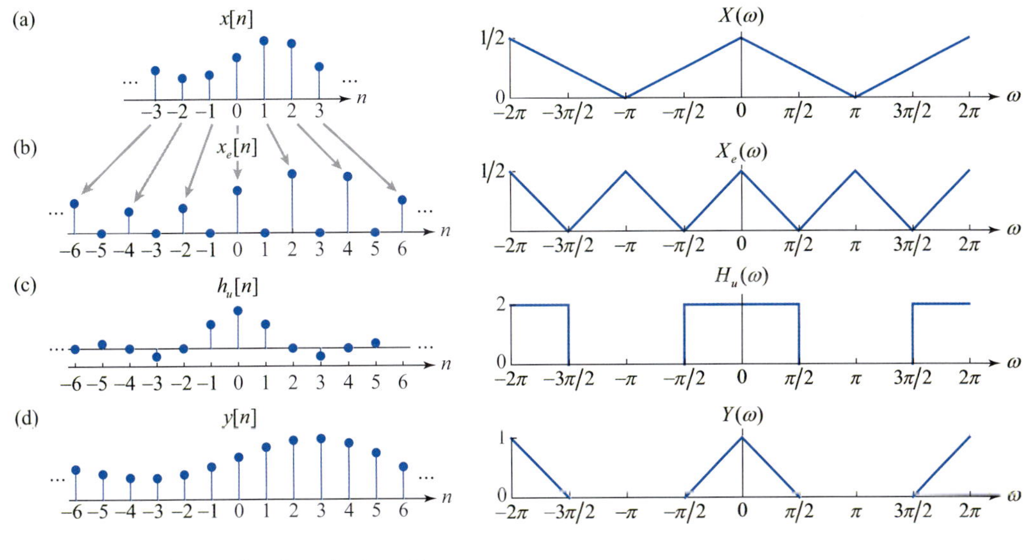

Figure 6.43 Upsampling by a factor of two

There are two steps necessary to interpolate $x[n]$ by a factor of U. First, create the **expanded sequence** $x_e[n]$ effectively by "inserting" $U - 1$ zeros between each pair of points of $x[n]$. Then, lowpass filter the expanded sequence to create $y[n]$. If everything goes according to plan, $y[n]$ should be identical to what we would have obtained had we sampled $x(t)$ at frequency Uf_s.

The left panel of **Figure 6.43b** shows the expanded sequence $x_e[n]$ for the example of $U = 2$. $x_e[n]$ is zero except at time points that are an integer multiple of U. We can write an expression that relates $x_e[n]$ to $x[n]$ using reasoning identical to that discussed in connection with the decimation operation, Equation (6.13),

$$x_e[n] = \sum_{k=-\infty}^{\infty} x[k]\delta[n - kU] = \begin{cases} x[n/U], & n = 0, \pm U, \pm 2U, \ldots \\ 0, & \text{otherwise} \end{cases}. \tag{6.16}$$

The spectrum for this sequence, derived in a manner identical to Equation (6.14), is

$$X_e(\omega) = \mathfrak{F}\{x_e[n]\} = \mathfrak{F}\left\{\sum_{k=-\infty}^{\infty} x[k]\delta[n-kU]\right\} = \sum_{k=-\infty}^{\infty} x[k]\mathfrak{F}\{\delta[n-kU]\} = \sum_{k=-\infty}^{\infty} x[k]e^{-jkU\omega}$$
$$= X(\omega U). \tag{6.17}$$

The relation between $X_e(\omega)$ and $X(\omega)$ is a simple compression of the frequency scale, as shown in the right panel of **Figure 6.43b**. Comparing the spectrum of the expanded signal, $X_e(\omega)$, with the spectrum of upsampled output, $Y(\omega)$ (in the right panel of **Figure 6.43d**), you can see that $X_e(\omega)$ comprises a baseband replica that is equal to $Y(\omega)$ (scaled in amplitude by a factor of U) plus image replicas of $Y(\omega)$ at frequencies $\pm 2\pi/U$. Accordingly, the second step required to produce the upsampled signal $y[n]$ is to filter the expanded sequence $x_e[n]$ with a discrete-time lowpass filter in order to remove the image replicas. Comparing the spectra of $Y(\omega)$ and $X_e(\omega)$ in **Figure 6.43b** and **Figure 6.43c**, we see that we can indeed derive $Y(\omega)$ from $X_e(\omega)$ by discrete-time lowpass filtering. Specifically, $Y(\omega) = X_e(\omega)H_U(\omega)$, where $H_U(\omega)$ is an ideal discrete-time lowpass filter with a cutoff frequency of π/U and a gain of U,

$$H_u(\omega) = \begin{cases} U, & |\omega| < \pi/U \\ 0, & \text{otherwise} \end{cases}. \tag{6.18}$$

The lowpass filter is sketched in **Figure 6.43c**. In the time domain, this filtering is achieved by convolving $x_e[n]$ with the impulse response of the lowpass filter, $h_u[n]$, as shown in the left panel of **Figure 6.43c**.

To get more insight into upsampling in both the time and frequency domains, let us look at a few examples of upsampling simple signals.

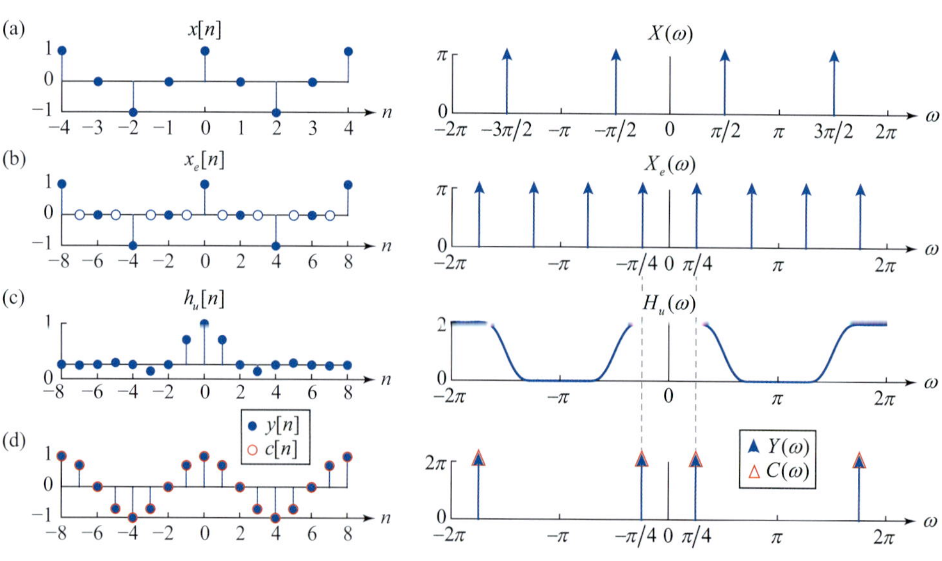

Figure 6.44 Upsampling of a cosine by a factor of two using Kaiser interpolating filter

6.5.5 ★ Upsampling a cosine

Consider an analog signal $x(t) = \cos(2\pi \cdot 1000t)$ that has been sampled at $f_s = 4000$ Hz to create the sequence $x[n] = \cos(n\pi/2)$ shown in the left panel of **Figure 6.44a**. The spectrum $X(\omega) = \mathfrak{F}\{x[n]\}$, shown in the right panel of **Figure 6.44a**, comprises impulses at $\omega = \pm\pi/2$ and, of course, replicas at all multiples of 2π. The rest of the figure shows the sequence of steps required to upsample $x[n]$ by a factor of $U = 2$ using a practical (Kaiser) discrete-time interpolating filter in order to form output sequence $y[n]$. This process should be effectively equivalent to sampling $x(t)$ at a rate of $\Omega_s = 2\pi \cdot 8000$.

The first step in upsampling by a factor of $U = 2$ is to create the expanded sequence $x_e[n]$ by "inserting" a zero between each pair of points of $x[n]$, as shown in the left panel of **Figure 6.44b**. The original points of $x[n]$ are shown with filled symbols, the inserted zero points are shown with open symbols. The spectrum of the expanded sequence $X_e(\omega)$ (right panel of **Figure 6.44b**) is just a frequency-scaled version of $X(\omega)$, namely $X_e(\omega) = X(2\omega)$, as expected from Equation (6.17). $X_e(\omega)$ thus contains impulses in the baseband at $\omega = \pm\pi/4$ as well as images centered at integer multiples of $2\pi/U = \pi$.

The second step necessary to create the upsampled sequence $y[n]$ from $x_e[n]$ is discrete-time lowpass filtering. As indicated by Equation (6.18), in order to preserve the baseband replica and remove the image replicas, we need a filter with a cutoff frequency of $\pi/U = \pi/2$ and a gain of $U = 2$. In this example, we will employ a simple finite-impulse-response (FIR) Kaiser lowpass filter of length $N = 17$, designed using the windowing technique that will be discussed in Chapter 7. The impulse response of this filter, $h_u[n]$, is shown in the left panel of **Figure 6.44c**. The frequency response of the filter, $H_u(\omega)$, shown on a linear amplitude scale in the right panel of **Figure 6.44c**, is adequate to filter out the unwanted frequency components of $X_e(\omega)$. Applying this filter to the expanded sequence means convolving $x_e[n]$ with $h_u[n]$ giving $y[n]$, as shown in the left panel of **Figure 6.44d** (filled blue circles). This panel also shows that the interpolated response using the FIR filter matches well with the desired result of the interpolation, $c[n] = \cos n\pi/4$ (open red circles), which is what we would have gotten had we sampled the original analog signal at $2f_s = 8000$ Hz. Incidentally, one interesting thing about our choice of the Kaiser image-rejection filter is that values of the interpolated sequence $y[n]$ at multiples of U (i.e., $n = 0, \pm U, \pm 2U, \cdots$) are exactly equal to the corresponding values of the original sequence, that is, $y[nU] = x[n]$ (see Problem 6-12).

Linear interpolation When interpolation is done using a Kaiser interpolating filter, the value of all the other interpolated points of $y[n]$ (i.e., $n \neq 0, \pm U, \pm 2U, \ldots$) is the weighted sum of multiple points of $x_e[n]$, and therefore of multiple points of $x[n]$. This interpolation scheme is much more sophisticated than a simple **linear interpolation** with which you are probably familiar. For linear interpolation with $U = 2$, the value of each interpolated point of $x_e[n]$ is just the average of the pair of adjacent points. For example,

$$y[1] = 0.5(x_e[0] + x_e[2]) = 0.5(x[0] + x[1]).$$

Linear interpolation of this sort can also be viewed as a discrete-time filtering operation, in which the interpolated sequence $y[n]$ is formed by convolving the expanded sequence $x_e[n]$ with a linear interpolating filter given by

$$h_u[n] = \begin{cases} 1 - |n|/U, & |n| < U \\ 0, & \text{otherwise} \end{cases}.$$

The following example compares linear interpolation with the interpolation using a Kaiser interpolating filter.

Example 6.3

Once again, take an analog signal $x(t) = \cos(2\pi \cdot 1000t)$ that has been sampled at $f_s = 4000$ Hz to create a sequence $x[n]$. Now use a linear interpolating filter to upsample $x[n]$ by a factor of $U = 2$ to form a sequence $x_u[n]$.

➤ Solution:

Figure 6.45 shows the sequence of steps required to upsample $x[n]$ by a factor of $U = 2$ using a linear interpolating filter.

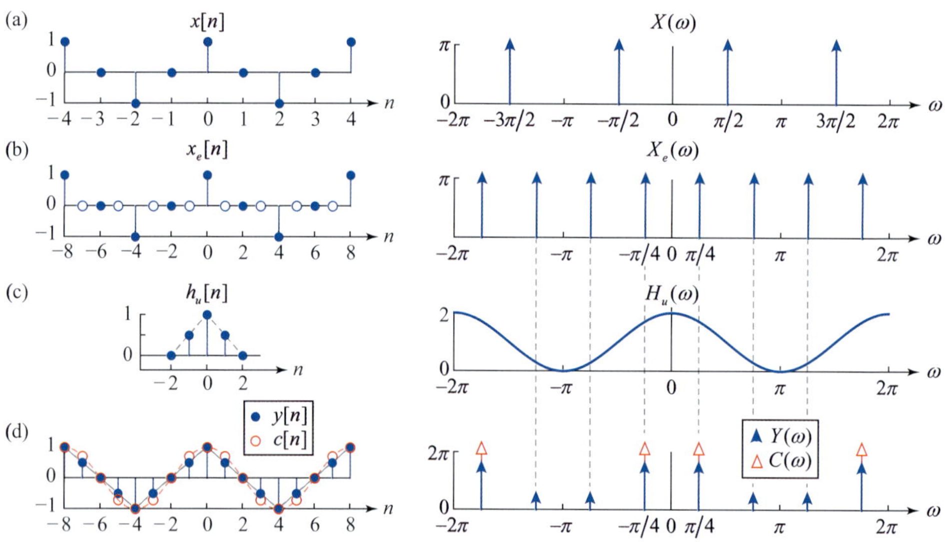

Figure 6.45 Interpolation of a cosine by a factor of two using linear interpolating filter

Figures 6.45a and **b** again show the time-domain and frequency-domain representations of $x[n] = \cos n\pi/4$ and the expanded sequence $x_e[n]$, respectively. The left panel of **Figure 6.45c** shows the impulse response of the linear interpolating filter $h_u[n]$ for $U = 2$, namely $h_u[n] = 0.5\delta[n+1] + \delta[n] + 0.5\delta[n-1]$. The frequency response of this filter, $H_u(\omega) = 1 + \cos\omega$, is shown in the right panel of **Figure 6.45c**. The left panel of **Figure 6.45d** shows the comparison of the interpolated response $y[n]$ (filled blue circles) with the ideal interpolated response $c[n]$ we would have expected had we sampled $x[n]$ at a rate of $2f_s = 8000$ Hz. The two sequences differ substantially. In fact, you can see that $y[n]$ is actually a triangle wave instead of a sine wave. The right panel of **Figure 6.45d** reveals that $Y(\omega)$ has spurious spectral components at $\omega = \pm 3\pi/4$, and also has roughly 15% lower amplitude at $\omega = \pm \pi/4$ compared to $C(\omega)$, the transform of the ideal interpolated sequence, $c[n] = \cos(n\pi/4)$.

Aliasing in upsampling There is no aliasing in upsampling. Why is that?

6.5.6 Upsampling D/A converter in a digital recording application

Having presented the details of upsampling, let us return again to the digital recording application we discussed in Section 6.4. The process of converting data stored on a CD back into an analog signal has issues similar to those we experienced with respect to A/D conversion. Remember that the main challenge was to design an appropriate analog reconstruction filter. As illustrated in **Figure 6.46**, there are basically two approaches.

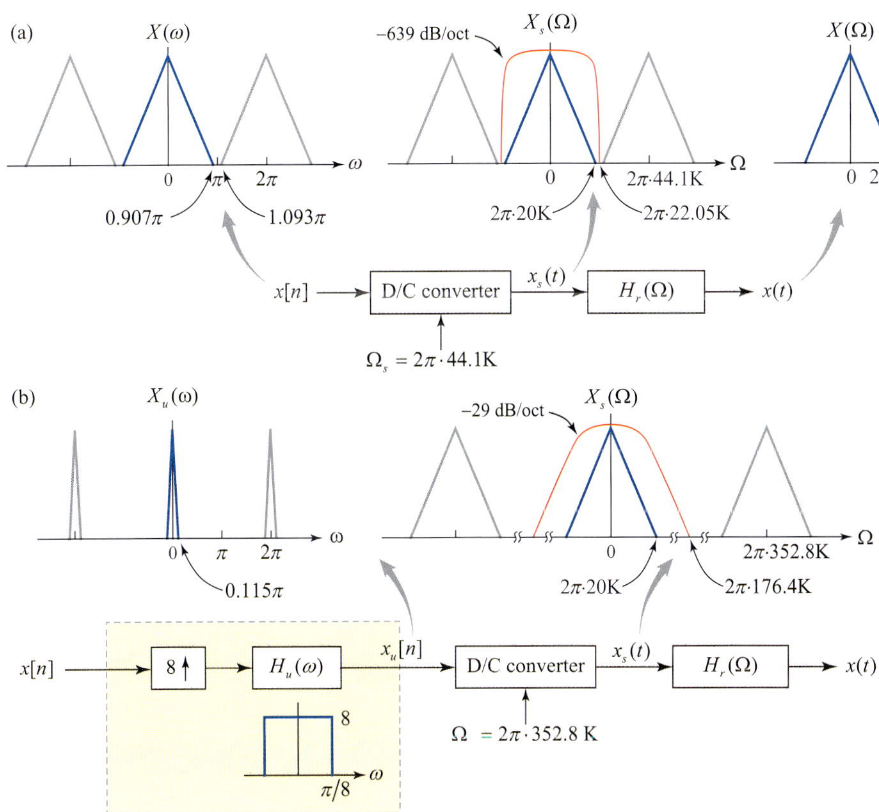

Figure 6.46 Use of upsampling in a D/A converter

Approach I: sharp analog reconstruction filter If the signal is reconstructed with a reconstruction rate equal to the sample rate, i.e., $\Omega_r = \Omega_s = 2\pi \cdot 44.1$ kHz, a very good analog reconstruction filter is necessary, as shown in **Figure 6.46a**. As we have mentioned, music recorded on the CD has a nominal maximum bandwidth of 20 kHz, and CD recording calls for a sample rate of $f_s = 44.1$ kHz, which is *just* greater than the Nyquist frequency (i.e., 40 kHz). Hence, the discrete-time signal on the CD, $x[n]$, has a bandwidth $2\pi(20/44.1) = 0.907\pi$, which is very close to π. The spectrum $X(\omega)$ is shown in the left plot of **Figure 6.46a**. The replicas centered at 0 and 2π are almost touching. So when the signal is reconstructed at 44.1 kHz, the replicas of the reconstructed signal are also almost touching. The transition band is very small, as shown by the middle plot of **Figure 6.46a**. In order to separate the baseband from the image bands, an analog reconstruction filter $H_r(\Omega)$ with a very sharp slope is necessary in the D/A converter. As we mentioned in Section 6.4.3, the design challenge here is identical to that of designing a sharp anti-aliasing filter in the A/D converter.

Approach II: oversampling D/A converter The way around the requirement for a superb reconstruction filter is to use an **oversampling D/A converter**. We first upsample $x[n]$ by a factor of U and then reconstruct the upsampled sequence with a D/A converter at rate $f_r = U \cdot f_s$, as indicated in **Figure 6.46b**. In this example, we upsample the sequence by a factor of $U = 8$ ("8× oversampling") to form $x_u[n]$, the spectrum of which, $X_u(\omega)$, is shown in the left panel. $x_u[n]$ is equivalent to a sequence that would have resulted from sampling the original analog signal at a frequency of $8 \times 44.1 = 352.8$ kHz. This signal is then reconstructed at 352.8 kHz. Now the job of the reconstruction filter in the D/A converter is easier. Clearly, $H_r(\Omega)$ does not have to have too sharp a cutoff to separate the baseband from the image replicas. The passband of the signal is still 20 kHz, but the transition band is now about three octaves wide. Hence, a relatively low-order filter will suffice to achieve the required amount of image rejection.

To summarize, in the A/D converter, we can use a higher sampling rate followed by downsampling to get around having to design an analog anti-aliasing filter with a sharp cutoff. Now, we have just seen that we can get around having to design a sharp analog reconstruction filter in a D/A converter by upsampling the discrete-time output sequence and using a higher reconstruction rate in the converter.

6.5.7 Resampling

So far we have considered upsampling and downsampling by integer factors of U and D respectively. However, we can achieve sample rate conversion by any rational factor of U/D by combining upsampling and downsampling in a process known as **sample-rate conversion** or **resampling**.[7] There are basically two approaches to resampling: downsampling by a factor of D followed by upsampling by a factor of U, as shown in **Figure 6.47a**, or upsampling followed by downsampling, as shown in **Figure 6.47b**.

These two ways of cascading the upsampling and downsampling operations are different, both practically and theoretically. From a practical standpoint, the first method (downsampling followed by upsampling) requires two separate lowpass filters, $H_u(\omega)$ and $H_d(\omega)$, located respectively at the input and output of the system. In the second method (upsampling followed by downsampling), the two lowpass filters occur in cascade, which means that their function can be combined into a single filter, $H_{ud}(\omega) = H_u(\omega)H_d(\omega)$, with a gain of U and a bandwidth that is the

[7] It is also possible to resample by non-rational factors, but that is a more advanced topic.

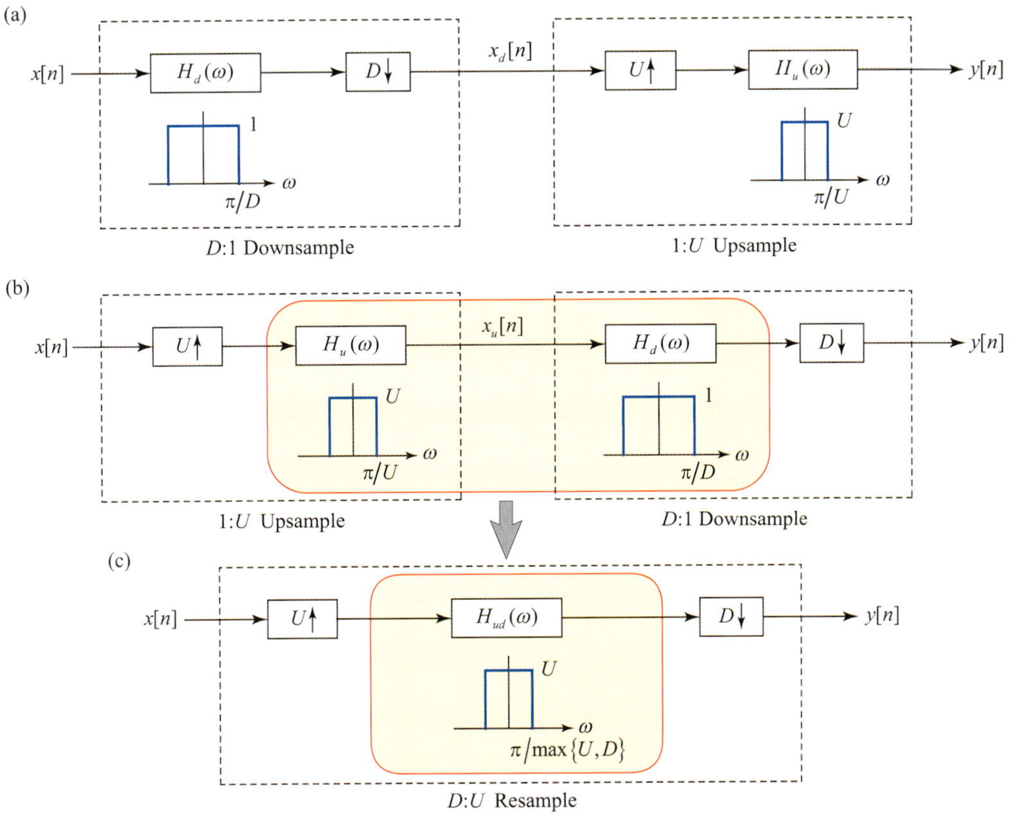

Figure 6.47 Resampling

minimum of π/U and π/D, as shown in **Figure 6.47c**. This D:U resampler has only one discrete-time filter, and therefore requires only one filtering operation to implement.

We can theoretically achieve any rational sample-rate conversion factor with a D:U resampler. For instance, to resample at 1.5× the original sample rate, we could upsample by $U = 3$ and then downsample by $D = 2$. In the resamplers shown in **Figures 6.47b** and **c**, the upsampling operation comes first and results in a reduction of the signal bandwidth by a factor of U. Subsequent downsampling by a factor of D expands the bandwidth by a factor of D. Therefore, as long as $U > D$, these resampler architectures guarantee that the spectrum of the resampled signal, $Y(\omega)$, is simply the spectrum of the input, $X(\omega)$, scaled in frequency and magnitude by a factor of U/D:

$$Y(\omega) = (U/D)X(\omega\, U/D).$$

Notice that this *is not* true for the resampler shown in **Figure 6.47a**, in which downsampling precedes upsampling. If the bandwidth of the input spectrum is greater than π/D, it will be truncated by the filter $H_d(\omega)$ in the downsampler. Even if $U > D$, the spectrum of the resampled signal will not be a simple scaled version of the input spectrum. Therefore, the two methods of cascading the upsampling and downsampling operations are not even theoretically equivalent. This is shown in the following example.

Example 6.4

Consider an input signal whose spectrum $X(\omega)$ has a bandwidth that spans the frequency range $-\pi < \omega < \pi$. Given $D = 2$ and $U = 3$,

(a) show that the resampler architecture shown in **Figure 6.48a**, in which the downsampler precedes the upsampler, results in a truncated output spectrum.

(b) show that the resampler architecture shown in **Figure 6.48c**, in which the upsampler precedes the downsampler, results in an output spectrum that is scaled in frequency and magnitude by a factor of $U/D = 3/2 = 1.5$.

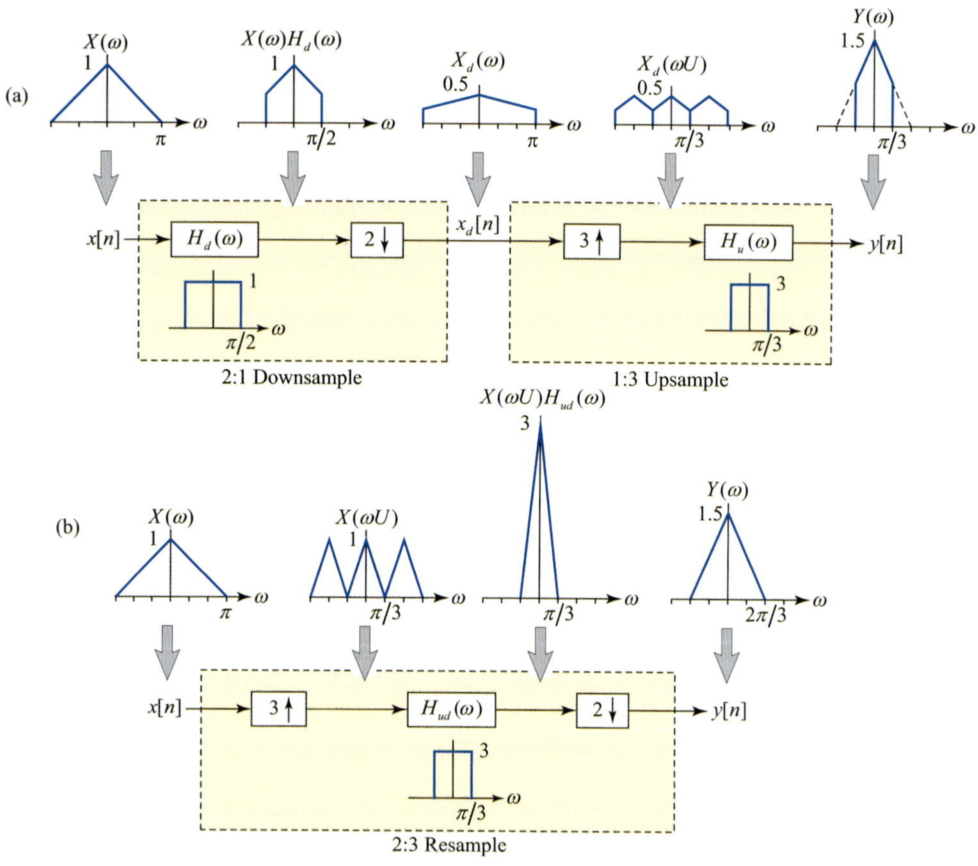

Figure 6.48 Comparison of resampler architectures.

➤ Solution:

(a) The input spectrum is truncated by filter $H_d(\omega)$ to $\pi/D = \pi/2$. The 2 : 1 decimation of the filter sequence expands the spectrum of $X_d(\omega)$ to π and halves the gain. Expansion of $x_d[n]$ by a factor of 1 : 3 produces three replicas, each of width $2\pi/3$. When filtered by $H_u(\omega)$, the result is a spectrum limited in frequency to $|\omega| \leq \pi/3$.

(b) Expansion of $x[n]$ by a factor of $1:3$ produces three replicas of $X(\omega)$, which when filtered by $H_{ud}(\omega)$ result in a single frequency scaled spectrum of width $|\omega| \le \pi/3$. Subsequent $2:1$ decimation results in an output spectrum $Y(\omega) = (U/D)X(\omega\, U/D) = 1.5X(1.5\omega)$.

There are additional practical issues if the ratio U/D has a very large numerator or denominator. For example, some professional digital audio recorders offer a sample rate of 48 kHz. In a recording application, we may require that music sampled at this rate eventually be resampled to 44.1 kHz in order to be encoded onto consumer CDs. This represents a resampling by a factor of $U/D = 44.1/48 = 441/480$. We can remove common factors from the numerator and denominator so that U/D can be expressed as the ratio of mutually prime factors: $U/D = 441/480 = 147/160$. This would seem to require upsampling by a factor of 147 followed by downsampling by a factor of 160, which means that the lowpass filter in **Figure 6.47c** would require a cutoff frequency of $\pi/160$. While theoretically this is not a problem, practically it is, because the complexity (i.e., order) of a discrete-time filter increases as the cutoff frequency decreases. For example, the length of the impulse response of a discrete-time Kaiser lowpass filter with a fixed cutoff slope and a cutoff frequency of π/N is directly proportional to N. Since the computational complexity of convolution is proportional to N^2 (or to $N\log_{10}N$ if convolution is implemented by means of FFTs), there is a big incentive to keep the upsample and downsample factors as small as possible. No worries, we can fix this. In this example, recognize that U/D can be factored:

$$\frac{U}{D} = \frac{147}{160} = \frac{3 \cdot 7 \cdot 7}{4 \cdot 5 \cdot 8} = \left(\frac{3}{4}\right) \cdot \left(\frac{7}{5}\right) \cdot \left(\frac{7}{8}\right).$$

Thus, we can implement this rational sample-rate reduction as a cascade of three resampling steps, as shown in **Figure 6.49**.

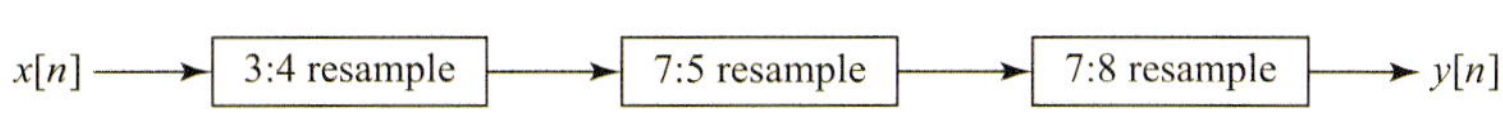

Figure 6.49 Resampling from 48 kHz to 44.1 kHz in a cascade of three stages

In this example, the first stage requires a filter with a bandwidth of only $\pi/4$. The remaining two stages have bandwidths of $\pi/7$ and $\pi/8$ respectively. That is a far cry from requiring a single filter with a bandwidth of $\pi/160$!

6.6 Matlab functions for sample-rate conversion

Matlab has several functions in the Signal Processing Toolbox that perform downsampling, upsampling and resampling.

Resample Matlab's `resample` function is your "go-to" function for performing resampling by a rational factor, as well as upsampling and downsampling by integer factors. The general syntax is

```
y = resample(x, U, D, [R] , [beta] )
y = resample(x, U, D, h),
```

where x is the input sequence, U is the upsampling factor, D is the downsampling factor. R and beta in the first syntax variant optionally specify the parameters of the lowpass anti-aliasing filter. (Specifically, this filter is designed using a least-square method (firls), windowed by a Kaiser filter (with parameter beta), whose length is determined by the cutoff frequency and parameter R (default 10): 1+2*R*max(U, D). Have no fear, we will discuss these filters in Chapter 7.) The second syntax variant allows you to specify h, the filter coefficients of the lowpass filter, yourself.

Internally, resample uses a polyphase resampling algorithm, upfirdn, implemented in a MEX file, so it is fast. It compensates for the signal delay of the lowpass filter so that the output and input are as aligned as possible. We will discuss polyphase resampling algorithms of this sort in Chapter 13.

Downsampling Matlab's decimate function performs downsampling. The general syntax is

```
y = decimate(x, D, ['FIR'])
y = decimate(x, D, [M], ['FIR']),
```

where x is the input sequence, D is the downsampling factor. M is the optionally specified order of the lowpass filter. By default, decimate uses an 8th-order Chebyshev IIR filter. (We discuss these filters in Chapter 8.) If 'FIR' is specified, the default is a 30th-order Hamming-window FIR. Internally, decimate performs zero-phase filtering on the input signal (see Section 7.4.1) before downsampling. Internally, decimate appears to be much slower than resample in performing downsampling, so unless you specifically want to control the filter order, I would suggest that you just use resample with U set to 1.

Upsampling Matlab's interp function performs by a factor of U. The general syntax is

```
y = interp(x, U, [N], [alpha]),
```

where x is the input sequence, U is the upsample factor. N and alpha are optionally specified parameters of the lowpass filter. The Matlab functions interp and resample use different designs for the image rejection filter; hence, the result of processing an input with these two methods will be somewhat different, though the difference is unlikely to be significant in most applications.

6.7 ★ Quantization

So far, life in this chapter has been fairly sweet. The process of A/D and D/A conversion described in the preceding sections is "ideal" in a number of ways. For example, both A/D and D/A conversion were characterized by ideal sampling in the time domain in which the amplitudes of both sampled and reconstructed signals were continuous, that is, unquantized. However, **quantization** of signals is an inherent feature of both A/D and D/A conversion. The A/D converter both *samples* the signal at intervals T and *quantizes* the amplitude of continuous input signals to discrete levels that can be expressed in binary words. The D/A converter takes quantized discrete-time signals and produces a continuous-amplitude output. Quantization represents a departure from the ideal behavior we have described so far, and one we have to understand. In this section, we will give a simplified theoretical model of the effect of quantization. Then in Section 6.8 we will present an overview of how A/D conversion, including quantization, is actually accomplished in hardware.

6.7.1 Model of quantization

Figure 6.50a shows a conceptual model of the full A/D conversion process including quantization. The portion of the model enclosed in the dotted box represents the ideal A/D sampler we have been discussing thus far. At each sample instant $t=nT$, which is a multiple of the sample period T, the ideal A/D converter converts the analog signal $x(t)$ into a discrete-time signal $x[n]=x(nT)$, which has a continuous, unquantized amplitude. Following the sampler is an N-bit **quantizer** that quantizes $x[n]$ into a finite number of levels, $L=2^N$, with N bits of precision, where the precision is determined by specifications of the particular converter. In a high-quality converter for audio signals, N can be 16, 24 or 32 bits. The result is the quantized discrete-time signal $\hat{x}[n]$. Finally, the encoder takes each quantized value and creates a digital output word that comprises N binary bits, $b[n]$. This output passes to subsequent parts of the system in either parallel or serial form, depending on the converter. We will now discuss quantization and encoding separately.

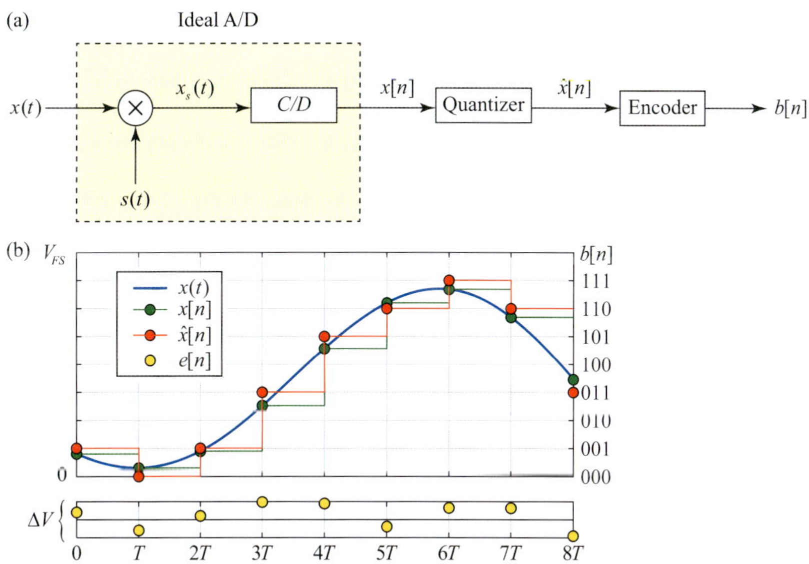

Figure 6.50 Model of A/D conversion with rounding quantization

Quantization **Figure 6.50b** shows how the quantization process affects signals for the hypothetical example of a converter with $N=3$ bits. For an N-bit converter, the **full-scale voltage** of the converter, V_{FS}, is divided into $L=2^N$ equal levels, each of which is an integer multiple of the **quantization voltage** ΔV,

$$\Delta V = \frac{V_{FS}}{L} = \frac{V_{FS}}{2^N} = V_{FS}2^{-N}. \tag{6.19}$$

So, for the 3-bit converter whose response is shown in **Figure 6.50b**, the quantization voltage is $\Delta V = V_{FS}/8$. At each sampling instant, the continuous-time input signal $x(t)$ (solid blue line) is sampled to create the unquantized discrete-time signal $x[n]$ (green dots). This particular

rounding quantizer rounds each sample to the nearest multiple of ΔV (red dots), creating quantized signal $\hat{x}[n]$. The difference between the quantized and unquantized voltages comprises an error signal

$$e[n] = \hat{x}[n] - x[n], \quad (6.20)$$

which is shown in the lower panel (yellow dots). Because the quantizer rounds (rather than truncates), the maximum absolute difference between $\hat{x}[n]$ and $x[n]$ cannot exceed $\Delta V/2$; hence,

$$-\Delta V/2 \leq e[n] < \Delta V/2.$$

The effect of the quantizer can be expressed by a nonlinear function that rounds the unquantized input to the nearest integer multiple of ΔV,

$$\hat{x}[n] = Q(x[n]).$$

Figure 6.51a shows the quantizer characteristics $Q(x)$ for two types of A/D converters, termed **unipolar** and **bipolar**, each with $L = 8$ levels, corresponding to $N = 3$ bits of quantization. Both are known as uniform, rounding (mid-tread) quantizers. The only difference between the two is the range of input voltage. For the unipolar converter (**Figure 6.51a**), the input range is nominally between a minimum value of 0 and a maximum voltage of V_{REF} (with qualifications to be noted in a moment), so the full-scale voltage is defined to be $V_{FS} \triangleq V_{REF} - 0 = V_{REF}$. For the bipolar converter (**Figure 6.51b**), the input range is nominally from $-V_{REF}$ to V_{REF}, so the full-scale voltage is $V_{FS} \triangleq V_{REF} - (-V_{REF}) = 2V_{REF}$. In our discussion below, we will focus on the unipolar converter, with the understanding that the bipolar converter differs only in the input range.

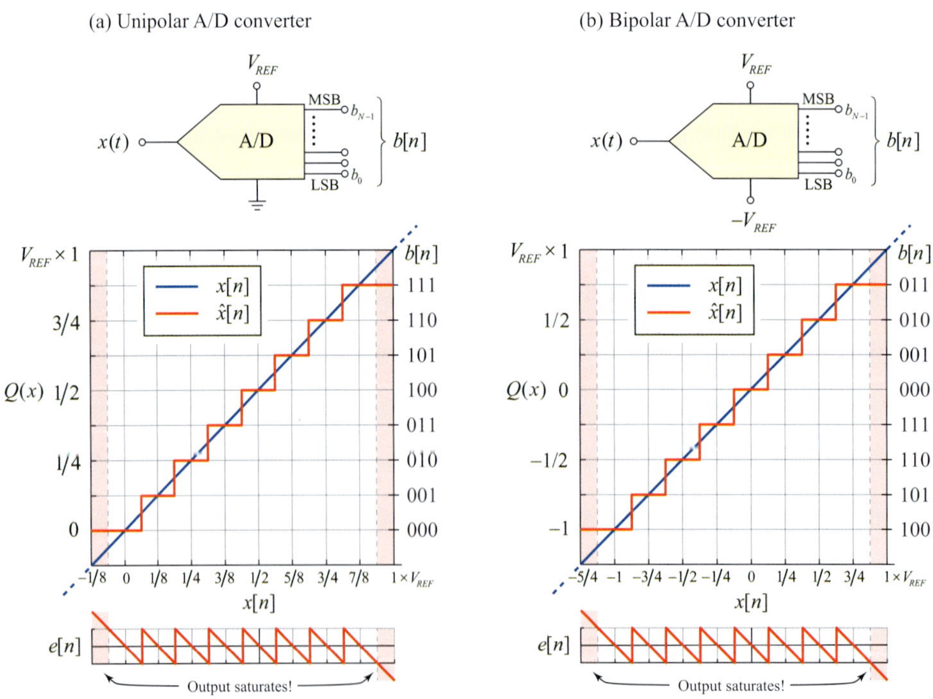

Figure 6.51 A/D converter quantizer characteristic

The blue lines indicate the unquantized input voltage $x[n]$, and the red "staircase" plot reflects the output of the quantizer $\hat{x}[n] = Q(x[n])$, which rounds the input voltage to L discrete output levels that are integer multiples of ΔV,

$$0, \Delta V, 2\Delta V, \ldots, (L-1)\Delta V.$$

The highest output level is not V_{REF}, but rather

$$(L-1)\Delta V = (2^N - 1)\frac{V_{REF}}{2^N} = V_{REF}(1 - 2^{-N}) = V_{REF} - \Delta V.$$

In the example of **Figure 6.51**, $L = 3$ so $\Delta V = V_{REF}/8$ and $Q(x)$ ranges between 0 and $V_{REF} - \Delta V = V_{REF}/8$.

Quantization is an inherently nonlinear and irreversible operation. Every flat "step" in the quantizer plot represents a mapping of a *range* of input values into a *single* output value. This is termed a **mid-tread quantizer** because each tread is centered around an output value. For example, every input value within the $\pm\Delta V/16$ range centered around $V_{REF}/2$ is mapped to the output value $V_{REF}/2$. For most of the input range, except the lowest and highest values, the treads are ΔV wide, meaning that the quantizer maps a ΔV range of the input to an output value. However, the quantizer **saturates** at both low and high input values. Every input value below $V_{REF}/16$ is mapped to the minimum output value, which is 0, and every input value above $13V_{REF}/16$ is mapped to the maximum output value, which is $V_{REF} - \Delta V = 7V_{REF}/8$.

The lower panel of **Figure 6.51a** shows the difference between the input and the output of the quantizer, namely the error $e[n] = \hat{x}[n] - x[n]$. For input values in the range $-\Delta V/2 \leq x[n] < V_{REF} - \Delta V/2$, $e[n]$ is in the range $-\Delta V/2 \leq e[n] < \Delta V/2$. However, for input values below $-\Delta V/2$ or values above $V_{REF} - \Delta V/2$, marked with the pink bars in **Figure 6.51a**, the output saturates at its minimum value (0) or maximum value $(7V_{REF}/8)$ and the absolute value of the error exceeds $\Delta V/2$. Hence, the practical range of input values for the unipolar converter is not $0 \leq x[n] < V_{REF}$ as implied in the introduction to this section, but actually $-\Delta V/2 \leq x[n] < V_{REF} - \Delta V/2$. However, the full-scale voltage of the converter is still

$$V_{FS} = (V_{REF} - \Delta V/2) - (-\Delta V/2) = V_{REF}.$$

The important moral here is that in any practical A/D application, it is critically important to make sure that the amplitude of the input signal is adjusted so that the output of the converter does not saturate.

Encoding As shown in **Figure 6.50b**, at each time point n, the quantized signal $\hat{x}[n]$ (red dots) is eventually encoded into an N-bit binary word $b[n]$, which has the form $b_{N-1}b_{N-2}\cdots b_1b_0$, where each of the N bits b_k, $0 \leq k < N$, has a value of either 0 or 1. The left-most bit of the word, b_{N-1}, is termed the **most significant bit (MSB)** and the right-most bit, b_0, is the **least significant bit (LSB)**. In a unipolar converter, $\hat{x}[n]$ is non-negative, so numbers corresponding to each of the L output levels are encoded using the **unsigned binary** representation, shown to the right of the plot in **Figure 6.51a**. In a bipolar converter, the output is both positive and negative, so numbers are generally encoded either as **two's-complement binary** (shown to the right of the plot in **Figure 6.51b**) or **offset binary**. See Appendix B for a detailed discussion of the different representations of fixed and floating-point numbers in digital systems.

6.7.2 Quantization error

Although quantization is a nonlinear transformation, under specific circumstances we will discuss below it is reasonable to model quantization as a source of additive **quantization noise**. Specifically, rewriting Equation (6.20), we get

$$\hat{x}[n] = x[n] + e[n].$$

Here, the quantization error $e[n]$ is viewed as a source of noise that is added to the "noise-free," unquantized signal $x[n]$ to give the quantized signal $\hat{x}[n]$.

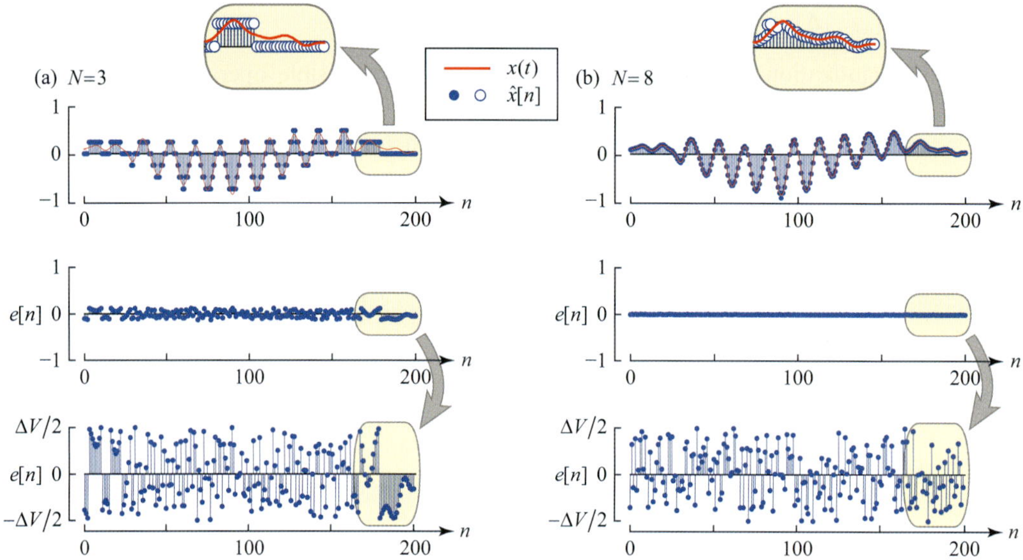

Figure 6.52 Quantization by an A/D converter

The top panel of **Figure 6.52a** shows the effect of quantizing a section of a continuous-time speech signal $x(t)$ (red trace) at a resolution of $N = 3$ bits, resulting in quantized points $\hat{x}[n]$ (blue circles). The middle panel shows the error signal $e[n] = \hat{x}[n] - x[n]$ on the same scale as the top panel and the bottom panel shows the same error signal expanded to fill the range $-\Delta V/2 \le e[n] < \Delta V/2$. For most of the waveform (except the beginning and end), $x(t)$ spans several quantizer levels. The error sequence $e[n]$ appears uncorrelated either with $\hat{x}[n]$ or with itself, and the values of $e[n]$ appear uniformly distributed in the range $-\Delta V/2 \le e[n] < \Delta V/2$. However, particularly for that portion of the waveform near the end ($n \ge 170$, shown in the inset of the top panel), $x(t)$ has a low amplitude and is only quantized into two converter levels: 0 or ΔV. The resulting error signal (middle and lower panels) appears highly correlated both with $\hat{x}[n]$ and with itself.

Figure 6.52b shows the effect of quantizing the same speech signal at a higher resolution, $N = 8$. Here, the number of quantization levels appears adequate to represent the signal accurately at all values of n, even when the signal has low amplitude. The important practical conclusion from this figure is that when designing or specifying an A/D system, the *largest* amplitude of the signal should to be adjusted to fill as much of the range of the converter as practicable ($0 - V_{REF}$ for a unipolar converter, $\pm V_{REF}$ for a bipolar converter). At any resolution of the converter (i.e., N), the absolute amplitude of the quantization noise is a constant independent of signal amplitude. Hence, N needs to be chosen to resolve the *smallest* features of the signal of interest.

If the quantization error $e[n]$ is uncorrelated either with $\hat{x}[n]$ or with itself, and $\hat{x}[n]$ varies in a relatively unsystematic manner such that quantization levels occur with roughly equal probability (e.g., speech or music), then the quantization noise of a rounding quantizer can be modeled as a random variable with a probability density function $p_e(e)$ that is uniformly distributed in the range $-\Delta V/2 \le e[n] < \Delta V/2$, as shown in **Figure 6.53**. (See Appendix D for a brief refresher of some of the relevant concepts of probability and random processes.)

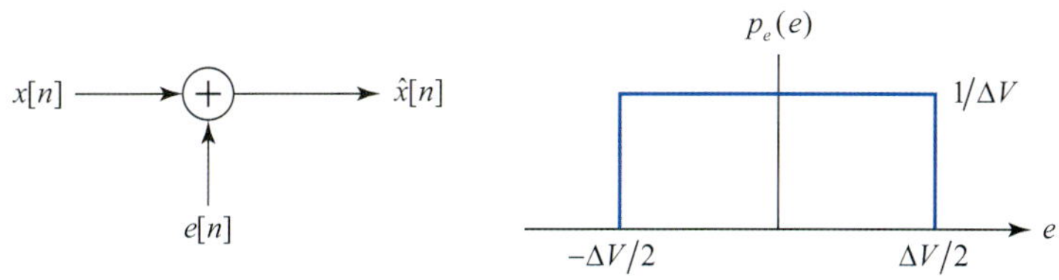

Figure 6.53 Probability density function of quantization error

The **noise power** of $e[n]$ is defined as the variance σ_e^2. For a rounding quantizer, $e[n]$ is a zero-mean process, so the variance is

$$\sigma_e^2 = \int_{-\Delta V/2}^{\Delta V/2} e^2 p_e(e)\, de = \frac{1}{\Delta V} \cdot \frac{1}{3} e^3 \Big|_{-\Delta V/2}^{\Delta V/2} = \frac{1}{\Delta V} \cdot \frac{2}{3} \left(\frac{\Delta V}{2}\right)^3 = \frac{\Delta V^2}{12}. \tag{6.21}$$

For an N-bit converter with full-scale voltage V_{FS}, Equation (6.19) relates ΔV to N and V_{FS}. Substituting this into Equation (6.21) gives

$$\sigma_e^2 = \frac{\Delta V^2}{12} = \frac{V_{FS}^2}{12 \cdot 2^{2N}}. \tag{6.22}$$

The **signal power** of a signal $x[n]$ is defined as the variance σ_x^2, and the **signal-to-noise ratio (SNR)** (expressed in decibels) is then proportional to the $\log_{10}$ of the ratio,

$$\mathrm{SNR} \triangleq 10 \log_{10} \frac{\sigma_x^2}{\sigma_e^2} = 10 \log_{10} \frac{12 \cdot 2^{2N} \sigma_x^2}{V_{FS}^2} = 6.02N + 10.79 + 20 \log_{10} \frac{\sigma_x}{V_{FS}}. \tag{6.23}$$

This expression shows that the SNR depends on three quantities: (1) the resolution of the converter (N), (2) the **root-mean-square (rms)** level of the signal (which is the standard deviation, σ_x) and (3) the maximum range of the converter (V_{FS}). Each additional bit of resolution increases the SNR by about 6 dB. In any practical application, the rms value of the input, σ_x, would be adjusted to be less than V_{FS} in order to prevent the converter from saturating. Hence, the ratio σ_x/V_{FS} will be less than one, which means that the last term of Equation (6.23) will be negative. The SNR will therefore be maximized by increasing the signal level of the signal to be as large as possible for a given V_{FS}, while taking care to keep the maximum amplitude of the signal within the range of the converter.

Example 6.5

Find the SNR of a unipolar A/D converter in which the signal is a sine wave that fills the range of the converter, namely that has an amplitude of $V_{FS}/2$.

➤ Solution:

The rms value of the sine wave of amplitude $V_{FS}/2$ is $\sigma_x = V_{FS}/2\sqrt{2}$, hence,

$$\text{SNR} = 6.02N + 10.79 + 20\log_{10}\frac{V_{FS}/2\sqrt{2}}{V_{FS}} = 6.02N + 1.76. \quad \textbf{(6.24)}$$

Equation (6.24) is often used as a "rule of thumb" to establish the **effective number of bits (ENOB)** that a converter must have to achieve a given signal-to-noise ratio for a sinusoid filling the dynamic range of the converter,

$$\text{ENOB} \triangleq \frac{\text{SNR} - 1.76}{6.02}.$$

Example 6.6

Find the effective number of bits for a unipolar converter to achieve a 96 dB signal-to-noise ratio.

➤ Solution:

$$\text{ENOB} = \left\lceil \frac{96 - 1.76}{6.02} \right\rceil = \lceil 15.65 \rceil = 16.$$

We have to round up to the nearest integer. In practice, we may have to increase the number of bits even further if an A/D converter meeting our specifications is not available at the minimum resolution specified by the ENOB.

6.7.3 Noise reduction by oversampling

Equation (6.23) indicates that the main way to increase the signal-to-noise ratio in A/D conversion is to increase the resolution, N. However, there are several other methods. In Section 6.5, we introduced the idea of the oversampled A/D converter, and showed that it offered a clever way to reduce the demands on the analog anti-aliasing filter that precedes the converter. We will now show that an N-bit oversampling converter also has a signal-to-noise ratio beyond that we would expect from Equation (6.23).

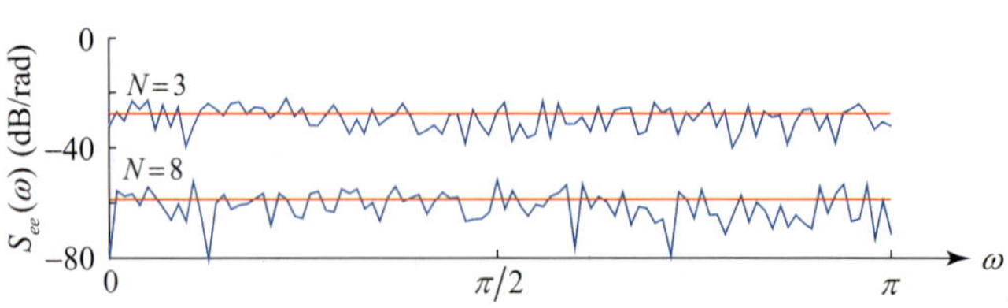

N	$10\log_{10}\sigma_e^2$
3	−27.5 dB
8	−58.5 dB

Figure 6.54 Power spectral density of quantization noise

Power spectral density of quantization noise In order to explain how oversampling can increase the SNR, we first need to understand a bit about the spectral representation of quantization noise. **Figure 6.54** shows the **power spectral density (PSD)** $S_{ee}(\omega)$ of the quantization noise $e[n]$ of the two signals that were shown in the lower panels of **Figure 6.52**, quantized with $N=3$ and $N=8$ bits.

Formally, the power spectral density $S_{ee}(\omega)$ is the Fourier transform of the autocorrelation function of $e[n]$ (see Appendix D and Chapter 14 for the full story). Essentially, the power spectral density specifies how the noise power σ_e^2 is distributed in frequency over the interval $-\pi \le \omega \le \pi$.[8] For an uncorrelated process such as $e[n]$, with a uniform probability density function $p_e(e)$, the noise power is uniformly distributed in frequency such that

$$\sigma_e^2 = \frac{1}{2\pi}\int_{-\pi}^{\pi} S_{ee}(\omega)\,d\omega.$$

Hence, the plot of $S_{ee}(\omega)$ vs. frequency is effectively flat with value (in dB) of $S_{ee}(\omega) = 10\log_{10}\sigma_e^2$, shown with the red lines in the figure for the two quantization levels $N=3$ and $N=8$. These values of σ_e^2 differ by about 30 dB, which is what we would expect from Equation (6.23) for a difference of five bits of resolution.

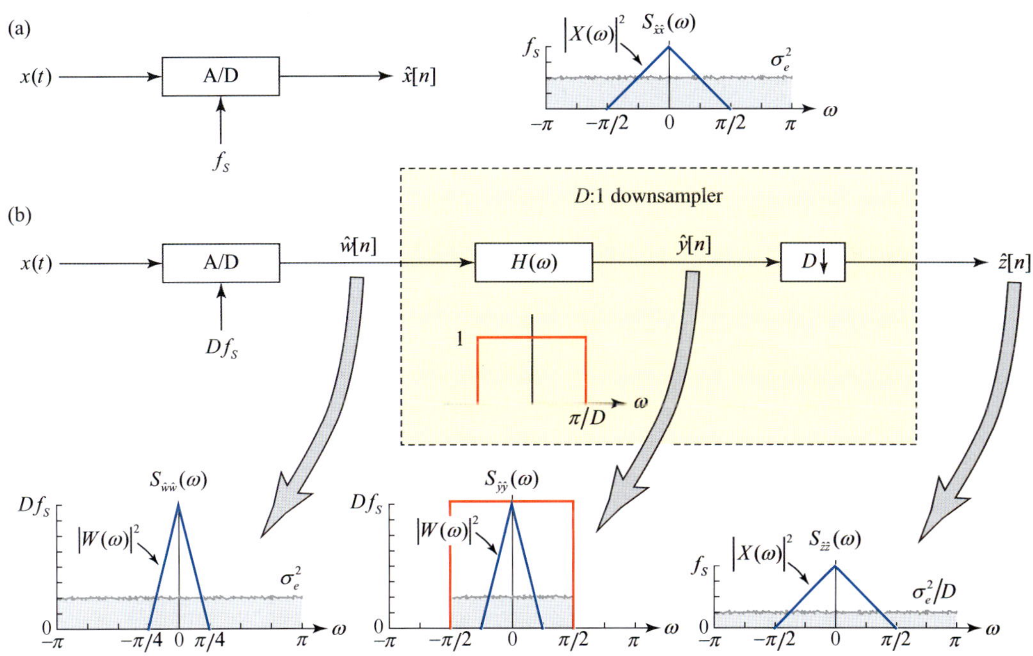

Figure 6.55 Oversampling A/D converter

Noise reduction in an oversampling A/D converter The key to understanding how over-sampling can reduce the effective quantization noise of an A/D converter – or, equivalently,

[8] The power spectral density computed by Matlab's `periodogram` function, used in Figure 6.54, is single-sided by default (i.e., it plots $2S_{ee}(\omega)$ for spectral frequencies ranging from $0 \le \omega \le \pi$). This is explained in Section 14.3.7.

increase the SNR – is that (at least theoretically) the amplitude and statistics of the quantization noise are *independent* of the sampling rate. **Figure 6.55a** shows the schematic representation of the output of an N-bit A/D converter sampling an analog signal $x(t)$ at rate f_s, yielding a sequence $\hat{x}[n]$ that is modeled as the sum of two components, an unquantized signal component $x[n]$, plus quantization noise $e[n]$, such that $\hat{x}[n] = x[n] + e[n]$. The power spectral density of the sum of these components (assuming they are uncorrelated) is

$$S_{\hat{x}\hat{x}}(\omega) \triangleq S_{xx}(\omega) + S_{ee}(\omega) = |X(\omega)|^2 + S_{ee}(\omega) = |X(\omega)|^2 + \sigma_e^2,$$

where $S_{xx}(\omega)$, the power spectral density of the signal $x[n]$, is simply the square of the magnitude of $X(\omega)$. The bandwidth of this signal component $|X(\omega)|^2$ (blue trace) is assumed to be limited to $|\omega| < \omega_B$, where in this example, $\omega_B = \pi/2$. In contrast, the noise power resulting from quantization, σ_e^2, is assumed to be spread uniformly over the range $-\pi \le \omega < \pi$, as shown by the (very exaggerated) grey area of **Figure 6.54a**.

Figure 6.55b shows the architecture of an **oversampling A/D converter**. Here, the same analog signal is oversampled at rate Df_s to form signal $\hat{w}[n]$, where $D = 2$ in this example. This signal also comprises two components, the unquantized oversampled signal component $w[n]$, plus a quantization noise component. The power density spectrum of the signal component, $S_{ww}(\omega) = |W(\omega)|^2$, has bandwidth ω_B/D ($\pi/4$ in this example) and an amplitude of Df_s. However – and this is the key point – because the quantization noise is independent of the sampling rate, the total noise power is still σ_e^2, and the power density spectrum of the noise is still uniformly distributed in the frequency range $-\pi \le \omega < \pi$. Hence, the power spectral density of the sum of these components, $S_{\hat{w}\hat{w}}(\omega)$, is

$$S_{\hat{w}\hat{w}}(\omega) \triangleq S_{ww}(\omega) + S_{ee}(\omega) = |W(\omega)|^2 + \sigma_e^2.$$

The quantized oversampled signal is now sent to a D:1 downsampler, which filters $\hat{w}[n]$ with a sharp discrete-time lowpass **decimation filter** $H(\omega)$ that has bandwidth π/D ($\pi/2$ in this example). The resulting output $\hat{y}[n]$ is a filtered random process that has power spectral density $S_{\hat{y}\hat{y}}$, given by

$$S_{\hat{y}\hat{y}}(\omega) = S_{\hat{w}\hat{w}}(\omega)|H(\omega)|^2 = (|W(\omega)|^2 + \sigma_e^2)|H(\omega)|^2 = |W(\omega)|^2 + \sigma_e^2|H(\omega)|^2.$$

(See Appendix D for a derivation of the power spectral density of a filtered random process.) Since the filter has a gain of Df_s in the frequency range of the signal, $-\pi/D \le \omega < \pi/D$, and a gain of zero outside this range, the effect of the filter is to pass the signal while reducing the total noise power by a factor of D. The second stage of the downsampler is the decimator, which expands the spectrum of *both* the signal and noise to cover the range $-\pi \le \omega < \pi$. Decimation reduces the amplitude of both signal and noise. The resulting quantized signal $\hat{z}[n]$ consists of the input $x[n]$ plus a quantization noise whose total power (i.e., the variance) has been reduced by a factor of D. Therefore, the power spectral density of the output of the oversampling A/D converter is

$$S_{\hat{z}\hat{z}}(\omega) = |X(\omega)|^2 + \sigma_e^2/D.$$

A good way to quantify the effect of noise reduction by the oversampling A/D converter is to express the increase in SNR in terms of an increase in the effective number of bits (ENOB) of the converter. First, it is useful to express D as a power of two, $D = 2^B$, where B is not necessarily an integer. Then, from Equation (6.22) the noise power is

$$\frac{\sigma_e^2}{D} = \frac{\Delta V^2}{12D} = \frac{2^{-2N} V_{FS}^2}{12 \cdot 2^B} = \frac{2^{-(2N+B)} V_{FS}^2}{12},$$

and Equation (6.23) becomes

$$\text{SNR} = 10 \log_{10} \frac{12\sigma_x^2}{2^{-(2N+B)} V_{FS}^2} = 6.02(N + 0.5B) + 10.79 + 20 \log_{10} \frac{\sigma_x}{V_{FS}}.$$

This equation says that each factor-of-two increase in oversampling adds the equivalent of half a bit of resolution, which is equivalent to increasing the effective SNR by roughly 3 dB. So, to gain the equivalent of one bit of resolution, you would have to increase the sampling rate by a factor of four. By itself, this particular approach to oversampling may not be a particularly useful way to increase the SNR, but it contains the germ of a very useful approach that we will describe in the next section that dramatically increases the SNR: the oversampling noise-shaping A/D converter.

6.7.4 ★ Noise-shaping A/D converters

The theory of the oversampling A/D converter shows that oversampling by a factor of D followed by decimation by D can reduce the noise power due to quantization, σ_e^2, by a factor of D, which corresponds to an increase in the SNR of roughly $3\log_2 D$ dB. This result assumes that the power density spectrum of the noise is uniformly distributed over the frequency range $-\pi \leq \omega < \pi$, and that it is independent of D. In this section, we will first show that we can do an even better job of noise reduction by "shaping" the quantization noise so that its power density spectrum is no longer uniform. By pushing a disproportional amount of the noise power out of the baseband towards higher frequencies, an **oversampling noise-shaping A/D converter** can achieve a significant increase in SNR. In Section 6.7.5, we will describe one variant of this converter – the one-bit **sigma-delta (Σ-Δ) converter** (also called delta-sigma converter) – that is the most popular converter for high-accuracy, high-resolution applications, such as audio signal processing.

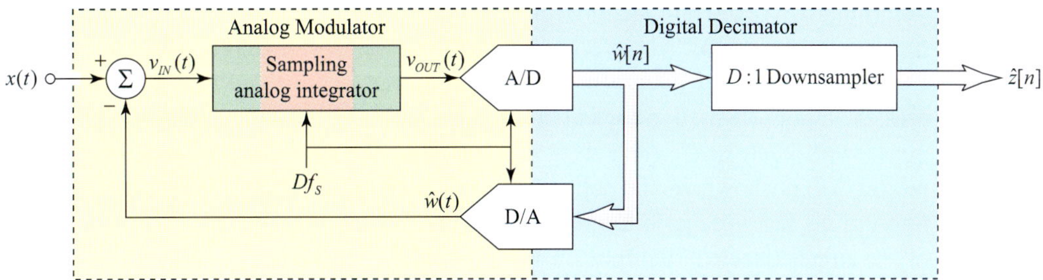

Figure 6.56 Noise-shaping A/D converter architecture

Figure 6.56 shows the overall architecture of a noise-shaping A/D converter. It comprises an analog section (called a **modulator**) and a digital section (called a **decimator**). The modulator is structured as a feedback loop that comprises a difference amplifier (schematized by Ⓢ), a

sampling discrete-time analog integrator, an A/D converter and a D/A converter. There is a lot going on here, so we shall describe this circuit in bite-size pieces and then put the pieces together to explain the whole thing.

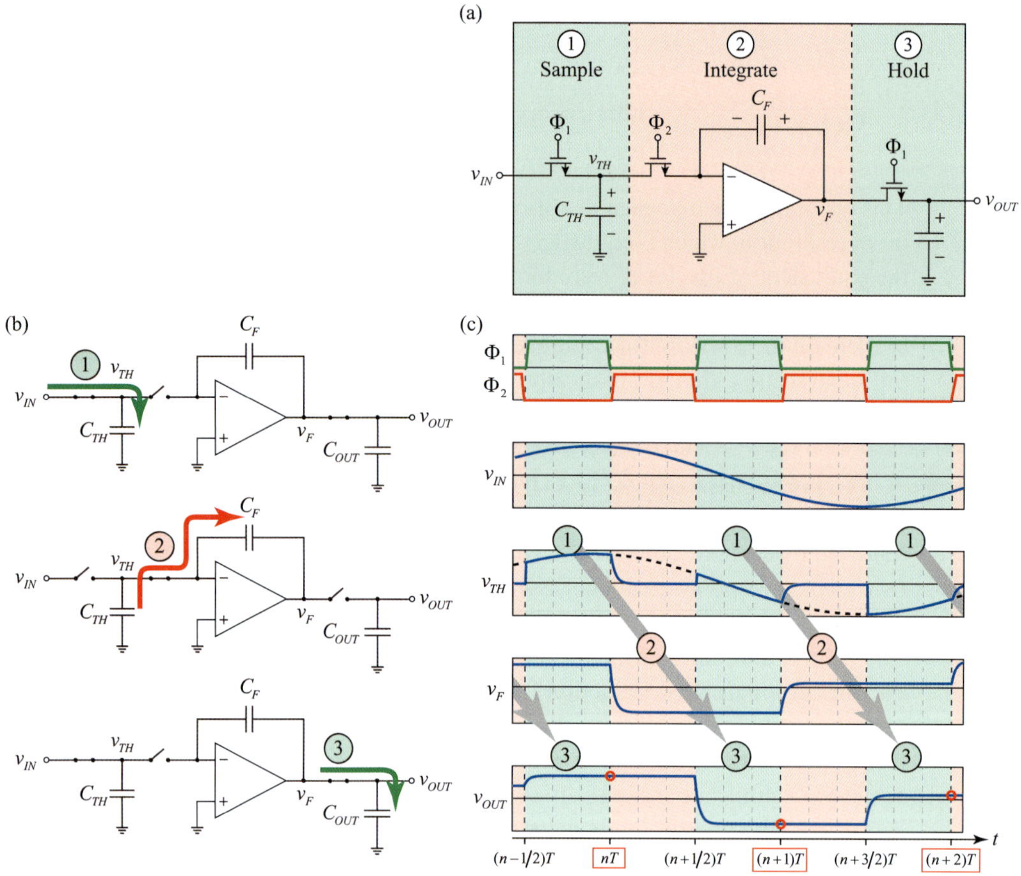

Figure 6.57 Sampling discrete-time analog integrator

Sampling discrete-time integrator The heart of the analog section is the **sampling discrete-time analog integrator**, whose schematic is shown in **Figure 6.57a**. This circuit has three parts – a sampler, an analog integrator and a hold buffer – connected by CMOS switches, which are turned on and off by two logical clock signals, Φ_1 and Φ_2. These clock signals operate out of phase with each other (i.e., $\Phi_2 = \bar{\Phi}_1$) at twice the sample frequency (i.e., half the sample period T), as shown in the top panel of **Figure 6.57c**. The other panels of this part of the figure show the voltages of important points in the circuit: v_{IN}, v_{TH}, v_F and v_{OUT}. The first part of the circuit (labeled "Sample" in **Figure 6.57a**) is a track-and-hold circuit. When clock Φ_1 is asserted (positive), the CMOS switch attaching input v_{IN} to capacitor C_{TH} is closed and current flows into the capacitor, as schematized in the top picture (labeled ①) of **Figure 6.57b**. At time nT, clock Φ_1 turns off, which disconnects the capacitor from the input. The capacitor voltage at the moment of disconnection is $v_{TH}(nT) = v_{IN}(nT)$, and its charge is therefore $C_{TH}v_{IN}(nT)$. Simultaneously, clock Φ_2 turns on, activating the CMOS switch that connects the capacitor to the input of the integrator. Because the negative terminal of the integrator acts like a virtual ground with effectively zero input impedance, when the capacitor discharges, the resulting current flows into the feedback capacitor C_F of the integrator, as schematized in the middle picture (labeled ②) of **Figure 6.57b**. At the end of this phase (i.e., at time $(n+1/2)T$), the charge on C_F has been reduced from its value at time nT by $C_{TH}v_{IN}(nT)$; that is,

$$C_F v_F((n+1/2)T) = C_F v_F(nT) - C_{TH} v_{IN}(nT).$$

Hence, the voltage at the output of the integrator is

$$v_F((n+1/2)T) = v_F(nT) - \left(\frac{C_{TH}}{C_F}\right) v_{IN}(nT).$$

At time $(n+1/2)T$, clock Φ_2 turns off, disconnecting C_{TH} from the integrator. Simultaneously, clock Φ_1 turns on again, activating the CMOS switch that connects the output of the integrator to capacitor C_{OUT}, as schematized in the bottom picture (labeled ③) of **Figure 6.57b**, where the voltage $v_{OUT}(t)$ is available to the rest of the circuit.

Because the output impedance of the integrator is essentially zero, the voltage across C_{OUT} at the end of this phase, $t=(n+1)T$, is equal to the voltage across v_F a half-cycle earlier:

$$v_{OUT}((n+1)T) = v_F((n+1)T) = v_F((n+1/2)T) = v_F(nT) - \frac{C_{TH}}{C_F} v_{IN}(nT).$$

Since $v_{OUT}(nT) = v_F(nT)$ we can write

$$v_{OUT}((n+1)T) = v_{OUT}(nT) - \frac{C_{TH}}{C_F} v_{IN}(nT),$$

or, equivalently,

$$v_{OUT}(nT) = v_{OUT}((n-1)T) - \frac{C_{TH}}{C_F} v_{IN}((n-1)T).$$

To be sure, there are many "real-world" issues – clock jitter and accuracy, parasitic capacitances, voltage "droop" of the capacitor, slew-rate of the integrator to name a few – that we have ignored. But the main point is that the output of the integrator at every discrete sampling time nT is the value of the integrator at the *previous* time $(n-1)T$, plus a constant times the value of the input at $(n-1)T$. Importantly, the values of the input and output are only updated at integer multiples of the sampling time T. This can therefore be viewed as a discrete-time sampling system with *unquantized* inputs and outputs related by difference equation

$$v_{OUT}[n] - v_{OUT}[n-1] = -\frac{C_{TH}}{C_F} v_{IN}[n-1].$$

If the circuit is designed such that $C_{TH} \simeq C_F$, and the input is buffered through an inverting amplifier, then the whole system reduces to the difference equation

$$v_{OUT}[n] - v_{OUT}[n-1] = v_{IN}[n-1].$$

Define the z-transform of this system to be $G(z)$,

$$G(z) \triangleq \frac{V_{OUT}(z)}{V_{IN}(z)} = \frac{z^{-1}}{1-z^{-1}},$$

as schematized in **Figure 6.58**.

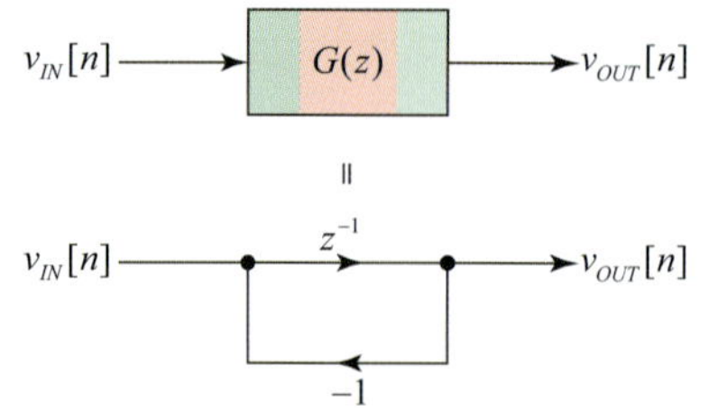

Figure 6.58 Discrete-time integrator

This is a sampling discrete-time analog integrator whose *unquantized* output $v_{OUT}[n]$ is updated every $T = 1/Df_s$ seconds.

Analysis of noise-shaping A/D converter Returning to the overall picture of **Figure 6.56**, the output of the sampling analog integrator $v_{OUT}[n]$ is an *unquantized* discrete-time signal. It is connected to an A/D converter whose output $\hat{w}[n]$ is a *quantized* N-bit binary word. Hence,

$$\hat{w}[n] = v_{OUT}[n] + e[n]. \tag{6.25}$$

The output of this A/D is then connected back to a D/A converter, which converts it into a quantized analog signal $\hat{x}(t)$. The input to the integrator, $v_{IN}(t)$, is the difference between input $x(t)$ and the quantized signal $\hat{w}(t)$,

$$v_{IN}(t) = x(t) - \hat{w}(t).$$

The analog integrator samples $v_{IN}(t)$ at an oversampled frequency of Df_s (period $T = 1/Df_s$) to produce

$$v_{IN}(nT) = x(nT) - \hat{w}(nT).$$

Recognize that $x(nT)$ and $\hat{w}(nT)$ correspond respectively to the unquantized and quantized oversampled signal components, so the equivalent discrete-time notation is

$$v_{IN}[n] = w[n] - \hat{w}[n]. \tag{6.26}$$

Figure 6.59 shows a revised block diagram for the noise-shaping A/D converter with all signals represented as discrete-time quantities operating at rate Df_s (except the output of the downsampler, which operates at rate f_s).

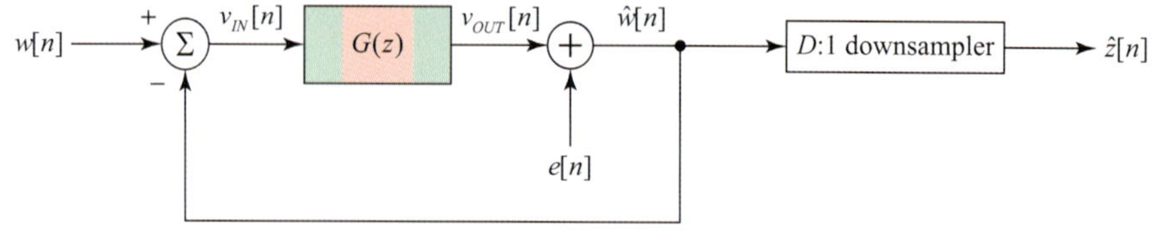

Figure 6.59 Block diagram of noise-shaping A/D converter

The key thing to note about the overall system model is that the quantization noise $e[n]$ is added *after* the integrator, as indicated in Equation (6.25). Putting it all together in terms of the z-transforms,

$$\hat{W}(z) = V_{OUT}(z) + E(z) = V_{IN}(z)G(z) + E(z) = (W(z) - \hat{W}(z))G(z) + E(z).$$

So,

$$\hat{W}(z) = \frac{G(z)}{1+G(z)}W(z) + \frac{1}{1+G(z)}E(z) = \underbrace{z^{-1}}_{F_1(z)} W(z) + \underbrace{(1-z^{-1})}_{F_2(z)} E(z) = F_1(z)W(z) + F_2(z)E(z),$$

where $F_1(z) = z^{-1}$ and $F_2(z) = 1 - z^{-1}$. Because the quantization due to the A/D converter happens after the integrator, the entire loop system filters the signal and quantization noise differently. In particular, since

$$\mathfrak{Z}^{-1}\{F_1(z)\} = \mathfrak{Z}^{-1}\{z^{-1}\} = \delta[n-1],$$

the signal is treated as if delayed by one sample with its magnitude unchanged,

$$|F_1(\omega)| = |e^{-j\omega}| = 1.$$

In contrast, the noise is filtered by a first-order highpass filter $F_2(z)$ which is, in fact, a differentiator. The magnitude of the frequency response of this filter in the range $0 \le \omega < \pi$ is

$$|F_2(\omega)| = |1 - e^{-j\omega}| = |e^{-j\omega/2}(e^{j\omega/2} - e^{-j\omega/2})| = |2je^{-j\omega/2}\sin\omega/2| = 2|\sin\omega/2| = 2\sin\omega/2.$$

The power spectral density of the output of the noise-shaping A/D converter, $S_{\hat{w}\hat{w}}(\omega)$, is then the sum of these two components: the filtered signal component $S_{xx}(\omega)$, and a filtered noise component $S_{ee}(\omega)$:

$$S_{\hat{w}\hat{w}}(\omega) = |F_1(\omega)|^2 S_{ww}(\omega) + |F_2(\omega)|^2 S_{ee}(\omega) = |W(\omega)|^2 + 4\sigma_e^2 \sin^2\omega/2.$$

Figure 6.60 shows the operation of the oversampling noise-shaping A/D converter, which you should compare to the oversampling converter without noise shaping shown in **Figure 6.55b**. The leftmost panel shows that the signal component $|W(\omega)|^2$ is unaffected by the filter $F_1(\omega)$, whereas the noise component is highpass filtered by $F_2(\omega)$ such that much of the noise power is moved out of the baseband to higher frequencies. The middle panel shows the effect of the lowpass decimation filter, which reduces the bandwidth of the noise to π/D while leaving the signal intact. Finally, the decimator expands the spectrum of both signal and noise to fill the range $-\pi \le \omega < \pi$. The power spectral density of the output of the oversampling noise-shaping A/D converter is

$$S_{\hat{z}\hat{z}}(\omega) = |X(\omega)|^2 + (4\sigma_e^2/D)\sin^2(\omega/2D).$$

The total noise power in the output is computed as the average of the integral of the power spectral density,

$$\frac{1}{2\pi}\int_{-\pi}^{\pi}(4/D)\sigma_e^2\sin^2\omega/(2D)\,d\omega = \frac{2\sigma_e^2}{\pi}\int_{-\pi/D}^{\pi/D}\sin^2\omega/2\,d\omega = \frac{\sigma_e^2}{\pi}\int_{-\pi/D}^{\pi/D}(1-\cos\omega)d\omega$$

$$= \frac{2\sigma_e^2}{\pi}(\pi/D - \sin\pi/D). \quad \textbf{(6.27)}$$

When $D = 1$, the total noise power is $2\sigma_e^2$, which is actually worse than if there had been no noise shaping at all! But, for any amount of oversampling, $D > 1$, things get better very quickly, because we are moving the preponderance of the noise out of frequency range of the baseband.

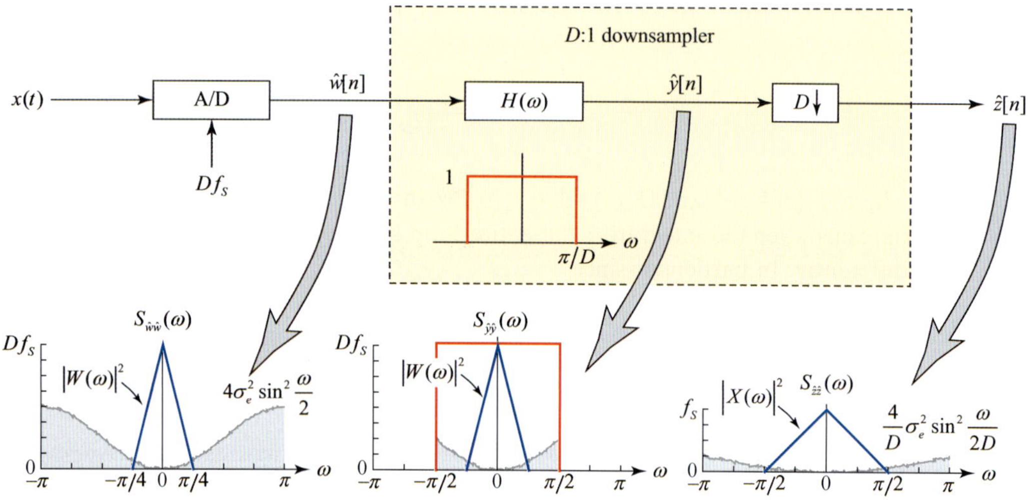

Figure 6.60 Oversampled noise-shaping A/D converter

Figure 6.61 shows the improvement in SNR that one can expect in an A/D converter as a result of oversampling by itself without noise shaping, as we described in conjunction with **Figure 6.55** (blue symbols) and for an oversampling converter with first-order noise shaping (green symbols), as just described in conjunction with **Figure 6.60**.

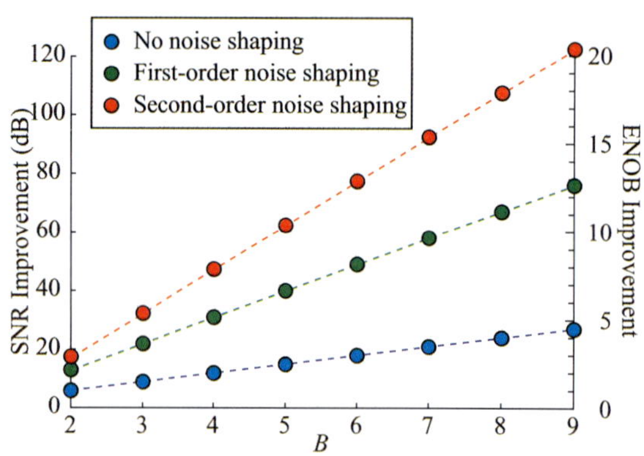

Figure 6.61 SNR improvement due to oversampling and noise shaping

The SNR improvement (left ordinate axis) is plotted against $B = \log_2 D$. For the converter with no noise shaping, the data are linear on this log–log scale so that every factor-of-two increase in D increases the SNR by about 3 dB, as shown in the first row of **Table 6.1**. From Equation (6.23), each improvement in SNR of 6.02 dB corresponds to one additional bit of resolution, so $3.01B$ dB is equivalent to a $0.5B$ bit improvement. The ENOB improvement is shown on the right axis of **Figure 6.60**. For example, oversampling by a factor of 256 ($B = 8$) results in an increase in

SNR of about 24 dB or four additional bits of resolution. For the converter with first-order noise shaping, the data for $B > 1$ are also well fit by a straight line.[9] Here, every factor-of-two increase in D increases the SNR by about 9 dB or 1.5 bits, as shown on the second line of **Table 6.1**. Higher-order noise-shaping networks can be constructed using similar principles and can improve the SNR even more. The red symbols in **Figure 6.61** are for a second-order noise-shaping network, created by putting two integrators in series in the modulator feedback loop (see Problem 7-28). Here, each factor-of-two increase in D increases the SNR by about 15 dB or the equivalent of 2.5 bits of resolution. That is a big deal!

Table 6.1 SNR and ENOB improvement of oversampling noise-shaping A/D converters

Noise shaping	SNR improvement (dB)	ENOB improvement
None	$3.01B$	$0.5B$
First-order	$9.03B - 5.17$	$1.5B - 0.86$
Second-order	$15.05B - 12.87$	$2.5B - 2.14$

Figure 6.62 shows simulated data for the noise-shaping A/D converter "designed" with an oversample factor of $D = 32$ and perfectly matched four-bit internal A/D and D/A converters.

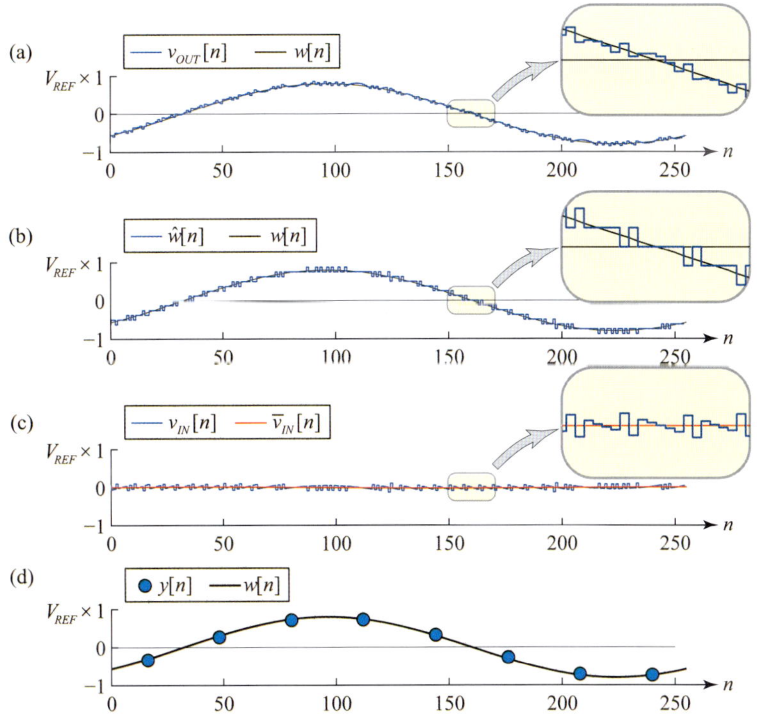

Figure 6.62 Simulation of oversampling A/D converter

[9] Straight line fits in this table are minimum-mean-square error fits of data in the range $4 \leq B \leq 12$.

Figure 6.62a shows the output of the integrator, $v_{OUT}[n]$ (blue trace), superimposed upon the input to the loop, $w[n]$ (thin black line), as a function of discrete-time intervals $t = nT$, where $T = 1/Df_s$, **Figure 6.62b** shows $\hat{w}[n]$, the N-bit quantized version of $v_{OUT}[n]$, and **Figure 6.62c** shows $v_{IN}[n]$, the difference between $w[n]$ and $\hat{w}[n]$, which forms the error signal that is the input to the integrator. The converter's feedback loop acts to keep $v_{IN}[n]$ oscillating around zero. To appreciate the operation of the loop, consider what happens at a particular moment if $w[n]$ exceeds $\hat{w}[n]$. Then, the error $v_{IN}[n] = w[n] - \hat{w}[n]$ goes positive and that difference is added to the output of the integrator at the next sample time, $v_{OUT}[n+1]$. The increased value of $v_{OUT}[n+1]$ causes $w[n+1]$ to increase, which reduces $v_{IN}[n+1]$. The end result is that $\hat{w}[n]$ tracks $w[n]$. **Figure 6.62d** shows the final output of the converter, $y[n]$, formed by the average of $w[n]$ over successive "frames" of $D = 16$ points.

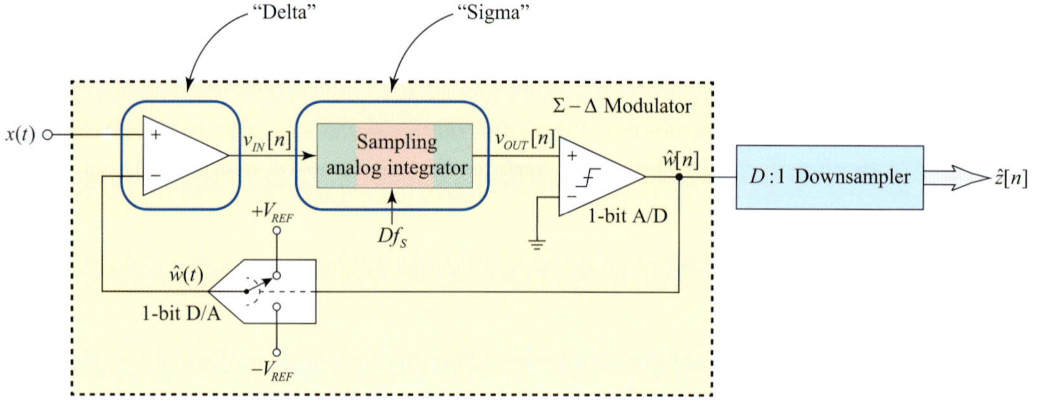

Figure 6.63 Sigma-delta converter

6.7.5 ★ Sigma-delta A/D converters

For a noise-shaping A/D converter, such as that discussed in the previous example and shown in **Figure 6.56**, the A/D and D/A converters in the loop filter can generally have N-bit resolution. However, one of the most popular and successful designs for many applications is the **sigma-delta ($\Sigma-\Delta$) A/D converter**, in which the A/D and D/A converters in the loop filter have a resolution of just one bit! How can this possibly work?

The revised block diagram for a sigma-delta converter with a first-order modulator is shown in **Figure 6.63**. The "sigma" is the sampling analog integrator and the "delta" is the difference amplifier, both of which we have seen before. However, the N-bit A/D converter in **Figure 6.56** has been replaced by a simple comparator, which is equivalent to a one-bit A/D that returns logic level 0 or 1. Similarly, the N-bit D/A has been replaced by a one-bit D/A that selects either a positive voltage $+V_{REF}$ or a negative voltage $-V_{REF}$, depending on the output of the one-bit A/D. A key advantage of the sigma-delta converter is that these one-bit A/D and D/A converters in the loop filter are incredibly simple to implement. They do not require the large

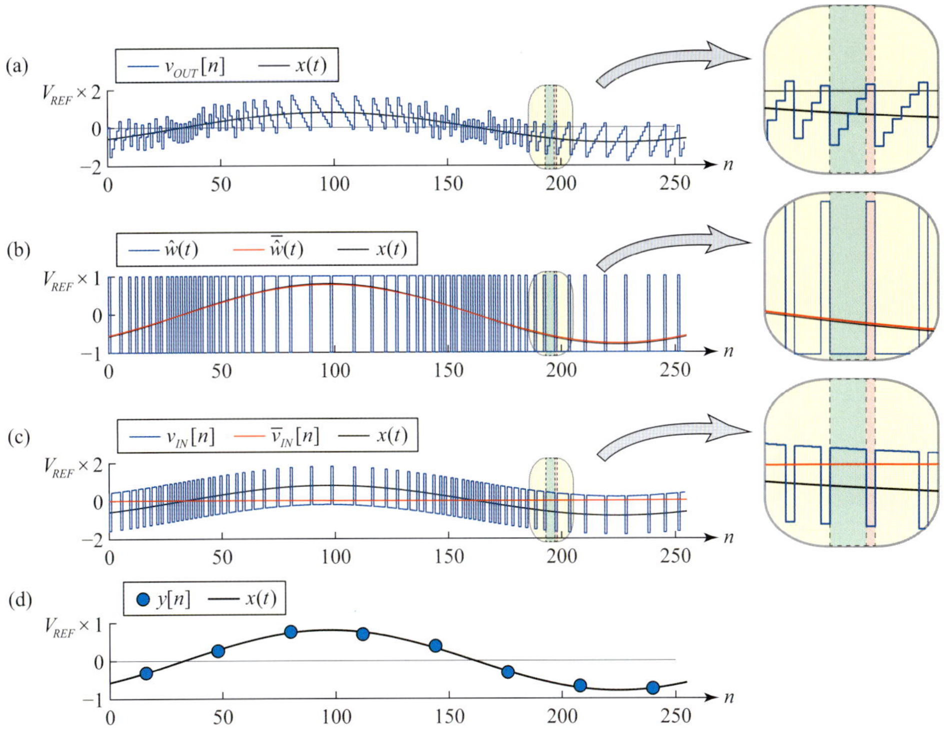

Figure 6.64 Simulation of sigma-delta A/D converter

input voltage range, linearity and precisely matched components of the N-bit A/D and D/A converters in the loop filter of the multi-bit oversampling converter. The simple architecture of the sigma-delta converter also scales well to very high resolutions, so much so that sigma-delta converters in resolutions exceeding 20 bits have become the converters of choice for high resolution, low-noise, relatively low-speed applications such as digital audio.

Figure 6.64 shows simulated data for various points in a sigma-delta A/D converter oversampled with a factor of $D = 32$. The highly jagged waveforms are somewhat daunting to look at, so in order to understand what is going on here, focus on the small sections of the traces shown magnified in the inset figures. Over this short time period, the input $x(t)$, plotted with the thin black line in **Figure 6.64a**, is less than zero, but greater than negative full scale: $-V_{REF} < x(t) < 0$. Five samples of $v_{OUT}[n]$ in the middle of the inset are highlighted in color. For the first four samples, highlighted in green, the output of the sampling A/D is less than zero: $v_{OUT}[n] < 0$. Hence the output of the A/D comparator is zero, $\hat{w}[n] = 0$, and the output of the one-bit D/A converter is full-scale negative, $\hat{w}[n] = -V_{REF}$, as shown in the inset of **Figure 6.64b**. The output of the difference amplifier, given by Equation (6.26), is therefore $v_{IN}[n] = w[n] - \hat{w}[n] = w[n] + V_{REF}$, which is greater than zero as shown in the inset of **Figure 6.64c**. $v_{IN}[n]$ is the input of the integrator, so on the next time point, the output of the integrator, $v_{OUT}[n]$, increases in a stepwise manner. This continues for each of the four samples shown in green, until on the fifth sample, highlighted in red, v_{OUT} exceeds zero. At this time, the comparator goes positive, which immediately makes $w[n] = +V_{REF}$, as shown in **Figure 6.64b**. This causes $v_{IN}[n]$ to go negative, $v_{IN}[n] = w[n] - \hat{w}[n] = w[n] - V_{REF}$, resulting in a big negative jump in the output of the integrator, $v_{OUT}[n]$, on the next time point. The essential point of the sigma-delta converter is that despite these large jumps in the voltages of the signals, the feedback loop of the sigma-delta feedback loop, like that of the loop with the N-bit converters, always acts to keep the *average value* of the input to the integrator, $\bar{v}_{IN}[n]$, zero, as shown by the thin red line in **Figure 6.64c**. As a consequence, the output of the one-bit A/D, $\hat{w}(t)$, of the sigma-delta A/D is effectively a pulse-width modulated signal whose average value $\overline{\hat{w}}(t)$ (plotted with the red line) in **Figure 6.64b**, closely tracks the input $x(t)$.

The output of the sigma-delta modulator, $\hat{w}[n]$, is a one-bit data stream at an oversampled frequency Df_s that can be in the multi-MHz range for sigma-delta converters targeted at audio applications. $\hat{w}[n]$ looks identical to $\hat{w}(t)$ in **Figure 6.64b** at values of $t = nT$, except it has binary values: zero or one. This data stream is passed to the final part of the sigma-delta converter, the D:1 downsampler, whose function is to reduce the sample rate to the Nyquist rate f_s, leaving the signal intact while truncating the spectrum of the output modulator to remove as much of the shaped quantization noise as possible. A number of ingenious methods have been developed that allow the downsampler in a sigma-delta converter not only to provide the decimation filter and reduce sample rate, but to do so in real time and with limited computational resources. The basic downsample task is to divide the one-bit data stream $\hat{w}[n]$ into frames of size D, as indicated in **Figure 6.65** for the case of a $D = 8$ downsample. The number of ones in each frame is the average value of $x[n]$ over the frame, which can range from 0 (i.e., all D bits 0) to 1 (i.e., all D bits 1). Given D as a power of two, each frame can be encoded as an N-bit word, where $N = \log_2 D$. In the example of **Figure 6.65**, each frame of eight bits is encoded into a three-bit word.

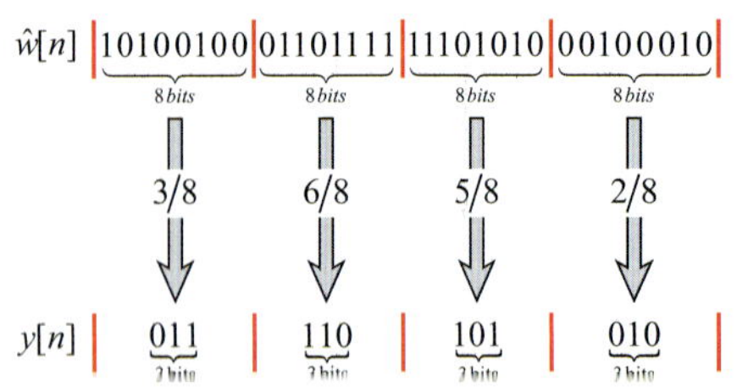

Figure 6.65 Downsampling from eight bits to three

The result of this downsampling is a lowpass output data stream at a rate of f_s, plotted with blue dots in **Figure 6.64d**.

6.8 ★ A/D converter architecture

In the preceding sections of Chapter 6, we discussed the theory of A/D conversion, including the effect of quantization. In supplementary material, we present a detailed discussion of some of the main hardware implementations of A/D converters. By way of introduction, **Figure 6.66a** shows a schematic representation of a typical A/D converter.

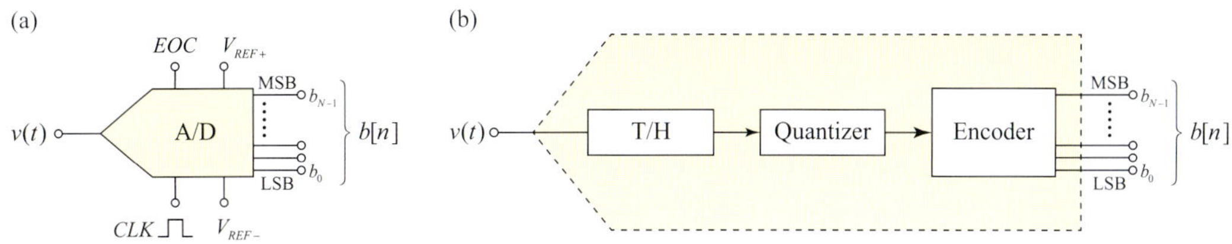

Figure 6.66 A/D converter architecture

This A/D converter converts a continuous-time analog signal $v(t)$ into an N-bit digital word $b[n]$. The range of the input signal is assumed to be in the range of $V_{REF-} < v(t) < V_{REF+}$. Most commonly, V_{REF+} is a positive supply voltage ranging from +1.9 to 5 volts for low-power circuits and up to +15 volts for circuits powered by standard split supplies (e.g., ±15 volts). V_{REF-} is usually 0 volts for single-supply circuits and equal to $-V_{REF+}$ for split supplies.

A/D conversion is triggered by the edge of a logical clock signal (CLK), which in most applications is periodic at the sampling period T. The end of each conversion is marked by a logical end-of-conversion (EOC) signal, which signals to the rest of the system that the converted data are available to be read. The data from each conversion is the N-bit digital word $b[n]$, which may be made available to the rest of the data acquisition system either in parallel form – through N parallel data paths, schematized in the figure as $b_0, \ldots, b_{N-1}$, where b_0 represents the least significant bit (LSB) and b_{N-1} is the most significant bit (MSB) – or in serial form, for example using the I^2C bus protocol for low-speed transfers (100 kbps to ~1 Mbps), the SPI protocol for medium speed (~10 Mbps) or the JESD204 high-speed (~3 Gbps) interface.

Figure 6.66b is a block diagram of the main conceptual elements of a practical A/D converter: a track-and-hold (T/H) circuit, a quantizer and an encoder. Variations of these basic elements appear in all the A/D architectures. Different A/D architectures are designed to balance the trade-off among performance characteristics such as speed, resolution and power consumption. In supplementary material, we provide a detailed description of a number of the most common of these A/D architectures – flash, successive approximation, subranging, pipelined, oversampling and sigma-delta A/D converters.

6.9 ★ D/A converter architecture

There are many different types of D/A converters, some of which share a lot of features with the corresponding A/D technologies. **Figure 6.67a** shows a schematic representation of a typical D/A converter. D/A converters are available with both serial and parallel inputs. In this instance, the input is the N-bit parallel digital word $b[n]$, where b_0 represents the least significant bit (LSB) and b_{N-1} is the most significant bit (MSB). D/A converters are available with either current or voltage output. In this example, the output is a voltage whose range, $V_{REF-} \leq v(t) \leq V_{REF+}$, is determined by externally applied output voltages V_{REF+} and V_{REF-}. **Figure 6.67b** shows the input–output relation between $b[n]$ and $v(t)$ for a three-bit D/A converter. Both **bipolar D/A converters** and **unipolar D/A converters** are available. Bipolar converters require split supplies and are typically used in applications requiring a symmetric voltage output range $-V_{REF} \leq v(t) \leq V_{REF}$. Unipolar converters are those in which $V_{REF-} = 0$ volts, so the output voltage range is $0 \leq v(t) \leq V_{REF}$. These converters are typically found in single-supply and embedded applications, where V_{REF} is generally in the range of 1.3 to 5 volts. A simple unipolar-to-bipolar circuit can also be designed to convert the output of a unipolar D/A converter to a desired bipolar output (see Problem 6-13). In supplementary material, we discuss the hardware architecture of several of the most commonly used D/A converters.

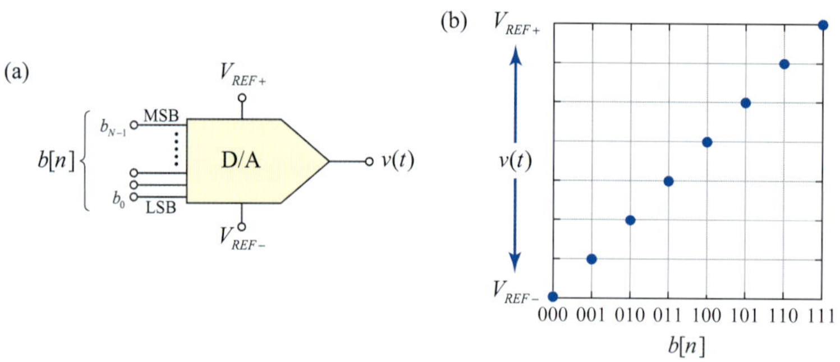

Figure 6.67 D/A converter architecture

SUMMARY

Analog-to-digital and digital-to-analog converters enable the power of digital signal processing algorithms to be applied to real-world problems. In this chapter, we first covered the basic theory of both A/D and D/A conversion and then discussed the elements of sample-rate conversion – upsampling, downsampling and resampling. In Chapter 13, we will extend basic notions of sample-rate conversion to **multirate systems**, which are discrete-time systems in which multiple sample rates operate simultaneously within the same system. These systems are designed to minimize the number of operations required to implement upsampling, downsampling, resampling and filtering.

PROBLEMS

Problem 6-1

The discrete-time filtering system shown in **Figure 6.68** comprises an A/D converter sampling at rate f_1, a discrete-time filter with frequency response $H(\omega)$ and an ideal D/A converter reconstructing at rate f_2. Ideal means that the converter contains an ideal lowpass reconstruction filter with a bandwidth of πf_2 and a gain of $1/f_2$. The spectrum of the input, $X(\Omega)$, is shown in **Figure 6.68**. Provide a fully labeled sketch of $X(\omega)$, $Y(\omega)$ and $Y(\Omega)$ for each of the following cases:

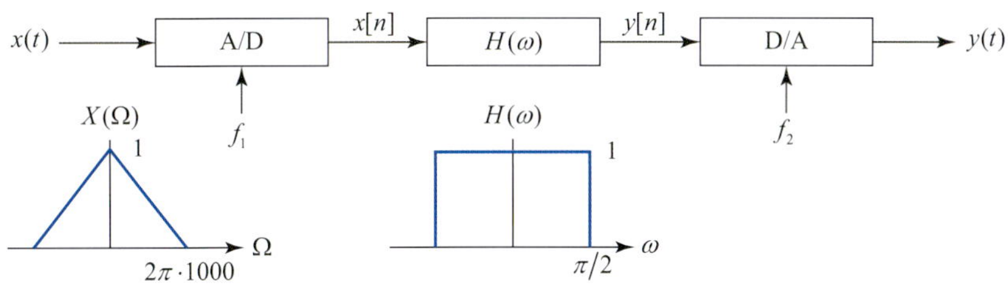

Figure 6.68

(a) $f_1 = f_2 = 4000$ Hz.
(b) $f_1 = f_2 = 2000$ Hz.
(c) $f_1 = 4000$ Hz, $f_2 = 2000$ Hz.
(d) $f_1 = 2000$ Hz, $f_2 = 4000$ Hz.

Problem 6-2

Given the discrete-time filtering system of **Figure 6.68** with $x(t) = \cos 2\pi \cdot 1000t$, provide a fully labeled sketch of $X(\omega)$, $Y(\omega)$ and $Y(\Omega)$ and find $y(t)$ for each of the following cases:
(a) $f_1 = f_2 = 4000$Hz.
(b) $f_1 = f_2 = 2000$ Hz.
(c) $f_1 = f_2 = 1333$ Hz.
(d) $f_1 = f_2 = 1000$ Hz.

Problem 6-3

Given the discrete-time filtering system of **Figure 6.68** with $f_1 = f_2 = 4000$Hz and the filter given by

$$H(\omega) = \frac{1 - \sqrt{2}e^{-j\omega}}{2 - \sqrt{2}e^{-j\omega}},$$

find the output $y(t)$ when the input is $x(t) = \cos 2\pi \cdot 500t$.

Problem 6-4

The discrete-time filtering system shown in **Figure 6.69** comprises an A/D converter sampling at rate f_1, a "bandpass upsampler" and an ideal D/A converter operating at rate f_2. "Ideal" means that the converter contains an ideal lowpass reconstruction filter with a bandwidth of πf_2 and a gain of $1/f_2$. The bandpass upsampler first expands sequence $x[n]$ by putting $U - 1$ zeros between adjacent points to produce intermediate output $x_e[n]$. Then, instead of the usual lowpass filter, the upsampler has a bandpass filter $H(\omega)$ with critical frequencies ω_1 and ω_2, and a gain of A. The spectra of the input and output of the complete system, $X(\Omega)$ and $Y(\Omega)$, are shown in the lower panel of the figure.

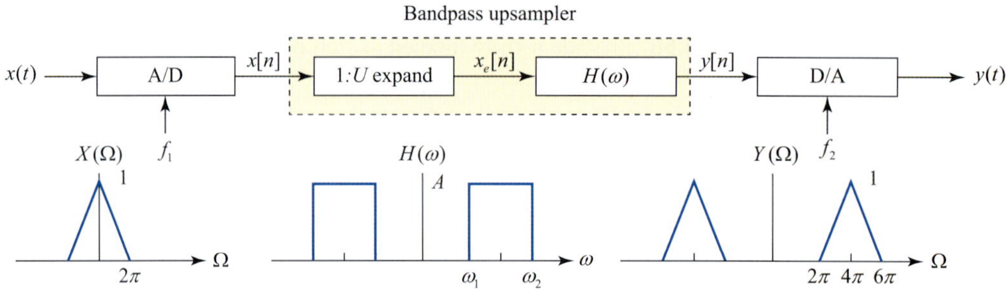

Figure 6.69

(a) Given $f_1 = 2$ Hz and $U = 2$, sketch $X(\omega)$, $X_e(\omega)$ and $Y(\omega)$, and find suitable values of f_2, ω_1, ω_2 and A such that the system performs as shown.

(b) Given $f_1 = 2$ Hz and $U = 3$, sketch $X(\omega)$, $X_e(\omega)$ and $Y(\omega)$, and find suitable values of f_2, ω_1, ω_2 and A such that the system performs as shown.

Problem 6-5

The discrete-time filtering system shown in **Figure 6.70** comprises an A/D converter sampling at rate f_1, a discrete-time filter with frequency response $H(\omega)$, a resampler that resamples at rate $D{:}U$ and an ideal D/A converter at rate f_2. "Ideal" means that the converter contains an ideal lowpass reconstruction filter with a bandwidth of πf_2 and a gain of $1/f_2$. Assume that the resampler is ideal (upsample by padding $y[n]$ with $U-1$ zeros, discrete-time filter with gain of U and bandwidth of $\pi/\max(U, D)$, downsample at D, tossing $D-1$ points). The spectrum of the input, $X(\Omega)$, is shown in the lower panel of the figure. For each of the following parts, plot the spectra $X(\omega)$, $Y(\omega)$, $Z(\omega)$ and $Z(\Omega)$.

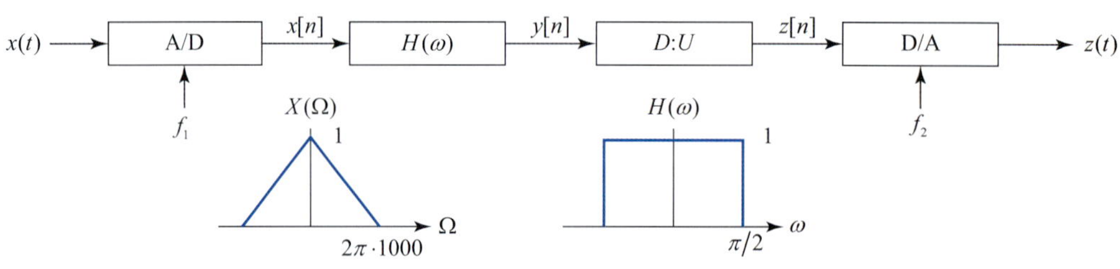

Figure 6.70

(a) $f_1 = 2000$ Hz, $f_2 = 1000$ Hz, $U = 1$, $D = 2$.
(b) $f_1 = 2000$ Hz, $f_2 = 4000$ Hz, $U = 2$, $D = 1$.

Problem 6-6

The discrete-time filtering system shown in **Figure 6.70** comprises an A/D converter sampling at rate $f_1 = 6000$ Hz, a filter with frequency response $H(\omega)$, as shown in the figure, a 2 : 1 downsampler and an ideal D/A converter reconstructing at rate $f_2 = 3000$ Hz. The input is $x(t) = 1 + \cos(2\pi \cdot 1000t) + \cos(2\pi \cdot 2000t)$. Provide a fully labeled sketch of $X(\omega)$, $Y(\omega)$, $Z(\omega)$ and $Z(\Omega)$.

Problem 6-7

Consider the discrete-time filtering system shown in **Figure 6.70**, except with the filter shown in **Figure 6.71**. Let $x(t) = \cos(2\pi \cdot 500t) + \cos(2\pi \cdot 1000t)$. For each of the following parts, sketch $X(\omega)$, $Y(\omega)$, $Z(\omega)$ and $Z(\Omega)$ and find $z(t)$.

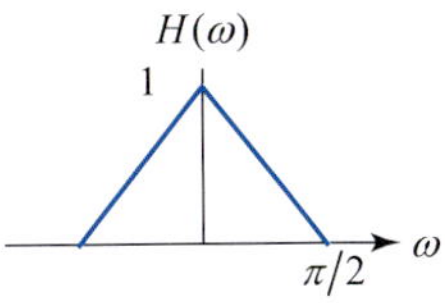

Figure 6.71

(a) $f_1 = f_2 = 4$ kHz and $U = D = 1$.
(b) $f_1 = 8$ kHz, $f_2 = 4$ kHz, $U = 1$ and $D = 2$.
(c) $f_1 = 8$ kHz, $f_2 = 16$ kHz, $U = 2$ and $D = 1$.

Problem 6-8

A discrete-time filtering system comprises an A/D converter sampling at rate f_1, decimation by a factor of D such that $y[n] = x[Dn]$, a discrete-time filter with frequency response $H(\omega)$ and bandwidth $3\pi/4$ and an ideal D/A converter operating at rate f_2, as shown in **Figure 6.72**. The "ideal" D/A converter contains an ideal lowpass reconstruction filter with a bandwidth of πf_2 and a gain of $1/f_2$. The input to the system is $x(t) = \cos(2\pi \cdot 1000t)$. For each of the following parts, make a detailed, accurate sketch of $X(\omega)$, $Y(\omega)$, $Z(\omega)$ and $Z(\Omega)$ and find $z(t)$.

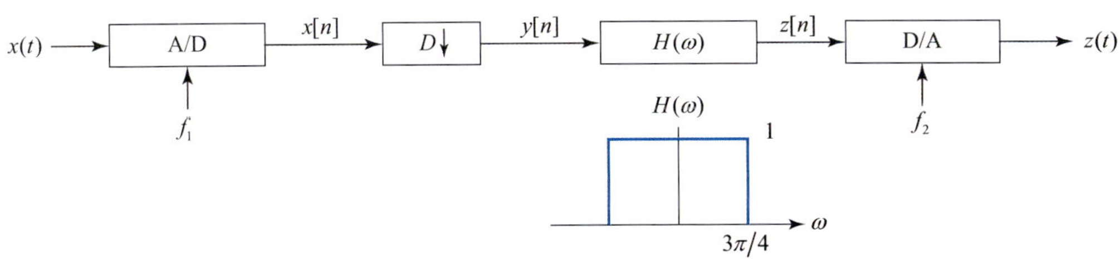

Figure 6.72

(a) $f_1 = 8$ kHz, $f_2 = 4$ kHz, $D = 2$.
(b) $f_1 = 4$ kHz, $f_2 = 2/3$ kHz, $D = 6$.

Problem 6-9

Same set-up as Problem 6-8, but with $x(t) = \cos(2\pi \cdot 1000t) + \cos(2\pi \cdot 2000t)$ and other parameters as follows:

(a) $f_1 = 10$ kHz, $f_2 = 5$ kHz and $D = 2$.
(b) $f_1 = 10$ kHz, $f_2 = 3333$ kHz and $D = 3$.

Problem 6-10

A discrete-time filtering system shown in **Figure 6.73** comprises an A/D converter sampling at rate $f_1 = 8000$ Hz, a 2:1 downsampler, a filter with frequency response $H(\omega)$ and an ideal D/A converter reconstructing at rate $f_2 = 4000$ Hz. "Ideal" means that the converter contains an ideal lowpass reconstruction filter with a bandwidth of πf_2 and a gain of $1/f_2$. Provide a fully labeled sketch of $X(\omega)$, $W(\omega)$, $Y(\omega)$, $Z(\omega)$ and $Z(\Omega)$ for each of the following cases:

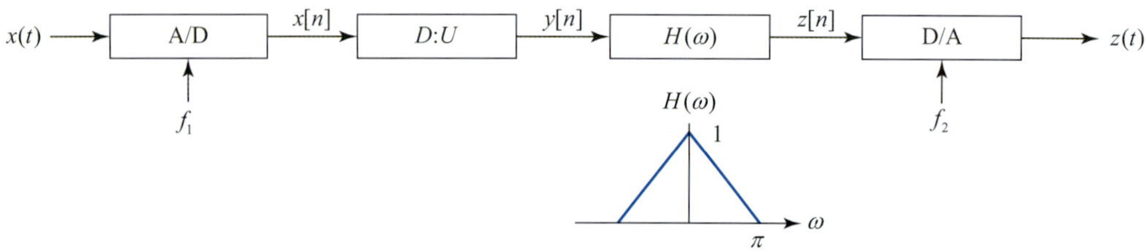

Figure 6.73

(a) $x(t) = \cos(2\pi \cdot 500t) + \cos(2\pi \cdot 1000t)$.
(b) $x(t) = \cos(2\pi \cdot 1500t) + \cos(2\pi \cdot 3000t)$.

Problem 6-11

A discrete-time demodulator shown in **Figure 6.74** comprises an A/D converter sampling at rate f_1, a discrete-time filter with response $H(\omega)$ and an ideal D/A converter reconstructing at rate f_2. The A/D converter has no anti-aliasing filter in it, so frequencies in the input greater than $2f_1$ are not attenuated. Given that the analog signal has an input spectrum, $X(\Omega)$, shown in the figure, with $\Omega_b = 2\pi \cdot 2$ kHz and $\Omega_c = 2\pi \cdot 10$ kHz, design a system such that the output spectrum of the analog signal is $Y(\Omega)$.

➤**Hint:** Think "undersampling."

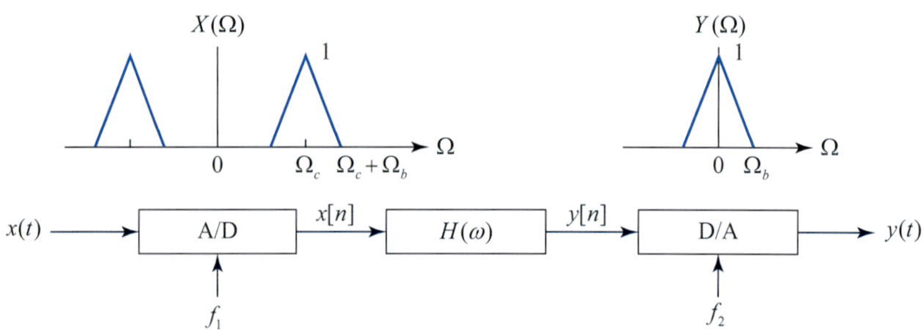

Figure 6.74

Problem 6-12

In Section 6.5.5, we discussed the procedure to upsample an input sequence $x[n]$ by a factor of U to create an output sequence $y[n]$. First, create the expanded sequence $x_e[n]$, Equation (6.16),

$$x_e[n] = \sum_{k=-\infty}^{\infty} x[k]\delta[n-kU] = \begin{cases} x[n/U], & n = 0, \pm U, \pm 2U, \ldots \\ 0 & \text{otherwise} \end{cases},$$

and then filter $x_e[n]$ with a discrete-time image-rejection filter that has impulse response $h_u[n]$. Consider the specific case of an FIR Kaiser filter of length N, specified by

$$h_u[n] = w[n]\,\mathrm{sinc}\,(n\pi/U), \quad |n| < (N-1)/2,$$

where $w[n]$ is a symmetric Kaiser window of length N. For the purposes of this problem, the exact shape of the Kaiser window does not matter, only that $w[0] = 1$. Using the expressions for $x_e[n]$, $h_u[n]$ and the fact that $y[n] = x_e[n] * h_u[n]$, show that values of the interpolated sequence $y[n]$ at multiples of U are exactly equal to the corresponding values of the original sequence; that is, $y[nU] = x[n]$.

Problem 6-13

A unipolar D/A converter shown in **Figure 6.75a** is assumed to have a voltage range of $0 \le v(t) \le V_{REF}$. The job of the unipolar-to-bipolar converter shown in **Figure 6.75b** is to convert the input voltage $v(t)$ into an output $\hat{v}(t)$, whose voltage spans the range $-\hat{V}_{REF} \le \hat{v}(t) \le +\hat{V}_{REF}$.

(a) Show that the voltage output of the converter, $\hat{v}(t)$, is

$$\hat{v}(t) = \hat{v}_1(t) + \hat{v}_2(t) = v(t)(1 + R_F/R_A + R_F/R_B) - V_{REF}R_F/R_B.$$

➤**Hint:** Consider $\hat{v}(t)$ as the sum of two terms, $\hat{v}(t) = \hat{v}_1(t) + \hat{v}_2(t)$. $\hat{v}_1(t)$ is the output of a non-inverting op-amp whose input $v(t)$ is the output of the unipolar D/A converter (with V_{REF} set to zero). $\hat{v}_2(t)$ is the output of an inverting op-amp due to voltage V_{REF} connected via a resistor, with the output of the unipolar D/A converter set to zero, $v(t) = 0$.

(b) Given a value of $V_{REF} = 2.5$ V and a nominal value of $R_B = 10$ kΩ, find values of R_A and R_F such that the bipolar converter has a full-scale voltage of $\pm\hat{V}_{REF} = \pm 10$ V.

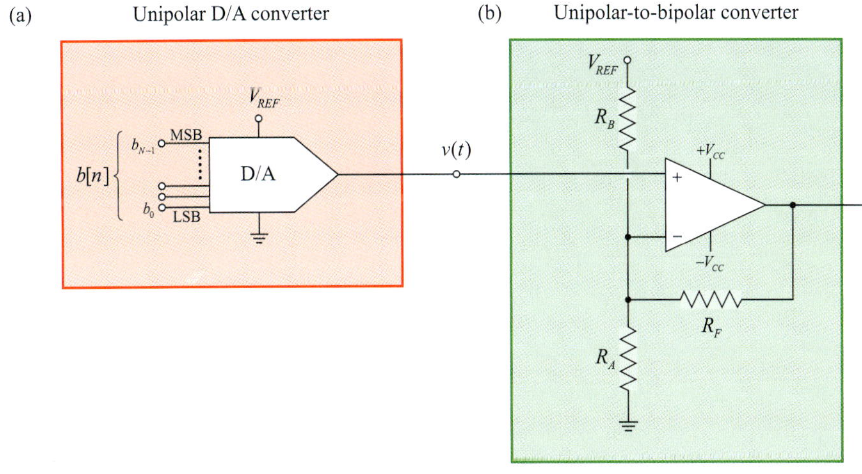

Figure 6.75

7 Finite impulse response filters

Introduction

As the name suggests, finite impulse response (FIR) filters have impulse responses that are of finite duration. For an FIR filter with impulse response $h[n]$, the output of the filter $y[n]$ is the finite-length convolution of input $x[n]$ with $h[n]$,

$$y[n] = \sum_{k=N_1}^{N_2} h[k]x[n-k],$$

where $h[n]$ is 0 outside the range $N_1 \leq n \leq N_2$.

FIR filters have many advantages that recommend them to the DSP designer. They are always stable and can be designed to fit practically any specified frequency-response profile. Most FIR filters are designed to have linear phase, something which is not possible with infinite impulse response (IIR) filters. FIR filters can be implemented by convolution in the time domain, or by multiplication of Fourier transforms in the frequency domain, which, when done using the fast Fourier transform (FFT), results in very fast computation times (see Chapter 11).

In this chapter, we will describe a number of methods of designing FIR filters to meet specific design criteria. These methods fall into four broad categories: **window-based** methods, **frequency-sampling** methods, **least-square-error** methods and **optimal design** methods. Window-based methods, which we cover first, are the easiest to understand and implement. If you only read one section of this chapter, it should be this one. The performance of window-based filters is high enough to meet the specifications of many common DSP tasks. However, these filters have a limited number of design parameters that can be altered, and the filter length necessary to meet given design criteria is not as low as can be achieved with other methods. Frequency sampling methods produce filters that match a desired frequency response at specific points. Least-square error and optimal design methods minimize some measure of the error between the frequency response of the designed filter and the desired frequency response over a range of frequencies. These methods are more mathematically involved but the filters they produce have smaller lengths than other methods.

Before we start, a word of encouragement. We are going to describe the details of the design algorithms for many different types of FIR filters. Most scientists and engineers will probably never have to implement these algorithms themselves from scratch because there exist a number of excellent filter-design packages (e.g., Matlab's Filter Design and Signal Processing Toolboxes) that can generate properly designed filters given a choice of filter type and a specification of input parameters. So, in this chapter we will explain how to use Matlab to design each type of filter as we go along, as well as how to design these filters from scratch without the Matlab toolboxes. But an important goal of this chapter is to make you aware of the properties and design trade-offs of the different filters – their strengths and weaknesses – so you can make an appropriate choice of filter. In order to do that, you need to understand what is going on under the hood of each of these algorithms. So, here goes!

7.1 Linear-phase FIR filters

All of the filters we are going to be discussing in this chapter have the property that their transforms have **linear phase**, a concept we introduced in Chapter 3 and expanded upon in Chapter 4. So, before we get down to the business of designing filters, we will now review how linear phase applies to filter design.

7.1.1 Types of linear-phase filters

As we discussed in Chapter 3, linear-phase FIR filters have impulse responses $h[n]$ that belong to one of four types as shown in **Figure 7.1**. The impulse responses of these causal filters can have either even or odd symmetry and have either even or odd length N.[1]

- A filter with even symmetry – which we will simply call a **symmetric filter** – has the property that $h[n] = h[N-1-n]$, $0 \le n \le N-1$, where N can be either odd or even, as shown in the examples of **Figures 7.1a** and **b** respectively.
- A filter with odd symmetry – which we will call an **antisymmetric filter** – has the property that $h[n] = -h[N-1-n]$, $0 \le n \le N-1$, as shown in the examples of **Figures 7.1c** and **d**.

The length and symmetry or antisymmetry of a filter plays a key role in determining the frequency response of each type of filter, which in turn determines the kind of filter (e.g., lowpass, highpass, bandpass or bandstop) for which each filter is best suited.

In order to formalize these ideas, we will derive results for a symmetric odd-length filter, also called a Type-I filter, such as the example shown in **Figure 7.1a**. Start by computing the DTFT of $h[n]$, and pulling a factor of $e^{-j\omega M}$ out of the summation

$$H(\omega) = \sum_{n=0}^{N-1} h[n]\, e^{-j\omega n} = e^{-j\omega(N-1)/2} \sum_{n=0}^{(N-1)/2} h[n]\, e^{-j\omega(n-(N-1)/2)} = e^{-j\omega M} \sum_{n=0}^{M} h[n] e^{-j\omega(n-M)},$$

where we define $M \triangleq (N-1)/2$.

For a symmetric filter, $h[n] = h[N-1-n] = h[2M-n]$, which allows us to express $H(\omega)$ as

[1] Let us remind ourselves here of the difference between filter *length* and filter *order*. The length of a filter, N, is the length of the impulse response. The order of the filter is $N-1$.

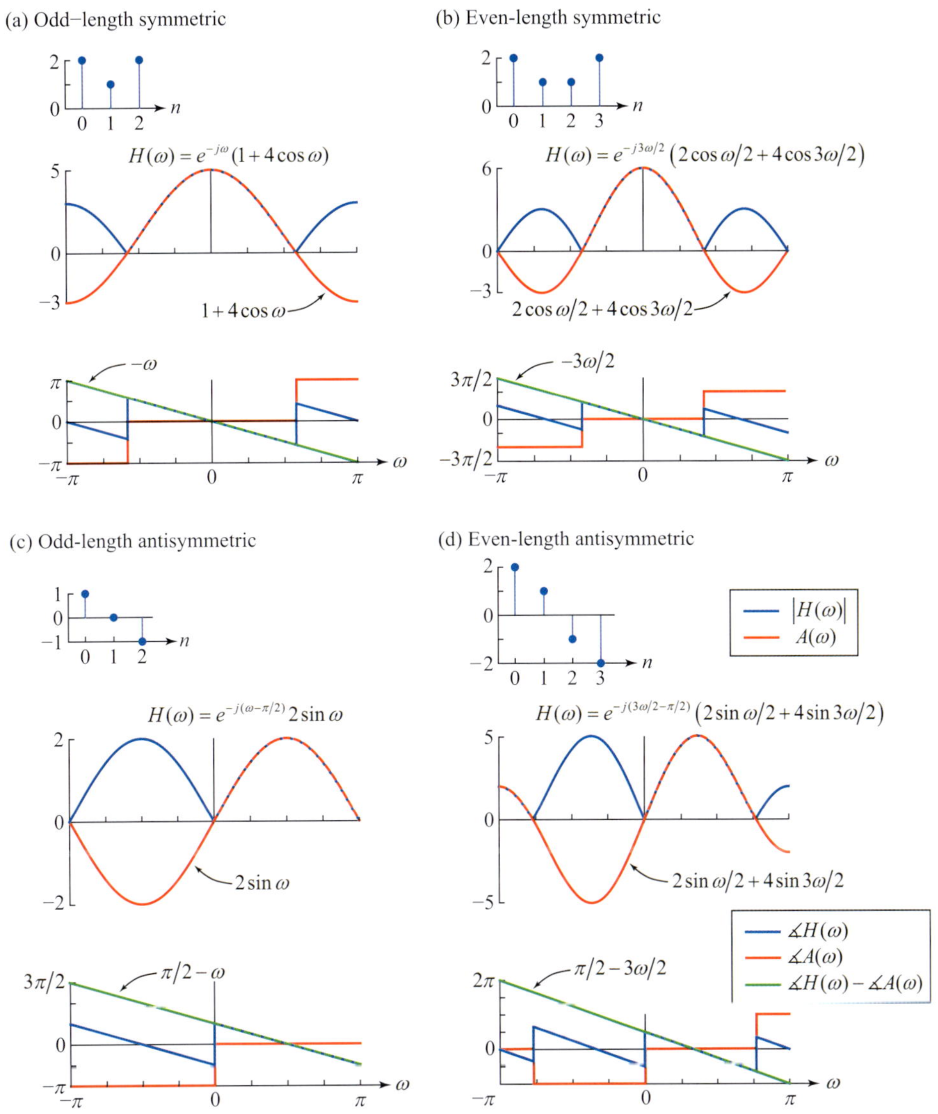

Figure 7.1 Frequency and amplitude responses of causal linear-phase filters

$$
\begin{aligned}
H(\omega) &= e^{-j\omega M}\sum_{n=0}^{M} h[n]e^{-j\omega(n-M)} \\
&= e^{-j\omega M}\left(\begin{array}{l} h[0]e^{j\omega M} + h[1]e^{j\omega(M-1)} + \cdots + h[M-1]e^{j\omega} + h[M] + h[M+1]e^{-j\omega} + \cdots \\ + h[2M-1]e^{-j\omega(M+1)} + h[2M]e^{-j\omega M} \end{array}\right) \\
&= e^{-j\omega M}\left(\begin{array}{l} h[0]e^{j\omega M} + h[1]e^{j\omega(M-1)} + \cdots + h[M-1]e^{j\omega} + h[M] + h[M-1]e^{-j\omega} + \cdots \\ + h[1]e^{-j\omega(M+1)} + h[0]e^{-j\omega M} \end{array}\right) \\
&= e^{-j\omega M}\left(\begin{array}{l} h[0](e^{j\omega M} + e^{-j\omega M}) + h[1](e^{j\omega(M-1)} + e^{j\omega(M-1)}) + \cdots \\ + h[M-1](e^{j\omega} + e^{j\omega}) + h[M] \end{array}\right)
\end{aligned}
$$

$$
\begin{aligned}
&= e^{-j\omega M}\left(\underbrace{h[M]}_{a[0]} + \underbrace{2h[M-1]}_{a[1]}\cos\omega + \cdots + \underbrace{2h[1]}_{a[M-1]}\cos\omega(M-1) + \underbrace{2h[0]}_{a[M]}\cos\omega M\right) \\
&= e^{-j\omega M}\underbrace{\sum_{m=0}^{M} a[m]\cos m\omega}_{A(\omega)} \\
&= e^{-j\omega M}A(\omega),
\end{aligned}
$$

where

$$A(\omega) = \sum_{m=0}^{M} a[m]\cos m\omega \tag{7.1}$$

and

$$a[m] = \begin{cases} h[M] = h[(N-1)/2], & m = 0 \\ 2h[M-m] = 2h[(N-1)/2 - m], & 1 \le m \le (N-1)/2 \end{cases}. \tag{7.2}$$

For a symmetric impulse response, $H(\omega)$ can therefore be expressed as the product of the purely real **amplitude term** $A(\omega)$ and a **linear phase-shift term** $e^{-j\omega(N-1)/2}$, where $(N-1)/2$ is an integer since N is odd. Because $e^{-j\omega(N-1)/2}$ has unity magnitude, the magnitude of the frequency response of $H(\omega)$ is equal to the magnitude of $A(\omega)$:

$$|H(\omega)| = |A(\omega)e^{-j\omega(N-1)/2}| = |A(\omega)|.$$

The phase of $H(\omega)$ is

$$\measuredangle H(\omega) = \measuredangle(A(\omega)e^{-j\omega(N-1)/2}) = \measuredangle A(\omega) + \measuredangle e^{-j\omega(N-1)/2} = \measuredangle A(\omega) - \omega(N-1)/2.$$

Since $A(\omega)$ is purely real, its phase $\measuredangle A(\omega)$ is either 0 (for values of ω for which $A(\omega)$ is positive) or $\pm\pi$ (for values of ω for which $A(\omega)$ is negative). Subtracting $\measuredangle A(\omega)$ from $\measuredangle H(\omega)$ yields a linear-phase term with slope $-(N-1)/2$,

$$\measuredangle H(\omega) - \measuredangle A(\omega) = -\omega(N-1)/2.$$

This explains why this filter is said to have linear phase.

Example 7.1

Find the amplitude function $A(\omega)$ and the slope of the linear-phase term for the Type-I filter of **Figure 7.1a**, which has an odd-length symmetric impulse response.

➤ Solution:

For this response, $h[n] = 2\delta[n] + \delta[n-1] + 2\delta[n-2]$, hence,

$$H(\omega) = 2 + e^{-j\omega} + 2e^{-2j\omega} = (2e^{j\omega} + 1 + 2e^{-j\omega})e^{-j\omega} = (1 + 4\cos\omega)e^{-j\omega}.$$

In this example, $H(\omega)$ has the form of Equation (7.1) with $N=3$. $A(\omega)$ satisfies Equation (7.1) with $a[0] = h[1] = 1$ and $a[1] = 2h[0] = 4$.

The magnitude $|H(\omega)|$ and phase $\measuredangle H(\omega)$ are shown in blue in **Figure 7.1a**. The amplitude function $A(\omega) = 1 + 4\cos\omega$ is shown in red. The phase of the amplitude term is 0 for $|\omega| \le 2\pi/3$ (where $A(\omega) \ge 0$) and $\pm\pi$ for $|\omega| > 2\pi/3$ (where $A(\omega) < 0$). Subtracting the phase of the amplitude term yields $\measuredangle H(\omega) - \measuredangle A(\omega)$, shown in green, which is linear in ω with a slope of -1. This corresponds to the observation that the impulse response $h[n]$ is symmetric when centered about $n=1$, or, equivalently, that a left shift of $h[n]$ by one would cause it to be centered about $n=0$.

7.1.2 Basic properties of linear-phase filters

Using derivations similar to those that lead to Equation (7.1) we can show (or rather, *you* can show in Problems 7-4 to 7-6) that the DTFT of each of the four filter types in **Table 7.1**, designated **Types I, II, III and IV**, can be expressed in the form

$$H(\omega) = A(\omega)e^{-j(\omega(N-1)/2-\beta)}, \tag{7.3}$$

where $A(\omega)$ is a real amplitude function comprising the sum of cosines (if $h[n]$ is symmetric) or sines (if $h[n]$ is antisymmetric) and $e^{-j(\omega(N-1)/2-\beta)}$ is a linear-phase term.

Table 7.1 DTFT of linear-phase filters

Type	Symmetry	Length	M	β	$A(\omega)$	Coefficients
I	Even	Odd	$(N-1)/2$	0	$\sum_{m=0}^{M} a[m]\cos m\omega$	$a[m] = \begin{cases} h[M], & m=0 \\ 2h[M-m], & 1 \le m \le M \end{cases}$
II		Even	$N/2$	0	$\sum_{m=1}^{M} b[m]\cos(m-1/2)\omega$	$b[m] = 2h[M-m],\ \ 1 \le m \le M$
III	Odd	Odd	$(N-1)/2$	$\pi/2$	$\sum_{m=1}^{M} c[m]\sin m\omega$	$c[m] = 2h[M-m],\ \ 1 \le m \le M$
IV		Even	$N/2$	$\pi/2$	$\sum_{m=1}^{M} d[m]\sin(m-1/2)\omega$	$d[m] = 2h[M-m],\ \ 1 \le m \le M$

For symmetric filters (i.e., Types I and II), $A(\omega)$ is the sum of cosines, and is therefore an even function of ω, $A(\omega) = A(-\omega)$. The phase $e^{-j\omega(N-1)/2}$ is linear with slope $-(N-1)/2$, and passes through 0 at $\omega = 0$. For the Type-I odd-length symmetric filter, the slope is an integer, which means that $h[n]$ can be centered about $n = 0$ by a left shift of $(N-1)/2$, as we saw in Example 7.1. For a Type-II even-length symmetric filter, $(N-1)/2$ is not an integer, so $h[n]$ cannot be shifted so as to be centered about $n = 0$.

Example 7.2

Find the amplitude function $A(\omega)$ and the slope of the linear-phase term for the Type-II filter of **Figure 7.1b**, which has an even-length, symmetric impulse response.

➤ Solution:

For this response, $h[n] = 2\delta[n] + \delta[n-1] + \delta[n-2] + 2\delta[n-3]$, so

$$\begin{aligned} H(\omega) &= 2 + e^{-j\omega} + e^{-2j\omega} + 2e^{-3j\omega} = (2e^{j3\omega/2} + e^{j\omega/2} + e^{-j\omega/2} + 2e^{-j3\omega/2})e^{-j3\omega/2} \\ &= (2\cos\omega/2 + 4\cos 3\omega/2)e^{-j3\omega/2}. \end{aligned}$$

By inspection, $H(\omega)$ satisfies Equation (7.3) with $A(\omega) = 2\cos\omega/2 + 4\cos 3\omega/2$, shown in red, and phase term $\measuredangle H(\omega) - \measuredangle A(\omega) = -3\omega/2$, shown in green, which is linear in ω with slope -1.5. The impulse response would be symmetric if we were able to center it about $n = 1.5$; however, non-integer shifts of a sequence are not possible.

For antisymmetric filters (i.e., Types III and IV), $A(\omega)$ is the sum of sines, and is therefore an odd function of ω, $A(\omega) = -A(-\omega)$. The phase $e^{-j(\omega(N-1)/2-\pi/2)}$ is linear with slope $-(N-1)/2$, and has an offset of $\pi/2$ at $\omega = 0$. So, we can also write

$$H(\omega) = A(\omega)e^{-j(\omega(N-1)/2-\pi/2)} = e^{j\pi/2}A(\omega)e^{-j\omega(N-1)2} = jA(\omega)e^{-j\omega(N-1)/2}.$$

Since $A(\omega)$ is a real and odd function of ω, $jA(\omega)$ is a purely imaginary and odd function, and $H(\omega)$ can therefore be viewed as a purely imaginary and odd function with a phase shift of $e^{-j\omega(N-1)/2}$. For a Type-III odd-length antisymmetric filter, the slope is an integer and $h[n]$ can be centered about $n=0$ by a left shift of $(N-1)/2$. For a Type-IV even-length antisymmetric filter, $(N-1)/2$ is not an integer, so $h[n]$ cannot be shifted so as to be centered about $n=0$.

Example 7.3

Find the amplitude function $A(\omega)$ and the slope and intercept of the linear-phase term for the Type-III filter of **Figure 7.1c**, which has an odd-length antisymmetric impulse response.

➤ Solution:

For this response, $h[n]=\delta[n]-\delta[n-2]$, so

$$H(\omega)=1-e^{-2j\omega}=(e^{j\omega}-e^{-j\omega})e^{-j\omega}=2j\sin\omega e^{-j\omega}=2\sin\omega e^{-j(\omega-\pi/2)}.$$

By inspection, $A(\omega)=2\sin\omega$. Hence, $\measuredangle H(\omega)-\measuredangle A(\omega)$, shown in green in **Figure 7.1**, is linear in ω with a slope of -1 and an offset of $\pi/2$. The impulse response is antisymmetric when centered about $n=1$.

Example 7.4

Find the amplitude function $A(\omega)$ and the slope and intercept of the linear-phase term for the Type-IV filter of **Figure 7.1d**, which has an even-length antisymmetric impulse response.

➤ Solution:

For this response, $h[n]=2\delta[n]+\delta[n-1]-\delta[n-2]-2\delta[n-3]$, so

$$\begin{aligned}H(\omega) &= 2+e^{-j\omega}-e^{-2j\omega}-2e^{-3j\omega}=(2e^{j3\omega/2}+e^{j\omega/2}-e^{-j\omega/2}-2e^{-j3\omega/2})e^{-j3\omega/2}\\ &= j(2\sin\omega/2+4\sin3\omega/2)e^{-j3\omega/2}=(2\sin\omega/2+4\sin3\omega/2)e^{-j(3\omega/2-\pi/2)}.\end{aligned}$$

By inspection, $A(\omega)=2\sin\omega/2+4\sin3\omega/2$, and $\measuredangle H(\omega)-\measuredangle A(\omega)$, shown in green, is linear in ω, with a slope of -1.5 and an offset of $\pi/2$.

Table 7.2 Four types of linear-phase FIR filters

Type	Must $A(0)=0$?	Must $A(\pi)=0$?	Lowpass?	Highpass?	Bandpass?	Bandstop?
I	✗	✗	✓	✓	✓	✓
II	✗	✓	✓	✗	✓	✗
III	✓	✓	✗	✗	✓	✗
IV	✗	✗	✓	✓	✗	✗

Table 7.2 summarizes important properties of $A(\omega)$ for the four types of linear-phase FIR filters. The important message contained in this table is that the behavior of the amplitude functions of these four types of filters at $\omega=0$ and $\omega=\pi$ determines the kind of filter for which each type of filter is most suitable. (For proofs, see Problems 7-13 to 7-16.)

- For a Type-I symmetric filter, the values of $A(\omega)$ at $\omega = 0$ and $\omega = \pi$ are not constrained, so this filter type is suitable for any of the four standard types of filters – lowpass, highpass, bandpass and bandstop.
- For a Type-II symmetric filter, $A(\omega)$ is constrained to be 0 at $\omega = \pi$, which makes this filter type particularly useful for lowpass or bandpass filters but unsuitable for highpass or bandstop filters, which require a non-zero response at $\omega = \pi$.
- For a Type-III antisymmetric filter, $A(\omega)$ is constrained to be 0 at both $\omega = 0$ and $\omega = \pi$, which makes this filter useful as a bandpass filter but unsuitable for use as a lowpass, highpass or bandstop filter.
- For a Type-IV antisymmetric filter, the response is only constrained to be 0 at $\omega = 0$, so it is useful for highpass or bandpass filters.

The key implication of this table is that the first step in FIR filter design should be to choose the right filter type to match the application. For example, if you wanted to design a bandstop filter, only a Type-I filter would do. When we discuss the design of optimal filters in Section 7.8, we will show an example of how a well-informed choice of filter type can reduce the length of the filter that is required to meet desired requirements.

Supplementary material available at www.cambridge.org/holton discusses some further symmetry properties of the amplitude functions $A(\omega)$ for the four types of linear-phase filters. Specifically, we show that whereas the magnitude functions $|H(\omega)|$ of all filters have a period of 2π, the amplitude function can have a period of either 2π or 4π.

7.1.3 ★ Time-aligned and zero-phase FIR filters

In Chapter 3, we showed several examples of a linear-phase filter whose output could be time-aligned with the input by shifting it by an integer number of samples. In supplementary material, we look a bit more closely at the notion of time-alignment from the point of view of the frequency domain, which leads us to introduce the notion of a **zero-phase filter**, one which has zero phase delay.

7.2 Preliminaries of filter design

Before we get down to the business of designing filters, let us start by defining the basic parameters of filter design, and introducing the ideal lowpass filter.

7.2.1 Specification of filter characteristics

In order to motivate the discussions of FIR filter design in this chapter, we need to define some basic terms. **Figure 7.2** shows the magnitude of the frequency response of a prototypical FIR lowpass filter for the frequency range $0 \le \omega \le \pi$. The magnitude of the response wiggles around the value $|H(\omega)| = 1$ at low frequencies and then drops at high frequencies. We can define three distinct frequency ranges of interest in the response of this filter:

- The **passband** is defined as the range of frequencies $0 \le \omega < \omega_p$ for which the response varies no more than a given amount δ_p from a nominal gain of one, as shown by the green region in **Figure 7.2**. The **passband frequency** ω_p defines the highest frequency of the passband. That is, $|1 - |H(\omega)|| < \delta_p,\ 0 \le \omega < \omega_p$. The variation of the response around $|H(\omega)| = 1$ is termed the **passband ripple**.

- The **stopband** is the range of frequencies $\omega_s \le \omega < \pi$ for which the response varies no more than a given amount δ_s from a nominal gain of 0, as shown by the red region in **Figure 7.2**. The **stopband frequency** ω_s defines the lowest frequency of the stopband. That is, $|H(\omega)| < \delta_p$, $\omega_s \le \omega < \pi$. The variation of the response around $|H(\omega)| = 0$ is termed the **stopband ripple**.
- The **transition band** is the range of frequencies $\omega_p \le \omega < \omega_s$ for which the magnitude of the response ranges between the highest magnitude of the stopband δ_s and the lowest magnitude of the passband $1-\delta_p$, as shown by the yellow region in **Figure 7.2**. That is, $\delta_s < |H(\omega)| < 1-\delta_p$, $\omega_p \le \omega < \omega_s$. The width of the transition band $\Delta\omega$ is defined as the difference between the stopband and passband frequencies, $\Delta\omega = \omega_s - \omega_p$.
- The **cutoff frequency** ω_c is defined as the frequency at which the magnitude of the frequency response drops by a given amount with respect to its nominal gain of 1. By analogy with the design of analog filters, we could define ω_c as the frequency at which the response magnitude drops by 3 dB. However, in the design of discrete-time FIR filters, ω_c is often taken to be the midpoint of the passband and the stopband frequencies, $\omega_c = (\omega_p + \omega_s)/2$.

Lowpass filter design – whether of FIR or IIR filters – starts by defining the important characteristics of the filter in the frequency domain. In general, the filter designer (i.e., you) is not necessarily interested in the details of the shape of the filter's response in the passband, transition band or stopband, only that the magnitude of the filter's response lies within the appropriate bands – that is, within the green, yellow and red rectangles in **Figure 7.2**. So, the designer only cares about specifying the parameters that control the filter's performance: ω_p, ω_s, δ_p and δ_s. For example, you generally do not care about the particular shape of the passband ripple, only that the amplitude of that ripple does not exceed a designated amount, which is controlled by the parameter δ_p. As we shall see, sometimes you cannot define all properties that you would like for a given filter type.

In this chapter, we will spend a lot of time designing practical, realizable filters that meet a given specification for ω_p, ω_s, δ_p and δ_s, so that the resulting response is "optimum" in some way. We will therefore have to define precisely what we mean by optimum as we go along. However, many FIR filter designs, specifically those we will discuss in Section 7.3 are based on approximating as closely as possible the response of the **ideal lowpass filter**, so let us discuss that first.

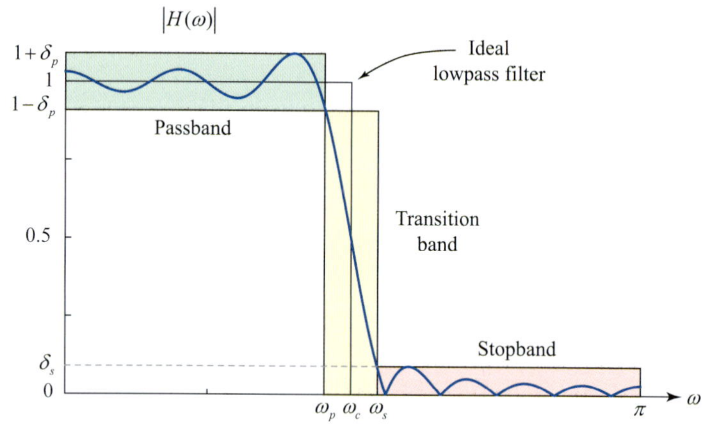

Figure 7.2 Specification of filter characteristics

7.2.2 The ideal lowpass filter

An ideal lowpass filter of bandwidth ω_c is defined as the filter that has frequency response

$$H_{ideal}(\omega) = \begin{cases} 1, & |\omega| < \omega_c \\ 0, & |\omega| > \omega_c \end{cases}.$$

Using the terminology of the last section, the ideal lowpass filter has a passband magnitude of exactly one (which means that $\delta_p = 0$), a stopband magnitude of 0 ($\delta_s = 0$) and a cutoff frequency of $\omega_c = \omega_p = \omega_s$, which means the filter's transition zone has zero width and the slope of the frequency response at the cutoff is infinite. This is a perfect filter! What's not to like?

The impulse response is "what's not to like." The impulse response of this marvelous filter is

$$h_{ideal}[n] = \frac{1}{2\pi}\int_{-\pi}^{\pi} H_{ideal}(\omega)e^{j\omega n}d\omega = \frac{1}{2\pi}\int_{-\omega_c}^{\omega_c} e^{j\omega n}d\omega = \frac{1}{2\pi jn}(e^{j\omega_c n} - e^{-j\omega_c n}) = \frac{1}{\pi n}\frac{e^{j\omega_c n} - e^{-j\omega_c n}}{2j}$$

$$= \frac{\omega_c}{\pi}\operatorname{sinc}\omega_c n. \tag{7.4}$$

The impulse response extends infinitely in time, $-\infty < n < \infty$. Therefore, it is unrealizable. The impulse response is also not absolutely summable, so the ideal lowpass filter is unstable as well (see Problem 7-30).

Figure 7.3 shows the impulse response and frequency response of a lowpass filter with bandwidths of $\omega_c = \pi/4$, $\pi/2$ and $3\pi/4$. As the bandwidth of the filter increases, the height of the main lobe of the impulse response, centered about $n = 0$, increases and the width of the main lobe decreases. In the limit, as the bandwidth approaches $\omega_c = \pi$, $H_{ideal}(\omega)$ approaches one and $h_{ideal}[n]$ approaches an impulse $h_{ideal}[n] = \delta[n]$.

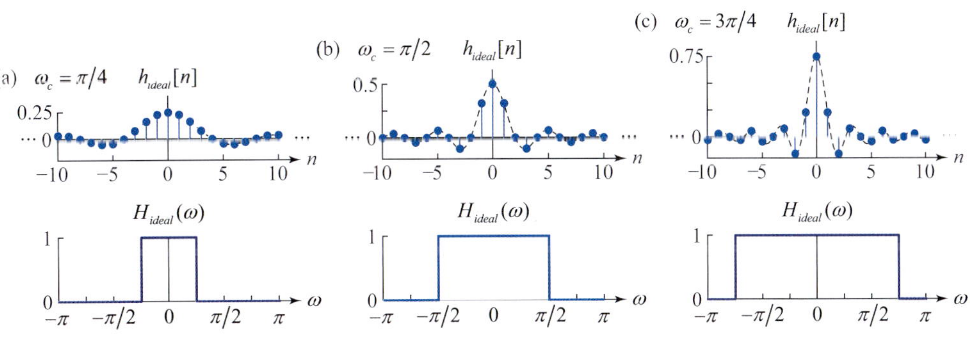

Figure 7.3 Time and frequency response of an ideal lowpass filter

7.2.3 The optimum least-square-error FIR filter

Since the ideal lowpass filter is not realizable, the question then becomes, "What is the impulse response $h[n]$ of a *realizable* (i.e., finite-length, causal, stable) filter that *best* approximates the ideal lowpass filter?" The answer depends on what we mean by "best." The general task of optimal filter design is to find the $h[n]$ that minimizes some measure of the difference (error) $E(\omega)$ between a desired frequency response $D(\omega)$ and the frequency response of a finite-length filter $H(\omega)$ that we might actually hope to design,

$$E(\omega) = D(\omega) - H(\omega).$$

In our case, the desired frequency response is that of the ideal lowpass filter, $D(\omega) = H_{ideal}(\omega)$. One very commonly used measure of optimality is the **least-square-error (LSE) criterion**. According to this criterion, we find $H(\omega)$ that minimizes the *square* of the error of $|E(\omega)|$ over a frequency range $\mathbb{W}$ that includes the passband and stopband:

$$\min\left(\int_{\omega\in\mathbb{W}} |E(\omega)|^2 d\omega\right). \tag{7.5}$$

The technique of minimizing the square-error between two quantities is frequently used in applied mathematics and engineering for several good reasons. For one thing, the square of the error is often directly related to a physically measurable quantity such as the noise power. Also, taking the square of the error weights deviations in $|E(\omega)|$ from zero much more strongly than would a simple distance measure (e.g., $E(\omega) = |D(\omega) - H(\omega)|$). Finally, the square-error criterion often leads to mathematically tractable equations that have closed-form solutions given the relevant parameters of several classes of filters, as we will show in Section 7.8.

For our purposes here, we will choose $\mathbb{W}$ to be the entire frequency range $-\pi \le \omega < \pi$, and further note that minimizing the integral of Equation (7.5) is equivalent to

$$\min\left(\frac{1}{2\pi}\int_{-\pi}^{\pi} |E(\omega)|^2 d\omega\right).$$

The integral of $|E(\omega)|^2$ is the energy of the error. The key to finding the impulse response $h[n]$ that minimizes this error is to use Parseval's theorem, discussed in Section 3.10.14, which relates energy in the time and frequency domains:

$$\frac{1}{2\pi}\int_{-\pi}^{\pi} |E(\omega)|^2 d\omega = \sum_{n=-\infty}^{\infty} |e[n]|^2 = \sum_{n=-\infty}^{\infty} e^2[n], \tag{7.6}$$

where in the last step we noted that for a real sequence, $|e[n]|^2 = e^2[n]$. Now, $e[n]$ is just the inverse transform of $E(\omega)$, so

$$e[n] = \mathfrak{F}^{-1}\{E(\omega)\} = \mathfrak{F}^{-1}\{D(\omega) - H(\omega)\} = d[n] - h[n].$$

In other words, the error sequence $e[n]$ is the difference between the impulse response of the desired filter, $d[n]$, and the impulse response of the filter we are trying to design, $h[n]$. In our case, the desired filter is the ideal lowpass filter, $d[n]=h_{ideal}[n]$, which is of infinite length, while $h[n]$ is assumed to be of finite length, having value only in the range $n\in\mathbb{N}$. Since $h[n]$ is 0 for values of $n\notin\mathbb{N}$, we can break the summation of Equation (7.6) into two parts, one for $n\in\mathbb{N}$, where both $h_{ideal}[n]$ and $h[n]$ have value, and one for $n\notin\mathbb{N}$, where only $h_{ideal}[n]$ has value:

$$\sum_{n=-\infty}^{\infty} e^2[n] = \sum_{n\in\mathbb{N}} (h_{ideal}[n] - h[n])^2 + \sum_{n\notin\mathbb{N}} h_{ideal}^2[n].$$

Both the terms on the right-hand side are summations of squared quantities, which must therefore be greater than or equal to 0. Because $h_{ideal}[n]$ is of infinite length, the second term is guaranteed to be greater than 0. Hence, the square-error will be minimized when the first summation term, for $n\in\mathbb{N}$, is 0. This occurs when $h[n]=h_{ideal}[n]$. Although the second summation cannot be 0 for $n\notin\mathbb{N}$, it will be minimized when the range $\mathbb{N}$ is symmetric around $n=0$. By this analysis, the least-square-error lowpass filter of odd length N is obtained by simply *truncating* the ideal lowpass filter to N points symmetrically about $n=0$,

$$h[n] = \begin{cases} \dfrac{\omega_c}{\pi}\operatorname{sinc}\omega_c n, & -(N-1)/2 \le n \le (N-1)/2 \\ 0, & \text{otherwise} \end{cases}. \qquad (7.7)$$

The result is a finite-length, real even impulse response. The residual error between $h[n]$ and $h_{ideal}[n]$ is the sum of $h_{ideal}[n]$ over the range $|n|>(N-1)/2$, an error that decreases monotonically as the length of $h[n]$ increases. As discussed in Chapter 3, because $h[n]$ is real and even, $H(\omega)$ is also real and even, with a magnitude $|H(\omega)|$ that is an even function of ω and a phase that is either 0 or $\pm\pi$.

7.2.4 Even- and odd-length causal filters

Although $h[n]$ in Equation (7.7) is of finite length, it is still not realizable because it is not causal. However, we can fix that by shifting the ideal response to the right by $(N-1)/2$ points,

$$h[n] = \begin{cases} \dfrac{\omega_c}{\pi}\operatorname{sinc}\omega_c (n-(N-1)/2), & 0 \le n \le N-1 \\ 0, & \text{otherwise} \end{cases}.$$

The result is shown in the top panel of **Figure 7.4a** for a value of $N=15$.

This is an example of a Type-I (symmetric odd-length) linear-phase filter. The top panel of **Figure 7.4b** shows the impulse response of a causal filter of even length $N=16$. This is a

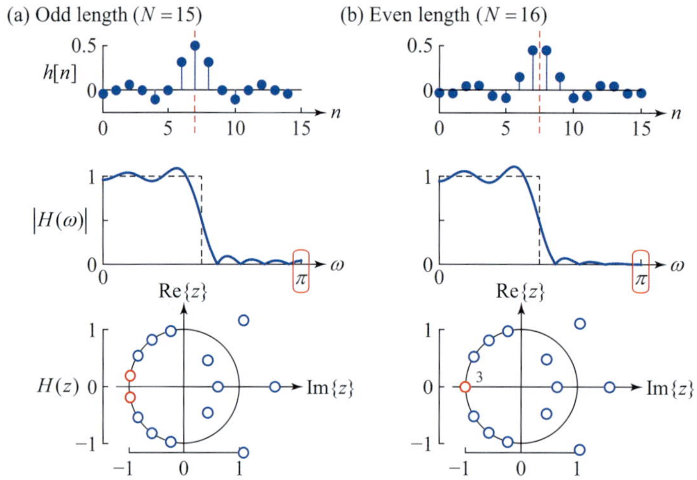

Figure 7.4 Even- and odd-length casual filters

Type-II (symmetric even-length) linear-phase filter. Unlike the Type-I filter, the z-transform of a Type-II filter has zero(s) at $z = -1$, the consequence of which is that the frequency response of the Type-II filter is guaranteed to go to 0 at $\omega = \pm\pi$. One advantage of an odd-length linear-phase filter is that its output can be time-aligned to compare with the input, as discussed in supplementary material.

7.3 Window-based FIR filter design

Many of the most commonly used FIR filters are designed based on modifying the response of the ideal lowpass filter by multiplying it with a window function $w[n]$. In this section, we will examine a few of the most common of these window-based filters. We will concentrate on designing odd-length filters that are symmetric about $n = 0$, with the understanding that both odd-length and even-length casual filters can be created in a manner similar to that shown in Section 7.2.4.

7.3.1 Rectangular window filter

As we have just shown in Section 7.2.3, the best approximation to the ideal filter in the least-square-error sense is truncation of the impulse response of the ideal lowpass filter. We will now look in detail at the frequency response of filters that result from this truncation. For a filter of odd length N, the truncation of $h_{ideal}[n]$ can be viewed as the multiplication of $h_{ideal}[n]$ by a symmetric rectangular window $w[n]$,

$$h[n] = h_{ideal}[n] \cdot w[n], \tag{7.8}$$

where

$$w[n] = \text{rect}_N[n] \triangleq \begin{cases} 1, & |n| \le (N-1)/2 \\ 0, & \text{otherwise} \end{cases}.$$

Since this rectangular window is central to our understanding of the properties of filters designed by windowing, we need to spend some time investigating its properties. For simplicity, we will consider only symmetric windows of odd length N, such as those shown in **Figure 7.5a** for $N = 5$, 9 and 13.

The DTFT of the rectangular window should look familiar to you:

$$\begin{aligned} W(\omega) &= \mathfrak{F}\{\text{rect}_N(n)\} = \sum_{n=-(N-1)/2}^{(N-1)/2} e^{-j\omega n} = \sum_{n=0}^{N-1} e^{-j\omega(n-(N-1)/2)} = e^{j\omega(N-1)/2} \sum_{n=0}^{N-1} e^{-j\omega n} \\ &= e^{j\omega(N-1)/2} \frac{1-e^{-j\omega N}}{1-e^{-j\omega}} = e^{j\omega(N-1)/2} \frac{e^{j\omega N/2}}{e^{j\omega/2}} \left(\frac{e^{j\omega N/2} - e^{-j\omega N/2}}{e^{j\omega/2} - e^{-j\omega/2}} \right) \\ &= \frac{\sin \omega N/2}{\sin \omega/2} = N \frac{\text{sinc}\, \omega N/2}{\text{sinc}\, \omega/2}. \end{aligned} \qquad \textbf{(7.9)}$$

In Chapter 3, we called $W(\omega)$ a **periodic sinc function** because it has a sinc-like shape but is periodic, as all DTFTs must be. It is shown in **Figure 7.5b** on a linear scale. The peak of the transform occurs at $\omega = 0$, and this peak value increases as the length of the window, N, increases; in fact, $W(0) = N$. The transform has $N - 1$ zero-crossings in the frequency range $-\pi < \omega < \pi$, which occur at multiples of $2\pi/N$; that is, at $\omega = 2\pi k/N$, for $0 < |k| \leq (N - 1)/2$. In **Figure 7.5c**, the same data are plotted on a dB scale, which we denote $|W(\omega)|_{dB}$, with its maximum value normalized to 0 dB over the range $0 \leq \omega \leq \pi$.

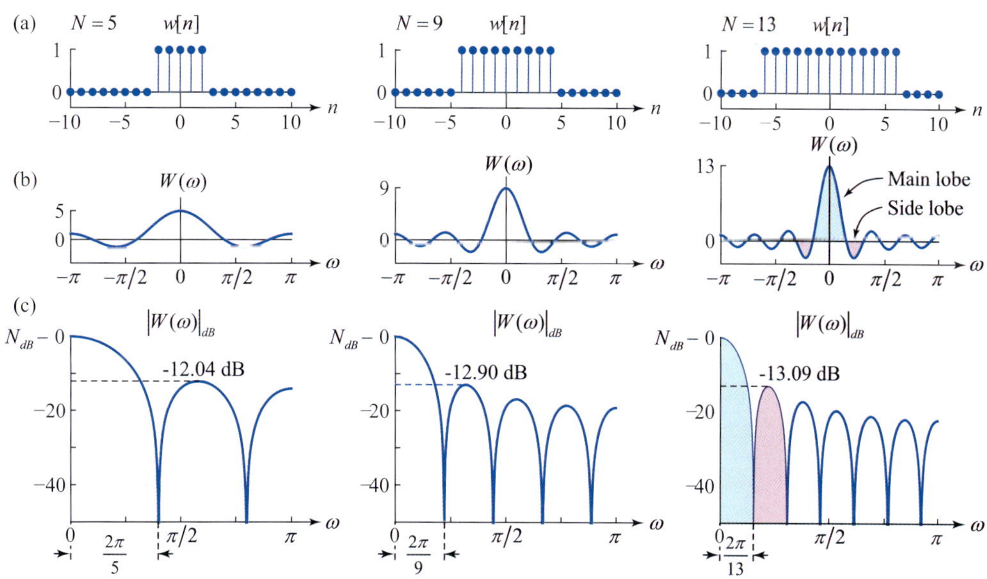

Figure 7.5 Time and frequency response of a rectangular window

We define the **main lobe** of $W(\omega)$ (cyan colored in **Figures 7.5b** and **c**) as the frequency range between the first zero-crossings of $W(\omega)$ around $\omega = 0$, namely for $|\omega| < 2\pi/N$. Hence, the main lobe has a width of $4\pi/N$. There are also a number of **side lobes** between subsequent zero-crossings. For example, the first side lobes on either side of the main lobe (magenta colored in **Figures 7.5b** and **c**) occur at frequencies $2\pi/N < |\omega| < 4\pi/N$. The width of the main lobe and all the side lobes decrease as N increases. The peak magnitude of the first side lobe is about a quarter (−13 dB) that of the main lobe, and the area of the first side lobe is about 13% of that of the main lobe. The attenuation of side lobes increases with frequency. However, neither the magnitude of the first side lobe nor the ratio of the area of the main lobe and the first side lobe change significantly as a function of the length of the window, N. All of these details will become important in a moment in helping us to understand why the rectangular filter, despite being optimal in the least-square sense, is actually a pretty poor filter, at least from the point of view of the level of attenuation the filter provides in the stopband.

Figure 7.6a illustrates the process of constructing a realizable lowpass filter with a finite-length impulse response by truncating the infinite impulse response of the ideal lowpass filter with a rectangular window of length $N = 21$.

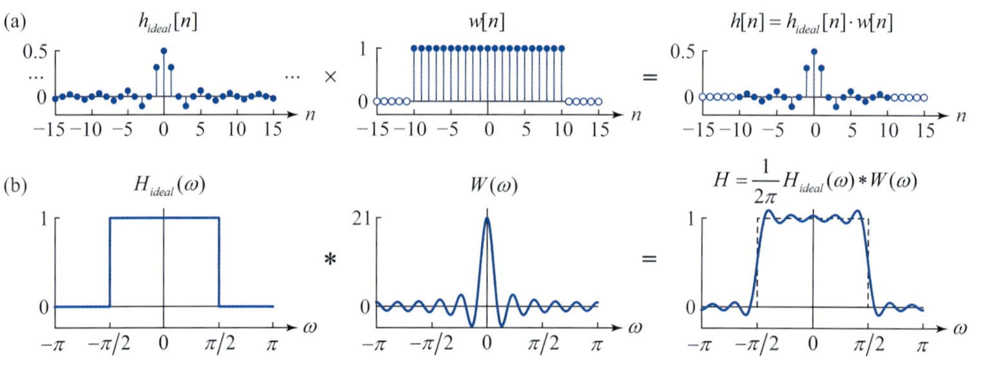

Figure 7.6 Windowing the impulse response of an ideal lowpass filter with a rectangular window

The ideal impulse response $h_{ideal}[n]$ (left panel) is multiplied by the finite-length rectangular window $w[n] = \text{rect}_N[n]$ (middle panel) to produce a finite-length impulse response $h[n] = h_{ideal}[n] \cdot w[n]$ (right panel). In the frequency domain, shown in **Figure 7.6b**, the frequency response of the filter $H(\omega)$ is the convolution of $H_{ideal}(\omega)$, the frequency response of the ideal lowpass filter, with $W(\omega)$, the frequency response of the rectangular window,

$$H(\omega) = \frac{1}{2\pi} H_{ideal}(\omega) * W(\omega) = \frac{1}{2\pi} \int_{-\pi}^{\pi} H_{ideal}(\xi) W(\omega - \xi) d\xi = \frac{1}{2\pi} \int_{-\omega_c}^{\omega_c} W(\omega - \xi) d\xi. \quad \textbf{(7.10)}$$

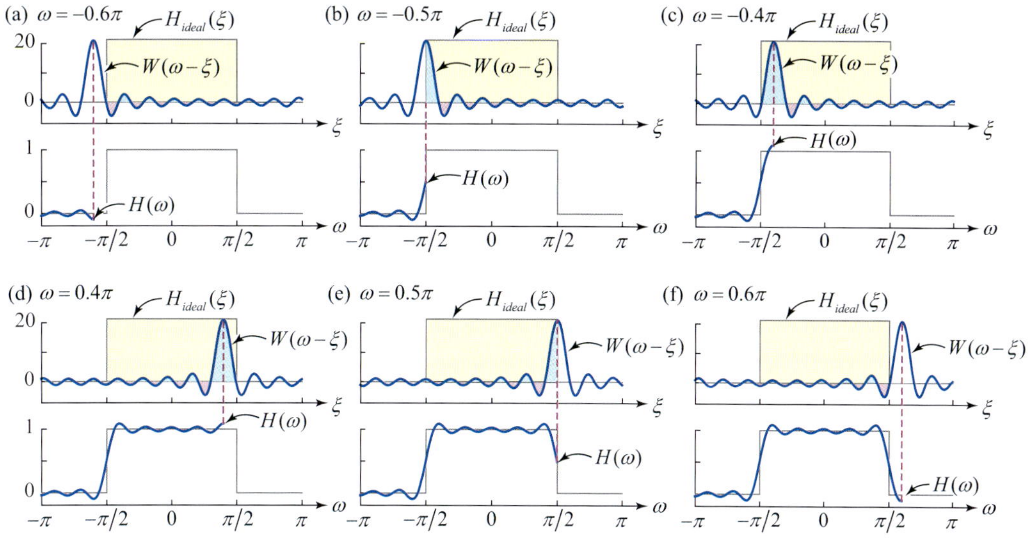

Figure 7.7 Convolution of $H_{ideal}(\omega)$ with $W(\omega)$

Figure 7.7 shows details of the convolution of $H_{ideal}(\omega)$ and $W(\omega)$ as an aid to understanding the shape of the frequency response of the resulting filter $H(\omega)$. The top panels of **Figure 7.7** show $H_{ideal}(\xi)$ and $W(\omega-\xi)$ for various values of ω at a fixed value of $N=21$. At any value of ω, Equation (7.10) states that $H(\omega)$ is equal to the integral of $W(\omega-\xi)$ over the frequency range $-\omega_c<\omega<\omega_c$. This corresponds to the sum of the positive and negative areas under the rectangle, shown graphically in the figure in cyan and magenta respectively. **Figure 7.7a** shows $H_{ideal}(\xi)$ and $W(\omega-\xi)$ in the upper panel and $H(\omega)$ in the lower panel for a value of $\omega=-0.6\pi$. At this value of ω, the largest contribution to the integral is the negative area of the first side lobe of $W(\omega-\xi)$; the resulting value of $H(\omega)$ reaches its minimum of $H(-0.6\pi)\simeq-0.09$; that is, it undershoots 0 by about 9%. At $\omega=-0.5\pi$ (**Figure 7.7b**), half of the main lobe of $W(\omega-\xi)$ falls inside the integral and the value of $H(-0.5\pi)\simeq0.5$. When $\omega=-0.4\pi$ (**Figure 7.7c**), the entire main lobe of $W(\omega-\xi)$ falls inside the integral and the value of $H(\omega)$ reaches its maximum of $H(-0.4\pi)\simeq1.09$; that is, it overshoots 1 by about 9%. The remaining panels of **Figure 7.7** show the details of the convolution at $\omega=0.4\pi$, 0.5π and 0.6π. The filter's response is clearly symmetric about $\omega=0$ (i.e., $H(\omega)$ is even), which is what we expect since the impulse response is even.

Figure 7.8a shows the magnitude of $H(\omega)$ on a linear scale over the range $0<\omega<\pi$ for the filter of **Figure 7.6** with $N=21$. The magnitude of the ideal lowpass filter with cutoff frequency $\omega_c=0.5\pi$ is also shown in a thin dotted black line. The frequency response of the finite-length filter $H(\omega)$ shown in **Figure 7.8a** departs from the response of the ideal lowpass filter in two important ways:

(1) There is a broad transition band in the response of $H(\omega)$, shown in yellow, around the cutoff frequency $\omega_c=0.5\pi$. As you can see by studying **Figure 7.7**, the width of this transition band is proportional to the width of the main lobe, which is in turn inversely proportional to N, the length of the impulse response of the filter. Hence, we would expect that increasing N would lead to a sharper filter – that is a reduced transition band.

(2) There are ripples in both the passband and stopband of $H(\omega)$ due to the fact that the filter's response overshoots $H(\omega)=1$ in the passband and undershoots $H(\omega)=0$ in the stopband,

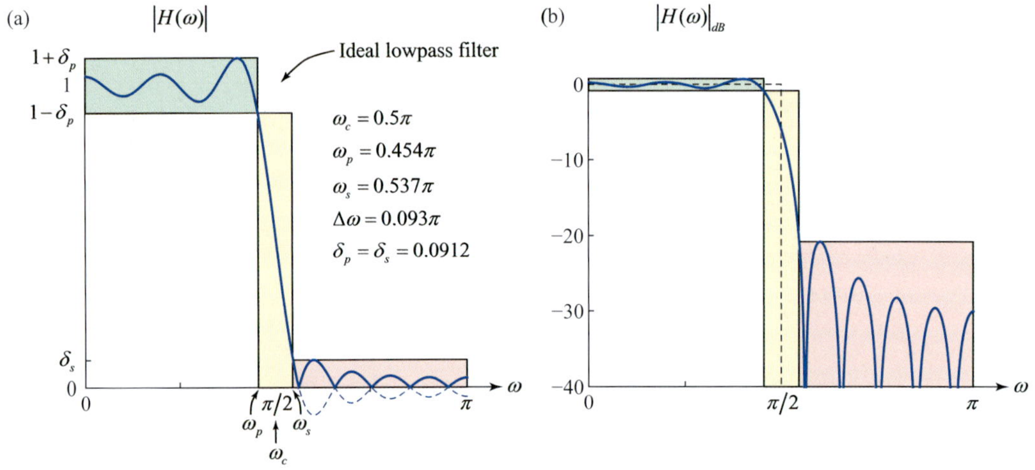

Figure 7.8 Frequency response of magnitude of an FIR filter on linear and log (dB) scales

> both by about 9%. As indicated in **Figure 7.7**, the magnitude of this overshoot and undershoot is proportional to the relative areas of the main lobe and first side lobe of $W(\omega)$. Because the ratio of the area of these lobes changes very little as a function of N, increasing N does not substantially change the amount of overshoot or undershoot in $H(\omega)$. Therefore, the passband or stopband tolerances of the filter do not significantly depend on N.

For the filter of **Figure 7.6**, the passband and stopband tolerances are equal, $\delta_p = \delta_s = 0.0912$, which corresponds to the 9% overshoot and undershoot of the response we found in **Figure 7.7**. The passband frequency is $\omega_p = \omega_c - 0.045\pi = 0.455\pi$, and the stopband frequency is $\omega_s = \omega_c + 0.045\pi = 0.545\pi$, giving a transition bandwidth of $\Delta\omega = \omega_s - \omega_p = 0.09\pi$.

Figure 7.8b shows the magnitude of $H(\omega)$ on a dB scale, which we denote $|H(\omega)|_{dB}$. On a log scale, the 9% overshoot with respect to a gain of 1 (i.e., $H(\omega) = 1 \pm 0.09$) corresponds to a small passband ripple (less than ± 1 dB), whereas the same 9% undershoot with respect to a gain of 0 (i.e., $H(\omega) = \pm 0.09$) corresponds to a stopband ripple whose peak value (i.e., minimum attenuation) is about −21 dB. In filter design, the minimum attenuation of the stopband determines the worst-case out-of-band rejection of a filter. In this example, the worst-case attenuation of −21 dB occurs at about $\omega = 0.6\pi$. Compared with other filters that we will study shortly, this is a very poor filter.[2]

In this example, we did not get to choose the filter's characteristic parameters – the passband tolerance, stopband tolerance or the width of the transition band: ω_p, ω_s, δ_p, and δ_s. These values simply "fell out" of our choice of the rectangular window with a particular length

[2] As an example of why this may not be a good filter, consider the sound filtering application in which the input signal is the sum of two tones, $x[n] = \cos(0.4\pi n) + \cos(0.6\pi n)$. If the goal of the filter is to remove frequencies above the equivalent of $\omega = 0.5\pi$, in this case it will only attenuate the second component by 21 dB. The "normal" human ear is exquisitely sensitive, having a dynamic range of about 120 dB, so this second component will be very clearly audible.

($N=21$). In the sections that follow, we will understand how the choice of window affects the filter's characteristic parameters.

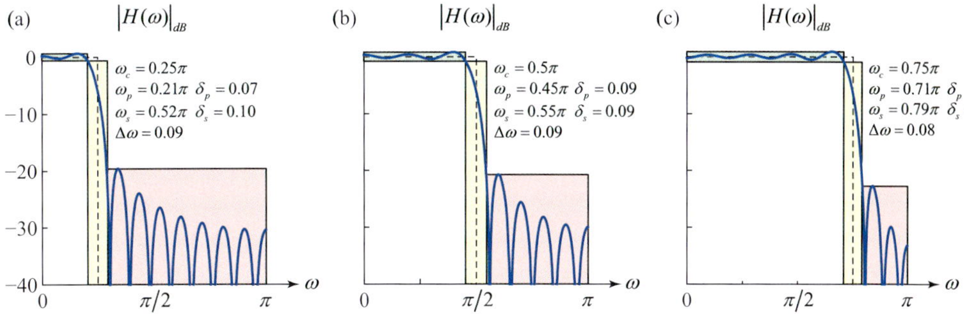

Figure 7.9 Rectangular-window filters with $N=21$, and (a) $\omega_C=\pi/4$, (b) $\omega_C=\pi/2$ and (c) $\omega_C=3\pi/4$

Characteristics of the rectangular-window FIR filter as a function of *N* and ω_c The rectangular-window FIR filter we "designed" in the previous section had only a couple of parameters that we could adjust: the cutoff frequency ω_c and the filter length N. However, as we will now show, some of the basic characteristics of the filter, for example the passband and stopband attenuation, do not change a lot as a function of cutoff frequency or filter length.

Figure 7.9 shows the magnitude of the frequency response of the filter as a function of the cutoff frequency ω_c for a filter of fixed length, $N=21$. The bottom line is that for a rectangular-window filter, the passband and stopband attenuation and the width of the transition band do not significantly change as a function of cutoff frequency.

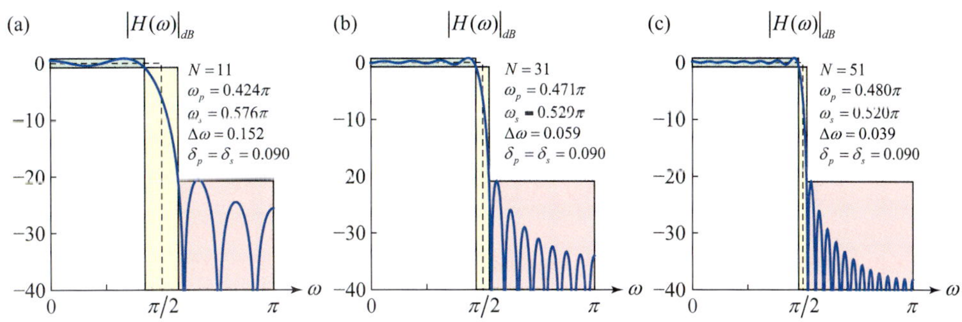

Figure 7.10 Rectangular-window filters with $\omega_C=\pi/2$, and (a) $N=11$, (b) $N=31$ and (c) $N=51$

Perhaps more interesting is the dependence (or independence) of filter characteristics on length N for a filter with a given cutoff frequency, as shown in **Figure 7.10**. Looking at these traces, there is good news and bad news. The good news is that the width of the transition band $\Delta\omega$ decreases as the length of the filter increases; that is, the filter gets "sharper," and this turns out to be true for a range of cutoff frequencies. The bad news is that the maximum magnitude of both the passband ripple and the stopband ripple remains constant at about 9%, essentially independent of filter length or cutoff frequency. This overshoot and undershoot is an example of the so-called **Gibbs phenomenon**, which you may have run across in your study of Fourier series approximations of continuous-time functions that have sharp discontinuities, such as a continuous-time square wave. The root of the problem, which we noted back in **Figure 7.5**, is that the relative areas of the main and side lobes of $W(\omega)$ are almost independent of N. As N increases, the width of the main lobe, $2\pi/N$, decreases but the magnitude of the lobe increases proportionally with N, so the area remains roughly constant at 1.18π. As shown in **Figures 7.7b** and **c**, the maximum overshoot in $H(\omega)$ corresponds approximately to half the total area of $W(\omega)$ (namely 0.5) plus the integrated area of the main lobe over 2π: $0.5 + 1.18\pi/2\pi \cong 1.09$. This passband overshoot of 9% persists as N approaches infinity even though the residual square error of the rectangular-window filter goes to 0!

The important point to take away from the preceding figures is this: increasing the length of the rectangular window makes the filter sharper by decreasing the width of the transition band, but the stopband tolerance, which determines the worst-case out-of-band attenuation of the filter, remains fixed at about −21 dB and cannot be improved by increasing N. Although we can in principle adjust the length of a rectangular window to match the desired width of the transition band, there is nothing we can do to change the maximum magnitude of the ripple in the passband and stopband. These are inherent characteristics of the rectangular window. If we want better passband and stopband characteristics of the filter, we will have to shop around for filters designed with other types of windows. So, that is just what we are now going to do.

Many types of simple FIR filters have been designed based on the same general principle as the rectangular-window filter; namely, shaping the infinite impulse response of an ideal LPF, $h_{ideal}[n]$, by multiplying it with a finite-length window $w[n]$ to produce a finite-length impulse response $h[n] = h_{ideal}[n] \cdot w[n]$. The challenge of filter design using this method lies in choosing $w[n]$ appropriately in order to retain some of the nice characteristics of the ideal LPF. The length and shape of the window will determine important characteristics of the filter, such as the width of the transition band, and the passband and stopband attenuations. We will now look at a few of the most common filters designed with non-rectangular windows.

7.3.2 Raised cosine window filters

Rectangular-window FIR filters have poor out-of-band attenuation because the rectangular window sharply truncates the impulse response of the ideal filter, $h_{ideal}[n]$, and these sharp transitions produce ripples in the frequency response, as we showed in **Figures 7.6** and **7.7**. The search for a better window starts with finding a shape of $w[n]$ that tapers at the edges of the window in order to prevent sharp transitions. Many different tapered windows have been proposed and implemented. Some of the most widely used FIR filters are based on **raised-cosine windows** formed from the sums of cosines. In this section, we will consider raised-cosine windows of odd length N, centered about $n = 0$. These windows all have the general form

$$w[n] = \begin{cases} \sum_{k=0}^{L-1} a_k \cos \frac{2\pi kn}{N-1}, & |n| \leq \frac{N-1}{2} \\ 0, & \text{otherwise} \end{cases} = \left(\sum_{k=0}^{L-1} a_k \cos \frac{2\pi kn}{N-1} \right) \text{rect}_N[n]. \quad \textbf{(7.11)}$$

Here, $w[n]$ is the sum of a constant (a_0) plus $L-1$ cosine terms at multiples of a common fundamental frequency $2\pi/(N-1)$, all truncated to the time range $-(N-1)/2 \le n \le (N-1)/2$. This family of windows includes the famous Hamming window and Hann window (sometimes called Hanning window), as well as other more esoteric windows, such as the Blackman window, all named for the people who proposed them. We will take a fairly detailed look at the Hamming window, and then mention some of the properties of the others.

Hamming window The Hamming window is perhaps the most popular raised-cosine window. It finds use in a variety of applications, including filtering and spectral analysis. It satisfies Equation (7.11) with only $L=2$ terms: a constant term of value $a_0=0.54$, plus one cycle of a cosine with amplitude $a_1=0.46$,

$$\text{hamming}_N[n] = 0.54 + 0.46\cos\frac{2\pi n}{N-1}, \quad |n| \le (N-1)/2, \tag{7.12}$$

for a window centered at $n=0$, and

$$\text{hamming}_N[n] = 0.54 - 0.46\cos\frac{2\pi n}{N-1}, \quad 0 \le n \le N-1,$$

for a causal window. (The minus sign in the causal formula results from a shift in the range of n by $(N-1)/2$ points with respect to the formula for the centered case.) The values of a_0 and a_1 for the Hamming window are chosen in order to minimize the magnitude of the first side lobe, a point we'll discuss in more detail below. Since a_0 and a_1 are fixed, the Hamming window is a function of only one parameter, the window length N. Matlab's `hamming` function can be used to design even- and odd-length causal Hamming windows. The equivalent do-it-yourself formula for the causal window is

```
hamming = @(N) 0.54-0.46*cos(2*pi*(0:(N-1))/(N-1));
```

Figure 7.11a shows centered Hamming windows, labeled $w[n]$, of three different lengths: $N=13$, 17 and 21. **Figures 7.11b** and **c** show their transforms, $W(\omega)$, on linear and dB scales, respectively.

A characteristic of the Hamming window is that the first side lobe of the transform is considerably attenuated compared to that of a rectangular window. In fact, all the side lobes of the Hamming window are relatively similar to each other in magnitude and all are substantially lower than the magnitudes of the side lobes of the rectangular window. However, the main lobe of the transform of the Hamming window is wider than that of the rectangular window of the same length.

Figure 7.12a makes this point more clearly by showing a comparison of a Hamming window of length $N=13$ (red circles) and a rectangular window (blue circles) of the same length. **Figures 7.12b** and **c** show the transform of the two windows on linear and dB scales, respectively, on a normalized magnitude scale.

The main lobe of the transform of the Hamming window is more than a factor of two wider than the transform of the rectangular window but the side lobes are at least 20 dB lower. In

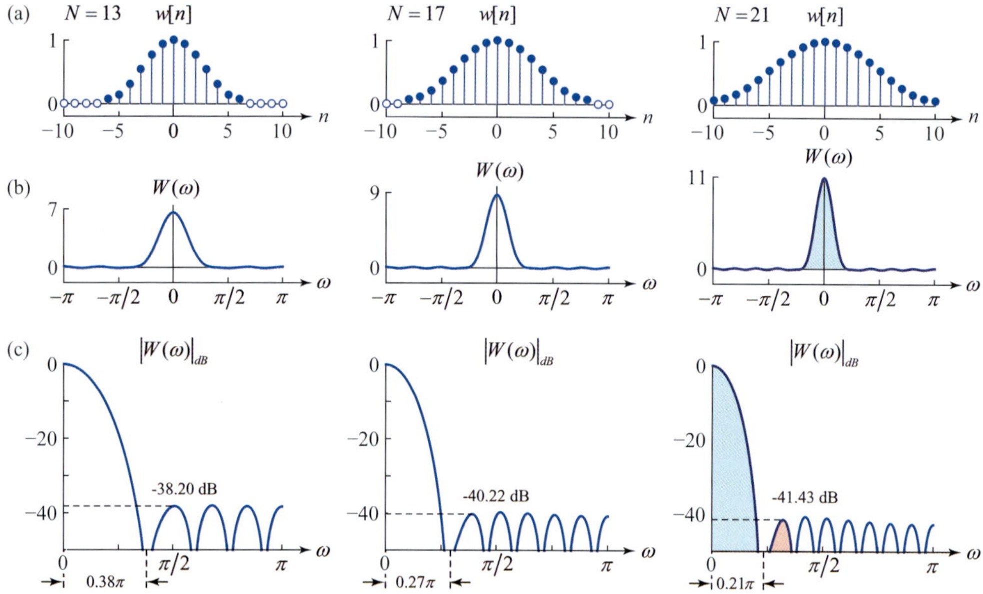

Figure 7.11 Time and frequency response of a Hamming window

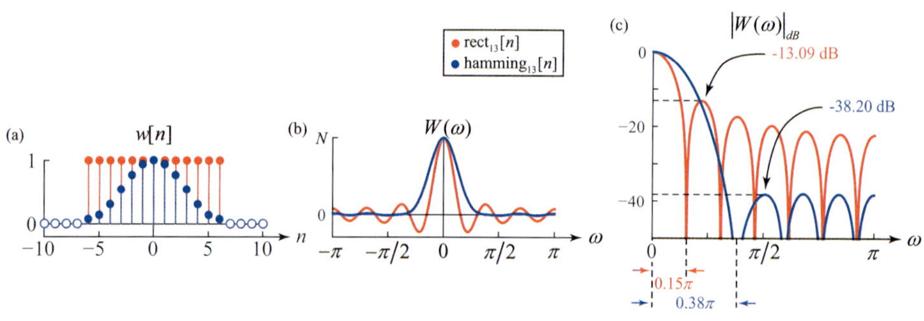

Figure 7.12 Comparison of Hamming and rectangular windows of length $N = 13$

particular, the peak magnitude of the first side lobe of the Hamming window is more than 25 dB below that of the first side lobe of the rectangular window.

Figure 7.13 shows why the magnitudes of the side lobes of the Hamming window are so attenuated compared to those of a rectangular window of comparable size, and also why the main lobe is so much wider. **Figure 7.13a** illustrates how to construct a Hamming window of length $N = 13$ from the sum of two components: a constant term, $0.54\,\text{rect}_N[n]$, plus a cosine term, $0.46\cos(2\pi n/(N-1))\,\text{rect}_N[n]$. The constant term (left panel of **Figure 7.13a**) is a rectangular pulse of length N centered at $n = 0$, with an amplitude of $a_0 = 0.54$. The cosine term (middle panel of **Figure 7.13a**) has frequency $2\pi/(N-1)$ and amplitude $a_1 = 0.46$, and is gated by a rectangular pulse of length N. It peaks at $n = 0$, and reaches a minimum at both the endpoints, $n = \pm(N-1)/2$. The sum of the constant and cosine terms is a raised cosine pulse, the Hamming window (right panel of **Figure 7.13a**). Since $\text{hamming}_N[\pm(N-1)/2] = 0.06$, the Hamming window has small discontinuities at its endpoints.

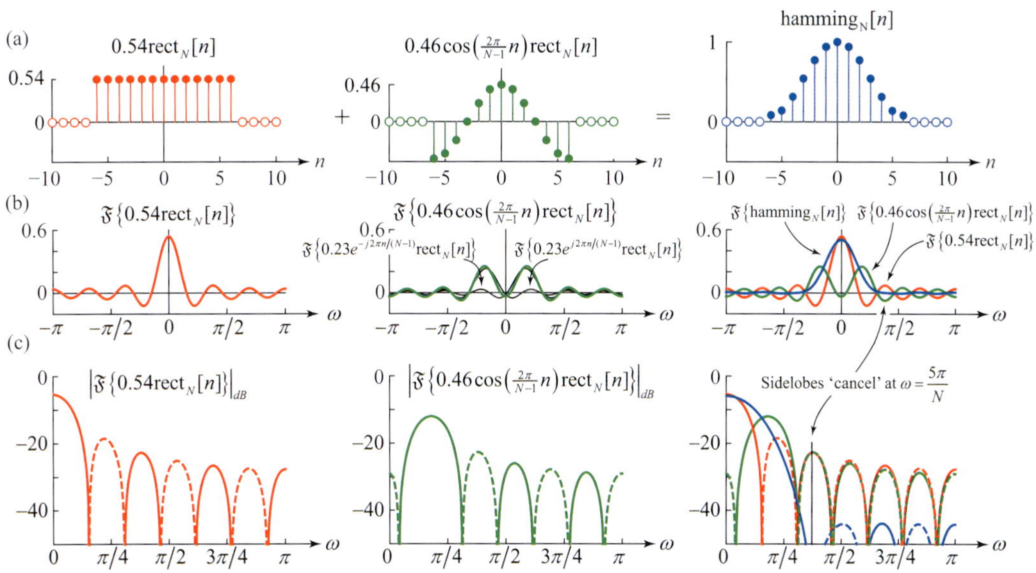

Figure 7.13 Components of the Hamming window

Figure 7.13b shows the transforms of the components in **Figure 7.13a**. The transform of the constant term, $\mathfrak{F}\{0.54\,\text{rect}_N[n]\}$, plotted in the left panel of **Figure 7.13b**, is a periodic sinc term centered at $\omega = 0$,

$$\mathfrak{F}\{0.54\,\text{rect}_N[n]\} = 0.54\frac{\sin \omega N/2}{\sin \omega/2}.$$

The transform of the cosine term, plotted in the middle panel of **Figure 7.13b**, is

$$\begin{aligned}\mathfrak{F}\{0.46\cos(2\pi/(N-1))\,\text{rect}_N[n]\} &= 0.46\cdot\mathfrak{F}\left\{\left(\tfrac{1}{2}e^{j2\pi n/(N-1)} + \tfrac{1}{2}e^{-j2\pi n/(N-1)}\right)\text{rect}_N[n]\right\}\\ &= 0.23\frac{\sin(\omega - 2\pi/(N-1))(N/2)}{\sin(\omega - 2\pi/(N-1))(1/2)} + 0.23\frac{\sin(\omega + 2\pi/(N-1))(N/2)}{\sin(\omega + 2\pi/(N-1))(1/2)}.\end{aligned}$$

This comprises the sum of two periodic sinc terms, as shown in thin black lines in the middle panel of **Figure 7.13b**: one term is centered on $\omega = +2\pi/(N-1)$; the other term is centered on $\omega = -2\pi/(N-1)$. The sum of these two translated spectral components is shown in the thicker green line.

Finally, the transform of the Hamming window $W(\omega)$, shown in blue in the right panel of **Figure 7.13b**, is the sum of the transforms of the two components: the transform of the constant term (shown in red), plus the transform of the cosine term (shown in green),

$$\begin{aligned}W(\omega) &= \mathfrak{F}\{0.54\,\mathrm{rect}_N[n]\} + \mathfrak{F}\{0.46\cos(2\pi n/(N-1))\,\mathrm{rect}_N[n]\} \\ &= 0.54\frac{\sin \omega N/2}{\sin \omega/2} + \left(0.23\frac{\sin(\omega - 2\pi/(N-1))(N/2)}{\sin(\omega - 2\pi/(N-1))(1/2)} + 0.23\frac{\sin(\omega + 2\pi/(N-1))(N/2)}{\sin(\omega + 2\pi/(N-1))(1/2)}\right).\end{aligned}$$

The key point about the Hamming window is that the positive-going side lobes of the cosine term, $\mathfrak{F}\{0.46\cos(2\pi n/(N-1))\,\mathrm{rect}_N(n)\}$, cancel to some extent the negative-going side lobes of constant component $\mathfrak{F}\{0.54\,\mathrm{rect}_N[n]\}$. This point is made more clearly in **Figure 7.13c**, which shows the magnitude of the transform components on a dB scale. The $\mathfrak{F}\{0.54\,\mathrm{rect}_N[n]\}$ component is again shown in red (left panel), and $\mathfrak{F}\{0.46\cos(2\pi n/(N-1))\,\mathrm{rect}_N(n)\}$ is shown in green (middle panel). Since both $\mathrm{rect}_N[n]$ and $\cos(2\pi n/(N-1))\,\mathrm{rect}_N(n)$ are real and even sequences, their transforms are purely real and even; hence, these components can have both positive-going and negative-going values, which are noted with solid and dotted traces respectively in **Figures 7.13b** and **c**. For frequencies below $4\pi/N \cong 0.31\pi$, which correspond to the main lobe and first side lobe of the transform of the rectangular component, the positive contribution of $\mathfrak{F}\{0.46\cos(2\pi n/(N-1))\,\mathrm{rect}_N(n)\}$ outweighs the negative contribution of $\mathfrak{F}\{0.54\,\mathrm{rect}_N[n]\}$, and the two components add; hence, the main lobe of the transform of the Hamming window is wider than that of a rectangular window of the same length. At higher frequencies, the side lobes of the two components become more similar in magnitude, but are of opposite sign, so they cancel to some extent. That is why the side lobes of the Hamming window are smaller than those of a rectangular window of the same length (see Problem 7-1 for further discussion of this point).

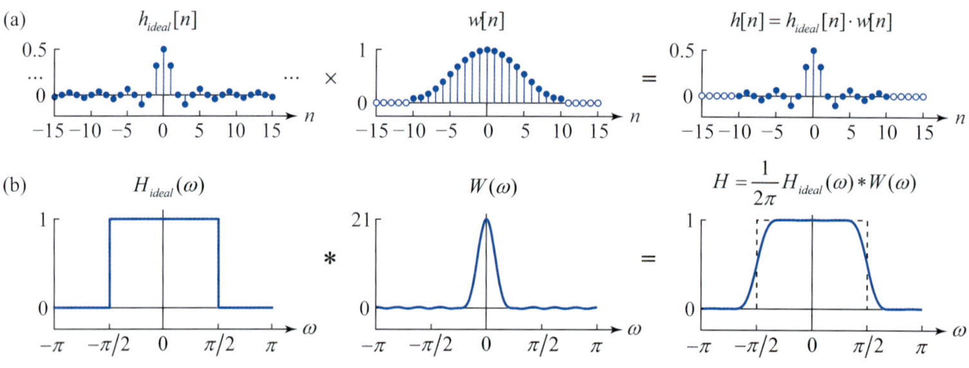

Figure 7.14 Time and frequency response of a Hamming-window filter

Given the smaller side lobes of the Hamming window, we might expect that an FIR filter designed using this window would have fairly good out-of-band attenuation, and we would be right! **Figure 7.14a** shows the construction of an FIR filter obtained by multiplying the response of the ideal lowpass filter $h_{ideal}[n]$ with bandwidth $\omega_c = 0.5\pi$, and a Hamming window $\mathrm{hamming}_N[n]$ of length $N = 21$. **Figure 7.14b** shows the transform of the Hamming window obtained as the convolution of the transforms of $h_{ideal}[n]$ and $\mathrm{hamming}_N[n]$. On a linear scale, the oscillations in the passband and stopband are not visible. So, we need to look on a dB scale.

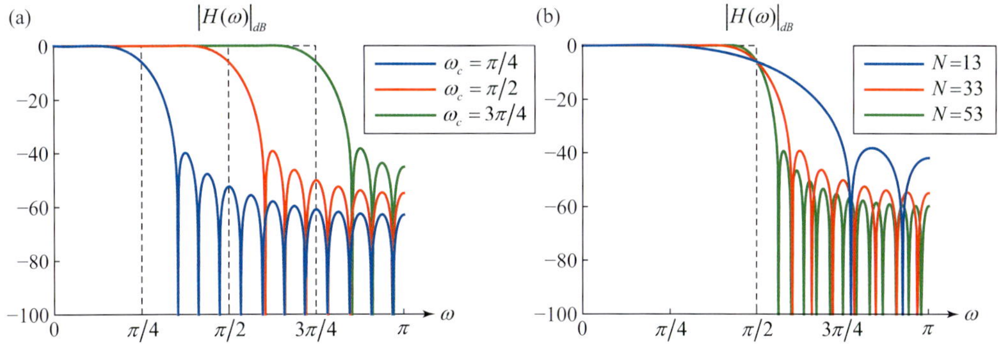

Figure 7.15 Hamming-window FIR filters as a function of ω_c and N

Figure 7.15a shows the magnitude on a dB scale of the response of a Hamming-window FIR filter of length $N = 33$, with cutoff frequencies of $\omega_c = \pi/4$, $\pi/2$ and $3\pi/4$. As with the rectangular-window filter (**Figure 7.9**), the basic characteristics of the Hamming-window filter – that is, the passband and stopband attenuation and the width of the transition band – do not change a lot as a function of ω_c. **Figure 7.15b** shows the effect of changing the length of the filter, N. Again, like the rectangular window filter, the width of the transition band, $\Delta\omega$, decreases as N increases, but the maximum magnitude of the stopband ripple, now about −40 dB, is a characteristic of the Hamming window that is essentially independent of the length of the filter.

The next few paragraphs discuss a couple of other common raised-cosine windows. The design principles of these windows are the same as those of the Hamming window.

Hann window The **Hann window** is another popular raised-cosine window that satisfies Equation (7.11) with only $L = 2$ terms. Only the coefficients a_0 and a_1 are different with respect to the Hamming window; in this case, $a_0 = a_1 = 0.5$. An odd-length Hann window, centered at $n = 0$, is defined as

$$\text{hann}_N[n] \triangleq 0.5 + 0.5\cos\frac{2\pi n}{N-1}, \quad |n| \leq (N-1)/2.$$

The equivalent causal window is

$$\text{hann}_N[n] \triangleq 0.5 - 0.5\cos\frac{2\pi n}{N-1}, \quad 0 \leq n \leq N-1.$$

The coefficients of the Hann window are chosen to create a roll-off of the side lobes of about −18 dB/octave. Like the Hamming window, the Hann window is a function of only one parameter, N. But unlike the Hamming window, the Hann window has no discontinuity at its endpoints; that is, $\text{hann}_N[\pm(N-1)/2] = 0$. There is also another definition of the centered window that avoids the 0 endpoints,

$$\text{hanning}_N[n] \triangleq 0.5 + 0.5\cos\frac{2\pi n}{N+1}, \quad |n| \leq (N-1)/2.$$

The equivalent causal window is

$$\text{hanning}_N[n] \triangleq 0.5 - 0.5\cos\frac{2\pi n}{N+1}, \quad 1 \leq n \leq N.$$

The Hann and Hanning windows are closely related in that $\text{hann}_{N+2}[n] = [0\ \text{hanning}_N[n]\ 0]$. The causal definitions also match Matlab's `hann` and `hanning` functions:[3]

```
hann = @(N) 0.5-0.5*cos(2*pi*(0:(N-1))/(N-1));
hanning = @(N) 0.5-0.5*cos(2*pi*(1:N)/(N+1));
```

Blackman window The Blackman window is our final example of a raised-cosine window. This window has $L = 3$ terms and satisfies Equation (7.11) with $a_0 = 0.42$, $a_1 = 0.5$ and $a_2 = 0.08$,

$$\text{blackman}_N[n] = 0.42 + 0.5\cos\frac{2\pi n}{N-1} + 0.08\cos\frac{4\pi n}{N-1}, \quad |n| < (N-1)/2,$$

for a window centered at $n = 0$. The equivalent causal window is

$$\text{blackman}_N[n] = 0.42 + \;-0.5\cos\frac{2\pi n}{N-1} + 0.08\cos\frac{4\pi n}{N-1}, \quad 0 \le |n| < N-1.$$

The three coefficients of the Blackman window produce a window that is, in some sense, a hybrid of the Hamming and Hann windows. Like the Hamming window, the magnitude of the first side lobe is low (about −60 dB); like the Hann window, the side lobes roll off with a slope of about −18 dB/octave. And, like the previous windows we have looked at, the Blackman window is a function of only one parameter, N. Matlab's causal `blackman` function can be used to design this window. Here is our equivalent definition:

```
blackman = @(N) 0.42-0.5*cos(2*pi*(0:(N-1))/(N-1))+ ...
        0.08*cos(4*pi*(0:(N-1))/(N-1));
```

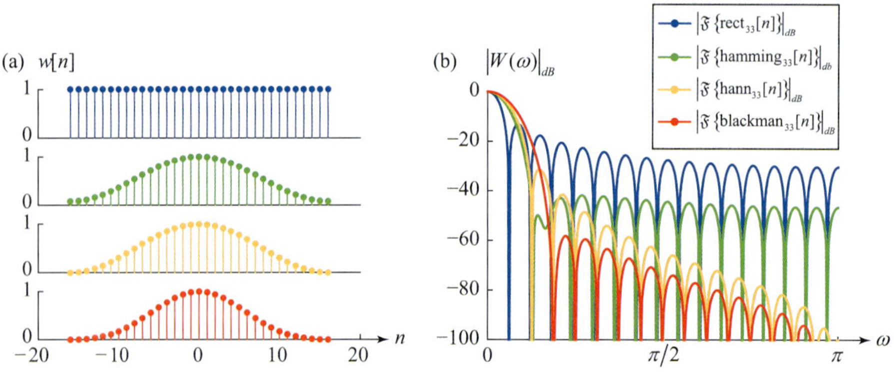

Figure 7.16 Comparison of rectangular, Hamming, Hann and Blackman windows

[3]The Hann window was named in honor of nineteenth-century Austrian meteorologist Julius von Hann. The terms Hann and Hanning are frequently used interchangeably to describe this window, but as you can see from the Matlab definitions, it is important to know exactly how the formula is defined or, better yet, create your own.

Comparison of raised-cosine windows **Figure 7.16a** shows a comparison of the rectangular window and the raised-cosine windows – Hamming window, Hann window and Blackman window – all of the same length, $N=33$. The rectangular and Hamming windows have discontinuities at their ends; the Hann and Blackman windows do not. **Figure 7.16b** shows a comparison of the transforms of these windows on a dB scale. The main observation is that there is an inverse relation between the width of the main lobe and the peak magnitude of the side lobes: the narrower the main lobe (e.g., rectangular window), the higher the peak magnitude of the side lobes. The rectangular window has the narrowest main lobe (0.12π) and highest peak side lobe (-13.2 dB). In contrast, the Blackman window has the widest main lobe (0.38π) and lowest side peak lobe (≤ -58.1 dB). The Hamming window (0.33π main lobe and side lobes ≤ -42.9 dB) and Hann window (0.24π main lobe and side lobes ≤ -33.4 dB) both satisfy Equation (7.11) with $L=2$ terms, and with only slightly different coefficients (Hamming: $a_0=0.54$, $a_1=0.46$. Hann: $a_0=a_1=0.5$). But their behavior is quite different. The Hamming window is specifically designed to minimize the peak magnitude of the first side lobe, but the discontinuity at the ends of $w[n]$ results in the other side lobes being higher than those of its cousin, the Hann window. The first side lobe of the Hann window has a higher magnitude than the Hamming window, but the magnitude of all subsequent side lobes falls fast due to the lack of discontinuity in $w[n]$.

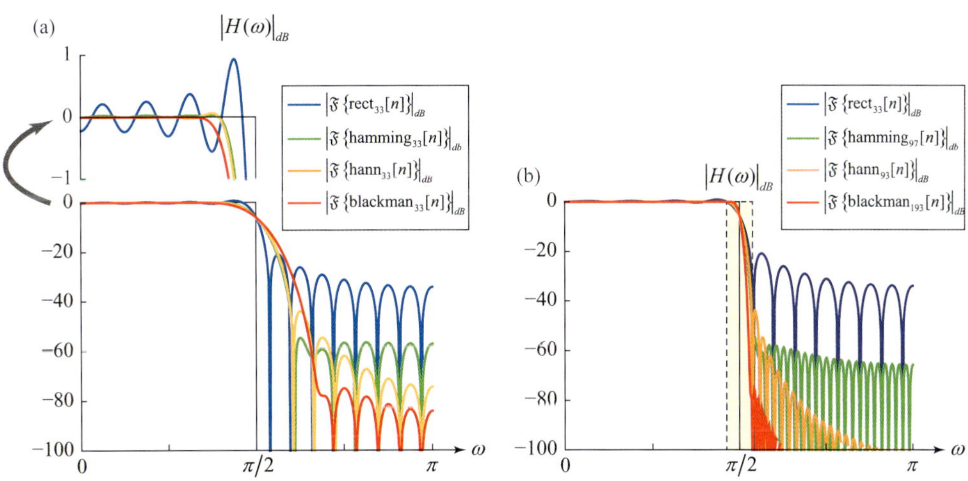

Figure 7.17 Comparison of FIR filters designed with rectangular, Hamming, Hann and Blackman windows

Figure 7.17a shows a comparison on a dB scale of the transform of lowpass FIR filters with a cutoff frequency, $\omega_c=0.5\pi$, designed using a rectangular window, Hamming window, Hann window and Blackman window of length $N=33$. As we expect from our discussion of **Figure 7.16**, for filters of the same length, the wider the transition zone of the window the lower the minimum stopband attenuation.

Design formulas for window-based FIR filters What do we do if we want to design a filter with both a small transition zone *and* good attenuation in the stopband? **Figure 7.15b** suggests an answer. It shows that for the Hamming window, the transition zone of the filter becomes narrower as the filter length increases, but the peak stopband attenuation does not appear to change appreciably. This result turns out to be generally true of all window-based filters, as shown in **Figure 7.18**.

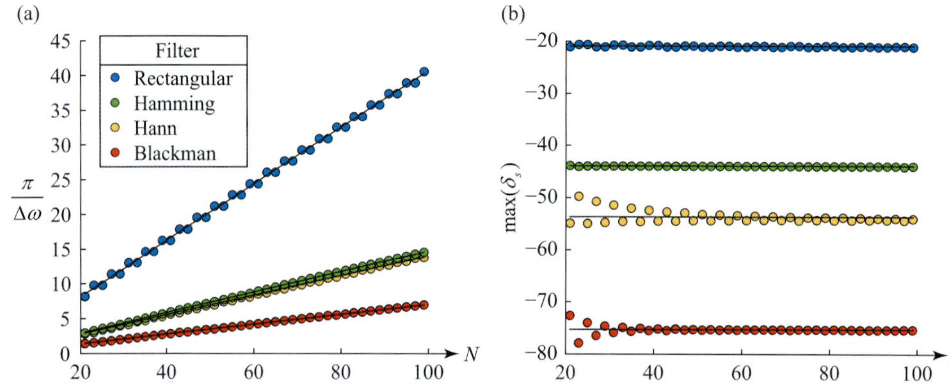

Figure 7.18 Filter transition bandwidth and peak stopband attenuation as a function of N

Figure 7.18a is a plot of the reciprocal of the width of the transition zone, $\pi/\Delta\omega$, as a function of filter length N for FIR filters based on all the windows we have studied so far. For each window, the data are well fit with a straight line whose equation is given in the middle column of **Table 7.3**. This table also gives data from a number of other filters we have not studied, but which Matlab has in its Signal Processing Toolbox. **Figure 7.18b** is a plot of the peak stopband magnitude, $\max(\delta_s)$, as a function of N. It shows that for each window, $\max(\delta_s)$ is a constant that is essentially independent of N (for large enough N). The value of this constant, given in the right column of **Table 7.3**, is characteristic of each window. All these filters have been designed with a cutoff frequency of $\omega_c = 0.5\pi$, but very similar results are obtained from filters designed with other values of ω_c. This figure and the table demonstrate that we *can* have both a small transition zone and good stopband attenuation; we just have to choose an appropriate window. Given a choice of window, the peak stopband magnitude is fixed, but the bandwidth can be adjusted to our specifications by changing N.

Table 7.3 Bandwidth and peak stopband attenuation for window-based filters

Window	$\pi/\Delta\omega$	$\max(\delta_s)$
Rectangular	$0.4069N$	−21
Hann	$0.1470N$	−44
Hamming	$0.1398N$	−54
Kaiser, $\beta = 5.874$	$0.1267N$	−62
Gaussian	$0.0891N$	−60
Blackman	$0.0704N$	−75
Nuttall	$0.0532N$	−111
Blackman Harris	$0.0528N$	−113
Chebyschev	$0.0635N$	−115

Figure 7.17b makes this point by showing a comparison of the frequency response of four filters whose transition regions have been chosen using the formulas in **Table 7.3** to match the transition region ($\Delta\omega = 0.074\pi$, highlighted in the yellow dashed box) of a rectangular window filter of length $N = 33$. This value of $\Delta\omega$ can be matched by a Hamming filter of length $N = 97$, a Hann filter of length $N = 93$ or a Blackman filter with $N = 193$. The stopband attenuations of all these other filters are better than that of the rectangular-window filter.

The moral of the story is that you can get the same transition width as the rectangular filter with better stopband attenuation, *but* at the cost of a higher filter length N. Of course, in a practical application, N matters. Among other things, a higher-order filter can mean increased computational cost. So, an effective approach to designing window-based filters is first to choose a filter window that provides the largest (i.e., least negative) peak stopband magnitude that meets your specifications for δ_s, and then select the length N that gives you the desired transition-zone width.

Example 7.5

Design a window-based lowpass filter with $\omega_c = 0.6$, $\Delta\omega = 0.15\pi$ and $\delta_s \leq -50$ dB.

➤ Solution:

Table 7.3 conveniently sorts the filter windows in order of the decreasing maximum peak magnitude, $\max(\delta_s)$. This allows us to rule out the rectangular and Hann windows, since their values of $\max(\delta_s)$ exceed -50 dB. That leaves a number of possible filters, everything from the Hamming window on down in the table. We choose the Hamming window, since it will achieve a specified value of $\Delta\omega$ at a lower value of N than the other windows in the table. That is because the slope of the equation relating $\pi/\Delta\omega$ to N in the second column of the table for the Hamming window (0.1424) is greater than that of any other filter with $\max(\delta_s) \leq -50$ dB. Using the formula in the table,

$$\frac{\pi}{\Delta\omega} = \frac{1}{0.15} = 0.1398N.$$

Solving gives $N = 47.69$, which we round up to the nearest odd integer, $N = 49$. Check for yourself the value of N that would have resulted had we chosen another filter instead, for example, the Blackman filter (Answer: $N = 95$).

Fractional bandwidth There is another practical consideration worth mentioning in designing filters, whether FIR or IIR. In many applications we want the transition bandwidth $\Delta\omega$ to decrease as the center frequency ω_c decreases. Specifically, we want the filter to have a constant **fractional bandwidth**[4]

$$\phi = \frac{\Delta\omega}{\omega_c}.$$

[4] This term comes from analog filter design and communication theory, where it is used as a figure of merit of bandpass and bandstop filters.

As an example, in the design of multirate discrete-time anti-aliasing and image-rejection filters for resampling (see Chapter 13), we will want $\Delta\omega = \phi\omega_c$, where $0.01 < \phi < 0.1$. Since the length of the filter, N, is inversely proportional to $\Delta\omega$, N will also be inversely proportional to ω_c.

7.3.3 Kaiser window

Raised-cosine windows are useful for many applications, but they only have one parameter that can be controlled: the length N. These windows do not provide independent control of important parameters such as the width of the transition band ($\Delta\omega$) or the magnitudes of the passband (δ_p) and stopband (δ_s). The Kaiser window is a very useful window that allows independent control over all these properties. For the Kaiser window, $w[n]$ has the form of a modified Bessel function of the first kind of order zero, $I_o(x)$. The Kaiser design equations accept two parameters provided by you, the filter designer: the ripple δ, and the desired bandwidth of the transition zone, $\Delta\omega$. For this filter, both the passband and the stopband ripple must be made the same, so δ would be set to be the smaller of δ_p and δ_s: $\delta \triangleq \min(\delta_p, \delta_s)$. For example, if we wanted a filter with a passband ripple of $\delta_p = 0.01$ and a stopband ripple of $\delta_s = 0.0001$, then $\delta = 0.0001$. This means that the passband would be "overdesigned," which is generally not a problem. Given values of δ and $\Delta\omega$, the Kaiser design equations allow us to compute the two important design parameters for the Kaiser filter: the filter order M and a shape parameter β. (Remember that the filter order is the filter length minus one: $M = N - 1$.) The smaller we wish to make $\Delta\omega$, the larger M will be. The shape parameter determines the ratio between the width of the main lobe and the peak side-lobe magnitude. Kaiser filters have the important property that the peak stopband magnitude is lower for a given width of the transition zone than that of other window-based filters.

Here is the procedure to compute β and M. First, express the value of δ in dB as the parameter A,

$$A = -20\log_{10}\delta.$$

Then use A to compute the shape parameter β according to the following formula:

$$\beta = \begin{cases} 0.1102(A - 8.7), & A > 50 \\ 0.5842(A - 21)^{0.4} + 0.07886(A - 21), & 21 \le A \le 50 \\ 0, & A < 21 \end{cases}.$$

Given values of A and $\Delta\omega$, the filter order M is

$$M = \left\lceil \frac{A - 7.95}{2.285\Delta\omega} \right\rceil,$$

where the notation $\lceil x \rceil$ means "largest integer not exceeding x," which is equivalent to Matlab's `ceil` function. If we wish to design a Type-I (odd-length symmetric) filter, then M has to be even so that N will be odd. If the design equations end up giving us an odd value of M, we need to increment it by one. Given M and β, the centered Kaiser window is

$$w[n] = \frac{I_o\left(\beta\left(1 - \left(\frac{2n}{M}\right)^2\right)^{\frac{1}{2}}\right)}{I_o(\beta)}, \quad |n| \le M/2.$$

Finally, the impulse response of the Kaiser lowpass filter is the product of the impulse response of the ideal lowpass filter and the Kaiser window,

$$h[n] = h_{ideal}[n]w[n].$$

A particularly attractive feature of the Kaiser filter is that we can change δ and $\Delta\omega$ independently using the design formulas. For example, **Figure 7.19** shows a family of Kaiser filters with $\omega_c = 0.5\pi$ and N and β adjusted to give a value of $\Delta\omega = 0.1\pi$ and values of $\delta = -40$ dB (blue), -50 dB (red) and -60 dB (green). They all have the same transition zone width, but different peak stopband attenuations.

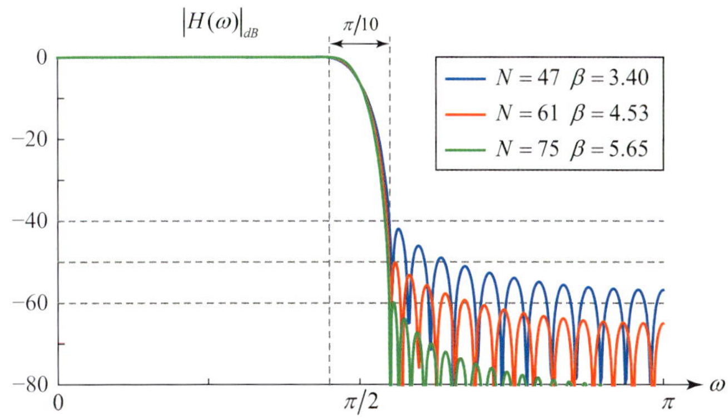

Figure 7.19 Kaiser filters

7.4 Highpass, bandpass and bandstop FIR filters

In the preceding sections, we have concentrated on the design of lowpass filters using windows. These lowpass filters can also be used to design simple **highpass**, **bandpass** and **bandstop** filters. There are several methods of designing these filters. One method uses the sum and difference of other responses of two filters. A second method is based on modulating the response of a prototype filter.

Complementary filters **Figure 7.20** shows a summary of a simple way of constructing highpass, bandpass and bandstop filters from a lowpass prototype by addition and subtraction of frequency responses.

Consider the process of designing a highpass filter from a lowpass prototype. Start with a lowpass filter $H_{LP}(\omega)$ with cutoff frequency ω_c, as shown in **Figure 7.20a**. **Figure 7.20b** shows a complementary highpass filter $H_{HP}(\omega)$ (red trace) with the same cutoff frequency ω_c, created by subtracting $H_{LP}(\omega)$ (green trace) from a constant frequency response (blue trace),

$$H_{HP}(\omega) = 1 - H_{LP}(\omega). \quad \textbf{(7.13)}$$

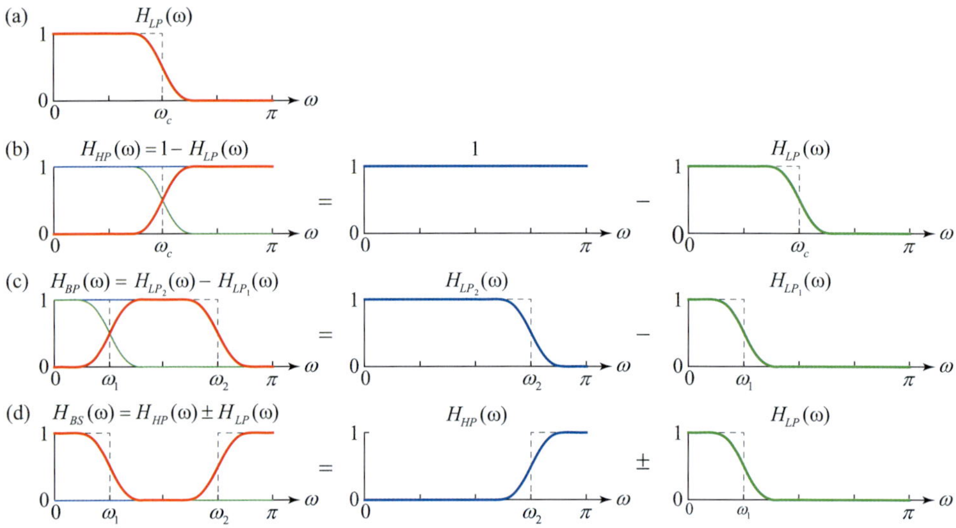

Figure 7.20 Design of highpass, bandpass and bandstop filters from a lowpass prototype

The inverse transform of Equation (7.13) gives the impulse response of the complementary filter in terms of the impulse response of the lowpass filter,

$$h_{HP}[n] = \delta[n] - h_{LP}[n]. \tag{7.14}$$

As an example, **Figure 7.21** shows the design of a highpass filter with $\omega_c = 0.25$ and length $N = 25$ using a rectangular-window lowpass prototype. Equation (7.14) applies when the impulse response $h_{LP}[n]$ is symmetric and non-causal, as is the case in **Figure 7.21**. To create a non-causal filter of odd length N, just delay everything by $(N-1)/2$,

$$h_{HP}[n-(N-1)/2] = \delta[n-(N-1)/2] - h_{LP}[n-(N-1)/2].$$

Example 7.6

Use Matlab to design a causal highpass filter with $\omega_c = 0.25$ and length $N = 25$ using a rectangular-window lowpass prototype.

➤ Solution:

Here is the code:

```
fc = 0.25;
N = 25;
n = -(N-1)/2:(N-1)/2;
hlp = fir1(N-1, fc, rectwin(N), 'noscale');
hhp = -hlp; % first negate response
hhp((N+1)/2) = hhp((N+1)/2) + 1; % then add one to center value
```

By default, the Matlab function `fir1` scales the coefficients so that the value of $H_{LP}(\omega = 0)$ is 0 dB. In this design, we want the coefficients to be unnormalized so that the frequency response of $|H_{LP}(\omega)|$

oscillates about 0 dB instead. Hence, we used `fir1` with the `noscale` option to design the prototype lowpass filter. Also, due to Matlab's ones-based indexing, the center tap of the filter is `hhp((N+1)/2)`, not `hhp((N-1)/2)`. The result is shown in **Figure 7.21**.

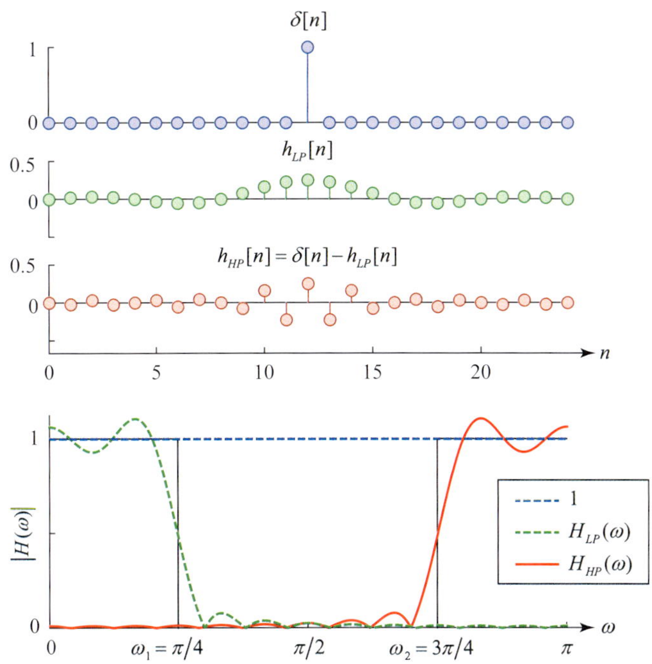

Figure 7.21 Design of a highpass filter from a lowpass prototype

This example indicates some of the issues connected with making a realizable filter. In constructing a derived filter such as this highpass filter, it is necessary to choose an appropriate prototype lowpass filter. According to **Table 7.2**, a linear-phase highpass filter must end up being either Type-I (odd-length symmetric) or Type-IV (even-length antisymmetric). Moreover, the highpass filter and the prototype lowpass filter must be the same length. Since a lowpass filter cannot be antisymmetric, only a Type-I prototype lowpass filter will do. You can also see from **Figure 7.21** that $h_{LP}[n]$ must be odd-length symmetric so that it can be subtracted from the impulse centered at $n = 0$. This example also indicates that the passband and stopband characteristics of the derived highpass filter are not necessarily exactly the same as those of the prototype filter. The exact shape of the side lobes in the stopband of the highpass filter depend on the shape of the ripples in the passband of the prototype lowpass filter, as illustrated in **Figure 7.7**.

Bandpass and bandstop filters Bandpass and bandstop filters can also be created from combinations of prototype lowpass filters. **Figure 7.20c** shows that a bandpass filter ($H_{BP}(\omega)$, red) with low cutoff ω_1 and high cutoff ω_2 can be created by subtracting the impulse response of a lowpass filter ($H_{LP_1}(\omega)$, green) with cutoff ω_1 from that of another lowpass filter ($H_{LP_2}(\omega)$, blue) with cutoff ω_2. In general,

$$H_{BP}(\omega) = H_{LP_2}(\omega) - H_{LP_1}(\omega),$$

so

$$h_{BP}[n] = h_{LP_2}[n] - h_{LP_1}[n].$$

A brief look at **Table 7.2** indicates that although a general linear-phase bandpass filter can be of any type (Types I–IV), a bandpass filter derived from a prototype lowpass filter can only be either Type-I or Type-II.

Example 7.7

Design a window-based bandpass filter with corner frequencies $\omega_1 = \pi/4$ and $\omega_2 = 3\pi/4$, and length $N = 25$, using rectangular-window lowpass prototypes.

➤ Solution:

The code is straightforward:

```
fc1 = 0.25;
fc2 = 0.75;
N = 25;n = -(N-1)/2:(N-1)/2;
hlp1 = fir1(N-1, fc1, rectwin(N), 'noscale');
hlp2 = fir1(N-1, fc2, rectwin(N), 'noscale');
hbp = hlp2 - hlp1;
```

The result is shown in **Figure 7.22**.

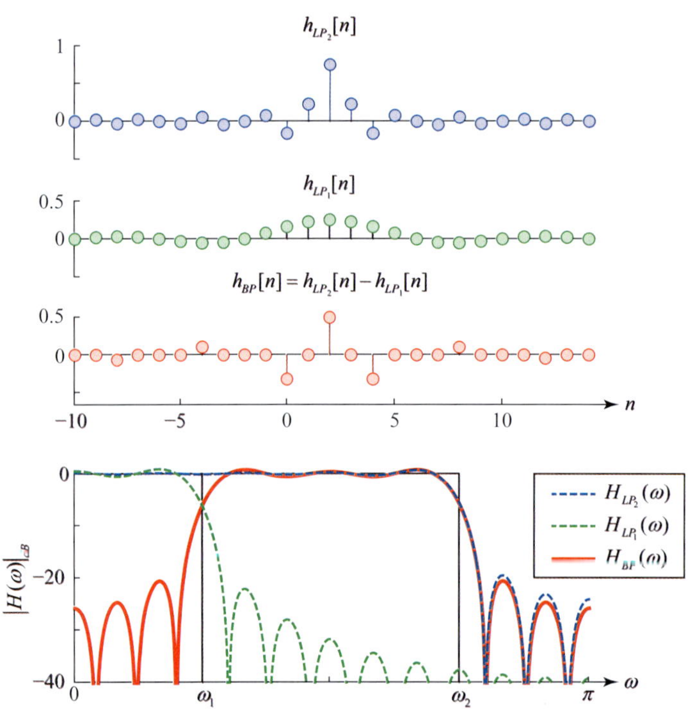

Figure 7.22 Design of a bandpass filter from lowpass prototypes

We could just as well have created this bandpass filter by subtracting $h_{LP_2}[n]$ from $h_{LP_1}[n]$: $h_{BP}[n] = h_{LP_1}[n] - h_{LP_2}[n]$. The magnitude of the frequency response would have been the same; the phase would differ by π.

Finally, **Figure 7.20d** shows a bandstop filter ($H_{BS}(\omega)$, red) with low cutoff ω_1 and high cutoff ω_2, created by subtracting the impulse response of a lowpass filter ($H_{LP}(\omega)$, green) with cutoff ω_1 from that of a highpass filter ($H_{HP}(\omega)$, blue) with cutoff ω_2. In general,

$$H_{BP}(\omega) = H_{HP}(\omega) \pm H_{LP}(\omega) = 1 - H_{LP_2}(\omega) \pm H_{LP_1}(\omega),$$

where we note that the highpass filter is itself created from a lowpass filter $H_{LP_2}(\omega)$ with cutoff ω_2. Hence,

$$h_{BP}[n] = h_{HP}[n] \pm h_{LP}[n] = (\delta[n] - h_{LP_2}[n]) \pm h_{LP_1}[n] = \delta[n] - h_{LP_2}[n] \pm h_{LP_1}[n].$$

As with the highpass filter, the filter length of the bandstop filter must be odd, so only a Type-I prototype lowpass filter can be used. While the bandstop filter can be created either by adding or subtracting the lowpass and highpass filter responses, the results differ, as shown in the next example.

Example 7.8

Design a window-based bandstop filter with corner frequencies $\omega_1 = \pi/4$ and $\omega_2 = 3\pi/4$, and length $N = 25$, using rectangular-window lowpass prototypes.

➤ Solution:

Figure 7.23a shows the impulse and frequency response of a bandstop filter (red) with cutoffs $\omega_1 = \pi/4$ and $\omega_2 = 3\pi/4$, created by *adding* the impulse responses of a lowpass rectangular-window filter ($H_{LP}(\omega)$, green) to the impulse response of a highpass filter ($H_{HP}(\omega)$, blue) with cutoff $\omega_2 = 3\pi/4$. **Figure 7.23b** shows the result of *subtracting* the same two filters.

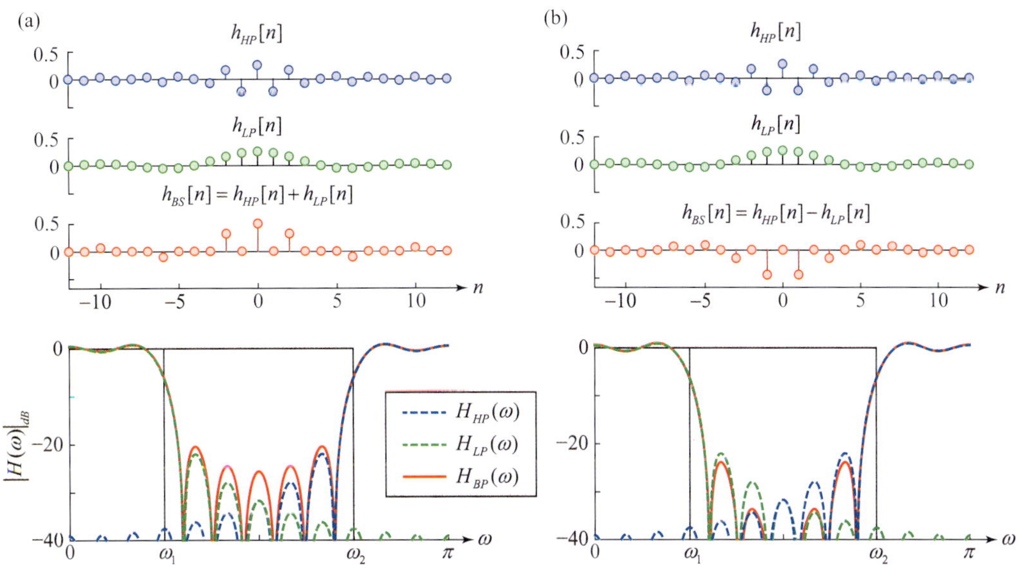

Figure 7.23 Bandstop rectangular window filters

In this particular example, the side lobes of the first filter add constructively and are higher than those of the constituent filters, whereas for the second filter they add destructively.

The preceding three examples emphasize an important point: for highpass, bandpass and bandstop filters created by the sum and difference of window-based filter responses, the characteristics of the passband and stopband are not simply the equal to those of either one of the prototype filters. You have to pay attention to the interaction of the side lobes and main lobes of the two filters, which change depending on the cutoff frequencies and window type of the filters.

Modulation method Another method of deriving highpass, bandpass and bandstop filters from a lowpass prototype uses the modulation property of the DTFT. Consider the example of designing a bandpass filter of width $\Delta\omega$, centered at ω_c. First, design a lowpass filter prototype with a cutoff frequency of $\Delta\omega/2$ and then translate that frequency response to be centered around $\pm\omega_c$ by multiplying the impulse response by $\cos\omega_c n$.

To create a highpass filter from a lowpass prototype, translate the spectrum to be centered at $\omega_c = \pm\pi$,

$$H_{hp}(\omega) = H_{lp}(\omega \pm \pi).$$

In the time domain, the modulation property of the DTFT gives

$$H_{hp}(\omega) = H_{lp}(\omega \pm \pi) = \sum_{n=-\infty}^{\infty} h_{lp}[n]e^{-j(\omega \pm \pi)n} = \sum_{n=-\infty}^{\infty} (h_{lp}[n]e^{\pm j\pi n})e^{-j\omega n}$$
$$= \sum_{n=-\infty}^{\infty} (h_{lp}[n](-1)^n)e^{-j\omega n} = \sum_{n=-\infty}^{\infty} h_{hp}[n]e^{-j\omega n}.$$

So, the highpass filter is created by multiplying the impulse response of the lowpass filter by $\cos\pi n$:

$$h_{hp}[n] = h_{lp}[n]\cos\pi n = h_{lp}[n](-1)^n.$$

Example 7.9

Use the modulation method to design a highpass filter with cutoff frequency $\omega_c = 0.25\pi$, $\Delta\omega = 0.25\pi$ and $\delta_s \leq -50$ dB.

➤ Solution:

To achieve the desired response, we first design a lowpass filter with cutoff frequency $\omega_c = 0.75\pi$, $\Delta\omega = 0.25\pi$ and $\delta_s \leq -50$ dB, and then multiply its impulse response $h_{LP}[n]$ by $\cos\pi n = (-1)^n$, which translates the frequency response by $\pm\pi$. Performing this multiplication is equivalent to negating every other value of $h_{LP}[n]$. From **Table 7.3**, we see that we can achieve the required stopband specification, $\delta_s \leq -50$, with a Hamming filter. The design formula in the table allows us to estimate the order required: $N = 29.44$. Rounding up to the nearest odd integer gives $N = 31$. **Figure 7.24a** shows the impulse response of a resulting Hamming lowpass filter, $h_{LP}[n]$. **Figure 7.24b** shows the sequence $\cos \pi n = (-1)^n$, and **Figure 7.24c** is the product, a Hamming highpass filter with impulse response $h_{HP}[n]$. Using Matlab, here is some code:

```
fc = 0.75;
df = 0.25;
N = ceil((1/df+0.18)/0.142); % find filter length from table
N = N + 1 - rem(N, 2);        % add one to the length if N is even
n = -(N-1)/2:(N-1)/2;
hlp = fir1(N-1, fc, hamming(N));
hhp = hlp .* (-1).^n;
```

Figure 7.24d shows the frequency responses $|H_{LP}(\omega)|$ (dotted blue line) and $|H_{HP}(\omega)|$.

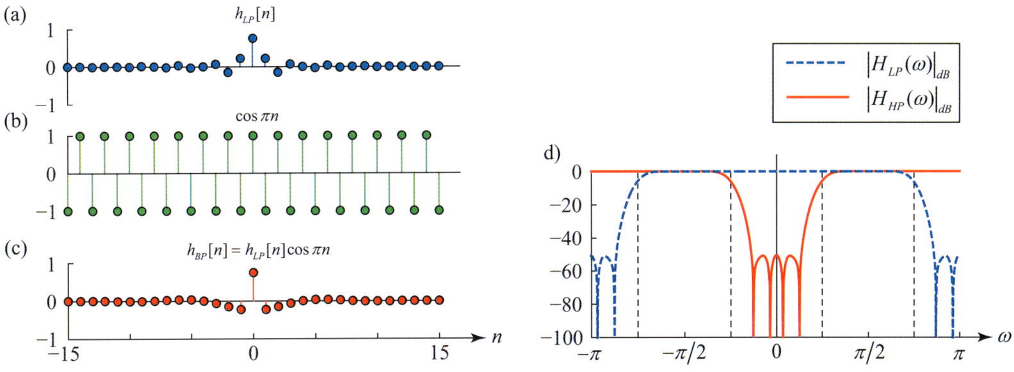

Figure 7.24 Construction of a highpass filter from a lowpass prototype using modulation

Figure 7.25 shows the construction of a bandpass filter of width $\Delta\omega = 0.5\pi$, centered at $\omega_c = 0.5\pi$. **Figure 7.25a** shows the impulse response $h_{LP}[n]$ (left panel) and frequency response magnitude $|H_{LP}(\omega)|_{dB}$ (right panel) of a Hamming lowpass filter with cutoff frequency $\omega_c = 0.25\pi$ and length $N = 31$. **Figure 7.25b** shows $\cos 0.5\pi n$, whose transform is two impulses at $\omega = \pm 0.5\pi$.

The left panel of **Figure 7.25c** is the product

$$h_{BP}[n] = \alpha h_{LP}[n] \cos \omega_c n,$$

whose transform is

$$\begin{aligned} H_{BP}(\omega) &= \mathfrak{F}\{\alpha h_{LP}[n] \cos \omega_c n\} = \alpha \frac{1}{2\pi} H_{LP}(\omega) * \mathfrak{F}\{\cos \omega_c n\} \\ &= \alpha \frac{1}{2\pi} H_{LP}(\omega) * (\pi\delta(\omega - \omega_c) + \pi\delta(\omega + \omega_c)) \\ &= \frac{\alpha}{2}(H_{LP}(\omega - \omega_c) + H_{LP}(\omega + \omega_c)). \end{aligned}$$

The value $\alpha = 1.9942$ has been chosen to make the magnitude of the frequency response exactly 0 dB at $\omega = \pm\omega_c$, as shown in the right panel of **Figure 7.25c**.

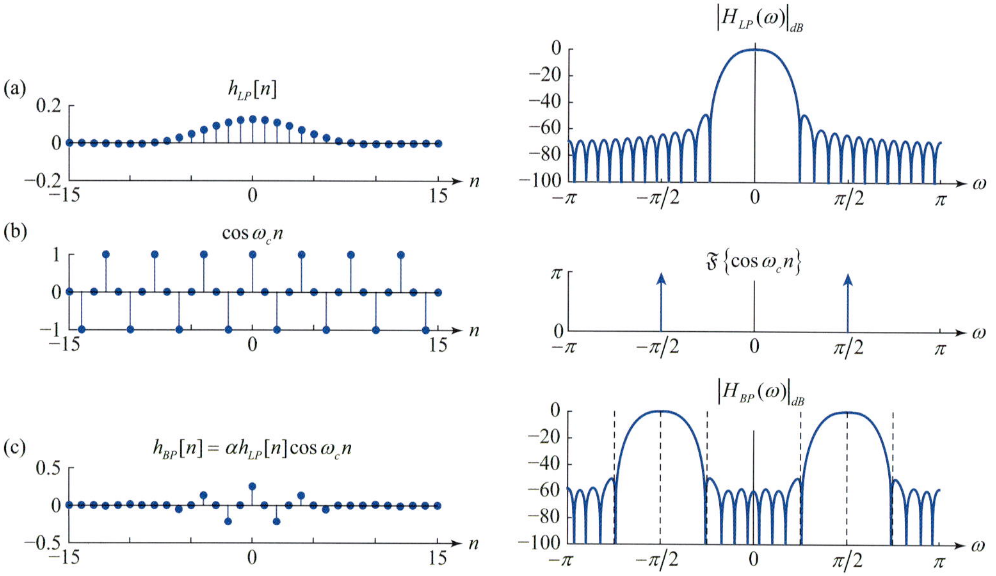

Figure 7.25 Construction of a bandpass filter by modulation of a lowpass prototype

7.5 Matlab implementation of window-based FIR filters

Matlab's Signal Processing Toolbox contains a number of functions that are useful in designing windows and window-based FIR filters. We have already mentioned individual functions for the windows we have discussed. `rectwin(N)`, `hamming(N)`, `hann(N)`, `blackman(N)` and `kaiser(N, beta)` all design windows of length N. There is also a more general window function `window(@windowname, N, [ options] )` that returns a window of length N with a window type given by the function handle `@windowname`, which can refer to one of 17 different types of windows. For example, for a Hamming window, you would use `@hamming`. Some windows, such as the Kaiser window, require the specification of options (e.g., β).

Designing an FIR filter with Matlab is easy if you have the Signal Processing Toolbox. It does the entire job for you using the `fir1` function, which implements window-based design of lowpass, highpass, bandpass and bandstop FIR filters (as well as multiband filters). The general syntax is

```
h = fir1(N-1, fc, 'type', window(N), 'noscale'),
```

where `N` is the length of the impulse response and `fc` is the cutoff frequency (or array of frequencies, in the case of bandpass and bandstop filters), normalized to π (not 2π).[5] The optional argument `type` specifies the type of filter (e.g., `'low'`, `'high'` or `'stop'`, with

[5]This point is important and can be confusing. In much of the DSP literature and common usage, the normalized frequency f is defined as angular frequency divided by 2π, $f \triangleq \omega/2\pi$, so that the maximum normalized frequency is one. In Matlab's signal processing functions, the normalized frequency is defined as $f \triangleq \omega/\pi$, so that the maximum normalized frequency is two.

'low' being the default). The optional argument window(N) indicates the type of window to be multiplied by the ideal impulse response to produce the response of the filter (e.g., blackman(N)), the default being a Hamming window. All of the 17 windows functions supported by Matlab are available for filter design. For example, to produce the response of **Figure 7.23a**, a bandstop rectangular-window filter of length $N = 25$ and cutoff frequencies $\omega_1 = \pi/4$ and $\omega_2 = 3\pi/4$, you would type

```
h = fir1(24, [0.25 0.75], 'stop', rectwin(25)).
```

The first argument of fir1 is the filter *order*, which is the length of the filter minus one. However, the correct filter *length* must appear as the argument of the window function (e.g., rectwin(25)). If you attempt to design a highpass or bandstop filter of even length, Matlab will either give an error (if you have specified a window function) or increase the length of the filter by one without telling you (if you are using the default Hamming window). The bandstop filter designed using the fir1 command, above, is actually the one shown in **Figure 7.23a**, where the side lobes add, not the one shown in **Figure 7.23b**, where they subtract.

By default, unless you use the noscale option, Matlab's fir1 function normalizes the magnitude of the filter's response to be exactly 0 dB at $\omega = 0$; that is, $H(0) = 1$. From the definition of the DTFT,

$$H(0) = H(\omega)|_{\omega=0} = \left(\sum_{n=-\infty}^{\infty} h[n] e^{-j\omega n} \right) \Bigg|_{\omega=0} = \sum_{n=-\infty}^{\infty} h[n].$$

Therefore, normalizing $H(0)$ to one is accomplished by dividing $h[n]$ by the sum of $h[n]$. The final optional argument of the fir1 function, 'noscale', indicates that you do not wish Matlab to perform this normalization. This option is important if you are trying to match Matlab's filter design with your own code. For example, to produce the response of the Hamming-window filter of length $N = 33$, shown in **Figure 7.17a**, you simply could use Matlab's fir1 function,

```
h = fir1(32, 0.5, hamming(33)).
```

If you do not have Matlab's Signal Processing Toolbox, you can create your own function for the lowpass case with just a few lines of code, using Equations (7.7), (7.8) and (7.12):

```
N = 33;
fc = 0.5; % corner frequency in fractions of pi: fc=wc/pi
n = -(N-1)/2:(N-1)/2;
h = (0.54+0.46*cos(2*pi*n/(N-1))).*(fc*sinc(fc*n));
h = h/sum(h); % normalize so that H(0)=1
```

Nota bene! Matlab's sinc function already includes a factor of π, so Equation (7.4), $h_{ideal}[n] = (\omega_c/\pi)\,\mathrm{sinc}\,\omega_c n$, translates into fc*sinc(fc*n), where fc is in fractions of π.

Matlab's kaiserord and kaiser functions simplify the task of designing Kaiser windows and filters. To design a Kaiser filter, you first use kaiserord to get values of M and β. The syntax of kaiserord is

```
[M, fc, beta, filtertype] = kaiserord(f, A, delta, [fs]),
```

where `f` is a vector of band-edge frequencies, normalized to the value of `fs`, which is 2 by default, such that a value of `f=1` corresponds to $\omega = \pi$. `A` is a vector of the associated band magnitudes, `delta` is a vector of the maximum allowable deviation in each band (i.e., δ) and `fs` is an optional sampling rate. This function allows you to design lowpass, highpass, bandpass and multiband filters, but for a simple lowpass filter, `f` is just a vector comprising $[\omega_p\ \omega_s]$, `A` is [1 0] and `delta` is $[\delta_p\ \delta_s]$. The output parameters of the function are the order of the filter `M`, an array of center frequencies `wc`, the Kaiser `beta` and the `filtertype` (e.g., '`low`'). Then, you plug these values into `fir1` using `kaiser` as the window function to get the impulse response:

```
h = fir1(M, fc, kaiser(M+1, beta)).
```

For example, to design the Kaiser filter in **Figure 7.19** with $\omega_c = 0.5\pi$, $\Delta\omega = 0.1\pi$ and $\delta = -40$ dB, we have $\omega_p = \omega_c - \Delta\omega/2 = 0.45\pi$, $\omega_s = \omega_c + \Delta\omega/2 = 0.55\pi$ and $\delta_p = \delta_s = \delta = 0.01$, so

```
[M, fc, beta] = kaiserord([0.45 0.55], [1 0], [0.01 0.01]).
```

This returns a value of filter order $M = 45$, so the filter length would be $N = M + 1 = 46$. However, since we wish to produce an odd-length symmetric filter, we would increment M by one to get $M = 46$ (i.e., $N = 47$). Of course, we can create our own Kaiser filter function and specialize it for the odd-length lowpass case:

```
function h = kaiser_lowpass(fc, df, delta)
    % KAISER_LOWPASS Impulse response of odd-length Kaiser lowpass filter
    %       h = KAISER_LOWPASS(fc, df, delta)
    %          h is the impulse response
    %          fc is cutoff (normalized by pi)
    %          df is transition bandwidth (normalized by pi)
    %          delta is a vector of the passband and stopband ripple
    delta = min(delta);
    A = -20*log10(delta);
    beta = 0.1102 * (A-8.7) .* (A>50) + ...
          (0.5842 * (A-21).^0.4 + 0.07886 * (A-21)) .* (A>=21 & A<=50);
    M = ceil((A - 7.95) / (pi * df * 2.285)); % don't forget the pi!
    M = M + mod(M, 2); % make sure M is even so N will be odd
    n = -M/2:M/2;
    w = besseli(0, beta * sqrt(1-(2*n/M).^2)) ./ besseli(0, beta);
    h = w .* (fc * sinc(fc * n));
    h = h / sum(h);
end
```

7.5.1 Matlab functions that implement FIR filtering

Matlab provides several methods of implementing FIR filtering of an input once the impulse response of the filter has been computed. Three of these are illustrated in **Figure 7.26**.

conv The input (blue trace)

$$x[n] = 3 + \cos(0.15\pi n) + \sin(0.25\pi n), \quad 0 \le n < 30$$

is of length $lx = 30$. The impulse response of the symmetric causal lowpass filter,

$$h[n] = \delta[n] + 2\delta[n-1\} + 3\delta[n-2] + 2\delta[n-3] + \delta[n-4],$$

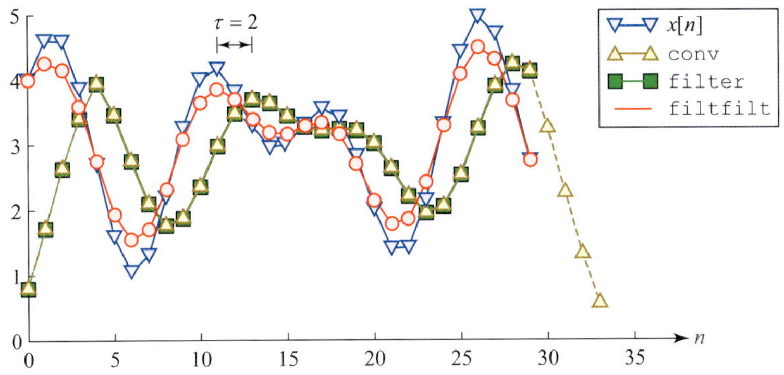

Figure 7.26 Comparison of FIR filtering functions in Matlab

is of length $lh = 5$. The output of this filter, `conv(x, h)` (yellow traces), is a response of length $lx + lh - 1 = 34$. The function implicitly pads both the input and output with $lh - 1 = 4$ zeros, so the first $lh - 1 = 4$ points of the convolution ramp up from 0 and the last $lh - 1 = 4$ points (both marked with dashed lines) ramp back down to 0. To see why this is done, consider the calculation of the first point of the output,

$$y[0] = \sum_{k=0}^{4} h[k]x[0-k].$$

The convolution sum requires values of $x[0]$, $x[-1]$, $x[-2]$, $x[-3]$ and $x[-4]$. Only $x[0]$ is a specified input value. The remaining four values are initial conditions that the function needs and implicitly sets to zero. Similar arguments apply to the calculation of $y[1]$, $y[2]$ and $y[3]$. Similar logic also applies to the end of the convolution. For example, the last point is

$$y[33] = \sum_{k=0}^{4} h[k]x[33-k].$$

The convolution sum requires values of $x[33]$, $x[32]$, $x[31]$, $x[30]$ and $x[29]$. Only $x[29]$ is a specified input value. The rest are implicitly set to zero. In the example shown in **Figure 7.26**, $h[n]$ is a causal Type-I linear-phase lowpass filter of length five. Hence, the output can be time-aligned with the input when it is shifted left by two samples.

`filter` FIR filtering can also be implemented with Matlab's general-purpose `filter` function (green trace). The syntax for FIR filtering is

```
[y, [yi]] = filter(h, 1, x, [xi]),
```

where `xi` is an optional array of the initial conditions for filtering of length $lh - 1$, which allows you to specify the first four values, $x[-1]$, $x[-2]$, $x[-3]$ and $x[-4]$. If these values are not specified, the first $lh - 1$ points are identical to those calculated by `conv` (dashed line). Unlike the `conv` function, the output of `filter` is the same length as the input, lx. The optional output array `yi` provides the last $lh - 1$ "partial" output points. If you want to make `filter` identical to `conv`, append $lx - 1$ zeros to the end of the input sequence and filter. The provision for including initial conditions in `filter` and generating partial output points is particularly useful in implementing the time-domain overlap-add method for long inputs, as described in Section 2.7.1. This is illustrated by the following code:

```
function y = overlap_add(x, h, blockSize)
    lx = length(x);
    lh = length(h);
    nBlocks = fix(lx / blockSize); % determine number of blocks
    xi = zeros(1, lh-1); % set initial conditions for first block
    y = zeros(1, lx);    % allocate output array

    % Main loop
    for n = 1:nBlocks
        indx = (n-1) * blockSize + (1:blockSize);  % index into arrays
        [y(indx), yi] = filter(h, 1, x(indx), xi);
       xi = yi; % use partial output points as initial conditions next time
    end

    % Last partial block (if necessary)
    indx = indx(end)+1:lx;
    y(indx) = filter(h, 1, x(indx), xi);
end
```

In each pass through the `for` loop, the partial output points generated by `filter` in processing one block of the input (i.e., `yi`) are used as the initial conditions for processing the next block (i.e., `xi`). This code is equivalent to `y = filter(h, 1, x)`. When using long impulse responses, the overlap-add algorithm can be made faster by doing the actual filtering by multiplication of transforms with the FFT instead of with `filter`. Matlab provides the `fftfilt` function for just this purpose. Its use will be described in Chapter 11.

`filtfilt` The Matlab function `filtfilt` can be used to implement zero-phase FIR filtering, as described in supplementary material. `filtfilt` is designed to work with both FIR and IIR filtering, but when employed with FIR filters the syntax is

```
y = filtfilt(g, 1, x),
```

where `g` is $g[n]$, as described in supplementary material. For the example shown in **Figure 7.26** (red trace), $h[n] = g[n] * g[-n]$, where $g[n] = \delta[n] + \delta[n-1] + \delta[n-2]$. Since the filter is zero-phase, the output of `filtfilt` is time-aligned with the input. `filtfilt` not only does zero-phase filtering; it also attempts to minimize the transients at the beginning and end of filtering by creating "reasonable" initial and ending conditions. The following code, `zerophasefilt_FIR`, provides output that is identical to `filtfilt`, but is faster and simpler to understand. It uses $h[n]$ instead of $g[n]$.

```
function y = zerophasefilt_FIR(x, h)
    lx = length(x);
    lh = length(h);

    % Mirror lh-1 points at the beginning and end of the sequence
    x_beg = 2 * x(1) - x(lh:-1:2);
    x_end = 2 * x(lx) - x((lx-1):-1:(lx-lh+1));
```

```
    % Prepend x_beg, append x_end and convolve
    xx = [x_beg x x_end];
    yy = conv(xx, h);

    % Select center lx points to time-align the output with x[n]
    off = 3*((lh-1)/2)+1;
    y = yy(off:end-off+1);
end
```

The idea here is to pad the beginning and ending points of the sequence with $lh-1$ points derived by mirroring the first and last $lh-1$ points of $x[n]$ respectively. For example, if the first $lh-1=4$ points of $x[n]$ form the ascending sequence [3 4 6 8 9], the function computes the cumulative difference between these elements (i.e., [1 3 5 6]), subtracts this from $x[0]=3$ and time-reverses the result to form array `x_beg` ([−3 −2 0 2]), which it then prepends to $x[n]$. The effect is to create a lead-in ascending sequence of coefficients that are not all zeros. The equivalent thing is done to the end of the sequence by appending sequence `x_end`. The augmented sequence `xx` is then convolved with the impulse response `h` to form `yy` and finally the center `lx` points of `yy` are selected to time-align the output.

7.6 ★ Spline and raised-cosine FIR filters

In Section 7.3, we showed that designing a lowpass filter using the rectangular window minimizes the least-square error with the ideal lowpass filter. The problem with the rectangular-window filter is that truncation of the impulse response of the ideal filter leads to poor behavior in the frequency domain: an overshoot in the passband of about 9%, and an undershoot of 9% in the stopband that corresponds to a peak attenuation of only about −21 dB. The approach of all window-based filters is to design window functions in the time domain that taper smoothly at the ends. Use of these windows leads to filters whose transforms have smaller passband and stopband ripples than the rectangular-window filter, though at the expense of a wider transition band.

Another fruitful approach to FIR filter design is to look at the problem in the *frequency domain*. In **Figure 7.7**, we showed that the poor frequency response of the rectangular window filter can be understood as the convolution in the frequency domain of the frequency response of the rectangular window, $W(\omega)$, with the step discontinuity in the frequency response of the ideal lowpass filter. We will now discuss two methods based on the idea of modifying the frequency response of the ideal lowpass filter by creating a transition region of finite width that removes the step discontinuity in the frequency response. The first method uses **splines** to specify the transition region; the second method uses a section of a **raised cosine**. Both these methods can be used to design Nyquist filters, discussed in Section 13.7, which are used in digital communication and filter-bank applications.

7.6.1 FIR filters designed using splines

Figure 7.27a shows the frequency response of an ideal lowpass filter $H_0(\omega)$ with cutoff frequency ω_c. We start by specifying a desired transition region of width $\Delta\omega$ centered on ω_c. **Figure 7.27b** shows the frequency response of a number of filters created by connecting the edge of the passband of the ideal filter (i.e., $H_0(\omega_c-\Delta\omega/2)=1$) to the edge of the stopband

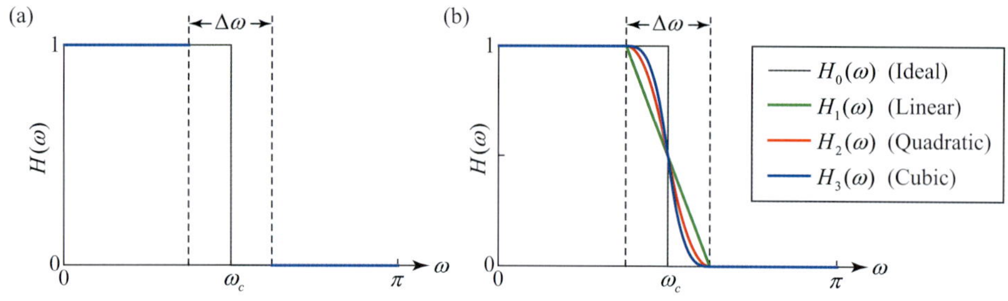

Figure 7.27 Spline filters

(i.e., $H_0(\omega_c + \Delta\omega/2) = 0$) with splines of different orders. A spline is just a polynomial function of ω. The simplest first-order spline is a straight line, resulting in a filter with a linear transition zone, $H_1(\omega)$, shown in green. **Figure 7.27b** also shows filters with frequency responses $H_2(\omega)$ and $H_3(\omega)$, whose transition zones are second-order and third-order splines respectively.

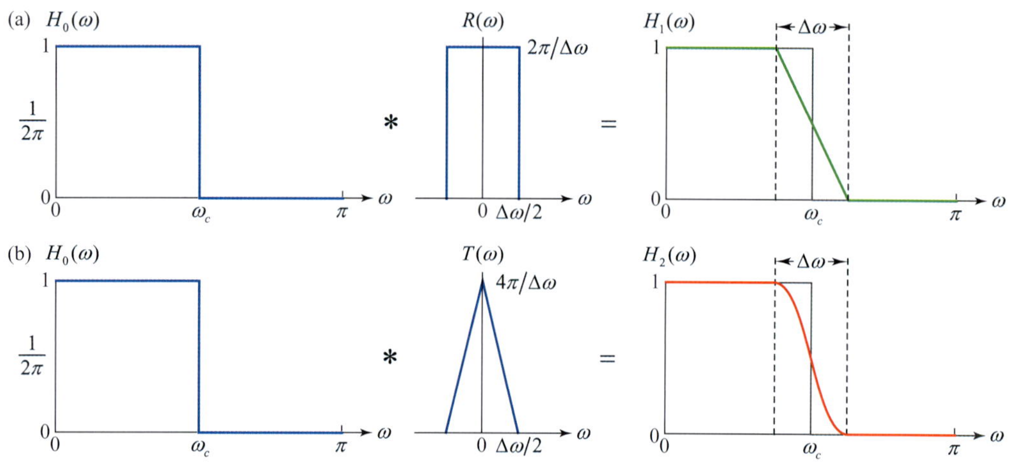

Figure 7.28 Construction of spline filters

Figure 7.28a illustrates a simple way of deriving the impulse response of a first-order spline filter, $h_1[n]$. First, write an expression in the frequency domain for the frequency response $H_1(\omega)$. To do so, create a rectangular function $R(\omega)$ centered at $\omega = 0$, with width $\Delta\omega$ and height $2\pi/\Delta\omega$. This function has a constant area of 2π,

$$R(\omega) = \begin{cases} \dfrac{2\pi}{\Delta\omega}, & -\Delta\omega/2 \le \omega \le \Delta\omega/2 \\ 0, & \text{otherwise} \end{cases}.$$

Then, $H_1(\omega)$ is the frequency-domain convolution of $H_0(\omega)$ with $R(\omega)$,

$$H_1(\omega) = \frac{1}{2\pi} H_0(\omega) * R(\omega).$$

The impulse response of the filter is $h_1[n] = h_0[n]r[n]$, where $h_0[n] = (\omega_0/\pi)\operatorname{sinc}\omega_c n$ is the impulse response of $H_0(\omega)$ and $r[n] = \operatorname{sinc} n\Delta\omega/2$ is the inverse transform of $R(\omega)$. Hence,

$$h_1[n] = \left(\frac{\omega_c}{\pi}\operatorname{sinc}\omega_c n\right)\operatorname{sinc}\frac{\Delta\omega}{2}n. \tag{7.15}$$

Figure 7.28b shows that the transform of a second-order spline filter,

$$H_2(\omega) = \frac{1}{2\pi}H_0(\omega) * T(\omega),$$

can be constructed in a similar manner (see Problem 7-31), as the convolution of the response of the ideal lowpass filter with a triangular response

$$T(\omega) = \begin{cases} \dfrac{4\pi}{\Delta\omega}\left(1 - \dfrac{2|\omega|}{\Delta\omega}\right), & 0 \leq |\omega| < \Delta\omega/2 \\ 0, & \text{otherwise} \end{cases}.$$

The impulse response of the second-order spline (Problem 7-31a) is

$$h_2[n] = \left(\frac{\omega_c}{\pi}\operatorname{sinc}\omega_c n\right)\operatorname{sinc}^2\frac{\Delta\omega}{4}n.$$

By extension, the impulse response of the Pth-order spline filter $H_P(\omega)$ can be shown to be (Problem 7-31c)

$$h_P[n] = \left(\frac{\omega_c}{\pi}\operatorname{sinc}\omega_c n\right)\operatorname{sinc}^P\frac{\Delta\omega}{2P}n. \tag{7.16}$$

$H_P(\omega)$ is continuous at frequencies $\omega \pm \Delta\omega/2$, as are its P derivatives (see Problem 7-32). The resulting impulse response $h_P[n]$ (Equation (7.16)) falls off as $1/(n+1)^P$. Despite the fact that we derived these spline filters using the frequency-domain behavior of their transition regions, these are all really just window-based filters, where the window in each case is a sinc function raised to a power instead of a Hamming or Kaiser window. Because $\operatorname{sinc} n$ falls off as $1/n$, a sinc window tapers the ideal impulse response $h_0[n]$ fairly dramatically.

In order to make any of the spline filters realizable, the impulse response of Equation (7.16) must be truncated (or multiplied by a window) to make it finite-length and shifted to make it causal. From Section 7.3, we have plenty of windows from which to choose. The resulting finite-length spline filter has at least five parameters that can be adjusted to change the shape of the filter – the cutoff frequency ω_c, the order of the spline, P, the width of the transition zone $\Delta\omega$, the length of the truncated impulse response, N, and the type of window used to shape or truncate the response. So, in principle, considerable flexibility in design is possible. Unfortunately, there are no standard design formulas for the spline filter, such as those for the Kaiser filter, to guide us in our choice of values of the parameters necessary to produce a desired response.

7.6.2 FIR filters designed using raised cosines

Like the spline filter, raised-cosine FIR filters are designed to have a "gentler" transition band in the frequency domain than the rectangular filter. These filters are used in digital communication systems to shape pulses that minimize **intersymbol interference**. It is unfortunate that the same words – raised cosine – are used in the literature to refer both to time-domain raised-cosine window filter designs, such as the Hamming and Hann filters we discussed in Section 7.3.2, and to the frequency-domain raised-cosine design we will describe in this section.

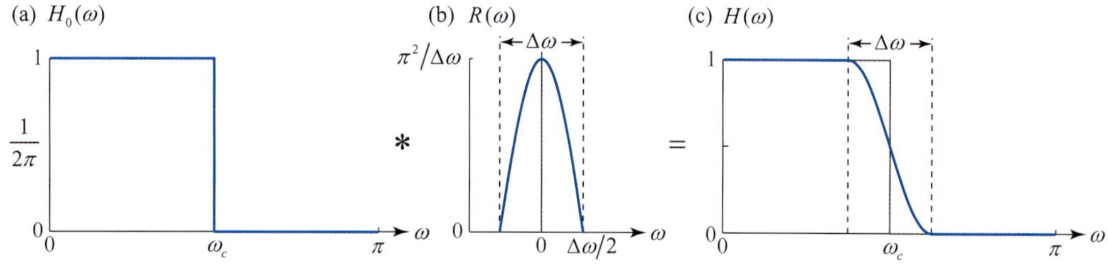

Figure 7.29 Construction of a raised-cosine filter

Figure 7.29 shows how the raised-cosine filter can be constructed in the frequency domain by the convolution of an ideal lowpass filter and a filter with a response comprising a half-cycle of a cosine. **Figure 7.29a** shows the frequency response of the ideal lowpass filter $H_0(\omega)$ with cutoff frequency ω_c. **Figure 7.29b** shows one half-cycle of a cosine $R(\omega)$, whose width in the frequency domain, $\Delta\omega$, is the only parameter,

$$R(\omega) = \begin{cases} \dfrac{\pi^2}{\Delta\omega} \cos \pi \dfrac{\omega}{\Delta\omega}, & -\Delta\omega/2 < \omega < \Delta\omega/2 \\ 0, & |\omega| > \Delta\omega/2 \end{cases}.$$

Figure 7.29c shows the convolution of $H_0(\omega)$ and $R(\omega)$ in the frequency domain. The result is the frequency response of the raised-cosine filter (see Problem 7-33),

$$H(\omega) = A(\omega)e^{-j(\omega(N-1)/2-\beta)}. \tag{7.17}$$

The response has a sinusoidal transition band of width $\Delta\omega$ centered at ω_c; it is equal to 1 for frequencies below the transition band and 0 above the band. The impulse response of the raised-cosine filter is

$$h[n] = h_0[n]r[n],$$

where $h_0[n] = (\omega_c/\pi) \operatorname{sinc} \omega_c n$ is the impulse response of the ideal lowpass filter and (see Problem 7-33)

$$r[n] = \frac{\pi}{4}\left(\operatorname{sinc}(\Delta\omega n/2 + \pi/2) + \operatorname{sinc}(\Delta\omega n/2 - \pi/2)\right).$$

So,

$$h[n] = h_0[n]r[n] = \frac{\omega_c}{4} \operatorname{sinc} \omega_c n(\operatorname{sinc}(\Delta\omega n/2 + \pi/2) + \operatorname{sinc}(\Delta\omega n/2 - \pi/2)).$$

Figure 7.30 gives an idea of how the spline filter and raised-cosine filter compare with some other filters. The figure shows (in red) the frequency response of second-order spline filters designed with $\omega_c = \pi/2$ and $\Delta\omega = \pi/5$, simply truncated to lengths $N = 31$ (**Figure 7.30a**) and $N = 35$ (**Figure 7.30b**). The response of a raised-cosine filter with a truncated impulse response is shown in green. For comparison, the responses of rectangular-window (black) and Hamming-window (blue) filters of the same cutoff frequency and lengths are shown.

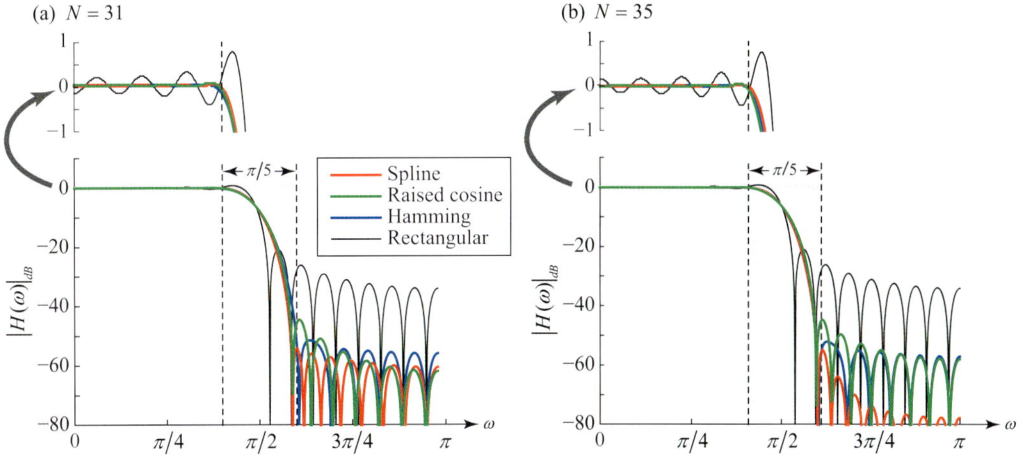

Figure 7.30 Spline and raised-cosine filters

With a careful selection of parameters, the performance of the spline and raised-cosine filters can be as good as (or better than) other popular window filters of the same order. For example, the stopband attenuation of the spline filters in the figure is roughly comparable to that of the Hamming-window filter, but the transition zones of the spline filters are narrower (i.e., the filters are "sharper"). The transition zones of the spline and Hamming-window filters are comparable, but the stopband attenuation of the spline filter might be judged to be superior. Compared with the rectangular-window filter, both spline and raised-cosine filters have better attenuation in the stopband as well as smaller oscillations in the passband (shown magnified on a ±1 dB scale in the top panels), though the transition zone of the rectangular-window filters is always sharper.

7.7 ★ Frequency-sampled FIR filter design

The window-based filters, spline filters and raised-cosine filters described in the previous sections are relatively simple to understand and to design. These filter design methods are all based on modifying the impulse response of the ideal filter with windows. While they have relatively few parameters that the designer can control, they are nonetheless well suited to many applications. However, there are other applications where one wishes to have greater control of the filter's frequency response. In this and the remaining sections of this chapter, we consider several frequency-domain methods of designing a linear-phase FIR filter that offer much finer control of the shape of the frequency response. These methods generally aim at finding a sequence $h[n]$ whose frequency response $H(\omega)$ matches a desired frequency response $H_d(\omega)$ at selected points (this section) or, alternately, so that the difference between $H(\omega)$ and $H_d(\omega)$ is minimized according to some criterion (Sections 7.8 and Section 7.9).

Filter design by **frequency sampling** is perhaps the simplest frequency-domain design technique to understand and implement, so we will discuss it first. The object is to create a filter of length N that matches a desired frequency response $H(\omega)$ at exactly N points, usually (though not always) equally spaced between $\omega = 0$ and $\omega = 2\pi$. The values of $H(\omega)$ at those N points form a sequence $H[k]$:

$$H[k] \triangleq H(\omega)\big|_{\omega = 2\pi k/N} = \sum_{n=0}^{N-1} h[n]e^{-j2\pi kn/N}, \quad 0 \le k \le N-1. \tag{7.18}$$

The impulse response of length N of the desired filter, $h[n]$, can then be obtained from $H[k]$ by one of several methods, which we shall now describe.

7.7.1 Inverse DFT

One simple and fast method of designing frequency-sampled filters with equally spaced points uses the discrete Fourier transform (DFT), which we will cover in detail in Chapter 10. If you are already familiar with the DFT, read on. If not, just skip to the next section.

Still here? Good. You should recognize that $H[k]$ in Equation (7.18) is just the DFT of $h[n]$, so $h[n]$ can be obtained by taking the inverse DFT,

$$h[n] = \frac{1}{N}\sum_{k=0}^{N-1} H[k]e^{j2\pi kn/N}, \quad 0 \le n \le N-1. \tag{7.19}$$

This approach is conceptually simple and also computationally fast since the inverse DFT can potentially be computed using the fast Fourier transform (FFT) algorithm (Chapter 11). With the frequency-sampling method, the frequency samples $H[k]$ can have any desired magnitude and phase. However, the design is particularly straightforward in the case of linear-phase filters. In this case, $H(\omega)$ can be expressed as the product of a real amplitude function $A(\omega)$ and the linear-phase term given in Equation (7.3) and **Table 7.1**,

$$H(\omega) = e^{-j(\omega(N-1)/2-\beta)}A(\omega).$$

The designer's job boils down to choosing the right filter type, specifying samples of $A(\omega)$, and then making sure the phase is correct. This is easiest to explain with an example.

Example 7.10

Design a symmetric (Type-I) filter of odd length $N = 17$ by uniformly sampling the frequency response of an ideal lowpass filter with a corner frequency of $\omega = \pi/2$.

➤ Solution:

Table 7.2 indicates that either a Type-I or a Type-II filter is suitable for a lowpass filter design. We will design a Type-I filter. As discussed in Section 7.1, we can express $H(\omega)$ as the product of a real amplitude function $A(\omega)$ and a linear-phase term $e^{-j\omega(N-1)/2}$:

$$H(\omega) = A(\omega)e^{-j\omega(N-1)/2} = \underbrace{\left(\sum_{m=0}^{(N-1)/2} a[m]\cos m\omega\right)}_{A(\omega)} e^{-j\omega(N-1)/2}. \quad \textbf{(7.20)}$$

In order to use the inverse DFT, we require $N = 17$ samples of the frequency response $H(\omega)$, equally spaced between $\omega = 0$ and $\omega = 2\pi$,

$$H[k] \triangleq H(\omega)|_{\omega=2\pi k/N} = e^{-j\omega(N-1)/2}A(\omega)|_{\omega=2\pi k/N} = e^{-j\pi k(N-1)/N}A[k], \quad 0 \le k \le N-1, \quad \textbf{(7.21)}$$

where $A[k]$ are the samples of the amplitude function. However, we can exploit the symmetry of $H(\omega)$ to simplify the problem. Because the impulse response $h[n]$ is real, the associated frequency response $H(\omega)$ is conjugate-symmetric: $H(\omega) = H^*(-\omega)$. $H(\omega)$ is also periodic with period 2π: $H(\omega) = H(\omega + 2\pi)$. Putting these two facts together yields $H(2\pi - \omega) = H^*(\omega)$. Sampling at $\omega = 2\pi k/N$ gives

$$H(2\pi - 2\pi k/N) = H(2\pi(N-k)/N) = H^*(2\pi k/N),$$

from which we conclude that

$$H[N-k] = H^*[k]. \quad \textbf{(7.22)}$$

Given an odd number of samples, N, we only need to compute $(N+1)/2$ values of $H[k]$, $0 \le k \le (N-1)/2$, using Equation (7.21). Then, the remaining values of $H[k]$ for $(N+1)/2 \le k < N$ can be found from Equation (7.22). The following piece of Matlab code shows the procedure:

```
N = 17;
wc = 0.5*pi;
k = 0:(N-1)/2;
w = 2*pi*k/N;
Ak = double(w<wc);
Hk = Ak .* exp(-j*pi*k*(N-1)/N);Hk = [Hk conj(Hk(end:-1:2))];
h = real(ifft(Hk));
```

First, we calculate ω and $A[k]$ for values $0 \le k \le (N-1)/2$. In the calculation of `Ak`, the expression `w<wc` produces a logical array with value 1 when $\omega < \omega_c$ and 0 otherwise. This array needs to be cast to double to be used in the calculation of $H[k]$. In the next line, $A[k]$ is multiplied by the appropriate phase, as specified in Equation (7.21), to give $H[k]$ for these values of k. The next line uses Equation (7.22) and Matlab's `conj` function to create $H[k]$ for all k. Then, Matlab's `ifft` command is used to compute the inverse DFT, Equation (7.19), which gives $h[n]$. Because $h[n]$ is real, it is not theoretically necessary to take the real part in the last line. However, doing so removes vestigial artifactual imaginary parts produced by the `ifft` computation. (Matlab's `ifft` command works regardless of the value of N. It need not be a power of two.)

Figure 7.31a shows $|H(\omega)|$, the magnitude of the frequency response of the resulting filter on a linear scale (red trace). **Figure 7.31b** shows the same frequency response on a log (dB) scale. In both panels, the samples of the amplitude function $A[k]$ are shown in red circles. There are $(N+1)/2 = 9$ points in the frequency range between $0 \le \omega \le \pi$ (red dots), corresponding to $0 \le k \le (N-1)/2$. Because this is an odd-length filter, there is no sample point at $\omega = \pi$. The frequency response passes through the design points, as it must, but you can see that the magnitude of the response in between those points is highly oscillatory. For example, the attenuation of the first side lobe in the stopband is less than −20 dB, which is not great. This is another example of the Gibbs phenomenon that results from the sharp transition of sample values around the corner frequency. In fact, the response of this frequency-sampled filter could be considered to be *worse* than that of a rectangular-window filter of the same length designed with the same corner frequency (blue line). Given our experience with the rectangular-window filter, you might expect that just changing the number of points, N, in the frequency-sampled filter will not fix the problem. One could fix this problem by multiplying $h[n]$ by a Hamming window or something similar. That would improve the out-of-band attenuation, but at the cost of a significantly broader transition zone. We will propose another solution in a moment, based on creating a less sharp transition region.

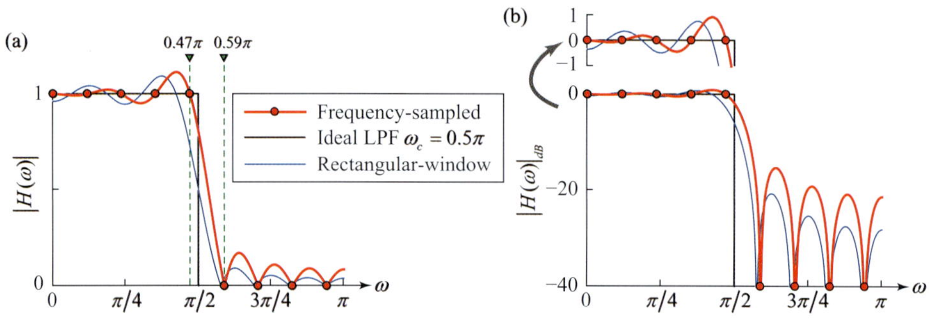

Figure 7.31 Frequency-sampled lowpass filter with $\omega_c = 0.5\pi$, and $N = 17$

This example brings up another problem with using frequency sampling to design a filter with a sharp transition. There is only a finite number of sampled points, in this case nine, spanning the range $0 \le \omega \le \pi$. Here, one point is to the left of the transition at $\omega = 0.47\pi$ and one is to the right at $\omega = 0.59\pi$, as indicated by the green dotted lines in **Figure 7.31a**. So, given a fixed value of the length of the filter, N, we would have designed exactly the same frequency-sampled filter to match any ideal lowpass filter with a corner frequency in the range $0.47\pi < \omega_c < 0.59\pi$. The solution is either to increase N and/or to introduce a transition region of finite width.

As a second example of the frequency-sampling method, we will design a bandpass filter. **Table 7.2** suggests that a Type-III filter is particularly suitable for a bandpass design, since $A(\omega)$ is already guaranteed to be zero at both $\omega = 0$ and $\omega = \pi$ for this filter type.

Example 7.11

Design a frequency-sampled filter of length $N = 17$ by sampling the frequency response of an ideal bandpass filter with a center frequency of $\omega_c = \pi/2$ and a width of $\Delta\omega = \pi/2$.

➤ Solution:

A quick check of **Table 7.2** suggests that we can use either a Type-I or Type-III filter. Let us choose the Type-III because it inherently guarantees that $A(0) = A(\pi) = 0$, which is what we want for a bandpass filter. Then, Equation (7.3) and **Table 7.1** give

$$H(\omega) = A(\omega)e^{-j(\omega(N-1)/2 - \pi/2)} = jA(\omega)e^{-j\omega(N-1)/2},$$

where

$$A(\omega) = \sum_{m=1}^{(N-1)/2} c[m] \sin m\omega.$$

We again sample $H(\omega)$ at $\omega = 2\pi k/N$ to form $H[k]$,

$$H[k] \triangleq jA[k]e^{-j\pi k(N-1)/N}.$$

As in the previous example, we only need to compute $H[k]$ at $(N+1)/2$ values of $A[k]$. (In point of fact, we only need $(N+1)/2 - 1 = (N-1)/2$ points because $A[0] = 0$.) We will again exploit the symmetry of $H(\omega)$ expressed by Equation (7.22) to provide $H[k]$ for all k, $0 \leq k \leq N-1$. The Matlab code is similar to that of Example 7.10, except that we define a variable `dw` for $\Delta\omega$, and change the way `Ak` and `Hk` are computed:

```
    ⋮
dw = 0.5*pi;
Ak = double(abs(w-wc)<=0.5*dw);
Hk = j * Ak .* exp(-j*pi*k*(N-1)/N);
    ⋮
```

Figure 7.32 shows the corresponding frequency response. The response matches the ideal response at the sampled frequencies, but the sharp transitions result in large variations in the response at other frequencies, similar to those we saw in the previous example.

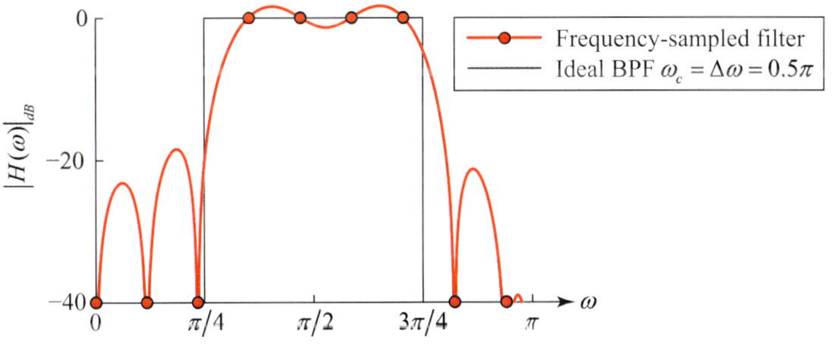

Figure 7.32 Frequency-sampled bandpass filter with $\omega_C = \Delta\omega = 0.5\pi$, and $N = 17$

Both of the previous examples resulted in filters that actually met the design goals (i.e., matched the frequency response of the ideal filter at the sampled frequencies), but both have poor out-of-band rejection due to the sharp transitions between sampled points.

One approach to mitigating the effect of the sharp transitions is to make the transition zone less abrupt, a trick we learned in designing spline filters. **Figure 7.33** shows (in red), on both linear and log scales, a frequency-sampled filter with $N = 17$ designed to match a lowpass filter with cutoff frequency $\omega_c = 0.5\pi$ with a linear (first-order spline) transition of width $\Delta\omega = 0.25\pi$ (shown in black).

Even though there are only a couple of points in the transition zone, the filter's out-of-band attenuation is at least 20 dB better than the filter with no transition zone shown in **Figure 7.31**.

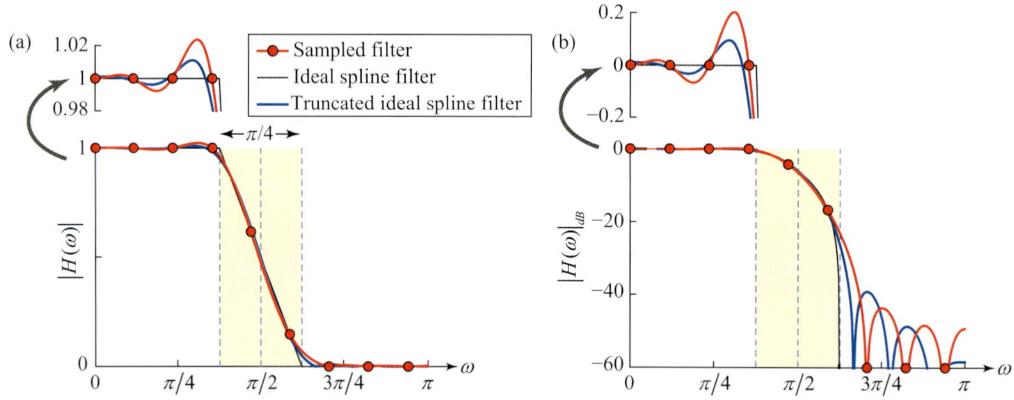

Figure 7.33 Frequency-sampled lowpass filter with spline transition

This figure also shows (blue line) the frequency response of a filter designed by truncating the response of the ideal first-order spline filter given in Equation (7.15) to $N = 17$ points. From our discussion in Section 7.2, the truncated ideal response provides the sharpest transition for a filter of this type and length. But, comparatively, the transition of the frequency-sampled filter is not too bad!

In the limit as the number of frequency sampling points, N, goes to infinity, the sampled filter's response approaches the desired response of the ideal filter in the sense that the mean-square error between the two responses goes to zero. However, the error does not necessarily go uniformly to zero as N increases. So, in practice it is important to try a range of N when designing a frequency-sampled filter.

7.7.2 Frequency sampling as interpolation

The frequency-sampling method can usefully be understood as an interpolation method in much the same way we understood the reconstruction of an analog signal from discrete samples in D/A conversion. The frequency response of the frequency-sampled filter $H(\omega)$ can be completely "reconstructed" from its samples $H[k]$ through the use of a periodic sinc interpolating function. To see this, substitute the equation for the inverse DFT, Equation (7.19), into the DTFT and turn the algebra crank a few times:

$$\begin{aligned} H(\omega) &= \sum_{n=0}^{N-1} h[n]e^{-j\omega n} = \sum_{n=0}^{N-1}\left(\frac{1}{N}\sum_{k=0}^{N-1} H[k]e^{j2\pi kn/N}\right)e^{-j\omega n} = \frac{1}{N}\sum_{k=0}^{N-1} H[k]\sum_{n=0}^{N-1} e^{-j(\omega-2\pi k/N)n} \\ &= \frac{1}{N}\sum_{k=0}^{N-1} H[k]\frac{1-e^{-j(\omega-2\pi k/N)N}}{1-e^{-j(\omega-2\pi k/N)}} = \frac{1}{N}\sum_{k=0}^{N-1} H[k]e^{-j(\omega-2\pi k/N)((N-1)/2)}\,\frac{\sin\left(\omega-\frac{2\pi k}{2}\right)\frac{N}{2}}{\sin\left(\omega-\frac{2\pi k}{2}\right)\frac{1}{2}}. \end{aligned}$$

For a Type-I filter, substitute Equation (7.21) for $H[k]$ to get an expression that relates $H(\omega)$ to samples of the amplitude response $A[k]$:

$$H(\omega) = \frac{1}{N}\sum_{k=0}^{N-1}\left(A[k]e^{-j\pi k(N-1)/N}\right)e^{-j(\omega-2\pi k/N)((N-1)/2)}\frac{\sin\left(\omega-\frac{2\pi k}{2}\right)\frac{N}{2}}{\sin\left(\omega-\frac{2\pi k}{2}\right)\frac{1}{2}}$$
$$= e^{-j\omega(N-1)/2}\frac{1}{N}\sum_{k=0}^{N-1}A[k]\underbrace{\left(\frac{\sin\left(\omega-\frac{2\pi k}{2}\right)\frac{N}{2}}{\sin\left(\omega-\frac{2\pi k}{2}\right)\frac{1}{2}}\right)}_{W(\omega-2\pi k/N)} = e^{-j\omega(N-1)/2}\frac{1}{N}\sum_{k=0}^{N-1}A[k]W\left(\omega-\frac{2\pi k}{2}\right), \quad \textbf{(7.23a)}$$

where the interpolation function,

$$W(\omega) \triangleq \frac{\sin \omega N/2}{\sin \omega/2} = N\frac{\operatorname{sinc} \omega N/2}{\operatorname{sinc} \omega/2}, \quad \textbf{(7.23b)}$$

is the periodic sinc function of Equation (7.9).

Equation (7.23a) has the form of Equation (7.20),

$$H(\omega) = A(\omega)e^{-j\omega(N-1)/2},$$

where $A(\omega)$ is a real amplitude function formed from the average of N interpolation functions $W(\omega)$, each one shifted in frequency by $2\pi k/N$ and scaled by $A[k]$,

$$A(\omega) = \frac{1}{N}\sum_{k=0}^{N-1}A[k]W(\omega - 2\pi k/N).$$

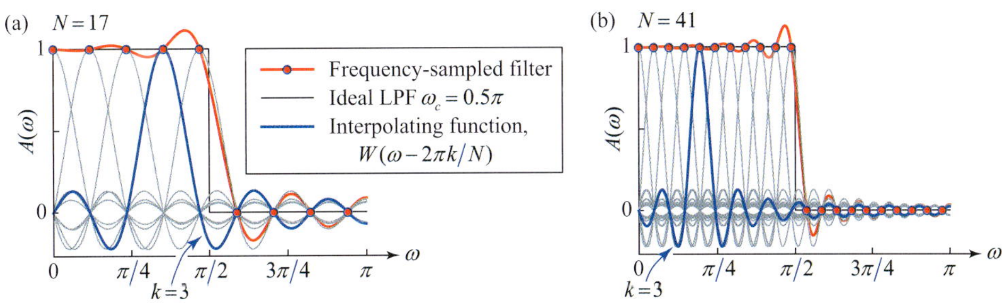

Figure 7.34 Interpolation functions for frequency-sampled filters

Figure 7.34a shows how the amplitude response of the lowpass filter example in **Figure 7.31**, with $N = 17$, can be interpreted in terms of this equation. Each of the blue curves is the shifted (scaled) interpolation function $A[k]W(\omega - 2\pi k/N)/N$ for one value of k. The curve for $k = 3$ is highlighted with a thicker line so you can see what an individual interpolation function looks like. The values of $A[k]$ are shown with the red dots. In this example, $A[k]$ only has eight non-zero values of k:

$$A[k] = \begin{cases} 1, & 0 \le k \le 4, 13 \le k \le 16 \\ 0, & 5 \le k \le 12 \end{cases}.$$

Five of these non-zero values are at frequencies $\omega < \pi/2$, and correspond to curves that have a peak amplitude of one in the figure, while the other three non-zero values are at frequencies $\omega > 3\pi/2$. The sum of all the scaled interpolating functions is the frequency response of the frequency-sampled filter, shown with the red curve.

This figure helps explain a few things. First, as we expect, $A(\omega)$ exactly matches the sample points $A[k]$ at each sample frequency $\omega = 2\pi k/N$. That is because only the peak of the kth interpolating function contributes to the response at that frequency. The functions at other values of k all pass through 0 at this frequency. Second, you can see why the response oscillates between sample frequencies; it is because the interpolating functions do not cancel at frequencies between sample points. These oscillations are, again, a manifestation of the Gibbs phenomenon and represent a key defect of the frequency-sampling method. Increasing the number of sample points, as shown in **Figure 7.34b**, reduces the oscillations between points but does not reduce the peak amplitude of the oscillations at the discontinuity. To do that, we will have to adopt a different filter design strategy, which we will do in subsequent sections.

7.7.3 Design formulas

An alternate approach to designing frequency-sampled filters is to use the design formulas shown in **Table 7.4**. These formulas are directly derived from the definition of the inverse DFT (Equation (7.19)), the relation of $H(\omega)$ and $A(\omega)$ (Equation (7.3)) and the symmetries of $A(\omega)$ about $\omega = 0$ and $\omega = \pi$. The results obtained by these formulas are identical to the inverse DFT approach, but they allow us to express $h[n]$ directly in terms of the sum of real sines or cosines. See Problems 7-17–7-20 to help you work through the derivations for yourself, if you are interested. Like the other methods we have discussed, these formulas only require samples of $A(\omega)$ for frequencies in the interval $0 \le \omega \le \pi$.

Table 7.4 Design formulas for $h[n]$

Type	Symmetry	Length	Points	$h[n]$
I	Even	Odd	$(N+1)/2$	$\frac{1}{N}\left(A[0] + 2\sum_{k=1}^{(N-1)/2} A[k]\cos\left(\frac{\pi k(N-1-2n)}{N}\right)\right)$
II	Even	Even	$N/2$	$\frac{1}{N}\left(A[0] + 2\sum_{k=1}^{N/2-1} A[k]\cos\left(\frac{\pi k(N-1-2n)}{N}\right)\right)$
III	Odd	Odd	$(N-1)/2$	$\frac{2}{N}\sum_{k=1}^{(N-1)/2} A[k]\sin\left(\frac{\pi k(N-1-2n)}{N}\right)$
IV	Odd	Even	$N/2$	$\frac{1}{N}\left(A[N/2]\sin \pi(N-1-2n)/2 + 2\sum_{k=1}^{N/2-1} A[k]\sin\left(\frac{\pi k(N-1-2n)}{N}\right)\right)$

7.7.4 Simultaneous equations

Frequency-sampled filters designed using the DFT method or the design formulas require the frequency samples to be uniformly distributed. We will now discuss an approach to designing these filters using simultaneous equations that does not have this restriction. Understanding this approach will also help us understand the design of filters using the least-square-error technique presented in Section 7.8.

Given N samples of the amplitude response $A[k]$, the formulas given in **Table 7.1** allow us to generate a series of N independent linear equations that can be solved for the impulse response $h[n]$ of length N. As an example, for a Type-I filter, the amplitude response is given by

$$A(\omega) = \sum_{m=0}^{M} a[m]\cos m\omega,$$

where $M = (N-1)/2$. To create $A[k]$, sample $A(\omega)$ at $M+1$ frequencies ω_k that span the interval $0 \le \omega < \pi$, not necessarily uniformly,

$$A[k] = \sum_{m=0}^{M} a[m]\cos m\omega_k, \quad 0 \le k \le M. \tag{7.24}$$

This results in a set of $M+1$ linearly independent equations in $M+1$ unknowns that can be solved for $a[n]$. Formulating this as a matrix problem, we have $\mathbf{A} = \mathbf{Ca}$,

$$\underbrace{\begin{bmatrix} A[0] \\ A[1] \\ \vdots \\ A[M] \end{bmatrix}}_{\mathbf{A}} = \underbrace{\begin{bmatrix} 1 & \cos\omega_0 & \cos 2\omega_0 & \cdots & \cos M\omega_0 \\ 1 & \cos\omega_1 & \cos 2\omega_1 & \cdots & \cos M\omega_1 \\ \vdots & \vdots & \vdots & \ddots & \vdots \\ 1 & \cos\omega_M & \cos 2\omega_M & \cdots & \cos M\omega_M \end{bmatrix}}_{\mathbf{C}} \underbrace{\begin{bmatrix} a[0] \\ a[1] \\ \vdots \\ a[M] \end{bmatrix}}_{\mathbf{a}}. \tag{7.25}$$

C is a square $(M+1)\times(M+1)$ matrix, whose inverse can be found by a number of standard methods (assuming it is well conditioned), leading to a unique solution for the parameter vector $\mathbf{a} = \mathbf{C}^{-1}\mathbf{A}$. Finally, all N values of $h[n]$ can be computed from $a[n]$ by "reversing" Equation (7.2),

$$h[n] = \begin{cases} a[M-n]/2, & 0 \le n \le M-1 \\ a[0], & n = M \\ a[n-M]/2, & M+1 \le n \le 2M \end{cases}. \tag{7.26}$$

In the specific case where the N samples of $A(\omega)$ are uniformly distributed in frequency, we have $\omega_k = k\omega_0$, where $\omega_0 \triangleq 2\pi/N$, so Equations (7.24) and (7.25) become

$$A[k] = \sum_{m=0}^{M} a[m]\cos mk\omega_0, \quad 0 \le k \le M,$$

and

$$\underbrace{\begin{bmatrix} A[0] \\ A[1] \\ \vdots \\ A[M-1] \\ A[M] \end{bmatrix}}_{\mathbf{A}} = \underbrace{\begin{bmatrix} 1 & 1 & 1 & \cdots & 1 \\ 1 & \cos\omega_0 & \cos 2\omega_0 & \cdots & \cos M\omega_0 \\ \vdots & \vdots & \vdots & \ddots & \vdots \\ 1 & \cos(M-1)\omega_0 & \cos 2(M-1)\omega_0 & \cdots & \cos M(M-1)\omega_0 \\ 1 & \cos M\omega_0 & \cos 2M\omega_0 & & \cos M^2\omega_0 \end{bmatrix}}_{\mathbf{C}} \underbrace{\begin{bmatrix} a[0] \\ a[1] \\ \vdots \\ a[M-1] \\ a[M] \end{bmatrix}}_{\mathbf{a}}.$$

Solving for **a**, and hence for $h[n]$, is a job for Matlab:

Example 7.12

Design the frequency-sampled filter of Example 7.10 using simultaneous equations.

➤ Solution:

The first five lines are essentially the same code as that of Example 7.10, except that w is a column vector:

```
N = 17;
wc = 0.5*pi;
k = 0:(N-1)/2;
w = 2*pi*k'/N;
Ak = double(w<wc);
C = cos(w*k);
a = C \ Ak;
h = [0.5*a(end:-1:2); a(1); 0.5*a(2:end)];
```

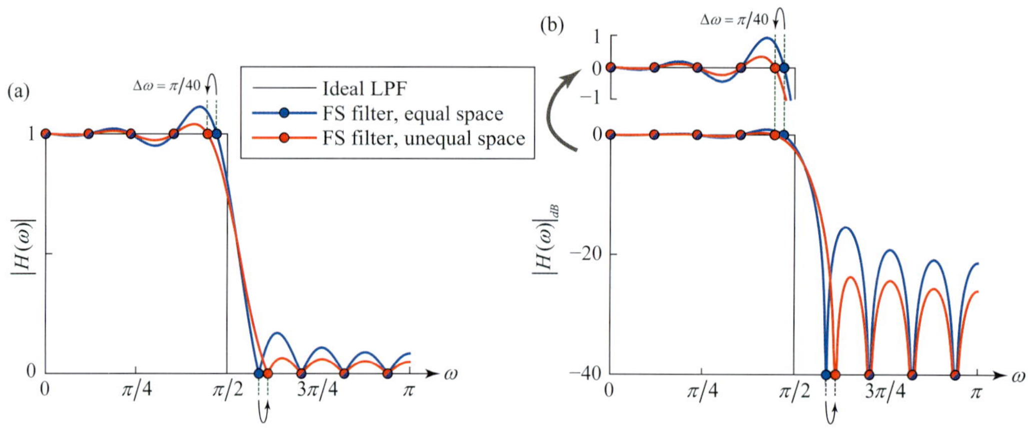

Figure 7.35 Frequency-sampled filter with an unequal frequency distribution

One advantage of using the method of simultaneous equations over the inverse DFT method or the design formulas is that the frequency samples of $A(\omega)$ can be non-uniformly distributed over the interval $0 \le \omega < \pi$. We can sample $A(\omega)$ at any $M+1$ points ω_k, $0 \le k \le M$, such that $0 \le \omega < \pi$. For example, **Figure 7.35** shows (in blue) the frequency response of the same frequency-sampled filter of Example 7.10, designed with uniform frequency samples. Shown in red is the response of a frequency-sampled filter in which only the two frequency samples nearest the discontinuity at $\omega_c = \pi/2$ have been moved away from their initial positions by just a tiny amount ($\Delta\omega = \pi/40$). As a result, the peak amplitude of the ripples in the passband and stopband is substantially attenuated.

Example 7.13

Design the bandpass filter of Example 7.11 using simultaneous equations.

➤ **Solution:**

This is a Type-III filter, so we implement the appropriate formula from **Table 7.1**. The code proceeds along the same lines as Example 7.12, except that the calculation of $A[k]$ is different, the **C** matrix uses a sine instead of a cosine and the "reverse" calculation of $h[n]$ uses the formula

$$h[n] = \begin{cases} c[M-n]/2, & 0 \le n \le M-1 \\ 0, & n = M \\ -c[n-M]/2, & M+1 \le n \le 2M \end{cases}.$$

Here is the code:

```
N = 17;
M = (N-1)/2;
wc = 0.5*pi;
dw = 0.5*pi;
k = 1:M;w = 2*pi*k'/N;
Ak = double(abs(w-wc)<=0.5*dw);
C = sin(w*k);
c = C \ Ak;
h = 0.5*[c(end:-1:1); 0; -c(1:end)];
```

7.7.5 Matlab implementation of frequency-sampled filters

Matlab's `fir2` function can be used to design frequency-sampled filters, although it actually does things in a somewhat different way than we have described here. In its basic form, the syntax is

```
h = fir2(N-1, w, A, npts, window),
```

where the length of the impulse response is `N`, and `w` is an array of critical break frequencies, normalized to π. The first frequency point must be 0 and the last point must be 1. `A` is an array of amplitudes that correspond to the `w` array. The `fir2` algorithm first internally interpolates the frequency and amplitude array over a large number of points `npts` (default 512), and creates a frequency-sampled filter with a long impulse response. This filter is then truncated to the desired number of points `N`. Finally, the algorithm multiplies the response by `window` (default `hamming(N)`) to form the output response `h`. Because `fir2` interpolates the critical frequencies and amplitudes over a large grid, the resulting filter approximates, in some sense, the impulse response of the ideal filter. As we discussed in Section 7.2, truncating the response to `N` points produces the optimum `N`-point least-square-error approximation of a filter of length `npts`.

7.8 ★ Least-square-error FIR filter design

In Section 7.7, we showed how the frequency-sampling method could be used to design a linear-phase filter whose frequency response $H(\omega)$ exactly matched that of a desired "model" filter

$H_d(\omega)$ at N points spaced uniformly in frequency over the interval $0 \le \omega < 2\pi$. We now turn to the more general topic of designing filters that minimize some measure of the square error between $H(\omega)$ and $H_d(\omega)$, either at a discrete number of frequencies (Section 7.8.1) or over a continuous range of frequency (Section 7.8.2).

7.8.1 Discrete least-square-error FIR filters

Consider first the problem of designing a filter of length N, whose frequency response $H(\omega)$ most closely matches the frequency response of a desired filter $H_d(\omega)$ at a specified number of points P that is *larger* than N, namely $P > N$. In this case, there is no unique, exact solution, so we will find the $H(\omega)$ that best matches $H_d(\omega)$, where "best" means that we will minimize the square error between $H_d(\omega)$ and $H(\omega)$ at those P discrete frequency points, ω_k, $0 \le k < P$.

We proceed in a manner similar to that of Section 7.7 by defining the error, $E(\omega_k)$, at frequency ω_k as

$$E(\omega_k) = H_d(\omega_k) - H(\omega_k).$$

If the frequency points are uniformly spaced in the interval $0 \le \omega < 2\pi$, such that $\omega_k = 2\pi k/P$, then we can write

$$E[k] = H_d[k] - H[k],$$

where $E[k] \triangleq E(\omega_k)$, $H_d[k] \triangleq H_d(\omega_k)$ and $H[k] \triangleq H(\omega_k)$ are the DFTs of P-point sequences $e[n]$, $h_d[n]$ and $h[n]$, respectively. Hence,

$$e[n] = h_d[n] - h[n].$$

The mean-square error of $E[k]$ is

$$\frac{1}{P}\sum_{k=0}^{P-1} |E[k]|^2 = \frac{1}{P}\sum_{k=0}^{P-1} |H_d(\omega_k) - H(\omega_k)|^2. \tag{7.27}$$

As in Section 7.2, Parseval's theorem allows us to relate the error in the frequency and time domains:

$$\frac{1}{P}\sum_{k=0}^{P-1} |E[k]|^2 = \sum_{n=0}^{P-1} |e[n]|^2 = \sum_{n=0}^{P-1} |h_d[n] - h[n]|^2. \tag{7.28}$$

In the case where both $H_d(\omega)$ and $H(\omega)$ are the frequency responses of symmetric (Type-1) lowpass FIR filters of odd length P we can write each one as the product of an amplitude response and a linear-phase term:

$$H_d(\omega) = D(\omega)e^{-j\omega(P-1)/2}$$
$$H(\omega) = A(\omega)e^{-j\omega(P-1)/2}.$$

Hence, Equation (7.27) becomes

$$\frac{1}{P}\sum_{k=0}^{P-1}|E[k]|^2 = \frac{1}{P}\sum_{k=0}^{P-1}|D(\omega_k) - A(\omega_k)|^2. \tag{7.29}$$

In the specific case where ω_k are uniformly distributed in frequency (i.e., $\omega_k = k\omega_0$, $0 \le k \le N-1$, where $\omega_0 = 2\pi/N$), then $D(\omega)$ and $A(\omega)$ will be real and even functions of ω, which correspond to real, even impulse responses $d[n]$ and $a[n]$, such that

$$\begin{aligned} h_d[n] &= d[n-(P-1)/2] \\ h[n] &= a[n-(P-1)/2]. \end{aligned}$$

Equation (7.28) then becomes

$$\begin{aligned} \frac{1}{P}\sum_{k=0}^{P-1}|E[k]|^2 &= \sum_{n=0}^{P-1}|h_d[n]-h[n]|^2 = \sum_{n=0}^{P-1}|d[n-(P-1)/2]-a[n-(P-1)/2]|^2 \\ &= \sum_{n=-(P-1)/2}^{(P-1)/2}|d[n]-a[n]|^2 = \sum_{n=-(P-1)/2}^{(P-1)/2}(d[n]-a[n])^2. \end{aligned} \tag{7.30}$$

The response $a[n]$ is symmetric about $n=0$, and is 0 outside the range $-(N-1)/2 \le n \le (N-1)/2$. Accordingly, we can rewrite Equation (7.30) as

$$\frac{1}{P}\sum_{k=0}^{P-1}|E[k]|^2 = \sum_{n=-(P-1)/2}^{(P-1)/2}(d[n]-a[n])^2 = \sum_{n\in\mathbb{N}}(d[n]-a[n])^2 + \sum_{n\notin\mathbb{N}}(d[n])^2,$$

where the range $n \in \mathbb{N}$ is $|n| \le (N-1)/2$ and the range $n \notin \mathbb{N}$ is $(N-1)/2 < |n| \le (P-1)/2$. As we argued in Section 7.2, the error is minimized when the first summation term is zero for $n \in \mathbb{N}$, which occurs when $a[n] = d[n]$. The residual error is then given by the second summation.

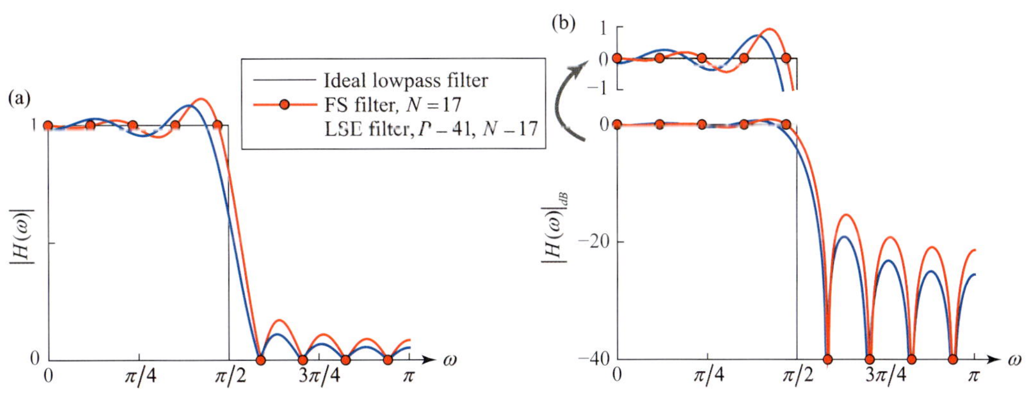

Figure 7.36 Discrete least-square-error filter

In summary, to design a filter of length N whose frequency response minimizes the square-error with respect to the frequency response of a desired filter at a larger number of points, $P > N$, first design the larger P-point frequency-sampled filter by any of the techniques in Section 7.7 and then truncate the response symmetrically to N samples. (This is essentially the same as what Matlab's `fir2` function, referenced above, does.)

Figure 7.36 shows an example of least-square-error design of a Type-I lowpass filter. The frequency-sampled filter of length $N = 17$ (shown in red) passes exactly through the sample points (red dots). This filter is equivalent to a least-square-error filter with $P = N = 17$. The least-square-error filter (shown in blue) is designed by first creating a frequency-sampled filter of length $P = 41$, and then truncating the impulse response to $N = 17$ points. The resulting filter has somewhat less pronounced oscillations in both the passband and the stopband than the frequency-sampled filter of length $N = 17$ (shown in red).

Matrix formulation of the discrete least-square-error filter The discrete least-square-error filter problem can also be formulated as a system of simultaneous equations. Using our treatment in Section 7.7 as a model, we seek to create a filter of length N that minimizes the square-error between a desired response $D(\omega)$ and an actual response $A(\omega)$ at $P > N$ frequencies ω_k, which are not necessarily uniformly distributed in the interval $0 \le \omega < 2\pi$. Taking again the case of the Type-I filter, we have

$$A(\omega_k) = \sum_{m=0}^{M} a[m] \cos \omega_k, \quad 0 \le k \le R,$$

where $M = (N-1)/2$ and $R = (P-1)/2$. This results in a set of $R+1$ equations in $M+1$ unknowns. In matrix form, we have $\mathbf{A} = \mathbf{Ca}$:

$$\underbrace{\begin{bmatrix} A(\omega_0) \\ A(\omega_1) \\ \vdots \\ A(\omega_R) \end{bmatrix}}_{\mathbf{A}} = \underbrace{\begin{bmatrix} 1 & \cos \omega_0 & \cos 2\omega_0 & \cdots & \cos M\omega_0 \\ 1 & \cos \omega_1 & \cos 2\omega_1 & \cdots & \cos M\omega_1 \\ \vdots & \vdots & \vdots & \ddots & \vdots \\ 1 & \cos \omega_R & \cos 2\omega_R & \cdots & \cos M\omega_R \end{bmatrix}}_{\mathbf{C}} \underbrace{\begin{bmatrix} a[0] \\ a[1] \\ \vdots \\ a[M] \end{bmatrix}}_{\mathbf{a}}. \quad (7.31)$$

If $R = M$, Equation (7.31) is exactly equal to Equation (7.25). In this case, the system is said to be **fully determined** (or **fully constrained**), and there is only one unique solution, $\mathbf{a} = \mathbf{C}^{-1}\mathbf{A}$. However, when $R > M$, this is an **overdetermined** (or **overconstrained**) system – the number of equations (rows) is greater than the number of unknowns (columns) – and there exists no exact solution for **a**. This is a classic problem that crops up in many disciplines. For example, it is the same issue one faces in trying to fit a polynomial with a small number of coefficients (e.g., two coefficients that define a straight line) to a large amount of data. A standard approach to this problem is to employ a least-square-error optimization method (see Appendix A). We define the square-error as in Equation (7.29), in terms of the difference between the desired and actual amplitude responses,

$$\sum_{k=0}^{R} |E[k]|^2 = \sum_{k=0}^{R} |D(\omega_k) - A(\omega_k)|^2.$$

Expressing this in matrix form, we define vector **E** as the difference between the desired amplitude response vector **D** and the actual frequency response vector **A** that results from a given choice of filter coefficients **a**,

$$\mathbf{E} = \mathbf{D} - \mathbf{A} = \mathbf{D} - \mathbf{Ca},$$

where,

$$\underbrace{\begin{bmatrix} E(\omega_0) \\ E(\omega_1) \\ \vdots \\ E(\omega_R) \end{bmatrix}}_{\mathbf{E}} = \underbrace{\begin{bmatrix} D(\omega_0) \\ D(\omega_1) \\ \vdots \\ D(\omega_R) \end{bmatrix}}_{\mathbf{D}} - \underbrace{\begin{bmatrix} A(\omega_0) \\ A(\omega_1) \\ \vdots \\ A(\omega_R) \end{bmatrix}}_{\mathbf{A}}.$$

The square-error is then

$$\sum_{k=0}^{R} |E(\omega_k)|^2 = \mathbf{E}^T\mathbf{E} = (\mathbf{D} - \mathbf{Ca})^T(\mathbf{D} - \mathbf{Ca}). \quad \textbf{(7.32)}$$

The task becomes to find the value of $\mathbf{a}$ that minimizes the square-error $\mathbf{E}^T\mathbf{E}$. We discuss the solution to this problem in detail in Appendix A, and show that the problem reduces to solving a system of so-called normal equations,

$$(\mathbf{C}^T\mathbf{C})\mathbf{a} = \mathbf{C}^T\mathbf{D}, \quad \textbf{(7.33)}$$

whose solution is

$$\mathbf{a} = (\mathbf{C}^T\mathbf{C})^{-1}\mathbf{C}^T\mathbf{D}. \quad \textbf{(7.34)}$$

There are a number of ways to use Matlab to solve the normal equations. For example, you could set up the $\mathbf{C}$ matrix and then blindly implement Equation (7.34):

```
a = inv(C'*C)*(C'*D).
```

However, it is preferable to avoid calculating the inverse $(\mathbf{C}^T\mathbf{C})^{-1}$ explicitly. Instead, you could use Matlab's matrix division operator (the backslash operator, "\ "):

```
a = (C'*C)\(C'*D).
```

Both these options have potential numerical issues. The better solution is to exploit the fact that Matlab's matrix division operator does double duty and can be used to solve both fully determined and overdetermined systems. For a fully determined system such as that of Equation (7.25), defined by $\mathbf{D} = \mathbf{Ca}$, where $\mathbf{C}$ is a non-singular square coefficient matrix, the Matlab command `a=C\D` gives the exact solution for $\mathbf{a}$. For an overdetermined system such as Equation (7.31), where $\mathbf{C}$ is a $P \times M$ matrix with $P > M$, `a=C\D` solves the normal equations to give the least-square-error solution for $\mathbf{a}$. This is the preferred solution method. Hence, for any fully determined or overdetermined system you can simply write

```
a = C\D.
```

Example 7.14

Design a least-square-error filter of length $N = 17$ with a corner frequency of $\omega = \pi/2$ that minimizes the square-error over $R = 41$ points uniformly in frequency.

➤ Solution:

The code to design this filter looks a lot like that of Example 7.12, except that the **C** matrix is not square:

```
N = 17;
M = (N-1)/2;
P = 41;
R = (P-1)/2;
wc = 0.5*pi;
k = 0:M;
l = 0:R;
w = 2*pi*l'/P;
D = double(w<wc);
C = cos(w*k);
a = C \ D;
h = [0.5*a(end:-1:2); a(1); 0.5*a(2:end)];
```

The resulting frequency response is shown in blue in **Figure 7.36**.

The only substantial difference between the frequency-sampled filter of Example 7.12 and the least-square-error filter of Example 7.14 is that, in the latter case, the **C** matrix has more rows than columns. If you were to set `P=17` in the code of Example 7.14, the result would be identical to the frequency-sampled filter. Accordingly, you can view the frequency-sampling filter as a special case of the discrete least-square-error filter in which the error is zero.

Discrete weighted least-square-error filter Least-square-error filter design can be made more useful by including a frequency dependent weighting factor $W(\omega_k)$ on the error. The weighting factor can be used to assign different weights to different frequency regions, for example the passband and stopband of a two-band filter. The weighted error is obtained by modifying Equation (7.32) to include the weighting factors:

$$\sum_{k=0}^{P} W(\omega_k)|D(\omega_k) - A(\omega_k)|^2 = \sum_{k=0}^{P} W(\omega_k)|E(\omega_k)|^2 = \mathbf{E}^T\mathbf{W}\mathbf{E}, \tag{7.35}$$

where **W** is a square weighting matrix whose diagonal is formed from the weighting factors $W(\omega_k)$,

$$\mathbf{W} = \begin{bmatrix} W(\omega_0) & 0 & \cdots & 0 \\ 0 & W(\omega_1) & \cdots & 0 \\ \vdots & \vdots & \ddots & \vdots \\ 0 & 0 & \cdots & W(\omega_p) \end{bmatrix}.$$

By a derivation similar to that which resulted in Equation (7.33), the least-square-error solution to this problem produces a set of modified normal equations,

$$(\mathbf{C}^T\mathbf{W}\mathbf{C})\mathbf{a} = \mathbf{C}^T\mathbf{W}\mathbf{D},$$

whose solution is

$$\mathbf{a} = (\mathbf{C}^T\mathbf{W}\mathbf{C})^{-1}\mathbf{C}^T\mathbf{W}\mathbf{D}. \tag{7.36}$$

Example 7.15

Design a discrete weighted least-square-error lowpass filter of length $N = 17$ with a corner frequency of $\omega = \pi/2$ that minimizes the square-error over $R = 41$ points uniformly in frequency, where the error of points in the stopband is weighted by a factor of $K = 10$.

➤ Solution:

The code is similar to that of Example 7.14, modified to include the weighting matrix:

```
N = 17;
R = 41;
P = (R-1)/2;
K = 10;
wc = 0.5*pi;
k = 0:(N-1)/2;
l = 0:P;
w = 2*pi*l'/R;
D = double(w<wc);
C = cos(w*k);
w_diag = ones(1, P+1);
w_diag(w>wc) = K;
W = diag(w_diag);
a = (C'*W*C) \ (C'*W*D); % better to use lscov(C, D, w_diag)
h = [0.5*a(end:-1:2); a(1); 0.5*a(2:end)];
```

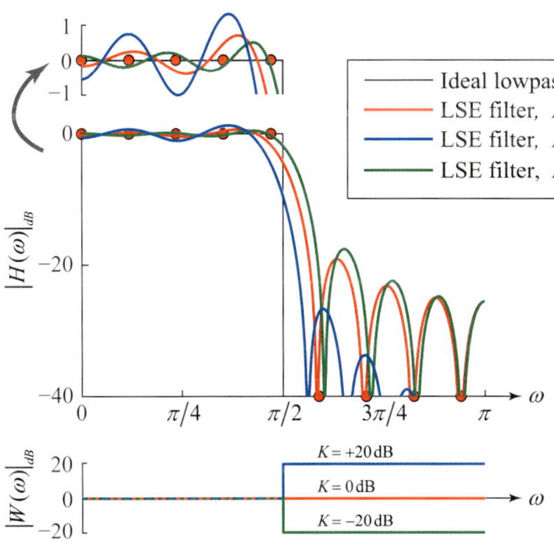

Figure 7.37 Discrete least-square-error filter with weighting

Here, `w_diag` is a $1 \times P+1$ vector of the weighting factors $[W(\omega_0)\ W(\omega_1)\ \cdots\ W(\omega_P)]$ that form the diagonal of the weighting matrix `W`. If `P` is not too large and the problem is well conditioned, you can probably get away with implementing Equation (7.36) directly, as I have done here,

```
a = (C'*W*C) \ (C'*W*D);
```

However, to assure numerical convergence, you can also use Matlab's `lscov` function to solve the normal equations with a weighting function:

```
a = lscov(C, D, w_diag);
```

In this particular example, the results are identical.

Figure 7.37 shows the responses of least-square-error filters designed with uniform weighting of all points (red curve) and with weighting of the stopband error by factors of 10 ($K=+20$ dB, blue curve) and 0.1 ($K=-20$ dB, green curve). When the error in the stopband is more heavily emphasized (i.e., $K=+20$ dB), then the stopband attenuation is significantly greater than the equally weighted case, while the opposite is true when the passband attenuation is more heavily weighted.

7.8.2 ★ Integral least-square-error FIR filter design

The least-square methods of the previous sections all have in common that they are based on minimizing the error between the desired and actual response over a finite number of points; for example, the discrete weighted least-square error (Equation (7.35) minimizes the sum of the weighted square error over P points,

$$\sum_{k=0}^{P} W(\omega_k)|D(\omega_k) - A(\omega_k)|^2.$$

Instead, it often makes sense to find the impulse response of the filter that minimizes the *integral* of the weighted square-error over one or more ranges of contiguous frequency in the interval $-\pi \le \omega < \pi$ rather than just as the sum of the error over a relatively small number of points:

$$\int_{-\pi}^{\pi} W(\omega)|D(\omega) - A(\omega)|^2 d\omega. \tag{7.37}$$

One of the important features of this design method is that it allows you to exclude portions of the frequency response from the integral error calculation by setting $W(\omega)=0$. That is useful in excluding frequencies around the transition region in which artifacts such as the Gibbs phenomenon would contribute heavily to the integral error. In supplementary material, we show how integral least-square-error filters are designed and develop some Matlab code that we apply to a number of examples.

7.9 ★ Optimal lowpass filter design

Chebyshev or **minimax optimization** is a very powerful method of FIR filter design that allows you to minimize the maximum error of the frequency response rather than the least-square error over a specified frequency range. Using this method, lowpass, highpass, bandpass, bandstop and multiband filters can be designed to meet almost arbitrary passband and stopband specifications. FIR filters designed using minimax optimization are also called **equiripple filters**, because the magnitude of the frequency response has ripples in both the passband and stopband and the ripples within any band are of identical amplitude.

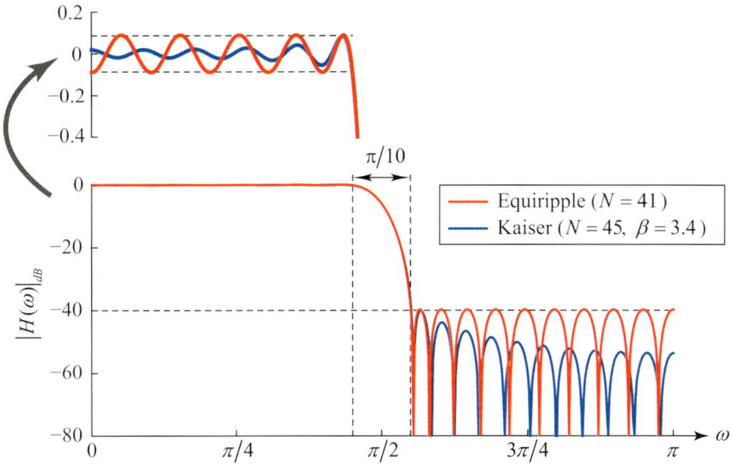

Figure 7.38 Equiripple and Kaiser filters designed with the same specifications

As an example, **Figure 7.38** shows the response of an equiripple lowpass filter designed with cutoff frequency $\omega_c = 0.5\pi$, transition bandwidth $\Delta\omega = 0.1\pi$ and with both passband and stopband attenuations set to $\delta = 0.01$ (red), which corresponds to −40 dB in the stopband and ±0.086 dB in the passband (shown with dashed lines). Also shown in the figure is the response of a Kaiser filter designed to meet approximately the same specifications. For the Kaiser filter, the largest ripple in the passband or stopband occurs at the boundaries of the transition zone (i.e., nearest ω_p and ω_s), and the size of the ripples decreases as ω moves away from $\omega = 0.5\pi$. In contrast, with the equiripple filter, all the ripples in the passband or stopband are of the same size. In the example of **Figure 7.38**, the length of the equiripple filter is $N = 41$, whereas the length of the Kaiser filter is $N = 45$. It turns out that the order of the equiripple filter necessary to meet a given set of passband and stopband criteria is always lower than that required by any other filter design. For this reason, the equiripple filter is also called the **optimal FIR filter**. The most commonly used design procedure for optimal FIR filters is the **Parks–McClellan algorithm**, named for the two researchers who developed and described it. In supplementary material, this method is derived in detail with a number of examples.

7.10 Multiband filters

A **multiband filter** has an amplitude profile that can be thought of as a combination of bandpass filters of different amplitudes, center frequencies and bandwidths. There are a number of design options to create a multiband filter. One simple approach is to superimpose window-based bandpass filters as designed in Example 7.7. However, several of the other filter design methods we have studied – the frequency-sampling method (Section 7.7), the least-square-error method (Section 7.8) and the optimum (Parks–McClellan) method (Section 7.16) – are all natively equipped to handle piecewise continuous multiband designs, as shown in the next example.

Example 7.16

Design a window-based two-band multiband filter with $\omega_c = \pi/4$ and $3\pi/4$, and $\Delta\omega = 0.2\pi$, on both bands using the sum of bandpass prototypes and compare with filters designed with the frequency-sampling, least-square-error and Parks–McClellan methods.

➤ Solution:

The code for the superposition of bandpass filters is essentially the same as that in Example 7.7:

```
fc = [0.15 0.35 0.65 0.85];
lfc = length(fc);
amp = [1 0.1];
N = 55;
n = -(N-1)/2:(N-1)/2;
h = zeros(1, N);
for i = 1:lfc/2
    h = h + amp(i)*(fir1(N-1, fc(2*i-1), kaiser(N, 4), 'noscale') - ...
         fir1(N-1, fc(2*i), kaiser(N, 4), 'noscale'));
end
```

Here, we allow the amplitude of each band to be set independently. The result is the blue curve in **Figure 7.39**. This response is compared with the response of filters designed with Matlab's `fir2`, `firls` and `firpm` methods. The `fir1` method also permits the design of multiband filters, but this method does not allow the amplitude of each band to be set independently.

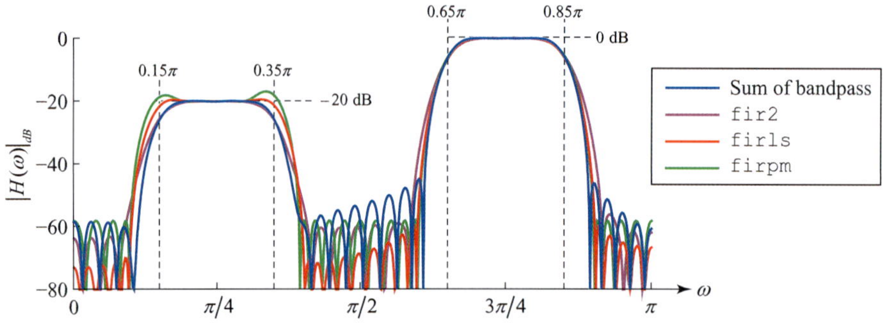

Figure 7.39 Multiband example

The remaining sections of this chapter cover the design of two special filters, differentiators and Hilbert transformers, and give examples of their use.

7.11 ★ Differentiator

An ideal discrete-time differentiator is a filter with transform

$$H(\omega) = j\omega$$

and impulse response (Problem 7-34)

$$h[n] = \begin{cases} \dfrac{1}{n}(-1)^n, & n \neq 0 \\ 0, & n = 0 \end{cases}. \tag{7.38}$$

Like the ideal lowpass filter discussed in Section 7.2.2, the ideal differentiator is unrealizable because it is non-causal, of infinite length and unstable.

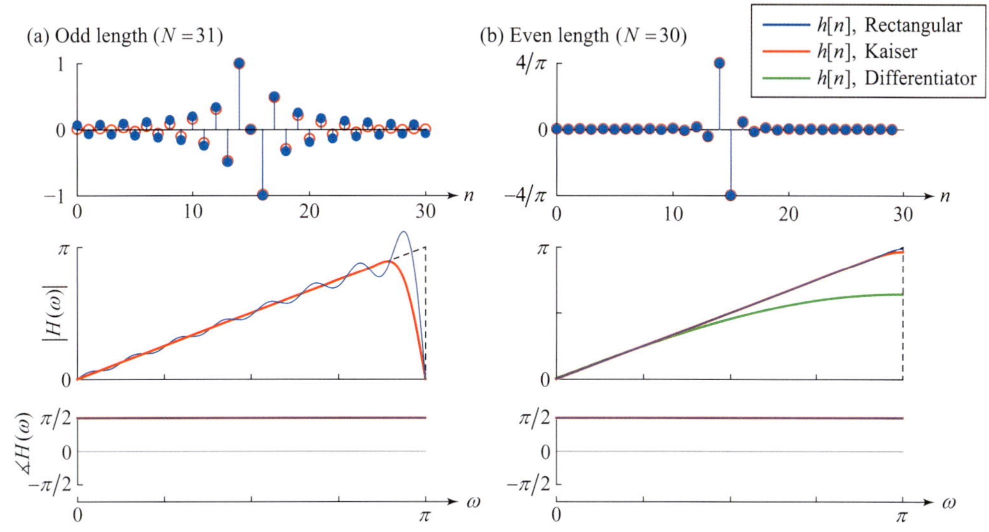

Figure 7.40 Impulse and frequency response of differentiator

However, there exist numerous designs of practical differentiators, both FIR and IIR. **Figure 7.40a** (top panel, blue symbols) shows the impulse response of a simple causal odd-length FIR differentiator produced by shifting and truncating the ideal response, Equation (7.38), to $N = 31$ points with a rectangular window, in much the same way we created a causal FIR lowpass filter in **Figure 7.4**,

$$h[n] = \begin{cases} \dfrac{(-1)^{n-(N-1)/2}}{(n-(N-1)/2)}, & 0 \le n \le N-1, n \neq (N-1)/2 \\ 0, & n = (N-1)/2 \end{cases}, \quad N \text{ odd.}$$

The phase of the frequency response of this filter, $\measuredangle H(\omega)$ (blue trace, bottom panel), has the desired value, $\pi/2$. However, the z-transform of this antisymmetric odd-length (Type-III) filter has zero(s) at $z = \pm 1$ that force the magnitude of the frequency response to 0 at $\omega = 0$ and $\omega = \pm\pi$. The zero at $z = +1$ is no problem since $H(\omega = 0) = 0$ in any event. However, the sharp discontinuity in the ideal response (middle panel, dashed black line) at $\omega = \pm\pi$ produces considerable oscillation in $|H(\omega)|$ (thin blue line) due to the Gibbs phenomenon. Fortunately, the effect of this discontinuity can be mitigated by multiplying $h[n]$ by one of the window functions we have studied, for example a Kaiser filter ($\beta = 5$, red trace), as we did with the lowpass filter discussed in Section 7.3.

Figure 7.40b shows the impulse response (top panel, blue symbols) and frequency response $|H(\omega)|$ (middle panel, thin blue trace) of a causal even-length ($N = 30$) differentiator (Problem 7-36),

$$h[n] = \frac{(-1)^{n-N/2+1}}{\pi(n-(N-1)/2)^2}, \; 0 \le n \le N-1.$$

The z-transform of this even-length antisymmetric (Type-IV) filter has zero(s) at $z = +1$, which are, again, no problem, since the response is 0 at $\omega = 0$. But, because there is no zero at $z = -1$, to force the frequency response magnitude to 0 at $\omega = \pm\pi$, this filter has a superior magnitude response compared with the odd-length filter and would be preferred in an application where it is not necessary to delay the output of the differentiator by an integer amount in order to align it precisely with the input. The Kaiser-filtered impulse response (red symbols and trace) confers no advantage in this case. For amusement, the green trace in the middle panel shows the frequency response of a simple first-order, backward-difference approximation of a differentiator,

$$h[n] = \delta[n] - \delta[n-1].$$

It is not really acceptable for inputs with frequencies above about $\omega = \pi/3$.

7.12 ★ Hilbert transformer

The **Hilbert transformer** is an important tool that enables many applications, particularly in communication and related areas. Applications include **single sideband (SSB)** modulation and demodulation, **amplitude modulation (AM)** and demodulation, **frequency (FM)** and **phase (PM)** modulated signals, **amplitude-shift keying (ASK)**, **phase-shift keying (PSK)**, automatic gain control and peak detection. In Section 7.12.1, we will develop the theory, leading to a description of some applications in Section 7.12.4.

7.12.1 Derivation of the Hilbert transformer

The concept of the Hilbert transformer is motivated by a simple observation. Consider a real signal $x_R[n]$ with transform $X_R(\omega)$, as shown in the top left panel of **Figure 7.41a**. This transform has information at both positive and negative frequencies. However, if the signal is real ($x_R[n] = x_R^*[n]$), then the transform exhibits conjugate symmetry (also called Hermitian symmetry), that is, $X_R(\omega) = X_R^*(-\omega)$. As a consequence, the information in the transform $X_R(\omega)$, at negative frequencies $-\pi \le \omega < 0$ can be derived from $X_R(\omega)$ at positive frequencies $0 \le \omega < \pi$. This redundancy of information leads us to ask, is there a signal $y[n]$ that has the same information as $x_R[n]$, but whose "one-sided" transform contains *only* the positive frequency components? There is and it is termed the **analytic signal**, the transform of which, $Y(\omega)$, is sketched on the top right panel of **Figure 7.41a**,

$$Y(\omega) = \begin{cases} 2X_R(\omega), & \omega > 0 \\ X_R(\omega), & \omega = 0 \\ 0, & \omega < 0 \end{cases}. \tag{7.39}$$

The factor of two assures that the analytic signal $y[n]$ has the same energy (or power) as the real signal $x_R[n]$ from which it is derived. As we shall soon show, $y[n]$ is complex, and is formed by adding a complementary imaginary component $x_I[n]$ to the real component $x_R[n]$:

$$y[n] = x_R[n] + jx_I[n]. \tag{7.40}$$

The imaginary part $x_I[n]$ is a real function[6] with transform $X_I(\omega)$, which must be designed so that the transform of the analytic signal,

$$Y(\omega) = X_R(\omega) + jX_I(\omega),$$

has only positive frequency components and satisfies Equation (7.39). To achieve this, $jX_I(\omega)$ must cancel the component of $X_R(\omega)$ at negative frequencies and add to the component at positive frequencies,

$$jX_I(\omega) = \begin{cases} +X_R(\omega), & \omega > 0 \\ 0, & \omega = 0 \\ -X_R(\omega), & \omega < 0 \end{cases}, \tag{7.41}$$

as shown in the middle panel of **Figure 7.41a**. The result is that the transform of the analytic signal is one-sided.

The Hilbert transformer can be viewed as an ideal phase shifter. To see this write Equation (7.41) as

$$X_I(\omega) = \begin{Bmatrix} -jX_R(\omega), & \omega > 0 \\ 0, & \omega = 0 \\ +jX_R(\omega), & \omega < 0 \end{Bmatrix} = X_R(\omega) \underbrace{\begin{Bmatrix} -j, & \omega > 0 \\ 0, & \omega = 0 \\ +j, & \omega < 0 \end{Bmatrix}}_{H_I(\omega)} = X_R(\omega) H_I(\omega), \tag{7.42}$$

where $H_I(\omega)$ has frequency response

$$H_I(\omega) \triangleq -j\,\mathrm{sgn}(\omega) = \begin{cases} -j, & \omega > 0 \\ 0, & \omega = 0 \\ +j, & \omega < 0 \end{cases}. \tag{7.43}$$

Because $H_I(\omega)$ is a purely imaginary and odd function of frequency, by the properties of the DTFT (see Section 3.5.4), the impulse response $h_I[n]$ is real and odd (Problem 7-35),

$$h_I[n] = \frac{\pi n}{2} \mathrm{sinc}^2 n\pi/2. \tag{7.44}$$

This filter is called the ideal **discrete-time Hilbert transformer**, and the output of the filter (i.e., the inverse transform of Equation (7.42)) is therefore also a real signal,

$$x_I[n] = x_R[n] * h_I[n].$$

The signals $x_R[n]$ and $x_I[n]$ together form a **Hilbert transform pair**. Equation (7.43) says that in order to produce $X_I(\omega)$, the transformer $H_I(\omega)$ induces a phase shift of one quadrant, namely $-\pi/2$, in $X_R(\omega)$ for $\omega > 0$ and $+\pi/2$ for $\omega < 0$. For this reason, $x_I[n]$ is also termed the **quadrature signal** of $x_R[n]$.

Figure 7.41b shows the conceptual steps necessary to construct the analytic signal $y[n]$ from $x_R[n]$ and the quadrature signal $x_I[n]$. The real signal $x_R[n]$ is passed through a filter with real impulse response $h_I[n]$ to produce a real signal $x_I[n]$. Then $x_I[n]$ is multiplied by j and added to $x_R[n]$ to form the complex analytic signal,

---

[6]Remember that given a complex number $z = a + jb$, both the real part a and the imaginary part b are *real* numbers. It is the j that combines them into a complex number.

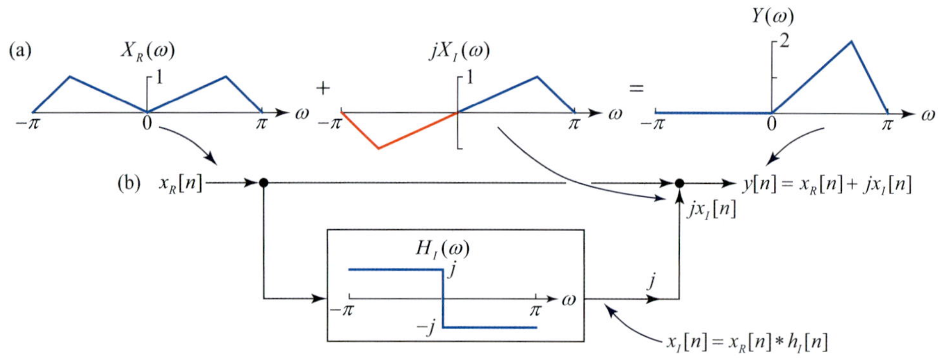

Figure 7.41 Hilbert transformer

$$y[n] = x_R[n] + jx_I[n].$$

This analytic signal can also be expressed in polar form,

$$y[n] = |y[n]|e^{j\measuredangle y[n]},$$

where

$$\begin{aligned} |y[n]| &= \sqrt{x_R^2[n] + x_I^2[n]} \\ \measuredangle y[n] &= \tan^{-1}(x_I[n]/x_R[n]) \end{aligned}. \tag{7.45}$$

In the polar form, $y[n]$ can be interpreted as a complex vector with instantaneous magnitude $|y[n]|$ and instantaneous phase $\measuredangle y[n]$, whose projections onto the real and imaginary axes are $x_R[n]$ and $x_I[n]$, respectively,

$$\begin{aligned} x_R[n] &= |y[n]| \cos \measuredangle y[n] \\ x_I[n] &= |y[n]| \sin \measuredangle y[n]. \end{aligned}$$

Example 7.17

Let $x_R[n] = A\cos\omega_o n$, with $\omega_0 > 0$. Find the quadrature signal $x_I[n]$ and the analytic signal $y[n] = x_R[n] + jx_I[n]$.

➤ Solution:

The transform of $x_R[n]$ consists of two impulses, one at $\omega = \omega_0$ and one at $\omega = -\omega_0$,

$$X_R(\omega) = A\pi(\delta(\omega + \omega_0) + \delta(\omega - \omega_0)).$$

The transform of $x_I[n]$ is given by

$$X_I(\omega) = H_I(\omega)X_R(\omega) = A\pi(j\delta(\omega + \omega_0) - j\delta(\omega - \omega_0)) = Aj\pi(\delta(\omega + \omega_0) - \delta(\omega - \omega_0)).$$

The inverse transform of $X_I(\omega)$ is the quadrature signal $x_I[n] = A\sin\omega_o n = A\cos(\omega_o n - \pi/2)$. From Equation (7.40), the analytic signal is then

$$y[n] = x_r[n] + jx_I[n] = A\cos(\omega_o n + \theta) + jA\sin(\omega_o n + \theta) = Ae^{j(\omega_o n + \theta)}, \tag{7.46}$$

whose transform is a single impulse at positive frequency $\omega = \omega_0$,

$$Y(\omega) = X_R(\omega) + jX_I(\omega) = 2\pi A e^{j\theta}\delta(\omega - \omega_0).$$

Figure 7.42 shows the transform pair $x_R[n]$ (blue trace) and $x_I[n]$ (red trace), from which the magnitude (black line) is computed,

$$|y[n]| = \sqrt{x_R[n] + x_I[n]} = \sqrt{A\cos^2\omega_0 n + A\sin^2\omega_0 n} = |A|.$$

In Section 7.12.4, we will see this same technique applied to the practical problem of demodulating amplitude modulated (AM) signals, where the amplitude A is a slowly varying function of time.

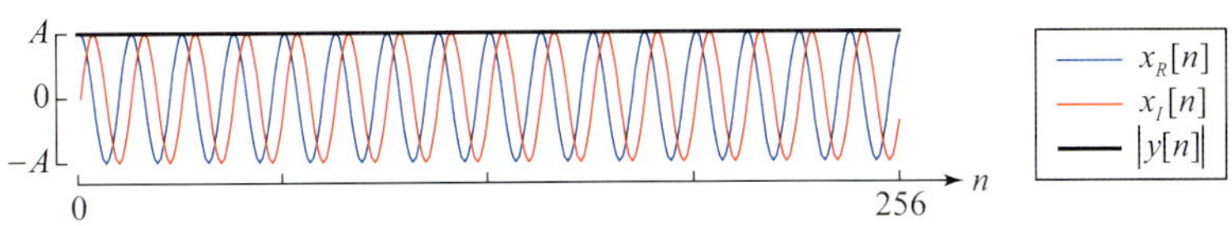

Figure 7.42 Hilbert transform pair for a cosine

7.12.2 FIR implementation of a Hilbert transformer

In attempting to design a practical Hilbert transformer, we encounter the same problem we faced in designing filters based on the ideal lowpass filter and the differentiator; namely that the impulse response of the ideal Hilbert transformer is non-causal, infinite in duration and it is also unstable since it is not absolutely summable. The solution is again to approximate the ideal response with a causal FIR filter derived from the ideal response. **Figure 7.43a** (top panel, blue symbols) shows the impulse response of a causal odd-length Hilbert transformer produced by shifting and truncating the ideal response, Equation (7.44), to $N = 31$ points,

$$h_I[n] = \frac{\pi(n - (N-1)/2)}{2}\,\mathrm{sinc}^2\pi(n - (N-1)/2)/2.$$

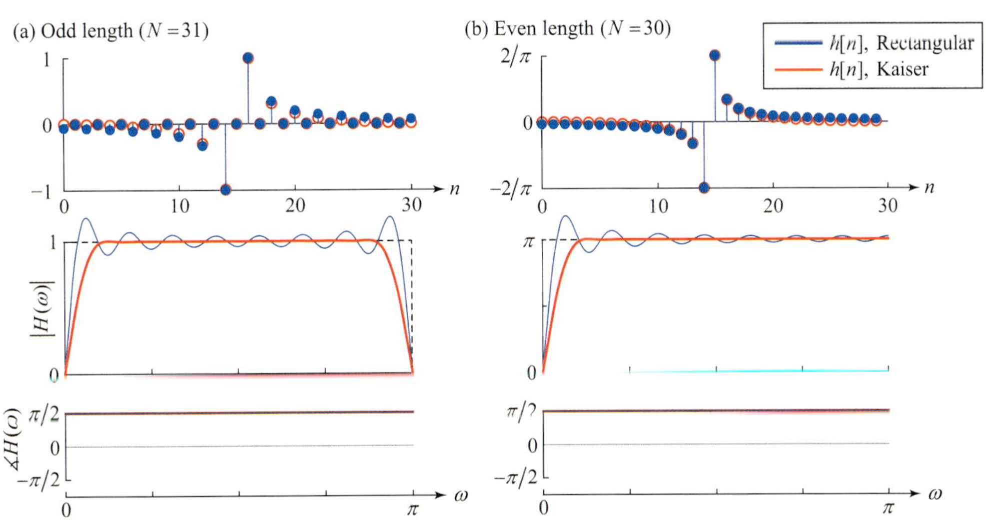

Figure 7.43 Impulse response and frequency response of the Hilbert transformer

As with the differentiator, zeros at $z = \pm 1$ in the z-transform of this Type-III filter force the magnitude of the response to 0 at $\omega = 0$ and $\omega = \pm \pi$, resulting in discontinuities that produce Gibbs oscillations in the response. The red symbols show the response when a 31-point Kaiser window (with $\beta = 4$) multiplies the response. The windowed Hilbert transformer has a flatter response and smaller oscillations around the discontinuities at the cost of a more gradual slope. **Figure 7.43b** shows the impulse response and frequency response of a Type-IV (odd-symmetric even-length) filter. Because the Hilbert transformer is usually applied to bandpass signals, either even- or odd-length filters would be acceptable. In practice the odd-length filter may be preferred (at least in the time-domain FIR implementation) since it can be implemented with fewer multiply and accumulate operations due to the fact that every other point in the impulse response is zero.

A window-based FIR Hilbert transformer can be simply implemented in Matlab using the `sinc` and `window` functions. For example, here is a code snippet that generates the odd-length Hilbert transformer using a Kaiser window in **Figure 7.43a**:

```
function h = hilb_h(N)
    n = 0:N-1; % assume N odd
    h = (pi* (n-(N-1)/2)/2) .* sinc((n-(N-1)/2)/2).^2 ...
      .* window(@kaiser, N, beta)';
end
```

The impulse response of this filter must be causal, so when using this function to help generate an analytic function, we have to account for the offset of the center point of the Matlab array, `h[(N+1)/2]`, which corresponds to the center point of the causal impulse response. The following example shows this.

Example 7.18

Generate the analytic signal for the cosine in Example 7.17 using Matlab.

➤ Solution:

```
N = 31;
w0 = pi/8;
xr = cos(w0*(0:256));
hi = hilb_h(N); % create the impulse response of the transformer
xh = filter(hi, 1, xr); % output is shifted
xi = [xh((N+1)/2:end) zeros(1, (N-1)/2)]; % re-center output and pad the end
y = xr + 1j * xi; % form analytic signal
```

The output of the `filter` function, `xh`, is delayed by $(N+1)/2$ with respect to the input `x`. The next line corrects for the delay of the causal filter by deleting the first $(N-1)/2$ points and pads the end with the same number of zeros so that the sizes of `xr` and `xi` will be equal. Here, I have used `filter` instead of `conv`, which would have required more offset. I leave it to you to plot `xr`, `xi`, `abs(y)` and `unwrap(angle(y))` and see if you get the expected results. In practice, you would most likely want to create a larger filter.

Because Hilbert transformers exhibit odd symmetry, they can also be created using Matlab methods such as `firpm` and `firls`, which have the ability to design Type-III (odd-length) or Type-IV (even-length) antisymmetric filters. See the Matlab documentation of these functions for more information.

7.12.3 FFT implementation of a Hilbert transformer

Equation (7.39) suggests an alternate method of implementing the Hilbert transformer in the frequency domain using the fast Fourier transform (FFT). This code snippet generates an analytic output `y` from an input `xr`.

```
function y = hilb_fft(xr)
    N = length(x);
    Xr = fft(xr(:), N); % assume N even
    Y = [Xr(1);2*Xr(2:N/2);Xr(N/2+1]; % double points except for w=0 and pi
    y = ifft(Y, N);
end
```

The N-point FFT `Xr` is equivalent to the DTFT $X_R(\omega)$ sampled at N points,

$$X_R[k] = X_R(\omega)|_{\omega = e^{j2\pi k/N}}.$$

The array `Y` corresponds to the positive frequencies of $X_R(\omega)$ in **Figure 7.41a**. All points except the first and last points, which correspond to $\omega = 0$ and π, are doubled. Then, we take the inverse transform of `Y` to produce the analytic signal `y`. The Matlab function `hilbert` gives the same output.

7.12.4 Applications of the Hilbert transformer

Figure 7.44 shows examples of two common applications that exploit the Hilbert transformer's ability to extract separate amplitude and phase information from an analytic signal.

AM demodulator The left-hand side of the figure shows how the Hilbert transformer can be used to demodulate AM signals by extracting the magnitude information from an analytic signal. **Amplitude modulation** is the technique that underlies AM and MW (European) radio transmission. It also is the basis of the quadrature amplitude modulation (QAM) method that is central to digital communication devices such as modems. In AM – either analog or digital – the **baseband signal** or **message** $m[n]$ (red trace in **Figure 7.44a**) represents the information we wish to transmit, for example, speech, music or a digital bit stream. The term "baseband" reflects the fact that the spectrum of the message, $M(\omega)$, is centered about frequency $\omega = 0$ (top (red) trace of the inset). In AM modulation, the baseband signal is used to modulate or multiply the amplitude of a **carrier**, normally a cosine $\cos \omega_0 n$ (blue trace in **Figure 7.44a**), to form the amplitude modulated signal $x_R[n]$ (thin green trace in **Figure 7.44b**),

$$x_R[n] = m[n] \cos \omega_0 n.$$

$x_R[n]$ is a bandpass signal. Its spectrum $X_R(\omega)$ (middle (green) trace in the inset) is two-sided; it has energy at both positive and negative frequencies. The carrier frequency ω_0 is designed to be high compared to the bandwidth of the message so that the waveform of the message $m[n]$ varies slowly compared to the waveform of the cosine.

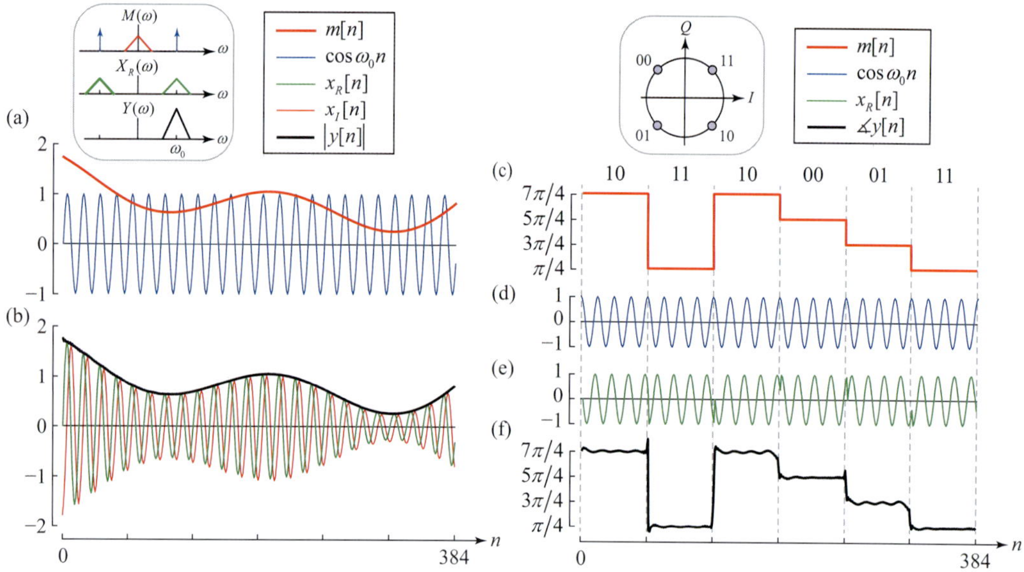

Figure 7.44 Demodulation of AM and PSK signals using Hilbert transformer

The Hilbert transformer can be used to demodulate the AM signal in order to recover the baseband message $m[n]$. The procedure is as follows: the real modulated signal $x_R[n]$ (thin green trace in **Figure 7.44b**) is processed through the Hilbert transformer to produce the quadrature signal $x_I[n]$ (thin red trace). As long as $m[n]$ is slowly varying compared with the cosine, it can be considered roughly constant, as in Example 7.17, and the quadrature component will be well approximated by

$$x_R[n] = m[n] \sin \omega_0 n.$$

The magnitude of the envelope signal, $|y[n]|$ (black trace in **Figure 7.44b**), is calculated from $x_R[n]$ and $x_I[n]$ using Equation (7.45),

$$\begin{aligned} |y[n]| &= \sqrt{x_R^2[n] + x_I^2[n]} = \sqrt{(m[n] \sin \omega_0 n)^2 + (m[n] \sin \omega_0 n)^2} \\ &= \sqrt{m^2[n]}\sqrt{\sin^2\omega_0 n + \cos^2\omega_0 n} \\ &= |m[n]|. \end{aligned}$$

The Matlab function `envelope` included in recent versions of the product returns the envelope of a real signal.

QPSK demodulator The right-hand side of **Figure 7.44** shows how the Hilbert transformer can be used to demodulate a phase-shift keyed (PSK) signal. In **phase-shift keying**, digital information in the message is encoded for transmission by modulating the *phase* of carrier signal, rather than its amplitude. The particular example in **Figure 7.44** illustrates a technique called **quadrature phase-shift keying (QPSK)**, which is extensively used in digital communication systems. In QPSK, the message comprises a series of two-bit "symbols," 00, 01, 10 and 11. Each symbol is encoded by shifting the phase of a constant-amplitude carrier by a different amount, $m[n] = 3\pi/4$, $5\pi/4$, $7\pi/4$ or $\pi/4$, for a few cycles of the carrier, as shown in **Figure 7.44c**. The mapping between the symbols and the amount of phase shift is also indicated by the **constellation diagram** in the inset at the top of the figure. The name QPSK refers to the fact that the phase shifts of symbols differ from each other by one quadrant, $\pi/2$.[7] **Figure 7.44d** shows the carrier, a cosine of constant amplitude and phase, $\cos\omega_0 n$. **Figure 7.44e** shows the transmitted QPSK signal $x_R[n]$ formed by phase shifting successive six-cycle intervals of the carrier by amount $m[n]$,

$$x_R[n] = \cos\underbrace{(\omega_0 n + m[n])}_{\phi[n]} = \cos\phi[n],$$

where the quantity $\phi[n] = \omega_0 n + m[n]$ is called the **instantaneous phase**. To demodulate the QPSK signal, $x_R[n]$ is filtered by a Hilbert transformer to produce the quadrature component

$$x_I[n] = \sin\phi[n].$$

Then, the instantaneous phase is extracted from the arctangent of the ratio, as specified in Equation (7.45),

$$\measuredangle y[n] = \tan^{-1}(x_I[n]/x_R[n]) = \tan^{-1}(\sin\phi[n]/\cos\phi[n]) = \phi[n],$$

from which the desired signal $m[n]$ is easily extracted. The result is shown in **Figure 7.44f**.

SUMMARY

In this chapter, we have explored a number of different practical FIR filter design techniques. All these filters are stable and have linear phase. The choice of filter depends on a number of factors. Design with window-based filters is exceptionally easy to understand and implement and is adequate for many applications, but there is a fundamental trade-off between the width of the transition zone, the amount of overshoot in the passband and the minimum attenuation in the stopband. These filters generally do not allow independent control of parameters such as passband and stopband frequencies and attenuations (the Kaiser filter being an exception). Spline filters offer better control of the transition zone. Frequency-sampled filters allow you to design complex filters in which the response amplitude can be precisely specified at specific sampled frequencies. However, these filters can have large deviations between the sampled frequency points, since the response is essentially interpolated by a sinc-like function. Discrete and integral least-square-error filters do not match the frequency response at precise frequencies, but allow you to set the total error between the desired and designed filter at selected frequency points or over selected frequency regions. Finally, the optimum filter design procedure allows you to design filters of minimum order that achieve specified passband and stopband requirements.

[7] The I and Q on the two axes of the constellation diagram stand for "in phase" and "quadrature," respectively.

PROBLEMS

Problem 7-1

The centered Hamming-window FIR filter is specified by the impulse response

$$\text{hamming}_N[n] = (0.54 + 0.46\cos(2\pi n/(N-1)))\,\text{rect}_N[n].$$

The values of the coefficients for this filter, 0.54 and 0.46, were supposedly chosen so that side lobes of the transform of the filter components $\mathfrak{F}\{0.54\,\text{rect}_N[n]\}$ and $\mathfrak{F}\{0.46\cos(2\pi/(N-1))\text{rect}_N[n]\}$ would cancel at frequency $\omega = \pm 4.5\pi/(N-1)$. This is almost, but not quite, true. In this problem, we will understand how these coefficients were chosen by examining the case of the continuous-time Hamming window of length T:

$$\text{hamming}_T(t) = (0.54 + 0.46\cos(2\pi t/T))\,\text{rect}_T(t).$$

Figure 7.45a shows a continuous-time symmetric raised-cosine window of length T defined as

$$w(t) = (\alpha + (1-\alpha)\cos(2\pi t/T))\,\text{rect}_T(t),$$

where α is a constant with $|\alpha| < 1$ and

$$\text{rect}_T(t) = \begin{cases} 1, & -T/2 \le t \le T/2 \\ 0, & \text{otherwise} \end{cases}.$$

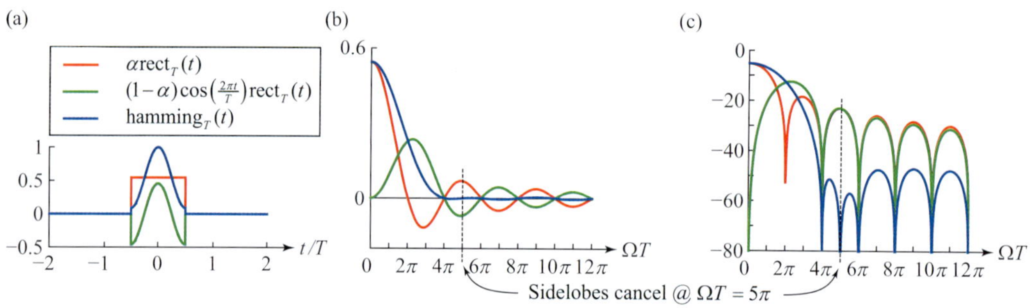

Figure 7.45

This window is the sum of a rectangular pulse of amplitude α (red trace), plus a gated cosine of amplitude $1-\alpha$ (green trace). **Figures 7.45b** and **c** show the transform of the two components and their sum on linear and dB scales. The purpose of this problem is to find the value of α that results in perfect cancelation of the transform components at frequency $\Omega = 5\pi/T$.

(a) Show that the continuous-time Fourier transform of $w(t)$ is

$$W(\Omega) = T\left(\alpha \operatorname{sinc} \Omega T/2 + \frac{1-\alpha}{2}(\operatorname{sinc}(\Omega T/2 - \pi) + \operatorname{sinc}(\Omega T/2 + \pi))\right).$$

(b) Show that the value of α necessary to achieve exact cancellation of the transform components at frequency $\Omega T = 5\pi$ is $\alpha = 25/46 \cong 0.5435$. When truncated to two significant figures, this results in $\text{hamming}_T(t) = (0.54 + 0.46\cos(2\pi t/T))\,\text{rect}_T(t)$.

Problem 7-2

The purpose of this problem is to show that the value of $\alpha = 25/46$, which achieves perfect cancelation for the continuous-time Hamming window in Problem 7-1, does not achieve cancelation for the discrete-time Hamming-window filter at frequency $\omega = 5\pi/N$. Use Matlab or other software to plot $|W(5\pi/N)|_{dB}$, vs. N for values of $13 \leq N \leq 53$, using $\alpha = 25/46$. Show that a minimum value of $|W(5\pi/N)|_{dB} = -64.22$ dB occurs for a value of $N = 17$.

Problem 7-3

A raised-cosine window FIR filter of two components is specified by the impulse response

$$w[n] = \left(\alpha + (1-\alpha)\cos\frac{2\pi n}{N-1}\right)\text{rect}_N[n].$$

The transform has two components:

$$\begin{aligned} W(\omega) &= \mathfrak{F}\{(\alpha + (1-\alpha)\cos(2\pi n/N-1))\,\text{rect}_N[n]\} \\ &= \alpha\mathfrak{F}\{\text{rect}_N[n]\} + (1-\alpha)\mathfrak{F}\{\cos(2\pi n/N-1)\,\text{rect}_N[n]\}. \end{aligned}$$

(a) Show that the value of α necessary to achieve perfect cancelation of two transform components at frequency $\omega = 5\pi/N$ is

$$\alpha = \frac{1}{2-\tan^2(\pi/(N-1))\cot^2(5\pi/2N)}.$$

(b) Verify this conclusion with Matlab for an odd value of N.

Problem 7-4

(a) Given a Type-II symmetric impulse response with even filter length N, show that the DTFT can be expressed as

$$H(\omega) = \sum_{n=0}^{N-1} h[n]e^{-j\omega n} = e^{-j\omega(N-1)/2}A(\omega),$$

where

$$A(\omega) = \sum_{m=1}^{M} b[m]\cos(m-1/2)\omega,$$

with $M = N/2$ and coefficients $b[m]$ given by

$$b[m] = 2h[M-m],\quad 1 \leq m \leq M.$$

(b) Show that

$$\cos(m-1/2)\omega = 2\cos(m-1)\omega\cos\omega/2 - \cos(m-3/2)\omega.$$

(c) Use the results of parts (a) and (b) to show that

$$A(\omega) = \cos\omega/2\sum_{m=0}^{M-1} r[m]\cos m\omega,$$

where

$$r[m] = \begin{cases} b[1] - r[1]/2, & m=0 \\ 2b[m+1] - r[m+1], & 1 \le m \le M-2 \\ 2b[M], & m = M-1 \end{cases}.$$

So,

$$r[m] = \begin{cases} 2h[M-1] - r[1]/2, & m=0 \\ 4h[M-1-m] - r[m+1], & 1 \le m \le M-2 \\ 4h[0], & m = M-1 \end{cases}.$$

➤**Hint:** Expand $A(\omega)$, and then substitute part (b) into each term starting at the end (i.e., $b[M]\cos(M-1/2)\omega$) and working backwards towards the first term.

Problem 7-5

(a) Given a Type-III antisymmetric impulse response with odd filter length N, show that the DTFT can be expressed as

$$H(\omega) = \sum_{n=0}^{N-1} h[n]e^{-j\omega n} = e^{-j\omega(M-\pi/2)}A(\omega),$$

where

$$A(\omega) = \sum_{m=1}^{M} c[m]\sin m\omega,$$

with $M = (N-1)/2$ and coefficients $c[m]$ given by

$$c[m] = 2h[M-m], \quad 1 \le m \le M.$$

(b) Show that

$$\sin(m-1/2)\omega = 2\cos(m-1)\omega\sin\omega/2 - \sin(m-3/2)\omega.$$

(c) Use the results of parts (a) and (b) to show that

$$A(\omega) = \sin\omega\sum_{m=0}^{M-1} r[m]\cos m\omega,$$

where

$$r[m] = \begin{cases} c[1] + r[2]/2, & m = 0 \\ 2c[m+1] + r[m+2], & 1 \le m \le M-3 \\ 2c[M-1], & m = M-2 \\ 2c[M], & m = M-1 \end{cases}.$$

So,

$$r[m] = \begin{cases} 2h[M-1] + r[2]/2, & m = 0 \\ 4h[M-1-m] + r[m+2], & 1 \le m \le M-3 \\ 4h[1], & m = M-2 \\ 4h[0], & m = M-1 \end{cases}.$$

➤**Hint:** Expand $A(\omega)$, and then substitute part (b) into each term starting at the end (i.e., $c[M]\sin M\omega$) and working backwards towards the first term.

Problem 7-6

(a) Given a Type-IV antisymmetric impulse response with N even, show that the DTFT can be expressed as

$$H(\omega) = \sum_{n=0}^{N-1} h[n]e^{-j\omega n} = e^{-j\omega(M-1/2-\pi/2)}A(\omega),$$

where

$$A(\omega) = \sum_{m=1}^{M} d[m]\sin(m-1/2)\omega,$$

with $M = N/2$ and coefficients $d[m]$ given by

$$d[m] = 2h[M-m], \quad 1 \le m \le M.$$

(b) Show that

$$\sin(m-1/2)\omega = 2\cos(m-1)\omega\sin\omega/2 + \sin(m-3/2)\omega.$$

(c) Use the results of parts (a) and (b) to show that

$$A(\omega) = \sin\omega/2 \sum_{m=0}^{M-1} r[m]\cos m\omega,$$

where

$$r[m] = \begin{cases} d[1] + r[1]/2, & m = 0 \\ 2d[m+1] + r[m+1], & 1 \le m \le M-2 \\ 2d[M], & m = M-1 \end{cases}.$$

So,

$$r[m] = \begin{cases} 2h[M-1] + r[1]/2, & m=0 \\ 4h[M-1-m] + r[m+1], & 1 \le m \le M-2\,. \\ 4h[0], & m = M-1 \end{cases}$$

➤Hint: Expand $A(\omega)$, and then substitute part (b) into each term starting at the end (i.e., $d[M]\sin(M-1/2)\omega$) and working backwards towards the first term.

Problem 7-7

(a) Show that an odd-length symmetric filter has the property that the amplitude function $A(\omega)$ is even about $\omega=\pi$; namely $A(\omega+\pi)=A(\omega-\pi)$.

(b) Show that an even-length symmetric filter has the property that the amplitude function $A(\omega)$ is odd about $\omega=\pi$; namely $A(\omega+\pi)=-A(\omega-\pi)$.

(c) Show that an odd-length antisymmetric filter has the property that the amplitude function $A(\omega)$ is even about $\omega=\pi$; namely $A(\omega+\pi)=A(\omega-\pi)$.

(d) Show that an even-length antisymmetric filter has the property that the amplitude function $A(\omega)$ is odd about $\omega=\pi$; namely $A(\omega+\pi)=-A(\omega-\pi)$.

Problem 7-8

For each of the following impulse responses, find and plot $|H(\omega)|$, $\measuredangle H(\omega)$, $A(\omega)$, $\measuredangle A(\omega)$ and $\measuredangle H(\omega)-\measuredangle A(\omega)$:

(a) Odd-length symmetric sequence: $h[n]=\delta[n]+\delta[n-1]+\delta[n-2]$.

(b) Even-length symmetric sequence: $h[n]=\delta[n]+\delta[n-1]+\delta[n-2]+\delta[n-3]$.

(c) Odd-length antisymmetric sequence: $h[n]=-\delta[n]+\delta[n-2]$.

(d) Even-length antisymmetric sequence: $h[n]=\delta[n]+\delta[n-1]-\delta[n-2]-\delta[n-3]$.

Problem 7-9

For a symmetric linear-phase filter (i.e., Type-I) of odd length N,

(a) show that $H(0)=0$ only if

$$\sum_{n=0}^{N-1} h[n] = 0,$$

or, equivalently,

$$\sum_{n=0}^{(N-1)/2-1} h[n] = -\tfrac{1}{2}h[(N-1)/2].$$

(b) show that $H(\pi)=0$ only if

$$\sum_{n=0}^{N-1} h[n](-1)^n = 0,$$

or, equivalently,

$$\sum_{n=0}^{(N-1)/2-1} h[n](-1)^n = -\tfrac{1}{2}(-1)^{(N-1)/2}h[(N-1)/2].$$

Problem 7-10

For a symmetric linear-phase filter (i.e., Type-II) of even length N,

(a) show that $H(0)=0$ only if

$$\sum_{n=0}^{N-1} h[n]=0,$$

or, equivalently,

$$\sum_{n=0}^{N/2} h[n]=0.$$

(b) show that $H(\pi)=0$.

Problem 7-11

For an antisymmetric linear-phase filter (i.e., Type-III) of odd length N, with impulse response $h[n]$,

(a) show that $H(0)=0$.

(b) show that $H(\pi)=0$.

Use the definition of the DTFT for $H(\omega)$ as your starting point.

Problem 7-12

For an antisymmetric linear-phase filter (i.e., Type-IV) of even length N, with impulse response $h[n]$,

(a) show that $H(0)=0$.

(b) show that $H(\pi)=0$, only if

$$\sum_{n=0}^{N-1} h[n](-1)^n=0,$$

or, equivalently,

$$\sum_{n=0}^{N/2} h[n](-1)^n=0.$$

Use the definition of the DTFT for $H(\omega)$ as your starting point.

Problem 7-13

For a symmetric linear-phase filter (i.e., Type-I) of odd length N, with $H(\omega)$ as defined in **Table 7.1**,

$$H(\omega) = e^{-j\omega(N-1)/2} A(\omega),$$

where $A(\omega)$ is the amplitude function

$$A(\omega) = \sum_{m=0}^{(N-1)/2} a[m] \cos m\omega,$$

(a) show that $A(\omega) = A(\omega + 2\pi)$, so that the amplitude function is periodic with period 2π.
(b) show that $A(\omega) = A(-\omega)$, so that the amplitude function is even.
(c) show that $A(\omega) = A(2\pi - \omega)$, or, equivalently, $A(\pi - \omega) = A(\pi + \omega)$, so that the amplitude function is symmetric about $\omega = \pi$.

Problem 7-14

For a symmetric linear-phase filter (i.e., Type-II) of even length N, with $H(\omega)$ as defined in **Table 7.1**,

$$H(\omega) = e^{-j\omega(N-1)/2} A(\omega),$$

where $A(\omega)$ is the amplitude function

$$A(\omega) = \sum_{m=1}^{N/2} b[m] \cos(m - 1/2)\omega,$$

(a) show that $A(\omega) = -A(\omega + 2\pi)$.
(b) show that $A(\omega) = A(\omega + 4\pi)$, so that the amplitude function is periodic with period 4π.
(c) show that $A(\omega) = A(-\omega)$, so that the amplitude function is even.
(d) show that $A(\omega) = -A(2\pi - \omega)$, or, equivalently, $A(\pi - \omega) = -A(\pi + \omega)$, so that the amplitude function is antisymmetric about $\omega = \pi$.
(e) show that $A(\pi) = 0$.

Problem 7-15

For an antisymmetric linear-phase filter (i.e., Type-III) of odd length N, with $H(\omega)$ as defined in **Table 7.1**,

$$H(\omega) = e^{-j(\omega(N-1)/2 - \pi/2)} A(\omega),$$

where $A(\omega)$ is the amplitude function

$$A(\omega) = \sum_{m=1}^{(N-1)/2} c[m] \sin m\omega,$$

(a) show that $A(\omega) = A(\omega + 2\pi)$.
(b) show that $A(\omega) = -A(-\omega)$, so that the amplitude function is odd.

(c) show that $A(\omega) = -A(2\pi - \omega)$, or, equivalently, $A(\pi - \omega) = -A(\pi + \omega)$, so that the amplitude function is antisymmetric about $\omega = \pi$.
(d) show that $A(0) = 0$.
(e) show that $A(\pi) = 0$.

Problem 7-16

For an antisymmetric linear-phase filter (i.e., Type-IV) of even length N, with $H(\omega)$ as defined in **Table 7.1**,

$$H(\omega) = e^{-j(\omega(N-1)/2 - \pi/2)} A(\omega),$$

where $A(\omega)$ is the amplitude function

$$A(\omega) = \sum_{m=1}^{N/2} d[m] \sin(m - 1/2)\omega,$$

(a) show that $A(\omega) = -A(\omega + 2\pi)$.
(b) show that $A(\omega) = A(\omega + 4\pi)$, so that the amplitude function is periodic with period, 4π.
(c) show that $A(\omega) = -A(-\omega)$, so that the amplitude function is odd.
(d) show that $A(\omega) = A(2\pi - \omega)$, or, equivalently, $A(\pi - \omega) = A(\pi + \omega)$, so that the amplitude function is symmetric about $\omega = \pi$.
(e) show that $A(0) = 0$.

Problem 7-17

Consider a symmetric linear-phase filter (i.e., Type-I) of odd length N, with $H(\omega)$ as defined in **Table 7.1**,

$$H(\omega) = e^{-j\omega(N-1)/2} A(\omega),$$

where $A(\omega)$ is the amplitude function

$$A(\omega) = \sum_{m=0}^{M} a[m] \cos m\omega,$$

and $M = (N-1)/2$. Choose N samples of frequency response $H(\omega)$ equally spaced between $\omega = 0$ and $\omega = 2\pi$,

$$H[k] \triangleq H(\omega)|_{\omega = 2\pi k/N}, \quad 0 \le k \le N-1.$$

(a) Use the properties of $A(\omega)$ derived in Problem 7.16 to show that $A[k] = A[N-k]$.
(b) Use the inverse DFT relation,

$$h[n] = \frac{1}{N} \sum_{k-0}^{N-1} H[k] e^{j2\pi kn/N},$$

and the result of part (a) to show that $h[n]$ can be derived from the samples of the amplitude function,

$$h[n] = \frac{1}{N}\left(A[0] + 2\sum_{k=1}^{(N-1)/2} A[k]\cos\left(\frac{\pi k(N-1-2n)}{N}\right)\right).$$

Problem 7-18

Consider a symmetric linear-phase filter (i.e., Type-II) of even length N, with $H(\omega)$ as defined in **Table 7.1**,

$$H(\omega) = e^{-j\omega(N-1)/2}A(\omega),$$

where $A(\omega)$ is the amplitude function

$$A(\omega) = \sum_{m=1}^{M} b[m]\cos(m-1/2)\omega,$$

and $M = N/2$. Choose N samples of frequency response $H(\omega)$ equally spaced between $\omega = 0$ and $\omega = 2\pi$,

$$H[k] \triangleq H(\omega)|_{\omega=2\pi k/N}, \quad 0 \le k \le N-1.$$

(a) Use the properties of $A(\omega)$ derived in Problem 7.16 to show that $A[k] = -A[N-k]$.
(b) Use the properties of $A(\omega)$ derived in Problem 7.16 to show that $A[N/2] = 0$.
(c) Use the inverse DFT relation,

$$h[n] = \frac{1}{N}\sum_{k=0}^{N-1} H[k]e^{j2\pi kn/N},$$

and the results of parts (a) and (b) to show that $h[n]$ can be derived from the samples of the amplitude function,

$$h[n] = \frac{1}{N}\left(A[0] + 2\sum_{k=1}^{(N-1)/2} A[k]\cos\left(\frac{\pi k(N-1-2n)}{N}\right)\right).$$

Problem 7-19

Consider an antisymmetric linear-phase filter (i.e., Type-III) of odd length N, with $H(\omega)$, defined in **Table 7.1**,

$$H(\omega) = e^{-j(\omega(N-1)/2-\pi/2)}A(\omega),$$

where $A(\omega)$ is the amplitude function

$$A(\omega) = \sum_{m=1}^{M} c[m]\sin m\omega,$$

and $M = (N-1)/2$. Choose N samples of frequency response $H(\omega)$ equally spaced between $\omega = 0$ and $\omega = 2\pi$,

$$H[k] \triangleq H(\omega)|_{\omega=2\pi k/N}, \quad 0 \le k \le N-1.$$

(a) Use the properties of $A(\omega)$ derived in Problem 7.16 to show that $A[k] = -A[N-k]$.
(b) Use the properties of $A(\omega)$ derived in Problem 7.16 to show that $A[0] = 0$.
(c) Use the inverse DFT relation,

$$h[n] = \frac{1}{N}\sum_{k=0}^{N-1} H[k]e^{j2\pi kn/N},$$

and the results of parts (a) and (b) to show that $h[n]$ can be derived from the samples of the amplitude function,

$$h[n] = \frac{2}{N}\sum_{k=1}^{(N-1)/2} A[k]\sin\left(\frac{\pi k(N-1-2n)}{N}\right).$$

Problem 7-20

Consider an antisymmetric linear-phase filter (i.e., Type-IV) of even length N, with $H(\omega)$ as defined in **Table 7.1**,

$$H(\omega) = e^{-j(\omega(N-1)/2-\pi/2)}A(\omega),$$

where $A(\omega)$ is the amplitude function

$$A(\omega) = \sum_{m=1}^{M} d[m]\sin(m-1/2)\omega,$$

and $M = N/2$. Choose N samples of frequency response $H(\omega)$ equally spaced between $\omega = 0$ and $\omega = 2\pi$,

$$H[k] \triangleq H(\omega)|_{\omega=2\pi k/N}, \quad 0 \le k \le N-1.$$

(a) Use the properties of $A(\omega)$ derived in Problem 7.16 to show that $A[k] = A[N-k]$.
(b) Use the properties of $A(\omega)$ derived in Problem 7.16 to show that $A[0] = 0$.
(c) Use the inverse DFT relation,

$$h[n] = \frac{1}{N}\sum_{k=0}^{N-1} H[k]e^{j2\pi kn/N},$$

and the results of parts (a) and (b) to show that $h[n]$, can be derived from the samples of the amplitude function,

$$h[n] = \frac{1}{N}\left(A[N/2]\sin\pi(N-1-2n)/2 + 2\sum_{k=1}^{N/2-1} A[k]\sin\left(\frac{\pi k(N-1-2n)}{N}\right)\right).$$

Problem 7-21

For each of the sequences $h[n]$ in **Figure 7.46**, determine whether $H(0) = 0$, and whether $H(\pi) = 0$.

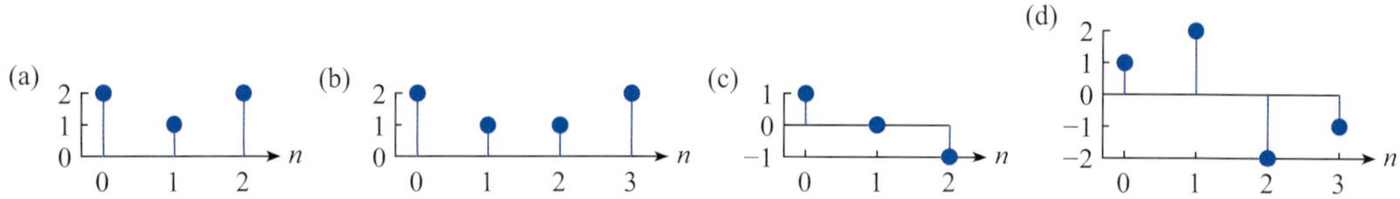

Figure 7.46

Problem 7-22

For each of the sequences $h[n]$ in **Figure 7.47**, determine whether $H(0) = 0$, and whether $H(\pi) = 0$.

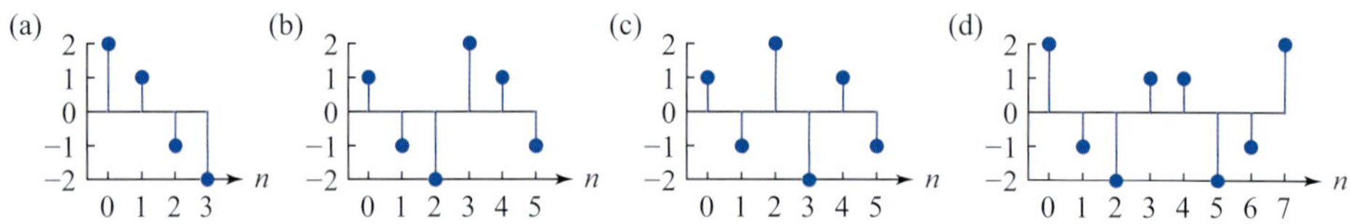

Figure 7.47

Problem 7-23

For each of the sequences $h[n]$ in **Figure 7.48**, determine whether $H(0) = 0$, and whether $H(\pi) = 0$.

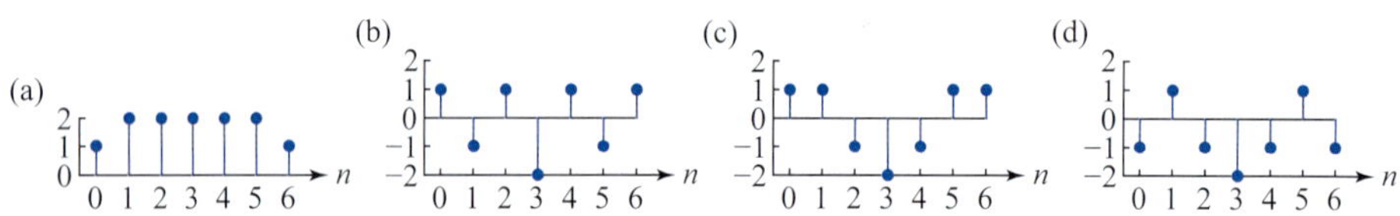

Figure 7.48

Problem 7-24

For each of the impulse responses $h[n]$ in Problem 7-21, express $H(\omega)$ in the form

$$H(\omega) = e^{-j(\omega(N-1)/2-\beta)}A(\omega),$$

where β and $A(\omega)$ are given by **Table 7.1**.

Problem 7-25

Repeat Problem 7-24 for each of the sequences in Problem 7-22.

Problem 7-26

Repeat Problem 7-24 for each of the sequences in Problem 7-23.

Problem 7-27

For each of the impulse responses $h[n]$ in Problem 7-21, express $H(\omega)$ in the form

$$H(\omega) = e^{-j(\omega(N-1)/2-\beta)}Q(\omega)A(\omega),$$

where β, $Q(\omega)$ and $A(\omega)$ are given by **Table 7.1**

Problem 7-28

Repeat Problem 7-27 for each of the sequences in Problem 7-22.

Problem 7-29

In Example 7.21, we used Matlab to design an optimal equiripple lowpass filter of Figure 7.58a with $\omega_p = 0.3150\pi$, $\omega_s = 0.4155\pi$ and $\delta_p = \delta_s = 0.1$. Check that the design meets the specifications, specifically that the values of δ_p and δ_s are correct.

Problem 7-30

Show that the ideal lowpass filter is unstable using the BIBO criterion. To do so, consider the output of the filter at time $n = 0$, when the input is $x[n] = \sin\omega_c n\, u[n]$, where ω_c is the cutoff frequency of the filter. Show that the resulting summation does not converge.

Problem 7-31

Figure 7.49 shows the frequency response of second-order spline filter, $H_2(\omega)$. This can be understood as the frequency-domain convolution of $H_0(\omega)$ with a triangular function $T(\omega)$ of width $\Delta\omega$ and area 2π,

$$H_2(\omega) = \frac{1}{2\pi}H_0(\omega) * T(\omega).$$

The figure also shows that $T(\omega)$ can be derived from the convolution of two identical rectangular functions of width $\Delta\omega/2$,

$$T(\omega) = \frac{1}{2\pi}R_2(\omega) * R_2(\omega),$$

where

$$R_2(\omega) = \begin{cases} \dfrac{4\pi}{\Delta\omega}, & -\Delta\omega/4 \le \omega \le \Delta\omega/4 \\ 0, & \text{otherwise} \end{cases}.$$

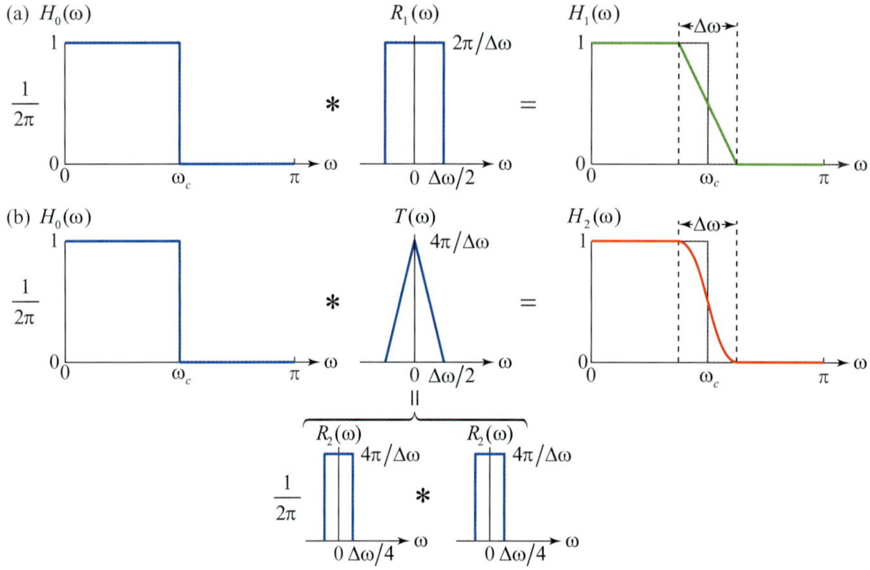

Figure 7.49

(a) Show that the impulse response of $T(\omega)$ is

$$t[n] = \mathrm{sinc}^2 \frac{\Delta\omega}{4} n,$$

and that the impulse response of the second-order spline filter is therefore

$$h_2[n] = h_0[n]r_2[n] = \left(\frac{\omega_c}{\pi} \mathrm{sinc}\, \omega_c n\right) \mathrm{sinc}^2 \frac{\Delta\omega}{2} n,$$

where $h_0[n]$ is the impulse response of $H_0(\omega)$.

(b) A general spline of order P can be constructed from the convolution of $H_0(\omega)$ with a function $W_P(\omega)$,

$$H_P(\omega) = \frac{1}{2\pi} H_0(\omega) * W_P(\omega),$$

where $W_P(\omega)$ is formed from the convolution of P identical rectangular functions,

$$W_P(\omega) = \frac{1}{2\pi} \underbrace{R_P(\omega) * \cdots * R_P(\omega)}_{P \text{ terms}}.$$

Find $R_P(\omega)$.

(c) Show that the impulse response of a pth-order spline filter is

$$h_P[n] = \left(\frac{\omega_c}{\pi} \mathrm{sinc}\, \omega_c n\right) \mathrm{sinc}^P \frac{\Delta\omega}{2P} n.$$

(d) Show that

$$H_P(\omega)=\begin{cases}1, & |\omega|<\omega_c-\Delta\omega/2\\ 1-\dfrac{1}{2}\left(1+\dfrac{2(\omega-\omega_c)}{\Delta\omega}\right)^{P+1}, & \omega_c-\Delta\omega/2<|\omega|<\omega_c\\ \dfrac{1}{2}\left(1-\dfrac{2(\omega-\omega_c)}{\Delta\omega}\right)^{P+1}, & \omega_c<|\omega|<\omega_c+\Delta\omega/2\\ 0, & |\omega|>\omega_c+\Delta\omega/2\end{cases}.$$

$$R_2(\omega)=\begin{cases}\dfrac{4\pi}{\Delta\omega}, & -\Delta\omega/4\le\omega\le\Delta\omega/4\\ 0, & \text{otherwise}\end{cases}.$$

Problem 7-32

A spline filter of order P is made by joining two endpoints with a continuous polynomial function of power P. Show that the P derivatives of the filter are 0 at the endpoints. To simplify matters, consider a continuous polynomial in ω of order P that has a transition at $\omega=0$,

$$S_P(\omega)=\omega^P u(\omega).$$

(a) Show that the first derivative of $S_P(\omega)$ is 0 at $\omega=0$ as long as $P\ge 1$.
(b) Show that the second derivative is 0 at $\omega=0$ as long as $P\ge 2$.
(c) Show that the third derivative is 0 at $\omega=0$ as long as $P\ge 3$.
(d) By inspection, argue that the nth derivative of $S_P(\omega)$ can be written as

$$\frac{d^n S_P(\omega)}{d\omega^n}=\sum_{k=0}^{n}\binom{n}{k}\frac{P!}{(P-(n-k))!}\omega^{P-(n-k)}\frac{d^k u(\omega)}{d\omega^k},$$

which will be 0 at $\omega=0$ as long as $P\ge n$.

Problem 7-33

The frequency response of a raised-cosine filter, $H(\omega)$, is created by the frequency-domain convolution of the response of the ideal lowpass filter $H_0(\omega)$ with cutoff ω_c, and a cosine filter $R(\omega)$ of width $\Delta\omega$:

$$H(\omega)=\frac{1}{2\pi}H_0(\omega)*R(\omega),$$

where

$$H_0(\omega)=\begin{cases}1, & |\omega|<\omega_c\\ 0, & |\omega|>\omega_c\end{cases},$$

and

$$R(\omega)=\begin{cases}\dfrac{\pi^2}{\Delta\omega}\cos\pi\dfrac{\omega}{\Delta\omega}, & -\Delta\omega/2<\omega<\Delta\omega/2\\ 0, & |\omega|>\Delta\omega/2\end{cases}.$$

(a) Show that

$$H(\omega)=\begin{cases}1, & |\omega|<\omega_c-\Delta\omega/2\\ \dfrac{1}{2}-\dfrac{1}{2}\sin\dfrac{\pi(\omega-\omega_c)}{\Delta\omega}, & \omega_c-\Delta\omega/2<|\omega|<\omega_c+\Delta\omega/2\\ 0, & |\omega|>\omega_c+\Delta\omega/2\end{cases}.$$

(b) Show that

$$r[n]=\frac{\pi}{4}(\operatorname{sinc}(n\Delta\omega/2+\pi/2)+\operatorname{sinc}(n\Delta\omega/2-\pi/2)).$$

(c) Show that an alternative form of $r[n]$ is

$$r[n]=\frac{1}{2\pi}\int_{-\pi}^{\pi}R(\omega)e^{j\omega n}d\omega=\begin{cases}\dfrac{\cos(n\Delta\omega/2)}{1-(n\Delta\omega/\pi)^2}, & n\neq\pm\pi/4\\ \dfrac{\pi}{4}, & n=\pm\pi/4\end{cases}.$$

(d) Show that the impulse response of the raised-cosine filter is

$$h[n]=\left(\frac{\omega_c}{4}\operatorname{sinc}\omega_c n\right)(\operatorname{sinc}(n\Delta\omega/2+\pi/2)+\operatorname{sinc}(n\Delta\omega/2-\pi/2)).$$

Problem 7-34

Show that the formula for the frequency response of the raised-cosine filter, $H(\omega)$, in Equation (7.17) is equivalent to the following formula found in the literature for $\omega>0$:

$$H(\omega)=\begin{cases}1, & \omega<\omega_c(1-\alpha)\\ \dfrac{1}{2}+\dfrac{1}{2}\cos\dfrac{\pi(\omega-\omega_c(1-\alpha))}{2\alpha\omega_c}, & \omega_c(1-\alpha)<\omega<\omega_c(1+\alpha)\\ 0, & \omega>\omega_c(1+\alpha)\end{cases},$$

where $\alpha=\Delta\omega/2\omega_c$ is called the **roll-off**.

Problem 7-35

Show that the impulse response of an ideal Hilbert transformer is

$$h_I[n] = \frac{1}{\pi n}(1 - \cos \pi n) = \frac{\pi n}{2}\mathrm{sinc}^2 n\pi/2.$$

Problem 7-36

Show that the impulse response of an ideal differentiator is

$$h[n] = \begin{cases} \frac{1}{n}(-1)^n, & n \neq 0 \\ 0, & n = 0 \end{cases}.$$

➤Hint: Use integration by parts to show that

$$h[n] = \frac{1}{2\pi}\int_{-\pi}^{\pi} u\,dv = \frac{1}{n}\cos \pi n - \frac{1}{\pi n^2}\sin \pi n, \tag{7.47}$$

where $u = j\omega$ and $dv = e^{j\omega n}d\omega$.

Problem 7-37

Show that the impulse response of a causal differentiator of length N is

$$h[n] = \begin{cases} \frac{(-1)^{n-(N-1)/2}}{(n - (N-1)/2)}, & N \text{ odd}, n \neq (N-1)/2 \\ 0, & N \text{ odd}, n = (N-1)/2 \\ \frac{(-1)^{n-N/2+1}}{\pi(n - (N-1)/2)^2}, & N \text{ even} \end{cases}.$$

➤Hint: Start with Equation (7.47) and substitute $n - (N - 1)/2$ for n.

Problem 7-38

Show that the frequency response of the Hilbert transformer goes to zero, $H_I(\omega) = 0$, at $\omega = 0$ and $\omega = \pm\pi$.

8 Infinite impulse response filters

Introduction

Infinite impulse response (IIR) filters are the second major class of discrete-time filters. These filters have a number of features that make them attractive to designers. The chief advantage of IIR filters is that an IIR filter designed to meet a given set of specifications (e.g., passband or stopband attenuation) generally has a lower order than an equivalently specified FIR filter. This means that IIR filters require fewer computational operations (multiplications and additions) and less memory to implement. IIR filters are also "faster" in that they have lower latency than FIR filters of equivalent specifications. However, IIR filter design is not suitable for every application. For example, unlike FIR filters, realizable IIR filters inherently have nonlinear phase. Also, unlike FIR filters, IIR filters can be unstable, and careful attention must be paid in implementation to avoid the effects of coefficient quantization and noise, topics we will cover in Chapter 9. However, IIR filters have proven their usefulness in many applications and are well worth our study.

There are several approaches to IIR filter design. The most common approach, which is the one covered in this chapter, is to base the design upon classical analog filter prototypes such as Butterworth, Chebyshev and elliptic filters. In this approach, the designer maps the specifications of the desired discrete-time filter – for example, the passband and stopband frequencies and attenuations – onto the parameters of an analog filter. Classical methods are generally limited to the design of lowpass, highpass, bandpass and bandstop filters. A second approach to IIR filter design, which is more advanced than we can cover here, uses an iterative parametric approach to design filters with arbitrary frequency responses.

Since we will be designing discrete-time IIR filters based on analog filter prototypes, we will first need to have a fairly thorough understanding of analog filter design. Accordingly, in Section 8.2 we will discuss the design of several commonly used types of analog lowpass filters. If you read only one part of this section, concentrate on Section 8.2.2, the design of the Butterworth filter. Although the Butterworth filter does not provide the lowest order filter for a given set of specifications, it is the simplest filter to understand, and the methods described in that section provide a good overview of the general approach we shall use to design all the analog filter prototypes covered in this chapter. In Sections 8.3 and 8.4, we show how to map the parameters of the desired discrete-time filter into the parameters of the analog prototype. There are two methods of doing this mapping, but the most useful is the bilinear transformation method discussed in Section 8.4. So again, if you have only limited time, that is the section you should read. Then, in Section 8.5, we will show how to transform a discrete-time lowpass filter prototype filter into highpass, bandpass and bandstop filters. As we mentioned, IIR filters inherently have nonlinear phase, but in Section 8.6 we will show how to design zero-phase IIR (and FIR) filters for non-real-time applications.

8.1 Definition of the IIR filter

Causal IIR filters are defined in the time domain by the general difference equation of the form

$$\sum_{k=0}^{N} a_k y[n-k] = \sum_{k=0}^{M} b_k x[n-k], \tag{8.1}$$

where $a_k \neq 0$ for at least some $k > 0$. Without loss of generality, let $a_0 = 1$ (which is equivalent to normalizing both sides of this equation by a_0). Then, reorganize Equation (8.1) to give

$$y[n] = \sum_{k=0}^{M} b_k x[n-k] - \sum_{k=1}^{N} a_k y[n-k]. \tag{8.2}$$

For an FIR filter, $a_k = 0$, $k > 0$, so the output $y[n]$ depends only upon scaled and shifted values of previous inputs, $x[n-k]$:

$$y[n] = \sum_{k=0}^{M} b_k x[n-k].$$

You will recognize this as nothing more than the convolution sum, where the coefficients b_k represent the impulse response of the filter. In contrast, the output of the general difference equation of Equation (8.2) depends both upon $x[n-k]$ and upon the scaled and shifted values of previous outputs, $y[n-k]$.

The z-transform of the difference equation, Equation (8.2), results in the transfer function of the IIR filter expressed as the ratio of two polynomials in z,

$$H(z) = \frac{Y(z)}{X(z)} = \frac{\sum_{k=0}^{M} b_k z^{-k}}{\sum_{k=0}^{N} a_k z^{-k}} = C \frac{\prod_{k=1}^{\max(M,N)} (z - z_k)}{\prod_{k=1}^{\max(M,N)} (z - p_k)},$$

where z_k are the zeros, p_k are the poles and C is a constant. As we discussed in Chapter 4, the poles of FIR filters, if they exist, can only occur at $z = 0$. In contrast, IIR filters have poles at non-zero locations in the z-plane, and it is the position of those poles that give IIR filters their essential qualities. It is also those non-zero poles that can lead to trouble. Unlike FIR filters, which are guaranteed to be stable, causal IIR filters can be unstable if their poles are located outside (or on) the unit circle. Furthermore, unlike FIR filters, which can be designed to have linear phase, the phase of a causal IIR filter is guaranteed to be nonlinear.

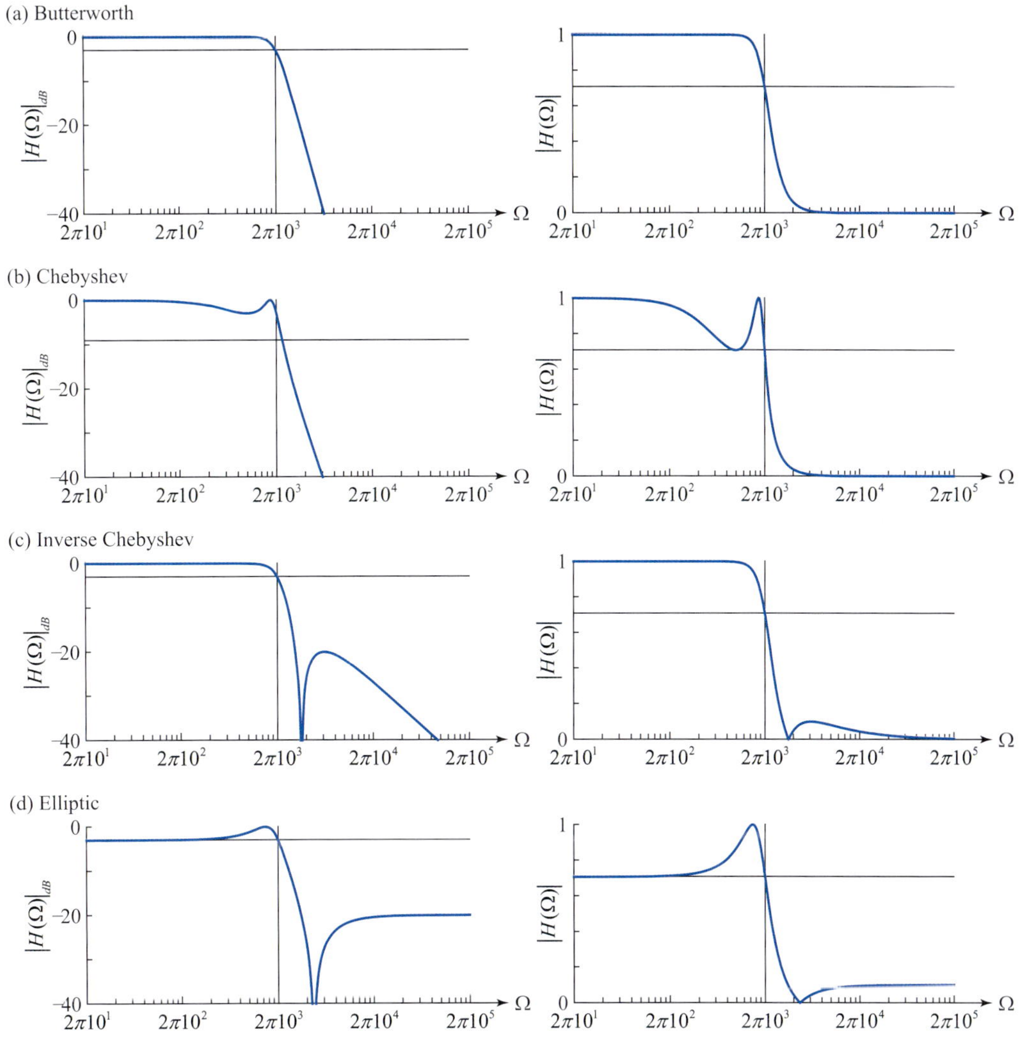

Figure 8.1 Butterworth, Chebyshev, Inverse Chebyshev and elliptic lowpass filters

To see why that is so, recall from Section 4.3 that the z-transform of a linear-phase filter must satisfy the relation

$$H(z) = z^{-(N-1)} H(z^{-1}), \tag{8.3}$$

meaning that the zeros of $H(z)$ must occur at conjugate-reciprocal positions in the z-plane. In particular, if there are zeros inside the unit circle, there will be an equal number outside the unit circle. In order for an IIR filter to be linear-phase and satisfy Equation (8.3), the poles of $H(z)$ inside the unit circle would have to have corresponding poles at conjugate-reciprocal positions outside the unit circle, which would make a causal filter unstable.

8.2 Overview of analog filter design

Analog filter design is a well-established art and many different designs of filters are in common use. **Figure 8.1** shows the response magnitude of lowpass filters designed from four of the most common analog prototypes – Butterworth, Chebyshev, inverse Chebyshev and elliptic filters – all plotted on a log-amplitude (dB) scale (left column) and a linear amplitude scale (right column). Other common filter prototypes exist, including Gaussian and Bessel filters, but the ones we have shown here are pretty representative of the types of filters used in many applications.

Each filter prototype has a characteristic magnitude and phase response that has its own particular advantages and disadvantages, as we will discuss below. Though the frequency response magnitude $|H(\Omega)|$ of each filter looks fairly complex, in each case it can be described by a function of frequency that is completely determined by a small number of free parameters. The filter designer's job is to determine the values of these parameters based on a detailed specification of what the filter is supposed to do; that is, what range of frequencies is the filter supposed to accentuate or attenuate and by how much. Our IIR designs will be based on specifying only the magnitude parameters of the desired discrete-time filter, $|H(\omega)|$, and matching them as best we can to the magnitude of the chosen analog filter, $|H(\Omega)|$. The phase of the filter will follow from the choice of filter, although we can subsequently modify the phase using an allpass filter.

8.2.1 Parameter definitions

The design of a practical analog filter starts with the selection of a class of filter – lowpass, bandpass or highpass – and a specification of important frequency and attenuation parameters of the filter. Initially, we will restrict our discussion to the design of lowpass filters, which are characterized by a **passband** and a **stopband**. The passband and stopband are defined by four parameters: the **passband frequency** Ω_p, **passband attenuation** δ_p, **stopband frequency** Ω_s and **stopband attenuation** δ_s, where δ_p and δ_s are measured here on a dB scale.[1] For example, the four prototype filters of **Figure 8.1** were all designed to have the same passband and stopband frequencies and attenuations. To illustrate this, the four responses have been plotted together in **Figure 8.2** with the passband and stopband parameters shown on a dB scale.

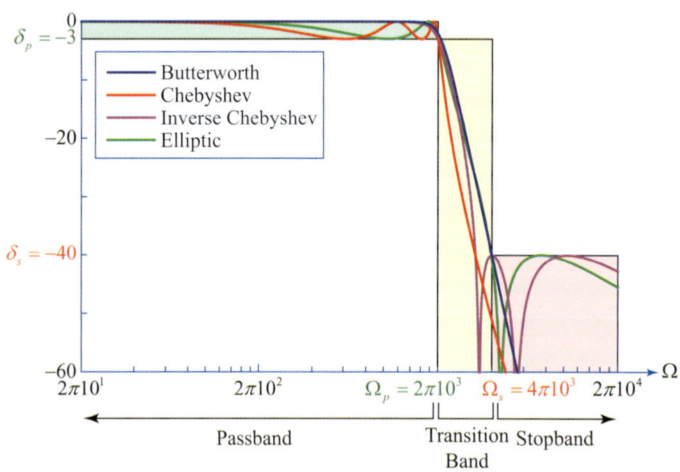

Figure 8.2 Schematic lowpass filters

[1] This is an opportune place to remind you of some notation: we use Ω for continuous-time frequency and ω for discrete-time frequency.

Passband The passband of the filter is defined as the frequency region below the passband frequency, $\Omega < \Omega_p$. In the passband, the magnitude of the filter's frequency response is greater than the passband attenuation,

$$|H(\Omega)|_{dB} \geq \delta_p, \quad |\Omega| \leq \Omega_p.$$

The four prototype filters were all designed to have a passband frequency of $\Omega_c = 2\pi \cdot 1000$ and a passband attenuation of $\delta_p = -3$ dB. Graphically, the passband is depicted by the green box in **Figure 8.2**. You can see that the response magnitudes of all filters lie within the box, with magnitudes ranging between −3 dB and 0 dB in the frequency range below 1 kHz. However, the shape of each particular response is different; the responses of the Butterworth and inverse Chebyshev filters are monotonically decreasing in the passband, whereas the responses of the Chebyshev and elliptic filters exhibit a ripple in the passband that is bounded by 0 dB on the high side and δ_p dB on the low side. In our designs, we will guarantee that filter's response exactly meets the passband specification δ_p at the passband frequency Ω_p. This means that the response curve will pass through the bottom right corner of the green box, at the point $|H(\Omega_p)|_{dB} = \delta_p$.

Stopband The stopband of the filter is the frequency region above the stopband frequency, $\Omega > \Omega_s$. In the stopband, the magnitude of the frequency response of the filter is guaranteed to be less than the stopband attenuation,

$$|H(\Omega)|_{dB} \leq \delta_s, \quad |\Omega| \geq \Omega_s.$$

The filters whose responses are shown in **Figure 8.2** were designed to have a stopband of $\Omega_s = 2\pi \cdot 2000$ and a stopband attenuation of $\delta_s = -40$ dB. The stopband is depicted by a red box in the figure. You can see that the response magnitudes of all filters lie within the box, with magnitudes below −40 dB in the frequency range above 2 kHz. However, the shape of each particular response is different: the responses of the Butterworth and Chebyshev filters are monotonically decreasing in the stopband, whereas the responses of the inverse Chebyshev and elliptic filters exhibit ripple in the stopband that is bounded by δ_s dB on the high side.

Transition band The frequency range between the passband and stopband frequencies, $\Omega_p < \Omega < \Omega_s$, is termed the **transition band**. The shape of the frequency response is not guaranteed to have any particular shape in the transition band. All we shall require in our filter designs is that the response magnitude stays in the region depicted in the yellow box in **Figure 8.2**, $-\infty \leq |H(\Omega)|_{dB} \leq \delta_p$, $\Omega_p \leq |\Omega| \leq \Omega_s$.[2]

Once the desired passband and stopband frequencies and attenuations have been selected, the filter will be designed by (1) choosing a filter prototype, (2) mapping the passband and stopband frequencies and attenuations that have been selected into the free parameters of the chosen filter and (3) finding the poles and zeros of the filter. We will now demonstrate this process by

[2]An alternative approach is to permit the transition band to include the range of response magnitudes, $-\infty \leq |H(\Omega)|_{dB} \leq 0$, $\Omega_p \leq |\ \Omega\ | \leq \Omega_s$. However, the consequence is that the effective passband will be wider than $0 \leq \Omega < \Omega_s$. That is, the response will no longer exactly pass through the bottom right corner of the green box at the point $|H(\Omega_p)|_{dB} = \delta_p$.

designing lowpass filters of each of the four prototypes to meet the passband and stopband specifications shown in **Figure 8.2**.

8.2.2 Butterworth filter

Filter definition The Butterworth filter is defined by the magnitude of the Fourier transform,

$$|H(\Omega)| = \frac{1}{\sqrt{1+(\Omega/\Omega_c)^{2N}}}. \tag{8.4}$$

The magnitude depends on only two parameters: the **filter order** N and the **cutoff frequency** Ω_c. Expressed on a decibel (dB) scale, the response magnitude of the Butterworth filter is

$$|H(\Omega)|_{dB} \triangleq 20\log_{10}|H(\Omega)| = 20\log_{10}\frac{1}{\sqrt{1+(\Omega/\Omega_c)^{2N}}} = -10\log_{10}\left(1+(\Omega/\Omega_c)^{2N}\right). \tag{8.5}$$

The distinguishing characteristics of the Butterworth filter are that the magnitude of the frequency response is unity at $\Omega=0$ (i.e., $|H(0)|=1$, or $|H(0)|_{dB}=0$ dB) and decreases monotonically with increasing frequency Ω for any value of N. For this reason, we say that the Butterworth is monotonic in both the passband and the stopband. Of all the common filter types, the Butterworth is the one with the flattest magnitude response. The response is termed **maximally flat**, meaning that the slope of the frequency response – in fact the first $2N-1$ derivatives of the frequency response – are zero at $\Omega=0$ (see Problem 8-2). The magnitude of the response is also closest to 0 dB in the passband in the least-square sense.

Mapping passband and stopband specifications to filter parameters **Figure 8.3** shows how the response magnitudes of the various Butterworth filter prototypes depend upon the filter's two free parameters: the order N and the cutoff (or corner) frequency Ω_c. **Figure 8.3a** shows the effect of increasing N while keeping Ω_c constant. For clarity of display, **Figure 8.3b** shows the same data on expanded frequency and amplitude scales. The frequency axes of both plots have been normalized to the cutoff frequency Ω_c, which is equivalent to setting $\Omega_c=1$. The first thing to note is that at the frequency $\Omega=\Omega_c$, the response magnitude is down by a factor of $\sqrt{2}$ from its peak value, independent of the value of N,

$$|H(\Omega_c)| = \frac{1}{\sqrt{1+(\Omega_c/\Omega_c)^{2N}}} = \frac{1}{\sqrt{2}},$$

or $|H(\Omega_c)|_{dB} = -10\log_{10}2 = -3.01$ dB. For this reason, the cutoff frequency is also termed the **−3-dB frequency** or **−3-dB point** of the filter. For a Butterworth filter, the cutoff frequency of the filter is also the frequency at which the output power is one half the input power (Problem 8-1). For this reason, the cutoff frequency is also called the **half-power frequency** or **half-power point** of the filter. Above the cutoff frequency, the slope of the response gets steeper as N increases.

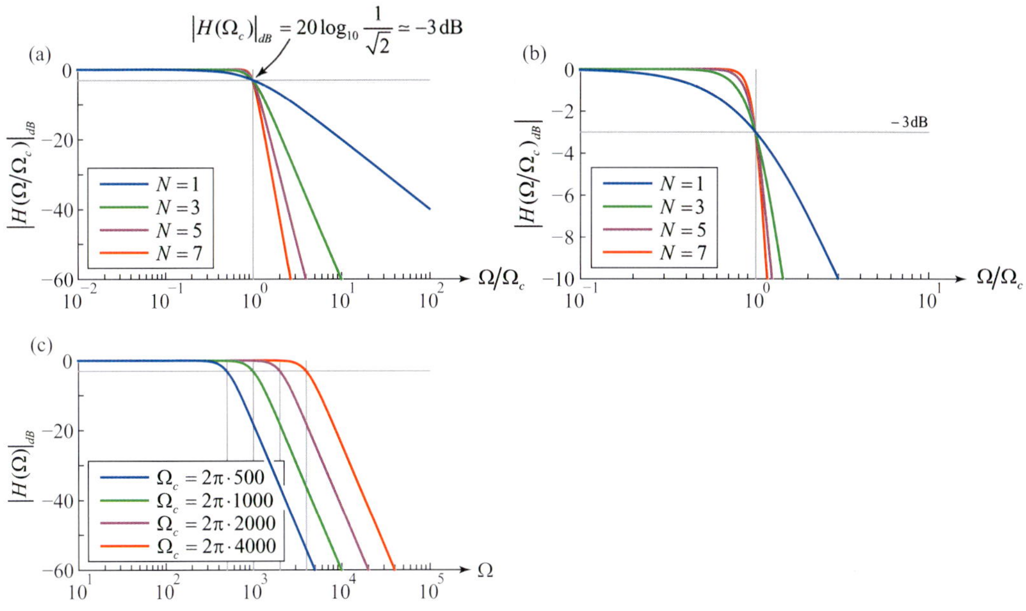

Figure 8.3 Dependence of the Butterworth filter on the filter parameters N and Ω_c

Figure 8.3c illustrates the effect of increasing the cutoff frequency, while keeping the order constant ($N = 5$). When the frequency response is plotted on a log–log scale, the effect of increasing Ω_c is to translate the entire plot to the right, while the steepness of the cutoff slope remains constant.

Given an understanding of how the frequency response of the Butterworth filter depends upon N and Ω_c, our task is now to determine values of N and Ω_c that best correspond to the passband and stopband specifications of a desired filter. The specifications require that the response magnitude be greater than or equal to δ_p dB in the passband ($|H(\Omega)|_{dB} \geq \delta_p$, $|\Omega| \leq \Omega_p$) and less than or equal to δ_s dB in the stopband ($|H(\Omega)|_{dB} \leq \delta_p$, $|\Omega| \geq \Omega_s$). Because the magnitude of the Butterworth filter's response decreases monotonically with increasing frequency, the lowest response magnitude within the passband will occur at the right edge of the passband, namely $\Omega = \Omega_p$. We will require that the magnitude of the response exactly meets the passband specifications, $|H(\Omega_p)|_{dB} = \delta_p$, or, from Equation (8.5),

$$-10\log_{10}(1 + (\Omega_p/\Omega_c)^{2N}) = \delta_p.$$

Similarly, the highest magnitude of the Butterworth filter's response in the stopband will occur at the band's lowest frequency, namely $\Omega = \Omega_s$. So, we require that $|H(\Omega_s)|_{dB} \leq \delta_s$, or

$$-10\log_{10}(1 + (\Omega_s/\Omega_c)^{2N}) \leq \delta_s.$$

Turning the algebra crank, we get two equations,

$$(\Omega_p/\Omega_c)^N = \underbrace{\sqrt{10^{-\delta_p/10} - 1}}_{\varepsilon_p}, \tag{8.6a}$$

and

$$(\Omega_s/\Omega_c)^N \geq \underbrace{\sqrt{10^{-\delta_s/10}-1}}_{\varepsilon_s}, \tag{8.6b}$$

where we have defined two constants that we will use extensively going forward, the **passband ripple**

$$\varepsilon_p \triangleq \sqrt{10^{-\delta_p/10}-1} \tag{8.7a}$$

and the **stopband ripple**

$$\varepsilon_s \triangleq \sqrt{10^{-\delta_s/10}-1}. \tag{8.7b}$$

Dividing Equations (8.6a) and (8.6b) gives

$$\underbrace{(\Omega_s/\Omega_p)^N}_{\xi} \geq \underbrace{\sqrt{\frac{10^{-\delta_s/10}-1}{10^{-\delta_p/10}-1}}}_{L}, \tag{8.8}$$

where we have defined two further constants, the **selectivity factor**

$$\xi \triangleq \Omega_s/\Omega_p \tag{8.9a}$$

and the **discrimination factor**

$$L \triangleq \frac{\varepsilon_s}{\varepsilon_p} = \sqrt{\frac{10^{-\delta_s/10}-1}{10^{-\delta_p/10}-1}}. \tag{8.9b}$$

The selectivity factor is a measure of the width of the transition band. The discrimination factor is a measure of the relative attenuation in the passband and stopband; L gets larger as the passband attenuation decreases (i.e., moves towards 0 dB) or the stopband attenuation increases (i.e., moves towards $-\infty$ dB). With these definitions, Equation (8.8) becomes

$$\xi^N \geq L,$$

which can be solved for N,

$$N = \left\lceil \frac{\log_{10} L}{\log_{10} \xi} \right\rceil = \left\lceil \frac{\log_{10}(\varepsilon_s/\varepsilon_p)}{\log_{10}(\Omega_s/\Omega_p)} \right\rceil = \left\lceil \frac{\log_{10}\sqrt{\frac{10^{-\delta_s/10}-1}{10^{-\delta_p/10}-1}}}{\log_{10}(\Omega_s/\Omega_p)} \right\rceil. \tag{8.10}$$

This is the **degree equation** that links all the important parameters of filter design. The ceiling brackets $\lceil \cdot \rceil$ indicate that N must be rounded up to the nearest integer, since the order of the filter has to be an integer. Note that N is inversely proportional to the log of the selectivity,

$$\log_{10}\xi = \log_{10}(\Omega_s/\Omega_p) = \log_{10}\Omega_s - \log_{10}\Omega_p,$$

which is just the width of the transition band expressed on a log scale. The sharper the filter (i.e., the narrower the transition band), the higher the filter order that is required. Once we have specified N, solving for Ω_c is just a matter of substituting N back into Equation (8.6a),

$$\Omega_c = \Omega_p \varepsilon_p{}^{-1/N} = \frac{\Omega_p}{\left(10^{-\delta_p/10} - 1\right)^{1/2N}}. \tag{8.11}$$

This value of Ω_c will ensure that the designed filter exactly meets the specifications of the passband,

$$|H(\Omega_p)|_{dB} = \delta_p,$$

so that the filter's response passes through the bottom right corner of the green box in **Figure 8.2**. We can then substitute Equation (8.11) into Equation (8.4) to arrive at a relation that defines the Butterworth filter in terms of Ω_p instead of Ω_c,

$$|H(\Omega)| = \frac{1}{\sqrt{1 + (\Omega/\Omega_c)^{2N}}} = \frac{1}{\sqrt{1 + \varepsilon_p^2(\Omega/\Omega_p)^{2N}}}. \tag{8.12}$$

Expressed this way, you can see that the passband ripple ε_p determines the maximum amount of attenuation in the passband, which occurs at frequency Ω_p,

$$|H(\Omega_p)| = \frac{1}{\sqrt{1 + \varepsilon_p^2}}.$$

At the stopband frequency Ω_s, the magnitude of the response is

$$|H(\Omega_s)| = \frac{1}{\sqrt{1 + (\Omega_s/\Omega_c)^{2N}}} = \frac{1}{\sqrt{1 + \varepsilon_p^2(\Omega_s/\Omega_p)^{2N}}} = \frac{1}{\sqrt{1 + \varepsilon_p^2 \xi^{2N}}}. \tag{8.13}$$

To appreciate this result, return for a moment to the degree equation, Equation (8.10). Once N is rounded up to an integer value, the degree equation can only be satisfied with equality if one or more parameters on the left side are altered. For example, if the passband parameters Ω_p and δ_p and the stopband frequency Ω_s are fixed, then the filter we design will exceed the stopband specifications, that is, the filter will have an actual, measured stopband attenuation of $\hat{\delta}_s = |H(\Omega_s)|$, given by Equation (8.13), that is lower (i.e., more negative) than δ_s. In a similar vein, we might ask what is the revised stopband frequency $\hat{\Omega}_s < \Omega_s$ at which the response *exactly* meets the stopband specification. Solving the degree equation with equality gives us a revised selectivity factor for this filter,

$$\hat{\xi} \triangleq \hat{\Omega}_s/\Omega_p = L^{1/N}, \tag{8.14a}$$

from which we can calculate the revised stopband frequency $\hat{\Omega}_s$,

$$\hat{\Omega}_s \triangleq \Omega_p L^{1/N}. \tag{8.14b}$$

This means that if we were to design a filter with passband frequency Ω_p and revised stopband frequency $\hat{\Omega}_s$, it would have an integer order and *exactly* meet both the passband and stopband attenuations; the response would pass through both the bottom right corner of the green box and the top left corner of the red box in **Figure 8.2**. In subsequent sections we will show how these same four parameters – ε_p, ε_s, ξ and L – can be used to characterize the other filters.

Example 8.1

(a) Design the Butterworth lowpass filter shown in **Figure 8.1**, whose passband is given by $\Omega_p = 2\pi \cdot 1000$ and $\delta_p = -3$ dB and whose stopband is given by $\Omega_s = 2\pi \cdot 2000$ with $\delta_s = -40$ dB.
(b) Determine the attenuation of the filter at the stopband frequency Ω_s.
(c) Determine the revised stopband frequency $\hat{\Omega}_s$ at which $|H(\hat{\Omega}_s)|_{dB} = \delta_s$.

➤ Solution:

(a) We can specify the filter in terms of Equation (8.4), which requires N and Ω_c, or in terms of Equation (8.12), which requires N and ε_p. To compute N, plug the specifications into Equation (8.10),

$$N \geq \frac{\log_{10}\sqrt{\dfrac{10^4 - 1}{10^{0.3} - 1}}}{\log_{10}(\Omega_s/\Omega_p)} = 6.647.$$

Since the order must be an integer, we round up to get $N = 7$. To solve for Ω_c, substitute into Equation (8.11),

$$\Omega_c = 2\pi \cdot 1000 \cdot \varepsilon_p^{-1/7} = 2\pi \cdot 1000.34,$$

where

$$\varepsilon_p = \sqrt{10^{0.3} - 1} = 0.9976.$$

Here is some code that calculates N and Ω_c:

```
Wp = 2*pi*1000;
Ws = 2*pi*2000;
dp = -3;
ds = -40;
ep = sqrt(10^-(dp/10)-1);
es = sqrt(10^-(ds/10)-1);
N = ceil(log10(es/ep) / log10(Ws / Wp));
Wc = Wp * ep^(-1/N); % Note: Matlab uses Wc = Ws * es^(-1/N)
```

The response of the resulting filter, shown in **Figure 8.4**, exactly passes through δ_p at frequency Ω_p; that is, $|H(\Omega_p)| = -3$ dB.

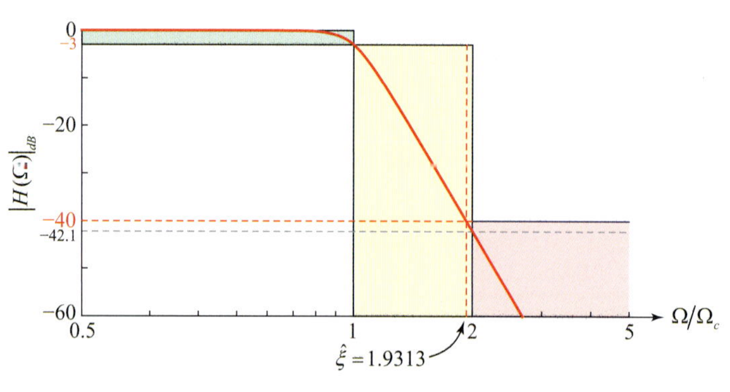

Figure 8.4 Analog Butterworth filter design example

(b) From Equation (8.12), the attenuation of the filter at the stopband frequency is

$$|H(\Omega_s)| = \frac{1}{\sqrt{1+\varepsilon_p^2(\Omega_s/\Omega_p)^{2N}}} = \frac{1}{\sqrt{1+\varepsilon_p^2\xi^{2N}}},$$

where $\xi = (\Omega_s/\Omega_p) = 2$. So,

$$|H(\Omega_s)|_{dB} = -10\log_{10}(1+\varepsilon_p^2\xi^{2N}) = -42.12\text{dB}.$$

This exceeds the desired stopband attenuation by a couple of dB, which is fine.

(c) Equation (8.14) gives $\hat{\Omega}_s = 2\pi\cdot 1931.34$ and $\hat{\xi} = 1.9313$, as shown by the dashed vertical red line. At that frequency, the stopband attenuation is exactly −40 dB, as shown by the dashed horizontal red line.

The Matlab function `buttord` is designed to find N and Ω_c. The syntax used in designing analog filters is

```
[N, Wc] = buttord(Wp, Ws, Rp, Rs, 's'),
```

where `Rp = -dp` and `Rs = -ds`. However, rather than using Equation (8.11) to calculate a value of $\Omega_c = 2\pi\cdot 1000.34$ such that $|H(\Omega_p)|$ meets the *passband* specifications exactly, as we have done, Matlab's function calculates a value of $\Omega_c = 2\pi\cdot 1035.90$ that meets the *stopband* specifications exactly, using

$$\Omega_c = \Omega_s \varepsilon_s^{-1/N}. \quad \textbf{(8.15)}$$

As a consequence, the filter's response magnitude at the passband frequency is $|H(\Omega_p)| = -2.07$ dB, which is above our specified values of δ_p. That means that the filter's response in the transition zone is no longer fully contained in the green box of **Figure 8.2**. For this filter, the passband frequency is $\Omega_p = \Omega_c = 2\pi\cdot 1035.90$, which exceeds the original design specifications.

The poles of the Butterworth filter The next step in analog filter design is to determine the filter's Laplace transform $H(s)$. Given our values of N and Ω_c, the poles and zeros of $H(s)$ can now be determined. The poles are the values of s in the finite s-plane at which $H(s)$ is infinite; the zeros are the values of s at which $H(s)$ is 0. We have defined the Butterworth filter only in terms of the magnitude of the Fourier transform, $|H(\Omega)|$, as given by Equation (8.4), but there is a simple relation between the Fourier transform and Laplace transform: the Fourier transform is the Laplace transform evaluated at values of s along the $j\Omega$-axis, namely $s = j\Omega$,

$$H(\Omega) = H(s)|_{s=j\Omega}.$$

To see the relation between $|H(\Omega)|$ and $H(s)$, write the magnitude-squared of the Fourier transform,

$$|H(\Omega)|^2 = H(\Omega)H^*(\Omega) = H(\Omega)H(-\Omega) = \frac{1}{1+(\Omega/\Omega_c)^{2N}}, \quad \textbf{(8.16)}$$

where we have used the fact that for a real filter, $H^*(\Omega) = H(-\Omega)$ (see Problem 8-13). Next, substitute $\Omega = s/j$ into Equation (8.16),

$$H(s)H(-s) = \frac{1}{1+(s/j\Omega_c)^{2N}}. \quad \textbf{(8.17)}$$

This system has $2N$ poles and no zeros. The poles are the roots of the denominator,

$$(s/j\Omega_c)^{2N} = -1.$$

Write $-1 = e^{j\pi(2k+1)}$, where k is an integer. Hence,

$$s = j\Omega_c e^{j\pi(2k+1)/2N} = e^{j\pi/2}\Omega_c e^{j\pi(2k+1)/2N} = \Omega_c e^{j\pi(2k+N+1)/2N}, \quad 0 \le k < 2N.$$

This says that the $2N$ poles of $H(s)H(-s)$ are equally spaced around a circle of radius Ω_c, as shown in **Figure 8.5** for various values of N.

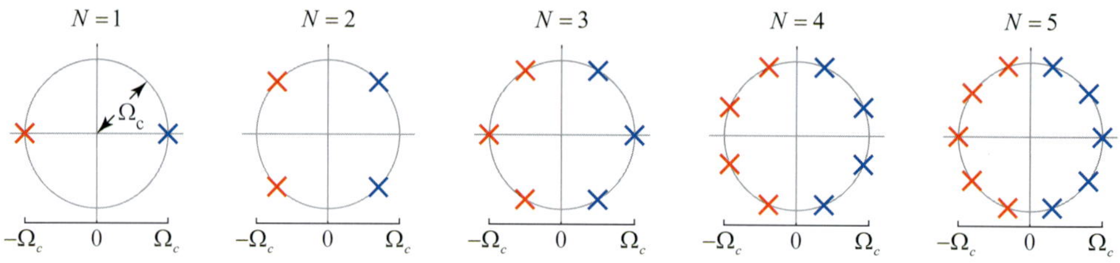

Figure 8.5 Poles of the Butterworth filter

Half of the poles are in the left half-plane and the other half are in the right half-plane. Since our Butterworth filter needs to be causal and stable, we choose the poles of $H(s)$ to be the N poles in the left half-plane on a circle of radius Ω_c, which we define as p_k, given by

$$p_k \triangleq \Omega_c e^{j\pi(2k+N+1)/2N}, \quad 0 \le k < N. \tag{8.18}$$

These poles are shown in red in **Figure 8.5**. The remaining N poles in the right half-plane, shown in blue, comprise $H(-s)$. Given the pole positions of Equation (8.18), we can now write $H(s)$ as a product of N pole terms,

$$H(s) = K \prod_{k=0}^{N-1} \frac{1}{s - p_k},$$

where K is a constant chosen to make $H(s=0) = H(\omega=0) = 1$, as required by Equation (8.4),

$$K = \prod_{k=0}^{N-1} (-p_k). \tag{8.19}$$

By plugging Equation (8.18) into Equation (8.19), you can show (Problem 8-7) that

$$K = \Omega_c^N.$$

Hence, we can write

$$H(s) = \prod_{k=0}^{N-1} \frac{-p_k}{s - p_k} = \Omega_c^N \prod_{k=0}^{N-1} \frac{1}{s - p_k} = \prod_{k=0}^{N-1} \frac{\Omega_c}{s - p_k}. \tag{8.20}$$

Normalized filter It is common practice to design a **normalized filter** first, one in which all frequencies are normalized to the cutoff frequency Ω_c, and then adapt that filter design to your desired value of Ω_c. Define the **normalized frequency** as

$$\tilde{\Omega} \triangleq \Omega/\Omega_c.$$

Then the definition of the Butterworth filter, Equation (8.4), becomes

$$|H(\tilde{\Omega})| = \frac{1}{\sqrt{1+\tilde{\Omega}^{2N}}}.$$

Apply this frequency normalization to the Laplace transform by dividing both the numerator and denominator of Equation (8.20) by Ω_c,

$$H(s) = \prod_{k=0}^{N-1} \frac{(\Omega_c/\Omega_c)}{(s/\Omega_c) - (p_k/\Omega_c)} = \prod_{k=0}^{N-1} \frac{1}{\tilde{s} - \tilde{p}_k},$$

where we have defined the normalized transform variable

$$\tilde{s} \triangleq s/\Omega_c = j\Omega/\Omega_c = j\tilde{\Omega},$$

and the normalized poles,

$$\tilde{p}_k \triangleq p_k/\Omega_c = e^{j\pi(2k+N+1)/2N},\ 0 \le k < N. \tag{8.21}$$

All the normalized poles lie along a circle of radius one. We will do an example in a moment.

Second-order sections As you can see from **Figure 8.5**, if the filter order N is even, $H(s)$ comprises $N/2$ pairs of complex-conjugate poles, so the poles p_k and $p_{(N-1)-k}$ are conjugates. If N is odd, then there are $(N-1)/2$ complex-conjugate pairs, plus one additional real pole at $s = -\Omega_c$. Hence, an analog Butterworth filter can be implemented by a cascade (series combination) of $N/2$ second-order (two-pole) filters, for example, the well-known Sallen–Key filter (see Problem 8-9), plus an additional first-order RC filter if the order is odd. For the normalized filter, each second-order stage would be of the form

$$\frac{1}{(\tilde{s}-\tilde{p}_k)(\tilde{s}-\tilde{p}_k^*)} = \frac{1}{\tilde{s}^2 - 2\mathrm{Re}\{\tilde{p}_k\}\tilde{s} + |\tilde{p}_k|^2} = \frac{1}{\tilde{s}^2 - 2\cos\left(\dfrac{2k+N+1}{2N}\pi\right)\tilde{s}+1},$$

and the first-order stage, for $k = (N-1)/2$, would be

$$\frac{1}{\tilde{s} - \tilde{p}_{(N-1)/2}} = \frac{1}{\tilde{s}+1}.$$

Putting it together, we have

$$H(\tilde{s}) = A(\tilde{s}) \prod_{k=0}^{\lfloor \frac{N}{2} \rfloor - 1} \frac{1}{\tilde{s}^2 - 2\cos\left(\dfrac{2k+N+1}{2N}\pi\right)\tilde{s}+1}, \tag{8.22a}$$

where

$$A(\tilde{s}) = \begin{cases} \dfrac{1}{\tilde{s}+1}, & N \text{ odd} \\ 1, & N \text{ even} \end{cases}. \tag{8.22b}$$

At this point, an example would probably be helpful.

Example 8.2

Find the transform of the Butterworth lowpass filter of Example 8.1 in both normalized form, $H(\tilde{s})$, and unnormalized form, $H(s)$.

➤ Solution:

For this filter, $\Omega_c = 2\pi \cdot 1000$ and $N = 7$. From Equation (8.21), the normalized poles are at $\tilde{p}_k = e^{j\pi(k+4)/7}$, $0 \le k < 7$. You can find these poles with a single line of Matlab code:

```
pk = exp(1j*pi*(2*(0:N-1)'+N+1)/2/N);
```

Alternatively, you can call Matlab's `buttap` function, which does exactly the same thing,

```
[z,p,k] = buttap(N);
```

To be compatible with Matlab's other filter design functions, `buttap` returns the values of the normalized zeros `z`, the poles `p`, and the constant `k`, necessary to make the value of the normalized filter unity at $\tilde{\Omega} = 0$. In the case of the Butterworth filter, `p` is equivalent to our `pk`, `z=[ ]` and `k=1`.

To implement the normalized filter using second-order sections, use Equation (8.22). For $N = 7$, there are three complex-conjugate pole pairs, $(\tilde{p}_0, \tilde{p}_0^*)$, $(\tilde{p}_1, \tilde{p}_1^*)$ and $(\tilde{p}_2, \tilde{p}_2^*)$ plus a real pole at $\tilde{p}_3 = -1$.

$$\begin{aligned} H(\tilde{s}) &= \left(\frac{1}{\tilde{s}^2 - 2\cos(4\pi/7)\tilde{s} + 1}\right) \cdot \left(\frac{1}{\tilde{s}^2 - 2\cos(5\pi/7)\tilde{s} + 1}\right) \cdot \left(\frac{1}{\tilde{s}^2 - 2\cos(6\pi/7)\tilde{s} + 1}\right) \cdot \left(\frac{1}{\tilde{s} + 1}\right) \\ &= \left(\frac{1}{\tilde{s}^2 + 0.4450s + 1}\right) \cdot \left(\frac{1}{\tilde{s}^2 + 1.2470s + 1}\right) \cdot \left(\frac{1}{\tilde{s}^2 + 1.8019s + 1}\right) \cdot \left(\frac{1}{\tilde{s} + 1}\right). \end{aligned} \tag{8.23}$$

The Matlab function `zp2sos` can calculate these coefficients for you,

```
sos = zp2sos(z, p, k).
```

For the unnormalized form, substitute $\tilde{s} = s/\Omega_c$ into Equation (8.23) with $\Omega_c = \Omega_p$ and turn the algebra crank, yielding

$$\begin{aligned} H(s) &= \frac{\Omega_c^2}{s^2 - 2\Omega_c \cos\left(\frac{4\pi}{7}\right)s + \Omega_c^2} \cdot \frac{\Omega_c^2}{s^2 - 2\Omega_c \cos\left(\frac{5\pi}{7}\right)s + \Omega_c^2} \cdot \frac{\Omega_c^2}{s^2 - 2\Omega_c \cos\left(\frac{6\pi}{7}\right)s + \Omega_c^2} \cdot \frac{\Omega_c}{s + \Omega_c} \\ &= \frac{39478000}{s^2 + 2796s + 39478000} \cdot \frac{39478000}{s^2 + 7835s + 39478000} \cdot \frac{39478000}{s^2 + 11322s + 39478000} \cdot \frac{6283}{s + 6283}. \end{aligned}$$

You can appreciate that the coefficients of the normalized filter are much more tractable.

8.2.3 ★ Chebyshev filter

Butterworth filters are characterized by a maximally flat, monotonic magnitude response in the passband and a monotonic, but relatively slow, roll-off in the stopband. The responses of the other filters we will now discuss are not monotonic; they have ripples in either the passband

(Chebyshev), the stopband (inverse Chebyshev) or both (elliptic). But, the benefit that they all offer is a sharper roll-off, for a given filter order, than the Butterworth filter.

The **Chebyshev filter**, also called the **Type-I Chebyshev filter**, is defined by the magnitude of the Fourier transform,

$$|H(\Omega)| = \frac{1}{\sqrt{1+\varepsilon_p^2 T_N^2(\Omega/\Omega_c)}}, \tag{8.24}$$

where $T_N(x)$ is the Chebyshev polynomial of order N, defined as

$$T_N(x) = \cos(N\cos^{-1}x), \tag{8.25}$$

and the passband ripple factor ε_p is identical to that defined in Equation (8.7a). Unlike the Butterworth filter, the Chebyshev filter depends on three parameters: the order N, the corner frequency Ω_c and the passband ripple factor ε_p. The distinguishing characteristic of the Chebyshev filter is that the magnitude of the frequency response exhibits characteristic **equiripple** behavior in the passband: the magnitude oscillates and the peak-to-peak amplitude of the oscillation is constant over the whole band. In the stopband, the response is monotonic. Of all the common filter types, the Chebyshev is the one that exhibits the smallest maximum error between the response magnitude in the passband and that of the ideal lowpass filter for a filter of a given order.

Chebyshev polynomials The equiripple response of the Chebyshev filter in the passband and monotonic response in the stopband result from the particular properties of the **Chebyshev polynomial** $T_N(x)$, so we need to invest a little time to understand this polynomial. First of all, Equation (8.25) certainly does not *look* like a polynomial, but it is easy to show that it can be expressed as one. For $N=0$, $T_0(x) = \cos(0\cdot\cos^{-1}x) = 1$ independent of x. For $N=1$, $T_1(x) = \cos(\cos^{-1}x) = x$. For larger values of N, the expression for $T_N(x)$ of Equation (8.25) can be described by a simple recursion relation (see Problem 8-16),

$$T_N(x) = 2xT_{N-1}(x) - T_{N-2}(x), \quad N > 1,$$

which can be solved to generate a series of Chebyshev polynomials, the first few of which are listed here:

$$T_N(x) = \begin{cases} 1, & N=0 \\ x, & N=1 \\ 2x^2-1, & N=2 \\ 4x^3-3x, & N=3 \\ 8x^4-8x^2+1, & N=4 \\ \vdots & \end{cases}.$$

Figure 8.6a shows the behavior of $T_N(x)$ as a function of x for different values of N over the range $0 \le x \le 1.1$. **Figure 8.6b** shows the behavior of $T_N^2(x)$ vs. x.

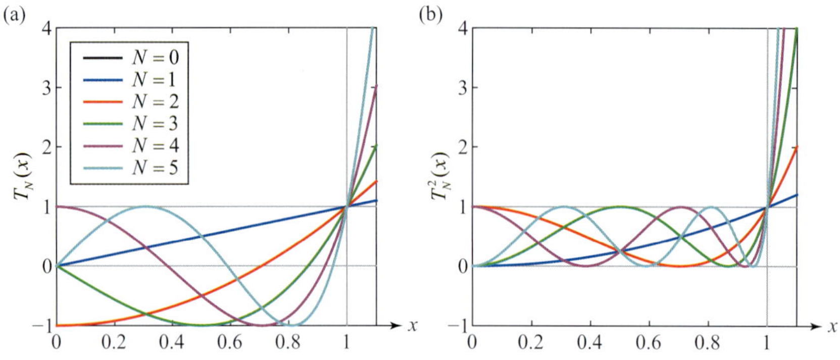

Figure 8.6 Chebyshev polynomials, $T_N(x)$

From the trigonometric form of the Chebyshev polynomial of Equation (8.25), you can see that $T_N(x)$ is basically just a cosine, $\cos\theta$, with a complicated angle argument $\theta = N\cos^{-1}x$. For values of x in the range $0 < x < 1$, $\cos^{-1}x$ ranges from $\pi/2$ to 0, so the argument $N\cos^{-1}x$ ranges from $N\pi/2$ to 0. This means that $T_N(x) = \cos(N\cos^{-1}x)$ is constrained to oscillate between -1 and 1 for values $0 \le x \le 1$, regardless of N. This is shown in **Figure 8.6a** for different values of N. As N increases, the number of oscillations within the range $0 \le x \le 1$ increases. This bounded oscillation is termed equiripple since the ripples are of equal maximum amplitude (± 1). $T_N^2(x)$ oscillates smoothly between 0 and 1, as shown in **Figure 8.6b**, and it turns out that this function has N maxima and minima within the range $0 \le x \le 1$, for $N > 0$ (see Problem 8-15).

At $x = 0$, $T_N(0) = \cos N\pi/2$, so $T_N^2(0)$ alternates between 0 (for N odd) and 1 (for N even). However, at $x = 1$, both $T_N(1)$ and $T_N^2(1)$ pass through 1, independent of N. For values of $x > 1$, $\cos^{-1}x$ is purely imaginary, so $T_N(x)$ has a very different, non-oscillatory behavior. Because $\cos^{-1}(jx) = \cosh x$, $T_N(x)$ can equivalently be written as

$$T_N(x) = \cosh(N\cosh^{-1}x). \tag{8.26}$$

This is a monotonically increasing function of x that is always greater than 1, as shown in **Figure 8.6** for $x > 1$. The slope of $T_N(x)$ at $x = 1$ is N^2 (Problem 8-17), so $T_N(x)$ grows rapidly as a function of N (see Problem 8-17). The distinctive behavior of Chebyshev polynomials – equiripple oscillations for $x \le 1$, and monotonically increasing values for $x > 1$ – gives the Chebyshev filter its characteristic frequency response. Returning to Equation (8.24), notice that the frequency response of the Chebyshev filter depends upon the behavior of $T_N(x)$ for $x = \Omega/\Omega_c$. The frequency range below the cutoff frequency, $0 < \Omega < \Omega_c$ (or equivalently, $0 \le \Omega/\Omega_c \le 1$), corresponds to the range $0 < x < 1$ in Equation (8.25). In this range, $T_N^2(\Omega/\Omega_c)$ oscillates between 0 and 1, giving rise to the equiripple passband. Above the cutoff frequency (i.e., in the frequency range $\Omega > \Omega_c$), $T_N(\Omega/\Omega_c)$ is greater than 1 and increases monotonically; hence, $|H(\Omega)|$ is monotonically decreasing in the stopband.

Mapping passband and stopband specifications to filter parameters

Figure 8.7 shows how the response of the Chebyshev filter depends upon the filter's three parameters: the order N, the corner frequency Ω_c and the passband ripple factor ε_p.

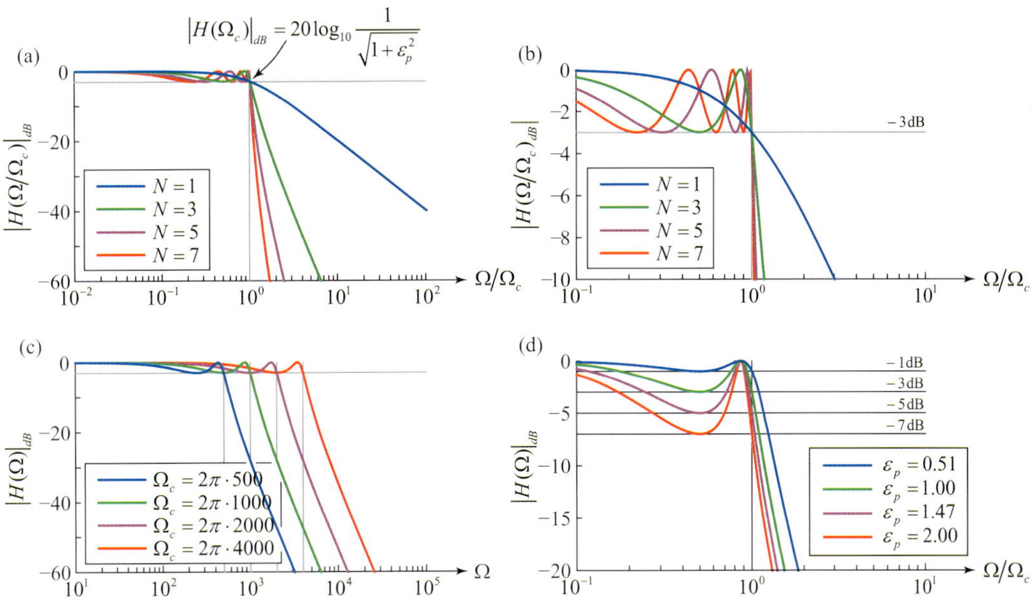

Figure 8.7 Dependence of Chebyshev filter parameters N, Ω_c and ε_p

Figure 8.7a shows the effect of increasing N while keeping the cutoff frequency and ripple factor ($\varepsilon_p = 1$) constant. **Figure 8.7b** shows the same data on expanded frequency and amplitude scale. Again, the frequency scales have been normalized to Ω_c. For the Chebyshev filter, the cutoff frequency corresponds to the frequency at which the response passes exactly through the value $|H(\Omega_c)|$,

$$|H(\Omega_c)| = \frac{1}{\sqrt{1+\varepsilon_p^2 T_N^2(\Omega_c/\Omega_c)}} = \frac{1}{\sqrt{1+\varepsilon_p^2 T_N^2(1)}} = \frac{1}{\sqrt{1+\varepsilon_p^2}},$$

where $T_N^2(1) = 1$, independent of N. On a dB scale,

$$|H(\Omega_c)|_{dB} = 20\log_{10}\frac{1}{\sqrt{1+\varepsilon_p^2}} = -10\log_{10}(1+\varepsilon_p^2)\,\text{dB}.$$

Below the cutoff frequency (i.e., $0 \le \Omega \le \Omega_c$), the magnitude of the frequency response oscillates in the range

$$\frac{1}{\sqrt{1+\varepsilon_p^2}} \le |H(\Omega)| \le 1,$$

hence, the response of the filter on a dB scale oscillates between a minimum value of $-10\log_{10}(1+\varepsilon_p^2)$ dB and a maximum value of 0 dB. The peak-to-peak amplitude of the ripple depends solely upon ε_p, and is independent of N. Most interesting from the perspective of the filter designer is that the minimum value of the ripple corresponds directly to a design parameter: the passband attenuation δ_p. This means that the cutoff frequency of the Chebyshev filter is always equal to the passband frequency, $\Omega_p = \Omega_c$. If you like, you can think of Ω_c as the "δ_p-dB point" of the filter. This is in contrast to the Butterworth filter, where Ω_c is always the -3-dB point of the filter. Hence, we can set the passband attenuation of the Chebyshev filter to any desired value simply by finding the appropriate value of the ripple factor,

$$\varepsilon_p = \sqrt{10^{-\delta_p/10} - 1}.$$

In the example of **Figure 8.7**, ε_p has been chosen to correspond to $\delta_p = -3$ dB, so,

$$\varepsilon_p = \sqrt{10^{0.3} - 1} \simeq 0.9976.$$

As shown in **Figures 8.7a** and **b**, above the cutoff frequency ($\Omega > \Omega_c$), all the curves fall monotonically. The cutoff slope above Ω_c is clearly a function of N; the larger the value of N, the steeper the slope. **Figure 8.7c** illustrates the effect of increasing the corner frequency Ω_c while leaving the other parameters fixed ($N = 5, \varepsilon \simeq 0.9976$). As with the Butterworth filter, increasing Ω_c translates the entire plot to the right on a log frequency scale. Neither the sharpness of the cutoff slope nor the peak-to-peak amplitude of the ripple is affected. Finally, **Figure 8.7d** shows what happens when we change the ripple factor ε_p while leaving the other parameters fixed ($N = 5, \Omega_c = 2\pi \cdot 1000$). As we expect, the peak-to-peak amplitude of the ripple increases with increasing ε_p, but notice that the slope of the response above the cutoff frequency is affected, too: it gets steeper as ε_p increases.

To design a Chebyshev filter, first calculate the value of ε_p necessary to meet given passband specifications exactly. As shown in **Figure 8.7**, the response of the filter at the cutoff frequency Ω_c, reaches the minimum value of δ_p dB, irrespective of N, the order of the filter; hence, we have $\Omega_c = \Omega_p$. That means

$$|H(\Omega_c)|_{dB} = |H(\Omega_p)|_{dB} = -10\log_{10}(1 + \varepsilon_p^2) = \delta_p,$$

from which we calculate the same ripple factor as in Equation (8.7a),

$$\varepsilon_p = \sqrt{10^{-\delta_p/10} - 1}.$$

Having specified Ω_c and ε_p, all that is left is to find the order N. We choose N to meet or exceed the specification for stopband attenuation, that is, $|H(\Omega_s)|_{dB} \leq \delta_s$, so

$$-10\log_{10}(1 + \varepsilon_p^2 T_N^2(\Omega_s/\Omega_p)) \leq \delta_s.$$

A little algebra yields the required relation,

$$T_N(\Omega_s/\Omega_p) \geq \sqrt{\frac{10^{-\delta_s/10} - 1}{10^{-\delta_p/10} - 1}}, \tag{8.27}$$

or, in terms of the selectivity ξ and discrimination factor L we defined in Equation (8.9b),

$$T_N(\xi) \geq L.$$

Since $\Omega_s/\Omega_p > 1$, we can express the Chebyshev function in the form of Equation (8.26) and solve Equation (8.27) for N,

$$N = \left\lceil \frac{\cosh^{-1} T_N(\Omega_s/\Omega_p)}{\cosh^{-1}\Omega_s/\Omega_p} \right\rceil = \left\lceil \frac{\cosh^{-1}\varepsilon_s/\varepsilon_p}{\cosh^{-1}\Omega_s/\Omega_p} \right\rceil = \left\lceil \frac{\cosh^{-1}\sqrt{\frac{10^{-\delta_s/10} - 1}{10^{-\delta_p/10} - 1}}}{\cosh^{-1}\Omega_s/\Omega_p} \right\rceil$$
$$= \left\lceil \frac{\cosh^{-1} L}{\cosh^{-1}\xi} \right\rceil, \tag{8.28}$$

where N is again rounded up to the nearest integer. This is the degree equation for the Chebyshev filter. Rounding the filter order up to an integer causes the designed filter to exceed the stopband attenuation. The exact attenuation at the stopband frequency will be

$$|H(\Omega_s)| = \frac{1}{\sqrt{1+\varepsilon_p^2 T_N^2(\Omega_s/\Omega_p)}} = \frac{1}{\sqrt{1+\varepsilon_p^2 T_N^2(\xi)}}. \tag{8.29}$$

From the degree equation, Equation (8.28), we could also calculate the revised discrimination factor at which the response exactly meets the stopband specification,

$$\hat{\xi} \triangleq \hat{\Omega}_s/\Omega_p = \cosh(\cosh^{-1}L/N), \tag{8.30a}$$

from which we would get the revised stopband frequency $\hat{\Omega}_s < \Omega_s$,

$$\hat{\Omega}_s \triangleq \Omega_p \cosh(\cosh^{-1}L/N). \tag{8.30b}$$

Example 8.3

Derive values of the filter parameters N, Ω_c and ε_p for the Chebyshev lowpass filter shown in **Figure 8.1**. The passband is given by $\Omega_p = 2\pi \cdot 1000$ with $\delta_p = -3$ dB and the stopband is given by $\Omega_s = 2\pi \cdot 2000$ with $\delta_s = -40$ dB.

➤ Solution:

We choose the cutoff frequency to equal the passband frequency, $\Omega_c = \Omega_p = 2\pi \cdot 1000$. To get the filter order, plug into Equation (8.28):

$$N \geq \frac{\cosh^{-1}\sqrt{\dfrac{10^4-1}{10^{0.3}-1}}}{\cosh^{-1}2} = 4.025.$$

Thus, $N = 5$ is the smallest integer that satisfies the condition. Equation (8.7a) gives the ripple factor,

$$\varepsilon_p = \sqrt{10^{0.3}-1} \simeq 0.9976.$$

The resulting response, with $N = 5$, $\varepsilon_p = 0.9976$ and $\Omega_c = 2\pi \cdot 1000$, is shown on a normalized frequency scale in **Figure 8.8**. Since $\Omega_p = \Omega_c$, the normalized passband frequency is $\Omega_p/\Omega_c = 1$ and the normalized stopband frequency is $\Omega_s/\Omega_c = \Omega_s/\Omega_p = \xi = 2$.

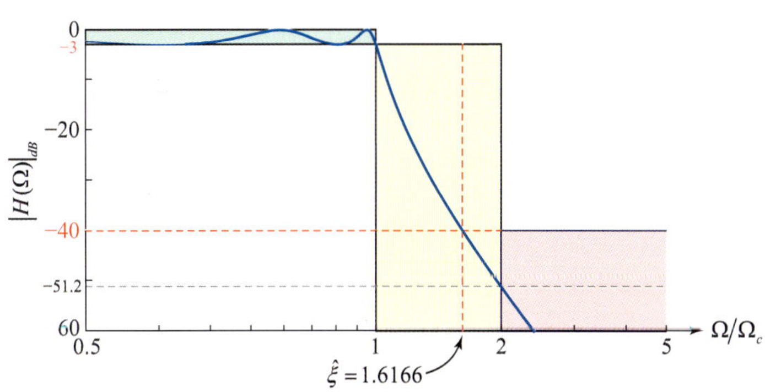

Figure 8.8 Analog Chebyshev filter design example

The filter's response is equiripple in the passband, as we expect. The peak-to-peak amplitude of the ripple on a dB scale is exactly $|\delta_p|$ dB, which was guaranteed since we chose the value of ε_p to match the passband tolerance exactly. The filter's response is monotonic in the transition and stopbands and exceeds the stopband requirement, $\delta_s = -40$, by quite a bit. In fact, from Equation (8.29) the value of $|H(\Omega_s)|_{dB} = -51.2$ dB. The figure also shows that the filter exactly matches the stopband (dotted red horizontal line) at a frequency (dotted red vertical line) of $\hat{\Omega}_s = 2\pi \cdot 1616.6$, which corresponds to a revised selectivity factor of $\hat{\xi} = \hat{\Omega}_s/\Omega_p = 1.6166$.

The Matlab function `cheb1ord` can be used to find N and Ω_c,

```
[N, Wc]=cheb1ord(Wp, Ws, Rp, Rs, 's'),
```

where `Rp = -dp` and `Rs = -ds`. The function uses an algorithm equivalent to the one we have presented here and gives identical results.

The poles of the Chebyshev filter Using the same approach we took to finding the roots of the Butterworth filter, it is (relatively) straightforward to show (see Problem 8-11) that the $2N$ poles of $H(s)H(-s)$ for the Chebyshev filter lie on an ellipse at locations

$$p_k = \Omega_c(\pm \sin\psi_k \sinh\varphi + j\cos\psi_k \cosh\varphi), \quad 0 \le k < N, \tag{8.31a}$$

where

$$\psi_k = (2k+1)\pi/2N \text{ and } \varphi = (1/N)\sinh^{-1}(1/\varepsilon_p). \tag{8.31b}$$

The major (real) axis of the ellipse is determined by $\cosh\varphi$ and the minor (imaginary) axis by $\sinh\varphi$. **Figure 8.9a** shows the poles of $H(s)$ in red and the poles of $H(-s)$ in blue for various values of N. **Figure 8.9b** shows the effect of changing the parameter ε_p for the case where $N = 5$.

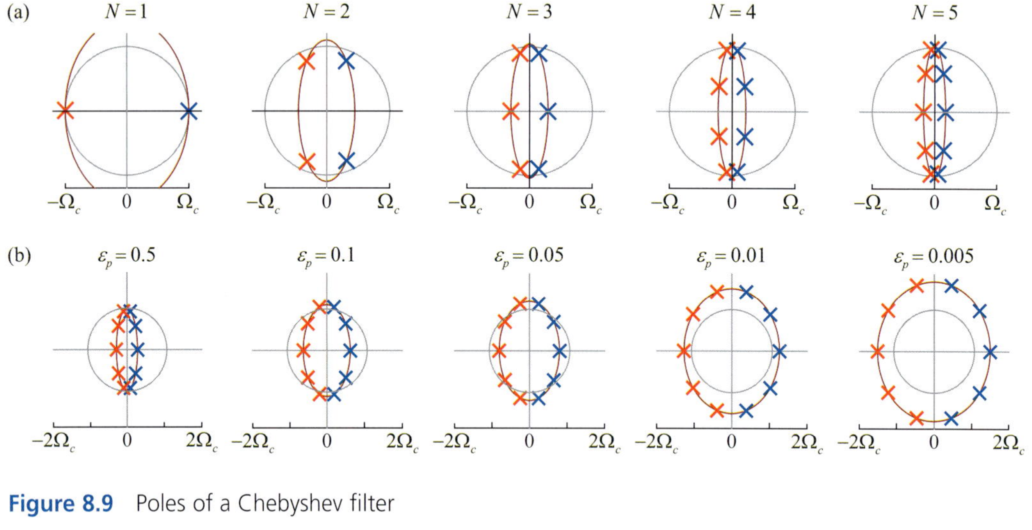

Figure 8.9 Poles of a Chebyshev filter

Like the Butterworth filter, $H(s)$ for the Chebyshev filter has N poles and no zeros in the finite s-plane, and can be written in normalized form as

$$H(\tilde{s}) = K \prod_{k=0}^{N-1} \frac{-\tilde{p}_k}{\tilde{s} - \tilde{p}_k}, \qquad \textbf{(8.32a)}$$

where $\tilde{p}_k = p_k/\Omega_c$ are the normalized poles, and the constant K is chosen to make $H(0) = 1$,

$$K = \begin{cases} 1/\sqrt{1+\varepsilon_p^2}, & N \text{ even} \\ 1, & N \text{ odd} \end{cases}. \qquad \textbf{(8.32b)}$$

For N even, $\tilde{p}_k$ and $\tilde{p}_{(N-1)-k}$ are conjugates, so $H(s)$ can be described as having $N/2$ pairs of complex-conjugate poles at positions

$$\tilde{p}_k = -\sin\psi_k \sinh\varphi \pm j\cos\psi_k \cosh\varphi,\ 0 \le k < N/2. \qquad \textbf{(8.33)}$$

For N odd, there is one additional real pole at $k = (N-1)/2$: $\tilde{p}_{(N-1)/2} = -\sinh\varphi$.

Having found the poles, $H(\tilde{s})$ can also be expressed as the product of second-order sections:

$$H(\tilde{s}) = B(\tilde{s}) \prod_{k=0}^{\lfloor \frac{N}{2} \rfloor - 1} \frac{|\tilde{p}_k|^2}{s^2 - 2\mathrm{Re}\{\tilde{p}_k\}s + |\tilde{p}_k|^2}, \qquad \textbf{(8.34)}$$

where

$$B(\tilde{s}) = \begin{cases} \dfrac{\sinh\varphi}{\tilde{s} + \sinh\varphi}, & N \text{ odd} \\ 1, & N \text{ even} \end{cases}.$$

Example, please!

Example 8.4

Find the transform of the Chebyshev lowpass filter of Example 8.3 in both normalized form, $H(\tilde{s})$, and unnormalized form, $H(s)$.

➤ Solution:

For this filter, $\Omega_c = 2\pi \cdot 1000$, $\varepsilon_p = \sqrt{10^{0.3} - 1} \simeq 0.9976$ and $N = 5$. From Equation (8.33), the normalized poles are located at

$$\tilde{p}_k = -\sin\psi_k \sinh\varphi \pm j\cos\psi_k \cosh\varphi,\ 0 \le k < 2,$$

where

$$\psi_k = (2k+1)\pi/10 \text{ and } \varphi = (1/5)\sinh^{-1}(1/\varepsilon_p).$$

If you do not want to perform that computation, Matlab's `cheb1ap` function can compute the poles for you:

```
[z,p,k] = chebap(N, -dp).
```

The Chebyshev filter has no zeros, so `z =[ ]`. The constant `k` that Matlab returns is equivalent to

$$K\prod_{k=0}^{N-1}(-\tilde{p}_k)$$

in Equation (8.32a). In this example, there are two complex-conjugate pole pairs, $(\tilde{p}_0, \tilde{p}_0^*)$ and $(\tilde{p}_1, \tilde{p}_1^*)$, and a real pole at $\tilde{p}_2$. Plugging this into Equation (8.34) yields the normalized transform,

$$H(s) = \left(\frac{0.9360}{\tilde{s}^2 + 0.1097\tilde{s} + 0.9360}\right)\cdot\left(\frac{0.3770}{\tilde{s}^2 + 0.2873\tilde{s} + 0.3770}\right)\cdot\left(\frac{0.1775}{\tilde{s} + 0.1775}\right).$$

Matlab's `[sos, g] = zp2sos` function will produce similar results, but will return a single gain term, `g`, that is the product of the numerator constants of the three sections.

8.2.4 ★ Inverse Chebyshev filter

The inverse Chebyshev filter, also known as a Type-II Chebyshev filter, is defined by the magnitude response

$$|H(\Omega)| = \frac{1}{\sqrt{1 + \varepsilon_s^2 T_N^{-2}(\Omega_s/\Omega)}},$$

where ε_s is the stopband ripple defined in Equation (8.7b). This filter is a modification of the Type-I Chebyshev filter that is monotonic in the passband and equiripple in the stopband. The motivating idea of the filter is that, all things being equal, we may prefer that the filter's response in the passband be maximally flat, like the Butterworth filter, while we may not care so much if the stopband has ripples, because the amplitude of the stopband is likely to be greatly attenuated anyway in any practical design.

Figure 8.10 shows how the response of the inverse Chebyshev filter depends upon the order N, the corner frequency Ω_c and the stopband ripple ε_s.

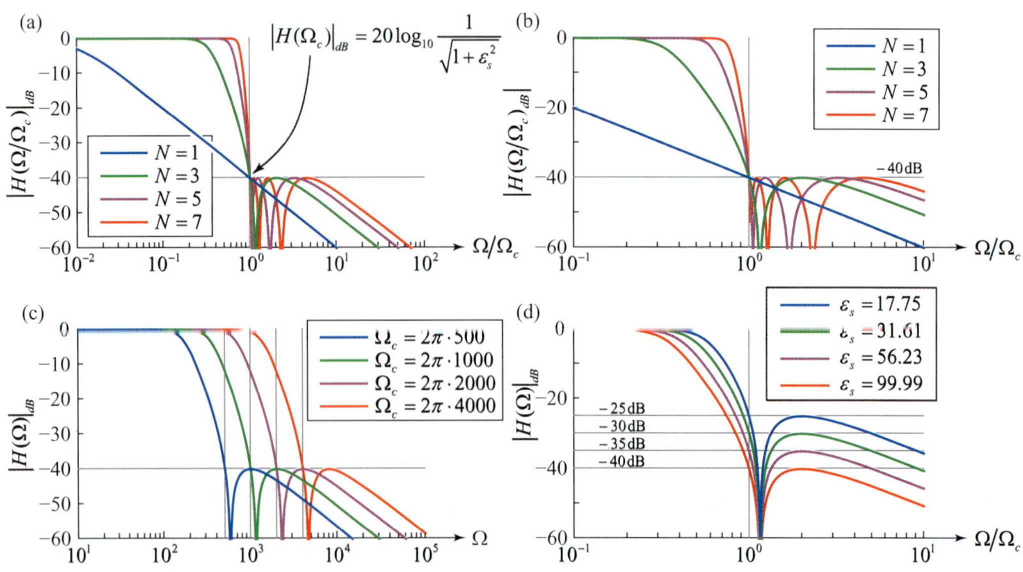

Figure 8.10 Dependence of the inverse Chebyshev filter on parameters N, Ω_c and ε_s

Figure 8.10a shows the effect of increasing N at a constant cutoff frequency. **Figure 8.10b** shows the same data on an expanded scale. For the inverse Chebyshev filter, the cutoff frequency Ω_c corresponds to the *stopband* frequency, $\Omega_c = \Omega_s$. That is because

$$|H(\Omega_s)| = |H(\Omega_c)| = \frac{1}{\sqrt{1+\varepsilon_s^2 T_N^{-2}(\Omega_c/\Omega_c)}} = \frac{1}{\sqrt{1+\varepsilon_s^2 T_N^{-2}(1)}} = \frac{1}{\sqrt{1+\varepsilon_s^2}}.$$

Expressed in dB,

$$|H(\Omega_c)|_{dB} = 20\log_{10}\left(1/\sqrt{1+\varepsilon_s^2}\right) = -10\log_{10}(1+\varepsilon_s^2) = \delta_s.$$

Below the cutoff frequency, $\Omega < \Omega_c$, the response increases monotonically from δ_s towards 1 as Ω decreases. Above the cutoff frequency, $\Omega > \Omega_c$, the response oscillates in the range

$$0 \le |H(\Omega)| \le \frac{1}{\sqrt{1+\varepsilon_s^2}},$$

and the peak-to-peak amplitude of this oscillation is independent of N. This figure also shows that, as with the other filters we have looked at so far, the slope of the inverse Chebyshev filter gets steeper with increasing N. **Figure 8.10c** shows that increasing the corner frequency Ω_c while leaving the other parameters fixed results in a translation of the entire plot to the right, as we saw with the other filters. Neither the sharpness of the cutoff slope nor the peak-to-peak amplitude of the ripple is affected. Finally, **Figure 8.7d** shows that the stopband attenuation is a function of ε_s; larger values of ε_s result in a greater stopband attenuation.

Mapping passband and stopband specifications to filter parameters To design an inverse Chebyshev filter, first find the value of ε_s that makes the magnitude of the filter's response meet the *stopband* specification

$$|H(\Omega_s)|_{dB} = -10\log_{10}(1+\varepsilon_s^2) = \delta_s,$$

which gives

$$\varepsilon_s = \sqrt{10^{-\delta_s/10} - 1},$$

the same formula for the stopband ripple as Equation (8.7b). Next, choose N to meet or exceed the *passband* specification $|H(\Omega_p)|_{dB} \ge \delta_p$, which means

$$-10\log_{10}(1+\varepsilon_s^2 T_N^{-2}(\Omega_s/\Omega_p)) \ge \delta_p.$$

Proceeding in a manner similar to that for the regular Chebyshev filter, N is the integer that satisfies

$$N - \left\lceil \frac{\cosh^{-1} T_N(\Omega_s/\Omega_p)}{\cosh^{-1}\Omega_s/\Omega_p} \right\rceil - \left\lceil \frac{\cosh^{-1}\sqrt{\dfrac{10^{-\delta_s/10}-1}{10^{-\delta_p/10}-1}}}{\cosh^{-1}\Omega_s/\Omega_p} \right\rceil$$

$$= \left\lceil \frac{\cosh^{-1}\varepsilon_s/\varepsilon_p}{\cosh^{-1}\Omega_s/\Omega_p} \right\rceil = \left\lceil \frac{\cosh^{-1} L}{\cosh^{-1}\xi} \right\rceil. \quad (8.35)$$

This degree equation is identical to that of the regular Chebyshev filter, Equation (8.28). Given the way we have calculated the values of Ω_c, ε_s and N, the filter's response will exactly satisfy the stopband specifications, $|H(\Omega_s)|_{dB} = \delta_s$, but will almost certainly exceed the passband tolerance, $|H(\Omega_p)|_{dB} > \delta_p$, meaning that the passband will be wider than our design specification for Ω_p. However, we can fix this. Since $\Omega_c = \Omega_s$ for the inverse Chebyshev filter, we just need to use the degree equation to calculate a revised stopband frequency $\hat{\Omega}_s$ at which the filter's response exactly meets the *passband* tolerance. Solving Equation (8.35) for the revised selectivity,

$$\hat{\xi} = \frac{\hat{\Omega}_s}{\Omega_p} = \cosh\left(\frac{1}{N}\cosh^{-1}L\right), \tag{8.36a}$$

gives the new corner frequency, which is equal to the stopband frequency for the inverse Chebyshev filter,

$$\hat{\Omega}_c = \hat{\Omega}_s = \Omega_p \cosh\left(\frac{1}{N}\cosh^{-1}L\right). \tag{8.36b}$$

To make this clear, let us design an inverse Chebyshev filter to meet the same specifications as the filter we designed in Example 8.3.

Example 8.5

Derive values of the parameters N, $\hat{\Omega}_c$ and ε_s for the inverse Chebyshev lowpass filter shown in **Figure 8.1**, with a passband given by $\Omega_p = 2\pi \cdot 1000$ with $\delta_p = -3$ dB and a stopband given by $\Omega_s = 2\pi \cdot 2000$ with $\delta_s = -40$ dB.

➤ Solution:

Consistent with our discussion, above, we initially choose the cutoff frequency to equal the stopband frequency, $\Omega_c = \Omega_s = 2\pi \cdot 2000$. Equation (8.7b) then gives

$$\varepsilon_s = \sqrt{10^4 - 1} \simeq 99.50,$$

and Equation (8.35) gives the identical result found in Example 8.3, $N \geq 4.025$, from which we conclude $N = 5$. The resulting response is shown in the blue trace of **Figure 8.11**. As you can see, the response exactly meets the stopband specifications at $\Omega = \Omega_s$ (i.e., passes through the top left corner of the red box), but exceeds the passband specifications in the frequency range $2\pi \cdot 1000 < \Omega < 2\pi \cdot 1237$. To make the filter match the passband specifications exactly, we use Equation (8.36b) to derive an adjusted value

$$\hat{\Omega}_c = \Omega_p \cosh\left(\frac{1}{N}\cosh^{-1}L\right) = 2\pi \cdot 1617.$$

On a log frequency scale, this results in translating the response curve (red trace) to the left so that it exactly meets the passband specifications while exceeding the stopband specifications.

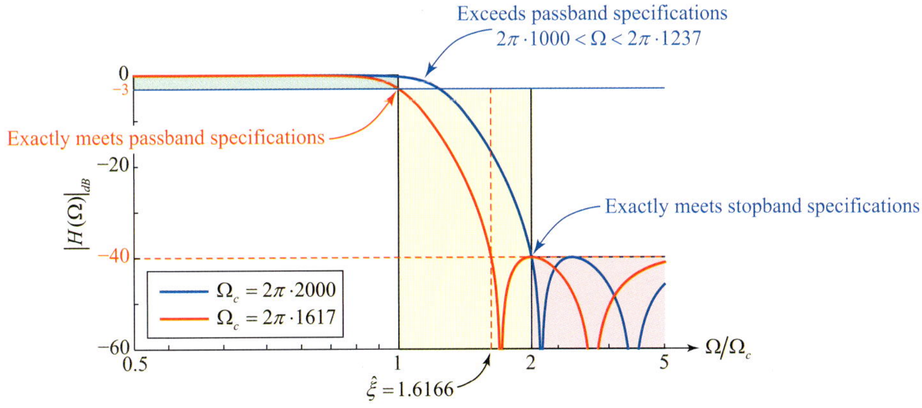

Figure 8.11 Analog inverse Chebyshev filter design example

The Matlab function `cheb2ord` finds N and Ω_c using an equivalent algorithm that makes the response exactly meet the requested passband specifications while exceeding the stopband specifications,

```
[N, Wc] = cheb2ord(Wp, Ws, Rp, Rs, 's'),
```

where `Rp = -dp` and `Rs = -ds`.

The poles and zeros of the inverse Chebyshev filter We find the poles and zeros of $H(s)H(-s)$ for the inverse Chebyshev filter using the same approach we used to find the poles of the Butterworth and Chebyshev filters (see Problem 8-12). Writing $H(s)H(-s)$ as

$$H(s)H(-s) = \frac{1}{1+\varepsilon_s^2 T_N^{-2}(j\Omega_c/s)} = \frac{\varepsilon_s^{-2} T_N^2(j\Omega_c/s)}{1+\varepsilon_s^{-2} T_N^2(j\Omega_c/s)},$$

it is clear that the inverse Chebyshev filter will have both poles (the roots of the denominator) and zeros (the roots of the numerator) in the finite s-plane. The N poles of $H(s)$ for the inverse Chebyshev filter lie in the left half-plane (see Problem 8-12) at locations

$$p_k = \Omega_c/(-\sin\psi_k \sinh\varphi + j\cos\psi_k \cosh\varphi), \quad 0 \le k < N, \tag{8.37}$$

where

$$\psi_k = (2k+1)\pi/2N \text{ and } \varphi = (1/N)\sinh^{-1}\varepsilon_s.$$

The zeros of $H(s)$ all lie on the $j\Omega$-axis at locations

$$z_k = j\Omega_c/\cos\psi_k, \quad 0 \le k < N. \tag{8.38}$$

It is these zeros of $H(s)$ that drive the frequency response $H(\Omega)$ to zero in the stopband, because $H(\Omega)$ is just $H(s)$ with s evaluated on the $j\Omega$-axis.

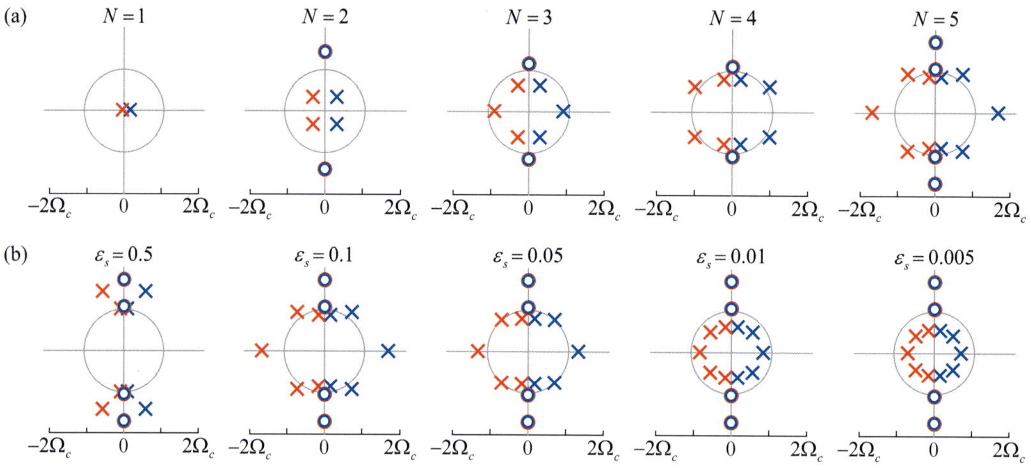

Figure 8.12 Poles of an inverse Chebyshev filter

Figure 8.12a shows the singularities of $H(s)$ in red and the singularities of $H(-s)$ in blue for various values of N. **Figure 8.12b** shows the effect of changing the parameter ε_s for the case where $N=5$. $H(s)$ has N poles and N zeros. However, for N odd, the zero at $k=(N-1)/2$ is located at ∞, so the inverse Chebyshev filter only has $2\lfloor N/2 \rfloor$ zeros in the finite s-plane,

$$H(s) = \frac{\prod\limits_{\substack{k=0 \\ k \neq (N-1)/2,\, k \text{ odd}}}^{N-1} (z_k - s)/z_k}{\prod\limits_{k=0}^{N-1} (p_k - s)/p_k}.$$

The poles of $H(s)$ can be grouped into $\lfloor N/2 \rfloor$ complex-conjugate pairs, plus one real pole when N is odd, which is located at $p_k = -(\Omega_c \sin \psi_k)^{-1}$ for $k=(N-1)/2$. Since $H(s)$ can only have an even number of zeros, they can also be grouped into $\lfloor N/2 \rfloor$ complex-conjugate pairs. Thus, the inverse Chebyshev filter can be represented as the cascade of second-order sections plus one first-order section if N is odd. Again, we choose to work with the normalized poles, zeros and transform variables,

$$\begin{aligned} \tilde{p}_k &= p_k/\Omega_c \\ \tilde{z}_k &= z_k/\Omega_c\,. \\ \tilde{s} &= s/\Omega_c \end{aligned} \tag{8.39}$$

Then, the normalized transform is

$$H(\tilde{s}) = C(\tilde{s}) \prod_{k=0}^{\lfloor N/2 \rfloor - 1} \left| \frac{\tilde{p}_k}{\tilde{z}_k} \right|^2 \frac{\tilde{s}^2 + |\tilde{z}_k|^2}{\tilde{s}^2 - 2\mathrm{Re}\{\tilde{p}_k\} s + |\tilde{p}_k|^2},$$

where

$$C(\tilde{s}) = \begin{cases} \dfrac{-\tilde{p}_{(N-1)/2}}{\tilde{s} - \tilde{p}_{(N-1)/2}}, & N \text{ odd} \\ 1, & N \text{ even} \end{cases}.$$

Example 8.6

Find $H(s)$ for the inverse Chebyshev lowpass filter of Example 8.5.

➤ **Solution:**

For this filter, $\Omega_c = 2\pi \cdot 1617$, $\varepsilon_s = \sqrt{10^4 - 1} \simeq 99.50$ and $N = 5$. From Equations (8.37) and (8.38), the normalized poles are located at

$$\tilde{p}_k = 1/(-\sin\psi_k \sinh\varphi + j\cos\psi_k \cosh\varphi),\ 0 \le k < 5,$$

where

$$\psi_k = (2k+1)\pi/10 \text{ and } \varphi = (1/5)\sinh^{-1}\varepsilon_s.$$

The normalized zeros are located at

$$\tilde{z}_k = j/\cos\psi_k,\ k = 0, 1, 3, 4.$$

Hence,

$$H(s) = \left(0.3595\frac{s^2+1.1056}{s^2+0.3118s+0.3975}\right)\cdot\left(0.1767\frac{s^2+2.8944}{s^2+1.0496s+0.5110}\right)\cdot\left(\frac{0.7878}{s+0.7878}\right).$$

Matlab's `cheb2ap` function can give the coefficients if you do not wish to calculate them yourself:

```
[z,p,k] = cheb2ap(N,-ds),
```

except that the constant `k` is the product of the constants of the individual sections.

8.2.5 ★ Elliptic filter

The **elliptic filter** is equiripple in both the passband and the stopband, and the amount of ripple in each band can be independently specified. Of all the filter types, this filter has the narrowest width of the transition zone for a given order of filter (or, equivalently, the lowest order for a given transition-zone width). The theory underlying this filter is also the most challenging of all the major filter types to understand, and is usually skipped in textbooks. However, being of adventurous spirit, we are going to try. In what follows, we will eschew many details of the theory in favor of giving a practical overview that highlights the features of this filter that are important for the designer.

Filter definition The elliptic filter is defined by the equation

$$|H(\Omega)| = \frac{1}{\sqrt{1+\varepsilon_p^2 R_N^2(\xi, \Omega/\Omega_c)}}, \tag{8.40}$$

where $R_N(\xi, \Omega/\Omega_c)$ is called the **elliptic rational function**. In contrast to the other filters we have studied, the design of the elliptic filter is determined by *four* separate parameters: the order N, the corner frequency Ω_c, the passband ripple factor ε_p and the selectivity factor ξ. This larger number of parameters gives us considerable flexibility in achieving our design objectives, as we will see when we get to the end of this section.

The elliptic rational function The classical definition of the elliptic rational function is daunting,

$$R_N(\xi, x) = \mathrm{cd}\left(\left(N\frac{K(1/L(\xi))}{K(1/\xi)}\right)\mathrm{cd}^{-1}(x, 1/\xi)\right), \tag{8.41}$$

where $L(\xi)$ is the **discrimination factor**, cd is the **Jacobi elliptic cosine function** and K is the **complete elliptic integral of the first kind**,

$$K(k) = \int_0^{\pi/2} \frac{d\theta}{\sqrt{1 - k^2 \sin^2\theta}}.$$

Do not despair! Equation (8.41) looks a little like our first definition of the Chebyshev polynomial function, Equation (8.25), in that it is defined in terms of a trigonometric function (in this case, cd) which is applied to an inverse function (cd^{-1}) that has been multiplied by a term having to do with the order N. And fortunately, like the Chebyshev polynomial function, the elliptic rational function $R_N(\xi, x)$ can be expressed in polynomial form, specifically as the ratio of polynomials in x,

$$R_N(\xi, x) = Ax^\alpha \prod_{i=1}^{\lfloor N/2 \rfloor} \frac{x^2 - z_i^2}{x^2 - p_i^2}, \tag{8.42a}$$

where $\pm z_i$ and $\pm p_i$ are, respectively, the zeros and poles of $R_N(\xi, x)$,

$$\alpha = \begin{cases} 1, & N \text{ odd} \\ 0, & N \text{ even} \end{cases}, \tag{8.42b}$$

and A is a constant calculated to make $R_N(\xi, 1) = 1$,

$$A = \frac{1}{\prod_{i=1}^{\lfloor N/2 \rfloor} \frac{1 - z_i^2}{1 - p_i^2}} = \prod_{i=1}^{\lfloor N/2 \rfloor} \frac{1 - p_i^2}{1 - z_i^2}. \tag{8.42c}$$

The poles and zeros of $R_N(\xi, x)$ determine the poles and zeros of the elliptic filter, and generate the filter's characteristic equiripple response, so let us spend a few moments understanding this function.

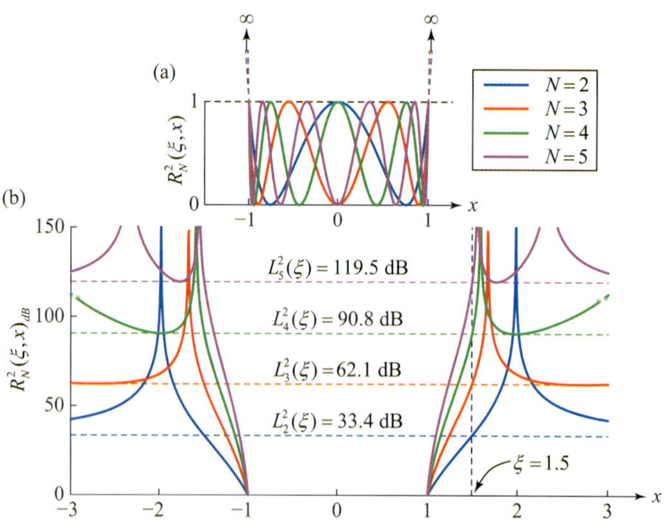

Figure 8.13 Elliptic rational function

Figure 8.13 illustrates the features of the elliptic rational function $R_N(\xi, x)$ that produce the equiripple response. **Figure 8.13a** shows $R_N^2(\xi, x)$ plotted on a linear scale over the range $|x| \le 1$ for a value of $\xi = 1.5$. In this range of x, $R_N(\xi, x)$ oscillates between values of -1 and 1, so $R_N^2(\xi, x)$ oscillates between 0 and 1. The range $|x| \le 1$ corresponds to the range of passband frequencies $|\Omega| < \Omega_c$ in Equation (8.40). Here, $|H(\Omega)|$ oscillates between 1 and $1/\sqrt{1+\varepsilon_p}$. That should remind you of the Chebyshev filter of Section 8.2.3, which has similar behavior. At exactly $x = 1$, $R_N^2(\xi, 1) = 1$, which corresponds to $\Omega = \Omega_c$, $|H(\Omega)| = 1/\sqrt{1+\varepsilon_p}$. That means that for the elliptic filter (like the Chebyshev filter), the cutoff frequency is equal to the passband frequency $\Omega_c = \Omega_p$. **Figure 8.13b** shows $R_N^2(\xi, x)$ plotted on a log (dB) scale for $|x| > 1$. Over the range $1 < |x| < \xi$, which corresponds to (at least part of) the transition band of the filter, $R_N^2(\xi, x)$ rises monotonically to ∞, which means $|H(\Omega)|$ falls monotonically to zero. At a value of $x = \xi = 1.5$, the magnitude of the elliptic filter is

$$|H(\Omega_s)| = \frac{1}{\sqrt{1+\varepsilon_p^2 R_N^2(\xi, \Omega_s/\Omega_p)}} = \frac{1}{\sqrt{1+\varepsilon_p^2 R_N^2(\xi, \xi)}} = \frac{1}{\sqrt{1+\varepsilon_p^2 L_N^2(\xi)}}, \tag{8.43}$$

where $L_N(\xi)$ is the discrimination factor that depends upon both the filter order N and the selectivity factor ξ. For values of $|x| > \xi$, $R_N^2(\xi, x)$ oscillates between the minimum value of $R_N^2(\xi, \xi)$ (dotted horizontal lines) and ∞. Hence, $|H(\Omega)|$ oscillates in the range

$$0 \le |H(\Omega)| \le \frac{1}{\sqrt{1+\varepsilon_p^2 R_N^2(\xi, \xi)}}.$$

These oscillations in $|H(\Omega)|$ for values of $|x| > \xi$ give rise to the equiripple behavior of the elliptic filter in the stopband. We will now show how the parameters of the elliptic rational function can be chosen to produce the desired response of the elliptic filter.

Mapping passband and stopband specifications to filter parameters The first step in the design of the elliptic filter is to determine the filter order N. The degree equation for this filter gives N as a ratio of terms involving the complete elliptic integral of $1/\xi$ and $1/L_N(\xi)$,

$$N = \left\lceil \frac{K(1/\xi^2)K(1-1/L_N^2(\xi))}{K(1-1/\xi^2)K(1/L_N^2(\xi))} \right\rceil.$$

Having determined N, the zeros and poles of $R_N(\xi, x)$ in Equation (8.42) are given by

$$\begin{aligned} z_k &= \mathrm{sn}\left(\left(\frac{2k-1}{N}+1\right)K(1/\xi^2), 1/\xi^2\right), \quad 1 \le k \le \lfloor N/2 \rfloor, \\ p_k &= \xi/z_k \end{aligned}$$

where sn is the Jacobi sine function evaluated with modulus $1/\xi$.

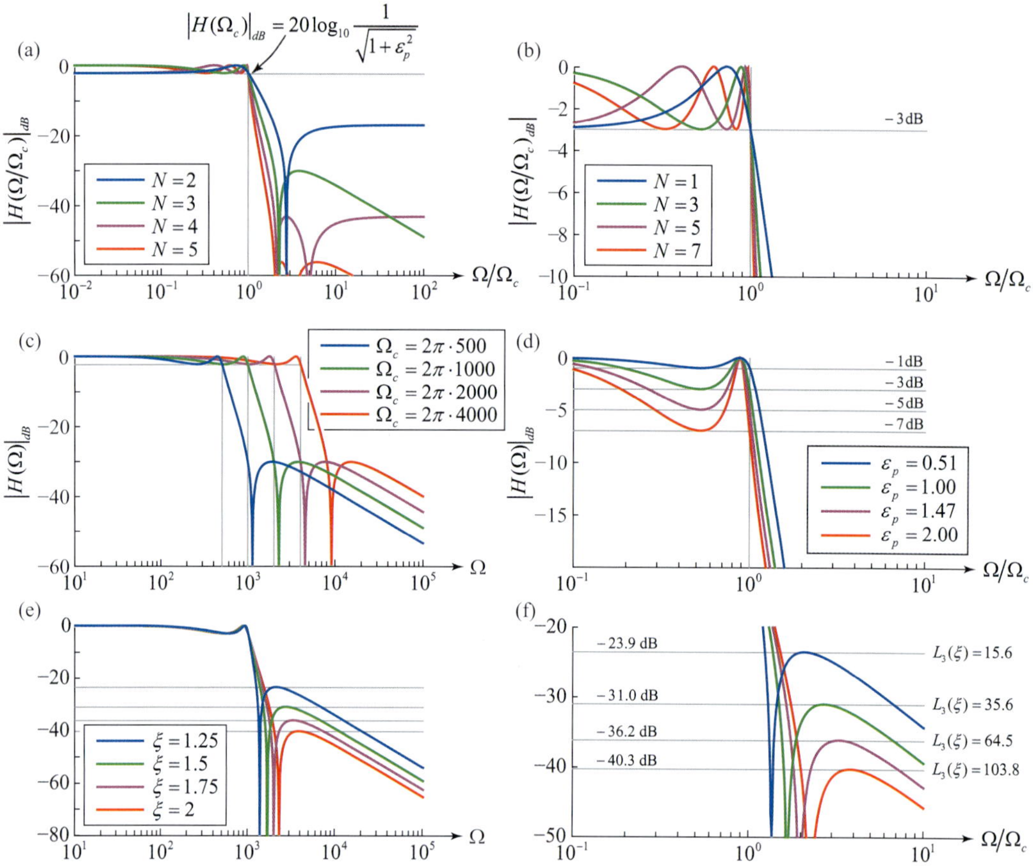

Figure 8.14 Dependence of elliptic filter on filter parameters N, Ω_c, ε_p and ξ

Matlab does not provide an elliptic rational function $R_N(\xi, x)$, but it does provide both the Jacobi sn function (`ellipj`) and the complete elliptic integral K (`ellipk`) , so we can write our own function. The following code computes $R_N(\xi, x)$, and also returns the poles and zeros of the function and constants A and α of Equation (8.42), all of which we will need when we come to design $H(\tilde{s})$, the transform of the filter, a bit later in this section.

```
function [R, zRn, pRn, A, alpha] = Rn(N, xi, x)
    lx = length(x);
    alpha = mod(N,2);
    M = floor(N/2);
    m = 1:M;
    zRn = ellipj(((2*m-1)/N+1)*ellipke(1/xi^2), 1/xi^2); % zeros of Rn
    pRn = xi./zRn; % poles of Rn
    A = 1 / prod((1-zRn.^2)./(1-pRn.^2)); % constant
    x2 = ones(M, 1)*x.^2; % this line and the next allow us to calculate Rn
    zRn2 = (zRn.^2).'*ones(1, lx); % when x is an array of values
    R = A * (x.^alpha).* prod((x2-zRn2)./ (x2-xi^2./zRn2));
end
```

Figure 8.14 shows how the response of the elliptic filter depends upon the filter's four parameters: the order N, the corner frequency Ω_c, the passband ripple factor ε_p and the selectivity factor ξ.

Figures 8.14a and **b** (on expanded frequency and magnitude scales) show the effect of increasing the order of the filter at a constant cutoff frequency and passband ripple ε_p. Unlike the other filters we have studied, both the width of the transition band and the minimum attenuation in the stopband vary with N. As we have seen in other filters, **Figure 8.14c** shows that the filter's response translates on a log frequency scale with changes in Ω_c. **Figure 8.14c** shows that increasing ε_p increases the passband ripple, as it does in Butterworth and Chebyshev filters. **Figures 8.14e** and **f** (on expanded scales) show the effect of changing the sensitivity factor, ξ, for a filter of order $N=3$. The width of the transition zone changes, but so does the maximum stopband attenuation $L_N(\xi)$. Because the elliptic filter has four design parameters, we can design multiple filters that meet both the passband and stopband specifications. We will show an example of that in a moment.

Poles and zeros of the elliptic filter As we did for the other filter types (for example, the Butterworth, Equation (8.17)) we now find the singularities (poles and zeros) of the transform,

$$H(s)H(-s)=\frac{1}{1+\varepsilon_p^2 R_N^2(\xi, s/j\Omega_c)}, \tag{8.44}$$

and then select those singularities that lie in the left half-plane to create $H(s)$. As we have with the other filters, we choose to work with the normalized poles, zeros and transform variables, so substituting Equations (8.39) and (8.42) into Equation (8.44), we get

$$\begin{aligned} H(\tilde{s})H(-\tilde{s}) &= \frac{1}{1+\varepsilon_p^2 A^2(\tilde{s}/j)^{2\alpha}\left(\prod_{i=1}^{\lfloor N/2\rfloor}\frac{(\tilde{s}/j)^2-\tilde{z}_i^2}{(\tilde{s}/j)^2-\tilde{p}_i^2}\right)^2} = \frac{1}{1+\varepsilon_p^2 A^2(-\tilde{s}^2)^{\alpha}\prod_{i=1}^{\lfloor N/2\rfloor}\left(\frac{\tilde{s}^2+\tilde{z}_i^2}{\tilde{s}^2+\tilde{p}_i^2}\right)^2} \\ &= \frac{\prod_{i=1}^{\lfloor N/2\rfloor}(\tilde{s}^2+\tilde{p}_i^2)^2}{\underbrace{\prod_{i=1}^{\lfloor N/2\rfloor}(\tilde{s}^2+\tilde{p}_i^2)^2}_{\text{Term due to poles of } R_N^2} + \underbrace{\varepsilon_p^2 A^2(-\tilde{s}^2)^{\alpha}\prod_{i=1}^{\lfloor N/2\rfloor}(\tilde{s}^2+\tilde{z}_i^2)^2}_{\text{Term due to zeros of } R_N^2}}. \end{aligned} \tag{8.45}$$

The zeros of $H(\tilde{s})H(-\tilde{s})$ are just the poles of $R_N^2(\xi, x)$ and are the product of terms of the form

$$(\tilde{s}^2+\tilde{p}_i^2)^2 = ((\tilde{s}+j\tilde{p}_i)(\tilde{s}-j\tilde{p}_i))^2 = (\tilde{s}+j\tilde{p}_i)^2(\tilde{s}-j\tilde{p}_i)^2.$$

Hence, there are $2\lfloor N/2\rfloor$ double zeros that lie on the imaginary axis of the $\tilde{s}$-plane at $\pm j\tilde{p}_i$. Finding the $2N$ poles of $H(\tilde{s})H(-\tilde{s})$ is a bit more of a challenge, because the denominator of the expression is the sum of terms derived from both the poles and the zeros of $R_N^2(\xi, x)$. Once the poles and zeros of the $H(\tilde{s})H(-\tilde{s})$ are found, we retain half of the double zeros – call them $\hat{z}_i$ – and the poles with a negative real part – call them $\hat{p}_i$ – to form $H(\tilde{s})$,

$$H(\tilde{s}) = C\frac{\prod_{i=1}^{2\lfloor N/2\rfloor}(\tilde{s}+\hat{z}_i^2)^2}{\prod_{i=1}^{N}(\tilde{s}^2+\hat{p}_i^2)^2},$$

where C is a constant chosen so that $H(\tilde{s})$ has the correct value at $\tilde{s}=0$. Since $H(\tilde{s}=0)=H(\Omega=0)$, from Equation (8.40), we have

$$|H(0)| = \frac{1}{\sqrt{1 + \varepsilon_p^2 R_N^2(\xi, 0)}}.$$

If you look at **Figure 8.13a**, you will see that $R_N^2(\xi, 0)$ is either 0 or 1, depending on N,

$$R_N^2(\xi, 0) = \left\{ \begin{array}{ll} 1, & N \text{ even} \\ 0, & N \text{ odd} \end{array} \right\} = 1 - \alpha,$$

where we have reused the value of α from Equation (8.42b). Putting the last few equations together, we get

$$C = \frac{1}{\sqrt{1 + \varepsilon_p^2 (1 - \alpha)}} \frac{\prod_{i=1}^{N} \left(1 + \hat{p}_i^2\right)^2}{\prod_{i=1}^{2\lfloor N/2 \rfloor} \left(1 + \hat{z}_i^2\right)^2}. \quad \textbf{(8.46)}$$

We certainly need an example to tie all this together.

Example 8.7

(a) Derive values of the filter parameters N, Ω_c, ε_p and ξ for an elliptic lowpass filter with $\Omega_p = 2\pi \cdot 1000$, $\delta_p = -3$ dB, $\Omega_s = 2\pi \cdot 2000$ and $\delta_s = -30$ dB.

(b) Compute the normalized poles and zeros of $H(\tilde{s})H(-\tilde{s})$, Equation (8.45).

(c) Compute the normalized poles and zeros of $H(\tilde{s})$.

➤ Solution:

(a) This is a job for Matlab.

```
% Input design parameters
Wp = 2*pi*1e3;
Ws = 2*pi*2e3;
dp = -3;ds = -30;

% Derived parameters
xi = Ws/Wp; % selectivity factor
ep = sqrt(10^-(dp/10)-1); % passband ripple factor
es = sqrt(10^-(ds/10)-1); % stopband ripple factor
L = es/ep; % (Initial) discrimination factor
N=ceil(ellipke(1/xi^2)*ellipke(1-1/L^2)/(ellipke(1-1/xi^2)*ellipke(1/L^2)));
```

If you wish, you can use Matlab's `ellipord` function to compute the order of the elliptic filter,

```
[N,Wc] = ellipord(Wp, Ws, Rp, Rs, 's'),
```

where `Rp = -dp` and `Rs = -ds`. The algorithm produces the same result as ours. In this case, `Wc = Wp`.

(b) First, obtain the poles and zeros of the elliptic rational function as well as the constants A and α from our code for the elliptic rational function, `Rn`,

```
% Poles and zeros of the elliptic rational function
[L, zRn, pRn, A, alpha] = Rn(N, xi, xi);
```

To find the poles and zeros of $H(\tilde{s})H(-\tilde{s})$, expand the numerator and denominator terms of the elliptic rational function:

$$\begin{aligned}(\tilde{s}^2+\tilde{z}_i^2)^2 &= ((\tilde{s}+j\tilde{z}_i)(\tilde{s}-j\tilde{z}_i))^2 = (\tilde{s}+j\tilde{z}_i)^2(\tilde{s}-j\tilde{z}_i)^2 \\ (\tilde{s}^2+\tilde{p}_i^2)^2 &= ((\tilde{s}+j\tilde{p}_i)(\tilde{s}-j\tilde{p}_i))^2 = (\tilde{s}+j\tilde{p}_i)^2(\tilde{s}-j\tilde{p}_i)^2\end{aligned}.$$

The denominator of $H(s)H(-s)$ in Equation (8.45) depends upon both the poles and the zeros of the elliptic rational function. Both these sets of roots can be expanded into a polynomial in $\tilde{s}$ with Matlab's `poly` function. If the order N is odd, this polynomial derived from the zeros must be multiplied by $-\tilde{s}^2$, which is accomplished by negating the array of polynomial coefficients and padding the end with a pair of zeros. The polynomial term, which is due to the poles of the elliptic rational function, must then also be padded with a pair of leading zeros so that the two polynomial terms can be added together. Then, the roots of the summed terms are calculated, yielding $2N$ poles.

```
% Compute poles and zeros of H(s)H(-s)
pR = ([1j;-1j;1j;-1j]*pRn); % expand denominator term
zR = ([1j;-1j;1j;-1j]*zRn); % expand numerator term
zers = zeros(2*alpha, 1); % if N odd, need to pad terms with zeros
rpH = roots([zers; real(poly(pR(:))')] + ...
      ((-1)^alpha)*(ep*A)^2*real(poly([zR(:); zers]))');
```

(c) The zeros of $H(\tilde{s})$ are just half the poles of $R_N^2(\xi,x)$. The poles of $H(\tilde{s})$ are the half of the poles of $H(\tilde{s})H(-\tilde{s})$ that have a negative real part. Finally, the constant C in Equation (8.46) is computed.

```
zH = reshape(pR([1 2], :), [], 1); % retain half the double zeros
pH = rpH(real(rpH)<0);    % retain only poles in the left half-plane
C = real(prod(-pH)/prod(-zH)) / sqrt(1+ep^2*(1-alpha));
```

The frequency response that results from this design is shown in the red curve of **Figure 8.15**. It clearly meets the design specifications: $|H(\Omega_p/\Omega_c)|_{dB} = |H(1)|_{dB} = \delta_p$ and $|H(\Omega_s/\Omega_c)|_{dB} = |H(\xi)|_{dB} = |H(2)|_{dB} =$ -40.3 dB, which is well below the stopband specification of -30 dB. Success, right?

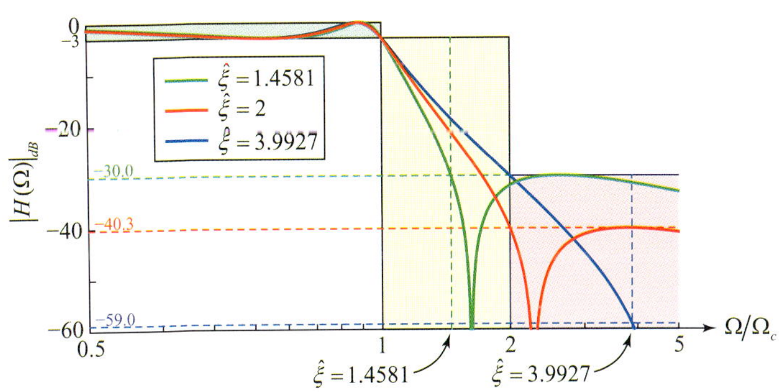

Figure 8.15 Analog elliptic filter design example

With great anticipation, we compare our results with those produced by either of Matlab's filter design functions for elliptic filters:

```
[z,p,k] = ellip(N,Rp,Rs,Wp,'s')
[b, a] = ellip(N,Rp,Rs,Wp,'s').
```

We get the curve shown in green in **Figure 8.15**. It also meets the design specifications, but it is different! What happened? A clue is provided by **Figures 8.14e** and **f**, which show the result of designing filters of order $N=3$, with various values of ξ. The red (lowest) traces in **Figures 8.14e** and **f** are the same as the "red" filter of **Figure 8.15**, both designed with $\xi=2$. At this value, Equation (8.43) gives the minimum stopband attenuation of $|H(\Omega_s/\Omega_c)|_{dB} = -40.3$. As the value of ξ decreases, the minimum stopband attenuation increases and the response becomes steeper in the transition band. Matlab's "green" filter of **Figure 8.15** was designed with a modified value of $\hat{\xi}=1.4581$, at which the minimum stopband attenuation is -30.0 dB. (This is close to the green curve in **Figures 8.14e** and **f**.) Just for completeness, the blue curve in **Figure 8.15**, corresponding to a filter designed with a value of $\hat{\xi}=3.9927$, *exactly* passes through both the passband and stopband design points: $|H(\Omega_p/\Omega_c)|_{dB} = -3$ and $|H(\Omega_s/\Omega_c)|_{dB} = -30.0$. The minimum stopband attenuation for this filter is -59.0 dB. This example shows that different elliptic filters of the same order N, designed with values of ξ in the range $1.4581 \le \xi \le 3.9927$, will all meet our specifications that $|H(\Omega_p/\Omega_c)|_{db} = \delta_p$ and $|H(\Omega_s/\Omega_c)|_{db} \le \delta_s$.

In order for us to match Matlab's result and produce the "green" curve, we just have to make a small modification to our code. Right after computing N in part (a), insert the following two lines:[3]

```
sd = 1/sqrt(1-1/L^2);
xi = 1/sqrt(1-1/Rn(N, sd, sd)^2);
```

Then, our values of the poles and zeros of $H(\tilde{s})$ will match those of Matlab exactly. What we are doing here is to find a revised value of $\hat{\xi} < \xi$ that corresponds to the value of $L_N(\hat{\xi}) = \varepsilon_s/\varepsilon_p$ for order N. This is similar to what we did with the Butterworth filter (i.e., Equation (8.14a)), the Chebyshev filter (i.e., Equation (8.30a)) and the inverse Chebyshev filter (i.e., Equation (8.36)), where we calculated a revised value of $\hat{\xi} < \xi$ that allowed us to meet the stopband specification exactly. The big difference is that with those previous filters, the revised $\hat{\xi}$ did not produce a different filter design, it only told us the value of the revised corner frequency $\hat{\Omega}_c$ at which the stopband specification would be met exactly. With the elliptic filter, the revised value of $\hat{\xi}$ creates a new filter, with different poles and zeros, that meets the design specifications exactly in an alternate manner.

8.2.6 Summary

The equations of the four prototype filters shown in **Figure 8.1** are listed in **Table 8.1**.

Table 8.1 Filter Prototypes

Filter Prototype (Passband/Stopband)	$\lvert H(\Omega)\rvert$	N	Ω_c
Butterworth (Monotonic/Monotonic)	$\frac{1}{\sqrt{1+(\Omega/\Omega_c)^{2N}}}$	$\left\lceil \frac{\log_{10} L}{\log_{10}\xi} \right\rceil$	$\Omega_p \varepsilon_p^{-1/N}$
Chebyshev (Equiripple/Monotonic)	$\frac{1}{\sqrt{1+\varepsilon_p^2 T_N^2(\Omega/\Omega_c)}}$	$\left\lceil \frac{\cosh^{-1} L}{\cosh^{-1}\xi} \right\rceil$	Ω_p
Inverse Chebyshev (Monotonic/Equiripple)	$\frac{1}{\sqrt{1+\varepsilon_s^2 T_N^{-2}(\Omega/\Omega_c)}}$	$\left\lceil \frac{\cosh^{-1} L}{\cosh^{-1}\xi} \right\rceil$	$\Omega_p \cosh\left(\frac{1}{N}\cosh^{-1} L\right)$
Elliptic (Equiripple/Equiripple)	$\frac{1}{\sqrt{1+\varepsilon_p^2 R_N^2(\xi,\Omega/\Omega_c)}}$	$\left\lceil \frac{K(1/\xi^2)K(1-1/L_N^2(\xi))}{K(1-1/\xi^2)K(1/L_N^2(\xi))} \right\rceil$	Ω_p

[3]A concise treatment of the design of elliptic filters and the different ways of meeting design specifications can be found in Dimopoulos, H.G. (2007).

8.3 ★ Impulse invariance

Once an analog filter prototype is chosen, the next step is to map the parameters of the desired discrete-time filter onto the design parameters of the chosen analog filter. There are several methods of doing this. In this section, we cover the impulse-invariance method. In Section 8.4, we discuss the bilinear-transformation method. The basic idea of the impulse-invariance method is an intuitively satisfying one: create a discrete-time filter whose impulse response is derived by sampling the sampled impulse response of an appropriately designed analog filter. It turns out that this method is only really applicable in the case of very bandlimited filters, but if you are curious why that is true, read on! Otherwise, skip ahead to the discussion of the bilinear transformation, Section 8.4.

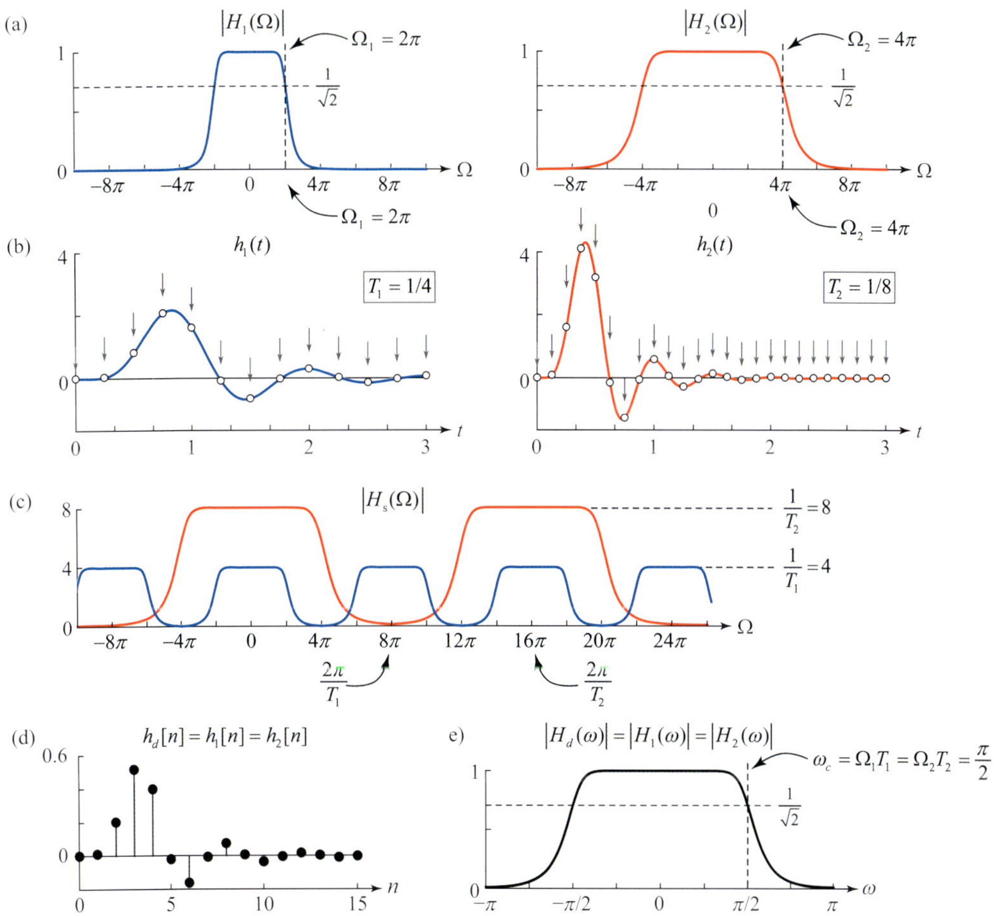

Figure 8.16 Impulse-invariance example

8.3.1 Impulse-invariance approach

The impulse-invariance method is based on performing a linear transformation of frequency between the continuous time and discrete-time domains. The essence of the approach is illustrated in **Figure 8.16**. The left panel of **Figure 8.16a** shows the magnitude of the frequency response of an analog Butterworth lowpass filter, $|H_1(\Omega)|$. The filter has been designed using the approach of Section 8.2.2 to meet specifications $\Omega_p = 2\pi$, $\delta_p = -3$ dB, $\Omega_s = 4\pi$ and $\delta_s = -40$ dB. Given these specifications, the design parameters of the filter given by Equations (8.10) and (8.11) are order $N = 7$ and corner frequency $\Omega_1 = 2\pi$. The left panel of **Figure 8.16b** shows the impulse response of this filter, $h_1(t)$. To form the impulse response of the corresponding discrete-time filter, we sample the analog response at a sampling period of $T_1 = 1/4$ s, as shown by the arrows in the figure (we will explain this choice of sampling period in a moment). As we discussed in Chapter 6, the transform of the sampled analog signal, $H_s(\Omega)$, comprises replicas of $H_1(\Omega)$ at multiples of the sampling frequency $\Omega_s = 2\pi/T_1 = 8\pi$ scaled by $1/T_1 = 4$, as shown by the red curve in **Figure 8.16c**,

$$H_s(\Omega) = \frac{1}{T_1} \sum_{k=-\infty}^{\infty} H_1(\Omega - k\Omega_s). \tag{8.47}$$

The discrete-time sequence is obtained by taking the sampled analog values and scaling them by the sampling period, $h_1[n] = T_1 h_1(nT_1)$, as shown in **Figure 8.16d**. The corresponding DTFT $H_1(\omega)$, shown in **Figure 8.16e**, is simply $H_s(\Omega)$ normalized in frequency by the sample rate so that $\omega = 2\pi\Omega/\Omega_s = \Omega T_1$, and also scaled in magnitude by T_1,

$$H_1(\omega) = T_1 H_s(\Omega)|_{\Omega = \omega/T_1}. \tag{8.48}$$

The corner frequency of the discrete-time filter $H_1(\omega)$ is just the corner frequency of the analog filter $H_1(\Omega)$ normalized by the sample rate: $\omega_1 = \Omega_1 T_1 = \pi/2$.

To the extent that the sampling produces no aliasing, the image replicas in Equation (8.47) (i.e., $H_1(\Omega - k\Omega_s)$ for $k \neq 0$) do not contribute to the sum. Then, the frequency response of the discrete-time filter becomes essentially equivalent to that of the analog filter with a frequency axis ω that is equal to Ω normalized by the sampling rate, as given by Equation (8.48). Thus, at least in theory, we can create a discrete-time filter by sampling the impulse response of an analog filter. We will soon see that this does not work so well in practice.

Before we go on, it is important to recognize that the transformation between the analog filter and the discrete-time filter in this example is not unique. To see this, consider a second Butterworth lowpass filter whose magnitude $|H_2(\Omega)|$ is shown in the right panel of **Figure 8.16a**. This filter, designed to meet specifications $\Omega_p = 4\pi$, $\delta_p = -3$ dB, $\Omega_s = 8\pi$ and $\delta_s = -40$ dB, again has order $N = 7$, but has a corner frequency of $\Omega_2 = 4\pi$, double the value of the previous filter. Hence, $H_2(\Omega) = H_1(\Omega/2)$. As expected from the properties of Fourier transforms, the impulse response of this new filter, $h_2(t)$, shown in the right panel of **Figure 8.16b**, is scaled by a factor of two in amplitude and a factor of one-half in time compared with $h_1(t)$: $h_2(t) = 2h_1(2t)$. If we now sample $h_2(t)$ with a period of $T_2 = 1/8$ s (in other words, half T_1), as shown by the arrows in **Figure 8.16b**, the transform of the resulting sampled analog signal comprises replicas of $H_2(\Omega)$ at multiples of the sampling frequency $\Omega_s = 2\pi/T_2 = 16\pi$, scaled by $1/T_2 = 16$, as shown by the red curve in

Figure 8.16c. Again, we can form a discrete-time sequence by taking the sampled analog values and scaling them by the sampling period: $h_2[n] = T_2 h_2(nT_2)$. This sampled sequence is the same as $h_1(t)$. Since $h_2(t) = 2h_1(2t)$ and $T_1 = 2T_2$, we have

$$h_2[n] = T_2 h_2(nT_2) = 2T_2 h_1(n2T_2) = T_1 h_1(nT_1) = h_1[n].$$

The DTFT $H_2(\omega)$ shown in **Figure 8.16e** is therefore the same as $H_1(\Omega)$. We can also arrive at the identical conclusion by normalizing the red curve in **Figure 8.16c** by the sampling frequency $\Omega_s = 16\pi$ and scaling the amplitude by $T_2 = 1/8$. The corner frequency of $H_2(\omega)$ is $\omega_2 = \Omega_2 T_2 = \pi/2$, which is the same as the value of ω_1 we found previously.

This example shows that neither the cutoff frequency of the analog filter Ω_c nor the sampling rate T matters by itself. All that matters is the *product* $\Omega_c T$. So, if we desire a discrete-time lowpass filter with a cutoff of $\pi/2$, we can sample the impulse response of an analog Butterworth filter with a cutoff frequency of $\Omega_c = 2\pi$ at $T = 1/4$ s, or sample the response of an analog filter with a cutoff frequency of $\Omega_c = 4\pi$ at $T = 1/8$ s, or any other combination of Ω_c and T such that $\omega_c = \Omega_c T = \pi/2$. In fact, we could arbitrarily decide to set $T = 1$, in which case, $\omega_c = \Omega_c$.

8.3.2 Impulse-invariance design procedure

With this example in mind, let us now lay out a systematic procedure for the design of a discrete-time filter using the impulse-invariance method and then do a few examples. The conceptual steps underlying the impulse-invariance design procedure are illustrated in **Figure 8.17**.

Step 1: Start with the critical parameters of the digital filter. Since we are designing a discrete-time filter, the first step in the design is to specify the important parameters of the filter *in the discrete-time domain*. For a lowpass filter, these parameters are the passband and stopband attenuations and frequencies: δ_p, δ_s, ω_p and ω_s.

Step 2: Choose an analog prototype and map the parameters of the digital filter to critical parameters of the analog filter. Next, choose an analog filter prototype and map the critical parameters of the discrete-time filter that were specified in Step 1 into the critical parameters of the analog filter. In keeping with our previous example, we will design a discrete-time filter by sampling the impulse response of an analog Butterworth filter. We therefore need to specify four parameters: δ_p, δ_s, Ω_p and Ω_s. From the discussion of **Figure 8.16**, the critical frequencies in the continuous- and discrete-time domains are related by a linear mapping $\omega = \Omega T$, where T is the sampling rate. Hence,

$$\Omega_p T = \omega_p \text{ and } \Omega_s T = \omega_s.$$

As we have discussed above, we do not have to specify a sampling parameter T, only the products $\Omega_p T$ and $\Omega_s T$.

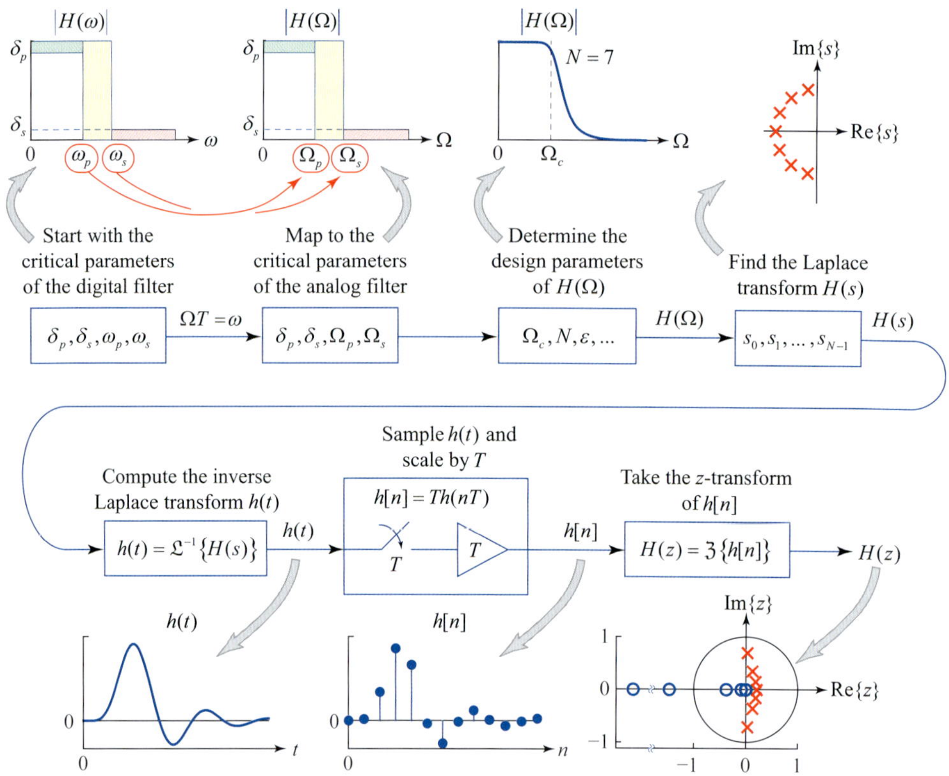

Figure 8.17 Impulse-invariance transformation design procedure

The transformation from the continuous-time to discrete-time domain amounts to a scaling of the frequency axis, but the amplitude axis is not affected by this transformation. Hence the values of the passband and stopband attenuations for the analog filter are the same as those of the digital filter.

Step 3: Determine the design parameters of the analog filter $H(\Omega)$. Having specified critical parameters of the analog filter use the results of the beginning of this chapter to design the analog filter $H(\Omega)$. For a Butterworth filter, we use the design formulas, Equations (8.10) and (8.11), to calculate the filter order N and corner frequency Ω_c. From Equation (8.10),

$$N = \left\lceil \frac{\log_{10}\left(\dfrac{10^{-\delta_s/10}-1}{10^{-\delta_p/10}-1}\right)}{2\log_{10}(\Omega_s/\Omega_p)} \right\rceil = \left\lceil \frac{\log_{10}\left(\dfrac{10^{-\delta_s/10}-1}{10^{-\delta_p/10}-1}\right)}{2\log_{10}(\omega_s/\omega_p)} \right\rceil.$$

Notice that the denominator of this expression depends only on the ratio of $\Omega_s/\Omega_p = \Omega_s T/\Omega_p T = \omega_s/\omega_p$, so the value of the sample rate T disappears from the calculation of N. From Equation (8.11),

$$\Omega_c = \Omega_p(10^{-\delta_p/10}-1)^{-1/2N}.$$

Multiplying both sides by T gives

$$\underbrace{\Omega_c T}_{\omega_c} = \underbrace{\Omega_p T}_{\omega_p} \cdot \left(10^{-\delta_p/10} - 1\right)^{-1/2N},$$

or, equivalently,

$$\omega_c = \omega_p \cdot (10^{-\delta_p/10} - 1)^{-1/2N}.$$

Since N, Ω_p and δ_p are all given or computed, we can compute ω_c. It would appear that we still need to specify a value of T if we wish to calculate the corner frequency of the analog filter Ω_c, but in a moment we will see that we do not need T, only the product $\omega_c = \Omega_c T$.

Step 4: Find the Laplace transform *H*(*s*). The poles of the Laplace transform are given by Equation (8.18),

$$p_k = \Omega_c e^{j\pi(2k+N+1)/2N}, \quad 0 \le k < N.$$

Equation (8.20) expresses the Laplace transform $H(s)$ as the product of terms, each of which contains one pole:

$$H(s) = \prod_{k=0}^{N-1} \frac{-p_k}{s - p_k}.$$

Step 5: Compute the inverse Laplace transform *h*(*t*). Find the impulse response of the analog filter $h_a(t)$ by taking the inverse Laplace transform of $H(s)$. The basic approach is first to express $H(s)$ as the sum of terms using the partial fraction decomposition,

$$H(s) = \sum_{k=0}^{N-1} \frac{R_k}{s - p_k},$$

where R_k are termed the **residues**. To aid in what follows, we define the normalized pole,

$$\hat{p}_k \triangleq p_k/\Omega_c = e^{j\pi(2k+N+1)/2N}, \quad 0 \le k < N,$$

which only depends on k and N. We can show (Problem 8-18) that the residues can be written as $R_k = \Omega_c \hat{R}_k$, where the normalized residue,

$$\hat{R}_k = \frac{e^{j\pi(2k+N+1)/2N}}{\prod_{\substack{l=0 \\ l \ne k}}^{N-1} (1 - e^{j\pi(k-l)/N})}, \tag{8.49}$$

only depends on k and N. The inverse Laplace transform of each term of $H(s)$ is

$$\mathfrak{L}^{-1}\left\{\frac{R_k}{s - p_k}\right\} = R_k e^{p_k t} u(t) = \Omega_c \hat{R}_k e^{\Omega_c \hat{p}_k t} u(t),$$

so,

$$h(t) = \Omega_c \sum_{k=0}^{N-1} \hat{R}_k e^{\Omega_c \hat{p}_k t} u(t).$$

Step 6: Sample $h(t)$ and scale by T. Now sample $h(t)$ at multiples of the sample period $t = nT$, and scale the result by T,

$$h[n] = Th(t)|_{t=nT} = Th(nT) = (\Omega_c T)\sum_{k=0}^{N-1} \hat{R}_k e^{n(\Omega_c T)\hat{p}_k} u[n] = \omega_c \sum_{k=0}^{N-1} \hat{R}_k e^{n\omega_c \hat{p}_k} u[n].$$

As we discussed in conjunction with **Figure 8.16**, $h[n]$ only depends on the product $\Omega_c T$, which is equal to the corner frequency of the discrete-time filter, $\omega_c = \Omega_c T$.

Step 7: Take the z-transform of $h[n]$ to get $H(z)$. Given $h[n]$, compute the z-transform $H(z)$, from which the appropriate difference equation can be derived. Each term of the $h[n]$ summation has z-transform

$$\mathfrak{Z}\{\hat{R}_k e^{n\omega_c \hat{p}_k} u[n]\} = \frac{\hat{R}_k}{1 - e^{\omega_c \hat{p}_k} z^{-1}},$$

so the whole summation is

$$H(z) = \sum_{k=0}^{N-1} \frac{\hat{R}_k}{1 - e^{\omega_c \hat{p}_k} z^{-1}}.$$

There is one more simplification we can make. Since only the product $\omega_c = \Omega_c T$ is relevant in our derivation, we could have simplified things at the outset by simply setting $T = 1$ so that $\Omega_c = \omega_c$. Making this substitution would mean that $\omega_c \hat{p}_k = \Omega_c \hat{p}_k = p_k$ and $\omega_c \hat{R}_k = \Omega_c \hat{R}_k = R_k$. Hence, we could just write

$$H(z) = \sum_{k=0}^{N-1} \frac{R_k}{1 - e^{p_k} z^{-1}}. \quad \textbf{(8.50)}$$

For values of $N > 1$, some or all of the residues R_k and the roots p_k will be complex. Thus, for any practical filtering application, $H(z)$ needs to be expressed in such a way that it does not contain complex terms. We can do this by combining terms of the summation so that $H(z)$ contains no complex quantities. This is best explained with a few progressively more complicated examples. Example 8.8 below presents a simple one-pole example. Supplementary material available at www.cambridge.org/holton contains further examples with multiple, complex poles.

Example 8.8

Use the impulse-invariance transformation to find the difference equation of a discrete-time lowpass filter with $\omega_p = 0.1\pi$, $\delta_p = -3$ dB, $\omega_s = 0.8\pi$ and $\delta_s = -15$ dB.

➤ Solution:

Given the specifications, we compute $N = 1$ and $\omega_c = 1.0024 \cdot \omega_p = 0.3149$. There is a single pole at $p_k = \Omega_c e^{j\pi} = -\Omega_c$, so $H(s)$ just has one term,

$$H(s) = \frac{\Omega_c}{s - p_k} = \frac{\Omega_c}{s + \Omega_c}.$$

The inverse of $H(s)$ is

$$h(t) = \Omega_c e^{-\Omega_c t} u(t).$$

The discrete-time sequence derived from sampling $h(t)$ at $t = nT$ and scaling by T is

$$h[n] = \Omega_c T e^{-\Omega_c T n} u[n] = \omega_c e^{-\omega_c n} u[n].$$

As expected, the term $\Omega_c T$ can be replaced by ω_c everywhere it appears. The z-transform is

$$H(z) = \frac{\omega_c}{1 - e^{-\omega_c} z^{-1}} = \frac{0.3149}{1 - 0.7299 z^{-1}}.$$

The difference equation for this simple first-order filter is $y[n] - 0.7299y[n-1] = 0.3149x[n]$.

Figure 8.18 shows a comparison of $H(\omega)$ and $H(\Omega)$ for a filter designed with the specifications given in this example.

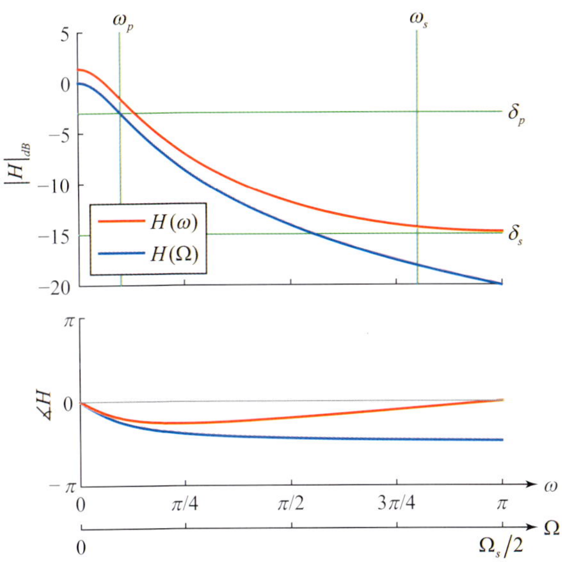

Figure 8.18 Impulse-invariance IIR filter with $N = 1$

The red curves show the magnitude (on a dB scale) and phase of $H(\omega)$, the frequency response of the discrete-time filter. The blue curves show $H(\Omega)$, the frequency response of the prototype analog filter, plotted with frequency normalized by the sampling frequency $\Omega_s = 2\pi/T$, so that a continuous-time frequency of $\Omega = \Omega_s/2$ corresponds to a discrete-time frequency of $\omega = 2\pi(\Omega_s/2)/\Omega_s = \pi$. The critical design parameters ω_p, δ_p, ω_s and δ_s are also shown with green lines. You can see that the frequency response $H(\omega)$ neither meets the design specifications, nor does it match $H(\Omega)$ well at any frequency. What is going on here?

The reason for the disparity in the frequency responses of the previous example is **aliasing**, which is an unavoidable artifact of the impulse-invariance method. **Figure 8.19** shows why this aliasing occurs.

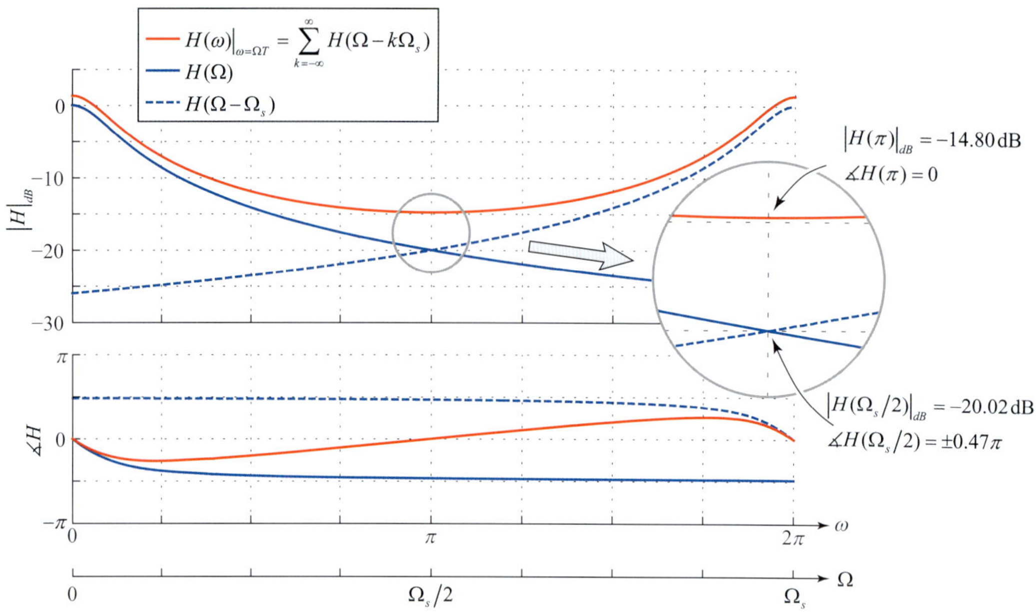

Figure 8.19 Aliasing of impulse-invariance IIR filter with $N = 1$

The red traces show the magnitude and phase of $H(\omega)$, the frequency response of the discrete-time filter. As we have discussed, $H(\omega)$ is equal to $H_s(\Omega)$, the transform of the sampled analog signal, with amplitude scaled by the sampling period T, and frequency scaled by the sampling frequency so that $\omega = \Omega T = 2\pi\Omega/\Omega_s$,

$$H(\omega)|_{\omega=\Omega T} = TH_s(\Omega) = \sum_{k=-\infty}^{\infty} H(\Omega - k\Omega_s). \tag{8.51}$$

$H_s(\Omega)$ comprises the sum of replicas of $H(\Omega)$ at multiples of the sampling frequency. The solid blue trace in **Figure 8.19** shows $H(\Omega)$ over the frequency range $0 < \Omega < \Omega_s$. The dashed blue trace shows $H(\Omega - \Omega_s)$, the image replica centered at $\Omega = \Omega_s$, over the same frequency range. If there were no aliasing in the sampling process, the replicas would be disjoint (i.e., non-overlapping in frequency). However, in this example there is a clear overlap between $H(\Omega)$ and image replicas centered at multiples of Ω_s, such as $H(\Omega - \Omega_s)$. That is aliasing. The biggest effect is at half the sampling frequency, $\Omega = \Omega_s/2$, which is equivalent to the discrete-time frequency $\omega = \pi$. Here, the value of the baseband replica $H(\Omega_s)$ adds to the value of the image replica centered at $\Omega = \Omega_s$, $H(-\Omega_s)$. If we assume that replicas at other multiples of Ω_s contribute negligibly to the sum, we get

$$H(\pi) = H_s(\Omega_s/2) \cong H(\Omega_s/2) + H(-\Omega_s/2).$$

In this example, you can show (Problem 8-21) that

$$|H(\pi)|_{dB} = -42.12 + 3.23 = -38.85\,\text{dB}.$$

The effect of aliasing is to reduce the attenuation of the lowpass filter by about 3 dB at frequencies near $\omega = \pi$. Since aliasing results from the overlap of replicas of $H(\Omega)$ centered at multiples of Ω_s, anything that increases the amount of overlap – for example, increasing the corner frequency ω_c or decreasing the slope N – makes aliasing worse.

Because this effect is the result of aliasing, it is tempting to say, "Just sample faster! Make T smaller!" But remember: we cannot choose the sampling parameter T independently of the corner frequency Ω_c. We can only choose the *product* $\Omega_c T$, and we have to choose that product so that it maps the desired specifications of the discrete-time filter into the specifications of the analog filter. In the current example, with the corner frequency of $\omega_c = \Omega_c T = \pi/2$, if we were to double the sampling rate, we would have to halve Ω_c to maintain the same value of the desired corner frequency.

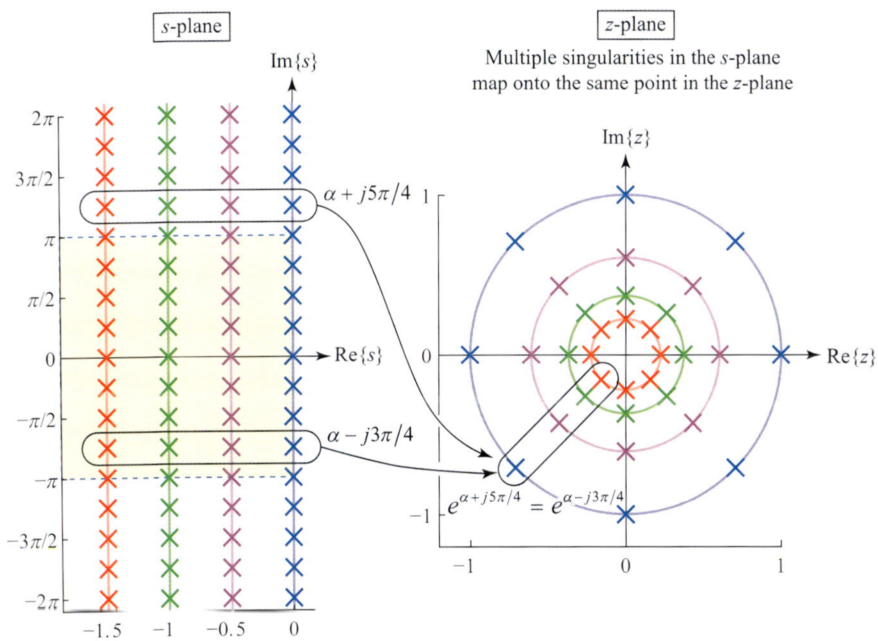

Figure 8.20 Impulse-invariance mapping the s-plane into the z-plane

8.3.3 Mapping of the s-plane to the z-plane

It is particularly useful to look at the impulse-invariance method as a general mapping between points in the s-plane and z-plane, as shown in **Figure 8.20**. From Equation (8.50), each point in the s-plane, $s = \alpha + j\Omega T$, maps into a point in the z-plane at position

$$z = e^s = e^\alpha \cdot e^{j\Omega T}.$$

The left panel of **Figure 8.20** shows lines in the s-plane, $s = \alpha + j\Omega T$, with fixed real part α (different colors), and imaginary part Ω in the range $-\infty < \Omega < \infty$. These lines map into circles in the z-plane, shown in the right panel. Hence, points on the $j\Omega$-axis of the s-plane (i.e., $\alpha = 0$, blue line and symbols) map onto the unit circle of the z-plane, and lines in the s-plane with real part α heading towards $-\infty$ (magenta, green and red lines and symbols) map to concentric circles in the z-plane with radii heading towards 0. The crux of the problem is that the mapping of points in the s-plane to the z-plane is not unique. Points in the s-plane with imaginary parts that differ by multiples of 2π, $s = \alpha + j(\Omega T + 2\pi k)$, map to the *same* point in the z-plane, because

$$z = e^{\alpha + j(\Omega T + 2\pi k)} = e^{\alpha + j\Omega T} \cdot e^{j2\pi k} = e^{\alpha + j\Omega T}.$$

This means that strips in the s-plane of width $2\pi j$, one of which is shown shaded in yellow in **Figure 8.20**, all map into the interior of the unit circle. For example, the figure shows that points in the s-plane at $\alpha + j5\pi/4$ map to the identical positions in the z-plane as points at $\alpha - j3\pi/4$,

$$H(z)\big|_{z=e^{sT}} = \sum_{k=-\infty}^{\infty} H(s - jk\Omega_s).$$

If $\alpha < 0$, the point is in the left half of the s-plane, and

$$|z| = |e^{\alpha}| < 1.$$

The impulse-invariance transformation does have the useful property that the left half of the s-plane maps into the interior of the unit circle of the z-plane, so any pole that is stable in the s-plane maps into a stable pole in the z-plane.

Figure 8.21 shows the mapping between the s- and z-planes for Example 8.8.

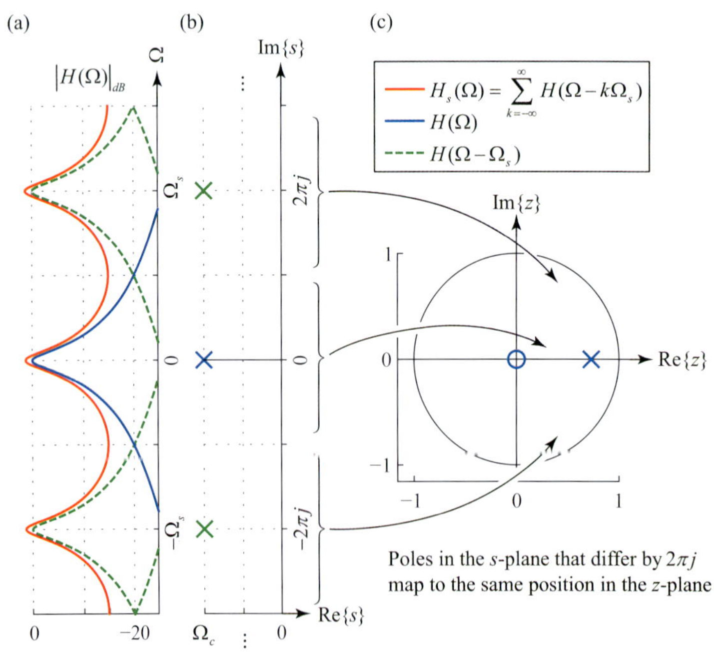

Figure 8.21 Mapping of the impulse-invariance IIR filter with $N = 1$

The blue curve in **Figure 8.21a**, centered around $\Omega = 0$, shows the frequency response $H(\Omega)$ of a first-order Butterworth filter with cutoff frequency Ω_c. This filter has a single real pole, shown in blue on the s-plane of **Figure 8.21b**. When an IIR filter is designed by the impulse-invariance method using this analog filter as a prototype, this pole maps into a pole in the z-plane at $z = e^s = e^{-\Omega_c T}$, as shown in **Figure 8.21c**. However, as we have discussed above, the impulse-invariance transformation actually maps the transform of the sampled analog response $H_s(\Omega)$ into the z-plane, so there are effectively image bands of poles in the s-plane at positions $s + j\Omega_s k$, $k \neq 0$, shown in green. In this figure, we have let $T = 1$ so $\Omega_s = 2\pi$. These bands of image poles correspond to the image components of the frequency response $H(\Omega - k\Omega_s)$, which are shown in green in **Figure 8.21a**. The image components add to the baseband to produce $H_s(\Omega)$, shown in red,

$$H_s(\Omega) = \sum_{k=-\infty}^{\infty} H(\Omega - k\Omega_s).$$

It is this transform, $H_s(\Omega)$, that is linearly related to $H(\omega)$ by $\omega = \Omega T = 2\pi\Omega/\Omega_s$, and it is the image components of the response that are ultimately responsible for the aliasing seen in $H(\omega)$.

As we just saw, the impulse-invariance transformation does have the nice property that the $j\Omega$-axis of the s-plane (i.e., $\alpha = 0$) maps onto the unit circle of the z-plane,

$$z = e^{j\omega} = e^{j\Omega T},$$

so $H_s(\Omega)$ is directly related to $H(\omega)$. But as we expect from our discussion of **Figure 8.20**, the mapping between ΩT and ω is not unique, as shown in **Figure 8.22**.

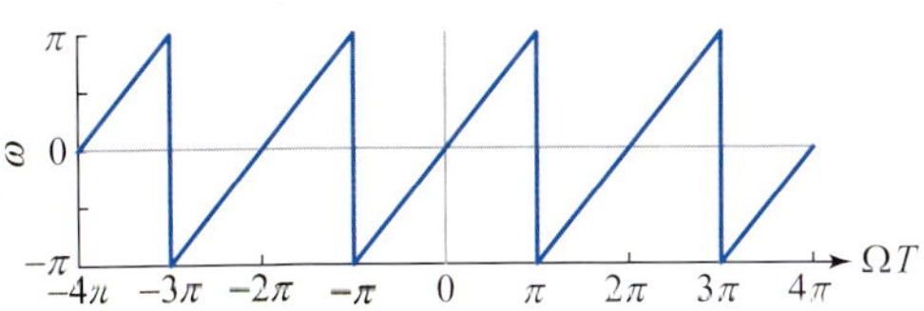

Figure 8.22 Mapping $j\Omega$ axis into the unit circle for the impulse-invariance transformation

The discrete-time frequency ω is linearly related to ΩT in the baseband range $-\pi \leq \Omega T < \pi$, as shown, but since $e^{j(\Omega T + 2\pi k)} = e^{j\Omega T}$, values of ΩT differing from each other by a multiple of 2π map into the same value of ω.

The effect of aliasing depends on the shape of the prototype filter. As the required corner frequency of a lowpass filter increases, the amount of aliasing increases. **Figure 8.23** shows the response of a first-order filter designed with the impulse-invariance transformation as a function of bandwidth ω_c. As $\omega_c \to \pi$, the amount of effect of aliasing increases dramatically, and noticeably affects the response at all frequencies.

Supplementary material continues the discussion of the impulse-invariance method with a number of examples.

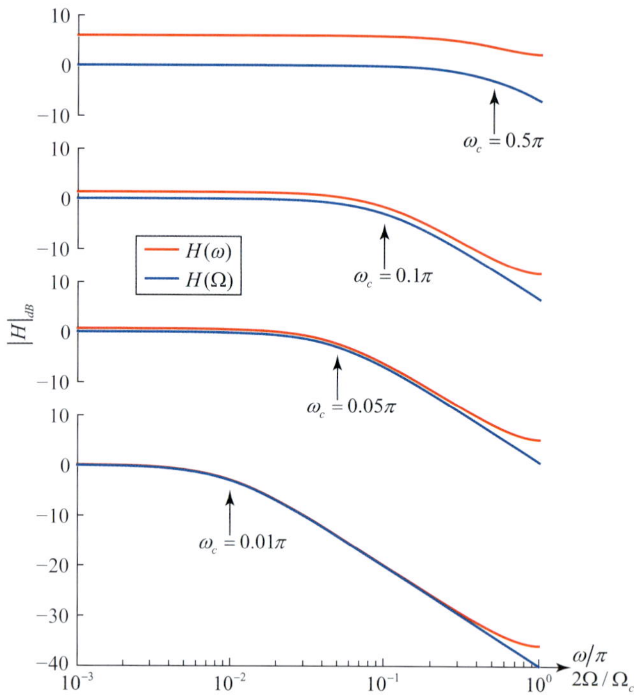

Figure 8.23 Aliasing of a first-order filter as a function of bandwidth

8.4 Bilinear transformation

There are other methods of mapping the s-plane into the z-plane, several of which can be derived from the numerical methods used to approximate continuous-time integration and differentiation. We will review a few potential candidates and show that one of them in particular, the **bilinear transformation** has some very nice properties.

To motivate the discussion that follows, consider the case where we are trying to design a discrete-time filter using a simple first-order analog Butterworth filter as a prototype, just as we did in Example 8.8. The Laplace transform of this filter is

$$H(s) = \frac{\Omega_c}{s + \Omega_c},$$

which corresponds to the differential equation

$$\frac{dy(t)}{dt} + \Omega_c y(t) = \Omega_c x(t).$$

In Section 8.3, we showed that we ran into trouble if we tried to "discretize" the impulse response by sampling. The approach we will take here is to investigate what happens if we "discretize" this differential equation instead, thereby producing a linear constant-coefficient difference equation (LCCDE). To understand the issues involved in converting a continuous-time differential equation to a discrete-time difference equation, let us start with a much simpler system, an analog differentiator defined by the input–output relation

$$\frac{dy(t)}{dt} = x(t), \tag{8.52a}$$

or the equivalent integral

$$y(t) = \int_{-\infty}^{t} x(\tau)d\tau. \tag{8.52b}$$

This system has Laplace transform

$$H(s) = \frac{Y(s)}{X(s)} = \frac{1}{s}. \tag{8.53}$$

We seek to find a discrete-time approximation of this system by sampling the input and output at period T, so that $t = nT$, $x(t) = x(nT)$ and $y(t) = y(nT)$. Then, Equation (8.52b) can be written as

$$y(nT) = \int_{-\infty}^{nT} x(\tau)d\tau = \int_{-\infty}^{(n-1)T} x(\tau)d\tau + \int_{(n-1)T}^{nT} x(\tau)d\tau = y((n-1)T) + \int_{(n-1)T}^{nT} x(\tau)d\tau,$$

or

$$y(nT) - y((n-1)T) = \int_{(n-1)T}^{nT} x(\tau)d\tau.$$

In words, the value of the integral at time point $t = nT$ (i.e., $y(nT)$) is equal to the value of the integral at the previous time point $t = (n-1)T$ (i.e., $y(n-1)T$), plus the area under the curve $x(t)$ from $(n-1)T < t < nT$. This area is shaded in yellow in **Figure 8.24a**.

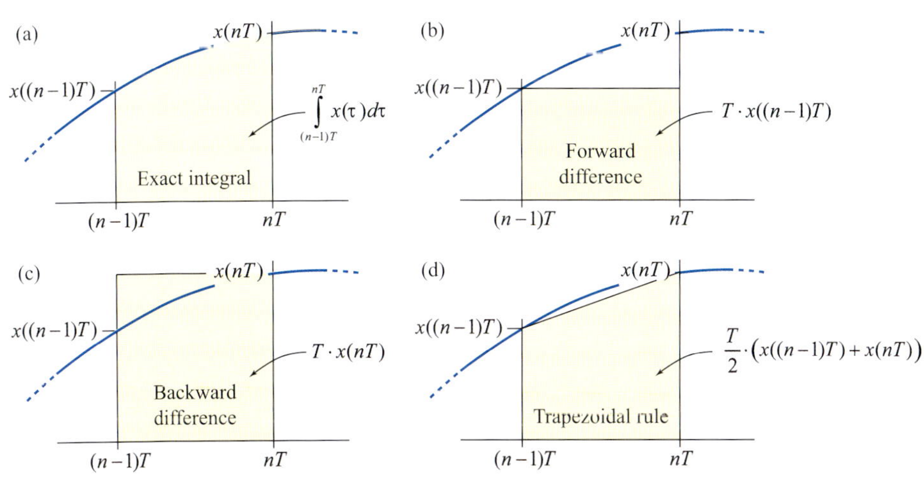

Figure 8.24 Discrete-time approximations of an integral

8.4.1 Forward-difference approximation

There are a several widely used approximations of this exact integral. One method, shown in **Figure 8.24b**, is to approximate the integral with the initial value of the interval, $x((n-1)T)$, multiplied by the length of the interval, T,

$$y(nT) - y((n-1)T) = T \cdot x((n-1)T).$$

Converting this to a difference equation yields $y[n] - y[n-1] = T \cdot x[n-1]$. Letting $n = n+1$ gives

$$x[n] = \frac{1}{T}(y[n+1] - y[n]).$$

We have effectively defined a discrete-time approximation of the derivative of Equation (8.52a) as the difference between a future value of the output $y[n+1]$ and the current value $y[n]$. This is called the **forward-difference approximation**. Taking the z-transform of both sides gives

$$H(z) = \frac{Y(z)}{X(z)} = T \cdot \frac{1}{z-1}. \tag{8.54}$$

Comparing Equations (8.53) and (8.54) leads to the conclusion that

$$H(z) = H(s)|_{s=\frac{1}{T}(z-1)}.$$

This equation therefore defines a mapping of the s-plane to the z-plane,

$$s = \frac{1}{T}(z-1).$$

This mapping is shown in **Figure 8.25** for a sample value of $T = 1$.

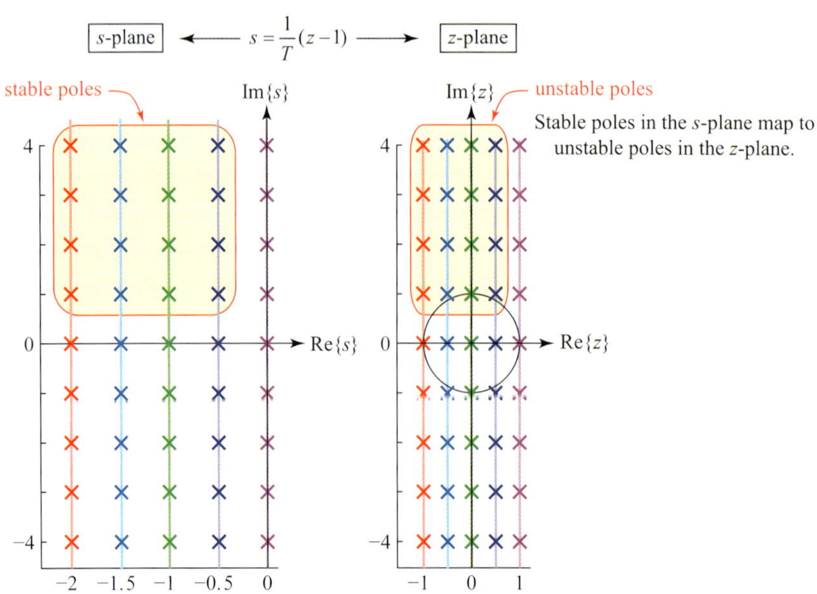

Figure 8.25 Mapping of the forward-difference approximation

There are a couple of big problems with this mapping, the biggest of which is that the left half of the s-plane does not map into the inside of the unit circle of the z-plane. Hence, stable poles in the s-plane do not necessarily map into stable poles in the z-plane, as shown in the figure. For this reason, we will not consider this approximation any further.

8.4.2 Backward-difference approximation

Another approach to approximating the exact integral over the interval $(n-1)T < t < nT$ is shown in **Figure 8.24c**. Here, the integral is approximated by the final value of the interval $x(nT)$, multiplied by the length of the interval T. Hence, $y(nT) - y((n-1)T) = T \cdot x(nT)$, which yields the difference equation

$$x[n] = \frac{1}{T}(y[n] - y[n-1]).$$

This is called the **backward-difference approximation**, since it is based on taking the difference between the current value of the output $y[n]$ and the past value $y[n-1]$. The z-transform is

$$H(z) = \frac{Y(z)}{X(z)} = T \cdot \frac{1}{1 - z^{-1}},$$

from which we conclude that the backward-difference approximation corresponds to the mapping between the s-plane and z-plane,

$$s = \frac{1}{T}(1 - z^{-1}),$$

as shown in **Figure 8.26**.

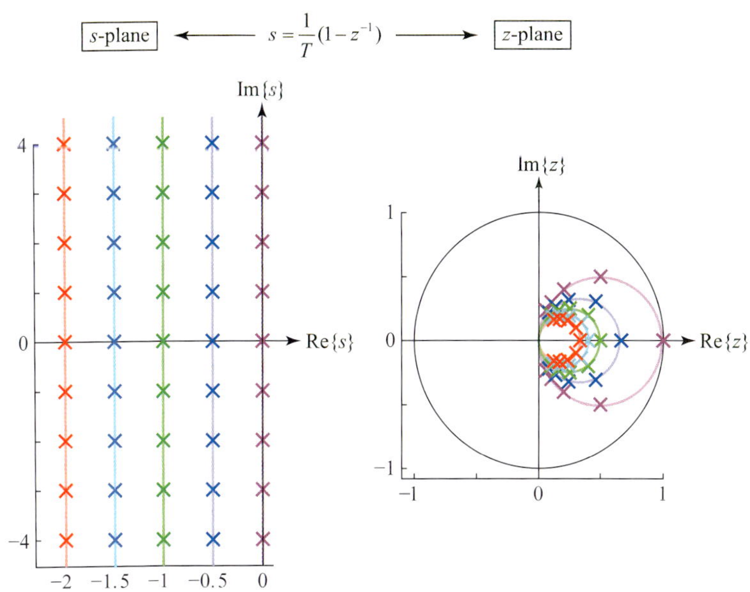

Figure 8.26 Mapping of the backward-difference approximation

In the backward-difference approximation, a line in the s-plane, $s = \alpha + j\Omega$, with fixed real part α and an imaginary part varying from $-\infty < \Omega < \infty$, maps into a circle in the z-plane with center $1/2(1+\alpha T)$ and radius $1/2(1+\alpha T)$ (see Problem 8-24). Points in the left half of the s-plane map entirely inside the unit circle (see Problem 8-23), so that stable poles in $H(s)$ map to stable poles in $H(z)$. That is good. However, as you can see, the backward-difference approximation does not map the *entire* left half-plane into the entire interior of the unit circle. Only a limited range of pole positions is possible in the z-plane. So, we dispense with this approximation as well.

8.4.3 Bilinear transformation

Figure 8.24d shows a more accurate approximation of the exact integral than either the forward-difference or backward-difference methods. Here, the integral is approximated with a trapezoid with vertices $x((n-1)T)$ and $x(nT)$, and base T. The area of the trapezoid is easily found to be

$$y(nT) - y((n-1)T) = \frac{T}{2}(x((n-1)T) + x(nT)),$$

which leads to the z-transform,

$$H(z) = \frac{T}{2}\frac{1+z^{-1}}{1-z^{-1}}.$$

So, the mapping between the s-plane and z-plane is given by

$$s = \frac{2}{T}\frac{1-z^{-1}}{1+z^{-1}} = \frac{2}{T}\frac{z-1}{z+1}, \tag{8.55}$$

as shown in **Figure 8.27**. This is called the **bilinear transformation**, and it has some nice properties that make it the most widely used of all the transformations for the design of IIR filters.

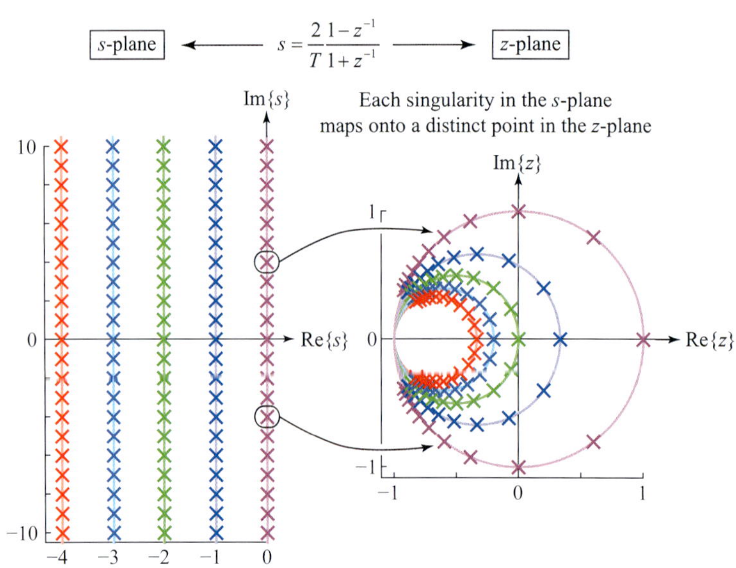

Figure 8.27 Bilinear transformation

In the bilinear transformation, lines in the s-plane, $s=\alpha+j\Omega$, with fixed real part α and imaginary part varying from $-\infty<\Omega<\infty$, map into circles in the z-plane (see Problem 8-26). Like the impulse-invariance transformation, the left half of the s-plane maps to the interior of the unit circle (see Problem 8-25), so stable poles in $H(s)$ map to stable poles in $H(z)$. Furthermore, this is a **complete mapping**, meaning that the entire left half of the s-plane maps into the entire inside of the unit circle. But, unlike the impulse-invariance transformation, the mapping of the bilinear transformation is *unique*, meaning that every point in the s-plane maps to a unique point in the z-plane. In particular, the bilinear transformation maps the entire $j\Omega$-axis of the s-plane onto the unit circle of the z-plane in a unique, invertible manner. This can easily be shown by substituting the equation of the unit circle, $z=e^{j\omega}$, into Equation (8.55),

$$\frac{2}{T}\frac{1-z^{-1}}{1+z^{-1}}\bigg|_{z=e^{j\omega}}=\frac{2}{T}\frac{1-e^{-j\omega}}{1+e^{-j\omega}}=\frac{2}{T}\frac{\cancel{e^{-j\omega/2}}}{\cancel{e^{-j\omega/2}}}\frac{e^{j\omega/2}-e^{-j\omega/2}}{e^{j\omega/2}+e^{-j\omega/2}}=j\frac{2}{T}\frac{\sin\omega/2}{\cos\omega/2}=j\frac{2}{T}\tan\omega/2=j\Omega.$$

Thus, the mapping between the discrete-time frequency ω and continuous-time frequency Ω is given by

$$\Omega T=2\tan\frac{\omega}{2}, \tag{8.56a}$$

or, equivalently, by the inverse,

$$\omega=2\tan^{-1}\frac{\Omega T}{2}. \tag{8.56b}$$

This means that the entire $j\Omega$-axis, from $-\infty\le\Omega<\infty$, maps into the unit circle from $-\pi\le\omega<\pi$, as shown in **Figure 8.28**.

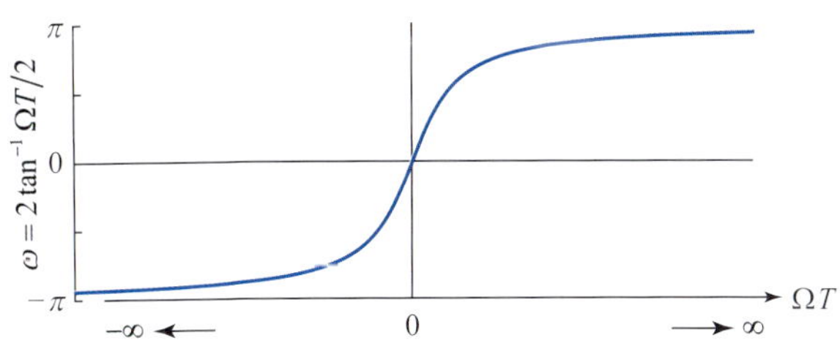

Figure 8.28 Mapping of the $j\Omega$-axis into the unit circle for the bilinear transformation

When $\Omega T=0$, then $\omega=0$. As $\Omega T\to\pm\infty$, then $\omega\to\pm\pi$. While the mapping between the $j\Omega$-axis and ω for the impulse-invariance transformation, shown in **Figure 8.22**, was piece-wise linear and non-unique, the mapping of the bilinear transformation is nonlinear and unique.

8.4.4 Bilinear-transformation procedure

The conceptual steps underlying the bilinear-transformation design procedure are illustrated in **Figure 8.29** for the Butterworth filter.

Step 1: Specify the critical parameters of the digital filter. The first step in the design of an IIR filter using the bilinear transformation is the same as the procedure for the impulse-invariance transformation: specify the critical parameters of the filter in the discrete-time domain.

Step 2: Prewarp the critical parameters of the analog filter. Choose an analog filter prototype and map the critical parameters of the discrete-time filter that were specified in Step 1 into

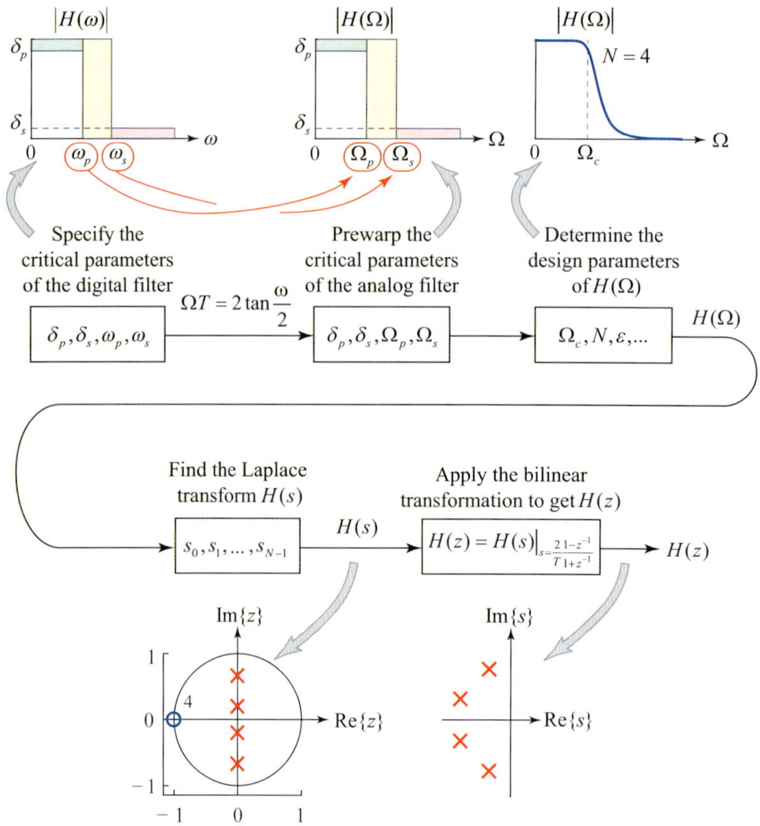

Figure 8.29 Bilinear-transformation design procedure

the critical parameters of the analog filter. In the impulse-invariance transformation, the relation between the continuous-time and discrete-time frequencies is simply linear: $\omega = \Omega T$. However, in the bilinear transformation, the relation between ω and Ω, given by Equation (8.56), is nonlinear. In order for the discrete-time filter to have a passband frequency ω_p, the analog passband frequency Ω_p must be chosen so that

$$\Omega_p T = 2\tan(\omega_p/2). \quad \textbf{(8.57a)}$$

Similarly, the analog stopband frequency Ω_s must be chosen so that

$$\Omega_s T = 2\tan(\omega_s/2). \quad \textbf{(8.57b)}$$

Alternately, if the design problem already specifies the discrete-time cutoff frequency ω_c then the analog cutoff frequency Ω_c is given by

$$\Omega_c T = 2\tan(\omega_c/2). \quad \textbf{(8.57c)}$$

This step is referred to as **prewarping** the critical frequencies. As in the impulse-invariance transformation, it is only necessary to specify the products $\Omega_p T$ and $\Omega_s T$ or $\Omega_c T$ rather than separate values of Ω_p, Ω_s, Ω_c and T. We will see why in a moment.

Step 3: Determine the design parameters of the analog filter $H(\Omega)$. Input the critical parameters of the analog filter, including the prewarped values of $\Omega_p T$ and $\Omega_s T$, into the design formulas

Equations (8.10) and (8.11) to determine the two design parameters of the analog Butterworth filter $H(\Omega)$,

$$N = \left\lceil \frac{\log_{10}\left(\dfrac{10^{-\delta_s/10}-1}{10^{-\delta_p/10}-1}\right)}{2\log_{10}(\Omega_s/\Omega_p)} \right\rceil = \left\lceil \frac{\log_{10}\left(\dfrac{10^{-\delta_s/10}-1}{10^{-\delta_p/10}-1}\right)}{2\log_{10}(\omega_s/\omega_p)} \right\rceil$$

and

$$\Omega_c T = \Omega_p T (10^{-\delta_p/10}-1)^{-1/2N}.$$

Step 4: Find the Laplace transform $H(s)$. By Equation (8.20), the Laplace transform is

$$H(s) = \prod_{k=0}^{N-1} \frac{-p_k}{s-p_k},$$

where the poles p_k are given by Equation (8.18),

$$p_k = \Omega_c e^{j\pi(2k+N+1)/2N}, \quad 0 \le k < N.$$

In the steps that follow, we only require the product $p_k T$, so write

$$p_k T = \Omega_c T e^{j\pi(2k+N+1)/2N}, \quad 0 \le k < N,$$

which is calculated using $\Omega_c T$ and N, as specified in Equations (8.10) and (8.11).

Step 5: Apply the bilinear transformation to get $H(z)$. Apply the bilinear transformation by substituting Equation (8.55) into $H(s)$ to give $H(z)$,

$$H(z) = H(s)\Big|_{s=\frac{2}{T}\frac{1-z^{-1}}{1+z^{-1}}} = \prod_{k=0}^{N-1} \frac{-p_k}{\dfrac{2}{T}\dfrac{1-z^{-1}}{1+z^{-1}} - p_k} = \prod_{k=0}^{N-1} \frac{-p_k T(1+z^{-1})}{2(1-z^{-1}) - p_k T(1+z^{-1})}$$

$$= \prod_{k=0}^{N-1} \left(\frac{-p_k T}{2-p_k T}\right) \frac{1+z^{-1}}{1-\left(\dfrac{2+p_k T}{2-p_k T}\right)z^{-1}}. \tag{8.58}$$

$H(z)$ has N identical zeros at $z = -1$ and N poles at

$$z = \frac{2+p_k T}{2-p_k T}.$$

To generalize this result, recognize that Butterworth, Chebyshev, inverse Chebyshev and elliptic filters all have transforms of the form

$$H(s) = K \frac{\prod_{k=0}^{M-1}(s-z_k)}{\prod_{k=0}^{N-1}(s-p_k)},$$

where p_k and z_k are the poles and zeros of $H(s)$ and K is an appropriately chosen constant. Application of the bilinear transformation, Equation (8.55), transforms $H(s)$ into $H(z)$, which is of the form

$$H(z)=\hat{K}(1+z^{-1})^{N-M}\frac{\prod_{k=0}^{M-1}(1-\hat{z}_k z^{-1})}{\prod_{k=0}^{N-1}(1-\hat{p}_k z^{-1})},$$

where the poles and zeros of $H(z)$ are given respectively by

$$\hat{p}_k=\frac{2+p_kT}{2-p_kT}\text{ and }\hat{z}_k=\frac{2+z_kT}{2-z_kT}.$$

For Butterworth and inverse Chebyshev filters, the constant $\hat{K}$ is chosen so that $H(0)=1$. For Chebyshev and elliptic filters, $\hat{K}$ is chosen so that

$$H(0)=\begin{cases}1/\sqrt{1+\varepsilon_p^2}, & N\text{ even}\\ 1, & N\text{ odd}\end{cases}.$$

To see how this process works, we will do a few examples.

Example 8.9

Use the bilinear transformation to find the difference equation of a discrete-time Butterworth lowpass filter with $\omega_p=0.1\pi$, $\delta_p=-3$ dB, $\omega_s=0.8\pi$ and $\delta_s=-15$ dB and compare the resulting filter to one designed using the impulse-invariance method.

➤ Solution:

First, prewarp the critical frequencies using Equation (8.57), which gives $\Omega_pT=2\tan\omega_p/2=0.3168$ and $\Omega_sT=2\tan\omega_s/2=6.1554$. Given these values, Equations (8.10) and (8.11) give $N=1$ and $\Omega_cT=1.0024\cdot\Omega_pT=0.3175$. There is a single pole at $p_k=\Omega_c e^{j\pi}=-\Omega_c$, so $H(s)$ just has one term,

$$H(s)=\frac{\Omega_c}{s-p_k}=\frac{\Omega_c}{s+\Omega_c}.$$

Making the bilinear transformation substitution, we get a first-order filter:

$$H_{bilin}(z)=H(s)\Big|_{s=\frac{2}{T}\frac{1-z^{-1}}{1+z^{-1}}}=\frac{\Omega_c}{\frac{2}{T}\frac{1-z^{-1}}{1+z^{-1}}+\Omega_c}=\frac{\Omega_cT(1+z^{-1})}{(\Omega_cT+2)+z^{-1}(\Omega_cT-2)}=\left(\frac{\Omega_cT}{\Omega_cT+2}\right)\frac{1+z^{-1}}{1-\left(\frac{2-\Omega_cT}{2+\Omega_cT}\right)z^{-1}}$$

$$=0.137\frac{1+z^{-1}}{1-0.727z^{-1}}.$$

As expected, the only parameter required in order to specify the filter is the prewarped corner frequency Ω_cT. The frequency response of this filter is shown in red in **Figure 8.30a**.

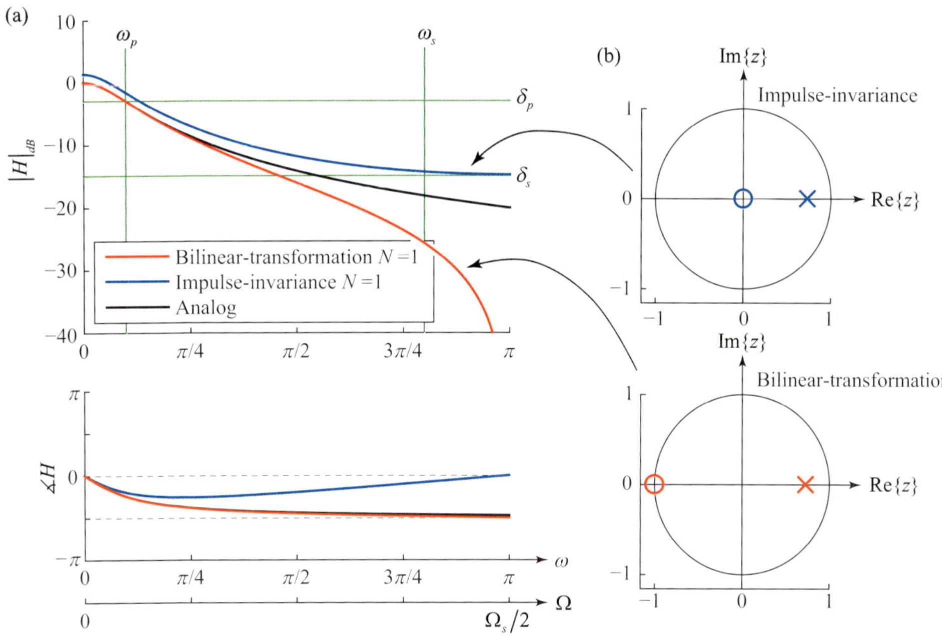

Figure 8.30 Comparison of bilinear-transformation and impulse-invariance methods

The critical design parameters ω_p, δ_p, ω_s and δ_s are shown with green lines. The filter meets the passband specification exactly ($|H_{db}(\omega_p)| = -3$ dB), and exceeds the stopband specification by a considerable amount ($|H_{db}(\omega_s)| = -25.7$ dB).

We can compare the frequency response of this filter with one designed using the impulse-invariance method to meet the same specifications. Using the methods outlined in Example 8.20, we get $\Omega_p T = \omega_p = 0.1\pi$ and $\Omega_s T = \omega_s = 0.8\pi$, from which we calculate $N = 1$ and $\omega_c = \Omega_c T = 1.0024 \cdot \Omega_p T = 0.3149$. This is also a first-order filter with z-transform

$$H_{impinv}(z) = \frac{\omega_c}{1 - e^{-\omega_c} z^{-1}} = \frac{0.3149}{1 - 0.7299 z^{-1}}.$$

The frequency response is shown in blue in **Figure 8.30a**. The figure also shows the frequency response of the first-order prototype analog filter $H(\Omega)$ from which H_{bilin} and $H_{impinv}(z)$ were derived. It is shown in black, plotted on a continuous frequency scale normalized by the sampling frequency $\Omega_s = 2\pi/T$, so that a continuous-time frequency of $\Omega = \Omega_s/2$ corresponds to the discrete-time frequency $\omega = \pi$. Due to aliasing, the filter designed by the impulse-invariance method does not meet either the passband ($|H_{db}(\omega_p)| = -1.63$ dB) or the stopband ($|H_{db}(\omega_s)| = -14.37$ dB) specifications. In fact, the response magnitude is greater than that of the prototype analog filter at all frequencies.

Lowpass filters designed using the bilinear transformation show no aliasing because there is a unique, albeit nonlinear, mapping of the $j\Omega$-axis to the unit circle such that $\omega \to \pm\pi$ as $\Omega \to \pm\infty$. Furthermore, the magnitude of the frequency response of lowpass filters designed using the bilinear transformation is guaranteed to head towards 0 ($-\infty$ on a dB scale) as $\omega \to \pm\pi$. To understand why this is so, look at the pole-zero plot of the filter shown in **Figure 8.30b**. When the bilinear transformation, Equation (8.55), is substituted into Equation (8.58), each pole in $H(s)$ produces a pole in $H(z)$ *plus* a zero at $z = -1$. In the present example, there is a single pole in $H(s)$, so there is a single zero at $z = -1$.

$$H(z) = \left.\frac{\Omega_c}{s+\Omega_c}\right|_{s=\frac{2}{T}\frac{z-1}{z+1}} = \frac{\Omega_c T(z+1)}{2(z-1)+\Omega_c T(z+1)} = \left(\frac{\Omega_c T}{\Omega_c T+2}\right)\frac{z+1}{z+\left(\frac{\Omega_c T-2}{\Omega_c T+2}\right)}.$$

This important result holds for all the lowpass filter prototypes we have discussed – Butterworth, Chebyshev, inverse Chebyshev and elliptic – of any order. For all these prototypes, an Nth-order filter has N zeros at $z = -1$.

Now, we will do an example with multiple ($N=4$) poles in which we use Matlab to help us make the appropriate computations.

Example 8.10

Design a discrete-time Butterworth lowpass filter using the bilinear-transformation method with $\omega_p = 0.5\pi$, $\delta_p = -3$ dB, $\omega_s = 0.7\pi$ and $\delta_s = -20$ dB.

➤ Solution:

The first steps are the same as with the impulse-invariance approach. Start with the critical parameters of the digital filter:

```
wp = 0.5 * pi;
ws = 0.7 * pi;
dp = -3;
ds = -20;
```

For the bilinear transformation, prewarp the critical frequencies before determining N and ω_c.

```
WpT = 2 * tan(wp / 2);
WsT = 2 * tan(ws / 2);
ep = sqrt(10^-(dp/10)-1);
es = sqrt(10^-(ds/10)-1);
N = ceil(log10(es/ep) / log10(WsT / WpT));
WcT = WpT * ep^(-1/N);
```

In this case $N=4$. Next, we need to find the Laplace transform

$$H(s) = \prod_{k=0}^{N-1} \frac{-p_k}{s-p_k},$$

where the poles are located at

$$p_k = \Omega_c e^{j\pi(2k+N+1)/2N}, \quad 0 \le k < N. \tag{8.59}$$

Substituting Equation (8.58) gives $H(z)$ as the product of N terms, which we then need to express as the ratio of numerator and denominator polynomials,

$$H(z) = H(s)\big|_{s=\frac{2}{T}\frac{1-z^{-1}}{1+z^{-1}}} = \prod_{k=0}^{N-1}\left(\frac{p_kT}{p_kT-2}\right)\frac{1+z^{-1}}{1-\left(\frac{2+p_kT}{2-p_kT}\right)z^{-1}} = \frac{\sum_{k=0}^{N-1} b_k z^{-k}}{\sum_{k=0}^{N-1} a_k z^{-k}}. \tag{8.60}$$

Matlab's prod and poly functions can be used to calculate the coefficients of the numerator and denominator polynomials b_k and a_k, respectively:

```
k = 0:N-1;
pkT = WcT * exp(1j*pi*(2*k+N+1)/2/N);
b = real(prod(pkT./(pkT-2))) * poly(-ones(1, N));
a = real(poly((2+pkT)./(2-pkT)));
```

The numerator and denominator coefficients are purely real; the real function just gets rid of trace imaginary parts that can result from the Matlab computation. The end result is

$$H(z) = 0.0941\frac{1+4z^{-1}+6z^{-2}+4z^{-3}+z^{-4}}{1+0.0015z^{-1}+0.4860z^{-2}+0.0003z^{-3}+0.0177z^{-4}}. \tag{8.61}$$

Designing a discrete-time Butterworth filter using Matlab's Signal Processing Toolbox functions is a two-step process. First, you can use the buttord function to compute the order N and cutoff frequency ω_c of the filter given values of parameters ω_p, ω_s, δ_p and δ_s:

```
[N, wc] = buttord(wp/pi, ws/pi, -dp, -ds).
```

The passband and stopband frequency arguments to buttord must be normalized by π and the passband and stopband attenuations must be positive. This function performs the bilinear transformation of discrete-time parameters into analog parameters, but as we have noted in Example 8.1, the function uses the stopband specifications (the equivalent to Equation (8.15)) rather than the passband specifications (the equivalent of Equation (8.11)) to compute wc. Given values of N and wc, use Matlab's butter function to compute the filter coefficients b_k and a_k,

```
[b, a] = butter(N, wc);
```

If we wish to use butter to compute b and a using the *passband* specifications, that is easily done since we have already calculated N and Ω_cT. By inverting Equation (8.57c), we get

$$\omega_c = 2\tan^{-1}(\Omega_cT/2), \tag{8.62}$$

which we can plug into butter to get the same result as Equation (8.61),

```
[b, a] = butter(N, 2*atan(WcT/2)/pi);
```

Figure 8.31a shows a comparison of frequency response of the discrete-time Butterworth filters designed using the bilinear-transformation and impulse-invariance methods to meet these identical specifications.

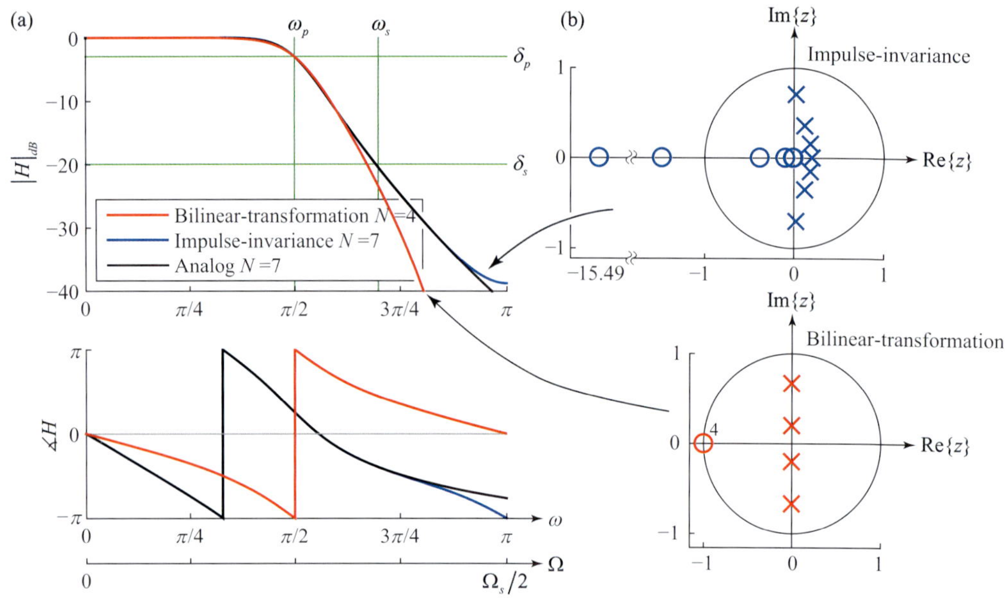

Figure 8.31 Comparison of bilinear-transformation and impulse-invariance Butterworth filters

Both the bilinear-transformation and impulse-invariance methods produce filters which meet or exceed the design specifications, but the bilinear-transformation method only requires a fourth-order filter ($N = 4$) to meet the specifications whereas the impulse-invariance method requires a seventh-order ($N = 7$) filter. It is characteristic of filters designed using bilinear transformation that they meet or exceed the design specifications using a lower order than the impulse-invariance method.

Notice that the filter designed using impulse-invariance matches both the magnitude and phase of the analog Butterworth template well beyond the stopband frequency, after which aliasing is evident. In contrast, the magnitude of the filter designed using bilinear-transformation shows no aliasing.

8.4.5 Cascade of second-order sections

In Chapter 9, we will discuss the architecture of FIR and IIR filters; that is, how they are actually implemented in software and hardware. While different implementations of a given filter (e.g., direct-form, cascade, parallel) are theoretically identical, practical considerations such as coefficient quantization favor some implementations over others. In particular, it turns out that it is preferable to implement IIR filters, particularly those of high order, as the cascade (product) of second-order filter sections rather than as a single section comprising the product of all the terms, as we did in Equation (8.61).

Equation (8.59) shows that the poles in the s-plane will be located at complex-conjugate positions, with

$$p_{N-1-k} = \Omega_c e^{j\pi(2(N-1-k)+N+1)/2N} = \Omega_c e^{-j\pi(2k+N+1)/2N} = p_k^*. \tag{8.63a}$$

If the filter length N is even, there are $N/2$ pairs of complex-conjugate poles. Each of these pairs will create one second-order section. If N is odd, then there is also a single pole located at $k = (N-1)/2$. But from Equation (8.63a), this pole is real since it is equal to its conjugate,

$$p_{N-1-(N-1)/2} = p_{(N-1)/2} = p^*_{(N-1)/2}. \tag{8.63b}$$

This pole will be used to create a first-order section. To proceed, note that Equation (8.60) can be written as the product of k terms,

$$H(z) = \prod_{k=0}^{N-1} H_k(z), \tag{8.64a}$$

where

$$H_k(z) = \left(\frac{p_k T}{p_k T - 2}\right) \frac{1 + z^{-1}}{1 - \left(\dfrac{2 + p_k T}{2 - p_k T}\right) z^{-1}}. \tag{8.64b}$$

Further for each k, $H_k(z)$ is effectively normalized so that $H_k(\omega) = 1$ at $\omega = 0$:

$$H(\omega)\big|_{\omega=0} = H(z)\big|_{z=1} = \left(\frac{p_k T}{p_k T - 2}\right) \frac{2}{1 - \left(\dfrac{2 + p_k T}{2 - p_k T}\right)} = \left(\frac{p_k T}{p_k T - 2}\right) \frac{2}{\left(\dfrac{2p_k T}{p_k T - 2}\right)} = 1.$$

Conjugate poles in the s-plane correspond to conjugate poles in the z-plane, so from Equations (8.63a) and (8.64b),

$$H_{N-1-k}(z) = H_k^*(z).$$

Hence, if N is even, Equation (8.64a) can be rewritten as the product of $N/2$ second-order terms,

$$H(z) = \prod_{k=0}^{N/2} H_k(z) H_{N-k}(z) = \prod_{k-0}^{N/2} H_k(z) H_k^*(z) = \prod_{k-0}^{N/2} |H_k(z)|^2.$$

If N is odd, the product has an additional term corresponding to a single real pole in the z-plane, given by Equation (8.64b) with $k = (N-1)/2$. Putting it all together, we have

$$H(z) = H_r(z) \prod_{k=0}^{\lfloor N/2 \rfloor} \left|\frac{p_k T}{p_k T - 2}\right|^2 \frac{1 + 2z^{-1} + z^{-2}}{1 + 2\mathrm{Re}\left\{\dfrac{p_k T + 2}{p_k T - 2}\right\} z^{-1} + \left|\dfrac{p_k T + 2}{p_k T - 2}\right|^2 z^{-2}}, \tag{8.65}$$

where

$$H_r(z) = \begin{cases} \left(\dfrac{p_{(N-1)/2} T}{p_{(N-1)/2} T - 2}\right) \dfrac{1 + z^{-1}}{1 - \left(\dfrac{2 + p_{(N-1)/2} T}{2 - p_{(N-1)/2} T}\right) z^{-1}}, & N \text{ odd} \\ 1, & N \text{ even} \end{cases}.$$

Example 8.11

Find the coefficients of the discrete-time Butterworth lowpass filter of Example 8.10, expressed as the product (cascade) of second-order sections.

➤ Solution:

The first few lines of the code are the same. After the computation of pkT we compute and display the coefficients of the sections. Here, we do not have to calculate the zeros, just the variable G, which is the gain of section.

```
N2 = floor(N/2); % number of second-order sections
p = pkT(1:N2)';
G = abs(p./(p-2)).^2; % gain of numerator of second-order sections
aa = [ones(N2, 1) 2*real((p+2)./(p-2)) abs((p+2)./(p-2)).^2];%section poles
if (rem(N, 2)) % it is odd -> need an additional real section
   G = [G; WcT/(WcT-2)];
   aa = [aa; [1 (WcT+2)/(WcT-2) 0]];
end

format long
disp('G                 a(2)                a(3)')
disp([G aa(:, [2 3])])

G                  a(2)              a(3)
0.361830352209086 0.000858646129096 0.446462762707250
0.260045830731828 0.000617105073088 0.039566217854224
```

So,

$$
\begin{aligned}
H(z) &= \frac{0.36183(1+2z^{-1}+z^{-2})}{1+0.00085865z^{-1}+0.44646z^{-2}} \cdot \frac{0.26005(1+2z^{-1}+z^{-2})}{1+0.00061711z^{-1}+0.039566z^{-2}} \\
&= \frac{0.36183+0.72366z^{-1}+0.36183z^{-2}}{1+0.00085865z^{-1}+0.44646z^{-2}} \cdot \frac{0.26005+0.52009z^{-1}+0.26005z^{-2}}{1+0.00061711z^{-1}+0.039566z^{-2}}.
\end{aligned}
$$

The Matlab function tf2sos can be used to convert from b_k and a_k in Equation (8.60) to the coefficient of second-order sections and gives the numerator and denominator coefficients of each section.

Example 8.12

Use the bilinear-transformation method to design a discrete-time (Type-I) Chebyshev lowpass filter with $\omega_p = 0.5\pi$, $\delta_p = -3$ dB, $\omega_s = 0.7\pi$ and $\delta_s = -40$ dB.

➤ Solution:

Proceed in a manner similar to the previous example, by using values of ω_p, δ_p, ω_s and δ_s to determine N and ω_c:

```
wp = 0.5 * pi;
ws = 0.7 * pi;
dp = -3;
ds = -40;
WpT = 2 * tan(wp / 2);
WsT = 2 * tan(ws / 2);
ep = sqrt(10^-(dp/10)-1);
es = sqrt(10^-(ds/10)-1);
x = acosh(es/ep);
N = ceil(x/acosh(WsT/WpT));
```

In this example, it turns out that $N = 5$. The Type-I Chebyshev filter is designed to meet the passband specifications exactly, so that the cutoff frequency Ω_c is equal to the passband frequency Ω_p. Hence, we set

```
WcT = WpT;
```

As with the Butterworth filter, the Chebyshev filter has N poles and no zeros. $H(s)$ is given by Equation (8.32), and the poles p_k are given by Equation (8.31). The bilinear transformation maps each pole in the s-plane to a pole in the z-plane located at $z = (s_k T + 2)/(s_k T - 2)$ and one zero located at $z = -1$,

$$\left.\frac{-p_k}{s - p_k}\right|_{s = \frac{2}{T}\frac{1 - z^{-1}}{1 + z^{-1}}} = \left(\frac{p_k T}{p_k T - 2}\right) \frac{1 + z^{-1}}{1 - \left(\dfrac{2 + p_k T}{2 - p_k T}\right) z^{-1}}.$$

In Matlab, we write

```
k = 0:N-1;
psi = (2*k+1)*pi/2/N;
phi = asinh(1/ep)/N;
pkT = WcT * (-sin(psi)*sinh(phi)+1j*cos(psi)*cosh(phi));
b = real(prod(pkT./(pkT-2))) * poly(-ones(1, N));
a = real(poly((2+pkT)./(2-pkT)));
```

The end result is

$$H(z) = 0.0156 \frac{1 + 5z^{-1} + 10z^{-2} + 10z^{-3} + 5z^{-4} + z^{-5}}{1 - 1.5097z^{-1} + 2.1610z^{-2} - 1.8229z^{-3} + 1.0800z^{-4} - 0.4083z^{-5}}.$$

Given values of the order N, passband attenuation δ_p and passband frequency ω_p, you can also use Matlab's `cheb1ord` and `cheby1` functions to compute the filter coefficients `b` and `a` of the Type-I filter. First,

```
[N, wc] = cheb1ord(wp/pi, ws/pi, -dp, -ds);
[b, a] = cheby1(N, -dp, wc);
```

Again, we can express this filter as the cascade of sections, using code similar to that of Example 8.11. In this case, there are two second-order sections and one first-order section:

$$H(z) = 0.0156 \frac{1 + 2z^{-1} + z^{-2}}{1 - 0.062544z^{-1} + 0.89273z^{-2}} \cdot \frac{1 + 2z^{-1} + z^{-2}}{1 - 0.74867z^{-1} + 0.6548z^{-2}} \cdot \frac{1 + z^{-1}}{1 - 69847z^{-1}}.$$

Example 8.13

Design a discrete-time inverse (Type-II) Chebyshev lowpass filter using the bilinear-transformation method using the same parameters as the previous example: $\omega_p = 0.5\pi$, $\delta_p = -3$ dB, $\omega_s = 0.7\pi$ and $\delta_s = -40$ dB.

➤ Solution:

We proceed in a similar manner to the previous example, using values of ω_p, δ_p, ω_s and δ_s to determine N and ω_c:

```
wp = 0.5 * pi;
ws = 0.7 * pi;
dp = -3;
ds = -40;
WpT = 2 * tan(wp / 2);
WsT = 2 * tan(ws / 2);
es = sqrt(10^-(ds/10)-1);
ep = sqrt(10^-(dp/10)-1);
x = acosh(es/ep);
N = ceil(x/acosh(WsT/WpT));
```

Here, again, $N=5$. Normally, the inverse Chebyshev filter would be designed to meet the stopband specifications exactly, so that the cutoff frequency Ω_c is equal to the stopband frequency Ω_s. In this case, we would just set

```
WcT = WsT;
```

If we wish to make $\Omega_c = \Omega_p$, we would apply Equation (8.36b) to recalculate Ω_c:

```
Wphat = WsT / cosh(x/N);
WcT = WsT * WpT / Wphat;
```

Unlike the Butterworth filter, the transform of the inverse Chebyshev filter has both poles and zeros,

$$H(s) = \frac{\prod\limits_{k=0,2} (z_k - s)/z_k}{\prod\limits_{k=0}^{2} (p_k - s)/p_k},$$

where the poles p_k are given by Equation (8.37) and the zeros z_k are given by Equation (8.38). In Matlab, we write

```
k = 0:N-1;
psi = (2*k+1)*pi/2/N;
phi = asinh(es)/N;
pkT = WcT ./ (-sin(psi)*sinh(phi)+1j*cos(psi)*cosh(phi));
psi(fix(N/2)+1) = []; % remove zero at infinity
zkT = 1j*WcT ./ cos(psi);
```

For an inverse Chebyshev filter with N odd, $H(s)$ has N poles, but only $N-1$ zeros in the finite s-plane. From Equation (8.38), the zero for $k=(N-1)/2$ is at infinity, since $z_{(N-1)/2} = j\Omega_c/\cos\psi_{(N-1)/2} = j\Omega_c/\cos\pi/2 = \infty$. It needs to be removed.

When the bilinear transformation is applied to $H(s)$, each pole in the s-plane maps to one pole in the z-plane located at $z = (2 + p_k T)/(2 - p_k T)$ and one zero located at $z = -1$, in a manner similar to that of the Type-I Chebyshev filter,

$$\left.\frac{-p_k}{s - p_k}\right|_{s=\frac{2}{T}\frac{1-z^{-1}}{1+z^{-1}}} = \left(\frac{p_k T}{p_k T - 2}\right) \frac{1 + z^{-1}}{1 - \left(\frac{2 + p_k T}{2 - p_k T}\right) z^{-1}}.$$

In our example, this produces three poles and three zeros. Similarly, the zeros of $H(s)$ also map to one pole in the z-plane located at $z = -1$ and one zero located at $z = (2 + z_k T)/(2 - z_k T)$,

$$\left(\frac{z_k T - 2}{z_k T}\right) \frac{1 - \left(\frac{2 + z_k T}{2 - z_k T}\right) z^{-1}}{1 + z^{-1}}.$$

In our example, this produces two poles and two zeros. The two poles at $z = -1$ cancel two of the zeros at $z = -1$, so $H(z)$ ends up with N poles and N zeros.

```
pls = (2+pkT)./(2-pkT);
zrs = [(2+zkT)./(2-zkT) -1];
b = real(prod((zkT-2)./zkT)*prod(pkT./(pkT-2))*real(poly(zrs)));
a = real(poly(pls));
```

The result, designed to meet the passband specifications, is

$$H(z) = \frac{0.1155 + 0.4048z^{-1} + 0.6923z^{-2} + 0.6923z^{-3} + 0.4048z^{-4} + 0.1155z^{-5}}{1 + 0.3171z^{-1} + 0.7907z^{-2} + 0.1975z^{-3} + 0.1083z^{-4} + 0.0115z^{-5}}. \quad (8.66)$$

Matlab's `cheb2ord` and `cheby2` functions can be used to compute the filter coefficients `b` and `a`,

```
[N, wc] = cheb2ord(wp/pi, ws/pi, -dp, -ds);
[b, a] = cheby2(N, -ds, wc);
```

However, unlike the analog form of `cheb2ord` discussed in Example 8.5, the digital form shown above computes `wc` based on the *stopband* rather than *passband* specifications. To get values of `b` and `a` that properly produce the $H(z)$ of Equation (8.66), you need to compute ω_c from $\Omega_c T$ using Equation (8.62) as in Example 8.10, and plug that value into `cheby2`. That is,

```
[b, a] = cheby2(N, -ds, 2*atan(WcT/2)/pi);
```

Figure 8.32a shows a comparison of the frequency response of a discrete-time inverse Chebyshev filter designed using the bilinear-transformation and impulse-invariance methods to meet these identical passband specifications. The analog filter is also shown on a normalized frequency scale. **Figure 8.32b** shows the pole-zero plots for these filters

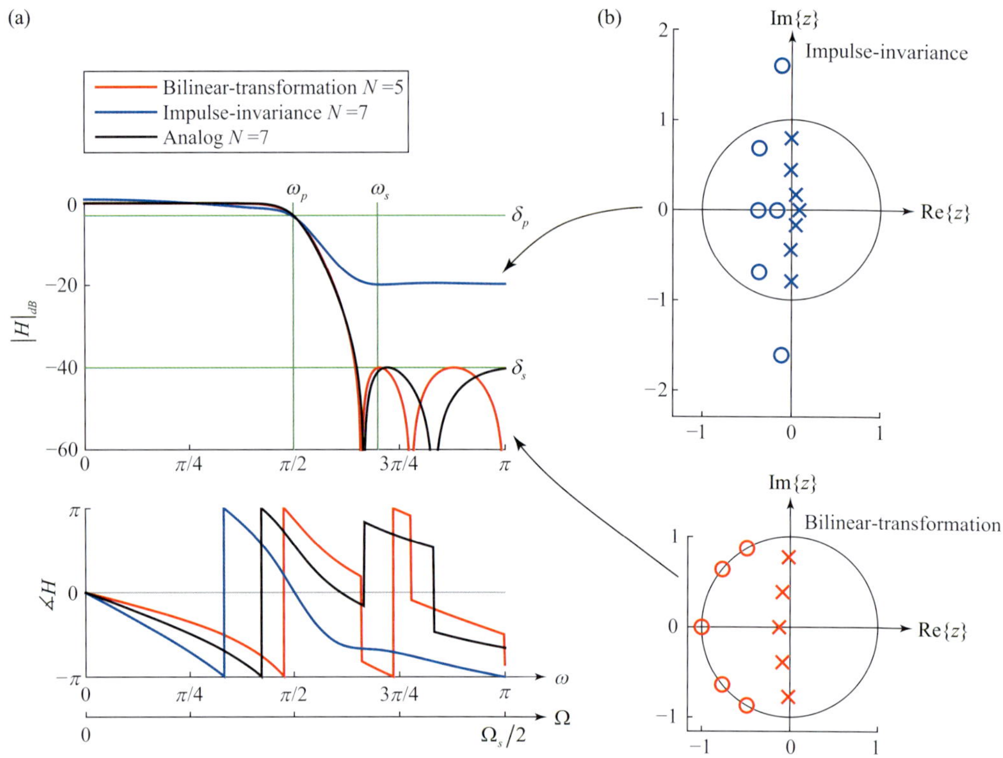

Figure 8.32 Comparison of bilinear-transformation and impulse-invariance inverse Chebyshev filters

The bilinear-transformation method produces a filter that meets or exceeds the design specifications (red traces). However, the impulse-invariance method does not even come *close* to meeting the design specifications; in fact, it is *horrible*. A clue to the problem is the position of the zeros of $H(z)$ shown in the pole-zero plots. For the analog Type-II Chebyshev filter (Equation (8.38)), all the zeros of $H(s)$ lie on the $j\Omega$-axis, as shown in **Figure 8.12**. The bilinear transformation maps the $j\Omega$-axis of $H(s)$ onto the unit circle of $H(z)$, so all the zeros of the bilinear-transformed discrete-time filter end up nicely disposed along the unit circle. Since the frequency response $H(\omega)$ is simply $H(z)$ for values of z along the unit circle, $|H(\omega)|$ is repeatedly driven to 0 at the position of the zeros, which has the effect of keeping it below the stopband value. In contrast, none of the zeros of the impulse-invariance filter are on the unit circle (see Problem 8-28). Their positions are determined by the values of both the zeros and the poles of $H(s)$. Hence, the response does not go to 0 as $\omega \to \pi$. This example carries an important lesson: the impulse-invariance method is not generally useful when $H(s)$ has zeros.

8.5 ★ Spectral transformations of IIR filters

Highpass, bandpass and bandstop IIR filters with specified characteristics can be created by first designing a prototype lowpass filter and then transforming it into the desired target lowpass, highpass, bandpass or bandstop filter by a process of **spectral transformation**. This transformation maps the singularities of the prototype filter into singularities of the target filter in a well-defined way with a minimum number of parameters to adjust.

There are essentially two approaches to spectral transformation: analog and digital. In the analog approach, you apply an analog transformation $\hat{s} = G(s)$, which maps the singularities of a prototype analog filter $H_a(\hat{s})$ in the s-plane into the singularities of the transformed analog lowpass, highpass, bandpass or bandstop filter $H(s) = H_a(G(s))$. Then, you apply the equivalent of the bilinear transformation to map the singularities of that analog filter from the s-plane into the singularities in the z-plane of the discrete-time filter $H(z)$. In the digital approach, you first transform the prototype analog filter into a digital prototype filter $H(\hat{z})$, using the bilinear transformation, exactly as we have done in Section 8.4, and then apply a digital transformation $\hat{z} = G(z)$ to map the singularities $H(\hat{z})$ into those of $H(z)$.[4] In this chapter, we shall adopt this second approach, because it builds upon the work that we have already done and because it is relatively easy to implement in computer code, as we shall see.

8.5.1 Lowpass-to-lowpass transformation

We will first consider the problem of designing a lowpass filter $H_{lp}(\omega)$ with a desired cutoff frequency of ω_c by transforming a prototype lowpass filter $H(\hat{\omega})$ that has a cutoff frequency of $\hat{\omega}_c$. This might seem like a pointless exercise, but it illustrates the approach we will use in other spectral transformations. The key idea is to specify a transformation

$$\hat{z} = G_{lp}(z)$$

that maps the z-transform of prototype lowpass filter $H(\hat{z})$ into the z-transform of the desired lowpass filter $H_{lp}(z)$; that is,

$$H_{lp}(z) = H(\hat{z}) = H(G_{lp}(z)). \tag{8.67}$$

The desired transformation $G_{lp}(z)$ should have the following properties:

- be a rational function of z,
- map values of z inside the unit circle (i.e., $|z| < 1$) to values of $\hat{z}$ inside the unit circle (i.e., $|\hat{z}| < 1$),
- map values of z outside the unit circle (i.e., $|z| > 1$) to values of $\hat{z}$ outside the unit circle (i.e., $|\hat{z}| > 1$),
- map the unit circle of z to the unit circle of $\hat{z}$.

These are the characteristics of an allpass transformation (see Section 5.7) of the form

$$G_{lp}(z) = \frac{z-\alpha}{1-\alpha z} = \frac{1-\alpha z^{-1}}{z^{-1}-\alpha}, \tag{8.68}$$

where $|\alpha| < 1$ (See Problem 8-29). This mapping converts the prototype filter $H(\hat{z})$ into a lowpass filter $H_{lp}(z)$ of the same order, but with a different corner frequency. To see this, consider the frequency responses of the transformed filter $H_{lp}(\omega)$ and the prototype filter $H(\hat{\omega})$. These are just their respective z-transforms evaluated for values of $z = e^{j\omega}$ and $\hat{z} = e^{j\hat{\omega}}$, respectively, on the unit circle, namely

$$\begin{aligned} H_{lp}(\omega) &= H_{lp}(z)\big|_{z=e^{j\omega}} \\ H(\hat{\omega}) &= H(\hat{z})\big|_{\hat{z}=e^{j\hat{\omega}}} \end{aligned}.$$

[4] These could rightly be called *Constantinides transformations* after the researcher who developed them in the late 1960s.

From Equation (8.67),

$$H_{lp}(\omega) = H(\hat{\omega}) = H(G_{lp}(\omega)),$$

where from Equation (8.68),

$$G_{lp}(\omega) = G_{lp}(z)\big|_{z=e^{j\omega}} = \frac{e^{j\omega} - \alpha}{1 - \alpha e^{j\omega}} = e^{j\omega}\frac{1 - \alpha e^{-j\omega}}{1 - \alpha e^{j\omega}}.$$

Expressing $G_{lp}(\omega)$ in magnitude and phase form,

$$G_{lp}(\omega) = |G_{lp}(\omega)|e^{\measuredangle G_{lp}(\omega)},$$

we find that the magnitude of $G_{lp}(\omega)$ is unity,

$$|G_{lp}(\omega)| = |e^{j\omega}|\left|\frac{1 - \alpha e^{-j\omega}}{1 - \alpha e^{j\omega}}\right| = 1,$$

as expected for an allpass type of transformation. This means that values of $z = e^{j\omega}$ on the unit circle map into values of $\hat{z} = e^{j\hat{\omega}}$ on the unit circle. Therefore, the transformation between ω and $\hat{\omega}$ is determined by the phase of $G_{lp}(\omega)$,

$$\begin{aligned}\hat{\omega} &= \measuredangle G_{lp}(\omega) = \measuredangle e^{j\omega} + \measuredangle(1 - \alpha e^{-j\omega}) - \measuredangle(1 - \alpha e^{j\omega}) \\ &= \omega + \measuredangle((1 - \alpha\cos\omega) + j\alpha\sin\omega) - \measuredangle((1 - \alpha\cos\omega) - j\alpha\sin\omega) \\ &= \omega + 2\tan^{-1}\left(\frac{\alpha\sin\omega}{1 - \alpha\cos\omega}\right)\end{aligned}. \quad \textbf{(8.69)}$$

$G_{lp}(\omega)$ is the nonlinear transformation or "warping" of the angle ω that is a function of one parameter, α. This transformation preserves the values $\measuredangle G(0) = 0$ and $\measuredangle G(\pm\pi) = \pm\pi$ irrespective of the value of α. In order to create the transformed filter, our job is to determine the precise value of α that transforms the corner frequency of the prototype filter, $\hat{\omega}_c$, into the desired corner frequency of the transformed filter, ω_c. In other words, we want $\measuredangle G(\omega_c) = \hat{\omega}_c$. Plugging into Equation (8.69) and solving for α (Problem 8-30a), we get

$$\alpha = \frac{\sin\left(\dfrac{\hat{\omega}_c - \omega_c}{2}\right)}{\sin\left(\dfrac{\hat{\omega}_c + \omega_c}{2}\right)}. \quad \textbf{(8.70)}$$

Figure 8.33 gives an overview of how this lowpass-to-lowpass transformation process works. **Figure 8.33a** shows $H(\hat{\omega})$, the frequency response of a prototype fourth-order Butterworth lowpass filter that has been designed to have a corner (−3-dB) frequency of $\hat{\omega}_c = \pi/2$ (open circle). **Figure 8.33b** shows the mapping of ω to $\hat{\omega}$, as specified by Equation (8.69) for seven values of α (colored curves). These values of α were chosen to transform $H(\hat{\omega})$ into the seven lowpass filters $H_{lp}(\omega)$ shown in **Figure 8.33c**, with corner frequencies (denoted with colored symbols) ranging from $\omega_c = \pi/8$ (dark blue) to $\omega_c = 7\pi/8$ (dark red). The thin grey arrows in the plot show how the value of $\hat{\omega}_c$ of the prototype filter (open circle in **Figure 8.33a**) is mapped by the transformation by Equation (8.69) at the appropriate value of α (colored circles in **Figure 8.33b**) into ω_c, the corner frequency of the prototype filter $H_{lp}(\omega)$ (colored circles in **Figure 8.33c**).

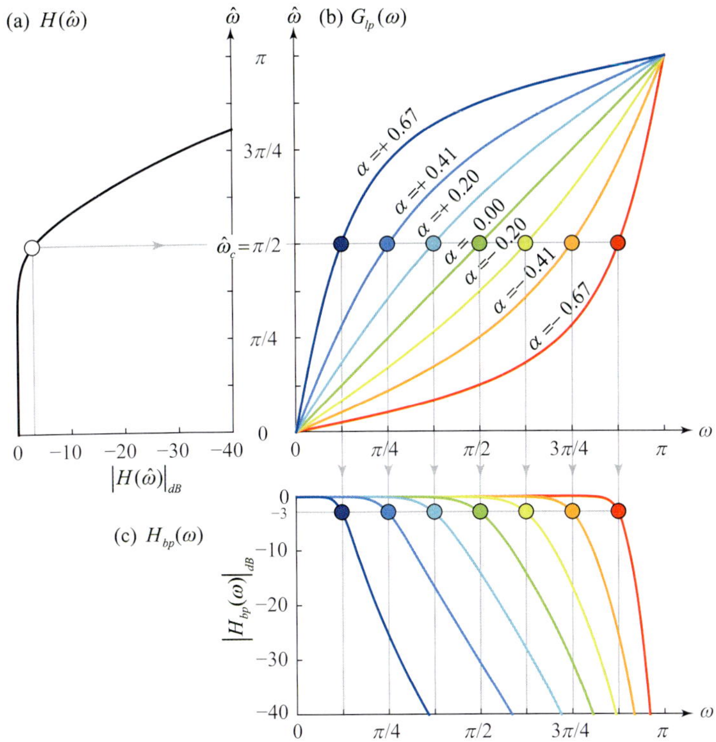

Figure 8.33 Lowpass-to-lowpass filter transformation

The transformation design parameter α generated by Equation (8.70) controls the shape of the transformation and guarantees that the desired corner frequency of the transformed filter, ω_c, will be exactly met. At a value of $\alpha = 0$, the transformation is a straight line ($\omega = \hat{\omega}$), so the transformed filter is identical to the prototype. For values of $\alpha > 0$, $\omega_c < \hat{\omega}_c$, and for values of $\alpha < 0$, $\omega_c > \hat{\omega}_c$.

As a specific example of the lowpass-to-lowpass transformation, consider $H(\hat{z})$, a prototype discrete-time Butterworth lowpass filter of arbitrary order and corner frequency, defined in Equation (8.58), and substitute $\hat{z} = G_{lp}(z)$ from Equation (8.68). After some uninspiring algebra (see Problem 8-31), the transformed lowpass filter can be expressed as

$$H_{lp}(z) = H(G_{lp}(z)) = \prod_{k=0}^{N-1} \left(\frac{p_k T}{p_k T - 2} \right) \frac{1 + z^{-1}}{1 - \left(\dfrac{2 + p_k T}{2 - p_k T} \right) z^{-1}}, \tag{8.71a}$$

where the poles of the transformed filter, p_k, are related to the poles of the prototype filter, $\hat{p}_k$, by

$$p_k = \hat{p}_k \left(\frac{1 - \alpha}{1 + \alpha} \right), \tag{8.71b}$$

and α is given by Equation (8.70). A comparison of Equation (8.71) with Equation (8.58) shows that the effect of the lowpass-to-lowpass transformation is to scale the poles of the original filter by a real factor $(1-\alpha)/(1+\alpha)$. You can show (Problem 8-31b) that a transformed filter $H_{lp}(z)$ with corner frequency ω_c is identical to a discrete-time filter designed from scratch to have corner frequency ω_c.

Example 8.14

Design a prototype fourth-order Butterworth lowpass filter with cutoff frequency $\hat{\omega}_c = \pi/2$, and transform it into a discrete-time lowpass filter with a corner frequency of $\omega_c = \pi/4$.

➤ Solution:

Here, we use Equations (8.70) and (8.71) and a little help from Matlab. First, prewarp $\hat{\omega}_c$ to form $\hat{\Omega}_c T = 2\tan\hat{\omega}_c/2$.

```
wc = pi / 4; % corner frequency of desired transformed filter
wc_hat = pi / 2; % corner frequency of prototype filter
WcT_hat = 2 * tan(wc_hat / 2); % bilinear transformation of prototype
N = 4; % filter order
```

Now, find the zeros and poles of the prototype filter. There are N zeros, all located at $z=-1$. The poles are located at

$$\hat{p}_k T = \hat{\Omega}_c T e^{j\pi(2k+N+1)/2N},\ 0 \le k < N.$$

Use Equation (8.70) to calculate $\alpha = 0.41$ and Equation (8.71b) to calculate the poles of the transformed filter, p_k.

```
k = 0:N-1;
pkT_hat = WcT_hat * exp(1j*pi*(2*k+N+1)/2/N); % poles of prototype filter
alpha = sin((wc_hat - wc) / 2) / sin((wc_hat + wc) / 2);
pkT = pkT_hat * (1 - alpha) / (1 + alpha); % poles of transformed filter
zers = -ones(N, 1);
pols = (2+pkT)./(2-pkT);
const = real(prod(pkT./(pkT -2)));
b = const * poly(zers(:));
a = real(poly(pols(:)));
```

Given the poles and zeros of $H_{lp}(z)$, the Matlab function `poly` creates the numerator and denominator polynomials. However, if we are too lazy to use `poly` ourselves, Matlab provides the function `zp2tf` that takes values of the poles, zeros and constant and produces the coefficients `b` and `a` of the transfer function:

```
[b, a] = zp2tf(zers, pols, const); % note: zers must be a column vector
```

Either way, the end result is

$$H(z) = \frac{0.0102 + 0.0408z^{-1} + 0.0613z^{-2} + 0.0408z^{-3} + 0.0102z^{-4}}{1 - 1.9684z^{-1} + 1.7359z^{-2} - 0.7245z^{-3} + 0.1204z^{-4}}.$$

As we discussed in Section 8.4.5, from a computational standpoint, it is usually a better idea to implement the filter as a cascade of second-order sections, so we are likely better off not calculating the transfer function, but simply calculating the poles and zeros, as we have done above, and using the cascade form of Equation (8.71) to calculate the second-order sections (see Problem 8-40). Matlab provides the function `zp2sos` for this purpose. Matlab also has a function `tf2sos`, which converts between the coefficients of the transfer function and the coefficients of the second-order section, but it is preferable to use `zp2sos`, since the first thing `tf2sos` does internally is to find the roots of the coefficients.

Figure 8.33a shows $|H(\hat{\omega})|_{dB}$, the frequency response of a prototype lowpass filter with $N=4$ and $\hat{\omega}_c=\pi/2$.

The second curve from the left in **Figure 8.33c** (corresponding to $\alpha=+0.41$) shows the frequency response $|H_{lp}(\omega)|_{dB}$ of the transformed lowpass filter with corner frequency $\omega_c=\pi/4$.

Matlab has a function `iirlp2lp` that can perform this entire transformation for you. This entire example can be done with the following two lines of code:

```
[b_prot, a_prot] = butter(4, 0.5); % create the prototype filter
[b, a] = iirlp2lp(b_prot, a_prot, 0.5, 0.25); % transform it
```

But where's the fun in that?

8.5.2 Lowpass-to-highpass transformation

We can employ a similar approach to transform a prototype lowpass filter into a highpass filter. As motivation, consider the simple transformation of a lowpass filter to a highpass filter shown in **Figure 8.34**. Let $H(z)$ be the transform of a lowpass filter with frequency response $H(\omega)$, shown in the blue trace. The red trace is the frequency response of a highpass filter $H(\pi-\omega)$ that results from the transformation $H(-z)$:

$$H(-z)|_{z=e^{j\omega}} = \sum_{n=-\infty}^{\infty} h[n](-e^{j\omega})^{-n} = \sum_{n=-\infty}^{\infty} h[n]e^{-j(\omega-\pi)} = H(\omega-\pi).$$

If the corner frequency of the lowpass filter is ω_c, then the corner frequency of the transformed highpass filter is $\pi-\omega_c$. Equivalently, if the corner frequency of the lowpass filter were $\pi-\omega_c$, then the corner frequency of the transformed highpass filter would be ω_c.

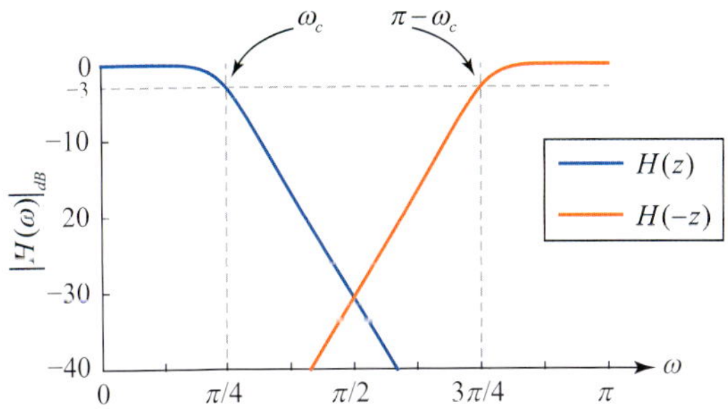

Figure 8.34 Relation of lowpass and transformed highpass filters with same corner frequency

This exercise suggests a conceptual two-step procedure to transform a prototype lowpass filter with corner frequency $\hat{\omega}_c$ into a highpass filter with corner frequency ω_c. First, use the lowpass-to-lowpass transformation procedure of Section 8.5.1 to transform the prototype lowpass filter $H(\hat{z})$ into a lowpass filter $H_{lp}(z)$ with corner frequency $\pi-\omega_c$. Then form the desired highpass filter with corner frequency ω_c by performing the discrete-time transformation $H_{hp}(z)=H_{lp}(-z)$. In summary, for a lowpass-to-highpass transformation,

$$\hat{z}=G_{hp}(z)=-\frac{z+\alpha}{1+\alpha z} \quad \text{or} \quad \hat{z}^{-1}=-\frac{z^{-1}+\alpha}{1+\alpha z^{-1}}.$$

The value of α required to transform $\hat{H}(\hat{\omega})$, the prototype lowpass filter with corner frequency $\hat{\omega}_c$, into $H_{hp}(\omega)$, a highpass filter of corner frequency ω_c, is (Problem 8-33)

$$\alpha=-\frac{\cos\left(\dfrac{\hat{\omega}_c+\omega_c}{2}\right)}{\cos\left(\dfrac{\hat{\omega}_c-\omega_c}{2}\right)}.$$

For the example of the transformed Butterworth lowpass prototype of Equation (8.71), we get

$$H_{hp}(z)=H_{lp}(-z)=\prod_{k=0}^{N-1}\left(\frac{p_kT}{p_kT-2}\right)\frac{1-z^{-1}}{1+\left(\dfrac{2+p_kT}{2-p_kT}\right)z^{-1}}, \tag{8.72a}$$

where p_k (the poles of the transformed highpass filter) are again related to $\hat{p}_k$ (the poles of the prototype lowpass filter) by

$$p_k=\hat{p}_k\left(\frac{1-\alpha}{1+\alpha}\right). \tag{8.72b}$$

The parameter α is given by Equation (8.70) with the substitution of $\pi-\omega_c$ for ω_c,

$$\alpha=\frac{\sin\left(\dfrac{\hat{\omega}_c-(\pi-\omega_c)}{2}\right)}{\sin\left(\dfrac{\hat{\omega}_c+(\pi-\omega_c)}{2}\right)}=\frac{\sin\left(\dfrac{\hat{\omega}_c+\omega_c}{2}+\dfrac{\pi}{2}\right)}{\sin\left(\dfrac{\hat{\omega}_c-\omega_c}{2}-\dfrac{\pi}{2}\right)}=-\frac{\cos\left(\dfrac{\hat{\omega}_c+\omega_c}{2}\right)}{\cos\left(\dfrac{\hat{\omega}_c-\omega_c}{2}\right)}. \tag{8.73}$$

You can also show (see Problem 8-33) that the lowpass-to-highpass transformation of Equation (8.72) is equivalent to directly applying the transformation $\hat{z}=G_{hp}(z)$ to the prototype lowpass filter, where

$$\hat{z}=G_{hp}(z)\triangleq G_{lp}(-z)=\frac{(-z)-\alpha}{1-\alpha(-z)}=-\frac{z+\alpha}{1+\alpha z} \quad \text{or} \quad \hat{z}^{-1}=-\frac{z^{-1}+\alpha}{1+\alpha z^{-1}}.$$

Figure 8.35 shows the lowpass-to-highpass transformation process. **Figure 8.35a** shows a prototype fourth-order Butterworth lowpass filter with a corner (–3-dB) frequency of $\hat{\omega}_c=\pi/2$ (the same one shown in **Figure 8.33a**). **Figure 8.35b** shows the mapping of ω to $\hat{\omega}$, as specified

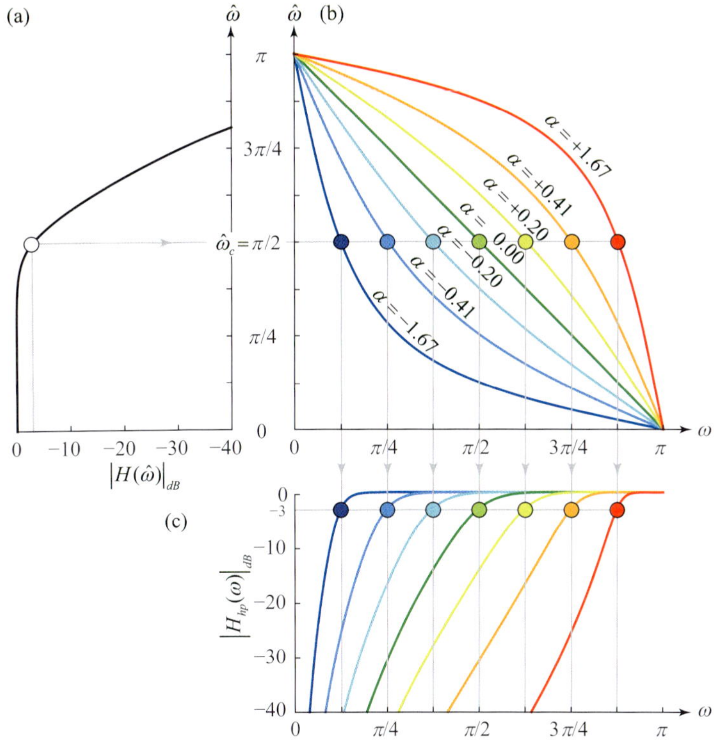

Figure 8.35 Lowpass-to-highpass filter transformation

by Equation (8.73) for seven values of α (colored curves), chosen to transform $H(\hat{\omega})$ into the seven highpass filters $H_{hp}(\omega)$ shown in **Figure 8.35c**, with corner frequencies (denoted with colored symbols) ranging from $\omega_c = \pi/8$ (dark blue) to $\omega_c = 7\pi/8$ (dark red).

Example 8.15

Transform a fourth-order Butterworth lowpass filter with cutoff frequency $\hat{\omega}_c = \pi/2$ into a discrete-time highpass filter with corner frequency $\omega_c = 3\pi/4$.

➤ Solution:

Proceed in a manner similar to Example 8.14:

```
wc = 3 * pi / 4; % corner frequency of transformed filter
wc_hat = pi / 2; % corner frequency of prototype filter
WcT_hat = 2 * tan(wc_hat / 2); % bilinear transformation of prototype
N = 4; % filter order
k = 0:N-1;
pkT_hat = WcT_hat * exp(1j*pi*(2*k+N+1)/2/N); % poles of prototype filter
alpha = -cos((wc_hat+wc)/2) / cos((wc_hat-wc)/2);
pkT = pkT_hat * (1 - alpha) / (1 + alpha); % poles of transformed filter
zers = ones(N, 1);
pols = -(2+pkT)./(2-pkT);
const = real(prod(pkT./(pkT -2)));
```

```
b = const * poly(zers(:));
a = real(poly(pols(:)));
```

The end result is

$$H(z)=\frac{0.0102-0.0408z^{-1}+0.0613z^{-2}-0.0408z^{-3}+0.0102z^{-4}}{1+1.9684z^{-1}+1.7359z^{-2}+0.7245z^{-3}+0.1204z^{-4}}.$$

Figure 8.35a shows $|H(\hat{\omega})|_{dB}$, the frequency response of a prototype lowpass filter with $N=4$ and $\hat{\omega}_c=\pi/2$.

The second curve from the right in **Figure 8.35c** (corresponding to $\alpha=+0.41$) shows the frequency response $|H_{lp}(\omega)|_{dB}$ of the transformed highpass filter with corner frequency $\omega_c=3\pi/4$.

Matlab has a function `iirlp2hp` that can perform this transformation for you. Here is the code:

```
[bb, aa] = butter(4, 0.5); % create the prototype filter with corner freq pi/2
[b, a]  = iirlp2hp(bb, aa, 0.5, 0.75); % transform it to 3pi/4
```

8.5.3 Lowpass-to-bandpass transformation

Using similar methods, you can show (Problem 8-35 and Problem 8-36) that you can directly transform a lowpass prototype into a bandpass filter using a lowpass-to-bandpass transformation given by

$$\hat{z}=G_{bp}(z)=-\left(\frac{z+\beta}{1+\beta z}\right)\bigg|_{z=z\left(\frac{z-\alpha}{1-\alpha z}\right)}=-\left(\frac{z\left(\frac{z-\alpha}{1-\alpha z}\right)+\beta}{1+\beta z\left(\frac{z-\alpha}{1-\alpha z}\right)}\right)=-\frac{z^2-\alpha(1+\beta)z+\beta}{\beta z^2-\alpha(1+\beta)z+1}, \tag{8.74}$$

where the constants α and β are given by

$$\alpha=\frac{\cos\left(\frac{\omega_2+\omega_1}{2}\right)}{\cos\left(\frac{\omega_2-\omega_1}{2}\right)}\text{ and }\beta=-\frac{\sin\left(\frac{\omega_2-\omega_1-\hat{\omega}_c}{2}\right)}{\sin\left(\frac{\omega_2-\omega_1+\hat{\omega}_c}{2}\right)}. \tag{8.75}$$

The lowpass-to-lowpass and lowpass-to-highpass transformations described in the previous sections were first-order in z. As a result, the order of the transformed filter was the same as the order of the prototype filter. In contrast, the lowpass-to-bandpass transformation of Equation (8.74) is quadratic. When this transformation is applied to a prototype filter, the order of the transformed filter will be double that of the prototype filter. To make the passband of the transformed bandpass filter $H_{bp}(\omega)$ extend over the desired frequency range $\omega_1<\omega<\omega_2$, the values of the constants α and β in Equation (8.75) are specified so that frequencies ω_1 and ω_2 both map to the cutoff frequency of the prototype lowpass filter $\hat{\omega}_c$ (see Problem 8-35).

In essence, α controls the center frequency of the filter and β controls the width. For example (see Problem 8-36), given values of α and β, a transformed Butterworth bandpass filter can be expressed as

$$H_{bp}(z) = \prod_{k=0}^{N-1} \frac{p_k T}{p_k T - 2} \frac{(1 - z^{-2})}{1 + \left(\frac{4\alpha}{p_k T - 2}\right) z^{-1} - \left(\frac{p_k T + 2}{p_k T - 2}\right) z^{-2}}, \tag{8.76a}$$

where p_k (the poles of the transformed filter) are related to $\hat{p}_k$(the poles of the prototype filter) by

$$p_k = \hat{p}_k \left(\frac{1 - \beta}{1 + \beta}\right). \tag{8.76b}$$

Figure 8.36 gives the outline of the lowpass-to-bandpass transformation process. Once again, we start with a prototype fourth-order Butterworth lowpass filter with a corner (−3-dB) frequency of $\hat{\omega}_c = \pi/2$ (**Figure 8.36a**). **Figures 8.36b** and **c** show how the frequency response of the bandpass filter depends on parameters α and β, respectively. The upper panel of **Figure 8.35b** shows the mapping of ω to $\hat{\omega}$, as specified by Equation (8.74) at a fixed value of $\alpha = 0$, and three values of β. The lower panel shows the magnitude of the resulting bandpass filters $|H_{bp}(\omega)|_{dB}$. At this value of α, all bandpass filters are centered on $\omega = \pi/2$, with a bandwidth $\Delta\omega = \omega_2 - \omega_1$ that is determined by β (e.g., $\Delta\omega = 3\pi/4$ when $\beta = -0.414$, and $\Delta\omega = \pi/4$ when $\beta = +0.414$). The upper panel of **Figure 8.35c** shows the mapping of ω to $\hat{\omega}$ at a fixed value of $\beta = +0.414$ and five values of α while the lower panel shows the magnitude of the resulting bandpass filters. Here, all the bandpass filters have the same bandwidth, $\Delta\omega = \pi/4$; the center frequencies of the filters vary from $\omega_c = \pi/4$ (when $\alpha = +0.765$) to $\omega_c = 3\pi/4$ (when $\alpha = -0.765$).

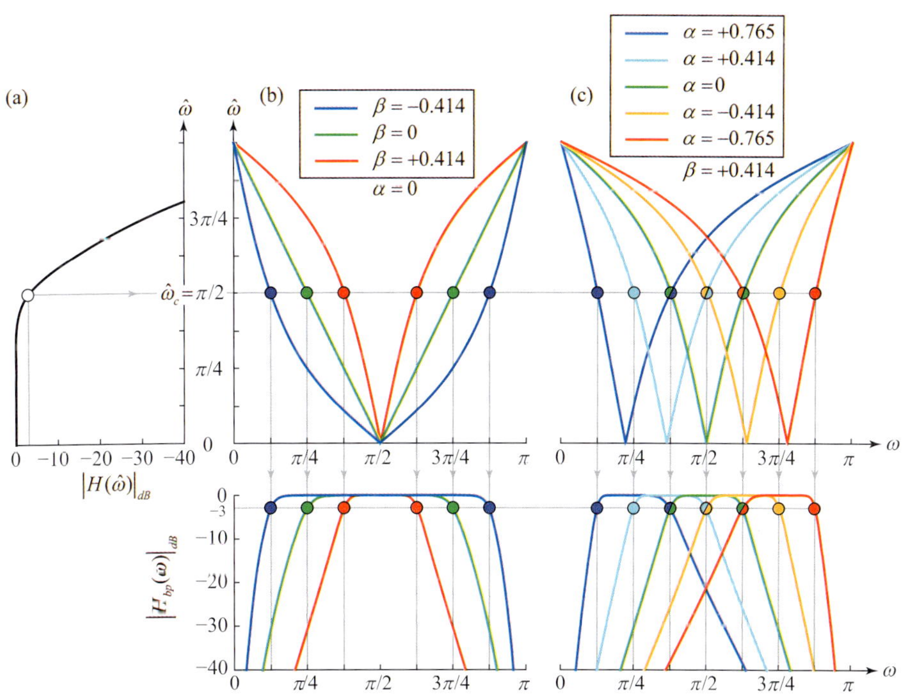

Figure 8.36 Lowpass-to-bandpass filter transformation

Example 8.16

Transform a fourth-order Butterworth lowpass filter with cutoff frequency $\hat{\omega}_c = \pi/2$ into a discrete-time bandpass filter with a -3 dB bandwidth of $5\pi/8 < \omega < 7\pi/8$.

➤ Solution:

This is a job for Matlab. But first, to make our code easier, we need to find the poles and zeros of each of the product terms of Equation (8.76a) so that we can use Matlab's `poly` function to create the numerator and denominator polynomials. The numerator term can be factored as $(1-z^{-2}) = (1-z^{-1})(1+z^{-1})$. Hence, each term has zeros at 1 and -1, which drives the frequency response to 0 at $\omega = 0$ and $\omega = \pm\pi$. The denominator can be factored (see Problem 8-36c) to give the poles,

$$\frac{\alpha \pm \sqrt{\alpha^2 - 1 + (p_k T/2)^2}}{1-(p_k T/2)}. \tag{8.77}$$

Then, we design a prototype filter as in Example 8.14, but with $\hat{\omega}_c = \pi/2$, and transform it:

```
wc_hat = pi / 2; % corner frequency of prototype
WcT_hat = 2 * tan(wc_hat / 2); % bilinear transformation of prototype
N = 4;
w = [5 7]*pi/8; % cutoff freqs of bandpass filter
dw = diff(w);
alpha = cos(sum(w)/2) / cos(dw/2);
beta = -sin((dw-wc_hat)/2) / sin((dw+wc_hat)/2);
k = 0:N-1;
pkT_hat = WcT_hat * exp(1j*pi*(2*k+N+1)/2/N);
pkT = pkT_hat * (1-beta)/(1+beta);
zers = [ones(1, N) -ones(1, N)];
pols = (alpha + [1;-1]*sqrt(alpha^2 - 1 + pkT.^2/4))./([1;1]*(1-pkT/2));
const = real(prod(pkT./(pkT-2)));
b = const * poly(zers(:));
a = real(poly(pols(:)));
```

The variable `pols` is a $2 \times N$ matrix of all the poles, as specified by Equation (8.77). The final answer is

$$H_{bp}(z) = \frac{0.0102(1-z^{-2})^4}{1+4.5680z^{-1}+9.9592z^{-2}+13.4991z^{-3}+12.4398z^{-4}+7.9500z^{-5}+3.4376z^{-6}+0.9231z^{-6}+0.1204z^{-7}}.$$

The right-most trace in the bottom panel of **Figure 8.36c** (corresponding to $\alpha = -0.765$) shows the frequency response $|H_{bp}(\omega)|_{dB}$ of the transformed bandpass filter. Again, this is a case where you would most likely want to implement this filter as a cascade of second-order sections (see Problem 8-41).

Matlab has a function `iirlp2bp` that can perform this transformation for you. Here is the code:

```
[bb, aa] = butter(4, 0.5); % create the prototype filter with wc=0.5pi
[b, a] = iirlp2bp(bb, aa, 0.5, [5 7]/8); % transform it to a bandpass
```

If you wish, you can then implement the filter as a cascade of second-order sections using the Matlab function `tf2sos`.

8.5.4 Lowpass-to-bandstop transformation

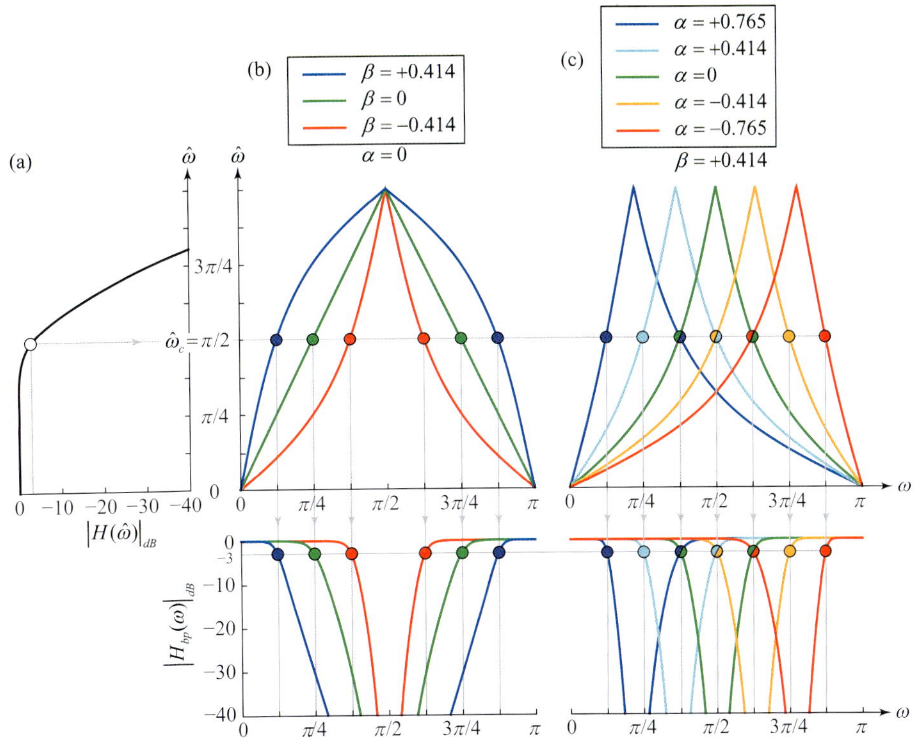

Figure 8.37 Lowpass-to-bandstop filter transformation

The lowpass-to-bandstop transformation is given by

$$\hat{z} = G_{bs}(z) = -\left(\frac{z+\beta}{1+\beta z}\right)\Bigg|_{z=-z\left(\frac{z-\alpha}{1-\alpha z}\right)} = -\left(\frac{-z\left(\frac{z-\alpha}{1-\alpha z}\right)+\beta}{1-\beta z\left(\frac{z-\alpha}{1-\alpha z}\right)}\right) = \frac{z^2-\alpha(1-\beta)z-\beta}{-\beta z^2-\alpha(1-\beta)z+1}.$$

Like the lowpass-to-bandpass transformation, the lowpass-to-bandstop transformation is quadratic, with the result that the transformed filter has an order double that of the prototype filter. The values of the constants α and β are again specified to make cutoff frequencies ω_1 and ω_2 map to the cutoff frequency of the prototype lowpass filter, $\hat{\omega}_c$,

$$\alpha = \frac{\cos\left(\frac{\omega_2+\omega_1}{2}\right)}{\cos\left(\frac{\omega_2-\omega_1}{2}\right)} \text{ and } \beta = -\frac{\cos\left(\frac{\omega_2-\omega_1+\hat{\omega}_c}{2}\right)}{\cos\left(\frac{\omega_2-\omega_1-\hat{\omega}_c}{2}\right)}.$$

For example (see Problem 8-38), given values of α and β, a transformed Butterworth bandstop filter can be expressed as

$$H_{bp}(z) = \prod_{k=0}^{N-1}\frac{p_kT}{p_kT-2}\frac{(1-2\alpha z^{-1}+z^{-2})}{1+\left(\frac{2\alpha p_kT}{p_kT-2}\right)z^{-1}+\left(\frac{p_kT+2}{p_kT-2}\right)z^{-2}}, \quad \textbf{(8.78a)}$$

where the poles of the transformed filter, p_k, are related to the poles of the prototype filter, $\hat{p}_k$, by

$$p_k = \hat{p}_k\left(\frac{1-\beta}{1+\beta}\right). \tag{8.78b}$$

Figure 8.37 gives the outline of the lowpass-to-bandstop transformation process. As with the lowpass-to-bandpass transformation, parameter β essentially controls the filter bandwidth and parameter α the center frequency.

Example 8.17

Transform a fourth-order Butterworth lowpass filter with cutoff frequency $\hat{\omega}_c = \pi/2$ into a discrete-time bandstop filter with a -3 dB bandwidth of $5\pi/8 < \omega < 7\pi/8$.

➤ Solution:

This is similar to Example 8.16, except the zeros of each of the product terms of Equation (8.78a) are located at

$$-\alpha \pm \sqrt{\alpha^2 - 1},$$

and the poles at

$$\frac{\alpha \pm \sqrt{\alpha^2 - 1 + (2/p_k T)^2}}{1 - (2/p_k T)}. \tag{8.79}$$

The design proceeds as in Example 8.16 with the following changes:

```
alpha = cos(sum(w)/2) / cos(dw/2);
beta = -cos((dw+wc_hat)/2) / cos((dw-wc_hat)/2);
k = 0:N-1;
pkT_hat = WcT_hat * exp(1j*pi*(2*k+N+1)/2/N);
pkT = pkT_hat * (1-beta)/(1+beta);
zers = (alpha + [1;-1]*sqrt(alpha^2 - 1))*ones(1, N);
b = real(prod(pkT./(pkT-2)) * poly(zers(:)));
pols = (alpha + [1;-1]*sqrt(alpha^2 - 1 + 4./pkT.^2))./([1;1]*(1-2./pkT));
a = real(poly(pols(:)));
```

The final answer is

$$H_{bp}(z) = \frac{0.3468 + 2.1236z^{-1} + 6.2632z^{-2} + 11.3465z^{-3} + 13.7369z^{-4} + 11.3465z^{-3} + 6.2632z^{-2} + 2.1236z^{-1} + 0.3468}{1 + 4.5680z^{-1} + 9.9592z^{-2} + 13.4991z^{-3} + 12.4398z^{-4} + 7.9500z^{-5} + 3.4376z^{-6} + 0.9231z^{-6} + 0.1204z^{-7}}.$$

The right-most trace in the bottom panel of **Figure 8.37c** (corresponding to $\alpha = -0.765$) shows the frequency response $|H_{bs}(\omega)|_{dB}$ of the transformed bandstop filter.

Matlab has a function, `iirlp2bs`, that can perform this transformation for you:

```
[bb, aa] = butter(4, 0.5); % create the prototype filter with wc=0.5pi
[b, a] = iirlp2bs(bb, aa, 0.5, [5 7]/8); % transform it to a bandstop
```

Summary of spectral transformations **Table 8.2** gives a summary of the transformations of a prototype lowpass filter with corner frequency $\hat{\omega}_c$ into a lowpass or highpass filter with corner frequency ω_c, as well as bandpass and bandstop filters with -3 dB points of ω_1 and ω_2.

Table 8.2 Spectral transformations of a prototype discrete-time lowpass filter

Transformation	$\hat{z} = G(z)$	Parameters
Lowpass	$\hat{z} = \dfrac{z-\alpha}{1-\alpha z}$	$\alpha = \dfrac{\sin\left(\dfrac{\hat{\omega}_c - \omega_c}{2}\right)}{\sin\left(\dfrac{\hat{\omega}_c + \omega_c}{2}\right)}$
Highpass	$\hat{z} = -\dfrac{z+\alpha}{1+\alpha z}$	$\alpha = -\dfrac{\cos\left(\dfrac{\hat{\omega}_c + \omega_c}{2}\right)}{\cos\left(\dfrac{\hat{\omega}_c - \omega_c}{2}\right)}$
Bandpass	$\hat{z} = -\dfrac{z^2 - \alpha(1+\beta)z + \beta}{\beta z^2 - \alpha(1+\beta)z + 1}$	$\alpha = \dfrac{\cos\left(\dfrac{\omega_2 + \omega_1}{2}\right)}{\cos\left(\dfrac{\omega_2 - \omega_1}{2}\right)}$, $\beta = -\dfrac{\sin\left(\dfrac{\omega_2 - \omega_1 - \hat{\omega}_c}{2}\right)}{\sin\left(\dfrac{\omega_2 - \omega_1 + \hat{\omega}_c}{2}\right)}$
Bandstop	$\hat{z} = \dfrac{z^2 - \alpha(1-\beta)z - \beta}{-\beta z^2 - \alpha(1-\beta)z + 1}$	$\alpha = \dfrac{\cos\left(\dfrac{\omega_2 + \omega_1}{2}\right)}{\cos\left(\dfrac{\omega_2 - \omega_1}{2}\right)}$, $\beta = -\dfrac{\cos\left(\dfrac{\omega_2 - \omega_1 + \hat{\omega}_c}{2}\right)}{\cos\left(\dfrac{\omega_2 - \omega_1 - \hat{\omega}_c}{2}\right)}$

8.6 ★ Zero-phase IIR filtering

In Chapter 7, we showed numerous examples of FIR filters that have linear phase. In contrast, the phase of the response of IIR filters is nonlinear; hence, the delay of each component in the filtered output at frequency ω is not a linear function of ω. However, it is possible to compensate for the nonlinear phase delay, at least in non-real-time applications where the input data are available for offline processing. The concept is indicated in **Figure 8.38**.

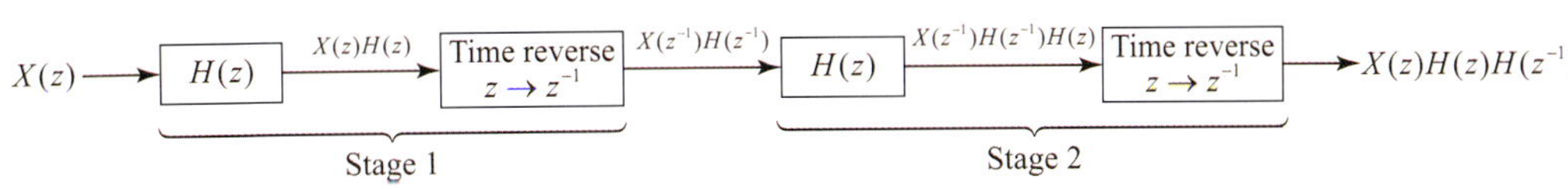

Figure 8.38 Zero-phase IIR filtering

The procedure comprises two stages: **forward filtering** and **backwards filtering**. In the first stage, an input signal $x[n]$ with z-transform $X(z)$ is filtered by a real causal stable IIR filter with impulse response $h[n]$ and z-transform $H(z)$ to form an output $y[n] = x[n] * h[n]$ and transform $Y(z) = X(z)H(z)$. The output is then time reversed to form $y[-n]$. Recall that the z-transform of the time-reversed signal is

$$\mathcal{Z}\{y[-n]\} = Y(z^{-1}) = X(z^{-1})H(z^{-1}).$$

In the second stage, the time-reversed signal is passed through the same filter again, resulting in a signal with transform $X(z^{-1})H(z^{-1})H(z)$. This signal is time-reversed a second time to yield an output signal with transform $X(z)H(z)H(z^{-1})$. The frequency response of the output is

$$X(z)H(z)H(z^{-1})|_{z=e^{j\omega}} = X(\omega)H(\omega)H(-\omega).$$

Since the impulse response of the IIR filter is real, the transform satisfies $H(-\omega) = H^*(\omega)$, so the transform of the output becomes

$$X(\omega)H(\omega)H^*(\omega) = X(\omega)|H(\omega)|^2.$$

The forward-backward filtering process is equivalent to filtering the input with transform $\hat{H}(\omega) \triangleq |H(\omega)|^2$. The transform $\hat{H}(\omega)$ is purely real and non-negative and thus has zero phase, but the magnitude is clearly the square of $|H(\omega)|$. Since $20\log_{10}|H(\omega)|^2 = 40\log_{10}|H(\omega)|$, the passband and stopband attenuations δ_s and δ_p are both doubled when expressed on a dB scale. This must be taken into account in designing $H(\omega)$.

Matlab's `filtfilt` function implements zero-phase filtering for both FIR and IIR filters. The syntax is

```
y = filtfilt(b, a, x),
```

where `x` and `y` are the input and output sequences, and `b` and `a` are arrays of coefficients that specify the z-transform,

$$H(z) = \frac{\sum_{k=0}^{M} b_k z^{-k}}{\sum_{k=0}^{N} a_k z^{-k}}.$$

In addition to the basic functionality just described, `filtfilt` also internally takes care of setting the initial conditions on the filter in order to minimize transients at the beginning and end of the response. The details of this function are discussed in Section 7.5.1.

Example 8.18

Compare the results of filtering a sequence $x[n] = 3 + \cos(0.15\pi n) + \sin(0.25\pi n)$ with a Butterworth IIR lowpass filter designed with $\omega_p = 0.2\pi$, $\delta_p = -3$ dB, $\omega_s = 0.4\pi$ and $\delta_s = -40$ dB, with and without using the zero-phase filtering technique.

➤ Solution:

Using the same procedure as Example 8.10, we design a sixth-order (non-zero-phase) filter $G(\omega)$ to meet the problem specifications. The magnitude and phase of this filter are shown in the blue traces in **Figure 8.39**.

Figure 8.40 shows the result of filtering $x[n]$ (green trace) with this non-zero-phase filter (blue trace) using Matlab's filter command: `y = filter(b, a, x)`. There are a couple of problems. First, there is a start-up transient; $y[n]$ starts at 0, and gradually ramps up to oscillate about the mean value of $x[n]$. This is due to the fact that the initial conditions for the difference equation that `filter` solves are taken to be 0 unless we explicitly specify them.

The second problem is that the two filtered frequency components of the output are not time-aligned either with each other or with the input. **Figure 8.39** indicates why this is so. The phase $\measuredangle G(\omega)$ of the component at $\omega = 0.15\pi$ is -0.99π, which corresponds to a delay of $0.99\pi/0.15\pi = 6.6$ samples. The phase of the component at $\omega = 0.25\pi$ is almost -2π, and corresponds to a delay of $(2\pi - 0.07\pi)/0.25\pi = 7.7$ samples. No single time adjustment of the output sequence will bring both of these components into time alignment with the input.

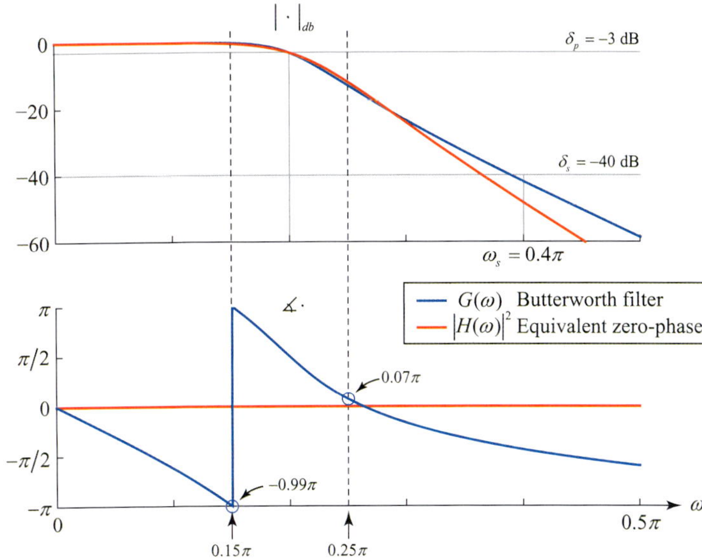

Figure 8.39 Frequency response of zero-phase filter

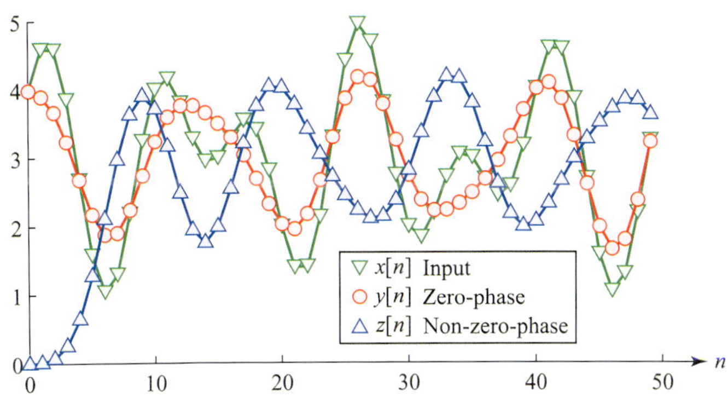

Figure 8.40 Zero-phase filtering example

In order to perform zero-phase filtering, a filter with response $H(\omega)$ must first be designed such that $|H(\omega)|^2$ matches $|G(\omega)|$ as closely as possible. That means the passband and stopband frequencies will be the same as those for $G(\omega)$ ($\omega_p = 0.2\pi$ and $\omega_s = 0.4\pi$), but the passband and stopband attenuations need to be half those of $G(\omega)$ on a dB scale: $\delta_p = -1.5$ dB and $\delta_s = -20$ dB. The effective frequency response of the resulting fourth-order filter, $|H(\omega)|^2$, is shown in the red trace in **Figure 8.39**. $|H(\omega)|^2$ exactly passes through -3 dB at $\omega_p = 0.15\pi$, and falls off somewhat more steeply than $|G(\omega)|$, which is to be expected since the filter order of $H(\omega)$ is more than twice that of $G(\omega)$. Because $|H(\omega)|^2$ is purely real and non-negative, the phase is identically 0 at all frequencies.

The red curve in **Figure 8.40** shows the result of filtering $x[n]$ with the zero-phase filter using Matlab's filter command: `y = filtfilt(b, a, x)`. The start-up transient is greatly minimized and the two filtered frequency components of the output are time-aligned both with each other and with the input.

SUMMARY

This chapter described the design of discrete-time filters that are based on analog filter prototypes. The first half of the chapter described the design of classical analog Butterworth, Chebyshev, inverse Chebyshev and elliptic filters. The chapter then described the two principal methods of converting from an analog prototype to a discrete-time realization – the impulse invariance and bilinear transformation – and argued that the bilinear transformation was preferred. The chapter concluded by developing discrete-time spectral transformations that convert a prototype discrete-time lowpass filter into a highpass, bandpass or bandstop filter.

PROBLEMS

Problem 8-1

Consider a Butterworth filter with input $x(t) = \cos \Omega t$ and output $y(t)$.

(a) Find the output $y(t)$ as a function of Ω.

(b) Show that the power transmitted by a filter, that is, the ratio of the output and input signals, is proportional to $|H(\Omega)|^2$.

(c) For a Butterworth filter, show that the -3 dB frequency of the filter is also the frequency at which the output power is one-half the input power. For this reason, the -3 dB frequency is also called the **half-power frequency** or **half-power point** of the filter.

Problem 8-2

Show that the Butterworth filter is maximally flat in the passband, namely that the first $2N-1$ derivatives of $|H(\Omega)|$ are 0 at $\Omega = 0$.

Problem 8-3

We have stated that Butterworth, Chebyshev and other analog filters that comprise multiple poles can be realized by a cascade of first-order and second-order sections. In this problem and the following two problems, we investigate how to choose the parameters of these sections to match the location of the poles derived from the design of these filters.

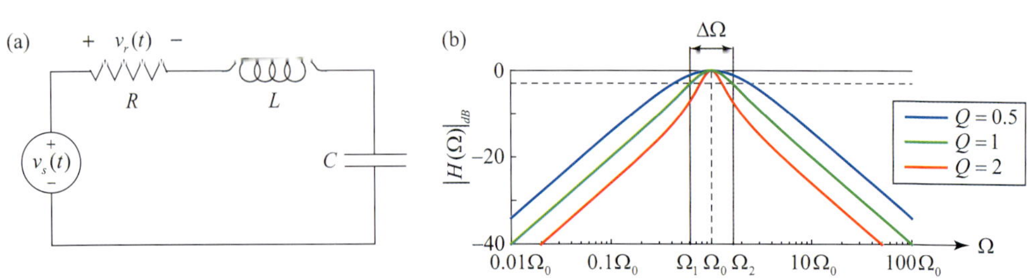

Figure 8.41

In this problem, we define the notions of **natural frequency** and ***Q*** for a second-order filter. **Figure 8.41a** shows the schematic of a second-order passive bandpass RLC filter.

The input is voltage $v_s(t)$ and the output is the voltage across the resistor, $v_r(t)$.

(a) Show that the transfer function of voltage for this filter, $H(\Omega)$, can be expressed as

$$H(\Omega) \triangleq \frac{V_r(\Omega)}{V_s(\Omega)} = \frac{1}{1 + jQ\left(\frac{\Omega}{\Omega_0} - \frac{\Omega_0}{\Omega}\right)}, \tag{8.80}$$

where

$$\Omega_0 = \frac{1}{\sqrt{LC}} \tag{8.81}$$

is termed the **natural frequency** or **resonant frequency** of the circuit, and

$$Q = \frac{1}{R}\sqrt{\frac{L}{C}} \tag{8.82}$$

is termed the **quality factor**. The Q of the filter is a measure of the sharpness of the magnitude of the transfer function. **Figure 8.41b** shows the magnitude of the transfer function for three values of Q plotted on a log-frequency axis (normalized to Ω_0) and log-magnitude (dB) scale. As Q increases, the response magnitude increases, which means that the filter is becoming more selective.

(b) Show that the peak of the magnitude of the transfer function occurs at $\Omega = \Omega_0$ and, at this frequency, the value of the transfer function is unity; that is, $|H(\Omega)| = 1$ or $|H(\Omega)|_{dB} = 0$ dB.

(c) An equivalent measure of the selectivity of the filter is $\Delta\Omega$, the -3 dB bandwidth of the transfer function. $\Delta\Omega$ is the difference between the frequencies Ω_1 and Ω_2 at which the magnitude of the transfer function falls by 3 dB (i.e., a factor of $1/\sqrt{2}$) from its peak value. In **Figure 8.41b**, Ω_1, Ω_2 and $\Delta\Omega = \Omega_2 - \Omega_1$ are marked to show the bandwidth of the transfer function for $Q = 1$.

Show that Q is equal to

$$Q = \frac{\Omega_0}{\Omega_2 - \Omega_1} = \frac{\Omega_0}{\Delta\Omega}.$$

(d) Find Ω_1, Ω_2 and $\Delta\Omega = \Omega_2 - \Omega_1$ (as a function of Ω_0) for $Q = 1$ and show that Ω_0 is the geometric mean of Ω_1 and Ω_2.

Problem 8-4

In Problem 8-3, we defined the natural frequency Ω_0 and quality factor of a second-order bandpass filter. In this problem, we turn our attention to the second-order filter whose circuit is shown in **Figure 8.42a**.

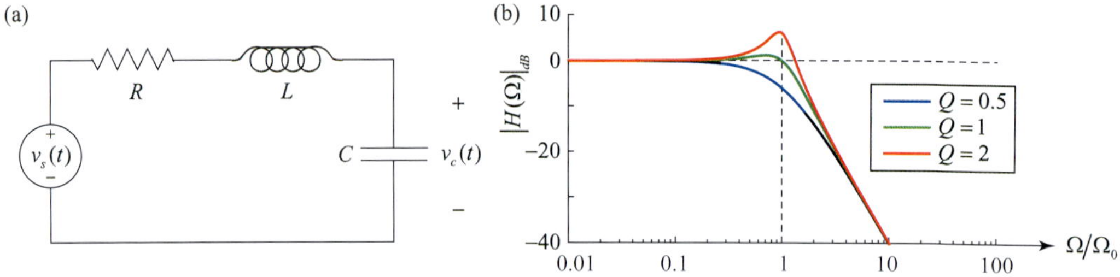

Figure 8.42

For this filter, the input is voltage $v_s(t)$, and the output is the voltage across the capacitor, $v_c(t)$. The frequency response of this filter is shown in **Figure 8.42b**.

(a) Show that the Laplace transform of this system is

$$H(s) \triangleq \frac{V_r(s)}{V_s(s)} = \frac{\Omega_0^2}{s^2 + (\Omega_0/Q)s + \Omega_0^2},$$

where

$$\Omega_0 = \frac{1}{\sqrt{LC}} \text{ and } Q = \frac{1}{R}\sqrt{\frac{L}{C}}.$$

(b) Show that the system has two poles, which occur at

$$s = \frac{\Omega_0}{2Q}\left(-1 \pm j\sqrt{4Q^2 - 1}\right).$$

(c) For values of $Q > 0.5$, the poles are complex and the system is said to be underdamped. Show that for an underdamped system, the poles form a complex-conjugate pair that lies in the left half-plane on a semicircle of radius Ω_0 at angles $\theta = \pm(\pi - \sec^{-1}Q)$, as diagrammed in **Figure 8.43**.

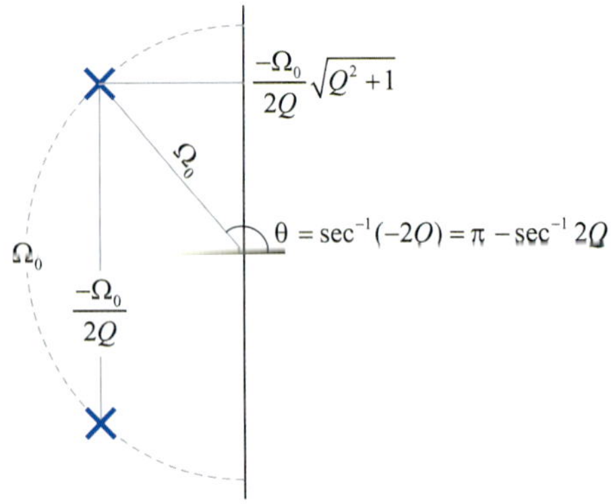

Figure 8.43

Problem 8-5

Given the bandpass filter shown in Problem 8-4,

(a) Show that the impulse response is

$$h(t) = \left(\frac{\Omega_0 Q}{\sqrt{4Q^2 - 1}}\right)(e^{s_1 t} - e^{s_2 t}),$$

where

$$s_{1,2} = \frac{\Omega_0}{2Q}\left(-1 \pm j\sqrt{4Q^2 - 1}\right).$$

(b) Using the impulse response invariance technique, let $t = nT$ and show that

$$h[n] = Th(nT) = \left(\frac{\omega_0 Q}{j\sqrt{4Q^2 - 1}}\right)\left(e^{(\omega_0/2Q)\left(-1+j\sqrt{4Q^2-1}\right)n} - e^{(\omega_0/2Q)\left(-1-j\sqrt{4Q^2-1}\right)n}\right),$$

where $\omega_0 = \Omega_0 T$.

(c) Show that

$$H(z) = \left(\frac{2\omega_0 Q}{\sqrt{4Q^2 - 1}}\right)\left(\frac{e^{-(\omega_0/2Q)}\sin\left((\omega_0/2Q)\sqrt{4Q^2 - 1}\right)z^{-1}}{1 - 2e^{-(\omega_0/2Q)}\cos\left((\omega_0/2Q)\sqrt{4Q^2 - 1}\right)z^{-1} + e^{-(\omega_0/Q)}z^{-2}}\right).$$

Problem 8-6

Given that the poles of the Butterworth filter of even order are

$$s_k = \Omega_c e^{j\pi(2k+n+1)/2N}, \quad 0 \le k < N,$$

show that the poles occur in complex-conjugate pairs such that $s_{(N-1)-k} = s_k^*$.

Problem 8-7

For a Butterworth filter of order N,

$$H(s) = K\prod_{k=0}^{N-1}\frac{1}{s - s_k},$$

where the poles are

$$s_k = \Omega_c e^{j\pi(2k+N+1)/2N}, \quad 0 \le k < N,$$

and the constant K is chosen such that $H(0) = 1$, show that

$$K = \prod_{k=0}^{N-1} -s_k = \Omega_c^N.$$

Problem 8-8

In each of the following parts, the specifications of the analog Butterworth filter are given in terms of the order N and center frequency Ω_c. For each filter, find the values of Ω_0 (Equation (8.81)) and Q (Equation (8.82)) of the cascade of second-order filters necessary to implement the filter.

(a) $N = 4$, $\Omega_c = 2\pi \cdot 1000$.
(b) $N = 5$, $\Omega_c = 2\pi \cdot 1000$.

➤ **Hint:** Use the results of Problem 8-4b.

Problem 8-9

Figure 8.44 shows the Sallen–Key circuit which implements a second-order lowpass filter with input $x(t)$ and output $y(t)$.

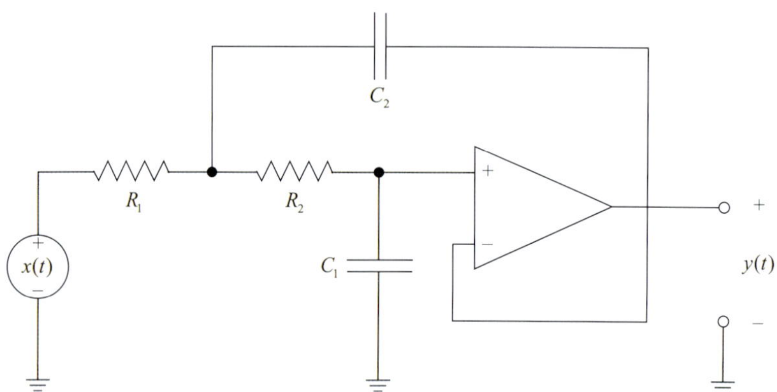

Figure 8.44

(a) Show that the analog transfer function of this system can be expressed as

$$H(s) = \frac{Y(s)}{X(s)} = \frac{\Omega_0{}^2}{s^2 + (\Omega_0/Q)s + \Omega_0{}^2}, \tag{8.85}$$

where the natural frequency Ω_0 and quality factor of the system are given in terms of the circuit elements as

$$\Omega_0 = \frac{1}{\sqrt{R_1 C_1 R_2 C_2}}$$

and

$$Q = \frac{\sqrt{R_1 C_1 R_2 C_2}}{(R_1 + R_2)C_1}.$$

(b) We wish to pick values of the circuit elements R_1, R_2, C_1 and C_2, to implement a filter with given design parameters Ω_0 and Q. There is no unique choice of the four circuit parameters to match the two design parameters. So, for simplicity, let $R_1 = R_2 \triangleq R$ and let $C_2 = \alpha^2 C_1$, where α is a constant. Show that $\alpha = 2Q$ and $RC_1 = (1/2)Q\,\Omega_0$. Hence, we can pick a nominal value of C_1 from which we can determine the values of the other circuit elements.

(c) Let $R_1 = R_2 \triangleq R$ and pick $C_1 = 100$ pF. Find reasonable values of R and C_2 such that the corner frequency is 1 kHz and $Q = 5$.

(d) Now assume that the resistors and capacitors of part (c) have $\pm 5\%$ tolerances. Find the range of Ω_0 and Q around the desired values.

Problem 8-10

In this problem, we will design a discrete-time filtering system to match the specifications of an analog filter. The analog filter is the Sallen–Key lowpass filter shown in **Figure 8.44**, which has a corner frequency of $f_0 = 1$ kHz and $Q = 0.7$. The equivalent digital system, whose architecture is shown in **Figure 8.45**, comprises an A/D and a D/A converter operating at sample rates of $f_s = f_r = 48$ kHz and a discrete-time IIR filter. Design a second-order IIR filter that matches the given specifications of the analog filter as closely as possible using the bilinear transformation.

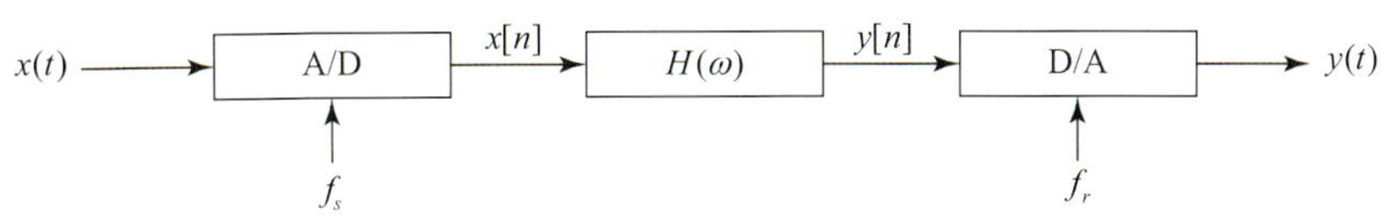

Figure 8.45

(a) Step 0: Find ω_0, the corner frequency of the discrete-time filter.

(b) Step 1: Map ω_0 into $\Omega_0 T$, the critical corner frequency of the second-order, continuous-time filter which will be mapped into the discrete-time, second-order filter.

(c) Step 2: Use the bilinear transformation to map $H(s)$, the transform of the second-order filter given in Equation (8.85), into $H(z)$, the transform of the equivalent discrete-time IIR filter.

(d) Plot the frequency response of the original Sallen–Key analog filter circuit, both magnitude and phase, and compare it with the frequency response of the complete digital filter circuit of **Figure 8.45**, including the IIR filter you just designed. Comment upon the differences.

Problem 8-11

(a) Show that the poles of $H(s)H(-s)$ for the Chebyshev filter are the roots of

$$1 + \varepsilon_p^2 T_N^2(s/j\Omega_c) = 0.$$

(b) Define $\cos\theta \triangleq s/j\Omega_c$, and show that the roots occur when

$$\theta = (1/N)\cos^{-1}(\pm j/\varepsilon_p),$$

so that the poles occur at

$$s = j\Omega_c \cos((1/N)\cos^{-1}(\pm j/\varepsilon_p)).$$

(c) Let

$$\cos^{-1}(\pm j/\varepsilon_p) = \alpha + j\beta,$$

where α and β are real. Then,

$$\pm j/\varepsilon_p = \cos(\alpha + j\beta).$$

By expanding this expression and equating real and imaginary parts on both sides, show that

$$\alpha = \pi/2 + k\pi \text{ and } \beta = \pm\sinh^{-1}(1/\varepsilon_p).$$

(d) Use the above results to show that the $2N$ roots of $H(s)H(-s)$ occur at

$$p_k = \Omega_c(\pm\sin\psi_k \sinh\varphi + j\cos\psi_k \cosh\varphi), \quad 0 \le k < N,$$

where

$$\psi_k = (2k+1)\pi/2N \text{ and } \varphi = (1/N)\sinh^{-1}(1/\varepsilon_p).$$

(e) Show that the roots of $H(s)$ are given by

$$p_k = \Omega_c(-\sin\psi_k \sinh\varphi + j\cos\psi_k \cosh\varphi), \quad 0 \le k < N,$$

and the roots of $H(-s)$ by

$$p_k = \Omega_c(-\sin\psi_k \sinh\varphi + j\cos\psi_k \cosh\varphi), \quad N \le k < 2N,$$

so that the roots of $H(s)H(-s)$ can be equivalently specified as

$$p_k = \Omega_c(-\sin\psi_k \sinh\varphi + j\cos\psi_k \cosh\varphi), \quad 0 \le k < 2N. \tag{8.86}$$

(f) Show that the roots of $H(s)H(-s)$ are disposed along an ellipse in the s-plane that has its major axis along the imaginary axis of length $2\cosh\varphi$ and minor axis along the real axis, of length $2\sinh\varphi$.

Problem 8-12

(a) Use the results of Problem 8-11 to show that the N poles of $H(s)$ for the inverse Chebyshev filter occur at

$$p_k = \Omega_c/(-\sin\psi_k \sinh\varphi + j\cos\psi_k \cosh\varphi), \quad 0 \le k < N,$$

where

$$\psi_k = (2k+1)\pi/2N \text{ and } \varphi = (1/N)\sinh^{-1}\varepsilon_s.$$

(b) Show that N zeros of $H(s)$ occur at

$$z_k = j\Omega_c/\cos((2k+1)\pi/2N), \quad 0 \le k < N.$$

Problem 8-13

Show that if a signal $x(t)$ is real (i.e., if $x(t) = x^*(t)$), then the Fourier transform is conjugate-symmetric (i.e., $X(\Omega) = X^*(-\Omega)$).

Problem 8-14

The Chebyshev trigonometric recursion relation states that

$$\cos \omega_0 n = 2 \cos \omega_0 n \cos \omega_0(n-1) - \cos \omega_0(n-2).$$

Prove this relation using trigonometry.

➤ **Hint:** Start by expressing

$$\cos\omega_0 n = \cos \omega_0((n-1)+1) = \cos \omega_0(n-1) \cos \omega_0 - \sin \omega_0(n-1) \sin \omega_0,$$

then expand the $\sin \omega_0(n-1)$ term in a similar manner.

Problem 8-15

In this problem, we show that the Chebyshev polynomial $T_N(x)^2$ has N maxima and minima within the range $0 < x < 1$ for $N > 0$.

(a) Show that

$$T_N^2(0) = \cos^2 N\pi/2,$$

so that it alternates in value between 0 and 1.

(b) Show that $T_N^2(1) = 1$, independent of N.

Problem 8-16

The Chebyshev polynomial of order N is defined as $T_N(x) = \cos(N\cos^{-1}x)$. Show that $T_N(x)$ can be equivalently defined by a simple polynomial recursion relation

$$T_N(x) = \begin{cases} 1, & N = 0 \\ x, & N = 1 \\ 2xT_{N-1}(x) - T_{N-2}(x), & N > 1 \end{cases}.$$

➤ **Hint:** Make the substitution, $x = \cos\theta$. Then, $T_N(\cos\theta) = \cos(N\cos^{-1}(\cos\theta)) = \cos(N\theta)$. Now use trigonometric relations such as

$$\cos(N\theta) = \cos(\theta + (N-1)\theta) = \cos\theta\cos(N-1)\theta - \sin\theta\sin(N-1)\theta$$

to prove the recursion relation.

Problem 8-17

Show that the slope of $T_N(x)$ at $x = 1$ is N^2 in the following ways:

(a) By taking the derivative of the polynomial recursion relation

$$T_N(x) = 2xT_{N-1}(x) - T_{N-2}(x).$$

(b) By taking the derivative of the definition $T_N(x) = \cos(N\cos^{-1}x)$.

Problem 8-18

We have shown that the Laplace transform of a Butterworth filter, $H(s)$, can be written as a product of N single-pole terms,

$$H(s) = \prod_{k=0}^{N-1} \frac{s_k}{s_k - s},$$

where the poles are disposed in the left-half plane on a circle of radius Ω_c given by

$$s_k = \Omega_c e^{j\pi(2k+N+1)/2N}, \quad 0 \le k < N.$$

(a) Show that $H(s)$ can also be written as the sum of N terms,

$$H(s) = \sum_{k=0}^{N-1} \frac{R_k}{s - s_k},$$

where the residues R_k are given by

$$R_k = -\frac{\prod_{l=0}^{N-1} s_l}{\prod_{\substack{l=0 \\ l \neq k}}^{N-1} (s_l - s_k)}.$$

(b) Show that R_k can be expressed as $R_k = \Omega_c \hat{R}_k$ where $\hat{R}_k$ is a factor that is independent of Ω_c,

$$\hat{R}_k = -\frac{s_k/\Omega_c}{\prod_{\substack{l=0 \\ l \neq k}}^{N-1} (1 - s_k/s_l)} = \frac{e^{j\pi(2k+N+1)/2N}}{\prod_{\substack{l=0 \\ l \neq k}}^{N-1} (1 - e^{j\pi(k-l)/N})}.$$

(c) Show that $s_{N-k+1} = s_k^*$ so that $R_{N-k+1} = R_k^*$.

(d) Show that for N odd, the pole at $k = (N-1)/2$ is real.

Problem 8-19

Given the results of Problem 8-18, design an Nth-order, discrete-time Butterworth filter using the impulse-invariance method.

(a) Start with the Laplace transform $H(s)$ of a Butterworth filter expressed as a sum of terms, as in Problem 8-18a. Show that, for $N > 1$,

$$H(s)=\sum_{k=0}^{N-1}\frac{R_k}{s-s_k}=\sum_{k=0}^{\lfloor N/2-1\rfloor}\left(\frac{R_k}{s-s_k}+\frac{R_k^*}{s-s_k^*}\right)+\begin{cases}\dfrac{R_{(N-1)/2}}{s-s_{(N-1)/2}}, & N \text{ odd}\\ 0, & N \text{ even}\end{cases}.$$

(b) Show that the inverse Laplace transform of $H(s)$ is

$$h(t)=\sum_{k=0}^{\lfloor N/2-1\rfloor}\left(R_k e^{s_k t}u(t)+R_k^* e^{s_k^* t}u(t)\right)+\begin{cases}R_{(N-1)/2}e^{s_{(N-1)/2}t}u(t), & N \text{ odd}\\ 0, & N \text{ even}\end{cases}.$$

(c) Sample $h(t)$ at period T and scale the result by T yielding $h[n]=Th(nT)$. Show that

$$\begin{aligned}h[n] &=\omega_c\sum_{k=0}^{\lfloor N/2-1\rfloor}\left(\hat{R}_k e^{\omega_c\hat{s}_k}n+\hat{R}_k^* e^{\omega_c\hat{s}_k^* n}\right)u[n]+\omega_c\begin{cases}\hat{R}_{(N-1)/2}e^{-\omega_c n}u[n], & N \text{ odd}\\ 0, & N \text{ even}\end{cases}\\ &=\sum_{k=0}^{\lfloor N/2-1\rfloor}\left(R_k e^{s_k n}+R_k^* e^{s_k^* n}\right)u[n]+\begin{cases}R_{(N-1)/2}e^{-\omega_c n}u[n], & N \text{ odd}\\ 0, & N \text{ even}\end{cases}.\end{aligned}$$

➤ **Hint:** Remember that $\omega_c=\Omega_c T$.

(d) Show that $H(z)$ can be expressed as

$$H(z)=2\omega_c\sum_{k=0}^{\lfloor N/2-1\rfloor}\left(\frac{\mathrm{Re}\{\hat{R}_k\}-z^{-1}\mathrm{Re}\{\hat{R}_k e^{s_k^* T_x}\}}{1-2\mathrm{Re}\{e^{s_k T_x}\}z^{-1}+e^{2\mathrm{Re}\{e^{s_k T}\}}z^{-2}}\right)+\omega_c\begin{cases}\dfrac{\hat{R}_{(N-1)/2}}{1-e^{-\omega_c}z^{-1}}, & N \text{ odd}\\ 0, & N \text{ even}\end{cases}.$$

Problem 8-20

Complete the steps of Example 8.19 to find the impulse response of a discrete-time filter with $\omega_p=0.3\pi$, $\delta_p=-3$ dB, $\omega_s=0.7\pi$ and $\delta_s-\ -20$ dB.

(a) Show that

$$H(s)=\sum_{k=0}^{N-1}\frac{R_k}{s-s_k}=\frac{R_0}{s-s_0}+\frac{R_1}{s-s_1}+\frac{R_2}{s-s_2},$$

where the residues are $R_0=-\frac{1}{\sqrt{3}}\Omega_c e^{j\pi/6}$, $R_1=\Omega_c$ and $R_2=-\frac{1}{\sqrt{3}}\Omega_c e^{-j\pi/6}$.

(b) Show that the z-transform, $H(z)$, can be written using real coefficients in parallel form as

$$H(z)=\frac{\omega_c}{1-e^{-\omega_c}z^{-1}}-\omega_c\left(\frac{1-z^{-1}\frac{2}{\sqrt{3}}e^{-\omega_c/2}\cos\left(\frac{\sqrt{3}}{2}\omega_c-\pi/6\right)}{1-2z^{-1}e^{-\omega_c/2}\cos\left(\frac{\sqrt{3}}{2}\omega_c\right)+z^{-2}e^{-\omega_c}}\right).$$

(c) Show that the z-transform $H(z)$ can be expressed as

$$H(z)=\omega_c\left(\frac{z^{-1}\left(-2e^{-\omega_c/2}\cos\left(\frac{\sqrt{3}}{2}\omega_c\right)+e^{-\omega_c}+\frac{2}{\sqrt{3}}e^{-\omega_c/2}\cos\left(\frac{\sqrt{3}}{2}\omega_c-\frac{\pi}{6}\right)\right)+z^{-2}\left(e^{-\omega_c}-\frac{2}{\sqrt{3}}e^{-3\omega_c/2}\cos\left(\frac{\sqrt{3}}{2}\omega_c-\frac{\pi}{6}\right)\right)}{1-z^{-1}\left(e^{-\omega_c}+2e^{-\omega_c/2}\cos\left(\frac{\sqrt{3}}{2}\omega_c\right)\right)+z^{-2}\left(e^{-\omega_c}+2e^{-3\omega_c/2}\cos\left(\frac{\sqrt{3}}{2}\omega_c\right)\right)-z^{-3}e^{-2\omega_c}}\right)$$

$$=\frac{0.0799z^{-1}+0.0526z^{-2}}{1-1.7824z^{-1}+1.1993z^{-2}-0.2843z^{-3}}.$$

Problem 8-21

In the example of **Figure 8.19**, we showed that the biggest effect of aliasing in the impulse-invariance method is for frequencies around half the sampling frequency, $\Omega=\Omega_s/2$, which is equivalent to the discrete-time frequency $\omega=\pi$. Here,

$$H(\pi)=H_s(\Omega_s/2)\cong H(\Omega_s/2)+H(-\Omega_s/2).$$

Show that, due to aliasing,

$$|H(\pi)|_{dB}=-42.12+3.23=-38.85\,\text{dB}.$$

Problem 8-22

In the previous example, the effect of aliasing is to make $|H(\omega)|_{dB}>|H(\Omega_s/2)|$; however, this is not always the case.

(a) Given $\omega_p=\pi/2$, $\delta_p=-3$ dB, $\omega_s=\pi$ and $\delta_s=-20$ dB, find N and ω_c for a Butterworth filter.

(b) Show that

$$|H(\Omega_s/2)|_{dB}=-24.08\,\text{dB and } |H(\pi)|_{dB}=-31.71\,\text{dB},$$

so that, in this case, the effect of aliasing is to reduce the response at $\omega=\pi$.

Problem 8-23

Show that for the backward-difference approximation, poles in the left half of the s-plane always map inside the unit circle of the z-plane.

Problem 8-24

Show that for the backward-difference approximation, a line in the s-plane $s = \alpha + j\Omega$, with fixed real part α and imaginary part varying from $-\infty < \Omega < \infty$, maps into a circle in the z-plane with center $1/2(1 + \alpha T)$ and radius $1/2(1 + \alpha T)$:

$$\frac{1}{2(1+\alpha T)}\left(1 + e^{j2\tan(\Omega/(1-\alpha T))}\right).$$

Problem 8-25

Show that for the bilinear transformation, points in the left half of the s-plane always map to points inside the unit circle of the z-plane.

Problem 8-26

Show that for the bilinear transformation, a line in the s-plane $s = \alpha + j\Omega$, with fixed real part and imaginary part varying from $-\infty < \Omega < \infty$, maps into a circle in the z-plane with center $\alpha T/(2 - \alpha T)$ and radius $2/(2 - \alpha T)$:

$$\frac{\alpha T + 2e^{j2\tan(\Omega T/(2-\alpha T))}}{2 - \alpha T}.$$

Problem 8-27

Design a discrete-time Butterworth filter using the bilinear transformation with $\omega_p = 0.3\pi$, $\delta_p = -3$ dB, $\omega_s = 0.7\pi$ and $\delta_s = -20$ dB.

Problem 8-28

The impulse-invariance method does not work well when the analog filter has zeros. Consider a simple analog filter with a zero at $s = 0$,

$$H(s) = \frac{s}{(s+1)(s+2)}.$$

Show that a discrete-time filter produced with the bilinear transformation has zeros on the unit circle at $z = \pm 1$, while a filter designed using the impulse-invariance method does not.

Problem 8-29

Given $|\alpha| < 1$, show that the transformation

$$\hat{z} = G(z) = \frac{z - \alpha}{1 - \alpha z}$$

maps values of z inside the unit circle ($|z|<1$) into values of $|\hat{z}|<1$ and values of z outside the unit circle (i.e., $|z|>1$) into values of $|\hat{z}|>1$.

Problem 8-30

For a lowpass-to-lowpass transformation of the IIR filter, we applied the mapping

$$\hat{z}=G_{lp}(z)=\frac{z-\alpha}{1-\alpha z},$$

where the parameter α determines the amount of frequency warping. For values of z on the unit circle (i.e., $z=e^{j\omega}$), we showed that the mapping corresponds to $\hat{z}=e^{j\hat{\omega}}$, where

$$\hat{\omega}=\omega+2\tan^{-1}\left(\frac{\alpha\sin\omega}{1-\alpha\cos\omega}\right).$$

(a) Show that the value of α necessary to map a given value of $\omega=\omega_c$ into a desired value of $\hat{\omega}=\hat{\omega}_c$ is

$$\alpha=\frac{\sin\left(\dfrac{\hat{\omega}_c-\omega_c}{2}\right)}{\sin\left(\dfrac{\hat{\omega}_c+\omega_c}{2}\right)}.$$

(b) From part (a) and the bilinear transformation, show that

$$\frac{1-\alpha}{1+\alpha}=\frac{2\tan\omega_c/2}{2\tan\hat{\omega}_c/2}=\frac{\Omega_c T}{\hat{\Omega}_c T}.$$

Problem 8-31

(a) Start with the equation for a prototype discrete-time Butterworth lowpass filter derived using the bilinear transformation,

$$H(\hat{z})=\prod_{k=0}^{N-1}\left(\frac{\hat{p}_kT}{\hat{p}_kT-2}\right)\frac{1+\hat{z}^{-1}}{1-\left(\dfrac{2+\hat{p}_kT}{2-\hat{p}_kT}\right)\hat{z}^{-1}},$$

where $\hat{p}_k$ are the poles of the filter. Apply a lowpass-to-lowpass transformation

$$\hat{z}=\frac{z-\alpha}{1-\alpha z},$$

and show that the z-transform of the transformed lowpass filter is

$$H_{lp}(z)=H(\hat{z})=\prod_{k=0}^{N-1}\left(\frac{p_kT}{p_kT-2}\right)\frac{1+z^{-1}}{1-\left(\dfrac{2+p_kT}{2-p_kT}\right)z^{-1}},$$

where the poles of the transformed filter are given by

$$p_k = \hat{p}_k\left(\frac{1-\alpha}{1+\alpha}\right).$$

So, the effect of the lowpass-to-lowpass transformation is equivalent to multiplying the radius of the poles of the original filter by a factor of $(1-\alpha)/(1+\alpha)$.

(b) Use the results of part (a) and Problem 8-30b to show that the filter $H_{lp}(z)$ with corner frequency ω_c, obtained by transforming the prototype filter $H(\hat{z})$ with corner frequency $\hat{\omega}_c$, is identical to a filter obtained by designing a filter of the same order, with corner frequency ω_c, from scratch. Specifically, show that the poles of the transformed filter are located at

$$p_k T = \Omega_c T e^{j\pi(2k+N+1)/2N}, \quad 0 \le k < N,$$

where, from the bilinear transformation,

$$\Omega_c T = 2\tan \omega_c/2.$$

Problem 8-32

(a) Design a prototype second-order Butterworth lowpass filter $H(\hat{z})$ with cutoff frequency $\hat{\omega}_c = \pi/2$ using the bilinear-transformation method. You should get

$$H(\hat{z}) = \frac{1}{2+\sqrt{2}} \frac{(1+\hat{z}^{-1})^2}{1+\left(\dfrac{2-\sqrt{2}}{2+\sqrt{2}}\right)\hat{z}^{-2}} = 0.2929\frac{(1+\hat{z}^{-1})^2}{1+0.1716\hat{z}^{-2}}.$$

(b) Transform this filter into a discrete-time lowpass filter $H_{lp}(z)$ with corner frequency $\omega_c = \pi/4$. You should get

$$H_{lp}(z) = 0.0976\frac{(1+z^{-1})^2}{1-0.9428z^{-1}+0.3333z^{-2}}.$$

(c) Plot $|H(\hat{\omega})|$, $G_{lp}(\omega)$ and $|H_{lp}(\hat{\omega})|$ in a manner similar to **Figure 8.33**.

Problem 8-33

In this problem, we derive the lowpass-to-highpass transformation directly, starting with the equation for a prototype discrete-time Butterworth lowpass filter,

$$H(\hat{z}) = \prod_{k=0}^{N-1}\left(\frac{\hat{p}_k T}{\hat{p}_k T - 2}\right)\frac{1+\hat{z}^{-1}}{1 - \left(\dfrac{2+\hat{p}_k T}{2-\hat{p}_k T}\right)\hat{z}^{-1}}.$$

(a) Show that the lowpass-to-highpass transformation is given by

$$\hat{z} = G_{hp}(z) = G_{lp}(-z) = -\frac{z+\alpha}{1+\alpha z},$$

where,

$$\alpha = \frac{\cos\left(\dfrac{\hat{\omega}_c + \omega_c}{2}\right)}{\cos\left(\dfrac{\hat{\omega}_c - \omega_c}{2}\right)}.$$

(b) Apply a lowpass-to-highpass transformation directly to the prototype lowpass filter and show that the transformed highpass filter is given by

$$H_{hp}(z) = H(G_{hp}(z)) = \prod_{k=0}^{N-1}\left(\frac{p_k T}{p_k T - 2}\right)\frac{1 - z^{-1}}{1 - \left(\dfrac{p_k T + 2}{p_k T - 2}\right)z^{-1}},$$

where

$$p_k = \hat{p}_k\left(\frac{1-\alpha}{1+\alpha}\right).$$

Problem 8-34

(a) Transform the prototype second-order Butterworth lowpass filter $H(\hat{z})$ with cutoff frequency $\hat{\omega}_c = \pi/2$, given by Equation (8.71), into a discrete-time highpass filter $H_{hp}(z)$ with corner frequency $\omega_c = 3\pi/4$. You should get

$$H_{hp}(z) = 0.2929\frac{(1 - z^{-1})^2}{1 + 2.8284z^{-1} + 3z^{-2}}.$$

(b) Plot $|H(\hat{\omega})|$, $G_{hp}(\omega)$ and $|H_{hp}(\hat{\omega})|$ in a manner similar to **Figure 8.33**.

Problem 8-35

In the lowpass-to-bandpass transformation of Equation (8.74),

$$\hat{z} = G_{bp}(z) = -\left(\frac{z+\beta}{1+\beta z}\right)\Bigg|_{z=z\left(\frac{z-\alpha}{1-\alpha z}\right)} = -\left(\frac{z\left(\dfrac{z-\alpha}{1-\alpha z}\right)+\beta}{1+\beta z\left(\dfrac{z-\alpha}{1-\alpha z}\right)}\right) = -\frac{z^2 - \alpha(1+\beta)z + \beta}{\beta z^2 - \alpha(1+\beta)z + 1},$$

the values of α and β specified in Equation (8.75) are designed so that, given a lowpass cutoff ω_1 and a highpass cutoff ω_2, when $z = e^{j\omega_1}$ we get $\hat{z} = e_c^{j\hat{\omega}}$ and when $z = e^{j\omega_2}$ we get $\hat{z} = e_c^{-j\hat{\omega}}$, where ω_c is the desired center frequency.

(a) Show that

$$\tan\frac{\hat{\omega}_c}{2} = \frac{\beta+1}{\beta-1}\,\frac{\cos\omega_1 - \alpha}{\sin\omega_1}$$

and

$$-\tan\frac{\hat{\omega}_c}{2} = \frac{\beta+1}{\beta-1}\frac{\cos\omega_2 - \alpha}{\sin\omega_2}.$$

➤ **Hint:**

$$\frac{1-e^{j\hat{\omega}_c}}{1+e^{j\hat{\omega}_c}} = \frac{-j\tan\hat{\omega}_c}{2}.$$

(b) Use the results of part (a) to show that

$$\alpha = \frac{\sin(\omega_2+\omega_1)}{\sin\omega_2 + \sin\omega_1} = \frac{\cos\left(\dfrac{\omega_2+\omega_1}{2}\right)}{\cos\left(\dfrac{\omega_2-\omega_1}{2}\right)}.$$

➤ **Hint:** Express

$$\omega_1 = \frac{\omega_2+\omega_1}{2} - \frac{\omega_2-\omega_1}{2} \text{ and } \omega_2 = \frac{\omega_2+\omega_1}{2} + \frac{\omega_2-\omega_1}{2}.$$

(c) Substitute the value of α into one of the first expressions in part (a) and show that

$$\beta = -\frac{\sin\left(\dfrac{\omega_2-\omega_1-\hat{\omega}_c}{2}\right)}{\sin\left(\dfrac{\omega_2-\omega_1+\hat{\omega}_c}{2}\right)}.$$

➤ **Hint:** Recognize that

$$\cos\left(\frac{\omega_2+\omega_1}{2}\right) = \cos\left(\frac{\omega_2-\omega_1}{2} + \omega_1\right).$$

Problem 8-36

Given the equation for a discrete-time Butterworth lowpass filter derived using the bilinear transformation,

$$H(\hat{z}) = \prod_{k=0}^{N-1}\left(\frac{\hat{p}_k T}{\hat{p}_k T - 2}\right)\frac{1+\hat{z}^{-1}}{1-\left(\dfrac{2+\hat{p}_k T}{2-\hat{p}_k T}\right)\hat{z}^{-1}},$$

(a) apply the lowpass-to-bandpass transformation,

$$\hat{z} = G_{bp}(z) = -\frac{z^2 - \alpha(1+\beta)z + \beta}{\beta z^2 - \alpha(1+\beta)z + 1},$$

and show that the transformed bandpass filter is given by

$$H_{bp}(z) = \prod_{k=0}^{N-1} \frac{p_k T}{p_k T - 2} \frac{(1 - z^{-2})}{1 + \left(\frac{4\alpha}{p_k T - 2}\right) z^{-1} - \left(\frac{p_k T + 2}{p_k T - 2}\right) z^{-2}}.$$

(b) show that the roots of the denominator of each of the product terms are located at

$$\frac{\alpha \pm \sqrt{\alpha^2 - 1 + (p_k T/2)^2}}{1 - (p_k T/2)}.$$

Problem 8-37

Design a prototype second-order Butterworth lowpass filter with cutoff frequency $\hat{\omega}_c = \pi/2$ and transform it into a bandpass filter with corner frequencies $\omega_1 = \pi/4$ and $\omega_2 = 3\pi/4$.

Problem 8-38

Given the equation for a discrete-time Butterworth lowpass filter derived using the bilinear transformation,

$$H(\hat{z}) = \prod_{k=0}^{N-1} \left(\frac{\hat{p}_k T}{\hat{p}_k} T - 2\right) \frac{1 + \hat{z}^{-1}}{1 - \left(\frac{2 + \hat{p}_k T}{2 - \hat{p}_k T}\right) \hat{z}^{-1}},$$

(a) apply the lowpass-to-bandstop transformation

$$\hat{z} = G_{bs}(z) = \frac{z^2 - \alpha(1 - \beta)z - \beta}{-\beta z^2 - \alpha(1 - \beta)z + 1},$$

and show that the transformed bandpass filter is given by

$$H_{bp}(z) = \prod_{k=0}^{N-1} \frac{p_k T}{p_k T - 2} \frac{(1 - 2\alpha z^{-1} + z^{-2})}{1 + \left(\frac{2\alpha p_k T}{p_k T - 2}\right) z^{-1} + \left(\frac{p_k T + 2}{p_k T - 2}\right) z^{-2}}.$$

(b) show that the roots of the denominator of the product terms are located at

$$\frac{\alpha \pm \sqrt{\alpha^2 - 1 + (p_k T/2)^2}}{1 - (p_k T/2)}.$$

Problem 8-39

Show that the bilinear transformation maps all the zeros of the Type-II Chebyshev filter from the $j\Omega$-axis in the a-plane onto the unit circle in the z-plane.

Problem 8-40

(a) Use Matlab and Equation (8.65) to find the coefficients for the lowpass-to-lowpass transformation in Example 8.14 as a cascade of second-order sections in the form

$$H_{bp}(z) = \prod_{k=0}^{M-1} C_k \frac{(1 + b_{1k}z^{-1} + b_{2k}z^{-2})}{1 + a_{1k}z^{-1} + a_{2k}z^{-2}}, \tag{8.87}$$

where C_k is a constant gain for each section k. The task here is to combine the N sections of Equations (8.71) into $M = N/2$ sections with real coefficients. In this example, $N = 4$ and so we will have $M = 2$ second-order sections.

(b) Check your work using Matlab's `zp2sos` function.

Problem 8-41

(a) Use Matlab to find the coefficients for the lowpass-to-bandpass transformation in Example 8.16 as a cascade of second-order sections in the form

$$H_{bp}(z) = C \prod_{k=0}^{M-1} \frac{(1 - b_{1k}z^{-1} + b_{2k}z^{-2})}{1 + a_{1k}z^{-1} - a_{2k}z^{-2}}, \tag{8.88}$$

where C is a constant gain. From Equation (8.76), a filter of order N has $2N$ poles and $2N$ zeros organized into N quadratic sections, each of which has complex coefficients in the denominator. The task here is to reorganize this into N sections with real coefficients in the denominator.

(b) Check your work using Matlab's `zp2sos` function.

9 Filter architecture

Introduction

In Chapters 7 and 8, we dealt with the theory of designing FIR and IIR filters to achieve particular specifications. However, creating the filter algorithms is only the first step. A designer also needs to make decisions on the precise manner in which the filter will be implemented, decisions that depend on such factors as the target processor and environment. Is the filter going to be implemented in software on a general-purpose microcomputer, or on a microcontroller, or on a special-purpose DSP chip, or on a VLSI circuit in a real-time environment? In software, we generally wish to maximize speed while minimizing code size and memory usage. In hardware, we want to minimize the number of memory elements, adders and multipliers while assuring maximum throughput. In both hardware and software implementations, we have to consider the fact that numbers, such as the coefficients that define the filter, are not represented precisely, but must be quantized to fit into the finite-precision fixed- or floating-point arithmetic.

Sections 9.1 through 9.6 represent the core material of this chapter. In Section 9.1, we start by introducing signal-flow graphs, which are an implementation-independent way of representing filter algorithms. Then, in Sections 9.2 and 9.3, we use these signal-flow graphs to represent the basic architecture of FIR and IIR filters. Sections 9.4 and 9.5 expand the basic discussion to cover two basic ways of implementing these filters: cascade and parallel configurations in both canonical and transpose variants. Section 9.6 shows a few special ways of implementing FIR filters. Section 9.7 introduces lattice and lattice-ladder filters, which are useful in a number of special applications, particularly in speech and audio signal processing. If you are planning to implement your designs in systems with less than full floating-point precision, then you need to consider how quantization of filter coefficients can affect the response characteristics of filters. This is covered in Section 9.8. Finally, in Section 9.9, we will consider a number of issues that arise in software and hardware implementations.

9.1 Signal-flow graphs

Signal-flow graphs are a powerful and economical way of representing and visualizing the flow of various signals in discrete-time signal processing systems such as filters. Originally developed to represent signal flow in continuous-time systems, particularly in control theory, they have proven useful in discrete-time systems to represent algorithms and their transformations theoretically and also to help designers implement these algorithms efficiently in hardware or software.

A signal-flow graph consists of a set of **nodes** (represented by open or closed circles), which represent signal inputs, output or connections. Nodes are connected to each other by one or more **branches** (represented by lines with arrowheads), which represent data paths that can also perform transformations of the data. Nodes come in several flavors, as shown in **Figure 9.1**.

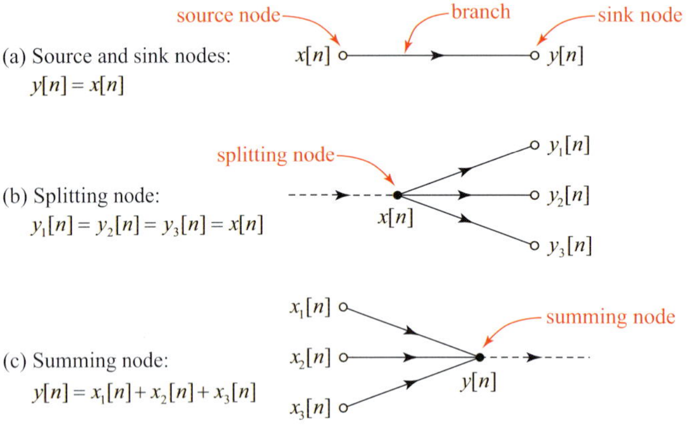

Figure 9.1 Signal-flow graph showing input, output, splitting and summing nodes

Source nodes and **sink nodes** are represented by open symbols. A source node represents an external input to the system, for example an input that is read from a data file or a real-time source such as an A/D converter. It is connected to the rest of the graph by a single branch. A sink node, connected to the rest of the graph by a single exiting branch, represents an output of the system, which may be written to a data file or sent to a D/A converter or to another part of the system. In this chapter, we will consider only systems with a single external input and a single output, so there will only be one source node and one sink node.

Figure 9.1a shows the world's simplest signal-flow graph comprising a single source node representing an input sequence $x[n]$, which is connected by a single branch to sink node representing an output sequence $y[n]$. In this case, the branch performs no transformation of the signal, so this signal-flow diagram represents the incredibly complicated algorithm $y[n] = x[n]$.

Splitting nodes and **summing nodes** are represented by closed circles in the signal-flow graphs. A splitting node (**Figure 9.1b**) replicates a signal entering the node from a single branch between two or more exiting branches. In this example, the signal-flow graph represents the equation $y_1[n] = y_2[n] = y_3[n] = x[n]$. This signal-flow graph indicates that the three output values $y_1[n]$, $y_2[n]$ and $y_3[n]$ change instantaneously, which would be possible to implement in a parallel hardware implementation such as an FPGA. However, in a software implementation on a general-purpose computer, the memory locations of the values of three output variables would most likely have to be set sequentially by separate instructions of code or microcode.

A summing node (**Figure 9.1c**) sums inputs from multiple entering branches into one output, for example, $y[n] = x_1[n] + x_2[n] + x_3[n]$. From an algorithmic point of view, the addition of multiple quantities is "instantaneous." However, in a software implementation, the **arithmetic logic unit (ALU)** responsible for addition most likely has two register inputs and produces one output, so the addition of N values would require $N-1$ sequential two-element additions by the ALU.

All branches in a signal-flow graph are unidirectional; the arrow at the center of the branch indicates the direction of the flow of information. A branch can perform a transformation on the input signal, and the symbol above or below the arrow head indicates the type of transformation. For the linear systems we are studying, there are two admissible transformations, multiplication by a constant (**Figure 9.2a**) or a one-unit time delay (**Figure 9.2b**).

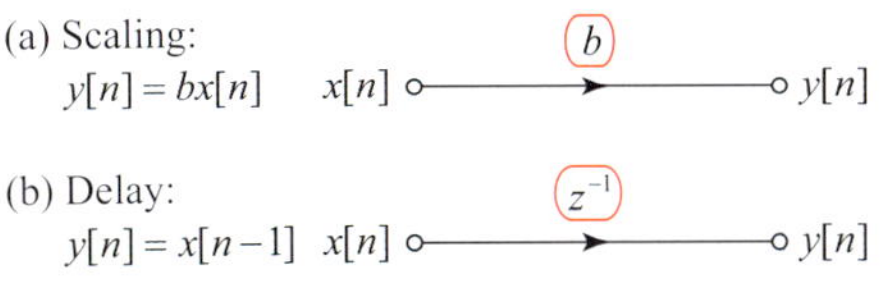

Figure 9.2 Branch transformations

A constant is denoted by a variable or a number, so in the example of **Figure 9.2a**, b is a constant. It is customary (though perhaps somewhat unfortunate) to denote the time delay by the symbol z^{-1}. So the example of **Figure 9.2b** corresponds to $y[n] = x[n-1]$. We emphasize that this notation does *not* mean that $y[n] = x[n]z^{-1}$, which would make no sense. A time delay in the signal-flow graph corresponds to a memory element in the hardware. To make a multiple-unit time delay, we concatenate multiple single-unit delays, as shown in **Figure 9.3** for the case of $y[n] = x[n-2]$.

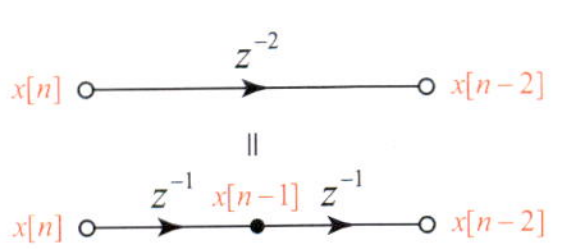

Figure 9.3 Multiple time delays

Here, there are two memory storage elements. It is important to understand how to translate the signal-flow graph into the appropriate sequence of software or hardware operations. In the case of a delay like $y[n] = x[n-2]$, three distinct operations need to be performed for each time point n:

1:	$y[n] \leftarrow x[n-2]$	Send stored value of $x[n-2]$ to output $y[n]$
2:	$x[n-2] \leftarrow x[n-1]$	Move contents of $x[n-1]$ to $x[n-2]$
3:	$x[n-1] \leftarrow x[n]$	Save current input value in memory.

Depending upon the implementation, delays of this sort may be implemented by updating pointers to memory addresses rather than physically moving memory contents (see Section 9.9.1).

9.2 Canonical filter architecture

A general Nth-order discrete time filter with one input, $x[n]$, and one output, $y[n]$, has the difference equation

$$\sum_{k=0}^{N} a_k y[n-k] = \sum_{k=0}^{M} b_k x[n-k]. \tag{9.1}$$

We can divide both sides by a_0, which is equivalent to making $a_0 = 1$. The difference-equation can then be written as

$$y[n] = \sum_{k=0}^{M} b_k x[n-k] - \sum_{k=1}^{N} a_k y[n-k].$$

As we have seen in Chapter 7, for an FIR filter, all the $a_k = 0$, $k > 0$, so the output $y[n]$ depends only upon scaled and shifted values of previous inputs $x[n-k]$.

9.2.1 First-order filters

Let us start by constructing signal-flow graphs of a few simple first-order filters, which are characterized by having only one memory element.

First-order FIR filter **Figure 9.4a** shows the signal-flow graph for a first-order FIR filter given by

$$y[n] = b_0 x[n] + b_1 x[n-1].$$

This is a **feed-forward configuration**, since the output depends only upon delayed and scaled versions of the input. For each time point n, the order of operations implied by this graph is

1:	$y[n] \leftarrow b_0 x[n] + b_1 x[n-1]$	Compute $y[n]$ using past and present input values
2:	$x[n-1] \leftarrow x[n]$	Save current input value in memory.

The first step, the computation of $y[n]$, itself involves multiple memory access and arithmetic operations, including two multiplications and an addition.

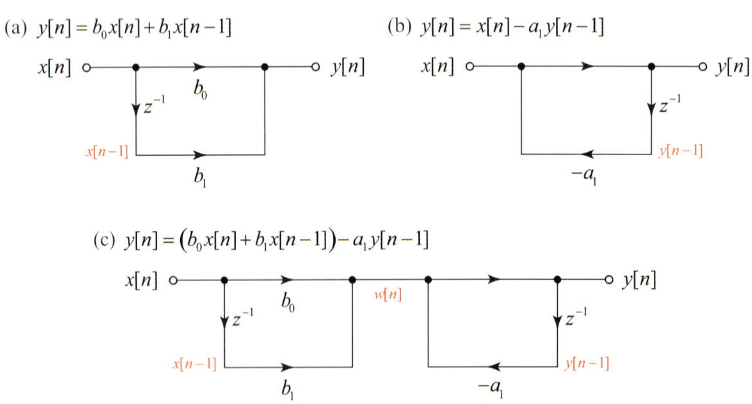

Figure 9.4 First-order FIR and IIR filters

First-order IIR filter **Figure 9.4b** shows the signal-flow graph for a first-order IIR filter given by

$$y[n] + a_1 y[n-1] = x[n],$$

or, equivalently,

$$y[n] = x[n] - a_1 y[n-1].$$

This is a **feedback configuration**, since the output depends on delayed and scaled versions of the output as well as the input. Notice that the sign of the scale factor becomes negative when the $y[n-1]$ term is moved to the right-hand side of the equation. The order of operations implied by this graph is

1:	$y[n] \leftarrow x[n] - a_1 y[n-1]$	Compute $y[n]$ using present input and past output
2:	$y[n-1] \leftarrow y[n]$	Save current output in memory.

Finally, **Figure 9.4c** shows the signal-flow graph for a first-order IIR filter with both feed-forward and feedback elements given by

$$y[n] + a_1 y[n-1] = b_0 x[n] + b_1 x[n-1].$$

This is represented by a **cascade** (a series concatenation) of the feed-forward and feedback elements. To analyze this configuration, it is useful to define the intermediate variable $w[n]$, which represents the output of the feed-forward path,

$$w[n] = b_0 x[n] + b_1 x[n-1].$$

Then the feedback operation can be expressed in terms of $w[n]$ as

$$y[n] = w[n] - a_1 y[n-1].$$

Hence, the order of operations is

1:	$w[n] \leftarrow b_0 x[n] + b_1 x[n-1]$	Compute $w[n]$ using past and present input values
2:	$y[n] \leftarrow w[n] - a_1 y[n-1]$	Compute $y[n]$ using present $w[n]$ and past output
3:	$y[n-1] \leftarrow y[n]$	Save current output value in memory
4:	$x[n-1] \leftarrow x[n]$	Save current input value in memory.

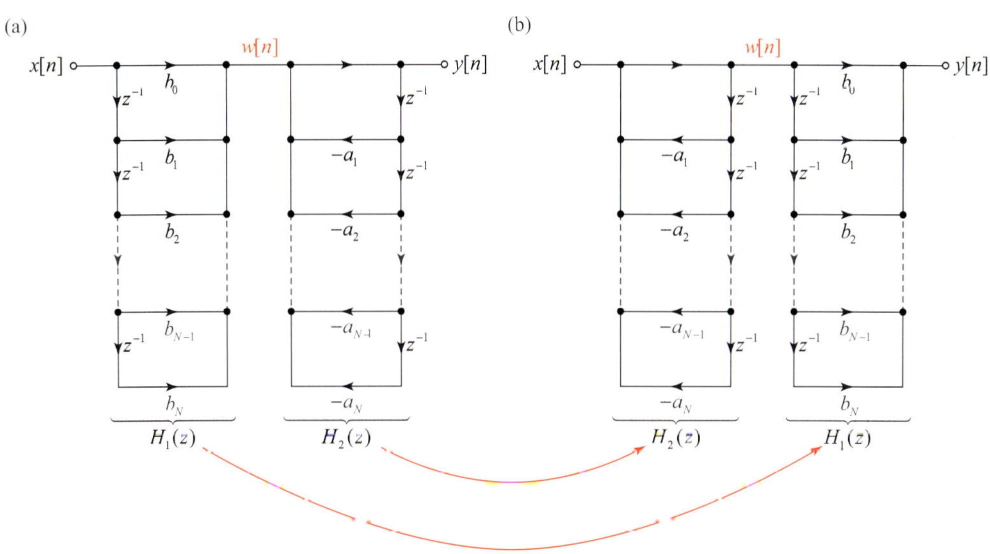

Figure 9.5 Signal-flow graph for Nth-order system

9.2.2 Canonical filter architecture

Figure 9.5a shows the signal-flow graph for a general Nth-order system. This corresponds to Equation (9.1) with $M = N$. If $M \neq N$, we can just set the appropriate values of a_k or b_k to zero. This architecture has $2N$ delays: N delays of $x[n]$ in the feed-forward path and N delays of $y[n]$ in the feedback path. Since each delay corresponds to a memory element in hardware or software, this architecture requires $2N$ memory elements. To aid in the analysis, define an intermediate value $w[n]$ to correspond to the output of the feed-forward stage, as we did in **Figure 9.4c**,

$$w[n] = \sum_{k=0}^{N} b_k x[n-k].$$

Taking the z-transform of this expression yields

$$\frac{W(z)}{X(z)} = \sum_{k=0}^{N} b_k z^{-k}.$$

The output of the filter is

$$y[n] = w[n] - \sum_{k=1}^{N} a_k y[n-k],$$

which leads to the z-transform

$$\frac{Y(z)}{W(z)} = \frac{1}{1 + \sum_{k=1}^{N} a_k z^{-k}}.$$

Then, the z-transform of the system function of Equation (9.1) is

$$H(z) = \frac{Y(z)}{X(z)} = \underbrace{\frac{W(z)}{X(z)}}_{H_1(z)} \cdot \underbrace{\frac{Y(z)}{\cancel{W(z)}}}_{H_2(z)} = \underbrace{\left(\sum_{k=0}^{N} b_k z^{-k}\right)}_{H_1(z)} \underbrace{\left(\frac{1}{1 + \sum_{k=1}^{N} a_k z^{-k}}\right)}_{H_2(z)}.$$

The z-transform factors into the product of two terms: $H_1(z) = W(z)/X(z)$ corresponds to the feed-forward path; $H_2(z) = Y(z)/W(z)$ corresponds to the feedback path. The architecture of **Figure 9.5a** directs us to compute the feed-forward path first, followed by the feedback path. However, from the commutative property of the z-transform we have

$$H(z) = H_1(z)H_2(z) = H_2(z)H_1(z),$$

so we can reverse the order of the feed-forward and feedback paths, as shown in **Figure 9.5b**. All the delay elements of both the feed-forward and feedback paths are now associated with the *same* intermediate variable $w[n]$, as shown in **Figure 9.6a**. Hence, we can simplify the signal-flow graph by joining the feed-forward and feedback paths together, as shown in **Figure 9.6b**. The resulting system, termed the **canonical filter architecture**, has only N delay elements, as well as $2N + 1$ multiplications by a constant and $2N$ two-branch additions.

At each time point n, the sequence of operations for the canonical filter is as follows:

$$
\begin{array}{lrl}
1: & w[n] & \leftarrow x[n] - a_1 w[n-1] - a_2 w[n-2] - \cdots - a_{N-1} w[n-(N-1)] - a_N w[n-N] \\
2: & y[n] & \leftarrow b_0 w[n] + b_1 w[n-1] + \cdots + b_{N-1} w[n-(N-1)] + b_N w[n-N] \\
3: & w[n-N] & \leftarrow w[n-(N-1)] \\
 & & \vdots \\
 & w[n-2] & \leftarrow w[n-1] \\
 & w[n-1] & \leftarrow w[n].
\end{array}
\tag{9.2}
$$

The first step is to compute a new value of $w[n]$ (the top node in **Figure 9.6b**) using the current input value $x[n]$, as well as the past N values of $w[n]$ (i.e., $w[n-1]$, $w[n-2]$, ..., $w[n-N]$), scaled by constants a_k. Then, a new value of $y[n]$ is computed from all $N+1$ values of $w[n]$, scaled by constants b_k. Finally, the N stored values of $w[n]$ are "shifted down," starting with the oldest value; that is $w[n-N] = w[n-(N-1)]$, until $w[n-1] = w[n]$.

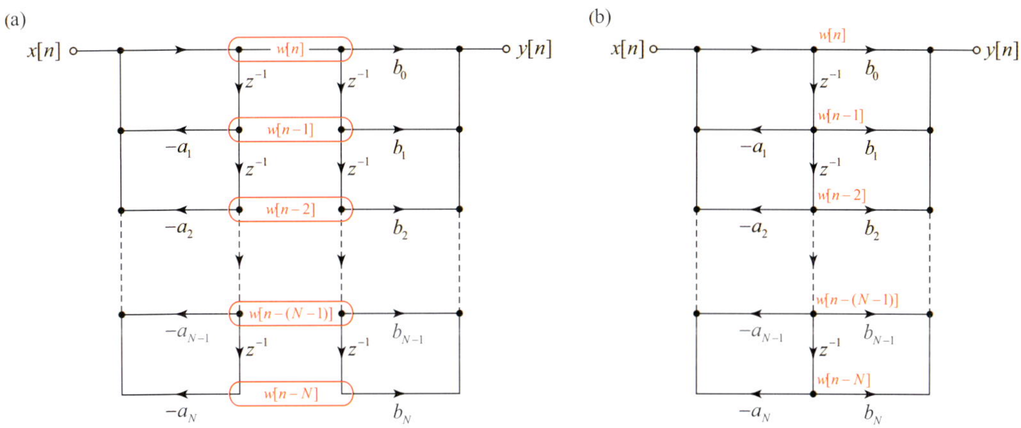

Figure 9.6 Canonical architecture

9.3 Transposed filters

There are multiple filter architectures, the relative advantages of which we will discuss as we go along. One of the most common is the **transposed canonical form**. To derive this, consider a canonical second-order filter with difference equation

$$y[n] + a_1 y[n-1] + a_2 y[n-2] = b_0 x[n] + b_1 x[n-1] + b_2 x[n-2], \tag{9.3}$$

whose implementation is shown in the top left panel of **Figure 9.7**.

The transposed filter is formed in three steps: (1) reverse all the signal paths so that summing nodes become splitting nodes, splitting nodes become summing nodes, the input node becomes the output node and the output node becomes the input node, (2) swap the input and the output variables $x[n]$ and $y[n]$ and (3) flip the entire graph left-right so the input is on the left side again. It is possible to show that the canonical filter and the transposed filter have identical system functions, and that this three-step procedure is generally applicable. But we can quickly justify

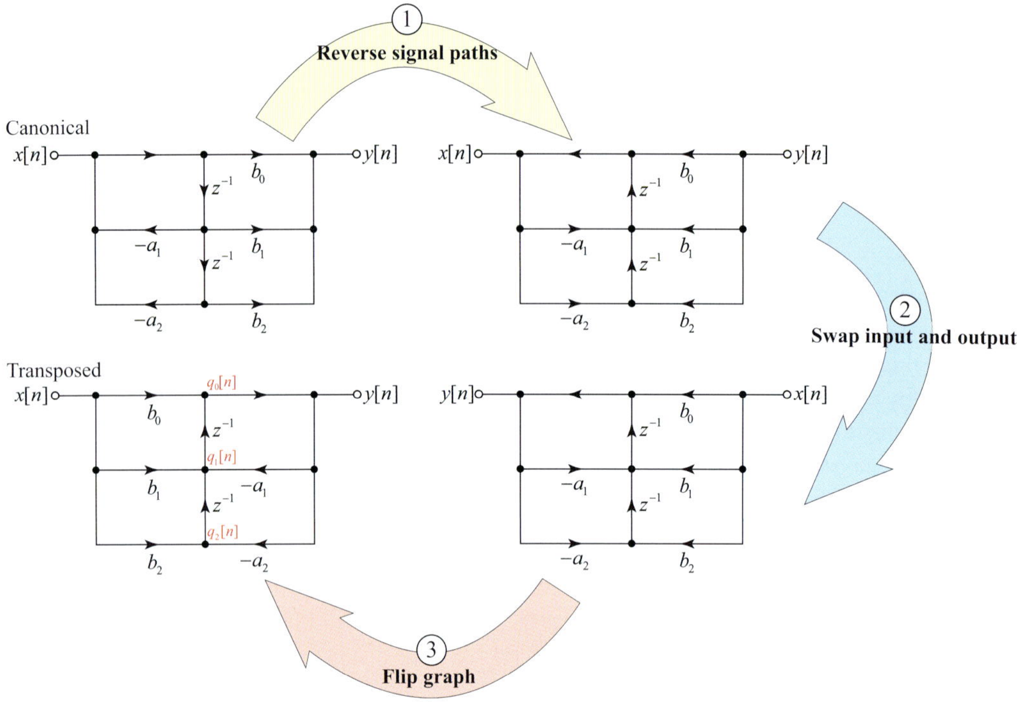

Figure 9.7 Transposed filter

the equivalence of the two filters in this example by a simple application of the z-transform. The z-transform of the second-order difference equation in Equation (9.3) is

$$Y(z)(1 + a_1 z^{-1} + a_2 z^{-2}) = X(z)(b_0 + b_1 z^{-1} + b_2 z^{-2}).$$

Rewrite this as a nested sum of terms, delayed by z^{-1},

$$\begin{aligned} Y(z) &= b_0 X(z) + z^{-1}(b_1 X(z) - a_1 Y(z)) + z^{-2}(b_2 X(z) - a_2 Y(z)) \\ &= \underbrace{b_0 X(z) + z^{-1} \underbrace{\left((b_1 X(z) - a_1 Y(z)) + z^{-1} \underbrace{(b_2 X(z) - a_2 Y(z))}_{Q_2(z)} \right)}_{Q_1(z)}}_{Q_0(z)} . \end{aligned}$$

Let $Q_2(z) \triangleq b_2 X(z) - a_2 Y(z)$. This corresponds to the difference equation $q_2[n] = b_2 x[n] - a_2 y[n]$, which is the output of the lower summing node in the signal-flow graph of the transposed filter, shown in red in the bottom left panel of **Figure 9.7**. Let $Q_1(z) \triangleq (b_1 X(z) - a_1 Y(z)) + z_{-1} Q_2(z)$. This corresponds to $q_1[n] = b_1 x[n] - a_1 y[n] + q_2[n-1]$, which is the output of the middle summing node of the transposed filter. Finally, let $Q_0(z) = b_0 X(z) + z^{-1} Q_1(z)$. This corresponds to $q_0[n] = b_0 x[n] + q_1[n-1]$, which is the output of the top summing node. It is also the output of the filter, because $y[n] = q_0[n]$. Since the system functions of the original and transposed filters are identical, their inputs and outputs are identical, too.

Figure 9.8 shows a comparison of Nth-order canonical and transposed filters.

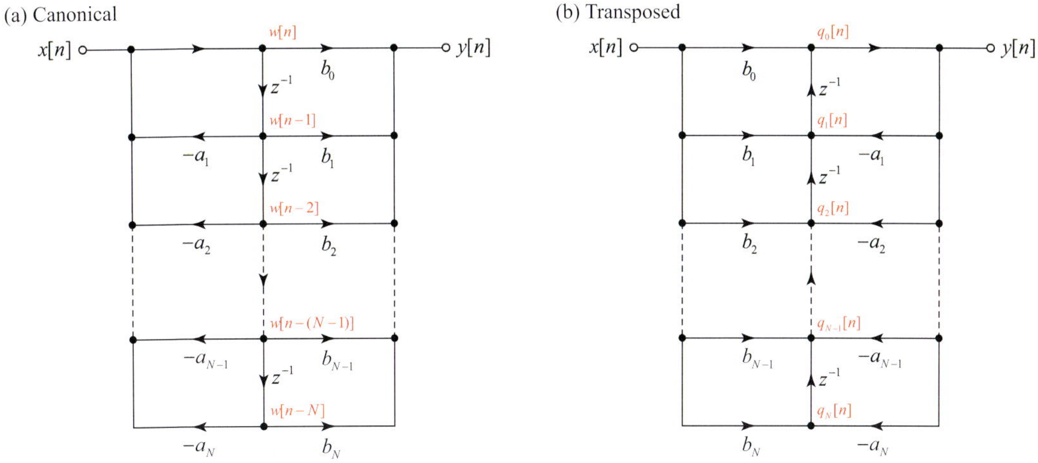

Figure 9.8 Canonical and transposed filters

At each time point n, the sequence of operations for the transposed canonical filter is as follows:

$$
\begin{aligned}
1:\quad & y[n] \leftarrow b_0 x[n] + q_0[n] \\
2:\quad & q_0[n] \leftarrow q_1[n] + b_1 x[n] - a_1 y[n] \\
& q_1[n] \leftarrow q_2[n] + b_2 x[n] - a_2 y[n] \\
& \quad\vdots \\
& q_N[n] \leftarrow b_N x[n] - a_N y[n]
\end{aligned}
$$

First, a new value of the output $y[n]$ is computed from the current input $x[n]$ and the stored value of a single intermediate value $q_0[n]$. Then, new values of all the intermediate values (i.e., $q_0[n]$, $q_1[n]$, …, $q_N[n]$) are computed in turn using values of $x[n]$ and $y[n]$ scaled by constants b_k and a_k, respectively.

Comparing the canonical filter of Section 9.2 with that of the transposed filter, you can show that the number of arithmetic operations – multiplications and additions – of the two architectures is identical, but the *sequence* of operations is significantly different. This has consequences for the **throughput** of the filter, which is the maximum speed at which the filter can process data. For the canonical architecture (Equations (9.2)), once an input point $x[n]$ is presented to the filter, we have to perform N multiplications and N additions in order to compute the top-node intermediate variable $w[n]$,

$$w[n] \leftarrow x[n] - a_1 w[n-1] - a_2 w[n-2] - \cdots - a_{N-1} w[n-(N-1)] - a_N w[n-N],$$

and then perform a further $N+1$ multiplications and N additions before we can produce the output point $y[n]$,

$$y[n] \leftarrow b_0 w[n] + b_1 w[n-1] + \cdots + b_{N-1} w[n-(N-1)] + b_N w[n-N].$$

Most modern general-purpose serial processors, including many microcontrollers, feature a variety of **multiplication and accumulate (MAC)** operations. The simplest MAC operation would perform the multiplication of two memory elements and then transfer or add the product to a register in one machine cycle. Here, computation of each time point for the canonical filter would require $2N+1$ cycles. In a hardware implementation of the canonical filter, all the multiplications (i.e., $a_k w[n-k]$ and $b_k w[n-k]$) could be performed in parallel, but if the additions need to be performed pair-wise, the maximum throughput would be essentially determined by $2N$ times the delay of the adder.[1] In contrast, for the transposed architecture, the computation of the output point requires only one MAC operation,

$$y[n] \leftarrow b_0 x[n] + q_0[n],$$

and the computation of each of the intermediate values (i.e., $q_0[n], q_1[n], \ldots, q_N[n]$) requires only two MAC operations. In a hardware implementation, these operations can be parallelized so that the maximum throughput of the filter is only two adder delays. We will elaborate on implementation issues such as this in Section 9.9.

9.4 Cascade architecture

A system with z-transform $H(z)$ that is the ratio of a numerator polynomial $N(z)$ of order M and a denominator polynomial $D(z)$ of order N can be expressed as the product of second-order terms $H_k(z)$:

$$H(z) = \frac{N(z)}{D(z)} = \frac{\sum_{k=0}^{M} b_k z^{-k}}{\sum_{k=0}^{N} a_k z^{-k}} = A \prod_{k=1}^{\lceil \max(M,N)/2 \rceil} H_k(z), \tag{9.4a}$$

where $A = b_0/a_0$ is a constant gain and

$$H_k(z) = \frac{1 + b_{1k} z^{-1} + b_{2k} z^{-2}}{1 + a_{1k} z^{-1} + a_{2k} z^{-2}}, \tag{9.4b}$$

with real coefficients b_{1k}, b_{2k}, a_{1k} and a_{2k}. Each of these second-order terms represented by $H_k(z)$ is called a **biquadratic (biquad) section** in recognition of the fact that both the numerator and denominator polynomials are quadratics. Most any practical filter can be realized with a cascade of biquadratic sections, and many microcontrollers have hardware/firmware that natively supports filtering using these sections. The total number of sections required is determined by the larger of the order of the numerator polynomial M and the order of the denominator polynomial N, namely $M = \lceil \max(M, N)/2 \rceil$, where $\lceil \cdot \rceil$ indicates rounding up to the largest integer. The signal-flow diagram for the cascade is shown in **Figure 9.9**.

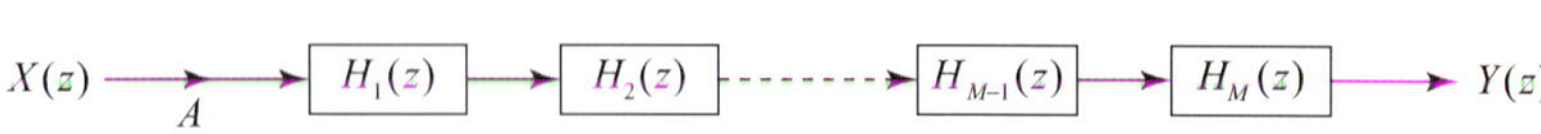

Figure 9.9 Cascade configuration

A few examples will lead us forward.

[1] It is possible to create so-called "pipelined" canonical IIR filters whose throughput is determined by two adder delays, but at a cost of larger **latency**, which is defined as the time delay between the input and output points.

Example 9.1

A second-order IIR system is described by the difference equation

$$y[n] + 0.75y[n-1] + 0.125y[n-2] = 2x[n] - 6x[n-1] + 4x[n-2].$$

Express $H(z)$ as a cascade of sections.

➤ Solution:

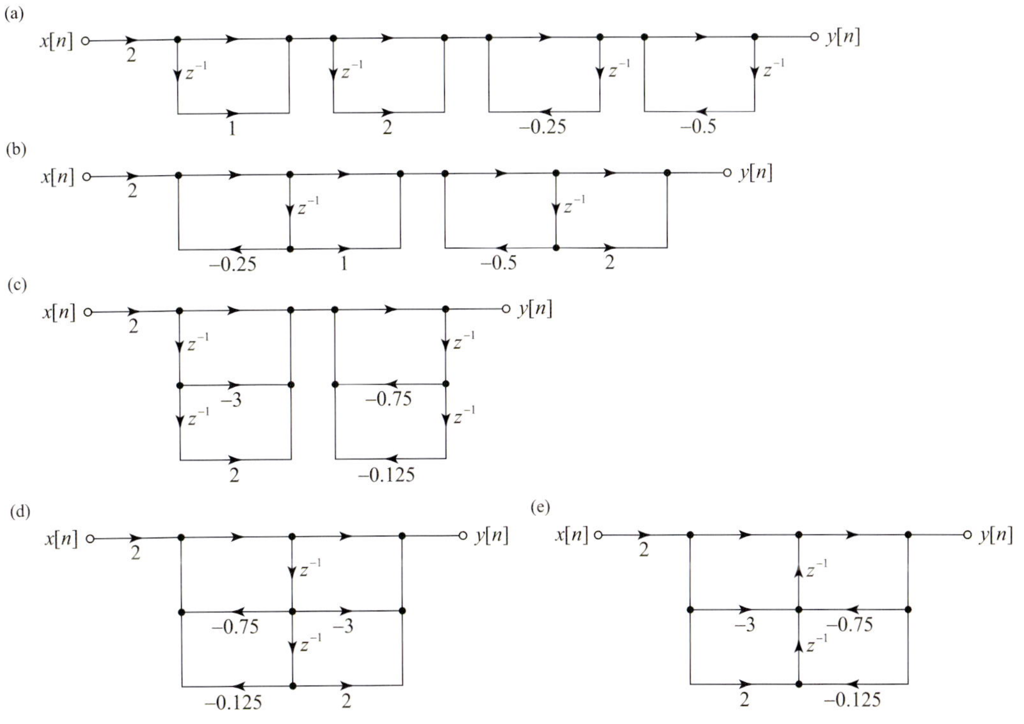

Figure 9.10 Cascade example

This can be realized as a product of terms in a number of ways, for example, as a cascade of first-order sections, shown in **Figure 9.10a**,

$$H(z) = 2(1 - z^{-1}) \cdot (1 - 2z^{-1}) \cdot \frac{1}{1 + 0.25z^{-1}} \cdot \frac{1}{1 + 0.5z^{-1}}.$$

The signs of the coefficients in the feedback sections of the signal-flow diagrams are the negative of those in the denominator terms (why?). The order of these sections can be altered, so there are 24 possible combinations. We can combine pairs of first-order sections into two canonical first-order sections, each of which comprises one numerator term and one denominator term, as shown in **Figure 9.10b**,

$$H(z) = 2\frac{1 - z^{-1}}{1 + 0.25z^{-1}} \cdot \frac{1 - 2z^{-1}}{1 + 0.5z^{-1}}.$$

The two numerator terms could be combined into one second-order section, and the two denominator terms into another second-order section, as shown in **Figure 9.10c**,

$$H(z) = 2(1 - 3z^{-1} + 2z^{-2}) \cdot \frac{1}{1 + 0.75z^{-1} + 0.125z^{-2}}.$$

Alternatively, we could have a second-order section and two first-order sections. You get the idea. Finally, we could represent the entire $H(z)$ with one canonical biquadratic section (**Figure 9.10d**), or its transposed equivalent (**Figure 9.10e**),

$$H(z) = \frac{2 - 6z^{-1} + 4z^{-2}}{1 + 0.75z^{-1} + 0.125z^{-2}} = 2\frac{(1 - z^{-1})(1 - 2z^{-1})}{(1 + 0.25z^{-1})(1 + 0.5z^{-1})}.$$

In Section 9.8.4, we will address the question of which of these possible combinations might be preferred in a practical implementation.

Here is an example in which $H(z)$ has complex poles and zeros.

Example 9.2

A system is described by the difference equation

$$y[n] + y[n-1] + 0.5y[n-2] + 0.125y[n-3] = 4x[n] - 14x[n-1] + 15x[n-2] - 7x[n-3] + 2x[n-4].$$

Express $H(z)$ as a cascade of sections.

➤ **Solution:**

$$H(z) = \frac{4 - 14z^{-1} + 15z^{-2} - 7z^{-3} + 2z^{-4}}{1 + z^{-1} + 0.5z^{-2} + 0.125z^{-3}} = 4\frac{(1 - 2z^{-1})(1 - z^{-1})(1 - 0.5e^{j\pi/3}z^{-1})(1 - 0.5e^{-j\pi/3}z^{-1})}{(1 + 0.5z^{-1})(1 - 0.5e^{j2\pi/3}z^{-1})(1 - 0.5e^{-j2\pi/3}z^{-1})}.$$

Because there are complex terms in both the numerator and denominator, it is not possible to express $H(z)$ as a cascade of first-order sections with purely real coefficients. However, because the complex terms occur in complex-conjugate pairs, they can be multiplied together to form second-order terms with real coefficients,

$$H(z) = 4\frac{(1 - 2z^{-1})(1 - z^{-1})(1 - 0.5z^{-1} + 0.25z^{-2})}{(1 + 0.5z^{-1})(1 + 0.5z^{-1} + 0.25z^{-2})}.$$

The resulting expression could be realized as the product of first- and second-order sections representing individual terms in the numerator and denominator, as we did in the previous example. However, a more efficient representation is obtained by combining pairs of real or complex-conjugate terms into pairs of canonical second-order filter sections,

$$H(z) = 4\left(\frac{1 - 3z^{-1} + 2z^{-2}}{1 + 0.5z^{-1}}\right)\left(\frac{1 - 0.5z^{-1} + 0.25z^{-2}}{1 + 0.5z^{-1} + 0.25z^{-2}}\right),$$

as shown in **Figure 9.11**.

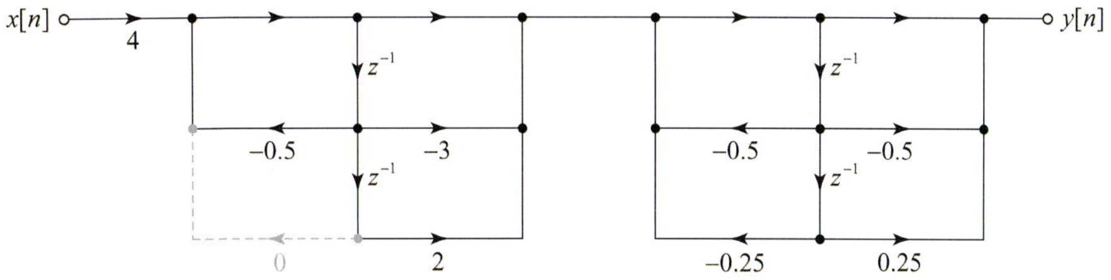

Figure 9.11 Second-order cascade example

The complete filter comprises a gain of four, followed by two second-order filter sections. The first section is "degenerate" in that it has a second-order numerator but only a first-order denominator. Hence, it only requires a first-order feedback path. We can still use a general-purpose second-order section, and simply set the gain of the lower feedback path to zero, as indicated by the dotted path in the signal-flow graph.

9.4.1 Allpass filters

The allpass filter discussed in Section 5.7 provides an example of canonical filter architecture. For a first-order allpass with real coefficients, the canonical filter section with transform

$$H_{ap}(z) = \frac{z^{-1} - a}{1 - az^{-1}}$$

is shown in **Figure 9.12a**, For the second-order filter, poles and zeros can occur in complex conjugate pairs, but the resulting filter coefficients must also be real,

$$H_{ap}(z) = \frac{\beta_2 z^2 + \beta_1 z + 1}{z^2 + \beta_1 z + \beta_2} = \frac{\beta_2 + \beta_1 z^{-1} + z^{-2}}{1 + \beta_1 z^{-1} + \beta_2 z^{-2}},$$

as shown in **Figure 9.12b**. The general (real) allpass filter with real and complex roots can be constructed as a cascade of first- and second-order sections.

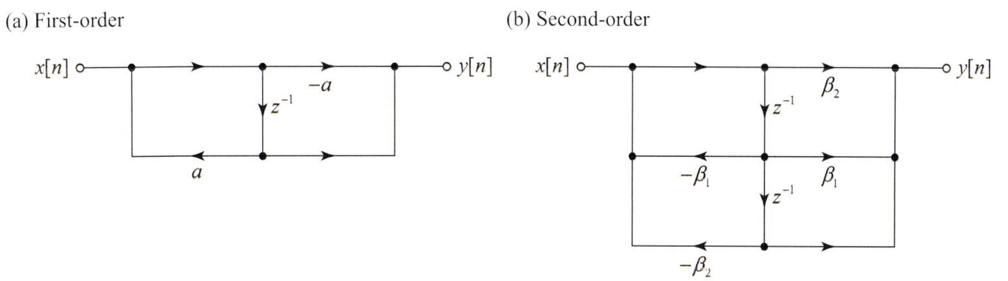

Figure 9.12 First-order and second-order allpass filter sections

In Section 9.7.4, we will show how allpass filters can be efficiently implemented using a lattice-filter architecture that makes them highly insensitive to coefficient quantization.

9.4.2 Using Matlab to design cascade filters

Matlab can help determine the coefficients of the cascade sections. Given the coefficients of the numerator and denominator polynomials a_k and b_k in Equation (9.4a), use Matlab's `roots` function to find the roots and then use the `poly` function to combine real pairs or complex-conjugate pairs of roots to form second-order sections. In the case of Example 9.2, we could do the following:

```
>> b = [4 -14 15 -7 2];
>> a = [1 1 0.5 0.125];
>> rb = roots(b)
        2.0000
        1.0000
        0.2500 + 0.4330i
        0.2500 - 0.4330i
>> ra = roots(a)
        -0.5000
        -0.2500 + 0.4330i
        -0.2500 - 0.4330i
>> b1 = poly(rb(1:2))
        1.0000    -3.0000    2.0000
>> a1 = poly(ra(1))
        1.0000     0.5000
>> b2 = poly(rb(3:4))
        1.0000    -0.5000    0.2500
>> a2 = poly(ra(2:3))
        1.0000     0.5000    0.2500
>> k = b(1)/a(1)
        4
```

Alternately, we could use Matlab's `tf2sos` or `zp2sos` functions, to which we referred in Chapter 8, to calculate the coefficients of the second-order sections for us:

```
>> [sos, k] = tf2sos(b, a)
sos =
      1.0000   -3.0000    2.0000    1.0000    0.5000         0
      1.0000   -0.5000    0.2500    1.0000    0.5000    0.2500
k =
    4
```

For a system, characterized by numerator and denominator polynomials of orders M and N, respectively, the number of rows is $\lceil \max(M, N)/2 \rceil$. The first three columns of each row are the b_k coefficients of the section; the second three columns are the a_k coefficients.

There are several reasons to favor filters made by the cascade of biquadratic sections, as opposed to some other order or mixture of orders. Sections can be no *less* than second-order, because that is the minimum necessary to represent a pair of complex-conjugate terms in the numerator or denominator. However, sections also need be no *more* than second-order, because a biquadratic section can represent fewer than two denominator and/or numerator terms simply by setting the appropriate coefficients to zero, as we did in Example 9.2. By concentrating on a single type of section, we can optimize hardware and/or software and reuse the same section architecture multiple times as needed with different values for the coefficients. Other reasons to concentrate on second-order sections are related to issues of stability and the quantization of coefficients that arises in fixed-point implementations. We will discuss those issues in Section 9.8.

9.5 Parallel architecture

For a causal filter, where the number of poles is equal to the number of zeros, $H(z)$ can also be expressed as the *sum* of second-order sections,

$$H(z) = \frac{\sum_{k=0}^{N} b_k z^{-k}}{1 + \sum_{k=1}^{N} a_k z^{-k}} = b_0 + \sum_{k=1}^{M} H_k(z), \tag{9.5a}$$

where

$$H_k(z) = \frac{b_{k1} + b_{k2} z^{-1}}{1 + a_{k1} z^{-1} + a_{k2} z^{-2}}, \tag{9.5b}$$

as shown in the signal-flow graph of **Figure 9.13**.

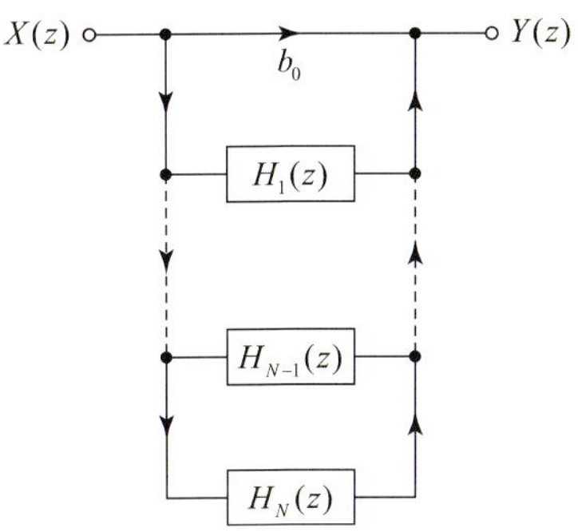

Figure 9.13 Parallel configuration

Example 9.3

A system is described by the difference equation

$$y[n] + y[n-1] + 0.5y[n-2] + 0.125y[n-3] = 4x[n] - 6x[n-1] + 3x[n-2] - x[n-3].$$

Express $H(z)$ as a parallel combination of sections.

➤ **Solution:**

$$\begin{aligned} H(z) &= \frac{4 - 6z^{-1} + 3z^{-2} - z^{-3}}{1 + z^{-1} + 0.5z^{-2} + 0.125z^{-3}} = -8 + \frac{12 + 2z^{-1} + 7z^{-2}}{1 + z^{-1} + 0.5z^{-2} + 0.125z^{-3}} \\ &= -8 + \frac{36}{1 + 0.5z^{-1}} + \frac{-12 - 2.3094j}{1 - 0.5e^{j2\pi/3}z^{-1}} + \frac{-12 + 2.3094j}{1 - 0.5e^{-j2\pi/3}z^{-1}} \\ &= -8 + \frac{36}{1 + 0.5z^{-1}} + \frac{-24 - 4z^{-1}}{1 + 0.5z^{-1} + 0.25z^{-2}}. \end{aligned}$$

First, split the product into the sum of first-order terms, some of which have complex-conjugate roots that cannot be implemented directly. Then, combine the terms with complex-conjugate roots into second-order sections with purely real coefficients. In this example, we end up with the sum of a direct term with gain

−8, a first-order section and a second-order section, as shown in **Figure 9.14a**. The transposed version of this filter is shown in **Figure 9.14b**.

As in Example 9.2, we choose to use a biquadratic section, Equation (9.5b), with appropriate coefficients set to zero in the place of a first-order section.

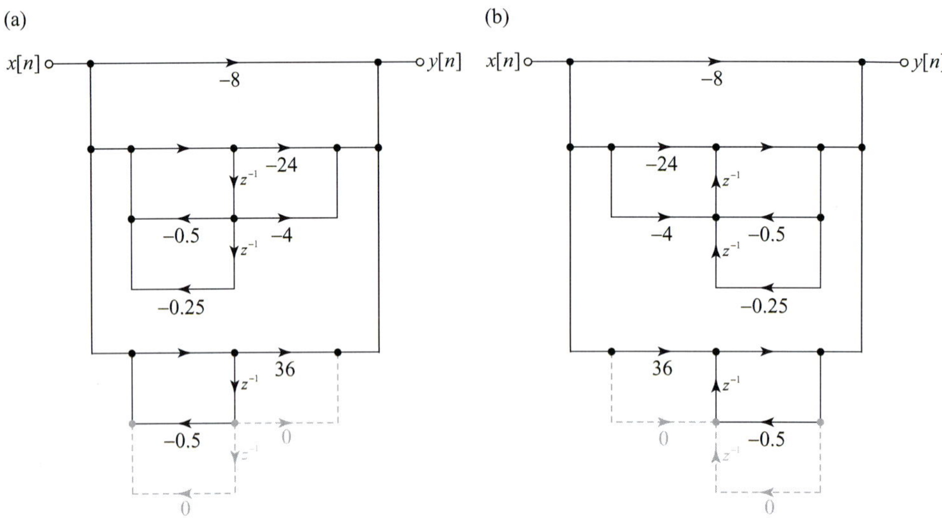

Figure 9.14 Second-order parallel example

9.5.1 Using Matlab to design parallel filters

Matlab's `residuez` function can be used to determine the coefficients of the sections of parallel filters. As described in Section 5.4.6, the `residuez` function performs partial fraction expansion that allows us to express $H(z)$ as the sum of a series of terms. The syntax is

```
[r, p, k] = residuez(b, a),
```

where `b` and `a` are the coefficients of $H(z)$ in powers of z^{-1}. Here is how we would go about determining the coefficients of the filter in Example 9.3:

```
>> b = [4 -6 3 -1];
>> a = [1 1 0.5 0.125];
>> [r, p, k] = residuez(b, a) % r = residues, p = poles, k = constant
r =
   36.0000
  -12.0000 - 2.3094i
  -12.0000 + 2.3094i
p =
   -0.5000
   -0.2500 + 0.4330i
   -0.2500 - 0.4330i
k =
    -8
```

Now use the `residuez` function "in reverse" to calculate the coefficients of $H_k(z)$ in Equation (9.5b), namely $b_{k1,2}$ and $a_{k1,2}$, given the residues, poles and direct terms:

```
[b, a] = residuez(r, p, k).
```

```
>> [b2, a2] = residuez(r(2:3), p(2:3), []) % recombine complex-conjugate terms
b2 =
  -24.0000    -4.0000
a2 =
    1.0000     0.5000     0.2500
```

While the cascade and parallel configurations are theoretically equivalent, there are instances where the parallel architecture can be exploited with multiple second-order sections in hardware to process data simultaneously. In fixed-point implementations where the filter coefficients are quantized, there can be reasons to favor cascade architecture, as we will show in Section 9.8.

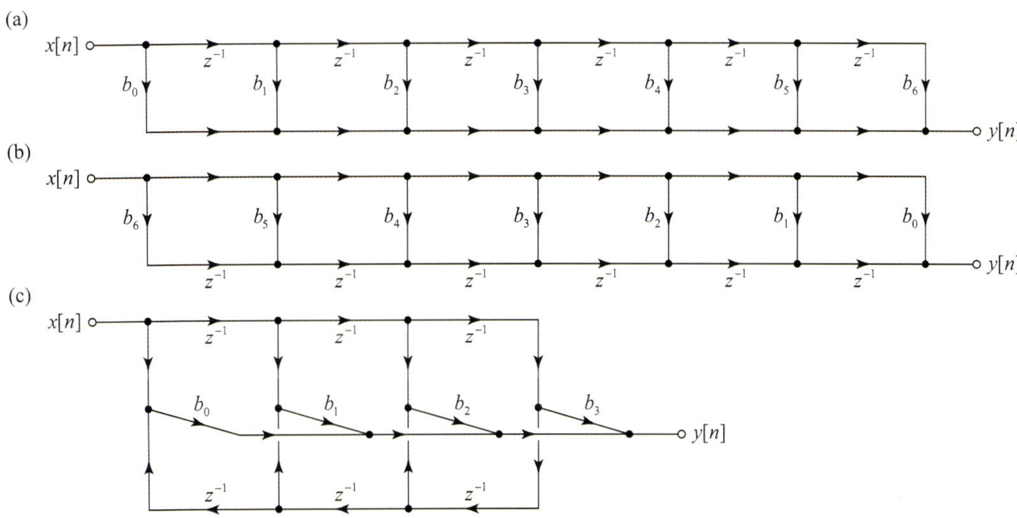

Figure 9.15 FIR architectures

9.6 FIR filters

A causal FIR filter of length N has LCCDE

$$y[n] = \sum_{k=0}^{N-1} b_k x[n-k].$$

Figure 9.15a shows the signal-flow graph for an example with $N = 7$. This filter requires N multiplications, $N - 1$ two-element additions and $N - 1$ delays.

Linear phase FIR filters, such as the Hamming and Kaiser filters discussed in Chapter 7, have either symmetric or antisymmetric impulse responses. For a symmetric filter, $b_k = b_{N-1-k}$; for an antisymmetric filter, $b_k = -b_{N-1-k}$. In such cases, the number of multiplications can be essentially cut in half. For example, a symmetric filter of even length can be written as

$$y[n]=\sum_{k=0}^{N-1} b_k x[n-k]=\sum_{k=0}^{N/2-1} b_k x[n-k]+\sum_{k=N/2}^{N-1} b_k x[n-k].$$

Letting $m=N-1-k$ in the second summation,

$$\begin{aligned} y[n]&=\sum_{k=0}^{N/2-1} b_k x[n-k]+\sum_{m=N/2-1}^{0} b_{N-1-m} x[n-(N-1-m)] \\ &=\sum_{k=0}^{N/2-1} b_k x[n-k]+b_{N-1-k} x[n-(N-1-k)] \qquad , N \text{ even.} \\ &=\sum_{k=0}^{N/2-1} b_k (x[n-k]+x[n-(N-1-k)]) \end{aligned}$$

Thus, an even-length impulse response requires $N/2$ multiplications and $N-1$ two-element additions. For N odd, the center point of the impulse response needs its own multiplier (unless the impulse response is antisymmetric, in which case $b_{(N-1)/2}=0$):

$$y[n]=b_{(N-1)/2} x[n-(N-1)/2]+\sum_{k=0}^{(N-3)/2} b_k (x[n-k]+x[n-(N-1-k)]), N \text{ odd.}$$

An odd-length response thus requires at most $(N+1)/2$ multiplications and $N-1$ additions, as shown in **Figure 9.15b** for $N=7$.

9.7 ★ Lattice and lattice-ladder filters

Lattice and **lattice-ladder** filters are specialized filter structures that can be used to implement both FIR and IIR filters. They have found extensive use in the design of allpass filters, adaptive filters, implementation of perfect reconstruction filter banks and implementation of linear-prediction algorithms for speech processing. In order to set up the discussion of this section, return to the general form of the LCCDE of Equation (9.1),

$$\sum_{i=0}^{N} a_i y[n-i]=\sum_{i=0}^{M} b_i x[n-i].$$

To simplify matters going forward, let $N = M$ so that both summations have the same limits, N. That is not a limitation; for example, if $N<M$, just let $a_i=0$, $M<i\leq N$. Also, let $a_0=1$ and $b_0=1$, which is equivalent to normalizing a_k by a_0, normalizing b_k by b_0/a_0 and applying a gain of b_0/a_0 to $x[n]$. Then, the z-transform of the system becomes

$$H(z)=\frac{Y(z)}{X(z)}=\frac{\sum_{i=0}^{N} b_i z^{-i}}{\sum_{i=0}^{N} a_i z^{-i}}=\frac{B_N(z)}{A_N(z)}, \tag{9.6}$$

where the z-transform of the numerator,

$$B_N(z)\triangleq\sum_{i=0}^{N} b_i z^{-i},$$

represents the zeros of the system, and the z-transform of the denominator,

$$A_N(z) \triangleq \sum_{i=0}^{N} a_i z^{-i},$$

represents the poles. The subscript N is the order of the system.

9.7.1 FIR lattice filters

In Section 9.6, we presented the architecture of a causal FIR filter, whose direct-form implementation is shown in **Figure 9.15**. This is an "all-zero" system whose z-transform $H(z)$ is equivalent to the z-transform of the coefficients b_i,

$$H(z) = \frac{Y(z)}{X(z)} = B_N(z) = \sum_{i=0}^{N} b_i z^{-i}. \tag{9.7}$$

Figure 9.16 shows the lattice architecture for FIR filters. An FIR lattice filter of order N comprises a cascade of N stages or sections. Each section has two inputs and two outputs. For the ith section, shown in the yellow box, the inputs (which are the outputs of the previous section) are $v_{i-1}[n]$ and $w_{i-1}[n]$, and the outputs (which are the inputs to the next section) are $v_i[n]$ and $w_i[n]$. By inspection of the figure, the inputs and outputs of the ith section are related through two recursion relations

$$\begin{aligned} v_i[n] &= v_{i-1}[n] + k_i w_{i-1}[n-1] \\ w_i[n] &= k_i v_{i-1}[n] + w_{i-1}[n-1], \end{aligned} \tag{9.8}$$

where k_i is a real constant, termed the **reflection coefficient** or ***k*-parameter**, that is associated with the ith section. The input to the first section is the sequence $x[n]$, so,

$$v_0[n] = w_0[n] = x[n]. \tag{9.9}$$

The final condition, on the top right end of the lattice, is

$$y[n] = v_N[n]. \tag{9.10}$$

An FIR filter of order N that is represented by the coefficients of the z-transform b_i, $1 \le i \le N$, can be equally well represented by a lattice filter of N sections with reflection coefficients k_i, $1 \le i \le N$. To show this, we need to relate the coefficients b_i to the reflection coefficients k_i. A quick example will indicate how this works.

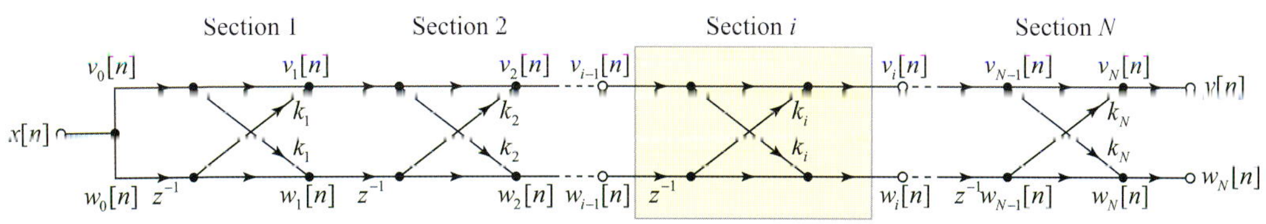

Figure 9.16 FIR lattice filter

Example 9.4

Find the reflection coefficients that correspond to the FIR filters described by each of the following difference equations:

(a) A first-order ($N = 1$) filter,

$$y[n] = x[n] + b_1 x[n-1].$$

(b) A second-order ($N = 2$) filter,

$$y[n] = x[n] + b_1 x[n-1] + b_2 x[n-2].$$

➤ Solution:

(a) Using the recursive relations of Equation (9.8), the initial conditions of Equation (9.9) and the final condition of Equation (9.10), write

$$\begin{aligned} y[n] = v_1[n] &= v_0[n] + k_1 w_0[n-1] = x[n] + k_1 x[n-1] \\ w_1[n] &= k_1 v_0[n] + w_0[n-1] = k_1 x[n] + x[n-1] \end{aligned}. \tag{9.11}$$

Comparing the first of these equations (for $y[n] = v_1[n]$) with the problem statement, we conclude that we need only one lattice filter section with $k_1 = b_1$.

(b) In this case, we have to go through the recursion equations twice, but we have already done the first recursion in part (a), so we can just use those results as the input to the second recursion:

$$\begin{aligned} y[n] = v_2[n] &= v_1[n] + k_2 w_1[n-1] = (x[n] + k_1 x[n-1]) + k_2(k_1 x[n-1] + x[n-2]) \\ &= x[n] + k_1(1+k_2)x[n-1] + k_2 x[n-2] \\ w_2[n] &= k_2 v_1[n] + w_1[n-1] = k_2(x[n] + k_1 x[n-1]) + k_1 x[n-1] + x[n-2] \\ &= k_2 x[n] + k_1(1+k_2)x[n-1] + x[n-1] \end{aligned}.$$

Again, comparing the first of these equations, for $y[n] = v_2[n]$, with the problem statement, we find that $b_1 = k_1(1+k_2)$ and $b_2 = k_2$. So, $k_1 = b_1/(1+b_2)$.

To gain further insight into the properties of the lattice filter, let us examine the architecture of the filter in the z-transform domain. Taking the z-transform of Equation (9.8) gives the two recursion relations of the ith section:

$$\begin{aligned} V_i(z) &= V_{i-1}(z) + k_i z^{-1} W_{i-1}(z) \\ W_i(z) &= k_i V_{i-1}(z) + z^{-1} W_{i-1}(z) \end{aligned}. \tag{9.12}$$

The initial and final conditions are,

$$\begin{aligned} V_0(z) &= W_0(z) = X(z) \\ V_N(z) &= Y(z) \end{aligned}. \tag{9.13}$$

Define the transfer function of the top and bottom branches of the ith section with respect to input $X(z)$ as

$$\begin{aligned} F_i(z) &\triangleq \frac{V_i(z)}{V_0(z)} = \frac{V_i(z)}{X(z)}, \quad 1 \le i \le N \\ G_i(z) &\triangleq \frac{W_i(z)}{W_0(z)} = \frac{W_i(z)}{X(z)}, \quad 1 \le i \le N \end{aligned}. \tag{9.14}$$

The recursive relations of Equation (9.12) become

$$\begin{aligned} F_i(z) &= F_{i-1}(z) + k_i z^{-1} G_{i-1}(z) \\ G_k(z) &= k_i F_{i-1}(z) + z^{-1} G_{i-1}(z), \end{aligned} \tag{9.15}$$

with initial conditions

$$\begin{aligned} F_0(z) &= V_0(z)/V_0(z) = 1 \\ G_0(z) &= W_0(z)/W_0(z) = 1. \end{aligned}$$

The transfer function of the entire FIR lattice filter of N sections is

$$H(z) \triangleq F_N(z) = \frac{V_N(z)}{V_0(z)} = \frac{Y(z)}{X(z)}.$$

If we repeat Example 9.4 from the point of view of the z-transform, we will discover something interesting.

Example 9.5

Find the lattice-filter transfer functions $H_i(z)$ and $G_i(z)$ for the difference equations in Example 9.4.

➤ Solution:

(a) For the first-order (N = 1) filter, we could apply the recursion relations, Equations (9.12)–(9.15), but let us take the shortcut of computing the z-transform of Equation (9.11):

$$\begin{aligned} V_1(z) &= X(z) + k_1 z^{-1} X(z) = X(z)(1 + k_1 z^{-1}) \\ W_1(z) &= k_1 X(z) + z^{-1} X(z) = X(z)(k_1 + z^{-1}) \end{aligned}.$$

Dividing both equations by $X(z)$ and applying Equation (9.15), we get

$$\begin{aligned} F_1(z) &= 1 + k_1 z^{-1} \\ G_1(z) &= k_1 + z^{-1} \end{aligned}.$$

Thus, $H(z) = F_1(z) = 1 + k_1 z^{-1}$. Notice that in this first-order example, $G_1(z) = z^{-1}(1 + k_1 z) = z^{-1} F_1(z^{-1})$.

(b) For a second-order (N = 2) filter, similar algebra yields the following:

$$\begin{aligned} F_2(z) &= 1 + k_1(1 + k_2) z^{-1} + k_2 z^{-2} \\ G_2(z) &= k_2 + k_1(1 + k_2) z^{-1} + z^{-2} \end{aligned}.$$

Here, $H(z) = F_2(z) = 1 + k_1(1 + k_2) z^{-1} + k_2 z^{-2}$. In this second-order example,

$$G_2(z) = z^{-2}(1 + k_1(1 + k_2) z + k_2 z^2) = z^{-2} F_2(z^{-1}).$$

You can show (Problem 9-6) that the results of this example generalize: for every section (e.g., the ith section) of the lattice filter, $G_i(z)$ and $F_i(z)$ are related by

$$G_i(z) = z^{-i} F_i(z^{-1}). \tag{9.16}$$

To see what this means, write out the z-transforms of $F_i(z)$ and $G_i(z)$ in terms of their coefficients, f_n and g_n:

$$F_i(z) = \sum_{n=0}^{i} f_n z^{-n}$$
$$G_i(z) = \sum_{n=0}^{i} g_n z^{-n}.$$

Recalling the time-reversal property of the z-transform, $F_i(z^{-1})$ corresponds to f_{-n}. Multiplication of $F_i(z^{-1})$ by z^{-i} to form $G_i(z) = z^{-i}F_i(z^{-1})$ corresponds to shifting f_{-n} by i to form $g_n = f_{i-n}$. In other words, the coefficients of $F_i(z)$ and $G_i(z)$ are the reverse of each other. In terms of the pole-zero plots, Equation (9.16) says that the zeros of $F_i(z)$ and $G_i(z)$ are at conjugate-reciprocal positions of each other with respect to the unit circle. So, if $z = re^{j\omega_0}$ is a zero of $F_i(z)$, then $z^{-1} = (1/r)e^{-j\omega_0}$ is a zero of $G_i(z)$.

The following example shows how to synthesize (derive) the reflection coefficients of an FIR lattice filter from the transfer function $H(z)$, as well as the inverse problem; that is, how to analyze an FIR lattice filter to find $H(z)$.

Example 9.6

Find the reflection coefficients that correspond to the transfer function

$$H(z) = 1 + 1.38z^{-1} + 1.34z^{-2} + 0.6z^{-3}.$$

➤ Solution:

This is a third-order function, so the lattice filter will comprise a cascade of three sections having transfer functions $F_1(z)$, $F_2(z)$ and $F_3(z)$. The way to do this synthesis problem is to start by finding the coefficient of the last section, $F_3(z)$, and work backwards iteratively to find $F_2(z)$ and finally $F_1(z)$. To see how this works, look back at Example 9.5. Notice that for a filter with N sections, the coefficient of the last section, $F_N(z)$, is associated with the highest (Nth) power of z^{-1}, namely $F_N(z) = 1 + \cdots + k_N z^{-N}$. In the present example, $F_3(z) = H(z)$, so $k_3 = 0.6$.

To find the remaining coefficients, massage Equation (9.15) to form a relation that expresses $F_{i-1}(z)$ in terms of $F_i(z)$ and $G_i(z)$,

$$\begin{aligned} F_{i-1}(z) &= F_i(z) - k_i z^{-1} G_{i-1}(z) \\ &= F_i(z) - k_i(G_i(z) - k_i F_{i-1}(z)). \\ &= F_i(z) - k_i G_i(z) + k_i^2 F_{i-1}(z) \end{aligned}$$

Hence, the reverse relation is

$$F_{i-1}(z) = \frac{F_i(z) - k_i G_i(z)}{1 - k_i^2}. \tag{9.17}$$

In this example, the coefficient for the second stage is found as follows:

$$\begin{aligned} F_2(z) &= \frac{F_3(z) - k_3 G_3(z)}{1 - k_3^2} = \frac{(1 + 1.38z^{-1} + 1.34z^{-2} + 0.6z^{-3}) - 0.6(0.6 + 1.34z^{-1} + 1.38z^{-2} + z^{-3})}{0.64} \\ &= 1 + 0.9z^{-1} + 0.8z^{-2}, \end{aligned}$$

from which we conclude that $k_2 = 0.8$. The transfer function for the first stage is

$$F_1(z) = \frac{F_2(z) - k_2 G_2(z)}{1 - k_2^2} = \frac{(1 + 0.9z^{-1} + 0.8z^{-2}) - 0.8(0.8 + 0.9z^{-1} + z^{-2})}{0.36} = 1 + 0.5z^{-1},$$

so $k_1 = 0.5$, and we are done. The final lattice is shown in **Figure 9.17**.

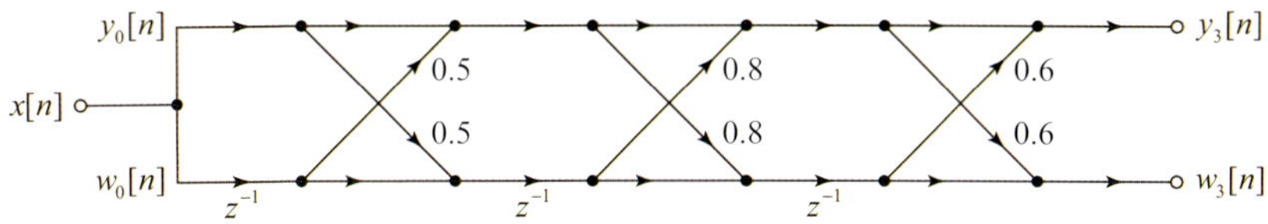

Figure 9.17 FIR lattice example

Matlab has a function `poly2rc` that computes the reflection coefficients k_i given the coefficients of $H(z)$. But we can easily make our own function by implementing Equation (9.17):

```
function k = tf2rc(f)
    f = f / f(1); % make sure f is normalized so that f(1)=1
    N = length(f)-1;
    k = zeros(1, N);
    for i = N:-1:1
        k(i) = f(i+1);
        f = (f(1:i) - k(i)*f(end:-1:2)) / (1-k(i)^2);
    end
end
>> k = tf2rc([1 1.38 1.34 0.6])
    k =
        0.5000    0.8000    0.6000
```

Notice that if $k_i = \pm 1$ for any k in any stage of the reverse iteration, then Equation (9.17) will blow up, and it will not be possible to find a lattice filter representation of the transfer function. Examples of transfer functions that fail include all linear-phase filters (see Problem 9-7), as well as some cases of filters whose transforms have zeros on the unit circle.

Example 9.7

Find the transfer function corresponding to the FIR lattice filter shown in **Figure 9.17**.

➤ Solution:

For the analysis problem, start at the first stage and work to the last stage using recursion, Equation (9.15).

$$F_1(z) = F_0(z) + k_1 z^{-1} G_0(z) = 1 + 0.5z^{-1}$$
$$G_1(z) = z^{-1} F_1(z^{-1}) = 0.5 + z^{-1}$$

$$F_2(z) = F_1(z) + k_2 z^{-1} G_1(z) = (1 + 0.5z^{-1}) + 0.8z^{-1}(0.5 + z^{-1}) = 1 + 0.9z^{-1} + 0.8z^{-2}$$
$$G_2(z) = z^{-1} F_2(z^{-1}) = 0.8 + 0.9z^{-1} + z^{-2}$$

$$\begin{aligned} F_3(z) &= F_2(z) + k_3 z^{-1} G_2(z) = (1 + 0.9z^{-1} + 0.8z^{-2}) + 0.6z^{-1}(0.8 + 0.9z^{-1} + z^{-2}) \\ &= 1 + 1.38z^{-1} + 1.34z^{-2} + 0.6z^{-3} \end{aligned}$$
$$G_3(z) = z^{-1} F_3(z^{-1}) = 0.6 + 1.34z^{-1} + 1.38z^{-2} + z^{-3}$$

So, $H(z) = F_3(z) = 1 + 1.38z^{-1} + 1.34z^{-2} + 0.6z^{-3}$. Matlab has an analysis function `rc2poly` to convert from the filter coefficients k_i to the coefficients of $H(z)$. Here is a simple function of our own that does the same thing:

```
function [f g] = rc2tf(k)
   M = length(k);
   f = 1;
   g = 1;
   for i = 1:N
       f = [f 0] + k(i)*[0 g];
       g = f(end:-1:1);
   end
end
>> f = rc2tf([0.5 0.8 0.6])
    f =
        1.0000    1.3800    1.3400    0.6000
```

9.7.2 Specialized FIR lattice filters

There are a number of specialized lattice filters that realize transfer functions for particular applications. In this section, we will highlight one example, the quadrature mirror filter (QMF) or conjugate quadrature filter (CQF) lattice shown in **Figure 9.18a**. This lattice can be used to implement a type of "perfect reconstruction" two-channel filter bank that we will be discussing in Chapter 13. A quadrature filter bank comprises an analysis section where an input signal is split into low- and high-frequency components by filters $F_N(z)$ and $G_N(z)$, respectively. $F_N(z)$ is a lowpass filter of order N, where N is odd, with a nominal bandwidth of $\omega_c = \pi/2$. $G_N(z)$ is the complementary highpass filter, also of order N, which is related to $F_N(z)$ as follows:

$$G_N(z) = z^{-(N-1)} F_N(-z^{-1}).$$

The QMF lattice shown in **Figure 9.18a** is just the FIR lattice of **Figure 9.16** with a couple of modifications: 1) the reflection coefficients of all even sections are zero and 2) in each section, one of the coefficients is positive and one is negative.

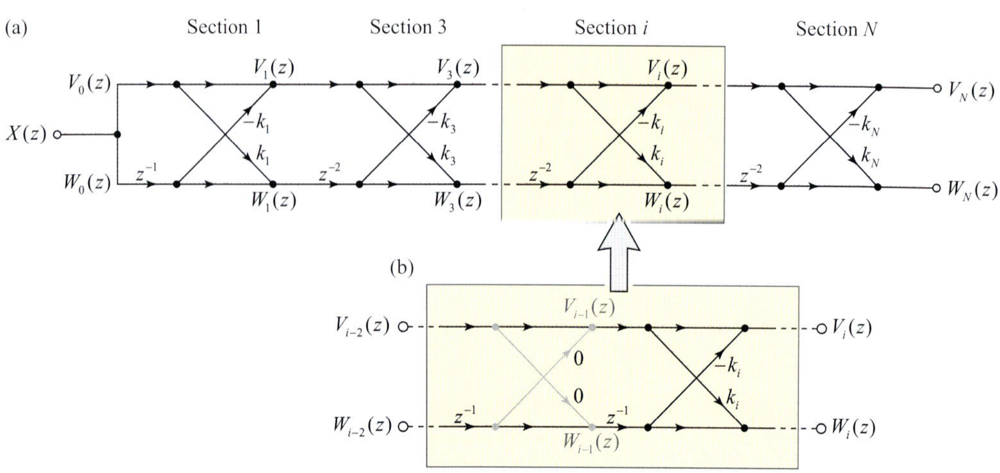

Figure 9.18 QMF lattice filter

Figure 9.18b shows the combination of a pair of sections, one even (greyed out) and one odd. Since the coefficients of the even section are zero, the only thing that section does is to delay the bottom signal path by an extra z^{-1}. The result is the QMF lattice of **Figure 9.18a**, which effectively comprises only odd sections with a delay of z^{-2}, with the exception of the first section, which has a delay of z^{-1}. For the QMF lattice, you can show (Problem 9-9) that

$$G_i(z) = z^{-i}F_i(-z^{-1}),$$

which means that this lattice "automatically" generates both $F_N(z)$ and $G_N(z)$.

9.7.3 IIR lattice filters

Figure 9.19 shows a lattice structure that can be used to implement an Nth-order difference equation for an "all-pole" IIR system,

$$\sum_{i=0}^{N} a_i y[n-i] = x[n], \tag{9.18}$$

where we take $a_0 = 1$. As with the FIR lattice, the IIR lattice comprises a cascade of sections, with reflection coefficients k_i. Compared with the FIR lattice, the IIR lattice has a reversed structure; the output of the $(i-1)$th section feeds back to the ith section. In each section, the sign of one of the reflection coefficients is positive and one is negative.

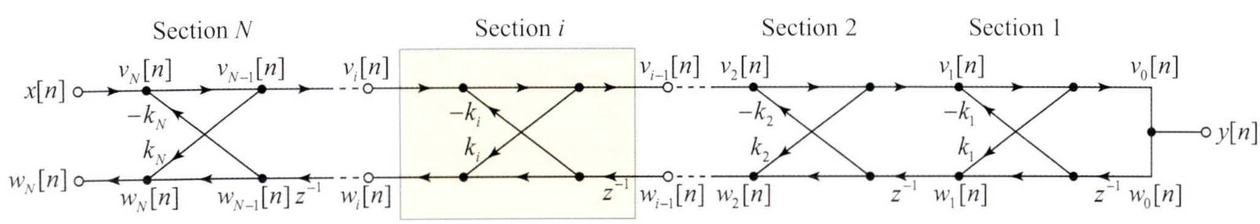

Figure 9.19 IIR lattice filter

The inputs and outputs of the ith section are related through two recursion relations:

$$\begin{aligned} v_{i-1}[n] &= v_i[n] - k_i w_{i-1}[n-1] \\ w_i[n] &= k_i v_{i-1}[n] + w_{i-1}[n-1], \end{aligned} \tag{9.19}$$

with initial and final conditions,

$$\begin{aligned} v_0[n] &= w_0[n] = y[n] \\ v_N[n] &= x[n]. \end{aligned} \tag{9.20}$$

Again, it is profitable to look at the IIR lattice in terms of the z-transform. The LCCDE of Equation (9.18) corresponds to the "all-pole" z-transform

$$H(z) = \frac{1}{\sum_{i=0}^{N} a_i z^{-i}} = \frac{1}{A_N(z)}, \tag{9.21}$$

where $A_N(z)$ is the denominator polynomial of order N that defines the poles.

From Equation (9.19), the z-transform relations for the IIR lattice are

$$\begin{aligned} V_i(z) &= V_{i-1}(z) + k_i z^{-1} W_{i-1}(z) \\ W_i(z) &= k_i V_{i-1}(z) + z^{-1} W_{i-1}(z), \end{aligned} \tag{9.22}$$

with initial and final conditions

$$\begin{aligned} V_0(z) &= W_0(z) = Y(z) \\ V_N(z) &= X(z). \end{aligned} \tag{9.23}$$

Define the transfer function of the ith IIR lattice section as

$$\begin{aligned} A_i(z) &= \frac{V_i(z)}{V_0(z)} = \frac{V_i(z)}{Y(z)}, \qquad 1 \le i \le M \\ B_i(z) &= \frac{W_i(z)}{W_0(z)} = \frac{W_i(z)}{Y(z)}, \qquad 1 \le i \le M, \end{aligned} \tag{9.24}$$

with

$$\begin{aligned} A_0(z) &= V_0(z)/V_0(z) = 1 \\ B_0(z) &= W_0(z)/W_0(z) = Y(z)/Y(z) = 1. \end{aligned}$$

The recursion relations of Equation (9.22) become

$$\begin{aligned} A_i(z) &= A_{i-1}(z) + k_i z^{-1} B_{i-1}(z) \\ B_i(z) &= k_i A_{i-1}(z) + z^{-1} B_{i-1}(z). \end{aligned} \tag{9.25}$$

The transfer function of a complete IIR lattice filter of N sections is then

$$H(z) \triangleq \frac{V_0(z)}{V_N(z)} = \frac{1}{A_N(z)}. \tag{9.26}$$

Example 9.8

Find the lattice-filter transfer functions $A_i(z)$ and $B_i(z)$ that correspond to the IIR filters described by each of the following difference equations.

(a) A first-order ($N = 1$) filter,

$$y[n] + a_1 y[n-1] = x[n].$$

(b) A second-order ($N = 2$) filter,

$$y[n] + a_1 y[n-1] + a_2 y[n-2] = x[n].$$

➤ Solution:

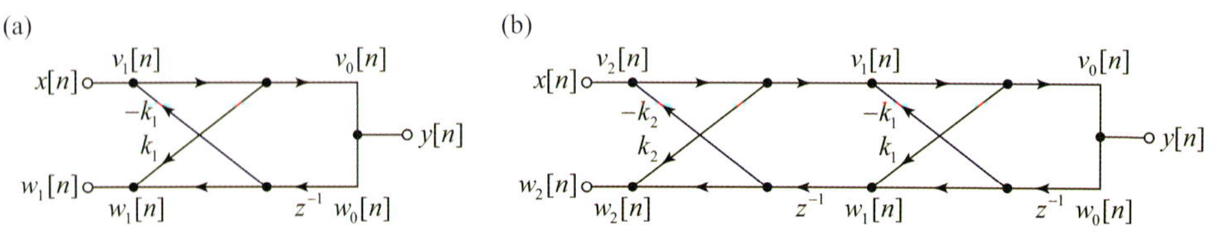

Figure 9.20 IIR lattice filter example

(a) A single-section lattice filter is shown in **Figure 9.20a**.

Using Equation (9.22) and the initial and final conditions of Equation (9.23), we get

$$V_1(z) = V_0(z) + k_1 z^{-1} W_0(z) = V_0(z)(1 + k_1 z^{-1})$$
$$W_1(z) = k_1 V_0(z) + z^{-1} W_0(z) = W_0(z)(k_1 + z^{-1}).$$

Applying Equation (9.24) gives

$$A_1(z) = \frac{V_1(z)}{Y(z)} = 1 + k_1 z^{-1}$$
$$B_1(z) = \frac{W_1(z)}{Y(z)} = k_1 + z^{-1}.$$

Hence,

$$H(z) = \frac{1}{A_1(z)} = \frac{1}{1 + k_1 z^{-1}} \tag{9.27}$$

has a pole at $z = -k_1$ and a zero at $z = 0$. As long as $|k_1| < 1$, the filter will be stable. Comparing Equation (9.27) with the transform of the LCCDE in the problem statement,

$$H(z) = \frac{Y(z)}{X(z)} = \frac{1}{A_1(z)} = \frac{1}{1 + k_1 z^{-1}} = \frac{1}{1 + k_1 z^{-1}},$$

we see that $k_1 = a_1$. Also, the coefficients of $B_1(z)$ and $A_1(z)$ are reversed so that $B_1(z) = z^{-1} A_1(z^{-1})$.

(b) The two-section lattice is shown in **Figure 9.20b**. The recursion relations and Equation (9.23) result in the following:

$$\begin{aligned} V_2(z) &= V_1(z) + k_2 z^{-1} W_1(z) \\ &= (V_0(z) + k_2 z^{-1} W_0(z)) + k_2 z^{-1} W_0(z)(k_1 + z^{-1}) \\ &= V_0(z)(1 + k_1(1 + k_2) z^{-1} + k_2 z^{-2}) \\ W_2(z) &= k_2 V_1(z) + z^{-1} W_1(z) \\ &= k_2 V_0(z)(1 + k_1 z^{-1}) + z^{-1} W_0(z)(k_1 + z^{-1}) \\ &= W_0(z)(k_2 + k_1(k_1 + k_2) z^{-1} + z^{-2}). \end{aligned}$$

So,

$$A_2(z) = \frac{V_2(z)}{V_0(z)} = \frac{X(z)}{Y(z)} = 1 + k_1(1 + k_2) z^{-1} + k_2 z^{-2}$$
$$B_2(z) = \frac{W_2(z)}{W_0(z)} = \frac{W_2(z)}{Y(z)} = k_2 + k_1(1 + k_2) z^{-1} + z^{-2} = z^{-2} A_2(z^{-1}).$$

Therefore,

$$H(z) = \frac{Y(z)}{X(z)} = \frac{1}{A_2(z)} = \frac{1}{1 + k_1(k_1 + k_2) z^{-1} + k_2 z^{-2}} = \frac{1}{1 + a_1 z^{-1} + a_2 z^{-2}},$$

from which we conclude that $a_1 = k_1(1 + k_2)$ and $a_2 = k_2$. So, $k_1 = a_1/(1 + a_2)$. Again, the coefficients of $B_2(z)$ and $A_2(z)$ are reversed so that in this case, $B_2(z) = z^{-2} A_2(z^{-1})$.

Comparing this IIR example with the FIR example of Example 9.5, you can see that the only difference is that the coefficients of the FIR lattice filter, k_i, map to the coefficients b_i of the numerator polynomial $B_N(z)$ of order N that define the zeros in Equation (9.7), whereas the reflection coefficients of the IIR lattice filter, k_i, map to the coefficients a_i of the denominator polynomial $A_N(z)$ that define the poles of $H(z)$ in Equation (9.21). Additionally, for the IIR filter, we can show that, for every section (e.g., the ith section) of the IIR lattice filter, $B_i(z)$ and $A_i(z)$ are related by

$$B_i(z) = z^{-i}A_i(z^{-1}),$$

so that

$$B(z) \triangleq B_N(z) = z^{-N}A_N(z^{-1}). \tag{9.28}$$

The recursion relations for the FIR lattice, Equation (9.12), and for the IIR lattice, Equation (9.25), are identical. Hence, the procedures for analyzing and synthesizing "all-pole" IIR lattices are identical to those we outlined in Examples 9.6 and 9.7 for FIR filters, and we can use the same Matlab programs. These IIR lattices have the same insensitivity to coefficient quantization as FIR lattices, something we will discuss in more detail in Section 9.8.5.

Example 9.9

Find the reflection coefficients that correspond to the transfer function

$$H(z) = \frac{1}{1 + 1.38z^{-1} + 1.34z^{-2} + 0.6z^{-3}}.$$

➤ Solution:

As with the analysis of the FIR filter in Example 9.6,

$$A_3(z) = 1 + 1.38z^{-1} + 1.34z^{-2} + 0.6z^{-3} = 1 + \cdots + k_3z^{-3}.$$

So, we identify k_3 as 0.6. The reverse recursion, which allows us to get $A_{i-1}(z)$ from $A_i(z)$, is equivalent to that in Equation (9.17),

$$A_{i-1}(z) = \frac{A_i(z) - k_iB_i(z)}{1 - k_i^2}.$$

Furthermore, all the algebra is identical:

$$\begin{aligned}
A_3(z) &= 1 + 1.38z^{-1} + 1.34z^{-2} + 0.6z^{-3} \\
B_3(z) &= 0.6 + 1.34z^{-1} + 1.38z^{-2} + z^{-3} \\
A_2(z) &= \frac{A_3(z) - k_3B_3(z)}{1 - k_3^2} = 1 + 0.9z^{-1} + 0.8z^{-2} \qquad \Rightarrow k_2 = 0.8 \\
B_2(z) &= z^{-2}A_2(z^{-1}) = 0.8 + 0.9z^{-1} + z^{-2} \\
A_1(z) &= \frac{A_2(z) - k_2G_2(z)}{1 - k_2^2} = 1 + 0.5z^{-1} \qquad \Rightarrow k_1 = 0.5 \\
B_1(z) &= z^{-1}A_1(z^{-1}) = 0.5 + z^{-1} \\
A_0(z) &= 1 \\
B_0(z) &= 1.
\end{aligned}$$

9.7.4 Allpass lattice filters

One important use of the lattice architecture is to implement allpass filters. **Figure 9.21** shows the architecture of an Nth-order allpass lattice filter. It is identical to the IIR lattice filter shown in **Figure 9.19**, except that the output $y[n]$ is now taken from the bottom left of the lattice.

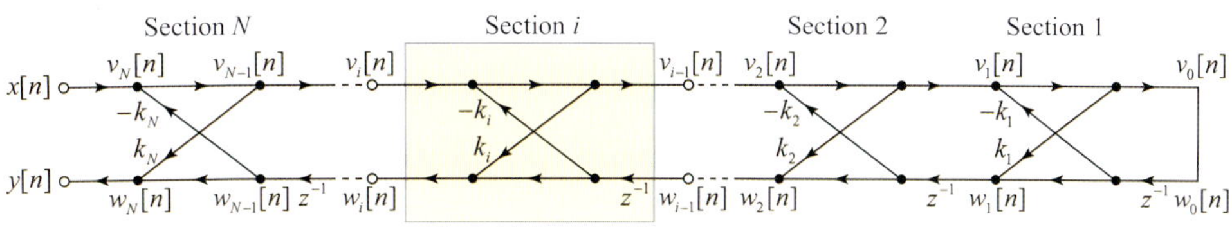

Figure 9.21 Allpass lattice filter

The recursion relations for each section are exactly the same as Equations (9.19) and (9.22), as are the initial conditions, Equations (9.20) and (9.23). The only difference is that for the allpass lattice, the final condition is now

$$Y(z) = W_N(z).$$

We are interested in the complete transfer function

$$H(z) = \frac{Y(z)}{X(z)} = \frac{W_N(z)}{V_N(z)},$$

but from Equations (9.24), (9.26) and (9.28), plus the fact that $W_0(z)/V_0(z) = 1$,

$$H(z) = \frac{Y(z)}{X(z)} = \frac{W_N(z)}{V_N(z)} = \underbrace{\frac{W_N(z)}{W_0(z)}}_{B_N(z)} \cdot \frac{W_0(z)}{V_0(z)} \cdot \underbrace{\frac{V_0(z)}{V_N(z)}}_{1/A_N(z)} = \frac{B_N(z)}{A_N(z)} = \frac{z^{-N}A_N(z^{-1})}{A_N(z)}.$$

This expression is exactly the definition of the allpass filter (see Section 5.7.4). Once again, a big selling point of allpass lattice filters is their insensitivity to coefficient quantization. Because the coefficients of the numerator polynomial $z^{-N}A_N(z^{-1})$ are the same as those of the denominator polynomial $A(z)$, just in reverse order, quantization of the coefficients affects both the numerator and the denominator equally.

9.7.5 Lattice-ladder IIR filters

The FIR lattice filters discussed in Section 9.7.1 implement all-zero systems with transforms given by Equation (9.7). The IIR lattice filters discussed in Section 9.7.3 implement all-pole systems with transforms given by Equation (9.21). In order to implement a system with both non-trivial poles and zeros, we can now modify the architecture of an IIR filter to produce the lattice-ladder filter shown in **Figure 9.22**.

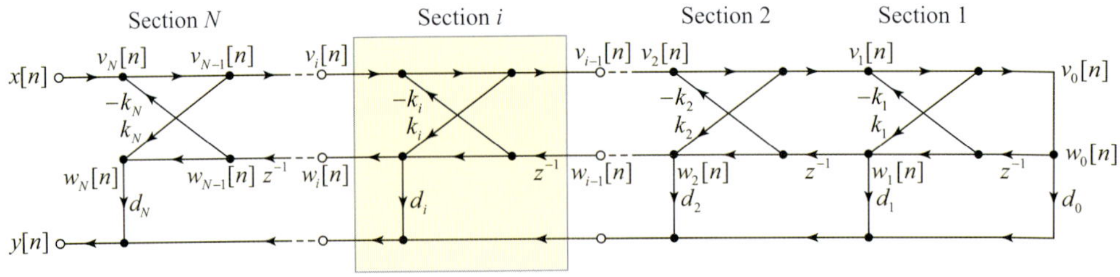

Figure 9.22 Lattice-ladder filter

The lattice-ladder is a hybrid, comprising a lattice, which is identical to the allpass filter of **Figure 9.21**, plus a ladder, which taps off the bottom of the lattice to generate the sum of the $w_i[n]$ terms, scaled by coefficients d_i, so that the output of the entire lattice-ladder filter is

$$y[n] = \sum_{i=0}^{M} d_i w_i[n]. \tag{9.29}$$

To analyze, take the transform of Equation (9.29) and use Equations (9.23) and (9.24):

$$H(z) = \frac{Y(z)}{X(z)} = \frac{1}{V_N(z)} \sum_{i=0}^{N} d_i W_i(z) = \sum_{i=0}^{N} d_i \underbrace{\frac{W_i(z)}{W_0(z)}}_{B_i(z)} \cdot \underbrace{\frac{W_0(z)}{V_0(z)}}_{1} \cdot \underbrace{\frac{V_0(z)}{V_N(z)}}_{1/A_N(z)}$$

$$= \frac{1}{A_N(z)} \underbrace{\sum_{i=0}^{M} d_i B_i(z)}_{D_N(z)} = \frac{D_N(z)}{A_N(z)},$$

where the numerator term is

$$D_N(z) \triangleq \sum_{i=0}^{N} d_i B_i(z) = \sum_{i=0}^{N} b_i z^{-i}. \tag{9.30}$$

The poles of the lattice-ladder filter $H(z)$ are completely determined by $A_N(z)$, and hence by the reflection coefficients of the lattice k_i, in exactly the same manner as they were for the IIR and allpass lattice filters described in Sections 9.7.3 and 9.8.4. The zeros of $H(z)$ are determined by the numerator term $D_N(z)$, which depends on the ladder coefficients d_i, as well as on the reflection coefficients of the lattice k_i that specify $B_i(z)$. So, given a general $H(z)$ of the form Equation (9.6), where the numerator is specified by coefficients b_i and the denominator by coefficients a_i, we first calculate the reflection coefficients by the techniques of Section 9.7.3, from which we simultaneously compute $B_i(z)$, $0 \le i < N$. Then, given $B_i(z)$ we solve Equation (9.30) iteratively to find the ladder coefficients d_i. This is easiest to explain with a specific example.

Example 9.10

Find the lattice-ladder filter coefficients that correspond to the transfer function

$$H(z) = \frac{1.13 + 1.664z^{-1} + 1.228z^{-2} + 0.6z^{-3}}{1 + 1.38z^{-1} + 1.34z^{-2} + 0.6z^{-3}}.$$

➤ Solution:

Since

$$H(z) = \frac{1.13 + 1.664z^{-1} + 1.228z^{-2} + 0.6z^{-3}}{1 + 1.38z^{-1} + 1.34z^{-2} + 0.6z^{-3}} = \frac{D_3(z)}{A_3(z)},$$

therefore

$$D_3(z) = \sum_{i=0}^{3} d_i B_i(z) = 1.13 + 1.664z^{-1} + 1.228z^{-2} + 0.6z^{-3}$$
$$A_3(z) = 1 + 1.38z^{-1} + 1.34z^{-2} + 0.6z^{-3}.$$

First, find the reflection coefficients. As with the analysis of the FIR filter in Example 9.6, the coefficient of $A_3(z)$ corresponding to the power z^{-3}, gives k_3:

$$A_3(z) = 1 + 1.38z^{-1} + 1.34z^{-2} + 0.6z^{-3} = 1 + \cdots + k_3 z^{-3}.$$

The reverse recursion that allows us to get $A_{i-1}(z)$ from $A_i(z)$ is equivalent to that in Equation (9.17):

$$A_{i-1}(z) = \frac{A_i(z) - k_i B_i(z)}{1 - k_i^2},$$

from which we can identify the remaining coefficients k_2 and k_1. All the algebra is identical to Example 9-6:

$$A_3(z) = 1 + 1.38z^{-1} + 1.34z^{-2} + 0.6z^{-3} \qquad \Rightarrow k_3 = 0.6$$
$$B_3(z) = 0.6 + 1.34z^{-1} + 1.38z^{-2} + z^{-3}$$

$$A_2(z) = \frac{A_3(z) - k_3 B_3(z)}{1 - k_3^2} = 1 + 0.9z^{-1} + 0.8z^{-2} \qquad \Rightarrow k_2 = 0.8$$
$$B_2(z) = z^{-2} A_2(z^{-1}) = 0.8 + 0.9z^{-1} + z^{-2}.$$

$$A_1(z) = \frac{A_2(z) - k_2 G_2(z)}{1 - k_2^2} = 1 + 0.5z^{-1} \qquad \Rightarrow k_1 = 0.5$$
$$B_1(z) = z^{-1} A_1(z^{-1}) = 0.5 + z^{-1}$$

$$A_0(z) = 1$$
$$B_0(z) = 1.$$

Now use $B_i(z)$, $0 \le i \le N$, from this analysis to find the ladder coefficients d_i, $0 \le i \le N$. Work recursively backwards from d_3. To do so, recognize that each $B_i(z)$ is an ith-order polynomial in z^{-1}. Specifically, $B_3(z)$ is the only polynomial with a z^{-3} term. Therefore, it is the only polynomial that could produce the $0.6z^{-3}$ term in $D_3(z)$,

$$D_3(z) = \sum_{i=0}^{3} d_i B_i(z) = 1.13 + 1.664z^{-1} + 1.228z^{-2} + 0.6z^{-3}.$$

That means $d_3 = 0.6$. Then, subtract $d_3 B_3(z)$ from $D_3(z)$ to get $D_2(z)$, and repeat:

$$\begin{aligned} D_2(z) &= \sum_{k=0}^{2} d_k B_k(z) = D_3(z) - d_3 B_3(z) \\ &= (1.13 + 1.664z^{-1} + 1.228z^{-2} + 0.6z^{-3}) - 0.6(0.6 + 1.34z^{-1} + 1.38z^{-2} + z^{-3}) \\ &= 0.77 + 0.86z^{-1} + 0.4z^{-2}. \end{aligned}$$

This gives $d_2 = 0.4$. Turn the algebra crank again:

$$D_1(z) = \sum_{i=0}^{1} d_i B_i(z) = D_2(z) - d_2 B_2(z)$$
$$= (0.77 + 0.86z^{-1} + 0.4z^{-2}) - 0.4(0.8 + 0.9z^{-1} + z^{-2})$$
$$= 0.45 + 0.5z^{-1}.$$

So, $d_1 = 0.5$. And, finally,

$$D_0(z) = D_1(z) - d_1 B_1(z)$$
$$= (0.45 + 0.5z^{-1}) - 0.5(0.5 + z^{-1})$$
$$= 0.2$$

which gives $d_0 = 0.2$. We are done.

It is not actually necessary to compute k_i and $B_i(z)$ for all i before starting the computation of d_i. We can compute k_i and d_i together. The following code is a modification of the `tf2rc` program of Example 9.6 that produces both sets of coefficients in a relatively efficient manner:

```
function [k, d] = tf2rclad(D, a)
    N = length(a)-1; % assume both D and a are the same length
    k = zeros(1, N); % initialize k and d arrays
    d = [zeros(1, N) D(end)];
    for i = N:-1:1
        k(i) = a(i+1);
        D = D(1:i) - d(i+1) * a(end:-1:2);
        d(i) = D(i);
        a = (a(1:i) - k(i) * a(end:-1:2)) / (1-k(i)^2);
    end
end
>> [k, d] = tf2rclad( [1.13 1.664 1.228 0.6], [1 1.38 1.34 0.6])
k =
    0.5000    0.8000    0.6000
d =
    0.2000    0.5000    0.4000    0.6000
```

Problem 9-8 shows how to use matrix methods to find coefficients.

9.7.6 Stability of IIR filters revisited

One of the key issues in filter design of IIR filters is determining whether a filter is stable. You may have designed a wonderful filter with floating-point coefficients $a[n]$ that define the denominator polynomial $A(z)$, but that filter may become unstable when the coefficients are quantized. We have already discussed several ways of determining the stability of IIR systems. For an LTI system, the bounded-input, bounded-output (BIBO) condition for stability in the time domain is equivalent to showing that the impulse response is absolutely summable. That is not a particularly practical test, since you would have to compute the impulse response. Another test is to find the location of poles, which are the roots of $A(z)$. For a causal stable system, all the poles must be inside the unit circle. However, root-finding algorithms can have numerical inaccuracies, particularly when tasked with finding the roots of polynomials of high order or when roots lie very close to the unit circle.

An alternative test of stability, called the **Schür–Cohn stability test**, can leverage the lattice filter representation we have developed in the previous section to determine the stability of IIR systems *without* explicitly finding the values of the roots. In brief, a necessary and sufficient test of stability of an Nth-order IIR filter is that the magnitudes of all the reflection coefficients of the IIR lattice representation must be less than one,

$$|k_i| < 1, \quad 1 \le i \le N.$$

That is easy to check using our `tf2rc` or Matlab's `poly2rc` functions. Here is a somewhat contrived example that compares tests of system stability using both the Schür–Cohn stability test and the position of the poles (i.e., the roots).

Example 9.11

Consider a stable system defined by $N = 18$ poles in the semi-circle in the right half-plane, very close to the unit circle,

$$A(z) = \prod_{k=1}^{N} (z - p_k) = \sum_{n=0}^{N} a[n] z^{-n},$$

where

$$p_k = 0.9999 e^{j\pi k/2N}, \quad -N/2 \le k \le N/2.$$

(a) Use Matlab to find $a[n]$.

(b) Determine whether the system is stable by using the Schür–Cohn stability test as well as by determining whether all the roots are inside the unit circle.

➤ Solution:

(a)

```
N = 18;
a = poly([0.9999*exp(1j*pi*(-N/2:N/2)/N/2)]); % compute a[n]
```

(b)

```
rc = poly2rc(a);    % compute reflection coefficients
ra = roots(a);      % compute roots using Matlab's roots function

>> all(abs(rc)<1) % Are abs(reflection coefficients) < 1 (Yes -> stable)
ans =
     1
>> all(abs(ra)<1) % Are all the roots inside the unit circle? (No -> wrong!)
ans =
     0
```

The magnitude of all the reflection coefficients is less than one, so the Schür–Cohn stability test correctly identified the system as stable, but Matlab's `roots` function incorrectly placed at least one root on or outside the unit circle.

The Schür–Cohn test can also be applied to determine if an FIR filter is minimum phase without calculating the roots of the z-transform. If all the reflection coefficients of an FIR lattice filter are less than one, $|k_i| < 1$, $1 \le i \le N$, then all the zeros are inside the unit circle, which means that the system is minimum phase.

9.8 ★ Coefficient quantization

So far, we have been assuming that the coefficients of the various filters, e.g., the a_k and b_k in Equation (9.1), are real numbers that can be expressed with arbitrary precision. However, in any real hardware or software implementation, coefficients are stored in a fixed amount of memory and are therefore represented as numbers that must be quantized in some manner. (Appendix B gives a comparative overview of the different ways of representing numbers in computers. It might be a good idea to review that material before proceeding with the rest of this section.) For example, a software program might represent filter coefficients as 64-bit, double-precision floating-point numbers while an embedded hardware implementation may represent the same coefficients as fixed-point fractional binary numbers with 8- or 16-bit precision. Designing filters essentially comes down to specifying the positions of the poles and zeros of $H(z)$ accurately. The effect of the quantization of coefficients is to constrain the precise placement of those poles and zeros, which can have dramatic and unwanted effects on the frequency response of the resulting filter. In the following paragraphs, we will first consider the effects of the quantization of coefficients in filters which have only poles or only zeros (except at $z = 0$), and then look at systems with both poles and zeros.

9.8.1 Systems with poles

An "all-pole" IIR system is characterized by the z-transform

$$H(z) = \frac{1}{\sum_{k=0}^{N} a_k z^{-k}} = \frac{1}{1 + \sum_{k=1}^{N} a_k z^{-k}},$$

where we have let $a_0 = 1$. This is termed an "all-pole" filter because it has only poles in the z-plane, except for N zeros located at $z = 0$. As an introductory example of the problem of coefficient quantization, consider a single-pole filter with z-transform

$$H(z) = \frac{1}{1 - az^{-1}} = \frac{z}{z - a}.$$

The pole is on the real axis at $z = a$, where for a causal stable filter, $|a| < 1$. Assume that the filter coefficient a has been stored as a fractional-binary signed-magnitude number spanning the range $-1 < a < 1$ with B fractional-binary bits of precision. So, if $B = 3$ then a can take on the $2^4 - 1 = 15$ distinct values between $1.111_2 = -0.875_{10}$ and $0.111_2 = +0.875_{10}$ with a quantization step size of $0.001_2 = +0.125_{10}$. The left panel of **Figure 9.23a** shows this quantization by vertical lines on the z-plane with small, blue × at values that correspond to the permissible pole positions.

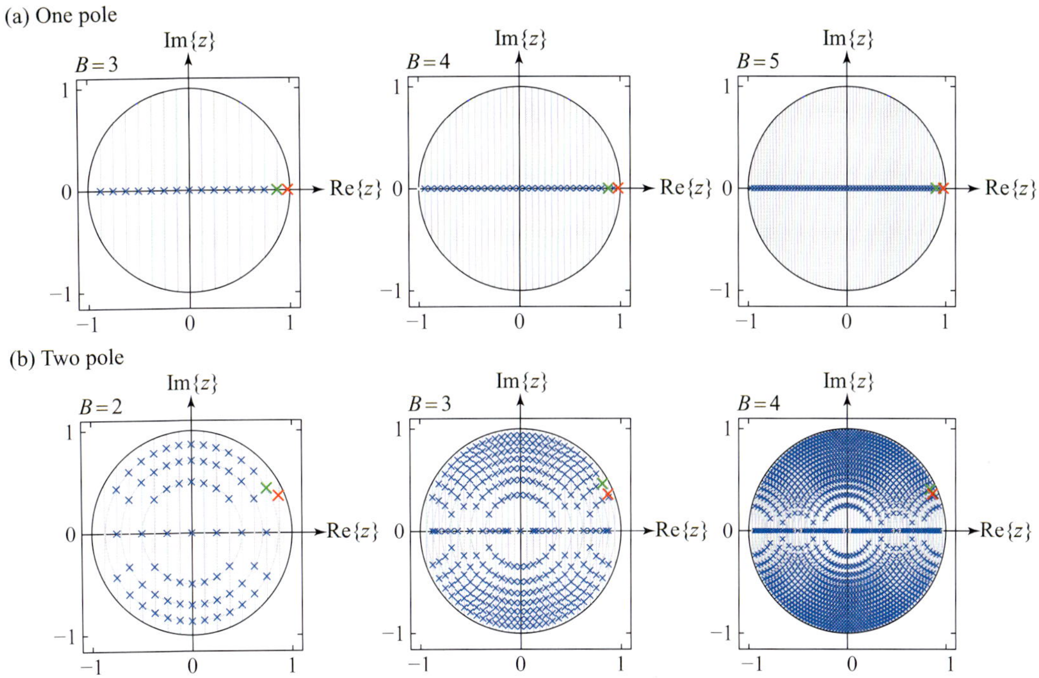

Figure 9.23 Pole locations of filters with quantized coefficients

As a specific example, let the actual (unquantized) pole position be $a = 0.98_{10} = 0.111110101110_2\ldots$, shown in the figure by a red ×. Since this value cannot be represented exactly in fractional binary, it will lose precision when quantized through rounding or truncation, as discussed in Appendix B. Quantization to $B = 3$ fractional-binary bits through rounding would make $a = 1.000_{10}$, which would lead to an unstable (or at best quasi-stable) system. Quantization through truncation reduces the value to $a = 0.111_2 = 0.875_{10}$, as shown by the green × in the figure. Because the quantization of the real axis is linear, each additional fractional bit decreases the quantization step-size by a factor of two, which brings the quantized pole closer to the unquantized value, as shown in **Figure 9.23a** for values of $B = 4$ and $B = 5$. **Figure 9.24a** shows the frequency response of this single-pole filter with a quantized from $B = 2$ to $B = 7$ bits.

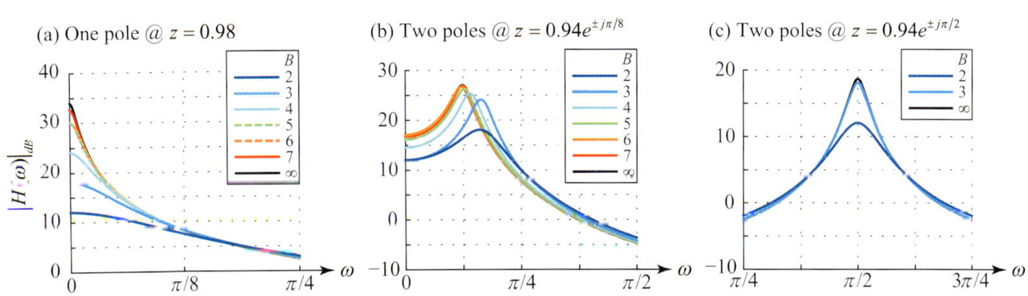

Figure 9.24 Frequency response of filters with quantized coefficients

As the number of fractional bits increases, the frequency response more closely approaches that of the unquantized pole ($B=\infty$). However, each additional bit of resolution does not necessarily reduce the error between the quantized and unquantized values of a, as shown in **Table 9.1**.

Table 9.1 Quantized pole position of single-pole filter

B	a (Binary)	a (Decimal)
2	0.11	0.75
3	0.111	0.875
4	0.1111	0.9375
5	0.11111	0.96875
6	0.111110	0.96875
7	0.1111101	0.97656
∞	0.111110101110...	0.98

In this example, the quantized values of a for values of $B=5$ and $B=6$ (shaded in the table) are the same because the additional bit of resolution only adds a 0 in the sixth fractional binary place.

Now, let us turn to the effect of the quantization of coefficients for a two-pole filter with z-transform

$$H(z)=\frac{1}{1+a_1z^{-1}+a_2z^{-2}}=\frac{z^2}{z^2+a_1z+a_2}=\frac{z^2}{(z-p_1)(z-p_2)}. \tag{9.31a}$$

The filter is described by two coefficients a_1 and a_2, both of which are assumed to be quantized to the same number of bits. Since a_1 and a_2 are real, the poles p_1 and p_2 are given by

$$p_{1,2}=\frac{-a_1\pm\sqrt{a_1^2-4a_2}}{2},$$

and either are both real (if the discriminant is non-negative, $a_1^2-4a_2\geq 0$) or occur as a complex-conjugate pair (if $a_1^2-4a_2<0$). Assuming complex roots, we can write $H(z)$ as

$$H(z)=\frac{z^2}{(z-p)(z-p^*)}=\frac{z^2}{z^2-(p+p^*)z+pp^*}=\frac{z^2}{z^2-2\mathrm{Re}\{p\}z+|p|^2}. \tag{9.31b}$$

A comparison of Equations (9.31a) and (9.31b) shows that $a_1=-2\,\mathrm{Re}\{p\}$ and $a_2=|p|^2$, or, equivalently, $\mathrm{Re}\{p\}=-a_1/2$ and $|p|=\sqrt{a_2}$. Hence, quantization of a_1 linearly affects the real part of the pole and the quantization of a_2 nonlinearly affects the magnitude of the pole. As an example, **Figure 9.23b** shows the permissible locations in the z-plane of the quantized poles of a second-order system with unquantized poles located at $z=0.94e^{\pm j\pi/8}$. The figure shows the results of $B=2$, 3 and 4 fractional-binary bits of quantization of coefficients a_1 and a_2. The $\times$ occur at the intersection of vertical lines representing $\mathrm{Re}\{p\}$ and circles representing $|p|$. For example, for $B = 2$, both a_1 and a_2 are quantized with a resolution of $0.01_2 = 0.25_{10}$. So, $\mathrm{Re}\{p\}$ is quantized by 0.125_{10} and can take on the 15 distinct values between -0.875_{10} and $+0.875_{10}$; $|p|^2$ can have four possible values, 0, 0.25, 0.5 or 0.75, so $|p|$ can be 0, 0.5, 0.707 or 0.866. The coverage of the z-plane is thus not uniform; the density of possible pole locations is higher for locations near $z = \pm j$ than for "sparse" locations near $z = 0$, $z = \pm 0.5$ or $z = \pm 1$. This means that filters with poles in those sparse locations will require finer quantization (i.e., larger B) than filters with poles near $z = \pm j$. In addition, the effects of coefficient quantization are particularly evident for pole locations near the unit circle, $|z| = 1$, because small changes in pole location will have a big effect on the frequency response. For example, **Figure 9.24b** shows the frequency response of a system when the poles are located at $z = 0.94e^{\pm j\pi/8}$, which is relatively close to $z = +1$. This filter requires at least $B = 7$ fractional bits of quantization for the frequency response of the quantized system to approximate reasonably well that of the unquantized one. In contrast, **Figure 9.24c** shows an example when the poles are at $z = 0.94e^{\pm j\pi/2}$, which is near $z = \pm j$. This filter requires only $B = 3$ fractional bits of quantization for the frequency responses of the quantized and unquantized systems to match reasonably well. One place this issue can become important is in a filter designed to process highly oversampled data. Recall that the discrete-time spectrum of the input, $H(\omega)$, is equivalent to the analog spectrum of the input with frequency normalized by the sampling frequency, f_s. As the sampling frequency increases, $H(\omega)$ is effectively compacted towards $\omega = 0$, which is equivalent to moving the singularities of the input data towards $z = 1$. Filters designed for oversampled data might therefore require a larger number of fractional bits of quantization. In cases like this, the filter designer needs to consider the trade-off of sampling rate and quantization level.

The effect of quantization also depends both on the method by which quantization is done (that is, through rounding or truncation), and on the way in which fractional numbers are represented (for example, signed magnitude or two's complement). **Figure 9.25** shows pole locations (×) and frequency responses of a second-order filter with unquantized pole locations at $z = 0.8e^{\pm j0.2\pi}$. The unquantized coefficients for this filter, $a_1 \sim -1.294427191$ and $a_2 = 0.64$, are given in the first row of **Table 9.2**.

Table 9.2 Effect of rounding and truncation on signed-magnitude filter coefficients

Quantization		a_1	a_2	z
Unquantized	×	$-1.2944\cdots$	0.64	$0.8e^{\pm j0.2\pi}$
Fix	×	-1.25	0.5	$0.71e^{\pm j0.17\pi}$
Round	×	-1.25	0.75	$0.87e^{\pm j0.24\pi}$
Floor	×	-1.5	0.5	1, 0.5
Nearest pole	×	-1.5	0.75	$0.87e^{\pm j0.17\pi}$

If the filter coefficients are represented as fractional-binary signed-magnitude numbers with infinite precision, they would be $a_1 = 11.0100101101011_2\cdots$ and $a_2 = 00.1010001111010_2\cdots$

For signed-magnitude numbers, quantization by truncation means that all fractional bits beyond B are simply removed. As described in Appendix B, the effect is to round both positive and negative numbers towards zero, which is equivalent to Matlab's `fix` operation. For example, if the number of fractional bits is $B = 2$, then $a_1 = 11.01_2 = -1.25_{10}$ and $a_2 = 00.10_2 = 0.5_{10}$. The poles of the resulting filter end up at $z = 0.71e^{\pm j0.17\pi}$, as shown in the figure in green (×). If the same coefficients are quantized by rounding to the nearest fractional-binary number, equivalent to Matlab's `round` operation, the results are $a_1 = 11.01_2 = -1.25_{10}$ and $a_2 = 00.11_2 = 0.75_{10}$, and the pole is at $z = 0.87e^{\pm j0.24\pi}$, shown in blue (×). Rounding towards $-\infty$, equivalent to Matlab's `floor` operation, yields $a_1 = 11.1_2 = -1.5_{10}$ and $a_2 = 00.1_2 = 0.5_{10}$, shown in brown (×). In this case, the filter has a pole at $z = +1$, and is therefore unstable (or at best quasi-stable). Finally, none of these truncation or rounding operations results in a system whose quantized poles are closest to the original poles in the sense of minimum distance in the z-plane. These are shown in orange (×).

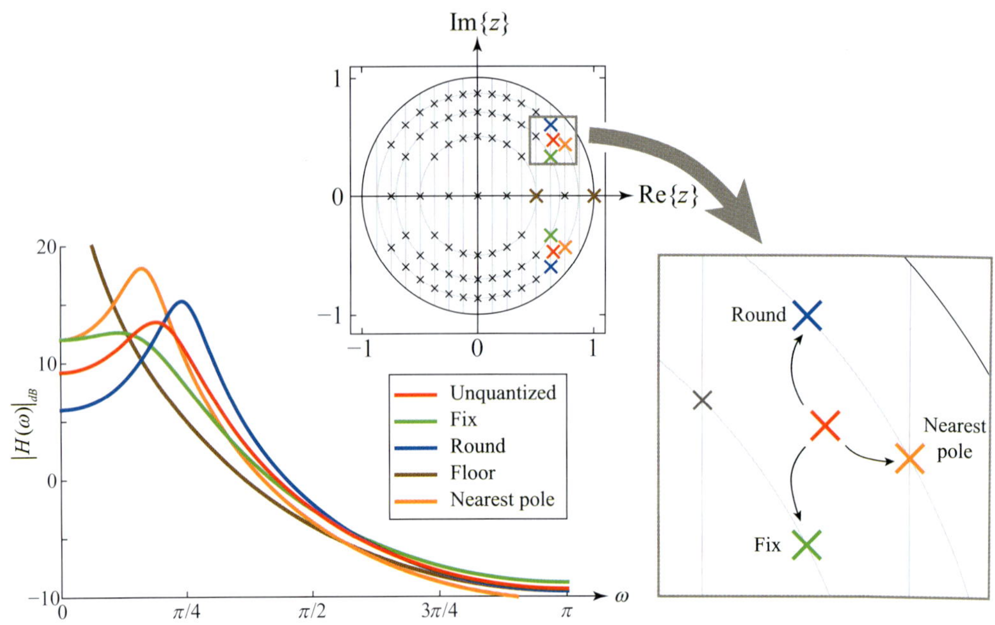

Figure 9.25 Effect of rounding and truncation on pole locations and frequency response

The preceding example brings up a point about the stability of systems with quantized coefficients. For a system to be stable, the poles must lie inside the unit circle. For a second-order system described by Equation (9.31a), it is easy to check the stability of a system without actually finding the roots or performing the Schür–Cohn stability test. In order to be stable, it is straightforward to show (Problem 9-4) that the filter coefficients must satisfy the relations $|a_2| < 1$ and $|a_1| < 1 + a_2$. This locus of a_1 and a_2 forms a so-called **stability triangle**, shown by the shaded regions of **Figure 9.26**.

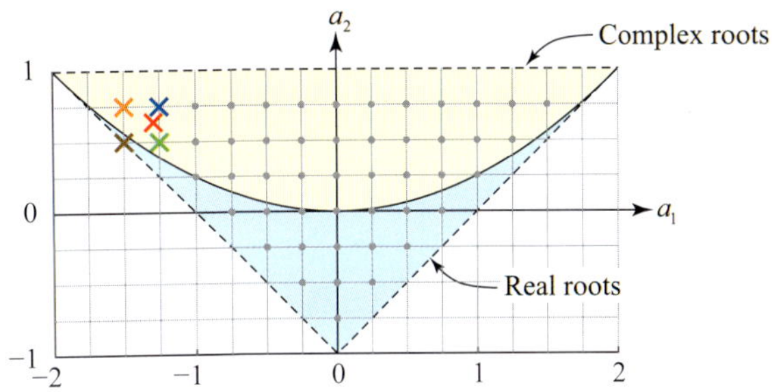

Figure 9.26 Stability triangle for a second-order filter

The lower-triangular blue shaded region corresponds to stable real roots. The upper parabolic yellow shaded region corresponds to stable complex roots, where the discriminant is positive, $a_1^2 - 4a_2 > 0$. The colored '**×**'s mark the coefficients of the filters for the example of **Figure 9.25**. The coefficients that result in the unstable root (**×**) lie directly on the margin of the stability triangle.

One of the advantages of making higher-order filters out of the cascade of second-order sections is the ease of checking the stability of these sections using simple algebra based strictly on the value of the coefficients. For higher-order systems, there also exist a variety of more mathematically sophisticated stability tests, such as the Schür–Cohn test, which we described in Section 9.7.6.

A cascade of second-order sections is also usually less sensitive to coefficient quantization than a single canonical filter of higher order. As an example, consider a fourth-order system comprising a pair of complex poles at $z = 0.94e^{\pm 0.1j\pi}$ and another pair at $z = 0.94e^{\pm 0.2j\pi}$. The filter can be realized as a single fourth-order expression with z-transform

$$\begin{aligned} H(z) &= \frac{1}{(1-0.94e^{+0.1j\pi}z^{-1})(1-0.94e^{-0.1j\pi}z^{-1})(1-0.94e^{+0.2j\pi}z^{-1})(1-0.94e^{-0.2j\pi}z^{-1})} \\ &= \frac{1}{1-3.3089z^{-1}+4.4866z^{-2}-2.9238z^{-3}+0.7807z^{-4}}, \end{aligned}$$

or as the product (cascade) of two second-order expressions, $H(z) = H_{C_1}(z) \cdot H_{C_2}(z)$,

$$\begin{aligned} H(z) &= \frac{1}{(1-0.94(e^{+0.1j\pi}+e^{-0.1j\pi})z^{-1}+0.94^2z^{-2})} \cdot \frac{1}{(1-0.94(e^{+0.2j\pi}+e^{-0.2j\pi})z^{-1}+0.94^2z^{-2})} \\ &= \underbrace{\left(\frac{1}{1-1.7880z^{-1}+0.8836z^{-2}}\right)}_{H_{C_1}(z)} \underbrace{\left(\frac{1}{1-1.5210z^{-1}+0.8836z^{-2}}\right)}_{H_{C_2}(z)}, \end{aligned}$$

or as the sum (parallel combination) of two second-order expressions, $H(z) = H_{P_1}(z) + H_{P_2}(z)$,

$$H(z) = \underbrace{\frac{6.6957-3.3089z^{-1}}{1-1.788z^{-1}+0.8836z^{-2}}}_{H_{P_1}(z)} + \underbrace{\frac{-5.6957+3.3089z^{-1}}{1-1.5210z^{-1}+0.8836z^{-2}}}_{H_{P_2}(z)}.$$

If the filter coefficients are not quantized, the three implementations are identical. The frequency response of the unquantized system is shown in black on the left of **Figure 9.27a** (labeled with center frequency $\omega_0 = 0.15\pi$).

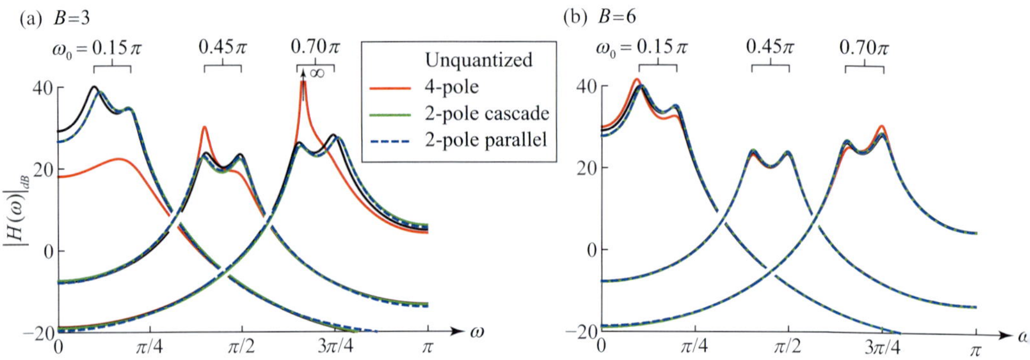

Figure 9.27 Comparison of quantized filter implementations

If $H(z)$ is realized as a single fourth-order filter with coefficients quantized by rounding to $B = 3$ bits of fractional resolution, the resulting frequency response (shown in red) does not match the unquantized response very well. If, instead, $H(z)$ is realized as the cascade or parallel combination of two second-order systems, each of which is quantized to $B = 3$ bits of resolution, the resulting frequency response (shown in green and blue, respectively) matches the desired response much better. **Figure 9.27a** also shows a comparison of the responses of fourth-order canonical filters and the equivalent cascade or parallel combination of second-order sections in cases where the center frequencies are $\omega_0 = 0.45\pi$ and $\omega_0 = 0.70\pi$. In each case, the quantized two-section filters match the unquantized response more accurately than the single fourth-order filter. In the case of $\omega_0 = 0.70\pi$, the quantized fourth-order filter is actually unstable, since it turns out that two of its poles are located on the unit circle at $z = e^{\pm j2\pi/3}$. **Figure 9.27b** shows frequency responses for filters quantized with $B = 6$ bits of resolution. As you might expect, the difference between the frequency responses of the quantized and unquantized filters decreases as the number of fractional bits increases.

While some of the examples in this section are contrived in that the quantization is unrealistically coarse (i.e., B is small), the main points are valid; designers of fixed-point systems need to be aware not only of the effect of quantization on pole position, but also of the way numbers are represented and rounded in hardware/software and of the different sensitivity to quantization of different filter architectures.

9.8.2 Systems with zeros

A causal FIR system characterized by the z-transform

$$H(z) = \sum_{k=0}^{N} b_k z^{-k}$$

is an "all-zero" filter in the sense that it has only zeros in the z-plane, except for poles located at $z = 0$. Quantization of the filter coefficients b_k results in the same issues we have just discussed in the preceding section for the all-pole filter, with a few important qualifications. Most obviously, though coefficient quantization can result in changes in the position of zeros, these changes can never make the filter unstable. **Figure 9.28** shows the frequency response of an all-zero fourth-order bandstop filter, which is the counterpart of the all-pole bandpass filter example of **Figure 9.27a**. There are complex zeros around $\omega_0 = 0.15\pi$ (i.e., $z = 0.94e^{\pm 0.1j\pi}$ and $z = 0.94e^{\pm 0.2j\pi}$), as well as around $\omega_0 = 0.45\pi$ and $\omega_0 = 0.70\pi$, all with coefficients quantized at $B = 3$ fractional-binary bits of resolution. Unsurprisingly, when plotted on a dB scale, the frequency responses of this filter are equivalent to those of **Figure 9.27a** flipped vertically about 0 dB. The quantized bandstop filter goes to zero ($-\infty$ dB) at the frequencies where the quantized bandpass filters went to ∞. Depending on the application, a filter designer may be more accepting of this variability in the bandstop region of the FIR filter than variability in the bandpass region of the IIR filter.

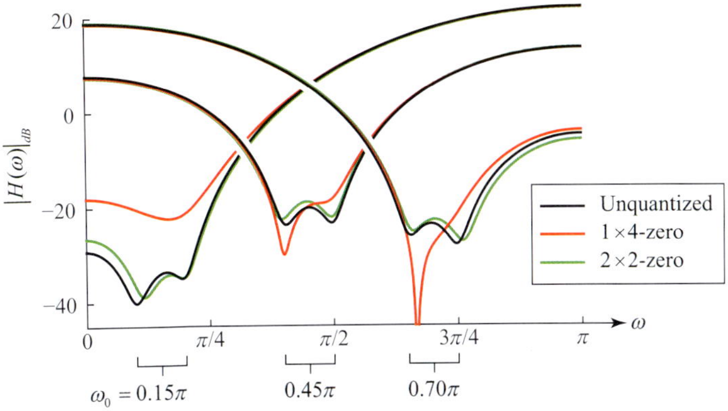

Figure 9.28 Comparison of quantized four-zero and two-zero cascade filter implementations

By extension of our discussion of all-pole filters, you might think that we would favor implementing FIR filters as a cascade of second-order filters instead of a single section, but that is not generally the case. For one thing, both special-purpose DSP processors and modern general-purpose CPUs feature hardware multiply-accumulate (MAC) operations, in which the product of two numbers is computed and added to an accumulator in one operation, which allows one to implement fast, efficient direct convolution algorithms. In addition, many important FIR filters – including all the designs such as Hamming and Kaiser, which are based on windowing – have impulse responses that are either symmetric or antisymmetric. As we showed in Chapter 4, the zeros of the z-transforms of these filters are constrained to occur in only a few configurations; for example, conjugate pairs on the unit circle or at conjugate-reciprocal positions off the unit circle. When subjected to quantization by rounding or truncation, the impulse responses of these window-based filters will retain their symmetry (or antisymmetry). Hence, the zeros of these quantized filters often retain the salient features of the unquantized filters.

As an example, **Figure 9.29** shows the pole-zero plot and frequency response for a lowpass Hamming-window filter of length $N = 13$, with a cutoff frequency of $\omega_c = 0.25\pi$, quantized at $B = 3$ (shown in red), compared with the unquantized filter (shown in red). The unquantized filter has six zeros on the unit circle that drive the frequency response to 0 for frequencies above $\omega = \pi/2$. Quantization of the coefficients moves these zeros, but they all still remain on the unit circle. The positions of the nulls in the frequency response change somewhat, but not the essential properties of the filter. The positions of the zeros at locations off the unit circle change as well, but remain at conjugate-reciprocal positions.

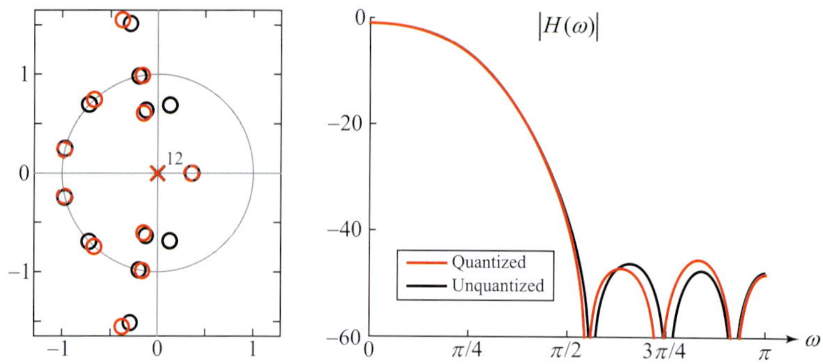

Figure 9.29 Effect of filter quantization on a Hamming-window filter

9.8.3 Systems with poles and zeros

The general form of a system with essential poles and zeros is

$$H(z) = \frac{\sum_{k=0}^{M} b_k z^{-k}}{\sum_{k=0}^{N} a_k z^{-k}}.$$

Such a system can have many implementations, the two most common of which are cascade and parallel. In **Figure 9.27**, we saw an example of an "all-pole" system for which there was very little difference between these two implementations, but, in systems with both poles and zeros, there can be substantial differences. As an example of what can happen, consider a notch filter with z-transform

$$\begin{aligned} H(z) &= \frac{1 - z^{-1} + z^{-2} - z^{-3} + z^{-4}}{1 - 0.94z^{-1} + 0.8836z^{-2} - 0.8306z^{-3} + 0.7807z^{-4}} \\ &= \frac{(1 - e^{j0.2\pi}z^{-1})(1 - e^{-j0.2\pi}z^{-1})(1 - e^{j0.6\pi}z^{-1})(1 - e^{-j0.6\pi}z^{-1})}{(1 - 0.94e^{j0.2\pi}z^{-1})(1 - 0.94e^{-j0.2\pi}z^{-1})(1 - 0.94e^{j0.6\pi}z^{-1})(1 - 0.94e^{-j0.6\pi}z^{-1})}. \end{aligned} \tag{9.32}$$

There are zeros on the unit circle at $z = e^{\pm j0.2\pi}$ and $z = e^{\pm j0.6\pi}$, and poles just off the unit circle at the same angles, $z = 0.94e^{\pm j0.2\pi}$ and $z = 0.94e^{\pm j0.6\pi}$, shown in black in **Figure 9.30a**. The zeros cause the frequency response of this filter to go to 0 ($-\infty$ dB) at $\omega = 0.2\pi$ and 0.6π, shown with black dotted lines in **Figure 9.30b**. This filter can be implemented as the cascade of two second-order sections, for example,

$$H(z) = \frac{(1-e^{j0.2\pi}z^{-1})(1-e^{-j0.2\pi}z^{-1})}{(1-0.94e^{j0.2\pi}z^{-1})(1-0.94e^{-j0.2\pi}z^{-1})} \cdot \frac{(1-e^{j0.6\pi}z^{-1})(1-e^{-j0.6\pi}z^{-1})}{(1-0.94e^{j0.6\pi}z^{-1})(1-0.94e^{-j0.6\pi}z^{-1})}$$

$$= \frac{1-1.6180z^{-1}+z^{-2}}{1-1.5210z^{-1}+0.8836z^{-2}} \cdot \frac{1+0.6180z^{-1}+z^{-2}}{1+0.5810z^{-1}+0.8836z^{-2}}.$$

Each cascade section is responsible for one pair of poles and one pair of zeros. The numerator polynomial of each section is of the form $1+\alpha z^{-1}+z^{-2}$. The zeros, which are the roots of these polynomials, lie on the unit circle at angles $\theta = \pm\cos^{-1}(-\alpha/2)$. When the coefficients of each section are quantized, they remain on the unit circle; only α, and hence the angles of the zeros, changes. The other coefficients of the numerator polynomials are 1, and are unchanged by quantization. In this example, when α is quantized with $B=3$ fractional-binary bits of resolution by rounding, the zeros of the two sections move from $\omega = \pm 0.2\pi$ to $\pm 0.1981\pi$ and from $\omega = \pm 0.6\pi$ to $\pm 0.6012\pi$, respectively; in other words, not a lot. As a consequence, the frequency response of the quantized cascade filter still goes to 0 very near the desired frequencies. The magnitude and angle of the pole locations also change due to quantization of the coefficients of the denominator polynomials, but very little. For example, the poles at $z=0.94e^{\pm j0.2\pi}$ move to $z=0.9354e^{\pm j0.2039\pi}$, as shown by the dashed green × in **Figure 9.30a**. Because the zeros remain on the unit circle, the response of the notch filter still goes to 0 near the designed frequencies, but since the angles of the poles and their associated zeros are no longer equal, the magnitude of the response is somewhat asymmetrical about the notch.

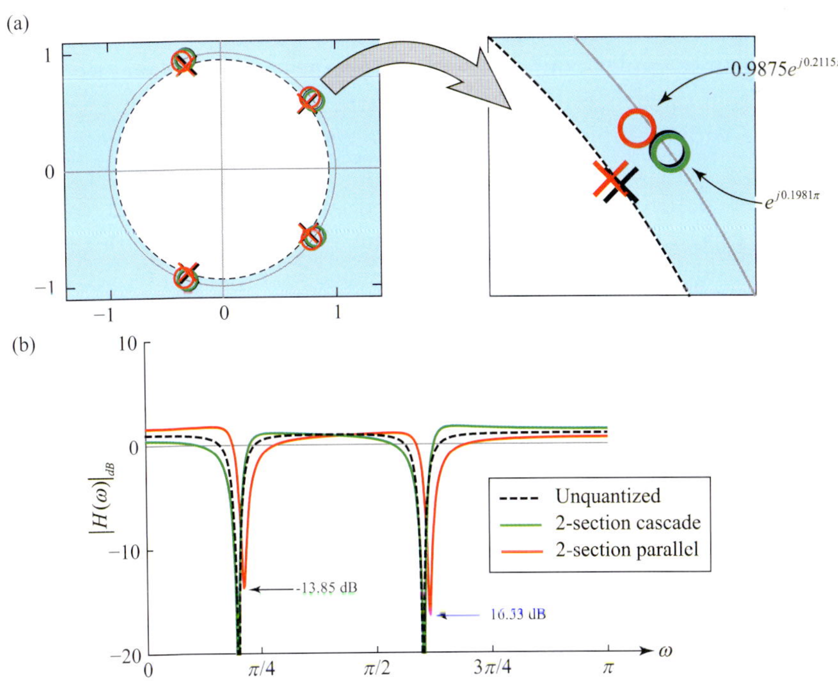

Figure 9.30 Comparison of quantized notch filters

When the filter is implemented as the parallel combination of two second-order sections, unfortunate things happen:

$$\begin{aligned} H(z) &= \frac{1 - z^{-1} + z^{-2} - z^{-3} + z^{-4}}{1 - 0.94z^{-1} + 0.8836z^{-2} - 0.8306z^{-3} + 0.7807z^{-4}} \\ &= 1.2808 + \frac{-0.1405 + 0.1077z^{-1}}{1 - 1.5210z^{-1} + 0.8836z^{-2}} + \frac{-0.1403 - 0.0354z^{-1}}{1 + 0.5810z^{-1} + 0.8836z^{-2}}. \end{aligned}$$

The unquantized denominator polynomials of the parallel filter are the same as those of the cascade filter, so when the coefficients are quantized, the poles of the two filters will be located at identical positions, shown in **Figure 9.30a** by dashed red **×**. However, the zeros of the parallel filter depend upon the cross-product of numerator and denominator polynomials of both sections. In this example, the result is that the zeros move off the unit circle to $z = 0.9875e^{\pm j0.2115\pi}$ and $z = 0.9906e^{\pm j0.6142\pi}$. As a consequence, the frequency response no longer goes to 0 at any frequency. In fact, it only reaches minima of -13.85 dB at $\omega = 0.2118\pi$ and -16.5334 dB at $\omega = 0.6143\pi$. The result is a not-very-good notch filter. Again, this is a fairly egregious example, since the quantization of coefficients is so coarse, but the main point is well taken: designers need to be careful when selecting the parallel implementation in filters with quantized coefficients when the locations of the zeros are critical.

9.8.4 Pairing poles and zeros

In constructing a filter with a cascade of biquadratic sections, we know that it is necessary to pair complex-conjugate zeros together in the numerator and complex-conjugate poles in the denominator so that all the coefficients of the section will be real; but in a system with multiple pairs of poles and zeros such as that shown in **Figure 9.30**, does it matter which zeros are paired with which poles? In theory, it does not, but in practice where coefficients are quantized and arithmetic can lead to overflows and underflows, it certainly can. One rule of thumb is to pair poles with the nearest zeros. To see why, consider again the notch filter specified by Equation (9.32). In this example, there are two ways of implementing the biquadratic sections, assuming we always keep complex-conjugate terms together. The left panel of **Figure 9.31a** shows the pole-zero plots for biquadratic sections created by matching each pole pair with the nearest pair of zeros, with the poles and zeros for each section shown in the same color. The top right panel of the figure shows the frequency responses of each of the two resulting biquadratic sections, with the color of the trace corresponding to the color of the associated section's poles and zeros. Because the pairs of poles and zeros are close to each other, the dynamic range of each section's response is small (except, of course, exactly at the frequency of the zero); the response does not "peak" a lot. **Figure 9.31b** shows the result of matching each pole pair with the farthest pair of zeros. In this case, the peak magnitude and dynamic range of response of each section is considerably greater than that of **Figure 9.31a**.

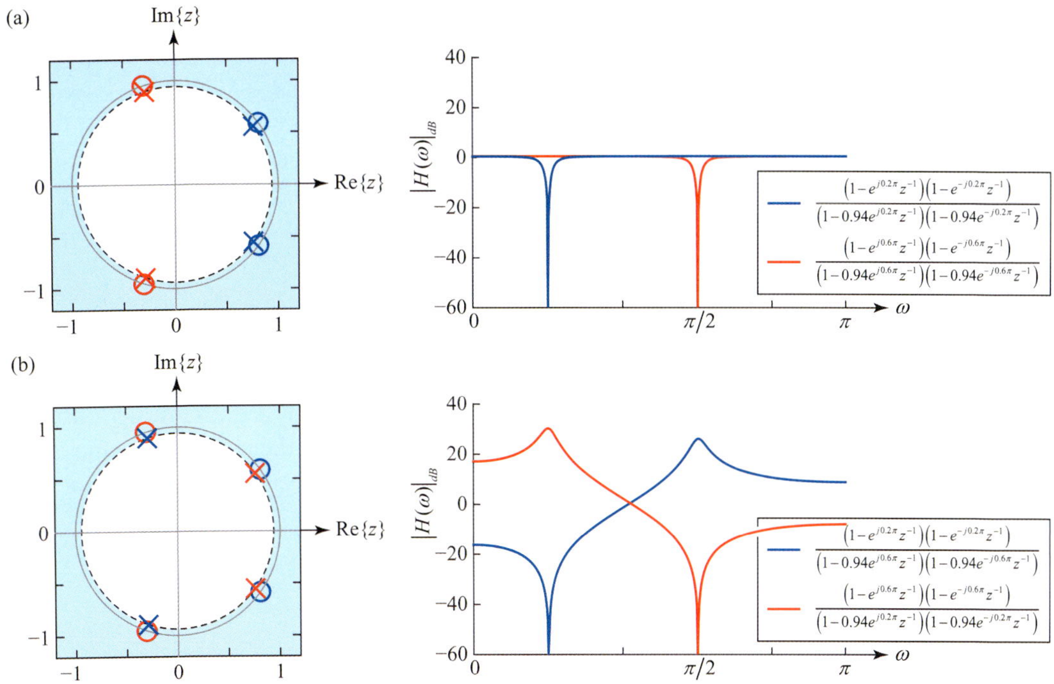

Figure 9.31 Pairs of poles and zeros in a cascade of biquadratic sections

While not shown by this example, the order in which filter sections are cascaded can also matter. The rule of thumb here is to put the section with the largest peak amplitude last in the cascade. This is often the section whose poles are closest to the unit circle.

9.8.5 Coefficient quantization of lattice filters

One of the salient characteristics of lattice filters is their relative insensitivity to the quantization of the reflection coefficients compared with direct-form implementations. Here is one simple example.

Example 9.12

Consider a filter with transform

$$H(z) = 1 + 1.38z^{-1} + 1.38z^{-2} + 0.6z^{-3}.$$

Let $H_Q(z)$ be the transform of a direct-form implementation of the system with coefficients whose fractional parts have been quantized to two fractional bits (e.g., 0.25) through rounding, and let $H_C(z)$ be the transform of the same system whose reflection coefficients have been quantized to the same two-bit precision. Find and plot both the pole-zero plot and frequency-response magnitudes of $H(z)$, $H_Q(z)$ and $H_C(z)$.

➤ Solution:

First, find the coefficients of $H_Q(z)$ and $H_C(z)$:

```
h = [1 1.38 1.38 0.6];
Hq = round(4*h)/4
Hq =
    1.0000    1.5000    1.5000    0.5000
Hc = rc2tf(round(4*tf2rc(h))/4)
Hc =
    1.0000    1.2500    1.1875    0.5000
```

Figure 9.32 shows the pole-zero plots and frequency responses $|H(z)|$ (blue trace), $|H_C(z)|$ (red trace) and $|H_Q(z)|$ (green trace). $H_Q(z)$ is formed by directly quantizing the coefficients of $H(z)$. It has two of its zeros on the unit circle, which results in a dramatically different frequency response from $H(z)$. $H_C(z)$, which results from quantizing the reflection coefficients, has a much smaller deviation from $H(z)$, in terms of both the pole-zero plot and the frequency response.

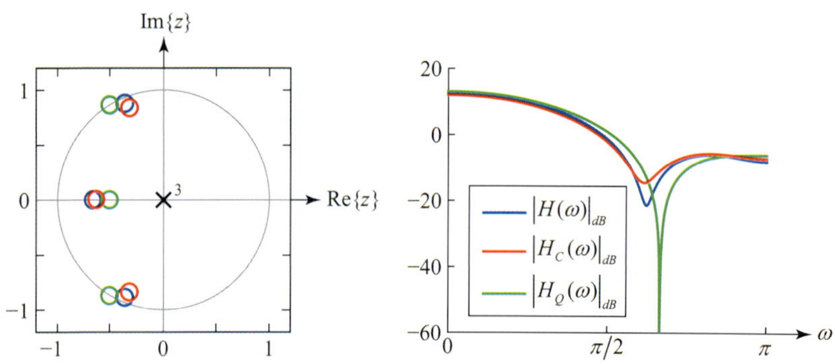

Figure 9.32 Effect of coefficient quantization in an FIR lattice filter

9.9 ★ Implementation issues

DSP algorithms can be implemented in hardware or software. The challenge to the designer is to tailor the implementation of the algorithms to the target, making use of the advantageous features while avoiding the limitations. The choices the designer will make are dependent on the particular hardware/software configuration, but we will briefly touch on some of these issues in the following sections.

9.9.1 Software implementation

Software implementations of discrete-time processing range from programs written in high-level languages that can run on a variety of processors to assembly language code targeted to a specific processor or processor family. Matlab and similar high-level simulation languages (e.g., Octave, Scilab, Rlab and Python, including the packages Numpy and Scipy) are generally interpreted (or byte-coded or just-in-time compiled) and are usually not designed for real-time operations. However, they are particularly well suited for rapid prototyping, development and testing of DSP algorithms. These languages natively handle complex numbers, allow vectorized operations and feature libraries of sophisticated signal processing functions that allow users to perform discrete-time filtering, fast Fourier transformation and other operations. Languages like C are still ubiquitous in science and engineering. Even though C is not expressly designed for signal processing, many libraries exist to aid the programmer, and excellent optimizing compilers exist for hardware platforms ranging from embedded microcontrollers to supercomputers. C code maps very efficiently to the underlying machine instructions, to the extent that C can often replace assembly programming in all but the most time-critical or memory-constrained applications.

A defining feature of most software implementations on serial processors is that computer code is ultimately translated into machine instructions that are executed sequentially, one at a time. Concurrent programming techniques, and instruction-level parallelism or pipelining, is also common in modern processors designed for general-purpose computing. However, the nature of DSP algorithms limits their effectiveness. To understand some of the issues involved in programming DSP algorithms, consider a software implementation of a real-time N-point FIR filter in which the input $x[n]$ is obtained one point at a time as the result of an interrupt from an A/D conversion that occurs at a constant rate. For each point of the input, the application is required to produce one point of the filtered output,

$$y[n] = \sum_{k=0}^{N-1} h[k]x[n-k]. \tag{9.33}$$

This is, of course, just convolution. To do a real-time convolution, we need to write an efficient routine, one that requires the minimum number of multiplications and additions as well as the minimum number of storage elements. In this section, we will discuss ways of making the basic convolution sum efficient. In Chapter 11, we will discuss using the fast Fourier transform (FFT) to do convolution.

Figure 9.33 shows a graphical interpretation of convolution for an example where $N = 3$.

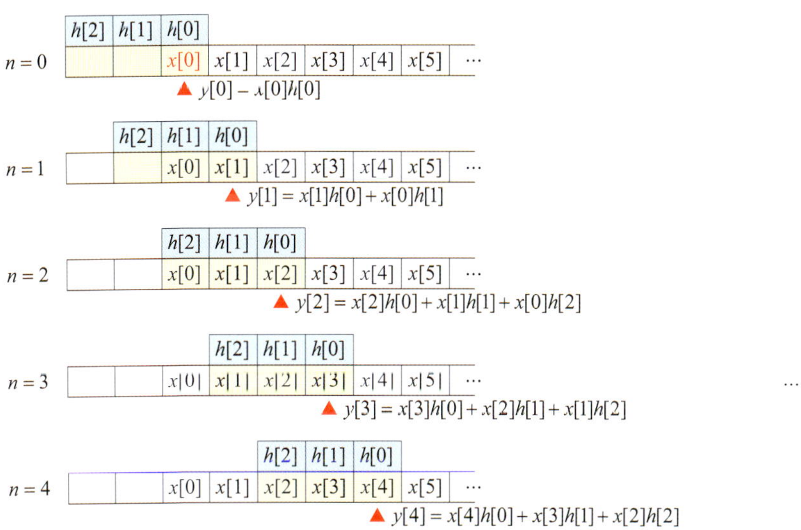

Figure 9.33 Convolution

To perform the convolution, the routine requires a buffer of N memory elements (shown shaded) for the input values. At each time point, one new value of $x[n]$ is added to the buffer, and the past $N-1$ values $x[n-k]$, $1 \le k < N$, are retained. Then, we have to do N multiplications and $N-1$ additions of Equation (9.33) to compute each point of the convolution sum $y[n]$. One option for managing this memory is to shift each existing element in the data buffer physically by one element, before adding the newest value, as diagrammed in the left column of **Figure 9.34** for a buffer of size $N = 3$.

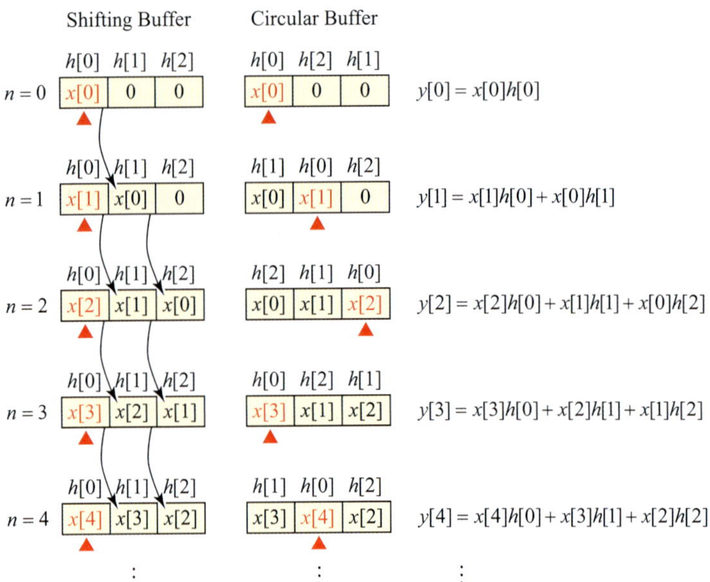

Figure 9.34 Implementing convolution using shifting and circular buffer

Before the convolution starts, all the elements of the data buffer (shown shaded in green) are initially set to zero. At the first time point (i.e., $n = 0$, first row), $x[0]$ is placed in the first (left-most) buffer element and multiplied by $h[0]$. Hence, the convolution sum is just $y[0] = x[0]h[0]$. At the next time point (i.e., $n = 1$), $x[0]$ is moved (i.e., written) into the second buffer position, and the new point $x[1]$ is written into the first position. The convolution sum is $y[1] = x[1]h[0] + x[0]h[1]$. For all subsequent points of the convolution, the oldest (right-most) element of the buffer is tossed, and the remaining $N-1$ elements of the buffer are shifted right. The newest value of $x[n]$ is then added to the initial position of the buffer (indicated by the red arrowhead). The central part of some sample C code might look something like this:

```
const int N = 3;
const float h[] = {1, 1, 1};
float *buff = (float*) calloc(N, sizeof(float));
do
{
  buff[0] = input();              // load newest value of x into buffer
  float sum = buff[0] * h[0];     // initialize sum
```

```
  for (int i = 1; i < N; ++i)      // remaining N-1 iterations
        sum += buff[i] * h[i];     // accumulate convolution sum
  output(sum);                     // output value of y[n]
  for (int i = N-1; i > 0; -i)     // shift data buffer right
        buff[i] = buff[i-1];
}
```

The code is pretty straightforward; just read the comments. The problem with the shifting buffer is that moving elements from one memory location to another in this manner is relatively expensive from a processor perspective. Each of the $N-1$ buffer shifts requires that the source and destination address be resolved, data read from the source and written to the destination.

The right panel of **Figure 9.34** shows a less processor-intensive approach to buffer design: the **circular buffer** (or **ring buffer**). For the first N points of the convolution (e.g., $n=0$, $n=1$ and $n=2$), the input data points populate the buffer from left to right, and the convolution is computed in a fashion similar to that of the shifting buffer. For all subsequent points of the convolution, the newest value of $x[n]$ replaces the oldest data point; for example, at $n=3$, $x[3]$ replaces $x[0]$. The remaining $N-1$ elements of the buffer remain unchanged. With the circular buffer, there are no memory moves, only one memory element gets written at each time point. In a high-level language such as C, a circular buffer is easily implemented by incrementing the index to the buffer modulo N:

```
const int N = 3;
const float h[] = {1, 1, 1};
float *buff = (float *) calloc(N, sizeof(float));
int k = 0;

do
{
   buff[k]= input();                 // load newest value of x[n] into buffer
   const float *hPtr = h;
   float sum = buff[k] * h[0];       // initialize sum with first value
   for (int i = 1; i < N; ++i)       // remaining N-1 iterations
   {
      k = ++k % N;                   // increment buffer index modulo N
      sum += buff[k] * h[i];         // accumulate convolution sum
   }
   *y++ = sum;    // output value of y[n]
}
```

While this is not the most efficient implementation, it is easy to understand. The routine creates an array `buff` that is the size of $h[n]$. Each new input point is written into the `k`th position of `buff`, multiplied by $h[0]$, and the product is used to initialize `sum`. The `for` loop calculates the remaining $N-1$ values of the sum. The key line is

```
k = ++k % N;
```

If `k`, the index of `buff`, exceeds the end of the buffer, the pointer is reset to the beginning of the buffer by division modulo N. In this implementation, the values of $h[n]$ are addressed so that the newest data point is always multiplied by $h[0]$. The minor downside of the circular buffer is that it requires a modulo arithmetic operation on the buffer index within the inner loop, but this costs next to nothing on modern processors and is pretty much guaranteed to be faster than moving $N-1$ memory elements in the shifting buffer.[2]

The most processor-intensive part of any convolution routine is the **multiply-and-accumulate (MAC)** calculation of the convolution sum, represented by this line in the inner loop,

```
sum += buff[k] * h[i];
```

For each pass through the loop, two numbers must be multiplied together and the product added to an accumulator. In the old days, such a MAC operation could take multiple clock cycles to access the two memory elements, add them together in the CPU's **arithmetic logic unit (ALU)** and write (or add) them to an output location. However, most modern computers can perform one or more MAC operations in a *single* clock cycle.

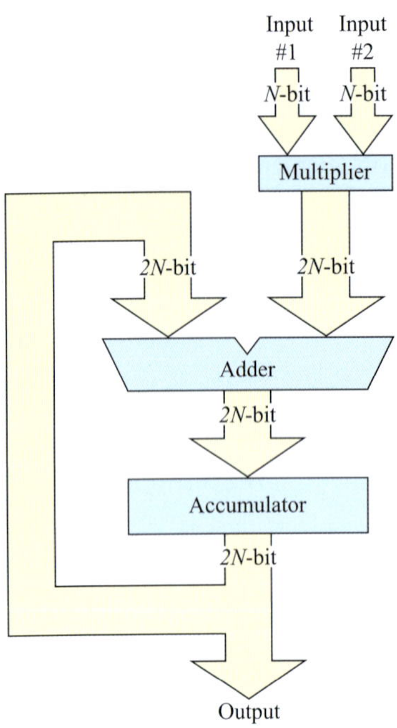

Figure 9.35 Multiply-accumulate (MAC) unit

Figure 9.35 shows the basic architecture of a simple MAC unit in a processor. Various flavors of floating and fixed-point MACs exist, but all of them include at least one multiplier, adder and accumulator. On each clock cycle the MAC fetches two numbers – in the example, two N-bit fixed-point numbers – multiplies them with full precision and adds the $2N$-bit product to an accumulator. The output of the accumulator is fed back to one input of the adder so that on each clock cycle, the accumulator maintains a running sum of the products. While originally

[2] Another approach to implementing the circular buffer is to initialize a pointer to the beginning of the data buffer. For each iteration of the loop, a new input data value is written into the dereferenced pointer position. Then the pointer is incremented and if it passes beyond the end of the buffer, it is reset to the beginning of the buffer. See Problem 1-5.

MAC operations were mostly found in special-purpose **DSP processors**, they are now a common feature of modern general-purpose processors – microprocessors and microcontrollers – as well as the GPUs found in graphics cards. These processors usually feature a variety of **single-instruction, multiple-data (SIMD) operations** that can implement multiple types of MAC operations. DSP processors (and DSP cores in modern general-purpose processors) have further specialized hardware that handles the circular buffer operations required for FIR filtering in a single clock cycle.

9.9.2 Hardware implementation

Increasingly, DSP algorithms for high-speed applications such as digital audio and video, communications applications including cellular telephony and cable and wireless modems, as well as many other areas are being implemented using techniques of **very-large-scale integration (VLSI)**, particularly using **application-specific integrated circuits (ASICs)** and **field-programmable gate arrays (FPGAs)**. ASICs offer better hardware customization, better performance and lower power consumption, but the FPGA approach has a lower cost to prototype than ASICs.

Designers specify their algorithms using a high-level **hardware description language (HDL)**, such as Verilog or VHDL. Writing in HDL is analogous to writing a computer program in a high-level language. A logic synthesizer "compiles" the high-level representations of a circuit into lower-level hardware representations at the gate-level and ultimately at the transistor-level in the case of ASICs, or a configuration bit-stream in the case of FPGAs. For example, variables in HDL correspond to registers or wires in the hardware. If-then-else constructs correspond to combinatorial and multiplexing logic. Arithmetic operations are implemented using **arithmetic logic units (ALUs)** and **floating-point units (FPUs)** that are derived from standard-cell "libraries" in the case of ASICs and look-up tables in the case of FPGAs.

A key feature of hardware implementations in VLSI is that highly parallel systems can easily be implemented. To see why that is important, consider again the Nth-order FIR filter we have been discussing,

$$y[n] = \sum_{k=0}^{N-1} b_k x[n-k],$$

where the filter coefficients are now labeled b_k. We will consider two ways of implementing this filter: the canonical architecture of Section 9.2 and the transposed architecture of Section 9.3. Although these two architectures are identical from an algorithmic standpoint, we will now show that they have very different performance when maximum throughput is considered.

Figure 9.36 shows the signal-flow graph of a canonical filter with $N = 3$. The hardware implementation of this filter has $N - 1$ delays which correspond to N memory elements (latches), designated by orange boxes (M_1, M_2 and M_3). There are N parallel multipliers, designated by the red arrowheads, and $N - 1$ two-input adders, designated by the green summing nodes. In the figure, the values of data at various points in the signal paths at time n are shown with the lozenges (e.g., $b_0x[n]$).

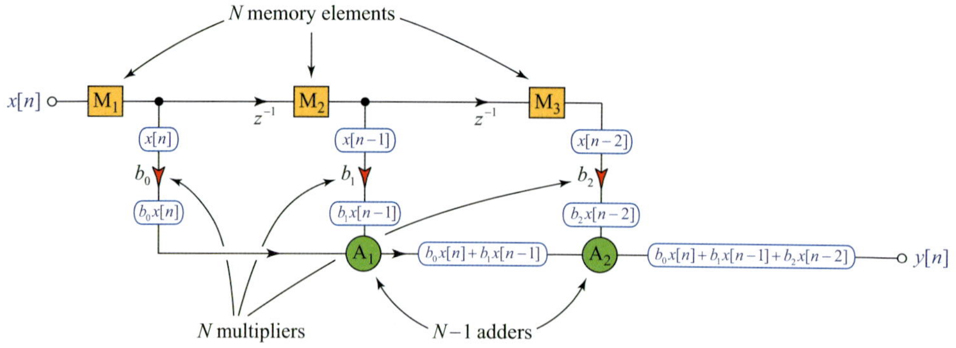

Figure 9.36 Canonical FIR filter

Figure 9.37 shows the state of the filter at times $n = 0$, 1, 2 and 3. The initial state of the memory registers, and the outputs of the multipliers and adders, are assumed to be 0. At $n = 0$, only the left-most memory element contains data, and the output of the filter is just $y[0] = b_0x[0]$. In the canonical architecture, the memory registers are essentially configured as a **serial-in, parallel-out (SIPO)** shift register. At each subsequent time point (i.e., each clock cycle), new input data are clocked into the left-most latch M_1, and data from previous time points are shifted right (diagrammed with the red arrows), multiplied by the appropriate constants and added to form the output. In order to understand what limits the speed of this filter, we need to look more carefully at the micro-structural sequence of events that begins when a new data point is presented to the input and culminates when the computed filter value is available at the output. **Figure 9.38** shows a detail of what happens in our canonical filter at time $n = 3$.

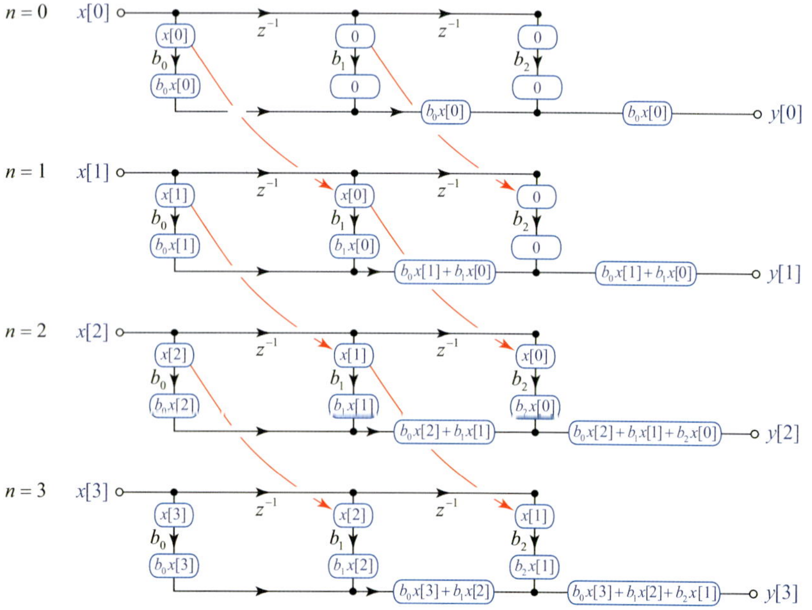

Figure 9.37 Canonical FIR filter

Figure 9.38a shows the state of the system just after the filter has produced an output for time point $n = 2$. At time $n = 3$, new input data are clocked into the left-most memory register M_1, and simultaneously data from previous time points are shifted right (**Figure 9.38a**). The contents of all the memory registers are available essentially instantaneously (with the propagation delay, T_D, of at most a couple of gates) to the input of the multipliers (**Figure 9.38b**). The time required for the multiplier to multiply the input data by a constant is T_M. Since the multipliers are in parallel, they simultaneously perform all the multiplications operations necessary to compute each output point (i.e., $b_0x[n]$, $b_1x[n-1]$, ..., $b_{N-1}x[0]$). Hence, valid data are available at the output of all the multipliers after the same delay of T_M (**Figure 9.38c**). The time required for each of the $N-1$ two-input adders to complete an addition is T_A. However, unlike multiplication, addition cannot happen in parallel. In the canonical architecture, the adders are daisy-chained together, meaning that the output of each adder is connected to one of the inputs of the next adder. Thus, every adder in the chain must wait for the result of the previous adder to become valid before it can perform a valid addition. In our example, there are only two adders. The first (left-most) adder completes its addition to form the partial sum $b_0x[3] + b_1x[2]$ with a delay of T_A (**Figure 9.38d**). Then the second adder requires another delay of T_A to add $b_2x[0]$ to the partial sum to form the final answer $y[3]$. The total time to generate this result is therefore $T = T_D + T_M + 2T_A$. Generalizing this result, the total time necessary to complete the calculation of a single output point in an Nth-order FIR filter is $T = T_D + T_M + (N-1)T_A$. This value of the total delay determines the maximum rate $f_{MAX} = 1/T$ at which data can be processed by the filter.

Now consider the hardware implementation of the FIR filter using a transposed architecture. Like the canonical filter, this transposed filter has N memory elements, N parallel multipliers and $N-1$ two-input adders, as shown in **Figure 9.39**.

Figure 9.40 shows the state of the filter at times $n = 0$, 1, 2 and 3. As expected, the output of the filter at each time point is the same as the canonical case, however, the microstructural details of what happens between time points is very different, and results in a higher rate at which data can be processed.

Figure 9.41 shows the series of events that follow the input of a new data point at time $n = 3$. At time $n = 3$, new input data appear at the output of latch M_1, with a propagation delay of T_D. Simultaneously, the outputs of multipliers b_1 and b_2 at the prior time point $n = 2$ are latched into M_2 and M_3, respectively (**Figure 9.41a**). The parallel multipliers require time T_M to multiply $x[3]$ by the filter coefficients b_1, b_2 and b_3 (**Figure 9.41c**). In the canonical filter, the adders are directly connected to each other, so that the output of each adder depends on the value of the preceding one. Hence, we need $(N-1)T_A$ for the result to ripple through all the adders. However, in the transpose filter, one of the inputs to each adder comes from a latch, not from another adder. While the time required for each adder to process its inputs is still T_A (**Figure 9.38d**), the output of the filter only depends on the output of the right-most adder A_2. So, the total time to generate each output point is $T = T_D + T_M + T_A$, and this is *independent* of the order of the filter. While it is possible to **pipeline** the canonical filter by imposing additional delays before the adders, this results in a more complex filter with a longer latency between input and output. The transposed filter is an example of a **self-pipelined** architecture.

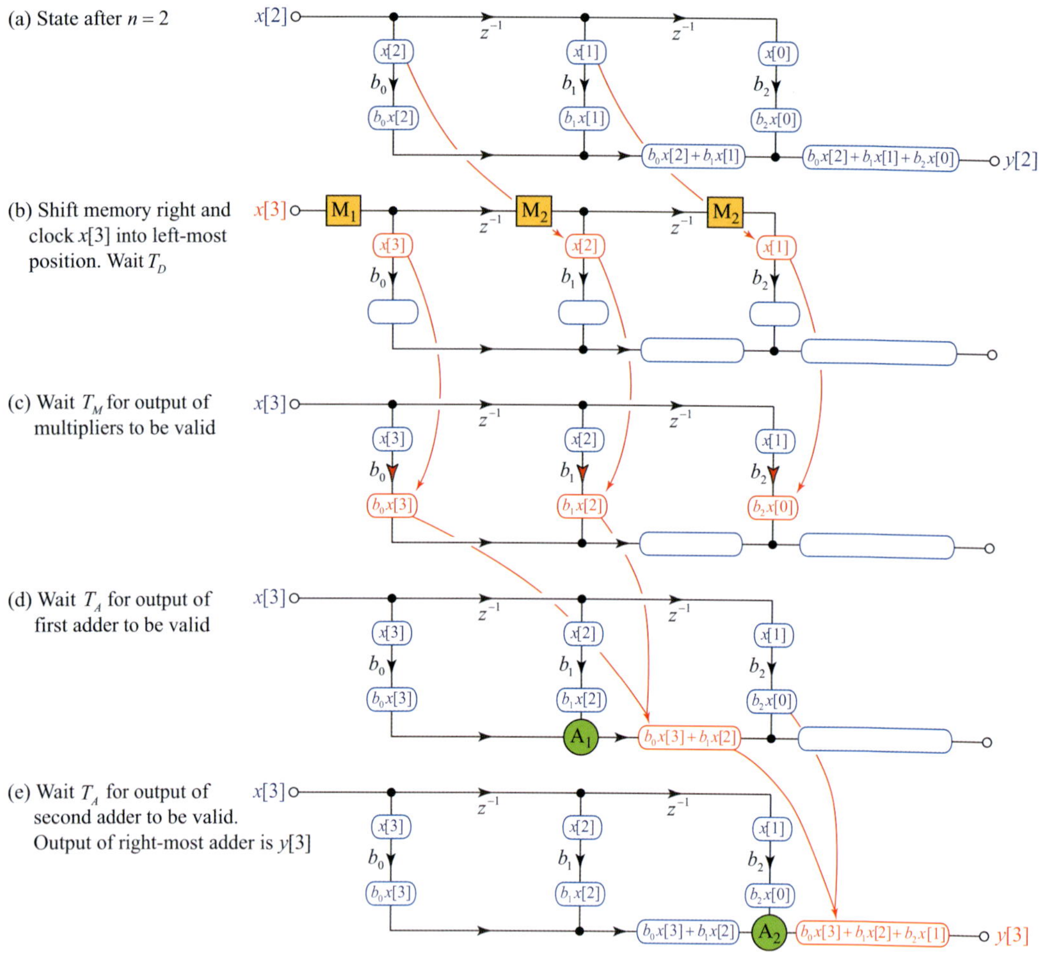

Figure 9.38 Canonical filter timing details

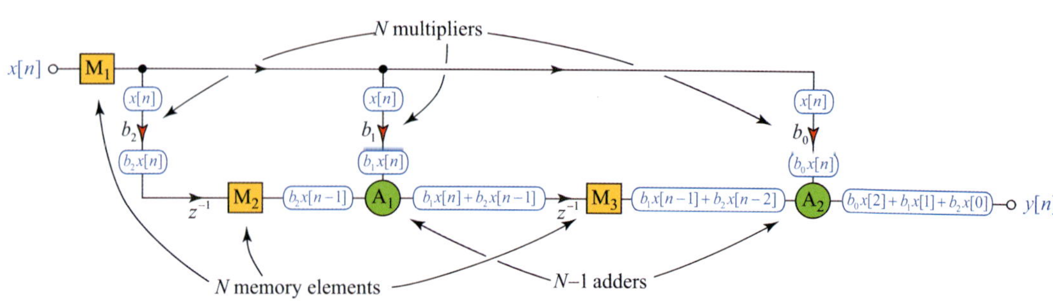

Figure 9.39 Transpose FIR filter

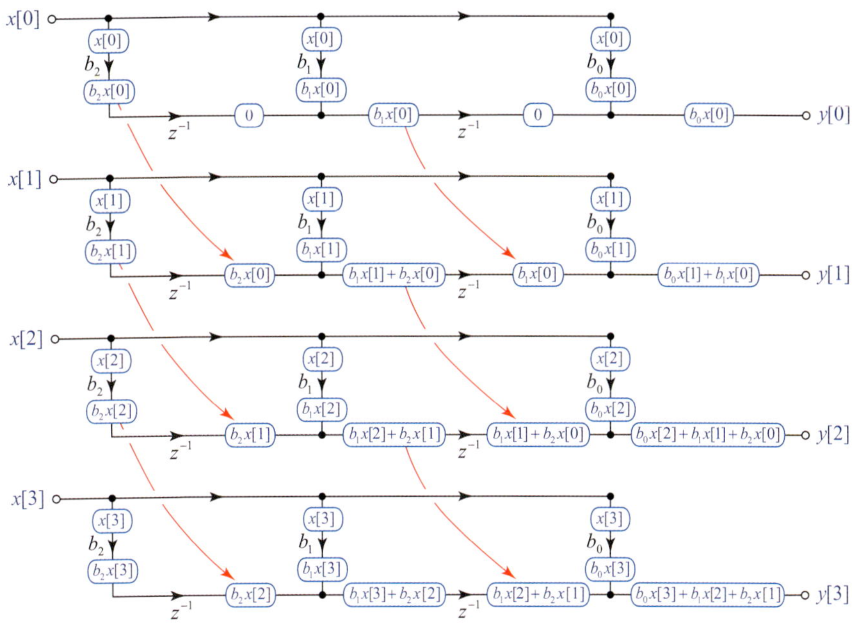

Figure 9.40 Transpose FIR filter

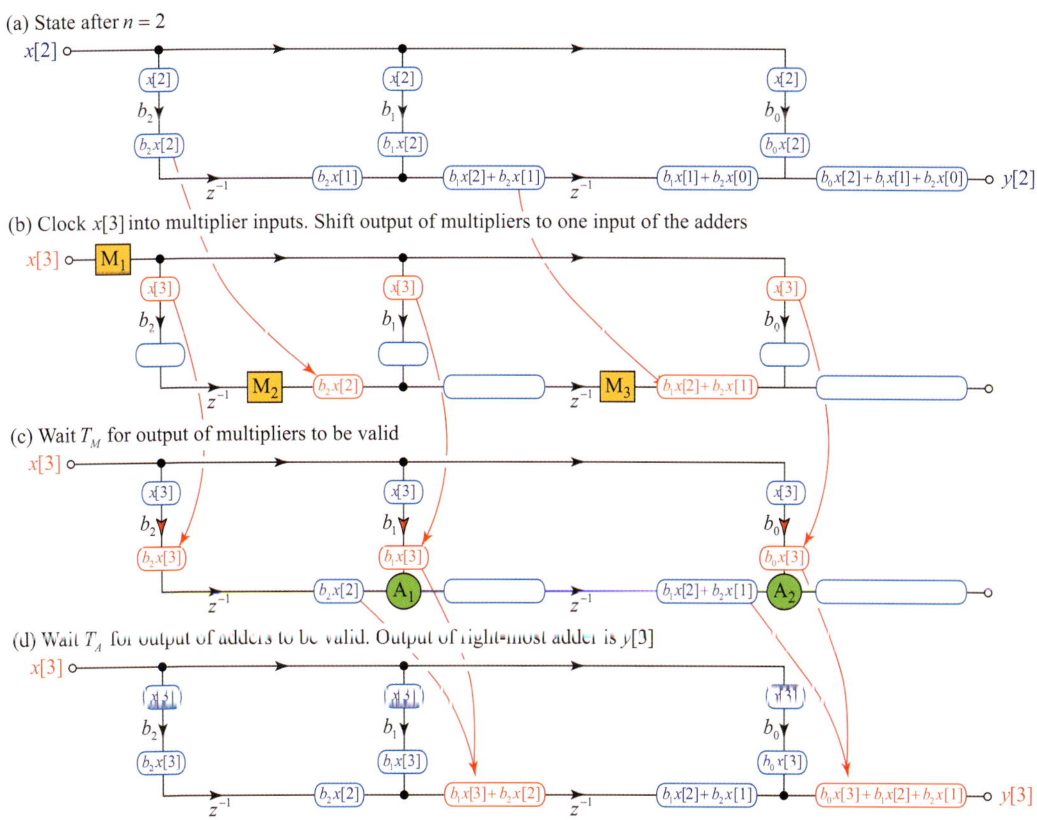

Figure 9.41 Transpose FIR filter timing details

SUMMARY

Designing FIR and IIR filters involves both designing the algorithms to achieve particular specifications, and making decisions about how the algorithms will be implemented. In this chapter, we have concentrated on the two most common filter architectures – cascade and parallel – as well as specialized lattice filters. We have highlighted particular areas of practical concern for the effect of coefficient quantization on the design of FIR and IIR filters.

PROBLEMS

Problem 9-1

Given a stable system with

$$H(z) = \frac{1 - 2z^{-1} + 2z^{-2}}{1 - \frac{1}{4}z^{-2}}.$$

(a) Express $H(z)$ as a cascade of one or more first-order sections or say why this cannot be done.

(b) Express $H(z)$ as a parallel combination of sections or say why this cannot be done.

Problem 9-2

Repeat Problem 9-1 with

$$H(z) = \frac{1 + \frac{1}{2}z^{-1} - \frac{1}{2}z^{-2}}{1 + \frac{7}{4}z^{-1} - \frac{1}{2}z^{-2}}.$$

Problem 9-3

Repeat problem Problem 9-1 with

$$H(z) = \frac{1 - z^{-1} + \frac{1}{2}z^{-2}}{1 + z^{-1} + \frac{1}{2}z^{-2}}.$$

Problem 9-4

A causal second-order filter has z-transform

$$H(z) = \frac{1}{1 + a_1 z^{-1} + a_2 z^{-2}} = \frac{z^2}{(z - p_1)(z - p_2)},$$

where the two coefficients a_1 and a_2 are real, so that the poles p_1 and p_2 are either both real or occur as a complex-conjugate pair.

(a) Show that, for a stable-filter, the following conditions on a_1 and a_2 apply:

$$\begin{aligned} 1 - a_2 &> 0 \\ 1 + a_2 &> 0 \\ 1 - a_1 + a_2 &> 0 \\ 1 + a_1 + a_2 &> 0. \end{aligned}$$

(b) Show that these inequalities are equivalent to $|a_2| < 1$ and $|a_1| < 1 + a_2$.

➤**Hint:** $a_1 = -(p_1 + p_2)$ and $a_2 = p_1 p_2$. Argue that $|p_1| < 1$ and $|p_2| < 1$. Use the quadratic formula to express $p_{1,2}$ in terms of a_1 and a_2.

Problem 9-5

In the program for convolution using the circular buffer, we stated that we could effectively rotate the buffer by initializing a pointer to the beginning of the data buffer. For each iteration of the loop, a new input point is written into the dereferenced pointer position. Here is the program minus a couple of key lines. Provide them.

```
const int N = 3;
const float h[]  = {1, 1, 1};
float * buff = (float *) calloc(N, sizeof(float)),
      * buffPtr = buff,
      * buffEnd = buff + N,
do
{
      * buffPtr = input();                // load x[n] into buffer
      const float *hPtr = h;
      float sum = *buffPtr * *hPtr++;     // initialize sum with first value
      for (int i = 1; i < N; ++i)         // remaining M-1 iterations
      {
            ** YOUR CODE HERE **          // rotate circular buffer
            ** YOUR CODE HERE **          // reset buffer if index overflow
            sum += *buffPtr * *hPtr++;    // accumulate convolution sum
      }
      output(sum);                        // output value of y[n]
}
```

Problem 9-6

The recursion relations for the FIR lattice filter in the z-transform domain are

$$\begin{aligned} F_k(z) &= F_{k-1}(z) + c_k z^{-1} G_{k-1}(z) \\ G_k(z) &= c_k F_{k-1}(z) + z^{-1} G_{k-1}(z). \end{aligned}$$

Use these relations to show the result:

$$G_k(z) = z^{-k}F_k(z^{-1}).$$

➤Hint: Substitute the result into both recursion relations so that both equations are expressed just in terms of $F_k(z)$ and $F_{k-1}(z)$, evaluate the bottom equation at $z = z^{-1}$ and compare the equations.

Problem 9-7

Show that lattice filters cannot directly represent linear-phase FIR filters.

➤Hint: Any linear-phase filter of order M can be expressed to within a constant gain factor as $H(z) = 1 + \cdots \pm z^{-M}$, where a "+" sign on the last term indicates a Type-I or Type-II (symmetric) filter and the "−" sign indicates a Type-III or Type-IV (asymmetric) filter.

Problem 9-8

In this problem, we use matrix methods to find the coefficients d_k of the lattice-ladder filter

$$H(z) = \frac{1.13 + 1.664z^{-1} + 1.228z^{-2} + 0.6z^{-3}}{1 + 1.38z^{-1} + 1.34z^{-2} + 0.6z^{-3}}.$$

In Example 9.9, we determined the reflection coefficients $c_1 = 0.5$, $c_2 = 0.8$ and $c_3 = 0.6$, as well as $B_k(z)$, $0 \le k \le 3$.

(a) Express $B_k(z)$ in matrix form $\mathbf{B} = \mathbf{Qz}$, where

$$\begin{aligned}\mathbf{B}^{\mathrm{T}} &= [B_0(z)\ \ B_1(z)\ \ B_1(z)\ \ B_2(z)]\\ \mathbf{z}^{\mathrm{T}} &= [1\quad z^{-1}\quad z^{-2}\quad z^{-3}],\end{aligned}$$

and the matrix $\mathbf{Q}$ comprises the coefficients of $B_k(z)$.

(b) Show that when expressed in matrix form,

$$\sum_{k=0}^{M} d_k B_k(z) = \sum_{k=0}^{M} b_k z^{-k}$$

becomes $\mathbf{d}^{\mathrm{T}}\mathbf{Q} = \mathbf{b}^{\mathrm{T}}\mathbf{z}$, where

$$\begin{aligned}\mathbf{d}^{\mathrm{T}} &= [d_1\quad d_2\quad d_3\quad d_4]\\ \mathbf{b}^{\mathrm{T}} &= [b_1\quad b_2\quad b_3\quad b_4].\end{aligned}$$

Hence $\mathbf{d}^{\mathrm{T}}\mathbf{Q} = \mathbf{b}^{\mathrm{T}}$, which can be solved to obtain $\mathbf{d}^{\mathrm{T}} = \mathbf{b}^{\mathrm{T}}\mathbf{Q}^{-1}$.

Problem 9-9

(a) Defining

$$F_k(z) \triangleq \frac{V_k(z)}{V_0(z)} = \frac{V_k(z)}{X(z)}, \qquad 1 \le k \le M$$

$$G_k(z) \triangleq \frac{W_k(z)}{W_0(z)} = \frac{W_k(z)}{X(z)}, \qquad 1 \le k \le M,$$

show that the recursion relations for the QMF lattice are given by

$$\begin{aligned} F_1(z) &= F_0(z) - c_1 z^{-1} G_0(z) = 1 - c_1 z^{-1} \\ F_1(z) &= c_1 F_0(z) + z^{-1} G_0(z) = c_1 + z^{-1}, \end{aligned} \qquad k = 1$$

$$\begin{aligned} F_k(z) &= F_{k-2}(z) - c_k z^{-2} G_{k-2}(z) \\ G_k(z) &= c_k F_{k-2}(z) + z^{-2} G_{k-2}(z), \end{aligned} \qquad 3 \le k \le M.$$

(b) Show that $G_k(z) = z^{-k} F_k(-z^{-1})$.

(c) Show that the reverse recursion relation is

$$F_{k-2}(z) = \frac{F_k(z) + c_k G_k(z)}{1 + c_k^2}, \; 3 \le k \le M.$$

Problem 9-10

(a) Write a Matlab function `c = tf2qmflat(f)` that takes transfer function coefficients `f`, and produces QMF lattice coefficients `c`. The test is as follows:

```
>> c = tf2qmflat([1 -0.5 -0.88 -0.56 -0.3 -0.6])
c =
    0.5000    0.8000    0.6000
```

(b) Write a Matlab function `f = tf2qmflat2tf(c)` that takes QMF lattice coefficients `c`, and produces transfer function coefficients `f`.

10 Discrete Fourier transform (DFT)

Introduction

In Chapter 3, we showed that there was a unique, invertible relation between a sequence $x[n]$ and its DTFT $X(\omega)$:

$$\begin{aligned} \text{Forward DTFT:} \quad & X(\omega) = \mathfrak{F}\{x[n]\} \triangleq \sum_{n=-\infty}^{\infty} x[n]e^{-j\omega n} \\ \text{Inverse DTFT:} \quad & x[n] = \mathfrak{F}^{-1}\{X(\omega)\} \triangleq \frac{1}{2\pi} \int_{\omega=0}^{2\pi} X(\omega)e^{j\omega n} d\omega. \end{aligned} \tag{10.1}$$

The DTFT is an essential tool for analyzing, understanding and visualizing the response of discrete-time systems. However, because the DTFT is a continuous function of frequency ω, its computation cannot be performed exactly on a digital computer. This limits the utility of the DTFT in many real-world signal-processing operations. As one important example, we showed that convolution of an input signal $x[n]$ and impulse response $h[n]$ could be implemented by multiplication of transforms in the frequency domain, as schematized here:

$$\begin{array}{ccccc} x[n] & * & h[n] & = & y[n] \\ \mathfrak{F}\downarrow & & \mathfrak{F}\downarrow & & \mathfrak{F}^{-1}\uparrow \\ X(\omega) & \cdot & H(\omega) & = & Y(\omega) \end{array}.$$

Specifically, we take the DTFT of $x[n]$ and $h[n]$, multiply the transforms together to form $Y(\omega)$, and then take the inverse DTFT to get $y[n]$. Each of these steps requires operations on the continuous variable ω.

In this chapter, we are going to address this limitation of the DTFT. Specifically, we shall show that if we restrict our consideration to a finite-length sequence of length N, $x[n]$, $0 \le n \le N-1$, then we can completely reconstruct $x[n]$ from only N values of the DTFT. The tool that makes this possible is the **discrete Fourier transform (DFT)**. The DFT of $x[n]$, which we denote with the symbol $\mathbb{F}_N\{x[n]\}$, is simply a sequence of N values, $X[k]$, $0 \le k \le N-1$, derived from sampling the DTFT of $x[n]$ at N frequencies equally spaced in the frequency range $0 \le \omega < 2\pi$,

$$X[k] = \mathbb{F}_N\{x[n]\} \triangleq X(\omega)|_{\omega = 2\pi k/N} = \sum_{n=0}^{N-1} x[n] e^{-j2\pi kn/N}, \quad 0 \le k \le N-1. \tag{10.2}$$

We will show that $x[n]$ can be completely reconstructed from $X[k]$ using the **inverse discrete Fourier transform (IDFT)**, denoted by the symbol $\mathbb{F}_N^{-1}\{X[k]\}$:

$$x[n] = \mathbb{F}_N^{-1}\{X[k]\} \triangleq \frac{1}{N} \sum_{k=0}^{N-1} X[k] e^{-j2\pi kn/N}, \quad 0 \le n \le N-1.$$

Because the formulas for the DFT and IDFT involve only the summation of a finite number of elements, all the operations necessary to convert time-domain sequences into the frequency domain, perform operations upon them and convert them back to the time domain can be done on digital hardware and software. For this reason, the DFT and IDFT – coupled with the efficient implementation using the fast Fourier transform (FFT) algorithms that we shall study in Chapter 11 – have provided the foundation for an enormous range of practical digital signal processing applications in spectral analysis, communication and sound processing, among other fields. To take one example, we will show that the convolution of two sequences $x[n]$ and $h[n]$ can be performed in the discrete frequency domain by taking the DFT of both sequences, multiplying the two resulting DFTs $X[k]$ and $H[k]$ to form $Y[k]$, and then taking the IDFT to form $y[n]$:

$$\begin{array}{ccccc} x[n] & * & h[n] & = & y[n] \\ \mathbb{F}_N\downarrow & & \mathbb{F}_N\downarrow & & \mathbb{F}_N^{-1}\uparrow \\ X[k] & \cdot & H[k] & = & Y[k] \end{array}.$$

In Chapter 11, we will further show that convolution implemented by this circuitous route using the FFT can be much faster than doing convolution in the time domain.

In the upcoming section, Section 10.1, we will present two ways of deriving and understanding the DFT. First, we will show that the DFT can be viewed in terms of the DTFT of periodic sequences. Then, we will approach the DFT from the viewpoint of sampling the DTFT. In Section 10.2, we will derive the DFT of some basic signals. In Section 10.3, we will present the important properties of the DFT, including the convolution property mentioned above, that enable a host of practical applications. You will find that while there is a lot of commonality between the properties of the DFT and the properties of the DTFT that you already know, there are also some critical differences that must be understood. In Section 10.5, we will explore some interesting applications of the DFT to upsampling in the time and frequency domains.

10.1 Derivation of the DFT

In this section, we present the key result: that a sequence $x[n]$ of length L can be uniquely represented in the frequency domain by $N \ge L$ values of its DTFT. A conceptual understanding of this result proceeds through a somewhat circuitous path, but please stick with it. To make

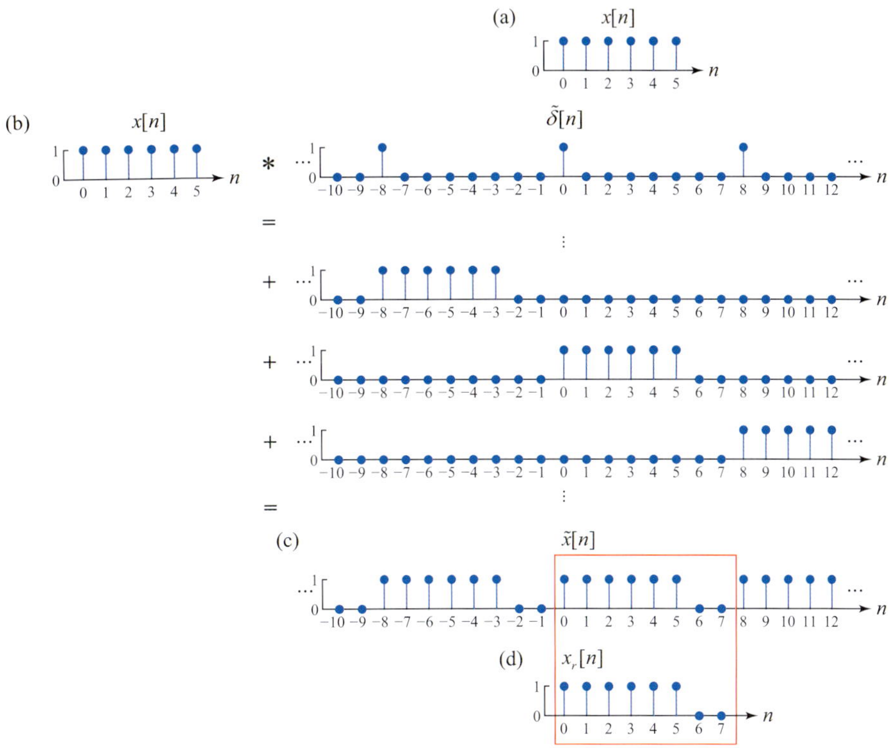

Figure 10.1 Construction of a periodic sequence with $N=8$

things easy to understand, let us work with a concrete example. Consider the finite-length sequence $x[n]$ of length $L=6$, shown in **Figure 10.1a**. Now create an infinite-length **periodic sequence** $\tilde{x}[n]$, comprising one period of $x[n]$ repeated with a period $N=8$, as shown in **Figure 10.1c**.[1]

Figure 10.1b shows that $\tilde{x}[n]$ can be constructed by convolving $x[n]$ with an infinite-length impulse train $\tilde{\delta}[n]$ that is periodic with period $N=8$,

$$\tilde{\delta}[n] \triangleq \sum_{k=-\infty}^{\infty} \delta[n-kN],$$

so that,

$$\tilde{x}[n] = x[n] * \tilde{\delta}[n] = x[n] * \sum_{k=-\infty}^{\infty} \delta[n-kN] = \sum_{k=-\infty}^{\infty} x[n-kN]. \tag{10.3}$$

Each impulse of $\tilde{\delta}[n]$ produces one replica of $x[n]$ in $\tilde{x}[n]$. Clearly, as long as the period of $\tilde{\delta}[n]$ is greater than or equal to the length of $x[n]$, $N \geq L$, then we can recover $x[n]$ from $\tilde{x}[n]$

[1] In this chapter, a tilde (~) on top of a variable indicates that it is periodic.

just by extracting the N values of $\tilde{x}[n]$ in the range $0 \le n < N$, as shown in the red box in **Figure 10.1d**.

The DTFT of $\tilde{x}[n]$ is the product of the transforms of $x[n]$ and $\tilde{\delta}[n]$,

$$\tilde{X}(\omega) = \mathfrak{F}\{x[n] * \tilde{\delta}[n]\} = \mathfrak{F}\{x[n]\} \cdot \mathfrak{F}\{\tilde{\delta}[n]\} = X(\omega)\tilde{\Delta}(\omega), \tag{10.4}$$

where $\tilde{\Delta}(\omega) \triangleq \mathfrak{F}\{\tilde{\delta}[n]\}$ is defined to be the DTFT of the impulse train. Since $x[n]$ is of finite length, $X(\omega)$ can be directly calculated from Equation (10.1) as a finite summation of elements,

$$X(\omega) = \sum_{n=-\infty}^{\infty} x[n]e^{-j\omega n} = \sum_{n=0}^{N-1} x[n]e^{-j\omega n}.$$

$X(\omega)$ is a continuous function of frequency and is periodic with a period of 2π. Its magnitude is shown in **Figure 10.2a**. To obtain the DTFT of the impulse train, $\tilde{\Delta}(\omega)$, we can use the fact that an impulse train of period N can be expressed as the sum of N complex exponential sequences that are multiples of a common fundamental frequency $\omega_s = 2\pi/N$,

$$\tilde{\delta}[n] = \frac{1}{N}\sum_{k=0}^{N-1} e^{jk\omega_s n} = \frac{1}{N}\frac{1 - e^{j2\pi \cancel{N} n/\cancel{N}}}{1 - e^{j2\pi n/N}} = \begin{cases} 1, & n = 0, \pm N, \pm 2N, \cdots \\ 0, & \text{otherwise} \end{cases}.$$

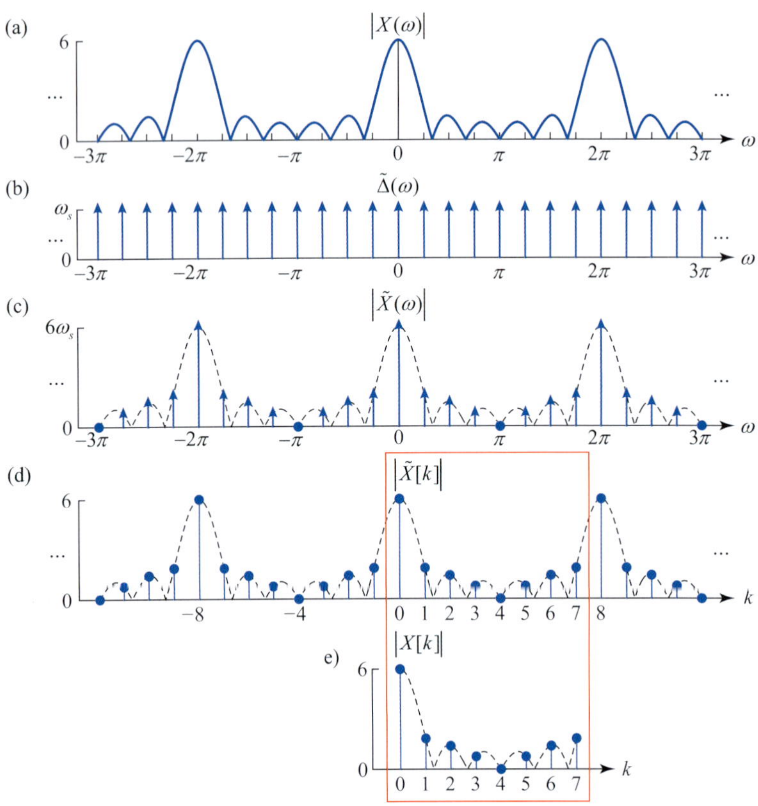

Figure 10.2 DTFT sampled at $N = 8$

Here, we have noted that the numerator of the fractional expression is always zero, and the denominator is non-zero except when n is a multiple of N, in which case the expression would appear to be indeterminate (i.e., 0/0). However, direct evaluation of the summation at these values of n is equal to 1 at these values. Therefore, the DTFT of $\tilde{\delta}[n]$ is

$$\tilde{\Delta}(\omega) \triangleq \mathfrak{F}\{\tilde{\delta}[n]\} = \mathfrak{F}\left\{\frac{1}{N}\sum_{k=0}^{N-1} e^{jk\omega_s n}\right\} = \frac{1}{N}\sum_{k=0}^{N-1}\mathfrak{F}\{e^{jk\omega_s n}\} = \frac{1}{N}\sum_{k=0}^{N-1} 2\pi\delta(\omega - k\omega_s)$$
$$= \omega_s \sum_{k=0}^{N-1} \delta(\omega - k\omega_s).$$

In words, $\tilde{\Delta}(\omega)$ comprises N impulses spaced at multiples of $\omega_s = 2\pi/N$ over the range $0 \le \omega < 2\pi$. However, since all DTFTs are periodic in frequency with a period of 2π, the limits of the summation need to be extended to $k = \pm\infty$,

$$\tilde{\Delta}(\omega) = \omega_s \sum_{k=-\infty}^{\infty} \delta(\omega - k\omega_s),$$

as shown in **Figure 10.2b**. Returning to Equation (10.4), $\tilde{\Delta}(\omega)$ multiplies (i.e., samples) $X(\omega)$ at multiples of $\omega_c = 2\pi/N$ and the product $\tilde{X}(\omega)$ is therefore a modulated impulse train,

$$\tilde{X}(\omega) = X(\omega)\tilde{\Delta}(\omega) = X(\omega)\cdot\omega_s \sum_{k-\infty 0}^{\infty} \delta(\omega - k\omega_s) = \omega_s \sum_{k=-\infty}^{\infty} X(k\omega_s)\delta(\omega - k\omega_s), \quad \textbf{(10.5)}$$

as shown in **Figure 10.2c**. In other words, whereas $\tilde{X}(\omega)$, the DTFT of the aperiodic sequence $x[n]$, contains energy over the entire frequency range, $\tilde{X}(\omega)$, the DTFT of the periodic sequence $\tilde{x}[n]$, has energy at only N unique values in any 2π range of frequency. If we now define $\tilde{X}[k]$ to be a periodic sequence comprising the values of $X(\omega)$ at multiples of the sampling frequency,

$$\tilde{X}[k] \triangleq X(k\omega_s), \quad \textbf{(10.6)}$$

then Equation (10.5) can be written in terms of $\tilde{X}[k]$ as

$$\tilde{X}(\omega) = \omega_s \sum_{k=-\infty}^{\infty} \tilde{X}[k]\delta(\omega - k\omega_s). \quad \textbf{(10.7)}$$

Since $X(\omega)$ is periodic with period 2π, the sequence $\tilde{X}[k]$ is also periodic, with period N, as shown in **Figure 10.2d**,

$$\tilde{X}[k+N] = X((k+N)\omega_s) = X(k\omega_s + N\cdot 2\pi/N) = X(k\omega_s) = \tilde{X}[k].$$

Finally, because $\tilde{X}[k]$ is periodic, we only actually require N contiguous values of k (for example, the values $0 \le k \le N-1$) to define the sequence completely for all k. Accordingly, we now define the **discrete Fourier transform (DFT)** $X[k]$ as the N values of $\tilde{X}[k]$, $0 \le k \le N-1$, shown in the red box in **Figure 10.2e**, which were obtained by sampling the DTFT at N equally spaced intervals in the frequency range $0 \leq \omega < 2\pi$:

$$X[k] \triangleq \tilde{X}[k], \quad 0 \leq k < N. \tag{10.8}$$

From Equations (10.6) and (10.8), we can finally write the DFT $X[k]$ directly in terms of $x[n]$:

$$\boxed{X[k] = \mathbb{F}_N\{x[n]\} \triangleq X(\omega)|_{\omega=2\pi k/N} = \sum_{n=0}^{N-1} x[n]e^{-j2\pi kn/N}, \quad 0 \leq k \leq N-1.} \tag{10.9}$$

10.1.1 The inverse discrete Fourier transform (IDFT)

The DFT is an invertible transform; that is, we can recover the N values of $x[n]$ directly from the N values of $X[k]$. First consider recovering the periodic sequence $\tilde{x}[n]$ from $\tilde{X}(\omega)$,

$$\tilde{x}[n] \triangleq \mathfrak{F}^{-1}\{\tilde{X}(\omega)\} = \frac{1}{2\pi}\int_{\omega=0}^{2\pi} \tilde{X}(\omega)e^{j\omega n}d\omega = \frac{1}{2\pi}\int_{\omega=0}^{2\pi}\left(\omega_s \sum_{k=-\infty}^{\infty} \tilde{X}[k]\delta(\omega - k\omega_s)\right)e^{j\omega n}d\omega.$$

Here, we have substituted the expression for $\tilde{X}(\omega)$ in terms of $\tilde{X}[k]$ using Equation (10.7). Because only the N impulses of $\delta(\omega - k\omega_s)$ for $0 \leq k < N$ fall inside the 2π range of frequency of the integral, the range of the summation can be reduced to $0 \leq k < N$,

$$\begin{aligned}
\tilde{x}[n] &= \frac{1}{2\pi}\int_{\omega=0}^{2\pi}\left(\omega_s \sum_{k=0}^{N-1} \tilde{X}[k]\delta(\omega - k\omega_s)\right)e^{j\omega n}d\omega \\
&= \frac{\omega_s}{2\pi}\sum_{k=0}^{N-1}\tilde{X}[k]\int_{\omega=0}^{2\pi}\delta(\omega - k\omega_s)e^{j\omega n}d\omega \\
&= \frac{1}{N}\sum_{k=0}^{N-1}\tilde{X}[k]e^{jk\omega_s n}.
\end{aligned}$$

Equation (10.8) states that over the range of $0 \leq k < N$, $\tilde{X}[k] = X[k]$, so

$$\tilde{x}[n] = \frac{1}{N}\sum_{k=0}^{N-1}\tilde{X}[k]e^{jk\omega_s n} = \frac{1}{N}\sum_{k=0}^{N-1}X[k]e^{j2\pi kn/N}. \tag{10.10}$$

Clearly, $x[n]$ can be recovered from $\tilde{x}[n]$, as shown with the red box in **Figure 10.1c**. We define the recovered sequence $x_r[n]$ as the points of $\tilde{x}[n]$ in the range $0 \leq n < N$,

$$x_r[n] \triangleq \tilde{x}[n](u[n] - u[n-N]) = \begin{cases} \tilde{x}[n], & 0 \leq n \leq N-1 \\ 0, & \text{otherwise} \end{cases}. \tag{10.11}$$

In the example, $x_r[n]$ is equivalent to $x[n]$ with $N - L = 2$ zeros appended to the end. Therefore, as long as $N \geq L$, $x[n]$ can be uniquely recovered from $\tilde{x}[n]$ and we can just state that $x[n] = x_r[n]$. Finally, Equations (10.10) and (10.11) can be combined to give $x[n]$ computed as a finite sum of terms,

$$\boxed{x[n] = x_r[n] = \mathbb{F}_N^{-1}\{X[k]\} = \frac{1}{N}\sum_{k=0}^{N-1} X[k]e^{j2\pi kn/N}, \quad 0 \le n \le N-1.} \tag{10.12}$$

This equation defines the **inverse DFT** or **IDFT**.

Our derivation of the DFT and IDFT has proceeded through a number of periodic intermediates – $\tilde{x}[n]$, $\tilde{X}(\omega)$ and $\tilde{X}[k]$ – shown in the colored box in **Figure 10.3**. In our discussion of the DFT and its properties in subsequent sections – particularly the shifting and convolution properties – it is going to be important to remember that although both $x[n]$ and $X[k]$ are finite-length sequences, they actually should be thought of as sequences of length N that have been produced windowing the periodic sequences $\tilde{x}[n]$ and $\tilde{X}[k]$, respectively, to the range $0 \le n \le N-1$.

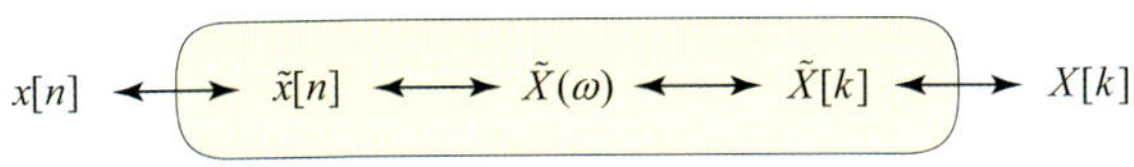

Figure 10.3 DFT and IDFT

Example 10.1

Calculate $X[k]$, the DFT of the sequence $x[n]$ shown in **Figure 10.1d**, and show that the IDFT of $X[k]$ is $x[n]$.

➤ Solution:

It is pointless to do this by hand. This is why we have computers. We can easily translate Equations (10.9) and (10.12) into a pair of Matlab functions that compute the DFT and IDFT:

```
function X = dft(x)
    N = length(x);
    X = exp(-1j*2*pi*(0:N-1)' * (0:N-1)/N) * x(:));
end

function x = idft(X)
    N = length(X);
    x = exp(1j*2*pi*(0:N-1)' * (0:N-1)/N) * X(:)/N;
end
```

Then,

```
>> X = dft([ones(1, 6) 0 0]);
>> x = idft(X);
>> disp('          X[k]                x[n]'); disp([X x])
          X[k]                x[n]
   6.0000 + 0.0000i    1.0000 + 0.0000i
  -0.7071 - 1.7071i    1.0000 - 0.0000i
   1.0000 - 1.0000i    1.0000 + 0.0000i
```

```
 0.7071 + 0.2929i   1.0000 + 0.0000i
 0.0000 - 0.0000i   1.0000 - 0.0000i
 0.7071 - 0.2929i   1.0000 - 0.0000i
 1.0000 + 1.0000i   0.0000 - 0.0000i
-0.7071 + 1.7071i   0.0000 - 0.0000i
```

In Chapter 11, we will show how the DFT can be implemented via the fast Fourier transform (FFT) algorithm. The Matlab functions `fft` and `ifft` provide extremely efficient implementations of the DFT and IDFT, respectively. The syntax of the `fft` and `ifft` functions is similar:

```
X = fft(x, [N])
x = ifft(X, [N])
```

The second, optional argument (`N`) of `fft` and `ifft` sets the desired size of the DFT or IDFT. If `N` is larger than the size of the input array, `fft` or `ifft` automatically pads the array with zeros to fill the array out to `N` points; so in Example 10.1, we could have done the following:

```
>> X = fft(ones(1, 6), 8);
>> x = ifft(X);
```

You can verify that the results are identical to those obtained using our home-brewed functions in Example 10.1, so for the remaining examples in the chapter, we will use `fft` or `ifft` whenever we need to compute a DFT or IDFT.

10.1.2 Orthogonality of complex exponential sequences

The complex exponential sequences $e^{-j2\pi kn/N}$ are **orthogonal** with respect to both k and n. The definition of orthogonality with respect to k is

$$\frac{1}{N}\sum_{n=0}^{N-1} e^{-j2\pi kn/N} e^{j2\pi ln/N} = \frac{1}{N}\sum_{n=0}^{N-1} e^{-j2\pi(k-l)n/N} = \frac{1}{N}\sum_{n=0}^{N-1}\left(\frac{1-e^{-j2\pi(k-l)}}{1-e^{-j2\pi(k-l)/N}}\right)$$
$$=\left\{\begin{array}{ll}1, & (k-l)=0,\ \pm N,\ \pm 2N,\ \ldots \\ 0, & \text{otherwise}\end{array}\right\} = \delta[(k-l)_N], \tag{10.13}$$

where $(k-l)_N$ means "$(k-l)$ modulo N." It is this orthogonality of complex exponential sequences that results in the invertibility of the DFT and IDFT. To show this, plug the candidate expression for the IDFT, $x[n]$, into the formula for the DFT and discover that you do, in fact, get the DFT $X[k]$:

$$\sum_{n=0}^{N-1} x[n]e^{-j2\pi kn/N} = \sum_{n=0}^{N-1}\left(\frac{1}{N}\sum_{l=0}^{N-1} X[l]e^{-j2\pi ln/N}\right)e^{-j2\pi k/N} = \sum_{l=0}^{N-1} X[l]\left(\frac{1}{N}\sum_{n=0}^{N-1} e^{-j2\pi(k-l)n/N}\right)$$
$$=\sum_{l=0}^{N-1} X[l]\delta[(k-l)_N] = X[k].$$

10.1.3 Periodicity of the DFT

Just as $X(\omega)$ is periodic with a period of 2π, so $X[k]$ is periodic with a period of N,

$$X[k+N]=\sum_{n=0}^{N-1}x[n]e^{-j2\pi(k+N)n/N}=\sum_{n=0}^{N-1}x[n]e^{-j2\pi kn/N}\cancel{e^{-j2\pi Nn/N}}=\sum_{n=0}^{N-1}x[n]e^{-j2\pi kn/N}$$
$$=X[k].$$

However, $X[k+N]$ actually exceeds the limits, $0\le k<N$, of the definition of the DFT in Equation (10.9). So, to be strictly accurate we should write

$$X[(k+N)_N]=X[k].$$

Again, the way to think about $X[(k+N)_N]$ is that it is equal to the sequence $\tilde{X}[k+N]$ that has been windowed to the values $0\le k<N$. Since $\tilde{X}[k]$ is defined to be periodic and to have value for *all* k, then $\tilde{X}[k+N]=\tilde{X}[k]$, and so, by Equation (10.8),

$$X[(k+N)_N]=\tilde{X}[k+N](u[k]-u[k-N])=\tilde{X}[k](u[k]-u[k-N])=X[k].$$

10.1.4 ★ Conditions for the reconstruction of a sequence from the DFT

In order to reconstruct $x[n]$ uniquely from $X[k]$ via the IDFT, the size of the DFT must be greater than or equal to the length of $x[n]$. In supplementary material available at www.cambridge.org/holton, we review the development of the DFT from the perspective of a problem of sampling the DTFT in the frequency domain and show that when the DTFT is undersampled, the result is **time-domain aliasing** of $x[n]$. Time-domain aliasing is a concept that will make a reappearance in Section 10.3.7, where we introduce the notion of circular convolution, and show that it can be interpreted in terms of time-domain aliasing.

10.2 DFT of basic signals

In order to gain some insight into the DFT, we now present the DFT of some basic signals.

10.2.1 DFT of an impulse

The N-point DFT of an impulse $x[n]=\delta[n-n_0]$, where $0\le n\le N-1$ and $0\le n_0\le N-1$, is

$$X[k]=\sum_{n=0}^{N-1}\delta[n-n_0]e^{-j2\pi kn/N}=e^{-j2\pi kn_0/N}.$$

So,

$$|X[k]|=1$$
$$\measuredangle X[k]=-2\pi kn_0/N.$$

Figure 10.4a shows $x[n]$ (left panel) and $X[k]$ (right panel) for the example of $n_0=2$ and $N=16$. The grey shaded region indicates the range of N over which the DFT is computed. The unfilled symbols remind us that the DFT is the windowed transform of a periodic sequence with period N. The thin red line shows the DTFT $X(\omega)$ over the frequency range $0\le\omega<2\pi$, so you can appreciate that the DFT is the DTFT sampled at N points.

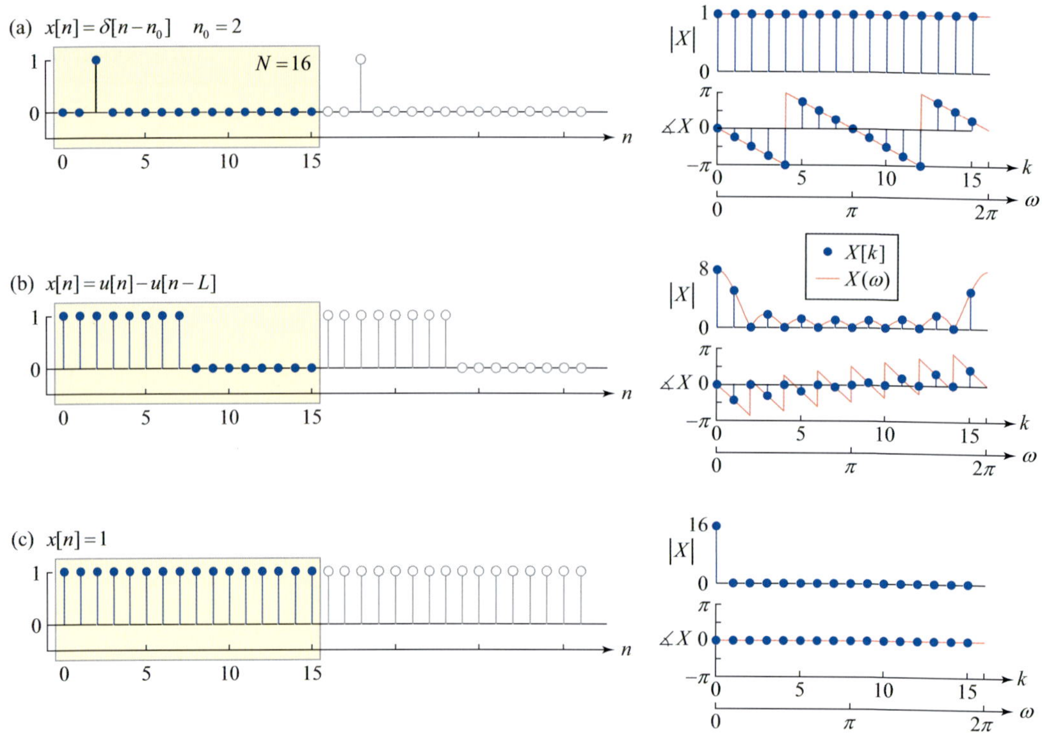

Figure 10.4 DFT of basic signals

10.2.2 DFT of a pulse

Define the DTFT of a pulse of length L,

$$\Psi_L(\omega) \triangleq \mathfrak{F}\{u[n]-u[n-L]\} = \sum_{n=0}^{L-1} e^{-j\omega n} = \frac{1-e^{-j\omega L}}{1-e^{-j\omega}} = \frac{e^{-j\omega L/2}}{e^{-j\omega/2}} \frac{e^{j\omega L/2}-e^{-j\omega L/2}}{e^{j\omega/2}-e^{-j\omega/2}}$$
$$= e^{j\omega(L-1)/2} \frac{\sin \omega L/2}{\sin \omega/2} \quad = e^{j\omega(N-1)/2} L \frac{\operatorname{sinc} \omega L/2}{\operatorname{sinc} \omega/2}. \tag{10.14}$$

This is the periodic sinc function that we first saw in Chapter 3, and which we will need in several places in this chapter. Like all DTFTs, this function is periodic with period 2π: $\Psi_L(\omega+2\pi) = \Psi_L(\omega)$.

Now use $\Psi_L(\omega)$ to find $X[k]$, the N-point DFT of a pulse of length $L < N$, $x[n] = u[n] - u[n-L]$, $0 \le n \le N-1$, which is shown in the left panel of **Figure 10.4b** for the example of $L = 8$. From Equation (10.14), the DFT $X[k]$ is obtained by sampling $\Psi_L(\omega)$ at values of $\omega = 2\pi k/N$,

$$X[k] = \Psi_L(\omega)|_{\omega=2\pi k/N} = e^{j\pi k(L-1)/N} L \frac{\operatorname{sinc} \pi kL/N}{\operatorname{sinc} \pi k/N}, \quad 0 \le k \le N-1, \tag{10.15}$$

as shown in the right panel.

10.2.3 DFT of a constant

For the case where $x[n]$ is a constant, $x[n]=1$, $0 \le n \le N-1$. We can view this as a pulse of length N,

$$\Psi_N(2\pi k/N) = \sum_{n=0}^{N-1} e^{-j2\pi kn/N} = e^{j\pi k(N-1)/N} N \frac{\operatorname{sinc} \pi k}{\operatorname{sinc} \pi k/N} = \begin{Bmatrix} N, & k=0 \\ 0, & k \ne 0 \end{Bmatrix} = N\delta[k], \qquad \textbf{(10.16)}$$

where we have recognized that when $k \ne 0$, $\operatorname{sinc} 0 = 1$. **Figure 10.4c** shows the response for the case $N = 16$.

10.2.4 DFT of a complex exponential sequence

Consider the N-point complex exponential sequence

$$x[n] = e^{j2\pi ln/N}, \quad 0 \le n \le N-1,$$

where l is an integer. The DFT of $x[n]$ follows from the orthogonality of complex exponentials, Equation (10.13),

$$X[k] = \sum_{n=0}^{N-1} e^{j2\pi ln/N} e^{-j2\pi kn/N} = \sum_{n=0}^{N-1} e^{-j2\pi(k-l)n/N} = N\delta[(k-l)_N], \quad 0 \le k \le N-1.$$

You can also get the same result from Equation (10.16) by viewing $X[k]$ as samples of the periodic sinc function $\Psi_N(\omega)$ at frequencies $\omega = 2\pi(k-l)/N$,

$$\begin{aligned} X[k] &= \Psi_N(2\pi(k-l)/N) = e^{-j\pi(k-l)(N-1)/N} N \frac{\operatorname{sinc} \pi(k-l)}{\operatorname{sinc} \pi(k-l)/N} \\ &= \begin{Bmatrix} N, & k-l=0, \pm N, \pm 2N, \ldots \\ 0, & \text{otherwise} \end{Bmatrix} = N\delta[(k-l)_N], \quad 0 \le k \le N-1, \end{aligned} \qquad \textbf{(10.17)}$$

10.2.5 DFT of a sinusoid

One of the most important practical applications of the DFT is **spectral analysis**, in which the energy of a signal is decomposed into its individual frequency components. We will devote an entire chapter (Chapter 14) to this topic, but, for now, let us look at just some of the issues that result from using the DFT to analyze a simple signal, a cosine of frequency f Hz and phase θ, that is sampled at sampling frequency f_s Hz: $x(t) = \cos(2\pi ft + \theta)$. First, create a sequence $x[n]$ by obtaining N points of $x(t)$, at a sampling period of $T = 1/f_s$,

$$x[n] = x(nT) = \cos(2\pi(f/f_s)n + \theta) = \cos(\omega_0 n + \theta), \quad 0 \le n \le N-1. \qquad \textbf{(10.18)}$$

To get some insight into some of the problems that arise from the selection of parameters, we will look at a couple of examples.

Example 10.2

A sequence $x[n]$ is created by sampling $N = 16$ points of a continuous-time signal $x(t) = \cos(2\pi ft + \theta)$. Find and plot the DFT $X[k]$ given the following parameters for $x(t)$:

(a) $f = 1000$ Hz, $\theta = -\pi/2$, $f_s = 8000$.

(b) $f = 1500$ Hz, $\theta = 0$, $f_s = 8000$.

➤ Solution:

(a) Equation (10.18) gives

$$x[n] = \cos(\omega_0 n + \theta) = \cos(\pi n/4 - \pi/2), \quad 0 \le n \le N-1.$$

The blue symbols in the yellow shaded area of the left panel of **Figure 10.5a** show the N points of $x[n]$. However, in this case, we can find the DFT directly using Equation (10.17), because ω_0 is an integer multiple of $2\pi/N$. For the general case, $\omega_0 = 2\pi l n/N$, where l is an integer, write

$$x[n] = \cos(2\pi l n/N + \theta) = \frac{1}{2}\left(e^{j(2\pi l n/N+\theta)} + e^{-j(2\pi l n/N+\theta)}\right) = \frac{1}{2}e^{j\theta}e^{j2\pi l n/N} + \frac{1}{2}e^{-j\theta}e^{-j2\pi l n/N}.$$

Then, by Equation (10.17),

$$X[k] = \frac{N}{2}e^{j\theta}\delta[(k-l)_N] + \frac{N}{2}e^{-j\theta}\delta[(k+l)_N]. \tag{10.19}$$

In this particular example, $\omega_0 = \pi/4 = 2\pi/8 = 2\cdot(2\pi/N)$, so $l = 2$ and $\theta = -\pi/2$, and Equation (10.19) gives

$$\begin{aligned} X[k] &= \frac{N}{2}e^{-j\pi/2}\delta[(k-2)_N] + \frac{N}{2}e^{j\pi/2}\delta[(k+2)_N] = \frac{N}{2}e^{-j\pi/2}\delta[k-2] + \frac{N}{2}e^{j\pi/2}\delta[k-(N-2)] \\ &= 8e^{-j\pi/2}\delta[k-2] + 8e^{j\pi/2}\delta[k-14] \end{aligned} \tag{10.20}$$

As shown in the right panel of **Figure 10.5a**, the DFT comprises only a pair of impulses of magnitude 8 and phase $\pm\pi/2$ (blue symbols), at $k = 2$ and 14.

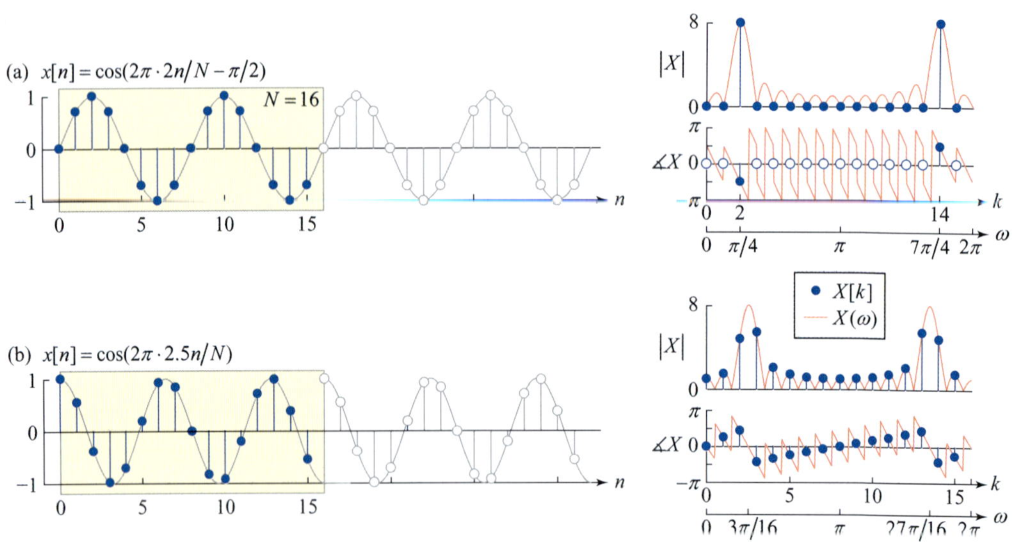

Figure 10.5 DFT of sinusoids

You might be thinking, "Of course, the transform of a cosine is a pair of impulses," but it is important to understand there is actually more going on here. The DFT in this example needs to be clearly understood as the *sampled* DTFT of a *windowed* cosine. Because we are only selecting N points of the sampled cosine, the actual signal we are sampling is NT seconds of a cosine,

$$x(t) = \cos(2\pi f t + \theta)\,(u(t) - u(t - NT)),$$

which leads to the N-point sequence of Equation (10.18), $x[n] = x(nT) = \cos(\omega_0 n + \theta),\ 0 \le n \le N-1$. You can see from the left panel of **Figure 10.5a** that because $\omega_0 = \pi/4 = 2\pi/8 = 2 \cdot (2\pi/N)$, we are sampling $x(t)$ at *exactly* an integer multiple (namely two) of $2\pi/N$. That means that the $N = 16$ points of $x[n]$ correspond to precisely two periods of $x(t)$. You can show (Problem 10-20a) that the DTFT of $x[n]$ is

$$X(\omega) = \frac{1}{2}(e^{j\theta}\Psi_{16}(\omega - \omega_0) + e^{-j\theta}\Psi_{16}(\omega + \omega_0)), \tag{10.21}$$

where $\Psi_{16}(\omega)$ is the periodic sinc function defined in Equation (10.14) evaluated at $N = 16$. In our example, with $\omega_0 = \pi/4$ and $\theta = -\pi/2$, the DTFT is shown in the thin red traces in the right panels of **Figure 10.5a**. Since the DFT $X[k]$ is the DTFT sampled at multiples of $2\pi/N$, you can show (Problem 10-20b) that Equation (10.21) becomes

$$\begin{aligned} X[k] &= X(\omega)|_{\omega = 2\pi k/N} = \frac{1}{2}(e^{-j\pi/2}\Psi_{16}(2\pi k/16 - \pi/4) + e^{j\pi/2}\Psi_{16}(2\pi k/16 + \pi/4)) \\ &= \frac{1}{2}e^{-j\pi/2}\Psi_{16}(2\pi(k-2)/16) + \frac{1}{2}e^{j\pi/2}\Psi_{16}(2\pi(k+2)/16) \\ &= 8e^{-j\pi/2}\delta[k-2] + 8e^{j\pi/2}\delta[k-14], \end{aligned}$$

which is identical to Equation (10.20). The key point here is that when the frequency of the cosine is an integer multiple of $2\pi/N$, namely $\omega_0 = 2\pi l/N$, then there are only two non-zero values of the DFT $X[k]$ corresponding to samples of $X(\omega)$ at $\omega_0 = 2\pi l/N$ and $2\pi(N-l)/N$.

Another useful way of understanding this result is to consider the DFT of a cosine $x[n]$ as N points of the DTFT of the infinite-length periodic sequence $\tilde{x}[n]$, which comprises identical periods of $x[n]$, as we did in Section 10.1. When ω_0 is an integer multiple of $2\pi/N$, then $\tilde{x}[n]$ is the pure cosine shown with both solid and open symbols in the left panel of **Figure 10.5a**, with no discontinuities between periods; hence, the transform comprises just two impulses.

(b) This example shows what happens when ω_0 is *not* an integer multiple of $2\pi/N$. The shaded portion of the left panel of **Figure 10.5b** shows the $N = 16$ points of $x[n]$ for the case $\omega_0 = 5\pi/16$ and $\theta = 0$. The DTFT $X(\omega)$ is shown in the thin red traces in the right panels of **Figure 10.5b**. The DFT $X[k]$ is now not simply two impulses; the samples of $X(\omega)$ that correspond to $X[k]$ have non-zero value of *all* values of k. If you interpret the DFT as the DTFT of an infinite-length periodic sequence $\tilde{x}[n]$, in this case there is a significant discontinuity between periods of $\tilde{x}[n]$ and, as a consequence, the transform has considerable energy at all N values of k.

10.2.6 Resolution and frequency mapping of the DFT

Example 10.2 brings up an important question: what is the relation between the DFT component and the actual frequency component of a sampled signal? In Section 10.1, we showed that the DFT samples the DTFT at N points, so that the kth DFT coefficient corresponds to $\omega = 2\pi k/N$. As discussed in Chapter 6, when a continuous-time signal of frequency f Hz (or $\Omega = 2\pi f$ rad/sec), is sampled at rate f_s (or $\Omega_s = 2\pi f_s$), the equivalent discrete-time frequency of the DTFT is $\omega = 2\pi\,\Omega/\Omega_s = 2\pi f/f_s$ rad. Equating these two, we get $\omega = 2\pi k/N = 2\pi\,\Omega/\Omega_s = 2\pi f/f_s$, or

$$f = k(f_s/N) \text{ or } \Omega = k(\Omega_s/N).$$

The **frequency resolution** of the DFT is the difference in frequency between adjacent coefficients of the transform,

$$\Delta f \triangleq f_s/N \text{ or } \Delta\Omega \triangleq \Omega_s/N. \tag{10.22}$$

Hence,

$$f = k\Delta f \text{ or } \Omega = k\Delta\Omega.$$

So, in Example 10.2, $\Delta f = 8000/16 = 50$ Hz and the coefficient $k = 2$ corresponds to the frequency 100 Hz. One of the key issues in spectral analysis is choosing a transform size adequate to obtain a required frequency resolution. We shall cover issues like this in detail in Chapter 14.

10.2.7 Summary (so far)

Let us pause a moment to summarize the results so far. We introduced the forward and inverse discrete Fourier transforms, the DFT and IDFT, respectively. For a finite-length sequence $x[n]$, the DFT coefficients are equal to N samples of the sequence's DTFT,

$$X[k] = X(\omega)|_{\omega = 2\pi k/N}.$$

A sequence of length $L \leq N$ can be completely reconstructed from $N \geq L$ DFT coefficients, yielding the relations

$$\begin{array}{|lll|} \text{DFT}: & X[k] = \mathbb{F}_N\{x[n]\} \triangleq \sum_{n=0}^{N-1} x[n]e^{-j2\pi kn/N}, & 0 \leq k < N \\ \text{IDFT}: & x[n] = \mathbb{F}_N^{-1}\{X[k]\} \triangleq \frac{1}{N}\sum_{n=0}^{N-1} X[k]e^{j2\pi kn/N}, & 0 \leq n < N \end{array}. \tag{10.23}$$

We have explicitly set the limits of the summation to $0 \leq n < N$ in the forward DFT to emphasize this fact. However, in thinking about the DFT, we emphasize yet again that it is important to keep in mind that $X[k]$ should be viewed as N values of the infinite-length periodic sequence $\tilde{X}[k]$, which has been windowed to keep only the values between $0 \leq k < N$,

$$X[k] = \tilde{X}[k], \quad 0 \leq k < N.$$

Both the DTFT and DFT provide unique and invertible transformations of finite-length sequences. If you compare the expressions for the DTFT, Equation (10.1), and the DFT, Equation (10.23), the practical significance of the DFT becomes clear. With the DTFT, the forward transform $X(\omega)$ is a continuous function of variable ω. Computation of the inverse transform from $X(\omega)$ requires the evaluation of an integral. In contrast, with the DFT, there are no continuous variables or integrals. The forward transform comprises a discrete finite-length sequence of coefficients $X[k]$. Computation of the inverse transform from $X[k]$ also only requires the evaluation of a finite summation. Both the DFT and IDFT are easily implemented on a computer. In Chapter 11, we will discuss a family of ingenious computational algorithms, the fast Fourier transform (FFT), that have been developed to implement the DFT efficiently in computer applications.

10.3 Properties of the DFT

There are many DFT properties, most of them easily derivable from the equivalent DTFT properties. We will only consider a few important properties here that will lead us to practically useful algorithms later.

10.3.1 Linearity

The linearity property states that the DFT of the scaled sum of inputs is the scaled sum of their DFTs:

$$\mathbb{F}_N\{ax_1[n]+bx_2[n]\}=a\mathbb{F}_N\{x_1[n]\}+b\mathbb{F}_N\{x_2[n]\}=aX_1[k]+bX_2[k],$$

where a and b are constants. You can easily derive this from Equation (10.23).

10.3.2 Complex conjugation

$$\begin{aligned}\mathbb{F}\{x^*[n]\} &=\sum_{n=0}^{N-1}x^*[n]e^{-j2\pi kn/N}=\sum_{n=0}^{N-1}\left(x[n]e^{j2\pi kn/N}\right)^*=\left(\sum_{n=0}^{N-1}x[n]e^{-j2\pi(-k)n/N}\right)^*\\ &=X^*[(-k)_N]=X^*[N-k], \qquad 0\le k\le N-1.\end{aligned} \tag{10.24}$$

We have noted that $X^*[-k]$ does not exist for negative k but, by the periodicity property,

$$X^*[-k]=X^*[(-k)_N]=X^*[N-k],\ \ 0\le k\le N-1.$$

10.3.3 Symmetry properties of the DFT

The symmetry properties of the DFT are important to understand, not just because they add insight into the way the DFT works, but because they lead directly to the development of fast algorithms such as the FFT. We are particularly interested in the properties that apply to real sequences.

Real sequences If $x[n]$ is real, then $x[n]=x^*[n]$. Hence, by the conjugation property, $\mathbb{F}_N\{x[n]\}=\mathbb{F}_N\{x^*[n]\}$, so by Equation (10.24),

$$X[k]=X^*[N-k],\ \ 0\le k\le N-1. \tag{10.25}$$

From this it also follows that

$$\begin{aligned}|X[k]|=|X^*[N-k]|=|X[N-k]|\\ \measuredangle X[k]=\measuredangle X^*[N-k]=-\measuredangle X[N-k]\end{aligned},\quad 0\le k\le N-1. \tag{10.26}$$

The magnitude is an even function of k and the phase is an odd function of k. Also,

$$\begin{aligned}\mathrm{Re}\{X[k]\}=\mathrm{Re}\{X^*[N-k]\}=\mathrm{Re}\{X[N-k]\}\\ \mathrm{Im}\{X[k]\}=\mathrm{Im}\{X^*[N-k]\}=-\mathrm{Im}\{X[N-k]\}\end{aligned}.$$

This set of results has some practical consequences. In general, the DFT of an arbitrary, possibly complex, N-point sequence $x[n]$ has N real values and N imaginary values, and therefore requires $2N$ storage locations. However, some interesting results fall out of the symmetry relations if we know that $x[n]$ is purely real. When $x[n]$ is real, Equation (10.25) gives $X[0]=X^*[N]$. But the periodicity property guarantees that $X^*[N]=X^*[0]$, so $X[0]=X^*[0]$, which means that the first DFT coefficient has to be real. Furthermore, if we choose N to be even (as we will do in many applications from here on), then from Equation (10.25),

$$X[N/2] = X^*[N - N/2] = X^*[N/2],$$

which means the "center point" of an even-length DFT must also be real. This center point corresponds to the DTFT frequency $\omega = (2\pi/N) \cdot (N/2) = \pi$. The remaining $N - 2$ points of the DFT (i.e., $X[1], \ldots X[N-1]$) are complex, but only half $(N/2 - 1)$ of these points $(X[1], \ldots X[N/2 - 1])$ contain unique information that is necessary to specify the entire DFT. The remaining $N/2 - 1$ values $(X[N/2 + 1], \ldots X[N-1])$ can be derived from the fact that $X[N-k] = X^*[k]$. This result has consequences for the amount of computer memory necessary to store the DFT. The DFT of an N-point real sequence, where N is even, can be completely specified by N elements of the total storage: two real coefficients, each of which requires one memory storage element, plus $N/2 - 1$ complex coefficients, each of which requires two memory storage elements, one for the real part and one for the imaginary part:

$$\underbrace{2}_{\substack{\text{\#real}\\\text{coefficients}}} \cdot \underbrace{1}_{\substack{\text{storage}\\\text{required}\\\text{for real}\\\text{coefficient}}} + \underbrace{(N/2 - 1)}_{\substack{\text{\#complex}\\\text{coefficients}}} \cdot \underbrace{2}_{\substack{\text{storage}\\\text{required}\\\text{for complex}\\\text{coefficient}}} = N.$$

If N is real and of odd length, $X[0]$ is still real, and we only need to compute half of the remaining $N - 1$ coefficients, since $X[N-k] = X^*[k]$ $1 \le k \le (N+1)/2$. Hence, for an odd real sequence, we have

$$\underbrace{1}_{\substack{\text{\#real}\\\text{coefficient}}} \cdot \underbrace{1}_{\substack{\text{storage}\\\text{required}\\\text{for real}\\\text{coefficient}}} + \underbrace{(N - 1)/2}_{\substack{\text{\#complex}\\\text{coefficients}}} \cdot \underbrace{2}_{\substack{\text{storage}\\\text{required}\\\text{for complex}\\\text{coefficient}}} = N.$$

This is a satisfying result: real sequences of length N, whether they are of even or odd length, require N storage elements in either the time or frequency domain; complex sequences need $2N$ elements in both domains.

The following example shows the relation of the DFT coefficients for a real sequence.

Example 10.3

Find the DFT coefficients $X[k]$ for the real sequence shown in **Figure 10.6**. Show that the symmetry relations apply.

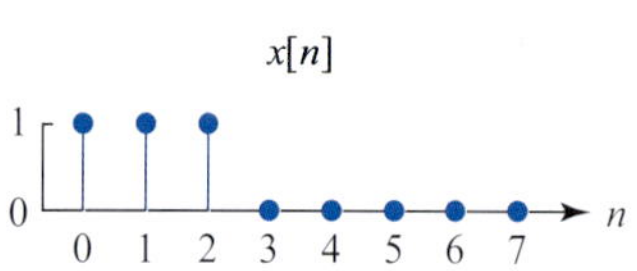

Figure 10.6

➤ **Solution:**

$$X[k] = \sum_{n=0}^{N-1} x[n]e^{-j2\pi kn/N} = \sum_{n=0}^{7} e^{-j\pi kn/4} = 1 + e^{-j\pi k/4} + e^{-j\pi k/2} = e^{-j\pi k/4}(e^{j\pi k/4} + 1 + e^{-j\pi k/4})$$
$$= e^{-j\pi k/4}(1 + 2\cos k\pi/4).$$

The values are as follows:

k	$X[k]$	$\lvert X[k]\rvert$	$\measuredangle X[k]$
0	3	3	0
1	$(1+\sqrt{2}/2)(1-j)$	$\sqrt{2}+1$	$-\pi/4$
2	$-j$	1	$-\pi/2$
3	$(1-\sqrt{2}/2)(1+j)$	$\sqrt{2}-1$	$\pi/4$
4	1	1	0
5	$(1-\sqrt{2}/2)(1-j)$	$\sqrt{2}-1$	$-\pi/4$
6	j	1	$\pi/2$
7	$(1+\sqrt{2}/2)(1+j)$	$\sqrt{2}+1$	$\pi/4$

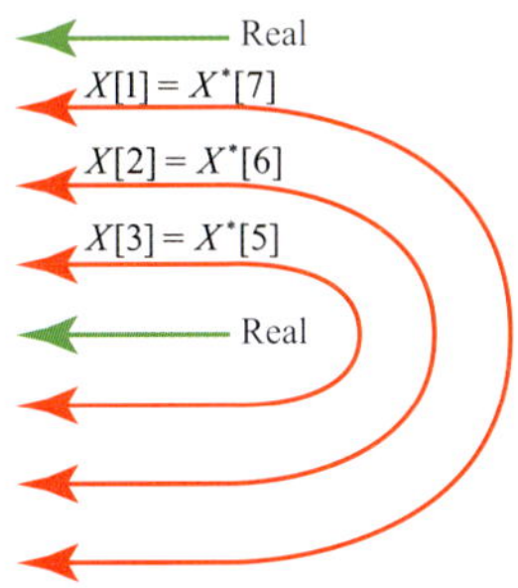

$$\lvert X[1]\rvert = \lvert X[7]\rvert, \measuredangle X[1] = -\measuredangle X[7]$$
$$\lvert X[2]\rvert = \lvert X[6]\rvert, \measuredangle X[2] = -\measuredangle X[6]$$
$$\lvert X[3]\rvert = \lvert X[5]\rvert, \measuredangle X[3] = -\measuredangle X[5]$$

As you can see, the coefficients $X[0]$ and $X[4] = X[N/2]$ are real. The remaining terms appear as complex-conjugate pairs, as indicated in Equation (10.25): $X[5] = X^*[3]$, $X[6] = X^*[2]$ and $X[7] = X^*[1]$. You can also see from the table that $\lvert X[k]\rvert = \lvert X[N-k]\rvert$ and $\measuredangle X[k] = -\measuredangle X[N-k]$, as indicated in Equation (10.26).

For a real sequence, the DFT coefficients have some intuitive meaning. To see this, express $x[n]$ as the IDFT of $X[k]$ and expand and regroup terms,

$$x[n] = \frac{1}{N}\sum_{k=0}^{N-1} X[k]e^{j2\pi kn/N}$$
$$= \frac{1}{N}\left\{ X[0]e^{j2\pi 0n/N} + X[N/2]e^{j2\pi(N/2)n/N} + \sum_{k=1}^{N/2-1} X[k]e^{j2\pi kn/N} + \sum_{k=N/2+1}^{N-1} X[k]e^{j2\pi kn/N} \right\}.$$

Let $l = N - k$ in the last summation:

$$\sum_{k=N/2+1}^{N-1} X[k]e^{j2\pi kn/N} = \sum_{l=N/2-1}^{1} X[N-l]e^{j2\pi(N-l)n/N} = \sum_{l=N/2-1}^{1} X[N-l]e^{j2\pi \cancel{N}n/\cancel{N}} \cdot e^{-j2\pi ln/N}$$
$$= \sum_{l=1}^{N/2-1} X^*[l]e^{-j2\pi ln/N} = \sum_{l=1}^{N/2-1} \left(X[l]e^{j2\pi ln/N} \right)^*,$$

where we have used the fact that $X[N-l]=X^*[l]$. Hence the entire IDFT is

$$\begin{aligned}
x[n] &= \frac{1}{N}\sum_{k=0}^{N-1} X[k]e^{j2\pi kn/N} \\
&= \frac{1}{N}\left\{ X[0]e^{j2\pi 0n/N} + X[N/2]e^{j2\pi(N/2)n/N} + \sum_{k=1}^{N/2-1} X[k]e^{j2\pi kn/N} + \sum_{k=1}^{N/2-1}\left(X[k]e^{j2\pi kn/N}\right)^* \right\} \\
&= \frac{1}{N}\left\{ X[0] + X[N/2](-1)^n + \sum_{k=1}^{N/2-1}\left(X[k]e^{j2\pi kn/N} + (X[k]e^{j2\pi kn/N})^*\right) \right\} \\
&= \frac{1}{N}\left\{ X[0] + X[N/2](-1)^n + 2\sum_{k=1}^{N/2-1} \operatorname{Re}\left\{X[k]e^{j2\pi kn/N}\right\} \right\} \\
&= \frac{1}{N}\left\{ X[0] + X[N/2](-1)^n + 2\sum_{k=1}^{N/2-1} \operatorname{Re}\left\{|X[k]|e^{j(2\pi kn/N + \measuredangle X[k])}\right\} \right\} \\
&= \frac{1}{N}\left\{ X[0] + X[N/2](-1)^n + 2\sum_{k=1}^{N/2-1} |X[k]|\cos(2\pi kn/N + \measuredangle X[k]) \right\}.
\end{aligned} \tag{10.27}$$

This expresses a real $x[n]$ as the sum of $N/2+1$ real terms. The $X[0]$ term is the constant ("DC") term that corresponds to the frequency $\omega=0: X[0]=X[0]\cos 0n$. The $X[N/2]$ term is the highest frequency term, and corresponds to the frequency $\omega=\pi: X[N/2](-1)^n = X[N/2]\cos \pi n$. The remaining $N/2-1$ terms, for $1\le k\le N/2-1$, are cosines of frequency $2\pi k/N$, with magnitude $(2/N)|X[k]|$, and phase $\measuredangle X[k]$.

Example 10.4

Show that the sequence $x[n]$ of Example 10.3 can be reconstructed from the IDFT as the sum of cosine terms using Equation (10.27) with $N=8$.

➤ Solution:

Figure 10.7a shows the sequence $x[n]$.

The colored dots in **Figure 10.7b** show the magnitude and phase of the DFT $X[k]$, computed with $N=8$. The magnitude and phase of the DTFT $X(\omega)$ are also shown superimposed in thin continuous lines where $k=0$ corresponds to $\omega=0$ and $k=8$ corresponds to $\omega=2\pi$. (A value of the DFT for $k=8$ is shown just to make the plot look symmetrical: $X[8]=X[0]$.)

Since $x[n]$ is real, the coefficients $X[0]$ (coded with the ● dot) and $X[4]$ (○) are real; therefore, their phase is 0. The remaining terms occur in complex-conjugate pairs, each of which is coded with the same color: $X[1]=X^*[7]$ (●), $X[2]=X^*[6]$ (○) and $X[3]=X^*[5]$ (●). You can see from the graph that $|X[k]|=|X[N-k]|$ and $\measuredangle X[k] = -\measuredangle X[N-k]$, as expected from Equation (10.26).

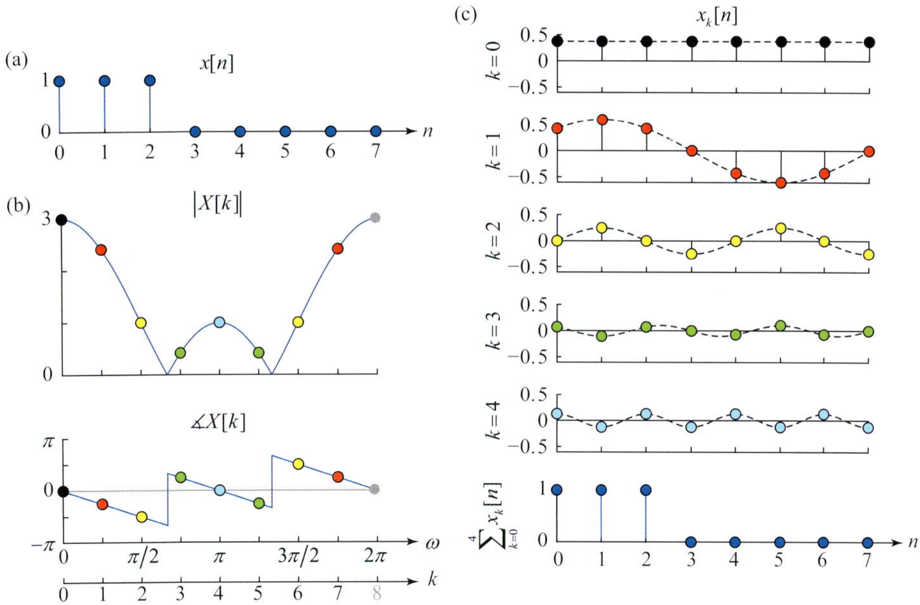

Figure 10.7

The IDFT for $N = 8$ can be expressed as the sum of five real terms using Equation (10.27):

$$\begin{aligned} x[n] &= \frac{1}{8}\left\{X[0] + X[4](-1)^n + 2\sum_{k=1}^{3}|X[k]|\cos(\pi kn/4 + \measuredangle X[k])\right\} \\ &= 0.375 + 0.125(-1)^n + 0.60\cos(\pi n/4 - \pi/4) + 0.25\cos(\pi n/2 - \pi/2) + 0.10\cos(3\pi n/4 - \pi/4) \\ &= \underbrace{0.375}_{x_0[n]} + \underbrace{0.125(-1)^n}_{x_1[n]} + \underbrace{0.60\cos(\pi(n-1)/4)}_{x_2[n]} + \underbrace{0.25\cos(\pi(n-1)/2)}_{x_3[n]} - \underbrace{0.10\cos(3\pi(n-1)/4)}_{x_4[n]}. \end{aligned}$$

Figure 10.7c shows each of these five terms coded in the appropriate color. The sum of these terms, in the bottom panel (●), is equal to the original $x[n]$ in **Figure 10.7a**.

★ General complex sequences For completeness, consider an arbitrary complex sequence with real part $x_r[n]$ and imaginary part $x_i[n]$,

$$x[n] = x_r[n] + jx_i[n],$$

where

$$\begin{aligned} x_r &= \tfrac{1}{2}(x[n] + x^*[n]) \\ x_i &= \tfrac{1}{2j}(x[n] - x^*[n]) \end{aligned} \quad (10.28)$$

Defining $X_r[k] \triangleq \mathbb{F}\{x_r[n]\}$ and $X_i[k] \triangleq \mathbb{F}\{x_i[n]\}$, and using the complex-conjugation property, Equation (10.24), the DFT of Equation (10.28), becomes

$$X_r[k] = \tfrac{1}{2}(X[k] + X^*[(-k)_N]) = \tfrac{1}{2}(X[k] + X^*[N-k])$$
$$X_i[k] = \tfrac{1}{2j}(X[k] - X^*[(-k)_N]) = \tfrac{1}{2j}(X[k] - X^*[N-k]) \qquad \textbf{(10.29)}$$

In Chapter 11 (Section 11.7), we shall use these relations to enhance the performance of the FFT in the computation of real sequences.

10.3.4 Circular time shifting

The shifting property has a "twist" (literally) compared to its DTFT counterpart. The differences between the DTFT and DFT are best shown by example. **Figure 10.8b** shows a sequence with three non-zero points, $x[n] = \delta[n] + 2\delta[n-1] + \delta[n-2]$.

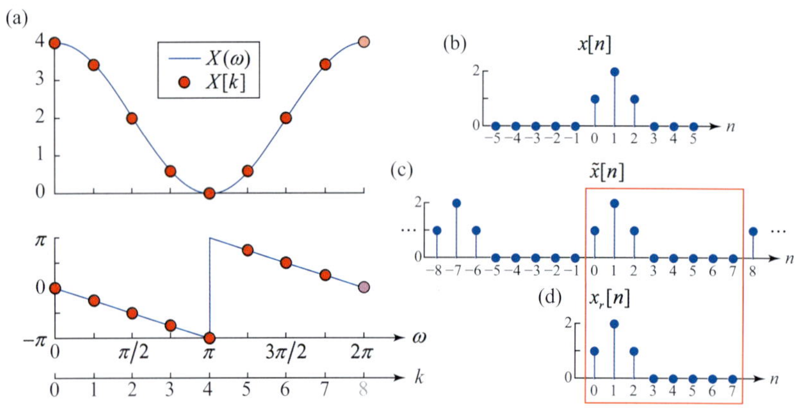

Figure 10.8

Figure 10.8a shows the DTFT of the sequence, $X(\omega)$ (thin blue curve), which has been sampled with $N=8$ to form the DFT coefficients $X[k]$, $0 \le k < N$ (red circles). As we have seen in previous examples, the DFT coefficients correspond to a periodic sequence $\tilde{x}[n]$ that comprises copies of $x[n]$ shifted by integer multiples of N and added together, as shown in **Figure 10.8c**. Then, the sequence $x_r[n]$ reconstructed from the IDFT is equal to the first N samples of $\tilde{x}[n]$, as shown in **Figure 10.8d**,

$$x_r[n] = \tilde{x}[n](u[n] - u[n-N]).$$

In this example, $x[n]$ is 0 outside the range $0 \le n < N$, so $\tilde{x}[n]$ is exactly equal to $x[n]$. So far, so good.

Now, recall the shifting property of the DTFT: if $X(\omega)$ is multiplied by $e^{-j\omega n_0}$, this corresponds to shifting $x[n]$ by n_0. Specifically, if $Y(\omega) = X(\omega)e^{-j\omega n_0}$, then $y[n] = x[n-n_0]$. For any ω,

$$|Y(\omega)| = |X(\omega)e^{-j\omega n_0}| = |X(\omega)|$$
$$\measuredangle Y(\omega) = \measuredangle X(\omega) - \omega n_0$$

Hence, $\measuredangle Y(\omega)$ differs from $\measuredangle X(\omega)$ by a continuous function $-\omega n_0$. The DFT $Y[k]$ of $y[n]$ is obtained by sampling $Y(\omega)$ as specified in Equation (10.2),

$$Y[k] = Y(\omega)|_{\omega = 2\pi k/N} = X(\omega)e^{-j\omega n_0}|_{\omega = 2\pi k/N} = X[k]e^{-j2\pi kn_0/N}, \qquad 0 \le k < N.$$

The magnitude and phase of $Y[k]$ are

$$\begin{aligned} |Y[k]| &= |X[k]e^{-j2\pi kn_0/N}| = |X[k]| \\ \measuredangle Y[k] &= \measuredangle X[k] - 2\pi kn_0/N \end{aligned}.$$

Hence, $\measuredangle Y[k]$ differs from $\measuredangle X[k]$ by an integer multiple of $2\pi k/N$. Now, the question is, "What is the sequence $y[n]$ that corresponds to the DFT $Y[k] = X[k]e^{-j2\pi kn_0/N}$?" A few examples will lead us forward.

Figure 10.9a shows the DTFT $Y(\omega) = X(\omega)e^{-j\omega n_0}$, and the DFT $Y[k] = X[k]e^{-j2\pi kn_0/N}$, when $n_0 = 1$. Comparing the DTFTs in **Figure 10.9a** and **Figure 10.8a**, you can see that the magnitudes are the same, $|Y(\omega)| = |X(\omega)|$, while the phases differ by a linear factor, $\measuredangle Y(\omega) = \measuredangle X(\omega) - \omega n_0 = \measuredangle X(\omega) - \omega$. For the DFTs, $|Y[k]| = |X[k]|$ and $\measuredangle Y[k] = \measuredangle X[k] - 2\pi kn_0/N = \measuredangle X[k] - k\pi/4$. The inverse transform of $Y(\omega)$ is the sequence $y[n] = x[n-1]$, as we might expect. The inverse transform of $Y[k]$ corresponds to the sequence $y_r[n]$ reconstructed from points $0 \le n < N$ of the periodic sequence $\tilde{y}[n]$, as shown in **Figures 10.8c** and **d**. Since $y[n]$ is zero outside the range $0 \le n < N$, $y_r[n]$ is exactly equal to $y[n]$. This suggests that the inverse transform of $X[k]e^{-j2\pi kn_0/N}$ is just $x[n - n_0]$. Life seems wonderful.

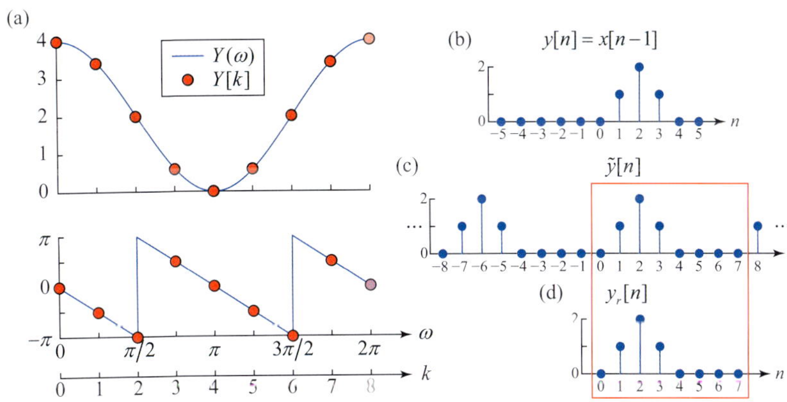

Figure 10.9

Figure 10.10 shows what happens when $n_0 = -1$. **Figure 10.10a** again shows the magnitude and phase of $Y(\omega)$. **Figure 10.10b** shows that the inverse transform of $Y(\omega)$ is the sequence $y[n] = x[n+1]$, as expected. However, the sequence $y_r[n]$, which corresponds to the inverse transform of $Y[k]$, is clearly not equal to $y[n]$. Instead, the relation between $y_r[n]$ and $x[n]$ can be summarized as

$$y_r[n] = x[(n - n_0)_N], \tag{10.30}$$

where the notation $(n - n_0)_N$ again means "$n - n_0$ modulo N." This is termed a **circular shift**.

The meaning of this term is made clearer in **Figure 10.11**. **Figure 10.11a** shows the sequence $x[n]$ and its shifted version $x[n+1]$ from **Figures 10.9** and **10.10**. **Figure 10.11b** shows the periodic sequences formed from the inverse DFT of $X[k]$ (top panel) and $Y[k] = X[k]e^{-j2\pi kn_0/N}$

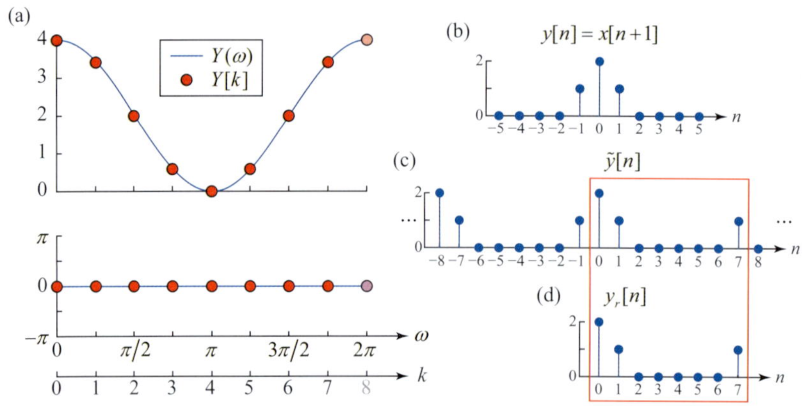

Figure 10.10

with $N = 8$ and a value of $n_0 = -1$ (bottom panel). Since $\tilde{x}[n]$ is periodic with period $N = 8$, when $\tilde{x}[n]$ shifts to the left to form $\tilde{y}[n] = \tilde{x}[n+1]$, the value $x_s[0]$ shifts out of the left edge of the red window, while the value $\tilde{x}[8]$ shifts into the right edge. Since $\tilde{x}[8] = \tilde{x}[0] = x[0]$, it is just as if the same value shifted out of the left edge of the window and back into the right edge. Formally, we would say

$$y_r[7] = x[(7+1)_8] = x[(8)_8] = x[0].$$

For this reason it is called a circular shift, as shown in **Figure 10.11c**. To summarize this discussion, we succinctly state the definition of the circular shift property:

$$\mathbb{F}\{x[(n-n_o)_N]\} = X[k]e^{-j2\pi k n_0/N}. \tag{10.31}$$

Circularly shifting $x[n]$ by n_0 corresponds to adding phase $-2\pi k n_0/N$ to $\measuredangle X[k]$.

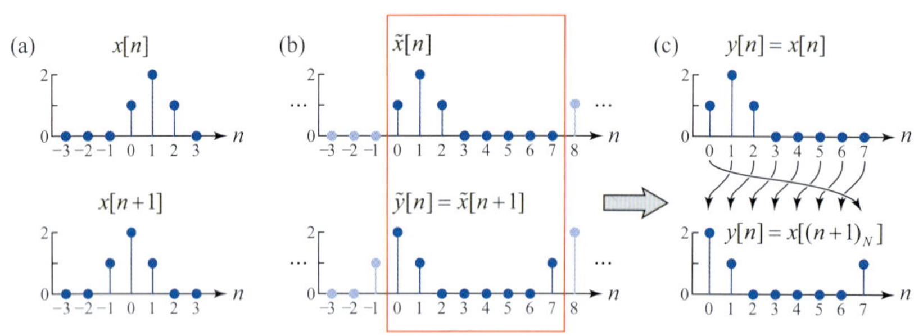

Figure 10.11

Example 10.5

Find the circular shift of the sequence $x[n] = \delta[n] + 2\delta[n-1] + \delta[n-2]$ in **Figure 10.11a** to form $y[n] = x((n+1)_N)$ when

(a) $N = 8$.

(b) $N = 4$.

➤ **Solution:**

(a) By direct calculation:

$$X[k] = \sum_{n=0}^{N-1} x[n]e^{-j2\pi kn/N} = 1 + 2e^{-jk\pi/4} + e^{-jk\pi/2}.$$

So,

$$\begin{aligned} Y[k] &= X[k]e^{-j2\pi kn_0/N} = X[k]e^{j\pi k/4} = e^{jk\pi/4} + 2 + e^{-jk\pi/4} = e^{jk\pi/4}{\cdot}e^{-jk2\pi} + 2 + e^{-jk\pi/4} \\ &= 1{\cdot}e^{-jk2\pi(7)/8} + 2{\cdot}e^{-jk2\pi(0)/8} + 1{\cdot}e^{-jk2\pi(1)/8}, \end{aligned}$$

and therefore,

$$y[n] = \mathbb{F}_N^{-1}\{Y[k]\} = 2\delta[n] + \delta[n-1] + \delta[n-7].$$

The Matlab function `circshift` can perform a circular shift of an array or matrix,

```
y = circshift(x, [r, c]),
```

where `r` and `c` are the amount of shift to be applied to the rows and columns, respectively. However, we can easily write a little Matlab function of our own to do the work, using Matlab's `mod` function:

```
function y = cshift(x, n0)
    N = length(x);
    y = x(1+mod((0:N-1)-n0, N));
end
```

Here, `n0` determines the amount of shift. For example,

```
>> cshift([1 2 1 0 0 0 0 0], -1)
ans =
    2    1    0    0    0    0    0    1
```

The `mod` function creates the proper index into the array:

```
>> 1+mod((0:N-1)-n0, N)
ans =
     2    3    4    5    6    7    8    1
```

We can also perform the circular shift using the DFT, by directly implementing Equation (10.31):

```
function y = cshift2(x, n0)
    N = length(x);
    X = fft(x);
    k = 0:N-1;
    Y = X .* exp(-1j*2*pi*k*n0/N);
    y = ifft(Y);
end
```

(b) By direct calculation:

$$X[k] = \sum_{n=0}^{N-1} x[n]e^{-j2\pi kn/N} = 1 + 2e^{-jk\pi/2} + e^{-jk\pi},$$

and

$$\begin{aligned} Y[k] &= X[k]e^{-j2\pi kn_0/N} = X[k]e^{j\pi k/2} = e^{jk\pi/2} + 2 + e^{-jk\pi/2} = e^{jk\pi/2}\cdot e^{-jk2\pi} + 2 + e^{-jk\pi/2} \\ &= 1\cdot e^{-jk2\pi(3)/4} + 2\cdot e^{-jk2\pi(0)/4} + 1\cdot e^{-jk2\pi(1)/4} \end{aligned}.$$

So,

$$y[n] = \mathbb{F}_N^{-1}\{Y[k]\} = 2\delta[n] + \delta[n-1] + \delta[n-4].$$

By Matlab:

```
>> cshift ([1 2 1 0], -1)
ans =
     2    1    0    1
```

10.3.5 Circular time reversal

Figure 10.12 shows what time reversal means from the point of view of the DFT. As we explained in the context of **Figure 10.1**, a sequence $x[n]$ whose N-point DFT is $X[k]$ is to be understood as N points of the periodic sequence $\tilde{x}[n]$, as shown in the example of **Figure 10.12a**. Consequently, the time-reversed sequence should be visualized as N points of the time-reversed periodic sequence $\tilde{x}[-n]$, as shown in **Figure 10.12b**. This portion of $\tilde{x}[-n]$ can be written as a time reversal modulo N,

$$x[(-n)_N] = \begin{cases} x[0], & n = 0 \\ x[N-n], & 1 \le n \le N-1 \end{cases}.$$

The DFT of the time-reversed sequence is

$$\mathbb{F}_N\{x[(-n)_N]\} = x[0] + \sum_{n=1}^{N-1} x[N-n]e^{-j2\pi kn/N}.$$

Let $\hat{n} = N - n$. Then,

$$\begin{aligned} x[0] + \sum_{\hat{n}=1}^{N-1} x[\hat{n}]e^{-j2\pi k(N-\hat{n})/N} &= x[0] + \sum_{n=1}^{N-1} x[n]e^{j2\pi kn/N} = x[0] + \left(\sum_{n=0}^{N-1} x[n]e^{j2\pi kn/N} - x[0]\right) \\ &= \sum_{n=0}^{N-1} x[n]e^{j2\pi kn/N} = \left(\sum_{n=0}^{N-1} x^*[n]e^{-j2\pi kn/N}\right)^*. \end{aligned}$$

So, the DFT of the time-reversed sequence is

$$\mathbb{F}_N\{x[(-n)_N]\} = (\mathbb{F}_N\{x^*[n]\})^*. \quad \textbf{(10.32)}$$

For a real sequence, $x^*[n] = x[n]$, and this just reduces to

$$\mathbb{F}_N\{x[(-n)_N]\} = (\mathbb{F}_N\{x[n]\})^* = X^*[k].$$

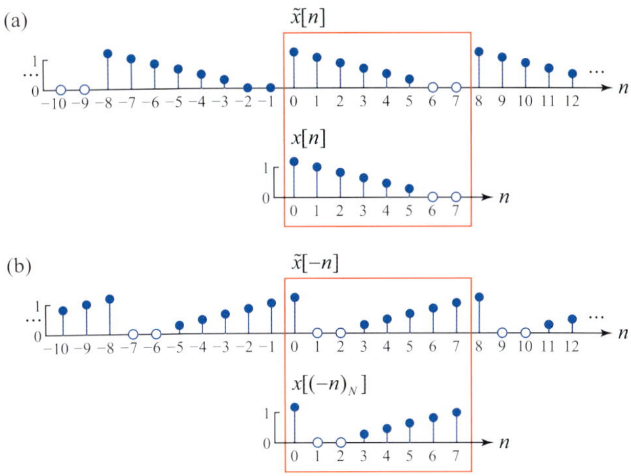

Figure 10.12 Time reversal by the DFT

Example 10.6

Use the circular time-reversal property of the DFT to find $x[(-n)_N]$ for the following sequences:

(a) $x[n] = [(1+j)\ (2+j)\ 0\ (6+j)]$.

(b) $x[n] = [1\ 2\ 3\ 0\ 6]$.

➤ Solution:

(a) From Equation (10.32), we have

$$x[(-n)_N] = \mathbb{F}_N^{-1}\left\{(\mathbb{F}_N\{x^*[n]\})^*\right\},$$

which translates to the Matlab expression, `ifft(fft(x')')`. Remember that in Matlab, the transpose operator performs the conjugate (Hermitian) transpose, which is exactly what we want in this instance:

```
>> x = [1+1j 2+1j 0 6+1j];
>> ifft(fft(x')')
ans =
    1.0000 + 1.0000i   6.0000 + 1.0000i   0.0000 + 0.0000i 2.0000 + 1.0000i
```

(b) In this case, x is real, so the conjugate is not strictly necessary, but we will leave it in so the final result is a row vector.

```
>> x = [1 2 3 0 6];
>> ifft(fft(x')')
ans =
    1.0000   6.0000   0   3.0000   2.0000
```

10.3.6 Circular frequency shift

This is the dual of the circular time-shift property. Given a sequence $x[n]$ of length N, let $y[n]$ be a sequence formed by multiplying $x[n]$ by a complex exponential at frequency $2\pi l/N$,

$$y[n] = x[n]e^{j2\pi ln/N}, \qquad 0 \le n \le N-1.$$

Then, the DFT $Y[k]$ is $X[k]$ shifted in frequency, modulo N,

$$\begin{aligned} Y[k] &= \sum_{n=0}^{N-1} y[n]e^{-j2\pi kn/N} = \sum_{n=0}^{N-1}\left(x[n]e^{-j2\pi ln/N}\right)e^{-j2\pi kn/N} = \sum_{n=0}^{N-1} x[n]e^{-j2\pi(k-l)n/N} \\ &= X[(k-l)_N], \quad 0 \le k \le N-1. \end{aligned} \tag{10.33}$$

Example 10.7

Let $x[n]$ be the sequence of Example 10.2, $x[n] = \cos(\pi n/4 - \pi/2)$, $0 \le n \le N-1$, where $N = 16$. Let $y[n] = x[n](-1)^n$. Use the circular frequency-shift property to find the DFT $Y[k]$ and $y[n]$.

➤ Solution:

From Equation (10.20), $X[k] = 8e^{-j\pi/2}\delta[(k-2)_{16}] + 8e^{j\pi/2}\delta[(k+2)_{16}]$. Since $N = 16$, write

$$y[n] = x[n](-1)^n = x[n]e^{j2\pi \cdot 8n/16};$$

so, from Equation (10.33),

$$\begin{aligned} Y[k] = X[(k-8)_N] &= 8e^{-j\pi/2}\delta[(k-2-8)_{16}] + 8e^{j\pi/2}\delta[(k+2-8)_{16}] \\ &= 8e^{-j\pi/2}\delta[(k-(16-6))_{16}] + 8e^{j\pi/2}\delta[(k-6)_{16}] \\ &= 8e^{-j\pi/2}\delta[(k+6)_{16}] + 8e^{j\pi/2}\delta[(k-6)_{16}] \end{aligned}$$

Hence, $y[n] = \cos(2\pi \cdot 6n/16 + \pi/2) = \cos(3\pi n/4 + \pi/2)$, $0 \le n \le N-1$.

10.3.7 Circular convolution

In Chapter 3, we discussed numerous properties of the DTFT that enhanced our ability to understand and analyze discrete-time systems. The convolution property was theoretically important because it allows us to perform convolution of sequences in the time domain by multiplication of transforms in the frequency domain. To perform the convolution of two finite-length sequences, $y[n] = x[n] * h[n]$, we would compute the transforms $X(\omega)$ and $H(\omega)$, respectively, multiply them together to form transform $Y(\omega) = X(\omega)H(\omega)$, and then compute the inverse DTFT to find $y[n]$, as indicated schematically in **Figure 10.13a**.

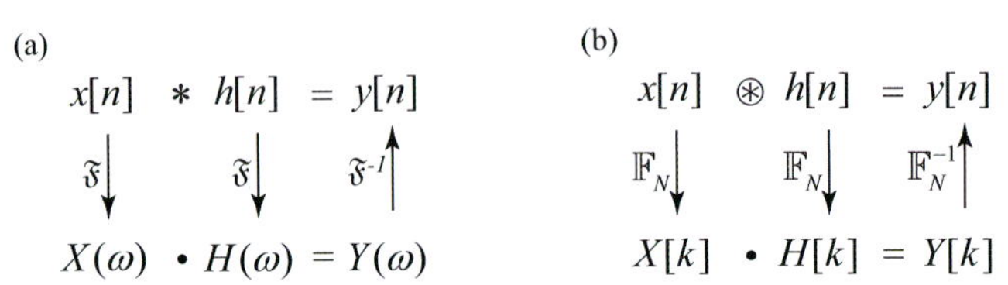

Figure 10.13

However, as we mentioned in the introduction to this chapter, implementing convolution via multiplication of DTFTs is not computationally practical. It would require the computation of $X(\omega)$ and $H(\omega)$, both of which are continuous functions of ω. Then the inverse transform of the product of these functions would have to be computed, which would require the computation of an integral. It is not clear how any of that could be done efficiently with a computer.

Because the DFT of a sequence is just the DTFT sampled at N equally spaced frequencies (Equation (10.2)), this suggests that we might be able to find a convolution property analogous to that shown in **Figure 10.13a** that would apply to the DFT. Given two N-point sequences $x[n]$ and $h[n]$, we propose to compute the N-point DFTs $X[k]$ and $H[k]$, multiply them together to form $Y[k]=X[k]H[k]$ and then take the inverse DFT to form a sequence $y[n]$. This entire process, shown schematically in **Figure 10.13b**, would require only discrete mathematical operations and should be possible to implement on a computer. In the same spirit that we defined the relation $X[k]e^{-jk2\pi n_0/N} \leftrightarrow x[(n-n_0)_N]$ to be the circular shift of $x[n]$, we shall define the inverse of the product of the DFT of two sequences as the **circular convolution** of $x[n]$ and $y[n]$:

$$Y[k]=X[k]H[k] \leftrightarrow y[n]=\frac{1}{N}\sum_{k=0}^{N-1} X[k]H[k]e^{j2\pi kn/N}=x[n]\circledast h[n]. \tag{10.34}$$

Circular convolution gets its own special symbol, $\circledast$, to distinguish it from the linear convolution of **Figure 10.13a**. Our task now is to determine under what circumstances (if any) circular and linear convolution are equivalent.

Before proceeding with the mathematical analysis, let us look at what happens when we attempt to perform circular convolution of two sequences using N-point DFTs at a couple of different values of N.

Example 10.8

Use Matlab to compute the circular convolution of the sequences shown in **Figure 10.14** with

(a) $N=8$.

(b) $N=4$.

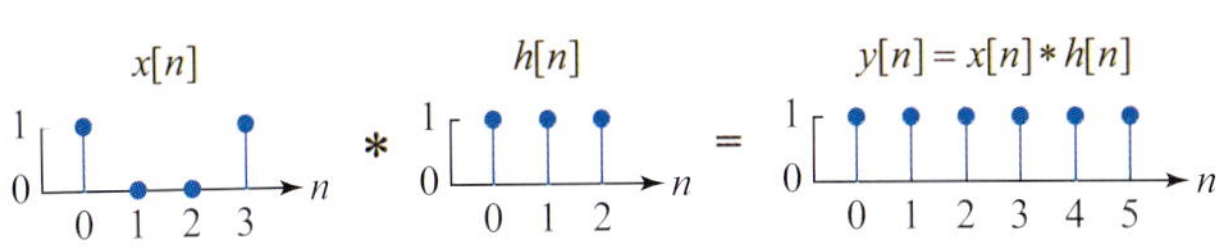

Figure 10.14

➤ Solution:

First, we will create a small function to implement circular convolution using Matlab's `fft` and `ifft` functions to implement the DFT and IDFT, respectively. The procedure is as follows:

1. Pad the end of $x[n]$ and $h[n]$ with zeros if necessary to fill the array out to N points;
2. Take N-point DFTs $x[n] \to X[k]$ and $h[n] \to H[k]$;
3. Form the product $Y[k] = X[k]H[k]$;
4. Take the N-point IDFT $Y[k] \to y[n]$.

```
function y = cconv(x, h, N)
    X = fft(x, N);
    H = fft(h, N);
    y = ifft(X.*H);
end
```

The second (N) argument to fft indicates the desired size of the DFT. If N is larger than the size of the input array, fft automatically pads the array with zeros to fill out N points. Here are the inputs:

```
x = [1 0 0 1];
h = [1 1 1];
```

(a)

```
>> cconv(x, h, 8)
     1   1   1   1   1   1   0   0
```

In this case, circular convolution is equal to the linear convolution $y[n] = x[n] * h[n]$, padded with two zeros at the end.

(b)

```
>> cconv(x, h, 4)
     2   2   1   1
```

Here, circular convolution is clearly not equal to linear convolution, nor could it be since the length of the linear convolution is $L = 6$ and the length of the IDFT is constrained to be $N = 4$.

In order to understand circular convolution analytically, and explain what just happened in this example, start by recalling that $y[n]$, the IDFT of $Y[k]$ in Equation (10.34), is defined as N points of the periodic sequence $\tilde{y}[n]$,

$$y[n] = \tilde{y}[n], \quad 0 \le n \le N-1, \tag{10.35}$$

where, in this case,

$$\tilde{y}[n] = \frac{1}{N}\sum_{k=0}^{N-1} Y[k]e^{j2\pi kn/N} = \frac{1}{N}\sum_{k=0}^{N-1} X[k]H[k]e^{j2\pi kn/N}.$$

Substitute expressions for $X[k]$ and $H[k]$ in terms of the periodic sequences $\tilde{x}[n]$ and $\tilde{h}[n]$, respectively:

$$\begin{aligned}\tilde{y}[n] &= \frac{1}{N}\sum_{k=0}^{N-1} X[k]H[k]e^{j2\pi kn/N} = \frac{1}{N}\sum_{k=0}^{N-1}\left(\sum_{m=0}^{N-1}\tilde{x}[m]e^{-j2\pi km/N}\right)H[k]e^{j2\pi kn/N} \\ &= \sum_{m=0}^{N-1}\tilde{x}[m]\underbrace{\left(\sum_{k=0}^{N-1}H[k]e^{j2\pi k(n-m)/N}\right)}_{\tilde{h}[n-m]} \\ &= \sum_{m=0}^{N-1}\tilde{x}[m]\tilde{h}[n-m].\end{aligned} \tag{10.36}$$

Over the limits of the summation, $0 \le m \le N-1$, $\tilde{x}[m]$ is always equal to the finite-length input sequence $x[m]$. By the circular-shift property, Equation (10.30), $\tilde{h}[n-m] = h[(n-m)_N]$. Hence, Equation (10.36) becomes

$$\tilde{y}[n] = \sum_{m=0}^{N-1} x[m]h[(n-m)_N].$$

Finally, by Equation (10.35), the circular convolution of $x[n]$ and $h[n]$ is just those points of $\tilde{y}[n]$ in the interval $0 \le n \le N-1$:

$$y[n] = x[n] \circledast h[n] \triangleq \sum_{m=0}^{N-1} x[m]h[(n-m)_N], \quad 0 \le n \le N-1. \tag{10.37}$$

While this equation provides a formal definition of circular convolution, it does not obviously explain how circular convolution is related to linear convolution. To gain a more intuitive understanding of this important point, return to Equation (10.36). Again, we have $\tilde{x}[m] = x[m]$ for $0 \le m \le N-1$. But now, using Equation (10.3), we can write

$$\tilde{h}[n] = \sum_{k=-\infty}^{\infty} h[n-kN] = h[n] * \sum_{k=-\infty}^{\infty} \delta[n-kN] = h[n] * \tilde{\delta}[n].$$

Accordingly, the shifted periodic sequence is $\tilde{h}[n-m] = h[n-m] * \tilde{\delta}[n]$. Equation (10.36) then becomes

$$\begin{aligned}\tilde{y}[n] &= \sum_{m=0}^{N-1} x[m](h[n-m] * \tilde{\delta}[n]) = \left(\sum_{m=0}^{N-1} x[m]h[n-m]\right) * \tilde{\delta}[n] \\ &= (x[n] * h[n]) * \sum_{k=-\infty}^{\infty} \delta[n-kN].\end{aligned}$$

Finally, by Equation (10.11), the circular convolution of $x[n]$ and $h[n]$ is the N points of $\tilde{y}[n]$, $0 \le n \le N-1$. So, formally we have

$$y[n] = x[n] \circledast h[n] = (x[n] * h[n]) * \sum_{k=-\infty}^{\infty} \delta[n-kN], \quad 0 \le n \le N-1. \tag{10.38}$$

Equation (10.38) is perhaps a more intuitively accessible definition of circular convolution than Equation (10.37). In words, it says that to perform the circular convolution of two N-point sequences $x[n]$ and $h[n]$ in the time domain (conceptually at least), you do the following:

1. Calculate the linear convolution of $x[n]$ and $h[n]$, $x[n] * h[n]$;
2. Form a periodic sequence $\tilde{y}[n]$ by summing replicas of the output of that convolution, each replica shifted by integer multiples of N;
3. Select the N points of $\tilde{y}[n]$, $0 \le n \le N-1$, which form the circular convolution $y[n]$.

Figure 10.15 demonstrates conceptually what happens if we perform the circular convolution using DFTs with $N=4$. We followed this exact procedure to explain the relation between a periodic and aperiodic sequence in **Figure 10.1**. **Figure 10.15** follows the identical steps using the linear convolution of Example 10.8, $x[n] * h[n]$ (**Figure 10.15a**). This sequence of length $L=6$ is convolved with a periodic impulse train $\tilde{\delta}[n]$ (**Figure 10.15b**), which has period $N=4$, to form a series of time-delayed replicas of $y[n]$ that add to form $\tilde{y}[n]$ (**Figure 10.15c**). Because the length of the linear convolution ($L=6$) is larger than the period ($N=4$), the last two points of each replica overlap with and add to the first two points of the succeeding replica which results in **time-domain aliasing**. Then, $N=4$ points of $\tilde{y}[n]$, $0 \le n \le N-1$, are selected to form the circular convolution $y[n]$. This exactly matches the result we got in Example 10.8b with $N=4$.

As shown in **Figures 10.15e–g**, we can visualize time-domain aliasing in circular convolution by breaking the linear convolution into chunks of length N (**Figure 10.15e**), time-aligning them (**Figure 10.15f**) and adding them to get the final N-point answer (**Figure 10.15g**). The result is identical to **Figure 10.15d**.

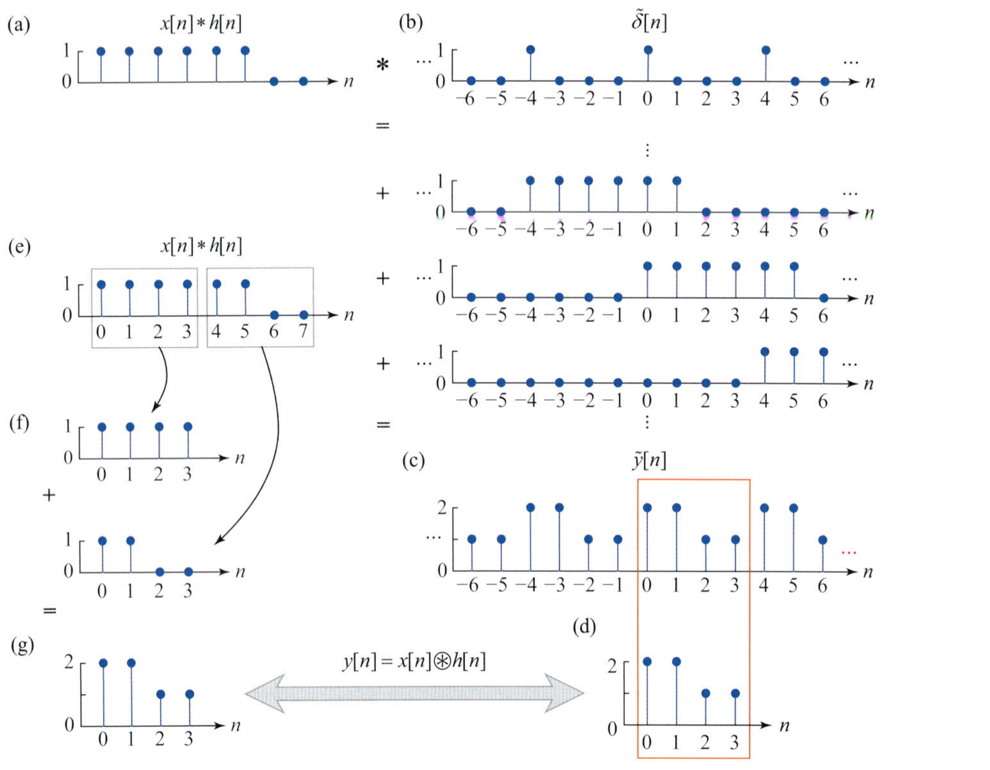

Figure 10.15 Circular convolution interpreted as linear convolution followed by time-domain aliasing

The key conclusion of the preceding analysis is that we can interpret circular convolution as linear convolution followed by time-domain aliasing. As long as the size of the DFT, N, is greater than or equal to the length of the linear convolution, L, then circular convolution is exactly equal to linear convolution. This means that instead of performing the convolution sum in the time domain, we can perform convolution of two finite-length sequences by multiplication of DFTs, as schematized in **Figure 10.13b**. If the two sequences have contiguous non-zero lengths L_1 and L_2, then all we have to do is to assure that the size of the DFTs is $N \geq L_1 + L_2 - 1$. While this procedure – taking the transforms of $x[n]$ and $h[n]$, multiplying them together, and then computing the inverse transform – seems like a very long trip to make to perform a convolution, in fact it turns out that it can involve substantially less computation than the direct computation of the convolution sum, principally because there exist highly efficient FFT algorithms for computing the DFT and IDFT, a topic we shall discuss in considerable detail in Chapter 11.

10.3.8 Multiplication

The multiplication property is the dual of the circular-convolution property. It states that the DFT of the product of two N-point sequences is the circular convolution of their DFTs. To show this, start with two sequences $x[n]$ and $w[n]$ and express their product $y[n]$ as follows:

$$y[n] = x[n]w[n] = \left(\frac{1}{N}\sum_{l=0}^{N-1} X[l]e^{j2\pi ln/N}\right)w[n] = \frac{1}{N}\sum_{l=0}^{N-1} X[l]\left(w[n]e^{j2\pi ln/N}\right), \quad 0 \leq n \leq N-1. \tag{10.39}$$

By the circular frequency-shifting property, Equation (10.33), the DFT of $w[n]e^{j2\pi ln/N}$ is $W[(k-l)_N]$; hence,

$$w[n]e^{j2\pi ln/N} = \mathbb{F}_N^{-1}\{W[(k-l)_N]\} = \frac{1}{N}\sum_{k=0}^{N-1} W[(k-l)_N]e^{j2\pi kn/N},$$

and Equation (10.39) becomes

$$\begin{aligned} y[n] &= \frac{1}{N}\sum_{l=0}^{N-1} X[l]\left(\frac{1}{N}\sum_{k=0}^{N-1} W[(k-l)_N]e^{j2\pi kn/N}\right) = \frac{1}{N}\sum_{k=0}^{N-1}\left(\frac{1}{N}\sum_{l=0}^{N-1} X[l]W[(k-l)_N]\right)e^{j2\pi kn/N} \\ &= \mathbb{F}_N^{-1}\left\{\frac{1}{N}\sum_{l=0}^{N-1} X[l]W[(k-l)_N]\right\}, \quad 0 \leq n \leq N-1. \end{aligned}$$

Taking the DFT of both sides gives the desired result,

$$Y[k] = \frac{1}{N}\sum_{l=0}^{N-1} X[l]W[(k-l)_N] = \frac{1}{N}X[k] \circledast W[k], \quad 0 \leq k \leq N-1.$$

Example 10.9

Use the multiplication property to find the DFT of the sequence

$$y[n] = \begin{cases} \cos 2\pi l n/N, & 0 \le n \le N/2 - 1 \\ 0, & N/2 \le n \le N-1 \end{cases}.$$

➤ Solution:

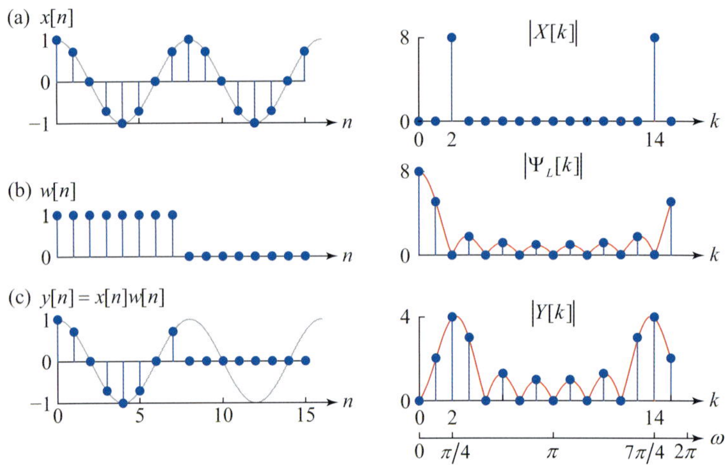

Figure 10.16

The first half of $y[n]$ is a cosine of frequency $2\pi l/N$, where l is an integer; the second half is 0. It can be represented as a product $y[n] = x[n]w_N[n]$, where $x[n]$ is a cosine of length N,

$$x[n] = \cos 2\pi l n/N, \qquad 0 \le n \le N-1,$$

and $w[n]$ is a "window" sequence

$$w[n] = \begin{cases} 1, & 0 \le n \le N/2 - 1 \\ 0, & N/2 \le n \le N-1 \end{cases}.$$

Because the frequency of the cosine is an integer multiple of $2\pi/N$, the DFT of $x[n]$ is given by Equation (10.19),

$$X[k] = \frac{N}{2}\delta[(k-l)_N] + \frac{N}{2}\delta[(k+l)_N].$$

The DFT of the window sequence of length L is given by Equation (10.15),

$$W[k] = e^{j\pi k(L-1)/N} L \frac{\operatorname{sinc} \pi k L/N}{\operatorname{sinc} \pi k/N} = e^{j\pi k(N/2-1)/N} \frac{N}{2} \frac{\operatorname{sinc} \pi k/2}{\operatorname{sinc} \pi k/N}.$$

Then, the DFT of $y[n]$ is given by the circular convolution of $X[k]$ and $W[k]$,

$$
\begin{aligned}
Y[k] &= \frac{1}{N} X[k] \circledast W[k] \\
&= e^{j\pi(k-l)(N/2-1)/N} \frac{N}{4} \frac{\operatorname{sinc} \pi(k-l)/2}{\operatorname{sinc} \pi(k-l)/N} + e^{j\pi(k+l)(N/2-1)/N} \frac{N}{4} \frac{\operatorname{sinc} \pi(k+l)/2}{\operatorname{sinc} \pi(k+l)/N}, \quad 0 \le k \le N-1.
\end{aligned}
$$

The left panels of **Figures 10.16a–c** show $x[n]$, $w[n]$ and $y[n]$ for example values of $N = 16$ and $l = 2$, and the right panels show the magnitude of their respective DFTs. (The thin red lines indicate $|W(\omega)|$ and $|Y(\omega)|$.) Truncation of the cosine in the time domain results in spectral spread of the DFT, something we also saw in Example 10.2. This topic will be discussed in detail in Chapter 14.

10.3.9 Parseval's theorem

The most commonly used variant of Parseval's theorem as it relates to the DFT equates the energy in a time sequence $x[n]$ of length N with the N corresponding DFT coefficients $X[k]$,

$$
\sum_{n=0}^{N-1} |x[n]|^2 = \sum_{n=0}^{N-1} x[n] x^*[n] = \sum_{n=0}^{N-1} x[n] \left(\frac{1}{N} \sum_{k=0}^{N-1} X[k] e^{j2\pi kn/N} \right)^* = \sum_{n=0}^{N-1} x[n] \left(\frac{1}{N} \sum_{k=0}^{N-1} X^*[k] e^{-j2\pi kn/N} \right)
$$

$$
= \frac{1}{N} \sum_{k=0}^{N-1} X^*[k] \left(\sum_{n=0}^{N-1} x[n] e^{-j2\pi kn/N} \right) = \frac{1}{N} \sum_{k=0}^{N-1} X^*[k] X[k] = \frac{1}{N} \sum_{k=0}^{N-1} |X[k]|^2.
$$

Example 10.10

Show that the sequence $x[n]$ and DFT $X[k]$ of Example 10.3 satisfy Parseval's theorem.

➤ **Solution:**

By calculation,

$$
\sum_{n=0}^{N-1} |x[n]|^2 = 3
$$

and

$$
\begin{aligned}
\frac{1}{N} \sum_{k=0}^{N-1} |X[k]|^2 &= \frac{1}{8} \left(|X[0]|^2 + |X[4]|^2 + 2(|X[1]|^2 + |X[2]|^2 + |X[3]|^2) \right) \\
&= \frac{1}{8} \left(3^2 + 1^2 + 2\left((\sqrt{2}+1)^2 + 1^2 + (\sqrt{2}-1)^2 \right) \right) \\
&= \frac{1}{8} \left(9 + 1 + 2\left((3+2\sqrt{2}) + 1 + (3-2\sqrt{2}) \right) \right) \\
&= \frac{1}{8} (9 + 1 + 2 \cdot 7) = 3.
\end{aligned}
$$

Of course, as we said before, this is what computers are for:

```
>> x = ones(1, 3);
>> X = fft(x, 8);
>> disp('    x^2    X^2');disp([x*x'  X*X'/8])
    x^2    X^2
     3      3
```

In Matlab, `X'` is the conjugate transpose $X^*[k]$, so $X[k]X^*[k] = |X[k]|^2$.

10.3.10 Summary of DFT properties

Table 10.1 gives a handy summary of the major properties covered in this section.

Table 10.1 DFT properties

Property	$x[n]$	$X[k]$
Linearity	$ax_1[n] + bx_2[n]$	$aX_1[k] + bX_2[k]$
Complex conjugation	$x^*[n]$	$X^*[(-n)_N]$
Circular time shift	$x[(n-n_0)_N]$	$X[k]e^{-j2\pi kn_0/N}$
Time reversal	$x^*[(-n)_N]$	$X^*[k]$
Circular frequency shifting	$x[n]e^{j2\pi ln/N}$	$X[(k-l)_N]$
Circular convolution	$x[n]\circledast h[n] = \sum_{m=0}^{N-1} x[m]h[(n-m)_N]$	$X[k]H[k]$
Multiplication	$x[n]w[n]$	$\frac{1}{N}X[k]\circledast W[k] = \frac{1}{N}\sum_{l=0}^{N-1} X[l]W[(k-l)_N]$
Parseval's theorem	$\sum_{n=0}^{N-1} \|x[n]\|^2$	$\frac{1}{N}\sum_{k=0}^{N-1} \|X[k]\|^2$
Symmetry	$x[n]$ real $(x[n] = x^*[n])$	$X[k] = X^*[(-k)_N]$ $\|X[k]\| = \|X[(-k)_N]\|$ $\measuredangle X[k] = -\measuredangle X[(-k)_N]$ $\mathrm{Re}\{X[k]\} = \mathrm{Re}\{X[(-k)_N]\}$ $\mathrm{Im}\{X[k]\} = -\mathrm{Im}\{X[(-k)_N]\}$

10.4 ★ Matrix representation of the DFT

The DFT and IDFT can be economically represented by matrix operations. To enable this discussion, we introduce a simple substitute notation for the complex exponential that is commonly found in the literature, and which we shall use extensively in Chapter 11 when we discuss the FFT,

$$W_N \triangleq e^{-j2\pi/N}.$$

(Note the minus sign in the exponent.) Then,

$$(W_N)^{kn} = W_N^{kn} = e^{-j2\pi kn/N}.$$

With this notation, the definitions of the DFT and IDFT, Equation (10.23), become

$$\text{DFT}: \quad \mathbb{F}_N\{x[n]\} \triangleq \sum_{n=0}^{N-1} x[n] W_N^{kn}, \qquad 0 \le k < N$$

$$\text{IDFT}: \quad \mathbb{F}_N^{-1}\{X[k]\} \triangleq \frac{1}{N}\sum_{n=0}^{N-1} X[k] W_N^{-kn}, \quad 0 \le n < N.$$

Now, the DFT can be written in matrix form. First, write

$$X[k] = \sum_{n=0}^{N-1} \underbrace{W_N^{kn}}_{\mathbf{W}[k,n]} x[n],$$

where $\mathbf{W}[k,n] = W_N^{kn}$ are the elements of a square $N \times N$ **DFT matrix** of complex exponentials, $\mathbf{W}$, with k and n being the row and column indices of the matrix, respectively. In matrix form,

$$\underbrace{\begin{bmatrix} X[0] \\ X[1] \\ \vdots \\ X[k] \\ \vdots \\ X[N-1] \end{bmatrix}}_{\mathbf{X}} = \underbrace{\begin{bmatrix} W_N^{00} & W_N^{01} & \cdots & W_N^{0n} & \cdots & W_N^{0(N-1)} \\ W_N^{10} & W_N^{11} & \cdots & W_N^{1n} & \cdots & W_N^{1(N-1)} \\ \vdots & \vdots & & \vdots & & \vdots \\ W_N^{k0} & W_N^{k1} & \cdots & W_N^{kn} & \cdots & W_N^{k(N-1)} \\ \vdots & \vdots & & \vdots & & \vdots \\ W_N^{(N-1)0} & W_N^{(N-1)1} & \cdots & W_N^{(N-1)n} & \cdots & W_N^{(N-1)(N-1)} \end{bmatrix}}_{\mathbf{W}} \underbrace{\begin{bmatrix} x[0] \\ x[1] \\ \vdots \\ x[n] \\ \vdots \\ x[N-1] \end{bmatrix}}_{\mathbf{x}}, \tag{10.40}$$

or just

$$\mathbf{X} = \mathbf{W}\mathbf{x}, \tag{10.41}$$

where $\mathbf{x}$ and $\mathbf{X}$ are column vectors of $x[n]$ and $X[k]$, respectively. The DFT matrix is square and symmetric, $\mathbf{W} = \mathbf{W}^T$. The IDFT is

$$x[n] = \frac{1}{N}\sum_{k=0}^{N-1} \underbrace{e^{j2\pi nk/N}}_{\mathbf{W}^*[n,k]} X[k].$$

In matrix form,

$$\underbrace{\begin{bmatrix} x[0] \\ x[1] \\ \vdots \\ x[n] \\ \vdots \\ x[N-1] \end{bmatrix}}_{\mathbf{x}} = \frac{1}{N} \underbrace{\begin{bmatrix} W_N^{-00} & W_N^{-10} & \cdots & W_N^{-k0} & \cdots & W_N^{-(N-1)0} \\ W_N^{-01} & W_N^{-11} & \cdots & W_N^{-k1} & \cdots & W_N^{-(N-1)1} \\ \vdots & \vdots & & \vdots & & \vdots \\ W_N^{-0n} & W_N^{-1n} & \cdots & W_N^{-kn} & \cdots & W_N^{-(N-1)n} \\ \vdots & \vdots & & \vdots & & \vdots \\ W_N^{-0(N-1)} & W_N^{-1(N-1)} & \cdots & W_N^{-k(N-1)} & \cdots & W_N^{-(N-1)(N-1)} \end{bmatrix}}_{(\mathbf{W}^*)^T} \underbrace{\begin{bmatrix} X[0] \\ X[1] \\ \vdots \\ X[k] \\ \vdots \\ X[N-1] \end{bmatrix}}_{\mathbf{X}}. \tag{10.42}$$

or

$$\mathbf{x} = \frac{1}{N}(\mathbf{W}^*)^T\mathbf{X} = \frac{1}{N}\mathbf{W}^*\mathbf{X}. \tag{10.43}$$

The DFT and IDFT matrices of Equations (10.40) and (10.42) use the same DFT matrix $\mathbf{W}$, except IDFT uses the conjugate $\mathbf{W}^*$. From Equations (10.41) and (10.43),

$$\mathbf{X} = \mathbf{W}\mathbf{x} = \frac{1}{N}\mathbf{W}\mathbf{W}^*\mathbf{X}.$$

Hence,

$$\frac{1}{N}\mathbf{W}\mathbf{W}^* = \mathbf{I}.$$

This is a statement of orthogonality of the transform matrix. The dot product of the kth row of $\mathbf{W}$ with the lth column of $\mathbf{W}^*$ gives the same orthogonality condition we saw in Equation (10.13):

$$\frac{1}{N}\sum_{n=0}^{N-1}\underbrace{e^{-j2\pi kn/N}}_{\mathbf{W}[k,n]}\underbrace{e^{j2\pi nl/N}}_{\mathbf{W}^*[n,l]} = \frac{1}{N}\sum_{n=0}^{N-1} W_N^{kn}(W_N^{nl})^* = \delta[k-l].$$

Also, by the definition of the matrix inverse, $\mathbf{W}\mathbf{W}^{-1} = \mathbf{I}$, so we can also state that the inverse of the DFT matrix is equal to the conjugate transpose to within a scale factor N.

$$\mathbf{W}^{-1} = \frac{1}{N}\mathbf{W}^*.$$

The first example in this chapter, Example 10.1, implicitly used these matrix operations. In our function `dft`, the matrix $\mathbf{W}$ is coded as

```
exp(-1j*2*pi*(0:N-1)' *(0:N-1)/N).
```

The array `x(:).'` represents the column vector $\mathbf{x}$. Hence, the product is $\mathbf{X} = \mathbf{W}\mathbf{x}$.

10.5 ★ Using the DFT to increase resolution in the time and frequency domains

In this section, we will show that through a simple manipulation – "zero-padding" – of the time sequence $x[n]$, we can economically increase the effective frequency resolution of the DFT $X[k]$. A corresponding method can be used to increase the time resolution of a sequence (i.e., upsample) by appropriate zero-padding of its transform $X[k]$.

10.5.1 Increasing frequency resolution by zero-padding in the time domain

Consider the simple sequence of length $L = 3$, $x[n] = \delta[n] + \delta[n-1] + \delta[n-2]$. What happens if N, the length of the DFT $X[k]$, exceeds L?

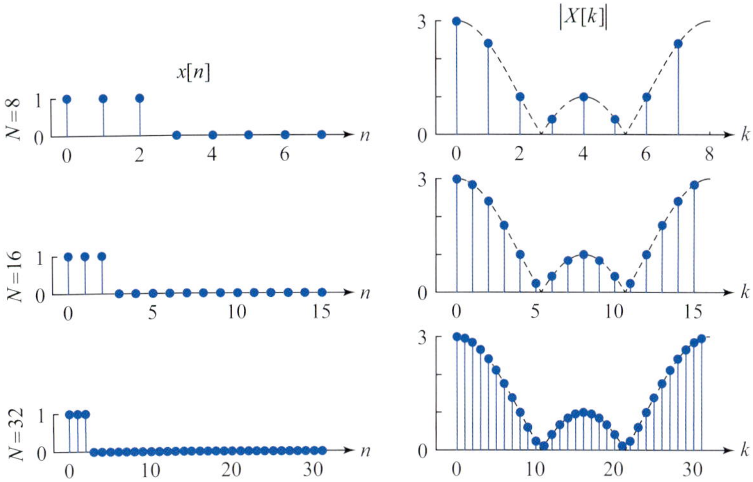

Figure 10.17 Increasing frequency resolution by zero-padding a sequence in the time domain

To find out, pad the end of $x[n]$ with $N-3$ zeros and take the transform:

$$X[k] = \sum_{n=0}^{L-1} x[l]e^{-j2\pi kn/N}, \quad 0 \leq k \leq N-1.$$

Because there are only L non-zero points of $x[n]$, that is all we include in the summation. However, the value of N determines the number of DFT coefficients, and therefore the frequency resolution of the transform. **Figure 10.17** shows the result of taking the DFT of this sequence with values of $N = 8$, 16 and 32. The right panels shows the DFTs for the different values of N. Superimposed on each graph is a thin line indicating $|X(\omega)|$ over the range $0 \leq \omega < 2\pi$, corresponding to values of $X[k]$ over the range $0 \leq k < N$. Since

$$X[k] = X(\omega)|_{\omega=2\pi k/N},$$

the effect of increasing N is to sample the DTFT more finely in frequency, at multiples of $2\pi/N$. This zero-padding "trick" is often employed in spectral analysis to increase the frequency resolution of the transform, since, by Equation (10.22), $\Delta f = f_s / N$.

Implementing time-domain zero-padding in Matlab is trivial using the `fft` function. You can either explicitly pad the input with zeros and take the FFT,

```
y = fft([x zeros(1, N-length(x))]),
```

or, easier still, just set the second (optional) argument `N` to the desired value,

```
y = fft(x, N),
```

in which case `fft` will automatically append the appropriate number of zeros to the end of the array `x`. Historically, input arrays were often padded to the nearest higher power of two, $N = 2^M$, because many of the most commonly used so-called "radix-2" FFT algorithms are specifically designed for inputs of that size. However, most modern FFTs – including Matlab's implementations – are optimized for many different radixes. (See Chapter 11 for details of the FFT.)

10.5.2 Upsampling in the time domain by zero-padding in the frequency domain

To understand how upsampling in the time domain can be achieved by zero-padding in the frequency domain, let us break things into a sequence of conceptual steps. Consider a concrete example: the sequence of even length $N=8$ shown with red symbols in the right panel of **Figure 10.18a**. It consists of a constant plus three cosine terms,

$$x[n] = \sum_{k=0}^{N/2-1} |x_k| \cos(2\pi kn/N + \measuredangle x_k), \quad 0 \le n \le N-1.$$

(The actual values are $x_0 = 0.5$, $x_1 = 2\measuredangle 0$, $x_2 = 2\measuredangle \pi/2$ and $x_3 = 2\measuredangle 0$.) The DFT $X[k]$, shown in the left panel of **Figure 10.18a**, consists of an impulse (blue symbol) at $k=0$ of value Nx_0 for the constant term, plus pairs of impulses for the cosine terms: one pair (green symbols) at $k=1$ with magnitude $(N/2)|x_1|$; one pair (red symbols) at $k=2$ with magnitude $(N/2)|x_2|$; and one pair (gold symbols) at $k=3$ with magnitude $(N/2)|x_3|$. This is an even-length sequence whose transform contains no spectral term at $k=N/2$. We will come back to that point later.

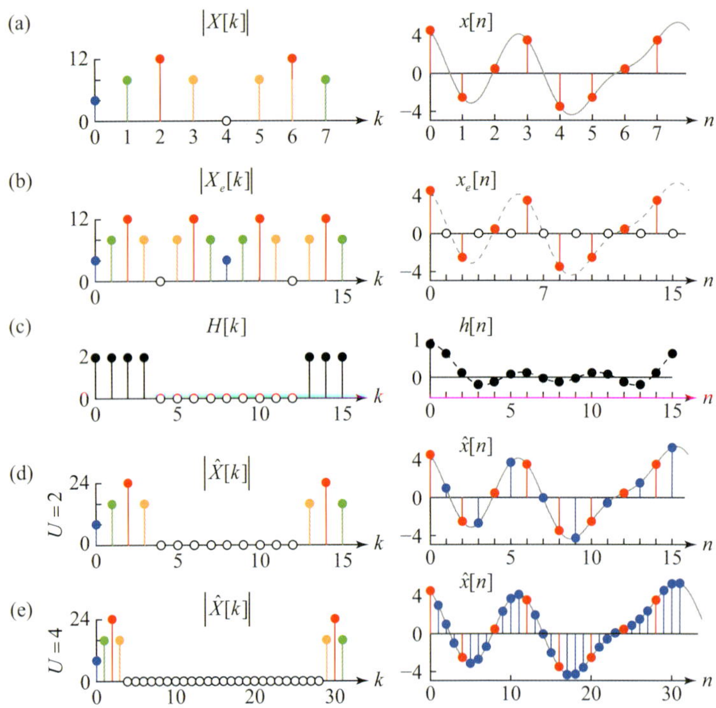

Figure 10.18 Increasing temporal resolution by zero-padding in the frequency domain

The first conceptual step is to create the expanded sequence $x_e[n]$ of length NU, effectively by placing $U-1$ zeros between each pair of points of $x[n]$:

$$x_e[n] = \sum_{l=0}^{N-1} x[l]\delta[n-lU], \quad 0 \le n \le NU-1. \tag{10.44}$$

We used this same technique in Chapter 6 to create an expanded sequence as part of an upsampling operation. The DFT of $x_e[n]$ is

$$\begin{aligned}X_e[k] &= \sum_{n=0}^{NU-1} x_e[n]e^{j2\pi kn/NU} = \sum_{n=0}^{NU-1}\left(\sum_{l=0}^{N-1} x[l]\delta[n-lU]\right)e^{j2\pi kn/NU}\\ &= \sum_{l=0}^{N-1} x[l]\sum_{n=0}^{NU-1}\delta[n-lU]e^{j2\pi kn/NU} = \sum_{l=0}^{N-1} x[l]e^{j2\pi klU/NU} = \sum_{l=0}^{N-1} x[l]e^{j2\pi kl/N}\\ &= X[(k)_N],\quad 0\le k\le NU-1.\end{aligned}$$

The only difference between $X_e[k]$ and $X[k]$ is that $X_e[k]$ is defined for k in the range $0\le k\le NU-1$, whereas $X[k]$ is only defined in the range $0\le k\le N$. Recall from our discussion of **Figure 10.2** that $X[k]$ is actually to be interpreted as one period of periodic sequence $\tilde{X}[k]$. Hence, $X_e[k]$ effectively comprises U periods of $\tilde{X}[k]$, as shown in **Figure 10.18b** for $U=2$.

To perform the equivalent of zero-padding in the frequency domain, multiply $X_e[k]$ by the DFT of the image-rejection filter $H[k]$, shown in the left panel of **Figure 10.18c**. The result is

$$\hat{X}[k] = X_e[k]H[k],$$

shown in the left panel of **Figure 10.18d**. The effect of this filtering operation is to remove the center (high-frequency) values of $X_e[k]$, leaving only the low-frequency components of $X[k]$ separated by zeros. Recognize that multiplication of $X_e[k]$ and $H[k]$ in the frequency domain corresponds to the circular convolution of $x_e[n]$ with $h[n]$ in the time domain, so

$$\hat{x}[n] = x_e[n]\circledast h[n],$$

where $h[n]$ is the impulse response of the image-rejection filter shown in the right panel of **Figure 10.18c** (see Problem 10-20 for a derivation),

$$h[n] - \frac{1}{N}\left(\frac{\sin(\pi(N-1)n/NU)}{\sin(\pi n/NU)}\right),\quad 0\le n\le NU-1. \tag{10.45}$$

For even values of N (see Problem 10-22), the final result is

$$\hat{x}[n] = \frac{1}{N}\sum_{l=0}^{N-1} x[l]\left(\frac{\sin(\pi(N-1)(n-lU)/NU)}{\sin(\pi(n-lU)/NU)}\right),\quad 0\le n\le NU-1, \tag{10.46}$$

shown in the right panel of **Figure 10.18d**. (See Problem 10-24 for odd values of N.) Equation (10.46) is an interpolation formula in which the interpolation function is the periodic sinc function of Equation (10.45). The interpolated values are shown with blue symbols. Interpolation is exact at values of n which are a multiple of U (shown with red symbols), that is values of $\hat{x}[lU]=x[l]$, for $0\le l\le N-1$ (see Problem 10-25).

In practice, we can implement upsampling directly in the frequency domain without going through all the steps of **Figure 10.18**. For even values of N, the DFT $\hat{X}[k]$ of

Figure 10.18d can be created directly from $X[k]$ by zero-padding the center of the transform with $NU-N+1$ zeros,

$$\hat{X}[k]=\begin{cases} UX[k], & 0\le k\le N/2-1 \\ 0, & N/2\le k\le NU-N/2 \\ UX[NU-k] & NU-(N/2-1)\le k\le NU-1 \end{cases}. \tag{10.47}$$

Then, $\hat{x}[n]$ is just the IDFT of $\hat{X}[k]$. An example should make things clear.

Example 10.11

Use Matlab to upsample the sequence shown in the right panel of **Figure 10.18a** by zero-padding the DFT, Equation (10.47), by a factor $U=4$.

➤ Solution:

Let us create a small function to do the work:

```
function xhat = zeropadf(x, U)
    N = length(x);              % Assume N is even
    X = fft(x);
    if (abs(X(1+N/2))>1e-6) % Check that center value of X[k]=0
        error('X[N/2] must be zero');
    end
    XX = zeros(1, U*N);
    XX(1:N/2) = X(1:N/2);
    XX(U*N:-1:U*N-N/2+2) = conj(XX(2:N/2)); % Assumes that x[n] is real
    xhat = U*ifft(XX);
end
```

Then,

```
N = 8;
n = 0:N-1;
x = 0.5 + 2*cos(2*pi*n/N) - 3*sin(2*pi*2*n/N) + 2*cos(2*pi*3*n/N);
y = zeropadf(x, 4);
```

The left panel of **Figure 10.18e** shows the DFT (XX in the code) and the right panel of the figure shows the resulting interpolation (y in the code), obtained by taking the IDFT.

The zero-padding approach to upsampling has certain limitations. In the time-domain upsampling method discussed in Chapter 6, the input signal $x[n]$ is first expanded by a factor of U, and the expanded signal is then filtered by an image-rejection filter whose specifications can be tailored by the designer to the demands of the problem. In the zero-padding approach via the DFT, the image-rejection filter is the periodic sinc function of Equation (10.45), whose characteristics are completely determined by the parameters N and U.

10.5.3 ★ Recovery of the DTFT from the DFT

The main utility of the discrete Fourier transform is that an N-point sequence can be completely reconstructed from only N samples of the DTFT, using the DFT coefficients $X[k]$ and the IDFT (Equation (10.12)). In supplementary material, we show that the entire DTFT $X_r(\omega)$ can be reconstructed for all frequencies ω from just these N DFT coefficients.

SUMMARY

The material in this chapter forms the theoretical bridge that connects the discrete-time Fourier transform (DTFT) to all practical applications of Fourier transformation using computers. The DFT is just the sampled DTFT, so most of the properties of the transforms are homologous. However, in working with the DFT, it is critical to keep in mind that both the N-point DFT sequence $X[k]$ and the IDFT sequence $x[n]$ need to be viewed as windowed N-point sections of the periodic sequences $\tilde{x}[n]$ and $\tilde{X}[k]$, respectively. It is this periodicity that underlies the most distinctive and practically important properties of the DFT: circular shift and circular convolution.

Due to the importance of the DFT in enabling a host of practical applications, a great deal of effort has been put into developing fast and efficient algorithms for computing the DFT. In Chapter 11, we will discuss a number of these algorithms, which are collectively called the fast Fourier transform (FFT).

In Chapter 12, we will build upon the material of this chapter and introduce a close relation of the DFT, the discrete cosine transform (DCT), which forms the core of many powerful applications, for example audio compression algorithms such as the well-known MP3 encoder for speech and music processing, and image compression algorithms such as the JPEG encoder that make possible efficient processing of still pictures and moving images.

PROBLEMS

Problem 10-1

Two 8-point sequences, $x[n]$ and $h[n]$, are shown in **Figure 10.19**.

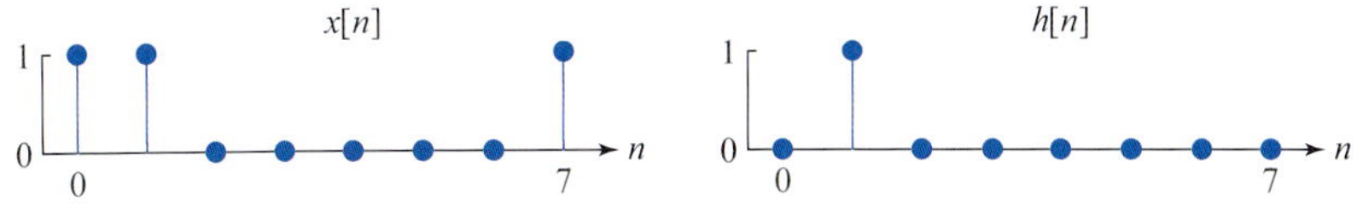

Figure 10.19

(a) Find the 8-point DFT of $x[n]$,

$$X[k] = \sum_{n=0}^{N-1} x[n]e^{-j2\pi kn/N},$$

by brute-force computation.

(b) Find $X[k]$ by recognizing that it can be obtained by sampling the transform of $x[n] = \delta[n+1] + \delta[n] + \delta[n-1]$ at eight points.
(c) Find the sequence $y[n]$ defined as the inverse DFT of $Y[k] = X[k]H[k]$. You *do not* have to compute $Y[k]$ to solve this problem.

Problem 10-2

A sequence $x[n]$ of length $N = 4$ has DFT $X[k] = \delta[k] + 2\delta[k-1] + 3\delta[k-2] + 4\delta[k-2]$. A new sequence is created, $y[n] = x[n]e^{-j1.5\pi n}$. Find $Y[k]$.

Problem 10-3

Given $x[n] = [1\ 2\ 3\ 4\ 5\ 0\ 0\ 6]$, find the circular shift $y[n] = x[(n - n_0)_8]$ for the following values of n_0:
(a) $n_0 = 2$.
(b) $n_0 = -2$.
(c) $n_0 = 1068$.

Problem 10-4

Find the circular convolution $y[n] = x[n] \circledast h[n]$ of the following two sequences,

$$x[n] = [1\ 3\ 0\ 2]$$
$$h[n] = [1\ 1\ 0\ 1],$$

given
(a) $N = 8$.
(b) $N = 6$.
(c) $N = 4$.

Problem 10-5

Find the circular convolution $y[n] = x[n] \circledast h[n]$ of the following two sequences,

$$x[n] = [1\ 2\ 0\ 0]$$
$$h[n] = [0\ 0\ 1\ 2],$$

given
(a) $N = 6$.
(b) $N = 4$.

Problem 10-6

Use the multiplication property of the DFT to compute the product of the following two sequences,

$$x[n] = [2\ \ 1\ \ 0\ \ 1]$$
$$w[n] = [1\ \ -1\ \ 1\ \ -1].$$

(a) Find the 4-point DFTs $X[k]$ and $W[k]$.
(b) Compute

$$Y[k]=\frac{1}{N}X[k]\circledast W[k].$$

(c) Find the IDFT $y[n]$, and verify that the result matches the multiplication of $x[n]$ and $w[n]$ in the time domain.

Problem 10-7

Two 4-point sequences, $x[n]$ and $h[n]$, are shown in **Figure 10.20**.

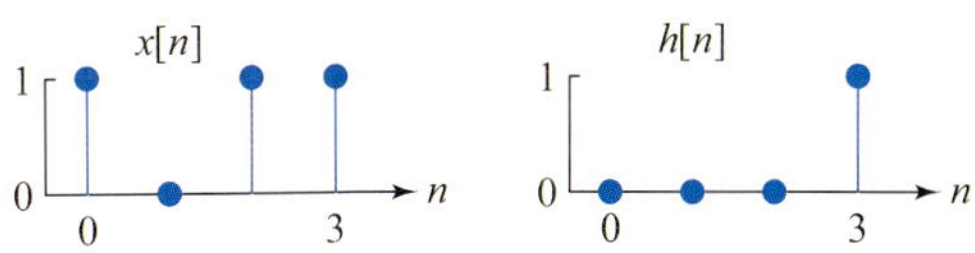

Figure 10.20

(a) Find $X[k]$ and $H[k]$, the 4-point DFTs of $x[n]$ and $h[n]$.
(b) The sequence $y[n]$ is defined as the inverse DFT of $Y[k]=X[k]H[k]$. Find $y[n]$ by computing $Y[k]$ and taking the inverse transform.
(c) Find $y[n]$ by periodic convolution (e.g., linear convolution followed by time-domain aliasing) and show that the results are the same.

Problem 10-8

Two 4-point sequences, $x[n]$ and y$[n]$, are shown in **Figure 10.21**.

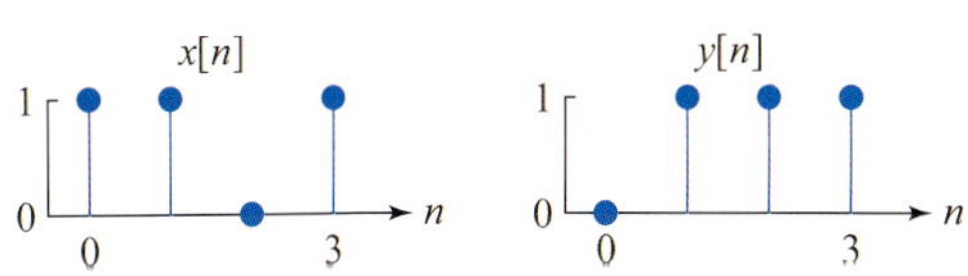

Figure 10.21

The sequence $y[n]$ results from the circular convolution of $x[n]$ with an unknown 4-point impulse response, $h[n]$. Find $h[n]$.

Problem 10-9

A signal $x[n]$ has DFT $X[k]=[1\ 2\ 3\ 4\ 5\ 6]$. Find $Y[k]$, the DFT of the sequence $y[n]=(-1)^n x[n]$.

Problem 10-10

Find the DFT of the sequence $x[n]=[2\ 0\ 2\ 0\ 2\ 0\ 2\ 0]$.

➤**Hint:** Express it as a constant plus a cosine.

Problem 10-11

For each of the following sequences $x[n]$ find the circular time-reversed sequence $x[(-n)_N]$.

(a) $x[n] = [4\ 3\ 2\ 1]$.

(b) $x[n] = [(1+j)\ (2+2j)\ (3+3j)\ 4\ (5+5j)]$.

Problem 10-12

For each of the sequences $x[n]$ shown in **Figure 10.22**, find and plot the magnitude and phase of the DFT, $|X[k]|$ and $\measuredangle X[k]$.

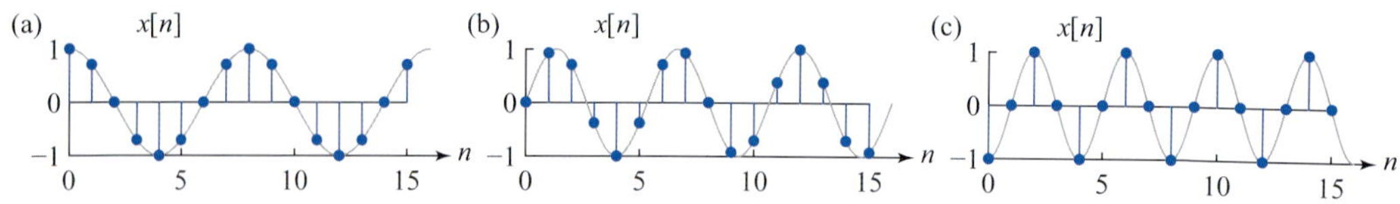

Figure 10.22

Problem 10-13

For each of the DFTs, $X[k]$, whose magnitude and phase are shown in **Figure 10.23**, find $x[n]$.

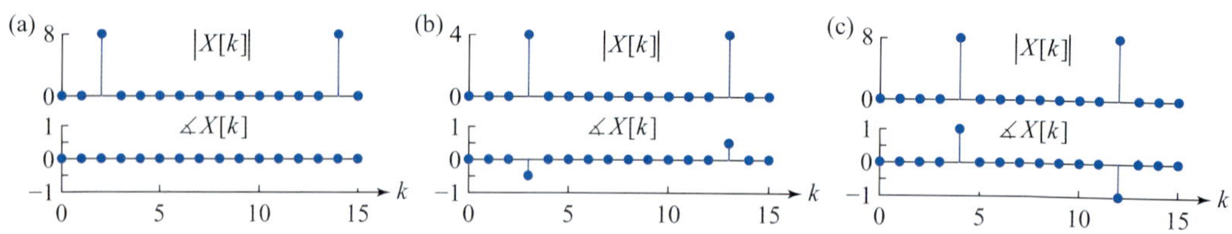

Figure 10.23

Problem 10-14

For each of the DFTs, $X[k]$, whose magnitude and phase are shown in **Figure 10.24**, find $x[n]$.

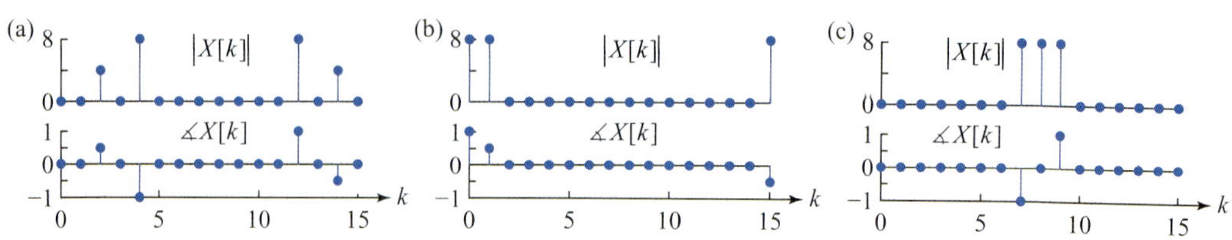

Figure 10.24

Problem 10-15

A signal $x[n] = [1\ 2\ 3\ 0\ 3\ 2]$ has DFT $X[k] = [11\ 0 - 4\ 3 - 4\ 0]$. Find $y[n]$, the sequence corresponding to the DFT $Y[k] = [3 - 4\ 0\ 11\ 0 - 4]$.

Problem 10-16

We define a Matlab function as follows:

```
function y = myfunc(x, h, N)
   XX = fft(x, N);
   HH = fft(h, N);
   YY = XX .* HH;
   y = ifft(YY, N);
end
```

(a) Given `x = [1 2 0 1]` and `h = [1 2 1]`, find `y = myfunc(x, h, 8)` without using Matlab.

(b) Given the same `x` and `h` as in part (a), determine `y = myfunc(x, h, 4)` without using Matlab.

(c) Given the same `x` as in part (a), find `h` such that `y = myfunc(x, h, 4)` yields `y = [1 1 2 0];`

Problem 10-17

We define a Matlab function as follows:

```
function y = myfunc2(x, n0)
   N = length(x);
   XX = fft(x, N);
   YY = XX .* exp(-1j*2*pi*n0*(0:N-1)/N);
   y = ifft(YY, N);
end
```

(a) Given `x = [1 2 0 1 0 0]`, find `y = myfunc2(x, 2)` without using Matlab.

(b) Given `x = [1 2 0 1 0 0]`, find `y = myfunc2(x, -2)` without using Matlab.

(c) Given `x = [1 2 0 1 0 0]`, find `y = myfunc2(x, 25)` without using Matlab.

Problem 10-18

We define a Matlab function as follows:

```
function y = myfunc3(x)
   X = fft(x);
   Y = [X(2:end) X(1)] + [X(end) X(1:end-1)];
   y = ifft(Y);
end
```

Given `x = cos((pi/4)*(0:7))`, find `y = myfunc3(x)` without using Matlab.

Problem 10-19

(a) Show that $X(\omega)$, the DTFT of the N-point sequence $x[n] = \cos(\omega_0 n + \theta)$, $0 \le n \le N-1$, is

$$X(\omega) = \frac{1}{2}(e^{j\theta}\Psi_N(\omega - \omega_0) + e^{-j\theta}\Psi_N(\omega + \omega_0)),$$

where,

$$\Psi_N(\omega) = e^{-j(N-1)\omega/2}\frac{\sin N\omega/2}{\sin \omega/2}.$$

(b) When ω_0 is an integer fraction, namely $\omega_0 = 2\pi l/N$, $0 \le l \le N-1$, use the results of part (a) and Equation (10.17) to show that the N-point DFT of $x[n]$ is

$$X[k] = \frac{N}{2}(\delta[(k-l)_N] + \delta[(k+l)_N]) = \frac{N}{2}(\delta[k-l] + \delta[k-(N-l)]).$$

Problem 10-20

In this problem, we derive Equation (10.45), the expression for $h[n]$, the impulse response of the image-rejection filter used to produce the sequence $\hat{x}[n]$ that results from interpolating a sequence $x[n]$ of even length by a factor U.

The impulse response can be calculated as follows:

$$h[n] = U \cdot \frac{1}{NU}\left(\sum_{n=0}^{N/2-1} e^{j2\pi kn/NU} + \sum_{n=NU-(N/2-1)}^{NU-1} e^{j2\pi kn/NU}\right).$$

In order to make the summations the same size, extend the limit of the second summation to NU and subtract $e^{j2\pi NUn/NU} = 1$. Then let $\hat{k} = NU - k$ in the second summation.

Problem 10-21

Show that the circular convolution of the expanded sequence $x_e[n]$, Equation (10.44), with the impulse response of the image-rejection filter $h[n]$, Equation (10.45), results in the interpolation formula found in Equation (10.46).

➤Hint: Use Equation (10.37).

Problem 10-22

In this problem, we provide an alternate derivation of Equation (10.46), the expression for the interpolated sequence $\hat{x}[n]$ resulting from zero-padding the DFT of a sequence $x[n]$ with zeros. Assume that N, the length of $x[n]$, is even, and that the length of $\hat{x}[n]$ is NU, where U is an integer. The algebra of this derivation can be pretty exhausting. You might consider starting with $\hat{x}[n]$ defined in terms of the IDFT of $\hat{X}[k]$ as follows:

$$\hat{x}[n] = \frac{1}{NU}\sum_{k=0}^{NU-1}\hat{X}[k]e^{j2\pi kn/NU}$$
$$= \frac{1}{NU}\left\{\sum_{k=0}^{N/2-1}\hat{X}[k]e^{j2\pi kn/NU} + \sum_{k=NU-(N/2-1)}^{NU}\hat{X}[k]e^{j2\pi kn/NU} - \hat{X}[NU]\right\}.$$

We are not allowing the center point of this DFT, $\hat{X}[NU/2]$, to have a non-zero value. The second summation has been extended to include an extra point, $\hat{X}[NU]$, so that this summation is the same length as the first summation. This extra point then needs to be subtracted. To proceed, recognize that

$$\hat{X}[k] = \hat{X}[NU-k] = UX[k],\ 0 \le k \le N/2-1.$$

Note that when $x[n]$ is upsampled by a factor of U, $X[k]$ needs to be scaled by U.

(a) Let $\hat{k} = NU - k$ in the second summation and remember that for real $x[n]$, $X[k] = X^*[N-k]$. After some manipulation, you should get

$$\hat{x}[n] = \frac{1}{N}\left\{2\,\mathrm{Re}\left\{\sum_{k=0}^{N/2-1}X[k]e^{j2\pi kn/NU}\right\} - X[0]\right\}.$$

(b) Then, for all instances of $X[k]$, substitute

$$X[k] = \sum_{m=0}^{N-1}x[m]e^{-j2\pi km/N}.$$

You should get

$$\hat{x}[n] = \frac{1}{N}\sum_{m=0}^{N-1}x[m]\left(2\mathrm{Re}\left\{\sum_{k=0}^{N/2-1}e^{j2\pi k(n-mU)/NU}\right\} - 1\right).$$

(c) Using the formula for a finite geometric sum, show that

$$\sum_{k=0}^{N-1}e^{jk\alpha} = e^{j\alpha(N-1)/2}\frac{\sin\alpha N/2}{\sin\alpha/2}.$$

Then, use this result to get

$$\hat{x}[n] = \frac{1}{N}\sum_{m=0}^{N-1}x[m]\left(\frac{2\sin(\pi(N/2)(n-mU)/NU)}{\sin(\pi(n-mU)/NU)}\cos(\pi(N/2-1)(n-mU)/NU) - 1\right).$$

(d) Use various trigonometric identities to get the final answer,

$$\hat{x}[n]=\frac{1}{N}\sum_{m=0}^{N-1}x[m]\left(\frac{\sin(\pi(N-1)(n-mU)/NU)}{\sin(\pi(n-mU)/NU)}\right). \tag{10.48}$$

Problem 10-23

In this problem, derive the expression for $\hat{x}[n]$, the interpolated sequence resulting from zero-padding the DFT of a sequence $x[n]$ of length N with zeros, assuming that N is *odd* and that the length of $\hat{x}[n]$ is NU, where U is an integer:

$$\hat{x}[n]=\frac{1}{N}\sum_{m=0}^{N-1}x[m]\left(\frac{\sin(\pi(n-mU)/U)}{\sin(\pi(n-mU)/NU)}\right).$$

Problem 10-24

A sequence $x[n]$ of even length N is interpolated by a factor U by zero padding its DFT $X[k]$ resulting in an interpolated sequence $\hat{x}[n]$. Assume that the center point of $X[k]$ is zero; that is,

$$X[N/2]=\frac{1}{N}\sum_{m=0}^{N-1}x[m]e^{-j2\pi(N/2)m/N}=\frac{1}{N}\sum_{m=0}^{N-1}x[m]e^{-j\pi m}=\frac{1}{N}\sum_{m=0}^{N-1}x[m](-1)^{-m}=0.$$

Use this fact and the interpolation formula for $\hat{x}[n]$ in terms of $x[n]$, Equation (10.48), to show that every Uth point of $\hat{x}[n]$ will be exactly equal to $x[n]$.

Problem 10-25

The right panel of **Figure 10.25a** shows $x[n]$, a sequence of even length $N=8$, whose DFT, $X[k]$, is shown in the left panel. The DFT has a non-zero spectral component at $X[N/2]$. In order to upsample $x[n]$ by a factor U, a DFT $\hat{X}[k]$ is created by zero-padding $X[k]$ to a length NU:

$$\hat{X}[k]=\begin{cases}UX[k], & 0\le k\le N/2-1\\ 0, & N/2\le k\le NU/2-1\\ UX[N/2], & k=NU/2\\ 0, & NU/2+1\le k\le NU-N/2\\ UX[NU-k] & NU-(N/2-1)\le k\le NU-1\end{cases}. \tag{10.49}$$

In order to maintain the symmetry of the transform, we set $\hat{X}[NU/2]=X[N/2]$.

(a) Show that the interpolated sequence $\hat{x}[n]$ corresponding to Equation (10.49) is

$$\hat{x}[n]=\frac{1}{N}\sum_{m=0}^{N-1}x[m]\left(\frac{\sin(\pi(N-1)(n-mU)/NU)}{\sin(\pi(n-mU)/NU)}+(-1)^{n-m}\right).$$

➤**Hint:** Modify Equation (10.48) using the fact that the middle point of $X[k]$ is

$$X[N/2] = \frac{1}{N}\sum_{m=0}^{N-1} x[m]e^{-j2\pi(N/2)m/N} = \frac{1}{N}\sum_{m=0}^{N-1} x[m]e^{-j\pi m} = \frac{1}{N}\sum_{m=0}^{N-1} x[m](-1)^{-m}.$$

(b) Use Matlab to show the plots for $\hat{X}[k]$ and $\hat{x}[n]$ for values of $U=2$ and $U=4$. Your answers should look like **Figure 10.25**.

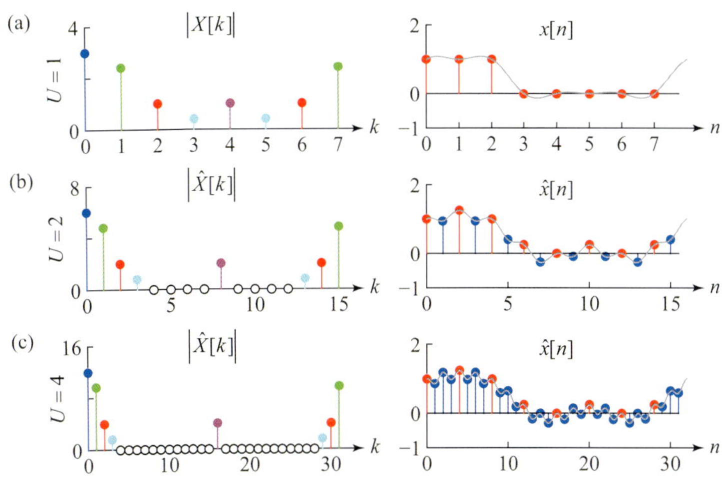

Figure 10.25

11 Fast Fourier transform (FFT)

Introduction

The **fast Fourier transform (FFT)** is a computationally efficient implementation of the DFT. It is based on a "divide-and-conquer" approach: dividing the DFT into smaller parts, and exploiting the periodicity and symmetry properties of the DFT to compute these smaller parts rapidly. The historical antecedents of the FFT date back two centuries, to 1801–1805, when the virtuoso German mathematician Carl Friedrich Gauss (1777–1855) developed a method equivalent to the modern decimation-in-frequency FFT as a solution to a harmonic analysis problem that involved predicting the orbit of the dwarf planet Ceres, which lies in the asteroid belt near Neptune. Over the next 160 years, a number of algorithms for efficiently performing harmonic analysis were proposed, but it was not until 1965 that J. W. Cooley and J. W. Tukey published the details of what we now know as the FFT algorithm. In the last half century, there has been an explosion of work on variants and refinements of the basic algorithm, some of which we will discuss in this chapter, making the FFT one of the most widely used and important numerical methods in science and engineering. The FFT has enabled a whole host of practical DSP applications in areas such as real-time spectral analysis, speech processing, communication, radar systems, data compression, data analysis and fast convolution. It is an indispensable tool in the DSP arsenal.

As with a number of other areas of DSP (filter design, for example), most scientists and engineers will probably never have to code their own FFT routine. Sophisticated, efficient software implementations of the algorithm are widely available for general-purpose machines and, increasingly, for microprocessors and even microcontrollers that include DSP cores with hard-coded FFTs. Nevertheless, a worker should always understand his or her tools. So, in this chapter, we will discuss a number of variants of the FFT and how to use them. In Section 11.1, we explain the simplest and most commonly implemented FFT algorithm, the radix-2 FFT. If you read only one section of this chapter, read that. Then, in the following three sections we discuss other transforms that offer some improvement in speed: radix 4, split radix and mixed-radix FFTs. In Section 11.4, we present the inverse FFT (the IFFT) and in Section 11.6 we show a couple of useful techniques that double the capacity or speed of the FFT and IFFT when used in the processing of real sequences. In Section 11.8, we discuss fast convolution of sequences using the multiplication of FFTs. Finally, at the end of the chapter, in Section 11.11, we talk about the implementation of these transforms. In the supplementary material available at www.cambridge.org/holton, we discuss additional FFT algorithms and their refinements.

11.1 Radix-2 FFT transforms

To understand the FFT, start with the expression for an N-point DFT,

$$X[k] = \sum_{n=0}^{N-1} x[n]e^{-j2\pi kn/N} = \sum_{n=0}^{N-1} x[n]W_N^{kn},$$

where we have defined $W_N \triangleq e^{-j2\pi/N}$ (note the minus sign in the exponent). The DFT is just a linear transformation that maps $x[n]$ into $X[k]$. We can diagram this transform as a "box" that takes N points of $x[n]$ as input and produces N points of $X[k]$ as output, as shown in **Figure 11.1**.

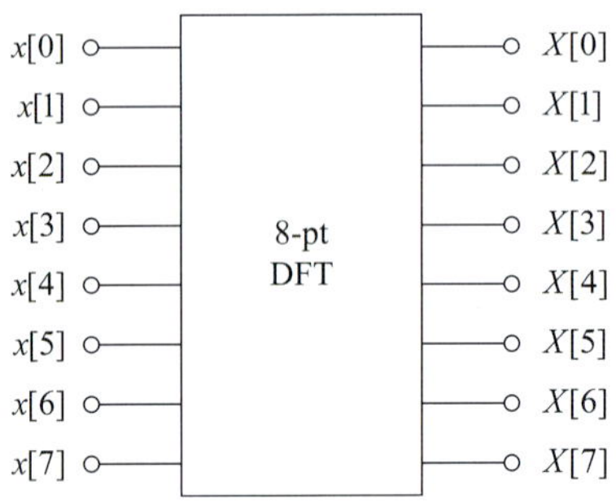

Figure 11.1 Eight-point DFT

The goal of the FFT is to evaluate the DFT fast. Traditionally, the number of multiplications – real or complex – in a given algorithm has been used as a measure of the execution time. The reason is that in the bad old days, when hardware multipliers were not common in computers, multiplication could be a rate-limiting step in the calculation. However, with modern processors that is no longer necessarily true. Furthermore, the speed of the transform is determined not only by the theoretical limits of algorithm chosen, but also, critically, by how it is implemented in practice, something we will talk about in Section 11.10. So, in this chapter, although we will continue to use the number of multiplications as a rough guide to the speed and efficiency of the algorithm, we are not going to count every single one.

The direct computation of an N-point DFT requires on the order of N^2 complex multiplications. So, to compute $X[k]$ for each of N values of k, we must compute and sum N values of the product $x[n]W_N^{kn}$, where in general $x[n]$ can be complex. The goal of FFT algorithms is to perform this same DFT in fewer than N^2 multiplications, preferably many fewer!

There are many different flavors of the FFT, some of which we will discuss in this chapter, but the essence of all FFT algorithms is the same: to partition the DFT recursively into groups of smaller DFTs, each of which requires a smaller number of complex operations. Many FFT transforms are classed by their **radix**. In a radix-r transform, the size of the DFT, N, is a power of r; that is, $N=r^M$, where $M=\log_r N$ is an integer. In this section, we will develop the easiest-to-understand FFTs, the radix-2 transforms. In a radix-2 transform, the DFT is recursively partitioned into two DFTs, each of half the original size. The size of the transformation, N, must therefore be a power of two (i.e., $N=2^M$), a restriction that does not turn out to be particularly onerous. There are two basic types of radix-2 transformations, the decimation-in-time FFT and the decimation-in-frequency FFT. We will first consider the decimation-in-time FFT.

11.1.1 Decimation-in-time FFT

In the **decimation-in-time (DIT)** FFT an N-point input sequence $x[n]$ is separated into two $N/2$-point subsequences $x_1[n]$ and $x_2[n]$, which represent $x[n]$ at the even and odd values of the index n, respectively. The DFT of $x[n]$ can then be expressed in terms of the transforms of $x_1[n]$ and $x_2[n]$:

$$\begin{aligned} X[k] &= \sum_{n=0}^{N-1} x[n]W_N^{kn} = \sum_{n=0}^{N/2-1} \underbrace{x[2n]}_{x_1[n]} W_N^{k2n} + \sum_{n=0}^{N/2-1} \underbrace{x[2n+1]}_{x_2[n]} W_N^{k(2n+1)} \\ &= \sum_{n=0}^{N/2-1} x_1[n]W_N^{k2n} + \sum_{n=0}^{N/2-1} x_2[n]W_N^{k(2n+1)}. \end{aligned} \tag{11.1}$$

Each of the two summations is half the size of the original. Now,

$$W_N^{k2n} = e^{-j2\pi k(2n)/N} = e^{-j2\pi kn/(N/2)} = W_{N/2}^{kn},$$

and

$$W_N^{k(2n+1)} = e^{-j2\pi k(2n+1)/N} = e^{-j2\pi k(2n)/N} e^{-j2\pi k/N} = W_{N/2}^{kn} W_N^k,$$

so Equation (11.1) becomes

$$X[k] = \underbrace{\sum_{n=0}^{N/2-1} x_1[n]W_{N/2}^{kn}}_{X_1[k]} + W_N^k \underbrace{\sum_{n=0}^{N/2-1} x_2[n]W_{N/2}^{kn}}_{X_2[k]} = X_1[k] + W_N^k X_2[k], \quad 0 \le k < N, \tag{11.2}$$

where $X_1[k]$ and $X_2[k]$ are the $N/2$-point DFTs of the even and odd points of $x[n]$, respectively:

$$X_1[k] = \sum_{n=0}^{N/2-1} x_1[n]W_{N/2}^{kn}, \quad 0 \le k < N$$

$$X_2[k] = \sum_{n=0}^{N/2-1} x_2[n]W_{N/2}^{kn}, \quad 0 \le k < N$$

Even though $X_1[k]$ and $X_2[k]$ are $N/2$-point DFTs, Equation (11.2) still requires that both be evaluated at all N values of k. Then, we assemble $X[k]$ from $X_1[k]$ and $X_2[k]$, by multiplying $X_2[k]$ by W_N^k and adding it to $X_2[k]$. The W_N^k factors are often called **twiddle factors**,[1] because

[1]Twiddle factors! Who says that engineers don't have a sense of humor?

the effect of multiplying $X_2[k]$ by a twiddle factor $W_N^k = e^{-j2\pi k/N}$ is to twiddle (i.e., change) just the phase by $-2\pi k/N$ before adding the result to $X_1[k]$.

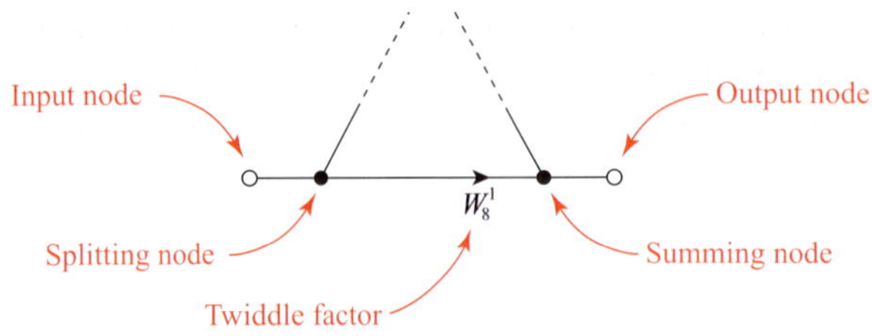

Figure 11.2 Signal-flow graph

It is easy to get bogged down in the algebra that underlies the algorithms of the FFT but, fortunately, most of the important concepts can be conveyed graphically. To aid in the discussion that follows, we will employ the signal-flow graph notation we introduced in Section 9.1 to represent filter architecture. As a reminder, **Figure 11.2** shows a portion of a signal-flow graph. Signals in this particular graph "flow" from left to right. Signals enter at an **input node** and exit at an **output node**, each of which is drawn with an open circle (○). **Splitting nodes** and **summing nodes** are represented by closed circles (•). At a splitting node, a signal entering from the left is split into two signals exiting to the right, where each path leaving the node has the same value. At a summing node, two signals entering from the left are summed into an output exiting the node to the right. An arrowhead (►) on a signal path represents a multiplication by a constant, which can be complex. In **Figure 11.2**, a multiplication by the complex twiddle factor W_8^1 is shown.

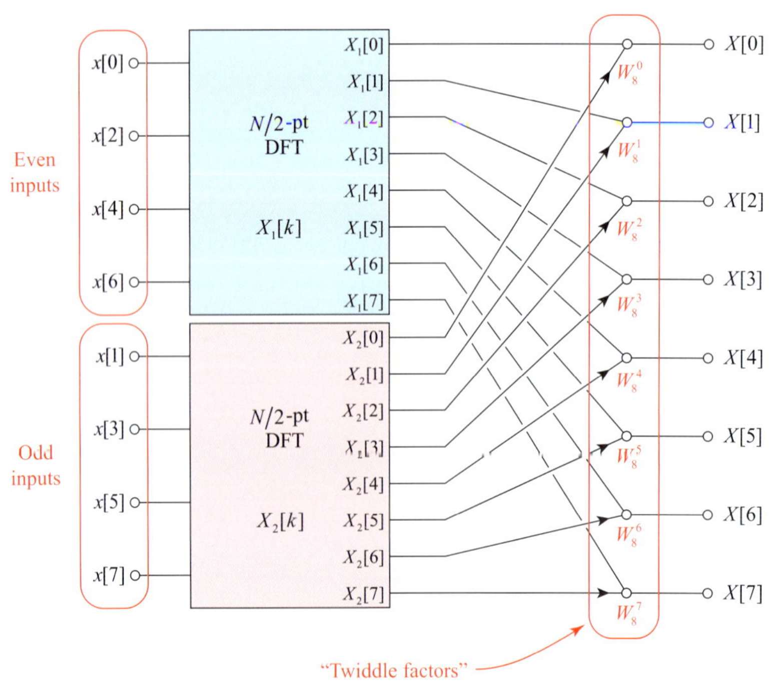

Figure 11.3 Intermediate stage of decomposition of an N-point DFT into two $N/2$-point DFTs

Figure 11.3 shows the signal-flow graph for Equation (11.2). The two boxes are the $N/2$-point DFTs $X_1[k]$ and $X_2[k]$. They each have $N/2 = 4$ inputs, corresponding to the even and odd input points respectively. However, each of these DFTs has $N = 8$ outputs, since they have to be evaluated at all values of k, $0 \leq k < N$. Each output value of $X_2[k]$ is then multiplied by the appropriate twiddle factor and added to the output of $X_1[k]$ to form $X[k]$.

Decomposing the N-point DFT into two $N/2$-point DFTs using Equation (11.2) does not, in itself, produce any savings in the number of complex multiplications. Each of the $N/2$-point DFTs requires $N^2/2$ complex multiplications ($N/2$ multiplications for each of N values of k) as well as $N/2 - 1$ additions. Also, each of the N values of $X_2[k]$ must be multiplied by a complex twiddle factor before being added to $X_1[k]$. So, to this point, the decomposition has actually *increased* the number of multiplications and additions required to compute the transform. The savings from the decomposition come from two important simplifying observations.

The first simplification is that every DFT is periodic. Specifically, an $N/2$-point DFT is periodic with a period of $N/2$. So, instead of having to compute N values of the DFTs $X_1[k]$ and $X_2[k]$ for $0 \leq k < N$, we only need to compute the first $N/2$ values, for $0 \leq k < N/2$, and then we can reuse these values to get the last $N/2$ values,

$$\begin{aligned} X_1[k+N/2] &= X_1[k] \\ X_2[k+N/2] &= X_2[k] \end{aligned}, \quad 0 \leq k < N/2. \tag{11.3}$$

The first $N/2$ points of $X[k]$ are given in Equation (11.2) as

$$X[k] = X_1[k] + W_N^k X_2[k], \quad 0 \leq k < N/2, \tag{11.4a}$$

and the last $N/2$ points of $X[k]$ are computed using the periodicity of the $N/2$-point DFT, Equation (11.3),

$$\begin{aligned} X[k+N/2] &= X_1[k+N/2] + W_N^{k+N/2} X_2[k+N/2] \\ &= X_1[k] + W_N^{k+N/2} X_2[k] \end{aligned}, \quad 0 \leq k < N/2. \tag{11.4b}$$

With $N = 8$, the signal flow graph representing Equations (11.4) is shown in **Figure 11.4**.

Each value of the $N/2$-point DFTs $X_1[k]$ and $X_2[k]$ is used *twice*, once to compute $X[k]$ and once to compute $X[k+N/2]$, both for $0 \leq k < N/2$. This gives the right side of the signal-flow graph the form of a series of $N/2$ "**butterflies**,"[2] one of which is highlighted in red. Each butterfly has two twiddle factors: a factor of W_N^k on the upper diagonal path and a factor of $W_N^{k+N/2}$ on the lower horizontal path.

[2] Butterflies! Who says that engineers don't have aesthetic sensibilities?

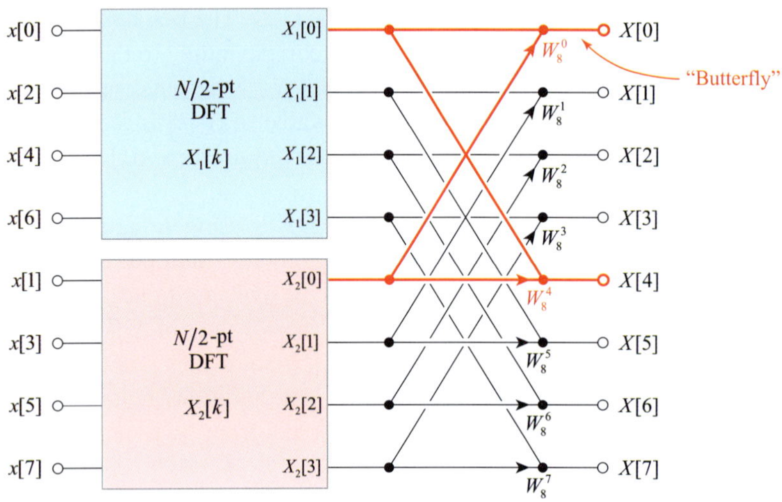

Figure 11.4 Decomposition of an N-point DFT into two-point DFTs

The second simplification of these equations and signal flow graph comes from the fact that the N twiddle factors on the butterflies are antisymmetric about $k = N/2$; that is,

$$W_N^{k+N/2} = W_N^k \cdot W_N^{N/2} = -W_N^k, \tag{11.5}$$

because $W_N^{N/2} = e^{-j2\pi(N/2)/N} = e^{-j\pi} = -1$. Using this result, Equations (11.4) become

$$\begin{aligned} X[k] &= X_1[k] + W_N^k X_2[k] \\ X[k+N/2] &= X_1[k] - W_N^k X_2[k] \end{aligned}, \quad 0 \le k < N/2. \tag{11.6}$$

Each butterfly in **Figure 11.4** has a twiddle factor of value $W_N^{k+N/2}$ in the middle of the bottom path and a twiddle factor of value W_N^k in the diagonal path. Using Equation (11.5), we recognize that $W_N^{k+N/2} = -W_N^k$, so we can replace both twiddle factors by a single twiddle factor of value W_N^k moved to the beginning of the bottom path, plus an additional multiplication by -1 in the middle of the bottom path, as shown in **Figure 11.5**. The effect of this simplification is to reduce the number of twiddle factors needed for each butterfly from two to one.

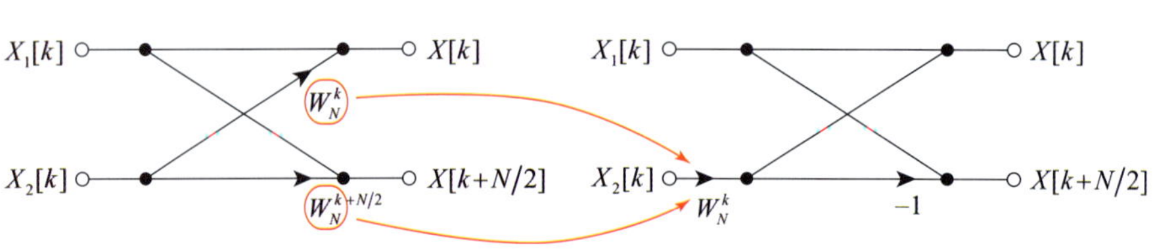

Figure 11.5 Simplifying a butterfly

If we are assessing the speed of an algorithm by the number of multiplications, then we have just done a very good thing. A multiplication by -1 can be achieved simply by changing the sign bit of a register and hence does not count as a multiplication.

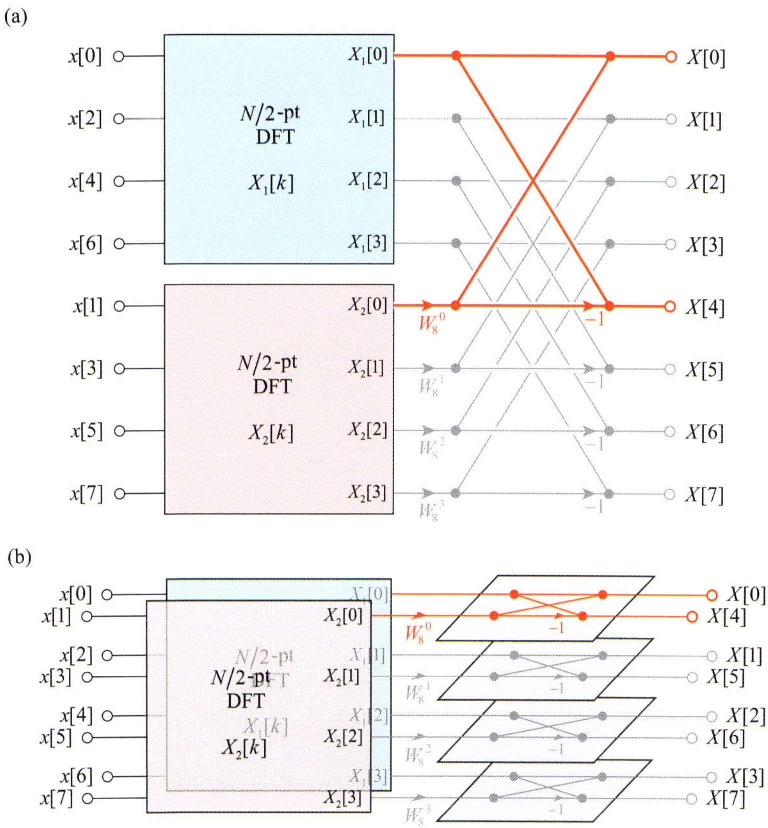

Figure 11.6 Decomposition of an N-point DFT with simplified butterflies

The signal-flow graph with the simplified butterflies is shown in **Figure 11.6a**. **Figure 11.6b** shows the same graph in "3-D." Each of the $N/2$-point DFTs requires only $N^2/4$ complex multiplications ($N/2$ multiplications for each of $N/2$ values of k) as well as $N/2 - 1$ additions. The butterflies themselves now only require $N/2$ complex multiplications since the number of twiddle factors has been cut in half and only the values of $X_2[k]$ must be multiplied by them. The butterflies also require N additions. After this stage of the decomposition, there are $2(N/2)^2 + N/2$ multiplications, which is less than N^2. So the number of complex multiplications has been reduced by the decomposition.

If N is a power of two, we can repeat this process again and decompose each of the $N/2$-point DFTs $X_1[k]$ and $X_2[k]$ into a pair of $N/4$-point DFTs plus associated twiddle factors. For example, $X_1[k]$ becomes

$$X_1[k] = \sum_{n=0}^{N/2-1} x_1[n] W_{N/2}^{kn} = \underbrace{\sum_{n=0}^{N/4-1} x_1[2n] W_{N/4}^{kn}}_{X_{11}[k]} + W_{N/2}^{k} \underbrace{\sum_{n=0}^{N/4-1} x_1[2n+1] W_{N/4}^{kn}}_{X_{12}[k]}$$

$$= X_{11}[k] + W_{N/2}^{k} X_{12}[k],$$

where $X_{11}[k]$ and $X_{12}[k]$ are $N/4$-point DFTs. Similarly, $X_2[k]$ can be decomposed into $X_{21}[k]$ and $X_{22}[k]$. The resulting signal-flow graph is shown in **Figure 11.7**.

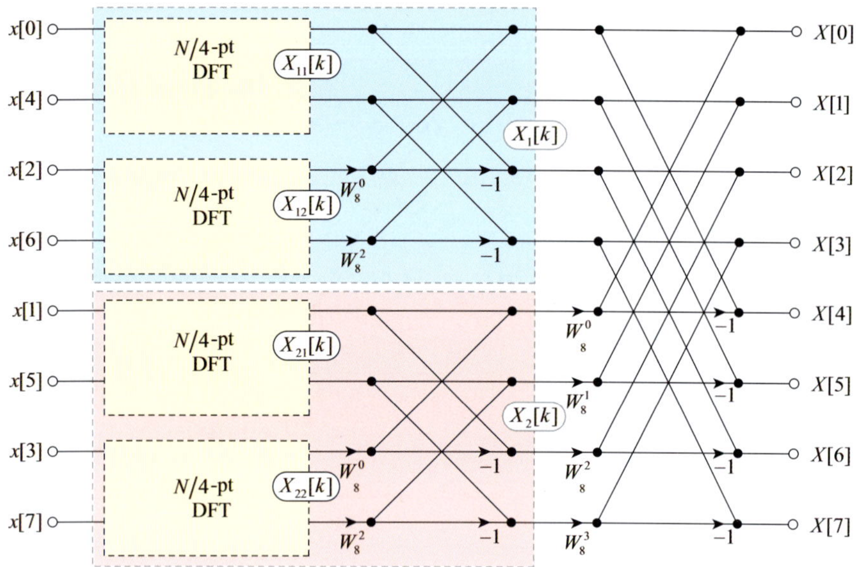

Figure 11.7 Second stage of decomposition of an eight-point DFT

We are left with four $N/4$-point DFTs. Since $N = 8$, each of these is simply a two-point DFT of the form

$$\hat{X}[k] = \sum_{n=0}^{N-1} \hat{x}[n]W_N^{kn} = \sum_{n=0}^{1} \hat{x}[n]W_2^{kn} = \hat{x}[0]W_2^0 + \hat{x}[1]W_2^k = \begin{cases} \hat{x}[0] + \hat{x}[1], & k = 0 \\ \hat{x}[0] - \hat{x}[1], & k = 1 \end{cases},$$

where $\hat{x}[n]$ and $\hat{X}[k]$ represent the inputs and outputs of the DFT, respectively. This simple butterfly has *no* multiplications, but for consistency with the other butterflies, we will include the twiddle factor $W_2^0 = 1$, as shown in **Figure 11.8**.

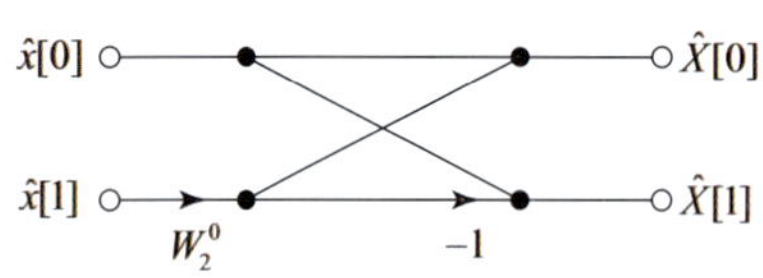

Figure 11.8 Simple two-point butterfly

The signal-flow graph of the completed transform is shown in **Figure 11.9**. The complete N-point FFT has a very simple and elegant form. The transform comprises $\log_2 N$ "stages" (in this case, $\log_2 8 = 3$), each of which contains $N/2$ butterflies. The only complex multiplications in the entire transform are the twiddle factors, one per butterfly. (All the twiddle factors have been converted to base-N to match the last stage. So, for example, W_4^1 in Stage 2 of **Figure 11.7** becomes W_8^2 in **Figure 11.9**.) All told, there are only a total of $(N/2)\log_2 N$ complex multiplications in an N-point FFT. This is substantially fewer than the N^2 multiplications required by direct calculation of the DFT.

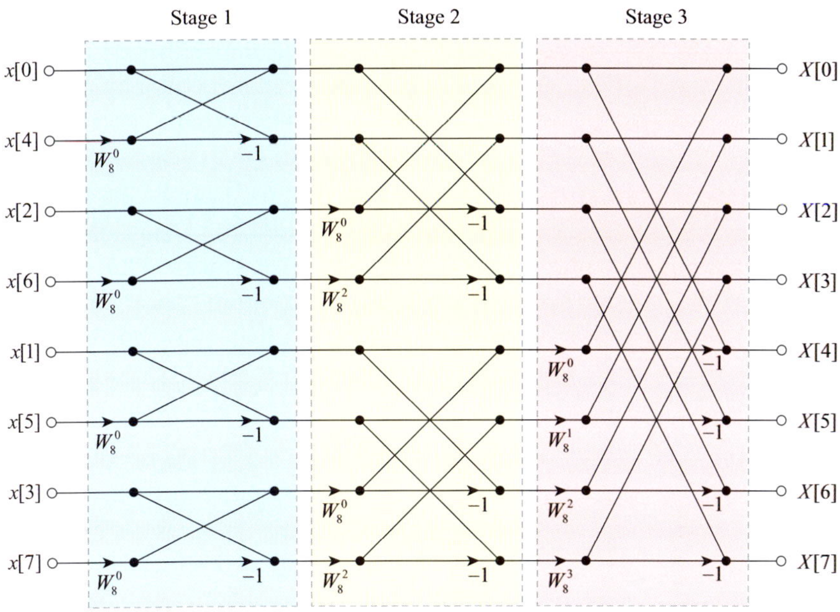

Figure 11.9 Eight-point DIT FFT

11.1.2 Computational gain

The theoretical speed advantage of the FFT over the direct computation of the DFT is enormous. **Table 11.1** shows the number of complex multiplications necessary to compute the DFT and the FFT of various sizes. We can define the **computational gain** as the ratio of the number of complex multiplications required to compute the DFT and FFT transforms. **Figure 11.10** plots the computational gain as a function of N on a log-log scale. Theoretically, an 8-point FFT is five times faster than the equivalent DFT and the speed advantage gets greater as N increases; for example, an 8192-point FFT is more than *three orders of magnitude* faster than the equivalent DFT. That is a huge improvement! You can understand how the invention (or perhaps re-invention) of the FFT algorithm by Cooley and Tukey in 1965, coupled with the cotemporaneous development of fast digital computers on which the algorithm could be implemented, can be said to have made a major part of modern, practical DSP possible.

The entire FFT only requires the computation of a total of $N/2$ twiddle factors: $W_N^k, 0 \leq k < N/2$. For example, for the 8-point FFT we only need to compute W_8^0, W_8^1, W_8^2 and W_8^3. Some of these values are used in more than one stage. Furthermore, not every twiddle factor actually represents a complex multiplication, and sophisticated transform implementations can use this fact to reduce further the number of required complex operations.[3] For example, the twiddle factor $W_N^0 = 1$ occurs $N-1$ times in an N-point FFT. Scaling a signal by 1 does not require any multiplications at all, so the FFT routine can be appropriately coded (or the signal path could be hard-wired in a hardware implementation) to avoid this factor completely. As

[3] A general complex multiplication of two numbers can be implemented by four real multiplications and two additions. In the specific case of the FFT, where the twiddle factors are pre-computed and stored, it is possible to perform a complex multiplication with a complex twiddle factor using only *three* real multiplications and *three* real additions (see Problem 11-4).

another example, the twiddle factor $W_N^{N/4} = -j$ occurs $N/2-1$ times in an N-point FFT. Multiplying a signal by j or $-j$ does not require any multiplications either. Multiplying a complex number $z = x + jy$ by j gives $jz = -y + jx$. This operation can be accomplished by swapping the real and imaginary parts and negating the (new) real part. Other twiddle factors can be coded so as to avoid the full computational cost of a complex multiplication. For example, multiplications by $W_N^{N/8} = e^{-j\pi/4} = (1-j)/\sqrt{2}$ and $W_N^{3N/8} = e^{-j3\pi/4} = -(1+j)/\sqrt{2}$ can be implemented with only two real multiplications and two real additions (see Problem 11-5). Using all these shortcuts, we could actually perform the entire eight-point transform of **Figure 11.9** with just four real multiplications. While trying to squeeze every last extraneous multiplication from the FFT can have costs in terms of hardware or program complexity that outweigh the benefits, we will discover that having an FFT "kernel" of such low computational cost can actually be useful when the FFT is implemented by a recursive method, as explained in Section 11.11. Also, transforms such as the radix-4 and split-radix transforms, which we will discuss below, use some of these simplifications to reduce the cost of the FFT by factors of 25% and 33%, respectively.

Table 11.1 Comparison of DFT and FFT operations

	Number of complex multiplications		
N	DFT (N^2)	FFT ((N/2) $\log_2 N$)	Computational gain of FFT
8	64	12	5.3
16	256	32	8.0
32	1024	80	12.8
64	4096	192	21.3
128	16384	448	36.6
256	65536	1024	64.0
512	262144	2304	113.8
1024	1048576	5120	204.8
2048	4194304	11264	372.4
4096	16777216	24576	682.7
8192	67108864	53248	1260.3

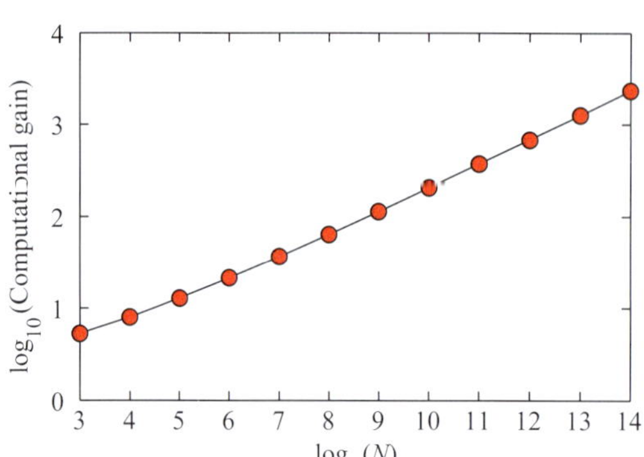

Figure 11.10 Computational gain

11.1.3 Bit reversal

Looking at the completed DIT transform, you can see that the output points $X[k]$ are in sequential or **normal order**: $X[0]$, $X[1]$, $X[2]$, . . . However, the input points $x[n]$ are scrambled: $x[0]$, $x[4]$, $x[2]$, $x[6]$, $x[1]$, $x[5]$, $x[3]$, $x[7]$. This ordering of the input sequence is termed **bit-reversed order**. Bit-reversed ordering occurs because at each stage, the DIT decomposition process sorts the input sequence into even and odd components. This process is illustrated in **Figure 11.11** for the eight-point FFT.

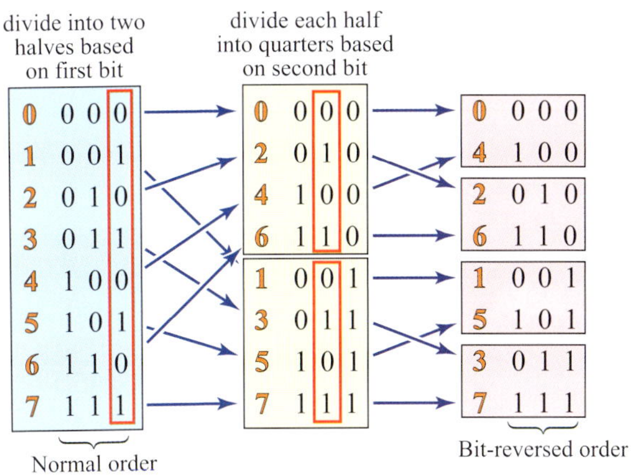

Figure 11.11 Bit reversal

The left-hand panel of the figure shows both the decimal and binary representation of the index n of the input sequence just before the first DIT decomposition. This corresponds to the situation depicted in **Figure 11.1**. In this first decomposition, the input into the DFT is divided into two halves: one representing the $N/2$-point DFT of the even points ($x[0]$, $x[2]$, $x[4]$ and $x[6]$), the other the DFT of the odd points ($x[1]$, $x[3]$, $x[5]$ and $x[7]$). This separation is equivalent to sorting the sequence into two halves based on the **least significant bit (LSB)** of the index: all the even indices end with 0; all the odd indices end with 1. In the second decomposition, each $N/2$-point DFT is further split into quarters: the "even" points of the first half are split into $x[0]$ and $x[4]$ in one quarter and $x[2]$ and $x[6]$ in the second quarter, as you can see from **Figure 11.7**. This separation is equivalent to sorting each half based on the second binary bit of the index. The end result of the DIT decomposition is shown in the right-hand panel. The sequence has indices whose binary representation is the reverse (i.e., mirror-image) of the original indices, hence the name: bit-reversed order indexing.

Depending on the way an FFT is implemented, the transform may need to provide hardware or software to bit-reverse the input. In terms of hardware, special-purpose DSP processors and many microcontrollers with DSP cores implement bit-reversed indexing in hardware. Traditional non-recursive software implementations of the FFT generally include code that bit-reverses the input data before the transform routine is called. A number of schemes have been developed to implement such bit reversal efficiently. A simple example of how bit reversal can be implemented at the machine level is shown in **Figure 11.12**.

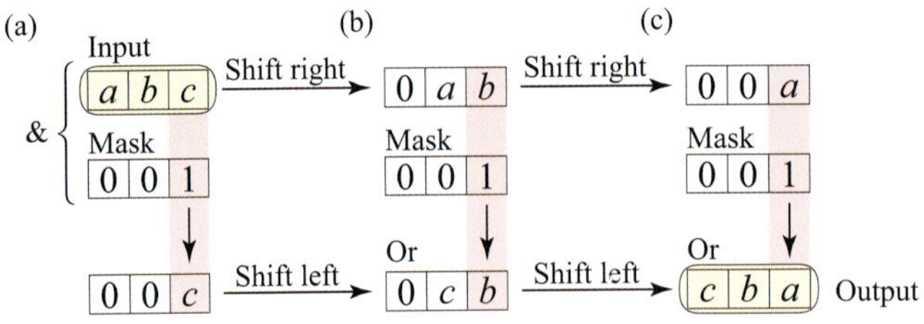

Figure 11.12 Bit reversal using shift registers

The top panel of **Figure 11.12a** shows a three-bit input word *abc*. The least significant bit of this word is selected by "anding" it with a mask, $abc\,\&\,001 = 00c$, and placing the result in an output register. Then the input register is shifted to the right and the new LSB is "anded" with the mask, $0ab\,\&\,001 = 00b$, and added to (or, equivalently, "ored" with) the output register that has been shifted to the left, to form $0cb$. The process is repeated once more and the result is the bit-reversed output, *cba*.

Matlab has a few functions that can bit-reverse the ordering of integer coefficients. The most relevant one in the Signal Processing Toolbox is

```
y = bitrevorder(x).
```

This function returns the bit-reversed data from any vector or matrix. It is functionally equivalent to

```
y = bin2dec(fliplr(dec2bin(x))).
```

Neither of these two routines is particularly fast, since they internally convert numerical data to characters! The following piece of code in Matlab is equivalent to the procedure shown in **Figure 11.12**.

```
function y = bitrev(x)
    r = floor(log2(max(x))); % How many bits do we need?
    y = zeros(size(x)); % initialize the output
    for k = 0:r
        y=bitshift(y, 1) +bitand(x, 1); % shift output left, add LSB of input
        x = bitshift(x, -1); % shift input right
    end
end
>> bitrev(0:7)
ans =
     0  4  2  6  1  5  3  7
```

In Section 11.11.2, we shall show that it is actually possible to write an efficient, recursive algorithm that performs bit-reversal automatically, without the need for any explicit bit-reversal code.

11.1.4 ★ Decimation-in-frequency FFT

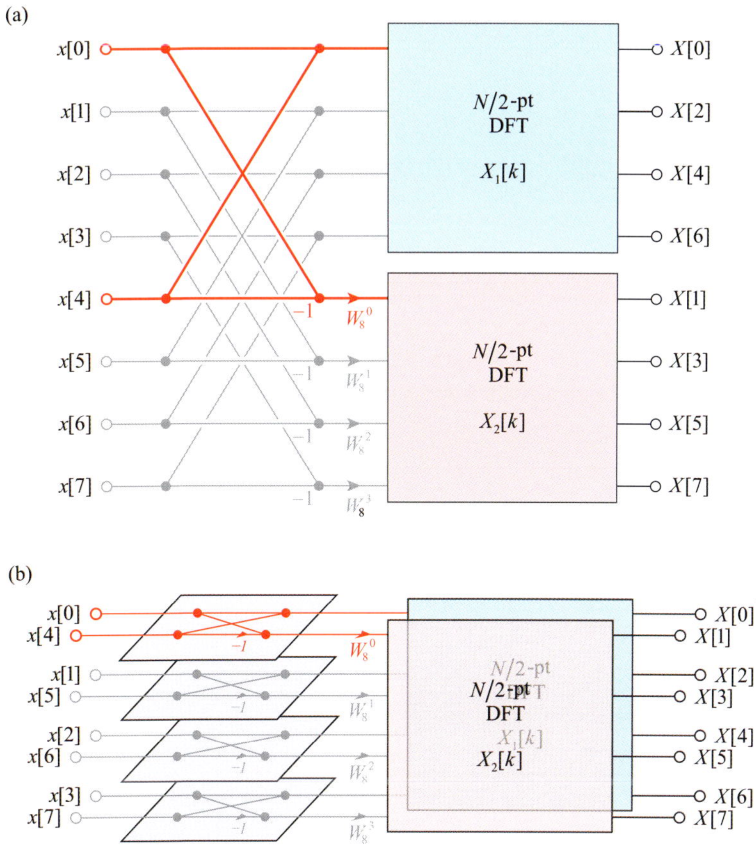

Figure 11.13 First stage of a DIF FFT

The decimation-in-time decomposition begins by separating the N-point sequence $x[n]$ into two $N/2$-point subsequences corresponding to the even and odd values of $x[n]$. However, there is another common decomposition, the **decimation-in-frequency (DIF)** decomposition, in which the DFT computation is split into two halves that represent the lower and upper values of $x[n]$:

$$
\begin{aligned}
X[k] &= \sum_{n=0}^{N-1} x[n] W_N^{kn} = \sum_{n=0}^{N/2-1} x[n] W_N^{kn} + \sum_{n=N/2}^{N-1} x[n] W_N^{kn} \\
&= \sum_{n=0}^{N/2-1} x[n] W_N^{kn} + \sum_{n=0}^{N/2-1} x[n+N/2] W_N^{k(n+N/2)} \\
&= \sum_{n=0}^{N/2-1} x[n] W_N^{kn} + W_N^{kN/2} \sum_{n=0}^{N/2-1} x[n+N/2] W_N^{kn} \\
&= \sum_{n=0}^{N/2-1} (x[n] + (-1)^k x[n+N/2]) W_N^{kn},
\end{aligned}
\quad 0 \le k \le N-1, \qquad (11.7)
$$

where we have recognized that $W_N^{kN/2} = (-1)^k$. This can be further simplified by considering $X[k]$ for k even and k odd separately:

$$
\begin{aligned}
k \text{ even: } X[2k] &= \sum_{n=0}^{N/2-1} (x[n] + (-1)^{2k} x[n+N/2]) W_N^{2kn} = \sum_{n=0}^{N/2-1} (x[n] + x[n+N/2]) W_{N/2}^{kn} \\
k \text{ odd: } X[2k+1] &= \sum_{n=0}^{N/2-1} (x[n] + (-1)^{(2k+1)} x[n+N/2]) W_N^{(2k+1)n} \\
&= \sum_{n=0}^{N/2-1} ((x[n] - x[n+N/2]) W_N^n) W_{N/2}^{kn}, \qquad 0 \le k \le N/2 - 1.
\end{aligned} \tag{11.8}
$$

The even points of $X[k]$ are just the $N/2$-point DFT of the sum of values of $x[n]$ from the lower and upper halves of the sequence: $x[n] + x[n+N/2]$. The odd points of $X[k]$ are the DFT of the difference of values of $x[n]$ from the lower and upper halves, which have been multiplied by twiddle factor W_N^n. This is shown in **Figure 11.13** for $N = 8$. Once again, we can repeat this decomposition to arrive at the final DIF FFT shown in **Figure 11.14**.

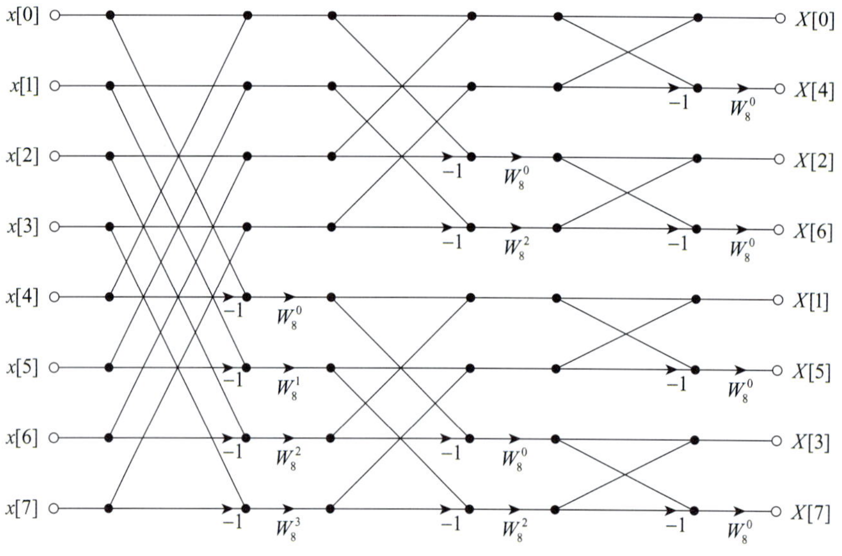

Figure 11.14 DIF FFT

As with the DIT FFT transform, in the DIF transform there are $(N/2)\log_2 N$ stages and $N/2$ butterflies per stage, each of which has one complex twiddle factor. Hence, the DIF transform also requires $(N/2)\log_2 N$ complex multiplications. A consequence of the DIF decimation is that the *inputs* to the transform are normal-ordered and the *outputs* are bit-reversed, just the opposite of the DIT FFT.

The DIT and DIF transforms are very similar. If you compare **Figure 11.9** with **Figure 11.14**, you will notice that the architecture of the two transforms is identical if you do the following: (1) reverse the signal paths by changing the direction of the arrow heads on the twiddle factors, (2) swap the output and input values (i.e., exchange $X[\cdot]$ and $x[\cdot]$) and (3) flip the entire graph so the input values are back on the left. In fact, these DIT and DIF transforms are *transposes* of each other, something we saw in discussing the architecture of FIR filters in Chapter 7.

While it is theoretically nice, what is the practical use of having multiple versions of the radix-2 FFT? We will suggest one answer after we have discussed the inverse FFT (IFFT) in Section 11.4.

11.2 ★ Radix-4 FFT

In a radix-2 transformation, a DFT with 2^M input points is successively divided into two smaller transforms in each of M stages. In a **radix-4 transformation**, a DFT with 4^M input points is successively divided into four smaller transforms. Each stage of the radix-4 transform essentially combines two successive stages of the radix-2 transformation and realizes simplifications that further reduce the number of multiplications required. The algebra is a bit messier, but the basic divide-and-conquer idea is the same.

11.2.1 The radix-4 decomposition

To understand the radix-4 transform, start by decimating the transform $X[k]$ in time into the sum of two $N/2$-point transforms, as in Equation (11.1), and then further decompose each of those $N/2$-point transforms so that we end up with the sum of four $N/4$-point transforms, $X_1[k]$, $X_2[k]$, $X_3[k]$ and $X_4[k]$, appropriately weighted by twiddle factors:

split into transform of even and odd values

split again into transform of ‘even’ and ‘odd’ values

$$\begin{aligned}
X[k] &= \sum_{n=0}^{N-1} x[n]W_N^{kn} \\
&= \sum_{n=0}^{N/2-1} x[2n]W_N^{2kn} + \sum_{n=0}^{N/2-1} x[2n+1]W_N^{k(2n+1)} \\
&= \sum_{n=0}^{N/4-1} \underbrace{x[4n]}_{x_1[n]}W_N^{4kn} + \sum_{n=0}^{N/4-1} \underbrace{x[4n+2]}_{x_3[n]}W_N^{k(4n+2)} + \sum_{n=0}^{N/4-1} \underbrace{x[4n+1]}_{x_2[n]}W_N^{k(4n+1)} + \sum_{n=0}^{N/4-1} \underbrace{x[4n+3]}_{x_4[n]}W_N^{k(4n+3)} \\
&= \underbrace{\sum_{n=0}^{N/4-1} x_1[n]W_{N/4}^{kn}}_{X_1[k]} + W_N^{2k}\underbrace{\sum_{n=0}^{N/4-1} x_3[n]W_{N/4}^{kn}}_{X_3[k]} + W_N^{k}\underbrace{\sum_{n=0}^{N/4-1} x_2[n]W_{N/4}^{kn}}_{X_2[k]} + W_N^{3k}\underbrace{\sum_{n=0}^{N/4-1} x_4[n]W_{N/4}^{kn}}_{X_4[k]} \\
&= X_1[k] + W_N^{2k}X_3[k] + W_N^{k}X_2[k] + W_N^{3k}X_4[k].
\end{aligned} \tag{11.9}$$

Each of the $N/4$-point transforms, $X_1[k]$, $X_2[k]$, $X_3[k]$ and $X_4[k]$, is periodic with period $N/4$, so each one only needs to be evaluated for $0 \le k < N/4$. Then, the output of each section of a radix-4 transform $X[k]$ can be expressed as four adjoining sections (see Problem 11-12),

$$\begin{aligned}
X[k] &= X_1[k] + W_N^{2k}X_3[k] + W_N^{k}X_2[k] + W_N^{3k}X_4[k] \\
X[k+N/4] &= X_1[k+N/4] + W_N^{2(k+N/4)}X_3[k+N/4] \\
&\quad + W_N^{(k+N/4)}X_2[k+N/4] + W_N^{3(k+N/4)}X_4[k+N/4] \\
X[k+N/2] &= X_1[k+N/2] + W_N^{2(k+N/2)}X_3[k+N/2] \\
&\quad + W_N^{(k+N/2)}X_2[k+N/2] + W_N^{3(k+N/2)}X_4[k+N/2] \\
X[k+3N/4] &= X_1[k+3N/4] + W_N^{2(k+3N/4)}X_3[k+N/4] \\
&\quad + W_N^{(k+3N/4)}X_2[k+N/4] + W_N^{3(k+3N/4)}X_4[k+N/4].
\end{aligned} \tag{11.10}$$

Equation (11.10) can be expressed graphically in the form of a radix-4 butterfly shown in **Figure 11.15a**. The radix-4 butterfly has only three twiddle factors: W_N^k, W_N^{2k} and W_N^{3k}. All the remaining multiplications within the dashed red box are by 1, -1, j or $-j$, and as we have argued previously, none of these counts as a multiplication. Accordingly, **Figure 11.15b** shows a graphically simplified radix-4 butterfly that only shows the twiddle factors.

Figure 11.16 shows the radix-4 decomposition of a 16-point DIT FFT using the simplified radix-4 butterflies. The four big radix-4 butterflies on the right side of the figure combine the outputs of four smaller $N/4$-point DFTs $X_1[k]$, $X_3[k]$, $X_2[k]$ and $X_4[k]$ (colored boxes). One of the large butterflies (for $k=1$) is highlighted in red. As with the radix-2 FFT, we can apply the same radix-4 decomposition to the $N/4$-point DFTs in the colored boxes to arrive at the complete radix-4 DIT FFT, as shown in **Figure 11.16a** in "2-D" and in **Figure 11.16b** in "3-D."

The number of points in this transform, $N=16$, is a power of four, so there are $\log_4 N=2$ stages of radix-4 decomposition. Each stage has $N/4=4$ butterflies, and each butterfly has three twiddle factors, so there are a total of $(3N/4)\log_4 N=(3N/8)\log_2 N$ twiddle factors in the radix-4 FFT. (We ignore the fact that many of the twiddle factors have value $W_{16}^0=1$.) Compared with a radix-2 decomposition of the same size, which has $(N/2)\log_2 N$ twiddle factors, the radix-4 transformation has only 75% as many complex multiplications.

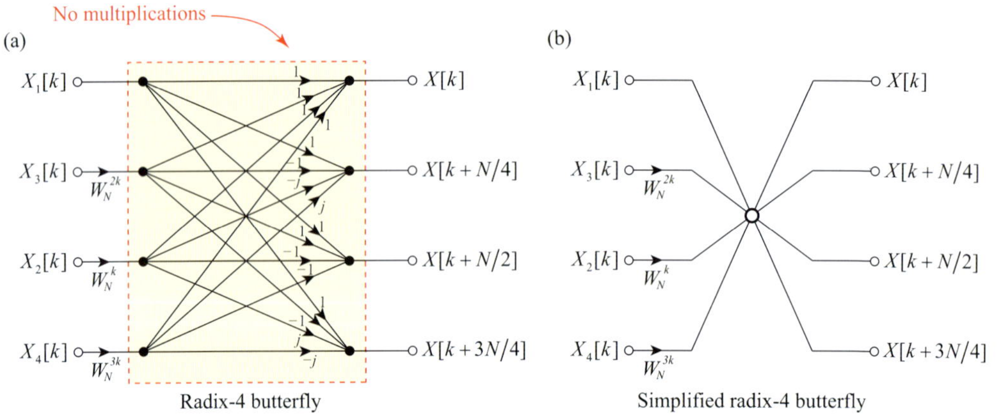

Figure 11.15 Radix-4 butterfly

A consequence of the way we set up the radix-4 butterflies (i.e., with the inputs permuted to be in the order $X_1[k]$, $X_3[k]$, $X_2[k]$ and $X_4[k]$) is that the inputs to the entire radix-4 transform end up being in bit-reverse order while the outputs are normal-ordered. Thus, this radix-4 transformation is "pin-for-pin compatible" with a radix-2 DIT transformation of the same size, except it requires fewer complex operations.

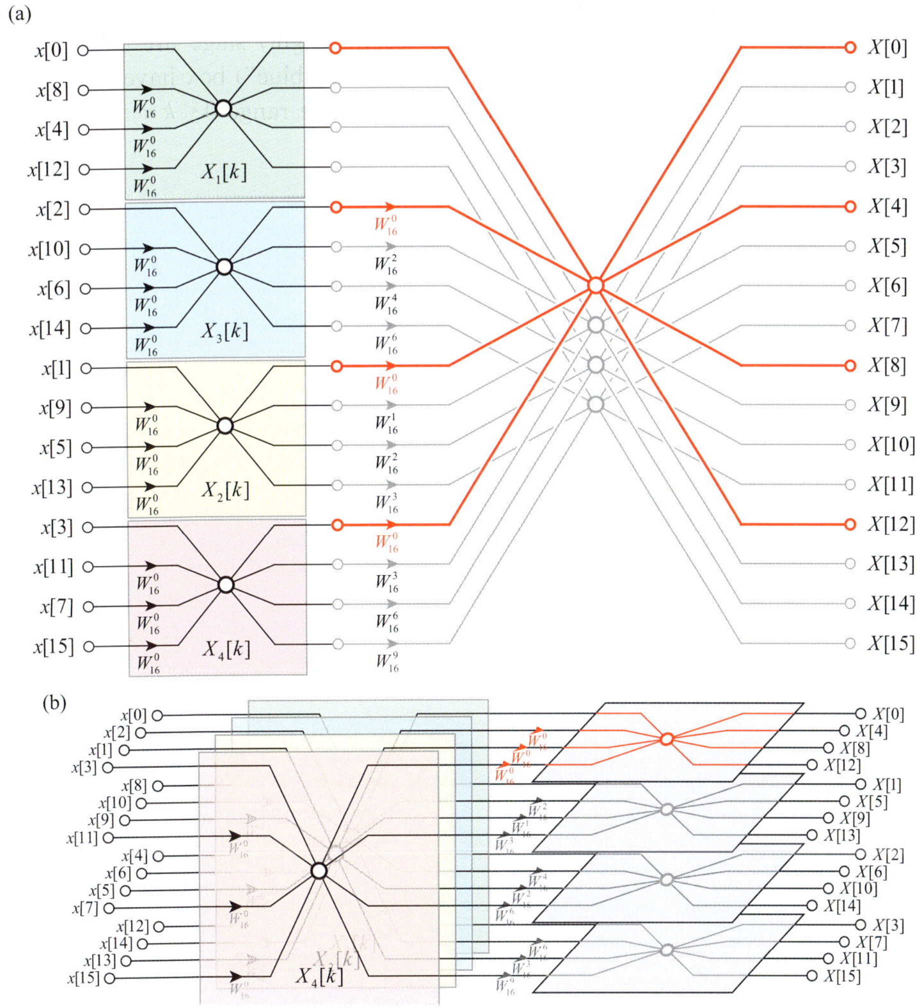

Figure 11.16 16-point radix-4 FFT

11.2.2 The radix-4 transform as a combination of radix-2 transforms

A revealing way to think about the radix-4 transform is as a combination of a pair of stages of radix-2 transformation. **Figure 11.17** shows an $N=16$-point radix-2 DIT FFT. There are $\log_2 N=4$ stages in the transform. Since N is a power of four, we can group the stages of this FFT in pairs, as shown by the dotted boxes. Each box contains $N/2$ "two-stage" butterflies, each of which comprises four radix-2 butterflies joined together. The figure shows a couple of examples of these two-stage butterflies in color.

Each of these two-stage butterflies has the same general pattern, which is shown in **Figure 11.18a**. The two-stage butterfly has four inputs and four outputs. There are four twiddle factors. The two associated with the first (left) stage are both of the form W_N^{2k}, where k is an integer. The two associated with the second (right) stage are W_N^k and $W_N^{k+N/4}$. In **Figure 11.17**, all the two-stage butterflies in the first ("blue") box have $k=0$. The two-stage butterflies in the second ("red") box have values in the range $0 \leq k < N/2$; for example, the butterfly highlighted in red has $k=3$.

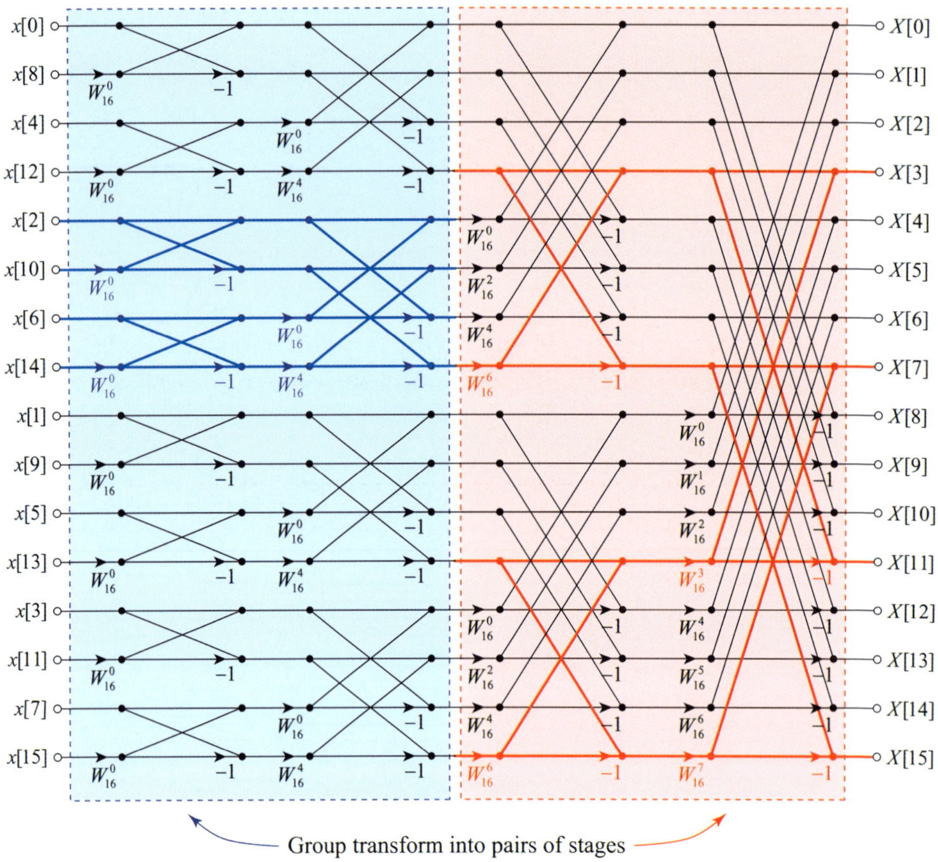

Figure 11.17 Understanding a radix-4 transform in terms of two radix-2 transforms

For each two-stage butterfly, the factor $W_N^{k+N/4}$ can be written as $W_N^{k+N/4} = W_N^k \cdot W_N^{N/4} = -jW_N^k$. A factor of W_N^k can be factored out of the middle and brought to the input (left) side of each two-stage butterfly, as shown in **Figure 11.18b**. The result, shown in **Figure 11.18c**, is a new two-stage butterfly that maps four inputs to four outputs using only three twiddle factors instead of four, a 25% decrease compared to the original radix-2 transformation. As always, none of the scaling operations in the dotted box (1, −1, j or $-j$) counts as a multiplication. Finally, combine the scaling operations in the dotted box to produce a single radix-4 butterfly, shown in **Figure 11.18d**, which is identical to the radix-4 butterfly of **Figure 11.15a**. Thus, a radix-4 FFT can be viewed as a combination of four radix-2 FFTs that has been modified to reduce the number of twiddle factors from four to three.

From a computational standpoint, there may actually be an advantage to implementing a radix-4 butterfly in two stages in some applications, as shown in **Figure 11.18c**, rather than in one stage, as shown in **Figure 11.18d**. The advantage lies in the number of complex additions required to compute the outputs. The two-stage solution requires only eight complex additions to compute all four outputs, whereas the one-stage solution requires 12. As we have mentioned before, further simplifications are possible by avoiding multiplications by the factors of $W_N^0 = 1$, which constitute the majority of the twiddle factors in the transform. As you might expect, there exist alternate versions of the radix-4 DIT transform as well as radix-4 DIF transforms.

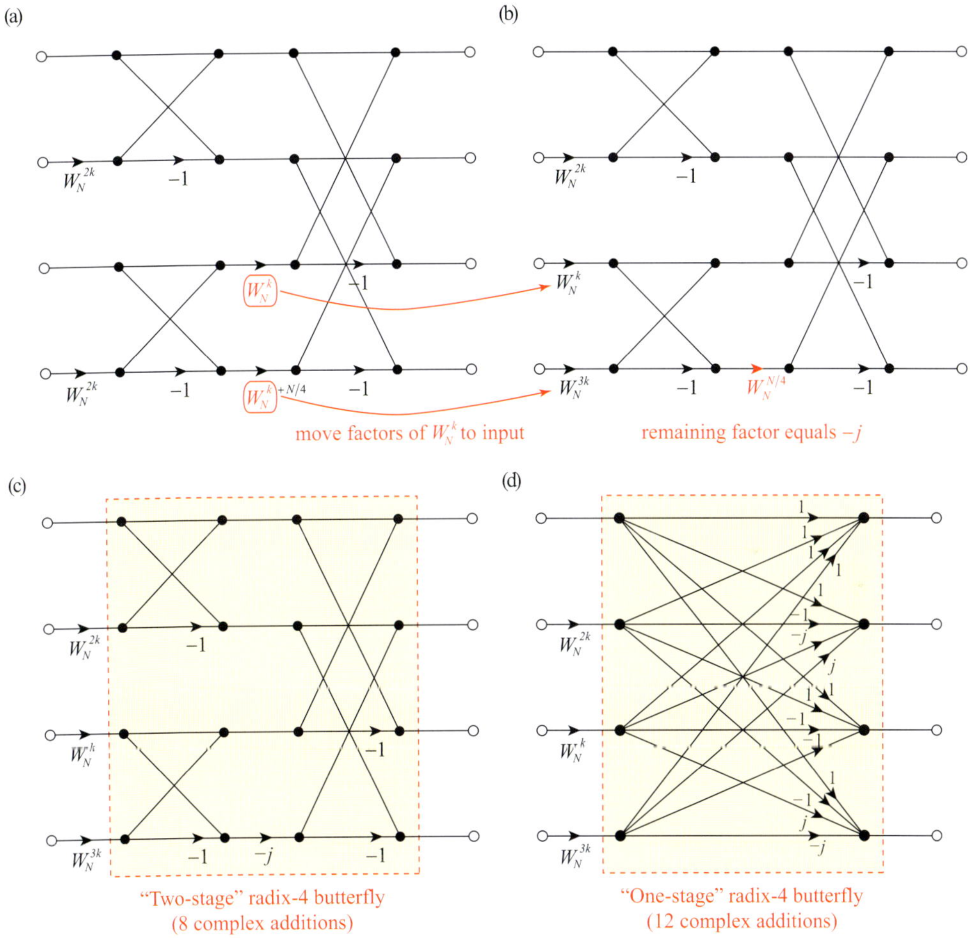

Figure 11.18 Deriving a radix-4 butterfly from radix-2 butterflies

11.3 ★ Composite (mixed-radix) FFT

The radix-2 and radix-4 DIT and DIF transforms are all implementations of radix-r algorithms in which the number of points in the transform, N, is a power of the radix r, such that $N = r^M$, where $M = \log_r N$. All these transforms can be viewed as special cases of a more general decomposition in which N is any **composite** size, where composite means that N is the product of two integer factors, $N = N_1 N_2$.

11.3.1 FFTs of composite size

If $N = N_1 N_2$ is a composite number, we can express the index of an N-point sequence $x[n]$ as $n = n_1 + n_2 N_1$, where $0 \le n_1 < N_1$ and $0 \le n_2 < N_2$. Then, we decompose the N-point DFT into N_1 transforms of length N_2:

$$\begin{aligned} X[k] &= \sum_{n=0}^{N-1} x[n] W_N^{kn} = \sum_{n_1=0}^{N_1-1} \sum_{n_2=0}^{N_2-1} x[n_1 + n_2 N_1] W_N^{k(n_1 + n_2 N_1)} \\ &= \sum_{n_1=0}^{N_1-1} W_N^{kn_1} \sum_{n_2=0}^{N_2-1} x[n_1 + n_2 N_1] W_N^{kn_2 N_1} \\ &= \sum_{n_1=0}^{N_1-1} W_N^{kn_1} \underbrace{\sum_{n_2=0}^{N_2-1} x[n_1 + n_2 N_1] W_{N_2}^{kn_2}}_{N_2\text{-pt DFT}}, \end{aligned} \tag{11.11}$$

where we recognize that

$$W_N^{kn_2 N_1} = W_{N/N_1}^{kn_2} = W_{N_2}^{kn_2}.$$

In Section 11.1, we found that splitting the DFT into the sum of smaller transforms did not in itself reduce the number of twiddle-factor multiplications. To reduce the number of twiddle factors, we also had to exploit the periodicity of the smaller transforms. That is true of transforms of composite size as well. Representing the index of the transform $X[k]$ as $k = k_2 + k_1 N_2$, where $0 \le k_1 < N_1$ and $0 \le k_2 < N_2$, Equation (11.11) becomes

$$\begin{aligned} X[k_2 + N_2 k_1] &= \sum_{n_1=0}^{N_1-1} W_N^{(k_2+N_2k_1)n_1} \sum_{n_2=0}^{N_2-1} x[n_1 + N_1 n_2] W_{N_2}^{(k_2+N_2k_1)n_2} \\ &= \sum_{n_1=0}^{N_1-1} W_N^{(k_2+N_2k_1)n_1} \sum_{n_2=0}^{N_2-1} x[n_1 + N_1 n_2] W_{N_2}^{k_2n_2} \\ &= \sum_{n_1=0}^{N_1-1} \left(\left(W_N^{k_2n_1} \left(\sum_{n_2=0}^{N_2-1} x[n_1 + N_1 n_2] W_{N_2}^{k_2n_2} \right) \right) W_{N_1}^{k_1n_1} \right), \end{aligned} \tag{11.12}$$

Compute $N_1 \times N_2$-pt DFTs

Multiply by $N_2 \times N_1$ twiddle factors

Compute $N_2 \times N_1$-pt DFTs

where we have recognized that

$$W_{N_2}^{(k_2+N_2k_1)n_2} = W_{N_2}^{k_2n_2} W_{N_2}^{N_2k_1n_2} = W_{N_2}^{k_2n_2},$$

and

$$W_N^{(k_2+N_2k_1)n_1} = W_N^{k_2n_1} W_N^{N_2k_1n_1} = W_N^{k_2n_1} W_{N/N_2}^{k_1n_1} = W_N^{k_2n_1} W_{N_1}^{k_1n_1}.$$

The last line of Equation (11.12), with the multicolored boxes, deserves a bit of scrutiny. It states that a composite DFT of length $N = N_1 N_2$ can be obtained by the following steps:

1. Cut the N-point input sequence $x[n]$ into N_1 chunks of length N_2 and take the N_2-point DFT of each chunk (the red box);

2. Scale the outputs of the DFTs by twiddle factors $W_N^{k_2 n_1}$ (orange box);
3. Group the scaled outputs into N_2 groups of length N_1 and take the N_1-point DFT of each group (green box).

Figure 11.19 shows a 3-D graphic representation of the composite DFT with $N_1 = 3$ and $N_2 = 4$.

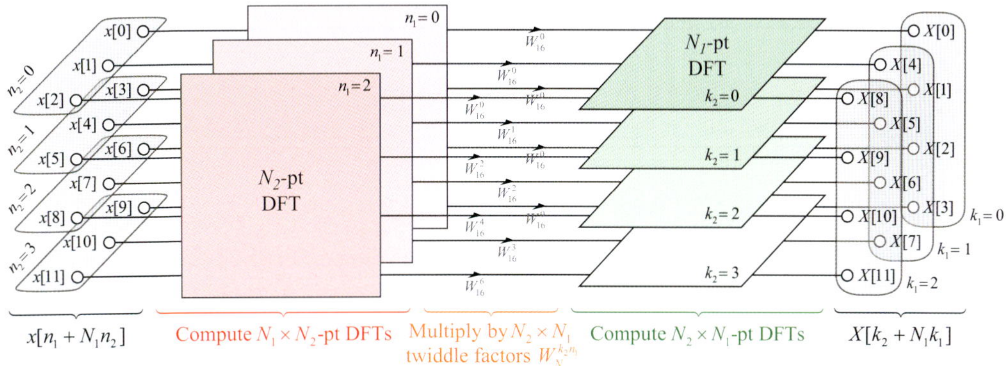

Figure 11.19 Composite $N_1 \times N_2$ DFT with $N_1 = 3$ and $N_2 = 4$

All the radix-2 and radix-4 transforms we have presented in Sections 11.1 and 11.2 have exactly the form of Equation (11.11), as indicated in the following examples.

Example 11.1

Show that the radix-2 DIT transform of Equation (11.6) is equivalent to the composite FFT of Equation (11.12) with $N_1 = 2$ and $N_2 = N/2$.

➤ Solution:

For the first decomposition of an N-point DIT radix-2 transform with $N_1 = 2$ and $N_2 = N/2$, we have $k = k_2 + k_1 N/2$ with $0 \le k_1 < 1$ and $0 \le k_2 < N/2$,

$$\begin{aligned} X[k_2 + k_1 N/2] &= \sum_{n_1=0}^{1} \left(W_N^{k_2 n_1} \sum_{n_2=0}^{N/2-1} x[n_1 + 2n_2] W_{N/2}^{k_2 n_2} \right) W_2^{k_1 n_1} \\ &= \underbrace{\sum_{n_2=0}^{N/2-1} x[2n_2] W_{N/2}^{k_2 n_2}}_{X_1[k_2]} + W_N^{k_2} W_2^{k_1} \underbrace{\sum_{n_2=0}^{N/2-1} x[2n_2+1] W_{N/2}^{k_2 n_2}}_{X_2[k_2]}, \\ &= X_1[k_2] + W_N^{k_2} (-1)^{k_1} X_2[k] \end{aligned}$$

where we recognize that $W_2^{k_2} = (-1)^{k_2}$. So,

$$\left. \begin{aligned} X[k_2] &= X_1[k_2] + W_N^{k_2} X_2[k_2] \\ X[k_2 + N/2] &= X_1[k_2] - W_N^{k_2} X_2[k_2] \end{aligned} \right\}, \quad 0 \le k_2 < N/2,$$

which is the same as Equation (11.6).

The composite DIF transform can be understood in exactly the same manner as the DIT transform, using the same equation, Equation (11.12). The only differences between the DIT and DIF transforms are the values of N_1 and N_2.

Example 11.2

Show that the radix-2 DIF transform of Equation (11.8) is equivalent to the composite FFT of Equation (11.12) with $N_1 = N/2$ and $N_2 = 2$.

➤ Solution:

With $0 \le k_1 < N/2$ and $0 \le k_2 < 1$, we have

$$X[k_2 + 2k_1] = \sum_{n_1=0}^{N/2-1} \left(W_N^{k_2 n_1} \sum_{n_2=0}^{1} x[n_1 + n_2 N/2] W_2^{k_2 n_2} \right) W_{N/2}^{k_1 n_1}$$

$$= \sum_{n_1=0}^{N/2-1} (W_N^{k_2 n_1}(x[n_1] + x[n_1 + N/2] W_2^{k_2})) W_{N/2}^{k_1 n_1}.$$

Evaluate separately for $k_2 = 0$ (i.e., k even) and $k_2 = 1$ (i.e., k odd):

$$k \text{ even:} \qquad X[2k_1] = \sum_{n_1=0}^{N/2} (x[n_1] + x[n_1 + N/2]) W_{N/2}^{k_1 n_1}$$

$$k \text{ odd:} \quad X[2k_1 + 1] = \sum_{n_1=0}^{N/2-1} ((x[n_1] - x[n_1 + N/2]) W_N^{n_1}) W_{N/2}^{k_1 n_1}.$$

This is the same as Equation (11.8).

Example 11.3

Show that the radix-4 DIT transform of Equation (11.9) is equivalent to the composite FFT of Equation (11.11) with $N_1 = 4$ and $N_2 = N/4$.

➤ Solution:

$$X[k] = \sum_{n_1=0}^{3} \sum_{n_2=0}^{N/4-1} x[n_1 + 4n_2] W_N^{k(4n_2+n_1)} = \sum_{n_1=0}^{3} W_N^{kn_1} \sum_{n_2=0}^{N/4-1} x[n_1 + 4n_2] W_N^{k4n_2}$$

$$= \underbrace{\sum_{n_2=0}^{N/4-1} x[4n_2] W_{N/4}^{kn_2}}_{X_1[k]} + W_N^k \underbrace{\sum_{n_2=0}^{N/4-1} x[4n_2+1] W_{N/4}^{kn_2}}_{X_2[k]} + W_N^{2k} \underbrace{\sum_{n_2=0}^{N/4-1} x[4n_2+2] W_{N/4}^{kn_2}}_{X_3[k]} + W_N^{3k} \underbrace{\sum_{n_2=0}^{N/4-1} x[4n_2+3] W_{N/4}^{kn_2}}_{X_4[k]}.$$

$$= X_1[k] + W_N^k X_2[k] + W_N^{2k} X_3[k] + W_N^{3k} X_4[k]$$

11.3.2 Mixed radix-2 and radix-4 transform

Take heart, we are almost done with our tour of basic and not-so-basic FFT architectures. One particularly useful example of a composite transform occurs when N can be written as $N = 2 \cdot 4^m$. In this case, we can enjoy the advantages of both the radix-4 and radix-2 butterflies. **Figure 11.20a** shows an $N = 8$ point **mixed-radix** DIT FFT. We can recognize that this can be arranged as two vertically stacked radix-4 butterflies (configured as two-stage butterflies, as in **Figure 11.18c**), which are connected by a set of radix-2 butterflies.

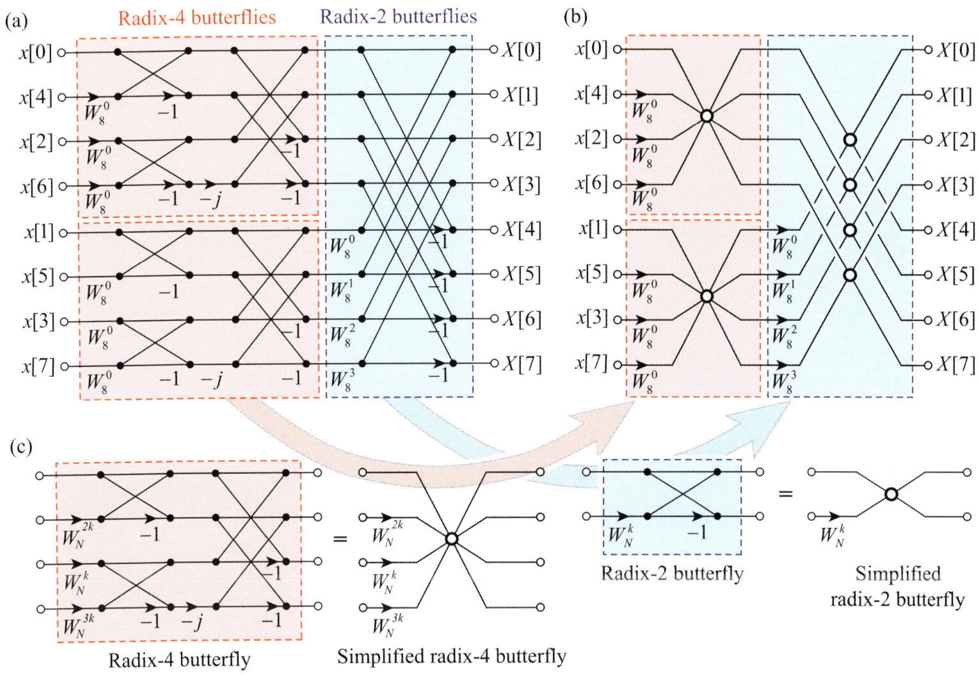

Figure 11.20 Mixed-radix FFT with radix-4 and radix-2 butterflies

Figure 11.20b shows a simplified version of this same mixed-radix transform where radix-4 and radix-2 butterflies have been replaced by graphically simpler versions (shown in **Figure 11.20c**). The resulting hybrid has ten twiddle factors instead of twelve for the 8-point radix-2 butterfly in **Figure 11.9**.

In addition to combinations of radix-2 and radix-4 transforms, there are a number of other transforms optimized for other radices, including radix-3, radix-5 and radix-8. There currently exist hardware DSP chips geared towards communication applications that can efficiently do mixed-radix transforms in which N is of the form $N = 2^k 3^l 4^m 5^n$.

11.3.3 Transposed and split-radix transforms

Sections 11.1–11.3 have introduced some of the major FFT algorithms in use, but there are plenty of other algorithms and variants, all aimed at reducing the complexity of the calculation in some way. Supplementary material available at www.cambridge.org/holton presents some of these variants, including **transposed forms** of the DIT and DIF transforms as well as a **split-radix transform** that further reduces the number of complex multiplications compared to either radix-2 or radix-4 transformations.

11.4 Inverse FFT

In many applications, we use the forward FFT to perform the transform from $x[n]$ to $X[k]$, do some processing in the frequency domain (e.g., filtering) and then convert the result back to the time domain with the **inverse FFT (IFFT)**. You will be relieved to hear that the IFFT can be implemented by simple modification of the FFT. In order to understand the modification, start with the definition of the N-point inverse DFT (the IDFT) of $X[k]$,

$$x[n] = \mathfrak{F}^{-1}\{X[k]\} = \frac{1}{N}\sum_{k=0}^{N-1} X[k]e^{j2\pi kn/N}. \tag{11.13}$$

This equation is very similar to the equation for the forward DFT,

$$X[k] = \mathfrak{F}\{x[n]\} = \sum_{n=0}^{N-1} x[n]e^{-j2\pi kn/N}.$$

The differences are:

- the DFT is a sum of N points of sequence $x[n]$ on index n, whereas the IDFT is a sum of N points of sequence $X[k]$ on index k;
- the DFT sum contains $e^{-j2\pi kn/N}$, whereas the IDFT contains the conjugate $e^{j2\pi kn/N}$;
- the IDFT is scaled by a factor of $1/N$.

This suggests that we can use the equation for the FFT to compute the IFFT. The strategy, diagrammed in **Figure 11.21a**, is as follows:

1. Take the complex conjugate of the coefficients $X[k]$ to form new coefficients $X^*[k]$ by negating the imaginary part.
2. Take the forward FFT of the sequence $X^*[k]$, i.e.,

$$\sum_{k=0}^{N-1} X^*[k]e^{-j2\pi kn/N}.$$

3. Take the complex conjugate of the result (by negating the imaginary part) and scale the result (both real and imaginary parts) by a factor of $1/N$,

$$\frac{1}{N}\left(\sum_{k=0}^{N-1} X^*[k]e^{-j2\pi kn/N}\right)^* = \frac{1}{N}\sum_{k=0}^{N-1} X[k]e^{j2\pi kn/N}.$$

Voila! The result is identical to Equation (11.13).

This result means that we do not have to write both forward and inverse FFT routines but can instead use a single FFT routine for both the forward and inverse transforms. Since the process of taking the conjugate of an array does not require any multiplications, the IFFT is no more computationally intensive than the FFT except that we do have to scale the output by a factor

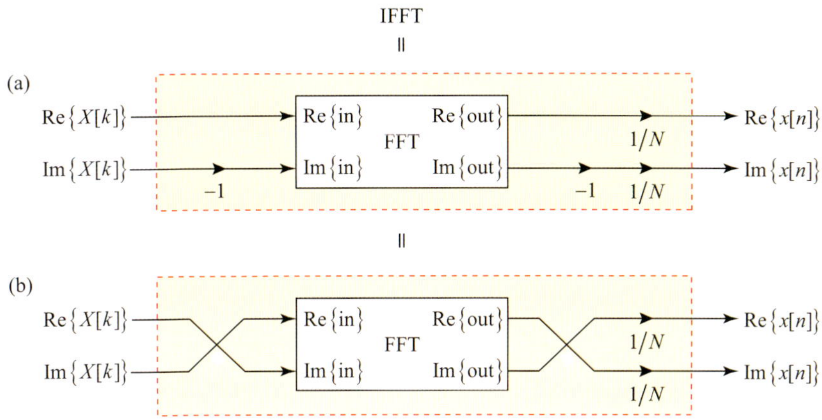

Figure 11.21 Performing the IFFT using the FFT

$1/N$. In many applications, the factor $1/N$ can be absorbed in some other part of the calculation. In block-floating-point implementations, scaling by this factor is trivial if N is a power of two.

Another simple way of computing the inverse DFT using the FFT is shown **Figure 11.21b**:

1. Swap the real and imaginary parts of $X[k]$. This is computationally equivalent to taking the conjugate of $X[k]$ and then multiplying by j:

$$\mathrm{Im}\{X[k]\} + j\,\mathrm{Re}\{X[k]\} = j(\mathrm{Re}\{X[k]\} - j\,\mathrm{Im}\{X[k]\}) = jX^*[k].$$

2. Take the forward FFT, so we have

$$\sum_{k=0}^{N-1} jX^*[k]e^{-j2\pi kn/N} = j\left(\sum_{k=0}^{N-1} X[k]e^{j2\pi kn/N}\right)^*.$$

3. Swap the real and imaginary parts of the result, which is again computationally equivalent to taking the conjugate of the above expression and multiplying by j. Then scale the result by $1/N$ to form $x[n]$:

$$\frac{1}{N}j\left(j\left(\sum_{k=0}^{N-1} X[k]e^{j2\pi kn/N}\right)^*\right)^* = \frac{1}{N}\sum_{k=0}^{N-1} X[k]e^{j2\pi kn/N}.$$

This strategy of forming the IFFT from the FFT by swapping the real and imaginary parts is particularly useful in that it is just a change in how data are addressed. For example, in an implementation in C, all one has to do is to swap the pointers for real and imaginary arrays to the FFT routine.

11.5 Matlab implementation

Matlab has two basic functions for computing the FFT and IFFT, unsurprisingly named `fft` and `ifft`. While these functions can operate on multi-dimensional matrices, the general syntax for one-dimensional arrays is

```
y = fft(x,[N])
x = ifft(y, [N]),
```

where the optional parameter N is the size of the transform. If N is unspecified, it is taken as the length of the input array. If N is specified and greater than the length of the input array, then the end of the array is padded with zeros. If N is specified and is less than the length of the input array, then the array is truncated to its first N points. Matlab's `fft` and `ifft` functions will work for any N. You can potentially increase the speed of execution by first running the Matlab function `fftw`, which experimentally determines the fastest algorithm for computing an FFT of a given size on your particular processor.

11.6 FFT of real sequences

Even if you do not code your own FFT, you can often realize significant computational savings by using them efficiently. For real sequences, you can exploit the symmetry properties of the DFT to reduce the number of computations necessary to perform the FFT by about a factor of two. In order to understand this, let us first review a few properties of the DFT (see Sections 10.3.2 and 10.3.3) that apply to real and imaginary sequences.

11.6.1 Properties of DFTs (revisited)

An arbitrary (complex-valued) sequence $x[n]$ has transform

$$X[k] = \mathfrak{F}\{x[n]\}.$$

The conjugate of the sequence, $x^*[n]$, has transform

$$\begin{aligned}\mathfrak{F}\{x^*[n]\} &= \sum_{n=0}^{N-1} x^*[n]e^{-j2\pi kn/N} = \left(\sum_{n=0}^{N-1} x[n]e^{j2\pi kn/N}\right)^* = \left(\sum_{n=0}^{N-1} x[n]e^{-j2\pi(-k)n/N}\right)^* \\ &= X^*[(-k)_N] = X^*[N-k].\end{aligned}$$

The periodicity of the DFT means that $X^*[(-k)_N] = X^*[N-k]$. Now, if the sequence is real, then $x^*[n] = x[n]$. Hence, $\mathfrak{F}\{x^*[n]\} = \mathfrak{F}\{x[n]\}$, so $X[k]$ is conjugate symmetric:

$$x[n] \text{ is real} \rightarrow X[k] = X^*[N-k]. \tag{11.14a}$$

In a similar manner, you can show that if $x[n]$ is purely imaginary, so that $x^*[n] = -x[n]$, then $X[k]$ is conjugate antisymmetric:

$$x[n] \text{ is imaginary} \rightarrow X[k] = -X^*[N-k]. \tag{11.14b}$$

In general, the transform of a real sequence will be complex valued, but $X[0]$ is guaranteed to be real. That is because, from Equations (11.14a) and the periodicity of the DFT, $X[0] = X^*[N] = X^*[0]$. The only way a number can be its own complex conjugate is if it is real. With N even, it is also true that $X[N/2]$ is real, since $X[N/2] = X^*[N/2]$. Thus, for a real sequence, we only need to compute two real values ($X[0]$ and $X[N/2]$) and $N/2-1$ complex values ($X[1]$, …, $X[N/2-1]$) to completely characterize $X[k]$, as shown in **Figure 11.22** (for $X[k]$ conjugate symmetric). The $N/2-1$ complex values ($X[N/2+1]$, …, $X[N-1]$) come from Equation (11.14a).

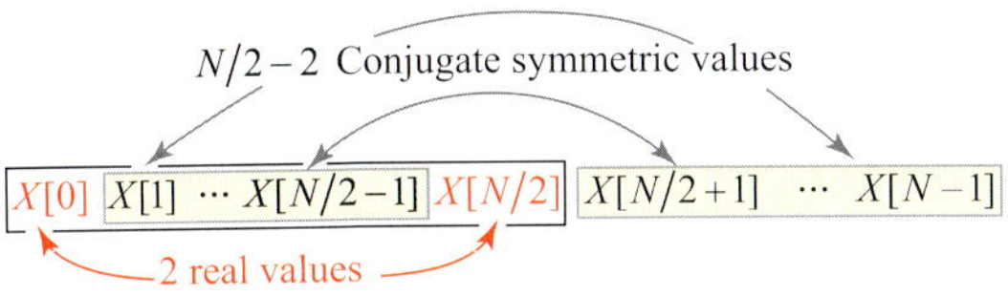

Figure 11.22 Conjugate symmetry of $X[k]$

Since each of the complex values comprises two real values, the total number of real values necessary to specify $X[k]$ is N:

$$\underbrace{2}_{X[0],\,X[N/2]} + \underbrace{2(N/2-1)}_{X[1],\,\cdots,\,X[N/2-1]} = N.$$

That is what we would expect from a linear transformation such as the DFT. It maps N real values of $x[n]$ into N real values of $X[k]$.

Example 11.4

Find the transform of a random sequence $x[n]$ and verify that $X[0]$ and $X[N/2]$ are real and $X[k]=X^*[N-k]$ for $0<k<N/2$.

➤ **Solution:**

Using Matlab:

```
>> disp(' k      X[k]'),disp([(0:7)' fft(randn(8, 1))])

   k       X[k]
   0       3.2983 ←——Real
   1       0.3949 - 3.2473i
   2       0.7360 + 1.0024i
   3      -1.1453 - 0.6272i
   4       3.2556 ←——Real        Complex conjugates
   5      -1.1453 + 0.6272i
   6       0.7360 - 1.0024i
   7       0.3949 + 3.2473i
```

11.6.2 N-point FFT of two real N-point sequences

The preceding results allow us to use a single complete N-point FFT to compute the transform of two real N-point sequences $x_1[n]$ and $x_2[n]$ simultaneously, which effectively doubles the capacity (or speed) of the transform. To perform the "double-real" FFT, first form a single N-point complex sequence from the two real N point sequences, $x[n]=x_1[n]+jx_2[n]$. This is a reversible operation, since $x_1[n]$ and $x_2[n]$ can easily be recovered from $x[n]$:

$$x_1[n] = \mathrm{Re}\{x[n]\} = \tfrac{1}{2}(x[n] + x^*[n])$$
$$x_2[n] = \mathrm{Im}\{x[n]\} = \tfrac{1}{2j}(x[n] - x^*[n]).$$

Similarly, the properties of the DFT can be used to recover the transform of $x[n]$ from the transforms of $x_1[n]$ and $x_2[n]$:

$$\begin{aligned} X_1[k] &\triangleq \mathfrak{F}\{x_1[n]\} = \mathfrak{F}\{\tfrac{1}{2}(x[n] + x^*[n])\} = \tfrac{1}{2}(X[k] + X^*[N-k]) \\ X_2[k] &\triangleq \mathfrak{F}\{x_2[n]\} = \mathfrak{F}\left\{\tfrac{1}{2j}(x[n] - x^*[n])\right\} = \tfrac{1}{2j}(X[k] - X^*[N-k]). \end{aligned} \tag{11.15}$$

Computing the FFT of two real sequences at the same time involves setting the real and imaginary parts of the FFT's input to $x_1[n]$ and $x_2[n]$, respectively, taking the complex FFT and then manipulating the real and imaginary parts of the output in a series of post-processing operations that implement Equation (11.15), as follows:

$$\mathrm{Re}\{X_1[k]\} = \tfrac{1}{2}(\mathrm{Re}\{X[k]\} + \mathrm{Re}\{X^*[N-k]\}) = \tfrac{1}{2}(\mathrm{Re}\{X[k]\} + \mathrm{Re}\{X[N-k]\})$$
$$\mathrm{Im}\{X_1[k]\} = \tfrac{1}{2}(\mathrm{Im}\{X[k]\} + \mathrm{Im}\{X^*[N-k]\}) = \tfrac{1}{2}(\mathrm{Im}\{X[k]\} - \mathrm{Im}\{X[N-k]\}),$$

and

$$\mathrm{Re}\{X_2[k]\} = \tfrac{1}{2}(\mathrm{Re}\{-jX[k]\} - \mathrm{Re}\{-jX^*[N-k]\}) = \tfrac{1}{2}(\mathrm{Im}\{X[k]\} + \mathrm{Im}\{X[N-k]\})$$
$$\mathrm{Im}\{X_2[k]\} = \tfrac{1}{2}(\mathrm{Im}\{-jX[k]\} - \mathrm{Im}\{-jX^*[N-k]\}) = -\tfrac{1}{2}(\mathrm{Re}\{X[k]\} - \mathrm{Re}\{X[N-k]\}).$$

The entire "double-real" FFT process is shown schematically in **Figure 11.23**.

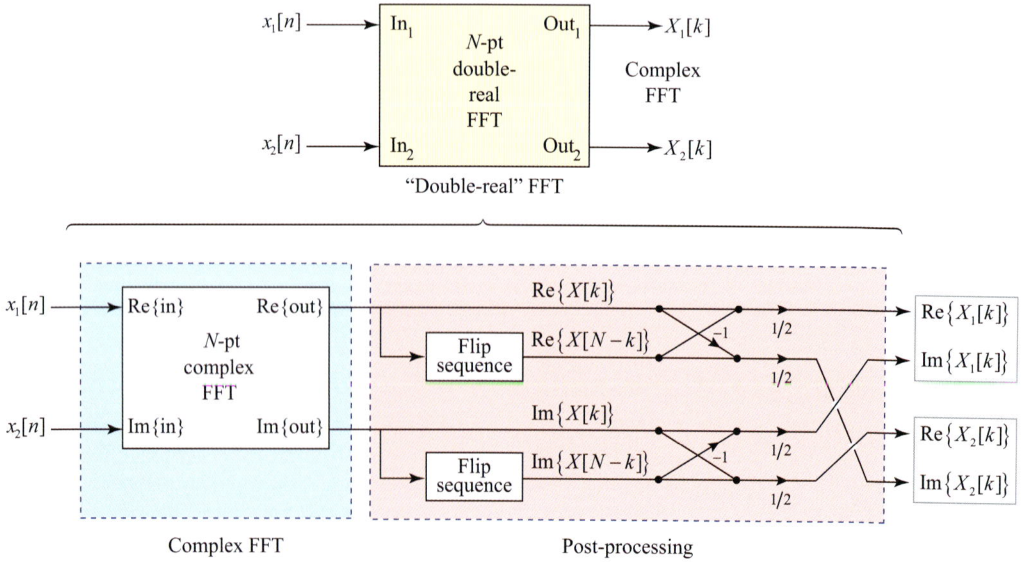

Figure 11.23 "Double-real" FFT

The post-processing stage requires $4(N-1)$ real additions, and $4N$ real scaling operations (i.e., the factors of 1/2). However, these additional operations are a small price to pay compared to the alternative of performing separate complex N-point FFTs for $x_1[n]$ and $x_2[n]$. A Matlab implementation of the double-real FFT is shown in the next example.

Example 11.5

Given the sequences $x_1[n] = [1\ 2\ 3\ 4]$ and $x_2[n] = [4\ 3\ 2\ 1]$, use a single 4-point FFT to find the transforms of the two sequences and verify that the answers are equivalent to taking the FFT of each sequence separately.

➤ Solution:

Let us first write a little Matlab function to perform the "double-real" FFT.

```
function [X1, X2] = doublerealfft(x1, x2)
%  DOUBLEREALFFT  Perform FFT of two real sequences x1[n] and x2[n]
%                 T. Holton
    x = x1+1j*x2;                          % x[n]
    X = fft(x);                            % perform complex FFT -> X[k]
    Xr_conj = conj([X(1) X(end:-1:2)]);    % flip & conjugate sequence -> X* [N-k]
    X1 = 0.5 * (X + Xr_conj);              % X1[k] derived from X[k]
    X2 = -0.5 * 1j * (X - Xr_conj);        % X2[k] derived from X[k]
end

% Compare FFT(x1[n]) with X1[k] derived from X[k]
x1 = [1 2 3 4];                        % x1[n]
x2 = [4 3 2 1];                        % x2[n]
[X1, X2] = doublerealfft(x1, x2);
disp('  fft(x1[n])             X1[k]'), disp([fft(x1).' X1.'])

  fft(x1[n])           X1[k]
  10.0000              10.0000
  -2.0000 + 2.0000i    -2.0000 + 2.0000i
  -2.0000              -2.0000
  -2.0000   2.0000i    -2.0000 - 2.0000i

% Compare FFT(x2[n]) with X2[k] derived from X[k]
disp('  fft(x2[n])             X2[k]'), disp([fft(x2).' X2.'])

  fft(x2[n])           X2[k]
  10.0000              10.0000
   2.0000 - 2.0000i     2.0000 - 2.0000i
   2.0000               2.0000
   2.0000 + 2.0000i     2.0000 + 2.0000i
```

11.6.3 *N*-point IFFT of two *N*-point transforms

It is equally easy to use a complex N-point IFFT to take the IFFT of two N-point transforms $X_1[k]$ and $X_2[k]$ simultaneously, as long as the inverses of those transforms, $x_1[n]$ and $x_2[n]$, are known to be real. First, form $X[k] = X_1[k] + jX_2[k]$. The inverse transform is $x[n] = x_1[n] + jx_2[n]$. So, $x_1[n] = \mathrm{Re}\{x[n]\}$ and $x_2[n] = \mathrm{Im}\{x[n]\}$.

Example 11.6

Given the transforms $X_1[k]$ and $X_2[k]$ found in Example 11.5, use a single 4-point IFFT to find the inverse transforms, i.e., the sequences $x_1[n]$ and $x_2[n]$, and verify that the answers are equivalent to those obtained by taking the IFFT of each transform separately.

➤ Solution:

```
function [x1, x2] = doubleifft(X1, X2)
% DOUBLEIFFT  Perform IFFT of two transforms X1[k] and X2[k],
%             whose inverses are real sequences x1[n] and x2[n]
%             T. Holton
    X = X1 + 1j * X2;
    x = ifft(X);
    x1 = real(x);
    x2 = imag(x);
end

% Compare x1[n] with Re{ifft(X[k])}
[xx1, xx2] = doubleifft(X1, X2);
disp('    x1[n] Re{x[n]}'), disp([x1' xx1'])
   x1[n] Re{ifft(X[k])}
     1      1
     2      2
     3      3
     4      4

% Compare x12[n] with Im{ifft(X[k])}
disp('    x2[n] Im{x[n]}'), disp([x2' xx2'])
    x2[n] Im{ifft(X[k])}
     4      4
     3      3
     2      2
     1      1
```

11.6.4 ★ *N*/2-point FFT of a **real** *N*-point sequence

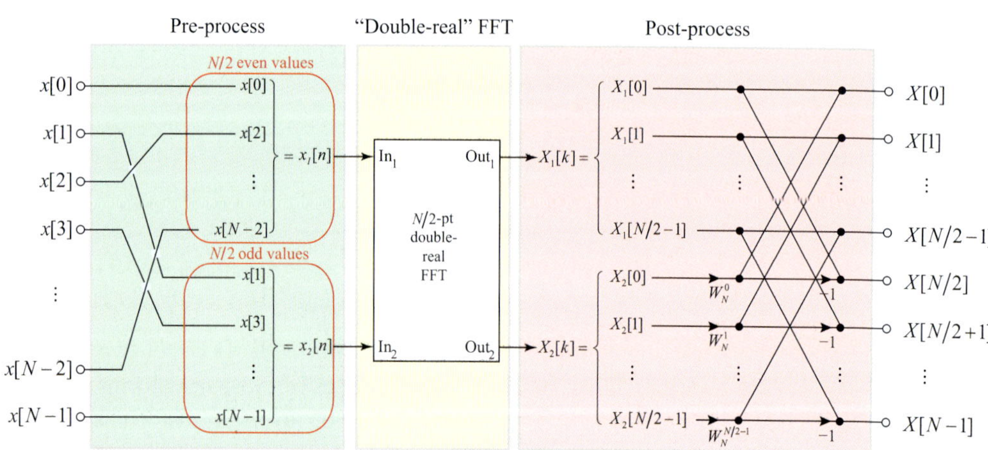

Figure 11.24 Pre- and post-processing of "double real" FFT

In the preceding section, we used an N-point double-real FFT to transform two real N-point sequences simultaneously. We can also use a double-real FFT in conjunction with pre-processing and post-processing steps to perform the FFT of a single real N-point sequence in roughly half the number of operations, as shown in **Figure 11.24**.

In the pre-processing step, a real N-point sequence $x[n]$ is split into even and odd subsequences of length $N/2$, $x_1[n]$ and $x_2[n]$, respectively. These subsequences are both real, so we can use an $N/2$-point double-real FFT to obtain the two $N/2$-point transforms $X_1[k]$ and $X_2[k]$ at the same time. What we have essentially done to this point is perform all but the last stage of an N-point DIT transform, as shown in **Figure 11.6a**. The post-processing step corresponds to the final stage of N butterflies that combines $X_1[k]$ and $X_2[k]$ into the N-point transform $X[k]$. This transform requires $(N/4)\log_2(N/2)$ complex multiplications for the $N/2$-point FFT plus an additional $N/2$ complex multiplications for the post-processing step.

Here is an example:

Example 11.7

Use a double-real 2-point FFT to find the transform of the 4-point real sequence $x[n] = [1\ 2\ 3\ 4]$, and verify that the answer is equivalent to that obtained using an 8-point FFT.

➤ Solution:

```
function X = doublerealfft2(x)
% DOUBLEREALFFT2  Perform double-real FFT of a N-point sequence
%                 using an N/2-point double real FFT
%                 T. Holton
    N = length(x);
    [X1, X2] = doublerealfft(x(1:2:N-1), x(2:2:N)); % even/odd pts
    X2 = X2 .* exp(-1j*2*pi*(0:N/2-1)/N); % do last butterfly
    X = [X1+X2 X1-X2];
end
x = [1 2 3 4];
X = doublerealfft2(x);
disp('  fft(x[n])            X[k]'), disp([fft(x).' X.'])
  fft(x[n])            X[k]
  10.0000              10.0000
  -2.0000 + 2.0000i    -2.0000 + 2.0000i
  -2.0000              -2.0000
  -2.0000 - 2.0000i    -2.0000 - 2.0000i
```

11.6.5 ★ $N/2$-point IFFT of an N-point transform

As you might expect, you can also use an $N/2$-point IFFT to perform the inverse transform of an N-point transform $X[k]$ when the inverse $x[n]$ is known to be real. The first step is to perform a decimation of the N-point IFFT of $X[k]$ into two $N/2$-point IFFTs using a derivation essentially equivalent to that leading to Equation (11.8). Then, recognize that, for a real sequence $x[n]$, $X[k]$ is conjugate symmetric, $X[k] = X^*[N-k]$, so we only need the $N/2$ values of $X[k]$, $0 \le k < N/2$, points. The end result (see Problem 11-1) is a pair of equations giving the even and odd points of $x[n]$,

$$\begin{aligned} n \text{ even:} \quad & x[2n] = \frac{1}{N}\sum_{k=0}^{N/2-1} \underbrace{(X[k]+X^*[N/2-k])}_{X_1[k]} W_{N/2}^{-kn} \\ n \text{ odd:} \quad & x[2n+1] = \frac{1}{N}\sum_{k=0}^{N/2-1} \underbrace{((X[k]-X^*[N/2-k])W_N^{-k})}_{X_2[k]} W_{N/2}^{-kn} \end{aligned}, \quad 0 \le n \le N/2-1. \tag{11.16}$$

Now, let

$$\begin{aligned} x_1[n] &= x[2n] \\ x_2[n] &= x[2n+1] \end{aligned}, \quad 0 \le n \le N/2-1,$$

and

$$\begin{aligned} X_1[k] &= X[k]+X^*[N/2-k] \\ X_2[k] &= (X[k]-X^*[N/2-k])W_N^{-k} \end{aligned}.$$

Here, $X_1[k]$ and $X_2[k]$ are the $N/2$-point IFFTs of $x_1[n]$ and $x_2[n]$, respectively, scaled by one-half, so

$$\begin{aligned} x_1[n] &= \frac{1}{2}\left(\frac{1}{N/2}\sum_{k=0}^{N/2-1} X_1[k]W_{N/2}^{-kn}\right) = \tfrac{1}{2}\mathrm{IFFT}_{N/2}(X_1[k]) \\ x_2[n] &= \frac{1}{2}\left(\frac{1}{N/2}\sum_{k=0}^{N/2-1} X_2[k]W_{N/2}^{-kn}\right) = \tfrac{1}{2}\mathrm{IFFT}_{N/2}(X_2[k]), \end{aligned} \quad 0 \le n < N/2.$$

Finally, use the same trick that we employed a few paragraphs ago to take the simultaneous IFFT of two sequences whose inverse transforms are known to be real; that is, form $X[k] = X_1[k] + jX_2[k]$, and take the inverse transform to obtain

$$\underbrace{\tfrac{1}{2}IFFT_{N/2}(X[k])}_{x[n]} = \underbrace{\tfrac{1}{2}IFFT_{N/2}(X_1[k])}_{x_1[n]} + \underbrace{j\tfrac{1}{2}IFFT_{N/2}(X_2[k])}_{jx_2[n]},$$

from which we get $x_1[n] = \mathrm{Re}\{x[n]\}$ and $x_2[n] = \mathrm{Im}\{x[n]\}$. An example, please!

Example 11.8

Given $X[k]$, which is the transform of the 8-point sequence $x[n] = [1\ 2\ 3\ 4\ 5\ 6\ 7\ 8]$, use a 4-point IFFT to find the inverse transform and verify that the answer is equal to $x[n]$.

➤ Solution:

```
function x = doubleifft2(X)
%  DOUBLEIFFT2  Perform N/2-point IFFT of N-point transform X[k]
%               whose inverse is real sequence x[n]
%               T. Holton
```

```
    N = length(X);
    X1 = X(1:N/2)+conj(X(N/2+1:-1:2));
    X2 = (X(1:N/2)-conj(X(N/2+1:-1:2))).*exp(1j*2*pi*(0:N/2-1)/N);
    XX = X1 + 1j * X2;
    xx = ifft(XX)/2;
    x1 = real(xx);
    x2 = imag(xx);
    x = [x1;x2];
    x = x(:)';
end

x = 1:8;
X = fft(x);
xx = doubleifft2(X);
disp('ifft(X[k]) x[n]'), disp([x' xx'])
ifft(X[k]) x[n]
     1     1
     2     2
     3     3
     4     4
     5     5
     6     6
     7     7
     8     8
```

11.7 FFT resolution

Taking the N-point DFT of an N-point sequence $x[n]$ is equivalent to measuring $X(\omega)$ at N equally spaced points in frequency $\omega = 2\pi k/N$, $0 \le k \le N-1$, which means that the frequency resolution of the DFT is $2\pi/N$ rad. How do you measure $X(\omega)$ with a lower or higher resolution in a computationally effective manner?

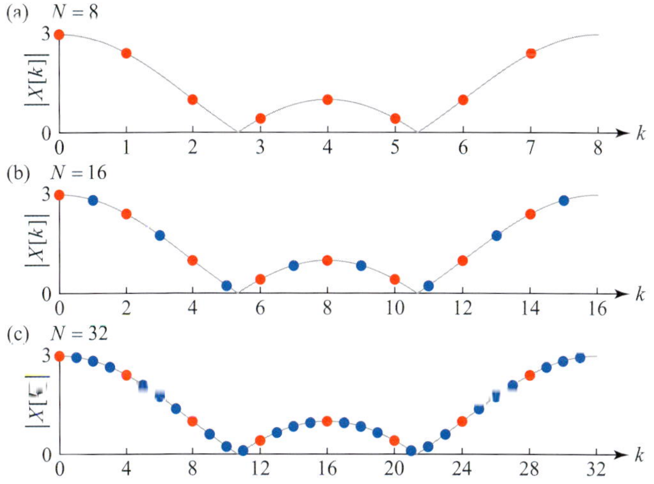

Figure 11.25 DFT of a sequence at three resolutions

11.7.1 Increasing the resolution of the FFT

As we discussed in Section 10.5.1, in order to measure $X(\omega)$ at a greater resolution with the DFT, we must effectively sample $X(\omega)$ at a larger value of N. For example, consider a sequence $x[n]$ of length M, where we wish to measure the DFT with frequency resolution $2\pi/N$ rad, where $N > M$. As we showed in Section 10.5.1, we accomplish this by padding $x[n]$ with $N-M$ zeros at the end and then taking the N-point DFT. For example, consider the sequence $x[n] = \delta[n] + \delta[n-1] + \delta[n-2]$, whose DTFT is $X(\omega) = e^{-j\omega}(1 + 2\cos\omega)$. **Figure 11.25** shows the magnitude of the N-point DFT $X[k]$ at three values of N that differ by a factor of two: $N = 8$, 16 and 32. For the 8-point DFT, each point corresponds to a multiple of $\omega = 2\pi/8 = \pi/4$; for the 16-point DFT, $\omega = 2\pi/16 = \pi/8$; and for the 32-point DFT, $\omega = 2\pi/32 = \pi/16$. Thus, every kth point of the 8-point DFT is equal to the $(2k)$th point in the 16-point DFT and the $(4k)$th point in the 32-point DFT, as indicated by the red points in the plots. These points correspond to the equivalent frequencies in the DTFT, which is shown in all plots by a thin gray line.

Increasing the resolution of the FFT in Matlab is just a matter of calling the `fft` function with the second argument `N` giving the length of the transform:

```
y = fft(x, N).
```

If `N` is greater than the length of `x`, the sequence is padded out with zeros. If `N` is smaller than the length of `x`, the sequence is truncated to `N` points.

11.7.2 Decreasing the resolution of the FFT

There are a number of algorithms, which fall under the general heading of **pruned FFT** algorithms, that are directed at reducing the computational requirements of the FFT either in cases where many of the input points of the transform are zero or in cases where you only want to compute a subset of the output points of the transform.

As a simple example, consider the problem where you have a 16-point sequence $x[n]$, and you only actually need to measure the eight points of the output of the DFT corresponding to frequencies $2\pi k/9$, $0 \le k < 8$. One approach that certainly *is not* going to work is to take the 8-point DFT of the first 8 points of $x[n]$. An approach that will work, but is not great, is suggested by looking at **Figure 11.25b**: take the 16-point DFT of $x[n]$ and then throw out every other output point. This works, but it is a pretty wasteful computational strategy, since we are calculating and throwing away an extra point for each one we use.

There is potentially another approach to computing the DFT of an N-point sequence with a resolution of $2\pi/M$ points, where $M < N$. To understand the approach, let us work out the specific example of $M = N/2$. The DFT of $x[n]$ is

$$X[k] = \sum_{n=0}^{N-1} x[n]e^{-j2\pi kn/N}.$$

What we really want is an $N/2$-point DFT $Y[k]$, defined by $Y[k] \triangleq X[2k]$, $0 \le k < N/2$:

$$Y[k] = X[2k] = \sum_{n=0}^{N-1} x[n]e^{-j2\pi kn/N} = \sum_{n=0}^{2(N/2)-1} x[n]e^{-j2\pi kn/(N/2)} = \sum_{n=0}^{2M-1} x[n]e^{-j2\pi kn/M}.$$

Split this into two summations of length M, and turn the algebra crank a few times to yield:

$$
\begin{aligned}
Y[k] &= \sum_{n=0}^{M-1} x[n]e^{-j2\pi kn/M} + \sum_{n=M}^{2M-1} x[n]e^{-j2\pi kn/M} = \sum_{n=0}^{M-1} x[n]e^{-j2\pi kn/M} + \sum_{n=0}^{M-1} x[n+M]e^{-j2\pi k(n+M)/M} \\
&= \sum_{n=0}^{M-1} x[n]e^{-j2\pi kn/M} + \sum_{n=0}^{M-1} x[n+M]e^{-j2\pi kn/M} \cancel{e^{-j2\pi kM/M}} = \sum_{n=0}^{M-1} \underbrace{(x[n]+x[n+M])}_{y[n]} e^{-j2\pi kn/M}.
\end{aligned}
$$

The final expression is just the M-point DFT of a sequence $y[n]$ of length $M = N/2$, which is formed by adding the two halves of $x[n]$. Since addition is computationally relatively cheap, this approach is much better for computing the DFT at a lower resolution. You can easily generalize this result to compute the M-point DFT of any sequence of length N, where N is an integer multiple of M (Problem 11-11),

$$
Y[k] = \sum_{n=0}^{M-1} \underbrace{\left(\sum_{m=0}^{N/M-1} x[n+mM] \right)}_{y[n]} e^{-j2\pi kn/M}. \tag{11.17}
$$

This is easy to implement in Matlab with just a few lines of preprocessing before using `fft`:

```
lx = length(x);
nc = ceil(lx/M); % number of rows of length M
ly = M * nc; % size of new array
newx = [x(:).' zeros(1, ly-lx)]; % may have to pad with zeros
x = sum(reshape(newx, M, nc)');
X = fft(x);
```

First, compute the number of rows, equivalent to N/M in Equation (11.17). If it is not an integral number, round up and pad $x[n]$ with zeros. All the heavy lifting is done by the `sum` and `reshape` functions, which create a matrix of rows and add them together.

11.8 Fast convolution using the FFT

One of the most important uses of the FFT is to perform fast convolution of input data with the impulse response of an FIR filter. In Chapter 10 (Section 10.3.7), we showed that the multiplication of transforms corresponds to circular convolution. In this section, we will show how this property can be used to do fast convolution.

11.8.1 Convolution of fixed-length input sequences

Figure 11.26 shows the general procedure for using transforms to convolve an input sequence $x[n]$ of length L_x with an impulse response $h[n]$ of length L_h to yield an output sequence $y[n] = x[n] * h[n]$ of length L_y. For a linear convolution $x[n] * h[n]$, the length of the output is $L_y = L_x + L_h - 1$. Hence, in order for circular convolution implemented by the multiplication of DFTs of length N to be equivalent to the linear convolution, we must ensure that $L_x + L_h - 1 \leq N$. To make sure that this condition is satisfied, the data in the $x[n]$ and $h[n]$ blocks are padded with $L_h - 1$ and $N - L_h$ zeros, respectively, so that each block is N points long, as shown in the top panel of **Figure 11.26**. To implement convolution by radix-2 transforms, the block size N would be chosen to be the next (nearest) higher power of two. To perform the circular convolution using FFTs, N-point FFTs of the input $x[n]$ and the impulse response $h[n]$

are obtained, yielding complex N-point transforms $X[k]$ and $H[k]$, respectively. These transforms are then multiplied together to form the transform $Y[k]$. Finally, the N-point IFFT of $Y[k]$ is taken to produce output $y[n]$, which is the circular convolution $x[n] \circledast h[n]$.

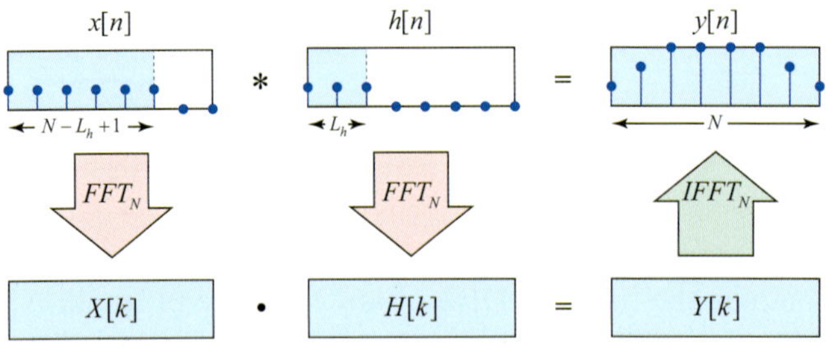

Figure 11.26 Convolution using the FFT and IFFT

Here is some Matlab code that implements the convolution of two fixed-length sequences using multiplication of FFTs:

```
function y = convfft(x, h)
%   CONVFFT Convolve two sequences using multiplication of FFTs
%           y = convfft(x, h) convolves sequences x and h
%           T. Holton
    Ly = length(x) + length(h) - 1;      % length of linear convolution
    N = 2^ceil(log2(Ly));                % size is next higher power of 2
    y = ifft(fft(x, N) .* fft(h, N));    % pad with zeros and convolve
    y = y(1:Ly);                         % truncate output
end
```

The first line of code calculates L_y, which is the required size of the linear convolution. The second line sets the actual transform size N to be the next higher power of two (you could also use `2^nextpow2(Ly)`). Then, the convolution is calculated using multiplication of transforms followed by an IFFT. The `fft` function uses the optional parameter `N` to pad both the `x` and `h` sequences out to `N` points. Finally, the result is truncated to L_y points.

Speed of convolution It is often – but not always – advantageous to use the multiplication of FFTs to compute the convolution, as opposed to simply implementing the convolution sum formula in the time domain. The advantage depends upon the parameters of the convolution, N and L_h, but it also depends critically on the implementation (hardware/software) and the design of the algorithms, as well as other factors such as memory utilization and overhead, some of which we will mention in Section 11.11. With those caveats in mind, it is still worth making a few back-of-the envelope calculations to get an idea of the sort of improvement we might expect. In what follows, we will just use the number of complex multiplications as an indication of the speed of convolution of each method. The FFT method requires that three N-point transforms be taken (two FFTs and one IFFT) per block of data, each of which requires on the order of $(N/2)\log_2 N$ complex multiplications, assuming a straightforward radix-2 transform. (In applications such as the FIR filtering of long sequences using the overlap-add or overlap-save methods described in the next section, the FFT of $h[n]$ can be pre-computed, which leaves two transforms to be done per block.) We also need N complex multiplications to form $X[k]H[k]$, for a total of $1.5N\log_2 N + N$ multiplications. In contrast, a direct convolution algorithm with possibly complex inputs should require no more than $L_h(N - L_h + 1)$ multiplications (Problem 11-8). The reason is that the leading and trailing $L_h - 1$ points of a convolution require fewer multiplications than the central $N - 2L_h + 2$ points. For small values of N or L_h, direct convolution is actually faster. For example, for the convolution shown in **Figure 11.27**, with $L_h = 3$ and $N = 8$, the FFT method would require 44 multiplications versus 18 for direct convolution. Similar calculations suggest that direct convolution could require fewer multiplications for any block size of $N < 32$, regardless of the length of the impulse response, L_h. But, for values of $N \geq 64$ and values of L_h greater than about 20, the advantage quickly swings to using the FFT method (see Problem 11-9). Of course, these simple calculations need to be taken with a large grain of salt. In practice, a number of other factors apart from the theoretical number of multiplications need to be considered, including the size and complexity of the code required for the FFT method, the possibility of round-off error in the calculation of floating-point FFTs and IFFTs, the relative overhead of the two methods and the issues of memory use in the computation of the FFT, some of which are discussed in Section 11.10. So, it is always important to test the two methods in your application before choosing.

11.8.2 Block convolution using the FFT

In Section 2.7, we discussed two methods – **overlap-add** and **overlap-save** – of implementing the convolution of a very long input sequence $x[n]$ with a finite impulse response $h[n]$. These methods of FIR filtering involve segmenting the input sequence into blocks, convolving each block with the impulse response to form an output block, and then stitching the output blocks together to form the complete output. In our previous treatment of these methods, convolution was implemented using the direct sum formula, so we did not require that these blocks be of any particular size, or even of the same size. However, now that we will be using multiplication of FFTs to do the convolution, data will need to be segmented (or padded out with zeros) into

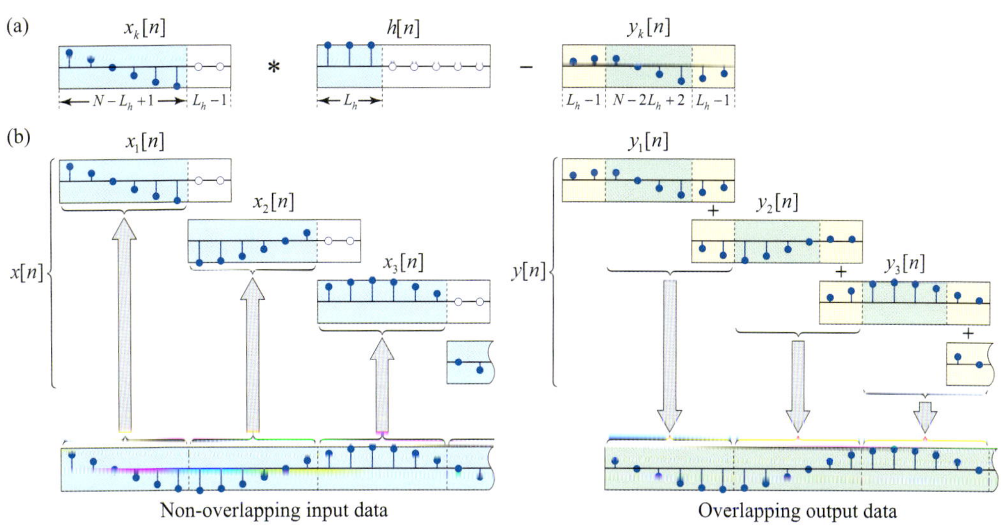

Figure 11.27 Convolution using the overlap-add method

blocks of the same length N. FFT methods are naturally suited to the "frame-based" processing of DSP processors whose architecture allows data to be accumulated and processed simultaneously. In many applications, particularly high-speed real-time applications, the DSP accumulates input data as it arrives into a frame buffer in memory. When the buffer is full, the DSP processes the data all at once (for example, by taking the FFT), while simultaneously beginning to accumulate the next frame of data.

Overlap-add method In the overlap-add method, diagrammed in **Figure 11.27**, the input data are segmented into *non-overlapping* chunks $x_1[n]$, $x_2[n]$, $x_3[n]$, ... , each of length $L_x = N - L_h + 1$. Each chunk $x_k[n]$ is padded with $L_h - 1$ zeros, so that the total block size is N, as shown in **Figure 11.27a**. The impulse response $h[n]$ of length L_h is also padded with zeros to form a block of length N. The output block, comprising a sequence $y_k[n]$, is created using the procedure we described in conjunction with **Figure 11.26**: the FFTs of $x_k[n]$ and $h[n]$ are computed and multiplied. The IFFT of the product is then equal to $y_k[n]$. In practice, the FFT of $h[n]$ is computed only once and used in the convolution of every block of data. Hence, there are only two transforms to be done per block. Because the total length of the output is $L_x + L_h - 1 \leq N$, the circular convolution that results from the multiplication of transforms is equivalent to linear convolution; there is no time-domain aliasing. The central $N - 2L_h + 2$ samples of $y_k[n]$ (colored in green in the figure) represent the portion of the output that relies completely on the convolution of data from input chunk $x_k[n]$ with $h[n]$. The leading and trailing $L_h - 1$ samples of each output chunk (colored in yellow in the figure) represent "partial results" that require the addition of data from either the previous or the following output chunk to complete the convolution. As described in Chapter 2, the series of output chunks $y_1[n]$, $y_2[n]$, $y_3[n]$, ... , overlap in such a way that when the last $L_h - 1$ samples of chunk $y_k[n]$ are added to the first $L_h - 1$ samples of the subsequent chunk $y_{k+1}[n]$, then that $(L_h - 1)$-point portion of the convolution is complete. After the addition, each output chunk has contributed $(N - 2L_h + 2) + (L_h - 1) = N - L_h + 1$ valid ("good") points to the output, as shown in **Figure 11.27b**.

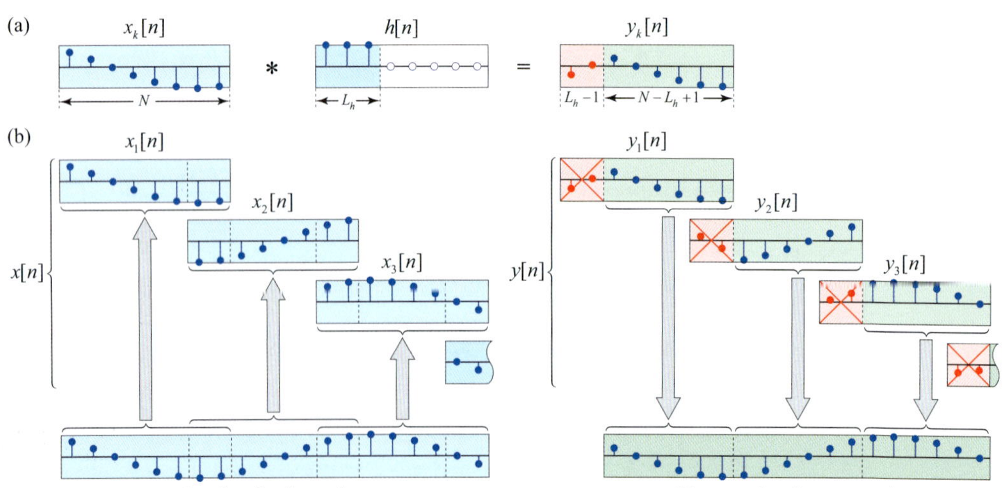

Figure 11.28 Convolution using the overlap-save method

Overlap-save method In the overlap-save method, diagrammed in **Figure 11.28a**, the input data is segmented into *overlapping* chunks $x_1[n]$, $x_2[n]$, $x_3[n]$, ... , each of length equal to the block size $L_x = N$. Now, because $L_x + L_h - 1 > N$, the circular convolution produced by the multiplication of transforms results in time-domain aliasing: the first $L_h - 1$ points of the output chunk $y_k[n]$, indicated in the figure by red shading, are corrupted by the wrap-around of the last $L_h - 1$ points of the convolution. In the overlap-save method, these "bad" points are tossed (ignored) and only the last $N - L_h + 1$ samples of $y_k[n]$ (colored in green in the figure), which represent the valid portion of the convolution, are "saved" (hence the name of the method). The "saved" points of successive output chunks $y_1[n]$, $y_2[n]$, $y_3[n]$, ... , do not overlap and contribute directly to the output, as shown in **Figure 11.28b**.

From **Figures 11.29** and **11.30**, you can see that the overlap-add and overlap-save methods have the same **throughput**; that is, each processed N-point output block contributes $N - L_h + 1$ completely valid points to the output. In terms of algorithmic complexity, both the overlap-add and overlap-save methods require the same number of multiplications per input data point when implemented with FFTs. The overlap-add method features simple, non-overlapping segmenting of the input data, but requires that the last $L_h - 1$ points of each output frame be saved and added to the first $L_h - 1$ points of the next output frame. The overlap-save features non-overlapping output segments and does not require summation of data from successive frames. However, it does require that the last $L_h - 1$ points of each input frame be saved and reused as the first $L_h - 1$ points of the next input frame. Based strictly on the number of total operations, the overlap-save method does not require the additional addition operations of the overlap-add method.

Choice of block size Choosing the appropriate block size of the FFT in the overlap-add or overlap-save methods is an exercise in balancing competing requirements. Larger FFT sizes allow one to process more data at once, but larger FFTs are also less efficient in the sense that doubling N more than doubles the number of multiplications that need to be performed (see Problem 11-8). In addition, because data are processed in blocks, there can be a delay of up to N samples between input and output, a delay that may be unacceptable in real-time applications (e.g., control). At least in principle, we can find the optimum data length L_x, and hence the optimum block length N, that minimizes the number of multiplications per processed input point for a given length of the impulse response L_h:

$$\min_{L_x}\left(\frac{(L_x + L_h - 1)\log_2(L_x + L_h - 1)}{L_x}\right).$$

Figure 11.29 shows the resulting relation between L_h and N. The plot suggests a block size in the range of four to eight times L_h over the range of L_h shown.

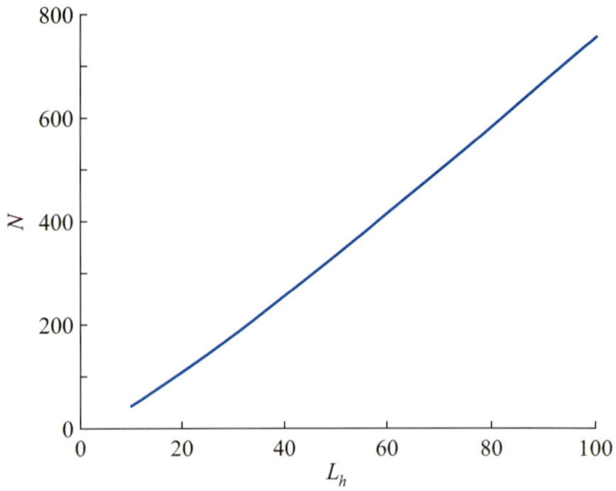

Figure 11.29 Optimum block size vs. length of impulse response

11.8.3 ★ Using both DIT and DIF transforms

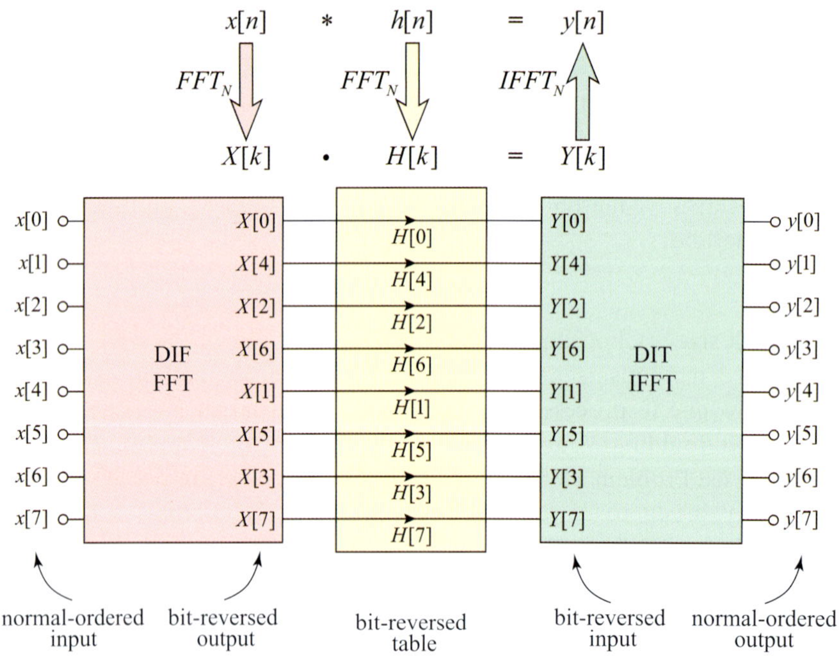

Figure 11.30 Use of both DIF and DIT FFTs

At the end of Section 11.1 we wondered why one might want to use both DIT and DIF transform implementations. One possible reason is shown in **Figure 11.30**. Consider a filtering operation in which frames of an input sequence $x[n]$ are convolved with an impulse response $h[n]$ to form frames of an output sequence $y[n]$. As we have just discussed, in the frequency domain the transforms $X[k]$ and $H[k]$ are multiplied to produce $Y[k]$ and then the inverse transformation is taken, yielding $y[n]$. In the implementation shown in **Figure 11.30**, the filter coefficients $H[k]$ are computed only *once* and stored in bit-reversed order. As each frame of the input $x[n]$ arrives, a DIF FFT is used to obtain the transform $X[k]$. Since the DIF transform naturally takes a normal-ordered input sequence and produces a bit reversed output, no bit reversal of the output is necessary. The transform $X[k]$ is then multiplied by $H[k]$ to form $Y[k]$. Since both $X[k]$ and $H[k]$ are in bit-reversed order, direct multiplication produces $Y[k]$ in bit-reversed order as well. Then a DIT IFFT takes the bit-reversed $Y[k]$ and produces a normal-ordered $y[n]$. By appropriate use of both DIF and DIT transforms, bit-reversal of the input or output has been completely avoided.

11.8.4 Matlab support for convolution using the FFT

Matlab's `fftfilt` function uses the overlap-add method to perform convolution of a data array or matrix with coefficients of an FIR filter. The syntax is

```
y = fftfilt(h, x, [N]),
```

where `h` is the impulse response of the FIR filter, `x` is the input and the optional parameter `N` determines the length of the FFT. `fftfilt` also supports parallel processing using graphic processing units (GPUs) via the Parallel Computing Toolbox.

11.9 ★ The Goertzel algorithm

The FFT in its many variants is the preferred way to compute an N-point DFT in cases where you need to compute $X[k]$ for many or all values of k, $0 \le k < N$. However, there are circumstances in which you only need one or a small subset of the DFT frequencies. For example, you may be interested in a spectral peak of a signal at a particular frequency or a small set of frequencies. In this case, it might be faster to compute the DTFT directly at the frequencies instead of computing the FFT in order to use only one or a few frequency bins.

The Goertzel algorithm is specifically designed to compute the discrete Fourier transform $X[k]$ at a single frequency value k more efficiently than direct computation of the DFT sum. To understand the algorithm, consider an N-point sequence $x[n]$, and write out the terms of $X[k]$ at a given value of k,

$$X[k] = \sum_{n=0}^{N-1} x[n] W_N^{kn} = x[0] + W_N^k x[1] + W_N^{2k} x[2] + \cdots + W_N^{k(N-2)} x[N-2] + W_N^{k(N-1)} x[N-1].$$

Multiply by $W_N^{-kN} = 1$:

$$X[k] = \sum_{n=0}^{N-1} x[n] W_N^{kn} \cdot W_N^{-kN} = \sum_{n=0}^{N-1} x[n] W_N^{-k(N-n)}$$
$$= W_N^{-Nk} x[0] + W_N^{-(N-1)k} x[1] + W_N^{-(N-2)k} x[2] + \cdots + W_N^{-2k} x[N-2] + W_N^{-k} x[N-1]. \quad \textbf{(11.18)}$$

The Goertzel algorithm casts the computation of $X[k]$ as a filtering problem in which the input is $x[n]$ and the output $y[n]$ is computed using a first-order difference equation

$$y[n] = W_N^{-k} y[n-1] + x[n]. \tag{11.19}$$

To see this, enumerate $y[n]$ in Equation (11.19) for $0 \le n \le N$, given the initial condition $y[-1]=0$:

$$\begin{aligned}
y[0] &= x[0] \\
y[1] &= W_N^{-k} y[0] + x[1] && = W_N^{-k} x[0] + x[1] \\
y[2] &= W_N^{-k} y[1] + x[2] && = W_N^{-2k} x[0] + W_N^{-k} x[1] + x[2] \\
&\vdots && \vdots \qquad \vdots \\
y[N-1] &= W_N^{-k} y[N-2] + x[N-1] && = W_N^{-(N-1)k} x[0] + W_N^{-(N-2)k} x[1] + \cdots + x[N-1] \\
y[N] &= W_N^{-k} y[N-1] + 0 && = W_N^{-Nk} x[0] + W_N^{-(N-1)k} x[1] + \cdots + W_N^{-k} x[N-1] = X[k].
\end{aligned}$$

If you compare the last line of this series of equations with Equation (11.18), you will see that $y[N]=X[k]$. The first N iterations of this filter require input data $x[n]$. But the last step, the computation of $y[N]$, is special in that there is no input. As presented so far, the recursive filter provides no computational savings; it still requires N complex multiplications for each value of $X[k]$. The computational savings come from observing that the z-transform of Equation (11.19) is

$$H(z) = \frac{Y(z)}{X(z)} = \frac{1}{1 - W_N^{-k} z^{-1}}.$$

Multiply both numerator and denominator by $1 - W_N^k z^{-1}$ to get

$$\begin{aligned}
H(z) &= \frac{Y(z)}{X(z)} = \frac{1}{1 - W_N^{-k} z^{-1}} \cdot \frac{1 - W_N^k z^{-1}}{1 - W_N^k z^{-1}} = \frac{1 - W_N^k z^{-1}}{1 - 2\cos(2\pi k/N) z^{-1} + z^{-2}} \\
&= \underbrace{\frac{1}{1 - 2\cos(2\pi k/N) z^{-1} + z^{-2}}}_{H_1(z)} \cdot \underbrace{(1 - W_N^k z^{-1})}_{H_2(z)}.
\end{aligned}$$

The filter has now been expressed as the cascade of two filters,

$$H(z) = \underbrace{\frac{Q(z)}{X(z)}}_{H_1(z)} \cdot \underbrace{\frac{Y(z)}{Q(z)}}_{H_2(z)} = H_1(z) \cdot H_2(z).$$

$H_1(z)$ is a second-order IIR filter with z-transform

$$H_1(z) = \frac{Q(z)}{X(z)} = \frac{1}{1 - 2\cos(2\pi k/N) z^{-1} + z^{-2}},$$

which corresponds to a difference equation with purely real coefficients,

$$q[n] = x[n] + 2\cos(2\pi k/N) q[n-1] - q[n-2]. \tag{11.20}$$

$H_2(z)$ is a first-order FIR filter with z-transform

$$H_2(z) = \frac{Y(z)}{Q(z)} = 1 - W_N^k z^{-1},$$

which corresponds to a difference equation that requires only one complex multiplication,

$$y[N] = q[N] - W_N^k q[N-1]. \tag{11.21}$$

To implement the Goertzel algorithm, calculate N values of Equation (11.20) and save only the final two values, $q[N]$ and $q[N-1]$. Each iteration of this difference equation requires only a single real multiplication. Then, perform one calculation of Equation (11.21) using the saved values of $q[N]$ and $q[N-1]$. This requires only a single multiplication of a real number by a complex number. Therefore, the computation of the Goertzel algorithm at a single frequency requires $N+2$ real multiplications. Here is some Matlab code that implements the algorithm:

```
function y = goertz(x, k)
% GOERTZ Goertzel algorithm
%           y = goertz(x, k)
%           Compute kth DFT coefficient of input array, x
%           at a single value of k
%           T. Holton
    N = length(x);
    W = exp(-1j* k* 2* pi/N);
    alpha = 2 * real(W);
    q1 = 0;
    q2 = 0;
    for i = 1:N
        q = x(i) + alpha * q1 - q2;
        q2 = q1;
        q1 = q;
    end
    q = alpha * q1 - q2;
    y = q - W * q1;
end
```

The complex quantity W_N^{-k} only needs to be computed at a single value of k, and is only used once at the end of the routine. Length N can have any integer value; it need not be a power of two, or a composite number. Furthermore, there is actually nothing that requires k to be an integer. That means we can efficiently calculate the value of the DTFT exactly at *any* frequency ω_0, simply by substituting $e^{j\omega_0}$ for W_N^{-k} and $\cos\omega_0$ for $\cos 2\pi k/N$ in the calculation.

DTMF tone decoding The Goertzel algorithm sometimes forms the basis of algorithms designed to decode which key was pressed on an old-school "Touch-Tone™" telephone keypad, schematized in **Figure 11.31**.

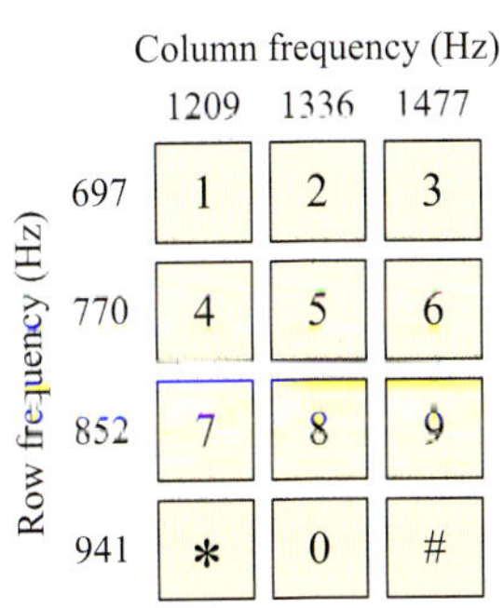

Figure 11.31 DTMF keypad

In **dual-tone multi-frequency (DTMF)** signaling, each press of a key produces the sum of tones at two frequencies, one that encodes the row and one that encodes the column. In a standard 4×3 keypad, there are four row tones (at 697, 770, 852 and 941 Hz) and three column tones (at 1209, 1336 and 1477 Hz).[4] For example, pushing the "*" key produces the sum of 941 and 1209 Hz tones. The task of DTMF decoding boils down to determining the location of the spectral peaks in a signal at seven known frequencies. If one uses the DFT/FFT to find the peaks, the size of the transform is determined by the minimum length of the DTMF tone (40 ms), and by the desire to have the seven tone frequencies correspond as closely as possible to frequencies of the transform bins. **Figure 11.32** shows a comparison of using the FFT and Goertzel algorithm to find the frequencies of the tones associated with the "*" key sampled at 8 kHz. The thin black trace shows the magnitude of the DTFT of a Hamming-windowed 32 ms segment of the DTMF tones for this key plotted on a dB scale. The two frequencies corresponding to this key occur at the peaks of the spectrum. The open blue circles show the spectral estimates obtained from a 256-point FFT of the recording and the solid red circles are the magnitude of the output of a Goertzel algorithm at the seven DTMF tone frequencies.

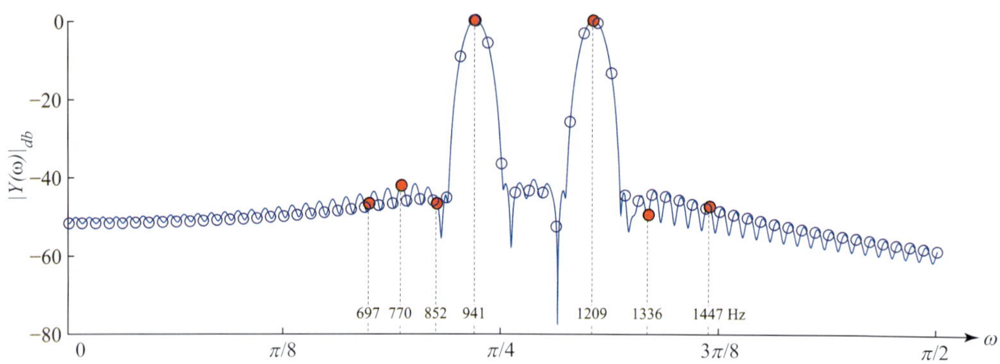

Figure 11.32 FFT of DTMF tones

While both techniques can be used successfully, the Goertzel algorithm is particularly well suited to real-time spectral measurements using small embedded processors, which often have low computational power and restricted program memory size. The output of the simple second-order IIR filter of the Goertzel algorithm (Equation 11.20) can be computed point by point as new data arrive. Then, only one calculation of the first-order FIR filter (Equation 11.21) is required when the last point of the frame arrives.

DTMF detection is an example of a spectral analysis problem in which we are primarily interested in the magnitude (or magnitude squared, $|Y[k]|^2$) of the DTFT at a particular frequency or frequencies. In this case, we can modify the last step of the Goertzel algorithm, Equation (11.21), to yield $|Y[k]|^2$ directly (see Problem 11-7):

$$|y[N]|^2 = |Y[k]|^2 = q^2[N] + q^2[N-1] - \cos(2\pi k/N)q[N]q[N-1].$$

[4]There are actually four columns in the full DTMF keypad, but the fourth column is almost never seen in home applications. The additional column, on the right of the keypad, has row buttons labeled "A," "B," "C" and "D." The frequency of the extra column tone is 1633 Hz.

In Chapter 14, we will discuss some of the principal techniques used in spectral analysis in detail.

11.10 Iterative and recursive implementations

There are two basic ways of implementing a fast FFT in software: **iterative** and **recursive** (or a hybrid of the two). The choice of implementation has consequences in terms of code complexity and memory efficiency. In this section and the next, we will look at some of the issues involved in these two approaches.

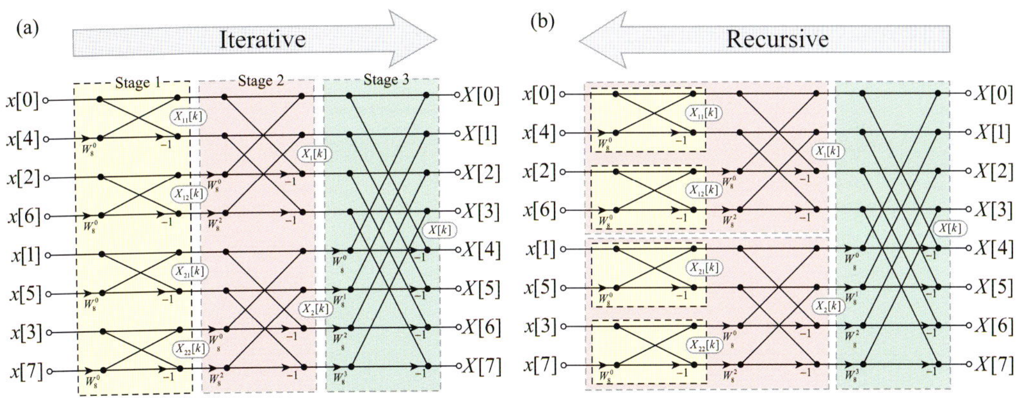

Figure 11.33 Iterative and recursive implementations of the FFT

11.10.1 Iterative implementation

Figure 11.33 shows the signal-flow graph of **Figure 11.9** interpreted in two different ways. Historically, most "textbook" FFT algorithms have been iterative or non-recursive, as schematized in **Figure 11.33a**. Iterative code starts with an input sequence and performs the computation of each stage of the transform, essentially working from the *left* side of **Figure 11.33a** towards the *right*, starting at the first stage of the decomposition (yellow box, with inputs $x[n]$) and proceeding sequentially to the last (green box, with outputs $X[k]$). In each of the $\log_2 N$ stages, all N points of data at the input to the stage are processed to produce N output points before proceeding to the subsequent stage. Each stage comprises $N/2$ butterflies; each butterfly takes two unique inputs and produces two unique outputs that go to the next stage, as shown in **Figure 11.5**. But once the computation of a given butterfly is completed, those two input values are no longer required in any subsequent stage, so the memory locations that held these values can be reused to hold the two output values of the computation. Computations that reuse the same memory locations in this manner are termed **in-place**. In-place computation means that an N-point FFT only requires memory storage for N input/output values plus $N/2$ twiddle factors (as well as some temporary storage). Furthermore, since each butterfly takes two unique inputs and produces two unique outputs, in principle all the butterflies in any given stage can be performed simultaneously.

An important issue with the in-place computation of the iterative approach is that each stage of the transform, comprising N input points, needs to be completed before calculations on the next stage can be completed; hence, all N points have to be available to complete the calculation of each stage. If N is large, chunks of memory likely have to be read in and out of cache, which will exact a substantial speed penalty compared to the recursive approach we will discuss below. However, the place where the iterative approach makes a lot of sense is in *parallel* implementations on GPUs, as well as implementations on special-purpose VLSI hardware, such as ASICs or FPGAs, which inherently support parallel designs. DSP chips uniformly use the iterative approach.

Here is a simple iterative DIT FFT in Matlab, which assumes that the input sequence has already been bit-reversed:

```
function x = fft_iterative(x)
    N = length(x);
    W = exp(-1j*2*pi*(0:N/2-1)/N); % set up the twiddle factors
    nStages = log2(N);
    nGroupsPerStage = N / 2;
    nFlysPerGroup = 1;
    for n = 1:nStages
        ibase = 1;
        iw = 1;
        for k = 1:nFlysPerGroup
            w = W(iw);   % compute twiddle factor for entire inner loop
            i0 = ibase; % index of input 0 of butterfly
            for m = 1:nGroupsPerStage
                i1 = i0 + nFlysPerGroup; % index of input 1 of butterfly
                y0 = x(i0);              % read input data into y0 and y1
                y1 = x(i1) * w;          % the only complex multiplication
                x(i0) = y0 + y1;         % write in-place into x(i0), x(i1)
                x(i1) = y0 - y1;
                i0 = i1 + nFlysPerGroup; % increment index 0 for next pass
            end
            iw = iw + nGroupsPerStage;
            ibase = ibase + 1;
        end
        nGroupsPerStage = nGroupsPerStage / 2;
        nFlysPerGroup = nFlysPerGroup * 2;
    end
end
```

While somewhat primitive, this code does illustrate some typical features of the iterative approach to a radix-2 DIT transform. There are `nStages` ($\log_2 N$) *stages*. In the nth stage, there are $N/2$ butterflies arranged in groups of `nGroupsPerStage` ($2^{nStages-n}$). Each group contains `nFlysPerGroup` (2^{n-1}) butterflies. Thus, the program comprises three nested `for` loops. The outer loop iterates through `nStages`, the middle loop through `nFlysPerGroup` and the inner loop through `nGroupsPerStage`. In the inner loop, where the program spends most of its time, the only multiplication is by the twiddle factor `w`. Temporary variables `y(i0)` and `y(i1)` are used so that the input data `x(i0)` and `x(i1)` are read only once. The results of the butterfly are written in-place back into `x(i0)` and `x(i1)`. The program sets up the twiddle factors in an array before entering the loops, so only `iw`, the index to the appropriate twiddle factor, has to be computed. The reason for the order of the middle and inner loops is that this minimizes the number of times that `iw` and `w` need to be computed. The twiddle factor determined in the middle loop is used for the entire inner loop. Also in this program, the computation of the indexes of the twiddle factor and the inputs to the butterflies (i.e., `i0` and `i1`) is done using only additions. In a language like C, factor-of-two scaling operations such as those of `nGroupsPerStage` and `nFlysPerGroup` at the end of the outer loop would be performed using the bit shifting assignment operators (<<= and >>=) rather than with multiplications or divisions by two.

11.10.2 Recursive implementation

Recursive algorithms are the simplest and most elegant in terms of code complexity and size and are particularly well suited to implementations of the FFT.[5] In a recursive implementation of the FFT, the code follows the signal-flow graph of **Figure 11.33b** from *right* to *left*. The rightmost stage (green box) shows that the final N-point transform $X[k] = X_1[k] + W_N^k X_2[k]$ is assembled from two $N/2$-point transforms $X_1[k]$ and $X_2[k]$ (red boxes), each of which is, in turn, assembled from two $N/2$-point transforms (yellow boxes),

$$\begin{aligned} X_1[k] &= X_{11}[k] + W_{N/2}^k X_{12}[k] \\ X_2[k] &= X_{21}[k] + W_{N/2}^k X_{22}[k] \end{aligned}.$$

In the recursive implementation, the program first calls for $X_1[k]$ to be computed, which in turn requires $X_{11}[k]$ and $X_{12}[k]$ to be computed in succession. Then $X_2[k]$ is computed, which requires $X_{21}[k]$ and $X_{22}[k]$ to be computed in that order. This is an example of an **out-of-place** computation. The most important advantage of this recursive code is its efficient utilization of memory. When implemented in general-purpose computers whose memory hierarchy includes various levels of cache, once the recursion gets to a point where a given section of input (e.g., the inputs necessary to compute $X_1[k]$, the upper red box in **Figure 11.33b**) is completely resident in cache, subsequent "deeper" recursive calls (e.g., to $X_{11}[k]$ and $X_{12}[k]$, the yellow boxes) use only a subset of the same memory in the cache. This can greatly increase the speed and efficiency of the routine, particularly when the size of the FFT is large, because cache memory accesses can be an order of magnitude or more faster than

[5] Recursive algorithms are sometimes considered to be less preferable than iterative ones due to concerns of storage efficiency or execution speed. While it is true that each recursive call requires the executable to create a new instance of the function and keep track of it through a call stack, the recursive algorithm has the key advantage of allowing efficient use of memory compared to the non-recursive algorithm. The speed gain due to this efficiency dwarfs the cost of implementing the recursion.

out-of-cache memory accesses.[6] Some of the fastest general-purpose FFT programs are highly optimized, "processor-aware" hybrids that use recursive calls in conjunction with relatively large kernels and careful attention to processor architecture and cache size. However, hardware vendors often offer FFT code (e.g., the Intel Math Kernel Library (MKL)) that is highly optimized for their particular processors and will run faster than general-purpose code in many applications.

Here is a bit of Matlab code that illustrates the elements of a recursive DIT FFT algorithm:

```
function X = fft_recursive(x)
    N = length(x);
    if (N == 1) % last pass -> no recursion
        X = x;
    else         % otherwise, recursion and apply twiddle factors to X2
        X1 = fft_recursive(x(1:2:N-1));
        X2 = fft_recursive(x(2:2:N)) .* exp(-1j*2*pi*(0:N/2-1)/N);
        X = [X1+X2 X1-X2];
    end
end
```

That's it! What we have done programmatically here is exactly what we showed graphically in **Figure 11.33b**. We start with an N-point sequence x, where we assume that N is a power of two, and recursively apply Equation (11.4). While in theory recursion is well suited to "divide-and-conquer" problems such as the FFT, this particular code exhibits a number of significant flaws, the worst of which is the wasteful computation of the twiddle factors at each stage of the recursion. At a minimum, the twiddle factors should be pre-computed and stored in a table, and the code should be written to index them appropriately. Another big flaw is that the last (N==1) pass of the FFT, the so-called **kernel**, in which the recursive routine spends a lot of time, does nothing useful; it just copies variable x into X. The efficiency of this recursive routine can be markedly improved by removing the computation of the twiddle factors and by making the kernel larger and hard-coding it to do some useful computational work. Here is a slightly more elaborate piece of code that addresses these issues:

```
function X = fft_recursive2(x)
    NN = length(x);
    W = exp(-1j*2*pi*(0:NN/2-1)/NN); % set up twiddle factors
    X = recurse(x, NN);

    function y = recurse(x, N)
       if (N == 4) % last pass -> bit-reversed radix-4 butterfly
           z = [x(1)+x(3) x(1)-x(3) x(2)+x(4) 1j*(x(2)-x(4))];
            y = [z(1)+z(3) z(2)-z(4) z(1)-z(3) z(2)+z(4)];
       else          % otherwise, recursion and apply twiddle factors to X2
           N2 = N / 2;
           X1 = recurse(x(1:2:N), N2);
           X2 = recurse(x(2:2:N), N2) .* W(1:NN/N:end);
```

[6] If an L1 cache reference requires on the order of 1 ns, an L2 cache reference would take ~4 ns, and a main memory (RAM) access ~100 ns. Big differences!

```
            y = [X1+X2 X1-X2];
        end
    end
end
```

The outer function. `fft_recursive2`, is a wrapper that sets up the twiddle factors `W`, and then calls the inner recursive function `recurse`. The inner recursion stops at a four-point, radix-4 kernel, which is implemented using the two-stage version shown in **Figure 11.18c** so that it has no complex multiplications (the `1j` does not count!). If you compare this code with the schematic in **Figure 11.18c**, you will see that the indexing of the inputs to this radix-4 butterfly have also been bit-reversed (i.e., `x(1)`, `x(3)`, `x(2)`, `x(4)`). This obviates the need for any further explicit bit-reversal code, and represents another significant advantage of this recursive method. Nice!

11.11 Implementation issues

Designing an FFT algorithm that is efficient on paper is just half the battle. How the algorithm is implemented in the target processor or application is critical to achieving anywhere close to the gains of speed and performance you predict from the theory. The nature of the hardware and software (e.g., language and compiler) has a first-order effect on the final speed of execution.

Multiplications and additions In our discussions so far, we have used the number of complex multiplications as the measure of the speed of the transform. But as we mentioned at the beginning of the chapter, that is because in the early days of computing, back when dinosaurs roamed the earth, multiplication was usually the rate-limiting step in the implementation of the algorithm. Accordingly, algorithm designers and programmers worried obsessively about reducing the number of multiplications, and the number of multiplications (and additions) became a convenient, easily calculable, method of comparing algorithms. If the processor uses a software/firmware implementation to perform multiplication, minimizing the number of multiplications would make sense. However, modern superscalar processors and DSPs feature multiple parallel instruction and data streams, and have pipelined arithmetic logic units that can perform one or more multiplication and accumulation (MAC) operations in a single machine cycle. So, the difference between the execution speed of a multiplication and an addition may be minimal.

Memory and cache The speed of MAC operations is not even necessarily the rate-limiting step in the computation of the FFT. A key set of issues – particularly in implementations of the FFT on general-purpose computers – revolves around the use of memory: loading and storing input and output data, allocating and deallocating memory, making sure memory reads and writes are word-aligned to physical memory and using available CPU registers and different levels of data cache effectively. The traditional iterative (non-recursive) implementations of the FFT discussed in Section 11.10.1 can be particularly problematic in this regard, since an efficient in-place computation of each stage of the transform would require that the entire data array be available in cache memory to be read and written. If the entire array does not fit in cache, then there will be "cache misses" – hardware interrupts that occur when a program is attempting to access data not currently located in the cache – that can exact a speed penalty of *hundreds of machine cycles*, many times more severe than the cost of a multiplication. The recursive or hybrid algorithms discussed in Section 11.10.2 can avoid some of these problems. As of this writing, probably the best collection of open-source FFT routines targeted at general-purpose processors is the wonderfully named "Fastest FFT in the West" (FFTW) library (www.fftw.org). FFTW code, which is written is ANSI C, is fast, portable and "processor aware," meaning that it supports a variety of hardware instructions set on general-purpose computers and can determine the fastest way of implementing a given size transform on a given computer. Matlab's FFT routines `fft` and `ifft` are all based on the FFTW library.

Fixed-point vs. floating-point Another important set of implementation issues revolves around the type of data the FFT is designed to process. FFTs can be implemented in hardware and software using either floating-point or fixed-point-integer arithmetic. At the heart of the FFT are the multiply/accumulate (MAC) operations in each butterfly, which involves the addition (or subtraction) of complex pairs of numbers, either preceded or followed by multiplication by complex twiddle factors. In fixed-point architectures, special attention has to be paid to the representation and scaling of these numbers to prevent either overflow or loss of precision. Consider a DIT FFT, such as that depicted in **Figure 11.9**, which is implemented using signed N-bit integer arithmetic; that is, the real and imaginary parts of the inputs and outputs of each butterfly as well as the twiddle factors are all stored as N-bit integers. To compute one of the output values, one N-bit input value is multiplied by an N-bit twiddle factor, which requires a $2N$-bit multiplier. Then the product is added to (or subtracted from) the other N-bit input number, which requires a $2N+1$-bit adder and the result is stored (in place) back as an N-bit output number. If the sum of this operation exceeds N bits, overflow will result.

One solution, implemented on fixed-point DSP processors, is to use so-called **block-floating-point (BFP)** arithmetic. BFP is essentially a fixed-point fractional representation (see Appendix B) in which the binary points of all the individual numbers in a block (for example all the butterflies comprising one stage of the FFT) are assigned a common value. Using this approach, the inputs to a stage can be scaled down conditionally by factor(s) of two to prevent overflow with the processor keeping track of the accumulated scaling. Conversely, inputs can be scaled up by factors of two so that they will fill the entire dynamic range of the multiply/accumulate processor.

Hardware implementations Finally, even though compilers for general-purpose computers have become extraordinarily powerful, the fastest FFT programs are still generally written directly in the assembly language of a target processor. General-purpose DSP chips (and special-purpose FFT processors) provide hardware specifically suited for the computation of FFTs, such as direct memory access, separate instruction and data caches, multiple multipliers and arithmetic units, and pipelined architecture that support instruction-level parallelism. Certain DSP machine instructions are also tailored to the FFT, such as the ability to perform bit-reverse indexing and compute radix-2 butterflies. The state of hardware changes rapidly. Many modern general-purpose processors have multiple cores, and most support some type of vectorized processing using **single-instruction multiple data (SIMD) instructions**, in which the processor can perform the same operation simultaneously on an array (vector) of data. Many embedded processors, particularly those aimed at the communications market (e.g., cellphones) now have multiple cores, including specialized DSP "cores." Fast FFTs have been implemented on **graphics processing units (GPUs)**, the graphics engines upon which video graphics cards are based. These GPUs feature a massively parallel, floating-point architecture. Extremely fast FFTs have also been implemented in special-purpose VLSI hardware, such as **custom application-specific integrated circuits (ASIC)** or **field-programmable gate arrays (FPGA)**.

SUMMARY

In this chapter, we derived a number of the most common variants of the fast Fourier transform, principally the radix-2, radix-4 and composite decimation-in-time (DIT) transforms. We found that computational savings can be achieved by exploiting the symmetry properties of the DFT in the calculation of real sequences. The chapter concluded with a discussion of implementation issues, particularly the role of cache memory in determining the speed of the algorithm, as well as a look at the relative merits of the recursive and non-recursive strategies for implementing the FFT.

PROBLEMS

Problem 11-1

Prove that the inverse transform of an N-point transform $X[k]$ when the inverse $x[n]$ is known to be real is given in Equation (11.16).

Problem 11-2

Derive the radix-4 decimation-in-frequency algorithm. Starting with $X[k]$, derive expressions for $X[4k]$, $X[4k+1]$, $X[4k+2]$ and $X[4k+3]$ for $0 \le k < N/4$.

Problem 11-3

The complex multiplication of two numbers, $z = xy$, where $x = a + jb$ and $y = c + jd$, conventionally requires four real multiplications and two real additions: $z = xy = (ac - bd) + j(ad + bc)$. Show that you can also perform a general complex multiplication using three real multiplications and five real additions.

➤**Hint:** Compute $p = (a + b)(c + d)$, $q = ac$ and $r = bd$.

Problem 11-4

In implementations of FFT algorithms, the twiddle factors are pre-computed. Consider the complex multiplication $z = xy$, where $x = a + jb$ is the input data and $y = c + jd$ is the twiddle factor. Assuming you pre-compute the quantities $q = c + d$ and $r = c - d$ once at the beginning of the program (so that we do not count those additions), show that you can perform the multiplication of x by y using only three real multiplications and three real additions.

➤**Hint:** Compute $p_1 = c(a+b)$, $p_2 = bq$ and $p_3 = ar$. This complex multiplication algorithm was discovered by Gauss!

Problem 11-5

Show that multiplications by $W_N^{N/8} = e^{-j\pi/4}$ and $W_N^{3N/8} = e^{-j3\pi/4}$ can be implemented with only two real multiplications and two real additions.

Problem 11-6

Determine the number of complex multiplications for the composite FFT represented by Equation (11.12).

Problem 11-7

Show that we can modify the last step of the Goertzel algorithm to yield the magnitude of the output directly:

$$|y[N]|^2 = |Y[k]|^2 = q^2[N] + q^2[N-1] - \cos\frac{2\pi k}{N} q[N]q[N-1].$$

Problem 11-8

In this problem, we investigate the relative computational cost of the convolution of two real sequences by two methods: direct convolution using the convolution sum and circular convolution implemented via the multiplication of FFTs. Consider two sequences $x[n]$ and $h[n]$ of lengths L_x and L_h, respectively. The convolution by either method yields a sequence $y[n]$ of length N. Hence, $N = L_x + L_h - 1$. Assume N is a power of two, which enables the use of radix-2 FFTs in the FFT method. Further assume that it takes four real multiplications per complex multiplication.

(a) For direct convolution, show that the minimum number of real multiplies required is $L_h(N - L_h + 1)$.

➤**Hint:** The first and last $L_h - 1$ points of the convolution do not require L_h multiplications.

(b) Show that the maximum number of real multiplies using direct convolution occurs when $L_h = (N+1)/2$.

(c) When N is an even number, the maximum value of L_h found in part (b) is not an integer. Show that the number of multiplies when $L_h = N/2$ is only 0.25 different from the number at $L_h = (N+1)/2$.

(d) For the FFT method, show that the double-real FFT method requires $N(\log_2(N/2) + 2)$ multiplies.

Problem 11-9

Using the results of Problem 11-8, use Matlab to show that direct convolution always requires the smallest number of multiplications for any $N < 32$.

Problem 11-10

Using the results of Problem 11-8, parts (a) and (d), use Matlab to find the smallest value of L_h such that the FFT method requires a smaller number of multiplications compared with direct convolution for values of $N = 32$, 64, 128, 256, 512, 1024 and 2048.

Problem 11-11

Show that $Y[k]$, the M-point DFT of any sequence of length N, where N is an integer multiple of M, can be expressed as the M-point DFT of a sequence $y[n]$ of length M,

$$Y[k] = \sum_{n=0}^{M-1} \underbrace{\left(\sum_{m=0}^{N/M-1} x[n+mM] \right)}_{y[n]} e^{-j2\pi kn/M} = \sum_{n=0}^{M-1} y[n] e^{-j2\pi kn/M}.$$

Problem 11-12

Derive the four outputs of the radix-4 transform $X[k]$ given in Equation (11.10).

Problem 11-13

Derive the four outputs of the split-radix transform $X[k]$ given in Equation (11.24).

12 Discrete cosine transform (DCT)

Introduction

In Chapter 10, we introduced the discrete Fourier transform (DFT), which is the basis of many practical applications, such as those concerned with spectral analysis (Chapter 14) and fast filtering via the FFT (Chapter 11). However, there are a number of important applications in which the DFT falls short, particularly in areas related to compression of signals such as speech, music and images. The **discrete cosine transform (DCT)** is a "cousin" of the DFT that addresses these applications.

In the first two sections of this chapter, we derive the DCT from the DFT and introduce the four principal variants of the transform as well as some of their important properties. The DCT is central to many applications that people see and interact with on a daily basis: their cellphones, music players and cameras. In Section 12.2, we discuss the **MP3** codec, which is nearly universally supported in music players, phones and streaming applications. Another variant of the DCT, the **two-dimensional (2D) DCT**, is at the heart of such image compression techniques as the **JPEG** standard for pictures, which we will discuss in Section 12.3. In supplementary material available at www.cambridge.org/holton, we will explain a couple of the DCT variants – the **modified DCT (MDCT)** and the **windowed MDCT** – which are employed in encoders that compress music and speech files.

12.1 The DCT

Figure 12.1 offers some motivation for our study of the DCT. **Figure 12.1a** shows a $2N$-point sequence $x[n]$ (solid blue dots), comprising the $N=5$ non-zero data points and N zeros, shown enclosed in the box. **Figure 12.1c** shows the energy of the DTFT of these points, $|X(\omega)|^2$ (blue trace), for frequencies $0 \le \omega \le \pi$. As we discussed in Chapter 10, when we compute the DFT $X[k]$ of this sequence, we are implicitly computing the DTFT of the periodic sequence $\tilde{x}[n]$, which consists of repeated periods of $x[n]$, a portion of which is plotted with open circles in **Figure 12.1a**. For a $2N$-point real sequence, the DFT comprises $2N$ coefficients, of which two are real-valued ($k=0$ and $k=N$) and the remainder occur as $N-1$ complex-conjugate pairs. The blue symbols in **Figure 12.1c** are $|X[k]|^2$, $0 \le k \le 5$. Because there is a sharp transient in $x[n]$ at the point $n = N - 1$, the DTFT has a substantial amount of energy at high frequencies; 95% of the energy in the transform is found at frequencies $|\omega| < 0.67\pi$, as indicated in the figure by the blue line. To represent 95% of the energy of the DFT, we need 90% of the DFT coefficients (the DC coefficient plus three complex conjugate pairs: $k = 0, 1, 2, 3, 7, 8$ and 9).

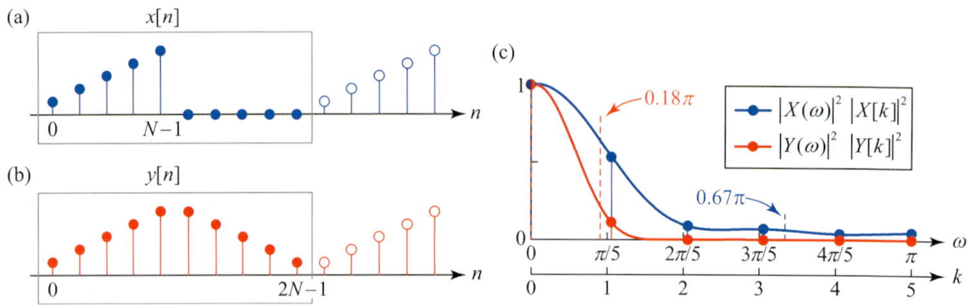

Figure 12.1 DFT of sequences with discontinuities

Figure 12.1b shows a $2N$-point sequence $y[n]$, which is formed as the **periodic extension** of $x[n]$ by taking the N non-zero points of $x[n]$, flipping them left-to-right and appending them to $x[n]$ in order to form the $2N$-point sequence $y[n]$,

$$y[n] = \begin{cases} x[n], & 0 \le n < N \\ x[2N - 1 - n], & N \le n \le 2N - 1 \end{cases}.$$

The DTFT and DFT of this sequence are shown in **Figure 12.1c** (the red trace and symbols respectively). The discontinuity at the edges of $y[n]$ (or between the periods of $\tilde{y}[n]$) is much smaller than that of $x[n]$ or $\tilde{x}[n]$. As a consequence, there is much less high-frequency energy in the transforms, $Y(\omega)$ and $Y[k]$. In fact, 95% of the energy of $Y(\omega)$ is found at frequencies $|\omega| < 0.18\pi$, as indicated in the figure by the red line; more than 95% of the energy of $Y[k]$ is contained in three of ten coefficients ($k = 0$, 1 and 9). Even though $y[n]$ was created from the same data as $x[n]$, its DFT has more energy at low frequencies than that of $x[n]$. This suggests a way of creating an invertible transform that potentially has better energy compaction properties than the DFT, particularly for sequences with discontinuities, such as the example in **Figure 12.1**. That transform is the DCT.

12.1.1 ★ Periodically extended sequences

The DCT of a sequence $x[n]$ can be conceptually understood as the DFT of a sequence formed from a **periodic extension** of the sequence $x[n]$ in a manner identical to that shown in **Figure 12.1b**. The top panel of **Figure 12.2** shows $x[n]$, and the remaining panels depict, enclosed in shaded boxes, the complete sequences ($x[n]$ plus its extension) that constitute the four principal variants of the DCT, called Type-I, -II, -III and -IV DCTs, labeled in the figure as DCT-I, DCT-II, DCT-III and DCT-IV.

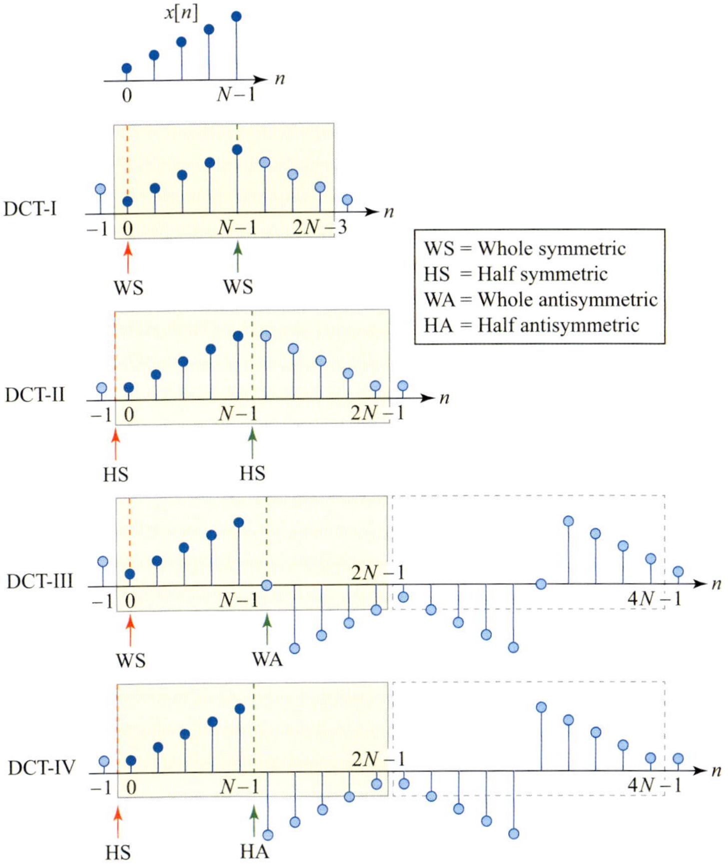

Figure 12.2 The four primary variants of the discrete cosine transform (DCT)

The main feature that determines the amount of energy compaction achieved by the DCT is the nature of the discontinuities at the left and right boundaries of a sequence and its extension (denoted respectively by the red and green arrows in the panels). Each boundary can be classed as either symmetric or antisymmetric, depending on whether the extension of $x[n]$ is even- or odd-symmetric with respect to $x[n]$ at that boundary; each boundary is further classified as either whole- or half-sample symmetric (or antisymmetric) depending on whether the point of symmetry or (antisymmetry) of $x[n]$ and its extension falls on a whole integer sample or on a point halfway between two samples. That gives four choices for each boundary and therefore 16 different periodic extensions. Of these, the DCT family comprises the eight variants whose left boundary is either whole- or half-sample symmetric. Of these eight, the four variants shown in **Figure 12.2** are the most commonly encountered.[1] That can be pretty confusing. To take a specific example, the periodic extension of the sequence labeled DCT-I is even-symmetric, formed by a right left flip of $x[n]$ about the right boundary, $x[N-1]$. The sequence and its extension share the point $x[N-1]$, so this boundary is classified as whole-sample symmetric (denoted "WS" in the figure). The left boundary of $x[n]$ is also symmetric about point $x[0]$,

[1]The eight variants that are antisymmetric on the left boundary form the basis of a related family of transforms, the **discrete sine transform (DST)**.

hence this is also labeled "WS." The sequence and its extension form a $(2N-2)$-point unit (shown in the box) that implicitly repeats periodically. The sequence labeled DCT-II is exactly the same as the sequence shown in **Figure 12.1b**. The periodic extension here is again even-symmetric, on the right boundary, but the point of symmetry lies halfway between the points $y[N-1]$ and $y[N]$. Hence this boundary is classified as half-sample symmetric (denoted "HS" in the figure). The left boundary of $x[n]$ is symmetric halfway between two points, $y[-1]$ and $y[0]$, hence this is also labeled "HS". The sequence and its extension form a $2N$-point unit.

The sequences that constitute the basis of DCT-III and DCT-IV are created from antisymmetric extensions by appending an inverted, flipped replica of $x[n]$ to its right boundary. DCT-III is whole-sample antisymmetric (WA) on the right boundary, but the only way a sample can be shared by $x[n]$ and its inverted extension is for that point to be zero. DCT-IV is half-sample antisymmetric (HA) on the right boundary. Once the extension has been created, the entire sequence plus its extension is copied, flipped and appended to the right end, which makes the complete extended sequence even-symmetric and $4N$ points long.

The bottom line is that all the fully extended sequences pictured in **Figure 12.2** are even-symmetric, the important consequence of which is that their transforms can all be expressed as sums of cosines (hence the name, discrete cosine transform). In the paragraphs that follow, we will derive in detail the forward and inverse transform formulas for the most commonly used DCT variant, the DCT-II, and indicate how this transform can be rapidly computed using the DFT/FFT. Then, we will briefly describe the other variants of the DCT. Because there is a considerable amount of algebraic grinding involved in these derivations, you will find detailed step-by-step derivations of the other variants, as well as a number of clever implementation techniques, presented as problems. Even if you do not work the problems, the problem statements themselves are designed to guide you through the main conceptual steps of the derivations, should you be interested.

12.1.2 "The" discrete cosine transform (DCT-II)

The most common of all the DCT variants is DCT-II, to the point that it is often simply referred to as "the DCT." The DCT-II is based on the DFT of a sequence $y[n]$, which has been formed from the symmetric expansion of a sequence $x[n]$ by gluing a flipped (mirror-image) copy of $x[n]$ to its right end, as shown in **Figure 12.2**,

$$y[n] = \begin{cases} x[n], & 0 \le n \le N-1 \\ x[2N-1-n], & N \le n \le 2N-1 \end{cases}. \tag{12.1}$$

The sequence $y[n]$ is of even length, $2N$, with a duplicate point at the center, namely $y[N] = y[N-1] = x[N-1]$. To find $Y[k]$, the $2N$-point DFT of $y[n]$, split the transform into two sums expressed solely in terms of $x[n]$,

$$\begin{aligned} Y[k] &= \sum_{n=0}^{2N-1} y[n]e^{-j2\pi kn/2N} = \sum_{n=0}^{N-1} y[n]e^{-j\pi kn/N} + \sum_{n=N}^{2N-1} y[n]e^{-j\pi kn/N} \\ &= \sum_{n=0}^{N-1} x[n]e^{-j\pi kn/N} + \sum_{n=N}^{2N-1} x[2N-1-n]e^{-j\pi kn/N}, \end{aligned}$$

where Equation (12.1) has been used to express $y[n]$ in terms of $x[n]$. Let $m = 2N - 1 - n$ in the second summation. Then,

$$\begin{aligned} Y[k] &= \sum_{n=0}^{N-1} x[n]e^{-j\pi kn/N} + \sum_{m=N-1}^{0} x[m]e^{-j\pi k(2N-1-m)/N} = \sum_{n=0}^{N-1} x[n]e^{-j\pi kn/N} + \sum_{m=0}^{N-1} x[m]e^{-j\pi k(2N-1-m)/N} \\ &= \sum_{n=0}^{N-1} x[n]\left(e^{-j\pi kn/N} + e^{-j\pi k(2N-1-n)/N}\right). \end{aligned}$$

Recognizing that $e^{-j\pi k(2N-1-n)/N} = e^{-j2\pi k}e^{j\pi k(n+1)/N} = e^{j\pi k(n+1)/N}$, we have

$$\begin{aligned} Y[k] &= \sum_{n=0}^{N-1} x[n]\left(e^{-j\pi kn/N} + e^{j\pi k(n+1)/N}\right) = e^{j\pi k/2N}\sum_{n=0}^{N-1} x[n]\left(e^{-j\pi k(n+1/2)/N} + e^{j\pi k(n+1/2)/N}\right) \\ &= e^{j\pi k/2N}\sum_{n=0}^{N-1} 2x[n]\cos\frac{\pi k(n+1/2)}{N} \\ &= e^{j\pi k/2N}\sum_{n=0}^{N-1} 2x[n]\cos\frac{\pi k(2n+1)}{2N}, \quad 0 \le k \le 2N-1. \end{aligned} \tag{12.2}$$

This equation seemingly requires that $Y[k]$ is to be computed at $2N$ values, $0 \le k \le 2N - 1$. However, if $x[n]$ is real then $y[n]$ is real, so we only actually need to compute N values of $Y[k]$, $0 \le k \le N - 1$. The reason is that if $y[n]$ is real, then $Y[k]$ is conjugate-symmetric, namely $Y[2N - k] = Y^*[k]$. The conjugate is

$$\begin{aligned} Y^*[k] &= \left(e^{j\pi k/2N}\sum_{n=0}^{N-1} 2x[n]\cos\frac{\pi k(2n+1)}{2N}\right)^* = e^{-j\pi k/2N}\left(\sum_{n=0}^{N-1} 2x[n]\cos\frac{\pi k(2n+1)}{2N}\right) \\ &= e^{-j\pi k/2N}\left(e^{-j\pi k/2N}\,Y[k]\right) = e^{-j\pi k/N}\,Y[k]. \end{aligned}$$

Hence,

$$Y[2N-k] = e^{-j\pi k/N}\,Y[k], \quad 0 \le k \le N-1. \tag{12.3}$$

Substituting $k = N$ into this equation, we find that $Y[N] = -Y[N]$, which can only happen if

$$Y[N] = 0. \tag{12.4}$$

So, we only need to calculate N unique values of $Y[k]$, $0 \le k \le N - 1$ using Equation (12.2). The remaining N values of $Y[k]$, $N \le k \le 2N - 1$, are provided by Equations (12.3) and (12.4).

We are almost done. We now define the **Type-II discrete cosine transform (DCT-II)**, $C_{II}[k]$, as

$$C_{II}[k] \triangleq \sqrt{\frac{2}{N}}\sum_{n=0}^{N-1} x[n]\beta[k]\cos\frac{\pi k(2n+1)}{2N}, \quad 0 \le k \le N-1, \tag{12.5}$$

where

$$\beta[k] = \begin{cases} 1/\sqrt{2}, & k = 0 \\ 1, & k \neq 0 \end{cases}.$$

This is one of a number of possible definitions of DCT-II, all of which differ from each other in the values of the scale factors (see, for example, Problem 12-11). The definition given in Equation (12.5) with the scale factors $\sqrt{2/N}$ and $\beta[k]$ is probably the most common, and is the definition that Matlab uses in its functions `dct` and `idct`. This choice of scale factors makes the basis vectors of the DCT orthonormal, as we will show in Section 12.1.6.[2] Since $C_{II}[k]$ is the sum of cosines, the DCT of a real sequence is *purely real*. Comparing Equations (12.5) and (12.2), you can see that the DCT $C_{II}[k]$ can be completely expressed in terms of the DFT $Y[k]$,

$$C_{II}[k] = \frac{1}{2}\sqrt{\frac{2}{N}}\beta[k]e^{-j\pi k/2N}Y[k] = \frac{1}{\sqrt{2N}}\beta[k]e^{-j\pi k/2N}Y[k], \quad 0 \leq k \leq N-1. \tag{12.6}$$

We will use this relation in Section 12.1.5 when we discuss ways of computing the DCT efficiently using the fast Fourier transform (FFT).

In a reciprocal manner, we can express $Y[k]$ in terms of $C[k]$ (see Problem 12-9), a fact we will exploit to compute the inverse DCT using the FFT,

$$Y[k] = \begin{cases} \dfrac{\sqrt{2N}}{\beta[k]}e^{j\pi k/2N}C_{II}[k], & 0 \leq k \leq N-1 \\ 0, & k = N \\ -\sqrt{2N}e^{j\pi k/2N}C_{II}[2N-k], & N+1 \leq k \leq 2N-1 \end{cases}, \tag{12.7}$$

or, equivalently (Problem 12-10),

$$\begin{aligned} Y[k] &= \frac{\sqrt{2N}}{\beta[k]}e^{j\pi k/2N}C_{II}[k], & 0 \leq k \leq N-1 \\ Y[N] &= 0 \\ Y[N+1+k] &= -j\sqrt{2N}e^{j\pi(k+1)/2N}C_{II}[N-1-k], & 0 \leq k \leq N-2. \end{aligned} \tag{12.8}$$

12.1.3 The inverse discrete cosine transform (IDCT)

The DCT is a truly invertible transform; that is, there exists an **inverse discrete cosine transform (IDCT)** that allows us to obtain $x[n]$ directly from $C_{II}[k]$. As shown in Equation (12.5), the DCT comprises the sum of even trigonometric basis functions,

[2] It might seem clumsy to put the factor of $\beta[k]$ inside the summation, but doing it this way emphasizes the symmetry of the formulas for the DCT and the inverse DCT, Equation (12.12).

$$\cos\frac{\pi k(2n+1)}{2N},\quad 0\le k\le N-1, \tag{12.9}$$

weighted by $x[n]$. The key to obtaining the IDCT is the fact that these cosine basis functions are orthogonal (see Problem 12-1 for a derivation), namely

$$\frac{1}{N}\sum_{n=0}^{N-1}\cos\frac{\pi k(2n+1)}{2N}\cos\frac{\pi l(2n+1)}{2N}=\begin{cases}1, & k=l=0\\ 1/2, & k=l\neq 0\\ 0, & k\neq l\end{cases}. \tag{12.10}$$

We can use this result to show that $x[n]$ can be expressed in terms of the orthogonal cosine basis functions. Specifically, we propose the definition of the IDCT to be

$$C_{II}^{-1}[n]\triangleq\sqrt{\frac{2}{N}}\sum_{k=0}^{N-1}C_{II}[k]\beta[k]\cos\frac{\pi k(2n+1)}{2N},\quad 0\le n\le N-1. \tag{12.11}$$

To verify that this is, in fact, the inverse, substitute this expression for $C_{II}^{-1}[n]$ back into the definition of the DCT, Equation (12.5), and turn the algebra crank a few times to show that the result is $C_{II}[k]$:

$$\begin{aligned}
&\sqrt{\frac{2}{N}}\sum_{n=0}^{N-1}\beta[k]C_{II}^{-1}[n]\cos\frac{\pi k(2n+1)}{2N}\\
&=\sqrt{\frac{2}{N}}\sum_{n=0}^{N-1}\beta[k]\left(\sqrt{\frac{2}{N}}\sum_{l=0}^{N-1}C_{II}[l]\beta[l]\cos\frac{\pi l(2n+1)}{2N}\right)\cos\frac{\pi k(2n+1)}{2N}\\
&=2\sum_{l=0}^{N-1}C_{II}[l]\beta[k]\beta[l]\left(\frac{1}{N}\sum_{n=0}^{N-1}\cos\frac{\pi l(2n+1)}{2N}\cos\frac{\pi k(2n+1)}{2N}\right)\\
&=2\sum_{l=0}^{N-1}C_{II}[l]\underbrace{\begin{Bmatrix}1/\sqrt{2}, & k=0\\ 1, & k\neq 0\end{Bmatrix}}_{\beta[k]}\underbrace{\begin{Bmatrix}1/\sqrt{2}, & l=0\\ 1, & l\neq 0\end{Bmatrix}}_{\beta[l]}\begin{Bmatrix}1, & k=l=0\\ 1/2, & k=l\neq 0\\ 0, & k\neq l\end{Bmatrix}\\
&=C_{II}[k],\quad 0\le k\le N-1.
\end{aligned}$$

To calculate the inverse DCT-II, use Equation (12.7) to get $Y[k]$, then take the inverse DFT to get $y[n]$ and select the first N points, as we will discuss in Section 12.1.5.

To summarize our work so far, here are the forward and inverse DCT-II relations:

$$C_{II}[k]=\sqrt{\frac{2}{N}}\sum_{n=0}^{N-1}x[n]\beta[k]\cos\frac{\pi k(2n+1)}{2N},\qquad 0\le k\le N-1$$

$$x[n]=\sqrt{\frac{2}{N}}\sum_{k=0}^{N-1}C_{II}[k]\beta[k]\cos\frac{\pi k(2n+1)}{2N},\quad 0\le n\le N-1. \tag{12.12}$$

$$\beta[k]=\begin{cases}1/\sqrt{2}, & k=0\\ 1, & k\neq 0\end{cases}$$

Example 12.1

(a) Find $C_{II}[k]$, the DCT-II of the input $x[n]$ shown in **Figure 12.3a**, and show that the IDCT of $C_{II}[k]$ is $x[n]$.

(b) Plot the cosine basis functions and indicate how their sum relates to $x[n]$.

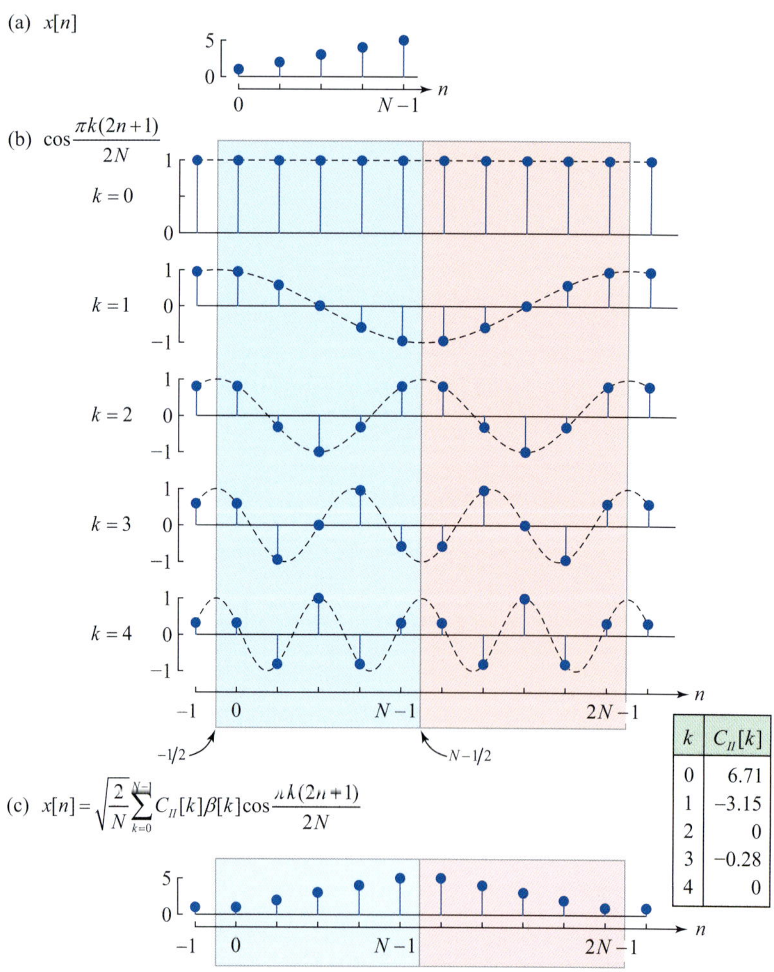

k	$C_{II}[k]$
0	6.71
1	−3.15
2	0
3	−0.28
4	0

Figure 12.3

➤ **Solution:**

(a) Use Matlab to implement Equation (12.12) directly using the sum of cosine terms. (We will discuss more computationally efficient implementations in Section 12.1.8.)

```
N = 5;
x = (1:N)';
M = sqrt(2/N)*cos(pi*(0:N-1)'*(2*(0:N-1)+1)/2/N); % matrix of cosines
M(1, :) = 1 / sqrt(N); % apply beta(k)
CII = M*x;    % DCT
xx = M'*CII; % IDCT
>> CII'
   6.7082   -3.1495   -0.0000   -0.2840   -0.0000

>> xx'
   1.0000    2.0000    3.0000    4.0000    5.0000
```

Here, we create a matrix of cosine basis functions,

$$M[k,n] = \sqrt{\frac{2}{N}}\cos\frac{\pi k(2n+1)}{2N}, \quad \textbf{(12.13)}$$

where k is the row index and n is the column index. Then we change the first row to reflect the product of the constants, $\sqrt{2/N}(\beta[0]\cos\pi 0(2n+1)/2N) = \sqrt{1/N}$. Next, compute the DCT $C_{II}[k]$ and the IDCT $x[n]$ by simple matrix multiplications.

(b) The DCT-II expresses a signal $x[n]$ of length N as the sum of N cosine basis functions that have a fundamental period of $2N$. **Figure 12.3b** shows the $N = 5$ cosine basis functions, Equation (12.13), for $0 \le k \le 4$. **Figure 12.3c** shows the IDCT-II, Equation (12.11): the sum of all N cosines, each weighted by $\sqrt{2/N}\beta[k]C_{II}[k]$, with the values of $C_{II}[k]$ for this particular $x[n]$ given in the small table in the figure.

The fundamental period of the basis functions for the DCT-II is $2N$, which corresponds to the length of $x[n]$ plus its extension. If non-integral values of n were allowed, each function would have a maximum at $n = -1/2$. As a consequence, each cosine is half-sample symmetric about $n = -1/2$ (i.e., at $n = 0$ and $n = -1$),

$$\cos\frac{\pi k(-1+1/2)}{N} = \cos\frac{\pi k(+1+1/2)}{N}.$$

It follows that the weighted sum $x[n]$ is also half-sample symmetric. A similar argument applies to values of the cosine and the sum around $n = N - 1/2$ and $n = 2N - 1/2$.

12.1.4 The four principal DCT variants

The four principal variants of the DCT are shown in **Figure 12.2**, of which we have discussed one, the DCT-II, in detail. The forward DCT and the inverse of all four variants are summarized in the first two columns of **Table 12.1**. You will find the derivations of DCT-I, DCT-III and DCT-V and their inverses in Problems 12-8, 12-18 and 12-19, respectively. Other forms of these four equations exist in the literature, which differ from those in Table 12-1 in the normalizing factors that are applied, but the particular normalizing factors shown in the table (e.g., $\sqrt{2/N}$) result in orthonormal transforms, and are the ones you will most commonly see in the literature. The last column of **Table 12.1** shows that the forward and inverse transforms of some of the variants are closely related. For example, if you swap variables n and k in the expression for the IDCT-II, you get DCT-III.

Table 12.1 The four principal variants of the DCT and IDCT

Type	DCT	IDCT	Relation
I	$C_I[k]=\sqrt{\frac{2}{N-1}}\sum_{n=0}^{N-1}\alpha[k]\alpha[n]x[n]\cos\frac{\pi kn}{N-1}$ $\alpha[k]=\begin{cases}1/\sqrt{2}, & k=0,N-1\\ 1, & \text{otherwise}\end{cases}$	$C_I^{-1}[n]=\sqrt{\frac{2}{N-1}}\sum_{k=0}^{N-1}\alpha[k]\alpha[n]C_I[k]\cos\frac{\pi kn}{N-1}$	$C_I^{-1}[k]=C_I[n]$
II	$C_{II}[k]=\sqrt{\frac{2}{N}}\sum_{n=0}^{N-1}x[n]\beta[k]\cos\frac{\pi k(2n+1)}{2N}$ $\beta[k]=\begin{cases}1/\sqrt{2}, & k=0\\ 1, & k\neq 0\end{cases}$	$C_{II}^{-1}[n]=\sqrt{\frac{2}{N}}\sum_{k=0}^{N-1}C_{II}[k]\beta[k]\cos\frac{\pi k(2n+1)}{2N}$	$C_{II}^{-1}[k]=C_{III}[n]$
III	$C_{III}[k]=\sqrt{\frac{2}{N}}\sum_{n=0}^{N-1}\beta[n]x[n]\cos\frac{\pi(2k+1)n}{2N}$	$C_{III}^{-1}[n]=\sqrt{\frac{2}{N}}\sum_{k=0}^{N-1}\beta[n]C_{III}[k]\cos\frac{\pi(2k+1)n}{2N}$	$C_{III}^{-1}[k]=C_{II}[n]$
IV	$C_{IV}[k]=\sqrt{\frac{2}{N}}\sum_{n=0}^{N-1}x[n]\cos\frac{\pi(2k+1)(2n+1)}{4N}$	$C_{IV}^{-1}[n]=\sqrt{\frac{2}{N}}\sum_{k=0}^{N-1}C_{IV}[k]\cos\frac{\pi(2k+1)(2n+1)}{4N}$	$C_{IV}^{-1}[k]=C_{IV}[n]$

12.1.5 Properties of the DCT

Before proceeding with some examples, let us briefly note some of the properties of all variants of the DCT and compare them with the DFT.

- **Linearity**: Like the DFT, all variants of the DCT are linear transforms, so,

$$\mathrm{DCT}\{\alpha x_1[n]+\beta x_2[n]\}=\alpha\,\mathrm{DCT}\{x_1[n]\}+\beta\,\mathrm{DCT}\{x_2[n]\},$$

  where DCT{ } represents any of the four variants of the DCT.
- **Real basis functions**: Both the DFT and the DCT are invertible transforms that represent signals as the sum of orthogonal basis functions. Whereas the DFT expresses an N-point signal $x[n]$ as the sum of N orthogonal, complex exponential functions of the form $e^{j2\pi kn/N}$, $0\le k\le N-1$, the DCT expresses $x[n]$ as the sum of N real orthogonal cosine functions of the form of Equation (12.9). The fact that the basis functions and the transform coefficients of the DCT (e.g., $C_{II}[k]$) require only real arithmetic make the design and implementation of certain computer algorithms easier. We will discuss some of these algorithms in Section 12.1.8, below.
- **Orthogonality**: The orthogonality of basis functions, Equation (12.10), is the property responsible for the invertibility of the DCT-II, but the DCT is not unique in that regard. The DFT is also an orthogonal transform, as are the discrete sine transform (DST) and others. What makes the DCT particularly useful is that its cosine basis functions turn out to be exceptionally well suited for the representation of one-dimensional signals with a lot of repeated structure, such as speech and music, as well as two-dimensional signals, such as images. We will discuss these applications of the DCT in Sections 12.2 and 12.3.
- **Energy conservation (Parseval's theorem):**
  Parseval's theorem, which relates the energy of signals in the time and frequency domains, applies to all four variants of the DCT (See Problem 12-17), as it does to the DFT,

$$\sum_{n=0}^{N-1} x^2[n] = \sum_{k=0}^{N-1} C_{II}^2[k].$$

- **Energy compaction:** Because the basis functions of both the DFT and the DCT are periodic, the transform of an N-point input $x[n]$ is equivalent to taking the transform of $x[n]$ that is periodically repeated. For the DFT, $x[n]$ is implicitly repeated with a period of N, with resulting discontinuities at the boundaries. In contrast, the different variants of the DCT correspond to periodically repeated extensions of $x[n]$, as shown in **Figure 12.2**. These extensions act to minimize the discontinuities at the boundaries for many signals. As a consequence, DFT and DCT allocate signal energy differently. Particularly in signals of interest to us, such as speech, music and images, the continuity of the boundaries of the DCT limits the spread of energy to high frequencies and results in the good energy compaction properties of the transform, something we hinted at in **Figure 12.1**. Section 12.1.7 gives more detail on this important property. Specifically we will show what happens when we take an N-point sequence $x[n]$, with DCT coefficients $C_{II}[k]$, $0 \le k \le N-1$, and construct a sequence $\hat{x}[n]$ using the IDCT of only a smaller number, $M < N$, of these DCT coefficients.

  It is worth mentioning that DCT is not theoretically the optimum transform for energy compaction. The **Karhunen–Loève transform (KLT)** is generally regarded as the optimum in the sense that the majority of the energy of the input is represented by the smallest number of transform coefficients. However, the basis functions of the KLT are *data dependent*: they must be computed based on the statistics (specifically, the covariance function) of the input signal. In contrast, the basis functions of the DCT are cosines that are *independent* of any feature of the input. The Type-II and Type-IV variants of the DCT turn out to be nearly optimal for compression of signals such as speech, music and images, as we will see when we discuss applications of the DCT in Sections 12.2 and 12.3.

12.1.6 ★ Matrix form of the DCT and IDCT

All four variants of the DCT and IDCT can be written in matrix form. For example, the DCT-II of Equation (12.12) can be written as,

$$C_{II}[k] = \sum_{n=0}^{N-1} \mathbf{M}[k,n]x[n] = \sum_{n=0}^{N-1} \underbrace{\left(\sqrt{\frac{2}{N}}\beta[k]\cos\frac{\pi k(2n+1)}{2N}\right)}_{\mathbf{M}[k,n]} x[n],$$

where $\mathbf{M}[k,n]$ are the elements of a square $N \times N$ matrix of cosines, $\mathbf{M}$, with k and n the row and column indices of the matrix, respectively. In matrix form,

$$\underbrace{\begin{bmatrix} C_{II}[0] \\ C_{II}[1] \\ \vdots \\ C_{II}[N-1] \end{bmatrix}}_{\mathbf{C}_{\mathrm{II}}} = \sqrt{\frac{2}{N}} \underbrace{\begin{bmatrix} 1/\sqrt{2} & 1/\sqrt{2} & \cdots & 1/\sqrt{2} \\ \cos\frac{\pi\cdot 1\cdot 1}{2N} & \cos\frac{\pi\cdot 1\cdot 3}{2N} & \cdots & \cos\frac{\pi\cdot 1\cdot(2N-1)}{2N} \\ \cos\frac{\pi\cdot 2\cdot 1}{2N} & \cos\frac{\pi\cdot 2\cdot 3}{2N} & \cdots & \cos\frac{\pi\cdot 2\cdot(2N-1)}{2N} \\ \vdots & \vdots & \cos\frac{\pi k(2n+1)}{2N} & \vdots \\ \cos\frac{\pi\cdot(N-1)\cdot 1}{2N} & \cos\frac{\pi\cdot(N-1)\cdot 3}{2N} & \cdots & \cos\frac{\pi\cdot(N-1)\cdot(2N-1)}{2N} \end{bmatrix}}_{\mathbf{M}} \underbrace{\begin{bmatrix} x[0] \\ x[1] \\ \vdots \\ x[N-1] \end{bmatrix}}_{\mathbf{x}},$$

or just

$$\mathbf{C}_{\mathbf{II}} = \mathbf{M}\mathbf{x}. \tag{12.14}$$

The IDCT is

$$x[n] = \sum_{k=0}^{N-1} \mathbf{M}^T[n,k]C_{II}[k] = \sum_{k=0}^{N-1} \underbrace{\left(\sqrt{\frac{2}{N}}\beta[k]\cos\frac{\pi k(2n+1)}{2N}\right)}_{\mathbf{M}^T[n,k]} C_{II}[k]. \tag{12.15}$$

The DCT-II and IDCT-II use the same matrix, **M**, but the IDCT-II uses the transpose, $\mathbf{M}^T$, since summation is over k for the IDCT as opposed to over n for the DCT,

$$\underbrace{\begin{bmatrix} x[0] \\ x[1] \\ \vdots \\ x[N-1] \end{bmatrix}}_{\mathbf{x}} = \underbrace{\sqrt{\frac{2}{N}}\begin{bmatrix} 1/\sqrt{2} & \cos\frac{\pi\cdot 1\cdot 1}{2N} & \cdots & \cos\frac{\pi\cdot(N-1)\cdot 1}{2N} \\ 1/\sqrt{2} & \cos\frac{\pi\cdot 1\cdot 3}{2N} & \cdots & \cos\frac{\pi\cdot(N-1)\cdot 3}{2N} \\ 1/\sqrt{2} & \cos\frac{\pi\cdot 1\cdot 5}{2N} & \cdots & \cos\frac{\pi\cdot(N-1)\cdot 5}{2N} \\ \vdots & \vdots & \cos\frac{\pi\cdot k\cdot(2n+1)}{2N} & \vdots \\ 1/\sqrt{2} & \cos\frac{\pi\cdot 1\cdot(2N-1)}{2N} & \cdots & \cos\frac{\pi\cdot(N-1)\cdot(2N-1)}{2N} \end{bmatrix}}_{\mathbf{M}^T} \underbrace{\begin{bmatrix} C_{II}[0] \\ C_{II}[1] \\ \vdots \\ C_{II}[N-1] \end{bmatrix}}_{\mathbf{C}_{\mathbf{II}}},$$

or just

$$\mathbf{x} = \mathbf{M}^T\mathbf{C}_{\mathbf{II}}. \tag{12.16}$$

Substituting Equation (12.16) into Equation (12.14) gives $\mathbf{C}_{\mathbf{II}} = \mathbf{M}\mathbf{x} = \mathbf{M}\mathbf{M}^T\mathbf{C}_{\mathbf{II}}$, from which we conclude that $\mathbf{M}\mathbf{M}^T = \mathbf{I}$. But, by definition, $\mathbf{M}\mathbf{M}^{-1} = \mathbf{I}$, so the transpose of **M** must be equal to its inverse, $\mathbf{M}^T = \mathbf{M}^{-1}$. A matrix whose transpose is equal to its inverse is an **orthogonal matrix**. The rows of **M** (or columns of $\mathbf{M}^T$) are the orthonormal **basis vectors** of the transform. Orthogonality means that each row (or column) vector is perpendicular to every other vector; the dot product of each row vector with itself is one and with any other row vector is zero. When you multiply **M** by an input vector **x**, the kth element of the resulting vector $\mathbf{C}_{\mathbf{II}}$ (namely $C_{II}[k]$) represents the projection of **x** onto the kth row vector of **M** (i.e., the dot product). Because the row vectors of **M** are orthogonal, the information in $C_{II}[k]$ for each value of k is "unique"; knowing the value of $C_{II}[k]$ at a particular value of k tells you nothing about the value of $C_{II}[l]$ for $k \neq l$.

Parseval's theorem in matrix form Parseval's theorem allows us to represent the total energy in the signal in terms of the matrix form of the DCT. Equation (12.16) gives the IDCT-II in matrix form as $\mathbf{x} = \mathbf{M}^T\mathbf{C}_{\mathbf{II}}$. So,

$$\sum_{n=0}^{N-1} x^2[n] = \mathbf{x}^T\mathbf{x} = (\mathbf{M}^T\mathbf{C}_{\mathbf{II}})^T\mathbf{M}^T\mathbf{C}_{\mathbf{II}} = \mathbf{C}_{\mathbf{II}}^T\underbrace{(\mathbf{M}\mathbf{M}^T)}_{\mathbf{I}}\mathbf{C}_{\mathbf{II}} = \mathbf{C}_{\mathbf{II}}^T\mathbf{C}_{\mathbf{II}} = \sum_{k=0}^{N-1} C_{II}^2[k]. \tag{12.17}$$

12.1.7 ★ Energy compaction of the DCT

To understand the energy compaction property of the DCT a bit better, let us go back to the IDCT, Equation (12.15), which says that an N-point sequence $x[n]$ can be exactly expressed as the sum of N orthogonal cosine functions $\mathbf{M}^T[n,k]$, each weighted by the appropriate DCT coefficient $C_{II}[k]$, $0 \le k \le N-1$. Now define the **compressed signal** $\hat{x}[n]$ as the IDCT of a subset $k \in \mathbb{M}$ of length $M < N$, selected from these same orthogonal cosine functions,

$$\hat{x}[n] = \sum_{k\in\mathbb{M}} \hat{\mathbf{M}}^T[n,k]\hat{C}_{II}[k] = \sum_{k\in\mathbb{M}} \underbrace{\left(\sqrt{\frac{2}{N}}\beta[k]\cos\frac{\pi k(2n+1)}{2N}\right)}_{\hat{\mathbf{M}}^T[n,k]} \hat{C}_{II}[k]. \quad \textbf{(12.18)}$$

From a matrix point of view, Equation (12.18) is

$$\hat{\mathbf{x}} = \hat{\mathbf{M}}^T\hat{\mathbf{C}}_{\mathrm{II}},$$

where $\hat{\mathbf{x}}$ is the $N\times 1$ vector of sequence values $\hat{x}[n]$; $\hat{\mathbf{M}}^T$ is the $N\times M$ matrix of cosines comprising all the rows of $\mathbf{M}^T$, but just the M columns that correspond to the subset $\mathbb{M}$; $\hat{\mathbf{C}}_{\mathrm{II}}$ is the $M\times 1$ vector of DCT coefficients. Now, how do we choose a smaller set of DCT coefficients $\hat{C}_{II}[k]$ so that $\hat{x}[n]$ will be as close as possible to $x[n]$ in the minimum square-error sense? This is an example of an **over-determined system** in which the number of equations (equal to N, the number of rows of $\hat{\mathbf{M}}^T$) is greater than the number of parameters (equal to M, the number of columns of $\hat{\mathbf{M}}^T$). Our task is to find the vector $\hat{\mathbf{C}}_{\mathrm{II}}$ that minimizes the square-error between $\hat{\mathbf{x}}$ and $\mathbf{x}$. The solution to this classic problem comes from the solution of the so-called **normal equations**, which is discussed at length in Appendix A. In this case, the solution is

$$\hat{\mathbf{C}}_{\mathrm{II}} = \left(\hat{\mathbf{M}}\hat{\mathbf{M}}^T\right)^{-1}\hat{\mathbf{M}}\mathbf{x}. \quad \textbf{(12.19)}$$

This looks daunting, but the solution is easy in this case because the column vectors of $\hat{\mathbf{M}}^T$ are just a subset of the full matrix, $\mathbf{M}^T$, and are therefore orthogonal to each other. Hence, $\hat{\mathbf{M}}\hat{\mathbf{M}}^T$ is a square $M\times M$ identity matrix, $\hat{\mathbf{M}}\hat{\mathbf{M}}^T = \mathbf{I}$, and Equation (12.19) reduces to

$$\hat{\mathbf{C}}_{\mathrm{II}} = \hat{\mathbf{M}}\mathbf{x}.$$

Comparing this with Equation (12.14) leads one immediately to the conclusion that $\hat{\mathbf{C}}_{\mathrm{II}}$ just consists of the M rows of $\mathbf{C}_{\mathrm{II}}$ that correspond to the values of $C_{II}[k]$, $k\in\mathbb{M}$. In other words, the best way (in the minimum square-error sense) to represent an N-point sequence $\hat{x}[n]$ with a selected set of orthogonal cosines $\mathbf{M}^T[n,k]$, $k\in\mathbb{M}$, is to reconstruct the sequence just using the corresponding DCT coefficients of $x[n]$, namely $\hat{\mathbf{C}}_{\mathrm{II}} = \mathbf{C}_{\mathrm{II}}[k]$, $k\in\mathbb{M}$. The square-error between the original signal $\mathbf{x}$ and the compressed signal $\hat{\mathbf{x}}$ is given by (Problem 12-20):

$$\sum_{n=0}^{N-1}(x[n]-\hat{x}[n])^2 = \sum_{k=0}^{N-1} C_{II}^2[k] - \sum_{k\in\mathbb{M}} C_{II}^2[k] = \sum_{k\notin\mathbb{M}} C_{II}^2[k]. \quad \textbf{(12.20)}$$

In words, the mean-square error is the sum of the squares of all the DCT coefficients that have been left out of subset $\mathbb{M}$; that is, $C_{II}[k]$, $k \notin \mathbb{M}$. In theory, one could sort the DCT coefficients $C_{II}[k]$ in order of decreasing absolute value, in order to determine the subset $\mathbb{M}$ that achieves the largest degree of energy compaction with the smallest square error for any value of M. An example might help.

Example 12.2

Given an N-point sequence $x[n]$, whose DCT coefficients $C_{II}[k]$, $0 \le k \le N-1$, were computed in Example 12.1, determine the energy in the compressed signal $\hat{x}[n]$ that follows from removing coefficients from the subset $\mathbb{M}$ one at a time, the smallest first.

➤ Solution:

Using Matlab, first sort $C_{II}^2[k]$ in order of decreasing size and determine their order.

```
>> [Csq, k]  = sort(CII'.^2, 'descend')
Csq =
    45.0000     9.9193     0.0807     0.0000     0.0000
k =
     1        2        4        5        3
```

Now find the cumulative energy of the coefficients, from biggest to smallest, normalized by the total energy $\mathbf{C}_{\mathrm{II}}^T\mathbf{C}_{\mathrm{II}}$:

```
>> 100*cumsum(Csq)/(CII'*CII)
ans =
81.8182   99.8534   100.0000   100.0000   100.0000
```

Reading this `ans` array and the `k` array from right to left, we see that removing coefficients 5 and 3 does not affect the energy at all (which should be obvious since they each have value zero). Removing coefficient 4 reduces the energy by only 0.15%. Roughly 82% of the energy is in the first coefficient.

In many applications, particularly those connected with music and pictures, which we will discuss in Sections 12.3 and 12.4, the characteristic of the signal is that the energy is naturally "front-loaded" in the earlier coefficients, so sorting of the coefficients is not necessary to achieve reasonable energy compaction. Instead, a common approach is to quantize the higher-energy coefficients more finely than those of lower energy.

12.1.8 ★ Implementation of the DCT-II and IDCT-II

In practice, it is inefficient to compute the DCT and IDCT using a sum of cosine terms. The literature is replete with clever and efficient ways to compute the various variants of the DCT and IDCT using the FFT and IFFT respectively. Here are a couple of relatively efficient ways of computing the DCT-II and IDCT-II.

Computation of the DCT-II Equation (12.6) allows us to obtain the DCT-II $C_{II}[k]$ from the $2N$-point DFT $Y[k]$. Here are the steps:

1. Create $y[n]$, a sequence of length $2N$ as the symmetric extension of $x[n]$, using Equation (12.1).

2. Calculate the $2N$-point DFT of $y[n]$ to form the $2N$-point transform $Y[k]$. Since $y[n]$ is of even length, a number of commonly available FFT algorithms, including those discussed in Chapter 11, are available to accomplish this task efficiently.
3. Derive the discrete cosine transform $C_{II}[k]$ from the first N points of $Y[k]$ using Equation (12.6).

This simple procedure does not exploit computational shortcuts that can be applied when $x[n]$ is of even length. Problems 12-15 and 12-16 present two faster algorithms that allow the DCT and IDCT of such $2N$-point sequences to be computed using only an N-point FFT. For example, you can also show (Problem 12-15), that if N is even, then you can calculate $C_{II}[k]$ with only an N-point DFT by following these steps:

1. Create an N-point sequence $y[n]$, the first half of which contains the even points of $x[n]$ and the second half the odd points in reverse order,

$$y[n] = \begin{cases} x[2n], & 0 \le n \le (N-1)/2 \\ x[2N-2n-1], & (N+1)/2 \le n \le N-1 \end{cases}.$$

2. Then take the DFT of $y[n]$ and form

$$C_{II}[k] = \sqrt{\frac{2}{N}}\beta[k]\,\mathrm{Re}\{e^{-j\pi k/2N}\,Y[k]\}.$$

3. Finally,

$$Y[k] = \sum_{n=0}^{N-1} y[n]e^{-j2\pi kn/N}.$$

Here is a simple implementation that uses the first procedure for odd-length sequences and the second one for even-length sequences. This is essentially identical to the algorithm used in Matlab's `dct` function.

```
function C = dct_II(x)
    N = length(x);
    b = exp(-1j*pi*(0:N-1)/2/N) / sqrt(2*N);
    b(1) = b(1) / sqrt(2);
  if (mod(N, 2)) % it is odd  -> length 2N sequence
      y = [x x(N:-1:1)];
  else                % it is even -> length  N sequence
      y = 2 * [x(1:2:N-1) x(N:-2:2)];
  end
  Y = fft(y);
  C = real(b .* Y(1:N));
end
```

Computation of the IDCT-II To compute the IDCT-II, you can perform a $2N$-point DFT using Equation (12.8):

1. Derive the $2N$-point DFT $Y[k]$ from $C_{II}[k]$ using Equation (12.8).
2. Take the inverse $2N$-point DFT of $Y[k]$ to form a $2N$-point sequence $y[n]$. Again, the FFT is a good way to do this.

3. Then $x[n]$ is the first N points of $y[n]$:

$$x[n] = \begin{cases} y[n], & 0 \le n \le N-1 \\ 0, & \text{otherwise} \end{cases}.$$

In the case of an even-length N-point DCT, you can show (Problem 12-16) that the IDCT can be computed using only an N-point IDFT, as follows:

1. Compute the following inverse transform of $C_{II}[k]$ to obtain the sequence $y[n]$,

$$y[n] = \sqrt{2N}\,\mathrm{Re}\{\mathbb{F}_N^{-1}\{\beta[k]C_{II}[k]e^{j\pi k/2N}\}\}, \quad 0 \le n \le N-1.$$

2. Then,

$$\begin{aligned} x[2n] &= y[n] \\ x[2n+1] &= y[N-1-n], \quad 0 \le n \le N/2-1. \end{aligned}$$

Here is some code that implements both options. It is equivalent to the procedure used in Matlab's `idft` function.

```
function x = idct_II(C)
    N = length(C);
    b = sqrt(2*N) * exp(1j*pi*(0:N-1)/2/N);
    if (mod(N, 2)) % it is odd  -> length 2N sequence
        b(1) = b(1) * sqrt(2);
        Y = [b .*C 0 -1j*b(2:N).*C(end:-1:2)];
        y = ifft(Y);
        x = real(y(1:N));
    else              % it is even -> length N sequence
        b(1) = b(1) / sqrt(2);
        Y = b .*C;
        y = real(ifft(Y));
        x = zeros(1, N);
        x(1:2:N-1) = y(1:N/2);
        x(N:-2:2) = y(N/2+1:N);
    end
end
```

The two functions that Matlab provides, `dct` and `idct`, compute the DCT-II and IDCT-II. At the time of writing, Matlab does not provide functions for computing the forward or inverse transforms of the other DCT variants.

12.1.9 ★ The modified discrete cosine transform (MDCT)

An important application of the DCT is the compression of sound and image files to make them smaller, while retaining as much of their perceived quality as possible. It is the energy compaction property of the DCT to which we have alluded in Section 12.1.7 that helps make this possible. **Figure 12.4a** shows a 256-point segment of a speech waveform $x[n]$ (thin black trace), plus the waveform of the IDCT $\hat{x}[n]$ reconstructed from the first $M = 15$ coefficients of the DCT (red trace). **Figure 12.4b** plots the cumulative square error between $x[n]$ and $\hat{x}[n]$ as a function of the number of DCT coefficients, M, used in the reconstruction. The square error drops very fast for the first few coefficients and then plateaus. The first 15 DCT coefficients account for over 92% of the energy in $x[n]$, or equivalently correspond to the square error between $x[n]$ and $\hat{x}[n]$ of less than 8%. This compaction of a preponderance of signal energy in a small number of early coefficients is the reason why the DCT is such a powerful compression tool. "Front-loading" of coefficients of this sort is typical of a periodic signal such as this section of speech.

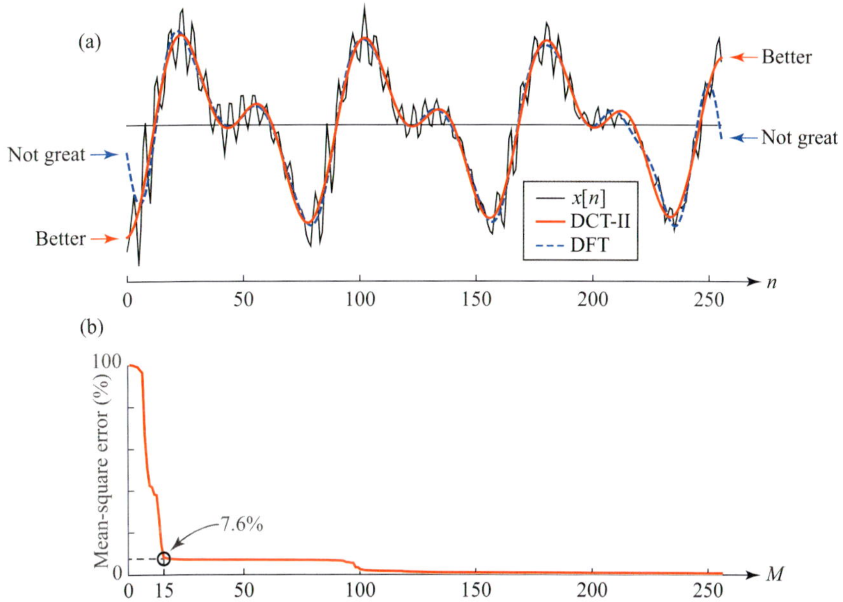

Figure 12.4 Compression and reconstruction of one frame of a waveform

Figure 12.4a also shows the reconstruction of the same segment of the waveform using $M = 29$ complex coefficients of the DFT (dashed blue trace), namely the DC coefficient plus 14 paired complex-conjugate coefficients. The fit is somewhat poorer at the ends of the segment than that of the DCT, reflecting the fact that the implicit periodic extension of the waveform by the DCT has a smaller discontinuity than by the DFT in this particular instance. However, the advantage of the DCT over the DFT is evident primarily in cases, such as those shown in **Figure 12.1**, where there is a sharp discontinuity at a frame boundary that can be ameliorated by periodic extension.

The DCT-II appears to do a reasonable job of representing the edges of a single frame of data, but there are problems that become obvious when we try to use the DCT-II and IDCT-II to compress and reconstruct signals comprising multiple frames or blocks of data.

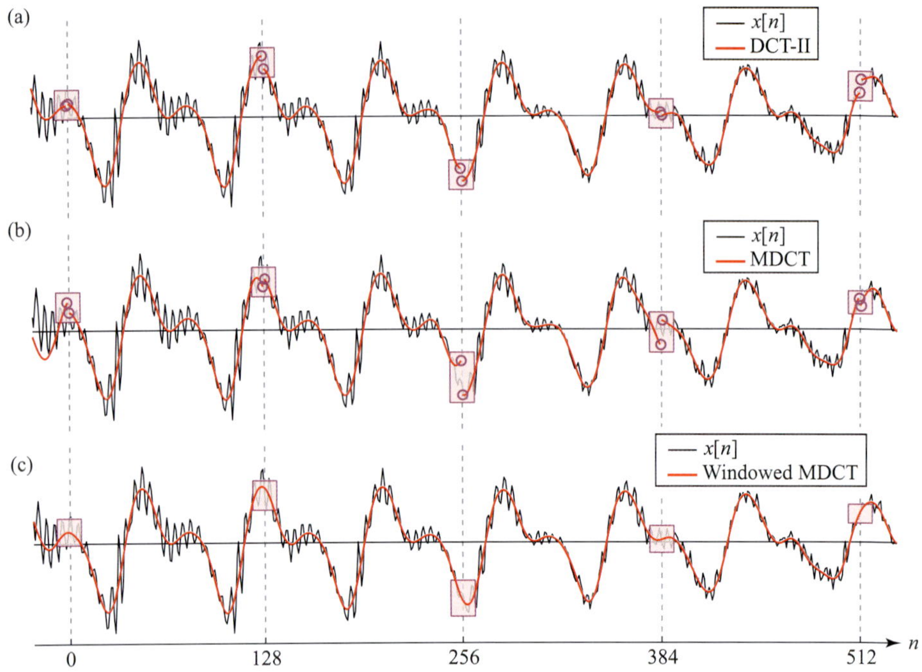

Figure 12.5 Compression and reconstruction of a multiple-frame waveform

Figure 12.5a shows a signal (thin black trace) comprising four frames of speech data, whose boundaries are indicated by the vertical dashed lines. The IDCT-II reconstruction of each frame with $M = 15$ DCT coefficients is also shown (red trace). Taken individually, the IDCT of each frame is a "pretty good" match to the original signal, but there are clear discontinuities between frames, highlighted in the pink boxes. The human auditory system is exquisitely sensitive to discontinuities such as these, which makes this approach of stitching together multiple frames a non-starter for important applications such as audio compression.

The solution to this problem starts with a modified version of DCT-IV, unsurprisingly called the **modified discrete cosine transform (MDCT)**. The MDCT is a **lapped transform**, meaning that it operates on frames of data of length $2N$ that *overlap* by N points so that the last half of each block overlaps with the first half of the succeeding block. Using the windowed version of the MDCT, the **windowed MDCT**, the artifacts at the ends of the blocks can be essentially eliminated through the process of **time-domain alias cancelation (TDAC)**, as shown in **Figure 12.5c**. Supplementary material describes in detail how the MDCT works and how it achieves TDAC.

12.2 MPEG audio compression

The MDCT is an important part of many audio compression encoders, such as those that produce the popular MP3 and AAC (Advanced Audio Coding) files. These encoders all use a combination of **lossy compression** based on an understanding of how humans process sounds and **lossless compression** techniques based on a purely algorithmic coding scheme. While a detailed description of the workings of any of these encoders is beyond what we can manage in a few paragraphs, it is still valuable to sketch the major features of one algorithm, the one used in the original MP3 codec, in order to give you an appreciation of the intricate manner in which psychoacoustic theory, DSP, and computer science techniques have been combined to produce a remarkably durable, practical algorithm.

To understand the technical need for audio compression, consider the data rate of music encoded on a compact disc (CD). Music on a standard CD is encoded as two channels of 16-bit linear, **pulse-code-modulated (PCM)** data at a sample rate of 44.1 kHz. That corresponds to $2 \times 16 \times 44.1 = 1411$ kilobits per second (kb/s). However, this is just the raw data rate. These data are then expanded by a digital modulation technique (**eight-to-fourteen modulation (EFM)**) that makes the data on the disk more resistant to errors caused by surface defects in the disk. EFM increases the size of each data word by 75%. Following this, considerable extra data are added to enable framing and synchronization. All told, each 16-bit data word ends up being expanded to 49 bits, which swells the total data rate to 4322 kb/s. Such high data rates are not compatible with many low-bandwidth applications, such as Internet and wireless communication. So the task is to find an audio compression technique that can reduce the data rate without affecting the perceived quality of the source material.

Historically, a variety of audio compression techniques have been developed that have been found to be acceptable for encoding speech at bit rates below 5 kb/s, for example, **linear predictive coding (LPC)** and **Code Excited Linear Prediction (CELP)** (see Chapter 14), but these techniques are not adequate to encode high-quality music. Beginning in the mid 1980s, Karlheinz Brandenburg, then a graduate student at the University Erlangen-Nuremberg, began working on the application of human psychoacoustics – that is, models of human auditory perception of sounds – to the development of high-quality digital audio coding algorithms. The results of his research, for which he received his Ph.D. in 1989, form the basis of the current MP3 algorithm as well as most other modern advanced codecs in use today.

The **MP3 algorithm** was the first international standard for compression of high-fidelity digital audio and is still among the most widely implemented, though it has been eclipsed by more recent encoders. MP3 is shorthand for **MPEG-1 Audio Layer III**, where MPEG stands for the Moving Pictures Expert Group, the international committee comprising experts from industry and academia in fields ranging from DSP to human auditory psychoacoustics that developed the original standard in 1992 (MPEG-1 Audio Layer I). The standard colloquially known as "MP3" is the third formal evolution of the standard (hence, "Layer III"). Interestingly, the MP3 standard does not dictate the specific algorithms by which compression is to be achieved; it only specifies the final characteristics of the output, such as the headers of the encoded material, so that MP3 decoders (music players and software) can do their job. As a consequence, a number of open-source MP3 algorithms (e.g., the LAME encoder) have been created that emulate the performance of the original algorithms.

Since the development of the original MP3 algorithm, a number of other advanced encoders have been developed to take its place, such as the **MPEG-4 Advanced Audio Coding (AAC)** standard introduced in 1997 (also known as .mp4a), Dolby AC-3, Ogg Vorbis and Windows

Media Audio (WMA). All these encoders use a combination of lossy and lossless techniques to discard data that are perceptually irrelevant and statistically redundant, and owe a large debt to the original MP3 encoder, so it is worth our time to give an overview of the MP3.

12.2.1 The MP3 encoder

The MP3 encoder achieves remarkable compression compared with PCM. The output bitrate ranges from 32 kb/s, which is generally acceptable only for speech, to 320 kb/s, which is the highest rate supported by the MP3 standard. At the default rate of 128 kb/s, the MP3 file represents a 12:1 reduction in data rate with respect to stereo 48 kHz, 16-bit PCM encoding with little to no perceptible decrease in sound quality.[3] The MP3 algorithms for encoding and decoding source material are asymmetrical, in that it takes a relatively large amount of processing power to encode source material, but rather less to decode it. This feature has allowed decoders to be implemented on many cheap hardware MP3 players (e.g., "iPods"), on cellphones and on software programs running on computers.

The overall structure of a simplified MP3 encoder is shown in **Figure 12.6**. The input to the MP3 encoder, $x[n]$, is a PCM audio signal at 32, 44.1 or 48 kHz (e.g., 16 bits $\times$48 kHz = 768 kb/s). The three main parts of the MP3 encoder are a **hybrid filter bank**, a **psychophysical model** and a **bit allocation stage** that brings both these parts together. The filter bank divides the audio signal into a number of **subbands** that are designed to match the "**critical bands**" that humans perceive as sound. Then, guided by a **psychoacoustic model**, each subband is quantized with a resolution chosen to minimize the quantization noise. Finally, the quantized outputs are encoded using **minimum entropy coding techniques** to achieve remarkable compression rates. In the following sections, we will briefly discuss each of these parts of the encoder.

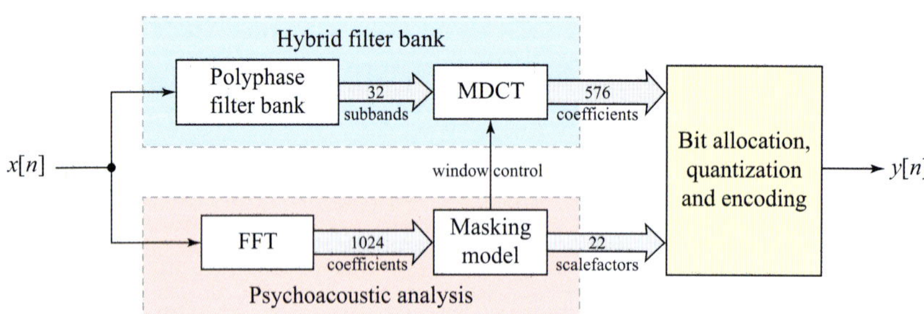

Figure 12.6 MP3 encoder

[3]Some audiophiles will vociferously take issue with this statement, claiming that MP3 – or really any digital recording technique – robs the music of a certain "musicality" that is only present in vinyl records or perhaps wax cylinder recordings. However, for the majority of listeners, the MP3 has proven to be an unalloyed technological marvel.

12.2.2 Hybrid filter bank and MDCT

The job of the hybrid filter bank is to perform a time-frequency analysis of the incoming signal. This is not an easy problem because the signal being analyzed – predominantly music – is non-stationary and its spectral content is highly variable as a function of time. As we will discuss in Chapter 14 (Spectral Analysis), there is a fundamental trade-off in Fourier analysis between resolution in the time and frequency domains. The broader the analysis window in the time domain, the finer the frequency resolution but the coarser the time resolution; conversely, the narrower the analysis window, the finer the time resolution but the coarser the frequency resolution. With music, you essentially need both types of analysis. The snap of a snare drum or other percussive instrument requires a narrow time window to capture its sharp onset. The long note from a wind or string instrument (e.g., an oboe or a violin) requires fine frequency resolution to capture the harmonic structure of the sound. Speaking or singing voices also present challenges. Speech comprises noisy elements with broad frequency distributions (e.g., fricatives and the initial plosive onset of many consonants) as well as more harmonically dense vowel segments, which are themselves made up of brief pitch pulses. The hybrid filter bank attempts to balance these conflicting requirements by doing the time-frequency analysis in two stages. The first stage is a 32-channel, maximally decimated, polyphase, pseudo QMF (PQMF) cosine-modulated filter bank of the sort that will be discussed in supplementary material. The filter bank's 32 channels divide the 48-kHz frequency range of the input into 750-Hz **subbands**, which results in a comparatively low frequency resolution, but good time resolution. The input frames are 1152 points in length for each of two stereo channels, which are then decimated by a factor of 32 by the filter bank into groups of 36 samples.

The second stage of the hybrid filter bank is a windowed MDCT whose purpose is to perform frequency analysis at a high frequency resolution. The output of each channel of the filter bank is transformed by the MDCT whose frame size can be switched adaptively under control of the psychoacoustic model between a long window (36 points long) and a short window (12 points long) depending on the nature of the signal; the short window is used to provide better temporal resolution for sharp transients. The MDCT has a 50% frame overlap and produces half as many output coefficients as input points. The result is that the output of each of the polyphase filter bank's 32 channels is subdivided into 18 finer frequency channels by the MDCT for the long window resulting in $32 \times 18 = 576$ spectral coefficients, each with a bandwidth of 48 kHz / 576 coefficients = 41.67 Hz. (For the short window, there are $32 \times 6 = 192$ spectral coefficients, each with a bandwidth of 48 kHz / 576 coefficients = 125 Hz.)

A key feature of the MP3 algorithm, and the one that is responsible for most of the compression, is the adjustment of the quantization by the MDCT based on the output of a psychoacoustic model that determines the maximum coarseness of quantization that can be applied to the output of the MDCT in order to remain perceptually inaudible. To explain how that works, we have to explain a bit about the psychoacoustic model.

12.2.3 The psychoacoustic model

MP3 and other advanced audio codecs use psychoacoustic models of human audio perception to reduce the bit rate of the encoded audio. These models seek to encapsulate an understanding of how the human auditory system perceives complex sounds. The ultimate purpose of the models is to analyze the input sound to determine the relative perceptual importance of information in different subband channels. The **subband coder** uses this information to allocate more bits to channels containing the more perceptually relevant information and to quantize more coarsely those subband channels that contain less perceptually important information.

In the early twentieth century, researchers, often associated with or funded by the telephone companies, amassed a body of detailed psychophysical experiments directed at determining the performance characteristics of the human auditory system.[4] Much of this work turned out to be useful in the development of perceptual encoding algorithms like MP3. In the next few paragraphs, we will look at three key psychophysical concepts that are responsible for most of the compression in the perceptual audio codecs: the **absolute threshold of hearing**, **critical bands** and **masking**.

The absolute threshold of hearing **The absolute threshold of hearing** (also called the **threshold of audibility**) is the minimum **sound pressure level (SPL)** of a pure tone that can be detected by a young healthy listener in a noiseless environment as a function of frequency. The units of sound pressure are **decibels (dB)** where 0 dB SPL is defined as 0.0002 dynes/cm^2 or 20 μPa (Micropascals). The decibel scale is logarithmic, given by the relation

$$\text{db (SPL)} = 20 \log_{10}(\text{SPL}/20\ \mu\text{Pa}).$$

Each factor of ten change in sound pressure corresponds to a 20 dB SPL. **Figure 12.7** shows a plot of the threshold of hearing as a function of frequency on a log scale. The frequency range of human hearing, again for a young, healthy adult, is conventionally given as 20 Hz–20 kHz. But people are most sensitive to sound in the 1–5 kHz range. The dynamic range of hearing is enormous: almost six log units or 120 dB SPL. The quietest sound an acute listener can hear, corresponding to minimum of the curve, is somewhat below 0 dB, and occurs near 3 kHz. By way of comparison, a whisper is about 30 dB, normal conversation in a quiet room is in the

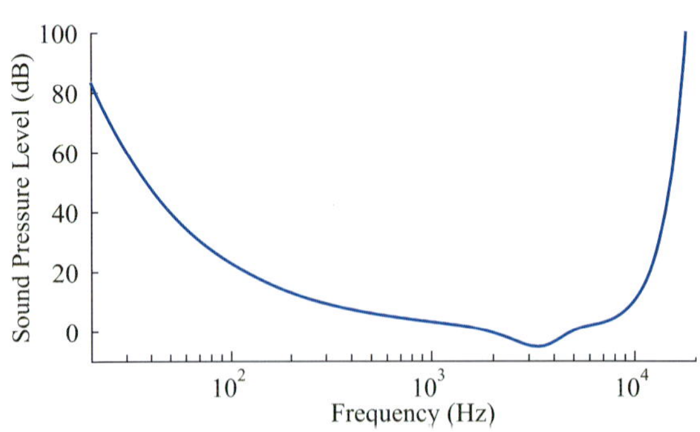

Figure 12.7 Absolute threshold of hearing

[4]Telephone companies have a keen desire to understand the characteristics of telephone apparatus and transmission systems necessary to support intelligible telephone conversations. Their interest is not purely altruistic. They are in the business of selling bandwidth. The smaller the bandwidth they can sell to you, the more bandwidth they have to sell to others.

60 dB range and the threshold of discomfort is about 120 dB. Because the threshold of hearing rises at both low and high frequencies, you can already see that it would be pointless to encode low amplitude information at those frequencies.

Critical bands The cochlea – the organ in the inner ear that is responsible for hearing – has been described as a filter bank of very sharp, frequency selective bandpass filters. Neurophysiologists have measured the frequency response of nerve fibers in the cochleas of animals and have determined that the effective bandwidths of these neural filters are non-uniform (asymmetric) and are both frequency and level dependent. Psychophysicists have found comparable results in psychoacoustic experiments using human listeners. One of the key concepts to come out of psychoacoustic research, which is essential to all perceptual audio codecs, is that of the **critical band**. A critical band is defined as a range of frequency within which the acoustic power of a group of tones (or a band-limited noise) is indistinguishable from a single tone of the same acoustic power.

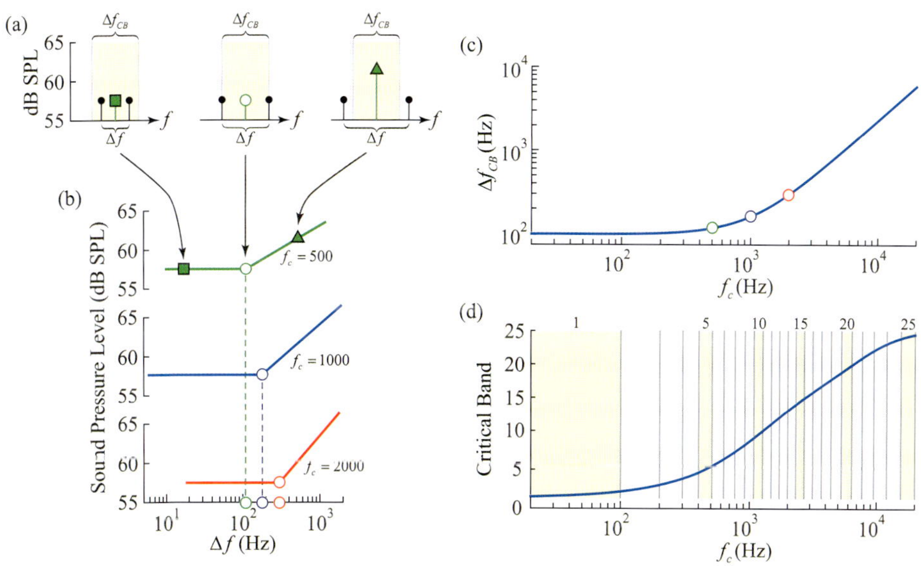

Figure 12.8 Critical bands

Figure 12.8a gives a schematic description of an experiment that establishes how a critical band is measured. In the experiment, a listener is presented with a couple of tones at frequencies that are Δf Hz apart, shown by the pair of solid black symbols in the left-most panel, centered on frequency f_c Hz. The listener is asked to adjust the level of a third (test) tone illustrated with the green square, until the perceived loudness of that tone matches the loudness of the tone pair. The experiment is then repeated as the distance between the tone pair is increased, as indicated by the two other small panels. The result of this experiment is that the perceived loudness of the tone pair remains constant, independent of Δf, until Δf reaches a critical value called the **critical bandwidth** Δf_{CB} (middle panel), after which the perceived loudness increases (right panel). **Figure 12.8b** shows the relation between the perceived loudness of the tone pair as a function of Δf for three values of the center frequency: $f_c = 500$ Hz (green), 1000 Hz (blue) and 2000 Hz (red).[5] The value of Δf that corresponds to the "knee" in the curve is the critical bandwidth Δf_{CB} at that frequency. **Figure 12.8c** shows that the critical bandwidth increases roughly logarithmically with the center frequency of the band, f_c, at higher values of f_c. The open green, blue and red symbols on the curve are the critical bandwidth values derived from the three experiments of **Figure 12.8b**.

The fact that tones within each critical band are perceptually equivalent has led researchers to propose a psychoacoustic frequency scale, called the **Bark scale**, which divides the range of audible frequencies into perceptually equal, contiguous critical bands, shown by the vertical lines in **Figure 12.8d**. Each of the 25 critical bands shown in the figure (six of which are highlighted) covers a perceptually equivalent frequency range. The blue curve shows the continuous mapping of the frequency scale to the Bark scale.

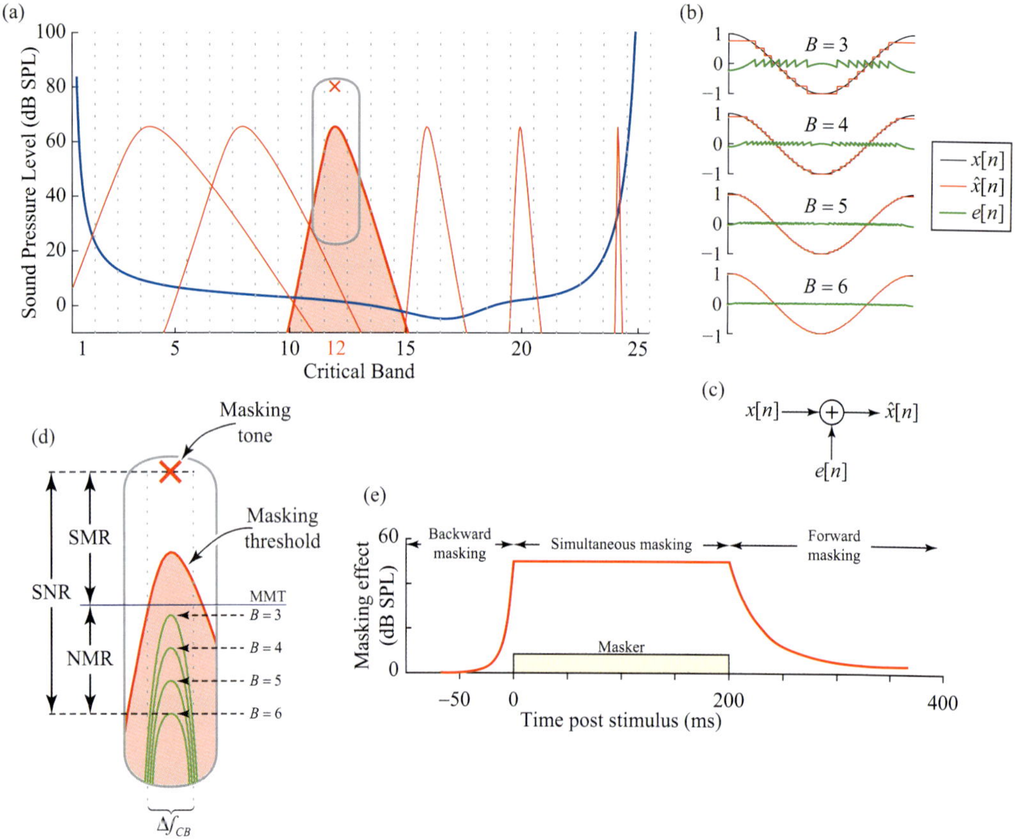

Figure 12.9 Masking and quantization

[5]These plots are derived from a paper published in 1957 by a team of psychophysicists at Harvard (Zwicker, E., Flottrorp, G. and Stevens, S. S. (1957)). There is a direct line between this experimental psychoacoustic work from the early and mid twentieth century and the current use of the results of that work in the development of perceptual audio encoders.

Masking The central concept that the audio codecs exploit is the psychoacoustic phenomenon of **auditory masking**. Masking refers to the fact that one sound – for example, a tone or bandlimited noise – can be rendered inaudible (i.e., "masked") by the presence of another tone or by noise, when these stimuli are presented either at the same time (**simultaneous masking**) or at different times (**temporal masking**). The blue curve in **Figure 12.9a** again shows the threshold of hearing, this time plotted on a frequency scale where each of the 25 linear divisions marked by the dotted lines is perceptually equivalent and corresponds to one critical band. Now perform the following experiment. Play a pure cosine tone, called the masking tone, or simply the **masker**, shown by the red × in the middle of critical band 12. This tone represents a cosine masker at frequency 1450 Hz and 80 dB SPL. Now play a second, test, tone and adjust that tone's frequency and level until it becomes just distinguishable from the masker. The locus of frequencies and levels at which the test tone is just distinguishable is called the **masking threshold**, and is shown by the solid red line below the ×. Sounds such as noise or other tones whose frequencies and amplitudes lie in the shaded area below the masking threshold are rendered inaudible by the masking tone. The masking threshold is a function of both the frequency and level of the masker. For example, **Figure 12.9a** also shows plots of masking threshold for 80 dB SPL tones at five other frequencies. At lower frequencies, the masking threshold curves spread over a number of critical bands.

Masking threshold is central to the design of perceptual encoders such as the MP3. The MP3 algorithm contains implementations of psychoacoustic models of two types of masking that are derived from psychoacoustic experiments: one in which the masker is a tone (a **tonal masker**, as shown in this figure) and a similar model in which the masker is a narrowband noise (a **noise masker**). The MP3 algorithms use these masking threshold curves to determine the minimum number of bits of quantization that are required to represent sound within each of the frequency subbands without allowing the noise that is introduced by quantization to become audible (see Section 6.7.2 for a discussion of quantization noise). For example, assume that our masking tone has been quantized with a resolution of B bits, which has the effect of dividing the full-scale amplitude of the waveform into 2^B levels. **Figure 12.9b** shows one cycle of an unquantized masking tone $x[n]$ (thin black traces), as well as waveforms of the quantized tone $\hat{x}[n]$ (red traces), for values of $B = 3$, 4, 5 and 6 bits. As we discussed in Section 6.7.2, the quantized waveform can be viewed as being equivalent to the unquantized signal plus an additive source of quantization noise $e[n]$ (green traces), as shown in **Figure 12.9c**,

$$\hat{x}[n] = x[n] + e[n]. \tag{12.21}$$

The finer the quantization (i.e., larger B), the smaller the quantization noise; each additional bit of resolution increases the signal-to-noise ratio (SNR) by about 6 dB. **Figure 12.9d** relates the spectrum of the quantization noise to the masking threshold within a critical band. The figure again shows the masking tone and masking threshold for band 12 enlarged from **Figure 12.9a**. The thin green curves represent the spectrum of the quantization noise of the masking tone within this critical band for the levels of quantization $B = 3$, 4, 5 and 6 bits. For a given level of masking tone, all sounds presented within the critical band with sound levels below a **minimum masking threshold (MMT)** (thin blue line) are likely to be inaudible. The difference in sound pressure between the level of the masking tone and the MMT is referred to as the **signal-to-**

masker ratio (SMR). The essential point here is that if the spectrum of the quantization noise $e[n]$ lies below the MMT, then the quantization noise will be inaudible. The **noise-to-masking ratio (NMR)** measures the "headroom" between the level of quantization noise and the minimum masking threshold. The codec estimates this value and adjusts the number of bits of quantization within each critical band to be as small as possible while maintaining adequate headroom. By continually adjusting the number of bits of quantization for each critical band so that the spectrum of the quantization noise remains below the MMT, the codec can achieve remarkable compression without introducing audible quantization noise.

The psychoacoustic model is implemented on a parallel path with the main hybrid filter bank. Each of the 1152-point frames of the input sound is divided in two, zero-padded, windowed and processed using a 1024-point FFT to produce an estimate of the power spectral density. The major task of the model is to identify the location in frequency and level of both tonal and noise maskers. A sample spectrum of music is shown using the blue trace in **Figure 12.10a**, with the location of the tonal and noise maskers marked with × and ○, respectively.[6] Maskers that occur too close together on a critical band scale are replaced by the strongest single masker, and any masker below the absolute threshold of hearing (dashed line) is discarded. Then, the individual (local) masking thresholds of each masker are estimated using prototype formulas as shown in **Figure 12.10b** (thin red lines), and these individual maskers are summed to form the **global masking threshold** (thick red line). If the remaining sections of the encoder keep the

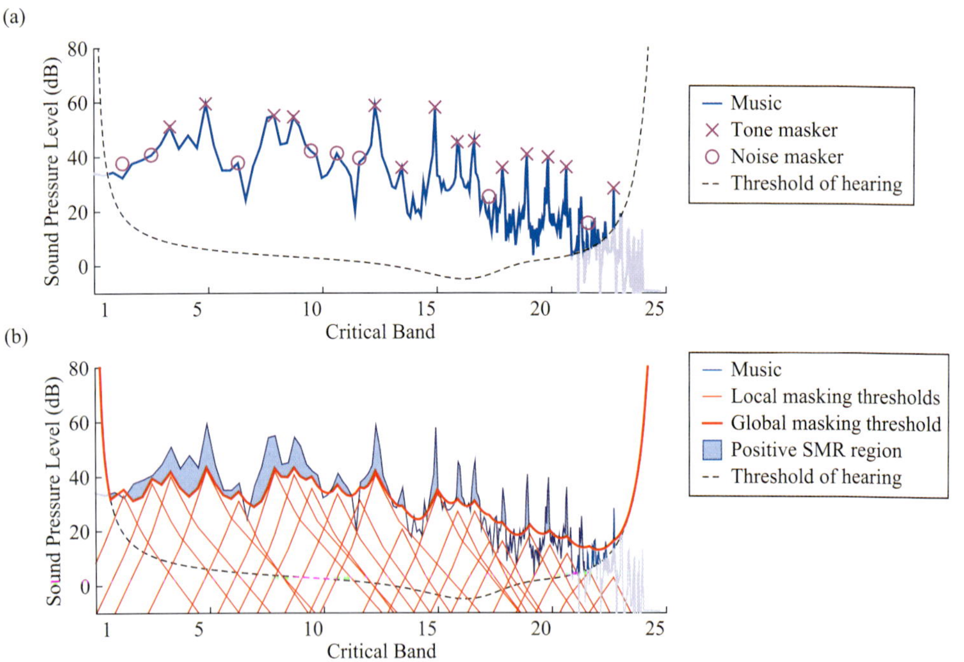

Figure 12.10 Implementation of the psychoacoustic model

[6]The sound sample is from the opening bars of Symphony No. 1 in E-flat Maj., K. 16, written by Wolfgang Amadeus Mozart (1756–1791) when he was eight years old. Contemplate that for a moment: *eight years old.*

quantization noise below the masking threshold in each critical band, then the compressed audio should be perceptually indistinguishable from the original input. The shaded blue region represents regions in which the SMR is positive. In essence, it is only the sound energy in this region that needs to be finely quantized by the next stage of the encoder, the quantization and bit allocation stage. In order to apply the information from the psychoacoustic model, each of the 576 frequency coefficients of the MDCT is assigned to one of 22 **scalefactor bands**, each of which corresponds roughly to one critical band. Each band is assigned a scale factor that corresponds to the SMR of the global masking threshold for that band.

Practical implementations of the MP3 codec also include temporal masking effects suggested by **Figure 12.9e**. When a masking tone is played, it does not only mask sounds occurring at the same time (simultaneous masking). The masking effect affects sounds for about 150 ms *after* the cessation of the masker (an effect termed forward masking), and also affects the perception of sounds for about 50 ms *before* the onset of the masker (backward masking)!

12.2.4 Bit allocation and quantization

The overall goal of the bit allocation, quantizing and encoding stage, shown in **Figure 12.6**, is to use the information from the psychoacoustic model, as represented by the scale factors, to encode the output of the MDCT at a given (generally fixed) number of bits per frame while maintaining the perceived sound quality as well as possible at any chosen bit rate. The approach is an iterative one, based on two nested loops, an inner loop known as the **rate control loop** and an outer loop known as a **noise (or distortion) control loop**. The purpose of the rate control loop is to assure that each output frame of MDCT data can be encoded in no more than the required number of bits. First, the output of the MDCT is non-uniformly quantized and encoded to a variable length code using a predetermined **Huffman table**. **Huffman coding**, which we will discuss in Section 12.2.5, below, assigns shorter code words to more frequent values, generally those with smaller quantization levels. This coding is considered "noiseless" in that it compresses the data in a lossless manner. If the total number of bits in a frame that results from this coding exceeds the available number of bits for that frame, then the global gain of the encoder, which determines the quantization step size of every coefficient, is increased in an iterative fashion until the overall rate of the coder is within the prescribed limit.

As you might imagine, increasing quantization globally can result in the quantization noise of some of the scalefactor bands becoming audible. It is the job of the noise control (outer) loop to deal with this problem. It does so by applying the scalefactors derived from the psychoacoustic model to reduce the quantization noise in each band to a perceptually inaudible level. The way this is done is to increase the number of quantization levels of the signal (i.e., B) in the affected bands, which results in a higher bit rate. If the overall resulting bit rate for all channels now exceeds the prescribed limit, the rate control (inner) loop has to be repeated, and the global gain changed. This noise-allocation process continues until the quantization noise in each scalefactor band is below the masking threshold for that band, and the total number of bits per frame is at the required value.

12.2.5 Minimum entropy coding

Both the MP3 encoder we are discussing and the JPEG encoder we will cover in Section 12.3 combine **lossy compression** algorithms that are based on "throwing away" perceptually irrelevant information with **lossless compression** techniques that are based on eliminating statistical redundancy of the resulting data. One of the most important of those lossless data compression techniques is **minimum entropy coding**.

Let us start by asking the general question: what is the optimum way – that is, the smallest number of bits necessary – to encode and transmit a group of N independent symbols, for example, the 1024 quantized levels of an A/D converter, or the 128 characters of the ASCII alphabet? If we use the familiar fixed-length binary code, we normally use $\log_2 N$ bits per symbol; so, we would need ten bits for the A/D converter or seven bits for each ASCII character. But what if we know that some symbols occur much more frequently than others? For example, it turns out that the letter E is 56 times more common than the letter Q, as measured in large corpus of English words. One might think that you might be able to devise a variable-length code that allocates fewer bits to the more common characters and more bits to the less common characters.[7]

There are really two related questions here: (1) What is the *theoretical* minimum number of binary bits we need per symbol? (2) Can we design a *practical* coding scheme that attains or approaches this minimum? The first question was answered by Claude Shannon[8] in 1948. As part of his quantitative analysis of the channel capacity of telephone lines, he formalized the idea that the **information** in a symbol or message is a measure of its **novelty** (or "**surprise**" as it is often called). A message that occurs rarely is considered very informative, whereas one that occurs frequently is not. So, "Tomorrow's weather will be just like today's" might be considered to have less "surprise" than, say, "Tomorrow, there will be a plague of locusts that will occlude the sky."

To quantify this notion, assume that your code consists of N symbols s_i, $1 \le i \le N$, each of which occurs with probability p_i. Shannon defined the amount of information of symbol s_i as

$$h_i \triangleq \log_2(1/p_i) = -\log_2 p_i.$$

Because the probability of any symbol cannot exceed one, $p_i \le 1$, the log function ensures that information conveyed by any symbol, $h_i = -\log_2 p_i$, is always non-negative. The negative sign

[7] Morse code, named in honor of Samuel F. B. Morse (1791–1872), an inventor of the telegraph, is an early example of just such a variable-length binary code, in which the lengths of symbol codes are related to their frequency. This code, dating from 1844 and still in occasional use today, uses an alternating sequence of short and long electrical signals, colloquially referred to as "dots" (.) and "dashes" (–) to form letters. The code for E is a single symbol (.); the code for Q is four symbols (– – . –).

[8] Claude E. Shannon (1916–2001) was a brilliant mathematician and engineer who essentially launched the field of information theory and also made fundamental contributions to numerous other fields, including communication theory, signal processing, artificial intelligence, cryptography and logic design. While still a graduate student at M.I.T., the 24-year-old Shannon showed (in his Master's thesis) that all Boolean algebra problems could be implemented by switching circuits, thereby laying the foundation of modern digital logic design. Working at Bell Laboratories during and after WWII, Shannon developed the mathematical ideas of information entropy and channel capacity that underlie all modern digital communication systems.

also means that the amount of information is inversely related to its probability of occurrence. A symbol that occurs with probability $p = 1$ gives $\log_2 1 = 0$ information. This makes sense since if there is only one message, then there is no "surprise" at all. The log function also has the nice property that if there are two independent symbols s_1 and s_2, with probabilities p_1 and p_2, respectively, then the information in the combined message $s_1 s_2$ is the sum of the information in each symbol:

$$-\log_2 (p_1 p_2) = -(\log_2 p_1 + \log_2 p_2) = h_1 + h_2.$$

Shannon then adapted the notion of entropy from statistical thermodynamics and defined the **entropy** of a code of N symbols as the expected value of the information,

$$H \triangleq -\sum_{i=1}^{N} p_i \log_2 p_i. \tag{12.22}$$

The entropy H is just the average amount of information in the entire code; it is the information in each symbol, $-\log_2 p_i$, multiplied by the probability of occurrence of that symbol, p_i, summed over all symbols.

Shannon's source coding theorem (also called the **theorem of noiseless coding**) established that the entropy H represents the lower bound to the average number of bits L that is required to encode messages using a set of N symbols. In fact, the source coding theorem states that

$$H \leq L < H + 1/N. \tag{12.23}$$

H therefore specifies the maximum amount of compression that can be achieved for data encoded with those symbols. Codes that lie within the range specified by Equation (12.23) are termed **minimum entropy codes**.

To take a specific example, assume we have four possible data symbols, A, B, C and D, each of which occurs with equal probability $p_i = 0.25$, $0 \leq i \leq 3$. The entropy is

$$H = -\sum_{i=1}^{4} 0.25 \log_2 0.25 = 2.$$

This says you need a minimum of two bits to encode each symbol. To generalize, if you had N unique data symbols of equal probability, $1/N$, you would assign each symbol a code with $\lceil \log_2 N \rceil$ binary bits. You will recognize this result as being exactly equivalent to standard binary coding. One can show that the entropy of a set of N symbols is actually maximum when the probabilities of the symbols are all equal, $p_i = 1/N$, so standard binary code is the best you can do if you know nothing about the actual probabilities of the symbols and are therefore implicitly assigning them all equal probability.

But now, consider what happens if you know that the probabilities of the symbols are not the same. For example, assume you know that the probabilities of the four symbols, A, B, C and D, are 0.5, 0.3, 0.15 and 0.05, respectively. From Equation (12.22), the minimum average number of bits per symbol required is

$$H = -\sum_{i=1}^{4} p_i \log_2 p_i = \underbrace{0.5 \cdot 1}_{\text{A}} + \underbrace{0.3 \cdot 1.737}_{\text{B}} + \underbrace{0.15 \cdot 2.737}_{\text{C}} + \underbrace{0.05 \cdot 4.322}_{\text{D}} = 1.648.$$

The source coding theorem tells us that it is theoretically possible to encode these four messages with fewer than two bits per symbol, but it does not tell us *how* to achieve this compression practically.

The **Huffman coding** algorithm, first published in 1952, was the first practical algorithm that showed how to achieve minimum entropy data compression.[9] It and its offshoots are still the most commonly used minimum entropy coding algorithms, finding application in codecs for sound, images and video as well as pure data compression. The Huffman coding algorithm has two phases. In the first phase, a binary tree is created using the symbols' known or estimated probabilities. Then, the symbols are assigned codes, with the smallest code lengths being assigned to the symbols whose probability of occurrence is highest. This is best explained by an example.

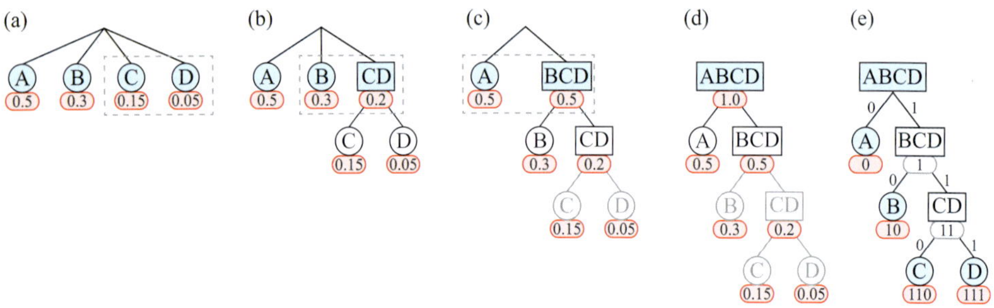

Figure 12.11 Huffman coding example

Figure 12.11 shows how a binary tree is built for our four symbols, A, B, C and D. Initially, each of the tree's top-level branches is one of the symbols, shown by the blue shaded circles in **Figure 12.11a**. The probability of each symbol is shown in the red oval below each symbol. Now, the two branches with the lowest probability – in this case, branches C and D – are combined and replaced with one branch, labeled CD, which has summed probability 0.2, as shown in the blue box in **Figure 12.11b**. Then, the two new top-level branches with the lowest probability – in this case, branches B and CD – are combined and their probabilities are

[9] In 1951, David A. Huffman (1925–1999) was a 25-year-old graduate student at M.I.T. enrolled in a course on information theory taught by Prof. Robert M. Fano, who was himself a student of Claude Shannon. Fano had given his students the choice between a final exam and a term paper, the topic of which was to find an optimum method of representing symbols in binary code, that is, a minimum entropy code. Fano craftily neglected to tell his students that he and Shannon had attempted this problem, but that their solution ("Shannon–Fano coding") was not optimum. Huffman chose the term paper, but after laboring on it futilely for months was just about to give up when the solution came to him. "It was the most singular moment of my life," he later recalled. "There was the absolute lightning of sudden realization." Huffman went on to a distinguished career as a mathematician and computer scientist, first as a professor at M.I.T. and later as the founder and chair of the computer science department at U.C. Santa Cruz, where he taught, among other things . . . DSP!

summed, to form branch BCD, as shown in **Figure 12.11c**. This process continues iteratively until the complete binary tree is formed, as shown in **Figure 12.11d**. In this example, the left branch (a circle) is a symbol and the right branch (a square) represents the remaining symbols, which have a lower or equal summed probability.

In the second phase of the Huffman algorithm, the tree is traversed from top to bottom and code values are assigned to the symbols in such a manner that the symbols with the highest probability get the shortest codes. In our example, **Figure 12.11e**, the symbol on the left branch, A, is given the unique one-bit code 0, and the right branch, representing all the other symbols, is assigned the prefix 1. At each subsequent level of the tree, the length of the symbol code increases by one bit, with the left branch getting a unique terminating suffix 0, and the remaining branches getting an additional 1 as prefix. So on the second level of our example, the symbol B gets the unique two-bit code 10, and symbols C and D get prefix 11. Finally, at the lowest level, C gets 110 and the lowest probability symbol, D, ends up with code 111. So, our complete variable-length code is as shown in **Table 12.2**:

Table 12.2 Huffman code example

Symbol	Code	Length	Probability
A	0	1	0.50
B	10	2	0.30
C	110	3	0.15
D	111	3	0.05

The average code length of this example is the sum of the word-length of each coded symbol, w_i, multiplied by its probability,

$$L = \sum_{i-1}^{N} w_i p_i = \underbrace{1 \cdot 0.5}_{w_A p_A} + \underbrace{2 \cdot 0.3}_{w_B p_B} + \underbrace{3 \cdot 0.15}_{w_C p_C} + \underbrace{3 \cdot 0.05}_{w_D p_D} = 1.7.$$

This is within the range of the minimum specified by Shannon's source coding theorem, Equation (12.23).

The Huffman code is an example of a **prefix code** (also confusingly called a **prefix-free code**), meaning that no symbol's entire code word is a prefix of any other code word. Hence, no symbol can be confused with any other symbol. This makes it possible to decode messages encoded with the Huffman algorithm unambiguously, even when the symbols' code words are strung together with no ancillary information to indicate the separation between symbols. For example, consider the message string: 01101110101001100. The only character that can start with 0 is A. Following that is 110 (C), 111 (D), 0 (A), and so on. So,

$$\underbrace{0}_{A} \mid \underbrace{110}_{C} \mid \underbrace{111}_{D} \mid \underbrace{0}_{A} \mid \underbrace{10}_{B} \mid \underbrace{10}_{B} \mid \underbrace{0}_{A} \mid \underbrace{110}_{C} \mid \underbrace{0}_{A}$$

Basically, whenever you see a terminal "0" or a "111," you know that you have decoded a symbol.

To create an optimal Huffman code, you have to build the Huffman tree, which means that either you must know the probabilities of the symbols *a priori*, or you have to have an estimate of them through measurement. In practice, many algorithms that bill themselves as using Huffman codes (including back-end algorithms for MP3 and JPEG) actually use pre-defined variable-length codes based on previously determined global averages of symbol probabilities that are then assigned to symbols using a look-up table. Encoders need to include these look-up tables in their headers so that the hardware or software doing the decoding can unambiguously decode the data.

Practical implementations of the MP3 codec have a lot of sophisticated features and complexities that we have not mentioned. For example, the encoder may use several Huffman tables for different frequency ranges of coefficients. It also takes advantage of the redundancy of stereo information in the encoding to reduce the bit rate of stereo material.

12.3 JPEG image compression

JPEG refers to a set of algorithms for the compression and coding of continuous-tone still images. These algorithms are found ubiquitously in devices that create or display digital images. If you have ever taken a photo with a digital camera or a cellphone, you have created "a JPEG." JPEG is actually an acronym for the Joint Photographic Experts Group, a committee of scientists and engineers drawn from international standards bodies and industry that was responsible for the creation of the original JPEG standard in the late 1980s and that continues to refine and extend the standards to this day. However, the word "JPEG" now colloquially refers to the highly compressed image files with a .jpg (or .jpeg) extension that have been created using these algorithms. Like the MP3 encoder discussed in Section 12.2, the JPEG encoder achieves substantial compression by using a remarkably ingenious combination of techniques from digital signal processing and computer science melded with an understanding of human visual psychophysics. An important part of the JPEG encoder is the DCT-II, so for that reason we will peek under the hood to see how the algorithm works.

Figure 12.12 gives a schematic overview of the process of forming a JPEG image.[10] There are essentially three stages: (1) color processing of the raw image, (2) lossy image compression and quantization using the DCT and, finally, (3) lossless encoding of the resulting coefficient data to form the JPEG file. In each of these steps, the algorithm takes advantage of redundancy in the data to achieve compression.

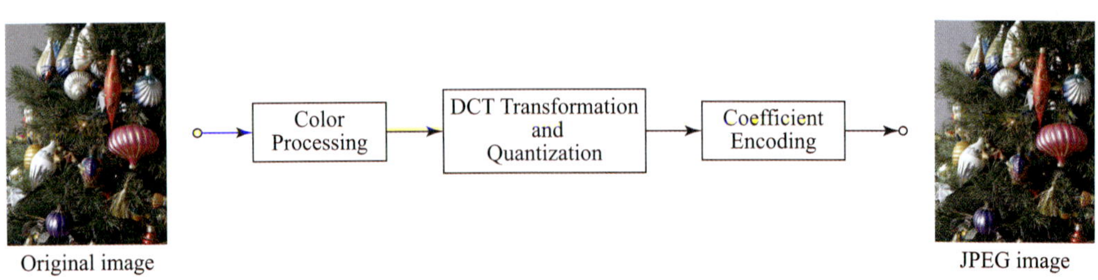

Figure 12.12 Overview of JPEG algorithms

[10]There many JPEG modes. What is outlined here is the simplest and most commonly used "baseline" mode.

Before we look at the algorithms for these three stages in some detail, let us look at the results. The original image on the left of the figure, obtained with a digital camera, is 1156×1536 pixels in size. Each pixel of the raw image encodes the three primary colors – red, green and blue – at 8 bits apiece for a total of 24 bits per pixel. Hence, the image comprises $1156 \times 1536 \times 24$ bits = 42,614,784 bits = 5,326,848 bytes. The image on the right represents the result of processing the original image using the JPEG encoder with a large amount of compression and then decoding it again to display. The JPEG standard allows one to adjust the amount of compression with a quality factor Q that ranges from 1 (lowest quality) to 100 (highest quality). The JPEG image in this figure was created with a miserly value of $Q = 10$, resulting in an image that has been encoded in a file of 214,865 bytes, representing about 4% the file size of the original image. Yet, even at this ridiculously low quality factor, the image is completely acceptable to the eye (well, to *my* eye) when viewed at full size on a computer monitor. At a more reasonable quality factor of $Q = 50$, the JPEG image requires only 7% the file size of the original image; at a quality factor of $Q = 80$, the JPEG image requires only 12% of the original image. At these higher quality factors, the JPEG images are essentially indistinguishable from the original, even under magnification by a persnickety observer. In the following paragraphs, we will provide a simplified treatment of the essentials of the JPEG methods to reveal how the magic is done.

12.3.1 Color processing

JPEG achieves its first level of compression by exploiting the fact that the human visual system is more sensitive to variations in the brightness within a scene than to variations in color. To understand how the JPEG encoder exploits this fact, first consider how color is represented in a picture from a digital camera at the sensor level. In the familiar **additive color model**, which is used by the sensors of essentially all electronic devices that capture or emit light (cameras, TVs, computer monitors, cellphones), all colors are represented as the sum of light in the three **primary colors**: red (R), green (G) and blue (B). These three colors can be pictorially represented as the vertices of orthogonal axes of the RGB color space "cube" (the smaller rotated cube inside the big cube in **Figure 12.13**). In a 24-bit RGB color representation, every available color in the color space is expressed as a tuple of eight-bit integer RGB values, $[R, G, B]$, each ranging from 0 to 255. The primary colors are at three of the cube's vertices: pure red ($R = [255, 0, 0]$), green ($G = [0, 255, 0]$) and blue ($B = [0, 0, 255]$). White and black are located diagonally at two other vertices. White is the sum of all the primaries ($255 \cdot R + 255 \cdot G + 255 \cdot B = [255, 255, 255]$), whereas black is the absence of all light ($0 \cdot R + 0 \cdot G + 0 \cdot B = [0, 0, 0]$). The three **secondary colors** – cyan, megenta and yellow – form the remaining vertices of the cube. Each secondary color is made by mixing two of the primaries:[11]

Cyan: $C = 0 \cdot R + 255 \cdot G + 255 \cdot B = [0, 255, 255]$
Magenta: $M = 255 \cdot R + 0 \cdot G + 255 \cdot B = [255, 0, 255]$
Yellow: $Y = 255 \cdot R + 255 \cdot G + 0 \cdot B = [255, 255, 0]$.

[11] An alternative color model, the **CMYK** (Cyan, Magenta, Yellow and Black) model is extensively used in four-color printing. In the CMYK model, all colors can be formed as the sum of the three secondary colors, which in the case of printing are the colors of the inks. Adding ink to the white paper absorbs (subtracts) light and what remains is reflected to the eye. Hence, the CMYK model is called a **subtractive color model**. White is represented by the absence of all color ink. Black is the sum of all secondary colors. Usually, pure black (K) ink is used instead of the sum, C+Y+M.

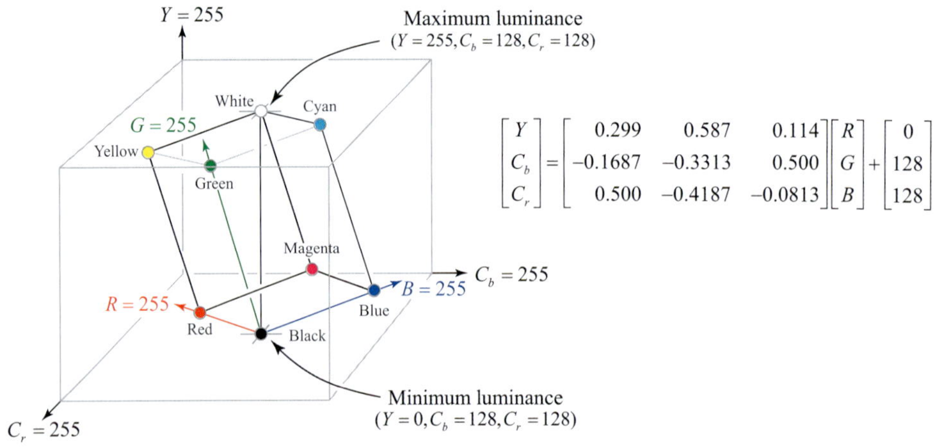

Figure 12.13 RGB and YCbCr Color spaces

Almost all consumer-grade digital cameras and cell phones initially record separate images of each RGB color plane as unsigned integers at a **bit depth** of at least eight bits per channel (more like 12 to 14 bits in better cameras). But by default, most cameras and cellphones process the RGB data into JPEGs in order to reduce the amount of on-camera storage required for each image, though higher-end cameras often provide the option of storing the "raw" RGB data for subsequent off-camera processing. The first step in the production of the JPEG – and the place where the first level of image compression occurs – is the mapping of RGB data into the so-called **YCbCr color space**,[12] represented by the outer cube in **Figure 12.13**. The YCbCr color space also has three axes: **luminance** (Y) plus two **chrominance** axes (C_b and C_r). The RGB color space is related to the YCbCr color space through a reversible matrix transformation

$$\begin{bmatrix} Y \\ C_b \\ C_r \end{bmatrix} = \begin{bmatrix} 0.299 & 0.587 & 0.114 \\ -0.1687 & -0.3313 & 0.500 \\ 0.500 & -0.4187 & -0.0813 \end{bmatrix} \begin{bmatrix} R \\ G \\ B \end{bmatrix} + \begin{bmatrix} 0 \\ 128 \\ 128 \end{bmatrix}. \tag{12.24}$$

An important effect of this transformation is to rotate the RGB cube so that the diagonal formed by the black and white vertices of the RGB cube is aligned with (i.e., in parallel with) the luminance (Y) axis of the YCbCr color space. As a result, luminance is effectively a measure of the "black-and-white" intensity of each pixel of the image. For example, if the values of R, G and B are all equal, say, $R = G = B = \alpha$, $0 \le \alpha \le 255$, then the transformation of Equation

[12] Even though we call YCbCr a color space, it is just a transformation of the RGB model, specified by Equation (12.41). There exist a number of mappings between the RGB and YCbCr spaces for different display devices. The one given in Equation (12.41) is from the original JPEG File Interchange Format specifications.

(12.24) gives a luminance value of $Y = \alpha$, and chrominance values $C_b = C_r = 128$, independent of α. The chrominance components C_b and C_r represent color difference information and are formed by subtracting the luminance from blue and red values respectively.

From a technical standpoint, the YCbCr color space transformation essentially decouples intensity information from color information. This is important because it turns out that the human visual system does something very similar; it treats brightness (the subjective perception of the intensity of illumination) and color separately. Moreover, the visual system is much more sensitive to changes in brightness across an image than to changes in color. The JPEG algorithms take advantage of these psychophysical facts by mapping the raw RGB data into the YCbCr space using the matrix calculation of Equation (12.24), and then compressing and encoding brightness (Y) and color (C_b and C_r) channels differently. We will now explain how that encoding is done and show the results.

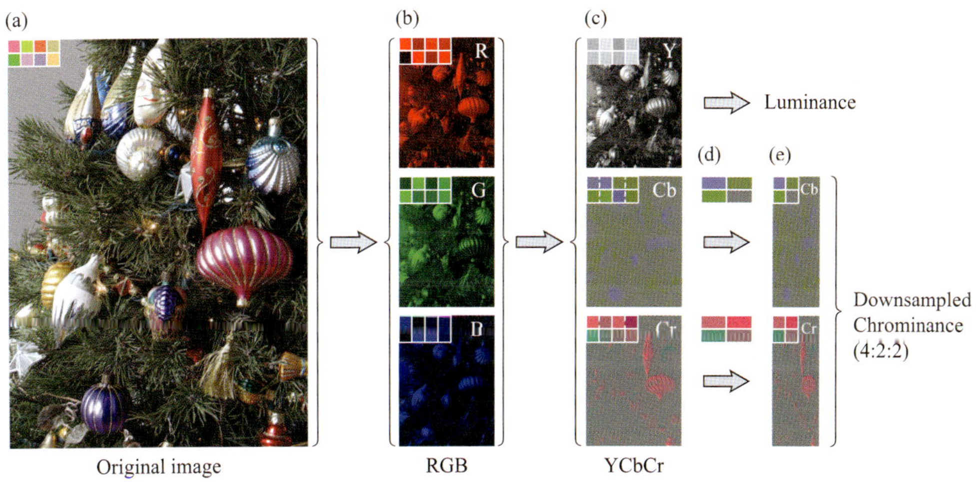

Figure 12.14 Color processing

Figure 12.14a shows the original scene before the camera, with the little inset of eight colors in the top left of the picture meant to indicate a typical range of colors in the image. The camera's sensors first record the three RGB image planes, denoted R, G and B in **Figure 12.14b**. The small insets in these panels track how the eight sample colors in the original image are represented in each plane. Next, the RGB images are mapped into the YCbCr color space, using the transformation of Equation (12.24). The Y plane represents the black-and-white intensity, whereas the Cb and Cr planes contain a mix of blue-green and red-green colors, respectively.

In order to take advantage of the relative insensitivity of the human visual system to chrominance, JPEG implementations almost always downsample both the chrominance planes by a factor of two or four. In the typical approach schematized in **Figure 12.14**, chrominance data are **subsampled** horizontally.[13] That is, each pair of horizontal pixels is first averaged, as shown in the insets of **Figure 12.14d**, and then downsampled to one pixel (**Figure 12.14e**), thereby halving the size of the chrominance data. The downsampling of chrominance data is a form of lossy compression, meaning that information is lost; the compression is irreversible. However, for most images, chrominance compression is imperceptible and well worth the trade-off obtained by the reduced file size that results.

12.3.2 DCT transformation and quantization

Most of the reduction of the JPEG file size takes place in the compression and quantization stage which follows the color processing stage. The principal tool responsible for the compression is the **two-dimensional DCT** – the 2-D DCT, which is based on the DCT-II we introduced in Section 12.1.2. The 2-D DCT of a matrix is a **separable transform** obtained by applying the DCT-II to each row of a matrix, and then applying the DCT again to each column. For simplicity, consider an $N \times N$ matrix of data, $f[n, m]$, $0 \le n \le N-1$, $0 \le m \le N-1$. By Equation (12.5), the DCT-II of the nth row is

$$C_n[k] = \sqrt{\frac{2}{N}} \sum_{m=0}^{N-1} f[n, m] \beta[k] \cos \frac{\pi k(2m+1)}{2M}, \quad 0 \le k \le N-1.$$

Then the DCT-II of $C_n[k]$ over all the rows is

$$\begin{aligned} C[k, l] &= \sqrt{\frac{2}{N}} \sum_{n=0}^{N-1} C_n[k] \beta[l] \cos \frac{\pi l(2n+1)}{2N} \\ &= \sqrt{\frac{2}{N}} \sum_{n=0}^{N-1} \left(\sqrt{\frac{2}{N}} \sum_{m=0}^{N-1} f[n, m] \beta[k] \cos \frac{\pi k(2m+1)}{2M} \right) \beta[l] \cos \frac{\pi l(2n+1)}{2N} \\ &= \frac{2}{N} \sum_{n=0}^{N-1} \sum_{m=0}^{N-1} f[n, m] \beta[k] \beta[l] \left(\cos \frac{\pi k(2m+1)}{2M} \cos \frac{\pi l(2n+1)}{2N} \right), \\ &\qquad 0 \le k \le N-1, \ 0 \le l \le N-1. \end{aligned} \tag{12.25}$$

For each value of k and l, the basis functions of the 2-D DCT form an $N \times N$ matrix of the product of "horizontal" and "vertical" cosines with frequencies $\pi k/N$ and $\pi l/N$, respectively,

$$\cos \frac{\pi k(2m+1)}{2N} \cos \frac{\pi l(2n+1)}{2N}, \quad 0 \le m \le N-1, \ 0 \le n \le N-1.$$

Since k and l can each take on N values, there are N^2 basis functions. Furthermore, because k and l are interchangeable, this matrix of basis functions is symmetric.

[13] This is called 4:2:2 subsampling, and is commonly used in many digital cameras. Another subsampling option found in devices like camera phones is 4:2:0, in which chrominance data are subsampled by a factor of two both horizontally and vertically, resulting in a factor-of-four compression. 4:4:4 means there is no horizontal or vertical chrominance subsampling at all. This is found in high-end cameras as well as in the JPEG output of image manipulation programs such as Adobe Photoshop at the higher quality settings.

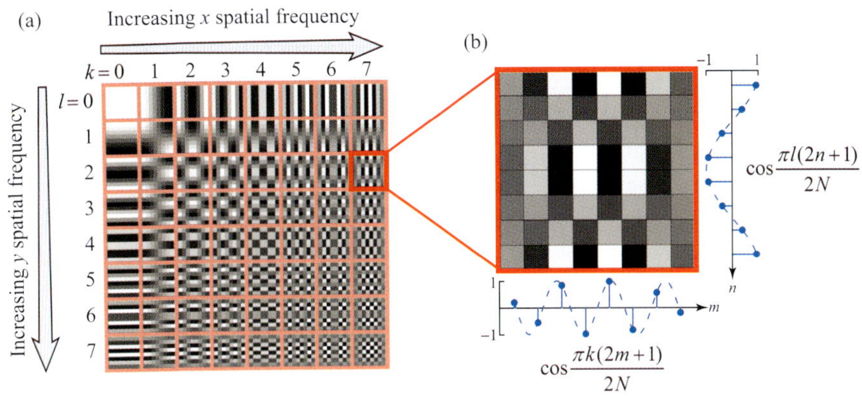

Figure 12.15 The basis functions of the 2-D DCT

Figure 12.15a shows the 64 basis functions of a 2-D DCT for the value $N=8$, displayed as an 8×8 grid with horizontal spatial frequency increasing from left to right with increasing k and vertical spatial frequency increasing from top to bottom with increasing l. **Spatial frequency** is the visual homologue of the frequency of sound. Consider the top row of **Figure 12.15a** ($l=0$). Each of the eight patterns ($0\le k\le 7$) consists of $N=8$ vertical bars whose brightness varies as a cosine as a function of horizontal position, $\cos\pi k(2m+1)/2N$, $0\le m\le N-1$. White corresponds to a value of one (maximum brightness) and black to zero. Spatial frequency is defined as the number of cycles of these bars that fall on the retina of an observer within one degree of visual angle, basically the number of bars per mm on the page (or screen) held a fixed distance from the eye. The basis function depicted in the top left-hand corner ($k=0$, $l=0$) represents the lowest spatial frequency, namely a constant brightness, $\cos 0\cdot\cos 0=1$, across all pixels, often called "DC." As k increases from left to right, the spatial frequency increases. Another example is the highlighted 8×8 pixel pattern corresponding to $k=7$ and $l=2$, shown enlarged in **Figure 12.15b**. Here, the higher frequency ($k=7$) cosine corresponds to the horizontal (m) axis and the lower frequency ($l=2$) cosine to the vertical (n) axis. Their product at each value of n and m creates this particular 64-pixel basis function. Like the 1-D DCT-II, the basis functions of the 2-D DCT are orthonormal (see Problem 12-3). Any $N\times M$ pattern can be represented by the sum of scaled basis functions, where the scaling factors are given by the 2-D DCT coefficients,

$$f[n,m]=\frac{2}{N}\sum_{k=0}^{N-1}\sum_{l=0}^{M-1}\beta[k]\beta[l]C_{xy}[k,l]\cos\frac{\pi k(2m+1)}{2M}\cos\frac{\pi l(2n+1)}{2N},$$
$$0\le n\le N-1,\quad 0\le m\le N-1. \tag{12.26}$$

Equations (12.25) and (12.26) are the expressions for the forward and inverse 2-D DCT, respectively

The DCT performs a transformation of images that is highly relevant to the way the human visual system processes visual information. The reason that the DCT is so well suited to image compression has to do with the relative insensitivity of the human visual system to high spatial frequencies in an image, with respect both to luminance and, especially, to chrominance. **Figure 12.16a** shows a sinusoidal grating of varying cycles of dark and light intensity whose spatial frequency increases from left to right across the picture, and whose **contrast** – the difference between the minimum (dark) and maximum (light) amplitudes of each cycle – increases from the top of the picture to the bottom. In perceptual experiments, the **contrast sensitivity function**, sketched in red, gives the minimum image contrast (ordinate) a person needs to detect the varying grating intensity of various spatial frequencies (abscissa). The conclusion from these experiments is that the human visual system has a bandpass characteristic that is relatively insensitive to higher spatial frequencies. Hence, this information is less important to preserve in compressed images. That means that DCT coefficients corresponding to lower horizontal and vertical spatial frequencies – namely, those with basis functions nearer the top left corner of **Figure 12.16b** – are more important to the visual system than those of higher spatial frequencies towards the bottom right.

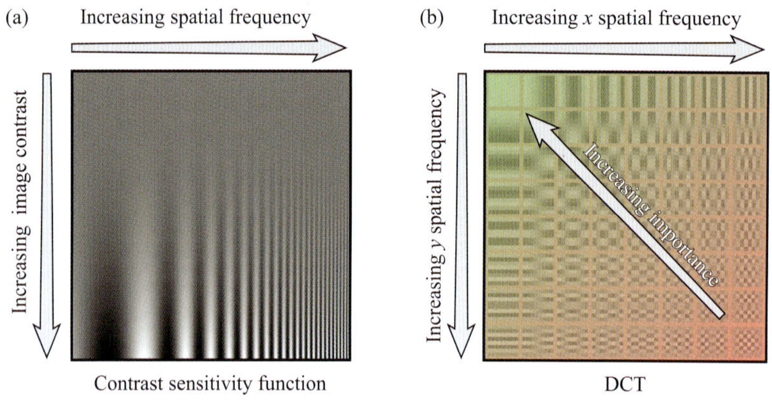

Figure 12.16 Contrast sensitivity function

With this in mind, let us now return to our JPEG example to see exactly how the 2-D DCT is used to compress information in pictures. Preparatory to applying the DCT, we transformed the digital picture into three YCbCr planes, and downsampled the Cb and Cr planes, as discussed in Section 12.3.1. Now, each plane is segmented into non-overlapping 8×8 pixel blocks. This particular size reflects an experimentally determined compromise between the amount of information in the block (which increases as the block size increases) and computational complexity of taking the DCT (which gets more burdensome as size increases).

Figure 12.17a shows a small 4096-pixel section of the original picture in **Figure 12.14** that has been divided into 64 non-overlapping 8×8 pixel blocks, one of which is shown expanded to the right. The top panel of **Figure 12.17b** shows the three 8×8 RGB color planes that comprise this small image, and below them the three 8×8 YCbCr planes that result from processing the RGB data using the matrix transformation of Equation (12.24). The three 8×8 matrices in **Figure 12.17c** give the numerical values of the data for these Y, Cb and Cr planes, ranging from 0 to 255, with the top left entry of each matrix corresponding to the top left corner of the corresponding image plane. In order to visualize the data more clearly, the data from each matrix are also presented as a 3-D false-color plot in the same row of **Figure 12.17c**, below the YCbCr images. While the Y (brightness) data vary considerably over the 64-pixel block, there is almost no variation in the value of the Cb and Cr data over the corresponding block.

We now apply the DCT to our image. The three matrices in **Figure 12.13d** represent the 2-D DCT of the Y, Cb and Cr data, rounded to the nearest integer. All the non-zero numbers have been highlighted in red. Almost all the information in the Y plane is compacted into the upper left triangle of the matrix, which corresponds to the lower spatial frequencies – the ones of most importance to the visual system. In this example, 99% of the energy is represented by only the 36 DCT coefficients within the region of the matrix shaded in yellow. With the Cb and Cr

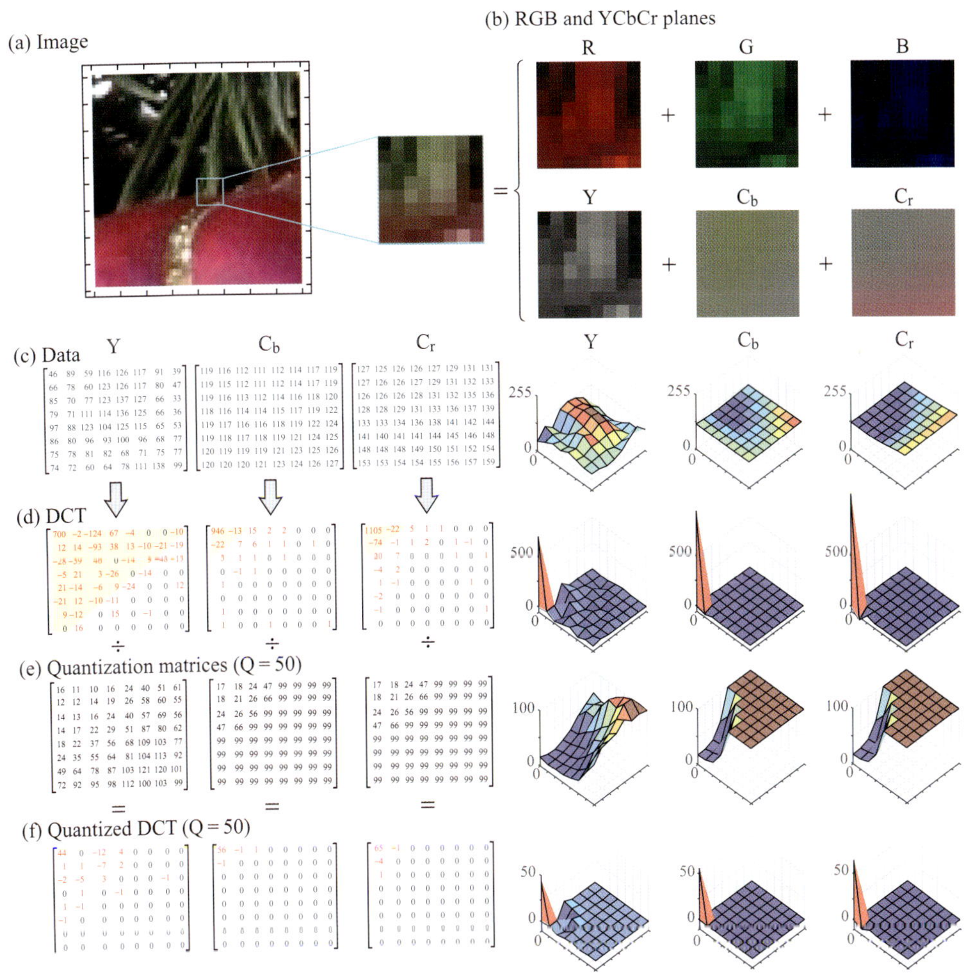

Figure 12.17 DCT compression of luminance and chrominance

planes, the result is even more dramatic; each DCT-transformed matrix comprises only a few non-zero coefficients near the top left of the matrix. 99% of the energy is contained in only three coefficients! The false-color 3-D plots of these matrices in the same row of **Figure 12.17d** emphasize the fact that by far the biggest coefficient in the DCT of all three planes is the constant, "DC," term.

Because the DCT transforms images in a manner similar to the way the human visual system processes visual information, we could achieve substantial compression simply by retaining some pre-determined fixed number of the most important coefficients and tossing the rest. However, most JPEG encoders take a more nuanced approach. Instead of keeping or retaining individual coefficients, each coefficient in the DCT matrix is quantized by an amount that reflects its perceived importance to the visual system, as determined by psychophysical experiments. Coefficients towards the top left corner of each matrix, corresponding to the more important lower spatial frequencies, are more finely quantized. Coefficients towards the bottom right corner, corresponding to higher spatial frequencies, are quantized more coarsely, which reflects their lower importance. To accomplish this differential quantization, every coefficient in the Y, Cb and Cr matrices is first divided by a value contained in a specifically designed **quantization matrix**. **Figure 12.17e** shows standard quantization JPEG matrices:[14] a **luminance quantization matrix** for Y and a **chrominance quantization matrix** that is used by both Cb and Cr. The 3-D visualizations of these matrices are shown on the right. You can see that values at higher spatial frequencies – particularly chrominance values – are more coarsely quantized. The quantization values of both these matrices can be varied algorithmically by adjustment of a single **quality factor parameter** Q, $1 \le Q \le 100$, to increase or decrease the amount of quantization, thereby trading image quality for file size. The quality factor ranges from $Q = 1$ (lowest quality, smallest file size) to $Q = 100$ (no quantization, largest file size).

Figure 12.18 shows a comparison of the small 64 × 64 pixel section of the original image with images reconstructed from the DCT coefficients quantized at quality factors of $Q = 80$, 50 and 10.

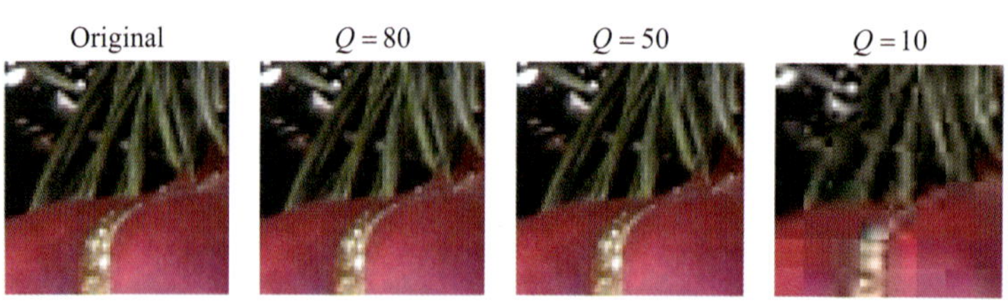

Compression (%)

Q	80	50	10
Y	69.1	80.2	97.0
C_b	95.9	97.8	99.1
C_r	96.7	98.3	99.1

Figure 12.18 Comparison of image quality and file compression

The numbers in the table at the right of the figure reflect the percent reduction in the number of bits in the three channels (Y, Cb and Cr) due to the combination of chrominance subsampling plus

[14] Nearly all digital cameras and software programs use their own proprietary quantization tables, which means that quality factors cannot be compared across devices or programs. However, all quantization tables are stored in the header of each JPEG file, which enables the image created by any device to be reconstructed by multiplying the DCT coefficients in the file by the included tables.

DCT transform and quantization. Even though DCT transformation and quantization are extremely lossy compression techniques (as was chroma subsampling), the effects of the compression are only really obvious at the lowest quality setting at high magnification.

12.3.3 Coefficient encoding

The last step in the formation of a JPEG file is lossless compression and encoding of the quantized DCT coefficients. **Figure 12.19** suggests how this is done. The left side of the figure shows a sample 8×8 pixel block of quantized coefficients – either luminance or chrominance data. Each pixel block comprises a **DC coefficient** in the top left corner (red dot in the figure) that represents the average value of all the pixels in the transformed image block, plus 63 coefficients, termed **AC coefficients**, that represent the other spatial frequencies. The DC and AC coefficients are encoded separately. In most pictures, the DC value is large, and the values of successive blocks are highly correlated. Because the DC value of the nth block, $DC[n]$, does not vary a great deal from the preceding block, only the difference between successive blocks, $\Delta DC[n]$, is encoded using (lossless) **differential pulse-code modulation (DPCM)** techniques,

$$\Delta DC[n] = DC[n] - DC[n-1].$$

The number of bits needed to represent $\Delta DC[n]$ is much smaller than that needed to represent either $DC[n]$ or $DC[n-1]$. Each of the DC difference values is assigned a **category code** (a Huffman code comprising variable numbers of zeros and ones) according to the number of bits needed to encode it. Then the category code is encoded followed by the actual data. This is another example of a lossless variable-bit-length coding algorithm similar to that we saw with the MP3 encoder.

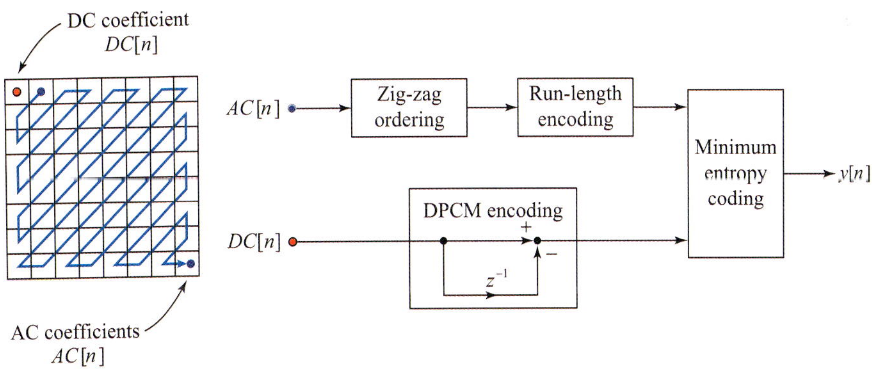

Figure 12.19 Coefficient encoding

The AC data are encoded using a different technique that takes advantage of both the sparseness of the quantized matrices and the fact that the quantized values are much smaller than the unquantized values. First the AC coefficients in each pixel block are ordered into a long sequence in zig-zag fashion, starting in the top left corner, shown by the blue line in **Figure 12.19**. This ordering puts the largest quantized coefficients at the front of the sequence followed by a lot of zeros, interspersed with the occasional coefficient, as you saw in the example of **Figure 12.17f**. Instead of encoding each of the resulting 63 individual values (most of them zero), this long sequence is **run-length encoded** as follows: the sequence is broken into subsequences each comprising 0–15 zeros followed by one non-zero coefficient. Then these data pairs are entropy-encoded using a Huffman-like coding scheme.[15] The number of zeros is encoded in a four-bit value followed by the non-zero coefficient, using essentially the same entropy coding technique described for the DC coefficient using a Huffman category code. The tables of Huffman codes that are used in the entropy encoding stage – two for the DC value and two for the AC values – are stored in the header of the JPEG file so that the file can be properly decoded by the target display device.

12.3.4 Implementation of 2D-DCT

The JPEG divides an image into small, non-overlapping 8×8 blocks. Because the pixel blocks do not overlap, they can be processed in parallel, using GPUs or purpose-built VLSI hardware using a variety of efficient DCT algorithms. In firmware implementations, it is common to use tightly coded, special-purpose DCTs to perform the necessary transforms. Most conventional algorithms for the 2D-DCT are based on the fact that the transform is separable into two one-dimensional DCT-IIs, so any efficient implementation of the DCT-II can be directly applied to the 2D-DCT. For example, Matlab's `dct` function can accept a matrix and performs a 1D-DCT on each column. So, given a matrix of frame data, `f`, in Matlab, we could take the column-wise DCT of a frame, perform the transpose and take the DCT again to get the row-wise DCT, and then perform a final transpose to return to the correct frame orientation:

```
dct(dct(f)')'
```

In fact, this is exactly what Matlab's `dct2` function does.

SUMMARY

In this chapter, we have introduced the discrete cosine transform and looked at a couple of its most important applications. The DCT has proven essential in signal compression applications. While the DCT is not theoretically an optimal transform for the purposes of compression, it turns out that the basis functions – cosines – make it particularly suitable to represent real, periodic signals such as speech and music. The modified DCT (MDCT), which is derived from the DCT-IV, is a lapped transform that is an essential part of most modern encoders for audio compression, including the MP3 and AAC. When overlapping frames of the output are added, the effects of time-domain aliasing are eliminated and edge discontinuities are mitigated. The basic DCT-II is at the heart of image compression techniques such as JPEG. The encoder uses the energy compaction properties of the 2D-DCT to remove perceptually irrelevant details in images.

[15] As we mentioned in Section 12.2.5, in a true Huffman coding scheme the codes would be based on the statistics of pattern frequency measured from the data. However, in most common JPEG applications, these category codes are predetermined and the look-up tables are included in the JPEG header.

PROBLEMS

Problem 12-1

Show that, for $a = 1$ or $a = 2$ and values $0 \leq |k| \leq 2N - 1$,

$$\frac{1}{N}\sum_{n=0}^{N-1}\cos\frac{\pi k(2an+1)}{2N} = \begin{cases} 1, & k = 0 \\ 0, & 1 \leq |k| \leq 2N-1 \end{cases}.$$

➤Hint: Express the sum of cosines as the real part of the sum of complex exponentials,

$$\frac{1}{N}\sum_{n=0}^{N-1}\cos\frac{\pi k(2n+1)}{2N} = \frac{1}{N}\mathrm{Re}\left\{\sum_{n=0}^{N-1} e^{j\pi k(2n+1)/2N}\right\}.$$

Then use the formula for the sum of a geometric series.

Problem 12-2

A different version of Problem 12-1 is necessary for an expanded range of k. Show that

$$\frac{1}{N}\sum_{k=0}^{N-1}\cos\frac{\pi(2k+1)r}{2N} = \frac{1}{N}\frac{\sin \pi r}{\sin \pi r/2N} = \begin{cases} (-1)^q, & r = 2Nq \\ 0, & \text{otherwise} \end{cases},$$

where q is an integer. In other words, the sum is zero unless r is a multiple of $2N$. Use the same hint as in Problem 12-1.

Problem 12-3

Many of the formulas for the DCT and IDCT are based on the orthogonality of cosines. Show that functions of the form

$$\cos\frac{\pi k(2an+1)}{2N},$$

where $a = 1$ or $a = 2$, are orthogonal with respect to k. Specifically, show that

$$\frac{1}{N}\sum_{n=0}^{N-1}\cos\frac{\pi k(2an+1)}{2N}\cos\frac{\pi l(2an+1)}{2N} = \begin{cases} 1, & k = l = 0 \\ 1/2, & k = l \neq 0 \\ 0, & k \neq l \end{cases},$$

for $0 \leq k \leq N - 1$ and $0 \leq l \leq N - 1$, when $a = 1$ or 2.

➤Hint: Use the results of Problem 12-1 and the trigonometric identity

$$\cos\theta_1\cos\theta_2 = \tfrac{1}{2}\cos(\theta_1 + \theta_2) + \tfrac{1}{2}\cos(\theta_1 - \theta_2).$$

Problem 12-4

The MDCT and IMDCT are constructed using basis functions of the form

$$\cos\frac{\pi(2k+1)(2n+1+N)}{4N}, \quad 0 \le k \le N-1, 0 \le n \le 2N-1.$$

Show that these functions are orthogonal with respect to k, namely that

$$\frac{1}{N}\sum_{n=0}^{2N-1}\cos\frac{\pi(2k+1)(2n+1+N)}{4N}\cos\frac{\pi(2l+1)(2n+1+N)}{4N} = \begin{cases} 1, & k=l \\ 0, & k \ne l \end{cases},$$

for $0 \le k \le N-1$ and $0 \le l \le N-1$. Hint: Use the trigonometric identity

$$\cos\theta_1\cos\theta_2 = \tfrac{1}{2}\cos(\theta_1+\theta_2) + \tfrac{1}{2}\cos(\theta_1-\theta_2),$$

plus the hint for Problem 12-1.

Problem 12-5

Show that, for $0 \le |k| \le 2N-1$,

$$\sum_{l=0}^{N-1}\frac{\cos\pi kl}{N} = \begin{cases} \frac{1}{2}(1-(-1)^k), & k \ne 0 \\ N, & k=0 \end{cases}.$$

Use the same hint as in Problem 12-1.

Problem 12-6

Show that functions of the form

$$\cos\frac{\pi(2k+1)(2n+1)}{4N}$$

are orthogonal with respect to n. Specifically, show that for $0 \le n \le N-1$ and $0 \le m \le N-1$,

$$\frac{1}{N}\sum_{k=0}^{N-1}\cos\frac{\pi(2k+1)(2n+1)}{4N}\cos\frac{\pi(2k+1)(2m+1)}{4N} = \begin{cases} 1/2, & n=m \\ 0, & n \ne m \end{cases}.$$

Use the same hint as Problem 12-4

Problem 12-7

Show that over the range $0 \le n \le 2N-1$, $0 \le m \le 2N-1$,

$$\frac{1}{N}\sum_{k=0}^{N-1}\cos\frac{\pi(2k+1)(2n+1+N)}{4N}\cos\frac{\pi(2k+1)(2m+1+N)}{4N}$$

$$=\frac{1}{4N}\frac{\sin\pi(n+m+1+N)}{\sin\pi(n+m+1+N)/2N}+\frac{1}{4N}\frac{\sin\pi(n-m)}{\sin\pi(n-m)/2N}$$

$$=\begin{cases}\frac{1}{2}, & m=n, & 0\le n\le 2N-1\\ \frac{1}{2}, & 3N-1-n, & N\le n\le 2N-1\\ -\frac{1}{2}, & m=N-1-n, & 0\le n\le N-1\\ 0, & \text{otherwise}\end{cases}.$$

➤**Hint:** Use the hints for Problem 12-4.

Problem 12-8

In this problem, we derive expressions for the Type-I DCT and IDCT. Consider a periodic $(2N-2)$-point symmetric extension of the N-point sequence $x[n]$ shown in **Figure 12.2**,

$$w[n]=\begin{cases}x[n], & 0\le n\le N-1\\ x[2N-2-n], & N\le n\le 2N-3\end{cases}.$$

In order to create an orthonormal transform, it turns out to be necessary to scale the points $x[0]$ and $x[N-1]$ by $\sqrt{2}$. Accordingly, define

$$\alpha[n]=\begin{cases}1/\sqrt{2}, & n=0, N-1\\ 1, & \text{otherwise}\end{cases}.$$

Then define a new sequence,

$$y[n]=\frac{w[n]}{\alpha[n]}=\begin{cases}\sqrt{2}x[0], & n=0\\ x[n], & 1\le n\le N-2\\ \sqrt{2}x[N-1], & n=N-1\\ x[2N-2-n], & N\le n\le 2N-3\end{cases}.$$

(a) Break $y[n]$ up into two sequences of length $N-2$, plus single points at $n=0$ and $n=N-1$, and show that $Y[k]$, the DFT of $y[n]$, is given by

$$
\begin{aligned}
Y[k] &= \sqrt{2}x[0] + \sqrt{2}x[N-1](-1)^k + \sum_{n=1}^{N-2} 2x[n]\cos\frac{\pi kn}{N-1} \\
&= \sum_{n=0}^{N-1} 2\alpha[n]x[n]\cos\frac{\pi kn}{N-1}, \quad 0 \le k \le 2N-3.
\end{aligned}
$$

(b) Show that $Y[k]$ is symmetric, namely that $Y[2N-2-k] = Y[k],\ 0 \le k \le N-1$.

(c) Define the Type-I DCT as

$$
C_I[k] \triangleq \frac{1}{\sqrt{2(N-1)}}\alpha[k]Y[k] = \sqrt{\frac{2}{N-1}}\sum_{n=0}^{N-1}\alpha[k]\alpha[n]x[n]\cos\frac{\pi kn}{N-1}, \quad 0 \le k \le N-1.
$$

Show that $Y[k]$ can be expressed in terms of $C_I[k]$,

$$
Y[k] = \frac{1}{\alpha[k]}\begin{cases} C_I[k], & 0 \le k \le N-1 \\ C_I[2N-2-k], & N \le k \le 2N-2 \end{cases}.
$$

(d) Show that the inverse Type-I DCT is

$$
x[n] = \sqrt{\frac{2}{N-1}}\sum_{k=0}^{N-1}\alpha[k]\alpha[n]C_I[k]\cos\frac{\pi kn}{N-1}, \quad 0 \le n \le N-1.
$$

➤**Hint:** Substitute the candidate expression for $x[n]$ into the definition of $C_I[k]$ to get

$$
\begin{aligned}
&\sqrt{\frac{2}{N-1}}\sum_{n=0}^{N-1}\alpha[k]\alpha[n]x[n]\cos\frac{\pi kn}{N-1} \\
&\quad = \frac{1}{N-1}\sum_{l=0}^{N-1}C_I[l]\alpha[k]\alpha[l]\left(2\sum_{n=0}^{N-1}\alpha^2[n]\cos\frac{\pi ln}{N-1}\cos\frac{\pi kn}{N-1}\right),
\end{aligned}
$$

and use the results of Problem 12-5 to show that

$$
\frac{2}{N-1}\sum_{n=0}^{N-1}\alpha^2[n]\cos\frac{\pi ln}{N-1}\cos\frac{\pi kn}{N-1} = \begin{cases} 2, & l=k=0, N-1 \\ 1, & l=k\ne 0, N-1 \\ 0, & l \ne k \end{cases}.
$$

Problem 12-9

Equation (12.2) gives the DFT of the extended sequence for the Type-II DCT $C_{II}[k]$:

$$
Y[k] = e^{j\pi k/2N}\sum_{n=0}^{N-1}2x[n]\cos\frac{\pi k(2n+1)}{2N}, \quad 0 \le k \le 2N-1.
$$

Show that this DFT can be obtained from the N values of $C_{II}[k]$,

$$
Y[k] = \begin{cases} \dfrac{\sqrt{2N}}{\beta[k]}e^{j\pi k/2N}C_{II}[k], & 0 \le k \le N-1 \\ 0, & k = N \\ -\sqrt{2N}e^{j\pi k/2N}C_{II}[2N-k], & N+1 \le k \le 2N-1 \end{cases}.
$$

➤**Hint:** Use Equation (12.5) to obtain the first N values of $Y[k]$ and Equations (12.3) and (12.4) to obtain the rest.

Problem 12-10

In deriving the DCT-II, show that Equation (12.7) can also be written as

$$\begin{aligned} Y[k] &= \frac{\sqrt{2N}}{\beta[k]} e^{j\pi k/2N} C_{II}[k], && 0 \le k \le N-1 \\ Y[N] &= 0 \\ Y[N+1+k] &= -j\sqrt{2N} e^{j\pi(k+1)/2N} C_{II}[N-1-k], && 0 \le k \le N-2. \end{aligned}$$

Problem 12-11

In Equation (12.5), we defined the orthonormal Type-II DCT of sequence $x[n]$ as

$$\begin{aligned} C_{II}[k] &= \sqrt{\frac{2}{N}} \sum_{n=0}^{N-1} x[n]\beta[k] \cos\frac{\pi k(2n+1)}{2N} \\ &= \sqrt{\frac{1}{2N}} \beta[k] e^{-j\pi k/2N} \underbrace{\left(e^{j\pi k/2N} \sum_{n=0}^{N-1} 2x[n] \cos\frac{\pi k(2n+1)}{2N} \right)}_{Y[k]} = \sqrt{\frac{1}{2N}} \beta[k] e^{-j\pi k/2N} Y[k], \end{aligned}$$

where the scale factors $\sqrt{1/2N}$ and $\beta[k]$ make this form of the DCT orthonormal. As indicated in this equation, $C_{II}[k]$ can be obtained directly from $Y[k]$, the DFT of the symmetrical expanded sequence $y[n]$. Another possibility is to define a non-orthonormal Type-II DCT as

$$\begin{aligned} \hat{C}_{II}[k] &= \frac{1}{N} \sum_{n=0}^{N-1} 2x[n] \cos\frac{\pi k(2n+1)}{2N} = \frac{1}{N} e^{-j\pi k/2N} \underbrace{\left(e^{j\pi k/2N} \sum_{n=0}^{N-1} 2x[n] \cos\frac{\pi k(2n+1)}{2N} \right)}_{Y[k]} \\ &= \frac{1}{N} e^{-j\pi k/2N} Y[k]. \end{aligned}$$

Given this definition, show that the IDCT is

$$x[n] = \sum_{k=0}^{N-1} \hat{\beta}[k] \hat{C}_{II}[k] \cos\frac{\pi k(2n+1)}{2N}, \quad 0 \le n \le N-1,$$

where

$$\hat{\beta}[k] = \begin{cases} 1/2, & k = 0 \\ 1, & k \ne 0. \end{cases}$$

➤**Hint:** Use the results of Problem 12-3.

Problem 12-12

In the derivation of the Type-II DCT, $C_{II}[k]$, we first created a sequence $y[n]$ of length $2N$ from the symmetric expansion of the N-point input sequence $x[n]$ shown in **Figure 12.2**,

$$y[n] = \begin{cases} x[n], & 0 \le n \le N-1 \\ x[2N-1-n], & N \le n \le 2N-1 \end{cases}.$$

In this problem, we show that the $C_{II}[k]$ is equivalent to the $4N$-point DFT of a sequence $q[n]$, whose odd values are given by $y[n]$ and whose even values are zero,

$$q[2n] = 0$$
$$q[2n+1] = y[n], \quad 0 \le n \le 2N-1.$$

(a) Express $q[n]$ in terms of $x[n]$.
(b) Find $Q[k]$, the $4N$-point DFT of $q[n]$, and show that

$$C_{II}[k] = \frac{1}{2}\sqrt{\frac{2}{N}}\beta[k]Q[k] = \sqrt{\frac{2}{N}}\sum_{n=0}^{N-1} x[n]\beta[k]\cos\frac{\pi k(2n+1)}{2N}, \quad 0 \le k \le N-1.$$

Problem 12-13

In this problem, we derive the inverse Type-II DCT from the IDFT. Start with the definition of the IDFT of $Y[k]$,

$$y[n] = \frac{1}{2N}\sum_{k=0}^{2N-1} Y[k]e^{j2\pi kn/2N},$$

where $y[n]$ is the $(2N-2)$-point even extension of sequence $x[n]$,

$$y[n] = \begin{cases} x[n], & 0 \le n \le N-1 \\ x[2N-1-n], & N \le n \le 2N-1 \end{cases}.$$

Substitute the equation that expresses $Y[k]$ in terms of $C_{II}[k]$ and show that

$$y[n] = \sqrt{\frac{2}{N}}\sum_{k=0}^{N-1}\beta[k]C_{II}[k]\cos\frac{\pi k(2n+1)}{2N}, \quad 0 \le n \le 2N-1,$$

where

$$\beta[k] = \begin{cases} 1/\sqrt{2}, & k=0 \\ 1, & k \ne 0 \end{cases}.$$

Then $x[n] = y[n]$, $0 \le n \le N-1$.

Problem 12-14

Show that the Type-II DCT of an N-point sequence $x[n]$ can be obtained by forming $\beta[n]x[n]$, where

$$\beta[n] = \begin{cases} 1/\sqrt{2}, & n=0 \\ 0, & n \ne 0 \end{cases},$$

and then taking the $2N$-point DFT of $\beta[n]x[n]$ appended with N zeros, multiplying the result by $\sqrt{2/N}e^{-jk\pi/2N}$ and finally taking the real part.

Problem 12-15

Show that you can calculate $C_{II}[k]$, the Type-II DCT of an even-length N-point sequence $x[n]$, using only an N-point DFT of a real sequence. Starting with the sequence

$$y[n] = \begin{cases} x[n], & 0 \le n \le N-1 \\ x[2N-1-n], & N \le n \le 2N-1 \end{cases},$$

separate $y[n]$ into its even and odd points,

$$\begin{matrix} e[n] = y[2n] \\ o[n] = y[2n+1] \end{matrix}, \quad 0 \le n \le N-1.$$

(a) Show that both $e[n]$ and $o[n]$ contain all the samples of $x[n]$. Specifically, show that the first half of $e[n]$ contains the even points of $x[n]$ and the second half contains the odd points in reverse order,

$$e[n] = \begin{cases} x[2n], & 0 \le n \le (N-1)/2 \\ x[2N-2n-1], & (N+1)/2 \le n \le N-1 \end{cases}.$$

(b) Show that $o[n]$ is just the reverse of $e[n]$,

$$o[n] = e[N-n-1], \;\; 0 \le n \le N-1.$$

(c) Show that the $2N$-point DFT of $Y[k]$ can be expressed as

$$Y[k] = e^{j\pi k/2N} \sum_{n=0}^{N-1} 2e[n] \cos \frac{\pi k(4n+1)}{2N}, \quad 0 \le k \le N-1,$$

and hence

$$C_{II}[k] = \sqrt{\frac{2}{N}} \beta[k] \sum_{n=0}^{N-1} e[n] \cos \frac{\pi k(4n+1)}{2N}, \quad 0 \le k \le N-1,$$

where

$$\beta[k] = \begin{cases} 1/\sqrt{2}, & k=0 \\ 1, & k \ne 0 \end{cases}.$$

➤**Hint:** Split $Y[k]$ into two summations, one for even points $e[n]$, and one for odd points $o[n]$.

(d) From the results of part (c), show that

$$C_{II}[k] = \sqrt{\frac{2}{N}} \beta[k] \,\mathrm{Re}\{e^{-j\pi k/2N} E[k]\},$$

where

$$E[k] = \sum_{n=0}^{N-1} e[n] e^{-j2\pi kn/N}$$

is the N-point DFT of the sequence $e[n]$.

Problem 12-16

In this problem we derive the even-length Type-II IDCT using an N-point IDFT.

(a) Given the even-length sequence $e[n]$ and its N-point DCT, as specified in Problem 12-15c,

$$C_{II}[k] = \sqrt{\frac{2}{N}}\beta[k]\sum_{n=0}^{N-1} e[n]\cos\frac{\pi k(4n+1)}{2N}, \quad 0 \le k \le N-1,$$

use the orthogonality property of cosines (Problem 12-3) to verify that the inverse DCT is

$$e[n] = \sqrt{\frac{2}{N}}\sum_{k=0}^{N-1}\beta[k]C_{II}[k]\cos\frac{\pi k(4n+1)}{2N}.$$

(b) Show that the result of part (a) can be rewritten as

$$\begin{aligned} e[n] &= \sqrt{2N}\,\mathrm{Re}\left\{\frac{1}{N}\sum_{k=0}^{N-1}(\beta[k]C_{II}[k]e^{j\pi k/2N})e^{j2\pi kn/N}\right\} \\ &= \sqrt{2N}\mathrm{Re}\{\mathbb{F}^{-1}\{\beta[k]C_{II}[k]e^{j\pi k/2N}\}\}. \end{aligned}$$

Hence, $e[n]$ can be recovered from the N-point IDFT of $C_{II}[k]$.

(c) Given $x[n]$, find $e[n]$.

Problem 12-17

Prove Parseval's theorem for DCT-II by brute force using the orthogonality property of cosines, Problem 12-3, namely that

$$\sum_{n=0}^{N-1} x^2[n] = C_{II}^2[k].$$

Problem 12-18

In this problem, we derive expressions for Type-III DCT and IDCT. Consider a periodic $4N$-point symmetric extension of an N-point sequence $x[n]$, as shown in **Figure 12.2**,

$$w[n] = \begin{cases} x[n], & 0 \le n \le N-1 \\ 0, & n = N \\ -x[2N-n], & N+1 \le n \le 2N \\ -x[n-2N], & 2N+1 \le n \le 3N-1 \\ 0, & n = 3N \\ x[4N-n], & 3N+1 \le n \le 4N-1 \end{cases}.$$

In order to create an orthonormal transform, scale the points $x[0]$ and $x[N-1]$ by $\sqrt{2}$. Accordingly, define

$$y[n] = \frac{w[n]}{\gamma[n]},$$

where,

$$\gamma[n] = \begin{cases} 1/\sqrt{2}, & n = 0, 2N \\ 1, & \text{otherwise} \end{cases}.$$

(a) Show that $Y[k]$, the DFT of $y[n]$, can be written as

$$\begin{aligned} Y[k] &= (1-(-1)^k)\left(x[0] + \sum_{n=1}^{N-1} 2x[n]\cos\frac{\pi kn}{2N}\right), \quad 0 \le k \le 4N-1 \\ &= \begin{cases} 0, & k \text{ even} \\ 2\left(x[0] + \sum_{n=1}^{N-1} 2x[n]\cos\frac{\pi kn}{2N}\right), & k \text{ odd} \end{cases}, \end{aligned}$$

hence,

$$\begin{aligned} Y[2k] &= 0 \\ Y[2k+1] &= 2\left(x[0] + \sum_{n=1}^{N-1} 2x[n]\cos\frac{\pi(2k+1)n}{2N}\right), 0 \le k \le 2N-1. \end{aligned}$$

➤**Hint:** Break $y[n]$ up into four sequences of length $N-1$, plus four single points, two of which have the value zero (for $n = N$ and $n = 3N$) and two of which have the value $\pm x[0]$ (for $n = 0$ and $n = 2N$).

(b) Show that $Y[k]$ is symmetric, namely that $Y[4N-k] = Y[k]$. Hence, the values of $Y[2k+1]$, $N \le k \le 2N-1$, are equal to $Y[2N-(2k+1)]$, $0 \le k \le N-1$.

(c) Define $C_{III}[k]$, the Type-III DCT, to be

$$C_{III}[k] \triangleq \sqrt{\frac{2}{N}} \sum_{n=0}^{N-1} \beta[n]x[n]\cos\frac{\pi(2k+1)n}{2N}, \quad 0 \le k \le N-1.$$

Use the results of parts (a) and (b) to show that $Y[k]$ can be expressed in terms of $C_{III}[k]$,

$$\begin{aligned} Y[2k] &= 0, \qquad 0 \le k \le 2N-1 \\ Y[2k+1] &= 4\sum_{n=1}^{N-1} \beta[n]x[n]\cos\frac{\pi(2k+1)n}{2N} \\ &= \begin{cases} 2\sqrt{2N}\,C_{III}[k], & 0 \le k \le N-1 \\ 2\sqrt{2N}\,C_{III}[2N-1-k], & N \le k \le 2N-1 \end{cases}. \end{aligned}$$

(d) By substituting the expression for $C_{III}[k]$, verify that the Type-III IDCT is

$$x[n] = \sqrt{\frac{2}{N}} \sum_{k=0}^{N-1} \beta[n] C_{III}[k] \cos\frac{\pi(2k+1)n}{2N}, \quad 0 \le n \le N \quad 1.$$

➤**Hint:** Use the results of Problem 12-3.

Problem 12-19

In this problem, we derive expressions for the Type-IV DCT and IDCT. Consider a periodic $4N$-point symmetric extension of an N-point sequence $x[n]$, as shown in **Figure 12.2**,

$$y[n] = \begin{cases} x[n], & 0 \le n \le N-1 \\ -x[2N-1-n], & N \le n \le 2N-1 \\ -x[n-2N], & 2N \le n \le 3N-1 \\ x[4N-1-n], & 3N \le n \le 4N-1 \end{cases}.$$

(a) Show that the DFT of $y[n]$ is

$$Y[k] = e^{j\pi k/4N}(1-(-1)^k)\sum_{n=0}^{N-1} 2x[n]\cos\frac{\pi k(2n+1)}{4N},$$

which can be written as

$$\begin{aligned} &Y[2k] = 0 \\ &Y[2k+1] = e^{j\pi(2k+1)/4N}\sum_{n=0}^{N-1} 4x[n]\cos\frac{\pi(2k+1)(2n+1)}{4N}, \qquad 0 \le k \le 2N-1. \end{aligned}$$

(b) Show that $Y[k]$ is conjugate-symmetric, namely that $Y[4N-k] = Y^*[k]$.

(c) Define the Type-IV DCT to be

$$C_{IV}[k] \triangleq \sqrt{\frac{2}{N}}\sum_{n=0}^{N-1} x[n]\cos\frac{\pi(2k+1)(2n+1)}{4N}, \quad 0 \le k \le N-1.$$

Show that

$$Y[2k] = 0, \quad 0 \le k \le 2N-1$$

$$Y[2k+1] = \begin{cases} 2\sqrt{2N}e^{j\pi(2k+1)/4N}C_{IV}[k], & 0 \le k \le N-1 \\ -2\sqrt{2N}e^{j\pi(2k+1)/4N}C_{IV}[N-k], & N \le k \le 2N-1 \end{cases}.$$

(d) By substituting the expression for $C_{IV}[k]$, verify that the Type-IV IDCT is

$$x[n] = \sqrt{\frac{2}{N}}\sum_{k=0}^{N-1} C_{IV}[k]\cos\frac{\pi(2k+1)(2n+1)}{4N}, \quad 0 \le n \le N-1.$$

➤**Hint:** Use the results of Problem 12-6.

Problem 12-20

In Section 12.1.7, we showed that, given an N-point sequence, $x[n]$, with DCT coefficients C_{II}, an N-point sequence, $\hat{x}[n]$, derived from a smaller set of orthogonal cosines, $\mathbf{M}^T[n,k]$, $k \in \mathbb{M}$, was optimum when the DCT of $\hat{x}[n]$ is $\hat{C}_{II} = C_{II}[k]$. Show that the square-error between $x[n]$ and $\hat{x}[n]$ is the sum of the energy in all the DCT coefficients not included in $\hat{C}_{II}$, as given by Equation (12.20),

$$\sum_{n=0}^{N-1}(x[n]-\hat{x}[n])^2=\sum_{k\notin\mathbb{M}}C_{II}^2[k].$$

(a) Show that

$$\sum_{n=0}^{N-1}(x[n]-\hat{x}[n])^2=\mathbf{x}^T\mathbf{x}-2\mathbf{x}^T\hat{\mathbf{x}}+\hat{\mathbf{x}}^T\hat{\mathbf{x}}.$$

(b) Show that the first term in part (a) is

$$\mathbf{x}^T\mathbf{x}=\sum_{k=0}^{N-1}C_{II}^2[k].$$

(c) Show that the middle term in part (a) is

$$2\mathbf{x}^T\hat{\mathbf{x}}=2\sum_{k\in\mathbb{M}}C_{II}^2[k].$$

➤**Hint:** A scalar is equal to its transpose.

(d) Show that the third term in part (a) is

$$\hat{\mathbf{x}}^T\hat{\mathbf{x}}=\sum_{k\in\mathbb{M}}C_{II}^2[k].$$

Adding the three terms together, we get the desired result.

Problem 12-21

Given $0\le n\le 2N-1$ and $0\le m\le 2N-1$, use the result of Problem 12-2 and the trigonometric identity

$$\cos\theta_1\cos\theta_2=\tfrac{1}{2}\cos(\theta_1+\theta_2)+\tfrac{1}{2}\cos(\theta_1-\theta_2),$$

to show that

$$\frac{1}{N}\sum_{k=0}^{N-1}\cos\frac{\pi(2k+1)(2n+1+N)}{4N}\cos\frac{\pi(2k+1)(2m+1+N)}{4N}$$
$$=\begin{cases}1/2, & n-m=0\\ -1/2, & n+m=N-1\\ 1/2, & n+m=3N-1\\ 0, & \text{otherwise}\end{cases}.$$

13 Multirate systems

Introduction

Multirate systems are systems in which multiple sample rates are present simultaneously. In Chapter 6, we presented some of the basic principles of sample-rate conversion – downsampling and upsampling of signals – and showed how these two operations could be used in combination to resample a signal by rational factors. Downsampling, upsampling and resampling are all examples of the basic operations of multirate systems. In this chapter, we will expand and build upon these important topics. This is an area of ongoing research and activity, and much of the material is quite advanced, at least compared with most of the other chapters in this book. However, multirate methods have become so important in practice that it is essential for us to cover at least the basics.

Multirate signal processing algorithms are extremely computationally efficient compared to their non-multirate counterparts. For this reason, they have proven essential to the cost-efficient implementation of a host of practical applications, particularly in areas such as communication systems, filtering and the compression, coding and transmission of data such as speech, music, images and video.

This chapter provides an introduction to a broad range of the topics in multirate and multistage signal processing. In Sections 13.1 and 13.2, we first review basic sample-rate conversion – upsampling, downsampling and resampling – and then show how the efficiency of sample-rate conversion can be dramatically increased by the use of **polyphase** techniques. In these sections we first develop an intuitive understanding of this material from the point of view of the time domain. Then, in Section 13.4, we analyze polyphase systems from the perspective of the z-transform, from which we develop the key multirate identities that we will need for the rest of the chapter. Section 13.5 shows how multirate techniques with more than one stage can improve downsampling and upsampling algorithms. Section 13.6 presents applications of multistage and multirate techniques to reduce the complexity of high-performance filter design. Section 13.7 introduces half-band filters and shows how they can be used to implement particularly simple downsampling and upsampling systems. Supplementary material available at www.cambridge.org/holton covers a number of more advanced multirate topics. One section describes how to implement polyphase sample-rate conversion algorithms in Matlab.

Supplementary material provides a brief introduction to the important topic of multirate filter banks, including quadrature mirror filters (QMF) and DFT-modulated filter banks, which are at the heart of digital signal processing applications in areas such as communication, speech and music processing.

13.1 Polyphase downsampling

13.1.1 Review of downsampling

Figure 13.1 shows a block diagram of downsampling by a factor of D, as we presented it in Chapter 6.

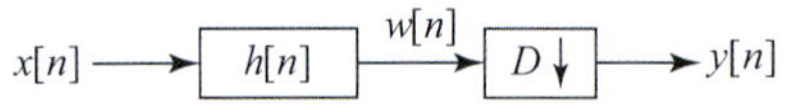

Figure 13.1 Downsampling block diagram

In order to prevent aliasing in downsampling, the input sequence $x[n]$ is first filtered by a lowpass filter with a cutoff frequency of π/D and impulse response $h[n]$ to form a filtered intermediate sequence

$$w[n] = x[n] * h[n] = \sum_{k=-\infty}^{\infty} x[k]h[n-k]. \tag{13.1}$$

This sequence is then decimated by a factor of D to form the output sequence

$$y[n] = w[Dn], \tag{13.2}$$

so,

$$y[n] = \sum_{k=-\infty}^{\infty} x[k]h[nD-k].$$

In what follows we will use the term **decimation** to refer to the $D\downarrow$ operation of Equation (13.2). **Downsampling** is the entire process diagrammed in **Figure 13.1**, comprising a discrete-time anti-aliasing lowpass filter followed by a decimation operation.

Figure 13.2 shows the steps of downsampling $x[n]$ by a factor of $D=3$ to form $w[n]$ and then $y[n]$. In this example, we have chosen $h[n]$ to be a symmetric non-causal FIR filter of odd length $N=15$. **Figure 13.2a** shows $x[k]$ and $h[k]$. **Figure 13.2b** shows $h[n-k]$ for D values of n; namely $n=0$, 1 and 2. The values of convolution $w[n]=x[n]*h[n]$ at these three values of n are shown in pale red, indicated by the arrows in **Figure 13.2c**. Then, every Dth value of $w[n]$ (shown in the grey lozenges) is retained to form $y[n]$, as shown in **Figure 13.2d**. Looking at this figure, it probably will occur to you that we are doing a lot of unnecessary calculation. We are only saving every Dth value of $w[n]$ to create $y[n]$, so why should we bother calculating the other $D-1$ points if we are just going to throw them out? Calculating a single point in the convolution of $w[n]$ requires N multiply-and-accumulate (MAC) operations. So, if we were to calculate every value of $w[n]$, we would need ND operations per output point, whereas if we calculate only every Dth value, we would only need N operations, which is equivalent to reducing the number of operations required to downsample by a factor of D. Much better, right?

Better, yes, but there is still a problem. Calculating every Dth value reduces the total *number* of MAC operations required per output point to N, but it does not change the *peak rate* at which the computation of each output point must be done. For each output point we are going to keep, we still need to perform N operations, and we need to do all these operations *immediately* as the point is required, in $1/f_s$ seconds, where f_s is the *input* sampling rate. That is true whether we compute every value of $w[n]$ and then throw out $D-1$ values or whether we are "clever" and only compute every Dth value. The peak number of operations per second remains the same, namely Nf_s operations per second. Because the speed and memory requirements for our processor have to be based on this peak rate, there would unfortunately appear to be little advantage to skipping points in the computation of $w[n]$. What we would really like to do is to distribute the N operations for each output point over all D sample points, thereby reducing the peak rate by a factor of D. In the next section, we are going to do exactly that by implementing a **polyphase downsampling** algorithm.

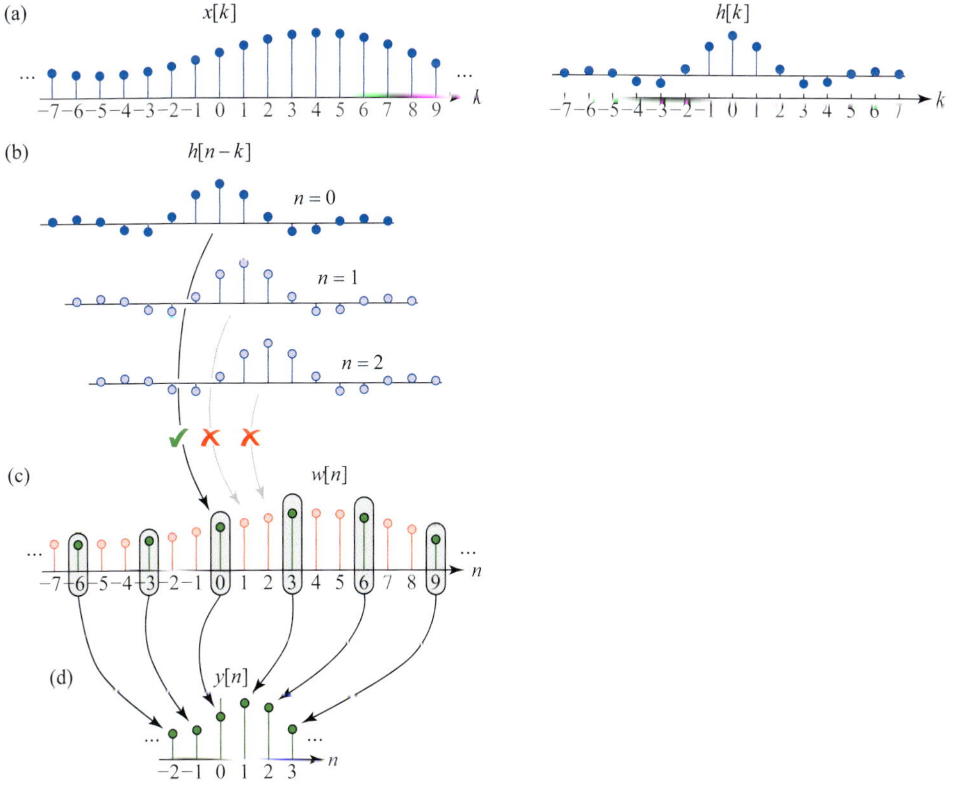

Figure 13.2 Downsampling example

13.1.2 Polyphase implementation of downsampling

The basic idea behind the polyphase implementation of downsampling is to split the convolution of Equation (13.1) into a number of smaller convolutions that can be done at a lower peak rate. **Figure 13.3** shows how this is accomplished for the case of $D=3$. The top panel of **Figure 13.3a** shows the input $x[n]$. In the remaining panels of **Figure 13.3a**, $x[n]$ has been decomposed into the sum of D interleaved subsequences $\hat{x}_d[n], 0 \leq d < D$, such that

$$x[n]=\sum_{d=0}^{D-1}\hat{x}_d[n]. \tag{13.3}$$

In this example, $D=3$, so $x[n]$ can be expressed as the sum of three interleaved subsequences,

$$x[n]=\hat{x}_0[n]+\hat{x}_1[n]+\hat{x}_2[n].$$

Each subsequence $x_d[n]$ comprises every Dth value of $x[n]$ offset by a shift of d, with d ranging from $0 \leq d \leq D-1$. The remaining values of each subsequence are zero. Formally, we can write

$$\hat{x}_d[n]=\sum_{k=-\infty}^{\infty}x[n]\delta[n-(kD-d)]=\begin{cases}x[n], & n=-d,\ -d\pm D,\ \ldots\\ 0, & \text{otherwise}\end{cases}. \tag{13.4}$$

The delta function inside the summation is only non-zero for values of $n=kD-d$; that is, for $n=-d,\ -d \pm D, \ldots$, and is zero otherwise.

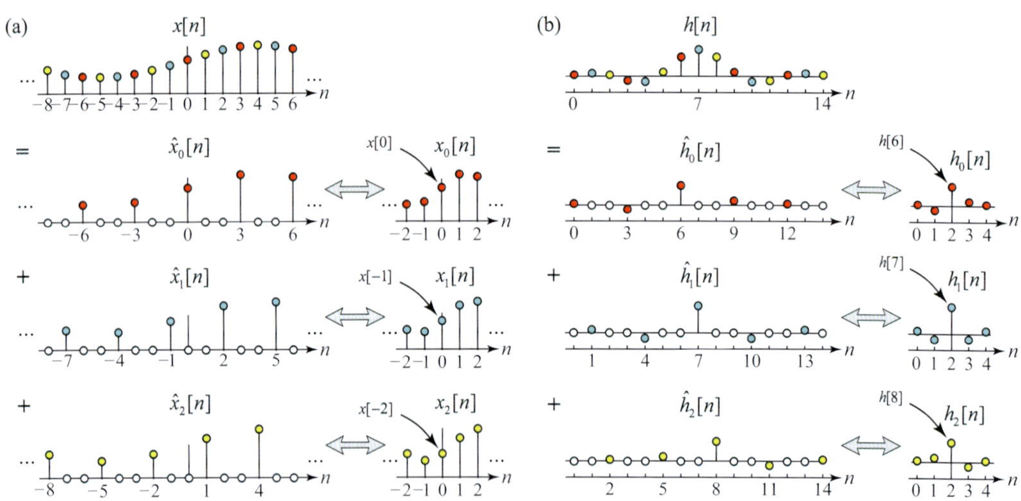

Figure 13.3 Polyphase representation of $x[n]$ and $h[n]$

Each subsequence $\hat{x}_d[n]$ can now be compacted into a "compressed" sequence $x_d[n]$ by removing the $D-1$ zeros between each pair of points, as shown in the right panels of **Figure 13.3a**. The values of $x_d[n]$ can be derived directly from $x[n]$,

$$x_d[n]=x[nD-d], \quad 0 \leq d < D. \tag{13.5}$$

Equation (13.5) says that $x_d[n]$ is obtained by delaying $x[n]$ by d samples and then decimating the resulting sequence by a factor of D. The formal relation between the "expanded" sequence

$\hat{x}_d[n]$ and the "compressed" sequence $x_d[n]$ is obtained by combining Equations (13.4) and (13.5):

$$\begin{aligned}\hat{x}_d[n] &= \sum_{k=-\infty}^{\infty} x[n]\delta[n-(kD-d)] = \sum_{k=-\infty}^{\infty} \underbrace{x[kD-d]}_{x_d[k]}\delta[n-(kD-d)] \\ &= \sum_{k=-\infty}^{\infty} x_d[k]\delta[n-(kD-d)], \quad 0 \le d \le D-1. \end{aligned} \tag{13.6}$$

The delta function inside the summation of Equation (13.6) will again only be non-zero for values of k and d such that $n = kD - d$. When this occurs, then the summation yields value $\hat{x}_d[n] = x_d[k] = x[kD - d]$. Otherwise, $x_d[n] = 0$. For example, to compute $\hat{x}_2[4]$ in **Figure 13.3a**,

$$\hat{x}_2[4] = \sum_{k=-\infty}^{\infty} x_2[k]\delta[4-(k\cdot 3-2)] = \begin{Bmatrix} x_2[2], & k=2 \\ 0, & k \ne 2 \end{Bmatrix} = x_2[2] = x[2\cdot 3-2] = x[4].$$

The shifted and decimated sequences $x_d[n]$, $0 \le d \le D-1$, are termed the **polyphase components** of $x[n]$. The term "polyphase" makes sense if you think about things from the perspective of the DTFT. The transform of each of the shifted, decimated components is

$$\mathfrak{F}\{x_d[n]\} = \mathfrak{F}\{x[nD-d]\} = \mathfrak{F}\{x[nD]\}e^{-j\omega d},$$

where $\mathfrak{F}\{x[nD]\}$ is the DTFT of the decimated sequence $x[nD]$. In words, the transform of each of the polyphase components has an added phase shift of $-\omega d$.

The top panel of **Figure 13.3b** shows an impulse response $h[n]$. We have made the impulse response causal, as it would be in an application. The lower panels of this figure show that $h[n]$ can also be decomposed into the sum of D expanded subsequences $\hat{h}_d[n]$, each comprising every Dth value of $h[n]$ offset by a shift of d, with d ranging from $0 \le d \le D-1$,

$$h[n] = \sum_{d=0}^{D-1} \hat{h}_d[n]. \tag{13.7}$$

Again, we can compact each of the $\hat{h}_d[n]$ into a "compressed" sequence $h_d[n]$ by removing the $D - 1$ zeros between each pair of points, as shown in the right panels of **Figure 13.3b**. These points can be derived directly from $h[n]$ by shifting $h[n]$ by d samples to the *left* and then decimating the resulting sequence by a factor of D,

$$h_d[n] \triangleq h[nD+d], \quad 0 \le d \le D-1. \tag{13.8}$$

Then, $\hat{h}_d[n]$ can be expressed in terms of $h_d[n]$ in a manner equivalent to Equation (13.6),

$$\begin{aligned}\hat{h}_d[n] &= \sum_{k=-\infty}^{\infty} h[n]\delta[n-(kD+d)] = \sum_{k=-\infty}^{\infty} \underbrace{h[kD+d]}_{h_d[k]}\delta[n-(kD+d)] \\ &= \sum_{k=-\infty}^{\infty} h_d[k]\delta[n-(kD+d)], \quad 0 \le d \le D-1. \end{aligned} \tag{13.9}$$

The $h_d[n]$ are called the **polyphase subfilters** of $h[n]$. In order for the polyphase algorithms we are about to describe to work, each subfilter has to be of the same length $L \triangleq \lceil N/D \rceil$. In the example of **Figure 13.3**, $N = 15$ and $D = 3$, so $L = 15/3 = 5$ and

$$\begin{aligned} h_0[n] &= h[0]\delta[n] + h[3]\delta[n-1] + h[6]\delta[n-2] + h[9]\delta[n-3] + h[12]\delta[n-4] \\ h_1[n] &= h[1]\delta[n] + h[4]\delta[n-1] + h[7]\delta[n-2] + h[10]\delta[n-3] + h[13]\delta[n-4]. \\ h_2[n] &= h[2]\delta[n] + h[5]\delta[n-1] + h[8]\delta[n-2] + h[11]\delta[n-3] + h[14]\delta[n-4] \end{aligned}$$

In this particular example, N is an exact integer multiple of D, so the length of each filter is just $L = N/D$. This is something we will assume in the derivations that follow. However, if N were not an integer multiple of D, then the length of $h[n]$ would have to be increased, either by padding the filter's impulse response with zeros or by creating a new filter of the larger length, so that all D subfilters would be of exactly the same length L, making N an exact integer multiple of D; that is, $N = LD$.

With this background, we are now ready to demonstrate the advantage of the polyphase implementation. Putting Equations (13.3) and (13.6) together yields an expression for $x[n]$ directly in terms of the polyphase components $x_d[n]$:

$$x[n] = \sum_{d=0}^{D-1} \hat{x}_d[n] = \sum_{d=0}^{D-1} \left(\sum_{k=-\infty}^{\infty} x_d[k]\delta[n + d - kD] \right).$$

From Equation (13.1), the filtered sequence $w[n]$ is

$$\begin{aligned} w[n] = x[n]*h[n] &= \sum_{d=0}^{D-1} \left(\sum_{k=-\infty}^{\infty} x_d[k]\delta[n - (kD - d)] \right) * h[n] \\ &= \sum_{d=0}^{D-1} \left(\sum_{k=-\infty}^{\infty} x_d[k]h[n - (kD - d)] \right). \end{aligned}$$

The final output sequence $y[n]$ is produced by decimating $w[n]$ by a factor of D, as specified in Equation (13.2),

$$y[n] = w[nD] = \sum_{d=0}^{D-1} \left(\sum_{k=-\infty}^{\infty} x_d[k]h[nD - (kD - d)] \right) = \sum_{d=0}^{D-1} \left(\sum_{k=-\infty}^{\infty} x_d[k]h[(n-k)D + d] \right).$$

From Equation (13.8) we recognize that $h[(n-k)D + d] = h_d[n-k]$, so the inner summation simplifies to the convolution of $x_d[n]$ and $h_d[n]$,

$$y[n] = w[nD] = \sum_{d=0}^{D-1} \left(\sum_{k=-\infty}^{\infty} x_d[k]h_d[n-k] \right) = \sum_{d=0}^{D-1} \underbrace{x_d[n]*h_d[n]}_{y_d[n]} = \sum_{d=0}^{D-1} y_d[n], \tag{13.10}$$

where we have defined $y_d[n]$ as the convolution of the dth polyphase component of $x[n]$ with the dth polyphase component of $h[n]$,

$$y_d[n] \triangleq x_d[n] * h_d[n], \quad 0 \leq d < D. \tag{13.11}$$

Finally, because each polyphase subfilter $h_d[n]$ is of finite duration $L = N/D$, we can rewrite $y_d[n]$ as a finite-length convolution,

$$y_d[n] = x_d[n]*h_d[n] = h_d[n]*x_d[n] = \sum_{k=0}^{L-1} h_d[k]x_d[n-k]. \tag{13.12}$$

Equation (13.10) says that to compute a single output point, we need to compute D output components $y_d[n]$, $0 \le d \le D$, where, in the example of **Figure 13.3**, $D = 3$. The essential advantage of the polyphase method is that these D components need not be computed *simultaneously*, but can instead be computed *in succession* as each new input point arrives. This point is illustrated in **Figure 13.4**.

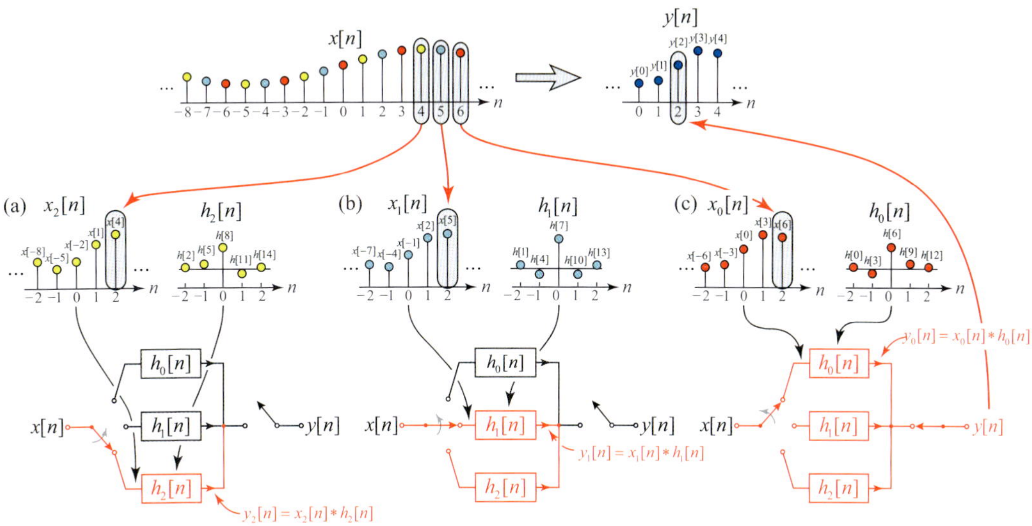

Figure 13.4 Example of polyphase downsampling for $D = 3$

This figure shows how polyphase downsampling would work in a "real-time" application with a downsampling factor of $D = 3$. The top left panel of the figure shows the input sequence $x[n]$. The figure shows the steps required to produce a single output point $y[2]$ of the downsampled sequence $y[n]$, which is shown in the top right panel. In this example, the impulse response $h[n]$ has a length of $N = 15$ and has been decomposed into $D = 3$ polyphase subfilters $h_d[n]$, $0 \le d \le 2$, each of length $L = N/D = 15/3 = 5$, according to Equation (13.8). These three subfilters, $h_2[n]$, $h_1[n]$ and $h_0[n]$, are shown in the top right panels of **Figures 13.4a**, **b** and **c** respectively (note the order of the filters). To compute the single output point $y[2]$, Equation (13.10) states that we need to compute and add together three output components,

$$y[2] = \sum_{d=0}^{D-1} y_d[2] = y_2[2] + y_1[2] + y_0[2].$$

In an efficient real-time application, each of these components would be computed in succession, one at a time, as a new input point arrives. This is schematized by the **commutator architecture** of the downsampler shown in the block diagrams in the bottom panels of the figure.[1] The order in which these components are computed matters; they must be computed in reverse order: first $y_2[2]$, then $y_1[2]$ and finally $y_0[2]$. In the example of **Figure 13.4a**, a new input point, $x[4]$, arrives and is used to form $x_2[2]$, since, by Equation (13.5), $x[4] = x[2 \cdot D - 2] = x_2[2]$. The figure shows the input

[1] A **commutator** is a switching device found in DC motors that sequentially sends current to a series of wire coils wound around a rotating armature, thereby creating an electromotive force that moves the rotor.

of the commutator connected to the filter $h_2[n]$, indicating that $x_2[n]$ is to be convolved with $h_2[n]$ to produce $y_2[2]$. From Equation (13.12),

$$y_2[2] = (h_2[n] * x_2[n])|_{n=2} = \sum_{k=0}^{L-1} h_2[k]x_2[2-k].$$

Because neither of the other two input subsequences, $x_1[n]$ or $x_0[n]$, contains input point $x[4]$, neither of the other two output components, $y_1[n]$ or $y_0[n]$, needs to be updated. Hence, they are shown "disconnected" from the input of the commutator in the block diagram. When the next input point, $x[5]$, arrives, it is used to update $x_1[n]$, since $x[5] = x[2 \cdot D - 1] = x_1[2]$. Then $x_1[n]$ is convolved with $h_1[n]$ to form $y_1[n]$, as indicated by the position of the commutator in **Figure 13.4b**. When $x[6]$ arrives, $x_0[n]$ is updated and $y_0[2]$ is computed by convolution with $h_0[n]$, as shown in **Figure 13.4c**. After all three components have been computed, the outputs of the subfilters are summed to form $y[2]$,

$$y[2] = \sum_{d=0}^{D-1} y_d[2] = y_0[2] + y_1[2] + y_2[2].$$

Computation of each of the D separate $y_d[n]$ components requires $L = N/D = 5$ MAC operations; therefore, computation of each output point $y[n]$ using the polyphase implementation still requires $DL = N$ operations, which is exactly the same as the non-polyphase implementation. So, what is the advantage of the polyphase approach? The answer is the *rate* at which operations need to be done. In the polyphase implementation, we have distributed those N operations over D input points. Computation of each of the $y_d[n]$ components occurs at the input data rate f_s, so the required rate of computation is Nf_s/D MAC operations per second. This is lower by a factor of D than the Nf_s rate of the non-polyphase approach.

13.1.3 Downsampling summary

Figure 13.5 shows signal-flow diagrams of non-polyphase (**Figure 13.5a**) and polyphase (**Figure 13.5b**) implementations of downsampling. The block diagram of the polyphase implementation comprises a ladder of D horizontal branches. Each branch is delayed by one sample (z^{-1} in z-transform notation), so the input to the dth branch is delayed by d samples, $x[n-d]$. Each branch is then decimated by D to form $x_d[n] = x[nD - d]$, as in Equation (13.5), and then convolved with polyphase filter $h_d[n]$ to form polyphase output component $y_d[n]$, as in Equation (13.11). Finally, the output components are summed to give $y[n]$, as in Equation (13.10).

It is useful to recognize the correspondence between the block diagram of **Figure 13.5** and the commutator architecture of **Figure 13.4**. In the block diagram, the delay and decimation operations at the input appear as separate elements, while the commutator essentially performs these two operations together by sequentially directing each new input point to a different filter, which is how you would design a program to implement polyphase downsampling in a real-time application. However, the two representations are algorithmically equivalent.

The essential difference between the polyphase and non-polyphase downsampling methods is the order of filtering and decimation. In the non-polyphase implementation (**Figure 13.5a**), filtering of the input data $x[n]$ by the anti-aliasing lowpass filter $h[n]$ precedes decimation and therefore occurs at the *input rate* f_s. Since $h[n]$ is of length N, the filter must process data at the rate Nf_s before being decimated.

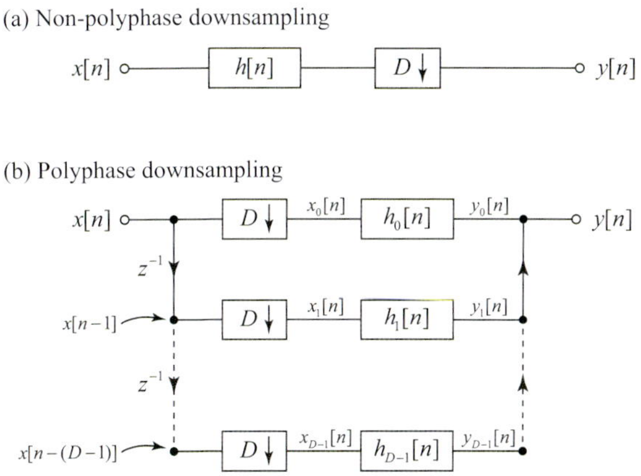

Figure 13.5 Block diagram of two implementations of downsampling

In the polyphase implementation (**Figure 13.5b**), the processes of decimation and filtering are **commuted** or swapped. Input data is first delayed and then decimated by a factor of D to form the polyphase components $x_d[n]$. These components are sequentially filtered by the subfilters $h_d[n]$, each of which is of length N/D. The result is that each filter processes data at a lower *output rate* Nf_s/D.

Example 13.1

Given the input $x[n]$ and impulse response $h[n]$ shown in **Figure 13.6**, compute the first four points of downsampled output $y[n]$ for $D=3$, using both the non-polyphase and polyphase methods.

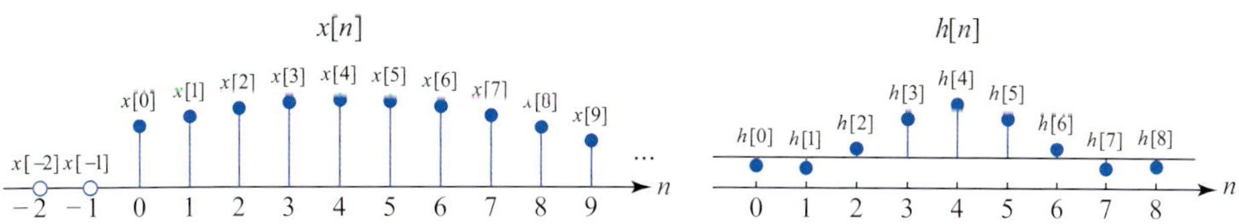

Figure 13.6 Input and impulse responses for downsampling

➤ Solution:

Non-polyphase downsampling

Figure 13.7 shows $x[k]$ and $h[n-k]$ for values of $n = 0$, 3, 6 and 9. The values of the filtered output, $w[n]=x[n] * h[n]$, are given in the green boxes of the figure for those values of n. Then the filtered output is decimated to yield $y[n]=w[Dn]=w[3n]$ for values of $n=0$, 1, 2 and 3, as follows:

$$y[0]=x[0]h[0]$$
$$y[1]=x[0]h[3]+x[1]h[2]+x[2]h[0]$$
$$y[2]=x[0]h[6]+x[1]h[5]+x[2]h[4]+x[3]h[3]+x[4]h[2]+x[5]h[1]+x[6]h[0]$$
$$y[3]=x[1]h[8]+x[2]h[7]+x[3]h[6]+x[4]h[5]+x[5]h[4]+x[6]h[3]+x[7]h[2]+x[8]h[1]+x[9]h[0].$$

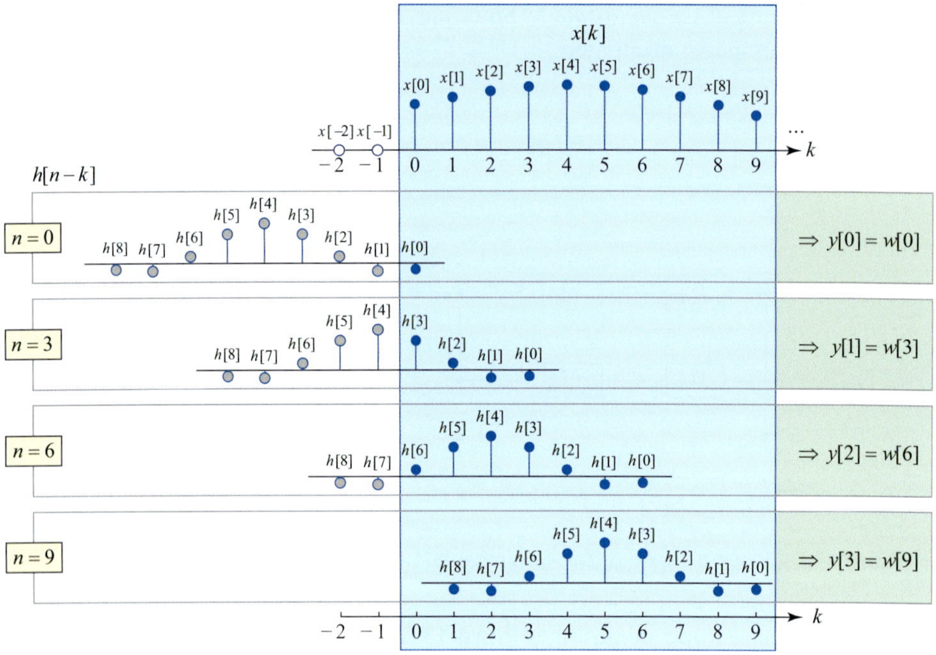

Figure 13.7 Non-polyphase downsampling by direct convolution

Polyphase downsampling

Figure 13.8 shows the polyphase decomposition of $x[n]$ and $h[n]$.

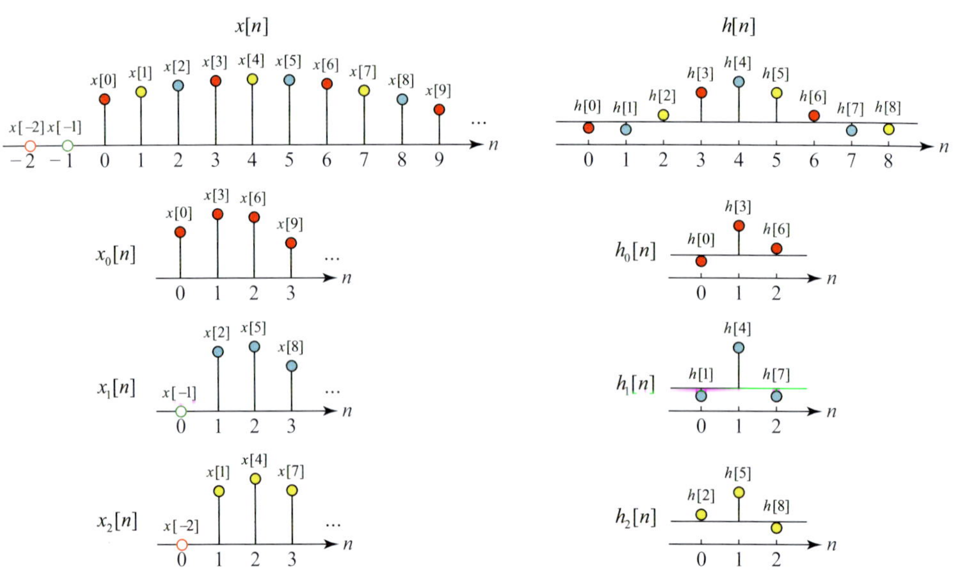

Figure 13.8 Polyphase decomposition of $x[n]$ and $h[n]$

The polyphase input is given by $x_d[n] = x[nD - d],\ 0 \le d < D$:

$$\begin{aligned} x_0[n] &= [x[0] \quad x[3] \quad x[6] \quad x[9] \quad \cdots] \\ x_1[n] &= [0 \quad x[2] \quad x[5] \quad x[8] \quad \cdots] \\ x_2[n] &= [0 \quad x[1] \quad x[4] \quad x[7] \quad \cdots]. \end{aligned}$$

Here, $x_d[n] = x[nD - d]$, so $x_1[0] = x[-1] = 0$ and $x_1[2] = x[-2] = 0$.

The polyphase subfilters are given by $h_d[n] = h[nD + d],\ 0 \le d < D$:

$$\begin{aligned} h_0[n] &= [h[0] \quad h[3] \quad h[6]] \\ h_1[n] &= [h[1] \quad h[4] \quad h[7]] \\ h_2[n] &= [h[2] \quad h[5] \quad h[8]]. \end{aligned}$$

Figure 13.9 shows the convolution of the polyphase input and subfilters, as given in Equation (13.12),

$$y_d[n] = h_d[n] * x_d[n] = \sum_{k=0}^{3} h_d[k] x_d[n-k].$$

The head of the three columns of the figure shows the polyphase filters $h_d[k]$ in *reversed* order; namely $h_2[k]$ (left column), $h_1[k]$ (center) and $h_0[k]$ (right). The order of these columns is important (at least in a real-time application), as we will see in a moment. The remaining four rows of each column are the polyphase subsequences $x_d[n-k]$ for output times $n = 0$ (top row), 1, 2 and 3 (bottom row). The columns correspond to values of $d = 3$, 2 and 1, reading left to right. To obtain the polyphase output component $y_d[n]$ for given values of d and n, you would sum the product of $h_d[k]$ and $x_d[n-k]$ in the appropriate shaded box.

In an application where you have access to all the data at once, such as a program that processes a file of data, the order in which you compute the output components $y_d[n]$ might not matter. However, as we have seen in conjunction with **Figure 13.4**, in an efficient real-time implementation you may need to process data as it arrives, and that means computing the output components in reverse order: $y_2[n]$, $y_1[n]$ and $y_0[n]$. To see this, note that each of the twelve grey shaded boxes in **Figure 13.9** has been numbered in the top left-hand corner with numbers from −2 to 9. These are the input times at which new data has arrived and been distributed to one of the three polyphase inputs in the order $x_2[n]$, $x_1[n]$ or $x_0[n]$. For example, the row highlighted in light blue ($n = 2$) shows the series of operations necessary to calculate output point $y[2]$. In the box numbered ④, data point $x[4]$ has just arrived and has been distributed to $x_2[n]$ (left column). As soon as this point is available, we can compute output subsequence $y_2[2] = x[1]h[5] + x[4]h[2]$. One clock tick later, in box ⑤, data point $x[5]$ arrives and is distributed to $x_1[n]$ (center column). Then we can compute $y_1[2] = x[2]h[4] + x[5]h[1]$. When $x[6]$ arrives and is distributed to $x_0[n]$ (right column), we are able to compute $y_0[2] = x[0]h[6] + x[3]h[3] + x[6]h[0]$. After we have computed all three output subsequences for any row in the order $y_2[n], y_1[n], y_0[n]$, we can sum them to give the downsampled output at the corresponding value of output time n, $y[n] = y_2[n] + y_1[n] + y_0[n]$. This is exactly the same as the process diagrammed with the input commutator in **Figure 13.4**. (Actually, we could compute the partial sum $y_2[n] + y_1[n]$ after the second point arrives, and add the last point, $y_0[n]$, to the partial sum to produce $y[n]$.) Here are all four downsampled output values for this example:

$$\begin{aligned} y[0] &= x[0]h[0] \\ y[1] &= x[0]h[3] + x[1]h[2] + x[2]h[1] + x[3]h[0] \\ y[2] &= x[0]h[6] + x[1]h[5] + x[2]h[4] + x[3]h[3] + x[4]h[2] + x[5]h[1] + x[6]h[0] \\ y[3] &= x[1]h[8] + x[2]h[7] + x[3]h[6] + x[4]h[5] + x[5]h[4] + x[6]h[3] + x[7]h[2] + x[8]h[1] + x[9]h[0]. \end{aligned}$$

Compare this output with the previous non-polyphase calculation in **Figure 13.7** and you will see that they are identical.

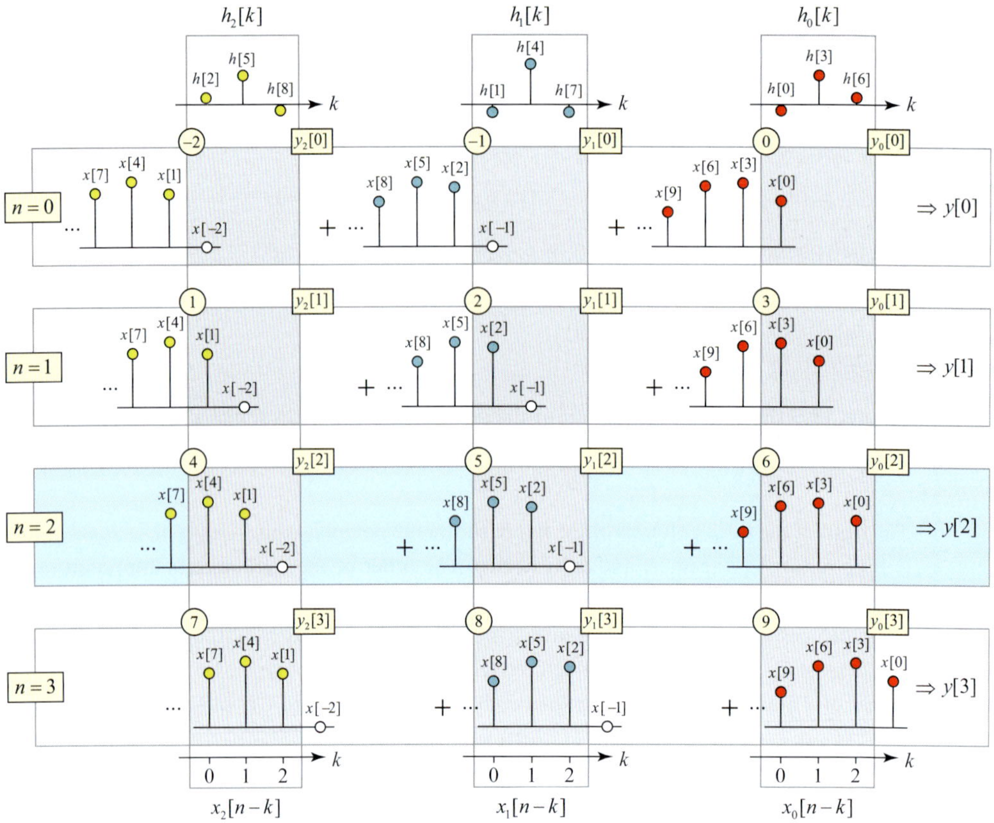

Figure 13.9 Convolution of polyphase input and subfilters in a downsampling example

The important conclusion of this example is that as each new input point arrives, it is distributed to the polyphase subsequences $x_d[n]$ in *reverse* order, $x_{D-1}[n], \ldots, x_0[n]$, exactly as the commutator did in **Figure 13.4**. Hence, that is the order in which the output subsequence $y_d[n] = h_d[x] * x_d[n]$ must be calculated. Then, when all D output subsequences have been computed, they are summed to create downsampled output point $y[n]$.

13.2 Polyphase upsampling

Not surprisingly, there is a polyphase implementation of upsampling as well.

13.2.1 Review of upsampling

By way of review, **Figure 13.10** shows a block diagram of the standard (non-polyphase) implementation of upsampling by a factor U, which we presented in Chapter 6.

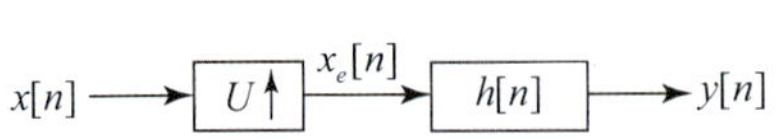

Figure 13.10 Upsampling block diagram

First, an expanded sequence $x_u[n]$ is created by inserting $U-1$ zeros between each pair of points of input $x[n]$:

$$x_u[n] = \sum_{k=-\infty}^{\infty} x[k]\delta[n-kU] = \begin{cases} x[n/U], & n=0, \pm U, \ldots \\ 0, & \text{otherwise.} \end{cases} \tag{13.13}$$

Then, the expanded sequence is filtered by a (causal) lowpass filter with impulse response $h[n]$, cutoff frequency π/U and gain U, to form the upsampled output sequence $y[n]$,

$$y[n] = x_u[n]*h[n] = \sum_{k=-\infty}^{\infty} x_u[k]h[n-k]. \tag{13.14}$$

In the paragraphs that follow, we shall use the words **expansion** to refer to the $U\uparrow$ operation of Equation (13.13) and **upsampling** to refer to the entire process diagrammed in **Figure 13.10**, comprising an expansion operation followed by discrete-time lowpass filtering.

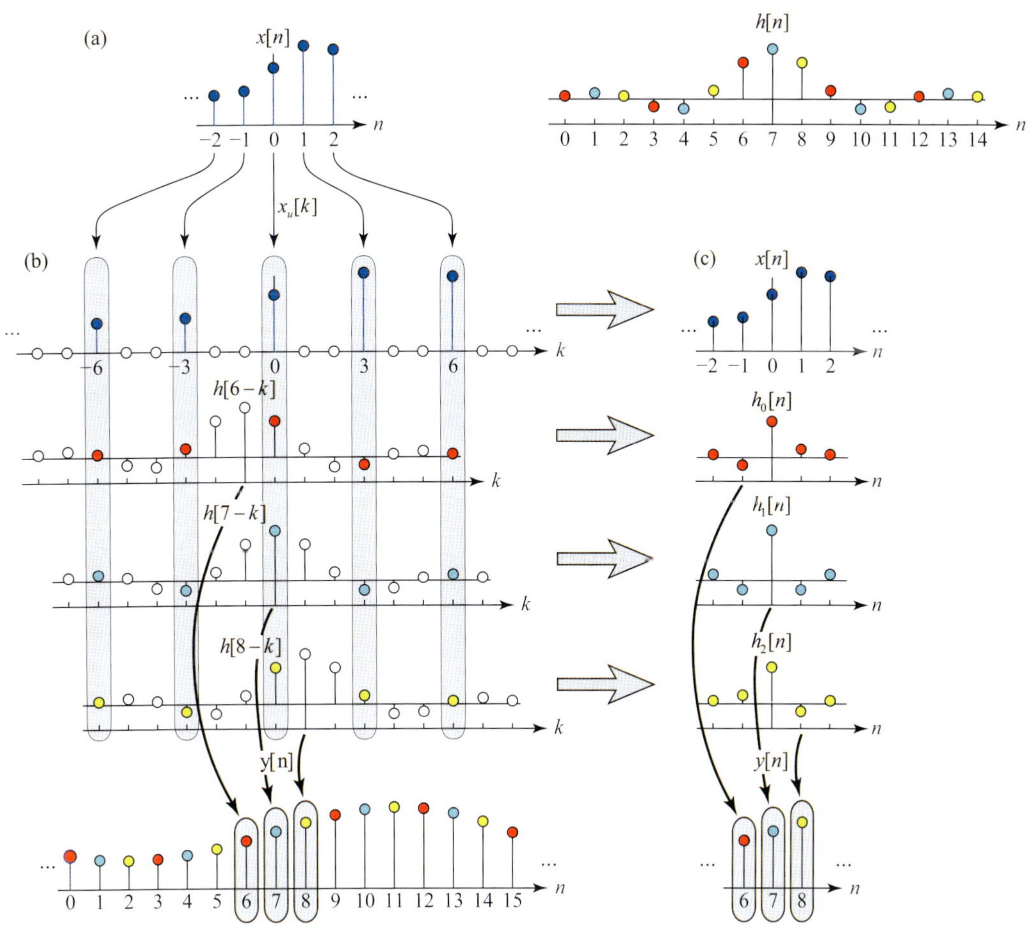

Figure 13.11 Polyphase upsampling

Figure 13.11 shows how the convolution of Equation (13.14) works for the example of $U=3$ and values of $n=6$, 7 and 8. The top left panel of **Figure 13.11a** shows a five-point section of $x[n]$, $-2 \le n \le 2$, and the top right panel shows the impulse response $h[n]$, which has length $N=15$. In

Figure 13.11b, the section of $x[n]$ is expanded by U to form the section of $x_u[k]$, shown in the second row. The following three rows of the figure show the flipped and shifted sequence $h[n-k]$ for values of shift $n=6$, 7 and 8. For each of these values of n, the computation of the convolution sum produces one value of the upsampled sequence; namely $y[6]$, $y[7]$ or $y[8]$, as shown in the bottom panel. Because $h[n]$ is causal and symmetrical with $N=15$, the upsampled output $y[n]$ is delayed by $(N-1)/2=7$ points with respect to $x_e[n]$.

The computation of each output point of $y[n]$ would nominally require N MAC operations if every point of $h[n]$ were multiplied by the corresponding point of $x_u[n]$. Given an input data rate of f_s, the filter must operate at the *output data rate* Uf_s, and therefore perform NUf_s MAC operations per second. However, it is clear from the figure that most of the values of $x_u[n]$ are zero; hence, $N(U-1)$ of these operations are unnecessary. By multiplying only the non-zero values of $x_e[n]$ with the appropriate points of $h[n-k]$, which are indicated by the colored points in the shaded lozenges, we can reduce the number of multiplications required to compute each output point by a factor of U. We will now show how to do just that using polyphase techniques.

13.2.2 Polyphase implementation of upsampling

The polyphase implementation of upsampling, shown in **Figure 13.11c** is an efficient method of avoiding the creation of the expanded sequence and the unnecessary multiplications that result from it. Using Equation (13.13), the convolution of Equation (13.14) becomes

$$y[n]=x_u[n]*h[n]=\left(\sum_{k=-\infty}^{\infty}x[k]\delta[n-kU]\right)*h[n]=\sum_{k=-\infty}^{\infty}x[k]h[n-kU]. \tag{13.15}$$

The sequences $h[n-kU]$ represent the shaded points of the filter's impulse response in **Figure 13.11b** for $n=6$, 7 and 8. By analogy with Equation (13.8), we recognize that these points correspond to U polyphase subfilters created by downsampling the impulse response by a factor of U after it has been shifted by n as shown in the middle three rows of **Figure 13.11c**.

For each single new input value of $x[n]$ that arrives, we compute U output points of the upsampled sequence $y[n]$. For example, in **Figure 13.11**, the same $N/U=5$ points of $x[n]$ shown in the top row of **Figure 13.11a** are used to compute each of the $U=3$ upsampled output points, $y[n]$, $n=6$, 7 and 8, shown in the grey box at the bottom of the figure. Importantly, computation of each of these output points requires only one of the three subfilters; hence, the computations can occur sequentially in time. To see this, define $y_u[n]$, $0\le u\le U-1$, to be the U upsampled output points starting at time nU,

$$y_u[n]\triangleq y[nU+u],\quad 0\le u\le U-1. \tag{13.16}$$

For example, the three output points shown in the bottom of **Figure 13.11b** are

$$\begin{aligned}y[6]&=y_0[2]\\ y[7]&=y_1[2].\\ y[8]&=y_2[2]\end{aligned}$$

Combining Equations (13.15) and (13.16) gives

$$\begin{aligned}y_u[n]&=\sum_{k=-\infty}^{\infty}x[k]h[(nU+u)-kU]=\sum_{k=-\infty}^{\infty}x[k]h[(n-k)U+u]=\sum_{k=-\infty}^{\infty}x[k]h_u[n-k]\\ &=x[n]*h_u[n],\end{aligned}$$

where $h_u[n]$ are the U polyphase subfilters of $h[n]$,

$$h_u[n] \triangleq h[nU+u]. \tag{13.17}$$

These filters $h_u[n]$ are identical to the polyphase subfilters $h_d[n]$ defined in Equation (13.8) for use in downsampling, except that $h[n]$ has been decimated by a factor of U instead of D. As was the case with polyphase downsampling, each $h_u[n]$ is of finite duration, $0 \le n \le L-1$, where $L \triangleq \lceil N/U \rceil$, so we can rewrite $y_u[n]$ as a finite-length convolution,

$$y_u[n] = x[n]*h_u[n] = h_u[n]*x[n] = \sum_{l=0}^{L-1} h_u[l]x[n-l]. \tag{13.18}$$

In what follows, we will again assume that N is an integer multiple of U, so that $L = N/U$. As we found in the polyphase implementation of downsampling, computing each of the output points $y_u[n]$, $0 \le u \le U-1$, only requires convolution with *one* of the subfilters $h_u[n]$. Hence, each convolution can be done separately and sequentially. All this is perhaps made clearer by the example of **Figure 13.12**.

The top left panel of the figure shows the same section of the input sequence $x[n]$ previously shown in **Figure 13.11a**, as well as a section of the upsampled sequence $y[n]$ in the top right panel. Given an impulse response $h[n]$ of length $N = 15$, and an upsampling factor of $U = 3$, the section of input $x[n]$ of length $N/U = 5$ generates $U = 3$ points of the output sequence, $y[6]$, $y[7]$ and $y[8]$, shown in the shaded lozenges in the top right panel. In a real-time application, these output points would be generated sequentially by convolving the section of $x[n]$ with the three polyphase subfilters $h_0[n]$, $h_1[n]$ and $h_2[n]$, *in that order*, as shown in **Figures 13.12a**, **b** and **c**, respectively. In these panels, upsampling is schematically described using a commutator architecture, where the commutator is on the output rather than the input (as it was in the case of downsampling in **Figure 13.4**). Note also that because the order of convolution is the reverse of that in **Figure 13.4**, the direction of the commutator is also reversed.

Polyphase upsampling is efficient. Each output point of $y[n]$ requires computation of just *one* of the $y_u[n]$ components, which in turn requires $L = N/U$ MAC operations. Given an input data rate

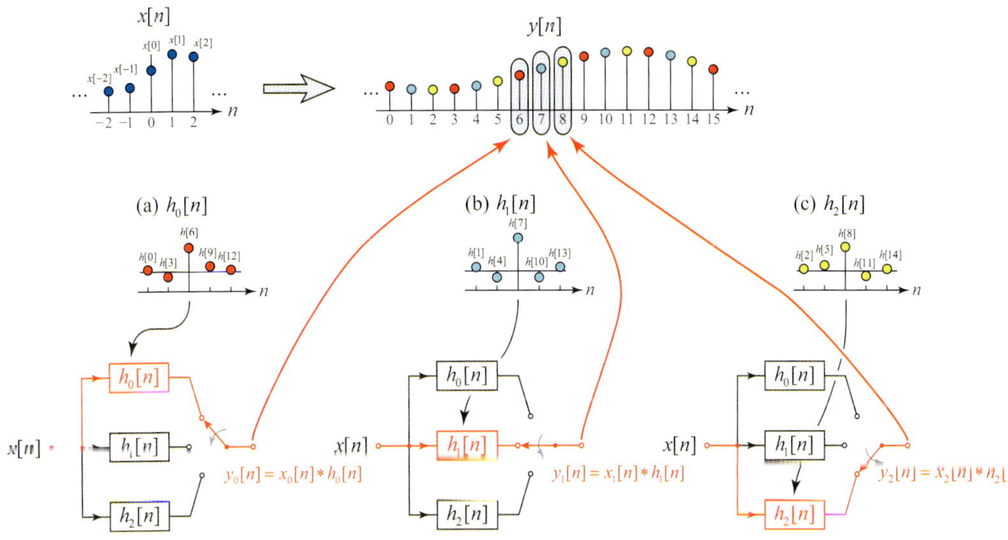

Figure 13.12 Example of polyphase upsampling for $U = 3$

f_s, this corresponds to an output data rate of Nf_s MAC operations per second, which is lower by a factor of U than the NUf_s output rate of the non-polyphase approach.

Finally, we can express $y[n]$ in terms of $y_u[n]$ as

$$y[n] = \sum_{u=0}^{U-1} \sum_{k=-\infty}^{\infty} y_u[k]\delta[n-(kU+u)] = \sum_{k=-\infty}^{\infty} y_u[k] \sum_{u=0}^{U-1} \delta[n-(kU+u)]. \tag{13.19}$$

This equation works by expanding $y_u[n]$ in a manner similar to Equation (13.13). For any value of n, the double summation picks only a single value of $y_u[k]$ such that k and u satisfy $n=kU+u$, with u constrained to be in the range $0 \le u \le U-1$. In our example above, with $U=3$, if $n=6$, then the values of $k=2$ and $u=1$ uniquely satisfy the equation. Hence, $y[6]=y[2\cdot U+0]=y_0[2]$. Similarly, $y[7]=y[2\cdot U+1]=y_1[2]$, $y[8]=y[2\cdot U+2]=y_2[2]$, and so on.

Example 13.2

Given input $x[n]$ and impulse response $h[n]$ shown in **Figure 13.13**, compute the first 12 points of upsampled output $y[n]$ for $U=3$ using both non-polyphase and polyphase methods.

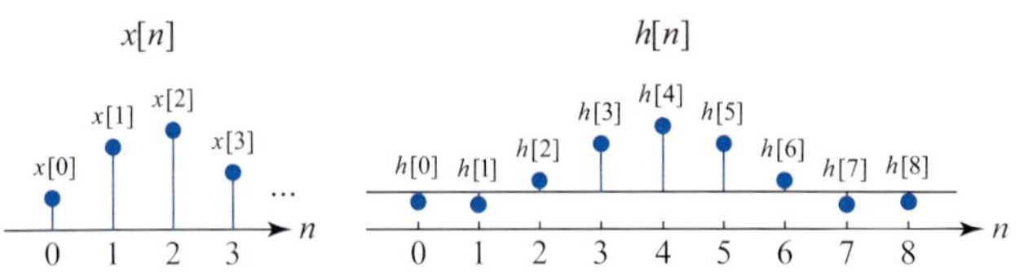

Figure 13.13 Sample input and impulse response for upsampling

➤ Solution:

Non-polyphase upsampling

The top panel of **Figure 13.14** shows the expanded sequence $x_e[k]$. The lower panels in the narrow boxes show $h[n-k]$ for sample values of $n=6$, 7 and 8. Each product of $x_e[k]h[n-k]$ has only non-zero values where $x_e[k]$ is non-zero. The output $y[n]=x_e[n]*h[n]$ is the sum of $y[6]$, $y[7]$ and $y[8]$. Actual values are shown in the bottom of the figure.

Polyphase upsampling

Figure 13.15 shows an example of upsampling a sequence by a factor of $U=3$ using polyphase filters. The task here is to compute the polyphase output components, as specified in Equation (13.18),

$$y_u[n] = h_u[n] * x[n] = \sum_{k=0}^{3} h_u[k]x[n-k],$$

where the polyphase subfilters $h_u[n]$, $0 \le u < 3$, created for upsampling by a factor $U=3$ are the same as the filters $h_d[n]$ we created for downsampling by a factor $D=3$, as shown in **Figure 13.8**. These subfilters are shown at the head of each column in the *normal* order: $h_0[k]$ (left), $h_1[k]$ (center) and $h_2[k]$ (right). Again, the order of these columns is important in an efficient application. Each horizontal row represents $x_u[n-k]$ for a single input time, either $n=0$ (top row), 1, 2 or 3 (bottom row). The columns correspond to values of $u=1$, 2 and 3. The sum of the product of $x[k]$ and $h_u[n-k]$ in each shaded box is

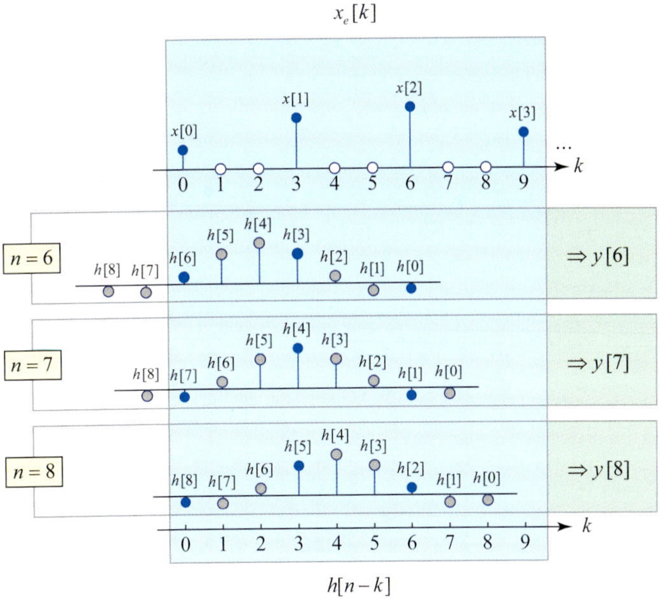

$$y[6] = x[0]h[6] + x[1]h[3] + x[2]h[0]$$
$$y[7] = x[0]h[7] + x[1]h[4] + x[2]h[1]$$
$$y[8] = x[0]h[7] + x[1]h[5] + x[2]h[2]$$

Figure 13.14 Non-polyphase upsampling by direct convolution

the convolution that calculates the output component $y_u[n]$. Hence each row generates $U = 3$ output components, each of which corresponds to one upsampled output point according to Equation (13.16), $y_u[n] = y[nU + u]$. The normal ordering of the subfilter ($u = 0$, 1, 2 and 3) assures that output sequences $y_u[n]$, and therefore the final output points $y[n]$, are generated in the proper order, exactly as we saw using the output commutator in **Figure 13.12**. For example, the third row produces $y[6] = y_0[2]$, $y[7] = y_1[2]$ and $y[8] = y_2[2]$. The bottom panel shows all 12 points of the upsampled sequence $y[n]$ generated from the four input points upsampled by $U = 3$.

$$\begin{aligned}
y[0] &= y_0[0] = x[0]h[0]\\
y[1] &= y_1[0] = x[0]h[1]\\
y[2] &= y_2[0] = x[0]h[2]\\
y[3] &= y_0[1] = x[0]h[3] + x[1]h[0]\\
y[4] &= y_1[1] = x[0]h[4] + x[1]h[1]\\
y[5] &= y_2[1] = x[0]h[5] + x[1]h[2]\\
y[6] &= y_0[2] = x[0]h[6] + x[1]h[3] + x[2]h[0]\\
y[7] &= y_1[2] = x[0]h[7] + x[1]h[4] + x[2]h[1]\\
y[8] &= y_2[2] = x[0]h[8] + x[1]h[5] + x[2]h[2]\\
y[9] &= y_0[3] = x[1]h[6] + x[2]h[3] + x[3]h[0]\\
y[10] &= y_1[3] = x[1]h[7] + x[2]h[4] + x[3]h[1]\\
y[11] &= y_2[3] = x[1]h[8] + x[2]h[5] + x[3]h[2].
\end{aligned}$$

The three rows highlighted in blue, above, are values of $y[n]$ at the same values of n as those obtained in **Figure 13.14** by direct convolution of the expanded sequence $x_e[n]$ with $h[n]$. This indicates that the two methods produce the same results.

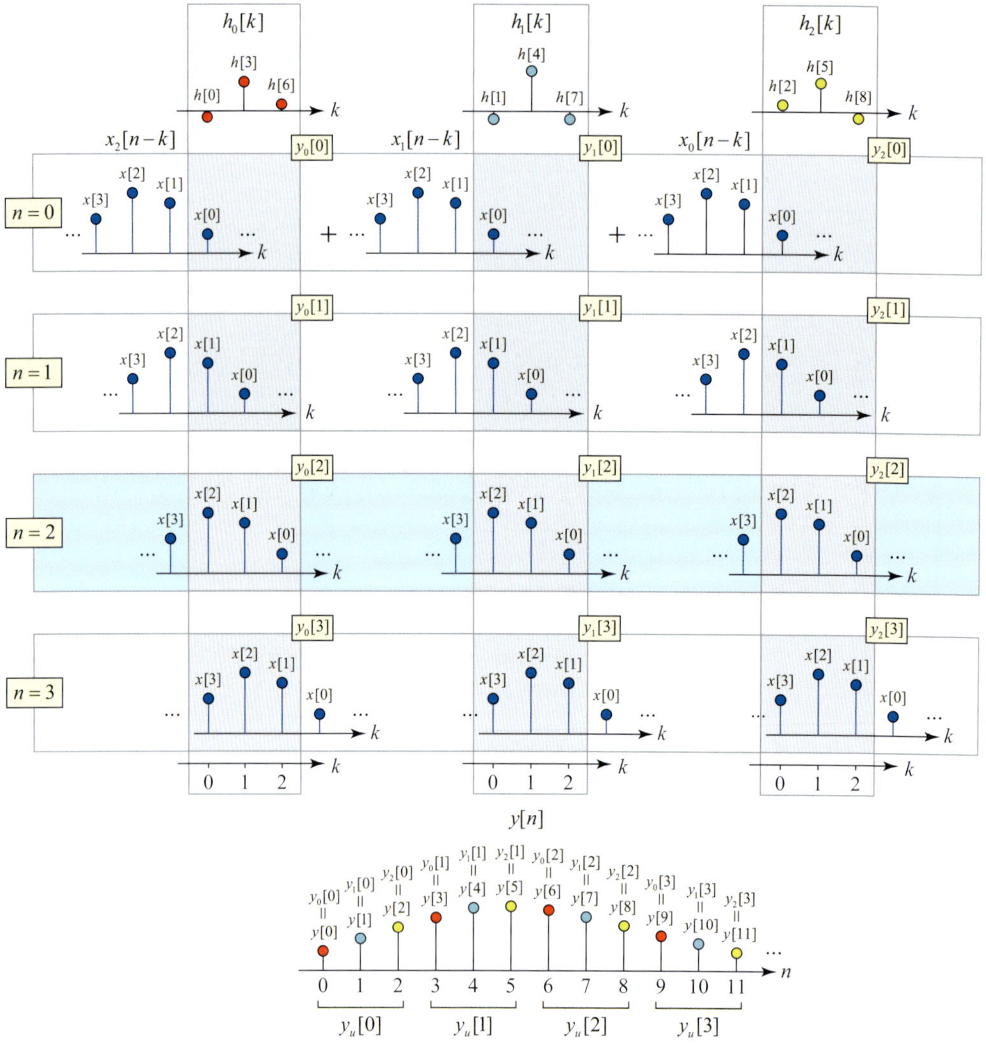

Figure 13.15 Convolution of input and polyphase subfilters in an upsampling example

13.2.3 Upsampling summary

Figure 13.16 shows signal-flow diagram representations of non-polyphase (**Figure 13.16a**) and polyphase (**Figure 13.16b**) implementations of upsampling. As with downsampling, the block diagram of the polyphase implementation comprises a ladder of U horizontal branches. Each branch filters the input $x[n]$ by a polyphase subfilter with impulse response $h_u[n]$, as given in Equation (13.17). The outputs of the uth branch are then expanded by U and delayed by u samples to form polyphase output components $y_u[n]$, as in Equation (13.18). Finally, the output components are summed to give $y[n]$, as in Equation (13.19).

As we discussed in conjunction with the summary of downsampling, the block diagram of **Figure 13.16** and the commutator architecture of **Figure 13.12** are algorithmically equivalent. In the block diagram, the delay and upsampling operations at the output appear as separate elements, while the commutator essentially performs these operations by sequentially directing each new input point to a different filter.

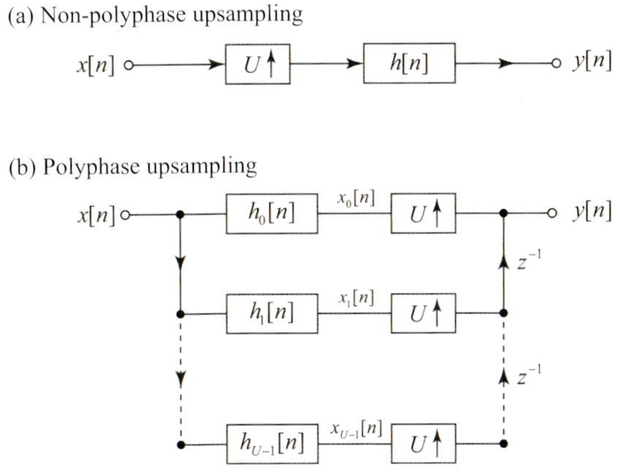

Figure 13.16 Block diagram of two implementations of upsampling

The important difference between the polyphase and non-polyphase upsampling methods is again the order of filtering and expansion. In the polyphase implementation (**Figure 13.16b**), filtering of the input data, $x[n]$, precedes expansion and therefore occurs at the *input* rate, f_s, whereas in the non-polyphase implementation (**Figure 13.16a**), filtering occurs after expansion and therefore occurs at the *output* rate, Uf_s.

13.3 ★ Polyphase resampling

Understanding how to implement an efficient polyphase approach to resampling in the time domain is not as straightforward as either the polyphase downsampling or upsampling approaches discussed in Sections 13.1 and 13.2. It is actually much simpler to understand resampling using the z-transform analysis of Section 13.4, so I recommend you head there. However, for completeness, and for diehard lovers of algebra, the outlines of one time-domain approach can be found in supplementary material.

13.4 Transform analysis of polyphase systems

Multirate systems are perhaps most easily understood from the perspective of the z-transform. In this section, we start by presenting some basic properties of systems with downsampling and upsampling components, as well as the important **multirate identities** that we will use to analyze polyphase systems in the z-transform domain, including combinations of decimation, expansion and filtering.

13.4.1 Basic decimation and expansion identities

Decimation ($D\downarrow$) and expansion ($U\uparrow$) operations are linear (but not time-invariant) transformations. **Figure 13.17** shows, in schematic form, some of the properties of decimation. Equivalent properties apply to expansion. Most of these properties are fairly self-evident.

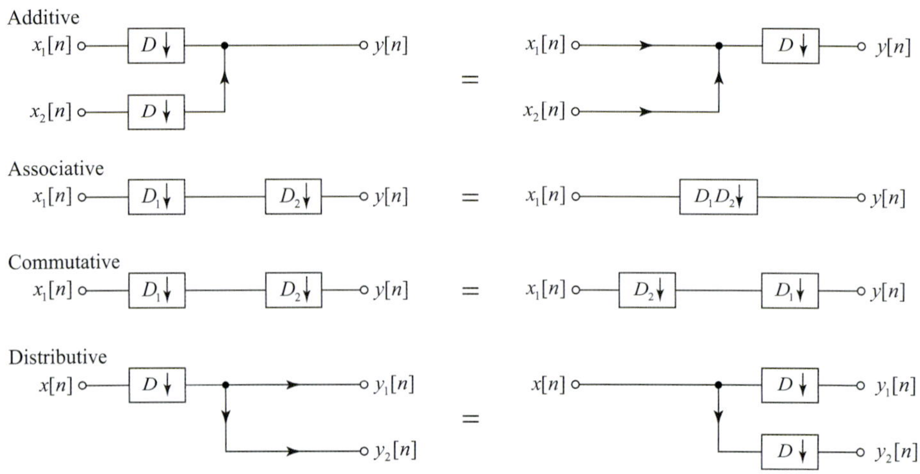

Figure 13.17 Decimation identities

Figure 13.18 shows the schematic description of one identity that is perhaps not as obvious as those shown in **Figure 13.17**; namely, that decimation and expansion commute: decimation by a factor of D followed by expansion by a factor of U is equivalent to expansion by U followed by decimation by D as long as the factors of D and U are **relatively prime** (also called **co-prime**), meaning that they have no common factors except 1.

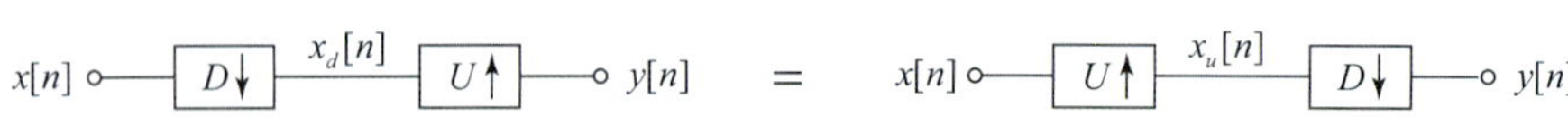

Figure 13.18 Interchanging upsampling and downsampling

This result is somewhat surprising since we might intuitively expect from our discussion of downsampling in Chapter 6 that decimation can lead to a loss of information (i.e., aliasing) that cannot be recovered by expansion. And in fact, this is true when D and U are not relatively prime (see Problem 13-8c). To see that the identity does hold when D and U are relatively prime, consider decimation followed by expansion, as indicated in the left panel of **Figure 13.18**. Decimating input $x[n]$ by a factor of D gives $x_d[n]=x[nD]$. Then, expanding $x_d[n]$ by a factor of U gives output $y[n]$,

$$y[n]=\sum_{k=-\infty}^{\infty} x_d[k]\delta[n-kU]=\sum_{k=-\infty}^{\infty} x[kD]\delta[n-kU].$$

The delta function is non-zero for values of n that are multiples of U, $n=kU$. For example, when $U=2$, the values of $k=n/U$ are integers, $k=n/U=\ldots,\ -2,\ -1,\ 0,\ +1,\ +2,\ \ldots$, so,

$$y[n]=x[kD]=\ldots,\ x[-2D],\ x[-D],\ x[0],\ x[D],\ x[2D],\ \ldots.$$

That is,

$$y[n]=\cdots+x[-2D]\delta[n+2U]+x[-D]\delta[n+U]+x[0]\delta[n]+x[D]\delta[n-U]+x[2D]\delta[n-2U]+\cdots. \tag{13.20}$$

The right panel of **Figure 13.18** shows the reverse situation, expansion followed by decimation. First, an expanded sequence $x_u[n]$ is created by inserting $U-1$ zeros between each pair of points of input $x[n]$,

$$x_u[n] = \sum_{k=-\infty}^{\infty} x[k]\delta[n-kU].$$

Then, $x_u[n]$ is decimated to form $y[n]$,

$$y[n] = x_u[nD] = \sum_{k=-\infty}^{\infty} x[k]\delta[nD-kU].$$

The delta function is non-zero for values of n such that $k = nD/U$ is an integer. If D and U are relatively prime, this occurs when n is again a multiple of U, namely $n = \ldots, -2U, -U, 0, +U, +2U, \ldots$. At these values of n, $k = nD/U = \ldots, -2D, -D, 0, +D, +2D, \ldots$, so

$$y[n] = x[k] = \ldots, x[-2D], x[-D], x[0], x[D], x[2D], \ldots.$$

That is,

$$y[n] = \cdots + x[-2D]\delta[n+2U] + x[-D]\delta[n+U] + x[0]\delta[n] + x[D]\delta[n-U] + x[2D]\delta[n-2U] + \cdots. \tag{13.21}$$

Equations (13.20) and (13.21) are identical, meaning that decimation and expansion operations can be interchanged (commuted) as long as D and U are relatively prime.

13.4.2 Multirate identities of downsampling and upsampling

We will now derive the z-transforms of the basic decimation and expansion operations. This will enable us to derive a pair of important multirate identities that will give us a very quick way of explaining polyphase downsampling, upsampling and resampling from the perspective of the z-transform.

z-transform of a decimated sequence

As an example, consider the sequence $x[n]$, shown in the left panel of **Figure 13.19**.

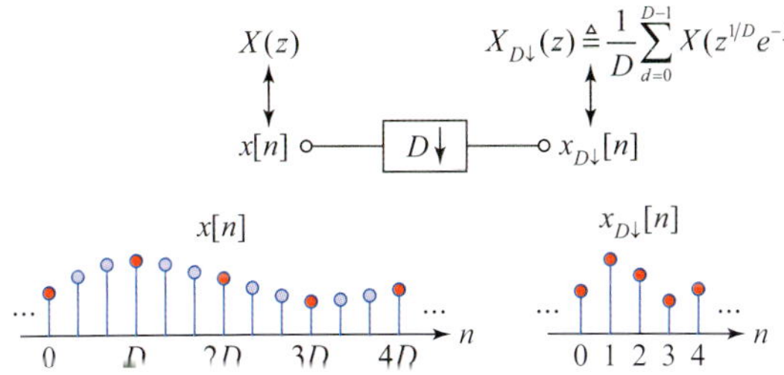

Figure 13.19 Decimated sequence

Define the decimated sequence, shown in the right panel, as

$$x_{D\downarrow}[n] \triangleq x[nD],$$

where $D\downarrow$ denotes the decimation operation. Similarly, define the z-transform of the decimated sequence as

$$X_{D\downarrow}(z) \triangleq \mathfrak{Z}\{x_{D\downarrow}[n]\}.$$

To find $X_{D\downarrow}(z)$, we have to be somewhat careful mathematically.[2] First, multiply $x[n]$ by an impulse train $s[n]$ that has a value of one at multiples of D and is zero otherwise,

$$s[n] = \sum_{k=-\infty}^{\infty} \delta[n-kD] = \begin{cases} 1, & n = 0, \pm D, \pm 2D, \ldots \\ 0, & \text{otherwise.} \end{cases}$$

To proceed, express $s[n]$ as a finite sum of complex exponentials,

$$s[n] = \frac{1}{D}\sum_{d=0}^{D-1} e^{j2\pi dn/D} = \frac{1}{D}\frac{1-e^{j2\pi n}}{1-e^{j2\pi n/D}} = \begin{cases} 1, & n = 0, \pm D, \pm 2D, \ldots \\ 0, & \text{otherwise.} \end{cases} \tag{13.22}$$

The expanded sequence (comprising the solid red points of $x[n]$ in **Figure 13.19**) is $x_e[n] = x[n]s[n]$. Then,

$$x_{D\downarrow}[n] \triangleq x_e[nD] = x[nD]s[nD].$$

So,

$$X_{D\downarrow}(z) = \sum_{n=-\infty}^{\infty} x_{D\downarrow}[n]z^{-n} = \sum_{n=-\infty}^{\infty} x[nD]s[nD]z^{-n}.$$

Let $m = nD$, and substitute Equation (13.22) for $s[m]$,

$$\begin{aligned} X_{D\downarrow}(z) &= \sum_{m=-\infty}^{\infty} x[m]s[m]z^{-m/D} = \sum_{m=-\infty}^{\infty} x[m]\left(\frac{1}{D}\sum_{d=0}^{D-1} e^{j2\pi dm/D}\right)z^{-m/D} \\ &= \frac{1}{D}\sum_{d=0}^{D-1}\left(\sum_{m=-\infty}^{\infty} x[m]\left(z^{1/D}e^{-j2\pi d/D}\right)^{-m}\right) = \frac{1}{D}\sum_{d=0}^{D-1} X\left(z^{1/D}e^{-j2\pi d/D}\right). \end{aligned} \tag{13.23}$$

The final expression may seem somewhat impenetrable, but by making the substitution $z = e^{j\omega}$, we arrive at a familiar expression, the DTFT of a decimated signal, which we have discussed extensively in Chapter 6 (Section 6.5.1),

$$X_{D\downarrow}(\omega) = X_{D\downarrow}(z)|_{z=e^{j\omega}} = \frac{1}{D}\sum_{d=0}^{D-1} X\left(\frac{\omega - 2\pi d}{D}\right).$$

The spectrum of the decimated signal, $X_{D\downarrow}(\omega)$, comprises D replicas of the input spectrum $X(\omega)$, scaled by $1/D$ in both frequency and amplitude.

z-transform of an expanded sequence

Now consider the sequence $x[n]$ shown in the left panel of **Figure 13.20**.

[2] Problem 11-6 shows how easily one can be led astray by a reasonable-looking, but wrong, derivation.

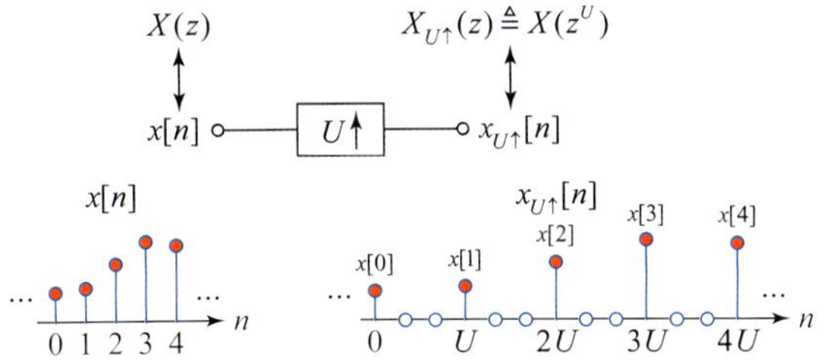

Figure 13.20 Expanded sequence

The expanded sequence $x_{U\uparrow}[n]$ in the right panel of the figure is formed by expanding $x[n]$ by a factor U by inserting $U-1$ zeros between each adjacent pair of points, where $U\uparrow$ represents the expansion operation. So,

$$\begin{aligned} x_{U\uparrow}[n] &= \cdots + x[0]\delta[n] + x[1]\delta[n-U] + x[2]\delta[n-2U] + x[3]\delta[n-3U] \\ &\quad + x[4]\delta[n-4U] + \cdots = \sum_{k=-\infty}^{\infty} x[k]\delta[n-kU]. \end{aligned} \tag{13.24}$$

The z-transform of $y[n]$ is

$$\begin{aligned} X_{U\uparrow}(z) &= \mathfrak{Z}\{x_{U\uparrow}[n]\} \triangleq \mathfrak{Z}\left\{\sum_{k=-\infty}^{\infty} x[k]\delta[n-kU]\right\} \\ &= \sum_{k=-\infty}^{\infty} x[k]\mathfrak{Z}\{\delta[n-kU]\} = \sum_{k=-\infty}^{\infty} x[k]z^{-kU} = \sum_{k=-\infty}^{\infty} x[k](z^U)^{-k} = X(z^U). \end{aligned} \tag{13.25}$$

With the basic expressions for the z-transform of decimated and expanded sequences in hand, let us now turn to the multirate identities of downsampling and upsampling.

Multirate downsampling identity

Consider the system diagrammed in the top panel of **Figure 13.21a**. The input $x[n]$ is first decimated by a factor of D to form $x_{D\downarrow}[n]$, with z-transform $X_{D\downarrow}(z)$, as indicated by Equation (13.23). The result is filtered with a filter with impulse response $h[n]$ and z-transform $H(z)$ to form $y[n]$,

$$y[n] = x[nD] * h[n] = \sum_{k=-\infty}^{\infty} h[k]x[(n-k)D].$$

The z-transform of $y[n]$ is the product of the z-transforms of $x[nD]$ and $h[n]$,

$$Y(z) = X_{D\downarrow}(z)H(z). \tag{13.26}$$

In the system schematized in the top panel of **Figure 13.21b**, $x[n]$ is first convolved with a filter with the expanded impulse response $h_e[n] = h_{D\uparrow}[n]$ to form the intermediate result $w[n]$,

$$w[n] = x[n] * h_{D\uparrow}[n],$$

where $h_{D\uparrow}[n]$ is formed by *expanding* the impulse response $h[n]$ by a factor of D in a manner similar to that given by Equation (13.24), as shown in **Figure 13.20**,

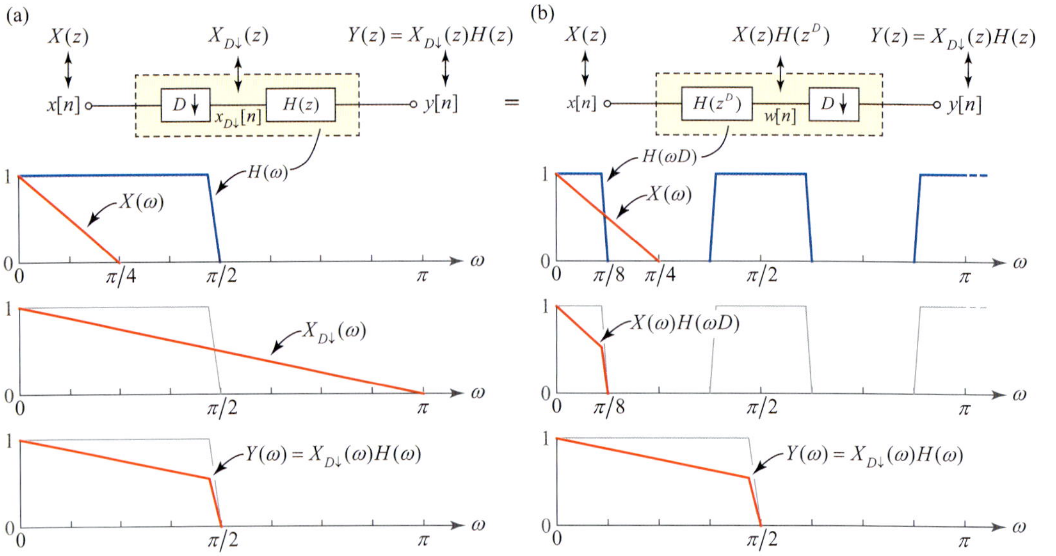

Figure 13.21 Multirate downsampling identity

$$h_{D\uparrow}[n] = \sum_{k=-\infty}^{\infty} h[k]\delta[n-kD].$$

By Equation (13.25), the z-transform of $h_{D\uparrow}[n]$ is $H_{D\uparrow}(z) = H(z^D)$, and the z-transform of $w[n]$ is

$$W(z) = X(z)H(z^D).$$

Then, $w[n]$ is decimated by a factor of D to form $y[n] = w[nD]$, so

$$Y(z) = W_{D\downarrow}(z) = (X(z)H(z^D))_{D\downarrow} = X_{D\downarrow}(z)H(z).$$

This is exactly the same as Equation (13.26), which indicates that the two systems are identical. In other words, decimating the input followed by filtering is equivalent to filtering the input with a filter with an expanded impulse response and then decimating the result. This is the **multirate downsampling identity**.

The spectral plots in the lower panels of the figure may help make this result more intuitive. The top spectral plot in **Figure 13.21a** shows an example of $X(\omega)$ and $H(\omega)$. When $x[n]$ is decimated by $D=4$ to form $x[nD]$, the resulting spectrum is $X_{D\downarrow}(\omega)$, shown in the middle panel. Because $X(\omega)$ is bandlimited to $\omega = \pi/4$ in this example, there is no aliasing in $X_{D\downarrow}(\omega)$. When this spectrum is filtered by $H(\omega)$, the result is $Y(\omega) = X_{D\downarrow}(\omega)H(\omega)$, shown in the bottom panel. The top spectral plot in **Figure 13.21b** again shows $X(\omega)$ as well as $H(\omega D)$, which is the spectrum of the expanded impulse response $h_e[n]$. The middle panel shows the filtered result $W(\omega) = X(\omega)H(\omega D)$. When $w[n]$ is decimated by a factor of D, the final result is again

$$Y(\omega) = (X(\omega)H(\omega D))_{D\downarrow} = X_{D\downarrow}(\omega)H(\omega),$$

as shown in the bottom panel of **Figure 13.21b**, which is the same as the bottom panel of **Figure 13.21a**.

Multirate upsampling identity

The **multirate upsampling identity** is the dual of the multirate downsampling identity.

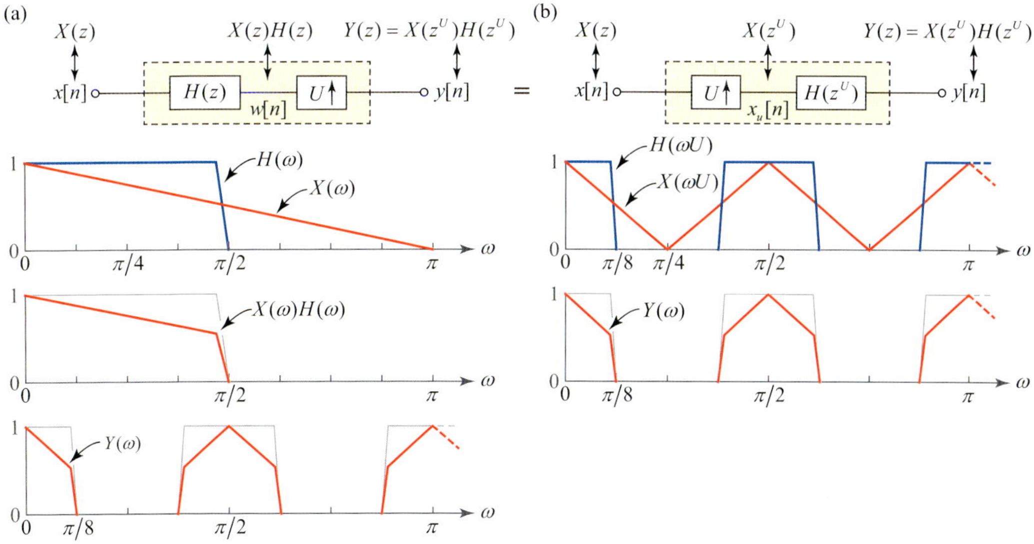

Figure 13.22 Multirate upsampling identity

Start with the system shown in the top panel of **Figure 13.22a**. The input $x[n]$ is first filtered by impulse response $h[n]$ to form the intermediate result,

$$w[n] = x[n] * h[n],$$

with transform

$$W(z) = X(z)H(z).$$

Then, $w[n]$ is expanded by a factor U to form the output $y[n]$. By Equation (13.25),

$$Y(z) = W(z^U) = X(z^U)H(z^U). \tag{13.27}$$

For the system shown in the top panel of **Figure 13.22b**, input sequence $x[n]$ is first expanded by a factor of U to form $x_{U\uparrow}[n]$, which has z-transform

$$X_{U\uparrow}(z) = X(z^U).$$

The expanded sequence is convolved with a filter with expanded impulse response $h_{U\uparrow}[n]$ to form output

$$y[n] = x_{U\uparrow}[n] * h_{U\uparrow}[n],$$

where $h_{U\uparrow}[n]$ is formed by expanding impulse response $h[n]$ by a factor of U as shown in **Figure 13.20**,

$$h_{U\uparrow}[n] = \sum_{k=-\infty}^{\infty} h[k]\delta[n - kU].$$

By Equation (13.25), the z-transform of $h_{U\uparrow}[n]$ is

$$H_{U\uparrow}(z) = H(z^U),$$

and the z-transform of $y[n]$ is therefore

$$Y(z) = X_{U\uparrow}(z)H_{U\uparrow}(z) = X(z^U)H(z^U),$$

which is the same as Equation (13.27). In words, the multirate upsampling identity states that filtering followed by expansion is equivalent to expansion followed by filtering with a filter that has an expanded impulse response.

The spectral plots in **Figure 13.22** help explain the identity. The top spectral plot in **Figure 13.22a** shows an example of $X(\omega)$ and $H(\omega)$. The middle panel shows the result of filtering $x[n]$ by $h[n]$ to form $w[n] = x[n] * h[n]$, with a resulting spectrum $X(\omega)H(\omega)$. When $w[n]$ is expanded by a factor of $U = 4$, the resulting spectrum, shown in the bottom panel, is $Y(\omega) = X(\omega U)H(\omega U)$. The top spectral plot in **Figure 13.22b** shows $X(\omega U)$, which is the spectrum of the expanded input $x_{U\uparrow}[n]$, as well as $H(\omega U)$, which is the spectrum of the expanded impulse response $h_{U\uparrow}[n]$. The middle panel shows the filtered result $Y(\omega) = X(\omega U)H(\omega U)$, which is the same as the bottom panel of **Figure 13.22a**.

Having developed these important multirate identities, we are now in a position to analyze polyphase downsampling, upsampling and resampling quickly and relatively intuitively from the perspective of the z-transform.

13.4.3 Transform analysis of polyphase downsampling

Figure 13.23 shows – step by step – how the basic and multirate downsampling identities described in the previous sections can be applied to reduce the rate of computation in a polyphase downsampling application.

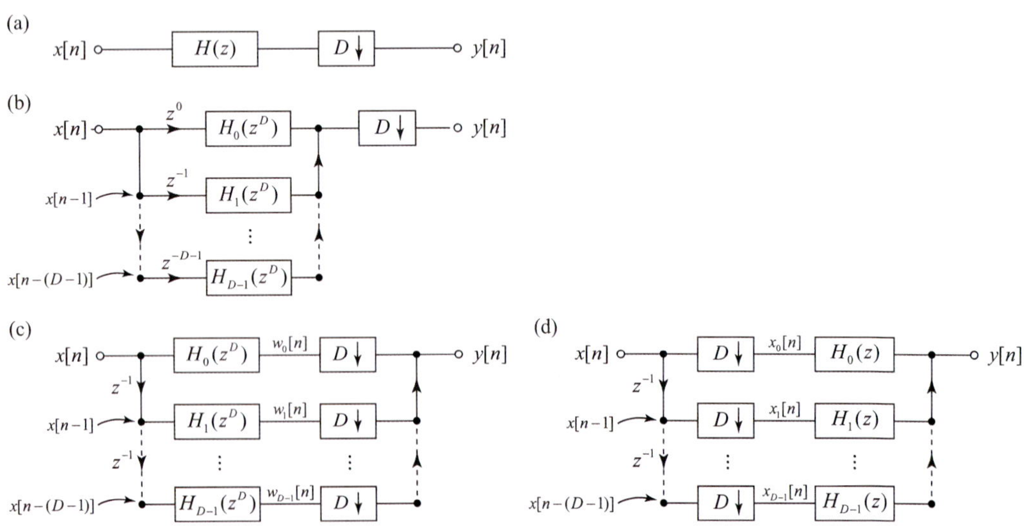

Figure 13.23 Polyphase downsampling

In non-polyphase downsampling, schematized in **Figure 13.23a**, samples of the input $x[n]$ are filtered by impulse response $h[n]$, with z-transform $H(z)$, and then decimated by a factor of D. As shown in **Figure 13.3b** and described in Equations (13.7)–(13.9), $h[n]$ can be decomposed into D expanded sequences $\hat{h}_d[k]$, $0 \le d \le D-1$, each shifted by time d,

$$h[n] = \sum_{d=0}^{D-1} \hat{h}_d[n],$$

where the expanded sequences are

$$\hat{h}_d[n] = \sum_{k=-\infty}^{\infty} h_d[k]\delta[n-(kD+d)], \qquad 0 \le d \le D-1,$$

and the polyphase components are

$$h_d[n] = h[nD+d], \qquad 0 \le d \le D-1. \tag{13.28}$$

The z-transform of $\hat{h}_d[n]$ is

$$\begin{aligned}\mathfrak{Z}\left\{\hat{h}_d[n]\right\} &= \mathfrak{Z}\left\{\sum_{k=-\infty}^{\infty} h_d[k]\delta[n-(kD+d)]\right\} = \sum_{k=-\infty}^{\infty} h_d[k]\mathfrak{Z}\{\delta[n-(kD+d)]\} \\ &= \sum_{k=-\infty}^{\infty} h_d[k]z^{-(kD+d)} = z^{-d}\sum_{k=-\infty}^{\infty} h_d[k](z^D)^{-k} \\ &= z^{-d}H_d(z^D),\end{aligned} \tag{13.29}$$

where $H_d(z)$ is the z-transform of polyphase component $h_d[n]$,

$$H_d(z) \triangleq \sum_{n=-\infty}^{\infty} h_d[n]z^{-n}.$$

Finally, from Equations (13.7) and (13.29),

$$H(z) = \mathfrak{Z}\left\{\sum_{d=0}^{D-1} \hat{h}_d[n]\right\} = \sum_{d=0}^{D-1} \mathfrak{Z}\left\{\hat{h}_d[n]\right\} = \sum_{d=0}^{D-1} z^{-d}H_d(z^D). \tag{13.30}$$

This is shown schematically in **Figure 13.23b**. Each horizontal branch of the figure corresponds to $z^{-d}H_d(z^D)$ for one value of d, $0 \le d \le D-1$. The responses of all the branches are then summed and the sum is decimated by a factor of D. In **Figure 13.23c**, we have exploited the fact (shown in **Figure 13.17**) that decimating the sum of responses by a factor D is equivalent to decimating each response separately and then adding the result. We have also replaced the delay of each branch by the product of z^{-1} delays: $z^{-d} = (z^{-1})^d$. Finally, in **Figure 13.23d** the multirate downsampling identity of **Figure 13.21** has been applied to commute the order of filtering and decimation in each branch of **Figure 13.23c**. This is exactly the same topology we derived by analysis in the time domain and presented in **Figure 13.5b**. For each value of d, the input signal is delayed by d samples to form $x[n-d]$, and then decimated by a factor of D to form $x_d[n]$. This sequence is filtered by the appropriate polyphase filter with impulse response $h_d[n]$ and z-transform $H_d(z)$, and the outputs of all branches are added to form $y[n]$. The central point of **Figure 13.23c** is that the filtering in each branch takes place at the input sampling rate f_s, whereas in **Figure 13.23d** filtering takes place at the output rate f_s/D, which is a factor D lower than the input rate.

13.4.4 Transform analysis of polyphase upsampling

Figure 13.24 shows how the multirate upsampling identity can reduce the rate of computation in a polyphase upsampling application. In non-polyphase upsampling, schematized in **Figure 13.24a**, samples of the input $x[n]$ are expanded by a factor U and then filtered by an impulse response $h[n]$ with z-transform $H(z)$. As we have previously seen, $h[n]$ can be split into U expanded sequences $\hat{h}_u[k]$, $0 \le d \le U-1$. The z-transform, developed in exactly the same fashion as Equation (13.30), is

$$H(z) = \mathfrak{Z}\left\{\sum_{u=0}^{U-1} \hat{h}_u[n]\right\} = \sum_{u=0}^{U-1} \mathfrak{Z}\left\{\hat{h}_u[n]\right\} = \sum_{u=0}^{U-1} z^{-u} H_u(z^U).$$

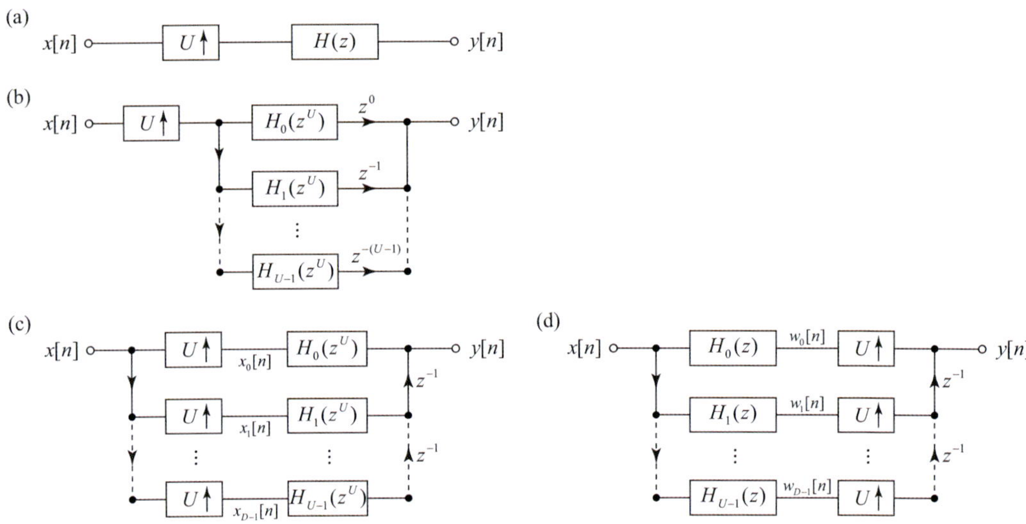

Figure 13.24 Polyphase upsampling

This is shown schematically in **Figure 13.24b**. Expanding the sum of responses by a factor of U is equivalent to expanding each response separately and then adding the result, as shown in **Figure 13.24c**. Finally, in **Figure 13.24d**, the multirate upsampling identity of **Figure 13.22** has been applied to commute the order of filtering and expansion in each branch of **Figure 13.24c**. As we have seen before, filtering in each branch in **Figure 13.24c** takes place at the output sampling rate Uf_s, whereas in **Figure 13.24d** filtering takes place at the input rate f_s.

13.4.5 Transform analysis of polyphase resampling

The multirate downsampling and upsampling identities can be combined to reduce the rate of computation in a polyphase resampling. **Figure 13.25** shows two ways of approaching polyphase resampling from the perspective of the z-transform, which we label Type-I and Type-II. Both approaches combine polyphase upsampling and downsampling, but the order in which they are applied differs.

The Type-I approach shown in the left column of **Figure 13.25a** starts by replacing the expansion and filtering operations with their polyphase equivalent using the multirate upsampling identity, exactly as described in conjunction with **Figure 13.24**. The approach shown in the right column of **Figure 13.25b** starts by replacing the filtering and decimation operations with their polyphase equivalents using the multirate downsampling identity as described in **Figure 13.23**. In the next few figures, we will describe the approach in the left column in detail; the approach shown in the right column is substantially similar.

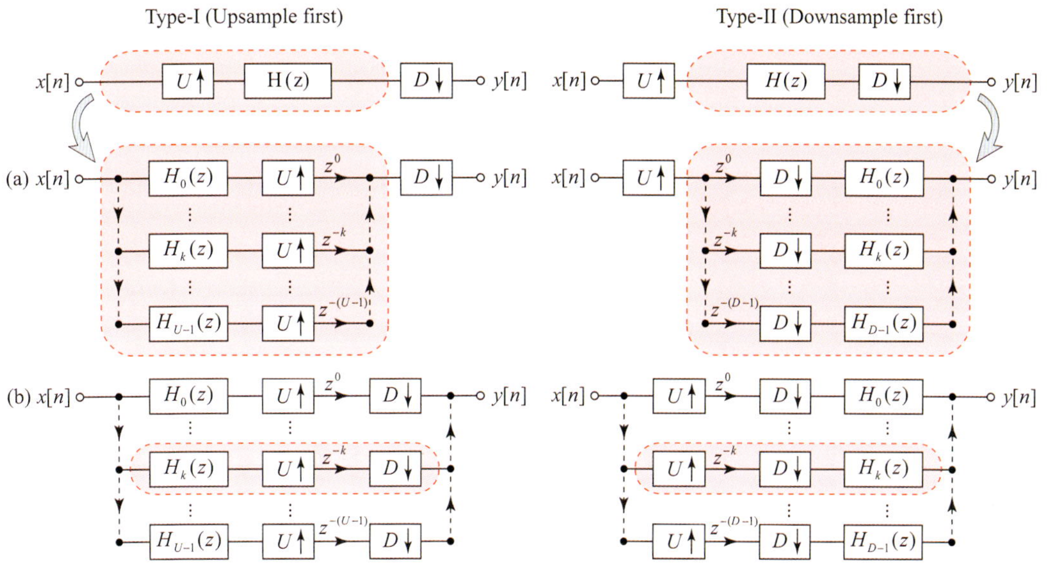

Figure 13.25 Polyphase resampling

In the Type-I approach of **Figure 13.25a**, the input $x[n]$ is processed by U parallel branches, $0 \le k < U$, each of which comprises a polyphase subfilter $H_k(z)$, followed by expansion by a factor of U. The output of each branch is then delayed by k samples (i.e., the transform is multiplied by z^{-k}), the delayed outputs are added together and the sum is decimated by D to form the output $y[n]$. In **Figure 13.25b**, the distributive identity of **Figure 13.17a** has been used to apply the decimation-by-D operation to each branch before the summation instead of after it.

Now, consider just the kth branch, highlighted in the pink lozenge in **Figure 13.25b**, which is also shown in the top left panel of **Figure 13.26**. This branch has a delay of z^{-k}, highlighted in **Figure 13.26a**, located between the expansion-by-U and decimate-by-D operations. The location of this delay is a roadblock to further progress because (looking ahead to **Figures 13.26f** and **g**) we will eventually want to apply polyphase downsampling to $H_k(z)$. The solution is to express z^{-k} as the product of two delays, one of which is a multiple of U, the other a multiple of D. To accomplish this, we invoke **Bézout's identity** from number theory.[3] In the context of our problem, Bézout's identity states that given relatively prime values of positive integers U and D, we can always find integers α and β such that $\alpha U + \beta D = 1$. Since both U and D are positive, either α or β (but not both) must be negative. Thus, we can express delay of the kth branch, z^{-k}, as

$$z^{-k} = z^{-k(\alpha U + \beta D)} = z^{-k\alpha U} z^{-k\beta D},$$

[3]Bézout's identity, named for French mathematician Étienne Bézout (1730–1783), states that the **greatest common divisor (gcd)** of two integers U and D can be expressed as the weighted sum of U and D, namely $\alpha U + \beta D = \gcd(U, D)$, where α and β are the two integer weights. When U and D are relatively prime, as they are in our application, then $\gcd(U, D) = 1$, and the identity reduces to $\alpha U + \beta D = 1$. This is a classic example of a linear Diophantine equation that has an infinite number of solutions for α and β (see Problem 11-9). This identity is, in turn, based on Euclid's algorithm for computing the gcd of two integers – one of the oldest algorithms in mathematics, dating from around 300 BCE.

as shown in **Figures 13.26b** and **c**. For example, if $U = 2$ and $D = 3$, then the identity is satisfied with $\alpha = -1$ and $\beta = 1$. So,

$$z^{-k} = z^{kU} z^{-kD} = z^{2k} z^{-3k}.$$

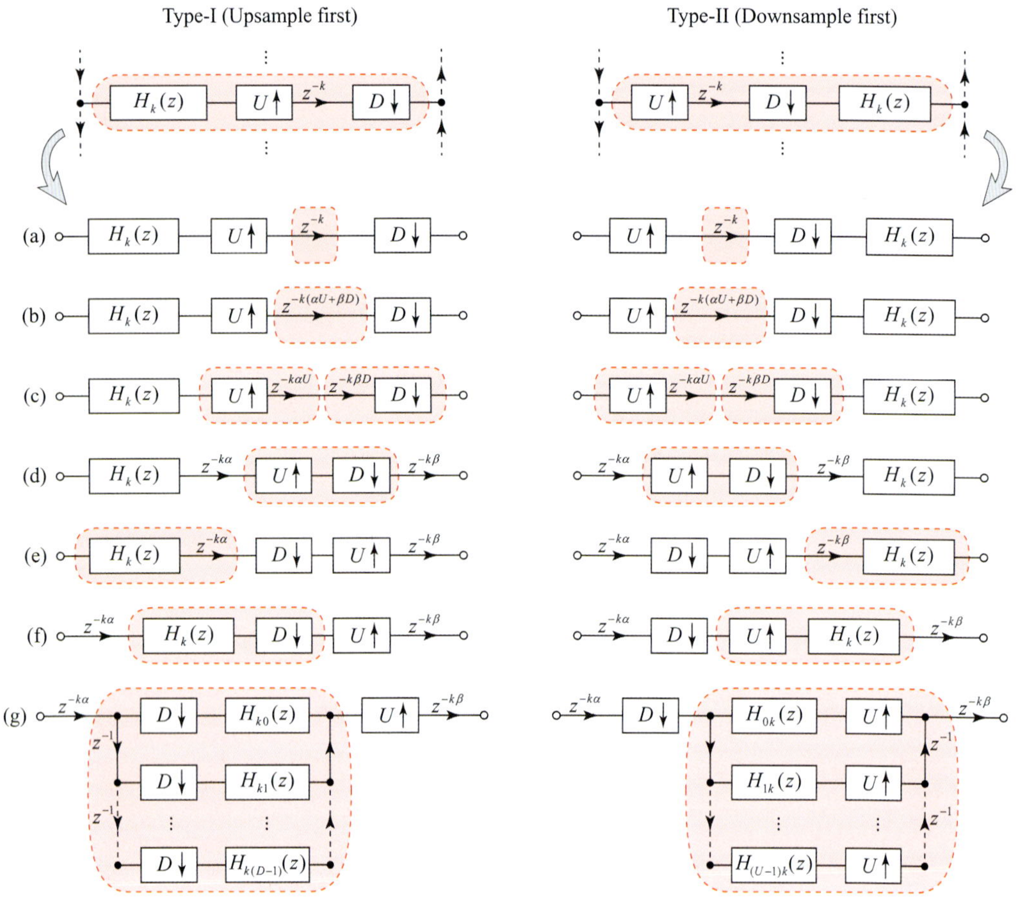

Figure 13.26 Detail of polyphase resampling

The next step is to exploit the polyphase downsampling and upsampling identities shown in **Figure 13.21** and **Figure 13.22**, respectively. The downsampling identity states that a delay of z^{-D} followed by a decimation by D is equivalent to decimation by D followed by a delay of z^{-1}, as shown in **Figure 13.27a**. Similarly, the upsampling identity states that expansion by U followed by a delay of z^{-U} is equivalent to a delay of z^{-1} followed by an expansion by U, as shown in **Figure 13.27b**.

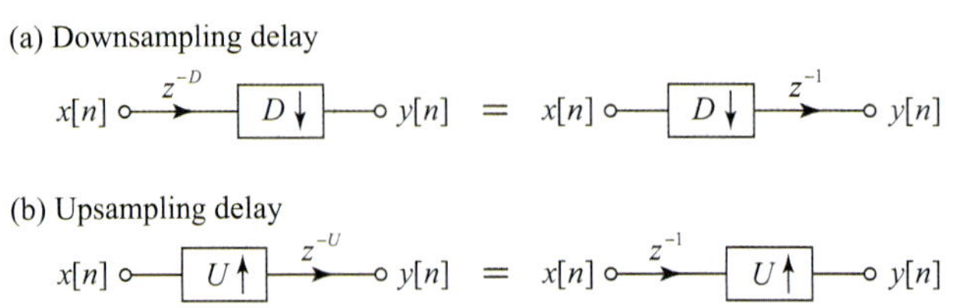

Figure 13.27 Multirate identities with delays

Making these substitutions results in the configuration shown in **Figure 13.26d**. The result is that the expansion and decimation blocks now adjoin one other. Since U and D are relatively prime, we can interchange the blocks, using the identity schematized in **Figure 13.18**, to arrive at the configuration shown in **Figure 13.26e**. Next, recognize that filtering (represented by $H_k(z)$) and delay (represented by $z^{-k\alpha}$) are both linear operations that can be commuted, as shown in **Figure 13.26f**. Finally, **Figure 13.26g** shows that the combination of filtering, $H_k(z)$, followed by decimation by D can be replaced by the polyphase equivalent schematized in **Figure 13.23**. The result is that the polyphase subfilters have been doubly decomposed, first into U subfilters, each of which is then further decomposed into D subfilters.

The Type-II column of the figure shows the result of the equivalent operations, replacing expansion by U followed by filtering with the polyphase equivalent.

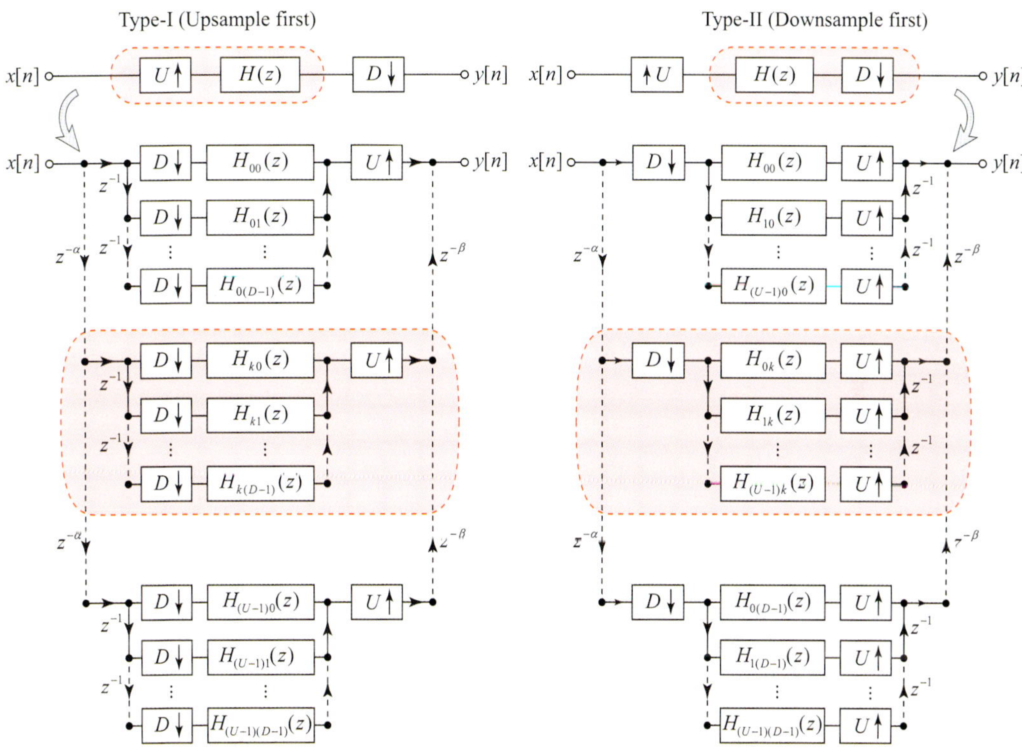

Figure 13.28 Polyphase resampling architectures

Figure 13.28 shows the final result of replacing each branch in **Figure 13.25** by the polyphase equivalent of **Figure 13.26**. The central benefit of these polyphase architectures is that they reduce both the number and the rate of computation compared with the non-polyphase architecture. In the non-polyphase architecture, with the lowpass filter located between expand-by-U and decimate-by-D blocks, the number and rate at which the lowpass filter must process data is determined by the larger of U and D. In the polyphase architectures, rate of computation is reduced by a factor of max([U D]).

The Type-I and Type-II polyphase implementations shown in **Figure 13.28** are, in fact, transposes of each other (something we discussed in Section 9.3). To obtain the schematic of either column, flip the schematic of the other column left–right, reverse the direction of all the arrowheads, interchange $D\downarrow$ and $U\uparrow$ and interchange $x[n]$ and $y[n]$. The two architectures have the same number of delays and the same number of subfilters, so under what circumstances might we prefer one vs. the other? A couple of examples will help us decide.

Example 13.3

Diagram Type-I and Type-II polyphase architectures for a resampled system with $U=2$ and $D=3$.

➤ Solution:

Figure 13.29a shows the two implementations, Type-I and Type-II, obtained just by substituting the appropriate numbers into **Figure 13.28**.

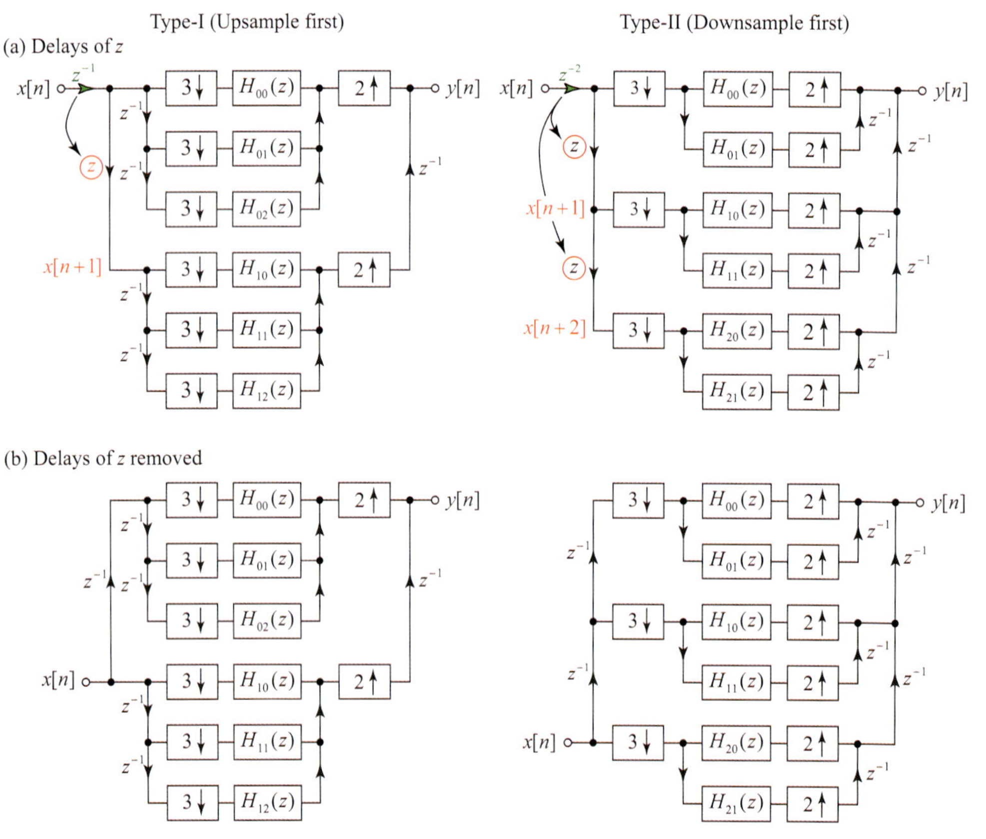

Figure 13.29 Type-I and Type-II resampling architectures for $U=2$ and $D=3$

Of particular importance are the values of the delay of the input branches, $z^{-\alpha}$, and output branches, $z^{-\beta}$. Recall that the values of α and β given by Bézout's identity are guaranteed to differ in sign; one of them must be positive and the other negative. In this example, $U = 2$ and $D = 3$, so the identity gives $\alpha = -1$ and $\beta = 1$. This means that the input to each of the lower branches in **Figure 13.29a** (circled in red) will be multiplied by z. This corresponds to a negative (non-causal) delay of $x[n]$; that is, $x[n+1]$, $x[n+2]$. That is not acceptable for any practically realizable system. The solution is to delay the input to the converter by an amount appropriate to cancel the positive powers of z in all branches. In this example, a delay of z^{-1} is necessary at the input of the Type-I system, whereas a delay of z^{-2} is required for the Type-II system, as shown in green in the figure. In **Figure 13.29b**, the diagrams have been appropriately modified to have only positive delays. We will have more to say about the details of implementing these delays in supplementary material, when we get down to the business of coding a practical algorithm for efficient polyphase resampling.

On to the question of which architecture – Type-I or Type-II – most efficiently realizes the sample-rate converter. Both architectures start with the same specification of lowpass filter $h[n]$, because the filter's bandwidth and therefore its length N are determined by max([U D]). When designing $h[n]$, it is practical to make N an integer multiple of UD. In the Type-I example, $h[n]$ is first decomposed into U subfilters, each of which is further decomposed into D subfilters. That means that we effectively end up with an array of $U \times D$ filters, in this example a 2×3 array, each of length $N/6$. In the Type-II example, the situation is reversed; $h[n]$ is first decomposed into D subfilters, each of which is further decomposed into U subfilters resulting in a $D \times U$ array of subfilters, in this example a 3×2 array, again of length $N/6$. So, in both Type-I and Type-II examples, the length of each subfilter is N/UD. No difference. The main distinction between the two architectures is the total number of expansion and decimation operations required. The Type-I system requires U expansion and UD decimation operations; the Type-II system requires D decimation and UD expansion operations. So, we would prefer Type-I when $U < D$ and Type-II when $U > D$.

Example 13.4

Diagram Type-I and Type-II polyphase architectures for a resampled system with $U = 3$ and $D = 2$.

➤ Solution:

Figure 13.30a shows the two implementations.

This example is similar to Example 13.3, except that $U = 3$ and $D = 2$, so Bézout's identity gives $\alpha = 1$ and $\beta = -1$. In this case, this results in a non-causal delay of the output $y[n]$; that is, $y[n+1]$, $y[n+2]$. Here, a solution is to delay the output to the converter to cancel the positive powers of z in all branches. In this example, a delay of z^{-2} is necessary at the output of the Type-I system, while a delay of z^{-1} is required for the Type-II system, as shown in green in the figure. **Figure 13.30b** shows the diagrams modified to have only positive delays. As in Example 13.3, the main distinction between the two architectures is the total number of upsampling and downsampling operations required. In this case $U > D$, so we would prefer the Type-II architecture.[4]

[4]The signs of α or β do not depend on whether $U > D$ or $U < D$, though one might be led to believe so based on the previous two examples. For example, when $U = 7$ and $D = 3$ (i.e., $U > D$), then $\alpha = 1$ and $\beta = -2$. But when $U = 7$ and $D = 4$ (still $U > D$), then $\alpha = -1$ and $\beta = 2$.

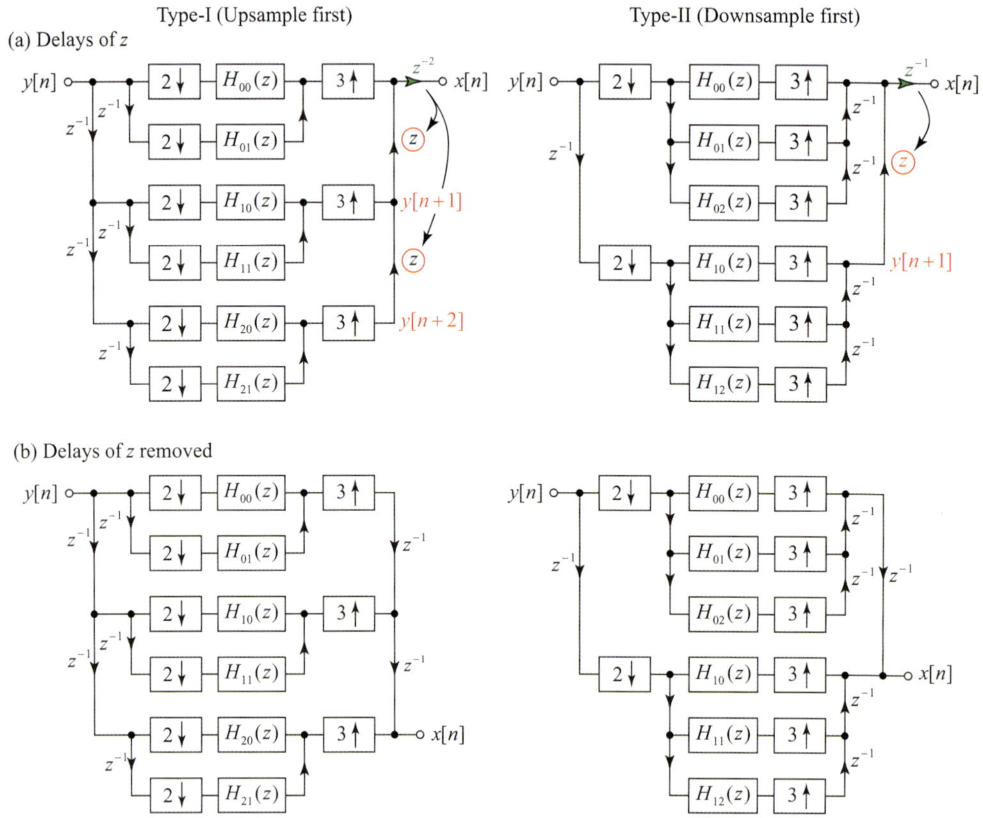

Figure 13.30 Type-I and Type-II resampling architectures for $U=3$ and $D=2$

13.4.6 ★ Matlab implementation of polyphase sample-rate conversion algorithms

Supplementary material provides a set of Matlab programs to implement polyphase downsampling, upsampling and resampling using the principles explained in Section 13.4. While the code presented there is not real-time, it does illustrate a few of the important issues that arise in translating the theory into practice.

13.5 Multistage systems for downsampling and upsampling

In this section, we explore applications of multirate techniques to improve the computational efficiency of algorithms for downsampling and upsampling. The main feature that determines the computational complexity of the algorithms is the design of the anti-aliasing or image-rejection filters, and specifically the order of these filters. In what follows, we will only consider linear-phase FIR filters, whose design parameters can be well estimated by relatively simple formulas. For these filters, the filter order M is inversely proportional to the transition bandwidth $\Delta\omega$:

$$M = \frac{K}{\Delta\omega},$$

where K is a constant that depends on the filter type and filter parameters, for example:

$$\text{Kaiser:} \quad M = \left\lceil \frac{-20\log_{10}(\min(\delta_p, \delta_s)) - 7.95}{2.285\Delta\omega} \right\rceil, \tag{13.31a}$$

$$\text{Parks–McClellan:} \quad M = \left\lceil -\frac{10\log_{10}(\delta_s\delta_p) + 13}{14.6\Delta\omega/2\pi} \right\rceil. \tag{13.31b}$$

The sharper the filter we require, the higher the order M. In the discussion that follows, we will use filter order as the basis of measures of the computational efficiency for multirate downsampling, upsampling and filtering.

13.5.1 Multistage downsampling

To motivate this discussion, consider the problem of downsampling a signal by a factor of D, where D is large. **Figure 13.31a** shows the schematic of downsampling using a single stage of anti-aliasing filter followed by decimation by D. **Figure 13.31b** shows a schematic response of the anti-aliasing filter $H(\omega)$ for downsampling by the values $D=2$ and $D=10$. The filter $H(\omega)$ has a passband ω_p that is determined by the highest frequency of valuable spectral content of the input signal. The stopband frequency ω_s is determined by D, and can be no greater than $\omega_s = \pi/D$ to ensure that there will be no aliasing in the output after decimation by D.

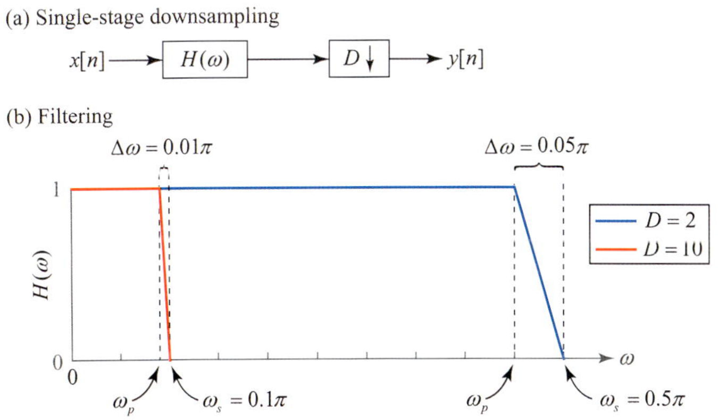

Figure 13.31 Filter bandwidth of anti-aliasing filter

In Section 7.3.2, we argued that the transition bandwidth $\Delta\omega$ of the discrete-time anti-aliasing and image-rejection filters for downsampling should be proportional to the corner frequency $\omega_c = \pi/D$; that is, these filters should have a constant **fractional bandwidth**, defined in Section 7.3.2 as

$$\phi = \frac{\Delta\omega}{\omega_c} = \frac{2\Delta\omega}{(\omega_s - \omega_p)}.$$

For example, the schematic filters depicted in **Figure 13.31b** both have a fractional bandwidth of $\phi = 1/9 = 0.11$. For filters with a constant fractional bandwidth, the order of the filter, M, ends up being directly proportional to D,

$$M = \frac{K}{\Delta\omega} = \frac{K}{\phi\omega_c} = \frac{K}{\phi\pi} D, \tag{13.32}$$

where K is a constant that depends on the type of filter employed. That is, if the transition bandwidth of the filter is small, the order of the filter will be large. In a downsampling algorithm, one simple measure of computational complexity, O, is the number of MAC operations necessary to process each *input* point. In a non-polyphase implementation, referring the number of calculations to the input (rather than to the output) makes sense because the sample rate of the system, f_s, is highest at the input, and hence the peak number of operations per second will be given by $\mathrm{O}f_s$. In a simple downsample-by-D algorithm, such as that diagrammed in **Figure 13.31a**, each input point is processed by an anti-aliasing filter of order M. Hence, O would just be proportional to M, which is in turn just proportional to D by Equation (13.32).[5] However, this is a pretty wasteful strategy. The process of decimation throws away $D-1$ of every D points, and we have already established that it would make more sense to compute only every Dth output of the filter. Then, the number of operations per input point would be lower by a factor of D,

$$\mathrm{O} \triangleq \frac{M}{D} = \frac{K}{\Delta\omega D} = \frac{K}{\phi\pi}, \tag{13.33}$$

which means that the computational complexity of downsampling would be effectively independent of D. (See Problems 13-2 and 13-3 for those results.)

Example 13.5

Determine the number of operations per input point for a single-stage downsampling system with $D = 12$ using a Kaiser anti-aliasing filter with $\omega_p = \pi/24$, $\omega_s = \pi/12$, $\delta_p = 0.1$ and $\delta_s = 0.001$.

➤ Solution:

Here, $\Delta\omega = \omega_s - \omega_p = \pi/24$. The order of filter is calculated using Equation (13.31a),

$$\begin{aligned} A &= -20\log_{10}(\min(\delta_p, \delta_s)) = 60 \\ M &= \frac{A - 7.95}{2.285\Delta\omega} = \frac{K}{\Delta\omega} = \frac{22.78}{\Delta\omega} = 174.0. \end{aligned}$$

Then, $\mathrm{O} = M/D = 14.5$. The order of the Kaiser filter depends on the minimum of δ_p and δ_s, so the value of δ_p is effectively ignored.

One way to reduce the computational requirements on the lowpass filter when D is large is to replace the single-stage downsampler, shown in **Figure 13.32a**, with a cascade of N stages, as shown in **Figure 13.32b**. Each stage comprises an anti-aliasing filter $H_k(\omega)$, followed by decimation by a smaller factor D_k, such that the product of all the decimation factors is $D = D_1 D_2 \cdots D_N$. Since the decimation factor of each stage is smaller than D, the bandwidth of

[5] O should really be proportional to the length of the filter, $N = M + 1$ but, for simplicity, we will use M.

$H_k(\omega)$, the associated filter, can be larger and therefore less computationally complex to implement.

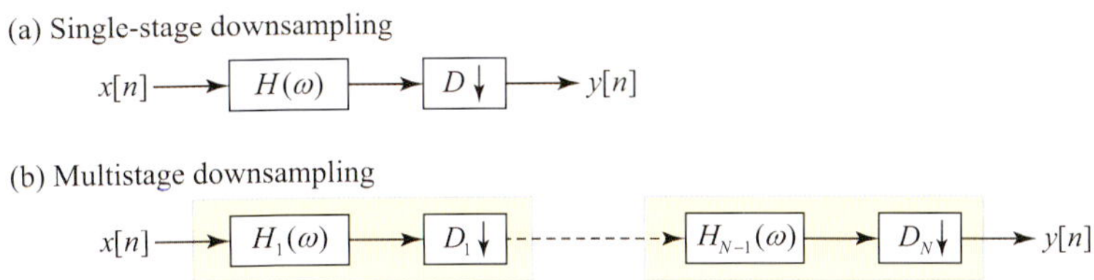

Figure 13.32 Single and multistage downsampling

The schematic of the procedure for the two-stage case is shown in **Figure 13.33a**. Assume that D can be expressed as the product of two integer factors, $D = D_1D_2$. **Figure 13.33b** shows the equivalent (desired) single-stage filter, with the response $H(\omega)$ that we want to result from application of the two-stage procedure. In the two-stage procedure, we first filter by a lowpass filter with response $H_1(\omega)$, as shown in the left column of **Figure 13.33c**. The passband of this filter, $\bar{\omega}_p = \omega_p$, is the same as that of the single-stage filter and is again determined by the highest frequency of interest of the input signal. The stopband of the filter is $\bar{\omega}_s = \pi/D_1$. There will be several (at least two) possible choices for D_1, depending on how many ways D can be factored into the product of integer factors, $D = D_1D_2$. The transition bandwidth of this filter (yellow shaded area) is $\Delta\omega_1 = \bar{\omega}_s - \bar{\omega}_p = \pi/D_1 - \bar{\omega}_p$. The output of the first filter is then decimated by a factor of D_1, with the result that both the passband and stopband are scaled by D_1, as shown in **Figure 13.33d** (blue trace). The second filter then has a passband of $\hat{\omega}_p = D_1\omega_p$ and a stopband of $\hat{\omega}_s = \pi/D_2$ with a resulting transition bandwidth of $\Delta\omega_2 = \hat{\omega}_s - \hat{\omega}_p = \hat{\omega}_s - \pi/D_2$. The result of the second filter stage is shown in **Figure 13.33e**, and the result of the entire two-stage process after the second decimation is shown in **Figure 13.33f**.

Before we do an example with some "real" numbers, note that in the above approach the stopband of the first filter, $\bar{\omega}_s = \pi/D_1$, is always set so that decimation by D_1 brings the frequency of the stopband edge to π. However, we can achieve computational savings by allowing the stopband of the first filter to be *greater* than π/D_1, as shown in the right-hand panels of **Figure 13.33**. **Figure 13.33c** shows what happens when $\bar{\omega}_s = \pi(2D_2 - 1)/D$, which is always greater than π/D_1 (Problem 13-4). To see how this value of $\bar{\omega}_s$ is arrived at, recognize that when decimated by D_1, a stopband edge that is greater than π/D_1 will move to $D_1\omega_s > \pi$, which means that aliasing will occur, as shown by the dotted lines in **Figure 13.33d**. Because $H_1(\omega)$ is periodic, there will be a replica of the aliased response centered at $\omega = 2\pi$ whose left edge extends down to $2\pi - D_1\bar{\omega}_s$. The stopband of the first filter, $\bar{\omega}_s$, must be specified so as to prevent this aliased signal from entering the transition band of the second filter, which has stopband $\hat{\omega}_s = \pi/D_2$. So,

$$\frac{\pi}{D_2} = 2\pi - D_1\bar{\omega}_s.$$

Solving for $\bar{\omega}_s$ gives

$$\bar{\omega}_s = \frac{(2D_2 - 1)\pi}{D_1D_2} = \frac{(2D_2 - 1)\pi}{D}.$$

Hence, the transition bandwidths of the two filters are

$$\Delta\omega_1 = \bar{\omega}_s - \bar{\omega}_p = \pi(2D_2 - 1)/D - \omega_p$$
$$\Delta\omega_2 = \hat{\omega}_s - \hat{\omega}_p = \pi/D_2 - D_1\omega_p = (\pi - D\omega_p)/D_2.$$

The result of the second stage of filtering is shown in **Figure 13.33e**, and the result of the second decimation is shown in **Figure 13.33f**. They are comparable to the results for filtering without allowing aliasing in the first stage.

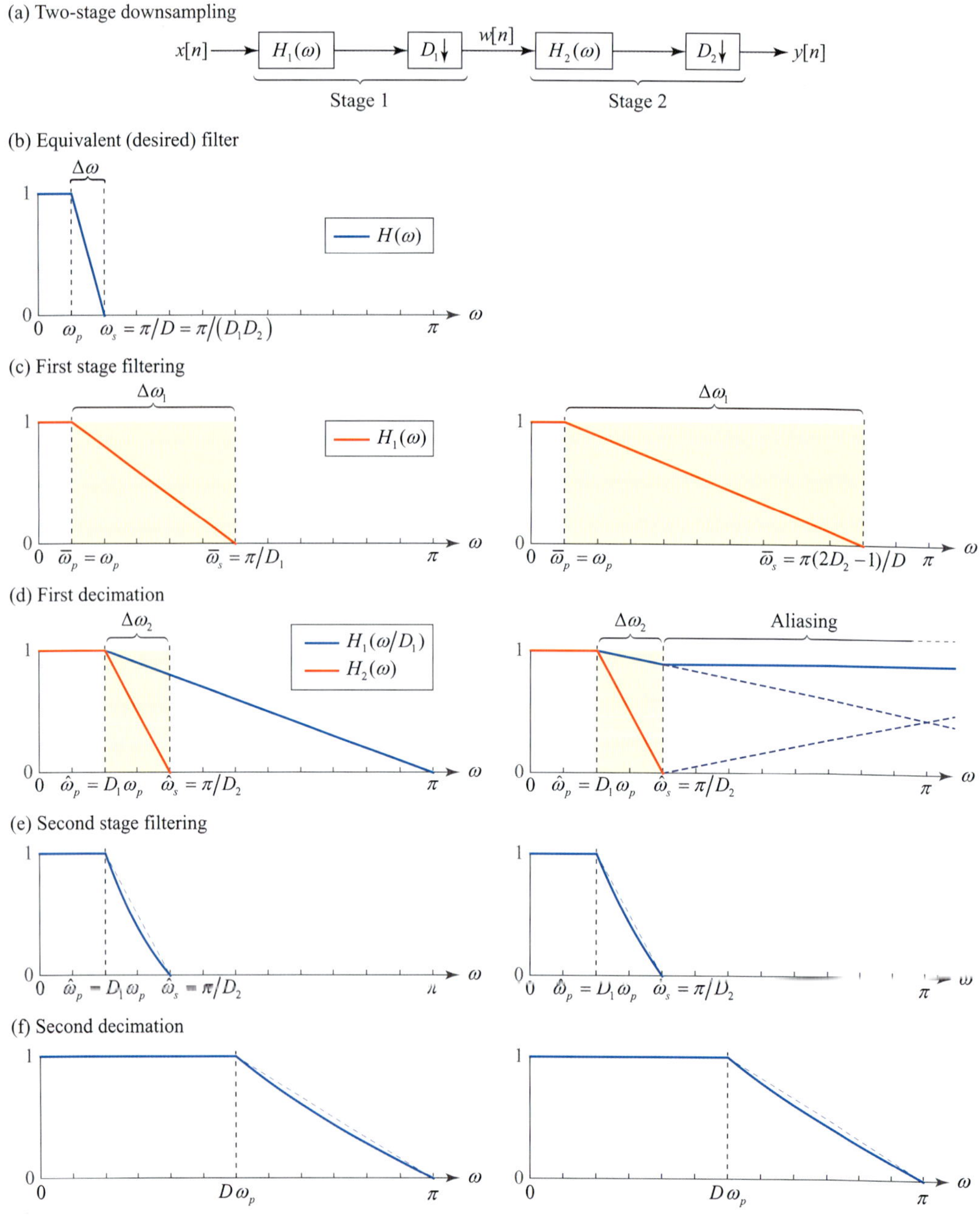

Figure 13.33 Two-stage downsampling

In addition to calculating the bandwidth of filters $H_1(\omega)$ and $H_2(\omega)$, we also need to determine their passband and stopband tolerances. These two filters are designed to do the job of a single equivalent filter $H(\omega)$ that has passband and stopband tolerances of δ_p and δ_s, respectively. While the decimation operations D_1 and D_2 of the multistage approach determine the frequency scaling of $H_1(\omega)$ and $H_2(\omega)$, they do not affect the passband and stopband tolerances of these two filters. From the point of view of determining the magnitude characteristics of the two-stage approach, these two filters effectively appear in cascade, $|H_1(\omega)H_2(\omega)| = |H(\omega)|$. Hence, we should require that

$$\begin{aligned} 1 - \delta_p \le |H_1(\omega)H_2(\omega)| &\le 1 + \delta_p, \qquad 0 \le \omega \le \omega_p \\ |H_1(\omega)H_2(\omega)| &\le \delta_s, \qquad \omega_s \le \omega \le \pi. \end{aligned}$$

In practice, one common approach is simply to specify the passband and stopband tolerances of $H_1(\omega)$ and $H_2(\omega)$ such that each filter has passband tolerance $\delta_p/2$, and then just to use δ_s as the stopband tolerance of both filters.

To determine the number of operations per input point for two-stage downsampling, consider each stage separately. Following the discussion leading to Equation (13.33), the number of operations per input point due to the first stage, O_1, is

$$\mathrm{O}_1 = \frac{K}{\Delta\omega_1 D_1}.$$

For the second stage, we have decimated twice, so the number of operations per input point is lower by a factor of $D_1 D_2 = D$,

$$\mathrm{O}_2 = \frac{K}{\Delta\omega_2 D_1 D_2} = \frac{K}{\Delta\omega_2 D}.$$

Then, the total number of operations per input point for two-stage downsampling is the sum

$$\mathrm{O} = \mathrm{O}_1 + \mathrm{O}_2. \tag{13.34}$$

In two-stage downsampling by a factor $D = D_1 D_2$, the choice of the factors D_1 and D_2 matters, as shown in the following example.

Example 13.6

Determine the total number of operations per input point for a two-stage downsampling system of Example 13.5 using a Kaiser filter with $D = 12$, with aliasing allowed in the first stage.

(a) $D_1 = 2$ and $D_2 = 6$.

(b) $D_1 = 6$ and $D_2 = 2$.

➤ Solution:

(a)

$$\begin{aligned} \Delta\omega_1 &= \pi(2D_2 - 1)/D - \omega_p = 11\pi/12 - \pi/24 = 21\pi/24 \\ \Delta\omega_2 &= \pi/D_2 - D_1\omega_p = \pi/6 - \pi/12 = \pi/12. \end{aligned}$$

So,

$$\mathrm{O}_1 = \frac{K}{\Delta\omega_1 D_1} = \frac{22.78}{(21\pi/24)2} = 4.1$$
$$\mathrm{O}_2 = \frac{K}{\Delta\omega_2 D} = \frac{22.78}{(\pi/12)/12} = 7.3,$$

and

$$\mathrm{O} = \mathrm{O}_1 + \mathrm{O}_2 = 11.4.$$

(b)

$$\Delta\omega_1 = \pi(2D_2 - 1)/D - \omega_p = \pi/4 - \pi/24 = 5\pi/24$$
$$\Delta\omega_2 = \pi/D_2 - D_1\omega_p = \pi/2 - \pi/4 = \pi/4.$$

So,

$$\mathrm{O}_1 = \frac{K}{\Delta\omega_1 D_1} = \frac{22.78}{(5\pi/24)6} = 5.8$$
$$\mathrm{O}_2 = \frac{K}{\Delta\omega_2 D} = \frac{22.78}{(\pi/4)12} = 2.4,$$

and

$$\mathrm{O} = \mathrm{O}_1 + \mathrm{O}_2 = 8.2.$$

(Incidentally, had we computed the computational complexity of a filter length of $N = M + 1$ instead of M, the answers to parts (a) and (b) would have been 12.0 and 8.5, respectively.)

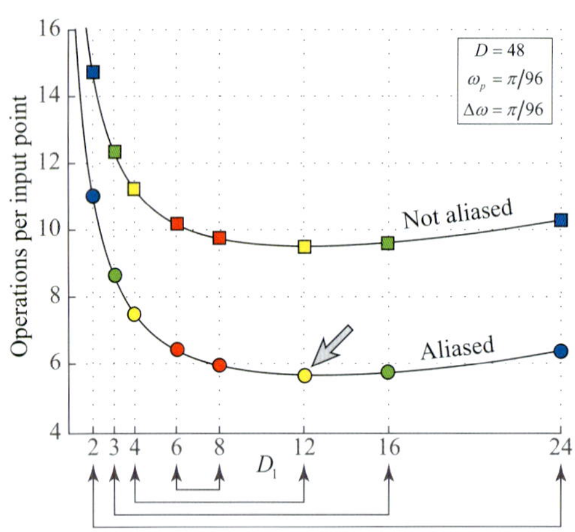

Figure 13.34 Total filter order vs. D_1

Example 13.6 demonstrates that the number of calculations necessary to implement the two-stage solution can be minimized by appropriate choice of D_1 and D_2. This is shown in **Figure 13.34**, which plots the total value of O for the two-stage solution vs. D_1 for aliased and non-aliased implementations when $D = 48$, $\omega_p = \pi/96$ and $\Delta\omega = \pi/96$. The values of O are calculated assuming that both stages are implemented using Kaiser filters designed with $\delta_p = \delta_s = 0.001$. The colored symbols on the graph represent the eight possible integer values of D_1: 2, 3, 4, 6, 8, 12, 16 and 24. Each value of D_1 has a corresponding value of $D_2 = D/D_1$ plotted in the same color, with pairs of D_1D_2 values connected by arrows on the abscissa. The graph shows several things. First, as one might expect, the aliased method always results in a lower total order for any D_1D_2 pair. The graph also shows that there is an optimum value of the D_1D_2 pair and, furthermore, that the order in which filtering is done matters: D_1 should be the larger of the two factors. For example, in this example, the optimum value of $D_1 = 12$ (grey arrow) and its "mate" $D_2 = 4$ results in a minimum total O = 5.7 (see Problem 13-3). Filtering with the same two values in the opposite order (i.e., $D_1 = 4$ and $D_2 = 12$) results in O = 7.5. By way of comparison, the single-stage solution using one Kaiser filter would require O = 696, so any of the two-stage solutions represents a considerable improvement.

The examples in this section have used Kaiser filters. However, other filters (e.g., the optimum "Parks–McClellan" filter) have a lower order for given specifications and can further lower the complexity of downsampling. These filters would also allow one to specify the passband and stopband tolerances separately.

13.5.2 Multistage upsampling

Multirate techniques can also be used to implement multistage upsampling. As with multistage downsampling, the idea is to replace single-stage upsampling by a factor of U, shown in **Figure 13.35a**, with a cascade of N stages, as shown in **Figure 13.35b**, each of which comprises an expansion by a factor of U_k, followed by an image-rejection filter $H_k(\omega)$, such that the product of all the upsampling factors is $U = U_1U_2 \cdots U_N$.

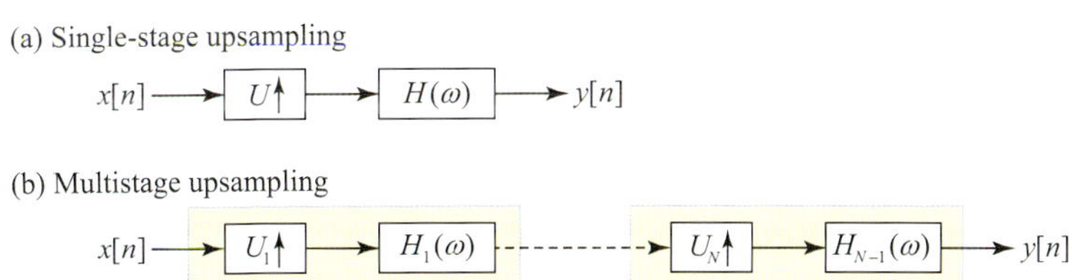

Figure 13.35 Single and multistage upsampling

Figure 13.36a shows a schematic of the procedure for the two-stage case. Assume that the frequency response of the input, $X(\omega)$, has essential frequencies up to the frequency ω_p, as shown in **Figure 13.36b**, and that we wish to upsample by an integer factor U. In the first stage, the input is upsampled by a factor of $U_1 < U$. The resulting spectrum $X(\omega U_1)$ (blue trace) comprises replicas of $X(\omega)$ scaled in frequency by $1/U_1$ and centered at multiples of $2\pi/U_1$, as shown in **Figure 13.36c**. The first image-rejection filter $H_1(\omega)$ (red trace) is designed to remove all except the baseband replica. Accordingly, the passband of the filter needs to match the passband of the replica, $\bar{\omega}_p = \omega_p/U_1$. The transition band of the filter extends from $\bar{\omega}_p$ on the right edge of the baseband to the equivalent point on the left edge of the first replica, which is located at $\bar{\omega}_s = 2\pi/U_1 - \omega_p/U_1 = (2\pi - \omega_p)/U_1$. The output of the first filter, with frequency response $W(\omega)$, shown in **Figure 13.36d**, is the result of the second expansion by a factor of U_2. This expansion results in scaled replicas of $W(\omega)$ at multiples of $2\pi/U_2$ (blue curve). The final image-rejection filter $H_2(\omega)$, with passband $\hat{\omega}_p = \bar{\omega}_p/U_2 = \omega_p/(U_1U_2) = \omega_p/U$ and stopband $2\pi/U_2 - \omega_p/U = (2\pi U_1 - \omega_p)/U$, produces the upsampled signal $Y(\omega)$.

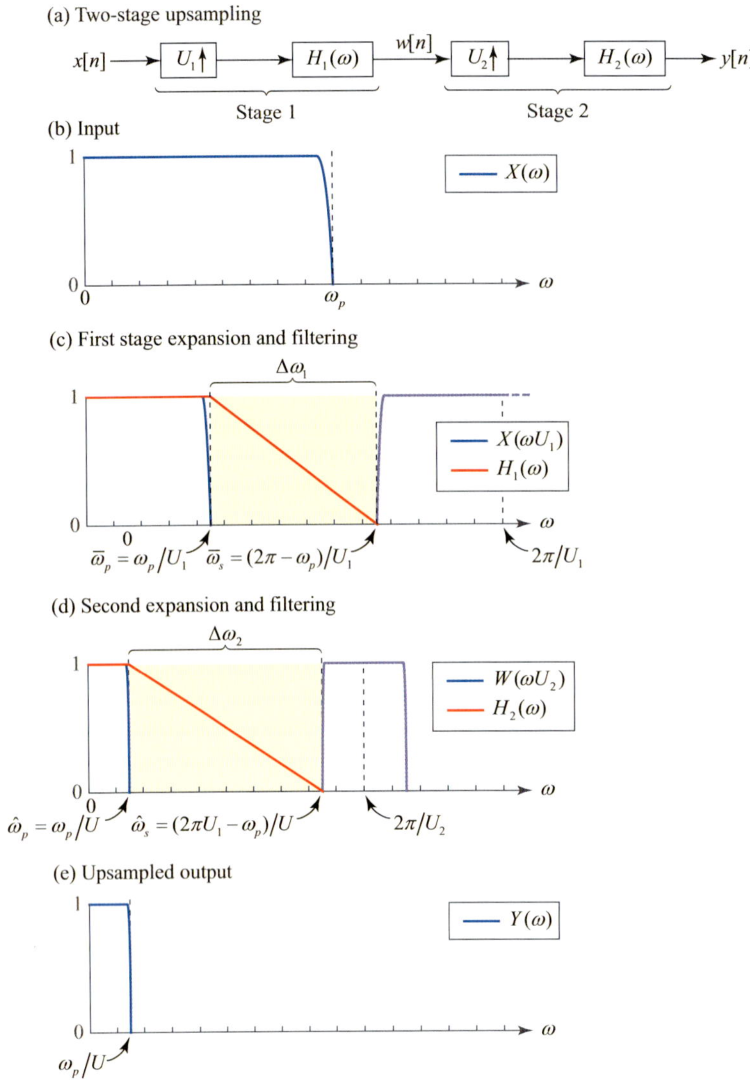

Figure 13.36 Two-stage upsampling

The computational complexity of upsampling algorithms is often defined in terms of the number of multiplications per *output* point. This makes sense since the peak number of operations per second of the system is determined by the output rate Uf_s. Starting from the output of a two-stage upsampler, the expansion in the second stage increases the number of points provided by the first stage ($\omega[n]$ in **Figure 13.36a**) by a factor of U_2. However, as we argued in conjunction with our discussion of decimation, $U_2 - 1$ of those points are zero and would be skipped by a well-designed system. Hence, the number of operations for the second stage should be divided by U_2:

$$\mathrm{O}_2 = \frac{K}{\Delta\omega_2 U_2}.$$

From the point of view of the output, we have upsampled the input twice, so the number of operations per output point is lower by a factor of $U_1 U_2$,

$$O_1 = \frac{K}{\Delta\omega_1 U_1 U_2} = \frac{K}{\Delta\omega_1 U}.$$

Then, the total number of operations per input point for a two-stage upsampling system is the sum

$$O = O_1 + O_2 = \frac{K}{\Delta\omega_1 U} + \frac{K}{\Delta\omega_2 U_2},$$

where the bandwidths of the two filters are

$$\Delta\omega_1 = \bar{\omega}_s - \bar{\omega}_p = (2\pi - \omega_p)/U_1 - \omega_p/U_1 = 2(\pi - \omega_p)/U_1$$
$$\Delta\omega_2 = \hat{\omega}_s - \hat{\omega}_p = (2\pi U_1 - \omega_p)/U - \omega_p/U = 2(\pi U_1 - \omega_p)/U.$$

Figure 13.37 shows the plot of the total filter order O vs. U_1 for a signal with bandwidth $\omega_p = \pi/2$, upsampled by a factor of $U = 48$. As with multistage downsampling, the order in which filtering is done matters. Here, U_1 should be the smaller of the two factors. The optimum value of U_1 in this example is $U_1 = 4$ (arrow) and $U_2 = 12$.

The ordinate values of **Figure 13.37** depend on the particular type of filter used to implement the upsampling, because O in Equation (13.34) is proportional to K, which in turn depends on the type of filter we select. However, the shape of the curve and the minimum value of U_1 do not depend on K.

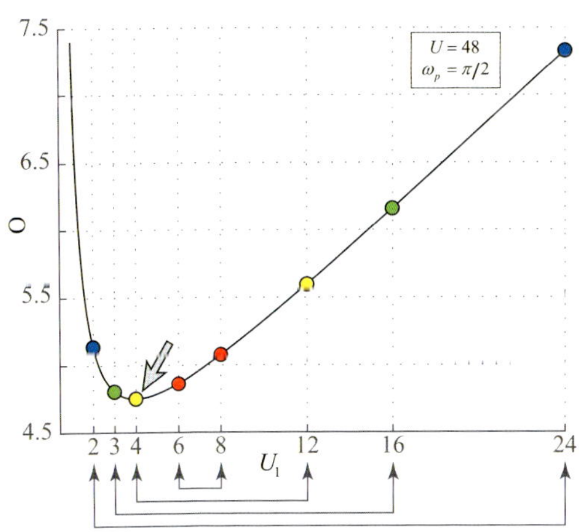

Figure 13.37 Total filter order vs. U_1

Example 13.7

Upsample the sequence $x[n] = 0.5\cos(n\pi/2) + 0.5\cos(n\pi/4)$ by a factor of $U = 48$, in two stages.

➤ Solution:

We will use a Kaiser filter. The code is pretty straightforward. We compute the parameters of the first Kaiser filter using $\bar{\omega}_p = \omega_p/U_1$ and $\bar{\omega}_s = (2\pi - \omega_p)/U_1$, and the parameters of the second Kaiser filter using $\hat{\omega}_p = \omega_p/U$ and stopband $\hat{\omega}_s = (2\pi U_1 - \omega_p)/U$. Then we create the input `x`, expand it by `U1` to form `xu`, filter it with `h1` to form `w`, expand that by `U2` to form `wu` and finally filter the result with `h2` to form the output `y`.

```
U = 48;
U1 = 4;
U2 = U / U1;
wp = 0.5*pi;
ds = 0.01;
fp = wp / pi;
kaiserf = (40-7.95)/2.285;

[M1,fc1, beta] = kaiserord([fp 2-fp]/U1, [1 0], ds*[1 1]);
[M2,fc2] = kaiserord([fp 2*U1-fp]/U, [1 0], ds*[1 1]);
h1 = U1 * fir1(M1, fc1, kaiser(M1+1, beta)); % same beta for both filters
h2 = U2 * fir1(M2, fc2, kaiser(M2+1, beta));

lx = 50;
x = 0.5*[1 1]*cos(wp*[0.5;1]*(0:lx-1)); % create input
xu = zeros(1, U1*lx);
xu(1:U1:end) = x;          % expand by U1
w = filter(h1, 1, xu);     % filter by H1
wu = zeros(1, U*lx);
wu(1:U2:end) = w;          % expand by U2
y = filter(h2, 1, wu);     % filter by H2
```

A portion of the resulting sequence $y[n]$ is shown in **Figure 13.38** (blue trace). Time-aligned points of $x[n]$ are shown with red circles for reference.

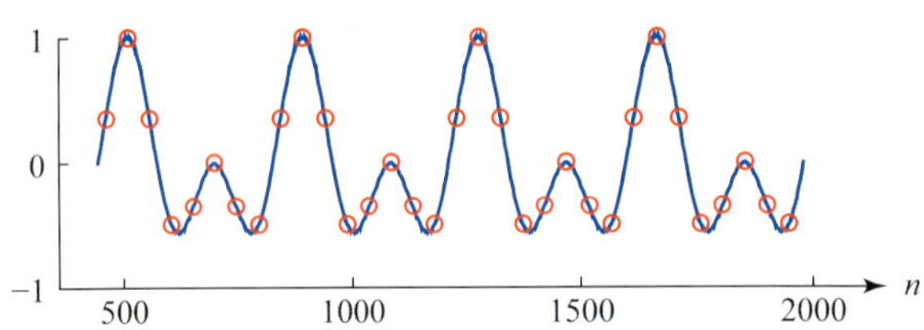

Figure 13.38 Two-stage upsampling example

For this two-stage example, O = 2.9, whereas if implemented with a single stage, O = 9.0.

13.6 Multistage and multirate filtering

Multirate techniques are the basis of a number of approaches aimed at improving the computational efficiency of filtering, particularly filters with low cutoff frequencies. In this section, we will highlight two approaches.

13.6.1 Multistage interpolated FIR (IFIR) filters

Multistage and multirate techniques are the basis of a particularly interesting class of filters called **interpolated finite impulse response (IFIR) filters**. Use of these techniques addresses the problem of designing a filter $H(\omega)$ that has a small passband ω_p and/or a sharp transition band $\Delta\omega$, such as that shown in **Figure 13.39b**.

The IFIR technique replaces this single FIR filter with a cascade of two filters of lower computational complexity. The process starts by designing a **model filter** $G(\omega)$, whose passband $\bar{\omega}_p = U\omega_p$ and transition band $\Delta\omega_1 = U\Delta\omega$ are both an integer factor U greater than those of the desired filter, as shown in **Figure 13.39c**. The left panel of the figure shows an example where $U=2$ and the right panel when $U=4$. The next step is to expand the model filter's impulse response $g[n]$ by the same factor U, by interposing $U-1$ zeros between each pair of points of $g[n]$, thereby forming an expanded impulse response $g_u[n]$. The response of the resulting filter $G(\omega U)$, shown in **Figure 13.39d** (blue trace), has a baseband component with the same

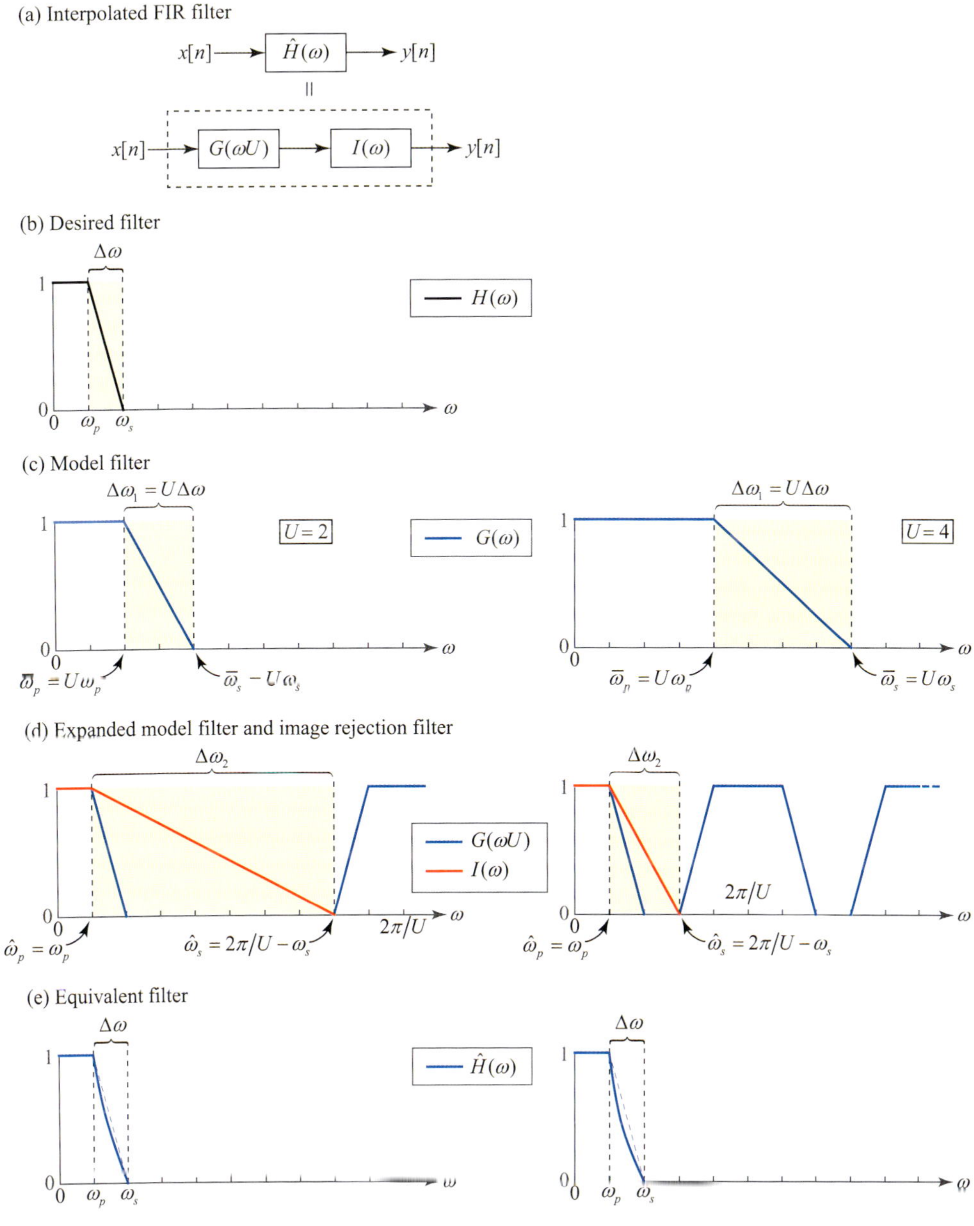

Figure 13.39 Interpolated FIR filters

bandwidth ω_p and transition bandwidth $\Delta\omega$ as $H(\omega)$, but it also has image replicas at multiples of $2\pi/U$. The second filter in the IFIR cascade is an image-rejection filter with response $I(\omega)$, whose function is to filter out these image replicas in much the same way as we discussed in conjunction with the two-stage upsampling technique in Section 13.5.2. This image-rejection filter (red trace) needs to have a passband $\hat{\omega}_p = \omega_p$ that matches that of $H(\omega)$, and a stopband $\hat{\omega}_s = 2\pi/U - \omega_s$ that attenuates all the image replicas.[6] The output of the cascade of two filters, $\hat{H}(\omega)$, is effectively equivalent to filtering by the desired filter $H(\omega)$, as shown in **Figure 13.39e**.

The particular choice of U determines the relative transition bandwidths of the $G(\omega U)$ and $I(\omega)$:

$$\Delta\omega_1 = U\omega_s - U\omega_p = U\Delta\omega$$
$$\Delta\omega_2 = 2\pi/U - \omega_s - \omega_p = 2\pi/U - (\omega_p + \Delta\omega) - \omega_p = 2(\pi/U - \omega_p) - \Delta\omega.$$

To compute the computational requirements of the IFIR filter, consider each filter separately. Even though the order of $G(\omega U)$, the filter with the expanded impulse response $g_u[n]$ is U times that of the model filter $G(\omega)$, the computational complexity of implementing $G(\omega U)$ should still be determined by $K/\Delta\omega_1$, the transition bandwidth of $G(\omega)$, because a reasonable implementation would avoid multiplying the input by the zero values of $g_u[n]$. The complexity of the image-rejection filter $I(\omega)$ should be no worse than $K/\Delta\omega_2$. Hence, for the IFIR filter, a reasonable measure of complexity is

$$\mathrm{O} = \frac{K}{\Delta\omega_1} + \frac{K}{\Delta\omega_2}.$$

Figure 13.39b shows that if $\Delta\omega_1$ is smaller, $\Delta\omega_2$ will be relatively larger (e.g., left panel) and vice versa (e.g., right panel). Hence, we suspect that given the parameters $\Delta\omega$ and ω_p of the desired filter, $H(\omega)$, there will be an optimum value of U that will minimize the computational complexity of the algorithm. To give an idea of the trade-off, **Figure 13.40** shows the relation between O and U for two Kaiser filters designed to emulate a single Kaiser filter with values of

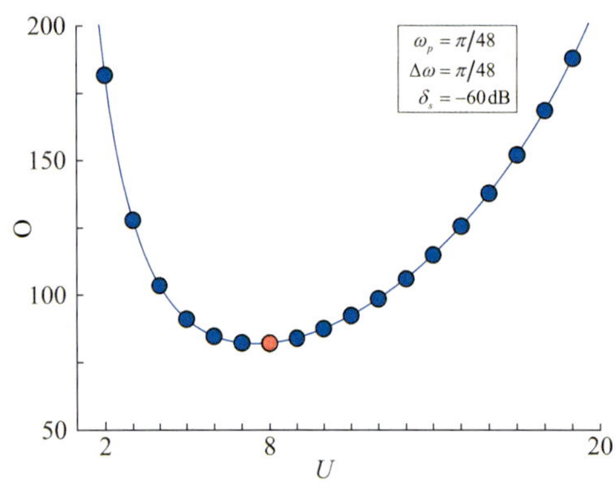

Figure 13.40 Total filter order vs. U

[6]While it is true that the filter does have to attenuate all the image replicas, it need not be a simple lowpass filter. See Example 11.8.

$\omega_p = \pi/48$, $\Delta\omega = \pi/48$ and $\delta_p = \delta_s = -60$ dB. The optimum value of U can be shown to be (Problem 13-1)

$$U = 1/(\sqrt{\Delta\omega/2\pi} + \omega_p/\pi + \Delta\omega/2\pi). \tag{13.35}$$

For the example above, Equation (13.35) gives $U = 7.5$, which we round to $U = 8$, as shown with the red symbol in the figure. Again, the ordinate of the curve in **Figure 13.40** depends upon K, which in turn depends upon the type of filter we select to implement the IFIR filter.

Example 13.8

Design an IFIR lowpass filter equivalent to a single optimal filter with passband $\omega_p = \pi/20$, transition bandwidth $\Delta\omega = \pi/40$, $\delta_p = 0.001$ and $\delta_s = 0.01$.

➤ Solution:

Given the parameters of the desired filter, we first use Equation (13.35) to give the optimum value of the upsampling factor that minimizes O, the sum of the orders of the two filters. From Equation (13.35), $U = 5.73$, which we round up to $U = 6$.

We can use Matlab functions `firpmord` and `firpm` to design optimal filters using the Parks–McClellan algorithm.

```
A = 40;
wp = pi/20;
dw = pi/40;

fp = wp / pi; % firpmord requires fractions of pi
df = dw / pi;
fs = fp + df;
dp = 0.001;
ds = 10^-(A/20);
U = round(1/(sqrt(dw/2/pi)+wp/pi+dw/2/pi)); % find optimum U

% Determine order of filters
[mh, fch, Ah, Wh] = firpmord([fp fs], [1 0], [dp ds]);               % H(w)
[mg, fcg, Acg, Wcg] = firpmord(U*[fp fs], [1 0], [dp/2 ds]);         % G(w)
[mi, fci, Aci, Wci] = firpmord([fp 2/U-fs], [1 0], [dp/2 ds]);  % I(w)

% Create impulse responses of filters
hh = firpm(mh, fch, Ah, Wh);       % H(w)
hg = firpm(mg, fcg, Acg, Wcg);     % G(w)
hgU = zeros(1, U*(mg+1));          % allocate upsampled response (length = mg+1)
hgU(1:U:end) = hg;                 % and create G(wU)
hi = firpm(mi, fci, Aci, Wci);     % I(w)
hhat = conv(hgU, hi);              % Hhat(w)
```

The results are plotted in **Figure 13.41**. **Figure 13.41a** shows the frequency response of the desired filter $H(\omega)$ (thin black trace), corresponding to `hh` in the code, with its transition region shaded in grey. The blue curve is the model filter $G(\omega)$, designed with passband and transition bandwidths (shown in yellow) that are a factor of $U = 6$ greater than those of $H(\omega)$. This filter corresponds to `hg` in the code. The thin blue trace in **Figure 13.41b** is $G(\omega U)$, the result of expanding the impulse response of $G(\omega)$ by a factor of U. This corresponds to `hgU` in the code. The red trace is the image-rejection filter $I(\omega)$, which

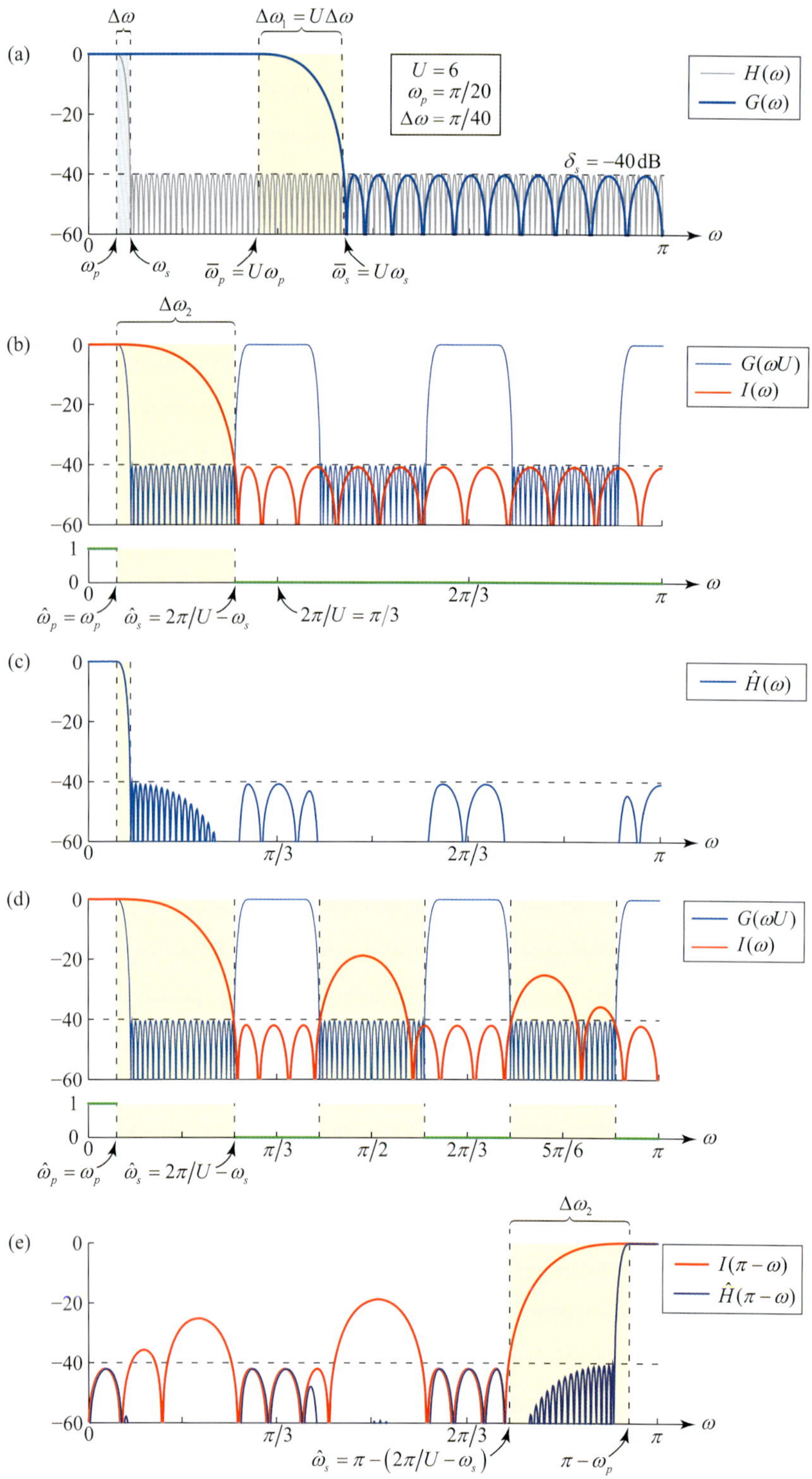

Figure 13.41 Design of IFIR lowpass and highpass filters

corresponds to `hi`. This filter is specified by the frequency vector `[fp 2/U-fs]` and amplitude vector `[1 0]`, which tell `firpmord` that the filter has a single passband from $0 \le |\omega| < \omega_p$ with a nominal amplitude of $1 \pm \delta_p$, and a single stopband that extends from $2\pi/U - \omega_s < |\omega| < \pi$. The green trace in the small panel below the main panel indicates the frequencies and amplitudes of these passband and stopband specifications. The yellow shaded region in the panel shows the transition region of the resulting filter.

The blue trace in **Figure 13.41c** shows $\hat{H}(\omega) = G(\omega U)I(\omega)$, the final result of applying $I(\omega)$ to the expanded response $G(\omega U)$, and corresponds to `hhat` in the code. In this example, the original filter $H(\omega)$ would have had order O = 209 if it had been implemented by a single optimal filter. The IFIR filters $G(\omega)$ and $I(\omega)$ have orders $O_1 = 37$ and $O_2 = 26$, respectively, for a total of $O = O_1 + O_2 = 63$, which represents considerable savings.

The image-rejection filter in **Figure 13.41b** has a single transition band and stopband, but this is not the only possibility. The red trace in **Figure 13.41d** shows the response of an image-rejection filter that has multiple transition and stopbands, which are diagrammed in the small lower panel. The passband is the same as in **Figure 13.41b**, but there are three stopbands located around the image frequencies and three "don't care" transition zones. The resulting filter, designed by specifying these parameters in `firpm`, has big deviations in the transition zone, but these do not matter to the final response $\hat{H}(\omega)$ because the product $G(\omega U)I(\omega)$ is still less than δ_s. As long as $I(\omega)$ appropriately rejects the images, $\hat{H}(\omega)$ will meet the specifications.

Finally, lowpass filters are not the only filters that can be designed with the IFIR method. **Figure 13.41e** shows an example of a highpass filter designed with this method. Highpass IFIR filters can be designed as long as U is even, in which case the upsampled response $G(\omega U)$ is a mirror image of the baseband replica located around $\omega = \pi$. To keep only this replica, all one has to do is create a frequency-reversed version of $I(\omega)$, namely $I(\pi - \omega)$, as shown in **Figure 13.41e** (red trace). To create $I(\pi - \omega)$ recall that

$$I(\pi - \omega) = \sum_{n=-\infty}^{\infty} i[n]e^{-j(\pi-\omega)n} = \sum_{n=-\infty}^{\infty} (i[n]e^{-j\pi n})e^{-j\omega n} = \sum_{n=-\infty}^{\infty} (i[n](-1)^n)e^{-j\omega n}.$$

So, we just negate every other point of `hi`. In Matlab,

```
hi_flipped = hi .* (-1).^(0:mi);
```

Applying this filter to $G(\omega U)$ results in the highpass response $H(\pi - \omega)$ in **Figure 13.41e** (blue trace).

13.6.2 Multirate lowpass filtering

The IFIR technique of the preceding section is a multistage technique, but the sample rate is the same in the cascade of the two filters. **Figure 13.42a** shows a multistage, multirate approach designing lowpass filters with both a steep transition region and a low cutoff frequency. The effective lowpass filter we want, $H(\omega)$, shown in **Figure 13.42b**, has passband ω_p and stopband ω_s. To achieve these specifications, the input is first downsampled by a factor of U to reduce the sample rate, then filtered at the lower rate by a so-called kernel filter $G(\omega)$, which has been designed with more relaxed specifications. Then, the result is upsampled by U to the original sample rate. The downsampling stage comprises an anti-aliasing filter $I(\omega)$, whose passband $\hat{\omega}_s$ is equal to the desired passband $\hat{\omega}_p = \omega_p$. Setting the stopband to $\hat{\omega}_s = 2\pi/U - \omega_s$ both removes components in the input signal that could alias when the signal is decimated and also allows the same filter design to be used for the image-rejection filter $I(\omega)$ in the upsampling stage. The passband and stopband of the kernel filter are designed with the specifications $\bar{\omega}_p = U\omega_p$ and $\bar{\omega}_s = U\omega_s$, respectively, as shown in **Figure 13.42c**, which means that this filter will have a lower order than $H(\omega)$. When the output of $G(\omega)$ is expanded by U, the baseband image of the resulting spectrum has the passband and stopband characteristics of the desired filter $H(\omega)$. The image-rejection filter at the end of the chain, which has response $I(\omega)$, attenuates all the spectral replicas of the output of the expander with the exception of the baseband.

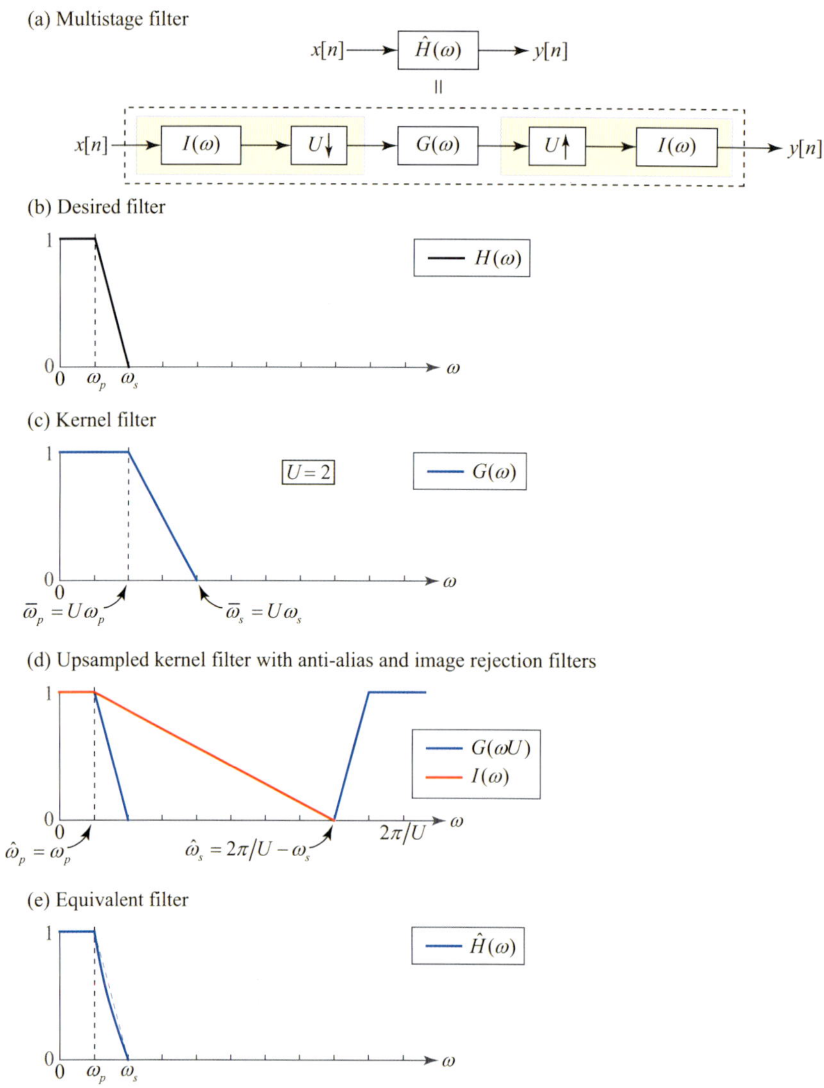

Figure 13.42 Multistage lowpass filtering

Comparing this method with the IFIR method, you can see that the kernel filter $G(\omega)$ of **Figure 13.42c** plays the same role as the model filter $G(\omega)$ of **Figure 13.39c**. An essential difference between these methods is that in the IFIR method, the input is processed at the input sample rate by a filter that has already been designed to have response $G(\omega U)$, whereas with the multirate method of this section, a filter with this response is effectively created by multirate processing of the input: by decimating the input by U, filtering by $G(\omega)$ and then upsampling the output by U. In both techniques, the final image-rejection filter $I(\omega)$ preserves only the baseband of the filtered output.

13.7 Special filters for multirate applications

This section describes two special types of FIR filters that we will find particularly useful in the multirate signal processing applications, such as the design of interpolators, as well as some high-efficiency filtering techniques, which are to be presented in supplementary material.

13.7.1 Half-band filters

As the name implies, half-band filters have a frequency response that passes half the frequency band, either the low-frequency half, $|\omega|<\pi/2$, or high-frequency half, $\pi/2<|\omega|<\pi$. The defining relation for a half-band filter in the frequency domain is

$$H(\omega)+H(\omega-\pi)=1, \tag{13.36}$$

or, equivalently, in terms of the z-transform,

$$H(z)+H(-z)=1. \tag{13.37}$$

Let $H_{LP}(\omega)\triangleq H(\omega)$ represent a half-band lowpass filter; then $H_{HP}(\omega)\triangleq H(\omega-\pi)$ is the corresponding half-band highpass filter. For example, **Figure 13.43a** shows the frequency response of an ideal, non-causal, half-band lowpass filter, $H_{LP}(\omega)$ (blue trace), and the corresponding half-band highpass filter, $H_{HP}(\omega)$ (red trace),

$$H_{LP}(\omega)=\begin{cases}1, & |\omega|<\pi/2\\ 0, & \pi/2<|\omega|<\pi\end{cases}$$

$$H_{HP}(\omega)=H_{LP}(\omega-\pi)=\begin{cases}0, & |\omega|<\pi/2\\ 1, & \pi/2<|\omega|<\pi.\end{cases}$$

By Equation (13.36), $H_{LP}(\omega)$ and $H_{HP}(\omega)$ form a pair of **complementary filters**, meaning that the sum of their frequency responses is unity,

$$H_{LP}(\omega)+H_{HP}(\omega)=H_{LP}(\omega)+H_{LP}(\omega-\pi)=1. \tag{13.38}$$

As we will now show, Equation (13.37) imposes significant conditions on the impulse responses of these half-band filters. If $H(z)$ is the transform of a half-band filter with the impulse response $h[n]$, then, $H(-z)$ is the transform of $h[n](-1)^n$, since

$$H(-z)=\sum_{n=-\infty}^{\infty}h[n](-z)^{-n}=\sum_{n=-\infty}^{\infty}(h[n](-1)^{-n})z^{-n}=\sum_{n=-\infty}^{\infty}(h[n](-1)^{n})z^{-n}.$$

Hence, the inverse transform of Equation (13.37) is

$$h[n]+h[n](-1)^n=h[n](1+(-1)^n)=\delta[n].$$

This means that every even point of $h[n]$ except the point at $n = 0$ must be zero, so

$$h[2n] = \begin{cases} 0.5, & n = 0 \\ 0, & n \neq 0 \end{cases}. \tag{13.39}$$

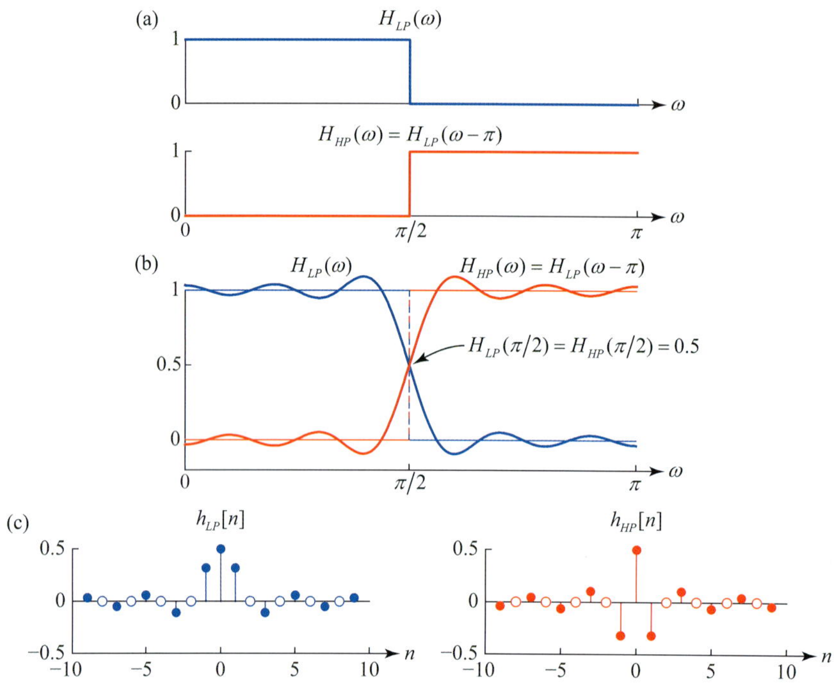

Figure 13.43 Half-band filters

You can show that Equation (13.39) is both a necessary and sufficient condition for a half-band filter; namely that any filter that satisfies Equation (13.39) also satisfies Equation (13.37) and is therefore a half-band filter (Problem 13-7). This important result points the way towards the design of practical half-band FIR filters. As a simple example, **Figure 13.43c** shows the impulse responses $h_{LP}[n]$ (blue) and $h_{HP}[n]$ (red), which have been created by truncating the infinite-length impulse response of the ideal half-band filters symmetrically to a finite length, N points, where N is odd,

$$\begin{aligned} h_{LP}[n] &= \tfrac{1}{2}\operatorname{sinc} n\pi/2, \qquad -(N-1)/2 \le n \le (N-1)/2 \\ h_{HP}[n] &= \delta[n] - h_{LP}[n] \\ &= \delta[n] - \tfrac{1}{2}\operatorname{sinc} n\pi/2, \quad -(N-1)/2 \le n \le (N-1)/2. \end{aligned}$$

The corresponding frequency responses $H_{LP}(\omega)$ and $H_{HP}(\omega)$ are shown in **Figure 13.43b**. Because Equation (13.39) is satisfied, these finite-length filters form a complementary pair. At the midpoint of frequency, $\omega_c = 0.5\pi$, Equation (13.36) gives $H_{LP}(\pi/2) = 1 - H_{LP}(\pi/2)$, so $H_{LP}(\pi/2) = 0.5$. Subtracting 0.5 from Equation (13.36) gives

$$H_{LP}(\omega) - 0.5 = 0.5 - H_{LP}(\pi - \omega).$$

This says that $H_{LP}(\omega)$ is symmetric in amplitude about a point of symmetry at $\omega_c = \pi/2$. So, if $H_{LP}(\omega)$ has passband frequency $\omega_p = 0.5\pi - \Delta\omega/2$, where $\Delta\omega$ is the width of the transition band, then the stopband frequency is $\omega_s = 0.5\pi + \Delta\omega/2$, such that $\omega_s + \omega_p = \pi$. Moreover, the passband and stopband attenuations are equal, $\delta_p = \delta_s$.

Looking back at the FIR filters presented in Chapter 7, you will see that all the filters of Sections 7.3–7.6 (e.g., Hamming, Kaiser, spline, raised-cosine) are essentially based on windowing or otherwise modifying the non-zero coefficients of the impulse response of the ideal lowpass filter $(\omega_c/\pi)\,\mathrm{sinc}\,\omega_c n$. Hence, the methods used to design these filters can also be used to design half-band filters when $\omega_c = \pi/2$. Other methods, such as the Parks–McClellan algorithm, can also be used to design half-band filters, as we will show in a moment.

The time-domain characteristics of half-band filters will turn out to be particularly useful in Section 13.11, where we show how half-band filters can be used to implement sharp multistage interpolators, filters and filter-banks in an efficient manner. For now, notice that nearly one-half of the points of an impulse response of a half-band filter of length N are zero, so only at most $(N+1)/2$ arithmetic operations (e.g., multiplications) are needed to implement the convolution of each point of an input. Actually, because the impulse response is also symmetric about $(N+1)/2$, we only need $\lfloor (N+1)/4 \rfloor + 1$ points. That is quite a saving!

As we have mentioned, the order of the half-band filter, M, must be even so that the length $N = M+1$ will be odd, as required for a Type-I filter. But, we also want to ensure that $M/2 = (N-1)/2$ will be odd as well. To see why that is the case, look at the impulse response of the half-band filter $h_{LP}[n]$ in the left panel of **Figure 13.43c**. This filter has order $M = 18$ (i.e., $N = 19$). Consider what would happen if $M = 20$ (i.e., $N = 21$). The two endpoints of the filter's response would then be zero: $h_{LP}[\pm(N-1)/2] = h_{LP}[\pm M/2] = 0$. Those zero values would contribute nothing to the filter's response. Reducing the filter order to $M = 18$ effectively removes those two points. Alternately, increasing the order to $M = 22$ would create the next larger filter whose endpoints are non-zero. All this means that we really want the order to be of the form

$$M = 4k - 2, \tag{13.40}$$

where k is an integer.

So far, we've assumed that the half-band filters $h_{LP}[n]$ and $h_{HP}[n]$ are non-causal, symmetric and centered about the point $n = 0$. Now, we can create causal half-band filters $g_{LP}[n]$ and $g_{HP}[n]$ of even order M simply by delaying the responses of the associated non-causal filters,

$$\begin{aligned} g_{LP}[n] &\triangleq h_{LP}(n - M/2) \\ g_{HP}[n] &\triangleq h_{HP}(n - M/2), \end{aligned} \tag{13.41}$$

so that

$$g_{LP}[n] + g_{HP}[n] = \delta[n - M/2].$$

In terms of the z-transform,

$$G_{LP}(z) = H_{LP}(z)z^{-M/2}$$
$$G_{HP}(z) = H_{HP}(z)z^{-M/2}.$$

Using Equation (13.38), we conclude that

$$G_{HP}(z) + G_{LP}(z) = (H_{HP}(z) + H_{LP}(z))z^{-M/2} = z^{-M/2}.$$

One final observation that we will employ a bit later is that

$$G_{HP}(z) = H_{HP}(z)z^{-M/2} = H_{LP}(-z)z^{-M/2} = -H_{LP}(-z)(-z)^{-M/2} = -G_{LP}(-z), \quad \textbf{(13.42)}$$

where we have noted that $H_{HP}(z) = H_{LP}(-z)$ and that $z^{-M/2} = -(-1)^{-M/2}z^{-M/2} = -(-z)^{-M/2}$. For the last step, we used the fact that M satisfies Equation (13.40), so $(-1)^{-M/2} = (-1)^{-(4k-2)/2} = (-1)^{-(2k-1)} = -1$, for all integers, k. An extended example will pull all these points together.

Example 13.9

Use Matlab to design a pair of half-band filters $h_{LP}[n]$ and $h_{HP}[n]$ using the Parks–McClellan method with $\Delta\omega = 0.1\pi$ and $\delta_p = \delta_s = 0.01$. Check that the even points of $h_{LP}[n]$ are zero and that $H_{LP}(\pi/2) = 0.5$.

➤ Solution:

We are going to do this using two methods, one straightforward and one clever. Both methods start by specifying the parameters of the filter in the frequency domain; namely,

$$\begin{aligned} \omega_p &= \omega_c - \Delta\omega/2 = 0.5\pi - 0.05\pi = 0.45\pi \\ \omega_s &= \omega_c + \Delta\omega/2 = 0.5\pi + 0.05\pi = 0.55\pi \end{aligned},$$

and $\delta_p = \delta_s = 0.01$.

Here is the code for this piece:

```
fc = 0.5;
df = 0.1;
fp = fc - 0.5 * df;
fs = fc + 0.5 * df;
dp = 0.01;
ds = 0.01;
```

Now we use `firpmord` to return the parameters the function `firpm` will use to create the filter: the order `M`, cutoff frequencies `f` and amplitudes `A`. (It is not necessary to include the fourth parameter returned by this function, the weights.)

```
[M, f, A] = firpmord([fp fs], [1 0], [dp ds]);
M = 4*ceil((M+2)/4)-2; % round order up to next higher odd value of M/2.
```

That last line results from the requirements of Equation (13.40). Given an initial value of M, we may need to round M up to

$$4\left\lceil \frac{M+2}{4} \right\rceil - 2. \quad \textbf{(13.43)}$$

In this example, the value of `M` that `firpmord` returns is 39, which gets rounded up to 42.

Straightforward method

All that remains is to call `firpm` to make the lowpass filter with the parameters computed by `firpmord` and then create the highpass filter by using the relation $g_{HP}[n] = \delta[n - M/2] - g_{LP}[n]$, which is the equivalent of $h_{HP}[n] = \delta[n] - h_{LP}[n]$. Here is the code:

```
glp = firpm(M, f, A);
ghp = -glp;
ghp(M/2+1) = ghp(M/2+1)+1;
```

The resulting amplitudes of the transforms of $H_{LP}(\omega)$ and $H_{HP}(\omega)$ are shown in **Figure 13.44**.

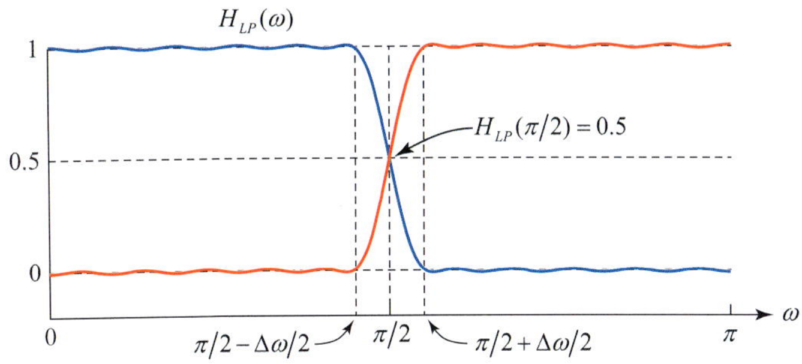

Figure 13.44 Half-band example

Now, let us check to make sure that the even points (except the center point) are zero:

```
>> max(abs(glp([2:2:M/2 M/2+3:2:end]))) % Are the even points really zero?
ans =
    1.4469e-005
```

To check that $H_{LP}(\pi/2) = 0.5$, we could compute the transform, but let's cheat with this one-liner:

```
>> real(ghp*((-j).^(-M/2:M/2)'))
ans =
    0.5000
```

What we have done here is evaluate $H_{LP}(\pi/2)$ directly,

$$H_{LP}(\pi/2) = \sum_{n=-(M-1)/2}^{(M-1)/2} h_{LP}[n]e^{-j\pi n/2} = \sum_{n=-(M-1)/2}^{(M-1)/2} h_{LP}[n](-j)^n.$$

The `real` just removes the vestigial imaginary part.

Clever method

The following method uses a clever "trick" to derive the impulse response of the lowpass half-band filter.[7] The method has more to recommend it than just cleverness (though that alone might be enough). It uses properties of FIR filters and upsampling to derive an impulse response for $h_{LP}[n]$ whose even points are *exactly* zero. Furthermore it directly gives us the filters necessary to implement the half-band filters in polyphase form, which we will use to advantage in Section 13.7.2.

[7] Vaidyanathan, P. P. and Nguyen, T. Q. (1987).

Here is the code:

```
g = firpm(M/2, [0 2*fp], [1 1]) ; % design the even filter
g2 = zeros(1, M+1);
g2(1:2:end) = g;                  % create the expanded filter
h = 0.5*g2;                       % scale it
h(M/2+1) = 0.5;                   % and add an impulse at n = M/2+1
```

We start with the same filter specifications as the straightforward method and derive the same value of M, but now create a full-band FIR filter with impulse response $g[n]$ that has *half* the order of the desired half-band filter, a passband frequency of $2\omega_p$ and a stopband frequency of $\omega_s = \pi$. Because the order $M/2 = 42/2 = 21$ is odd, the result will be an even length ($N = M/2 + 1 = 22$) symmetric (Type-II) FIR filter with passband and stopband tolerances of $2\delta_p$ and $2\delta_s$, respectively, as shown in **Figure 13.45a** (blue trace). In Chapter 3, we established that the amplitude response of this type of filter is zero at $\omega = \pi$ and antisymmetric about $\omega = \pi$ (i.e., $A(\pi - \omega) = -A(\pi + \omega)$, as shown in **Figure 13.45b** (blue trace, labeled $G(\omega)$).

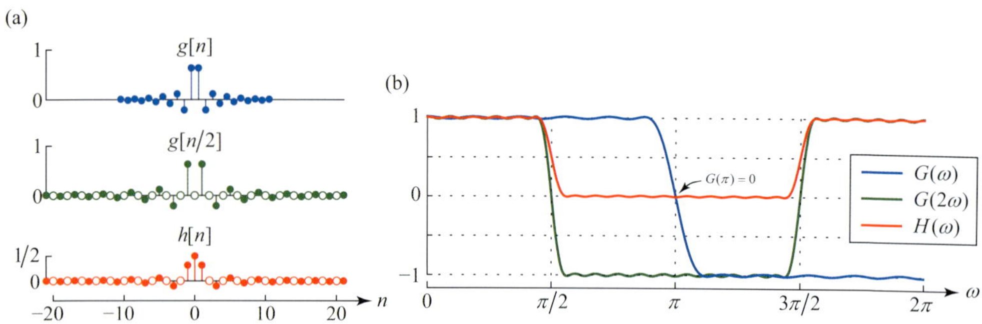

Figure 13.45 Half-band filter "trick"

The next step is to expand $g[n]$ by a factor of two, by inserting zeros between each point of $g[n]$, resulting in the odd-length, ($N = M + 1 = 43$) sequence (g2 in the code) shown in **Figure 13.45a** (green trace), whose frequency response $G(2\omega)$ is "contracted" by a factor of two (**Figure 13.45b**, green trace). The passband frequency of $G(2\omega)$ is therefore moved back from $2\omega_p$ to ω_p. The frequency response of the desired half-band filter $H(\omega)$ is obtained by scaling the amplitude of $G(2\omega)$ by one-half and adding one-half: $H(\omega) = 0.5G(2\omega) + 0.5$ (**Figure 13.45b**, red trace). In the time domain, this is achieved by scaling g2 by one-half and adding an impulse of height one-half to the center position (since $\mathfrak{F}\{0.5\delta[n]\} = 0.5$), as shown in **Figure 13.45b** (red trace). The final filter has one-half the passband and stopband tolerances of the full-band filter, namely δ_p and δ_s, as required.

Matlab's Filter Design Toolbox has a function, firhalfband, that can design half-band filters for you given appropriate specifications.

13.7.2 Polyphase downsampling and upsampling using half-band filters

Half-band filters are frequently used in downsampling and upsampling applications because they can be implemented very efficiently. The impulse response of the half-band lowpass filter $h_{LP}[n]$ is symmetric, and almost half the points are zero. When used in a downsampling-by-two or upsampling-by-two application, half-band filters cry out for an efficient polyphase implementation. To show this, consider $G_{LP}(z)$, the z-transform of a causal half-band lowpass filter of even order M:

$$\begin{aligned} G_{LP}(z) &= \sum_{n=0}^{M} g_{LP}[n]z^{-n} \\ &= g_{LP}[0]z^{0} + 0\cdot z^{-1} + g_{LP}[2]z^{-2} + \cdots + g_{LP}[M/2]z^{-M/2} + 0\cdot z^{-(M/2+1)} \\ &\quad + g_{LP}[M/2+2]z^{-(M/2+2)} + \cdots + g_{LP}[M]z^{-M}. \end{aligned}$$

Pull the center point, $g_{LP}[M/2] = 0.5$, out of the summation and compact the remaining sum by removing the zero points:

$$G_{LP}(z) = \tfrac{1}{2}z^{-M/2} + \sum_{n=0}^{M/2} \underbrace{g_{LP}[2n]}_{g_0[n]} z^{-2n} = \tfrac{1}{2}z^{-M/2} + \sum_{n=0}^{M/2} g_0[n]z^{-2n},$$

where $g_0[n] = g_{LP}[2n]$ is a decimated impulse response of even length $M/2+1$, comprising all the non-zero points of the causal response $g_{LP}[n]$ except the center point $g_{LP}[M/2]$. This $g_0[n]$ is *exactly* the same sequence `g` that we generated by the "clever method" of Example 13.9. Given that the z-transform of $g_0[n]$ is

$$G_0(z) \triangleq \sum_{n=0}^{M/2} g_0[n]z^{-n},$$

$G_{LP}(z)$ becomes

$$G_{LP}(z) = G_0(z^2) + \tfrac{1}{2}z^{-M/2}. \tag{13.44}$$

Incidentally, in Example 13.9, `h = 0.5*g2` is $G_0(z^2)$ and `h(M/2+1) = 0.5` is $(1/2)z^{-(M-1)/2}$. Equation (13.44) can now be expressed in polyphase form:

$$\begin{aligned} G_{LP}(z) &= G_0(z^2) + \tfrac{1}{2}z^{-M/2} \\ &= G_0(z^2) + z^{-1}\left(\tfrac{1}{2}z^{-M/2+1}\right) \\ &= G_0(z^2) + z^{-1}\left(\tfrac{1}{2}z^{-(M-2)/2}\right) \\ &= G_0(z^2) + z^{-1}\underbrace{\left(\tfrac{1}{2}(z^2)^{-(M-2)/4}\right)}_{G_1(z^2)} \\ &= G_0(z^2) + z^{-1}G_1(z^2), \end{aligned} \tag{13.45}$$

where we define $G_1(z) \triangleq (1/2)z^{-(M-2)/4}$. Use Equation (13.42) to derive the complementary highpass half-band filter,

$$\begin{aligned} G_{HP}(z) &= -G_{LP}(-z) = -G_0((-z)^2) - \tfrac{1}{2}(-z)^{-M/2} = -G_0(z^2) - (-1)^{-M/2}\tfrac{1}{2}z^{-M/2} \\ &= -G_0(z^2) + \tfrac{1}{2}z^{-M/2}. \end{aligned}$$

The "+" sign on the last term arises because, from Equation (13.40), we require that $M = 4k - 2 = 2(2k-1)$, where k is an integer, so the quantity $(-1)^{-M/2} = (-1)^{-(2k-1)} = -1$. Putting this in polyphase form in the same manner as Equation (13.45), we get

$$G_{HP}(z) = -G_0(z^2) + z^{-1}G_1(z^2). \tag{13.46}$$

Compared with Equation (13.45), the only difference is the sign of the first term.

Half-band filters are useful in applications in which we downsample by two (i.e., $D = 2$) or upsample by two ($U = 2$), as schematized in **Figure 13.46**. In downsampling (**Figure 13.46a**, top panel), the lowpass half-band filter $H_{LP}(z)$ is replaced by the polyphase form, Equation (13.45) (middle panel). Then, the order of decimation and filtering is interchanged using the multirate downsampling identity (shown in **Figure 13.21**) leading to an efficient architecture (bottom panel) with two polyphase filters, each of which operates at one-half the input sample rate. The first polyphase term, $G_0(z)$, is an even-length filter with $M/2+1$ symmetrical coefficients, so we only need half the multipliers, that is,

$$\begin{aligned} G_0(z) &= \sum_{n=0}^{M/2} g_0[n]z^{-n} = \sum_{n=0}^{(M-2)/4} \left(g_0[n]z^{-n} + g_0[M/2-n]z^{-(M/2-n)}\right) \\ &= \sum_{n=0}^{(M-2)/4} g_0[n]\left(z^{-n} + z^{-(M/2-n)}\right). \end{aligned}$$

The second polyphase term, $G_1(z)$, is just the transform of a shifted and scaled impulse $g_1[n] = (1/2)\delta[n-(m-2)/4]$. The total number of multiplications per input point (decimation) or output point (interpolation) is thus $(M/2+1)/2+1$.

Exactly the same arguments apply to the case of polyphase upsampling using half-band filters. In upsampling (**Figure 13.46b**), the lowpass half-band filter $G_{LP}(z)$ is again replaced by the

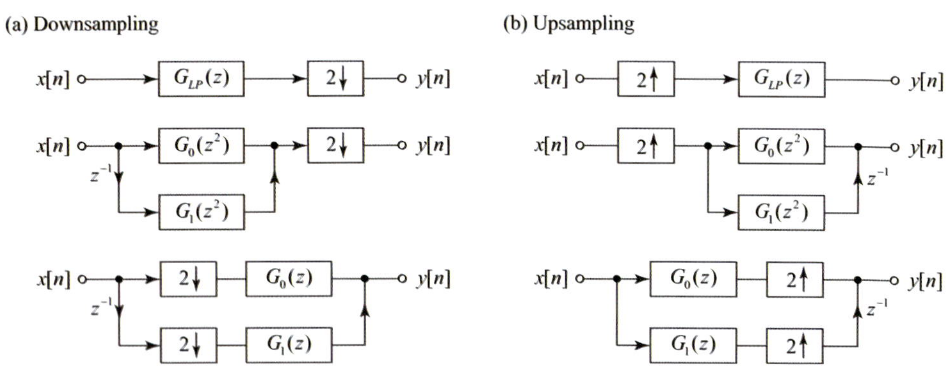

Figure 13.46 Polyphase half-band lowpass filters in downsampling and upsampling

polyphase form and the order of expansion and filtering is interchanged using the multirate upsampling identity (shown in **Figure 13.22**) leading to an efficient architecture with two polyphase filters, each of which operates at one-half the output sample rate.

Example 13.10

Use Matlab to design an upsampling system with $U = 2$ using polyphase half-band lowpass filters for the input $x[n] = \cos(n\pi/4)$.

➤Solution:

The filter g, which we developed in Example 13.9, is $2G_0(z)$, the first polyphase filter component scaled by two. Since the image rejection filter needs a gain of two, this is just what we need.

```
npts = 100;
x = cos(pi*(0:npts-1)/4); % Make an input

% ... Here we would make filter 'g' as in Example 13.9

y = zeros(1, 2*npts);              % allocate the expanded output
y(1:2:end) = filter(g, 1, x);      % filter input with 2*G0(z) and upsample
st = (M-2)/2+2;                    % compute the appropriate delay for x[n]
y(st:2:end) = x(1:npts-st/2+1);    % add delayed and upsampled x[n]
```

Filtering with $2z^{-1}G_1(z)$ and expanding is a matter of computing the appropriate index into the output array y, and adding the unscaled input x to every other point of y. The results are shown in **Figure 13.47**. The upsampled output is shown in blue circles. By using a half-band filter, every other point of the output is identical to original points of the input, shown with red ×'s.

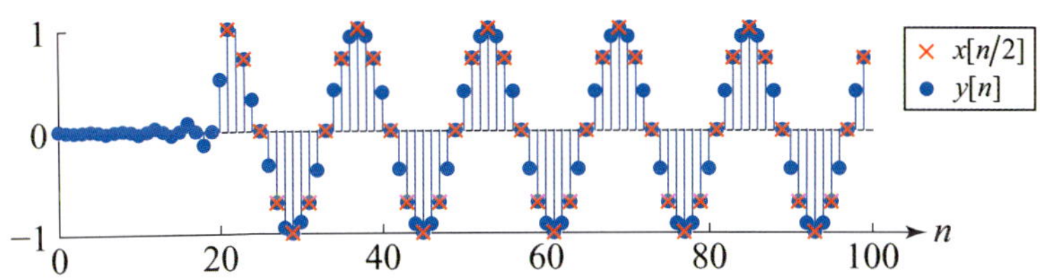

Figure 13.47 Half-band upsampling example

13.7.3 *L*-band (Nyquist) filters

***L*-band filters**, also called ***M*-band** or **Nyquist filters**, are Type-I FIR filters in which every Lth point of the impulse response is zero. For a non-causal filter, the defining relation is

$$h[nL] = \begin{cases} 1/L, & n = 0 \\ 0, & n \neq 0 \end{cases}.$$

For the odd-length causal filter,

$$g[n] = h[n - (N-1)/2].$$

Comparing these with Equations (13.39) and (13.41), respectively, you can see that the half-band filters we have been studying are just a special case of the L-band filter, where $L = 2$.

Figure 13.48a shows symmetric impulse responses for lowpass rectangular L-band filters of the form

$$h[n] = \frac{1}{L}\text{sinc}\, n\pi/L, \qquad -(N-1)/2 \le n \le (N-1)/2$$

where $L = 2$, 3 and 4 and the number of points is $N = 21$. As indicated by the open symbols in the figure, every Lth point of the impulse response is zero except the center point. **Figure 13.48b** shows the associated frequency responses $H(\omega)$, each of which has its corner frequency at $\omega_c = \pi/L$.

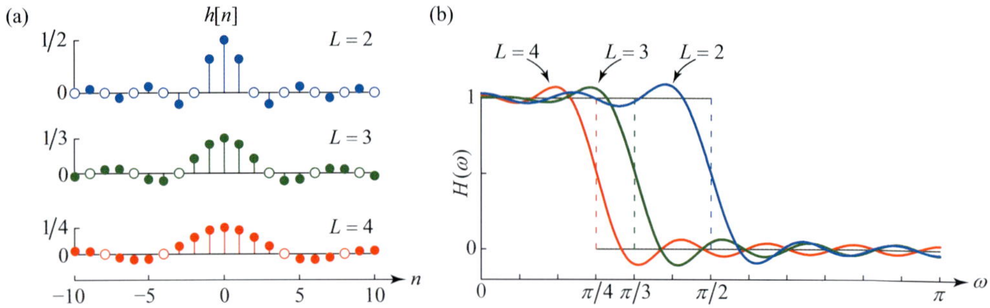

Figure 13.48 L-band filters

In the frequency domain, L-band filters satisfy the relation (Problem 13-11)

$$\sum_{k=0}^{L-1} H(\omega - 2\pi k/L) = 1,$$

which is a generalization of Equation (13.38). **Figure 13.49** shows the L-band filters $H(\omega - 2\pi k/L)$ for $L = 4$. Here, $H(\omega)$ is the same filter whose response is shown in red in **Figure 13.48**. The sum of the responses of all the filters is one.

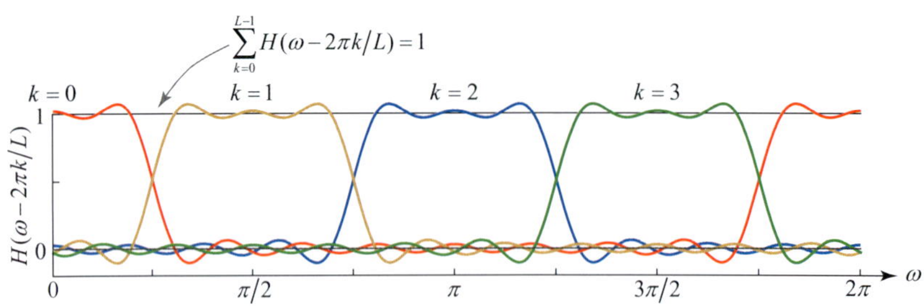

Figure 13.49 Sum of L-band filters

L-band filters are useful in upsampling and downsampling by a factor of L and are well suited to polyphase applications. One of the attractive features of an L-band filter in an upsampling application is that every Lth point of the output is exactly equal to the appropriate input point $y[n] = x[n/L]$. (Why?) We just saw this in Example 13.10 for the case $L = 2$. Matlab's `interp` function uses a Nyquist filter to retain every Uth point in an upsample-by-U application.

13.8 ★ Multirate filter banks

One of the key uses of multirate techniques is in the design and implementation of **multirate filter banks**, which are used extensively in applications involving signal communication and compression. Supplementary material contains an extended discussion of these filter banks. We first show that many of the issues encountered in the use of filter banks with multiple channels can be understood by analyzing a simple **two-channel filter bank**. This leads to a discussion of a couple of filter-bank architectures that have found wide implementation: **quadrature mirror filters (QMF)** and **conjugate quadrature filters (CQF)**. Many important applications, such as those in communications and audio signal processing, require that the signal be partitioned into multiple subbands. So, supplementary material includes a discussion of **tree-structured** and **octave filter banks** and concludes with a treatment of **complex-modulated filter banks (DFT filter banks)** and **cosine-modulated filter banks**, which are found in applications such as the audio codecs (e.g., the MP3 codec) discussed in Chapter 12.

SUMMARY

This chapter has covered a lot of ground. We have explored the essential topics of polyphase downsampling, upsampling, resampling and filtering in some detail from the point of view of both the time-domain and the frequency domain. The development of the multirate identities provided the framework that allowed us to reduce the computational burden of resampling. We then discussed the design of multistage and multirate filters with low cutoff frequencies and/or narrow transition zones.

The remainder of the chapter introduced the half-band and L-band filters and showed how they could be used to design efficient downsamplers and upsamplers. Supplementary material extends the discussion to include quadrature mirror filters and indicates some of the complexities involved in designing "perfect reconstruction" two-channel filter banks. That material also includes a brief tour of filter banks with more than two channels, which have become of central importance in many applications of digital signal processing.

PROBLEMS

Problem 13-1

A lowpass IFIR filter is designed to have passband ω_p and bandwidth $\Delta\omega$. The filter comprises a model filter $G(\omega)$ with bandwidth $\Delta\omega_1$ and an image-rejection filter $I(\omega)$ with bandwidth $\Delta\omega_2$.

(a) Show that the sum of the required orders of the two filter stages necessary to implement the IFIR is

$$M = \frac{K}{\Delta\omega_1} + \frac{K}{\Delta\omega_2} = K\left(\frac{1}{\Delta\omega U} + \frac{U}{2(\pi - \omega_p U) - \Delta\omega U}\right).$$

The length of the first filter has been increased by a factor U by interpolation with respect to the impulse response of the model filter.

(b) Show that the upsample factor required to minimize the combined order of the filters is the integer closest to

$$U = \frac{1}{\sqrt{\Delta\omega/2\pi + \omega_p/\pi + \Delta\omega/2\pi}}.$$

Problem 13-2

Consider a two-stage downsampler, where the two stages downsample an input by factors of D_1 and D_2 respectively *without aliasing*, such that the total downsampling factor is $D \triangleq D_1 D_2$.

(a) Show that the total number of operations per input point is

$$\mathrm{O} = KD_2\left(\frac{1}{\pi D_2 - D\omega_p} + \frac{1}{\pi D - D^2\omega_p}\right).$$

(b) Show that the value of D_2 required to minimize the combined order of the filters is the integer closest to

$$D_2 = \frac{D\omega_p}{\pi}\left(1 + \sqrt{\frac{\pi}{\omega_p} - D}\right).$$

Problem 13-3

Repeat Problem 13-2 but this time *allow aliasing* in the first stage of filtering.

(a) Show that the total number of operations per input point is

$$\mathrm{O} = KD_2\left(\frac{1}{\pi(2D_2 - 1) - D\omega_p} + \frac{1}{\pi D - D^2\omega_p}\right).$$

(b) Show that the value of D_2 required to minimize the combined order of the filters is the integer closest to

$$D_2 = \frac{1}{2} + \frac{D\omega_p}{2\pi}\left(1 + \sqrt{D\left(\left(\frac{\pi}{D\omega_p}\right)^2 - 1\right)}\right).$$

Problem 13-4

The stopband of the filter in the first stage of a two-stage decimation by $D = D_1 D_2$ is $\bar{\omega}_s = \pi(2D_2 - 1)/D$ for a system where there is aliasing. Show that this frequency is always larger than the stopband of a first-stage filter which does not allow aliasing.

Problem 13-5

Consider a two-stage interpolator, where the two stages interpolate an input with passband ω_p by factors of U_1 and U_2 respectively.

(a) Show that the total number of operations per output point is

$$\mathrm{O} = \frac{KU_1}{2U}\left(\frac{1}{\pi - \omega_p} + \frac{U}{\pi U_1 - \omega_p}\right).$$

(b) Show that the value of U_1 that minimizes O is

$$U_1 = \sqrt{U(\omega_p/\pi)(1 - \omega_p/\pi)} + \omega_p/\pi.$$

Problem 13-6

In Section 13.4.2, we derived the z-transform of the decimated sequence $y[n] = x[nD]$ by multiplying $x[n]$ by an impulse train

$$s[n] = \sum_{k=-\infty}^{\infty} \delta[n - kD] = \frac{1}{D}\sum_{d=0}^{D-1} e^{j2\pi dn/D},$$

and taking the z-transform of the product, yielding

$$Y(z) = \frac{1}{D}\sum_{d=0}^{D-1} X\left(z^{1/D}e^{-j2\pi d/D}\right).$$

Here is another approach to the solution, which seems reasonable but leads to the wrong answer. To form the output $y[n]$, sample every Dth point of the input $x[n]$,

$$\begin{aligned} y[n] &= \cdots + x[0]\delta[n] + x[D]\delta[n-1] + x[2D]\delta[n-2] + x[3D]\delta[n-3] + x[4D]\delta[n-4] + \cdots \\ &= \sum_{k=-\infty}^{\infty} x[kD]\delta[n-k]. \end{aligned}$$

The z-transform is

$$Y(z) \triangleq \mathfrak{Z}\{y[n]\} = \mathfrak{Z}\left\{\sum_{k=-\infty}^{\infty} x[kD]\delta[n-k]\right\} = \sum_{k=-\infty}^{\infty} x[kD]\mathfrak{Z}\{\delta[n-k]\} = \sum_{k=-\infty}^{\infty} x[kD]z^{-k}.$$

Let $n = kD$, then

$$Y(z) = \sum_{n=-\infty}^{\infty} x[n]z^{-n/D} = \sum_{n=-\infty}^{\infty} x[n]\left(z^{1/D}\right)^{-n} = X(z^{1/D}).$$

This is incorrect. Why?

Problem 13-7

Let $h[n]$ be the impulse response of a filter with z-transform $H(z)$.

(a) Split $h[n]$ into two sequences for even and odd values of n, and show that $H(z)$ can be expressed in terms of two polyphase components:

$$H(z) = H_0(z^2) + z^{-1}H_1(z^2),$$

where $H_0(z)$ is the transform of the even values,

$$H_0(z) = \sum_{n=-\infty}^{\infty} h[2n]z^{-n},$$

and $H_1(z)$ is the transform of the odd values,

$$H_1(z) = \sum_{n=-\infty}^{\infty} h[2n+1]z^{-n}.$$

(b) Now consider the impulse response of a half-band filter in which every even element is zero except $h[0] = 0.5$, namely

$$h[2n] = \begin{cases} 0.5, & n = 0 \\ 0, & n \neq 0 \end{cases}.$$

Use this fact and the result of part (a) to show that $H(z) + H(-z) = 1$.

Problem 13-8

Consider two multirate systems shown in **Figure 13.50**. The system on the left has a decimator (by a factor of D) followed by an (by a factor of U). The system on the right has an expander followed by a decimator. If D and U are relatively prime then the outputs of the two systems are equivalent; that is $x_{du}[n] = x_{ud}[n]$. If D and U are not relatively prime, they are not equivalent.

Figure 13.50

(a) Consider an input $x[n] = \cos(2\pi n/16)$, $0 \leq n < 16$, with $D = 2$ and $U = 3$. Make plots for $x[n]$, $x_d[n]$, $x_{du}[n]$, $x_u[n]$ and $x_{ud}[n]$. Verify that $x_{du}[n] = x_{ud}[n]$.

(b) Repeat part (a) for $D = 3$ and $U = 2$ and verify that $x_{du}[n] = x_{ud}[n]$.

(c) Repeat part (a) for $D = 2$ and $U = 2$ and verify that $x_{du}[n] \neq x_{ud}[n]$.

Problem 13-9

Bézout's identity states that, given relatively prime values of U and D, there exist integers u and d such that $uU + dD = 1$. Show that, given one solution, u_0 and d_0, an infinite number of solutions exist.

Problem 13-10

A real, symmetric Type-I half-band lowpass filter is defined by $\omega_c = \pi/2$ and

$$H_{LP}(\omega) = 1 - H_{LP}(\pi - \omega).$$

(a) Show that $H_{LP}(\omega)$ is symmetric in amplitude about $\omega_c = \pi/2$.
(b) Show that if $H_{LP}(\omega)$ has passband frequency $\omega_p = 0.5\pi - \Delta\omega/2$, where $\Delta\omega$ is the width of the transition band, then the stopband frequency is $\omega_s = 0.5\pi + \Delta\omega/2$ such that $\omega_s + \omega_p = \pi$.
(c) Show that the passband and stopband attenuations are equal, $\delta_p = \delta_s$.

Problem 13-11

In this problem we prove that an L-band filter with impulse response $h[n]$ and transform $H(\omega)$ satisfies the frequency-domain relation

$$\sum_{k=0}^{L-1} H(\omega - 2\pi k/L) = 1.$$

Figure 13.51 suggests how to accomplish this. The defining relation of the L-band filter in the time domain is that every Lth point of the impulse response $h[n]$ is zero except for $h[0] = 1/L$, as shown in **Figure 13.51a** for the example of $L = 4$. Consider the impulse train $s[n]$ shown in **Figure 13.51b**. We have shown that $s[n]$ can be expressed as the sum of a finite number of complex exponentials,

$$s[n] = \sum_{k=-\infty}^{\infty} \delta[n - kL] = \frac{1}{L}\sum_{k=0}^{L-1} e^{-j2\pi kn/L}.$$

Use the fact that the product of $h[n]s[n] = (1/L)\delta[n]$ to prove the frequency relation in the problem statement.

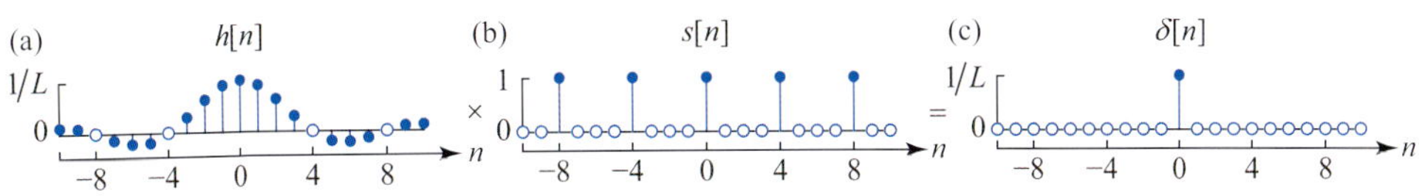

Figure 13.51

Problem 13-12

The condition for a perfect reconstruction of quadrature mirror FIR filters is that

$$H_0^2(z) - H_0^2(-z) = z^{-n_0},$$

or, in terms of the frequency response,

$$H_0^2(\omega) - H_0^2(\omega - \pi) = e^{-j\omega n_0}.$$

The purpose of this problem is to show that the only filters that satisfy this criterion are of the form

$$h_0[n] = \tfrac{1}{2}(\delta[n - n_1] + \delta[n - n_2]),$$

or, in terms of the z-transform,

$$H_0(z) = \tfrac{1}{2}(z^{-n_1} + z^{-n_2}),$$

where n_1 is even, n_2 is odd and $n_0 = n_1 + n_2$, which is therefore also odd.

(a) Show that $H_0^2(z) - H_0^2(-z)$ can be expressed as the product of two terms,

$$H_0^2(z) - H_0^2(-z) = E_1(z)E_2(z),$$

where

$$e_1[n] = \mathcal{Z}^{-1}\{E_1(z)\} = \begin{cases} 2h_0[n], & n \text{ even} \\ 0, & n \text{ odd} \end{cases}$$

$$e_2[n] = \mathcal{Z}^{-1}\{E_2(z)\} = \begin{cases} 2h_0[n], & n \text{ odd} \\ 0, & n \text{ even} \end{cases}.$$

(b) Show that neither $E_1(z)$ nor $E_2(z)$ has any zeros and that either $E_1(z)$ or $E_2(z)$ (or both) must have a pole at $z = 0$ and nowhere else.

(c) From part (b), show that $e_1[n] = \delta[n - n_1]$ and $e_2[n] = \delta[n - n_2]$ such that $n_1 + n_2 = n_0$.

(d) From parts (a) and (c), show that the impulse response, $h_0[n]$, can only consist of two impulses,

$$h_0[n] = \tfrac{1}{2}(\delta[n - n_1] + \delta[n - n_2]),$$

where n_1 is even and n_2 is odd.

(e) Show that $n_0 = n_1 + n_2$, which is therefore also odd.

Problem 13-13

Show that the Haar filter pair

$$H_0(z) = \tfrac{1}{2}(1 + z^{-1})$$

$$H_1(z) = \tfrac{1}{2}(1 - z^{-1})$$

satisfies the criterion for perfect reconstruction,

$$H_0(z)H_1(-z) - H_0(-z)H_1(z) = z^{-n_0},$$

with $n_0 = 1$.

Problem 13-14

Show that the Haar filter pair

$$H_0(z) = \tfrac{1}{2}(1 + z^{-1})$$
$$H_1(z) = \tfrac{1}{2}(1 - z^{-1})$$

is power complementary:

$$|H_0(\omega)|^2 + |H_0(\omega - \pi)|^2 = 1.$$

Problem 13-15

Show that the filter pairs

$$H_0(z) = \tfrac{1}{2}(z^{-n_1} + z^{-n_2})$$
$$H_1(z) = \tfrac{1}{2}(z^{-n_1} - z^{-n_2}),$$

where n_0 is even and n_1 is odd, are power complementary:

$$|H_0(\omega)|^2 + |H_0(\omega - \pi)|^2 = 1.$$

Problem 13-16

We have shown that if the prototype filter $H_0(z)$ of a two-channel QMF filter bank is chosen to be a linear-phase filter of length N, then $H_0(z)$ can be expressed as the product of an amplitude function $A(z)$ and a linear-phase term $z^{-(N-1)/2}$:

$$H_0(z) - A(z)z^{-(N-1)/2}.$$

Show that N must be even by showing that, otherwise, the QMF criterion for perfect reconstruction, $F_0(z) = H_0^2(z) - H_0^2(-z) = z^{-n_0}$, cannot be satisfied at all frequencies.

➤Hint: $\omega = \pi/2$.

Problem 13-17

The zeros of a valid half-band filter $T(z)$ of even order M can only occur as quartets at conjugate-reciprocal positions in the z-plane (two complex-conjugate pairs with reciprocal magnitudes), as pairs at conjugate positions on the unit circle or as pairs on the real axis. If there are zeros at $z = \pm 1$, they must occur in even numbers (e.g., 0, 2, . . .). The conjugate mirror filter $H_0(z)$ is then created by spectral factorization of $T(z) = H_0(z)H_0(z^{-1})$ by assigning half (i.e., $M/2$) the zeros of $T(z)$ to $H_0(z)$ and the remainder to $H_0(z^{-1})$. Argue that any assignment of zeros to $H_0(z)$ that uses half the complex-conjugate pairs (but not both pairs from the same quartet) plus half of any real pairs will result in the same frequency response magnitude $|H_0(\omega)|$.

Problem 13-18

The prototype half-band analysis filter $H_0(z)$ can be decomposed into its polyphase components $H_{00}(z)$ and $H_{10}(z)$, as shown in **Figure 13.52**, such that

$$\begin{aligned} H_0(z) &= H_{00}(z^2) + z^{-1}H_{01}(z^2) \\ H_1(z) &= H_{00}(z^2) - z^{-1}H_{01}(z^2) \end{aligned}.$$

(a) Show that the remaining analysis and synthesis filters can be expressed in terms of $H_{00}(z)$ and $H_{10}(z)$, as follows:

$$\begin{aligned} G_0(z) &= G_{00}(z^2) + z^{-1}G_{01}(z^2) = 2(H_{00}(z^2) + z^{-1}H_{01}(z^2)) \\ G_1(z) &= G_{10}(z^2) + z^{-1}G_{11}(z^2) = 2(-H_{00}(z^2) + z^{-1}H_{01}(z^2)) \end{aligned}.$$

(b) Derive the polyphase realization of the analysis and synthesis filter bank as shown in Figure 13.52.

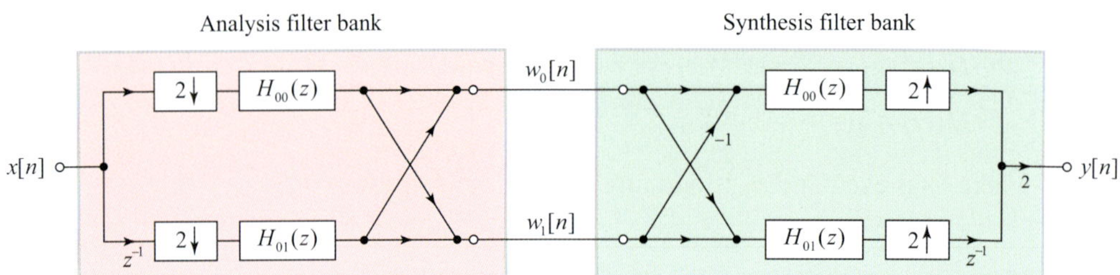

Figure 13.52

Problem 13-19

(a) Show that CQF analysis and synthesis filters are related through

$$\begin{aligned} G_0(z) &= 2z^{-(N-1)}H_0(z^{-1}) \\ G_1(z) &= 2z^{-(N-1)}H_1(z^{-1}) \end{aligned}.$$

(b) Show that $G_0(\omega)$ and $H_0(\omega)$ are related through an allpass filter, as are $G_1(\omega)$ and $H_1(\omega)$, so that

$$\begin{aligned} |G_0(\omega)| &= 2|H_0(\omega)| \\ |G_1(\omega)| &= 2|H_1(\omega)| \end{aligned}.$$

14 Spectral analysis

Introduction

Spectral analysis is one of the essential applications of DSP. It is the process of measuring, estimating and characterizing the frequency content of signals – both deterministic and probabilistic. The techniques of spectral analysis that we will discuss in this chapter are widely applied in fields as diverse as astrophysics, where they are used to measure the velocity of galaxies and the expansion of the universe, medicine, where they are used in a variety of instruments such as ultrasound imaging machines, geology, where they are used to analyze ground movement and seismic events, chemistry, where they are used in mass spectrograms to determine the molecular and chemical constituents of materials, and engineering and computer science, where they are used as the basis for speech processing and communication systems.

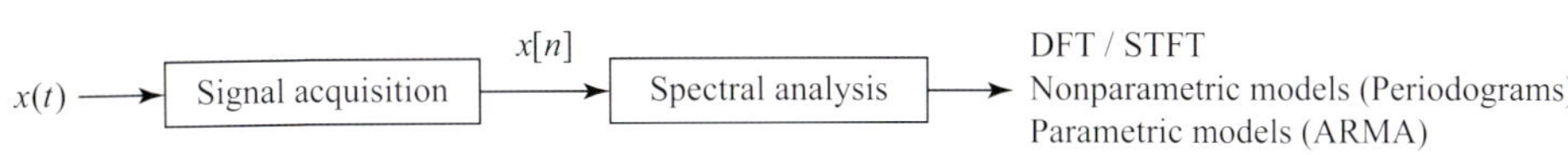

Figure 14.1 Overview of spectral analysis system

Figure 14.1 shows a schematic overview of a spectral analysis system. The majority of spectral analysis problems start with data in continuous-time domain (e.g., $x(t)$), which are then sampled by a signal acquisition system comprising an analog front-end (amplifier, filter) and an A/D converter in order to produce a digital signal $x[n]$. In Chapter 6, we discussed a number of the parameters of the signal acquisition system, including the effect of sample rate and signal quantization. After the signal is digitized, it is subjected to spectral analysis. The technique(s) of analysis that are used depend on the goals of spectral analysis, which vary widely depending on the application. These goals include estimating the frequencies and amplitudes of particular spectral components of the signal (e.g., in molecular spectroscopy), visualizing the spectrum of a complex time-varying signal (e.g., speech or music) or estimating parameters of a model of the signal from the measured spectra (e.g., linear predictive coding for telephony).

In this chapter, we will consider all these problems. In Section 14.1, we will consider the frequency analysis of a simple pure-tone signal in the absence of noise. In Section 14.2, we will see how to measure and display the frequency content of a time-varying signal such as speech. Section 14.3 presents some of the techniques of nonparametric spectral analysis, which do not depend on a particular model of the signal being analyzed. Finally, in Sections 14.4 and 14.5, we discuss parametric models of spectral estimation and their application to extracting psychophysically important features, such as formant frequencies, from speech.

14.1 Basics of spectral analysis

In this section, we look at two fundamental issues that affect the practical measurement of the frequency spectrum of both deterministic and probabilistic signals: **data windowing** and **frequency sampling**.

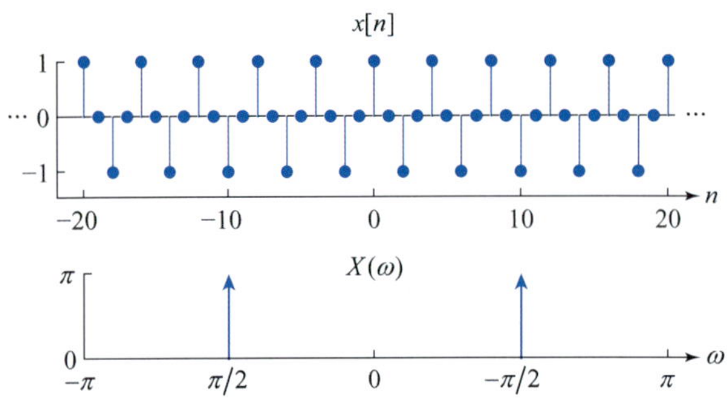

Figure 14.2 Spectrum of pure cosine

In principle, the task of analyzing the frequency content of a pure tone – a single sinusoid of constant frequency – is easy. Consider an infinite-length sinusoid $x(t) = \cos \Omega_0 t$ that is sampled at continuous-time frequency Ω_s to form the discrete-time infinite-length sequence $x[n] = \cos \omega_0 n$, where $\omega_0 = 2\pi(\Omega_0/\Omega_s)$. This signal has DTFT $X(\omega) = \pi(\delta(\omega - \omega_0) + \delta(\omega + \omega_0))$. **Figure 14.2** shows the picture of $x[n]$ and its transform $X(\omega)$ for a frequency of $\omega_0 = \pi/2$.

The characterization of this signal in the frequency domain is easy for two reasons: (1) a spectrum of a cosine of infinite length has all its power concentrated at only two frequencies, ω_0 and $-\omega_0$; (2) we have a theoretical tool – the DTFT – that allows us to analyze this signal with infinite precision in frequency. However, in any practical application neither of these conditions applies. An infinite length of signal is not available for analysis; we have to choose a segment of the signal to analyze. Furthermore, in practice we cannot use a continuous-time measure of frequency analysis such as the DTFT; instead we must rely on the sampled transform, the DFT/FFT. These two issues, data windowing and frequency sampling, are the two central sources of spectral ambiguity that need to be addressed in a discussion of spectral analysis, even of this simple signal. We will take these issues in turn in the next two subsections.

14.1.1 Spectral effects of windowing

The first departure from ideal behavior stems from the fact that any signal we measure is finite in duration. We are most often selecting a portion of a long signal for analysis, and the selected portion is created by truncating the longer signal to a given length or perhaps by explicitly multiplying it with a tapered window of finite length. In Section 3.10.10, we established the theoretical result that such a time-limited signal cannot be bandlimited. This fact influences important decisions we must make about the signals we measure. For example, how long a signal must we obtain in order to distinguish features of the frequency spectrum that are of interest to us? How do we measure the spectral content of time-varying signals effectively?

To formalize this discussion, a finite-length signal $\hat{x}[n]$ can be viewed as a signal of infinite length, $x[n]$, multiplied by a window $w_N[n]$ of length N that truncates it or shapes it in some way,

$$\hat{x}[n] = x[n]w_N[n].$$

We have encountered exactly this situation in Chapter 7, where we constructed FIR filters by truncation of the infinite impulse response of an ideal lowpass filter using windows (e.g., Hamming, Hann and Kaiser). As an example, **Figure 14.3** shows the result of windowing a cosine by truncating it with a rectangular window in the time domain (top panels) and frequency domain (bottom panels).

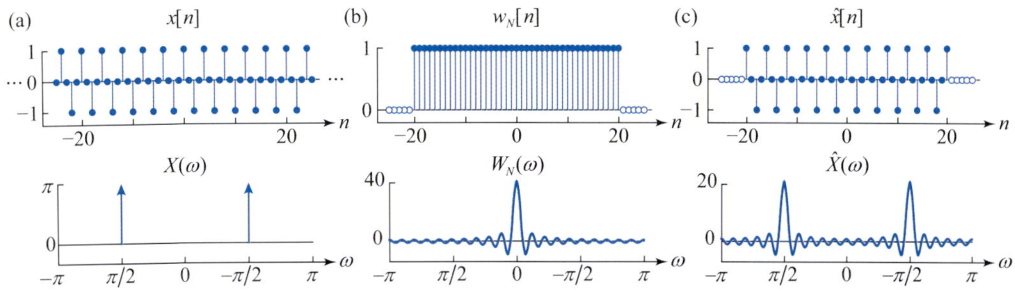

Figure 14.3 Spectrum of a rectangular windowed cosine

Figure 14.3a shows a portion of an infinite-length cosine $x[n]$ (top) and its transform $X(\omega)$ (bottom). **Figure 14.3b** shows a rectangular window $w_N[n]$ of length $N = 41$ and its transform, the periodic sinc function (bottom),

$$W_N(\omega) = \frac{\sin \omega N/2}{\sin \omega/2} = N\frac{\text{sinc}\, \omega N/2}{\text{sinc}\, \omega/2}.$$

Figure 14.3c shows the windowed cosine $\hat{x}[n] = x[n]w[n]$ (top) and its spectrum $\hat{X}(\omega)$ (bottom). In the frequency domain, the effect of truncating or windowing $x[n]$ by $w_N[n]$ is to convolve $X(\omega)$ with $W_N(\omega)$,

$$\begin{aligned}\hat{X}(\omega) &= \frac{1}{2\pi}X(\omega) * W_N(\omega)\\ &= \frac{1}{2\pi}\underbrace{\pi(\delta(\omega+\omega_0)+\delta(\omega-\omega_0))}_{X(\omega)} * \underbrace{N\frac{\text{sinc}(\omega N/2)}{\text{sinc}(\omega/2)}}_{W_N(\omega)}\\ &= \underbrace{\frac{N}{2}\frac{\text{sinc}((\omega+\omega_0)N/2)}{\text{sinc}((\omega+\omega_0)/2)}}_{\hat{X}_1(\omega)} + \underbrace{\frac{N}{2}\frac{\text{sinc}((\omega-\omega_0)N/2)}{\text{sinc}((\omega-\omega_0)/2)}}_{\hat{X}_2(\omega)}.\end{aligned} \quad (14.1)$$

Whereas the power of the infinite-length cosine is concentrated at only two frequencies, ω_0 and $-\omega_0$, the energy of the spectrum of the windowed cosine is spread over the entire frequency range $-\pi \leq \omega < \pi$. In particular, the spectrum of the windowed cosine, $\hat{X}(\omega)$, is the sum of the two terms, $\hat{X}_1(\omega)$ and $\hat{X}_2(\omega)$, with broad peaks, centered at ω_0 and $-\omega_0$, respectively, plus substantial energy at other frequencies.

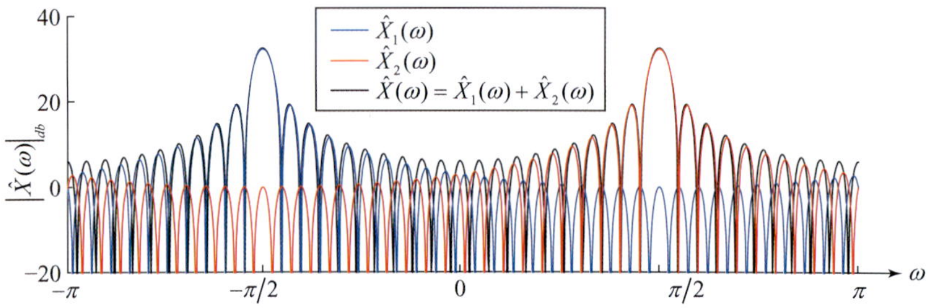

Figure 14.4 Spectrum of a rectangular windowed cosine (dB scale)

Figure 14.4 shows the spectrum of the rectangular-windowed cosine from **Figure 14.3c** plotted on a dB scale. In this picture, the two components of the resulting spectrum, $\hat{X}_1(\omega)$ and $\hat{X}_2(\omega)$, are each plotted in a different color, corresponding to the colors in Equation (14.1). Each term contributes energy to all frequencies in $\hat{X}(\omega)$, and the energy of the two terms sums to form $\hat{X}(\omega)$ (thin black line). The effect of windowing a signal is to introduce ambiguity into the frequency-domain representation. This ambiguity can be mitigated either by increasing the length of the window or by choosing a more appropriate shape of the window. However, each of these solutions has trade-offs, as we shall see.

14.1.2 Effect of window choice

The amount of spectral ambiguity depends on the shape and length of the window $w_N[n]$. As an example, **Figure 14.5** shows the result of multiplying a cosine with a Hamming window of length $N = 41$. The top panels (reading from left to right) show the un-windowed sequence $x[n] = \cos \omega_0 n$ (**Figure 14.5a**), the Hamming window $w_N[n]$ (**Figure 14.5b**) and the Hamming-windowed cosine sequence $\hat{x}[n]$ (**Figure 14.5c**). The bottom panels show the transforms of these time functions on a linear scale.

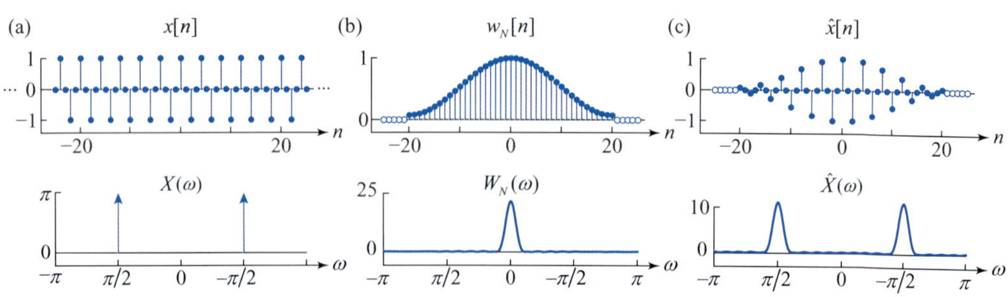

Figure 14.5 Spectrum of Hamming-windowed cosine

Compared with the rectangular-windowed cosine (**Figure 14.3**), the main lobes of the Hamming-windowed cosine centered around ω_0 and $-\omega_0$ are wider, but the peak amplitude of the ripples in the transform is lower.

In our discussion of window-based FIR filters in Chapter 7, we showed how the choice of window and its parameters (i.e., length) affected the frequency response of the filter. The same principles apply to the selection of the window that we will now use to multiply data preparatory to calculating the DTFT. **Figure 14.6** shows a comparison of the magnitude of the transform, $|W_N(\omega)|_{dB}$, of four of the most popular spectral data windows – rectangular (blue), Hamming (green), Hann (yellow) and Blackman (red) – all of the same order ($N = 31$) and all plotted on a log (dB) scale normalized to a maximum amplitude of one $\left(\text{i.e., } |W_N(0)|_{db} = 0\right)$. The main lobe of each of these transforms (the lobe centered about $\omega = 0$) has a width $\Delta\omega$ that varies inversely with window length N, as shown in the table in the figure. Both the absolute amplitude of the main lobe and the magnitude of the sidelobes increase with increasing N, but the *ratio* of the amplitude of the largest sidelobe to the main lobe (ΔA_{MAX} in the figure) is roughly independent of window size for large enough N. As we saw in Chapter 7, the transforms of Hamming, Hann and Blackman windows all have lower maximum sidelobe amplitudes compared with the rectangular window, but this comes at the expense of greater main-lobe width $\Delta\omega$. Each of the windows has its own particular character, which may be useful in different spectral estimation problems, as we will see in examples to follow. The rectangular window has the narrowest main lobe ($4\pi/N$), but the highest sidelobes (-13 dB). The sidelobes "decay" with a slope of about $sl = -6$ dB/octave. The Hamming window is specifically designed to minimize the amplitude of the first sidelobe. The remaining sidelobes are about 20 dB below those of the rectangular window and also decay with a slope of about $sl = -6$ dB/octave. The main lobe of the Hann window is the same width as the Hamming window; the maximum sidelobe amplitude (-32 dB) is much higher than that of the Hamming window, but the sidelobes decay much faster ($sl = -18$ dB/octave). The Blackman window has the lowest sidelobes of the four windows shown here (-58 dB), and the sidelobes decay faster ($sl = -18$ dB/octave), but the main lobe is the widest ($12\pi/N$).

	$\Delta\omega$	ΔA_{MAX} (dB)	sl (dB/oct)
Rectangular	$4\pi/N$	−13	−6
Hamming	$8\pi/N$	−43	−6
Hann	$8\pi/N$	−32	−18
Blackman	$12\pi/N$	−58	−18

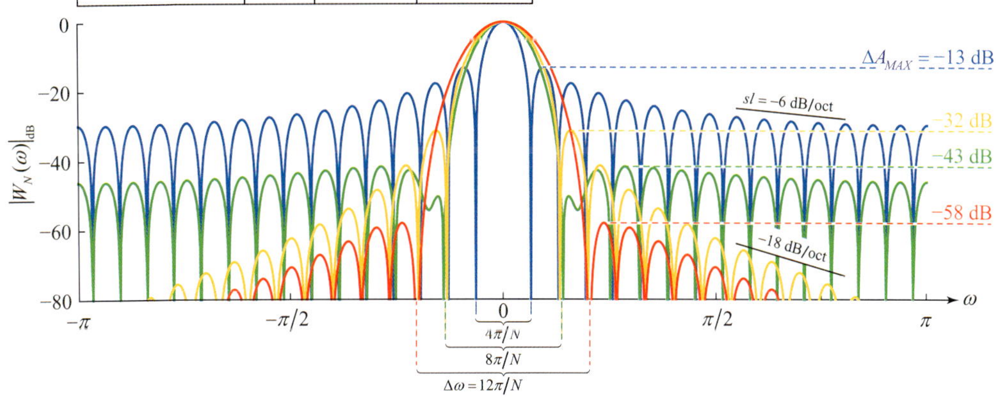

Figure 14.6 Spectral data windows

Figure 14.7 shows the effect of multiplying a pure cosine with each of the windows shown in **Figure 14.6**. In the transform domain, this is equivalent to convolving each of the spectra with a pair of impulses, $\pi(\delta(\omega-\omega_0)+\delta(\omega+\omega_0))$, that represent the cosine. The out-of-band energy, which we define as the energy in the sidelobes of the transform, is decreased by using Hamming, Hann or Blackman windows, though this decrease comes at the expense of a wider main lobe around $\omega=\pm\omega_0$.

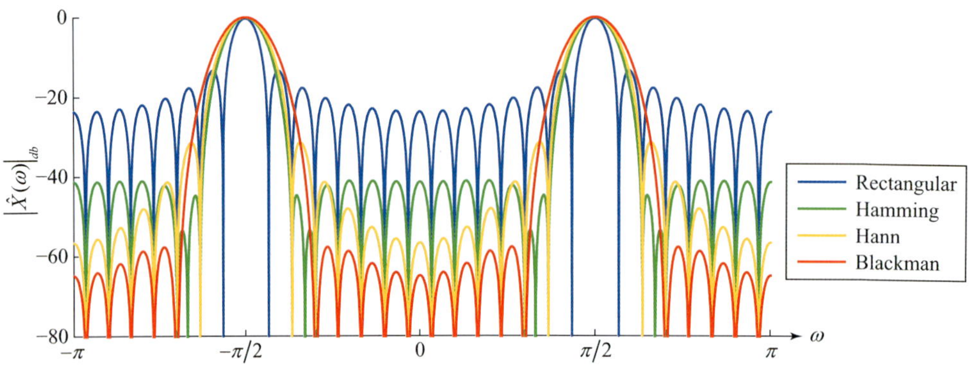

Figure 14.7 Comparison of windowed cosines

14.1.3 Spectral spread and leakage

Both the width of the main lobe and the out-of-band energy of the other lobes pose problems for accurate spectral estimation. There are two related issues here: **spectral spread** and **spectral leakage**. Spectral spread refers to the effect that the width of the main lobe has on limiting the accurate resolution of spectral components that are close together in frequency. Spectral leakage refers to the energy of out-of-band spectral components – the sidelobes – that create a "noise floor" that effectively limits the measurement of low-amplitude spectral components.

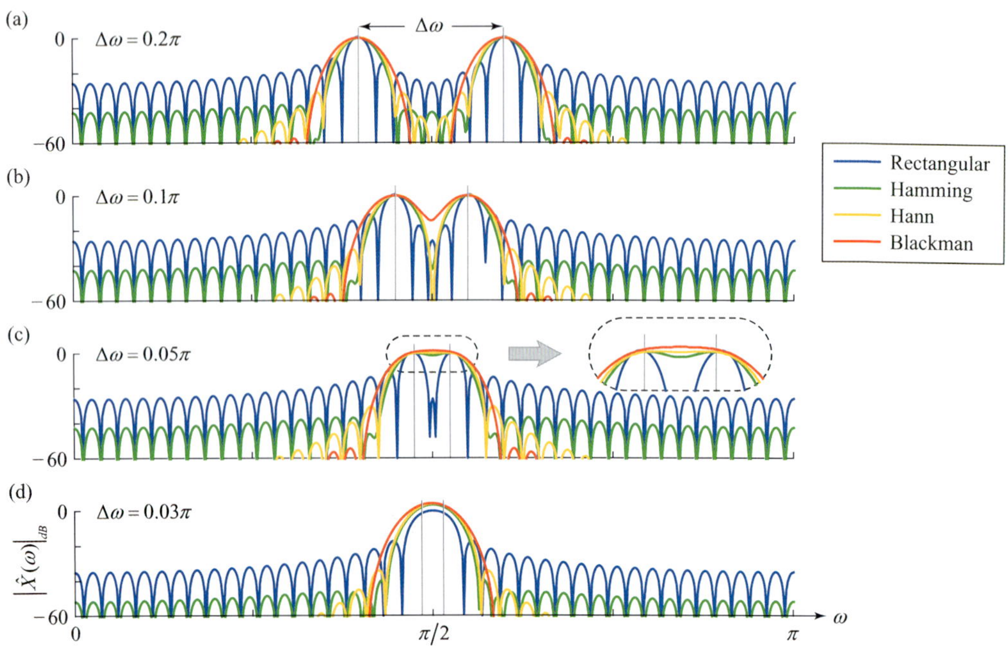

Figure 14.8 Effect of frequency separation on frequency ambiguity

Effect of main lobe width on spectral resolution As an example of the first problem, consider a signal $x[n]$, comprising the sum of two pure cosines of equal amplitude separated by frequency $\Delta\omega$,

$$x[n] = \cos(\omega_0 - \Delta\omega/2)n + \cos(\omega_0 + \Delta\omega/2)n.$$

Then, $x[n]$ is multiplied by a spectral window $w_N[n]$ of length $N = 81$ to form $\hat{x}[n] = x[n]w_N[n]$. **Figure 14.8** plots the spectrum of this signal, $|\hat{X}(\omega)|$, for different values of the separation of the two tones, $\Delta\omega$. When $\Delta\omega = 0.2\pi$ (**Figure 14.8a**), it is easy to distinguish the two peaks in the spectrum, regardless of the spectral window used. However, as $\Delta\omega$ decreases (remaining panels), the lobes centered at the two frequencies overlap and merge, and the individual spectral components cannot be resolved. Since the amount of overlap depends principally on the width of the main lobe of the window, the **spectral resolution** of two closely spaced frequency components – the minimum separation of tones that can be resolved in the DTFT – depends on the choice of window. In the case of the sum of two cosine tones, the spectral peaks appear to be well separated when $\Delta\omega$ is greater than the width of the main lobe, as given in the table of **Figure 14.6**. For example, for a Hamming window of length $N = 81$, the main lobe is $\Delta\omega = 8\pi/81 \sim 0.1\pi$. **Figure 14.8b** confirms that the main lobes of the Hamming-windowed spectrum are, indeed, well separated when $\Delta\omega = 0.1\pi$. In general, the conventional advice is to avoid the rectangular window due to the fact that the amplitudes of the sidelobes are so much higher than those of the other windows. But in this instance, you might want to select this window because the width of the main lobe is the smallest. For a rectangular window of length $N = 81$, the main lobe is $\Delta\omega = 4\pi/81 \sim 0.05\pi$, and **Figure 14.8c** shows that the spectral peaks of the rectangular-windowed data (blue traces) are distinct when the separation of cosine frequencies is $\Delta\omega = 0.05$, whereas the peaks of the spectra of data windowed with windows with wider main lobes have essentially merged. In order to separate these peaks in the spectrum, you would have to use a higher-order window, which means taking more points of $x[n]$.

Figure 14.9 shows the spectrum of a single cosine truncated with a Hamming window of length $N = 21$ (blue), 41 (green) and 61 (red), normalized to a peak amplitude of one (i.e., 0 dB).

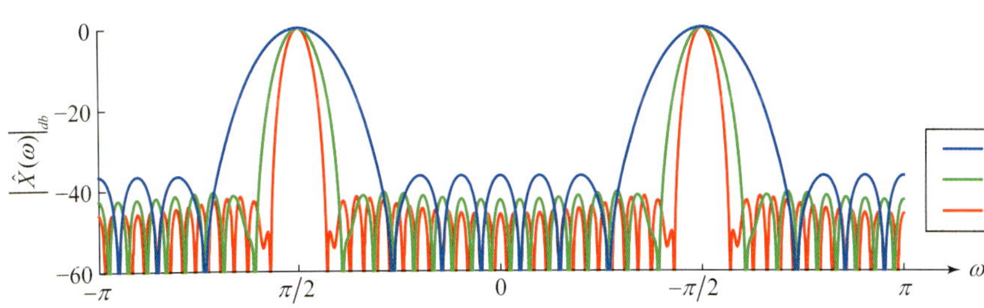

Figure 14.9 Effect of window length on spectrum of cosine

As the length of the window increases by a factor of three, the width of the main lobe of the spectrum gets proportionally narrower but the amplitude of the sidelobes does not decrease very much.

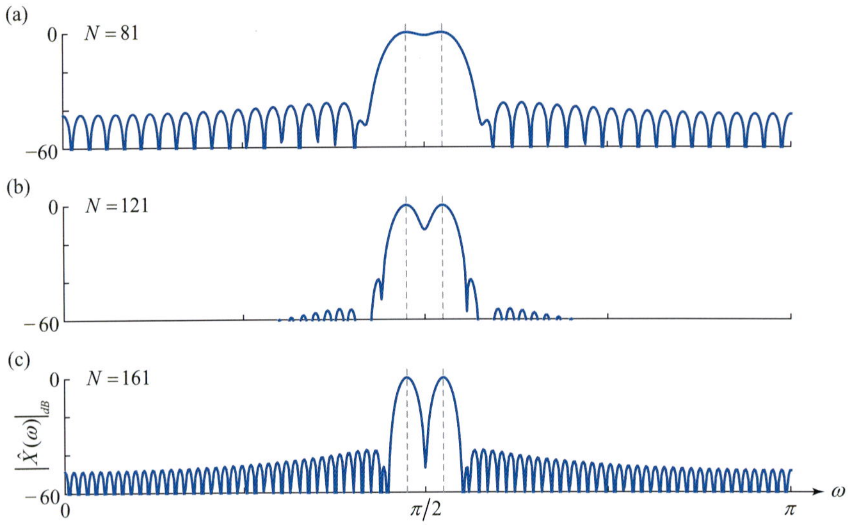

Figure 14.10 Effect of window length on spectral resolution

Figure 14.10 shows what happens when Hamming windows of different sizes N are applied to a sum of cosines, $x[n]$, with a fixed frequency separation of $\Delta\omega = 0.05\pi$. At a relatively low value of $N = 81$ (**Figure 14.10a**), the spectral peaks have almost merged (as we saw previously in **Figure 14.8c**) and it is difficult to distinguish between them but, as N increases, the lobes representing the two frequencies become increasingly distinct. At a value of $N = 161$ (**Figure 14.10c**) the frequency separation of the cosines corresponds to the width of the main lobe of the Hamming window, $8\pi/161 \cong 0.05\pi$, and the peaks are well resolved. In spectral analysis problems where one has to deal with noise and the presence of other signals, it may be necessary to choose a window whose length corresponds to more than the main lobe width.

Effect of spectral leakage on spectral resolution In addition to the issue of the width of the main lobe, the other inevitable consequence of windowed signals is **spectral leakage**: the distribution of energy from spectral components of large energy into the sidelobes. This out-of-band energy essentially acts as a noise floor for the spectral measurement; it sets the lowest level at which spectral components can be resolved. For example, **Figure 14.11** shows the spectrum of the sum of two pure cosines of unequal amplitude separated by $\Delta\omega = 0.5\pi$,

$$x[n] = \cos(\omega_0 - \Delta\omega/2)n + A\cos(\omega_0 + \Delta\omega/2)n.$$

$x[n]$ is then windowed by a Hamming window $w_N[n]$ of length $N = 81$ to form $\hat{x}[n]$.

The amplitude A_{dB} of the second tone in dB at frequency $\omega = 3\pi/4$ ranges from 0 dB (a gain of one, top panel) to -60 dB (a gain of 0.001, bottom panel) with respect to the tone at $\omega = \pi/4$. Spectral leakage from the tone at $\omega = \pi/4$ produces sidelobes that form an effective noise floor between -40 and -50 dB for the measurement of other spectral components, depending on frequency. In this example, when the spectral amplitudes of two cosine components differ by more than about –60 dB, the smaller component becomes almost indistinguishable from this sidelobe "noise." The spectral analysis of signals such as this, which comprise components

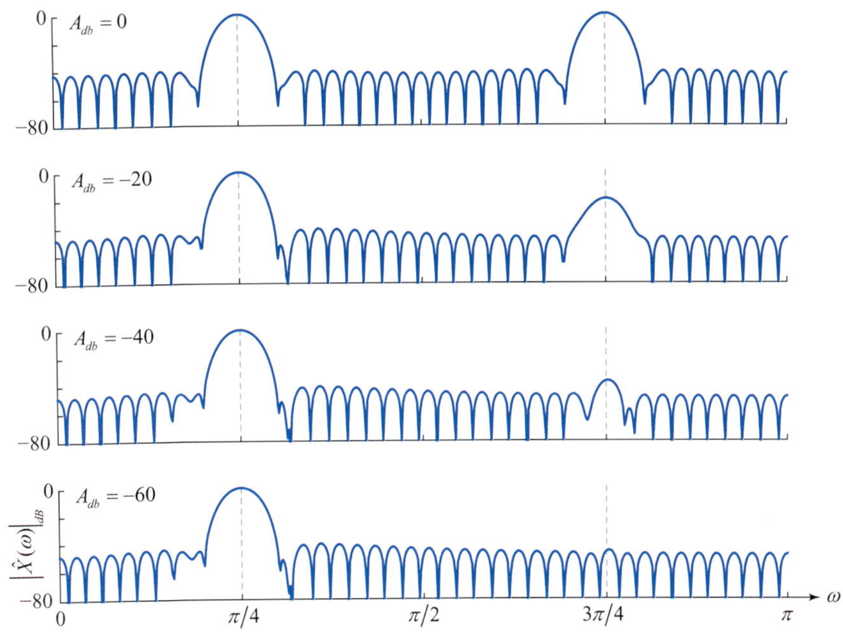

Figure 14.11 Effect of Hamming window on amplitude resolution

which vary widely in amplitude, poses particular problems. An example of this sort of problem would be using spectral analysis to identify the presence of a trace element that presents a very faint spectral signature at a particular frequency in the presence of a very strong component at another frequency.

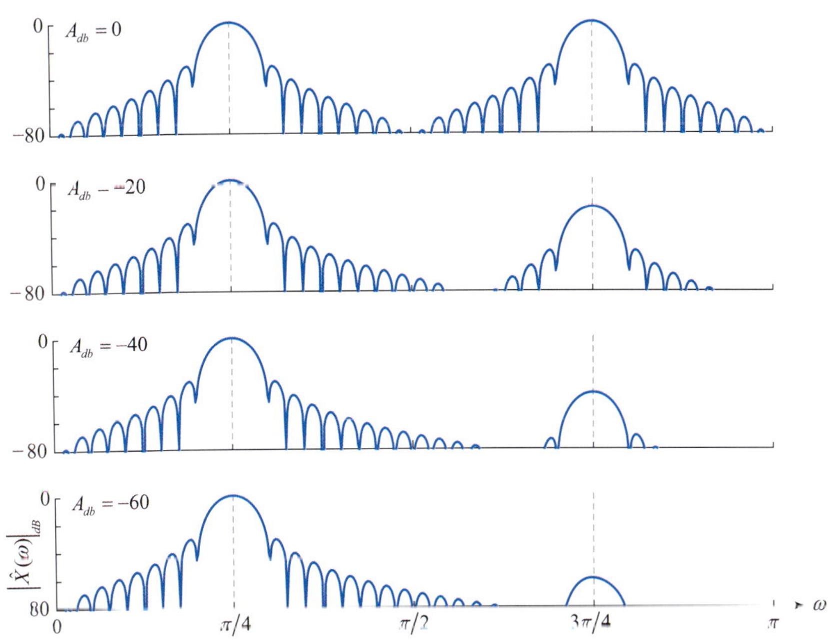

Figure 14.12 Effect of Hann window on amplitude resolution

As with the example of **Figure 14.8**, the problem of the out-of-band energy can be ameliorated by the choice of a different window. **Figure 14.12** shows the effect of reducing the amplitude using a Hann window of the same length, $N=81$. Even though the maximum sidelobe attenuation of the Hann window is higher than that of the Hamming window (−32 dB vs. −43 dB), the faster decay of the sidelobes for the Hann window (−18 dB/oct vs. −6 dB/oct) means that the amount of spectral leakage from the dominant spectral component is considerably reduced, and this particular spectral component of lower amplitude can be resolved.

14.1.4 Spectral effect of sampling

In addition to the issues of spectral estimation introduced by the *windowing* of the sequence, a second set of issues derives from the fact that practical analysis techniques usually require *sampling* the spectrum at discrete frequencies using a transform such as the DFT, rather than using a transform of continuous frequency resolution such as the DTFT. Sampling imposes its own set of constraints on frequency measurements with which we have to contend.

In most spectral estimation problems, a continuous-time signal $x(t)$ is sampled at rate f_s to form a sequence $x[n]$. That sequence is windowed to form $\hat{x}[n]$, from which the spectrum is calculated. In the preceding discussion, we have taken the spectrum to be the DTFT of the windowed signal, $\hat{X}(\omega)$, which is a continuous function of frequency ω. However, in practice we do not measure $\hat{X}(\omega)$, but instead $\hat{X}[k]$, the N-point DFT of the windowed input sequence $\hat{x}[n]$,

$$\hat{X}[k]=\sum_{n=0}^{N-1}\hat{x}[n]e^{-j2\pi kn/N}=\sum_{n=0}^{N-1}x[n]w[n]e^{-j2\pi kn/N}.$$

In Chapter 10, we showed that the DFT can be viewed as the DTFT sampled in frequency at multiples of $2\pi/N$,

$$\hat{X}[k]=\hat{X}(\omega)\big|_{\omega=2\pi k/N},$$

and in Chapter 11 we showed that the DFT could be computed efficiently using the fast Fourier transform (FFT). All the issues with the estimation of spectral frequency due to windowing of the sequence we discussed in Section 14.1.3 – the spreading of the spectral peaks and the noise floor due to spectral leakage – still apply to the case of sampling of the DTFT by the DFT. But in addition the DFT introduces a new problem to spectral estimation: the size of the transform, N, determines the **frequency resolution** of the transform, $\Delta\omega$, which is the difference in spectral frequency (in radians) between two adjacent points in the transform,

$$\Delta\omega-2\pi/N.$$

For spectral estimation problems that start with a continuous-time signal $x(t)$, the frequency resolution Δf, expressed in Hz, is determined both by the transform size N and by the sampling frequency f_s. By the sampling theorem, f_s maps to the discrete-time frequency 2π, so

$$\Delta f=(f_s/2\pi)\Delta\omega=f_s/N.$$

As an example, consider the signal $x(t)=\cos 2\pi f_0 t$, where $f_0=1143$ Hz. This signal is sampled at rate $f_s=8000$ Hz to form sequence $x[n]=\cos\omega_0 n$, where $\omega_0=2\pi f_0/f_s=0.2858\pi$. Then, a 16-point segment of $x[n]$, $0 \le n \le 15$, is selected. This is equivalent to multiplying $x[n]$ by a rectangular window of length $N = 16$. The result is the finite-length sequence $\hat{x}[n]$ shown in the left panel of **Figure 14.13a**. As we have previously shown (e.g., **Figure 14.12**), the application of a time window results in a broad peak in the DTFT $\hat{X}(\omega)$, shown in the right panel (dashed black line), plotted vs. frequency, $0 \le \omega \le \pi$. The actual frequency of the cosine, ω_0, is indicated with a thin red line in each panel. Superimposed on the plot is the magnitude of the N-point DFT, $|\hat{X}[k]|$, as a function of k, $0 \le k \le 8$ (filled symbols).

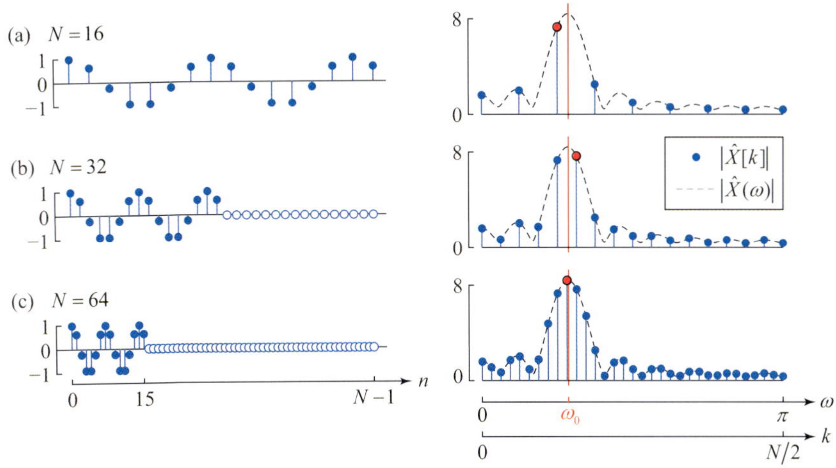

Figure 14.13 Spectrum of a cosine as a function of increasing DFT length

At a value of $N = 16$, the frequency resolution is coarse, $\Delta\omega = \pi/8$ ($\Delta f = 500$ Hz). The peak of the DFT occurs at $k_{\max} = 2$ (highlighted red symbol), which corresponds to a frequency of $\hat{\omega}_0 \triangleq k_{\max}\Delta\omega = 0.25\pi$ ($\hat{f}_0 \triangleq k_{\max}\Delta f = 1000$ Hz). In order to increase the frequency resolution of the DFT, the same 16-point data sample can be padded with $N - 16$ zeros, as shown in the left panels of **Figure 14.13b** (for $N = 32$) and **Figure 14.13c** (for $N = 64$), before the DFT is taken. Appending zeros to $x[n]$ does not change the DTFT $\hat{X}(\omega)$, but it does increase the number of points in the DFT, and therefore the frequency resolution. N is most often chosen to be a power of two, because radix-2 FFTs are commonly used to implement the DFT, as discussed in Chapter 11. With each factor of two increase in N, the frequency resolution of the DFT improves by a factor of two (i.e., $\Delta\omega$ gets smaller), so that the maximum difference between the "true" frequency, ω_0, and the frequency corresponding to the peak of the DFT, $\hat{\omega}_0$, is guaranteed to be no greater than $\Delta\omega/2 = \pi/N$. When $N = 32$, $\hat{\omega}_0 = 5\Delta\omega = 0.3125\pi$ ($\hat{f}_0 = 1250$ Hz), and when $N = 64$, $\hat{\omega}_0 = 9\Delta\omega = 0.2813\pi$ ($\hat{f}_0 = 1125$ Hz).

In the example of **Figure 14.13**, the number of sampled data points was fixed at $M = 16$ points, and the size of the transform, N, was adjusted. This separated the effect of *windowing* the data from the effect of *sampling* the DTFT at discrete frequencies by taking the DFT at different values of N. **Figure 14.14** shows an example of the converse situation in which the size of the DFT is fixed at $N = 64$ points, and the size of the data window, $M \le N$, is changed. The analog signal here is the sum of two cosines closely spaced in frequency, $x(t) = \cos 2\pi f_1 t + \cos 2\pi f_2 t$, where $f_1 = 914$ Hz and $f_1 = 1143$ Hz. This signal is then sampled at rate $f_s = 8000$ Hz to form the discrete-time signal $x[n] = \cos \omega_1 n + \cos \omega_2 n$, where $\omega_1 = 0.23\pi$ and $\omega_2 = 0.29\pi$; this means that the separation of spectral peaks is only $\omega_2 - \omega_1 = 0.06\pi$. $x[n]$ is then truncated to various lengths M, padded with $N - M$ zeros to form $\hat{x}[n]$ and the 64-point DFT is taken to produce $\hat{X}[k]$. Since the frequency resolution of the DFT is fixed at $\Delta\omega = 2\pi/64 = 0.03\pi$, we can investigate the effect of the data windowing as M changes. The task here is to resolve both spectral peaks of $\hat{X}[k]$. **Figure 14.14a** shows the spectrum of the DTFT ($|\hat{X}(\omega)|$, dashed black line) and the DFT ($|\hat{X}[k]|$, solid symbols) of a section of $\hat{x}[n]$ of length $N = 16$. When the windowed duration of $\hat{x}[n]$ is relatively short, the effective spectral resolution of the DTFT is determined by the size and nature of the data window, as previously shown in **Figure 14.8**. Because the separation of spectral peaks is so small, individual peaks cannot be resolved in the spectrum of the DTFT $|\hat{X}(\omega)|$ at a data window size of $M = 16$ (**Figure 14.14a**). So, increasing the size of the DFT alone by further zero-padding the sequence, as in **Figure 14.13**, would be insufficient to allow us to resolve the two spectral peaks at *any* value of N. The remaining panels of the figure show the result of taking the transform of larger sections of $x[n]$: $M = 32$ (**Figure 14.14b**) and $M = 64$ (**Figure 14.14c**). As the size of the data window increases, the two spectral peaks become increasingly distinct.

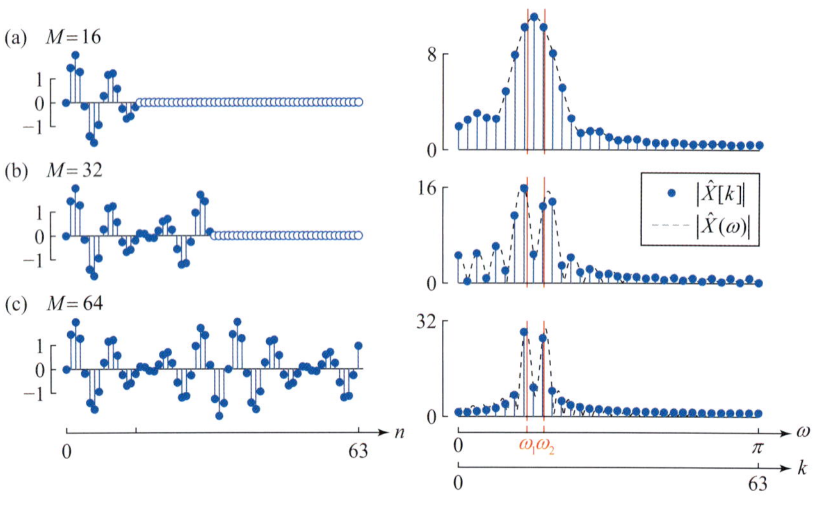

Figure 14.14 Spectrum of a signal with two tones

This result suggests that one should use the largest data window and DFT possible when analyzing the spectrum of a signal, consistent with any computational constraints. However, this approach only makes sense if the main spectral features of the signal – that is, the frequencies and amplitudes of the spectral components of $x[n]$ – are constant for the entire duration of the window. In many cases of the most practical interest, such as the analysis of speech and music, the signals are inherently time-varying and we need a different approach to perform effective spectral analysis, which we describe in the next section.

14.2 The short-time Fourier transform (STFT)

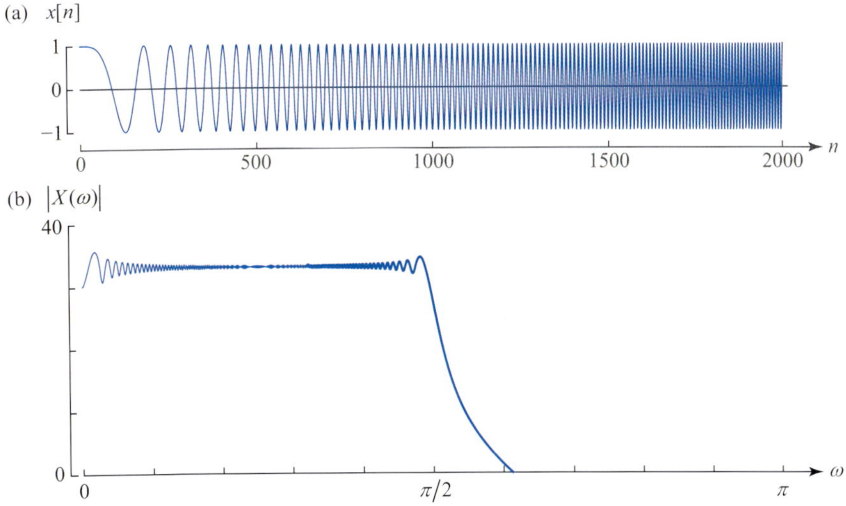

Figure 14.15 DTFT of a chirp

In order to motivate the discussion of the short-time Fourier transform, consider the time-varying signal called a **linear FM sweep** or **chirp**,

$$x[n] = \cos 2\pi(n/128)^2, \tag{14.2}$$

shown in **Figure 14.15a**. The instantaneous frequency of this signal, given by

$$\frac{d}{dn} 2\pi(n/128)^2 = 4\pi n/128^2, \tag{14.3}$$

varies linearly with time n, from 0 when $n = 0$ to $\pi/2$ when $n = 2048$.

Figure 14.15b shows $|X(\omega)|$, the DTFT of the entire signal. The fact that the instantaneous frequency of $x[n]$ varies with time is not at all revealed in the transform of the entire signal. A solution is to create what amounts to a time-varying DTFT. The idea is to break $x[n]$ into a series of smaller subsequences in which the instantaneous frequency is more or less constant, then take the DTFT "snapshots" of each subsequence, and paste these snapshots together to obtain a panoramic picture of spectral frequency as a function of time.

To formalize these ideas, we define a **frame** as a subsequence formed by multiplying $x[n]$ by a finite-length window function $w_L[n]$, such as a rectangular, Hamming or Hann window of **frame length** L,

$$x[n]w_L[n - mD],$$

where the parameter D is the **frame offset** and m is the **frame index**. The product mD determines the shift of the window with respect to the input, so that $x[n]w_L[n - mD]$ represents the short subsequence of $x[n]$ of length L to be analyzed. Now, the DTFT of each frame, m, is computed,

$$X_L(\omega, m) \triangleq \mathfrak{F}\{x[n]w_L[n - mD]\} = \sum_{n=-\infty}^{\infty} (x[n]w_L[n - mD])e^{-j\omega n}.$$

The result, $X_L(\omega, m)$, is the discrete-time **short-time Fourier transform (STFT)**. Whereas the DTFT $X(\omega)$ is a function of only one parameter, ω, the properties of the STFT $X_L(\omega, m)$ depend both on ω and on the parameters of the window $w_L[n]$: its shape, length L and offset D. We will now investigate how the STFT depends on these factors. We will be particularly interested in the conditions under which the STFT is a true, invertible transform.

14.2.1 Constant overlap-add criterion

Given an appropriate choice of the window $w_L[n]$, and the parameters L and D, we will now first show that the total input $x[n]$ can be reassembled from the sum of the smaller windowed subsequences,

$$x[n] = \sum_{m=-\infty}^{\infty} x[n]w_L[n - mD]. \tag{14.4}$$

This will not work for just any choice of window parameters. To see what is required to assemble $x[n]$ from the sum of the windowed subsequences, consider a few possible choices for the window parameters L and D. If $D > L$, then $w_L[n - mD]$ cuts $x[n]$ into disjoint frames of length L with gaps of length $L - D$ between them so we certainly cannot reassemble $x[n]$ from the pieces. If $D = L$, then $w_L[n - mD]$ cuts $x[n]$ into a series of frames that exactly adjoin one another. Equation (14.4) can be satisfied, but only if $w_L[n]$ is a rectangular window. Of particular interest is the case when $1 \le D < L$. Then, subsequent frames of $x[n]w_L[n - mD]$ overlap by $L - D$ points. We will now show that if the window type $w_L[n]$ and parameters L and D satisfy the **constant overlap-add (COLA)** criterion,

$$\sum_{m=-\infty}^{\infty} w_L[n - mD] = 1,$$

then Equation (14.4) will be satisfied and STFT will be a truly invertible transform. To understand the COLA criterion, consider an example in which $w_L[n]$ is a causal, discrete-time Hann window of odd length $M = 17$,

$$w_L[n] = 0.5 - 0.5\cos\frac{2\pi n}{L-1}, \quad 0 \le n \le L - 1,$$

as shown in **Figure 14.16a**.

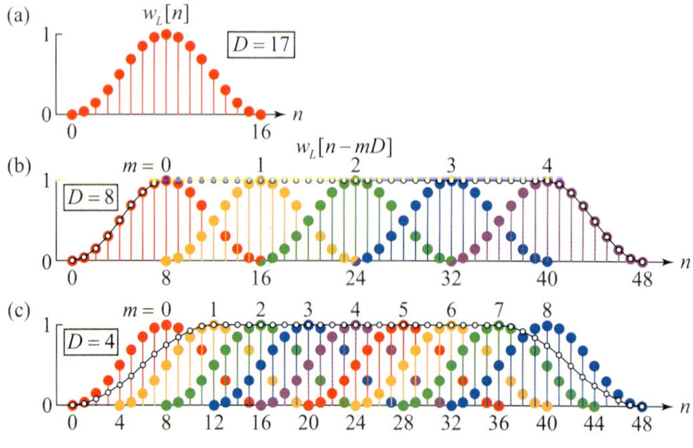

Figure 14.16 Overlapping windows for the STFT

When the offset is $D = (L-1)/2 = 8$ samples, as shown in **Figure 14.16b**, then successive windows overlap by 50%. With this choice of window type, length and overlap, the sum of overlapping windows (thin black line and open symbols) is a constant equal to one (Problem 14-1a),

$$\sum_{m=-\infty}^{\infty} w_L[n-mD] = \sum_{m=-\infty}^{\infty} w_L[n-m(L-1)/2] = 1, \tag{14.5}$$

except for the points at the beginning and end of the entire sum. The Hann window is not unique in satisfying the COLA criterion. Other raised-cosine windows such as the Hamming window also have (or can be modified to have) this property (see Problem 14-1c).

If $w_L[n]$ satisfies the COLA criterion, then the sum of all frames of the STFT is equal to the DTFT of the entire signal:

$$\sum_{m=-\infty}^{\infty} X_L(\omega,m) = \sum_{m=-\infty}^{\infty}\left(\sum_{n=-\infty}^{\infty}(x[n]w_L[n-mD])\right)e^{-j\omega n} = \sum_{n=-\infty}^{\infty} x[n]e^{-j\omega n}\underbrace{\left(\sum_{m=-\infty}^{\infty} w_L[n-mD]\right)}_{1}$$

$$= \sum_{n=-\infty}^{\infty} x[n]e^{-j\omega n} = X(\omega).$$

The COLA criterion will generally be satisfied for odd-length windows and integer values of overlap $D = (L-1)/R$, where R is an integer **overlap factor**. In this case, the fractional frame overlap is $(R-1)/R$ and the sum formula of Equation (14.5) generalizes to

$$\sum_{m=-\infty}^{\infty} w_L[n-mD] = \frac{2}{R}\sum_{m=-\infty}^{\infty} w_L[n-m(L-1)/R] = 1.$$

For example, **Figure 14.16c** shows that when $R = 4$, then $D = (L-1)/4 = 4$, and successive windows overlap by 75%.

14.2.2 The spectrogram

One key application of the STFT is the **spectrogram**: a two-dimensional graphical representation of the frequency spectrum of a time-varying signal. The spectrogram is extensively used to characterize the spectral content of speech and music, but it is also used to analyze signals from fields as diverse as geology (e.g., signals generated by earthquakes and volcanic eruptions), astronomy (e.g., measurements of electromagnetic radiation from space in optical, infrared and radio bands) and medicine (e.g., analysis of electrocardiographic (EKG) signals from the heart, electroencephalographic signals (EEG) from the brain and electromyographic signals (EMG) from muscle). Spectrographic displays are also common in many object detection and classification applications (e.g., radar, sonar and fish-finding(!)).

Spectrogram of a chirp **Figure 14.17** shows how a spectrogram is constructed for the simple chirp signal we introduced in **Figure 14.15a**.

Figure 14.17b shows the chirp $x[n]$ given by Equation (14.2). **Figure 14.17a** shows a series of 16 overlapping Hann windows $w_M[n-mL]$ (thin grey traces) of length $M = 241$ that span the

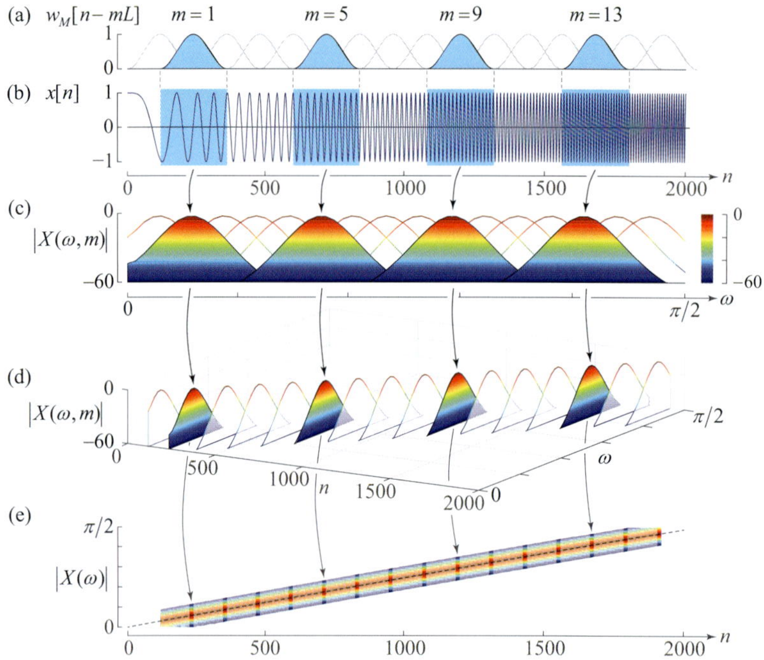

Figure 14.17 Construction of the spectrogram of a chirp

time range of $x[n]$. The windows are created with an overlap of 50% (i.e., an overlap factor of $R=2$), giving a frame offset of $L=(M-1)/R=120$ points. Four of the windows that are evenly spaced in time ($m=1$, 5, 9 and 13) are highlighted in blue, and correspond to the highlighted regions of $x[n]$ shown in **Figure 14.17b**. **Figure 14.17c** shows the magnitude of the STFT, $|X(\omega, m)|$, of all 16 windowed segments $x[n]w[n-mL]$, $0 \le m \le 15$, plotted on a dB scale vs. frequency ω. The traces are plotted in false color so that the highest-magnitude portions of the trace (near 0 dB) are dark red and the lowest-magnitude portions (at or below −60 dB) are dark blue. A color bar to the right of the plot shows the range of colors and their relation to magnitude. The STFTs of the four highlighted time segments in **Figure 14.17a** are plotted filled with color. The spectra of all the frames have a similar shape and are equally spaced apart in frequency, with peaks corresponding to the instantaneous frequency of the center of their respective frames. In order to appreciate fully how the STFT frames depend on time, **Figure 14.17d** shows all of these STFTs in three dimensions as spectral "slices" ($|X(\omega, m)|$ vs. ω) plotted as a function of the time n of the center of each windowed segment. Finally, because we have represented the spectral magnitude in this plot both as the z-axis of the plot and as the color of the trace, the z-axis of the plot is actually redundant, since at any time-frequency point in the plot, the color tells us everything we need to know about the magnitude. In order to produce a more compact, two-dimensional spectral representation, we rotate the 3-D plot so that we are effectively looking on it from above. The result (**Figure 14.17e**) is the conventional spectrogram, which is a plot of spectral frequency (ordinate) vs. time (abscissa) with magnitude encoded in color. In order to show the spectrogram with finer time resolution, this plot was created using an overlap factor of $R = 16$, so that the number of spectral slices is a factor of eight greater than **Figure 14.17d**. The thin dotted black line superimposed on the plot is the instantaneous frequency given in Equation (14.3).

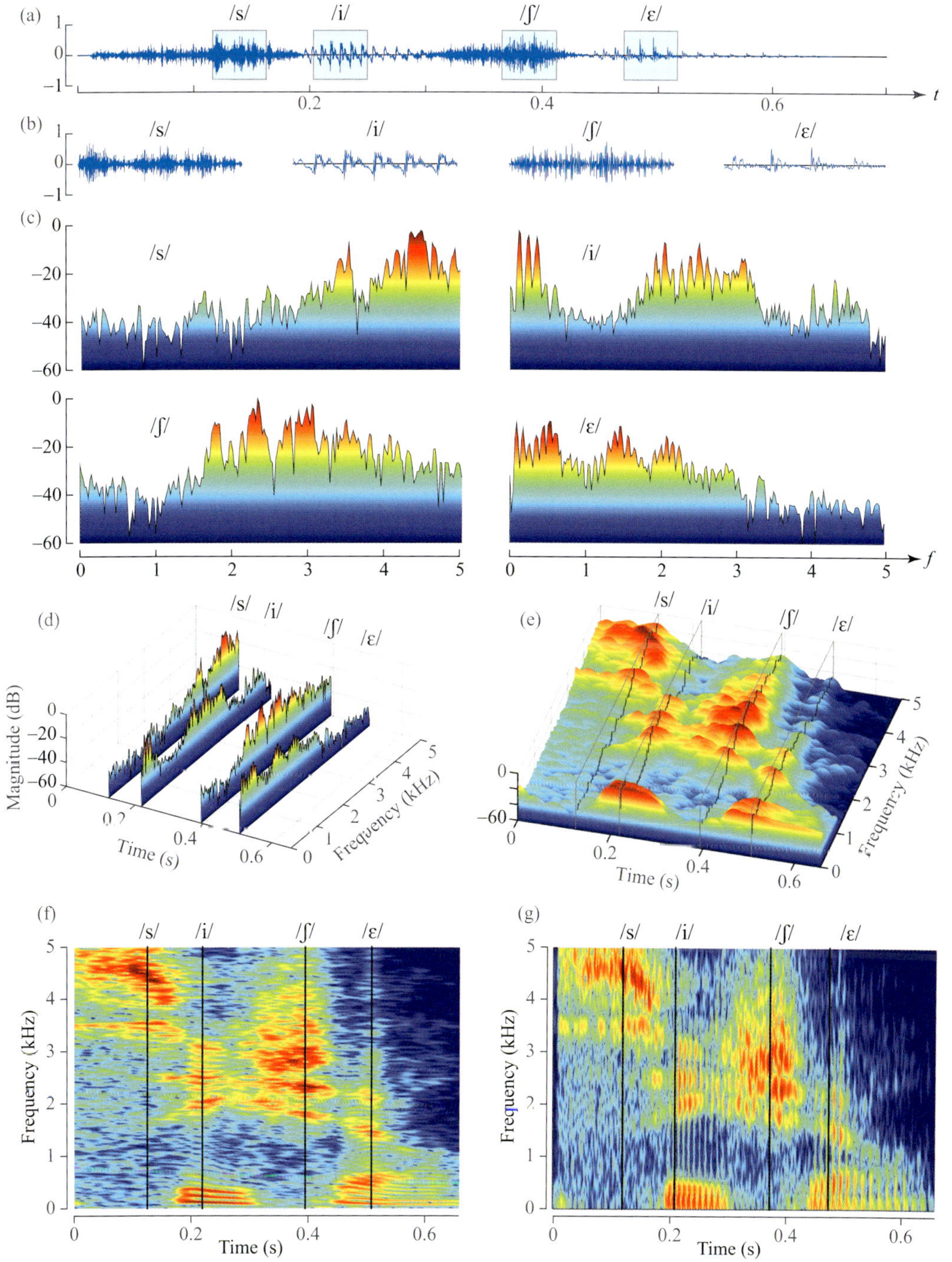

Figure 14.18 Spectrogram of the word "seashell"

Spectrogram of speech So far, all the signals we have considered in this chapter have been comparatively simple: a continuous tone or a constant-amplitude chirp. However, most of the signals we wish to describe or analyze in real-world applications are considerably more complex. Consider the problem of applying Fourier analysis to speech. **Figure 14.18a** shows a plot of a 700 ms sample of speech: the word "seashell," spoken by a male speaker (your author) digitized at a sample rate of 11025 samples/s. Like the chirp, speech is time-varying, but, unlike the chirp, the time waveform is not a constant-amplitude linear sweep of frequency. Rather, the amplitude of the utterance changes markedly with time. Like most speech utterances, this word is a concatenation of **consonants** and **vowels** occurring in alternation. Highlighted in light blue are four 46 ms segments of the word, corresponding to the consonant /s/ (="s") and the vowel /i/ (="ee") in "sea," followed by the consonant /ʃ/ (="sh") and the vowel /ɛ/ (="eh") in "shell." **Figure 14.18b** shows plots of these four highlighted segments at higher resolution. Clearly these segments look different in the time domain. The vowels /i/ and /ɛ/ have a clear periodic structure, while the consonants /s/ and /ʃ/ both look "noisy," that is, not obviously periodic.

Figure 14.18c shows the magnitude of the STFT spectrum of each of the four segments of **Figure 14.18b** plotted from 0 to 5 kHz, which corresponds to the frequency range in which most of the energy in speech lies. The magnitude is plotted on a dB scale in "false color," as in **Figure 14.17c**. The spectra of the consonants and vowels are markedly different. For example, the spectrum of the vowel /i/ comprises a series of peaks; there is a big peak of energy at about 90 Hz, which corresponds to the fundamental pitch frequency of the vowel, and there are also peaks roughly at multiples of this pitch frequency. Most of the energy in this spectrum is below about 3 kHz, which is very characteristic of the spectrum of vowels. In contrast, the spectrum of the consonant /s/ does not have a lot of structure. It has a broad peak above 4 kHz, and very little energy below 1 kHz. You can observe a similar dichotomy between the vowel /ɛ/ and consonant /ʃ/. **Figure 14.18d** replots these four spectra as time "slices" on a 3-D plot, in a manner similar to **Figure 14.17d**. **Figure 14.18e** shows a plot of the spectra of 165 slices in time (46.4 ms long segments of the utterance, each offset from the previous segment by 5.8 ms, corresponding to an overlap factor of $R = 8$). Here again, the x-axis is time, the y-axis is frequency and the z-axis is magnitude. The plot appears to be a continuous surface that resembles a topographic map where the red mountain peaks correspond to high spectral magnitude and the blue valleys represent the low-energy portions of the spectrum. While beautiful, the z-axis of the plot is redundant, so we can rotate this 3-D spectral plot to form the 2-D spectrogram (**Figure 14.18f**).

Spectrograms such as this are an essential tool in the graphical representation of the spectra of time-varying waveforms such as speech. In this example of "seashell" you can clearly see the alternation of distinct spectral patterns that correspond to the consonants and vowels. The /s/ of "sea" is a **sibilant fricative**, which is a fancy way of saying "a hissy consonant." On the spectrogram, the sibilant displays as a relatively diffuse blob of high-frequency energy above 4 kHz. The vowel /i/ comprises energy predominately in three frequency bands: near 400Hz, 2 kHz and 3 kHz. These bands are termed the **formants** of the vowel. The formants correspond to the resonant peaks in the vocal tract that characterize this vowel. You can see that there is actually considerable microstructure in the spectrum of these vowels. Each formant is made of a series of striations, which are spaced at multiples of the pitch period of the utterance, in this case about 90 Hz, which correspond to the peaks we saw in the spectral slice of /i/ in

Figure 14.18c. Following the vowel is another sibilant consonant, /ʃ/ of "shell." This consonant again displays as a diffuse high-frequency blob, but in this case the centroid of the spectrum is of somewhat lower frequency than the /s/ in "sea." Finally, the /ε/ of "shell" is another vowel whose formants are at a different frequency from the vowel /i/.[1]

In Section 14.5.2, we will continue our discussion of speech when we introduce a model of speech production that is the basis of a family of signal processing applications for speech coding and analysis called **linear predictive coding**.

Time resolution vs. frequency resolution of the spectrogram The spectral representation of the spectrogram is critically dependent on the choice of parameters. In particular, there is an inextricable trade-off between the time and frequency resolution of the STFT that results from the choice of the windowing function and its length, issues we have also covered in Section 14.1. A large time window gives finer frequency resolution but coarser time resolution. Conversely, a small time window gives better time resolution at the expense of poorer frequency resolution. The choice of window is highly dependent on the nature of the signal being analyzed. In the example of "seashell" in **Figure 14.18f**, the data window of the STFT is 46 ms long, resulting in a so-called **narrowband transform**. This provides good localization in frequency; the horizontal striations show the detailed formant structure of vowels. In contrast, the spectrogram of "seashell" in **Figure 14.18g** was calculated with a data window only 11.6 ms long, resulting in a **wideband transform**. Here, we have given up fine frequency resolution in exchange for fine time resolution. Although we can no longer see the formant structure of the vowels, the narrow analysis window allows us to discern the individual pitch periods of the vowels as vertical striations in the display of /i/ and /ε/.

14.2.3 ★ Implementation of the discrete STFT in Matlab

The **discrete STFT** is the time-varying equivalent of the DFT, formed by sampling the STFT at N points in frequency,

$$X[k,m] = X(\omega,m)|_{\omega=2\pi k/N}.$$

In practice, the FFT (Chapter 11) is used to calculate the DFT. The following code demonstrates how the spectrogram of **Figure 14.17e** was created. First, produce a 2048-point chirp using Equation (14.2).

```
% Create chirp
lx = 2048;
n = 0:lx-1;
x = cos(2*pi*(n/128).^2)';
```

[1] These two vowels and two sibilant consonants are just a few of the many sounds, or **phones**, that make up the building-blocks of speech. There are many vowels, not all of which are found in every language. There are also many different consonants, including **nasals** (e.g., /m/, /n/), **plosives** (/b/, /d/, /g/), **voiceless** (/s/) and **voiced** (/z/) sibilant fricatives and voiceless (/ϕ/="th") and voiced (/v/) non-sibilant fricatives, among others. The combination of these speech sounds makes up human language. The study of these speech sounds – their production and perception– is called **acoustic phonetics** (from the Greek φωνή, *phone* = sound/voice).

Next, the size of the FFT, Nfft, is set at 256 and a Hann window h is calculated and scaled to make the maximum value of the STFT be close to 0 dB. The length of the data window, M, needs to be odd and smaller than Nfft. Here the length of each frame, M, has been chosen to be 241 points, which results in the largest possible number of frames consistent with an overlap factor of $R = 16$.

```
% Create w[n]
Nfft = 256;    % size of DFT
R = 16;        % overlap factor
L = Nfft-R+1; % frame size (length of w[n])
h = hann(L);   % make w[n]
h = 2 * h / sum(h); % normalize
```

Next, each frame m of the STFT is a column vector in the matrix, Xwm(:, m), which is computed by multiplying a frame of $x[n]$ by the window and taking the FFT. We are actually calculating $x[n+mD]w_L[n]$ here, which is equivalent to $x[n]w_L[n-mD]$. The second argument of the FFT function, Nfft, automatically pads the frame with zeros.

```
% Compute STFT
D = (L-1)/R; % frame overlap
nFrames = floor((lx-L)/D+1);   % number of frames of STFT data
Xwm = zeros(Nfft, nFrames);    % allocate STFT ...
for m = 1:nFrames              % ... and calculate it
    indx = (m-1)*D + 1;
    Xwm(:, m) = fft(x(indx:indx+L-1).*h, Nfft); % this is x[n]w[n-mD]
end
```

Finally, we plot the spectrogram. Since the maximum instantaneous frequency of the chirp is $\pi/2$, we only need one-quarter of the calculated points of the FFT. Matlab's surf function creates the three-dimensional plot, where the x-axis is time, the y-axis is frequency and the z-axis is spectral magnitude on a dB scale. A time offset toff of half the frame size, $(L-1)/2$, has been added to x to align the time of each spectral "slice" of the spectrogram with the center of the associated time window. The view command, here incorporated in set(gca,...), rotates the 3-D plot so we are looking on it from above, thereby creating the 2-D spectrogram of x (time) vs. y (frequency). The ZLim and CLim arguments respectively set the range of magnitude to be displayed and set the color map to span that range.

```
% Plot spectrogram
wmax = 0.5; % max freq = pi/2
npts = Nfft*wmax/2+1; % select part of FFT

x = (0:D:(nFrames-1)*D);
y = linspace(0, wmax, npts);
z = db(abs(Xwm(1:npts, :)));
noff = (L-1)/2; % offset of center of time window
surf(x+noff, y, z, 'EdgeColor', 'none');
hold on;
plot([0 lx-1], 4*[0 lx-1]/128^2, 'k')
zl = -50; % minimum magnitude to display
set(gca, 'View', [0 90], 'XLim', [0 2000], 'ZLim', [zl 0], 'CLim', [zl 0])
```

The Matlab function `spectrogram` provides the basic functionality to compute and display spectrograms such as this. It allows the user to specify the window function, the frame overlap and the size of the FFT, among other parameters.

14.3 ★ Nonparametric methods of spectral estimation

Most of the chapters in this book have been concerned with **deterministic** systems. The defining characteristic of a deterministic system – for example, an FIR filter – is that each time a given input is applied to the system, it produces the same output. In contrast, most of the systems encountered in the real world are, to a greater or lesser degree, random. The defining characteristic of a random or **probabilistic** system is that its output is uncertain or unpredictable. Randomness may arise intrinsically – for example, the variation in sound waveforms of a speaker saying the same thing – or it may arise extrinsically from noise added to a signal or system due to environmental, electronic or computational factors.

For the remainder of this chapter, we turn to the spectral analysis of random signals. We are generally interested in estimating some characteristics of a signal (which may itself be random) in the presence of noise. A simple example of this task would be to estimate the frequency of a single sinusoid with random phase in the presence of white noise. A more complicated example would be to estimate the formant frequencies of a segment of speech. We shall look at both of those problems.

There are two main categories of spectral estimation techniques: nonparametric and parametric. **Nonparametric spectral methods**, which we discuss in this section, are methods that do not depend on knowing any specific information of the signal being analyzed; they assume no model or particular structure to the data. In contrast, **parametric spectral methods**, which we will discuss in Section 14.4, are based on estimating the parameters of models that have been specifically designed to represent important features of the process being analyzed, such as speech.

Much of the material in this section assumes that you have some familiarity with the basic ideas of probability and random processes. Appendix D provides a brief review of these concepts and also introduces the notation we will be using for our derivations in the paragraphs to follow.

14.3.1 The periodogram

In practical applications for spectral estimation, the task is to estimate features of a signal based on finite-length segments of data. The periodogram provides the critical link between quantities that you *can* measure directly from the finite-length signal (i.e., the DFT) and those that you *cannot*, namely the power spectral density of the underlying random process.

Definition Let $x[n]$ be an infinitely long, zero-mean, wide-sense stationary (WSS) discrete-time random process whose autocorrelation function $r_{XX}[l]$ is absolutely summable,

$$\sum_{l=-\infty}^{\infty} |r_{XX}[l]| < \infty.$$

(See Appendix D for a definition of all these terms.) The **power spectral density (PSD)** of this process, $S_{XX}(\omega)$, is defined as the DTFT of the autocorrelation function,

$$S_{XX}(\omega) \triangleq \mathfrak{F}\{r_{XX}[l]\} = \sum_{l=-\infty}^{\infty} r_{XX}[l]e^{-j\omega l}. \tag{14.6}$$

However, in any real problem we encounter, we do not have access to an infinite amount of data $x[n]$, nor do we know the underlying **probability density function** $p_X(x)$, which would be necessary to calculate the "true" autocorrelation function $r_{XX}[l]$ and its transform, the "true" power spectral density $S_{XX}(\omega)$. The task of nonparametric spectral estimation therefore comes down to estimating $S_{XX}(\omega)$ given limited data.

Assume that we only have access to a portion of $x[n]$ of length N, $0 \le n \le N-1$, and wish to develop an **estimator** of the autocorrelation function for this finite-length process. As discussed in Appendix D, for a process that is **autocorrelation-ergodic**, we initially propose defining the estimator as

$$\frac{1}{N}\sum_{n=0}^{N-1} x[n]x[n-l]. \tag{14.7}$$

However, in order to insure that $x[n-l]$ is only evaluated in the range $0 \le n-l \le N-1$, it is first useful to define a sequence $x_N[n]$ that is equal to $x[n]$ multiplied by a rectangular window $w_N[n]$ that has a value of one within the range $0 \le n \le N-1$ and zero outside of it,

$$x_N[n] \triangleq x[n]w_N[n], \tag{14.8}$$

where

$$w_N[n] = \begin{cases} 1, & 0 \le n \le N-1 \\ 0, & \text{otherwise} \end{cases}. \tag{14.9}$$

Then an estimator of the autocorrelation function $\hat{r}_{XX}[l, N]$, termed the **correlogram**, can be defined in terms of $x_N[n]$ (see Appendix D.2.6),

$$\hat{r}_{XX}[l, N] \triangleq \frac{1}{N}\sum_{n=-(N-1)}^{N-1} x_N[n]x_N[n-l] = \frac{1}{N}\sum_{n=-\infty}^{\infty} x_N[n]x_N[n-l] = \frac{1}{N}x_N[l] * x_N[-l]. \tag{14.10}$$

The argument N in $\hat{r}_{XX}[l, N]$ makes it expressly clear that this estimator depends on the length of the sequence, N. The infinite limits on the second summation in Equation (14.10) reflect the fact that the product $x_N[n]x_N[n-l]$ will be zero for $|n| > N-1$, so we can expand the summation's limits to cover all time. The fact that the product $x_N[n]x_N[n-l]$ is zero for $|l| > N-1$ means that $\hat{r}_{XX}[l, N]$ will be zero for $|l| > N-1$. Now, by analogy with Equation (14.6), we define the estimator of the power spectral density $\hat{S}_{XX}(\omega, N)$ to be the Fourier transform of the correlogram,

$$\hat{S}_{XX}(\omega, N) \triangleq \mathfrak{F}\{\hat{r}_{XX}[l, N]\} = \sum_{l=-(N-1)}^{N-1} \hat{r}_{XX}[l, N]e^{-j\omega l}. \tag{14.11}$$

$\hat{S}_{XX}(\omega, N)$ is termed the **periodogram**, and it too depends on the length N of the underlying sequence, $x_N[n]$. Because $x[n]$ is a random process, each measurement of N points will be different, which means that the estimator $\hat{S}_{XX}(\omega, N)$ is itself a random process.

The periodogram can also be defined directly in terms of the Fourier transform of the sequence $x[n]$. To see this, first substitute Equation (14.10) into Equation (14.11),

$$\hat{S}_{XX}(\omega, N) = \sum_{l=-(N-1)}^{N-1} \hat{r}_{XX}[l, N]e^{-j\omega l} = \mathfrak{F}\left\{\frac{1}{N}x_N[l] * x_N[-l]\right\} = \frac{1}{N}\mathfrak{F}\{x_N[l]\} \cdot \mathfrak{F}\{x_N[-l]\}. \tag{14.12}$$

Let $X_N(\omega)$ be the DTFT of $x_N[n]$. Then, from Equation (14.8),

$$X_N(\omega) \triangleq \mathfrak{F}\{x_N[n]\} = \sum_{n=-\infty}^{\infty} x_N[n]e^{-j\omega n} = \sum_{n=-\infty}^{\infty} x[n]w_N[n]e^{-j\omega n} = \sum_{n=0}^{N-1} x[n]e^{-j\omega n}. \tag{14.13}$$

The DTFT of $x_N[-n]$ is

$$\mathfrak{F}\{x_N[-n]\} = X_N(-\omega) = X_N^*(\omega). \tag{14.14}$$

Here, we have used the time-reversal property of the DTFT as well as the fact that $x_N[n]$ is real, which mean that $X_N(\omega)$ is conjugate symmetric, $X_N(-\omega) = X_N^*(\omega)$. Putting Equations (14.12)–(14.14) together, we get an expression for $\hat{S}_{XX}(\omega, N)$ in terms of the DTFT of $x[n]$,

$$\begin{aligned}\hat{S}_{XX}(\omega, N) &= \frac{1}{N}\mathfrak{F}\{x_N[l]\} \cdot \mathfrak{F}\{x_N[-l]\} = \frac{1}{N}X_N(\omega)X_N^*(\omega) = \frac{1}{N}|X_N(\omega)|^2 \\ &= \frac{1}{N}\left|\sum_{n=0}^{N-1} x[n]e^{-j\omega n}\right|^2.\end{aligned} \tag{14.15}$$

The periodogram $\hat{S}_{XX}(\omega, N)$, as defined in Equation (14.15), is an estimator of the "true" power spectral density $S_{XX}(\omega)$, based on N measured samples of $x[n]$. But how good an estimator is it?

Bias of the periodogram As discussed in Appendix D, there are two measures that are most commonly used to assess the "goodness" of an estimator: **bias** and **consistency**. Bias is the difference between the expected value (i.e., the mean) of the estimate and the "true" value of the quantity that it purports to estimate. The bias of $\hat{S}_{XX}(\omega, N)$ is therefore

$$E(\hat{S}_{XX}(\omega, N)) - S_{XX}(\omega).$$

The expected value of $\hat{S}_{XX}(\omega, N)$ is

$$E(\hat{S}_{XX}(\omega, N)) = E\left(\sum_{l=-(N-1)}^{N-1} \hat{r}_{XX}[l, N]e^{-j\omega l}\right) = \sum_{l=-(N-1)}^{N-1} E(\hat{r}_{XX}[l, N])e^{-j\omega l}. \tag{14.16}$$

In Appendix D, we show that the sample autocorrelation function $\hat{r}_{XX}[l, N]$ is an **asymptotically unbiased estimator** of $r_{XX}[l]$, with expectation value

$$E(\hat{r}_{XX}[l,N]) = \underbrace{\left(\frac{N-|l|}{N}\right)}_{b_{2N-1}[l]} r_{XX}[l] = b_{2N-1}[l] r_{XX}[l], \tag{14.17}$$

where $b_{2N-1}[n]$ is a symmetric triangular (Bartlett) window of odd length $2N-1$ that has a maximum value one at $l = 0$ and tapers linearly to zero when $|l| = N$,

$$b_{2N-1}[l] = \frac{1}{N} w_N[n] * w_N[-n] = \begin{cases} \dfrac{N-|l|}{N}, & 0 \le |l| \le N-1 \\ 0, & \text{otherwise} \end{cases},$$

where $w_N[n]$ is the rectangular window defined in Equation (14.9).

Substituting Equation (14.17) into Equation (14.16) and using the multiplication (windowing) property of the DTFT, we get

$$\begin{aligned} E(\hat{S}_{XX}(\omega,N)) &= E\left(\sum_{l=-(N-1)}^{(N-1)} \hat{r}_{XX}[l,N] e^{-j\omega l}\right) = \sum_{l=-\infty}^{\infty} (b_{2N-1}[l] r_{XX}[l]) e^{-j\omega l} = \mathfrak{F}\{b_{2N-1}[l] r_{XX}[l]\} \\ &= \frac{1}{2\pi} \mathfrak{F}\{b_{2N-1}[l]\} * \mathfrak{F}\{r_{XX}[l]\} = \frac{1}{2\pi} B_{2N-1}(\omega) * S_{XX}(\omega), \end{aligned} \tag{14.18}$$

where $B_{2N-1}(\omega)$ is the DTFT of the triangular window $b_{2N-1}[n]$ (see Example 3.35),

$$\begin{aligned} B_{2N-1}(\omega) &= \mathfrak{F}\left\{\frac{1}{N} w_N[n] * w_N[-n]\right\} = \frac{1}{N} W_N(\omega) W_N^*(\omega) = \frac{1}{N} |W_N(\omega)|^2 \\ &= \frac{1}{N}\left(\frac{\sin \omega N/2}{\sin \omega/2}\right)^2. \end{aligned} \tag{14.19}$$

Here, we have used the fact that the triangular window is the convolution of two rectangular windows in the frequency domain, hence the transform is the square of the periodic sinc given by

$$W_N(\omega) = \frac{\sin \omega N/2}{\sin \omega/2} = N \frac{\operatorname{sinc} \omega N/2}{\operatorname{sinc} \omega/2}.$$

Equation (14.18) indicates that $\hat{S}_{XX}(\omega, N)$ is a biased estimator of $S_{XX}(\omega)$, since it is not equal to $S_{XX}(\omega)$. However, in the limit as $N \to \infty$, $b_{2N-1}[n]$ approaches a constant, $b_{2N-1}[n] = 1$, and $B_{2N-1}(\omega)$ approaches an impulse at frequency $\omega = 0$,

$$\lim_{N\to\infty} B_{2N-1}(\omega) = \lim_{N\to\infty} \frac{1}{N}\left(\frac{\sin \omega N/2}{\sin \omega/2}\right)^2 = 2\pi\delta(\omega).$$

Thus, in the limit as $N \to \infty$, $E(\hat{S}_{XX}(\omega, N))$ approaches $S_{XX}(\omega)$,

$$\lim_{N\to\infty} E(\hat{S}_{XX}(\omega,N)) = \lim_{N\to\infty} \frac{1}{2\pi} B_{2N-1}(\omega) * S_{XX}(\omega) = \frac{1}{\cancel{2\pi}} \cancel{2\pi}\delta(\omega) * S_{XX}(\omega) = S_{XX}(\omega). \tag{14.20}$$

For this reason, $\hat{S}_{XX}(\omega, N)$ is termed an **asymptotically unbiased** estimator of the power density spectrum.

Equations (14.11) and (14.15) provide two definitions of the periodogram, one in terms of the Fourier transform of the autocorrelation function, one in terms of the Fourier transform of $x[n]$. Putting these two equations together with the result of Equation (14.20) and the definition of $S_{XX}(\omega)$ in Equation (14.6), we arrive at a key summary result:

$$\lim_{N\to\infty} E(\hat{S}_{XX}(\omega, N)) = \lim_{N\to\infty} E\left(\frac{1}{N}\left|\sum_{n=0}^{N-1} x[n]e^{-j\omega n}\right|^2\right) = S_{XX}(\omega) = \mathfrak{F}\{r_{XX}[l]\} = \sum_{l=-\infty}^{\infty} r_{XX}[l]e^{-j\omega l}.$$

In words, in the limit as the length of the signal increases, the average magnitude-square of the DTFT of a sequence converges to the power spectral density. This result is known as the **Wiener–Khinchin theorem**. It provides the key link between the quantity we want to estimate (the power spectral density) and a quantity we can actually hope to measure (the DTFT, or its sampled version, the DFT).

Consistency of the periodogram An estimator is said to be **consistent** if its variance approaches zero as $N\to\infty$. That means that the estimates get "better"; that is, the scatter of repeated estimates of the process about the sample mean decreases as the amount of data in the periodogram increases. It turns out that as $N\to\infty$, the variance of the periodogram approaches a constant that depends on the "true" value of the spectral density,

$$\lim_{N\to\infty} E(\hat{S}_{XX}(\omega, N)) \approx S_{XX}^2(\omega).$$

This can be seen in **Figure 14.19** for the example of a periodogram measured for a zero-mean Gaussian random process with a variance of $\sigma_X^2 = 1$.

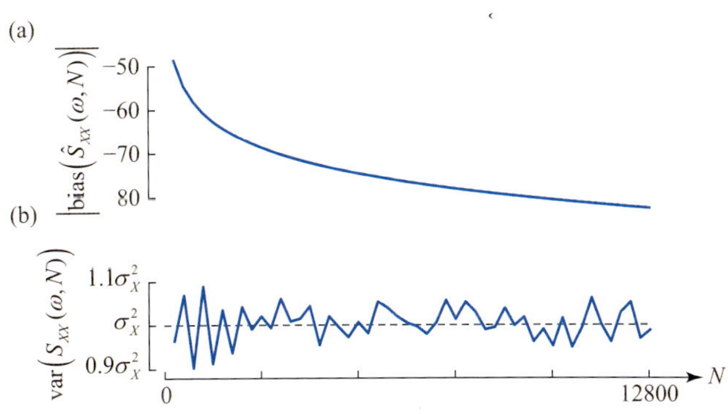

Figure 14.19 Bias and variance of the periodogram

The magnitude of the bias, expressed on a dB scale (**Figure 14.19a**), decreases monotonically for a series of periodograms, whose lengths range from $N = 256$ to 12800 in multiples of a 256-point frame size. However, the variance (**Figure 14.19b**) does not decrease systematically. So, although $\hat{S}_{XX}(\omega, N)$ is an asymptotically unbiased estimator of the PSD, it is *not* a consistent estimator.

In the sections that follow, we will discuss a number of practical implementations of the periodogram that seek to provide *both* unbiased and consistent estimates of the power spectral density.

14.3.2 Bartlett's method

The periodogram described in Section 14.3.1 is asymptotically unbiased but not consistent. **Bartlett's method** is a method of finding an estimator that is both unbiased and consistent. It is motivated by the fact that if you average a set of M uncorrelated measurements of a random variable to form the **sample mean**, the variance of this sample mean decreases in proportion to M. Assume that we have a sequence of random data $x[n]$ of length N. In Bartlett's method, we first divide $x[n]$ into M contiguous *non-overlapping* frames $x[n+mL]$, $0 \le m < M$, each of length $L = N/M$, such that the entire length of the sequence is $N = LM$. Then calculate the periodogram of each of the M frames and average these periodograms together to form the **averaged periodogram** $P_B(\omega, L, M)$,

$$P_B(\omega, L, M) \triangleq \frac{1}{M}\sum_{m=0}^{M-1}\left(\frac{1}{L}\left|\sum_{n=0}^{N-1} x[n+mL]e^{-j\omega n}\right|^2\right) = \frac{1}{N}\sum_{m=0}^{M-1}\left(\left|\sum_{n=0}^{N-1} x[n+mL]e^{-j\omega n}\right|^2\right). \tag{14.21}$$

The expected value of the averaged periodogram $P_B(\omega, L, M)$ is the same as that of the periodogram of one frame,

$$E(P_B(\omega, L, M)) = E(\hat{S}_{XX}(\omega, L)),$$

hence, the averaged periodogram is also asymptotically unbiased as $L \to \infty$ (Problem 14-4a). On the assumption that periodograms of different frames are statistically uncorrelated and have equal variance, the variance of the sum of M periodograms is $1/M$ the variance of the periodogram of each frame; hence, the variance of the averaged periodogram is

$$\mathrm{var}(P_B(\omega, L, M)) = \frac{1}{M}\mathrm{var}(\hat{S}_{XX}(\omega, L)),$$

which tends to zero as $M \to \infty$ (Problem 14-4b). Taken together, this means the averaged periodogram is both asymptotically unbiased *and* a consistent estimator of the power spectral density. However, nothing good comes for free in engineering, as we will now see.

Periodogram of random signal in noise Consider the example of using the periodogram and averaged periodogram to estimate the power spectral density of a WSS process comprising a cosine of constant frequency $\omega_0 = 0.541\pi$ in the presence of noise,

$$x[n] = A\cos(\omega_0 n + \theta) + v[n], \tag{14.22}$$

where the phase of the cosine θ is a random variable uniformly distributed over the interval $-\pi \le \omega < \pi$ with probability density function (PDF)

$$p_\Theta(\theta) = \frac{1}{2\pi}, \quad -\pi \le \theta < \pi, \tag{14.23}$$

and $v[n]$ is a zero-mean white-noise process with variance $\sigma_V^2 = 1$. (In Appendix D, we show that the random-phase cosine is a WSS and ergodic process.)

First let us look at the case of the "ideal" power spectral density that results when $x[n]$ is of *infinite length*. Because the cosine and white noise are uncorrelated, the ideal power spectral density $S_{XX}(\omega)$ is the sum of two terms (Problem 14-2): the power density spectra of the cosine, which we denote $S_{CC}(\omega)$, and the power density spectrum of the noise, which is just a constant equal to its variance σ_V^2,

$$S_{XX}(\omega) = S_{CC}(\omega) + \sigma_V^2 = \underbrace{\frac{A^2\pi}{2}(\delta(\omega - \omega_0) + \delta(\omega + \omega_0))}_{S_{CC}(\omega)} + \sigma_V^2.$$

$S_{CC}(\omega)$ comprises two impulses at frequencies $\omega = \pm\omega_0$. The vertical red impulse at $\omega_0 = 0.541\pi$ in each panel of **Figure 14.20** denotes the positive frequency term of $S_{CC}(\omega)$, that is, $A^2\pi\delta(\omega - \omega_0)/2$. The thin horizontal red line shows the power density spectrum of the noise, which is a constant, σ_V^2.

Now, consider the example in which $\hat{S}_{XX}(\omega, L)$ represents the periodogram of a signal of length L. The expectation value of $\hat{S}_{XX}(\omega, L)$ is the sum of the expectations of the two components,

$$E(\hat{S}_{XX}(\omega, L)) = E(\hat{S}_{CC}(\omega, L)) + \sigma_V^2,$$

where $\hat{S}_{CC}(\omega, L)$ is the power spectral density of a fixed-length segment of the random-phase cosine of length L. From Equations (14.18) and (14.19), $E(\hat{S}_{CC}(\omega, L))$ is calculated by convolving each impulse of $S_{CC}(\omega)$ with the transform of a Bartlett window, $B_{2L-1}(\omega)$, of length $2L - 1$,

$$\begin{aligned} E(\hat{S}_{CC}(\omega, L)) &= \frac{1}{2\pi} B_{2L-1}(\omega) * S_{CC}(\omega) = \frac{1}{2\pi} B_{2L-1}(\omega) * \left(\frac{A^2\pi}{2}(\delta(\omega - \omega_0) + \delta(\omega + \omega_0))\right) \\ &= \frac{A^2}{4L}\left(\frac{\sin(\omega - \omega_0)L/2}{\sin(\omega - \omega_0)/2}\right)^2 + \frac{A^2}{4L}\left(\frac{\sin(\omega + \omega_0)L/2}{\sin(\omega + \omega_0)/2}\right)^2. \end{aligned}$$

The grey trace shaded in pink in each panel of **Figure 14.20** shows the positive frequency term of $E(\hat{S}_{CC}(\omega, L))$ as we change the length of the signal from $L = 2500$ (top panel) to $L = 50$ (bottom panel), with $A = 1$. Data are plotted on a log magnitude scale for frequencies $0 \le \omega < \pi$. The effect of the window is to spread the impulsive spectrum of the cosine over the entire frequency range, something we discussed in detail in Section 14.1.1. The resolution of the periodogram is then limited by the spectral spread due to the main lobe of $B_{2L-1}(\omega)$, which has width $4\pi/L$ (the same as that of a rectangular window of length L). The larger the value of L, the narrower the main lobe of $B_{2L-1}(\omega)$ and the finer the resolution of the periodogram will be.

Now, consider an example in which $x[n]$ is of *fixed* length $N = 2500$, and the signal is then divided into M frames of length L such that $N = ML$. Because N is fixed, the number of frames must be inversely proportional to their length; the frame size L is set to $L = N/M$ so that the product ML will be a constant equal to the total number of available points, $ML = N = 2500$. Each of the four panels of **Figure 14.20** shows the averaged periodogram $P_B(\omega, L, M)$ (blue trace), computed for a number of averaged periodograms ranging from $M = 1$ (top panel) to $M = 50$ (bottom panel) using Equation (14.21). The top panel of the figure, with values of $M = 1$

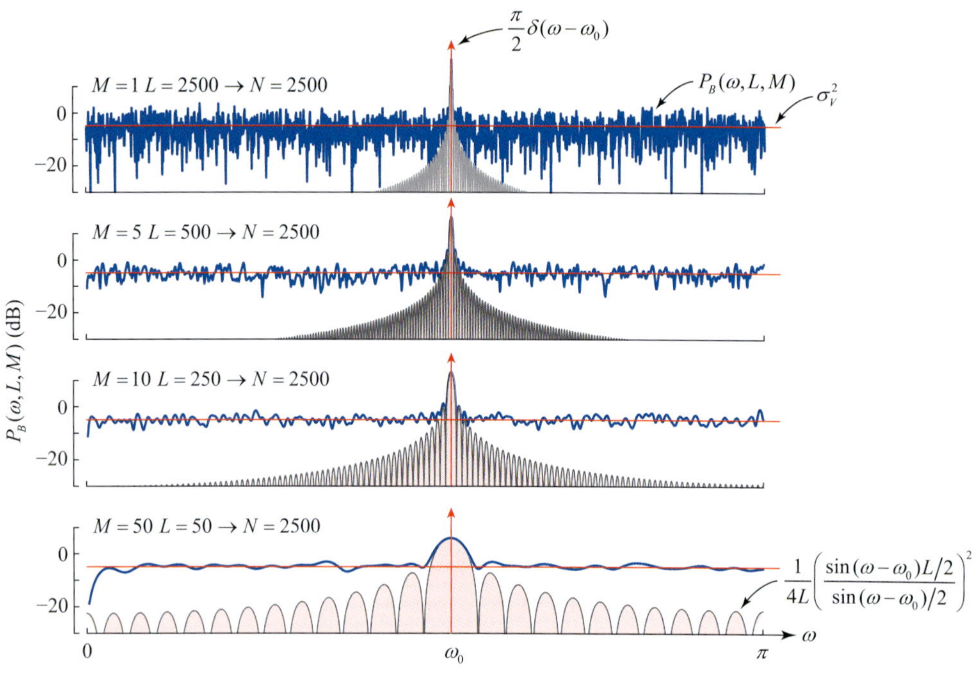

Figure 14.20 Averaged periodogram, constant N

and $L=2500$, is the unaveraged periodogram of all $L=N=2500$ points. The width of the main lobe is very narrow in this case, $\Delta\omega=4\pi/2500=0.0016\pi$, so the periodogram has a sharp peak at ω_0. However, the variance is large. In the remaining panels of the figure, as the number of periodograms averaged, M, increases and their size, L, decreases, the spectral resolution gets worse (as we expect due to the decreasing size of the data window), but the variance gets smaller. This shows the essential trade-off that exists for averaged periodograms of a fixed size signal: there is an inverse relation between good spectral resolution (large L) and low variance (large M).

To help parse the effects of spectral resolution and variance, **Figure 14.21** shows the effect on the periodogram of averaging different numbers of frames, M, each of constant length $L=100$. The spectral resolution of the average is determined by L, with a main lobe width $\Delta\omega\simeq 4\pi/L=0.04\pi$. The top panel of the figure, with a value of $M=1$, is the unaveraged periodogram. Because the periodogram comprises only $N=100$ points, the variance here is large. The remaining panels of the figure show that as the number of periodograms averaged, M, increases from $M=1$ to $M=50$ (bottom panel), the amount of data increases from $N=100$ points (top panel) to $N=5000$ points (bottom panel) and the variance decreases in proportion to M, as we might expect.

To complete this example, **Figure 14.22** shows the performance of the averaged periodogram $P_B(\omega, L, M)$ (blue trace), with $L=100$ and $M=50$, for a cosine in white noise as the level of the noise increases from $\sigma_V^2=0.01$ (top panel) to $\sigma_V^2=4$ (bottom panel). The horizontal red lines that correspond to levels of the noise $\sigma_V^2=0.01$, 0.25, 1.0 and 4 are plotted at $10\log_2(\sigma_V^2/\pi)=-25.0$, -11.0, -5.0 and 1.0 dB. The results are relatively well predicted by

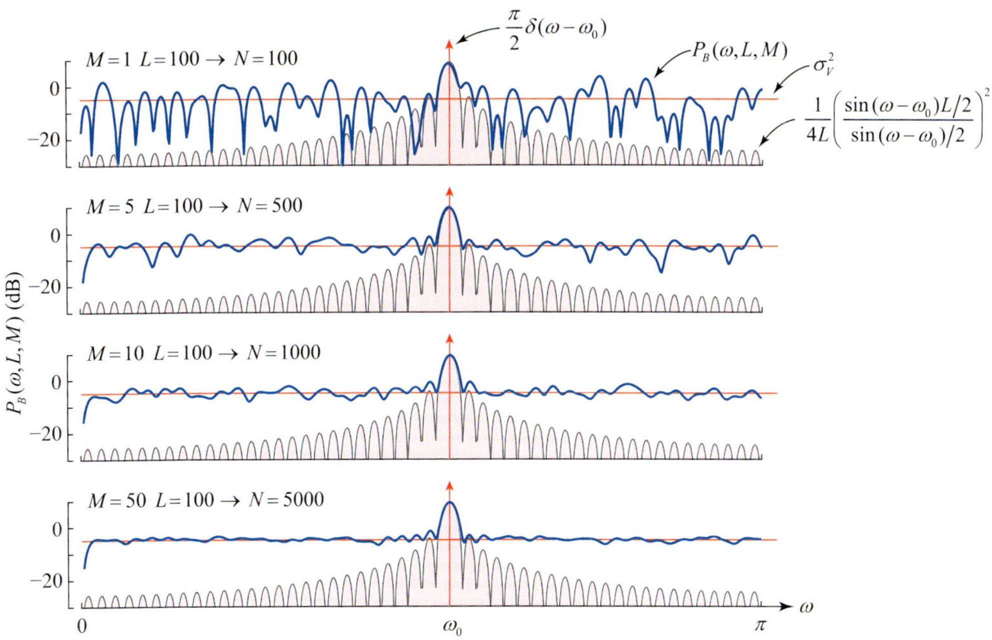

Figure 14.21 Averaged periodograms, constant L

the theory. At low levels of added noise, the spectral leakage due to the sidebands acts as the dominant source of noise. However, as the level of the added noise increases, the sidebands get "submerged" until only the peak is visible.

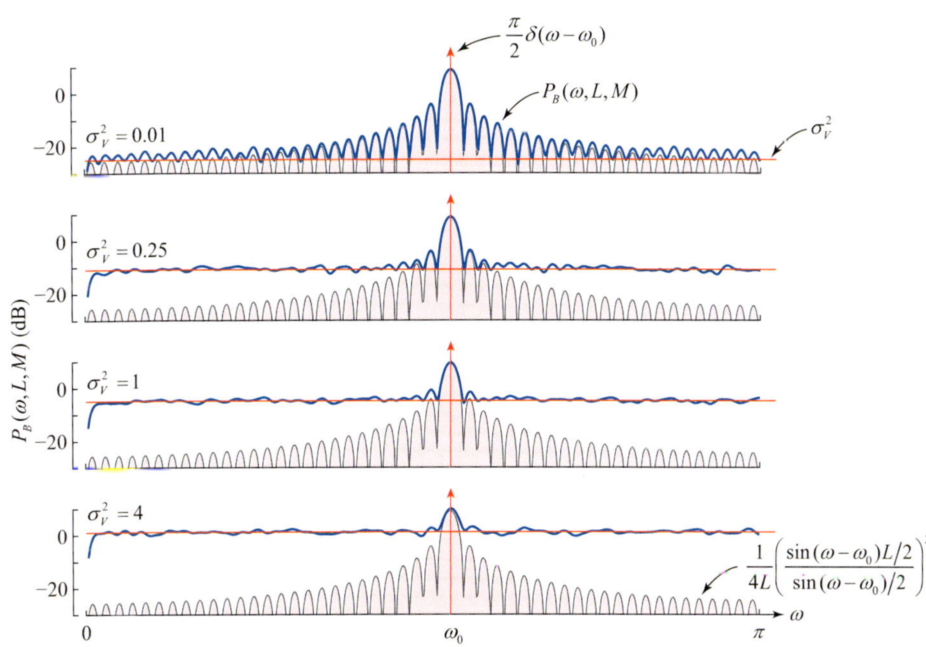

Figure 14.22 Random-phase cosine in noise

14.3.3 The modified periodogram

The periodogram was defined in Equation (14.15) as the sum of the magnitude squared of the DTFT of $x_N[n]$. This sequence of N points of $x[n]$ was created by multiplying $x[n]$ by a rectangular window $w_N[n]$ (Equation (14.8)). Now let $w_N[n]$ be a general window of length N, such as a Hamming or Hann window. Starting with Equations (14.13), we tentatively define the **modified periodogram** $P_M(\omega, N)$ as

$$\begin{aligned}
P_M(\omega, N) &\triangleq \frac{1}{N}\left|\sum_{n=-\infty}^{\infty} x_N[n]e^{-j\omega n}\right|^2 = \frac{1}{N}\left|\sum_{n=-\infty}^{\infty} x[n]w_N[n]e^{-j\omega n}\right|^2 \\
&= \frac{1}{N}\left(\sum_{n=-\infty}^{\infty} x[n]w_N[n]e^{-j\omega n}\right)\left(\sum_{m=-\infty}^{\infty} x[m]w_N[m]e^{-j\omega m}\right)^* \\
&= \frac{1}{N}\sum_{n=-\infty}^{\infty}\sum_{m=-\infty}^{\infty} x[n]x[m]w_N[n]w_N[m]e^{-j\omega(n-m)}.
\end{aligned}$$

The expected value is

$$\begin{aligned}
E(P_M(\omega, N)) &= \frac{1}{N}\sum_{n=-\infty}^{\infty}\sum_{m=-\infty}^{\infty} E(x[n]x[m])w_N[n]w_N[m]e^{-j\omega(n-m)} \\
&= \frac{1}{N}\sum_{n=-\infty}^{\infty}\sum_{m=-\infty}^{\infty} r_{XX}[n-m]w_N[n]w_N[m]e^{-j\omega(n-m)},
\end{aligned} \tag{14.24}$$

where we have recognized that for a WSS process, $E(x[n]x[m]) = r_{XX}[n-m]$. Let $l = n - m$. After changing the limits on the second sum to l, Equation (14.24) becomes

$$\begin{aligned}
E(P_M(\omega, N)) &= \frac{1}{N}\sum_{n=-\infty}^{\infty}\sum_{l=-\infty}^{\infty} r_{XX}[l]w_N[n]w_N[n-l]e^{-j\omega l} \\
&= \frac{1}{N}\sum_{l=-\infty}^{\infty} r_{XX}[l]\underbrace{\left(\sum_{n=-\infty}^{\infty} w_N[n]w_N[n-l]\right)}_{w_N[n]*w_N[-n]} e^{-j\omega l} \\
&= \frac{1}{N}\sum_{l=-\infty}^{\infty}(r_{XX}[l](w_N[n]*w_N[-n]))e^{-j\omega l} \\
&= \frac{1}{N}\mathfrak{F}\{r_{XX}[l](w_N[n]*w_N[-n])\} = \frac{1}{2\pi N}\mathfrak{F}\{r_{XX}[l]\}*\mathfrak{F}\{w_N[n]*w_N[-n]\} \\
&= \frac{1}{2\pi N}S_{XX}(\omega)*|W_N(\omega)|^2.
\end{aligned} \tag{14.25}$$

If $w_N[n]$ is a rectangular window of length N given by Equation (14.9), then Equation (14.25) is identical to Equation (14.18); hence, the regular periodogram of Section 14.3.1, $\hat{S}_{XX}(\omega, N)$, can be considered as a special case of the modified periodogram with a rectangular window.

Equation (14.25) indicates that the modified periodogram is a biased estimator. To be asymptotically unbiased, the expected value of $P_M(\omega, N)$ needs to approach $S_{XX}(\omega)$ as $N \to \infty$,

$$\lim_{N\to\infty} E(P_M(\omega, N)) = S_{XX}(\omega).$$

This means that $|W_N(\omega)|^2$ in Equation (14.25) has to approach an impulse with area $2\pi N$ as $N \to \infty$,

$$\frac{1}{2\pi}\int_{-\pi}^{\pi} |W_N(\omega)|^2 d\omega = N.$$

By Parseval's theorem,

$$\frac{1}{2\pi}\int_{-\pi}^{\pi} |W_N(\omega)|^2 d\omega = \sum_{n=0}^{N-1} w_N^2[n].$$

So, to be asymptotically unbiased, $w_N[n]$ needs to be normalized so that its total energy is equal to N, or, equivalently,

$$\frac{1}{N}\sum_{n=0}^{N-1} w_N^2[n] = 1.$$

For this reason, we now formally define the **modified periodogram** $P_M(\omega, N)$ as

$$P_M(\omega, N) \triangleq \frac{\left|\sum_{n=0}^{N-1} x[n]w_N[n]e^{-j\omega n}\right|^2}{\sum_{n=0}^{N-1} w_N^2[n]}, \qquad -\pi \le \omega < \pi. \tag{14.26}$$

If $w_N[n]$ is a rectangular window of length N, then

$$\sum_{n=0}^{N-1} w_N^2[n] = N,$$

and the modified periodogram, Equation (14.26), becomes the same as the regular periodogram $\hat{S}_{XX}(\omega, N)$, Equation (14.15).

14.3.4 Averaged modified periodogram

The modified periodogram $P_M(\omega, N)$ is not a consistent estimator since its variance does not go to zero as $N \to \infty$. However, we can easily extend the definition of averaged periodogram, Equation (14.21), to include sequences windowed with other than rectangular windows. The result is the **averaged modified periodogram**,

$$P_{AM}(\omega, L, M) \triangleq \frac{1}{M}\sum_{m=0}^{M-1}\left(\frac{\left|\sum_{n=0}^{N-1} x[n+mL]w_N[n]e^{-j\omega n}\right|^2}{\sum_{n=0}^{N-1} w_N^2[n]}\right) = \frac{\frac{1}{M}\sum_{m=0}^{M-1}\left(\left|\sum_{n=0}^{N-1} x[n+mL]w_N[n]e^{-j\omega n}\right|^2\right)}{\sum_{n=0}^{N-1} w_N^2[n]}. \tag{14.27}$$

As an example of the effect of different windows, **Figure 14.23** shows the averaged modified periodogram for a random process comprising the sum of two random-phase cosines in the presence of a zero-mean white noise,

$$x[n] = A_1\cos((\pi-\Delta\omega)n/2+\theta_1) + A_2\cos((\pi+\Delta\omega)n/2+\theta_2) + v[n],$$

where $A_1 = A_2 = 1$ and $v[n]$ has variance $\sigma_V^2 = 1$ (see Problem 14-3). The frequencies of the cosines are separated by $\Delta\omega$ around a center frequency of $\pi/2$. The phases of the cosines, θ_1 and θ_2, are each described by a uniformly distributed random variable over the interval $-\pi \leq \omega < \pi$, with PDF given by Equation (14.23). **Figures 14.23a** and **b** show measured periodograms $P_{AM}(\omega, L, M)$ (blue traces) on a log magnitude scale, for the process $x[n]$ windowed with a rectangular window and a Hamming window, respectively. The frequency separation of the cosines ranges from $\Delta\omega = 0.02\pi$ (top panel) to $\Delta\omega = 0.16\pi$ (bottom panel). Each trace is the average of 50 periodograms, each comprising $L = 100$ points of $x[n]$. The thin red trace is the theoretically expected value $E(P_{AM}(\omega, L, M))$.

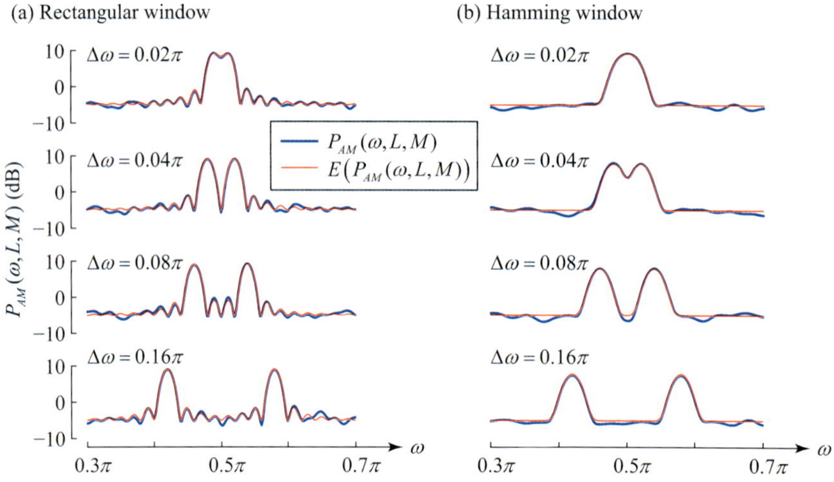

Figure 14.23 Average modified periodogram of two cosines with random phase in noise

The results are consistent with our discussion of the effect of window shape in Section 14.1.2. The spectral spread due to the rectangular window is smaller than that of the Hamming window, but the effect of spectral leakage is greater.

14.3.5 Welch's method

Welch's method is a widely used technique of averaging periodograms in an effort to reduce the variance. This method is essentially equivalent to the averaged modified periodogram except that, in this method, the sequence of random data, $x[n]$, is divided into M frames of length L that *overlap* with each other by D points. For each frame, data are multiplied by a window, $w_L[n]$, in the manner described for the averaged modified periodogram and periodograms of each frame are computed and averaged together to form Welch's periodogram, $P_W(\omega, L, D, M)$:

$$P_W(\omega, L, D, M) = \frac{\dfrac{1}{M}\sum_{m=0}^{M-1}\left(\left|\sum_{n=0}^{L-1} x[n+mD]w_L[n]e^{-j\omega n}\right|^2\right)}{\sum_{n=0}^{L-1} w_L^2[n]}. \tag{14.28}$$

The periodogram computed by Welch's method is an unbiased and consistent estimator of the power spectral density. On the assumption that segments overlap by about 50% (i.e., $D = L/2$), the variance of this method can be shown to be about half that of the averaged modified periodogram described in Section 14.3.7. Welch's method is essentially a superset of all the previously defined periodogram measures in the sense that with appropriate choice of parameters L, D and M, Equation (14.28) can emulate the defining relations of the periodogram (Equation (14.15)), Bartlett's method (Equation (14.21)), the modified periodogram (Equation (14.26)) and the averaged modified periodogram (Equation (14.27)).

14.3.6 Discrete-time periodograms

All the formulas for periodograms that have been presented so far in this section have been functions of continuous frequency ω. In practical applications, periodograms are computed at discrete frequencies using the DFT, with the corresponding formulas obtained by sampling ω at L points. For example, the formula for Welch's periodogram, Equation (14.28), becomes

$$P_W[k,L,D,M] = P_W(\omega,L,D,M)|_{\omega=2\pi k/L}$$
$$= \frac{\frac{1}{M}\sum_{m=0}^{M-1}\left(\left|\sum_{n=0}^{L-1}x[n+mD]w_L[n]e^{-j2\pi kn/L}\right|^2\right)}{\sum_{n=0}^{L-1}w_L^2[n]} = \frac{\frac{1}{M}\sum_{m=0}^{M-1}\left(|\text{DFT}_L(x[n+mD]w_L[n])|^2\right)}{\sum_{n=0}^{L-1}w_L^2[n]}.$$

The inner summation of the numerator is just the DFT of a windowed section of $x[n]$ of length L.

14.3.7 Matlab implementation of the periodogram functions

Matlab has two basic functions, `periodogram` and `pwelch`, that compute and plot periodograms. The `periodogram` function computes periodograms and modified periodograms as described in Sections 14.3.1 and 14.3.3. The `pwelch` function implements Welch's method, as described in Section 14.3.5, and can therefore compute averaged periodograms of either overlapping or non-overlapping windowed segments. Here are some details relating to the implementation of both of these Matlab functions:

- Matlab computes discrete-time periodograms $P_{XX}[k]$ using an L-point DFT, which has the effect of sampling the periodogram at L points. Either a two-sided or a one-sided periodogram can be computed. For a **two-sided periodogram**, all $2L$ points are retained, corresponding to the entire frequency range $0 \le \omega < 2\pi$. However, because the power spectral density is an even function of ω (i.e., $S_{XX}(\omega) = S_{XX}(-\omega)$), Matlab's default mode is to compute only a **single-sided periodogram**, in which only $L/2+1$ points are retained, corresponding to one-half the frequency range, $0 \le \omega \le \pi$. The one-sided periodogram includes both endpoints; the first point $P_{XX}[1]$ corresponds to $\omega = 0$, and the last point $P_{XX}[L/2+1]$ corresponds to $\omega = \pi$.
- In order to assure that the total power of the single-sided and two-sided periodograms are equal, Matlab doubles the value of points in the single-sided periodogram array that are duplicated in the two-sided array, that is, the points $P_{XX}[k]$, $2 \le k \le L/2$. The first and last points of the array, $P_{XX}[1]$ and $P_{XX}[L/2+1]$, are not doubled because they are not duplicated in the two-sided periodogram.

- The Matlab periodogram computes the total power divided by the entire frequency range of the distribution, namely 2π.

The syntax of the Matlab functions is

```
Pxx = periodogram(x, window, Nfft),
Pxx = pwelch(x, window, noverlap, Nfft),
```

where `x` is the input vector, `window` is an optional array that specifies a window used to compute the modified periodogram and `Nfft` is an optional parameter that allows one to increase the resolution of the periodogram beyond the default value, which is the greater of 256 or the next higher power of two larger than the length of `x`. Whereas the `periodogram` function computes a single periodogram of length `Nfft`, the `pwelch` function computes an average of windowed periodograms, where the `noverlap` parameter determines the number of samples that successive frames overlap. If `periodogram` or `pwelch` is called without an output argument (e.g., `Pxx`), the function plots the periodogram. The implementation of `periodogram` and `pwelch` functions is best shown by examining some bare-bones code of our own that provides the same functionality as both of these Matlab functions:

```
function out = mypsd(x, L, D, wind, Nfft)
    lx = length(x);
    w = window(wind, L);
    M = 1 + floor((lx-L)/D);
    X = ones(L, 1);
    for m = 1:M
        indx = 1+(m-1)*D;
        X = X + abs(fft(x(indx+(0:L-1))' .* w, Nfft)).^2 / sum(w.^2);
    end
    P = [X(1); 2*X(2:Nfft/2); X(Nfft/2+1)]/M/2/pi;
    if (nargout)
        out = P;
    else
        plot(linspace(0, 1, Nfft/2+1), 10*log10(P))
        xlabel('Frequency (rad /\pi)')
        ylabel('PSD (dB/rad)')
        grid on
    end
end
```

In our code, `wind` is a handle to a data window (e.g., `@rectwin`) rather than a data array. In keeping with the default behavior of the `periodogram` and `pwelch` functions, our code produces single-sided periodograms. An `Nfft`-point FFT is computed spanning the equivalent range of frequencies $0 \leq \omega < 2\pi$, and then normalized by the window energy to form array `X`. Then, `Nfft/2+1` points of `X` are retained, spanning the range $0 \leq \omega \leq \pi$. All points, except the two endpoints are doubled and then the array is divided by 2π. For a standard periodogram (i.e., no averaging), you would call this function as

```
Pxx = mypsd(x, length(x), length(x), @rectwin, Nfft).
```

For an averaged periodogram, you would specify a value of the frame size `L`,

```
Pxx = mypsd(x, L, length(x)/L, @hamming, Nfft).
```

For Welch's method, you would specify values of both `L` and `D`,

```
Pxx = mypsd(x, L, D, @hamming, Nfft).
```

When our `mypsd` function is called without an output argument, it plots the energy on a log scale, as do Matlab's implementations.

14.4 ★ Parametric methods of spectral estimation

The nonparametric spectral estimation techniques described in the previous section – the STFT and periodograms – make no assumptions about the nature of the signals being analyzed. They rely solely on computation of standard measures such as the DFT. However, in many important cases, we have a model of the random signal whose parameters we wish to derive. The single random-phase cosine in noise in Section 14.3.2 is an example of a process whose model is effectively given by Equation (14.22). The parameters we wish to estimate here would be A and ω_0. A more complex case, which we shall discuss in depth in Section 14.5.2, is an all-pole model of speech whose parameters reflect important attributes of speech production.

The essential idea of **parametric spectral estimation** is to represent a complicated random signal with a model that has a limited number of parameters that can be efficiently measured. In this section, we will develop two related parametric modeling approaches, the **autoregressive moving-average (ARMA) model** and **linear prediction modeling**, and show how they can be applied to some example problems.

14.4.1 The ARMA model

In Chapters 3 and 4, we saw that useful deterministic systems could be well described by linear constant-coefficient difference equations (LCCDEs) in the time domain and the ratio of rational z-transforms in the frequency domain. In much the same way, in the model-based approach to spectral estimation, a discrete-time WSS random process $y[n]$ can be expressed as the output of a discrete-time linear filter that is excited by an input random process $x[n]$. Such a parametric model can be generally defined by a difference equation

$$\sum_{i=0}^{N} a_N[i]y[n-i] = \sum_{i=0}^{M} b_M[i]x[n-i], \tag{14.29}$$

where M and N define the model order and $a_N[i]$ and $b_N[i]$ are the model coefficients. The coefficients are subscripted with the model orders to emphasize the fact that models of different orders will generally have different coefficients; that is, $b_N[i] \neq b_{N+1}[i]$. Let $a_N[0] = 1$ and rewrite to give

$$y[n] = -\sum_{i=1}^{N} a_N[i]y[n-i] + \sum_{i=0}^{M} b_M[i]x[n-i]. \tag{14.30}$$

Equation (14.30) defines the general form of the **autoregressive moving-average (ARMA) model**. The first summation defines the "autoregressive" (AR) part of the model, in which the present value of $y[n]$ depends on values of the N past outputs $y[n-i]$, $1 \leq i \leq N$, weighted by coefficients $a_N[i]$. The second summation defines the "moving average" (MA) part of the model, which has the form of a finite-length convolution of $M+1$ values of $x[n-i]$, $0 \leq i \leq M$, with coefficients $b_M[i]$. The two major assumptions of this model are (1) that the system defined by the model parameters is causal and stable, and (2) that the present value of the input process $x[n]$ is uncorrelated with past values of the output $y[n]$, that is,

$$E(x[n]y[n-i]) = 0, \quad i > 0.$$

The ARMA model can be treated as a problem of filtering random processes, which is discussed in Appendix D.2.5. To show this, rewrite Equation (14.29) as

$$y[n] + \sum_{i=1}^{N} a_N[i]y[n-i] = \sum_{i=0}^{M} b_M[i]x[n-i].$$

The model can then be represented as a filter with z-transform

$$H(z) = \frac{B_M(z)}{A_N(z)} = \frac{\sum_{i=0}^{M} b_M[i]z^{-i}}{1 + \sum_{i=1}^{N} a_N[i]z^{-i}}. \tag{14.31}$$

Then (see Appendix D.3.2), given an input with PSD $S_{XX}(\omega)$, the PSD of the output is

$$S_{YY}(\omega) = S_{XX}(\omega)|H(\omega)|^2 = S_{XX}(\omega)\left|\frac{B_M(\omega)}{A_N(\omega)}\right|^2, \tag{14.32}$$

where

$$H(\omega) = H(z)|_{z=e^{j\omega}} = \left.\frac{B_M(z)}{A_N(z)}\right|_{z=e^{j\omega}}.$$

In the event that the input is a zero-mean white-noise process $X[n]$, with variance σ_X^2, then the power spectral density of the input is flat, $S_{XX}(\omega) = \sigma_X^2$, and the power density spectrum output is determined entirely by parameters of the filter,

$$S_{YY}(\omega) = \sigma_X^2 \left|\frac{B_M(\omega)}{A_N(\omega)}\right|^2. \tag{14.33}$$

The transform $H(z)$ of the general ARMA model of Equation (14.31) has both poles and zeros. There are two important special cases of this model. When $a_N[i] = 0$, $1 \leq i \leq N$, and at least some $b_M[i] \neq 0$, then Equation (14.31) becomes a pure MA model, with an "all-zero" filter z-transform,

$$H(z) = \sum_{i=0}^{M} b_M[i]z^{-i}.$$

Alternatively, when $b_M[0] \neq 0$ and $b_M[i] = 0,\ 1 \leq i \leq M$, and at least some $a_N[i] \neq 0$, then Equation (14.31) becomes a pure AR model, with an "**all-pole**" filter z-transform,[2]

$$H(z) = \frac{b_M[0]}{A_N(z)} = \frac{b_M[0]}{1 + \sum_{i=1}^{N} a_N[i] z^{-i}}. \tag{14.34}$$

All three variants of the model – MA, AR and full ARMA – have found application in signal processing, but the AR model has proven particularly popular. The theory is well developed, tractable and relatively simple to understand, and there are important applications, such as speech processing, in which the all-pole AR model turns out to be naturally well suited to the underlying process being estimated, as we shall show in the sections that follow.

Autoregressive (AR) model In this section, we will analyze an **autoregressive (AR) model** of order N, described in Equation (14.34). To simplify the analysis, let $b_M[0] = 1$, so we have

$$H(z) = \frac{1}{A_N(z)} = \frac{1}{1 + \sum_{i=1}^{N} a_N[i] z^{-i}}. \tag{14.35}$$

Then, Equation (14.30) can be written as

$$y[n] = x[n] - \sum_{i=1}^{N} a_N[i] y[n-i]. \tag{14.36}$$

A fundamental assumption of the AR model is that $y[n]$ represents a zero-mean, wide-sense stationary (WSS) random process that is completely specified by its autocorrelation function $r_{YY}[k]$. The goal of our analysis is to determine the values of the coefficients $a_N[i],\ 1 \leq i \leq N$, that describe the poles of the model based on measurements of the autocorrelation function of the output, $\hat{r}_{YY}[k]$, on the assumption that the input $x[n]$ is a zero-mean white-noise process. In what follows, we will further assume that our estimate of the autocorrelation function is sufficiently unbiased and consistent that we can take $\hat{r}_{YY}[k] = r_{YY}[k]$.

To compute $r_{YY}[k]$, multiply both sides of Equation (14.36) by $y[n-k]$ and take the expected value:

$$\begin{aligned} r_{YY}[k] &= E(y[n]y[n-k]) = E\left(x[n]y[n-k] - \sum_{i=1}^{N} a_N[i] y[n-i] y[n-k] \right) \\ &= E(x[n]y[n-k]) - \sum_{i=1}^{N} a_N[i] E(y[n-i]y[n-k]) \\ &= r_{XY}[k] - \sum_{i=1}^{N} a_N[i] E(y[n]y[n-(k-i)]) = r_{XY}[k] - \sum_{i=1}^{N} a_N[i] r_{YY}[k-i], \\ &\qquad 0 \leq k \leq N. \end{aligned} \tag{14.37}$$

[2] Of course, a causal all-zero model will have at least M poles at the origin ($z = 0$), and a causal all-pole model can have up to M zeros at the origin. However, these singularities at $z = 0$ do not affect $|H(\omega)|$.

In the last step, we used the fact that $y[n]$ is a WSS random process whose autocorrelation function is independent of absolute time, so $E(y[n]y[n-(k-i)]) = r_{YY}[k-i]$. To find the cross-correlation $r_{XY}[k] = E(x[n]y[n-k])$, evaluate Equation (14.36) at time $n-k$, then multiply both sides by $x[n]$ and compute the expected value,

$$\begin{aligned} r_{XY}[k] &= E(x[n]y[n-k]) = E(x[n]x[n-k]) + \sum_{i=1}^{N} a_N[i]E(x[n]y[(n-k)-i]) \\ &= r_{XX}[k] + \sum_{i=1}^{N} a_N[i]r_{XY}[k+i]. \end{aligned} \tag{14.38}$$

The assumptions of the AR model are that the system is causal (and stable) and that the input process $x[n]$ is not correlated with the past values of $y[n]$, namely

$$r_{XY}[l] = E(x[n]y[n-l]) = 0, \quad l > 0.$$

Given these assumptions, all terms in the summation of Equation (14.38) are zero for values of $k \geq 0$. Finally, if $x[n]$ is zero-mean white-noise, then

$$r_{XX}[k] = \begin{Bmatrix} \sigma_X^2, & k = 0 \\ 0, & k \neq 0 \end{Bmatrix} = \sigma_X^2 \delta[k],$$

so Equation (14.38) reduces to

$$r_{XY}[k] = r_{XX}[k] = \sigma_X^2 \delta[k],$$

and Equation (14.37) finally becomes

$$r_{YY}[k] = \sigma_X^2 \delta[k] - \sum_{i=1}^{N} a_N[i] r_{YY}[k-i]. \tag{14.39}$$

The Yule–Walker equations Equation (14.39) can now be rewritten as a pair of equations called the **Yule–Walker equations**:

$$r_{YY}[k] = \begin{cases} \sigma_X^2 - \sum_{i=1}^{N} a_N[i] r_{YY}[-i], & k = 0 \\ -\sum_{i=1}^{N} a_N[i] r_{YY}[k-i], & k > 0 \end{cases}. \tag{14.40}$$

The quantity $r_{YY}[0] = \sigma_{YY}^2$ is the variance of the output of the model. Define ε_N as the variance (the **residual mean-square error**) of the model estimate in terms of the N model parameters $a_N[i]$ and autocorrelation coefficients $r_{YY}[i]$, $1 \leq i \leq N$,

$$\varepsilon_N \triangleq r_{YY}[0] + \sum_{i=1}^{N} a_N[i] r_{YY}[-i], \quad N > 0, \tag{14.41}$$

where

$$\varepsilon_0 \triangleq r_{YY}[0] = \sigma_X^2.$$

We have defined the residual error ε_N with the subscript to make explicit the fact that the error depends on the model order N. When $k > 0$, Equation (14.40) generates a system of N linear equations that can be solved to produce an estimate of the N unknown model parameters $a_N[1], \ldots, a_N[N]$,

$$r_{YY}[k] = -\sum_{i=1}^{N} a_N[i] r_{YY}[k-i], \quad 1 \le k \le N, \quad N > 0. \tag{14.42}$$

To proceed, it is useful to write Equation (14.42) in matrix form,

$$\underbrace{\begin{bmatrix} r_{YY}[0] & r_{YY}[-1] & \cdots & r_{YY}[1-N] \\ r_{YY}[1] & r_{YY}[0] & \cdots & r_{YY}[2-N] \\ \vdots & \vdots & \ddots & \vdots \\ r_{YY}[N-1] & r_{YY}[N-2] & \cdots & r_{YY}[0] \end{bmatrix}}_{\boldsymbol{\Gamma}_N} \underbrace{\begin{bmatrix} a_N[1] \\ a_N[2] \\ \vdots \\ a_N[N] \end{bmatrix}}_{\mathbf{a}_N} = -\underbrace{\begin{bmatrix} r_{YY}[1] \\ r_{YY}[2] \\ \vdots \\ r_{YY}[N] \end{bmatrix}}_{\mathbf{r}_N}, \tag{14.43}$$

or

$$\boldsymbol{\Gamma}_N \mathbf{a}_N = -\mathbf{r}_N, \tag{14.44}$$

where

$$\boldsymbol{\Gamma}_N = \begin{bmatrix} r_{YY}[0] & r_{YY}[1] & \cdots & r_{YY}[N-1] \\ r_{YY}[1] & r_{YY}[0] & \cdots & r_{YY}[N-2] \\ \vdots & \vdots & \ddots & \vdots \\ r_{YY}[N-1] & r_{YY}[N-2] & \cdots & r_{YY}[0] \end{bmatrix} \quad \mathbf{a}_N = \begin{bmatrix} a_p[1] \\ a_p[2] \\ \vdots \\ a_p[N] \end{bmatrix} \quad \mathbf{r}_N = \begin{bmatrix} r_{YY}[1] \\ r_{YY}[2] \\ \vdots \\ r_{YY}[N] \end{bmatrix}. \tag{14.45}$$

Because the autocorrelation function is symmetric, $r_{YY}[k] = r_{YY}[-k]$, we have expressed all the elements of the $N \times N$ autocorrelation matrix $\boldsymbol{\Gamma}_N$ in Equation (14.45) solely in terms of elements with positive indices, $r_{YY}[k]$, $k \ge 0$. If $\boldsymbol{\Gamma}_N$ is well conditioned, the Yule–Walker equations, Equation (14.44), can be solved by standard matrix inversion techniques (e.g., Gauss–Jordan elimination, see Appendix A) to give the vector of model coefficients,[3]

$$\mathbf{a}_N = -\boldsymbol{\Gamma}_N^{-1} \mathbf{r}_N. \tag{14.46}$$

Since $r_{YY}[i] = r_{YY}[-i]$, the residual error, Equation (14.41), can be written in vector-matrix form as

$$\varepsilon_N = r_{YY}[0] + \sum_{i=1}^{N} a_N[i] r_{YY}[i] = r_{YY}[0] + \mathbf{a}_N^T \mathbf{r}_N = r_{YY}[0] + \mathbf{r}_N^T \mathbf{a}_N. \tag{14.47}$$

The computational complexity of general-purpose matrix inversion methods, assessed by the number of multiply-accumulate (MAC) operations, scales as N^3, i.e., $O(N^3)$. However, the fact that the autocorrelation matrix $\boldsymbol{\Gamma}_p$ has the form of a symmetric **Toeplitz matrix** is the basis of a computationally efficient, $O(N^2)$, algorithm for solving for the coefficients, the Levinson–Durbin algorithm, which we will now explain.

14.4.2 The Levinson–Durbin algorithm

The **Levinson–Durbin algorithm** is a computationally efficient, recursive procedure for solving for $\mathbf{a}_N$, the vector of the parameters of a model of order N, using $\mathbf{a}_{N-1}$, the vector of the parameters

[3] The covariance matrix, $\boldsymbol{\Gamma}_N$, can be shown to be positive definite such that $\mathbf{a}_N$ in Equation (14.46) is the unique solution.

of a model of order $N-1$. The procedure uses only simple algebra and does not explicitly require inverting matrices. The key to the Levinson–Durbin algorithm is the observation that $\mathbf{\Gamma}_N$, the symmetric Toeplitz matrix of order N, can be partitioned into another symmetric Toeplitz submatrix $\mathbf{\Gamma}_{N-1}$, of lower order $N-1$, augmented by an extra row and column,

$$\mathbf{\Gamma}_N = \left[\begin{array}{cccc|c} r_{YY}[0] & r_{YY}[1] & \cdots & r_{YY}[N-2] & r_{YY}[N-1] \\ r_{YY}[1] & r_{YY}[0] & \cdots & r_{YY}[N-3] & r_{YY}[N-2] \\ \vdots & \vdots & \ddots & \vdots & \vdots \\ r_{YY}[N-2] & r_{YY}[N-3] & \cdots & r_{YY}[0] & r_{YY}[1] \\ \hline r_{YY}[N-1] & r_{YY}[N-2] & \cdots & r_{YY}[1] & r_{YY}[0] \end{array}\right] = \begin{bmatrix} \mathbf{\Gamma}_{N-1} & \tilde{\mathbf{r}}_{N-1} \\ \tilde{\mathbf{r}}_{N-1}^T & r_{YY}[0] \end{bmatrix}.$$

We will use the convention that a tilde on the top of a vector quantity signals reversed index order. So, the extra column vector $\tilde{\mathbf{r}}_{N-1}$ of length $N-1$ has the same elements as $\mathbf{r}_{N-1}$, but in the reversed order,

$$\tilde{\mathbf{r}}_{N-1} \triangleq [r_{YY}[N-1]\ r_{YY}[N-2]\ \cdots\ r_{YY}[1]]^T.$$

Due to the symmetry of the $\mathbf{\Gamma}_N$ matrix about the diagonal, the extra row vector is just the transpose of the extra column vector $\tilde{\mathbf{r}}_{N-1}^T$. Now, also express the vectors $\mathbf{a}_N$ and $\mathbf{r}_N$ of Equation (14.45) in partitioned form,

$$\mathbf{a}_N = \left[\begin{array}{c} a_N[1] \\ a_N[2] \\ \vdots \\ a_N[N-1] \\ \hline a_N[N] \end{array}\right] = \begin{bmatrix} \hat{\mathbf{a}}_N \\ a_N[N] \end{bmatrix} \quad \mathbf{r}_N = \left[\begin{array}{c} r_{YY}[1] \\ r_{YY}[2] \\ \vdots \\ r_{YY}[N-1] \\ \hline r_{YY}[N] \end{array}\right] = \begin{bmatrix} \mathbf{r}_{N-1} \\ r_{YY}[N] \end{bmatrix},$$

where the vector $\hat{\mathbf{a}}_N$ comprises the first $N-1$ coefficients of $\mathbf{a}_N$. Using these partitioned matrices and vectors, Equation (14.43) can be economically expressed as

$$\mathbf{\Gamma}_N \mathbf{a}_N = \begin{bmatrix} \mathbf{\Gamma}_{N-1} & \tilde{\mathbf{r}}_{N-1} \\ \tilde{\mathbf{r}}_{N-1}^T & r_{YY}[0] \end{bmatrix} \begin{bmatrix} \hat{\mathbf{a}}_N \\ a_N[N] \end{bmatrix} = -\begin{bmatrix} \mathbf{r}_{N-1} \\ r_{YY}[N] \end{bmatrix}. \tag{14.48}$$

The matrix multiplication in Equation (14.48) results in two equations:

$$\mathbf{\Gamma}_{N-1}\hat{\mathbf{a}}_N + \tilde{\mathbf{r}}_{N-1} a_N[N] = -\mathbf{r}_{N-1} \tag{14.49a}$$

$$\tilde{\mathbf{r}}_{N-1}^T \hat{\mathbf{a}}_N + r_{YY}[0] a_N[N] = -r_{YY}[N]. \tag{14.49b}$$

We are going to show that these equations can be solved recursively in an efficient manner but, before doing so, we require one more pair of relations. Because $\mathbf{\Gamma}_N$ is a symmetric Toeplitz matrix, we can show (Problem 14-7) that the solution of the Yule–Walker equations to the index-reversed vector of correlation coefficients $\tilde{\mathbf{r}}_N$ is the index-reversed vector of model parameters $\tilde{\mathbf{a}}_N$. That is, we can show that the index-reversed equivalent of Equation (14.46) is

$$\tilde{\mathbf{a}}_N = -\mathbf{\Gamma}_N^{-1}\tilde{\mathbf{r}}_N. \tag{14.50}$$

For the same reason, the index-reversed equivalent of the error equation, Equation (14.47), is

$$\varepsilon_N = r_{YY}[0] + \tilde{\mathbf{a}}_N^T \tilde{\mathbf{r}}_N = r_{YY}[0] + \tilde{\mathbf{r}}_N^T \tilde{\mathbf{a}}_N. \tag{14.51}$$

Now, putting it all together, solve Equation (14.49a) for model parameters $\hat{\mathbf{a}}_N$ with the aid of Equations (14.46) and (14.50),

$$\hat{\mathbf{a}}_N = -\boldsymbol{\Gamma}_{N-1}^{-1}\mathbf{r}_{N-1} - \boldsymbol{\Gamma}_{N-1}^{-1}\tilde{\mathbf{r}}_{N-1}a_N[N] = \mathbf{a}_{N-1} + \tilde{\mathbf{a}}_{N-1}a_N[N]. \tag{14.52}$$

Substitute Equations (14.51) and (14.52) into Equation (14.49b),

$$\begin{aligned} -r_{YY}[N] &= \tilde{\mathbf{r}}_{N-1}^T\hat{\mathbf{a}}_N + r_{YY}[0]a_N[N] = \tilde{\mathbf{r}}_{N-1}^T(\mathbf{a}_{N-1} + \tilde{\mathbf{a}}_{N-1}a_N[N]) + r_{YY}[0]a_N[N] \\ &= \tilde{\mathbf{r}}_{N-1}^T\mathbf{a}_{N-1} + (r_{YY}[0] + \tilde{\mathbf{r}}_{N-1}^T\tilde{\mathbf{a}}_{N-1})a_N[N] = \tilde{\mathbf{r}}_{N-1}^T\mathbf{a}_{N-1} + \varepsilon_{N-1}a_N[N]. \end{aligned} \tag{14.53}$$

Now, solve for $a_N[N]$, the final coefficient of $\mathbf{a}_N$,

$$\begin{aligned} a_N[N] &= -\frac{r_{YY}[N] + \tilde{\mathbf{r}}_{N-1}^T\mathbf{a}_{N-1}}{\varepsilon_{N-1}} = -\frac{r_{YY}[N] + \mathbf{r}_{N-1}^T\tilde{\mathbf{a}}_{N-1}}{\varepsilon_{N-1}} \\ &= -\frac{r_{YY}[N]a_{N-1}[0] + \sum_{i=1}^{N-1}a_{N-1}[N-i]r_{YY}[i]}{\varepsilon_{N-1}} = -\frac{\sum_{i=1}^{N}a_{N-1}[N-i]r_{YY}[i]}{\varepsilon_{N-1}}. \end{aligned} \tag{14.54}$$

Here, we have used the fact that $\mathbf{r}_{N-1}^T\tilde{\mathbf{a}}_{N-1} = \tilde{\mathbf{r}}_{N-1}^T\mathbf{a}_{N-1}$ and that $a_{N-1}[0] = 1$. The final coefficient, $a_N[N]$, is called the **reflection coefficient** of the Nth-order filter and is given its own symbol,

$$k_N \triangleq a_N[N].$$

The reflection coefficient is the key to the recursion. For a given value of N, Equation (14.54) shows that k_N can be computed using the previously computed model parameters $\mathbf{a}_{N-1}$, plus the residual error ε_{N-1} and the autocorrelation $r_{YY}[N]$. Once we have computed k_N, we substitute it back into Equation (14.52) to generate all the remaining coefficients of the vector $\hat{\mathbf{a}}_N$ (which are the same as the first $N-1$ coefficients of $\mathbf{a}_N$) from the previous model parameters $\mathbf{a}_{N-1}$,

$$\hat{\mathbf{a}}_N = \mathbf{a}_{N-1} + k_N\tilde{\mathbf{a}}_{N-1}. \tag{14.55}$$

Finally, reassemble the complete vector $\mathbf{a}_N$ from $\hat{\mathbf{a}}_N$ and $a_N[N]$,

$$\mathbf{a}_N = \begin{bmatrix} \hat{\mathbf{a}}_N \\ a_N[N] \end{bmatrix} = \begin{bmatrix} \hat{\mathbf{a}}_N \\ k_N \end{bmatrix} = \begin{bmatrix} \mathbf{a}_{N-1} \\ 0 \end{bmatrix} + k_N\begin{bmatrix} \tilde{\mathbf{a}}_{N-1} \\ 1 \end{bmatrix}. \tag{14.56}$$

Equation (14.56) is called the **order update equation**. This equation shows that a single reflection coefficient k_N is sufficient to generate the entire model of order N given the coefficients $\mathbf{a}_{N-1}$ of the model of order $N-1$. Hence, all that is necessary to generate a family of models, namely $\mathbf{a}_m$, $1 \le m \le N$, is to compute one reflection coefficient for each model order k_m, $1 \le m \le N$. The order update equation for a model of order N is then computed by applying Equation (14.56) recursively, starting with the initial condition, $\mathbf{a}_0 = \tilde{\mathbf{a}}_0 = 1$.

To calculate the updated residual error ε_N in terms of the previous error ε_{N-1}, substitute values of $\mathbf{a}_N$ and $\mathbf{r}_N$ into Equation (14.47),

$$\begin{aligned}\varepsilon_N &= r_{YY}[0]+\mathbf{r}_N^T\mathbf{a}_N = r_{YY}[0]+[\mathbf{r}_{N-1}^T \ r_{YY}[N]]\begin{bmatrix}\hat{\mathbf{a}}_N\\ k_N\end{bmatrix} = r_{YY}[0]+\mathbf{r}_{N-1}^T\hat{\mathbf{a}}_N + k_N r_{YY}[N]\\ &= r_{YY}[0]+\mathbf{r}_{N-1}^T(\mathbf{a}_{N-1}+k_N\tilde{\mathbf{a}}_{N-1})+k_N r_{YY}[N]\\ &= \underbrace{r_{YY}[0]+\mathbf{r}_{N-1}^T\mathbf{a}_{N-1}}_{\varepsilon_{N-1}} + \underbrace{k_N(r_{YY}[N]+\mathbf{r}_{N-1}^T\tilde{\mathbf{a}}_{N-1})}_{-k_N\varepsilon_{N-1}}\\ &= \varepsilon_{N-1} - k_N(k_N\varepsilon_{N-1}) = \varepsilon_{N-1}(1-k_N^2), \ N>0.\end{aligned} \tag{14.57}$$

The last lines used Equations (14.51) and (14.53). Equation (14.57) is called the **error update equation**. By applying the error update equation recursively, we can express the error ε_N of a model of order N in terms of the initial value $\varepsilon_0 = \sigma_X^2$ and all the reflection coefficients k_i, $1 \le i \le N$,

$$\begin{aligned}\varepsilon_1 &= \varepsilon_0(1-k_1^2)\\ \varepsilon_2 &= \varepsilon_1(1-k_2^2) = \varepsilon_0(1-k_1^2)(1-k_2^2)\\ &\vdots\\ \varepsilon_N &= \varepsilon_0\prod_{i=1}^{N}(1-k_i^2) = \sigma_X^2\prod_{i=1}^{N}(1-k_i^2), \ 1\le i\le N.\end{aligned} \tag{14.58}$$

Equation (14.58) leads to some interesting conclusions:

1. **The reflection coefficients are bounded by unity magnitude.**
   Because the error ε_N must be non-negative for all orders N, the factors of $(1-k_i^2)$ in Equation (14.58) must all be positive. That means $k_i^2 \le 1$ for all i. Hence, the magnitude of every reflection coefficient must be less than one,

   $$|k_i| \le 1, \ 1\le i\le N.$$

2. **The error is monotonically decreasing.**
   Because the maximum magnitude of each reflection coefficient is one, each of the factors of $(1-k_i^2)$ in Equation (14.58) must be a positive number between zero and one,

   $$0 \le (1-k_i^2) \le 1.$$

   Because ε_N in Equation (14.58) is proportional to the product of these factors, the error must either decrease or remain constant with increasing model order N,

   $$\varepsilon_i \le \varepsilon_{i-1}, \ 1\le i\le N.$$

3. **The Levinson–Durbin recursion is computationally efficient.**
   For a model of order N, there are N iterations of the algorithm. In each iteration, the main work is computing the reflection coefficient, Equation (14.54), which requires N MAC operations. Computing the remaining $N-1$ coefficients using Equation (14.52) requires only additions. Hence, the Levinson–Durbin algorithm has a complexity of $O(N^2)$.

An AR model of order N is often termed an **AR(N) model**. Here are a couple of simple examples of the recursion as applied to AR(1) and AR(2) models.

Example 14.1

Determine the AR model coefficients for
(a) a first-order AR model, AR(1).
(b) a second-order AR model, AR(2).

➤ Solution:

(a) Let $N=1$ in Equation (14.36):

$$y[n]=x[n]+a_1[1]y[n-1].$$

There is only one parameter to compute, $a_1[1]$, which is the same as reflection coefficient k_1. To start the recursion, let $\mathbf{a}_0=1$, $\mathbf{r}_0=0$ and solve for the reflection coefficient using Equation (14.54) and the resulting error ε_1 using the error update equation, Equation (14.57):

$$\begin{aligned} k_1 &= a_1[1] = -(a_0[0]r_{YY}[1])/\varepsilon_0 = -r_{YY}[1]/\sigma_X^2 \\ \varepsilon_1 &= \varepsilon_0(1-k_1^2). \end{aligned}$$

(b) Let $N=2$ in Equation (14.36):

$$y[n]=x[n]+a_2[1]y[n-1]+a_2[2]y[n-2].$$

We shall use the value of $k_1=a_1[1]$ that we computed in part (a) to compute the parameters of $\mathbf{a}_2$. First compute the final coefficient of $\mathbf{a}_2$, which is reflection coefficient $k_2=a_2[2]$, and then use it to compute the remaining coefficient of $\mathbf{a}_2$, namely $a_2[1]$:

$$\begin{aligned} k_2 &= a_2[2] = -(a_1[0]r_{YY}[2]-a_1[1]r_{YY}[1])/\varepsilon_1 = -(r_{YY}[2]-k_1 r_{YY}[1])/\varepsilon_1 \\ a_2[1] &= \hat{\mathbf{a}}_2 = \mathbf{a}_1 + a_2[2]\tilde{\mathbf{a}}_1 = a_1[1]+a_2[2]a_1[1] = a_1[1](1+a_2[2]) = k_1(1+k_2) \\ \varepsilon_2 &= \varepsilon_0(1-k_1^2)(1-k_2^2). \end{aligned}$$

14.4.3 Matlab implementation of the Levinson–Durbin algorithm

Matlab provides several functions that implement the Levinson–Durbin algorithm. The basic algorithm is `levinson`,

```
[a, eta, rcs] =levinson(r, [N]),
```

where the input `r` is an array of autocorrelation coefficients, with the first one being $r_{XX}[0]$. The optional input parameter `N` tells the function to take only the first $N+1$ autocorrelation coefficients, which corresponds to computing an Nth-order model. The output parameter `a` is the array of model coefficients, `eta` is the prediction error ε_N, and `rcs` is an array of the reflection coefficients. Matlab also provides two other functions, `lpc` and `aryule`, which produce the `N` coefficients, `a`:

```
a=lpc(x, N)
[a, eta, rcs] = aryule(x, N).
```

These two functions accept the sequence `x` and the desired model order `N` as input. They then calculate the required autocorrelation sequence internally and solve the Yule–Walker equations using the Levinson–Durbin recursion. Here is our bare-bones version of `aryule`:

```
function [a, eta, rcs] = yule(x, N)
    % Compute Levinson-Durbin recursion on sequence x
    %     Returns Model parameters (a)
    %             Array of residual error (eta)
    %             Array of reflection coefficients (rcs)
    %             T. Holton
    rxx = xcorr(x(:), N)/length(x); % compute 'biased' autocorrelation
     rxx = rxx(N+1:end);             % make column vector of necessary lags
    a = 1;                           % initialize a(0)
    eta = rxx(1);                    % and initialize eta array
    rcs = [];                        % array of reflection coefficients
    for i = 2:N+1
        atilde = a(end:-1:1);
        rc = -a(end:-1:1)*rxx(2:i)/eta(end);
        rcs = [rcs; rc];
        a = [a 0]+[0 rc*atilde];
        eta = [eta; eta(end)*(1-rc^2)];
    end
end
```

To calculate the model coefficients, we first calculate `rxx`, the "biased" autocorrelation function, namely Equation (14.7), from the input sequence `x`. Matlab's autocorrelation function `xcorr(x, N)` generates a two-sided, $2N+1$-point autocorrelation, of which we choose only the last $N+1$ coefficients, which correspond to $r_{YY}[i]$, $0 \leq i \leq N$, as required by the Yule–Walker equations.[4] After initializing the outputs, the `for` loop computes the reflection coefficient `rc` corresponding to $a_N[n]$ recursively using Equation (14.54). Parameter `a` is updated on every cycle using Equation (14.56).

The following example shows an application of the Levinson–Durbin recursion to a problem of autoregressive modeling. Here, the task is to determine the parameters of an all-pole filter, namely $a_N[i]$, $1 \leq i \leq N$, that produces an output autoregressive process that has the same statistics as the input process that we are trying to model when excited with white noise. In this example, we will use our `yule` function to determine the coefficients for autoregressive models of different orders to a filtered noise signal.

Example 14.2

Let $H(z)$ be an all-pole filter with five poles,

$$\begin{aligned} H(z) &= \frac{1}{1-1.7043z^{-1}+0.6509z^{-2}+0.9931z^{-3}-1.1350z^{-4}+0.3572z^{-5}} \\ &= \frac{1}{(1+0.9z^{-1})(1-0.9e^{-j\pi/4}z^{-1})(1-0.9e^{j\pi/4}z^{-1})(1-0.7e^{-j\pi/10}z^{-1})(1-0.9e^{j\pi/10}z^{-1})}, \end{aligned} \quad (14.59)$$

[4] Here we are replacing the "true" autocorrelation coefficients $r_{XX}[l]$ with the sample autocorrelation function $\hat{r}_{XX}[l, N]$.

and let $y[n]$ be the autoregressive output of this filter when the input $v[n]$ is a zero-mean, Gaussian noise process with variance $\sigma_V^2 = 1$. Determine the AR model coefficients $a_N[i]$, $1 \le i \le N$, the poles and zeros of $H(z) = 1/A_N(z)$, for AR models of order $N = 1$, 3 and 5.

➤ Solution:

The following code generates the filter coefficients $a_N[n]$ for a model of order N:

```
v = randn(1, 1000);           % create Gaussian noise
v = (v - mean(v)) / std(v);   % make sure it is zero-mean, unity variance
aa = poly([-0.9 0.9*exp(1j*0.25*pi*[1 -1]) 0.7*exp(1j*0.1*pi*[1 -1])]);
yy = filter(1, aa, v);        % filter the noise by H(z)
npts = 500;
y = yy(end-npts+1:end);       % last points avoid start-up transient
[a, eta, rcs] = yule(y, N); % compute model coefficients
```

The filter $H(z)$ in Equation (14.59) is specified by its pole coefficients `aa`. Gaussian white noise is filtered by the model and the last `npts` points of the resulting time sequence are retained to form the autoregressive process to be analyzed, `y`. Next, the model coefficients `a` are computed using `yule`, described above (or using the Matlab function `lpc`).

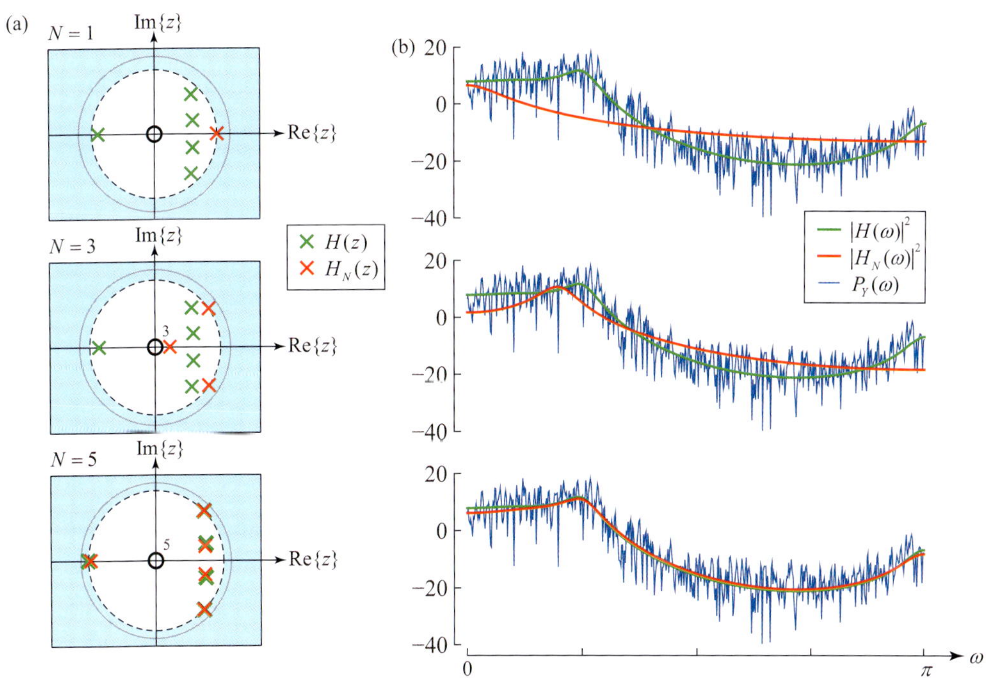

Figure 14.24 AR analysis of all-pole model driven by noise

Figure 14.24a shows pole-zero plots of AR(1) ($H_1(z)$, top panel), AR(3) ($H_3(z)$, middle panel) and AR(5) ($H_5(z)$, bottom panel) models fit to filter $H(z)$ driven by zero-mean Gaussian noise. The poles of each model are shown in red (×). The five poles of $H(z)$ are also shown in green (×). **Figure 14.24b** shows the power spectral density $P_Y(\omega)$ of the autoregressive input process $y[n]$ computed using Matlab's `periodogram` function (or our equivalent `mypsd`, as described in Section 14.3.7) and plotted on a dB power scale (i.e., $10\log_{10}P_Y(\omega)$). In order to compare the fit of the model to the input process, we use the result of Equation (14.33) that the power spectral density of a filtered white-noise process is proportional to the square of the filter magnitude, in this case,

$$S_{YY}(\omega) = \sigma_V^2 |H_N(\omega)|^2,$$

where $\sigma_V^2 = 1$. Accordingly, superimposed on the plots of **Figure 14.24b** are the frequency responses of $H(\omega)$ (green traces) and $H_N(\omega)$ (red traces). As you can see, for a model of order $N = 5$, the pole-zero plots and power spectral density plots of $H_5(z)$ fit the autoregressive input $H(z)$ quite well.

14.5 Linear prediction

Linear prediction is a powerful technique for modeling stochastic processes that has found wide use in multiple areas including digital communication, adaptive filtering and identification of signals in noise and speech processing, where it is an integral part of applications for speech recognition, compression and synthesis. As we will now show, the theory underlying linear prediction is closely related to the theory of AR modeling in Section 14.4.1, and in fact, leads to the same Yule–Walker equations and to their solution via the Levinson–Durbin algorithm. Then we will demonstrate the application of AR modeling and linear prediction to a couple of practical problems.

Let $y[n]$ be a WSS random process. The purpose of linear prediction is to predict the output of this process, $\hat{y}_N[n]$, at time n using a linear combination of the N previous values of the process,

$$\hat{y}_N[n] \triangleq -\sum_{i=1}^{N} a_N[i]y[n-i] = -\mathbf{y}_N^T \mathbf{a}_N,$$

where $\mathbf{y}_N^T[n]$ is the vector of previous values of $y[n]$ and $\mathbf{a}_N$ is the vector of **linear predictor coefficients**,

$$\begin{aligned} \mathbf{y}_N^T &= [y[n-1]\ y[n-2]\ \cdots\ y[n-N]] \\ \mathbf{a}_N &= [a_N[1]\ a_N[2]\ \cdots\ a_N[N]]^T. \end{aligned}$$

This is termed **forward linear prediction**, since we are attempting to use past values of the process ($y[n-i]$, $i = 1, \cdots, N$) to look forward in time and predict the value of $\hat{y}_N[n]$. The **predictor order** N is the equivalent of the AR model order in Equation (14.35). The subscript N in the linear prediction coefficients $a_N[i]$ and the predicted value $\hat{y}_N[n]$ both emphasize the fact that these quantities depend on the predictor order, so $\mathbf{a}_{N-1}$ and $\mathbf{a}_N$ represent two different sets of linear prediction coefficients. We define the **prediction error** as the difference between the predicted value $\hat{y}_N[n]$ and the actual value of the autoregressive process $y[n]$ at time n,

$$e_N[n] = y[n] - \hat{y}_N[n] = y[n] + \sum_{i=1}^{N} a_N[i]y[n-i] = y[n] + \mathbf{y}_N^T \mathbf{a}_N. \tag{14.60}$$

The prediction error, which depends on the order of the predictor, is also called the **innovation** of the process,[5] because it describes the component of $y[n]$ that is "novel," that is, not completely described by the Nth-order model $\hat{y}_N[n]$.

[5]The root of the word "innovate" is the Latin *novare*, which is literally "to make new."

The solution to the linear prediction problem can be cast as a problem to determine the value of the set of linear prediction coefficients $a_N[i]$, $1 \le i \le N$, that will minimize the expected **mean-square prediction error** ε_N,

$$\varepsilon_N = E(e_N^2[n]). \tag{14.61}$$

Since $e_N[n]$ is taken to be a WSS random process, its expected value ε_N is independent of time. To calculate ε_N, substitute Equations (14.60) into Equation (14.61):

$$\begin{aligned}
\varepsilon_N = E(e_N^2[n]) &= E((y[n] + \mathbf{y}_N^T\mathbf{a}_N)^T(y[n] + \mathbf{y}_N^T\mathbf{a}_N)) \\
&= E\big((y^T[n] + \mathbf{a}_N^T\mathbf{y}_N)(y[n] + \mathbf{y}_N^T\mathbf{a}_N)\big) \\
&= E(y^T[n]y[n]) + E(\mathbf{a}_N^T\mathbf{y}_N y[n]) + E(y[n]\mathbf{y}_N^T\mathbf{a}_N) + E(\mathbf{a}_N^T\mathbf{y}_N\mathbf{y}_N^T\mathbf{a}_N) \\
&= E(y^2[n]) + \mathbf{a}_N^T E(\mathbf{y}_N y[n]) + E(y[n]\mathbf{y}_N^T)\mathbf{a}_N + \mathbf{a}_N^T E(\mathbf{y}_N\mathbf{y}_N^T)\mathbf{a}_N \\
&= r_{YY}[0] + \mathbf{a}_N^T\mathbf{r}_N + \mathbf{r}_N^T\mathbf{a}_N + \mathbf{a}_N^T\boldsymbol{\Gamma}_N\mathbf{a}_N \\
&= r_{YY}[0] + 2\mathbf{r}_N^T\mathbf{a}_N + \mathbf{a}_N^T\boldsymbol{\Gamma}_N\mathbf{a}_N.
\end{aligned} \tag{14.62}$$

Here, we have used the notation of Equation (14.45) to express $E(\mathbf{y}_N y[n]) = \mathbf{r}_N$, $E(y[n]\mathbf{y}_N^T) = \mathbf{r}_N^T$ and $E(\mathbf{y}_N\mathbf{y}_N^T) = \boldsymbol{\Gamma}_N$. Equation (14.62) is the matrix form of the so-called **normal equations**, a version of which we have seen before in connection with least-square error FIR filters in Chapter 7 (Section 7.8.1) and also in conjunction with the problem of energy compaction of the DCT in Chapter 12 (Section 12.1.7). Solution of these normal equations gives the values of the linear prediction coefficients $\mathbf{a}_N$ that minimize the error ε_N. Appendix A gives a full discussion of the solution of these equations in both over-constrained and fully constrained systems of equations. In this case, the system is fully constrained since $\boldsymbol{\Gamma}_N$ is a square $N \times N$ matrix (assumed to be non-singular) with inverse $\boldsymbol{\Gamma}_N^{-1}$, and the solution to these normal equations reduces to

$$\mathbf{a}_N = -\boldsymbol{\Gamma}_N^{-1}\mathbf{r}_N. \tag{14.63}$$

This is exactly the same as the solution to the Yule Walker equations for the parameters of the AR model, given in Equation (14.46)! Substituting Equation (14.63) back into Equation (14.62) gives the minimum square-error,

$$\begin{aligned}
\varepsilon_N &= r_{YY}[0] + 2\mathbf{r}_N^T\mathbf{a}_N + \mathbf{a}_N^T\boldsymbol{\Gamma}_N\mathbf{a}_N \\
&= r_{YY}[0] + 2\mathbf{r}_N^T\mathbf{a}_N - \mathbf{a}_N^T\boldsymbol{\Gamma}_N(\boldsymbol{\Gamma}_N^{-1}\mathbf{r}_N) \\
&= r_{YY}[0] + 2\mathbf{r}_N^T\mathbf{a}_N - \mathbf{a}_N^T\mathbf{r}_N \\
&= r_{YY}[0] + \mathbf{r}_N^T\mathbf{a}_N,
\end{aligned}$$

which is identical to the residual error of the AR model, Equation (14.47). The transfer function of the difference equation, Equation (14.60),

$$H_N(z) = \frac{Y_N(z)}{E_N(z)} = \frac{1}{A_N(z)} = \frac{1}{1 + \sum_{i=1}^{N} a_N[i]z^{-i}}, \tag{14.64}$$

is also identical to Equation (14.35).

Figure 14.25 shows two complementary ways of interpreting linear prediction, as defined by Equation (14.60). **Figure 14.25a** shows linear prediction as an *analysis* technique. The entire block in light blue can be viewed as a **prediction error filter** $A_N(z)$, whose output is the prediction error. The predictor block predicts the current value of the output $y[n]$ from a linear combination of the previous values of the output. The predictor extracts linear prediction coefficients $a_N[i]$, $1 \le i \le N$ (or, equivalently, the reflection coefficients $k_N[i]$, $1 \le i \le N$). The prediction error $e_N[n] = y[n] - \hat{y}[n]$ represents that part of $y[n]$ that is not predicted. In the event that the predictor is "perfect," the prediction error will be white-noise with a flat power spectral density. **Figure 14.25b** interprets linear prediction as a *synthesis* technique. Here, the block in light green is the filter $H_N(z) = 1/A_N(z)$. The innovation process $e_N[n]$ is considered to be the excitation of the filter whose output $y[n]$ is the sum of $e_N[n]$ plus a linear combination of the N previous values of the output. In the event that the predictor is perfect and that the input innovation process $e_N[n]$ is white noise, the output of this synthesis will be $y[n]$.

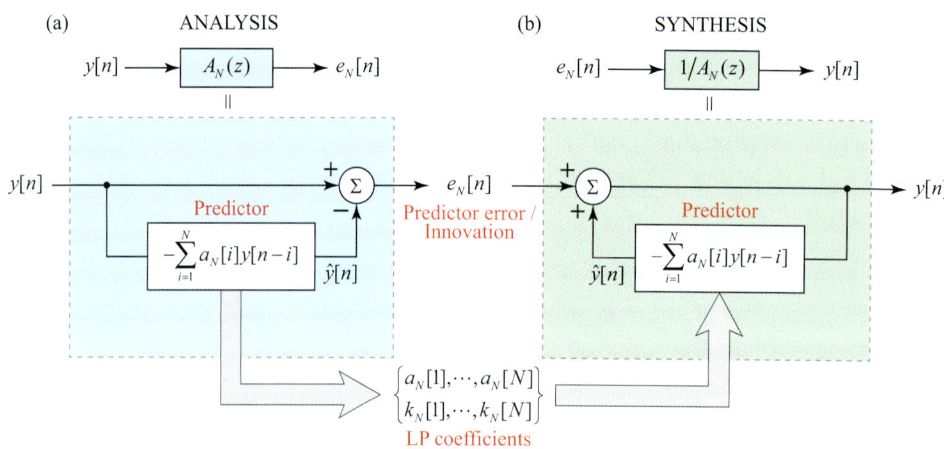

Figure 14.25 Linear prediction model

14.5.1 Prediction error and the estimation of model order

In Example 14.2, we generated (synthesized) a random process by filtering a white-noise input with an all-pole filter of known order $N = 5$, and then (unsurprisingly) determined that the model coefficients for an AR model of order $N = 5$ fit the data "quite well." This raises two important questions: (1) What does "quite well" mean? (2) How do we determine the appropriate order of an AR model that fits a general random process? As a point of departure, let us cast Example 14.2 as a linear prediction problem and examine the prediction error $e_N[n]$ in the time domain as a function of the predictor (model) order N. As suggested in our discussion of **Figure 14.25**, if the predictor were "perfect," the (residual) prediction error would be a white-noise process $e_N[n] = \sigma_X^2 \delta[n]$.

Example 14.3

Plot the prediction error $e_N[n]$ for the three models of Example 14.2.

➤ Solution:

The following lines of code generate the prediction error $e_N[n]$ of a model of order N:

```
en = filter(a, 1, y);      % compute prediction error
en = en(end-npts+1:end); % and select last points
```

We perform the inverse filter using `filter` and the predictor coefficients `a` generated in Example 14.2 to produce the prediction error array `en`, and select points corresponding to the same time range as `y`.

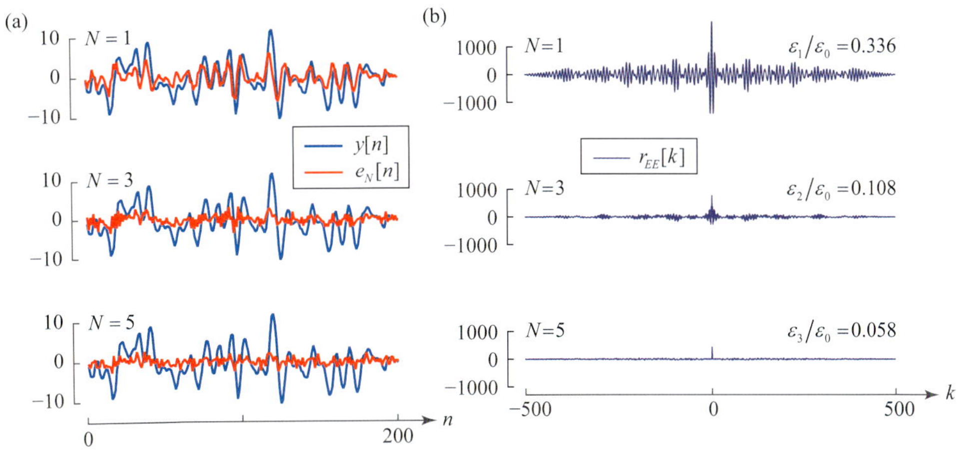

Figure 14.26 Prediction error of AR models

Figure 14.26a shows the input to the predictor `y` (blue traces), and the resulting prediction error `en` (red traces), for the three models of Example 14.2. For the model of order $N = 1$, the prediction error appears highly correlated with the input signal, indicating that the AR(1) model is not a good fit. For the model of order $N = 5$, the error appears less correlated with the input indicating a better fit. **Figure 14.26b** shows the autocorrelation function $r_{EE}[k]$ computed from the prediction error $e_N[n]$ for the three models. The autocorrelation of the error of the AR(5) model is approximately an impulse, which indicates that the prediction error in this case is well approximated by a white-noise process with a flat spectrum, as one might expect for a near-perfect fit. In this example, the linear prediction filter is said to act as a **whitening filter**, since its output, the prediction error, is effectively white noise.

Figure 14.27 suggests two (related) ways to estimate the appropriate model order based on looking at the residual mean-square prediction error ε_N. **Figure 14.27a** shows the normalized mean square error $\varepsilon_N/\varepsilon_0$, plotted on a log scale. As we expect from the discussion of Section 14.4.2, the error is monotonically decreasing with increasing N, but, in this example, there is no change in the fourth decimal place of the error after the first five coefficients, suggesting that number of model coefficients is sufficient to describe the process.

The **partial correlation (PARCORR) coefficients** of the process, shown in **Figure 14.27b**, provide a relatively objective method for selecting the model order. Recall that the

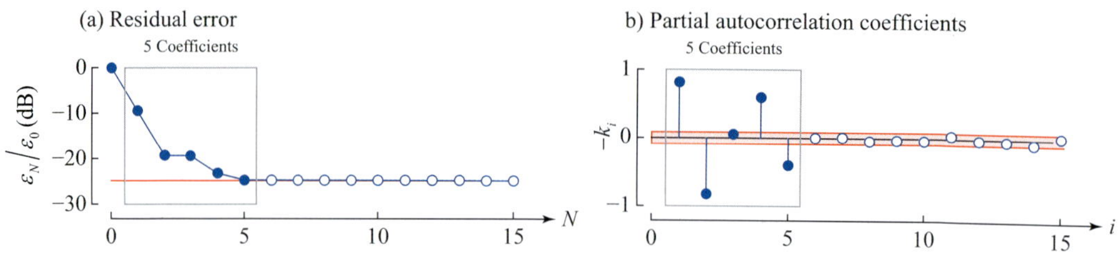

Figure 14.27 Estimation of model order using partial correlation coefficients

autocorrelation function of a process, $r_{YY}[l]$, measures the degree of correlation between two samples, $y[n]$ and $y[n-l]$, where l is the lag. However, $y[n]$ and $y[n-l]$ are related not only directly, but also *indirectly* through the correlation of intermediate values $y[n-1], \ldots, y[n-l+1]$. That is to say, $y[n]$ is correlated with $y[n-1]$, which is in turn correlated with $y[n-2]$, and so on to $y[n-l]$. **Partial correlation** is a measure of the direct correlation of $y[n]$ and $y[n-l]$ after the indirect correlation due to the intervening terms has been removed. Since linear prediction is based on using a linear combination of the N previous values of $y[n]$ to predict the current value, it makes sense to choose N to be the maximum lag for which the partial correlation is significant. While the derivation of partial correlation is beyond the scope of our presentation, it can be shown that the partial correlation coefficients are equal to the negative of the reflection coefficients, $-k_i$, which is particularly fortuitous because our `yule` function generates the entire array of reflection coefficients `rcs`. These are shown in **Figure 14.27** with a confidence interval of 95% shown with red lines. Here again, it appears that five coefficients are adequate.

14.5.2 Linear predictive coding (LPC)

Linear prediction is the basis of **linear predictive coding (LPC)**, a family of methods that is widely used in signal processing applications for speech coding, speech recognition and verification, speech synthesis and speech compression. The common feature of all of these methods is their reliance on a parametric model of speech production or synthesis that allows a complex

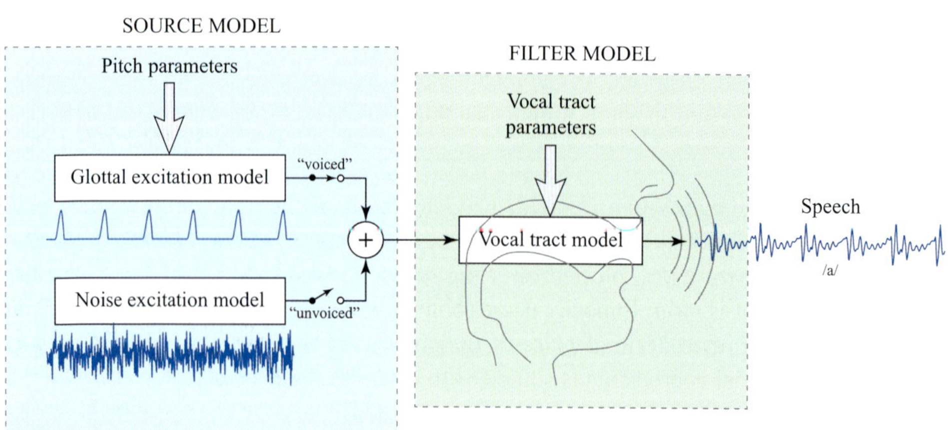

Figure 14.28 The source-filter model

acoustic signal to be represented by a relatively small number of parameters. The original LPC approach was developed in the 1970s and the many successful refinements and variants of this method have followed.

14.5.3 The source-filter model

The foundation of the LPC approach is the **source-filter model** of speech production diagrammed in **Figure 14.28**. To understand the model, first consider how a **voiced** speech sound, such as the vowel (e.g., /a/, as in "arm"), is produced. Air expelled from the lungs passes up through the trachea and excites the vocal cords in the **glottis** (part of the **pharynx**). The vocal cords open and close periodically at the **pitch frequency** (generally in the range of 50 Hz–400 Hz) and modulate the flow of air resulting in a series of **glottal pulses** of air pressure. The glottal pulses pass through the upper vocal tract, which acts as an adjustable resonant cavity whose resonances are controlled by the position of the **articulators** – the jaw, tongue, lips and soft palette (**velum**). The spectral peaks of the vocal tract resonances are called the **formants**. As we saw in our discussion of the spectrogram in Section 14.2.2, different vowels are characterized by distinct frequencies and bandwidths of these formants. Finally, sound escapes the mouth or the nose (depending on the position of the velum). **Unvoiced sounds**, such as some **fricatives** (e.g., /s/ "sea" and /ʃ/ in "shell"), are produced when air passes through the glottis without exciting the vocal folds; the resulting turbulent airflow, modified by constrictions or restrictions of the vocal tract and lips, generates the characteristic hissing sounds.

The source-filter model of **Figure 14.28** models the complex biophysics of speech production as a cascade of two independent components: a source of sound excitation followed by a filter with variable frequency parameters. The **source model**, shown in the light blue box, comprises two elements, a **glottal excitation model**, which models the glottal pulse train that produces voiced sounds, and a **noise model**, which models the turbulent air flow that produces unvoiced sounds. The source model permits excitation of the vocal tract by either glottal pulses, turbulent noise or both (e.g., in the case of voiced postalveolar fricatives such as /ʒ/ (as in "genre" or "leisure"). The **filter model**, shown in the light green box, models the resonances of the upper vocal tract by an **all-pole model**, namely Equation (14.34). Such an all-pole model is predicted by the physics of a simplified model of the vocal tract. a hard-walled tube, closed at one end, comprising a series of cylindrical sections of varying diameters. Despite the fact that none of these conditions actually apply to the real vocal tract, the all-pole model works remarkably well in fitting the spectra of speech sounds. The wide adoption of the LPC approach to speech processing is due to the fact that it provides an effective and computationally economical method of computing the parameters of an all-pole vocal-tract model excited by both voiced and unvoiced sources.

14.5.4 Linear predictive coding architecture

The object of the linear predictive coding (LPC) is to encode speech at a low bit-rate for transmission over a communications channel, such that the speech decoded at the other end of the channel is of high-enough quality to be acceptable for the intended application (e.g., wireless telephony). The combination of an **enc**oder and **dec**oder is called a **codec**. There are many variants of LPC codecs, most of which share the same basic architecture.

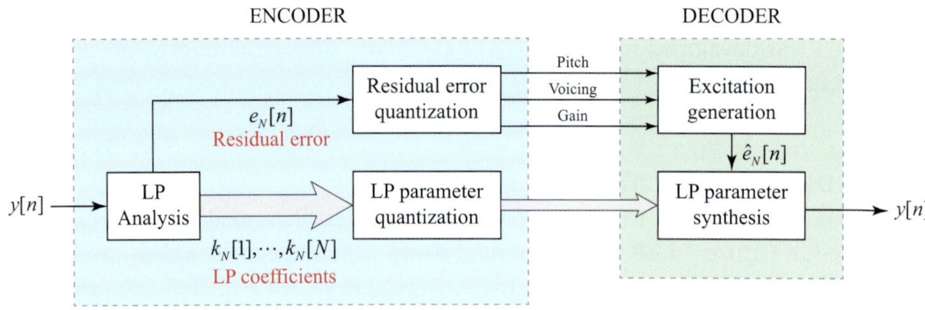

Figure 14.29 Basic LPC codec architecture

Basic LPC architecture **Figure 14.29** schematized the architecture of the basic **LPC-10** codec that has many of the elements of more modern encoders. In the encoder (blue box), speech is segmented into 22.5 ms frames (i.e., 44.4 frame/s). For each frame, a decision is made whether the speech segment is voiced or unvoiced. Linear prediction analysis is performed on each frame and the reflection coefficients $k_N[n]$ and prediction error sequence $e_N[n]$ are extracted. Part of the art of LPC design is finding an efficient representation for both the reflection coefficients and the prediction error sequence. For voiced segments, the LPC-10 encoder extracts $N = 10$ coefficients (hence, the name, LPC-10) and quantizes them from two to five bits each for a total of 41 bits. (For unvoiced segments, only four coefficients are obtained and quantized at five bits each for a total of 20 bits.) The prediction error signal is processed separately to derive an additional 13 bits of pitch, voicing and gain information, for a total of 54 bits/frame. Accordingly, the LPC-10 encoder operates at $44.4 \times 54 = 2.4$ kbps. When decoded (green box), the resulting speech is intelligible, but has a relatively unnatural, robotic quality.

Many of the more sophisticated and natural-sounding modern varieties of LPC codecs have been designed to model and encode the residual error more accurately in order to improve perceived speech quality. The most widely used of these are the **code excited linear prediction (CELP)** family of codecs, variants of which are found in cellphones, among other common communications applications. In CELP encoders, linear prediction coefficients and the predictor error are extracted, essentially as we have described above. The predictor error is then compared to one or more libraries of excitation sequences called **codebooks**, and the index of the codebook sequence that minimizes the distortion of a perceptual model is encoded along with the linear prediction coefficients and other parameters. These codecs operate at a variety of bit rates, with rates of about 13 kbps and above having high perceived speech quality.

Example of using linear prediction in encoding and decoding **Figure 14.30** shows an example of the application of linear prediction to the analysis and synthesis of a voiced segment of speech. **Figure 14.30a** shows $y[n]$, a generous 50 ms voiced section of the vowel /a/. The sound was recorded at $f_s = 11.025$ kHz, so $\omega = \pi$ corresponds to 5.513 kHz. The speech waveform comprises a train of pulses that occur at multiples of the **pitch period** $T_0 \simeq 11.3$ ms. The high-frequency oscillations in the waveform following each pulse represent the vocal tract resonances that have been excited by each glottal pulse. The thin blue traces in each of the four panels of **Figure 14.30d** show the measured power spectral density $P_Y(\omega)$ of this vowel on a log (dB) scale. The rapidly oscillating peaks in the spectrum occur at multiples of the **pitch frequency** $f_0 = 1/T_0$. The broad peaks correspond to the vocal tract resonances, the formants. Following the procedure described in Example 14.2, we extract the linear prediction coefficients $a_N[i]$, $1 \leq i \leq N$, for models of order N ranging from 1 to 30. The red traces in **Figure 14.30d** show the spectra computed from the parameters for four of these models, with $N = 4$, 8, 16 and 24. The fit of the model to the data gets better as N increases. To get an idea of the model order required to represent this sound, **Figure 14.30b** plots the normalized residual error $\varepsilon_N/\varepsilon_0$, and **Figure 14.30c** shows the partial correlation coefficients, as a function of model order N, with the values of error for the four models whose spectra are shown in **Figure 14.30d** plotted with red symbols. As expected, the error decreases monotonically and the partial correlation generally becomes less significant as the model order increases. These data, which are fairly typical of voiced speech segments, indicate that you get diminishing returns for models of order greater than the mid-teens, though the actual error curves depend on a number of factors, such as sample rate and speaker (i.e., male or female). However, just as we discussed in Chapter 12 in conjunction with the MP3 and JPEG encoders, it is important to point out that ultimately, the minimum number of coefficients to be retained needs to be judged not by arbitrary measures such as the minimum mean-square error, but by an assessment of the voice quality of the decoded speech by actual humans.

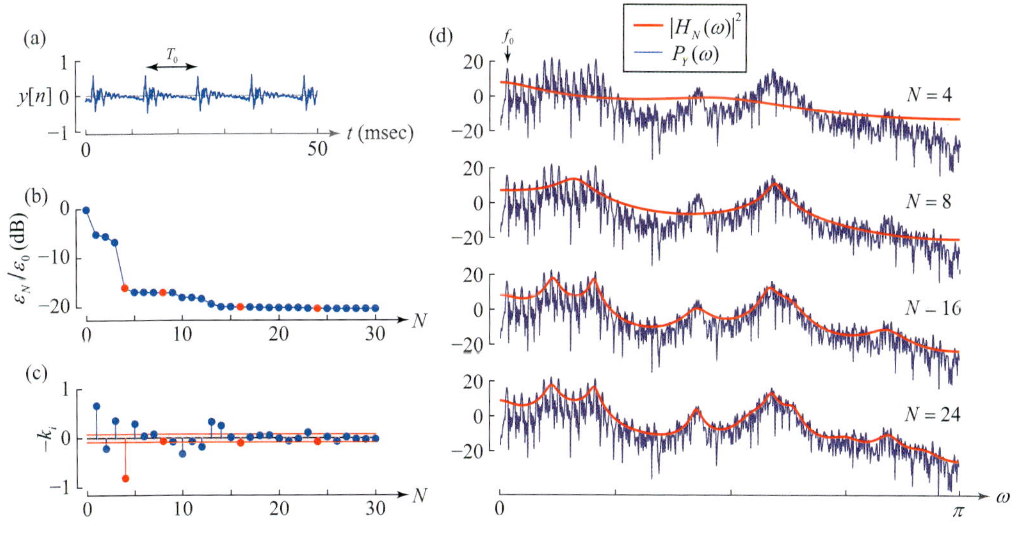

Figure 14.30 LPC analysis of a voiced speech segment

Figure 14.31 shows how the vowel can be decoded – resynthesized – in a codec using the linear prediction coefficients and the prediction error for a model of order $N = 16$. The blue and red traces in **Figure 14.31a** are replotted from **Figure 14.30b** for reference. The prediction error $e_N[n]$ is plotted in blue in **Figure 14.31b**. Unlike the prediction error in Example 14.2, which was essentially white noise, the prediction error of this speech example consists of a series of pulses, which reflects the fact that the vocal tract is driven not by white noise (as in Example 14.2), but by the glottal pulse train. In LPC encoders, considerable effort goes into modeling and encoding this residual signal. However, in our simple example, we have just estimated the times of pitch pulses using a simple peak-picking algorithm (grey arrows) and then modeled each pitch pulse by a three-point sequence (shown in the inset in orange). To resynthesize the vowel, we used the pitch-pulse times and the model pitch pulse to create a synthetic pitch-pulse train $\hat{e}_N[n]$ (shown in orange). This pulse train is input into the inverse filter $1/A_N(z)$ using the extracted linear prediction coefficients, as suggested in **Figure 14.25b**. The resulting output $\hat{y}[n]$ is shown in **Figure 14.31c** (orange trace). The waveform of the resynthesized output (orange trace) looks relatively similar to that of the original vowel input (shown in blue). However, as we mentioned above, the real test of similarity is not how it looks, or how low the mean-square error difference is between these waveforms, but how it sounds to a person. (It sounds pretty natural to me.)

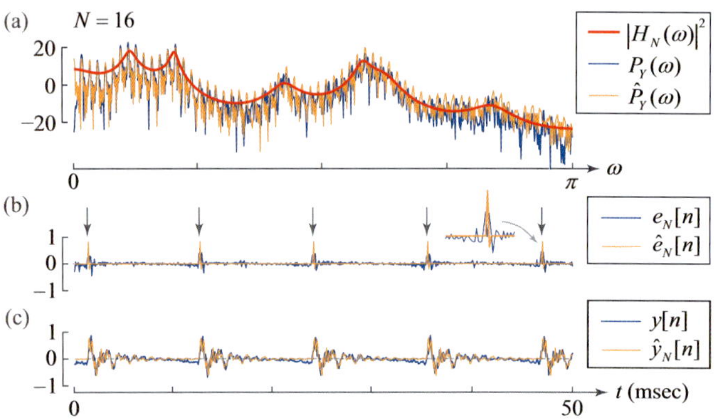

Figure 14.31 Re-synthesis of vowel from linear prediction coefficients

14.5.5 ★ Alternative formulations of linear prediction equations

Supplementary material available at www.cambridge.org/holton presents two alternate representations of the linear prediction coefficients – **lattice filter** and **line spectral frequencies** – that are more numerically stable, more perceptually relevant or both.

SUMMARY

This chapter has covered a wide range of topics relating to the spectral analysis of deterministic and probabilistic systems, starting with an explanation of two fundamental issues that affect the practical measurement of the frequency spectrum of signals: data windowing and frequency sampling. We explored the effects of the choice of data window and its parameters for signals, and then introduced the short-term Fourier transform (STFT) for the many cases where signals are inherently time-varying, such as speech and music. The remaining sections of the chapter covered spectral measurement and estimation of probabilistic systems, both nonparametric and parametric. The nonparametric spectral methods we discussed include the periodogram and its modifications, including Bartlett's method and Welch's method. For parametric methods, we examined the autoregressive (AR) model in some detail, derived the Yule–Walker equations and solved them using the Levinson–Durbin algorithm. The chapter concluded with a discussion of linear prediction and its application to encoding and decoding speech sounds using the source-filter model of speech production.

PROBLEMS

Problem 14-1

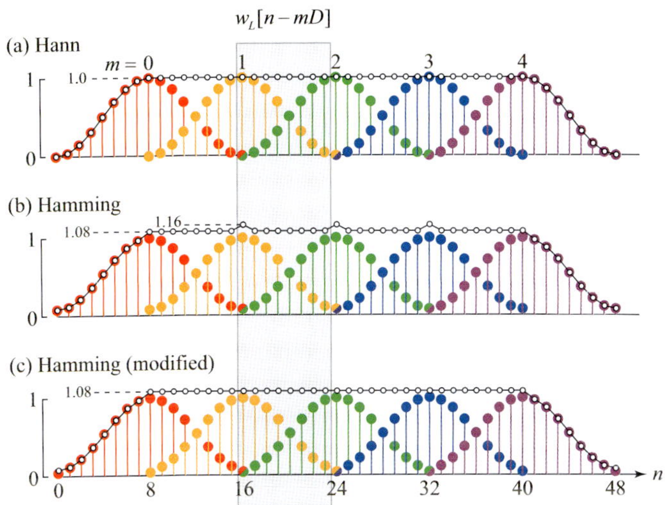

Figure 14.32 Overlapping STFT windows

Figure 14.32 shows three different STFT overlapping windows: Hann (**Figure 14.32a**), Hamming (**Figure 14.32b**) and modified Hamming (**Figure 14.32c**). Each window, $w_L[n-mD]$, is of length L, and overlaps subsequent windows by 50%, which is $D=(L-1)/2$ points (highlighted with a grey box). Let $s_L[n-mD]$ be the sum of the points in the mth overlapping section of D points (excluding the ramp-up and ramp-down), shown with the thin black line and open symbols. The sequence $s_M[n]$ comprises the sum of points from three windows, the last point of window $m-1$ (e.g., the "red" window, $w_L[L-1]$), plus the second half of window m (the "orange" window, $w_L(n+D)$, $0 \le n \le D-1$), plus the first half of the "green" window ($w_L(n)$, $0 \le n \le D-1$):

$$s_L[n] = \begin{cases} w_L[L-1] + w_L(D) + w_L(0), & n=0 \\ w_L(n+D) + w_L(n), & 1 \le n \le D-1 \end{cases}. \tag{14.65}$$

(a) A Hann window of length L (**Figure 14.32a**) is defined as

$$w_L[n] = 0.5 - 0.5\cos\frac{2\pi n}{L-1}, \quad 0 \le n \le L-1.$$

Show that the sum of Hann windows that overlap by 50% is a constant, one, over the overlapping range; namely, that

$$s_L[n] = 1, \quad 0 \le n \le D - 1 .$$

(b) A Hamming window of length L (**Figure 14.32b**) is defined as

$$w_L[n] = 0.54 - 0.46 \cos \frac{2\pi n}{L-1}, \qquad 0 \le n \le L - 1.$$

Show that the sum of Hamming windows that overlap by 50% is *not* constant over the overlapping range, but rather,

$$s_L[n] = \begin{cases} 1.16, & n = 0 \\ 1.08, & 1 \le n \le D - 1 \end{cases}.$$

(c) A modified symmetric Hamming window of length L (**Figure 14.32c**) is created by dividing the endpoints of the standard Hamming window by two,

$$w_L[n] = \begin{cases} 0.04, & n = 0 \\ 0.54 - 0.46 \cos \dfrac{2\pi n}{L-1}, & 1 \le n \le L - 2 \\ 0.04, & n = L - 1 \end{cases}.$$

Show that the sum of modified Hamming windows that overlap by 50% *is* a constant over the overlapping range,

$$s_L[n] = 1.08, \quad 0 \le n \le D - 1.$$

Problem 14-2

Given a random process comprising a cosine with random phase plus white noise,

$$x[n] = A \cos(\omega_0 n + \theta) + v[n],$$

where the phase of the cosine, θ, is uniformly distributed between $-\pi \le \theta < \pi$,

$$p_\Theta(\theta) = \frac{1}{2\pi}, \qquad -\pi \le \theta < \pi,$$

and $v[n]$ is a zero-mean, white-noise process with a variance σ_V^2 that is uncorrelated with the cosine:

(a) show that the mean of the process is $E(x[n]) = 0$.

(b) show that the autocorrelation function is

$$r_{XX}[l] = \frac{A^2}{2} \cos \omega_0 l + \sigma_V^2 \delta[l].$$

(c) show that the power density spectrum is

$$S_{XX}(\omega) = \frac{A^2 \pi}{2} (\delta(\omega - \omega_0) + \delta(\omega + \omega_0)) + \sigma_V^2.$$

Problem 14-3

Given a random process comprising the sum of two cosines with random phase plus white noise,

$$x[n] = A_1 \cos(\omega_1 n + \theta_1) + A_2 \cos(\omega_2 n + \theta_2) + v[n],$$

where the phase of the cosine, θ, is uniformly distributed between $-\pi \le \theta < \pi$,

$$p_\Theta(\theta) = \frac{1}{2\pi}, \quad -\pi \le \theta < \pi,$$

and $v[n]$ is a zero-mean, white-noise process with variance σ_V^2 that is uncorrelated with the cosine:

(a) show that the mean of the process is $E(x[n]) = 0$.

(b) show that the autocorrelation function is

$$r_{XX}[l] = \frac{A_1^2}{2} \cos \omega_1 l + \frac{A_2^2}{2} \cos \omega_2 l + \sigma_V^2 \delta[l].$$

(c) show that the power density spectrum is

$$S_{XX}(\omega) = \frac{A_1^2 \pi}{2}(\delta(\omega - \omega_1) + \delta(\omega + \omega_1)) + \frac{A_2^2 \pi}{2}(\delta(\omega - \omega_2) + \delta(\omega + \omega_2)) + \sigma_V^2.$$

Problem 14-4

(a) Show that the expectation of the averaged periodogram, $E(P_B(\omega, L, M))$, defined by Equation (14.21) is the same as the expectation of the periodogram of one frame, namely,

$$E(P_B(\omega, L, M)) = E(\hat{S}_{XX}(\omega, L)),$$

so that the averaged periodogram is asymptotically unbiased.

(b) Show that the variance of the averaged periodogram tends to zero as $M \to \infty$,

$$\text{var}(P_B(\omega, L, M)) = \frac{1}{M} \text{var}(\hat{S}_{XX}(\omega, L)).$$

Problem 14-5

The general parametric ARMA model of Equation (14.29) can be expressed as

$$\sum_{i=0}^{N} a_N[i] y[n-i] + \sum_{i=0}^{M} b_M[i] x[n-i],$$

with $a_N[0]=1$. Show that the power spectral density of the output is given by Equation (14.32),

$$S_{YY}(\omega)=S_{XX}(\omega)\left|\frac{B_M(\omega)}{A_N(\omega)}\right|^2,$$

where $B_M(\omega)$ and $A_N(\omega)$ are the transforms of the model coefficients $b_M[n]$ and $a_N[n]$, respectively. It might be useful to break the problem into two parts, as suggested by **Figure 14.33**. Find the relation between $S_{XX}(\omega)$ and $S_{WW}(\omega)$ and between $S_{WW}(\omega)$ and $S_{YY}(\omega)$.

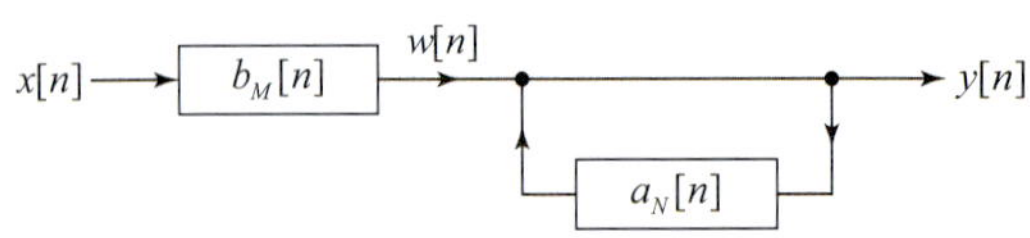

Figure 14.33 Schematic of ARMA model

Problem 14-6

A reversal matrix, $\mathbf{J}$, is a row/column-reversed version of the identity matrix that has a value of one along the anti-diagonal,

$$\mathbf{J}=\begin{bmatrix} 0 & 0 & \cdots & 0 & 1 \\ 0 & 0 & \cdots & 1 & 0 \\ \vdots & \vdots & \ddots & \vdots & \vdots \\ 0 & 1 & \cdots & 0 & 0 \\ 1 & 0 & \cdots & 0 & 0 \end{bmatrix}.$$

(a) If $\mathbf{J}$ and $\mathbf{\Gamma}_N$ are $N\times N$ matrices, what is the result of the matrix multiplication $\mathbf{\Gamma}_N\mathbf{J}$?
(b) What is the result of $\mathbf{J}\mathbf{\Gamma}_N$?
(c) If a_N is an $N\times 1$ column vector, what is the result of $\mathbf{J}\mathbf{a}_N$?

Problem 14-7

Show that if $\mathbf{\Gamma}_N$ is a symmetric Toeplitz matrix, the solution of the Yule–Walker equations to the order-reversed vector of correlation coefficients $\tilde{\mathbf{r}}_N$ is the order-reversed vector of model parameters $\tilde{\mathbf{a}}_N$. That is, show that if $\mathbf{\Gamma}_N\mathbf{a}_N=-\mathbf{r}_N$, then $\mathbf{\Gamma}_N\tilde{\mathbf{a}}_N=-\tilde{\mathbf{r}}_N$.

➤**Hint:** Use the results of Problem 14-6.

Problem 14-8

Equation (14.68) shows the results of applying the definition of the z-transform of $a_N[n]$, Equation (14.67), to the pair of equations in Equation (14.66). Using this definition,

(a) show that

$$\sum_{i=1}^{N-1} a_N[i]z^{-i} = A_N(z) - 1 - k_N z^{-N}.$$

(b) show that the z-transform of $a_N[i] = a_{N-1}[i] + k_N a_{N-1}[N-i]$ is $A_N(z) = A_{N-1}(z) + k_N z^{-N} A_{N-1}(z^{-1})$.

➤Hint: Multiply both sides of the equation by z^{-i}, sum over $1 \le i \le N-1$ and use the results of part (a).

(c) show that

$$\sum_{i=1}^{N-1} a_N[N-i]z^{-i} = z^{-N}(A_N(z^{-1}) - 1) - k_N.$$

(d) Use the results of parts (a) and (b) to show that the z-transform of $a_N[N-i] = a_{N-1}[N-i] + k_N a_{N-1}[i]$ is $z^{-N}A_N(z^{-1}) = z^{-N}A_{N-1}(z^{-1}) + k_N A_{N-1}(z)$.

Problem 14-9

Show that the error sequence $e_N[n]$ is orthogonal to $y[n-k]$, $1 \le k \le N$. That is, show that $E(e_N[n]y[n-k]) = 0$, $1 \le k \le N$.

➤Hint: Remember Equation (14.42).

A. Linear algebra

There are multiple topics in this book – convolution, FIR filter design, DFT, FFT, DCT, MDCT, STFT, AR modeling, Levinson–Durbin algorithm – in which matrix formulations provide an economical and powerful way of representing and solving problems. This appendix provides a brief summary of some of the important concepts of linear algebra.

A.1 Systems of linear equations

One of the most effective uses of matrices is in the solution of systems of linear equations. Consider the linear transformation of a set of M input variables x_k, $1 \leq k \leq M$, to a set of N output variables y_k, $1 \leq k \leq N$. This transformation can be expressed as a system of N linear equations,

$$\begin{aligned} y_1 &= a_{11}x_1 + a_{12}x_2 + \cdots + a_{1M}x_M \\ y_2 &= a_{21}x_1 + a_{22}x_2 + \cdots + a_{2M}x_M \\ &\vdots \\ y_N &= a_{N1}x_1 + a_{N2}x_2 + \cdots + a_{NM}x_M, \end{aligned}$$

where a_{nm} are the constant coefficients that define the transformation from x to y. This system of equations can be compactly expressed in matrix form as

$$\underbrace{\begin{bmatrix} y_1 \\ y_2 \\ \vdots \\ y_N \end{bmatrix}}_{\mathbf{y}} = \underbrace{\begin{bmatrix} a_{11} & a_{12} & \cdots & a_{1M} \\ a_{21} & a_{22} & \cdots & a_{2M} \\ \vdots & \vdots & \ddots & \vdots \\ a_{N1} & a_{N2} & \cdots & a_{NM} \end{bmatrix}}_{\mathbf{A}} \underbrace{\begin{bmatrix} x_1 \\ x_2 \\ \vdots \\ x_M \end{bmatrix}}_{\mathbf{x}},$$

or more compactly as

$$\mathbf{y} = \mathbf{A}\mathbf{x}, \tag{A.1}$$

where

$\mathbf{y}$ is an $N \times 1$ vector of values;
$\mathbf{x}$ is an $M \times 1$ vector of values;
$\mathbf{A}$ is an $N \times M$ **transformation matrix** of coefficients.

If the input **x** and coefficients of the transformation matrix **A** are known, determining **y** is a simple matter of matrix multiplication. However, frequently we need to find the value or values of **x** that satisfy Equation (A.1) for known values of **y** and **A**. When $N > M$, there are more equations than unknowns and the system is said to be **overconstrained** or **overdetermined**, and no exact solution (i.e., values of **x**) can be found. If $N < M$, the system is **underconstrained** or **underdetermined**. We will deal with those two cases later. However, when $N = M$, the system is **fully determined**, "just right," like the Goldilocks fairy tale, in that an exact solution (or solutions) to the equations can be obtained. We will now investigate the solution of Equation (A.1) in two cases of fully determined systems. When $\mathbf{y} \neq 0$, the system of equations is termed **inhomogeneous**. When $\mathbf{y} = 0$, the system of equations is termed **homogeneous**.

A.2 Solution of an inhomogeneous system of equations

Two of the classic methods for solving a system of linear equations are **Gaussian elimination** and its refinement, **Gauss–Jordan elimination**, which we will describe here. These methods are easily applied when the system is expressed in matrix form. The first step is to form an **augmented matrix** by appending the $N \times 1$ column vector **y** to the $N \times N$ coefficient matrix **A** to form an $N \times (N+1)$ matrix that encapsulates both the left and right sides of the linear equations. A vertical line is generally drawn to indicate the partition between **A** and **y**:

$$[\mathbf{A}|\mathbf{y}] = \left[\begin{array}{cccc|c} a_{11} & a_{12} & \cdots & a_{1N} & y_1 \\ a_{21} & a_{22} & \cdots & a_{2N} & y_2 \\ \vdots & \vdots & \ddots & \vdots & \vdots \\ a_{N1} & a_{N2} & \cdots & a_{NN} & y_N \end{array}\right].$$

There are three possible outcomes in the solution of an inhomogeneous system of equations:

(1) there is one unique solution;
(2) there is an infinite number of solutions;
(3) there is no solution.

We will demonstrate each of these cases with an example.

A.2.1 Unique solution

Consider the following system of equations:

$$\begin{aligned} -x_1 + x_2 - 3x_3 &= -6 \\ 2x_1 + 3x_2 + x_3 &= 2 \\ 3x_1 + x_2 - 2x_3 &= 3. \end{aligned}$$

In matrix notation this is

$$\underbrace{\begin{bmatrix} -1 & 1 & -3 \\ 2 & 3 & 1 \\ 3 & 1 & -2 \end{bmatrix}}_{\mathbf{A}} \underbrace{\begin{bmatrix} x_1 \\ x_2 \\ x_3 \end{bmatrix}}_{\mathbf{x}} = \underbrace{\begin{bmatrix} -6 \\ 2 \\ 3 \end{bmatrix}}_{\mathbf{y}},$$

and the augmented matrix is

$$\left[\begin{array}{ccc|c} -1 & 1 & -3 & -6 \\ 2 & 3 & 1 & 2 \\ 3 & 1 & -2 & 3 \end{array}\right].$$

The Gauss–Jordan elimination process uses **elementary row operations** to transform the augmented matrix into a form – termed **reduced row-echelon** or **canonical** form – from which the result (i.e., $\mathbf{x}$) can be read by inspection. The following three operations can be applied to the system of equations (or, equivalently, to the augmented matrix) without changing the solution:

(1) multiplying any row by a non-zero constant;
(2) adding any two rows together. We can combine this operation with the preceding one, adding a scaled multiple of any row to another row;
(3) interchanging any two rows.

The Gauss–Jordan elimination process uses these elementary row operations to transform the augmented matrix into reduced row-echelon form. In reduced row-echelon form,

(1) the leading entry of each row, defined as the left-most non-zero entry in that row, is in a column to the left of the leading entry of the row below it;
(2) each row has been normalized so that its leading entry is one;
(3) any rows that have all zero elements are at the bottom of the matrix.

To put our example into reduced row-echelon form, we start with the first row and scale it by -1 so that the leading entry is one:

$$\left[\begin{array}{ccc|c} -1 & 1 & -3 & -6 \\ 2 & 3 & 1 & 2 \\ 3 & 1 & -2 & 3 \end{array}\right] \xrightarrow{-1} \left[\begin{array}{ccc|c} 1 & -1 & 3 & 6 \\ 2 & 3 & 1 & 2 \\ 3 & 1 & -2 & 3 \end{array}\right]$$

Add the top row to the second and third rows, scaled by -2 and -3 respectively:

$$\left[\begin{array}{ccc|c} 1 & -1 & 3 & 6 \\ 2 & 3 & 1 & 2 \\ 3 & 1 & -2 & 3 \end{array}\right] \xrightarrow{-2 \times \text{row 1} \,+\, \text{row 2}} \left[\begin{array}{ccc|c} 1 & -1 & 3 & 6 \\ 0 & 5 & -5 & -10 \\ 3 & 1 & -2 & 3 \end{array}\right]$$

$$\left[\begin{array}{ccc|c} 1 & -1 & 3 & 6 \\ 0 & 5 & -5 & -10 \\ 3 & 1 & -2 & 3 \end{array}\right] \xrightarrow{-3 \times \text{row 1} \,+\, \text{row 3}} \left[\begin{array}{ccc|c} 1 & -1 & 3 & 6 \\ 0 & 5 & -5 & -10 \\ 0 & 4 & -11 & -15 \end{array}\right]$$

This "clears" the first (left-most) column so that there is only one non-zero entry in that column, and that is in the first row. We proceed in an analogous fashion with the second row. Normalize this row by its leading entry (i.e., divide by 5):

$$\left[\begin{array}{ccc|c} 1 & -1 & 3 & 6 \\ 0 & 1 & -1 & -2 \\ 0 & 4 & -11 & -15 \end{array}\right].$$

Add this row to the first and third rows, scaled by 1 and -4 respectively, which clears the second column:

$$\left[\begin{array}{ccc|c} 1 & 0 & 2 & 4 \\ 0 & 1 & -1 & -2 \\ 0 & 0 & -7 & -7 \end{array}\right].$$

Normalize the third row by its leading entry, -7,

$$\left[\begin{array}{ccc|c} 1 & 0 & 2 & 4 \\ 0 & 1 & -1 & -2 \\ 0 & 0 & 1 & 1 \end{array}\right],$$

and add appropriately scaled replicas to the first two rows to clear the final column. The result is in reduced row-echelon form:

$$\left[\begin{array}{ccc|c} 1 & 0 & 0 & 2 \\ 0 & 1 & 0 & -1 \\ 0 & 0 & 1 & 1 \end{array}\right].$$

This augmented matrix corresponds to the system of linear equations

$$\begin{aligned} 1x_1 + 0x_2 + 0x_3 &= 2 \\ 0x_1 + 1x_2 + 0x_3 &= -1 \\ 0x_1 + 0x_2 + 1x_3 &= 1, \end{aligned}$$

so,

$$\begin{aligned} x_1 &= 2 \\ x_2 &= -1 \\ x_3 &= 1, \end{aligned}$$

and we are done.

When the system of linear equations has a unique solution, the row-echelon procedure reduces the augmented matrix $[\mathbf{A} \mid \mathbf{y}]$ to the form $[\mathbf{I} \mid \mathbf{x}]$; with an $N \times N$ identity matrix to the left of the partition and a non-zero $N \times 1$ vector to the right. Another way to find $\mathbf{x}$ in this case is to find the inverse of the coefficient matrix and then compute $\mathbf{x} = \mathbf{A}^{-1}\mathbf{y}$. In fact, applying the Gauss–Jordan method to the augmented matrix is exactly equivalent to multiplying this matrix by $\mathbf{A}^{-1}$,

$$\mathbf{A}^{-1}[\mathbf{A} \mid \mathbf{y}] = [\mathbf{A}^{-1}\mathbf{A} \mid \mathbf{A}^{-1}\mathbf{y}] = [\mathbf{I} \mid \mathbf{x}],$$

except that we never have to calculate the inverse explicitly. Another test that a system of linear equations has a unique solution is that the inverse $\mathbf{A}^{-1}$ exists, which requires that the determinant be non-zero, $|\mathbf{A}| \neq 0$. This is equivalent to stating that all the equations are linearly independent. The number of linearly independent equations of a matrix is termed the **rank** of a matrix, and is simply the number of non-zero rows of a matrix when it has been reduced to canonical form. Our example, above, has rank three. A square matrix with only non-zero rows is said to be of **full rank**. These matrices are non-singular and invertible, as we saw in this section. For a system defined by a square coefficient matrix of full rank, $\mathbf{A}$, the Matlab command

```
x=A\y
```

provides the exact solution of $\mathbf{Ax} = \mathbf{y}$.

A.2.2 Infinite number of solutions

Not every system of linear equations has a unique solution. Consider this example:

$$\begin{aligned} x_1 + 2x_2 - x_3 &= 1 \\ 3x_1 + 7x_2 - 6x_3 &= 2 \\ 2x_1 + 2x_2 + 4x_3 &= 4. \end{aligned}$$

Express these equations as an augmented matrix and then massage into reduced row-echelon form:

$$\left[\begin{array}{ccc|c} 1 & 2 & -1 & 1 \\ 3 & 7 & -6 & 2 \\ 2 & 2 & 4 & 4 \end{array}\right] \longrightarrow \left[\begin{array}{ccc|c} 1 & 2 & -1 & 1 \\ 0 & 1 & -3 & -1 \\ 0 & -2 & 6 & 2 \end{array}\right] \longrightarrow \left[\begin{array}{ccc|c} 1 & 0 & 5 & 3 \\ 0 & 1 & -3 & -1 \\ 0 & 0 & 0 & 0 \end{array}\right].$$

The last row of the final matrix becomes all zeros because the second and third rows of the middle matrix are identical, except for a scale factor of −2. An all-zero row indicates that this system of equations is not linearly independent and that any one of the equations can be derived from the other two. You can work out the details yourself, but by reducing the augmented matrix to row echelon form, we have effectively shown that the bottom equation is eight times the top equation minus two times the middle equation. Therefore, our original system of three equations in three unknowns is actually equivalent to a system of two linearly independent equations in three unknowns – that is, it is an underconstrained system:

$$\begin{aligned} x_1 - 5x_3 &= 3 \\ x_2 - 3x_3 &= -1. \end{aligned}$$

This system has an infinite number of solutions. If any one of the variables is specified, then the others can be determined in terms of that variable. For example, if x_3 is specified, then,

$$\begin{aligned} x_1 &= 5x_3 + 3 \\ x_2 &= 3x_3 - 1. \end{aligned}$$

In this example, the matrix has rank two, since there was one row of zeros at the bottom of the matrix. Consider another example, this one with four equations in four unknowns:

$$\begin{aligned} x_1 + 2x_2 + 3x_3 - x_4 &= -1 \\ 4x_1 + 7x_2 + 9x_3 - 2x_4 &= -1 \\ 2x_1 + 2x_2 + 2x_4 &= 4 \\ 3x_1 + 2x_2 - 3x_3 + 5x_4 &= 9. \end{aligned}$$

The reduced row-echelon form has only two non-zero rows, which indicates that there are only two linearly independent equations:

$$\left[\begin{array}{cccc|c} 1 & 2 & 3 & -1 & -1 \\ 4 & 7 & 9 & -2 & -1 \\ 2 & 2 & 0 & 2 & 4 \\ 3 & 2 & -3 & 5 & 9 \end{array}\right] \longrightarrow \left[\begin{array}{cccc|c} 1 & 2 & 3 & -1 & -1 \\ 0 & -1 & -3 & 2 & 3 \\ 0 & -2 & -6 & 4 & 6 \\ 0 & -4 & -12 & 8 & 12 \end{array}\right] \longrightarrow \left[\begin{array}{cccc|c} 1 & 0 & -3 & 3 & 5 \\ 0 & 1 & 3 & -2 & -3 \\ 0 & 0 & 0 & 0 & 0 \\ 0 & 0 & 0 & 0 & 0 \end{array}\right]$$

While is not possible to find a unique solution to an underconstrained system, it is often possible to find an "optimal" solution to a set of equations that matches a given criterion. We will have more to say about underconstrained systems below.

A.2.3 No solution

A system of equations can have no solution. For example:

$$\begin{aligned} x_1 + 2x_2 - x_3 &= 1 \\ 3x_1 + 7x_2 - 6x_3 &= 2 \\ 2x_1 + 2x_2 + 4x_3 &= 5. \end{aligned}$$

Converting to reduced row-echelon form yields

$$\left[\begin{array}{ccc|c} 1 & 2 & -1 & 1 \\ 3 & 7 & -6 & 2 \\ 2 & 2 & 4 & 5 \end{array}\right] \longrightarrow \left[\begin{array}{ccc|c} 1 & 0 & 5 & 3 \\ 0 & 1 & -3 & -1 \\ 0 & 0 & 0 & 1 \end{array}\right] \text{ Oops!}$$

The bottom row states the impossible condition that

$$0x_1 + 0x_2 + 0x_3 = 1,$$

or simply that $0 = 1$. This set of equations is insoluble and is termed **inconsistent**. The indication of inconsistency is that the last row of a matrix in reduced row-echelon form is all zeros except for the final element, which is non-zero.

A.3 Solution of a homogeneous system of equations

A homogeneous system of linear equations is described by the matrix equation

$$\mathbf{Ax} = \mathbf{0},$$

where $\mathbf{0}$ is a $N \times 1$ vector of zeros. This could be viewed as a special case of the inhomongeneous equation with $\mathbf{y} = \mathbf{0}$. However, there are only two possible outcomes for a homogeneous system of equations:

(1) there is one unique solution, $\mathbf{x} = \mathbf{0}$, termed the **trivial solution**;
(2) there is an infinite number of solutions.

A.3.1 Trivial solution

A homogeneous system of equations *always* has at least one solution, the trivial solution $\mathbf{x} = \mathbf{0}$, since $\mathbf{Ax} = \mathbf{A0} = \mathbf{0}$. This solution is obviously independent of $\mathbf{A}$. Consider this example:

$$\begin{aligned} x_1 + 3x_2 + 2x_3 &= 0 \\ 2x_1 + 4x_2 + 6x_3 &= 0 \\ -x_1 + x_2 - x_3 &= 0. \end{aligned}$$

The augmented matrix looks like $[\mathbf{A} \mid \mathbf{0}]$ and the reduced row-echelon procedure yields the following:

$$\left[\begin{array}{ccc|c} 1 & 3 & 2 & 0 \\ 2 & 4 & 6 & 0 \\ -1 & 1 & -1 & 0 \end{array}\right] \longrightarrow \left[\begin{array}{ccc|c} 1 & 3 & 2 & 0 \\ 0 & -2 & 2 & 0 \\ 0 & 4 & 1 & 0 \end{array}\right] \longrightarrow \left[\begin{array}{ccc|c} 1 & 0 & 5 & 0 \\ 0 & 1 & -1 & 0 \\ 0 & 0 & 5 & 0 \end{array}\right] \longrightarrow \left[\begin{array}{ccc|c} 1 & 0 & 5 & 0 \\ 0 & 1 & -1 & 0 \\ 0 & 0 & 1 & 0 \end{array}\right] \longrightarrow \left[\begin{array}{ccc|c} 1 & 0 & 0 & 0 \\ 0 & 1 & 0 & 0 \\ 0 & 0 & 1 & 0 \end{array}\right].$$

In other words, the solution is $x_1 = x_2 = x_3 = 0$. The end result can be expressed as $[\mathbf{I} \mid \mathbf{0}]$ and is the indication that this system of homogeneous equations has only a trivial solution. As we saw before, applying the row-reduction procedure to the augmented matrix $[\mathbf{A} \mid \mathbf{0}]$ is equivalent to multiplying this matrix by the inverse of the coefficient matrix, $\mathbf{A}^{-1}$:

$$\mathbf{A}^{-1}[\mathbf{A} \mid \mathbf{0}] = [\mathbf{A}^{-1}\mathbf{A} \mid \mathbf{A}^{-1}\mathbf{0}] = [\mathbf{I} \mid \mathbf{0}].$$

Stated another way, if the inverse exists (which means that the determinant of $\mathbf{A}$ is non-zero, $|\mathbf{A}| \neq 0$), then the trivial solution is the unique solution. A non-zero determinant also means that the equations are linearly independent, so we can also state that if the equations are linearly independent, only the trivial solution exists.

A.3.2 Infinite number of solutions

If the equations are linearly dependent, that is, if the determinant of $\mathbf{A}$ is zero (i.e., $|\mathbf{A}| = 0$), then there is an infinite number of solutions to the homogeneous system of equations. For example,

$$\begin{aligned} x_1 + 3x_2 + x_3 &= 0 \\ -x_1 - x_2 + x_3 &= 0 \\ 2x_1 + 4x_2 + 2x_3 &= 0 \end{aligned}$$

gives

$$\left[\begin{array}{ccc|c} 1 & 3 & 3 & 0 \\ -1 & -1 & 1 & 0 \\ 2 & 4 & 2 & 0 \end{array}\right] \longrightarrow \left[\begin{array}{ccc|c} 1 & 3 & 3 & 0 \\ 0 & 2 & 4 & 0 \\ 0 & -2 & -4 & 0 \end{array}\right] \longrightarrow \left[\begin{array}{ccc|c} 1 & 0 & -3 & 0 \\ 0 & 1 & 2 & 0 \\ 0 & 0 & 0 & 0 \end{array}\right].$$

The rank is two, so there are only two independent equations in three unknowns, and hence there is an infinite number of solutions. The augmented matrix corresponds to two equations:

$$\begin{aligned} x_1 - 3x_3 &= 0 \\ x_2 + 2x_3 &= 0. \end{aligned}$$

Hence, given a value of one of the variables, one can find values of the other two. For example, if we set $x_3 = 1$, then $x_1 = 3$ and $x_2 = -2$.

A.4 Least-square-error optimization

In Section A.1, we said that a system defined by N linear equations in M unknown parameters, where $N > M$, is said to be overconstrained. A classic example of this occurs if we wanted to fit a straight line to a data set. In this case, there are $M = 2$ parameters to be determined, the slope m and intercept b of the line. If the data set has only two points (x_1, y_1) and (x_2, y_2), then clearly we can find an exact solution to the equation of the line. There are $N = 2$ equations to solve:

$$\begin{aligned} y_1 &= mx_1 + b \\ y_2 &= mx_2 + b, \end{aligned}$$

or, in matrix form,

$$\begin{bmatrix} y_1 \\ y_2 \end{bmatrix} = \begin{bmatrix} x_1 & 1 \\ x_2 & 1 \end{bmatrix} \begin{bmatrix} m \\ b \end{bmatrix}.$$

As long as $x_1 \neq x_2$, the unique solution is

$$\begin{bmatrix} m \\ b \end{bmatrix} = \begin{bmatrix} x_1 & 1 \\ x_2 & 1 \end{bmatrix}^{-1} \begin{bmatrix} y_1 \\ y_2 \end{bmatrix} = \frac{1}{x_1 - x_2} \begin{bmatrix} 1 & -1 \\ -x_2 & x_1 \end{bmatrix} \begin{bmatrix} y_1 \\ y_2 \end{bmatrix} = \frac{1}{x_1 - x_2} \begin{bmatrix} y_1 - y_2 \\ x_1 y_2 - y_1 x_2 \end{bmatrix}.$$

However, if we want to fit a straight line to $N > 2$ data points, we have

$$\underbrace{\begin{bmatrix} y_1 \\ y_2 \\ \vdots \\ y_N \end{bmatrix}}_{\mathbf{y}} = \underbrace{\begin{bmatrix} x_1 & 1 \\ x_2 & 1 \\ \vdots & \vdots \\ x_N & 1 \end{bmatrix}}_{\mathbf{C}} \underbrace{\begin{bmatrix} m \\ b \end{bmatrix}}_{\mathbf{p}}.$$

This is an overdetermined system – the number of equations (rows) is greater than the number of unknowns (columns) – and an exact solution for the two parameters m and b cannot be determined. For an overdetermined system with N equations in M unknowns, we can cast the problem in the form $\mathbf{y} = \mathbf{Cp}$ where the coefficient matrix $\mathbf{C}$ is an $N \times M$ matrix of coefficients formed from the abscissa values of the data, $\mathbf{p}$ is the vector of parameters to be determined (i.e., m and b) and $\mathbf{y}$ is the $M \times 1$ vector of data to be fit. Although we cannot solve $\mathbf{y} = \mathbf{Cp}$ exactly for the parameter vector $\mathbf{p}$, we can find a value of $\mathbf{p}$ that minimizes some measure of the difference, or error, between $\mathbf{y}$ and $\mathbf{Cp}$,

$$\mathbf{e} = \mathbf{y} - \mathbf{Cp},$$

where $\mathbf{e}$ is an $N \times 1$ vector of values e_k, $1 \leq k \leq N$. A tractable approach to this problem is to find $\mathbf{p}$ that minimizes a scalar function of $\mathbf{e}$. The most common such function is the **square-error**, which is just the sum of the squares of the e_k's:

$$E = \sum_{k=1}^{N} e_k^2.$$

Expressing E in matrix form, and doing a bit of manipulation, gives

$$\begin{aligned} E &= \sum_{k=1}^{N} e_k^2 = \mathbf{e}^T\mathbf{e} = (\mathbf{y} - \mathbf{Cp})^T(\mathbf{y} - \mathbf{Cp}) \\ &= (\mathbf{y}^T - \mathbf{p}^T\mathbf{C}^T)(\mathbf{y} - \mathbf{Cp}) \\ &= \mathbf{y}^T\mathbf{y} - \mathbf{y}^T\mathbf{Cp} - \mathbf{p}^T\mathbf{C}^T\mathbf{y} + \mathbf{p}^T\mathbf{C}^T\mathbf{Cp} \\ &= \mathbf{y}^T\mathbf{y} - 2\mathbf{p}^T\mathbf{C}^T\mathbf{y} + \mathbf{p}^T\mathbf{C}^T\mathbf{Cp}. \end{aligned}$$

The last step comes from recognizing that $\mathbf{p}^T\mathbf{C}^T\mathbf{y} = \mathbf{y}^T\mathbf{Cp}$. That is because $\mathbf{p}^T\mathbf{C}^T\mathbf{y}$ is a scalar. It is the product of a $1 \times M$ vector ($\mathbf{p}^T$) by an $M \times 1$ vector ($\mathbf{C}^T\mathbf{y}$), which is itself formed from the product of an $M \times N$ matrix ($\mathbf{C}^T$) with an $N \times 1$ vector ($\mathbf{y}$). Since $\mathbf{p}^T\mathbf{C}^T\mathbf{y}$ is scalar, it is its own transpose: $(\mathbf{p}^T\mathbf{C}^T\mathbf{y})^T = \mathbf{y}^T\mathbf{Cp}$, from which the last step follows.

We want to minimize the square error E with respect to $\mathbf{p}$,

$$\min_{\mathbf{p}} (E) = \min_{\mathbf{p}} (\mathbf{y}^T\mathbf{y} - 2\mathbf{p}^T\mathbf{C}^T\mathbf{y} + \mathbf{p}^T\mathbf{C}^T\mathbf{Cp}).$$

It turns out that E is a non-negative quadratic function of $\mathbf{p}$, hence the value of $\mathbf{p}$ at which the square error is a minimum can be obtained by taking the derivative of E with respect to $\mathbf{p}$

and setting it to zero. E comprises three terms. The first term, $\mathbf{y}^T\mathbf{y}$, does not depend on $\mathbf{p}$ so its derivative with respect to $\mathbf{p}$ is zero. As we showed above, the second term, $-2\mathbf{p}^T\mathbf{C}^T\mathbf{y}$, is a scalar. To find its derivative with respect to $\mathbf{p}$, first let $\mathbf{b} \triangleq -2\mathbf{C}^T\mathbf{y}$. This is an $M \times 1$ vector with elements b_k, $1 \leq k \leq M$. Since $\mathbf{p}^T$ is a $1 \times M$ vector with elements x_k, $1 \leq k \leq M$, we can write

$$-2\mathbf{p}^T\mathbf{C}^T\mathbf{y} = \mathbf{p}^T(-2\mathbf{C}^T\mathbf{y}) = \mathbf{p}^T\mathbf{b} = \sum_{k=1}^{M} p_k b_k.$$

The derivative of this term with respect to $\mathbf{p}$ is a vector of partial derivatives with respect to p_k,

$$\frac{d}{d\mathbf{p}}(\mathbf{p}^T\mathbf{b}) = \left[\frac{\partial}{\partial p_1}(\mathbf{p}^T\mathbf{b}) \quad \frac{\partial}{\partial p_2}(\mathbf{p}^T\mathbf{b}) \quad \cdots \quad \frac{\partial}{\partial p_M}(\mathbf{p}^T\mathbf{b})\right]^T,$$

but the kth partial is

$$\frac{\partial}{\partial p_k}(\mathbf{p}^T\mathbf{b}) = \frac{\partial}{\partial p_k}\left(\sum_{k=1}^{M} p_k b_k\right) = b_k;$$

hence,

$$\frac{d}{d\mathbf{p}}(\mathbf{p}^T\mathbf{b}) = [b_1 \quad b_2 \quad \cdots \quad b_M]^T = \mathbf{b} = -2\mathbf{C}^T\mathbf{y}.$$

The third term of the error E is $\mathbf{p}^T\mathbf{C}^T\mathbf{C}\mathbf{p}$. Define $\mathbf{M} \triangleq \mathbf{C}^T\mathbf{C}$. Since $\mathbf{C}$ is an $N \times M$ matrix, $\mathbf{M}$ is a square symmetric $M \times M$ matrix with elements m_{ij}, $1 \leq i \leq M$, $1 \leq j \leq M$. Then, $\mathbf{Mp}$ is an $M \times 1$ column vector whose elements are

$$\mathbf{Mp} = \left[\sum_{j=1}^{M} m_{1j}p_j \quad \sum_{j=1}^{M} m_{2j}p_j \quad \cdots \quad \sum_{j=1}^{M} m_{Mj}p_j\right]^T,$$

and $\mathbf{p}^T\mathbf{C}^T\mathbf{C}\mathbf{p}$ is a scalar whose value is

$$\mathbf{p}^T\mathbf{C}^T\mathbf{C}\mathbf{p} = \mathbf{p}^T\mathbf{M}\mathbf{p} = \sum_{i=1}^{M} p_i \sum_{j=1}^{M} m_{ij}p_j.$$

The derivative of $\mathbf{p}^T\mathbf{C}^T\mathbf{C}\mathbf{p} = \mathbf{p}^T\mathbf{M}\mathbf{p}$ with respect to $\mathbf{p}$ is a vector of partial derivatives with respect to p_k, each of which is found using the chain rule,

$$\begin{aligned}\frac{\partial}{\partial p_k}(\mathbf{p}^T\mathbf{C}^T\mathbf{C}\mathbf{p}) &= \frac{\partial}{\partial p_k}\left[\sum_{i=1}^{M} p_i \sum_{j=1}^{M} m_{ij}p_j\right] = \sum_{i=1}^{M} p_i\left(\frac{\partial}{\partial p_k}\sum_{j=1}^{M} m_{ij}p_j\right) + \sum_{i=1}^{M}\frac{\partial p_i}{\partial p_k}\left[\sum_{j=1}^{M} m_{ij}p_j\right] \\ &= \sum_{i=1}^{M} p_i m_{ik} + \sum_{j=1}^{M} m_{kj}p_j = 2\sum_{j=1}^{M} m_{kj}p_j = 2\mathbf{Mp} = 2\mathbf{C}^T\mathbf{C}\mathbf{p}.\end{aligned}$$

The combining of the two sums in the last step follows because $\mathbf{M}$ is a symmetrical matrix, so that $m_{jk} = m_{kj}$. Putting it all together, the derivative of E is

$$\frac{dE}{d\mathbf{p}} = -2\mathbf{C}^T\mathbf{y} + 2\mathbf{C}^T\mathbf{C}\mathbf{p}.$$

Setting the derivative to zero gives a system of so-called **normal equations**, which can be solved for the value of $\mathbf{p}$ that minimizes the square error,

$$\mathbf{C}^T\mathbf{C}\mathbf{p} = \mathbf{C}^T\mathbf{y}. \quad \text{(A.2)}$$

The solution of the normal equations can be found directly, at least in theory, by multiplying both sides of Equation (A.2) by $(\mathbf{C}^T\mathbf{C})^{-1}$,

$$\mathbf{p} = (\mathbf{C}^T\mathbf{C})^{-1}\mathbf{C}^T\mathbf{y}. \quad \text{(A.3)}$$

The quantity $(\mathbf{C}^T\mathbf{C})^{-1}\mathbf{C}^T$ is termed the **pseudoinverse** of $\mathbf{C}$, and is given the symbol $\mathbf{C}^+$,

$$\mathbf{C}^+ \triangleq (\mathbf{C}^T\mathbf{C})^{-1}\mathbf{C}^T.$$

Hence, the solution of the normal equations reduces to calculating

$$\mathbf{p} = \mathbf{C}^+\mathbf{y}. \quad \text{(A.4)}$$

The pseudoinverse has the property that

$$\mathbf{C}^+\mathbf{C} = ((\mathbf{C}^T\mathbf{C})^{-1}\mathbf{C}^T)\mathbf{C} = (\mathbf{C}^T\mathbf{C})^{-1}(\mathbf{C}^T\mathbf{C}) = \mathbf{I}. \quad \text{(A.5)}$$

For the special case of a fully constrained system, where there are N equations in N unknowns, $\mathbf{C}$ will be a square $N \times N$ matrix. If this matrix is non-singular, it is invertible with inverse $\mathbf{C}^{-1}$. Then, by Equation (A.5), the pseudoinverse is identical to the regular inverse,

$$\mathbf{C}^+ = \mathbf{C}^{-1}.$$

and the solution to the normal equations, Equation (A.4), simply becomes

$$\mathbf{p} = \mathbf{C}^{-1}\mathbf{y}. \quad \text{(A.6)}$$

For these reasons, the pseudoinverse $\mathbf{C}^+$ can be viewed as a generalization of $\mathbf{C}^{-1}$ for overdetermined systems, which reduces to the regular inverse when the system is fully determined.

In practice, it turns out that the matrix $\mathbf{C}^T\mathbf{C}$ is frequently ill-conditioned, so that attempting to solve the normal equations directly using Equation (A.3) can be numerically unstable. However, there exist a number of matrix methods (e.g., **QR factorization**) that provide an efficient and numerically stable solution. The Matlab backslash operator ("\") does double duty, allowing one to solve both fully determined and overdetermined systems. For a fully determined system defined by the non-singular square $N \times N$ coefficient matrix $\mathbf{C}$, the Matlab command

```
p=C\y
```

provides the exact solution of Equation (A.6). For an overdetermined system defined by the $N \times M$ coefficient matrix $\mathbf{C}$, with $N > M$, the Matlab command `p=C\y` gives the least-square solution for $\mathbf{p}$. As an example, consider the fit of a straight line $y = mx + b$ to the following data set:

x	y
0	0.49
1	1.48
2	2.24
3	3.09
4	4.07
5	4.98

In matrix form, $\mathbf{y} = \mathbf{C}\mathbf{p}$:

$$\underbrace{\begin{bmatrix} 0.49 \\ 1.48 \\ 2.24 \\ 3.09 \\ 4.07 \\ 4.98 \end{bmatrix}}_{\mathbf{y}} = \underbrace{\begin{bmatrix} 0 & 1 \\ 1 & 1 \\ 2 & 1 \\ 3 & 1 \\ 4 & 1 \\ 5 & 1 \end{bmatrix}}_{\mathbf{C}} \underbrace{\begin{bmatrix} m \\ b \end{bmatrix}}_{\mathbf{p}}.$$

From Matlab, `p=C\y` gives

$$\mathbf{p} = \begin{bmatrix} m \\ b \end{bmatrix} = \begin{bmatrix} 0.888 \\ 0.506 \end{bmatrix},$$

and the square error is $E = (\mathbf{y} - \mathbf{C}\mathbf{p})^T(\mathbf{y} - \mathbf{C}\mathbf{p}) = 0.017$. The data (red dots) and the straight-line fit are plotted below in **Figure A.1**.

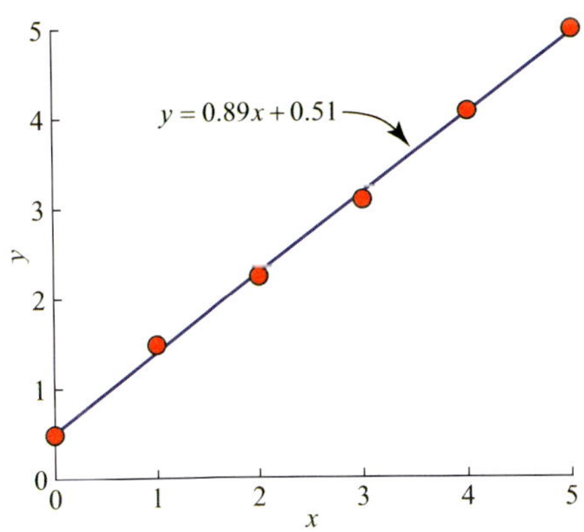

Figure A.1

B. Numeric representations

Practical signal processing involves translating algorithms into a form that can be implemented on a hardware platform, which might be a mainframe, a general-purpose microprocessor, an embedded processor, a special-purpose DSP processor or a custom VLSI solution. To implement algorithms effectively, a designer has to understand how numbers are represented in a given processor and the nature of the arithmetic that is used to operate on those numbers. In this appendix, we will give a brief introduction to how integers and non-integers are represented in hardware and software, including processors and A/D and D/A converters.

B.1 Integer representation

A **byte** is defined as eight binary bits. It is conventionally the smallest independently addressable element in computer memory. Memory words and registers of computers and microcontrollers are almost always multiples of eight-bit bytes in size, though hardware such as A/D and D/A converters commonly have word sizes ranging from 6 to 32 bits. A variety of representations are used in various devices to represent signed and unsigned numbers. In this section, we will discuss some of the most common ones.

B.1.1 Unsigned binary

An N-bit digital word can be used to represent 2^N integers. The **unsigned binary** or **unipolar binary** representation represents non-negative integers between 0 and 2^N-1. An unsigned N-bit integer binary number x can be written as

$$x = \sum_{k=0}^{N-1} b_k 2^k,$$

where each of the N bits b_k, $0 \le k < N$, has a value of either 0 or 1. For convenience, we can also write this number in a form that shows all the bits,

$$x \triangleq b_{N-1} b_{N-2} \cdots b_1 b_0,$$

where the left-most bit b_{N-1} is termed the **most significant bit (MSB)** and the right-most bit b_0 is the **least significant bit (LSB)**. The range of the unsigned binary representation is $0 \le x \le (2^N-1)$, as shown on the left side of **Table B.1** for $N=4$.

Table B.1 Unsigned, signed-magnitude, two's-complement and offset-binary integer numbers

Decimal	Unsigned binary
15	1111
14	1110
13	1101
12	1100
11	1011
10	1010
9	1001
8	1000
7	0111
6	0110
5	0101
4	0100
3	0011
2	0010
1	0001
0	0000

Decimal	Signed magnitude	Two's complement	Offset
7	0111	0111	1111
6	0110	0110	1110
5	0101	0101	1101
4	0100	0100	1100
3	0011	0011	1011
2	0010	0010	1010
1	0001	0001	1001
+0	0000	0000	1000
−0	1000	—	—
−1	1001	1111	0111
−2	1010	1110	0110
−3	1011	1101	0101
−4	1100	1100	0100
−5	1101	1011	0011
−6	1110	1010	0010
−7	1111	1001	0001
−8	—	1000	0000

Much of the time, we need to represent signed numbers: those that have both positive and negative values. Common representations of signed numbers are **signed magnitude**, **two's complement** and **offset binary**. (Another representation, **one's complement**, which was used in early computer systems, is much less commonly found today and will not be described.)

B.1.2 Signed-magnitude binary

In the signed-magnitude binary representation, the most significant bit indicates whether the number is positive (0) or negative (1). The remaining $N-1$ bits code the magnitude, interpreted as an unsigned binary number. This representation has a range of $-(2^{N-1}-1) \leq x \leq (2^{N-1}-1)$, as shown in **Table B.1** for $N=4$. Notice that this representation has a redundant representation of zero: both +0 (0000) and −0 (1000) exist.

B.1.3 Two's-complement binary

The most common hardware representation of signed numbers is two's-complement binary. As with the signed-magnitude representation, the most significant bit indicates whether the number is positive (0) or negative (1). Positive numbers range from $0 \leq x \leq (2^{N-1}-1)$, and are coded in the same way as signed magnitude and unsigned binary. Negative numbers range from $-2^{N-1} \leq x < 0$. They are calculated by taking the one's complement of the positive number (i.e., inverting each bit) and then adding one to the least significant bit, ignoring any carry-out. For example, 6_{10} is coded as 0110_2. The one's complement is 1001_2. Adding one to the least significant bit gives 1010_2, which is -6_{10}. The two's-complement representation has a unique value for zero (e.g., 0000). The maximum value is the same as that of the signed magnitude representation

$(2^{N-1}-1)$, but the minimum value is one lower (i.e., more negative), -2^{N-1}. An N-bit two's-complement number can be converted into an $(N+M)$-bit number, thereby increasing the range of numbers that can be represented by 2^M, by **sign extension**: appending M identical copies of the sign bit to the left of the binary string. So, the N-bit number $b_{N-1}b_{N-2}\cdots b_1b_0$ would be sign-extended to become the $(N+M)$-bit number

$$\underbrace{b_{N-1}b_{N-1}\cdots b_{N-1}}_{M\text{ bits}}\,\underbrace{b_{N-1}b_{N-2}\cdots b_1b_0}_{N\text{ bits}}.$$

For example, 6_{10} expressed as an eight-bit number would be 00000110_2, where the four leading zeros are extensions of the (0) sign bit. Similarly, -6_{10} would be converted to an eight-bit number by appending four 1s to the left-hand side, yielding 11111010_2. The corollary of sign extension would be reducing the range of representation by removing one or more identical sign-extension bits from the left of a number. So, $\cancel{000}010_2 = 010_2 = 2_{10}$ and $\cancel{111}110_2 = 110_2 = -2_{10}$. Be careful not to remove the sign bit itself!

A significant advantage of the two's-complement representation is that the arithmetic operations addition and subtraction are handled identically in the hardware. Subtraction of two operands x and y is just performed by adding the two's complement of y to x. For example, $5_{10}-7_{10}$ corresponds to $0101_2+1001_2=1110_2$, which is -2_{10}. Notice that $5_{10}+7_{10}$ corresponds to $0101_2+0111_2=1100_2$, which is -4_{10}. Hey wait a minute, that's not right! What has happened, of course, is that the sum overflows the capacity of the four-bit signed-binary representation. Similarly, $-5_{10}-7_{10}$ yields an underflow, $1011_2+1001_2=0100_2$, which is 4_{10}. Fortunately, overflow and underflow conditions are easily detected in two's-complement arithmetic hardware. One of the ways this is done is for the **arithmetic logic unit (ALU)**, the piece of hardware that does the addition, to sign-extend both operands by one bit to $N+1$ bits before performing addition, as shown in **Table B.2**.

Table B.2 Overflow and underflow detection in two's complement arithmetic

Decimal	N-bit binary	$(N+1)$-bit binary
$5_{10}-7_{10}$	$0101_2+1001_2=1110_2 \rightarrow -2_{10}$ ✓ O.K.	$00101_2+11001_2=\boxed{11}110_2 \rightarrow -2_{10}$; XOR → 0 ✓ O.K.
$5_{10}+7_{10}$	$0101_2+0111_2=1100_2 \rightarrow -4_{10}$ ✗ Overflow!	$00101_2+00111_2=\boxed{01}100_2$; XOR → 1 ✗ Error!
$-5_{10}-7_{10}$	$1011_2+1001_2=0100_2 \rightarrow 4_{10}$ ✗ Underflow!	$11011_2+11001_2=\boxed{10}100_2$; XOR → 1 ✗ Error!
$7_{10}-5_{10}$	$0111_2+1011_2=0010_2 \rightarrow 2_{10}$ ✓ O.K.	$00111_2+11011_2=\boxed{00}010_2 \rightarrow 2_{10}$; XOR → 0 ✓ O.K.

The first column shows the desired decimal operation, the second column is the result of a four-bit two's-complement addition without sign extension. Overflow and underflow occurs whenever the sum of the operands exceeds the signed four-bit range, which is greater than +7 (0111_2) or less than $-8(1000_2)$. Sign-extending the operands to $N+1$ bits, as shown in the right column of the figure, doubles the range of the representation and guarantees that the sum will not overflow. If the two most significant bits of the resulting addition are the same (either 00 or 11), we can safely truncate the result back to the desired N bits without overflow or underflow. However, if the two most significant bits are not the same (either 01 or 10), then the result cannot be truncated to N bits, and we know that overflow or underflow has occurred. Testing for an error condition can be done by a simple logical XOR operation of the two most significant bits of the sign-extended sum.

One important property of the two's-complement representation is that the two's-complement addition of a series of numbers will not result in an overflow as long as the final sum is within range, even if intermediate steps in the additions overflow. This is best explained with an example. Consider this sum: $5_{10}+7_{10}-8_{10}=4_{10}$. The summation proceeds in two steps, first $5_{10}+7_{10}$ are added together to form an overflowing intermediate result, -4_{10}. Then, -8_{10} is added to obtain the final result, 4_{10}. When expressed in four-bit binary, we get the following:

$$
\begin{array}{rcrl}
 & 00101_2 \longrightarrow & 5_{10} & \\
+ & 00111_2 \longrightarrow & 7_{10} & \\
\hline
 & \boxed{01}100_2 \longrightarrow & -4_{10} & \times\ \text{Intermediate result overflows!} \\
+ & 01000_2 \longrightarrow & -8_{10} & \\
\hline
 & \boxed{00}100_2 \longrightarrow & 4_{10} & \checkmark\ \text{Final result O.K.}
\end{array}
$$

Even though the first addition results in an overflow, the final result is in the range of the binary representation and is thus accurate.

B.1.4 Offset binary

The final integer binary representation of interest to us is offset (or biased) binary. This representation is similar to unsigned binary, except the numbers are offset by half the range, 2^{N-1},

$$x=\left(\sum_{k=0}^{N-1} b_k 2^k\right)-2^{N-1}.$$

So, the smallest binary number (0000_2) represents the most negative decimal number, -2^{N-1}, and the largest binary number (1111_2) represents the most positive decimal number, $(2^{N-1}-1)$. Therefore, the range of offset binary is the same as that of two's complement. Offset binary is used in coding the exponent of floating-point numbers, as we will see in Section B.3. It is also sometimes used in DSP hardware such as unipolar A/D and D/A converters, which measure or output positive voltages over the range $0-V_{REF}$ volts. As an example, one can use a unipolar A/D converter to measure bipolar signals that range from $-V_{REF}/2$ to $+V_{REF}/2$ volts by adding an offset of $+V_{REF}/2$ volts to the input. Then, an input voltage of $+V_{REF}/2$ volts is represented in offset binary by all ones ($11\ldots111_2$), an input voltage of 0 volts is represented by a 1 in the most significant bit followed by all zeros ($10\ldots000_2$) and an input voltage of $-V_{REF}/2$ volts is represented all zeros ($00\ldots000_2$).

B.1.5 Converting between binary formats

It is easy to convert between bipolar binary formats, with either hardware or software. The required operations are shown in **Table B.3**:

Table B.3 Conversion between bipolar binary representations

From \ To	Signed magnitude	Two's complement	Offset binary
Signed magnitude		If MSB = 1 Complement other bits and add 1.	If MSB = 0 Complement MSB else Complement all bits and add 1
Two's complement	If MSB = 1 Complement other bits and add 1		Complement MSB
Offset binary	If MSB = 1 Complement MSB else Complement all bits and add 1	Complement MSB	

B.2 Fixed-point (fractional) representation

There are a number of ways in which fractional numbers can be represented using N bits, as shown in **Table B.4**. All are essentially reinterpretations of the schemes we described for representing integers. A **fractional unsigned** N-bit binary number x is defined as

$$x = \sum_{k=-B}^{M-1} b_k 2^k = \left(\sum_{k=0}^{N-1} b_{k-B} 2^k \right) \Big/ 2^B,$$

where each of the N bits b_k has a value of 0 or 1. Here again, we can define a notation

$$x \triangleq \underbrace{b_{M-1} b_{M-2} \cdots b_1 b_0 \cdot b_{-1} b_{-2} \cdots b_{-B}}_{N \text{ bits}},$$

where the dot ($\cdot$) is called the **binary point**. The M bits to the left of the binary point are the integer part of x, formed from the weighted sum of positive powers of 2^k for $0 \le k \le M-1$. The $B = N-M$ bits to the right of the point are the fractional part, formed from the weighted sum of negative powers of 2^k for $-B \le k \le -1$. It is worth noting that there is no hardware "binary point"; it is just a useful notation to help us represent fractional numbers. It is up to the programmer or hardware designer to keep track of the binary point so they will know how to interpret results. For example, assume a hardware register holds four bits: 1011. If $B=0$, we interpret these same bits as an integer,

$$1011_2 = 1 \cdot 2^3 + 0 \cdot 2^2 + 1 \cdot 2^1 + 1 \cdot 2^0 = 11_{10},$$

whereas if $B = 2$, we interpret these same four bits as fractional unsigned binary,

$$10.11_2 = 1 \cdot 2^1 + 0 \cdot 2^0 + 1 \cdot 2^{-1} + 1 \cdot 2^{-2} = 2.75_{10}.$$

Fractional unsigned binary can just be viewed as integer unsigned binary divided by 2^B; namely $10.11_2 = 1011_2/2^2 = 11/4 = 2.75_{10}$. For a given choice of N and B, the range of numbers that can be expressed using the fractional unsigned format is between 0 and $(2^N - 1)/2^B = 2^{N-B} - 2^{-B} = 2^M - 2^{-B}$ with a resolution (i.e., the increment between successive numbers) determined by the least significant bit, 2^{-B}.

The left side of **Table B.4** shows an example with $N = 4$ and $B = 2$. The range is $0 \leq x \leq 3.75$, and the resolution is 0.25.

Table B.4 Unsigned, signed-magnitude and two's-complement fractional-binary numbers

Number	Unsigned binary
3.75	11.11
3.50	11.10
3.25	11.01
3.00	11.00
2.75	10.11
2.50	10.10
2.25	10.01
2.00	10.00
1.75	01.11
1.50	01.10
1.25	01.01
1.00	01.00
0.75	00.11
0.50	00.10
0.25	00.01
0.00	00.00

Number	Signed magnitude	Two's complement
1.75	01.11	01.11
1.50	01.10	01.10
1.25	01.01	01.01
1.00	01.00	01.00
0.75	00.11	00.11
0.50	00.10	00.10
0.25	00.01	00.01
0.00	00.00	00.00
−0.00	10.00	—
−0.25	10.01	11.11
−0.50	10.10	11.10
−0.75	10.11	11.01
−1.00	11.00	11.00
−1.25	11.01	10.11
−1.50	11.10	10.10
−1.75	11.11	10.01
−2.00	—	10.00

The two most common representations of signed fractional numbers are **fractional signed magnitude** and **fractional two's complement**, which can be simply understood as their integer equivalents described in Section B.1, divided by 2^B. In the fractional signed-magnitude representation, the most significant bit is again the sign bit and the remaining bits code the magnitude, interpreted as fractional unsigned binary. This representation therefore has a range of $-(2^{M-1} - 2^{-B}) \leq x \leq (2^{M-1} - 2^{-B})$. The right part of **Table B.4** shows an example of signed magnitude for $N = 4$ and $B = 2$. As with the integer case, there is a redundant representation of zero: both $+0.00_{10}(00.00_2)$ and $-0.00_{10}(10.00_2)$ exist.

Fractional two's-complement numbers range from the maximum positive value, $2^{M-1}-2^{-B}$, to the minimum negative value, -2^{M-1}, and the resolution is 2^{-B}. Negative numbers are formed by taking the one's complement of the positive number and adding one to the least significant bit (i.e., 2^{-B}). The range of fractional two's-complement numbers can be increased by sign extension: appending copies of the sign bit to the left of the number. The resolution of the representation (and to a small extent, the range) can be increased by adding zeros to the right of the number. For example, $-1.75_{10} = 10.01_2 = 1110.0100_2$. For a fixed word length (i.e., N), the range and resolution of a number trade off. Increasing the range by a factor of two means moving the binary point one digit to the right, which concomitantly reduces the resolution of representation by a factor of two.

B.2.1 Rounding and truncation

Only rational numbers that are integer multiples of 2^{-B} (called **dyadic numbers**) can be exactly represented in fractional binary with B bits of precision. For example, $1.4375_{10} = 23 \times 2^{-4}$ is exactly 01.0111_2 with $B = 4$ fractional bits of precision. All other numbers will lose precision when expressed in fractional binary through the processes of **rounding** and **truncation**. Understanding these processes is central to working with fixed- and floating-point numbers. Arithmetic logic units (e.g., multipliers) and floating point processors of computers do their calculations internally at high precision and deliver rounded results. In addition, as discussed in Chapter 9, the quantization of the coefficients of discrete-time filters through rounding and truncation can have a significant effect on the position of the filter's singularities (i.e., its poles and zeros).

There are numerous methods of rounding in use, both in software and in hardware. For example, the **IEEE-754 standard** implemented by many software and VHDL compilers specifies five different rounding methods. The precise effect of these methods depends on the way numbers are represented (e.g., unsigned, signed magnitude or two's complement).

Truncation is a common rounding method, and the easiest of all to implement in hardware. Truncation to B bits simply means that only B fractional bits are retained and the rest discarded. The effect of truncation depends on how fractional-binary numbers are represented. The left side of **Table B.5** shows the truncation of positive and negative four-bit signed-magnitude numbers to $B = 2$ bits. When positive numbers are truncated, they are effectively rounded down (i.e., towards 0, indicated by ⬇). For example, $1.4375_{10} = 01.0111_2$ becomes $01.01_2 = 1.25_{10}$, which produces a difference (error) of 0.1875 between the input and truncated output. When negative signed-magnitude numbers are truncated, they are effectively rounded up (i.e., towards 0, indicated by ⬆). For example, $-1.4375_{10} = 11.0111_2$ becomes $11.01_2 = -1.25_{10}$. So, truncation of signed-magnitude numbers is equivalent to rounding towards 0, and can be modeled by Matlab's `fix` function (specifically, `fix(x*2^B)/2^B`).

Table B.5 Truncation of signed-magnitude and two's-complement fractional-binary numbers with $B=2$

Signed-magnitude input	Truncated output Fix		Error
$1.4375_{10}=01.0111_2$	$01.01_2=1.25_{10}$	↓	0.1875
$1.3750_{10}=01.0110_2$	$01.01_2=1.25_{10}$	↓	0.1250
$1.3125_{10}=01.0101_2$	$01.01_2=1.25_{10}$	↓	0.0625
$1.2500_{10}=01.0100_2$	$01.01_2=1.25_{10}$	↓	0.0000
0	0		0
$-1.2500_{10}=11.0100_2$	$11.01_2=-1.25_{10}$	↑	−0.0000
$-1.3125_{10}=11.0101_2$	$11.01_2=-1.25_{10}$	↑	−0.0625
$-1.3750_{10}=11.0110_2$	$11.01_2=-1.25_{10}$	↑	−0.1250
$-1.4375_{10}=11.0111_2$	$11.01_2=-1.25_{10}$	↑	−0.1875
			$\Sigma=0$

Signed-magnitude input	Truncated output Fix		Error
$1.4375_{10}=01.0111_2$	$01.01_2=1.25_{10}$	↓	0.1875
$1.3750_{10}=01.0110_2$	$01.01_2=1.25_{10}$	↓	0.1250
$1.3125_{10}=01.0101_2$	$01.01_2=1.25_{10}$	↓	0.0625
$1.2500_{10}=01.0100_2$	$01.01_2=1.25_{10}$	↓	0.0000
0	0		0
$-1.2500_{10}=11.0100_2$	$11.01_2=-1.25_{10}$	↓	−0.0000
$-1.3125_{10}=11.0101_2$	$11.01_2=-1.25_{10}$	↓	−0.0625
$-1.3750_{10}=11.0110_2$	$11.01_2=-1.25_{10}$	↓	−0.1250
$-1.4375_{10}=11.0111_2$	$11.01_2=-1.25_{10}$	↓	−0.1875
			$\Sigma=0$

Truncation is an unbiased rounding method for signed-magnitude numbers, meaning that the truncation error averaged over all possible positive and negative inputs is zero.

The right side of **Table B.5** shows the truncation of two's-complement numbers to $B=2$ bits. Since positive two's-complement numbers are the same as positive signed-magnitude numbers, truncation acts in an identical manner: they are rounded down. However, when negative two's-complement numbers are truncated, they are also rounded down. For example, $-1.4375_{10}=10.1001_2$ becomes $10.10_2=-1.50_{10}$. In other words, truncation of two's-complement numbers is equivalent to rounding towards $-\infty$, which is described by Matlab's `floor` function (i.e., `floor(x*2^B)/2^B)`. Truncation is a biased rounding method for two's-complement numbers, since the average error for all possible inputs is non-zero.

Three of the most common methods of **rounding** are shown in **Table B.6** (for signed-magnitude representation) and **Table B.7** (for two's-complement representation). You will note that the results are independent of the representation.

Table B.6 Rounding modes for signed-magnitude numbers with $B=2$

Signed-magnitude input	Round ties to nearest even		Round ties away from zero		Hardware rounding	
	`convergent`		`round`		`round`	
$1.4375_{10}=01.0111_2$	$01.10_2=1.50_{10}$	↑	$01.10_2=1.50_{10}$	↑	$01.10_2=1.50_{10}$	↑
$1.3750_{10}=01.0110_2$	$01.10_2=1.50_{10}$	↑	$01.10_2=1.50_{10}$	↑	$01.10_2=1.50_{10}$	↑
$1.3125_{10}=01.0101_2$	$01.01_2=1.25_{10}$	↓	$01.01_2=1.25_{10}$	↓	$01.01_2=1.25_{10}$	↓
$1.1250_{10}=01.0010_2$	$01.00_2=1.00_{10}$	↓	$01.01_2=1.25_{10}$	↑	$01.01_2=1.25_{10}$	↑
0	0		0		0	
$-1.1250_{10}=11.0010_2$	$11.00_2=-1.00_{10}$	↑	$11.01_2=-1.25_{10}$	↓	$11.01_2=-1.25_{10}$	↓
$-1.3125_{10}=11.0101_2$	$11.01_2=-1.25_{10}$	↑	$11.01_2=-1.25_{10}$	↑	$11.01_2=-1.25_{10}$	↑
$-1.3750_{10}=11.0110_2$	$11.10_2=-1.50_{10}$	↓	$11.10_2=-1.50_{10}$	↓	$11.10_2=-1.50_{10}$	↓
$-1.4375_{10}=11.0111_2$	$11.10_2=-1.50_{10}$	↓	$11.10_2=-1.50_{10}$	↓	$11.10_2=-1.50_{10}$	↓

Table B.7 Rounding modes for two's-complement numbers with $B=2$

Two's-complement input	Round ties to nearest even		Round ties away from zero		Hardware rounding	
	`convergent`		`round`		`round`	
$1.4375_{10}=01.0111_2$	$01.10_2=1.50_{10}$	↑	$01.10_2=1.50_{10}$	↑	$01.10_2=1.50_{10}$	↑
$1.3750_{10}=01.0110_2$	$01.10_2=1.50_{10}$	↑	$01.10_2=1.50_{10}$	↑	$01.10_2=1.50_{10}$	↑
$1.3125_{10}=01.0101_2$	$01.01_2=1.25_{10}$	↓	$01.01_2=1.25_{10}$	↓	$01.01_2=1.25_{10}$	↓
$1.1250_{10}=01.0010_2$	$01.00_2=1.00_{10}$	↓	$01.01_2=1.25_{10}$	↑	$01.01_2=1.25_{10}$	↑
0	0		0		0	
$-1.1250_{10}=10.1110_2$	$11.00_2=-1.00_{10}$	↑	$11.01_2=-1.25_{10}$	↓	$11.01_2=-1.25_{10}$	↓
$-1.3125_{10}=10.1011_2$	$11.01_2=-1.25_{10}$	↑	$11.01_2=-1.25_{10}$	↑	$11.01_2=-1.25_{10}$	↑
$-1.3750_{10}=10.1010_2$	$11.10_2=-1.50_{10}$	↓	$11.10_2=-1.50_{10}$	↓	$11.10_2=-1.50_{10}$	↓
$-1.4375_{10}=10.1001_2$	$11.10_2=-1.50_{10}$	↓	$11.10_2=-1.50_{10}$	↓	$11.10_2=-1.50_{10}$	↓

"Round ties to nearest even" is equivalent to Matlab's `convergent` function and "round ties away from zero" is equivalent to Matlab's `round` function. Both are unbiased rounding methods (considered over the entire range of positive and negative numbers), though the first one is also unbiased for positive and negative numbers considered as separate subsets. The only difference between these methods is how they deal with "ties," namely values of the input exactly between two possible output values. For example, looking at **Table B.7** with $B=2$, the value $1.1250_{10}=01.0010_2$ is exactly halfway between the available output values $01.00_2=1.00_{10}$ and $01.01_2=1.25_{10}$. Since $1.1250_{10}=4.5/4$, rounding this "tie" to the nearest even gives an output of $1.00_{10}=4.0/4$; rounding away from zero gives $1.25_{10}=5.0/4$. In contrast, the input value $1.3750_{10}=01.0110_2$ is halfway between output values $01.01_2=1.25_{10}$ and $01.10_2=1.50_{10}$. Since $1.3750_{10}=5.5/4$, both rounding methods give an output of $1.50_{10}=6.0/4$.

B.2.2 Arithmetic of fractional numbers

The procedure for adding and subtracting fractional two's-complement numbers is identical to that described for integer binary numbers in Section B.1.3, as long as the range and resolution (i.e., the binary point) of the two operands are the same. If the range and resolution of the operands differ, they could be adjusted, at least in principle, to have the same size and resolution so they can be added exactly without a loss of either range or resolution. For example, consider two four-bit operands with the same size but different binary points,

$$2.5_{10}+0.875_{10}\rightarrow 010.1+0.111_2=010.100_2+000.111_2=011.011_2\rightarrow 3.375_{10}.$$

The first operand has $M=3$ and $B=1$; the second has $M=1$ and $B=3$. To add them exactly using fixed-point arithmetic, both have to be adjusted to have the same size and resolution, $M=3$ and $B=3$. Unfortunately, in practice word lengths are fixed, so the size and resolution of the operands cannot be arbitrarily adjusted. When adding numbers with different ranges and resolutions, something has to give. If we adjust the second operand to match the range of the first (by rotating it two bits to the right), it loses two bits of resolution. Instead of $0.111_2=0.875_{10}$, we have $000.1_2=0.5_{10}$ and the sum becomes

$$010.1+000.1_2=011.0_2\rightarrow 3_{10}.$$

If instead we adjust the first operand to match the resolution of the second (by rotating it two bits to the left), things are much worse. Instead of $010.1_2=2.5_{10}$, we have $0.100_2=0.5_{10}$ and the sum becomes

$$0.100_2+0.111_2=1.011_2\rightarrow 1.375_{10}.$$

B.3 Floating-point representation

The floating-point binary representation allows numbers to be represented in binary scientific notation,

$$-1^S\times M\times 2^E,$$

where the **sign bit** S can be 0 or 1, the **mantissa** M is a fractional unsigned binary number and the **exponent** E is an integer. Unsigned fractional-binary numbers can be economically represented in binary floating point in **normalized form**, as shown in the examples of **Table B.8**.

Table B.8 Unsigned-binary and floating-point numbers

Number	Unsigned-binary	Floating-point
6.00	110.000	1.1×2^2
3.00	011.000	1.1×2^1
1.50	001.100	1.1×2^0
0.75	000.110	1.1×2^{-1}
0.375	000.011	1.1×2^{-2}

Normalized means that in each case the exponent has been adjusted so that the mantissa is always of the form $1.xxx\cdots xxx$; that is, an integer part of 1 followed by the binary point followed by the fractional remainder of the mantissa.

The most common floating-point representation used in computers is the IEEE-754 standard, which comes in several flavors, the most common of which are single precision (32-bit) and double precision (64-bit). The bit allocation for single-precision is shown in **Figure B.1**.

Figure B.1 Bit allocation of IEEE-754 single-precision floating-point numbers

A floating-point word comprises three fields: a sign field (one bit), an exponent field (eight bits) and a mantissa field (23 bits). A sign bit of 0 denotes a positive number; 1 denotes a negative number. (In IEEE-754 double-precision floating-point, there is one sign bit, the exponent field is 11 bits and the mantissa field has 52 bits.) The integer exponent of single-precision numbers ranges from 0 to $2^8-1=255$. To allow both positive and negative exponents to be represented, the exponent field is coded in a variant of the integer offset-binary code we described in Section B.1.4. To obtain the decimal-equivalent exponent, subtract a bias of $2^{N-1}-1=127_{10}$. So the smallest exponent is $00000000_2 \rightarrow -127_{10}$ and the largest exponent is $11111111_2 \rightarrow +128_{10}$. The exponents -127_{10} and $+128_{10}$ are actually reserved for special numbers, as we will discuss in a moment, so the real range of exponents is -126_{10} to $+127_{10}$. Numbers in this range are termed **normalized** because the mantissa is interpreted as an unsigned fractional-binary number in normalized form; namely, $1.xxx\cdots xxx$. However, since the leading digit is always 1, the mantissa field only expressly encodes the 23 bits of the fractional part of the mantissa; the integer part (the leading 1) is implicit. Normalized floating-point numbers therefore have a full $23+1=24$ bits of precision. For example, 1.4_{10} expressed as a single-precision floating-point number is effectively truncated to 1.399993896484375_{10}. The resolution (that is, the difference between subsequent numbers) is not fixed but depends on the exponent: 2^{E-31}. The smallest expressible single-precision floating-point normalized numbers are $\pm 2^{-126}$ (when the mantissa is all zeros) and the largest are $\pm(2-2^{-23}) \times 2^{127}$ (when the mantissa is all ones). The range of normalized double-precision floating-point numbers is $\pm 2^{-1022}$ to $\pm(2-2^{-52}) \times 2^{1023}$.

When the exponent field is set to all zeros, the resulting floating-point numbers are referred to as **denormalized**. In this case the mantissa is interpreted as $0.xxx\cdots xxx$, that is, with a leading 0 instead of a leading 1. Also (confusingly), the exponent is set to -126_{10}. Thus, the smallest (non-zero) expressible single-precision floating-point denormalized numbers are $\pm 2^{-149}$ (mantissa is all zeros except for the least significant bit) and the largest are $\pm(1-2^{-23}) \times 2^{-126}$ (the mantissa is all ones). Double-precision denormalized numbers range from $\pm 2^{-1074}$ to $\pm(1-2^{-52}) \times 2^{-1022}$. However, the resolution of denormalized numbers gets smaller as the numbers get smaller. The number 0 is a denormalized number formed by setting the exponent and mantissa fields to 0. Both ± 0 can be represented.

The IEEE-754 standard allows some special numbers to be represented when the exponent field is set to all ones. Infinities ($\pm\infty$) are coded by setting the mantissa field to 0 and setting the sign bit appropriately. Arithmetic operations with infinity are well defined in the standard. For example, $1 \pm \infty = \pm\infty$ and $1/\pm\infty = \pm 0$. The standard also allows the floating-point processor (or software emulator) to return special codes when indeterminate or invalid operations are requested. **NaN** (Not a Number) is coded by setting the mantissa field to a non-zero value, where bit 22 is set to 1 if the operation is indeterminate (termed a "quiet *NaN*" or **qNaN**) and set to 0 if the operation is invalid (a "signaling *NaN*" or **sNaN**). A *qNaN* is generated by an arithmetic operation whose result is not mathematically defined, such as $\pm 0/\pm 0$ or $\pm\infty/\pm\infty$. Since arithmetic operations with *qNaN* are well defined (e.g., $NaN + 1 = NaN$), they do not raise an exception (i.e., they are quiet) and are propagated through to the output for the program to deal with. *sNaN*s result from operations, such as overflow and underflow, that need to be signaled immediately to the program by means of a raised exception.

Comparing floating-point to fixed-point for a same word size, the range of floating-point numbers is greater than that of fixed-point, though the precision is lower.

B.4 Computer representation of numbers

There are several "standard" versions of the C language that have evolved from the original versions developed at Bell Labs in the 1970s, of which the most common for general-purpose computers are ANSI 1989 (a.k.a. C89) and ISO 1999 (a.k.a. C99), though there are several more recent iterations of the language as well (C11 from 2011 and C18 from 2018). There are also numerous extensions to C for embedded processors which support fixed-point arithmetic, interrupts and I/O operations. The variable types for number representation fall into two categories: **integer** and **floating point**. The two basic integer types are `char` and `int`, to which the optional prequalifiers `short` (or `long`), `signed` (or `unsigned`) can be attached, as indicated in **Table B.9**. The common floating-point types are `float` and `double`.

Table B.9 Numerical data types in C

Data category	Data type		Minimum number of bits, N	Minimum value	Maximum value
Integer	`char`	`unsigned char`	depends on processor (usually $N = 8$)	0	$2^N - 1$ $= 255$
		`signed char`		$-2^{N-1} =$ -128	$2^{N-1} - 1 =$ 127

	`int`	`short`
		`short int`
		`signed short`
		`signed short int`
			16	$-2^{N-1} =$
				-32768
					$2^{N-1} - 1 =$
					32767

		`unsigned short` `unsigned short int`		$-2^{N-1} =$ -32768	$2^{N-1} - 1 =$ 32767
		`int` `signed int`		$-2^{N-1} =$ -32768	$2^{N-1} - 1 =$ 32767
		`unsigned int` `unsigned int`		$-2^{N-1} =$ -32768	$2^{N-1} - 1 =$ 32767

		`long`
		`long int`
		`signed long`
		`signed long int`
			32	$-2^{N-1} =$
				-2147483648
					$2^{N-1} - 1 =$
					2147483647

		`unsigned long` `unsigned long int`		0	$2^{N-1} - 1 =$ 4294967295
		`long long`[1] `long long int`[1] `signed long long`[1] `signed long long int`[1]	64	$-2^{N-1} =$ $-(\text{a lot})$	$2^{N-1} - 1 =$ $+(\text{a lot})$
		`unsigned long long`[1] `unsigned long long int`[1]		0	$2^{N-1} - 1$
Floating point	`float`		32	$\pm 2^{-149}$	$\pm(2 - 2^{-23}) \times 2^{127}$
	`double`		64	$\pm 2^{-1074}$	$\pm(2 - 2^{-52}) \times 2^{1023}$
	`long double`[1,2]		128	$\pm 2^{-16494}$	$\pm(2 - 2^{-112}) \times 2^{16383}$

1. C99 only
2. The definition of long double is compiler dependent. This table gives what is sometimes called quad floating point.

The C language standard only specifies the minimum number of bits that must be allocated to each of the data types, and that `sizeof(long int)≥sizeof(int)≥sizeof(short int)≥sizeof(char)`. The precise number of bits for each data type depends on the specific compiler and the hardware for which the compiler is written. Different implementations can exceed the minimum requirements. So, for example, the C compiler for an embedded processor might define `int` to be 16 bits (the minimum specified by the standard), whereas the compiler for a general-purpose processor might define it to be 32 or 64 bits. The C header files `limits.h` and `float.h` give specific information on the size of each of the integer and float data types that are implemented by a particular compiler. Check them carefully!

Matlab offers a number of data types that basically map to the standard C data types, as shown in **Table B.10**.

Table B.10 Numerical data types in Matlab

Data category		Matlab data type	C data type
Integer	signed	`int8`	`signed char`
		`int16`	`signed int`
		`int32`	`signed long`
		`int64`	`signed long long`
	unsigned	`uint8`	`unsigned char`
		`uint16`	`unsigned int`
		`uint32`	`unsigned long`
		`uint64`	`unsigned long long`
Floating point		`single`	`float`
		`double`	`double`

By default, Matlab stores all numeric values as double-precision (64-bit) floating point. However, you can initialize or cast a variable to one of the other types. Matlab's Fixed Point Toolbox provides support for fixed-point data types and operations.

C. Matlab tutorial

C.1 Introduction to Matlab

This is a brief tutorial on Matlab. The purpose is not to teach you to program, but to indicate features that Matlab has in common with other languages you may already know – C, Java, Python – and also to highlight those special features that make Matlab especially suited to the processing of arrays, which is at the heart of DSP. If you know any programming, Matlab is easy to pick up on your own and there are many resources available to help you.

- Matlab's online `help` and `doc` functions are always there for you. Typing "`doc Matlab`" at the command prompt will get you to the top level of Matlab's excellent help system.
- An extensive selection of tutorials is available online, at the MathWorks website and on YouTube.

C.1.1 What is Matlab?

Matlab is a program and development environment for scientific applications. It is a calculator, a programming language, an integrated development environment, and a sophisticated system with hundreds of specialized functions for analyzing and displaying data. It is well suited to DSP operations and is used extensively in academia and industry to develop and prototype sophisticated algorithms and applications. There is extensive built-in documentation as well as a large web-based user community to help with solutions to problems.

Matlab's roots are in the 1970s. It was initially developed as an interactive front-end to the LINPACK and EISPACK libraries for matrix and eigenvalue computations that were written in FORTRAN (hence the name: Matlab = Matrix Laboratory). Matlab's core strengths remain its speedy and robust algorithms for matrix manipulation and numerical analysis, many of which are now based on the open-source libraries designed for modern computer architectures, such as LAPACK and FFTW. However, some of Matlab's core deficiencies also derive from its FORTRAN roots. Specifically, addressing of arrays in Matlab still follows the ones-based indexing convention of FORTRAN, meaning the first element of an array `s` is `s(1)`, not `s(0)`, as it would be in C, C++, Java or other modern languages with zero-based indexing. This can be a source of much programming sadness, and is therefore something all Matlab programmers need to keep in mind as they code. Other issues that frequently raise the ire of Matlab users are its awkward handling of strings, multi-dimensional matrices and matrices with mixed data types (e.g., `cell`). MathWorks, the makers of Matlab, updates the software semiannually. This introduces new features, but also frequently redefines old features. However, despite its flaws, Matlab continues to dominate the market and is well worth your effort to learn.

C.1.2 What is Matlab *not*?

Matlab is not cheap – at least the full-blown commercial versions. It is a relatively expensive, proprietary, commercial product that comprises the core Matlab product plus a number of application-specific toolboxes of particular interest to practitioners of DSP, such as the Signal Processing Toolbox and the Image Processing Toolbox. However, there is a reasonably priced Student Edition of Matlab available for those who qualify. The basic price gets you the core Matlab product, and you can add a number of the most important toolboxes for a nominal cost.

As an alternative to Matlab, there exist a number of free, open-source Matlab-like clones, chief among which are Scilab and GNU Octave. These have a syntax similar to that of Matlab and can perform many of Matlab's basic functions. Octave, in particular, contains many functions that are important for signal processing, including `fft`, `ifft` and `filter`. There is also a growing community of programmers, particularly in academia, who use Python in combination with additional packages (e.g., Numpy, Scipy and Matplotlib) to do scientific computing. Because the basic syntax of Matlab and Python is quite similar, it should be relatively easy for an experienced programmer to translate code in this book from Matlab to Python. Depending on your application, one of these open-source alternatives might suffice for your needs. However, all the examples in this book are written in basic Matlab.

C.2 The elements of Matlab

Matlab features an impressive collection of basic and advanced functions. Here are a few highlights.

C.2.1 Calculator functions

At its simplest, Matlab is a calculator. After the prompt (>>), you type the expression you wish to evaluate:

```
>> 1+1
ans =
     2
```

More complex calculations can be done:

```
>> (1+2^3) / (6-3 * (4/2-1))
ans =
     3
```

The usual rules of precedence apply in Matlab to expressions such as this; exponentiation is done first, multiplication and division are next (and equal) in precedence and are done left to right. Addition and subtraction are lowest in precedence. Parentheses can be used to override these usual rules of precedence.

C.2.2 Variables

In addition to using Matlab as a calculator, you can set and manipulate variables.

Variable basics At its simplest, you can set variables from the command line:

```
>> x = 3
x =
     3

>> y = 2
y =
     2

>> z = x + y
z =
     5
```

Matlab echoes your work each time you type return. You can suppress this tedious echoing by terminating each line with a semicolon:

```
>> x = 3;
>> y = 2;
>> z = x + y;
>> z
z =
     5
```

The last line shows that just by typing the variable name, Matlab gives you the value of the requested variable. You can also get Matlab to display a variable's value less verbosely:

```
>> disp(z)
     5
```

Matlab allows multiple commands on a single line, both on the command line and in the editor, for example,

```
>> x = 3; y = 2; z = x + y;
```

However, I recommend that you do not do this. Putting multiple commands on a single line makes code harder for you (and others) to read and harder to debug.

Matlab keeps all your variables around until you clear them with the `clear` command. You can clear all variables at once or just specific variables. Here is how you can query all your variables in the current (base) Matlab workspace:

```
>> who
Your variables are:
ans  x  y  z
```

You can also get a more involved display:

```
>> whos
Name       Size          Bytes  Class
ans        1x1               8  double array
x          1x1               8  double array
y          1x1               8  double array
z          1x1               8  double array
Grand total is 3 elements using 24 bytes
```

`ans` is a built-in Matlab variable that gives the result of the last computation. You can use `ans` just like any other variable, and it is replaced with each successive computation

```
>> 1+1
ans =
     2
>> ans + 2
ans =
     4
```

There are other “sort-of” reserved Matlab variables, such as `i` and `j` (see below) and `pi`. We will have more to say about the “sort-of” qualification, below.

```
>> pi
ans =
    3.1416
```

By default, all numeric variables and calculations upon them are done in double-precision floating point and are displayed in a format specified by the Matlab `format` command. The default format is short floating point. That is why the value of `pi`, above, was a bit truncated. That is easy to fix:

```
>> format long
>> pi
ans =
    3.14159265358979
```

Matlab is inherently capable of working with complex numbers. The variables `i` and `j` are used by Matlab to denote the imaginary part (mathematicians and physicists tend to use `i` while engineers use `j`). All operations (i.e., multiplication, division, addition and subtraction) work transparently with complex numbers:

```
>> x = 1 + 2 * j
x =
  1.0000+ 2.0000i

>> y = 2 - j
y =
  2.0000- 1.0000i

>> z = x + y
z =
  3.0000+ 1.0000i
```

All Matlab’s arithmetic functions also work with complex numbers:

```
>> exp(z)
ans =
  10.8523 +16.9014i
```

However, watch out! You can use `j` (or any other built-in Matlab variable) as a variable in your own program by reassigning its value, but then it will no longer denote the imaginary operator:

```
>> j = 2;
>> x = 1 + 2 * j
x =
     5
```

Ooops! This is really pretty terrible. Matlab should not allow this, but it does. You can reset a built-in Matlab variable such as `j` to its default value by clearing it:

```
>> clear j
>> j
ans =
     0 + 1.0000i
```

Better yet, use the *truly* reserved built-in variable `1j` instead of using `j` to make your syntax unambiguous:

```
>> x = 1 + 2 * 1j
x =
     1.0000 + 2.0000i
```

The previous example shows that you can use `clear` to delete some or all variables and functions that you have defined:

```
>> who
Your variables are:
ans        x          y          z
>> clear y z
>> who
Your variables are:
ans        x
```

"Reserved" variables As we have just seen, Matlab does not prevent you from redefining important predefined variables, so it is incumbent upon *you* to know what they are. Here are the most important ones:

`ans`	Default variable name returned by a function with no specified output argument
`pi`	π
`i` or `j`	j
`inf` or `Inf`	∞. This occurs naturally when attempting to do a division such as 1/0. You can also use it as an argument; for example, `atan(Inf)`
`nan` or `NaN`	Not a number. Returned by a calculation such as 0/0 or sin(inf).
`eps`	The smallest distinguishable positive double floating-point number, 2^{-52}.

Vectors and matrices Matlab's most powerful feature is its ability to express vectors and matrices compactly and operate upon them efficiently. Vectors (or arrays) in Matlab are declared using square brackets:

```
>> x1 = [1 2 3 4 5]
x1 =
     1    2    3    4    5
```

x1 is a row vector. Putting a semicolon between elements creates a column vector:

```
>> x2 = [1; 2; 3; 4; 5]
x2 =
     1
     2
     3
     4
     5
```

The transpose operator converts row vectors to column vectors:

```
>> x3 = x2'
x3 =
     1   2   3   4   5
```

Be careful when taking the transpose of complex quantities, since the transpose operator creates the Hermitian (the conjugate transpose):

```
>> xc = [1+1j 2+1j 3+1j]
xc =
   1.0000 + 1.0000i 2.0000 + 1.0000i 3.0000 + 1.0000i

>> xc'
ans =
     1.0000 - 1.0000i
     2.0000 - 1.0000i
     3.0000 - 1.0000i
```

To get a non-conjugate transpose, use the dot-transpose operator:

```
>> xc.'
ans =
     1.0000 + 1.0000i
     2.0000 + 1.0000i
     3.0000 + 1.0000i
```

Another source of error is white space. Some programmers put a lot of white space between characters to aid in readability (e.g., `x = 1 + 2 * 1j`), whereas other folks like a more compact style (e.g., `x=1+2*1j`). In Matlab, adding white space does not matter... until it does. Look at this:

```
>> xc = [1 +1j 2+1j 3+1j]
xc =
   1.0000 + 0.0000i 0.0000 + 1.0000i 2.0000 + 1.0000i 3.0000 + 1.0000i
```

Oops! The white space between the `1` and the `+1j` led to an array of four numbers, not three.

Matlab provides an easy way of generating sequences like x1, above:

```
>> x1 = 1:5
x1 =
     1   2   3   4   5
```

You can increment or decrement by integers or fractions,

```
>> 1:2:5
ans =
     1   3   5
```

Or backwards,

```
>> 9:-2:3
ans =
    9   7   5   3
```

Or in non-integer increments or decrements,

```
>> 5:-0.5:3
ans =
    5.0000 4.5000 4.0000 3.5000 3.0000
```

Matrix arithmetic with Matlab is easy, since all arithmetic operators are vectorized. For example, to add two vectors of the same dimension, you simply use the + sign:

```
>> y = x1 + x3
y =
  2   4   6   8   10
```

Vector multiplication is also simple. Given `x1` and `x2` from above,

```
>> z = x1 * x2
z = 55
```

Of course, this is different from

```
>> z = x2 * x1
ans =
      1   2    3    4    5
      2   4    6    8   10
      3   6    9   12   15
      4   8   12   16   20
      5  10   15   20   25
```

Matrices (and arrays) of compatible dimensions can be concatenated. Matrices with the same number of rows can be horizontally concatenated by using `[m1 m2 ...]`; matrices with the same number of columns can be vertically concatenated using `[m1; m2; ...]`, for example,

```
>> [x1 x3]
ans =
     1    2    3    4    5    1    2    3    4    5

>> [x1; x3]
ans =
     1    2    3    4    5
     1    2    3    4    5
```

Adding or subtracting vectors or matrices with differing dimensions in Matlab will generally produce an error (try it). However, the newest versions of Matlab allow what is called "implicit arithmetic expansion." Prior to version R2016b, if you had a matrix like `z`, above, and you wanted to add numbers `1:5` to each column respectively, you had to add a matrix that was exactly the same size as the original one, something like this:

```
>> z + ones(5, 1)*(1:5)
```

In newer versions of Matlab, you can just do this

```
>> z + (1:5);
```

This new syntax could be useful, but it can also lead to unexpected errors if used improperly.

C.2.3 Matlab functions

All Matlab's arithmetic functions accept vectors and matrices as arguments, for example:

```
>> cos(2*pi*(0:7)/8)
ans =
    1.0000   0.7071   0.0000   -0.7071   -1.0000   -0.7071   -0.0000   0.7071
```

In a language such as C, you would have to use a `for` loop to obtain the same result.

All of Matlab's elementary functions also accept complex arguments, which means that complicated expressions can be implemented very economically. This is something we frequently want to do in DSP programs. For example, consider the N-point discrete Fourier transform (DFT) of an N-point sequence $x[n]$ defined by

$$X[k] = \sum_{n=0}^{N-1} x[n]e^{-j2\pi kn/N}, \qquad 0 \leq k < N.$$

All N values of $X[k]$ can be efficiently computed with a single line of code:

```
>> N = 8;
>> x = [1 1 1 1 1 0 0 0];
>> k = 0:N-1;
>> n = 0:N-1;
>> X = exp(-1j*2*pi*k'*n/N)*x' % Compute DFT of x[n]
X =
   5.0000 + 0.0000i
   0.0000 - 2.4142i
   1.0000 + 0.0000i
   0.0000 - 0.4142i
   1.0000 + 0.0000i
  -0.0000 + 0.4142i
   1.0000 + 0.0000i
  -0.0000 + 2.4142i
```

It is worth understanding what happened here. A matrix multiplication such as `k'*n` creates a $k \times n$ matrix, where each row corresponds to a value of index k. This entire matrix is then exponentiated in one fell swoop to form the $k \times n$ matrix $e^{-jk2\pi n/N}$. When this matrix is multiplied by the $n \times 1$ column vector corresponding to $x[n]$, the result is the $1 \times k$ row vector that corresponds to $X[k]$. Writing vectorized code like this is extremely efficient. Computing this DFT without vectorization would require a pair of nested `for` loops and would be almost an order of magnitude slower. While Matlab's vectorized operators are powerful, they also allow you to write code that is as dense as a neutron star, and so obtuse as to be almost unintelligible to others and possibly even to yourself when you come back to it after a while. Accordingly, it is a good idea to comment byzantine code if you use it.

Accessing array values You can access any value of an array or vector with subscripts of the general form, `z(row, column)`:

```
>> z(2, 3)
ans =
     6
```

It is important to remember that Matlab uses ones-based indexing. Trying to declare or access array values with non-positive integers produces an error:

```
>> y(0)
??? Index into matrix is negative or zero.
```

This is really unfortunate for DSP, since we would like to be able to manipulate arrays with zero or negative indices. Fortunately, we will find a way around this restriction in the laboratory for Chapter 1, by creating our own `sequence` class, which will allow us to index with both positive and negative integers.

In Matlab, you can use integer arrays as subscripts as well:

```
>> y(1:3)
ans =
     2   4   6

>> z(2:4, :)
ans =
     2   4    6    8   10
     3   6    9   12   15
     4   8   12   16   20
```

In the preceding example, the first argument to `z` says that you want rows 2 through 4. The second argument (the colon) is shorthand that tells Matlab you want all the columns. You can do very sophisticated things with the array indices, such as replacing or manipulating portions of vectors and matrices. Here we replace rows 2 to 4 in columns 3 and 5 with the scalar value 99:

```
>> z(2:4, [3 5]) = 99
z =
    1    2    3    4    5
    2    4   99    8   99
    3    6   99   12   99
    4    8   99   16   99
    5   10   15   20   25
```

Like its forebear, FORTRAN, Matlab stores data internally using column-wise ordering or indices, so you can access a multi-dimensional array with a single index if you like:

```
>> z(3:8)
ans =
     3        4        5        2        4        6
```

Particularly with large arrays, indexing data by columns first is faster than indexing by rows.

Matlab gives you the ability to produce arrays and matrices with all zeros or ones using the `zeros` or `ones` commands. For example, here we produce an array of four columns:

```
>> zeros(1, 4)
ans =
      0      0      0      0

>> ones(1, 4)
ans =
      1      1      1      1
```

One of the uses of the `zeros` command is to preallocate a data array when you are performing a calculation in a `for` loop that adds an element to an array, particularly a large array. Consider the following fragment:

```
y = [];
for n = 1:10000
    y = [y n];
end
```

Here, Matlab has to resize, move and copy the array for `y` repeatedly as new elements are added on each pass through the loop. This could lead to slow execution. Do this instead:

```
y = zeros(1, 10000);
for n = 1:10000
    y(n) = n;
end
```

Here, we first preallocate the entire array. Then, each pass through the loop just replaces one of the 0 values with the desired value.

Size of matrices and arrays You can determine the size of a matrix with the size command, which has several forms:

```
>> sz = size(ones(4, 5))
sz =
     4     5

>> [nrows, ncols] = size(ones(4, 5))
nrows =
     4
ncols =
     5

>> nrows = size(ones(4, 5), 1)
nrows =
     4

>> ncols = size(ones(4, 5), 2)
ncols =
     4
```

For arrays (i.e., column or row vectors) you can use `length`, but watch out: the function will return the larger of the two dimensions.

```
>> len = length(zeros(2, 4))
len =
     4

>> len = length(zeros(4, 2))
len =
     4
```

The reserved keyword `end` can be used in the index of arrays. For example

```
>> x = 1:5;
>> x(3:end)
ans =
     3     4     5
```

`end` can be used to flip arrays:

```
>> x(end:-1:1)
ans =
     5     4     3     2     1
```

You can also use the Matlab commands `fliplr` and `flipud`, to flip arrays and matrices. However, this brings up an important point. In Matlab, there are literally hundreds and hundreds of functions. You can spend a good chunk of your life looking for *just* the right function to perform your particular task. In some cases, finding the right function (e.g., `fft`) is essential, because `fft` is a complicated, highly optimized function you really do not want to program yourself. In other simpler cases (e.g., `fliplr`), it may be easier just to use basic Matlab language functions. The fewer special functions you call, the easier your code will be for others to understand and to port to other languages.

Data types Matlab is not a strongly typed language, in the sense that C is. That is, you do not have to declare variables as floating point or integer or character. However, Matlab does have a number of data types available in order to express floating-point and integer numbers, characters and logical quantities (see Appendix B.4). By default, all numerical expressions are double-precision floating point (eight bytes), but they can be cast (i.e., converted) to single-precision (four bytes) with the operator `single` and back to double-precision by `double`. Operators `int` and `uint` with appropriate suffixes cast numerical quantities to integers. This is useful to implement fixed-point algorithms. For example, `uint8(x)` casts `x` to an eight-bit (one-byte) unsigned integer; `int16(x)` casts `x` to a signed 16-bit (two-byte) integer. When dealing with integer arithmetic, watch out for overflows and underflows; for example,

```
>> uint8(-1)
ans =
   0

>> int8(257)
ans =
  127
```

Strings in Matlab are arrays of alphabetic characters enclosed in single quotes, for example,

```
>> str = 'hello';
```

Individual characters of the string can be addressed and concatenated just like elements of numerical arrays:

```
>> ['j' str(2:end)]
ans =
jello
```

Strings are of type `char`. String arrays can be cast to integers (or double) and integer arrays representing ASCII codes can be cast to strings:

```
>> x = int8(str)
x =
  104   101   108   108   111

>> char(x)
ans =
hello
```

Logical data arise from Matlab's logical operations and have values of either 0 or 1. A number of Matlab's functions and operations yield logical output.. These include comparison operators on numbers (e.g., <, ==, >, `all`, `any` and `find`), logical operations on logical data (e.g., `|` and `&`) and a number of string operators (e.g., `strcmp`, `strfind` and `strmatch`). Judiciously used, these logical operators can save a lot of programming. For example,

```
>> x = 1:10;
>> indx = (x>3)&(x<7)
indx =
     0     0     0     1     1     1     0     0     0     0
```

Here, `indx` is logical data that can be used to select the parts of the numerical data array corresponding to logical 1:

```
>> x(indx)
ans =
     4     5     6
```

Notice that the only time you can use zero indices in an array is if they are of the logical data type. You can also do the entire operation in one line:

```
>> x((x>3)&(x<7))
ans =
     4     5     6
```

You can cast logical data to double. For example, here is a trivial way to make a step function $u[n]$ in Matlab:

```
function y = u(n)
   y = double(n >= 0);
end
```

Structures Structures in Matlab are just like structures in C. They are containers that allow you to group together a variety of data under the umbrella of one variable name. The form of a structure is `variableName.field`. For example:

```
>> x.data = [1 2 3 4 5];
>> x.offset = -1;
>> x
x =
      data: [1 2 3 4]
    offset: -1
```

We have declared a variable `x` with two fields. The `data` field contains an array; the `offset` field contains another number. Structures of this type can be useful for us in expressing sequences in Matlab. As we mentioned previously, Matlab always assumes that the index of the first element of an array is 1. So, when we just say

```
>> x = [1 2 3 4 5];
```

it means that `x[1] =1`, `x[2] =2`, and so on. But, what if we want to express a sequence `x[-1] =1`, `x[0] =2`, ... ? We obviously need to provide the user with some additional information in addition to the array data, namely an offset. An offset of −1 tells the user that "we want you to interpret the first point in the Matlab array as corresponding to `x[-1]` ." We can use structures for this purpose. We will have a bit more to say about structures when we discuss classes, below.

C.3 Programming in Matlab

With Matlab, you can create applications using both procedural and object-oriented programming techniques. As a procedural language, you can create your own scripts and functions.

C.3.1 Scripts

A **script** is just a sequence of operations that Matlab executes one after the other, exactly as if you had entered the same operations in the command window one after another. Create a script using Matlab's editor by typing `edit` on the command line. When you finish entering your script, save it as a .m file. Make sure you save the file in your home directory, or in a directory that is on Matlab's path (see documentation for the `path` command). For example, here is a file `myscript.m`:

```
% MYSCRIPT A little script to add two numbers
%
% T. Holton
x = 3;
y = 5;
z = x + y
```

Now, when you type `myscript` following the Matlab prompt, you get the answer of this very complex calculation:

```
>> myscript
z =
     8
```

The first few lines of the script start with the % character, and by default Matlab colors them green in the editor. This denotes a comment. It is good practice to comment your programs, giving the name of the program, the author and perhaps the revision history. If you put comments like this in the first few lines of the program, then your script is automatically accessible from the Matlab's help function:

```
>> help myscript
MYSCRIPT A little script to add two numbers

T. Holton
```

C.3.2 Functions

All the variables that you define in a script are available in the base Matlab workspace. This may be an advantage in some situations, but most often you would like the script to work as an independent unit, perhaps one that takes one or more inputs and produces one or more outputs and does not litter your workspace with variables. This is precisely what Matlab's **functions** are for. Matlab's user-defined functions are like scripts except that all variables declared in the function and all calculations take place in a separate workspace (like those in other languages such as C). Functions look just like scripts except for the first and last lines:

```
function out = myfunction(in1, in2)
    % MYFUNCTION A little function to add two numbers
    %
    % T. Holton
    temp = 2 * in1;
    out = temp + in2; % this is a comment
end
```

The first line of the function specifies the input and output variables. In this case there is one output variable out, and two input variables in1 and in2. The last line of the function is the end statement. By default, the Matlab editor colors this and other recognized commands in blue. Now, call this function:

```
>> clear
>> y = myfunction(3, 4)
y =
    10
>> who
Your variables are:
y
```

Because all lines in myfunction are terminated with semi-colons, the printing of all calculations from within the function has been suppressed. Matlab did all the calculations required by the function in a separate workspace that is normally not accessible to you. None of the variables that we created in myfunction will show up in the calling (base) workspace. Also, since the input variables are dummy variables, if you ask for them in the function caller's workspace, you will be out of luck:

```
>> in1
??? Undefined function or variable 'in1'.
```

If you call a function from within another function, each function operates in its own workspace. There are ways of sharing variables between workspaces in Matlab if absolutely necessary; specifically, variables can be declared `global` or one can use Matlab's built-in `assignin` function. However, extensive use of global variables is generally considered a hallmark of bad program design. It defeats the purpose of using functions, which is to isolate variables in different workspaces so that they cannot interact with variables in the main workspace or other workspaces in unforeseen ways.

Function handles The "`@`"sign is a Matlab operator that indicates a function handle, which is equivalent to a function pointer in C. Matlab functions have function handles that are just the function name preceded with the "`@`" sign; e.g., `@sin` and `@cos`. One can evaluate a function in a couple of ways, for example, by using the `feval` function with appropriate arguments:

```
feval(@functionhandle, arguments)
```

So for example, `feval(@sin, pi)` is equivalent to `sin(pi)`. You can also get the handle to any existing function. The syntax is

```
f_handle = @function_name
```

So,

```
>> f_handle = @cos;
>> f_handle(pi)
ans =
     -1
```

Use of function handles is not restricted to Matlab's existing functions. Any function you have written is fair game. For example, in the previous subsection, we defined `myfunction(in1, in2)`, a function of two input variables and one output variable. Using the function handle syntax, we could call the function this way,

```
f_handle = @my_function;
fhandle(in1, in2)
```

or just

```
feval(@myfunction, in1, in2)
```

Anonymous functions One useful application of function handles is to define an **anonymous function**. An anonymous function is a function with a single output variable and one or more input variables that can be defined on a single line. Anonymous functions can be defined on the command line, in scripts or inside other functions. The syntax of the anonymous function is

```
output_variable = @(input_variables) function_of_input_variables;
```

For example, a single-line function that calculates the DFT could be defined as follows:

```
y = @(x) exp(-1j*2*pi*(0:length(x)-1)'*(0:length(x)-1)/length(x))*x(:);
```

The output variable is y. In this example, there is a single input variable x, prefaced by the "@" sign and enclosed in parentheses. Following this is the definition of the function. As is the case with regular functions, the input variable is a dummy variable. Also, any variable used in the function definition must either appear in the list of input variables or have been previously defined. Incidentally, the syntax x(:) assures that the x will be a column vector, which it needs to be in this case.

C.3.3 Conditionals and loops

Like all reasonable programming languages, Matlab has syntax that enables conditional and repeated execution of code. Conditionals are of the form

```
if (condition)  statements
elseif (condition)
    statements
else (condition)
    statements
end
```

where else and elseif are optional. The parentheses around condition statements are also not necessary, but I recommend them to aid in the readability of the code.

Loops either have the form of a while loop or of the familiar for loop:

```
while (condition)
    statements
end

for index = values
    statements
end
```

break allows for the premature termination of a for or while loop, for example:

```
for k = 1:10
    if (k > 4)
        break
    end
end
```

By default, the editor displays all recognized function names in your programs in blue.

C.3.4 Classes

Beyond simple scripts and functions, Matlab also offers a relatively full-featured **object-oriented programming (OOP)** language that permits you to create more complex applications with user-defined classes. Purists may complain that Matlab's OOP implementation lacks a number of features (e.g., templates, strong typing) found in other OOP languages. But in recent versions, Matlab's OOP syntax and implementation has evolved to look similar enough to other "standard" OOP languages that if you are familiar with object-oriented languages like C++, Java or Python, you will feel right at home with Matlab. If you have not run across object-oriented programming and plan to spend any time professionally with Matlab, this approach to programming is well worth learning.

In brief, an **object** is essentially a software "container." The basic container holds data variables, termed **properties**, and the functions, termed **methods**, that define the permissible ways the world can access or operate on the data. Objects are members of **classes** and are defined by a **class definition file**. For example, floating-point variables in Matlab are members of the `double` class; logical variables are members of the `logical` class; string variables are members of the `char` class and so on.

You can use Matlab's `class` and `isa` operators to determine the class of objects:

```
>> y = 6;
>> class(y)
ans =
    double

>> isa(y, 'double')
ans =
     1
```

There are a *lot* of methods in Matlab that allow us to operate on numeric data. To see a list of all the methods for the `double` class, type `methods('double')`. For example, in the list you will see `plus` and `minus`. Matlab's `plus` method returns the sum of two numeric variables, which is also of the `double` class:

```
>> plus(1, 2)
ans =
     3
```

Of course, ordinarily, we never call commonly used functions like `plus` and `minus` directly, since Matlab offers the familiar shortcut notation:

```
>> 1 + 2
ans =
     3
```

What Matlab has done behind the scenes here is to note that the "+" operator has been called with two numeric arguments and then applied the `plus` function of the numeric class to arrive at the answer.

A fundamental feature of object-oriented programming is that we can create our own classes and define how operators, even existing operators such as `plus` and `minus`, will work on our class properties. This reuse of the same operators in different classes is called **operator overloading**, and is one of the key features of object-oriented programming.

As an example, let us create a `sequence` class to allow us to define and manipulate sequences. The class comprises a data structure that describes the sequence, plus a collection of methods that allow us to operate on one or more sequences, for example to add two sequences together using the "+" operator or to plot a sequence.

To define a class in Matlab, we create an .m file with the name of the class, in this case, `sequence.m`. The class definition as well as all the methods that operate on sequence data, including those methods such as plus and minus that override (i.e., redefine) Matlab operations, live in this file. Here is what the beginning of the file looks like:

```
classdef sequence
    properties
        data
        offset
    end

    methods
        function s = sequence(data, offset)
            % SEQUENCE Sequence object
            %          S = SEQUENCE(DATA, OFFSET) creates sequence, s
            %          using DATA and OFFSET
            %
            %          T. Holton

            s.data = data;
            s.offset = offset;
        end

        function s = plus(s1, s2)
          % Add two sequences, s1 and s2, and provide an output sequence, s
          data = s1.data + s2.data;        % <- bogus
          offset = s1.offset + s2.offset;
          s = sequence(data, offset);
        end
    end
end
```

The code starts with classdef sequence, which tells Matlab that we are defining a class called sequence. Following this is a properties block terminated by an end statement, which enumerates the properties of the class. In this case, only two properties are necessary to define a discrete-time sequence completely: data and offset. There follows a methods block, also terminated by an end statement, which includes all the functions that are necessary to construct the class and allow us to operate on its data. The first function in the block is required to be the class **constructor**:

```
function s = sequence(data, offset)
```

The constructor creates and returns an **instance** of the sequence class: an **object** s is a structure that has the data of the sequence in the data member and the offset (i.e., the first index) in the offset element. For example, to create an object for sequence $\delta[n+1]+2\delta[n]+3\delta[n-1]+4\delta[n-2]$, we write

```
>> s = sequence([1 2 3 4], -1)
s =
  sequence handle

  Properties:
      data: [1 2 3 4]
    offset: -1
```

Matlab tells us, in a very verbose fashion, that we have just created a sequence object. The object `s` is not a member of the numeric class; it is a member of our newly defined `sequence` class.

```
>> class(s)
ans =
    sequence
```

or,

```
>> isa(x, 'sequence')
ans =
     1
```

We can now define functions – such as addition, subtraction and plotting – that work on objects of the `sequence` class. In the code fragment above, I have defined a `plus` method that adds two sequences `s1` and `s2` with different offsets and data, and generates a new output sequence `s`. The particular algorithm I have used to compute the output sequence is bogus, but the last line of the method is relevant. Once you have properly computed the parameters of the output sequence (e.g., `data` and `offset`), you call the constructor, which returns a new sequence object with proper values of `data` and `offset`.

In the laboratory exercise accompanying Chapter 1, you will write a non-bogus `plus` method that can do the following:

```
>> x = sequence([1 2 3 4], -1);
>> h = sequence([1 1 1], -2);
>> x + h
ans =
      data: [1 2 3 3 4]
    offset: -2
```

Here, we are using the same operator ("+") for the sequences that we do for numeric data. When you call the `plus` operator either explicitly (e.g., `plus(x, h)`) or implicitly (e.g., `x+h`) with objects of the sequence class as arguments, Matlab will look in the `sequence.m` file to find the overloaded `plus` method, and then use it. You can also overload functions such as Matlab's `conv` function to deal with sequences, which you will do in the laboratory exercise.

```
>> conv(x, h)
ans =
      data: [1 3 6 9 7 4]
    offset: -3
```

Pretty handy! All the overloaded functions for your sequence class such as `plus`, `minus` and `conv` must live in the methods block of the `sequence.m` file, which is where Matlab expects to find them.

C.4 Matlab help

Matlab has an excellent help system. To get help on anything else in Matlab, use the `help` or `doc` functions. `help` lists information in the command window; `doc` opens Matlab's documentation browser in a separate window and provides more comprehensive information and often examples of usage. For example, to get help on any given function, type

```
>> help sin
>> doc sin
```

C.5 Plotting

One of Matlab's best features is its comprehensive set of powerful and easy-to-use plotting functions. The most useful for us are the basic `plot` and `stem` functions. Here is a small example that demonstrates a few of the available features:

```
set(gcf, 'Unit', 'inch', ...
    'Position', [1 1 4 2], ...
    'PaperPositionMode', 'auto')
n = 0:31;
y = sin(2*pi*n/16);
plot(n, y, 'o-', ...
    'MarkerSize', 6, ...
    'MarkerFaceColor', 'r')
grid on
axis tight
xlabel('n')
ylabel('x')
title('My little plot')
print('MyLittlePlot', '-dpng') % output to a .png file
```

The editor displays strings in purple. **Figure C.1** shows the result of running the code:

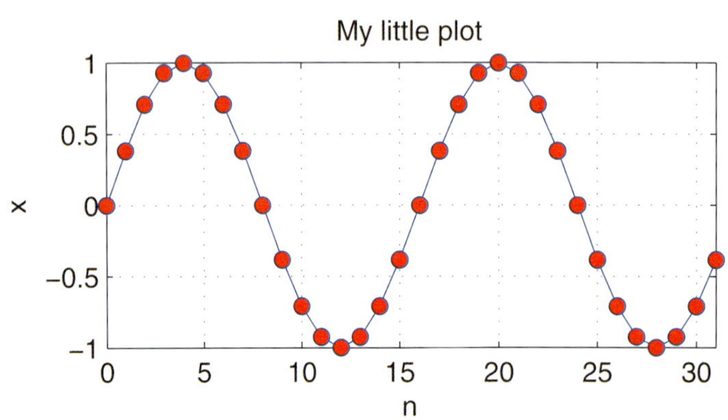

Figure C.1

The `set` command specifies the units and position of the current plot on the screen (`gcf` means 'get current figure'). Setting `PaperPositionMode` to `auto` is necessary only if you wish to maintain the aspect ratio of the plot you see on the screen when you print your figures to an output device or to a file using Matlab's `print` command, as I have done here.

Relatively sophisticated control of plot attributes, such as colors, line styles, symbols, text and axes, as well as multiple and overlaid plots are possible with Matlab's plotting functions.

C.6 The Matlab environment

Matlab's integrated development environment (IDE) makes the work of writing, debugging and running code easy. The Matlab IDE has a number of components, the most important of which are the Command window and the Editor.

C.6.1 Command window and editor

Command window When you launch Matlab for the first time, a large window opens that contains a number of panels. Depending on the version of Matlab you have installed, these can include the *Command Window*, *Current Directory*, *Workspace* and *Command History*. My personal preference is to close all of these windows except the Command Window so that you will have the whole window to enter commands and view results.

Editor Typing "`edit`" at the command prompt (">>") in the command window launches Matlab's *Editor*. In the editor, you can write, edit, run and debug code. While the editor may initially be docked inside the main Matlab window, you will most likely find it convenient to undock it to its own window on your desktop. The IDE has many parameters that are controlled via the *Preferences* panel, which is accessed through the File menu of the Main Window in older versions of Matlab, and through the Environment tab in more recent versions. In order to make your code easier to read, allow me to recommend that you set the editor to use "Emacs-style" tabs, as shown in **Figure C.2a**, and require it to indent all functions, as shown in **Figure C.2b**.

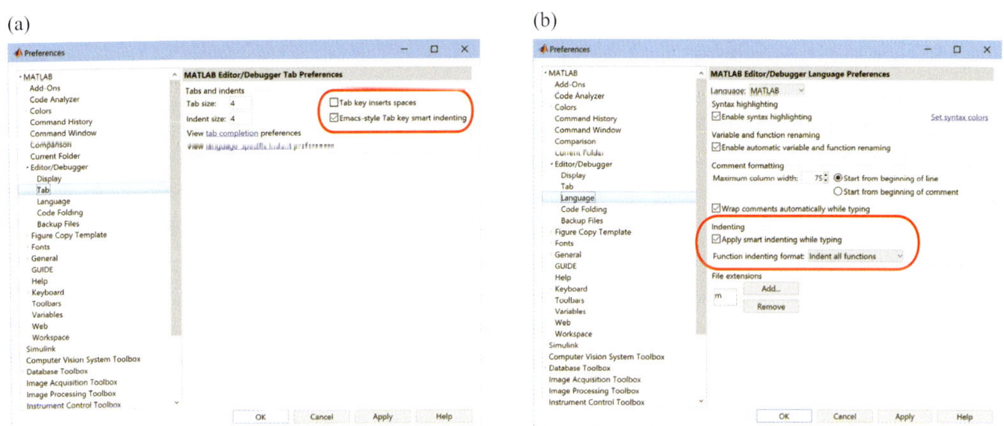

Figure C.2

With these settings, when you select all your code (e.g., by typing "Ctrl-a") and push the tab key, your code will be properly indented, which improves readability and helps in debugging.

C.6.2 Debugging and writing "clean" code

With the Matlab IDE, you can set absolute or conditional breakpoints, step through code line-by-line or run it in sections. You can also interrogate workspace variables. In the example in **Figure C.3**, I have set a breakpoint in the editor at line 6 by clicking in the margin at the location of the red dot. When the program is run, it will stop at that line. By hovering the mouse over any variable in the editor, you will see its value.

Matlab acts like an interpreted language, in that you can enter commands directly in the command window and it will execute them immediately, but it is actually a just-in-time (JIT) compiler. That means that while you are writing a piece of code in the editor, each time you add a character or terminate a line, Matlab is furiously recompiling your program in the background. Matlab includes a valuable tool to check your code called *M-lint*. M-lint is Matlab's version of the C-language "lint" program. It is built into the editor and enabled by default. It continuously combs through your code looking for incorrect, ambiguous, inefficient or deprecated syntax and usage, unused or uninitialized variables and more. You may not find it necessary to use M-lint for short programs, but if/when you write longer programs, it can be valuable in identifying potential problems before they cause you headaches. In the screenshot of the short program shown in **Figure C.3**, the M-lint notifications are indicated on the right window border. Here, M-lint has noticed something minor, namely that one statement has not been terminated with a semicolon, which means that this line is going to produce unwanted output in the command window when the function is called. It has flagged the offending line with a thin yellow bar and has issued a warning that you can see by hovering your mouse over the line. You can choose to have M-lint fix the problem or ignore the warning, and you can disable M-lint entirely by unchecking a box in the preference panel of the editor if its punctiliousness annoys you.

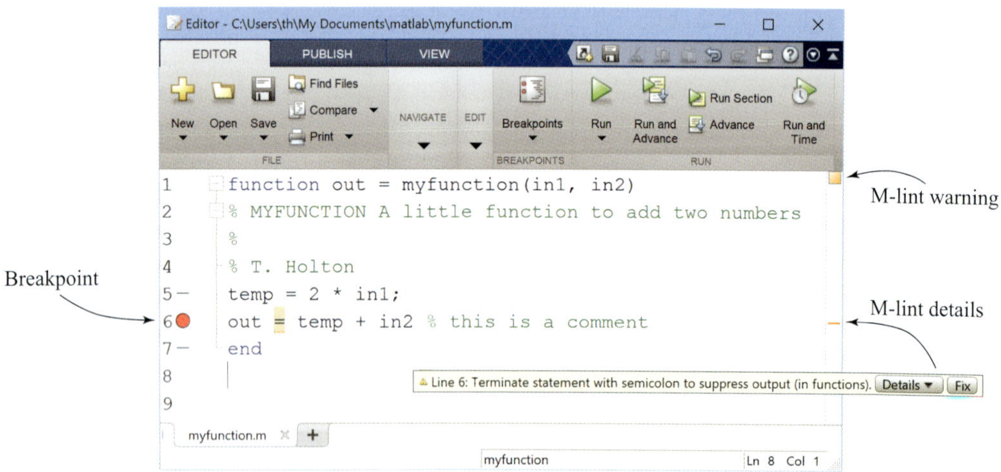

Figure C.3

D. Probability and random processes

D.1 Probability distribution and density functions

Most of the systems encountered in the real world are, to a greater or lesser degree, **probabilistic** or random. The defining characteristic of a probabilistic system is that its output is uncertain or unpredictable. Randomness may arise intrinsically – for example, the variation in sound waveforms of a speaker saying the same thing – or it may arise extrinsically from noise added to a signal or engendered in the system due to environmental, electronic or computational factors. In this section, we briefly review some of the essentials of probability theory as it applies to the two main types of probabilistic system, *discrete* and *continuous*, and investigate the ways to characterize them.

D.1.1 Discrete probability density function

Consider a simple **discrete probabilistic system**: flipping a coin. Each flip of the coin can be considered an **experiment** or **trial**. Each trial results in an **outcome**, which is a unique, measurable result of the experiment. In a discrete system, there are a countable (finite or infinite) number of discrete possible outcomes. In the case of the coin, there are only two possible outcomes: heads or tails, which together comprise the **sample space** of the experiment. A **random variable** X is defined as a function that maps each possible outcome of the experiment to a real number. In what follows, we adopt the common convention that the random variable is written in upper case (e.g., X), and a particular value of the random variable is denoted in lower case (e.g., x). In our coin-flip experiment, X will be assigned the value of 1 for heads and 0 for tails. The **probability distribution function**[1] $\Pr_X(x)$ of the discrete random variable X assigns the probability that X is associated with a given outcome x. So, in this example, if the coin is "fair" (e.g., equally weighted), the probability assigned to each outcome is equal, namely 0.5, and we would write

$$\Pr_X(x) = \begin{cases} 0.5, & x = 0 \\ 0.5, & x = 1 \end{cases}.$$

[1] For discrete probabilistic events such as this, the term **probability mass function** is often preferred.

D.1.2 Continuous probability density function

A **continuous probabilistic system** is characterized by outcomes that can take on an infinite number of values that span a finite or infinite range. For example, consider a system in which we spin a dial and record the angle of the pointer, ω. Each spin of the dial is a trial, and each trial results in an outcome, which in this case is the angle of the pointer. In this experiment, the sample space comprises infinitely many angles spanning the range $-\pi \le \omega < \pi$. Here, we will choose the random variable X to be ω/π, so that the range of values is $-1 \le X < 1$. Because there are an infinite number of possible outcomes, the probability that a given spin of the dial will result in any particular, exact, value, say $x = 0.000\cdots$, is zero! However, the probability that the pointer lands within a specified *range* of values is finite and can be calculated. To do so, we define the **cumulative distribution function (CDF)** $P_X(x)$ to be the probability that X will be less than or equal to a particular value x,

$$P_X(x) = \Pr(X \le x).$$

Figure D.1a shows the CDF of our spinning dial. Because X cannot be less than -1, the cumulative probability for $x \le -1$ is $P_X(x) = 0$, as indicated by the dashed line to the left of the x axis. Similarly, because X cannot exceed 1, $P_X(x) = 1$, $x \ge 1$. By definition, the CDF is cumulative and is therefore always a monotonically increasing function of x. If our dial is evenly weighted, the cumulative probability that X is less than or equal to any given value x will be proportional to x, that is,

$$P_X(x) = \begin{cases} 0, & x < -1 \\ (x+1)/2, & -1 \le x \le 1, \\ 1, & x > 1 \end{cases}$$

as shown in **Figure D.1a**.

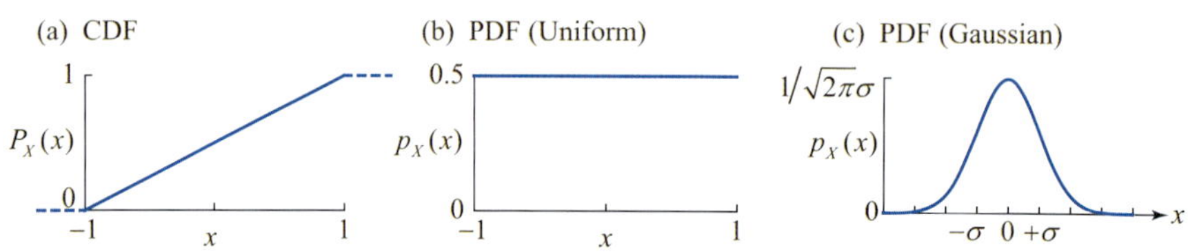

Figure D.1 Cumulative probability distribution function and probability density function

The **probability density function (PDF)** $p_X(x)$ of a continuous random variable is defined as the derivative of the CDF,

$$p_X(x) = \frac{dP_X(x)}{dx},$$

or conversely, the CDF is defined as the integral of the PDF,

$$P_X(x) = \int_{-\infty}^{x} p_X(\xi)d\xi.$$

Uniform PDF The PDF that describes our spinning dial is a **uniform probability distribution**,

$$p_X(x) = \frac{d(x+1)/2}{dx} = \begin{cases} 0.5, & -1 \le x < 1 \\ 0, & \text{otherwise} \end{cases},$$

shown in **Figure D.1b**. It is important to emphasize that probability *density* is not the same as probability. The probability that X is any particular, exact value, e.g., $x = 0.000\cdots$, is still zero, even though $p_X(0) = 0.5$. What the PDF *does* allow us to do is to calculate the probability that an outcome lies within a specified *range* of values $x_0 \le X \le x_1$. We do this by integrating the PDF over the desired range,

$$\Pr(x_0 \le X < x_1) = \int_{x_0}^{x_1} p_X(x)dx = \int_{-\infty}^{x_1} p_X(x)dx - \int_{-\infty}^{x_0} p_X(x)dx = P_X(x_1) - P_X(x_0). \quad \textbf{(D.1)}$$

In our example, $\Pr(x_0 \le X < x_1) = 0.5(x_1 - x_0)$. The uniform PDF is extensively used in engineering in general and DSP in particular to model physical processes where the outcomes have an equal likelihood of occurrence within a limited range of values, such as our spinning dial example. In Chapter 6, we use the uniform PDF to model the quantization error that arises in A/D conversion as a result of rounding or truncation.

Gaussian PDF The **Gaussian distribution** or **normal distribution** is perhaps the most widely used distribution to model probabilistic phenomena in mathematics, biology, engineering, economics, finance and many other fields. For example, **Gaussian white noise** is very commonly used to model electronic (thermal) noise in electronic equipment, such as the analog front end of an A/D measuring system. The Gaussian or normal PDF is given by

$$p_X(x) = \frac{1}{\sqrt{2\pi}\sigma_X} e^{-\frac{1}{2}\left(\frac{x-\mu_X}{\sigma_X}\right)^2}, \quad \textbf{(D.2)}$$

where μ_X is the **mean** of the distribution and σ_X is the **standard deviation**. The normal distribution is the familiar bell-shaped curve, a symmetrical curve centered on the mean value μ_X, with a width that is proportional to σ_X. **Figure D.1c** shows an example of a normal distribution that has a zero mean. The Matlab function `randn` produces outcomes from a normal PDF.

Unlike the uniform distribution, the range of values of the Gaussian distribution is infinite, $-\infty < X < \infty$, which means that it is theoretically possible for an outcome to have any value. However, the bell-shaped nature of the distribution means that most outcomes will be near the mean value. For example, the probability that an outcome lies within one standard deviation of the mean is

$$\Pr(-\sigma_X \le X \le +\sigma_X) \triangleq \int_{-\sigma_X}^{+\sigma_X} p_X(x)\,dx = 0.6827.$$

The probability that the outcome lies within four standard deviations (the range shown in **Figure D.1c**) is $\Pr(-4\sigma_X \leq X \leq 4\sigma_X) = 0.9999$. These values are tabulated and can be calculated by the Matlab `cdf` function.

D.1.3 Joint, marginal and conditional probability distributions

Often we are interested in the relation between two or more random events. To do so, we introduce the notion of a **joint PDF**. The joint distribution of more than one random variable is termed **multivariate**, and the specific case of a joint distribution of two random variables is termed **bivariate**. Given random variables X and Y, the joint (bivariate) PDF is written as $p_{XY}(x, y)$. The probability of the joint event that X is in the range $x_0 \leq X < x_1$ *and* Y is in the range $y_0 \leq Y < y_1$ can be written as a double integral of the joint PDF in a manner similar to Equation (D.1),

$$\Pr(x_0 \leq X < x_1, y_0 \leq Y < y_1) = \int_{x_0}^{x_1}\int_{y_0}^{y_1} p_{XY}(x, y)dxdy.$$

If a bivariate PDF is integrated over all values of one of the random variables, the result is termed the **marginal PDF** of the other variable,

$$\begin{aligned} p_X(x) = \Pr(x, -\infty < Y < \infty) &= \int_{-\infty}^{\infty} p_{XY}(x, y)dy \\ p_Y(y) = \Pr(-\infty < X < \infty, y) &= \int_{-\infty}^{\infty} p_{XY}(x, y)dx. \end{aligned} \tag{D.3}$$

Two random variables are said to be **independent** if their joint PDF is equal to the product of the marginal PDFs,

$$p_{XY}(x, y) = p_X(x)p_Y(y). \tag{D.4}$$

The **conditional probability distribution** of X given Y is defined as the joint PDF of X and Y over the marginal PDF of Y,

$$p_{X|Y}(x|y) = \frac{p_{XY}(x, y)}{p_Y(y)}. \tag{D.5}$$

The conditional probability allows one to calculate the probability of an event range on X given an event range on Y,

$$\Pr(x_0 \leq X < x_1 | y_0 \leq Y < y_1) = \frac{\int_{x_0}^{x_1}\int_{y_0}^{y_1} p_{XY}(x, y)dxdy}{\int_{y_0}^{y_1} p_Y(y)dy}.$$

If two random variables X and Y are independent, the conditional PDF is equal to the marginal PDF. For example, substituting Equation (D.4) into Equation (D.5) gives

$$p_{X|Y}(x|y) = \frac{p_{XY}(x, y)}{p_Y(y)} = \frac{p_X(x)\cancel{p_Y(y)}}{\cancel{p_Y(y)}} = p_X(x).$$

In other words, if two random variables are independent, then knowing the probability of occurrence of one of them gives you no information about the other.

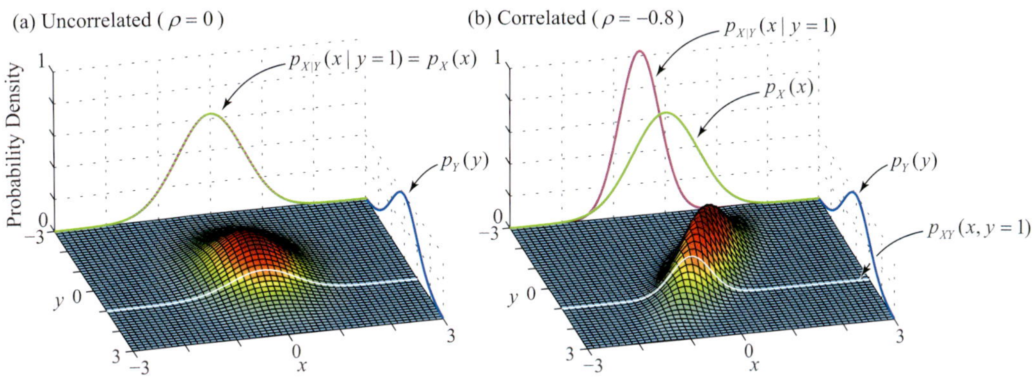

Figure D.2 Uncorrelated and correlated random variables

As an example illustrating the preceding ideas, let X and Y be two jointly distributed Gaussian random variables that represent measurements of the manufacturing tolerance of two different parts coming off an assembly line. The general form of the bivariate Gaussian distribution is

$$p_{XY}(x,y) = \frac{1}{2\pi\sigma_X\sigma_Y\sqrt{1-\rho^2}} e^{-\frac{1}{2(1-\rho^2)}\left(\left(\frac{x-\mu_X}{\sigma_X}\right)^2 - 2\rho\left(\frac{x-\mu_X}{\sigma_X}\right)\left(\frac{y-\mu_Y}{\sigma_Y}\right) + \left(\frac{y-\mu_Y}{\sigma_Y}\right)^2\right)},$$

where μ_X and μ_Y are the means of X and Y, σ_X and σ_Y are the variances and ρ is called the **correlation coefficient**. In our example, let $\mu_X = \mu_Y = 0$, $\sigma_X = \sigma_Y = 0.6$.

Figure D.2a shows a three-dimensional surface plot of the joint PDF $p_{XY}(x, y)$, as a function of x and y for the case where the correlation coefficient is $\rho = 0$, which means that the random variables X and Y are **uncorrelated**, a concept we will explain further in Section D.1.5. When $\rho = 0$, $p_{XY}(x, y)$ is separable into the product of two expressions, $p_X(x)$ and $p_Y(y)$:

$$\begin{aligned} p_{XY}(x,y) &= \frac{1}{2\pi\sigma_X\sigma_Y} e^{-\frac{1}{2}\left(\left(\frac{x-\mu_X}{\sigma_X}\right)^2 + \left(\frac{x-\mu_X}{\sigma_X}\right)^2\right)} = \left(\frac{1}{\sqrt{2\pi}\sigma_X} e^{-\frac{1}{2}\left(\frac{x-\mu_X}{\sigma_X}\right)^2}\right)\left(\frac{1}{\sqrt{2\pi}\sigma_Y} e^{-\frac{1}{2}\left(\frac{y-\mu_Y}{\sigma_Y}\right)^2}\right) \\ &= p_X(x)p_Y(y). \end{aligned} \quad \text{(D.6)}$$

This means that, for the bivariate Gaussian, uncorrelated random variables are also independent. Because the two random variables are independent, the three-dimensional surface looks symmetric about both the x and y axes. Integrating $p_{XY}(x, y)$ over x yields the marginal PDF, $p_Y(y)$ (blue curve on the "wall" of the plot); integrating $p_{XY}(x, y)$ over y gives $p_X(x)$ (green curve). The white curve superimposed on the surface is the joint PDF evaluated at the single value of $y = 1$, $p_{XY}(x, 1)$. Because the two random variables are independent in this example, the conditional PDF $p_X(x|y)$ at this value of y (or any other value for that matter) is exactly equal to the marginal PDF $p_X(x)$, as shown by the dotted purple curve that exactly superimposes on the green curve,

$$p_{X|Y}(x|1) = \frac{p_X(x)\, \cancel{p_Y(1)}}{\cancel{p_Y(1)}} = p_X(x).$$

Figure D.2b shows an example in which the two random variables X and Y are correlated ($\rho = -0.8$) and not independent. The marginal PDFs $p_Y(y)$ (blue curve) and $p_X(x)$ (green curve) are the same as in **Figure D.2b**, since Equation (D.3) applies regardless of whether X and Y are independent or not. However, in this example, the conditional PDF (purple curve) is not equal to the marginal PDF for X (green curve), $p_{X|Y}(x|\,1) \neq p_X(x)$.

D.1.4 Expected value and moments

The **expected value** or **mean** of a random variable X is computed by summing all possible values of X weighted by their probability of occurrence. For a discrete random variable with N possible outcomes, $x_1, x_2, \ldots, x_N$, the expected value of X, defined as $E(x)$, is calculated as the sum of the value of each outcome x_i multiplied by its probability $P_X(x_i)$,

$$E(x) \triangleq \sum_{i=1}^{N} x_i P_X(x_i).$$

As a simple example, consider the experiment in which X represents the number that comes up when you throw a fair, six-sided die. The die has only six faces, so the sample space of the experiment comprises exactly six values, 1, 2, . . ., 6, each of which has an equal probability of occurring, namely 1/6. Hence, the expected value,

$$E(x) = \sum_{i=1}^{6} x_i P_X(x_i) = 1\cdot\frac{1}{6} + 2\cdot\frac{1}{6} + 3\cdot\frac{1}{6} + 4\cdot\frac{1}{6} + 5\cdot\frac{1}{6} + 6\cdot\frac{1}{6} = \frac{21}{6} = 3.5,$$

is just the simple average of the face numbers. Now, assume that you have stumbled into a gambler's den and the die is "loaded" so that the probability of numbers 1–4 is 0.1, the probability of number 5 is 0.2 and the probability of number 6 is 0.4. Then, the expected value is the weighted average of the numbers,

$$E(x) = \sum_{i=1}^{6} x_i P_X(x_i) = 1\cdot0.1 + 2\cdot0.1 + 3\cdot0.1 + 4\cdot0.1 + 5\cdot0.2 + 6\cdot0.4 = 4.4.$$

For a continuous random variable, the expected value (mean) of X is defined as μ_X,

$$\mu_X \triangleq E(x) = \int_{-\infty}^{\infty} x p_X(x) dx. \tag{D.7}$$

In the example of the uniform distribution, the mean value is

$$\mu_X = \int_{-1}^{1} 0.5x dx = 0.$$

For the Gaussian PDF, you can show (Problem D-1a) that the parameter μ_X in the definition, Equation (D.2), is in fact the mean value,

$$E(x) = \int_{-\infty}^{\infty} x p_X(x) dx = \int_{-\infty}^{\infty} x \frac{1}{\sqrt{2\pi}\sigma_X} e^{-((x-\mu_X)/\sigma_X)^2/2} dx = \mu_X.$$

More generally, one can compute the expectation of any function of the random variable $f(x)$ as

$$E(f(x)) = \int_{-\infty}^{\infty} f(x) p_X(x) dx.$$

For the particular case $f(x) = x^N$, the expectation $E(x^N)$ is termed the **Nth moment** of X,

$$E(x^N) = \int_{-\infty}^{\infty} x^N p_X(x) dx.$$

The two most widely used moments are the **first moment** $E(x)$, which is the **mean**, and the **second moment** $E(x^2)$. An important related quantity is the **Nth central moment**, which is formed by taking the expected value of a random variable that has had its mean subtracted, raised to the Nth power, $f(x) = (x - E(x))^N$. The second central moment is the most important of these. It is universally known as the **variance**, and given the symbol σ_X^2,

$$\begin{aligned}\sigma_X^2 &\triangleq E((x - E(x))^2) = \int_{-\infty}^{\infty} (x - E(x))^2 p_X(x) dx \\ &= \int_{-\infty}^{\infty} x^2 p_X(x) dx - 2E(X) \int_{-\infty}^{\infty} x p_X(x) dx + E(x)^2 \int_{-\infty}^{\infty} p_X(x) dx \\ &= E(x^2) - 2E(x)^2 + E(x)^2 = E(x^2) - E(x)^2.\end{aligned}$$

As an example, the variance of the uniform distribution in **Figure D.1a** is

$$\sigma_X^2 = \int_{-\infty}^{\infty} (x - E(x))^2 p_X(x) dx = \frac{1}{2} \int_{-1}^{1} x^2 dx = \frac{1}{6} x^3 \Big|_{-1}^{1} = \frac{1}{3}.$$

The **standard deviation** σ_X is then defined as the square root of the variance. It is a measure of the dispersion of a distribution about the mean, and is always positive. You can show (Problem D-1b) that the variance of the Gaussian PDF is, in fact, σ_X^2. Several important probabilistic processes, including the monovariate Gaussian and uniform distributions, are completely characterized by their mean and variance.

D.1.5 Covariance and correlation

The **covariance** σ_{XY} is a measure of how two random variables X and Y vary with respect to each other. It is defined as

$$\begin{aligned}\sigma_{XY} &= \sigma_{YX} \triangleq E((x - E(x))(y - E(y))) = E(xy) - 2E(x)E(y) + E(x)E(y) \\ &= E(xy) - E(x)E(y)\end{aligned}, \qquad \textbf{(D.8)}$$

where

$$E(xy) = \int_{-\infty}^{\infty}\int_{-\infty}^{\infty} xy\, p_{XY}(x,y)dxdy. \quad \textbf{(D.9)}$$

Note that although the standard deviation σ_X is always positive, the covariance σ_{XY} can be positive or negative.

A special case of covariance occurs when $Y = X$. Then, the covariance is equal to the variance,

$$\sigma_{XX} = E(x^2) - E(x)^2 = \sigma_X^2. \quad \textbf{(D.10)}$$

If random variables X and Y are independent, then their covariance is zero, $\sigma_{XY} = 0$. The reason is that, by Equation (D.4), $p_{XY}(x,y) = p_X(x)p_Y(y)$, so

$$E(xy) = \int_{-\infty}^{\infty}\int_{-\infty}^{\infty} xy\, p_X(x)p_Y(y)dxdy = \left(\int_{-\infty}^{\infty} xp_X(x)dx\right)\left(\int_{-\infty}^{\infty} yp_Y(y)dy\right) = E(x)E(y).$$

Hence,

$$\sigma_{XY} = E(xy) - E(x)E(y) = E(x)E(y) - E(x)E(y) = 0.$$

However, the converse is *not* generally true; random variables that have a covariance of 0 are not necessarily independent. The classic counterexample is to let X be a zero-mean, symmetric PDF (such as the uniform or zero-mean Gaussian distribution shown in Figure D.1), so that $p_X(x) = p_X(-x)$, and let $Y = X^2$. These distributions are clearly not independent, but their covariance is zero:

$$\sigma_{XY} = E(xy) - E(x)E(y) = E(x^3) - E(x)E(x)^2 = 0. \quad \textbf{(D.11)}$$

The first term is zero because

$$E(x^3) = \int_{-\infty}^{\infty} x^3 p_X(x)dx = \int_{-\infty}^{0} x^3 p_X(x)dx + \int_{0}^{\infty} x^3 p_X(x)dx = \int_{\infty}^{0} (-x)^3 p_X(-x)d(-x) + \int_{0}^{\infty} x^3 p_X(x)dx$$

$$= -\int_{0}^{\infty} x^3 p_X(x)dx + \int_{0}^{\infty} x^3 p_X(x)dx = 0.$$

The second term of Equation (D.11) is zero because $E(x) = 0$.

The bivariate Gaussian PDF is an important case where the distribution is guaranteed to be independent as long as the covariance of the distribution is zero. That will prove significant in Section D.2.4, when we consider the characteristics of the Gaussian white-noise process.

An important property of the covariance is that the values of σ_{XY} are bounded,

$$|\sigma_{XY}| \le \sigma_X \sigma_Y. \quad \textbf{(D.12)}$$

The proof derives from the famous Cauchy–Schwarz inequality (see Problem D-2).

Finally, the **correlation** is defined as the covariance normalized by the product of the standard deviations,

$$\rho_{XY} = \frac{\sigma_{XY}}{\sigma_X \sigma_Y}.$$

By Equation (D.12), $|\sigma_{XY}| \leq \sigma_X \sigma_Y$, so the correlation is also bounded: $|\rho_{XY}| \leq 1$, or explicitly,

$$-1 \leq \rho_{XY} \leq 1.$$

Because the covariance of independent random variables is zero, $\sigma_{XY} = 0$, their correlation is zero as well.

D.2 Random processes

A random signal is unpredictable, meaning that we cannot determine in advance the value at any given time instant. However, the *statistics* of a random signal – for example, its mean and variance – *can* be determined precisely from a specification of the underlying probabilistic process (e.g., uniform or Gaussian). One of the key problems in signal processing, which we address in Chapter 14, is to estimate parameters of interest of random signals. In this section, we explore some of the issues involved in characterizing signals that are either inherently random or have a random component (such as additive noise).

As an example, the blue symbols in **Figure D.3a** are a plot of 16 points of the deterministic signal $w[n] = A\cos(\omega_0 n + \theta)$, where $A = 1$, $\omega_0 = \pi/8$ and $\theta = 0$. At any point in time n, the value of $w[n]$ is completely determined by the parameters of cosine function: A, ω_0 and θ. In contrast, the red symbols in **Figure D.3a** are a plot of cosine $\hat{w}[n]$ to which a random "noise" $x[n]$ has been added at each time point. This random value may have resulted from a sampled physical process – for example electronic (thermal) noise in a measuring circuit, or environmental noise incident on a microphone – or it might represent quantization noise of an A/D converter, as discussed in Section 6.7. Regardless of the source of the noise, the result is a random signal, $\hat{w}[n] = w[n] + x[n]$, whose values are not completely predictable simply given values of the cosine parameters and time n. **Figure D.3b** shows just the random component of the signal.

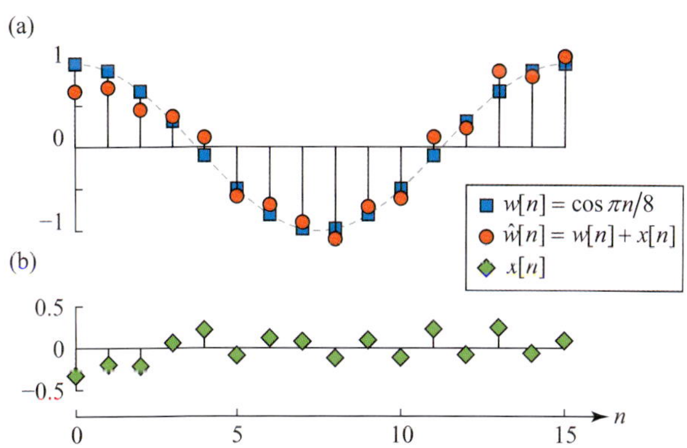

Figure D.3 Deterministic and random signals

In the examples of Section D.1, each random variable characterized an experiment (roll of the die, spin of the dial) that produced a single real value. Now consider what happens if each experiment instead produces a finite or infinite sequence of values as a function of time n, such as the purely random signal $x[n]$ shown in **Figure D.3b**. **Figure D.4a** shows examples of repeated measurements of the signal $x[n]$, labeled $x_1[n]$, $x_2[n]$, ... Each measurement of $x[n]$, for example $x_k[n]$, represents the outcome of a single experiment, and is called a **sample function** (also called a **realization**) of the process. For example, $x[n]$ might represent repeated measurements of a noisy signal waveform. The **sample space** for the experiment is the (possibly infinite) **ensemble** of all these sample functions. The values of this set of functions at a single moment in time n_1 are called a **time sample** and are characterized by **a random variable** $X[n_1]$ that has PDF $p_X(x, n_1)$. At a different moment in time, say n_2, the PDF of random variable $X[n_2]$, namely $p_X(x, n_2)$, may or may not be the same as $p_X(x, n_1)$. The ensemble of all these random variables over all the time points $X[n] \in \{X[n_0], X[n_1], \ldots, X[n_N]\}$, where N could be infinite, is called a **random process**, also called a **stochastic process**. When the time index n is discrete, as it is in the example of **Figure D.4a**, this is called **discrete-time random process**, and is characterized by an ensemble of joint PDFs $p_X(x, n) = p_X(x_1, x_2, \ldots, n_1, n_2, \ldots)$ that expresses the probability of the entire collection of values that make up the process at all moments of time.

In order to characterize a random process fully in accordance with the discussion in the preceding paragraph, we would require the value of *every* possible sample function in the sample space at *every* moment of time, clearly an impossible task. Fortunately, in many cases of practical interest to us, such as those discussed in Chapter 14, the random process can be characterized by a small number of standard statistical measures that are relatively easy to compute. We now briefly describe those measures.

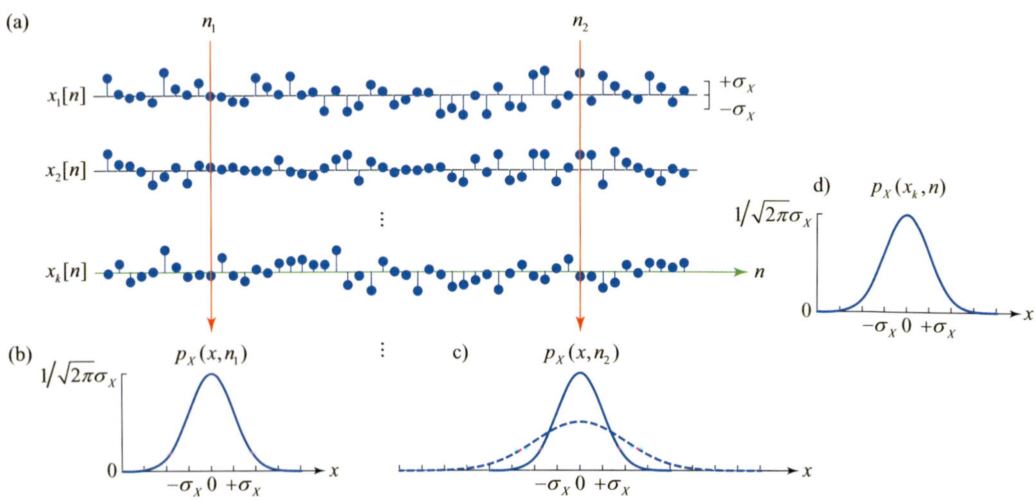

Figure D.4

D.2.1 Statistics of a random process

Since a discrete-time random process can be viewed as a family of random variables $X[n]$ indexed on time n, we can apply our treatment of the statistics of continuous random variables in Sections D.1.4 and D.1.5 to random processes.

Mean The expected value of random process $X[n]$ is defined in a manner similar to the expected value of the continuous random variable X in Equation (D.7), except that since $X[n]$ depends on time n its expected value is also a function of n,

$$\mu_X[n] \triangleq E(x[n]) \triangleq \int_{-\infty}^{\infty} x[n]p_X(x,n)dx. \tag{D.13}$$

Here, the notation $p_X(x,n)$ makes explicit the fact that the PDF of $X[n]$ depends on n.

Autocorrelation The **autocorrelation function** (often just called the **correlation function**) of a random process $r_{XX}[n_1,n_2]$ is homologous to $E(x^2)$ for a random variable defined in Equation (D.9). It measures the correlation between the random variables of the same process at two times n_1 and n_2. Define these random variables to be $X_1 = X[n_1]$ and $X_1 = X[n_2]$. Then,

$$r_{XX}[n_1,n_2] = E(x[n_1]x[n_2]) = \int_{-\infty}^{\infty}\int_{-\infty}^{\infty} x[n_1]x[n_2]p_X(x_1,x_2,n_1,n_2)dx_1dx_2, \tag{D.14}$$

where $p_X(x_1,x_2,n_1,n_2)$ is the joint PDF of the random variables X_1 and X_2 at times n_1 and n_2. The autocorrelation function is symmetric with respect to n_1 and n_2:

$$r_{XX}[n_1,n_2] = r_{XX}[n_2,n_1].$$

When $n \triangleq n_1 = n_2$, the autocorrelation is just the (time-dependent) second moment,

$$r_{XX}[n,n] = E(x^2[n]). \tag{D.15}$$

Crosscorrelation In Chapter 14, we will be considering the relation between two random processes $X[n]$ and $Y[n]$. In order to characterize the relation between these two processes fully, we would require the joint PDF of $X[n]$ and $Y[n]$. However, in the cases of primary concern to us, we will only be interested in the relation between the joint *statistics* of these two processes – in particular, the first and second moments . To that end, we now define a measure — the crosscorrelation — that can be viewed as an extension of the autocorrelation defined in the preceding section.

The **crosscorrelation function** of the random variables of the two processes at two times $X[n_1]$ and $Y[n_2]$ is defined as

$$r_{XY}[n_1,n_2] \triangleq E(x[n_1]y[n_2]) = \int_{-\infty}^{\infty}\int_{-\infty}^{\infty} x[n_1]y[n_2]p_X(x,y,n_1,n_2)dxdy. \tag{D.16}$$

The symmetry relation for the crosscorrelation function is

$$r_{XY}[n_1,n_2] = r_{YX}[n_2,n_1]. \tag{D.17}$$

(Note the swapping of the subscripts of r.)

Matrix form of correlation In Chapter 14, we will use the matrix form of the autocorrelation. Define the column vector of random variables at times n_1, n_2, …, n_N as,

$$\mathbf{x} = [x[n_1]\ \ x[n_2]\ \ \cdots\ \ x[n_N]].$$

Then, the vector of mean values is defined as

$$\boldsymbol{\mu}_X = E(\mathbf{x}) = [u_X[n_1]\quad u_X[n_1]\quad \cdots \quad u_X[n_N]]. \tag{D.18}$$

The $N \times N$ autocorrelation matrix $\boldsymbol{\Gamma}_{\mathbf{XX}}$ that expresses the autocorrelation of pairs of variables at times n_1 and n_2 is then defined in matrix form as

$$\boldsymbol{\Gamma}_{XX} = E(\mathbf{x}\mathbf{x}^T) = \begin{bmatrix} r_{XX}[n_1,n_1] & r_{XX}[n_1,n_2] & \cdots & r_{XX}[n_1,n_N] \\ r_{XX}[n_2,n_1] & r_{XX}[n_2,n_2] & \cdots & r_{XX}[n_2,n_N] \\ \vdots & \vdots & \ddots & \vdots \\ r_{XX}[n_N,n_1] & r_{XX}[n_N,n_2] & \cdots & r_{XX}[n_N,n_N] \end{bmatrix}. \tag{D.19}$$

Similarly, an $N \times N$ crosscorrelation matrix, $\boldsymbol{\Gamma}_{XY}$, is defined as

$$\boldsymbol{\Gamma}_{XY} = E(\mathbf{x}\mathbf{y}^T) = \begin{bmatrix} r_{XY}[n_1,n_1] & r_{XY}[n_1,n_2] & \cdots & r_{XY}[n_1,n_N] \\ r_{XY}[n_2,n_1] & r_{XY}[n_2,n_2] & \cdots & r_{XY}[n_2,n_N] \\ \vdots & \vdots & \ddots & \vdots \\ r_{XY}[n_N,n_1] & r_{XY}[n_N,n_2] & \cdots & r_{XY}[n_N,n_N] \end{bmatrix}.$$

D.2.2 Stationary random processes

Figure D.4b shows the time sample at a particular time n_1 (red line). The values of the sample are distributed according to a continuous zero-mean Gaussian random variable $X[n_1]$, with standard deviation σ_X and PDF $p_X(x, n_1)$. **Figure D.4c** shows the time sample at another time n_2, characterized by random variable $X[n_2]$. As we have mentioned, in general $X[n_1]$ and $X[n_2]$ need not have the same PDFs. For instance, the characteristics of the measured noise could change with time. As a consequence, $p_X(x, n_2)$ (dashed blue curve curve) could be different from $p_X(x, n_1)$ (solid blue curve).

A random process is said to be **stationary** if some or all of its probabilistic properties do not change with time. The process is termed **strict-sense stationary** if the PDF $p_X(x, n)$ does not depend on absolute time n, as suggested by the two identical "blue" Gaussian PDFs in **Figures D.4b** and **c**. In a process that is strict-sense stationary, all the moments of the process are independent of absolute time.

A less stringent criterion, **wide-sense stationarity** (often abbreviated **WSS**), is very frequently used in models of random processes of physical systems. A process is said to be wide-sense stationary if the mean and autocorrelation of the process are independent of absolute time. A process that is strict-sense stationary is also wide-sense stationary, but the converse is not generally true, since the higher moments of the process may depend on absolute time. Important simplifications occur in the equations for mean, autocorrelation and covariance of WSS processes, as we will now show.

Mean The mean of a WSS process, $\mu_X[n]$, is independent of n, so from Equation (D.13), it is a constant,

$$\mu_X[n] = \mu_X.$$

Autocorrelation The autocorrelation function of a WSS process, $r_{XX}[n_1, n_2]$, does not depend on absolute time, but only on the *difference* between times n_1 and n_2, which we define as the **lag** $l \triangleq n_1 - n_2$. To see this, let $n = n_1$ and $n_2 = n_1 - l = n - l$. Then the autocorrelation in Equation (D.14) becomes $r_{XX}[n, n-l]$. Since the autocorrelation is independent of absolute time n, we can define it solely as a function of the lag parameter,[2]

$$r_{XX}[l] \triangleq r_{XX}[n, n-l]. \tag{D.20}$$

The autocorrelation function is a symmetric, real function of l since

$$r_{XX}[-l] \triangleq r_{XX}[n, n+l] = r_{XX}[n-l, (n-l)-l] = r_{XX}[n-l, n] = r_{XX}[n, n-l] = r_{XX}[l]. \tag{D.21}$$

From Equations (D.15) and (D.20), $r_{XX}[0]$ is the second moment, which is guaranteed to be non-negative,

$$r_{XX}[0] = E(x^2[n]) \geq 0.$$

The observation that $r_{XX}[0] \geq 0$ follows from the fact that both $x^2[n]$ and the PDF $p_X(x[n])$ are non-negative, so their integral, which is the expectation, is also non-negative.

For a WSS process, a derivation similar to that which resulted in Equation (D.12) leads one to the conclusion that the magnitude of the autocorrelation of a WSS process has its maximum value at a lag of zero (see Problem D-3),

$$|r_{XX}[l]| \leq r_{XX}[0].$$

In the case of a WSS random process, the matrix form of the autocorrelation function, Equation (D.19), is considerably simplified. The elements of the autocorrelation matrix, $r_{XX}[n_1, n_2]$, are replaced by $r_{XX}[l]$. Because the autocorrelation function of a WSS process is symmetric, by Equation (D.21), $r_{XX}[n_2, n_1]$ is replaced by $r_{XX}[-l]$, yielding

$$\mathbf{\Gamma_{XX}} = E(\mathbf{xx}^T) = \begin{bmatrix} r_{XX}[0] & r_{XX}[1] & \cdots & r_{XX}[N-1] \\ r_{XX}[-1] & r_{XX}[0] & \cdots & r_{XX}[N-2] \\ \vdots & \vdots & \ddots & \vdots \\ r_{XX}[1-N] & r_{XX}[2-N] & \cdots & r_{XX}[0] \end{bmatrix}$$

This has the form of a symmetric Toeplitz matrix in which all the elements of each diagonal have the same value. This matrix makes an important appearance in Chapter 14.

[2] The usual convention is to define lag such that a positive lag implies that $n_1 > n_2$.

Example D.1

Consider a random process comprising a cosine with random phase,

$$x[n] = A\cos(\omega_0 n + \theta),$$

where the phase of the cosine, θ, is uniformly distributed between $-\pi \le \theta < \pi$,

$$p_\Theta(\theta) = \frac{1}{2\pi}, \quad -\pi \le \theta < \pi.$$

Determine whether this process is WSS.

➤ **Solution:**

The mean of the process is

$$\mu_X[n] = E(x[n]) = E(A\cos(\omega_0 n + \theta)) = \int_{-\pi}^{\pi} A\cos(\omega_0 n + \theta) p_\Theta(\theta) d\theta = \frac{A}{2\pi}\int_{-\pi}^{\pi} \cos(\omega_0 n + \theta) d\theta = 0.$$

The autocorrelation is

$$\begin{aligned} r_{XX}[n,l] &= E(x[n]x[n-l]) = A^2 E(\cos(\omega_0 n + \theta)\cos(\omega_0(n-l) + \theta)) \\ &= \frac{A^2}{2}\cos\omega_0 l + \frac{A^2}{2} E(\cos(\omega_0(2n-l) + 2\theta)) \\ &= \frac{A^2}{2}\cos\omega_0 l + \left(\frac{A^2}{2}\int_{-\pi}^{\pi} \frac{1}{2\pi}\cos(\omega_0(2n-l) + 2\theta) d\theta\right) = \frac{A^2}{2}\cos\omega_0 l, \end{aligned}$$

where we have used the trigonometric identity

$$\cos\varphi_1 \cos\varphi_2 = \tfrac{1}{2}\cos(\varphi_1 - \varphi_2) + \tfrac{1}{2}\cos(\varphi_1 + \varphi_2),$$

and recognized that the integral of the cosine evaluates to zero because the angle is uniformly distributed between $-\pi \le \theta < \pi$. Because both $\mu_X[n]$ and $r_{XX}[n,l]$ are independent of time n, the process is WSS.

Crosscorrelation We are frequently interested in the statistical relation between two random processes $X[n]$ and $Y[n]$. If both are WSS such that their first- and second-order statistics only depend on the lag, then from Equation (D.16) we define the **crosscorrelation function** as

$$r_{XY}[l] \triangleq r_{XY}[n, n-l].$$

Using Equation (D.17), the symmetry relation for the crosscorrelation becomes

$$r_{XY}[-l] = r_{XY}[n, n+l] = r_{XY}[n-l, n] = r_{YX}[n, n-l] = r_{YX}[l]. \qquad \textbf{(D.22)}$$

Like their counterparts discussed in Section D.1.4, the measures discussed in this section that are calculated from the underlying random processes are either values (in the case of the mean and variance) or deterministic functions (in the case of the correlations).

D.2.3 White noise

A discrete-time **white noise** is a zero-mean, wide-sense stationary process defined by autocorrelation and covariance function,

$$r_{XX}[l] = \begin{cases} \sigma_X^2, & l = 0 \\ 0, & l \neq 0 \end{cases} = \sigma_X^2 \delta[l]. \tag{D.23}$$

Importantly, Equation (D.23) states only that samples are serially uncorrelated – that is, each sample of the process is not correlated with any sample except itself. It does not specify the PDF of the process, namely the distribution of amplitudes of the noise. Perhaps the most common white-noise PDF is the Gaussian distribution, which has proven to be a good model for many physical systems.

D.2.4 Filtered random processes

In Chapter 14, we will be considering systems in which an input random process $X[n]$ is filtered through an LTI system with impulse response $h[n]$ to yield an output random process $Y[n]$.

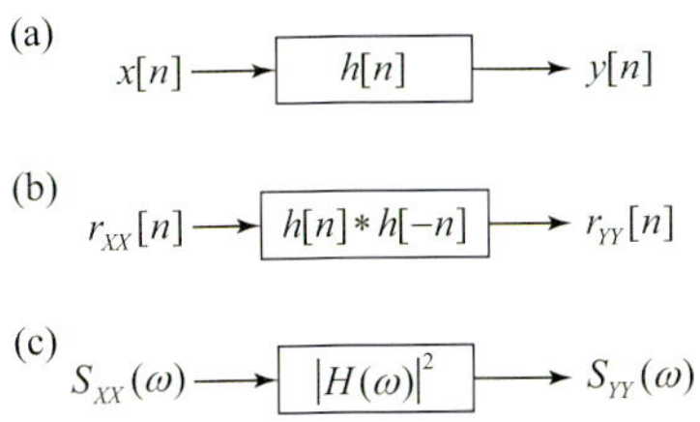

Figure D.5

We will now derive an important result; namely, that if $X[n]$ is a WSS process, then $Y[n]$ is also a WSS process. To do so, we will show that both the mean and autocorrelation function of $Y[n]$ are independent of absolute time n.

First, recognize that the input process comprises a set of input sample functions $x[n]$. When a given sample function passes through the filter, it produces an output sample function $y[n]$, as shown in **Figure D.5a**,

$$y[n] = x[n] * h[n] = \sum_{k=\infty}^{\infty} h[k]x[n-k].$$

The expected value (mean) of the output process Y is a constant,

$$\mu_Y = E(y[n]) = E\left(\sum_{k=\infty}^{\infty} h[k]x[n-k]\right) = \sum_{k=\infty}^{\infty} h[k]E(x[n-k]) = u_X \sum_{k=\infty}^{\infty} h[k].$$

Here, we used the fact that since $X[n]$ is WSS, its mean is a constant independent of absolute time, $E(x[n-k]) = \mu_X$. Also, if the filter is stable, then the impulse response is absolutely summable, so μ_Y will be finite.

In order to find the autocorrelation of Y, first compute the crosscorrelation of Y and X,

$$r_{YX}[n,n-l] = E(y[n]x[n-l]) = E\left(\left(\sum_{k=\infty}^{\infty} h[k]x[n-k]\right)x[n-l]\right) = \sum_{k=\infty}^{\infty} h[k]E(x[n-k]x[n-l])$$
$$= \sum_{k=\infty}^{\infty} h[k]E(x[n]x[n-(l-k)]) = \sum_{k=\infty}^{\infty} h[k]r_{XX}[l-k] = h[l]*r_{XX}[l].$$

In the second line, we have again recognized that since $X[n]$ is WSS, the autocorrelation function depends only on the lag $l-k$. The last summation is just the convolution of two functions of lag l. Hence, crosscorrelation is only a function of l,

$$r_{YX}[l] = r_{XX}[l] * h[l]. \tag{D.24}$$

In a similar fashion, the autocorrelation of the output is

$$r_{YY}[n,n-l] = E(y[n]y[n-l]) = E\left(\left(\sum_{k=\infty}^{\infty} h[k]x[n-k]\right)y[n-l]\right) = \sum_{k=\infty}^{\infty} h[k]E(x[n-k]y[n-l])$$
$$= \sum_{k=\infty}^{\infty} h[k]E(x[n]y[n-(l-k)]) = \sum_{k=\infty}^{\infty} h[k]r_{XY}[l-k] = h[l] * r_{XY}[l]. \tag{D.25}$$

Putting Equations (D.24) and (D.25) together and using the symmetry properties, Equations (D.21) and (D.22), yields

$$r_{YY}[l] = h[l] * r_{XY}[l] = h[l] * r_{YX}[-l] = h[l] * (h[-l] * r_{XX}[-l]) = r_{XX}[-l] * \underbrace{(h[l] * h[-l])}_{\hat{h}[l]}$$
$$= r_{XX}[l] * \hat{h}[l], \tag{D.26}$$

where $\hat{h}[n]$ is the (deterministic) autocorrelation function of the filter's impulse response $h[n]$,

$$\hat{h}[n] \triangleq h[n]*h[-n] = \sum_{k=-\infty}^{\infty} h[k]h[n+k]. \tag{D.27}$$

Because both the mean μ_Y and the autocorrelation $r_{YY}[l]$ are independent of absolute time, the filtered WSS random process $Y[n]$ is also a WSS process.

Of particular interest to us is the case in which the input to the filter is a zero-mean white noise, as discussed in Section D.2.3, with an autocorrelation function given by Equation (D.23), $r_{XX}[l] = \sigma_X^2 \delta[l]$. In this case, the output of the filter is

$$r_{YY}[l] = \sigma_X^2 \delta[l] * \hat{h}[l] = \sigma_X^2 \hat{h}[l].$$

In other words, the output of the filter is a WSS process whose autocorrelation function is determined to within a multiplicative constant (σ_X^2) by the characteristics of the filter. In theory, this means that we can design a random noise process with an autocorrelation function that we specify using $\hat{h}[n]$. Alternatively, we can model a given autocorrelation function $r_{YY}[l]$, as the result of a white noise process that has been filtered by the filter with impulse response $\hat{h}[n]$.

D.2.5 The Wold decomposition

The Wold decomposition theorem, diagrammed in **Figure D.6**, is a central result in the study of random processes. It states that any WSS random process $Z[n]$, can be expressed as the sum of two components, a *random* component $Y[n]$, and a *deterministic* component $D[n]$,

$$Z[n] = D[n] + Y[n].$$

The random component $Y[n]$ is derived by filtering a white-noise process $X[n]$ with a filter described by

$$y[n] = \sum_{k=0}^{\infty} h_k x[n-k]$$

and z-transform

$$H(z) = \sum_{k=0}^{\infty} h_k z^{-k}, \tag{D.28}$$

where h_k are the filter parameters. The deterministic component $D[n]$ is assumed to be uncorrelated with $Y[n]$, that is, $E(y[n]d[n]) = 0$, and $Y[n]$ is assumed to be uncorrelated with the filter constants h_k.

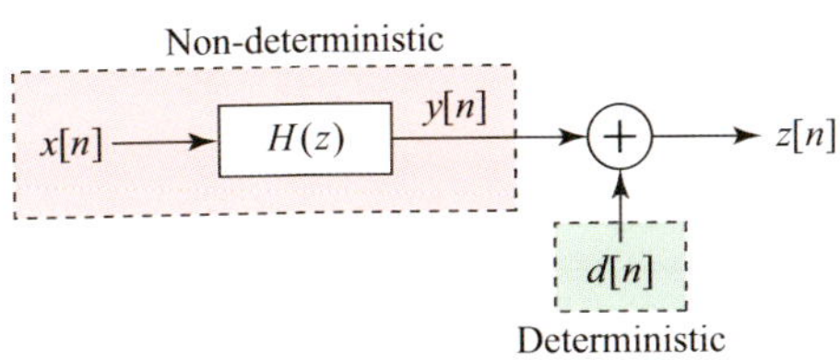

Figure D.6 Wold decomposition

The Wold decomposition provides the framework for the parametric modeling of random processes in Chapter 14. Given a description of $D[n]$, and a white-noise input process $X[n]$ with autocorrelation, $r_{XX}[l] = \sigma_X^2 \delta[l]$, a complete specification of the autocorrelation function for process $Z[n]$ would require knowing all the filter coefficients h_k, of which there may be an *infinite* number. That is unfortunate, since the entire purpose of parametric modeling is to find an *economical* representation of the process – that is, a description of the process that requires only a few model parameters. The key is to approximate the infinite-order polynomial $H(z)$ of Equation (D.28) by the ratio of two polynomials of finite order, a numerator polynomial of order M and a denominator polynomial of order N,

$$H(z) \cong \frac{B(z)}{A(z)} = \frac{\sum_{k=0}^{M} b_k z^{-k}}{\sum_{k=0}^{N} a_k z^{-k}}.$$

This $H(z)$ corresponds to the linear constant-coefficient difference equation (LCCDE),

$$\sum_{k=0}^{N} a_k y[n-k] = \sum_{k=0}^{M} b_k x[n-k],$$

or, letting $a_0 = 1$,

$$y[n] = -\sum_{k=1}^{N} a_k y[n-k] + \sum_{k=0}^{M} b_k x[n-k].$$

This is called the **autoregressive moving-average (ARMA) model**. The task of parametric modeling is then to specify the model parameters and their values. In Chapter 14, we discuss several approaches, and explore one popular one – **autoregressive (AR) modeling** – in considerable detail.

D.2.6 Estimators

In the entire discussion to this point, we have been assuming that we know the underlying probability distributions that allow us to calculate measures such as the mean, variance, correlation and covariance. However, in practice, we do not have access to these probability density functions. What we must do instead is to attempt to *estimate* the desired measures based on a finite number of observations of the process. In this section, we will discuss some of these sample measures.

Sample mean The **sample mean** $\hat{\mu}_X[N]$ is defined as the average of N observations of the process,

$$\hat{\mu}_X[N] \triangleq \frac{1}{N}\sum_{n=0}^{N-1} x[n]. \tag{D.29}$$

Because $x[n]$ is a random variable, each time we measure and average N values, we are likely to get a different value of $\hat{\mu}_X[N]$. For this reason, the sample mean is *itself* a random variable! $\hat{\mu}_X[N]$ is termed an **estimator** of the mean of the underlying random process μ_X and each measurement of $\hat{\mu}_X[N]$ with N values is an **estimate** of μ_X. The Matlab function `mean` performs this function.

Sample autocorrelation In a similar fashion, the **sample autocorrelation** $\hat{r}_{XX}[l, N]$ might be defined as

$$\hat{r}_{XX}[l, N] = \frac{1}{N}\sum_{n=0}^{N-1} x[n]x[n-l].$$

There is a bit of a problem with this definition, as values of $x[n-l]$ only exist for values n and l such that $0 \le n - l \le N$. We will deal with this by defining a sequence $x_N[n]$ that is equal to $x[n]$ within the range $0 \le n \le N-1$ and zero outside of it,

$$x_N[n] = \begin{cases} x[n], & 0 \le n \le N-1 \\ 0, & \text{otherwise} \end{cases}.$$

Then the sample autocorrelation $\hat{r}_{XX}[l, N]$ can be expressed in terms of $x_N[n]$,

$$\hat{r}_{XX}[l, N] \triangleq \frac{1}{N}\sum_{n=-(N-1)}^{N-1} x_N[n]x_N[n-l]. \tag{D.30}$$

By this definition, $\hat{r}_{XX}[l,N]=0$ for values of lag, $|l|\geq N$.

Estimators such as the sample mean $\hat{\mu}_X[N]$, and the sample autocorrelation $\hat{r}_{XX}[l,N]$ approximate the underlying quantities μ_X and $r_{XX}[l]$, respectively, but how good are these estimators? The two measures that are most commonly used to assess the "goodness" of an estimator are **bias** and **consistency**.

Bias Bias is defined as the difference between the expected value (i.e., the mean) of the estimator and the "true" value of the quantity that it purports to estimate. If this difference is zero, then the estimator is said to be **unbiased**. For example, consider $\hat{\mu}_X[N]$, which is an estimator of μ_X. The expected value of $\hat{\mu}_X[N]$ is

$$E(\hat{\mu}_x[N])=E\left(\frac{1}{N}\sum_{n=0}^{N-1}x[n]\right)=\frac{1}{N}\sum_{n=0}^{N-1}E(x[n])=\frac{1}{N}N\mu_X=\mu_X,$$

since, by definition, the expectation $E(x[n])=\mu_X$. Hence, the bias of this estimator is

$$E(\hat{\mu}_x[N])-\mu_X=0,$$

and we conclude that the sample mean of a WSS process is an unbiased estimator of the actual mean.

Now consider the bias of sample autocorrelation $\hat{r}_{XX}[l,N]$,

$$E(\hat{r}_{XX}[l,N])-r_{XX}[l].$$

To compute the expected value of $\hat{r}_{XX}[l]$, we need to calculate the expectation of Equation (D.30) carefully, making sure to respect the fact that the limits of the sum are different for positive and negative values of lag, l, due to the need to exclude values of $x[n]$ that lie outside the range, $0\leq n\leq N-1$:

$$\begin{aligned}E(\hat{r}_{XX}[l,N])&=\frac{1}{N}\sum_{n=-(N-1)}^{N-1}x_N[n]x_N[n-l]\\&=\begin{cases}\frac{1}{N}\sum_{n=l}^{N-1}E(x[n]x[n-l])=\frac{1}{N}\sum_{n=l}^{N-1}r_{XX}[l]=\frac{N-l}{N}r_{XX}[l], & 0\leq l\leq N-1\\ \frac{1}{N}\sum_{n=0}^{N-1+l}E(x[n]x[n-l])=\frac{1}{N}\sum_{n=0}^{N-1+l}r_{XX}[l]=\frac{N+l}{N}r_{XX}[l], & -(N-1)\leq l\leq -1\end{cases}\\&=\frac{N-|l|}{N}r_{XX}[l],0\leq |l|\leq N-1.\end{aligned} \tag{D.31}$$

Since $E(\hat{r}_{XX}[l,N])-r_{XX}[l]\neq 0$, $\hat{r}_{XX}[l,N]$ is a **biased estimator** of the autocorrelation function. However, as N gets large, the ratio $(N-|l|)/N$ approaches one for a fixed value of l, and the estimator is said to be **asymptotically unbiased**. The estimator of $\hat{r}_{XX}[l,N]$ given by Equation (D.30) is not unique. Other estimators have other properties (see Problem D-4).

Consistency An estimator is said to be **consistent** if its variance approaches zero as $N\to\infty$. Consistency is a desirable quality for an estimator to have since it means that the scatter of estimated values about the sample mean decreases as you take more data (i.e., larger N). For example, the variance of the sample mean is

$$\mathrm{var}(\hat{\mu}_X[N]) = E(\widehat{\mu}_X^2[N]) - E(\hat{\mu}_X[N])^2 = E(\widehat{\mu}_X^2[N]) - \mu_X^2[N].$$

The second moment is

$$E(\widehat{\mu}_X^2[N]) = E\left(\left(\frac{1}{N}\sum_{n=0}^{N-1} x[n]\right)\left(\frac{1}{N}\sum_{m=0}^{N-1} x[m]\right)\right) = \frac{1}{N^2}\sum_{n=0}^{N-1}\sum_{m=0}^{N-1} E(x[n]x[m])$$
$$= \frac{1}{N^2}\sum_{n=0}^{N-1}\sum_{m=0}^{N-1} r_{XX}[n-m].$$

This assumes that $x[n]$ and $x[m]$ are samples of a WSS process whose autocorrelation therefore only depends on the difference of time indices $n-m$. Then,

$$\mathrm{var}(\hat{\mu}_X[N]) = \frac{1}{N^2}\sum_{n=0}^{N-1}\sum_{m=0}^{N-1} r_{XX}[n-m] - \mu_X^2[N]. \tag{D.32}$$

If the process is not zero-mean (i.e., $\mu_X[N] \neq 0$), then the variance will not go to 0 as $N \to \infty$. However, if the process is zero-mean and $r_{XX}[n-m]$ is assumed to be bounded, the variance will approach zero, so that the sample mean will be a consistent estimator. A particular case of interest is white noise (see Section D.2.3), a zero-mean process whose observations $x[n]$ and $x[m]$ are uncorrelated for $n \neq m$,

$$r_{XX}[n-m] = \left\{\begin{matrix} r_{XX}[0] = \sigma_X^2, & n = m \\ 0, & n \neq m \end{matrix}\right\} = \sigma_X^2 \delta[n-m].$$

In this case, Equation (D.32) becomes

$$\mathrm{var}(\hat{\mu}_X[N]) = \frac{1}{N^2}\sum_{n=0}^{N-1} \sigma_X^2 = \frac{1}{N}\sigma_X^2.$$

The variance of the autocorrelation function $\hat{r}_{XX}[l]$, is relatively complicated to calculate, but for large N can be shown to approach

$$\mathrm{var}(\hat{r}_{XX}[l]) \cong \frac{1}{N}\sum_{n=-\infty}^{\infty} (r_{XX}^2[n] + r_{XX}[n-l]r_{XX}[n+l]).$$

Since the variance approaches 0 as $N \to \infty$, $\hat{r}_{XX}[l]$ is a consistent estimator of $r_{XX}[l]$.

D.2.7 Ergodic processes

A complete characterization of a random process essentially requires knowing the joint PDFs of the ensemble of random variables $X[n]$ that constitute the process as a function of time, or equivalently, the joint statistics of the ensemble of every order. If we know (or assume) that the process is WSS, then we only need to know the first- and second-order statistics (e.g., mean, correlation, covariance) of the ensemble of random variables as a function of time. That is still an impossibility, since computing those statistics requires access to every sample function in the sample space.

An **ergodic** random process is one in which we can estimate the statistics of the *ensemble of random variables* $X[n]$ from the statistics of the time-average of *a single sample function* (e.g., a measurement) of the process. For a WSS random process, only the mean and second moments of the ensemble average need be the same as those derived from a time sample. We have seen that the ensemble mean of a WSS process is not a function of absolute time, $E(x[n]) = \mu_X[n] = \mu_X$, and the autocorrelation function is only a function of the time difference (lag) between two time points, $r_{XX}[n, n-l] = r_{XX}[l]$. If the sample mean of a WSS process, $\hat{\mu}_X[N]$, converges to the ensemble mean μ_X in the mean-square sense,

$$\lim_{N\to\infty} E((\hat{\mu}_X[N] - \mu_X)^2) = 0,$$

then the process is said to be **mean-ergodic**, so

$$\lim_{N\to\infty} \frac{1}{N} \sum_{n=0}^{N-1} x[n] = \mu_X.$$

It can be shown that a necessary and sufficient condition for mean-ergodicity is that the process be asymptotically unbiased and consistent.

If the sample autocorrelation of a WSS process, $\hat{r}_{XX}[l, N]$, converges to the ensemble autocorrelation $r_{XX}[l]$ in the mean-square sense,

$$\lim_{N\to\infty} E((\hat{r}_{XX}[l, N] - r_{XX}[l])^2) = 0,$$

then the process is said to be **auto-correlation-ergodic**, so,

$$\lim_{N\to\infty} \frac{1}{N} \sum_{n=0}^{N-1} x[n]x[n-l] = r_{XX}[l].$$

Example D.2

Determine whether the random-phase cosine process of Example D.1, $x[n] = A\cos(\omega_0 n + \theta)$, is mean-ergodic and autocorrelation-ergodic.

➤ Solution:

By Example D.1, the ensemble statistics are $\mu_X = 0$ and $r_{XX}[l] = (A^2/2)\cos\omega_0 l$. The sample mean is

$$\lim_{N\to\infty} \frac{1}{N} \sum_{n=0}^{N-1} x[n] = A \lim_{N\to\infty} \frac{1}{N} \sum_{n=0}^{N-1} \cos(\omega_0 n + \theta) = 0,$$

so the process is mean-ergodic. Similarly,

$$\begin{aligned}\lim_{N\to\infty} \frac{1}{N} \sum_{n=0}^{N-1} x[n]x[n-l] &= \lim_{N\to\infty} \frac{A^2}{N} \sum_{n=0}^{N-1} \cos(\omega_0 n + \theta)\cos(\omega_0(n-l) + \theta) \\ &= \lim_{N\to\infty} \frac{A^2}{2N} \sum_{n=0}^{N-1} \cos\omega_0 l + \cancel{\lim_{N\to\infty} \frac{A^2}{2N} \sum_{n=0}^{N-1} \cos(\omega_0(2n-l) + 2\theta)} = \frac{A^2}{2}\cos\omega_0 l,\end{aligned}$$

so the process is also autocorrelation-ergodic.

The Gaussian random process is a case of special interest, because the process is completely defined by the first and second moments. For example, **Figure D.4d** shows $p_X(x_k, n)$, the PDF of a single time sample $x_k[n]$ of a Gaussian process. This is equivalent to the PDF of the ensemble, $p_X(x, n_1)$, for a stationary ergodic process since the Gaussian PDF is completely characterized by its mean and variance.

D.3 Power spectral density

The DTFT is an invaluable tool that provides a frequency domain interpretation of deterministic sequences and systems. The problem with attempting to apply this method to random signals is that the Fourier transform of a random signal is random! However, WSS random processes are completely characterized by their autocorrelation functions $r_{XX}[l]$ and $r_{YY}[l]$, and their crosscorrelation function $r_{XY}[l]$, which are *deterministic* functions of index l. Considerable insight into the nature of a wide-sense stationary random process can be gained by looking at the frequency spectrum of these functions. Because their transforms are deterministic, all the properties of DTFTs we studied in Chapter 3 apply.

D.3.1 Definition

The **power spectral density (PSD)** is defined as the DTFT of the autocorrelation function $r_{XX}[l]$,

$$S_{XX}(\omega) \triangleq \sum_{l=-\infty}^{\infty} r_{XX}[l]e^{-j\omega l}.$$

The sufficient condition for the existence of the PSD is the same as that for transforms of deterministic sequences, namely that the sequence $r_{XX}[l]$ is absolutely summable,

$$\sum_{l=-\infty}^{\infty} |r_{XX}[l]| < \infty.$$

The inverse DTFT recovers $r_{XX}[l]$ from $S_{XX}(\omega)$,

$$r_{XX}[l] = \frac{1}{2\pi}\int_{-\pi}^{\pi} S_{XX}(\omega)e^{j\omega l}d\omega.$$

The PSD represents the *distribution of power* of signal $r_{XX}[l]$ as a function of frequency. The integral of $S_{XX}(\omega)/2\pi$ over the entire frequency range $-\pi \le \omega < \pi$ is equal to the variance of the process, which is the **total power** of the signal,

$$\sigma_X^2 = E(x^2) = r_{XX}[0] = \frac{1}{2\pi}\int_{-\pi}^{\pi} S_{XX}(\omega)d\omega. \quad \textbf{(D.33)}$$

Because the $r_{XX}[l]$ is real and symmetric (even), its transform, the PSD, is also real and symmetric,

$$S_{XX}(\omega) = S_{XX}(-\omega).$$

Furthermore, it can be shown[3] that the PSD is always non-negative, that is,

$$S_{XX}(\omega) \ge 0.$$

[3] An elegant, indirect way to show this is due to Papoulis, A. (1965). p. 397. Imagine that $S_{XX}(\omega)$ *were* negative, $S_{XX}(\omega) < 0$, over some contiguous frequency range $\omega_0 < |\omega| < \omega_1$. Now, filter this process through a bandpass filter that only passes this frequency range,

$$H(\omega) = \begin{cases} 1, \omega_0 < |\omega| < \omega_1 \\ 0, \text{otherwise} \end{cases}.$$

Therefore, by Equation (D.34), $S_{YY}(\omega) = S_{XX}(\omega)|H(\omega)|^2$ is either negative or zero over the entire frequency range $-\pi \le \omega < \pi$. By Equation (D.33), the integral of $S_{YY}(\omega)$ is proportional to the variance of the output process σ_Y^2, which must be non-negative, $\sigma_Y^2 > 0$. Hence, $S_{YY}(\omega)$ must also be non-negative. Since $|H(\omega)|^2 \ge 0$, $S_{XX}(\omega)$ must also be non-negative, $S_{XX}(\omega) \ge 0$.

D.3.2 Power spectral density of a filtered random process

An important set of results relates to the PSD of filtered random processes. If the filter's impulse response $h[n]$ is real with transform $H(\omega)$, then by the properties of the DTFT, the transform of $h[-n]$ is $H(-\omega) = H^*(\omega)$ due to conjugate symmetry. Hence, the transform of $\hat{h}[n] = h[n] * h[-n]$ in Equation (D.27) is $H(\omega)H^*(\omega) = |H(\omega)|^2$. Then, from Equations (D.24) and (D.26), we obtain the transforms of $r_{XY}[l]$ and $r_{YY}[l]$, the PSD of the filtered crosscorrelation and autocorrelation functions, respectively:

$$\begin{aligned} S_{XY}(\omega) &= S_{XX}(\omega)H(\omega) \\ S_{YY}(\omega) &= S_{XX}(\omega)|H(\omega)|^2 \end{aligned}. \tag{D.34}$$

D.3.3 Power spectral density of noise

From Equation (D.23), the autocorrelation function of a white-noise process is $r_{XX}[l] = \sigma_X^2 \delta[l]$. Hence, the PSD of a white-noise process is

$$S_{XX}(\omega) = \sum_{l=-\infty}^{\infty} \sigma_X^2 \delta[l] e^{-j\omega l} = \sigma_X^2.$$

In words, the defining characteristic of white noise in the frequency domain is that the PSD is flat (constant) over the whole frequency range $-\pi \leq \omega < \pi$.

So-called "pink-noise" is created by passing white noise through a filter with frequency response $H(\omega)$. Because the PSD of white noise is a constant with respect to frequency, by Equation (D.34) the PSD of the output of the filter is determined by the magnitude of the filter,

$$S_{YY}(\omega) = \sigma_X^2 |H(\omega)|^2.$$

D.4 Matlab functions

Matlab has a number of basic functions that implement various PDFs and statistical operations.

D.4.1 Random number generators

Uniform random variables The Matlab function `rand` produces an array of N outcomes of a uniform PDF, with values distributed between 0 and 1,

```
x = rand(1, N).
```

Gaussian random variable The Matlab function `randn` produces an array of `N` outcomes of a zero-mean Gaussian PDF of unit variance,

```
x = randn(1, N).
```

Random number generator The Matlab function `rng` provides the "seed" of the random number generator for `rand` and `randn`,

```
rng(sd).
```

Calling `rng` with a non-negative integer `sd` resets the random number generator to a defined state so that subsequent calls to `rand` and `randn` will produce "predictably random" results every time a program is run.

D.4.2 Autocorrelation and crosscorrelation

The function `xcorr` produces an estimate of the autocorrelation or crosscorrelation of random variables.

```
r = xcorr(x, [y], [options])
```

If `x` and `y` are both specified, then the crosscorrelation is computed. If only `x` is specified and is a vector of length N, the autocorrelation is computed. In this case, `xcorr` essentially implements the "raw" sample autocorrelation, equivalent to

```
r = conv(x, x(end:-1:1)).
```

The resulting vector `r` is a symmetrical array of length $2N-1$, with the value of $r_{XX}[0]$ located in the middle of the array. If the optional parameter `N` is used only the center $2N+1$ coefficients will be computed. If the optional parameter `"biased"` is appended to the arguments, then the autocorrelation is divided by `length(x)`. See the documentation for further arguments.

PROBLEMS

Problem D-1

(a) Show that the parameter μ_X in the definition of the univariate Gaussian PDF, Equation (D.2), is in fact the mean value, $E(x)=\mu_X$.

(b) Show that $E((x-E(x))^2)=\sigma_X^2$ is the variance.

Problem D-2

In this problem, we show that the covariance of two random variables X and Y is bounded,

$$|\sigma_{XY}| \leq \sigma_X \sigma_Y.$$

Define a new random variable z as a combination of x and y,

$$z=(x-E(x))-(y-E(y))\frac{\sigma_{XY}}{\sigma_Y^2}.$$

The expectation of any squared quantity must be non-negative, so $E(z^2) \geq 0$. Expand z^2 and use the fact that the expectation is a linear operator; that is, the expected value of the sum of random variables weighted by constants is equal to the weighted sum of their individual expected values, regardless of whether the random variables are independent. So,

$$0 \leq E(z^2)= \cdots = \sigma_X^2 - \frac{\sigma_{XY}^2}{\sigma_Y^2}.$$

Fill in the remaining steps of the derivation to conclude that $\sigma_{XY}^2 \leq \sigma_X^2 \sigma_Y^2$, from which Equation (D.12) follows.

Problem D-3

In this problem, we show that the maximum value of the autocorrelation function of a WSS random process, $X[n]$, occurs at a lag of $l=0$,

$$|r_{XX}[l]| \leq r_{XX}[0].$$

Define a new random process as

$$z[n_1, n_2] = x[n_1] \pm x[n_2],$$

where $n_2 = n_1 - l$. The expectation of any squared quantity must be non-negative, so $E(z^2[n_1, n_2]) \geq 0$. Expand z^2 and show that

$$0 \leq E(z^2[n_1, n_2]) = \cdots = 2(r_{XX}[0] \pm r_{XX}[l]).$$

Fill in the remaining steps of the derivation to conclude that $|r_{XX}[l]| \leq r_{XX}[0]$.

Problem D-4

We can create an unbiased estimator $\hat{r}_{XX}[l]$ of the autocorrelation function by modifying Equation (D.30) as follows:

$$\hat{r}_{XX}[l] = \begin{cases} \dfrac{1}{N-l} \displaystyle\sum_{n=l}^{N-1} x[n]x[n-l], & 0 \leq l \leq N-1 \\ \dfrac{1}{N+l} \displaystyle\sum_{n=0}^{N-1+l} x[n]x[n-l], & -(N-1) \leq l \leq -1. \end{cases}$$

Show that this estimator is, in fact, unbiased,

$$E(\hat{r}_{XX}[l]) = r_{XX}[l], \quad 0 \leq |l| \leq N-1.$$

Unfortunately, the variance of this estimator turns out to be

$$\mathrm{var}(\hat{r}_{XX}[l]) \cong \frac{N}{N-|l|} \sum_{n=-\infty}^{\infty} (r_{XX}^2[n] + r_{XX}[n-l]r_{XX}[n+l]).$$

Hence, the variance of the estimate increases markedly as $|l| \to N$.

Problem D-5

A random process has autocorrelation function

$$r_{XX}[l] = \alpha^{|l|}, \quad |\alpha| < 1.$$

Show that the power spectral density is

$$S_{XX}(\omega) = \frac{1-\alpha^2}{1-\alpha(e^{j\omega}+e^{-j\omega})+\alpha^2}.$$

REFERENCES

Basic linear systems theory

The books by Lathi (2000) and Oppenheim and Willsky (1996) are representative of modern approaches to teaching both continuous-time and discrete-time signal processing together at a sophomore or junior level. The books by Bracewell (1986) and Papoulis (1960) are classics. They used to be standard texts for signal processing, and are still full of insight for those who want to delve deeply into the topic.

R. N. Bracewell. *The Fourier Transform and its Applications*. McGraw-Hill, 1986

B. P. Lathi. *Signal Processing and Linear Systems*. Oxford University Press, 2000

A.V. Oppenheim and A. S. Willsky. *Signals and Systems*, 2nd edition. Pearson, 1996

A. Papoulis. *The Fourier Transform and its Applications*. McGraw Hill, 1960

DSP textbooks

Of the many DSP textbooks currently available, Oppenheim and Schafer (2009) is probably the most popular, and for good reason. The coverage overlaps with and extends the material in my book and is often used for graduate courses. The book by Lyons (2010) is notable not only for its light-hearted approach, but for the many practical tips and tricks offered by this practicing engineer. Porat (1997) is a concise book at a more advanced level than some others, but would repay your inspection after you've mastered the basics at the level of this book. Rabiner and Gold (1975) is not a basic textbook by any means, but, after 40+ years, it still has much to offer to the DSP practitioner.

R. G. Lyons. *Understanding Digital Signal Processing*, 3rd edition. Prentice Hall, 2010

S. K. Mitra. *Digital Signal Processing*, 3rd edition. McGraw-Hill, 2006

A. V. Oppenheim and R. W. Schafer. *Discrete-Time Signal Processing*, 3rd edition. Prentice Hall, 2009

B. Porat. *A Course in Digital Signal Processing*. Wiley, 1997

J. G. Proakis and D. G. Manolakis. *Digital Signal Processing*, 4th edition. Prentice Hall, 2007

L. R. Rabiner and B. Gold. *Theory and Application of Digital Signal Processing*. Prentice Hall, Inc., 1975

A/D and D/A conversion

The basics of theory of A/D and D/A conversion are well handled by almost any of the basic DSP textbooks. The paper by Hauser (1991) is an excellent tutorial. For an understanding of the hardware and practical applications of A/D and D/A converters, your best bet is often the tutorials and application notes written (often anonymously) by the companies that produce these devices, e.g., Analog Devices, Maxim, Burr-Brown and Texas Instruments. A few of these are cited below. In particular, the book by Kester (2004) is considered a bible by practicing engineers.

M. W. Hauser. "Principles of Oversampling A/D Conversion." *Journal of the Audio Engineering Society*, 39 (1/2). 3–26, 1991

W. Kester, ed. *Data Conversion Handbook*. Newnes, 2004

"Which ADC Architecture Is Right for Your Application?" *Analog Dialogue*, 39 (6), 2005: www.analog.com/analogdialog

Maxim Integrated. "Understanding SAR ADCs: Their Architecture and Comparison with Other ADCs." Tutorial 1080

"Demystifying Delta-Sigma ADCs." Tutorial 1070

"A Simple ADC Comparison Matrix." Tutorial 2094

"Understanding Pipelined ADCs." Tutorial 1023

"Understanding Flash ADCs." Tutorial 810

S. Pavan, R. Schreier, and G. C. Temes. *Understanding Delta-Sigma Data Converters*, 2nd edition. Wiley-IEEE Press, 2017

Probability and statistics

Kay (2005) is a reasonable book for engineers and includes Matlab. By modern standards, Papoulis (1965) is somewhat dense, but there's no one who explains things more clearly.

S. Kay. *Intuitive Probability and Random Processes using MATLAB*. Springer, 2005

A. Papoulis. *Probability, Random Variables, and Stochastic Processes*. McGraw-Hill, 1965

FIR and IIR filtering

Antoniou (2018) is a textbook that is particularly strong in IIR filter design. The books by Jackson (1995) and Parks and Burrus (1987) are useful once one has mastered the basics. Most of the remaining references are primary papers.

A. Antoniou. *Digital Filters: Analysis, Design, and Signal Processing Applications*. McGraw-Hill, 2018

C. S. Burrus, A. W. Soewito, and R. A. Gopinath. "Least Squared Error FIR Filter Design with Transition Bands." *IEEE Transactions on Signal Processing*, 40 (6): 1327–1340, 1992

A. G. Constantinides. "Spectral Transformations for Digital Filters." *Proceedings of the IEEE*, 117 (8): 1585–1590, 1970

H. G. Dimopoulos. "Optimal Use of Some Classical Approximations in Filter Design." *IEEE Transactions Circuits and Systems II: Express Briefs*, 54 (9): 780–784, 2007

f. j. harris. "On the Use of Windows for Harmonic Analysis with the Discrete Fourier Transform." *Proceedings of the IEEE*, 66 (1): 51–83, 1978

L. B. Jackson. *Digital Filters and Signal Processing*, 3rd edition. Kluwer Academic Publishers, 1995

L. J. Karam, J. H. McClellan, E. W. Selesnick, and C. S. Burrus. "Digital Filtering." In *Digital Signal Processing Handbook*, ed. V. K. Madisetti and D. B. Williams. CRC Press, 1999

T. W. Parks and C. Burrus. *Digital Filter Design*. Wiley, 1987

T. W. Parks and J. McClellan. "Chebyshev Approximation for Nonrecursive Digital Filters with Linear Phase." *IEEE Transactions on Circuit Theory*, 19 (2): 189–194, 1972

A. D. Poularikas. "Windows." In *The Handbook of Formulas and Tables for Signal Processing*, ed. A. D. Poularikas. CRC Press, 1999

S. C. Roy. "A Simple Derivation of the Spectral Transformations for IIR Filters." *IEEE Transactions of Education*, 48 (2): 274–278, 2005

P. P. Vaidyanathan and T. Q. Nguyen. "A 'Trick' for the Design of FIR Half-Band Filters." *IEEE Transactions on Circuits and Systems*, 34 (3): 297–300, 1987

"A Simple Proof of the Alternation Theorem." *Conference Record of the Forty-First Asilomar Conference on Signals, Systems and Computers*, 1111–1115, 2007

DFT/FFT

The book by Smith (2011) is particularly accessible (and is freely available on the Internet), whereas that of Rao et al. (2010) is more suited for a graduate level. After more than 30 years, the later sections of the book by Brigham (1988) dealing with the FFT and its applications still have a lot of useful detail (though the programs are written in FORTRAN).

E. O. Brigham. *Fast Fourier Transform and Its Applications*. Prentice Hall, 1988.

C. S. Burrus, M. Frigo, and S. G. Johnson. *Fast Fourier Transforms*, www.cnx.org, 2019

J. Cooley and J. Tukey. "An Algorithm for the Machine Calculation of Complex Fourier Series." *Mathematics of Computation*, 19 (90): 297–301, 1965

P. Duhamel and H. Hollmann. "Split-Radix FFT Algorithm." *Electronics Letters*, 20 (1): 14–16, 1984

P. Duhamel and M. Vetterli. "Fast Fourier Transforms: A Tutorial Review and State of the Art." *Signal Processing*, 19: 259–299, Apr. 1990

M. Frigo and S. G. Johnson. "The Design and Implementation of FFTW3." *Proceedings of the IEEE*, 93 (2): 216–231, 2005

G. Goertzel. "An Algorithm for the Evaluation of Finite Trigonometric Series." *American Mathematics Monthly*, 65 (1): 34–35, 1958.

M. T. Heideman, D. H. Johnson and C. S. Burrus. "Gauss and the History of the Fast Fourier Transform." *IEEE Acoustics, Speech, and Signal Processing Magazine*, 1: 14–21, 1984

K. R. Rao, D. N. Kim, and J. J. Hwang. *Fast Fourier Transform – Algorithms and Applications*. Springer, 2010

J. O. Smith. *Mathematics of the Discrete Fourier Transform (DFT): With Audio Applications*, 2nd edition. W3K Publishing, 2011

DCT

The review by Painter and Spanias (1997) is a useful guide to the complexities of the MP3 algorithm. Most of the rest are primary papers.

G. Bonnerot and M. Bellanger, "Odd-Time Odd-Frequency Discrete Fourier Transform for Symmetric Real-Valued Series." *Proceedings of the IEEE*, 64 (3): 392–393, 1976

M. Bosi and R. E. Goldberg. *Introduction to Digital Audio Coding and Standards*. Kluwer Academic Publishers, 2003

K. Brandenburg and G. Stoll. "ISO/MPEG-1 Audio: A Generic Standard for Coding of High-Quality Digital Audio." *Journal of the Audio Engineering Society*, 42 (10): 780–792

V. Britanak. "A Survey of Efficient MDCT Implementations in MP3 Audio Coding Standard: Retrospective and State-of-the-Art." *Signal Processing*, 91 (4): 624–672, 2011

P. Duhamel, Y. Mahieux, and J. P. Petit. "A Fast Algorithm for the Implementation of Filter Banks Based on 'Time Domain Aliasing Cancellation.'" *IEEE Proceedings of the International Conference on Acoustics, Speech, and Signal Processing (ICASSP)*: 2209–2212, 1991

R. C. Gonzalez and R. E. Woods. *Digital Image Processing*, 4th edition. Pearson, 2017

A. Nasir, T. Natarajan and K. R. Rao. "Discrete Cosine Transform." *IEEE Transactions on Computers*, C-23 (1): 90–93, 1974

T. Painter and A. Spanias. "A Review of Algorithms for Perceptual Coding of Digital Audio Signals." *Proceedings of 13th International Conference on Digital Signal Processing*, 1: 179–208, 1997

J. P. Princen, A. W. Johnson, and A. B. Bradley. "Subband/transform Coding Using Filter Bank Designs Based on Time Domain Aliasing Cancellation." *IEEE Proceedings of the International Conference on Acoustics, Speech, and Signal Processing (ICASSP)*: 2161–2164, 1987

K. R. Rao and P. C. Yip. *The Transform and Data Compression Handbook*. CRC Press, 2000

N. Roma and L. Sousa. "A Tutorial Overview on the Properties of the Discrete Cosine Transform for Encoded Image and Video Processing." *Signal Processing*, 91 (11): 2443–2464, 2011

C. E. Shannon. "A Mathematical Theory of Communication." *Bell System Technical Journal*, 27 (3): 379–423, and 27 (4): 623–666, 1948

S. S. Stevens and H. Davis. *Hearing: Its Psychology and Physiology*. Acoustical Society of America, 1983

G. Stix. "Profile: David A. Huffman: Encoding the 'Neatness' of Ones and Zeroes." *Scientific American*, 265 (3): 54–61, 1991

T. K. Truong, P. D. Chen, and T. C. Cheng. "Fast Algorithm for Computing the Forward and Inverse MDCT in MPEG Audio Coding." *Signal Processing*, 86 (5): 1055–1060, 2006

E. Zwicker and H. Fastl. *Psychoacoustics: Facts and Models*. Springer Verlag, 1999

E. Zwicker, G. Flottrorp, and S. S. Stevens. "Critical Bandwidth in Loudness Summation." *Journal of the Acoustical Society of America*, 29 (5): 548–557, 1957

Multirate systems

The books by Fliege (1994) and Vaidyanathan (1993) provide similar, complete treatments of many aspects of multirate signal processing. The book by harris (2004) is a good complement to these.

R. E. Crochiere and L. R. Rabiner. *Multirate Digital Signal Processing*. Prentice Hall, 1983

N. Fliege. *Multirate Digital Signal Processing*. Wiley, 1994

f. j. harris. *Multirate Signal Processing for Communication Systems*. Prentice Hall, 2004

E. Hogenauer. "An Economical Class of Digital Filters for Decimation and Interpolation." *IEEE Transactions on Acoustics, Speech, and Signal Processing*, 29 (2): 155–162, 1981

A. Mertins and D. A. Mertins. *Signal Analysis: Wavelets, Filter Banks, Time-Frequency Transforms and Applications*. Wiley, 1999

L. Milić. *Multirate Filtering for Digital Signal Processing*. Information Science Reference, 2008

Y. Neuvo, Dong Cheng-Yu, and S. Mitra. "Interpolated Finite Impulse Response Filters." *IEEE Transactions on Acoustics, Speech, and Signal Processing*, 32 (3): 563–570, 1984

P. P. Vaidyanathan. "Multirate Digital Filters, Filter Banks, Polyphase Networks, and Applications: A Tutorial." *Proceedings of the IEEE*, 78 (1): 56–93, 1990

Multirate Systems and Filter Banks. Prentice Hall, 1993

Spectral analysis

The book by Hayes (1996) is exceptionally clear and thorough. The book by Smith (2011) is particularly useful for its concentration on audio applications. The tutorial by Vaidyanathan (2008) is an accessible introduction to linear prediction.

J. R. Deller, Jr., J. G. Proakis, and J. H. Hansen. *Discrete-Time Processing of Speech Signals*. Macmillan, 1993

M. H. Hayes. *Statistical Digital Signal Processing and Modeling*. John Wiley & Sons, 1996

P. Ladefoged. *Elements of Acoustic Phonetics*. University of Chicago Press, 1996

J. Makhoul. "Linear Prediction: A Tutorial Review." *Proceedings of the IEEE*, 63 (4): 561–580, 1975

J. D. Markel and A. H. Grey. *Linear Prediction of Speech*. Springer, 1976

T. F. Quatieri. *Discrete-Time Speech Signal Processing*. Prentice Hall, 2002

L. R. Rabiner and R. W. Schafer. *Theory and Application of Digital Speech Processing*. Prentice Hall, 2010

J. O. Smith. *Spectral Audio Signal Processing*. W3K Publishing, 2011

K. N. Stevens. *Acoustic Phonetics*. MIT Press, 1998

P. Stoica and R. Moses. *Introduction to Spectral Analysis*. Prentice Hall, 1997

P. P. Vaidyanathan. *The Theory of Linear Prediction*. Morgan & Claypool, 2008

P. D. Welch. "The Use of Fast Fourier Transform for the Estimation of Power Spectra: A Method Based on Time Averaging over Short, Modified Periodograms." *IEEE Transactions on Audio and Electroacoustics*, AU-15: 70–73, 1967

INDEX